Lecture Notes in Civil Engineering

559

Lecture Notes in Civil Engineering (LNCE) publishes the latest developments in Civil Engineering—quickly, informally and in top quality. Though original research reported in proceedings and post-proceedings represents the core of LNCE, edited volumes of exceptionally high quality and interest may also be considered for publication. Volumes published in LNCE embrace all aspects and subfields of, as well as new challenges in, Civil Engineering. Topics in the series include:

- Construction and Structural Mechanics
- Building Materials
- Concrete, Steel and Timber Structures
- Geotechnical Engineering
- Earthquake Engineering
- Coastal Engineering
- Ocean and Offshore Engineering; Ships and Floating Structures
- Hydraulics, Hydrology and Water Resources Engineering
- Environmental Engineering and Sustainability
- Structural Health and Monitoring
- Surveying and Geographical Information Systems
- Indoor Environments
- Transportation and Traffic
- Risk Analysis
- Safety and Security

To submit a proposal or request further information, please contact the appropriate Springer Editor:

- Pierpaolo Riva at pierpaolo.riva@springer.com (Europe and Americas);
- Swati Meherishi at swati.meherishi@springer.com (Asia—except China, Australia, and New Zealand);
- Wayne Hu at wayne.hu@springer.com (China).

All books in the series now indexed by Scopus and EI Compendex database!

Bao-Jie He · Deo Prasad · Li Yan ·
Ali Cheshmehzangi · Gloria Pignatta
Editors

International Conference on Urban Climate, Sustainability and Urban Design

 Springer

Editors
Bao-Jie He
School of Architecture and Urban Planning
Chongqing University
Chongqing, China

Deo Prasad
School of Built Environment
University of New South Wales Sydney
Kensington, NSW, Australia

Li Yan
School of Civil Engineering and Architecture
Southwest University of Science
and Technology
Mianyang, China

Ali Cheshmehzangi ⓘ
School of Architecture, Design and Planning
University of Queensland
Brisbane, QLD, Australia

Gloria Pignatta ⓘ
School of Built Environment
University of New South Wales
Sydney, NSW, Australia

ISSN 2366-2557   ISSN 2366-2565 (electronic)
Lecture Notes in Civil Engineering
ISBN 978-981-97-8400-4   ISBN 978-981-97-8401-1 (eBook)
https://doi.org/10.1007/978-981-97-8401-1

This Springer imprint is published by the registered company Springer Nature Singapore Pte Ltd.
The registered company address is: 152 Beach Road, #21-01/04 Gateway East, Singapore 189721, Singapore

If disposing of this product, please recycle the paper.

Bao-Jie He dedicates this book to all his collaborators and colleagues at Centre for Climate-resilient and Low-Carbon Cities. Also dedicated to all invited speakers, moderators, and participants of the International Conference of Urban Climate and Urban Design.

Deo Prasad dedicates this book to the collaborators and colleagues working on Net-Zero Carbon Built Environment, and to all members in the NSW Decarbonisation Innovation Hub, previous CRC Low Carbon Living, and School of Built Environment (UNSW Sydney). A special dedication to all PhD and Masters graduated from his research team.

Li Yan dedicates this book to the researchers in the field of urban climate and urban design, dedicated to the partners and friends who have made tremendous efforts for the organization of this conference and the publication of the proceedings.

Ali Cheshmehzangi dedicates this book to his team at his previous institute, Qingdao City University, with whom he worked extensively on establishing new platforms and creating innovative opportunities. In particular, Center of Innovation for Education and Research (CIER), Internationalisation and Global Partnership Team, College of Architecture, and the Excalibur team.

Gloria Pignatta dedicates this book to all members of her Applied Material and Building Physics (AMBP) research group working at the University of New South Wales, Australia, and internationally, and to all collaborators and participants of the International Conference of Urban Climate and Urban Design. A special dedication to her family, particularly her newborn Alexander Michael, whose love and support have been the cornerstone of her academic journey.

Foreword

Extreme weather events have been increasingly frequent all over the world in recent years. How to respond to climate change, reduce climate risks, and pursue a sustainable development pathway with climate resilience is a common challenge for all of humanity. It is also closely related to the fate of every country, city, industry, and each individual. Urban planning and design, as an important means to address climate change, can help us protect natural resources, reduce carbon emissions, and improve urban sustainability.

Based on the Mianyang Science and Technology City (MSTC) Talent Summit and the series conference of Urban Climate and Urban Design, we launched the 2023 International Conference on Urban Climate, Sustainability, and Urban Design at Mianyang, Sichuan, China. It is of great significance for the construction of low-carbon cities and sustainable urban development in China to have globally renowned scholars in the field of urban planning and design contribute their ideas. The MSTC is the only national science and technology city approved for construction by the State Council and the State Council in September 2000. It is an important region for the comprehensive innovation and experiments of the national system, and it shoulders the important mission of exploring the path and accumulating experience for the construction of new urbanization.

We invited academicians, senior experts, and young scholars from universities and research institutions to deliver speeches and talks, exchange the latest research results, and discuss how to alleviate the challenges and problems caused by local and global climate change through urban planning and design. The conference is expected to generate suggestions and strategies for climate change mitigation and adaptation, high-quality urban development, rural revitalization, and promote the development of disciplines related to urban climate and urban design.

Locally, by holding an internationally influential conference, we aim to make high-level talents from around the world know Mianyang, seek paths for the high-quality development of the city, and create excellent urban and rural spatial quality. The efficient and convenient collaboration among academia, government, and enterprise provides a sound environmental foundation for attracting talents, while the excellent urban and rural spatial quality provides a fundamental environment. By actively organizing such a high-level academic conference, we provide talents with a livable, business-friendly, innovative, enjoyable, and sustainable urban and rural environment, which is an important tool for retaining talents and promoting the innovative and sustainable development of Mianyang City.

Preface

Human beings are under the risks and threats of many challenges. Climate change, associated with greenhouse (GHG) emissions and anthropogenic activities, is the most severe one with significantly environmental, economic, and social consequences. Climate-related disasters and impacts are non-linear, where the impacts are unbearable once the warming level exceeds 1.5 °C. However, we are standing on a warming level of 1.1-1.2 °C which is very close to the critical threshold by 2030-2035. There is an urgent to address climate change before our human beings are locked into the irreversible circumstances.

However, many other challenges such as urbanization, environmental deterioration, economic growth, future development, and pandemic are also on, bringing prominent threats to human health, safety, comfort, and wellbeing. Many of those challenges including climate change can be interlinked. For instance, climate-related disasters such as extreme heat and intense precipitation evolve into urban overheating and urban flooding. It is the unsustainable urban planning and design that extensively integrate hard surfaces into cities and reduce climate-proofing capacity. The release of air pollutants during urban production and operation causes airborne pollution, while the increase of building height and density, the reduction of urban greening, and replacement of open spaces reduce the capacity of pollutant dispersion.

Given this, the 1st International Conference of Urban Climate, Sustainability, and Urban Design was held in November 2023, with the attendance of experts and researchers from academia and industries in the field of urban planning and design, architecture, atmospheric environment, ecology, geography, and economy to share the ideas, thoughts, technologies, and practices on how to address climate change and enhance sustainable development. The conference covers eight themes, including

- Technical strategies for the regulation of the urban heat island effect
- Urban anthropogenic heat exhaustion and environmental pollution
- Low-carbon urban development and climate-environmental correlation assessment
- Urban design methods based on thermal comfort evaluation
- Big data processing for urban climate and environment management
- Resilient city construction based on climate change assessment
- Performance-oriented urban renewal strategies
- Other related contents of urban design and urban planning.

A total of 70 submissions with new findings and outstanding contributions are accepted for publication in this conference proceeding after critical peer review. This conference proceeding consists of seven main parts, including

- Climate Predication and Impact Assessment
- Decarbonization and Carbon Neutral Solutions
- Sustainable Land Use and Management
- Urban Heat Mitigation and Adaptation

- Urban Environmental Quality Assessment and Improvement
- Sustainable Urban and Rural Planning
- Resilience to Climate and Health Challenges.

Overall, this edited conference proceeding is expected to enable more experts and researchers to understand the methods and theories on the climate-induced impact and assessment, climate change mitigation and adaptation, urban environmental sustainability improvement, sustainable urban-rural planning and design, and urban resilience to emerging challenges.

Acknowledgments

This collection of papers was able to be published successfully thanks to the support and assistance of many organizations and individuals.

Li Yan and Bao-Jie He appreciate the sponsorship of the Organization Department of Mianyang Government, Southwest University of Science and Technology, and Sichuan Provincial Meteorological Bureau for this International Conference on Urban Climate, Sustainability and Urban Design Conference.

We appreciate the support from the School of Civil Engineering and Architecture at Southwest University of Science and Technology, Mianyang Meteorological Bureau, and School of Architecture and Urban Planning at Chongqing University for their assistance in organizing the conference.

We appreciate the support from the Meteorological Impact and Risk Research Center of China Meteorological Administration, School of Architecture and Urban Planning of Chongqing University, Mianyang Land and Space Planning Association, and People's Government of Pingwu County for their strong support for conference propaganda and promotion.

We appreciate Scientia Professor Deo Prasad AO, FTSE at UNSW Sydney, Prof. Bao-Jie He at Chongqing University, and Dr. Li Yan at Southwest University of Science and Technology for chairing the conference. Many thanks go to keynote and invited speakers, international scientific organizing committee members, local organizing committee members, and reviewers all over the world.

Special thanks go to Dr. Yangli Li, Ms. Lu Che, Ms. Erli Zeng, Mr. Mingqiang Yin, Ms. Yisha Liu, and Mr. Xiaowei Shang for their tremendous contributions to funding application, conference organization, paper collection, and feedback collection.

Scientific Organizing Committee

Conference Chair

Deo Prasad — Scientia Professor, FTSE, University of New South Wales, Australia

Academic Chairs

Bao-Jie He — Professor, PhD, Chongqing University, China
Li Yan — Lecturer, PhD, Southwest University of Science and Technology, China

Scientific Organizing Committee Members

Amirhosein Ghaffarianhoseini — Professor, PhD, Auckland University of Technology, New Zealand
Amos Darko — Assistant Professor, PhD, University of Washington, United States
Andreas Matzarakis — Professor, PhD, University of Freiburg, Germany
Attia Shady — Professor, PhD, University of Liège, Belgium
Ayyoob Sharifi — Professor, PhD, Hiroshima University, Japan
Beta Paramita — Associate Professor, PhD, Universitas Pendidikan Indonesia, Indonesia
Federica Rosso — PhD, DICEA-Sapienza Università di Roma, Italy
Gloria Pignatta — Scientia Lecturer, PhD, University of New South Wales, Australia
Jack Ngarambe — PhD, Kyung Hee University, South Korea
Jihui Yuan — Associate Professor, PhD, Osaka Metropolitan University, Japan
Jinda Qi — Assistant Professor, PhD, National University of Singapore, Singapore
Joni Jupesta — PhD, University of United Nations, Japan
Karam Al-Obaidi — Senior Lecturer, PhD, Sheffield Hallam University, UK
Minal Pathak — Associate Professor, PhD, Ahmedabad University, India

Nasrin Aghamohammadi	Associate Professor, PhD, Curtin University, Australia
Regiane Relva Romano	Professor, PhD, Centro Universitário Fieo, Brazil
Rongrong Cheacharoen	PhD, Chulalongkorn University, Thailand
Samad Sepasgozar	Associate Professor, PhD, University of New South Wales, Australia
Shamila Haddad	PhD, University of Sydney, Australia
Siliang Yang	PhD, Leeds Beckett University, UK
Tobi Morakinyo	Assistant Professor, PhD, University College Dublin, Ireland
Waqas Ahmed Mahar	Professor, PhD, BUITEMS, Pakistan
Zhikang Bao	Assistant Professor, PhD, Heriot-Watt University, UK

Local Organizing Committee Members

Bin Jia	Professor, PhD, Southwest University of Science and Technology, China
Chuntao Zhang	Professor, PhD, Southwest University of Science and Technology, China
Dachuan Wang	Asso/Professor, PhD, Southwest University of Science and Technology, China
Mingying Zeng	Professor, PhD, Southwest University of Science and Technology, China
Xiaowei Shang	Lecturer, Southwest University of Science and Technology, China
Yangli Li	Lecturer, PhD, Southwest University of Science and Technology, China
Yisha Liu	Lecturer, Southwest University of Science and Technology, China
Yu Wang	Asso/Professor, PhD, Southwest University of Science and Technology, China

Student Organizing Committee Members

Mingqiang Yin	Chongqing University, China
Lu Che	Southwest University of Science and Technology, China
Erli Zeng	Southwest University of Science and Technology, China; University of Tokyo, Japan

Reviewers

Regiane Relva Romano	Smart Campus Facens, Facens University, Sorocaba, Brazil
Shamila Haddad	School of Architecture, Design and Planning, The University of Sydney, Australia
Siliang Yang	School of Built Environment, Engineering and Computing, Leeds Beckett University, UK; Mott Macdonald, UK
Simei Wu	School of Management, Xi'an University of Architecture and Technology, China
Siqi He	College of Landscape Architecture, Nanjing Forestry University, China
Waqas Ahmed Mahar	Faculty of Applied Sciences, Université de Liège, Belgium; Faculty of Architecture and Town Planning (FATP), Aror University of Art, Architecture, Design and Heritage, Pakistan
Weicong Fu	College of Landscape Architecture, Fujian Agriculture and Forestry University, China
Wen Liu	School of Architecture and Urban Planning, Shenzhen University, China
Wenlong Li	School of Architecture and Urban Planning, Beijing University of Civil Engineering and Architecture, China
Xiao Liu	School of Architecture, South China University of Technology, China
Xiao Ouyang	Hunan Institute of Economic Geography, Hunan University of Finance and Economics, China
Xiaohong Chen	Energy Accelerator Programme, Build Planet Zero Ltd, UK
Xin Dong	School of Architecture, South China University of Technology, China
Xuan Ma	School of Architecture, Chang'an University, China
Xuecheng Fu	School of Architecture, South China University of Technology, China
Yanan Liu	College of Architecture and Urban Planning, Chongqing Jiaotong University, China
Yang Liu	Zhongnan Hospital of Wuhan University, Wuhan University, China; School of Economics and Management, Wuhan University, China
Yangli Li	School of Civil Engineering and Architecture, Southwest University of Science and Technology, China

Keynote and Invited Speakers

Keynote Speakers

Deo Prasad, Scientia Professor, AO FTSE
School of Built Environment, University of New South Wales, Australia
Moving beyond sustainability in the built environment—Decarbonisation, resilience & regenerative

Bin Lyu, Professor
Department of Urban and Regional Planning, School of Environment, Peking University, China
Exploration of multi scale spatial adaptability design paths for climate change

Min Zhao, Professor
College of Architecture and Urban Planning, Tongji University, China
Urban renewal and design in the stock era

Tian Chen, Professor
School of Architecture, Tianjin University, China
Climate adaptive ecological city design and disaster prevention planning

Wen Hu, Professor
School of Architecture and Urban Planning, Chongqing University, China
Building a front-end architectural design strategy for urban ventilation and energy conservation

Invited Speakers

Ali Cheshmehzangi, Professor
Center for Innovation in Teaching, Learning, and Research, Qingdao City University, China;
School of Architecture, Design and Planning, University of Queensland, Australia
Sustainable urban design pathways in facing climate change impacts on cities

Samad Sepasgozar, Associate Professor
School of Architecture and Environment, University of New South Wales, Australia
AI agents and digital twin applications for shaping sustainable and smart buildings and cities

Yonghong Liu, Professor
Center for Numerical Forecasting of Earth System, China Meteorological Administration, China
Construction of urban ventilation corridors and local climate effects

Wei Zhao, Professor
School of Architecture and Environment, Sichuan University, China
Exploration of urban design from the perspective of fine governance of urban space

Ziyu Tong, Professor
School of Architecture and Urban Planning, Nanjing University, China
Urban form and climate at the micro scale

Linglan Bi, Professor
School of Architecture, Southwest Jiaotong University, China
Construction of park city design technology system to cope with climate change

Yu Dong, Professor
School of Architecture, Harbin Institute of Technology, China
Health risks of urban and rural community residents under climate change

Long Zhou, Associate Professor
Graduate School, City University of Macau, Macau, China
Urban form and climate resilience—research and reflection based on Macau, China

Mingxuan Chen, Senior Research Scientist
Institute of Urban Meteorology, China Meteorological Administration, China
Refined forecasting technology system and observation experiments for the megacity of Beijing

Mingying Zeng, Professor
School of Civil Engineering and Architecture, Southwest University of Science and Technology, China
Spatiotemporal characteristics and planning response strategies of urban and rural resilience in climate change sensitive areas

Jianwei Li, Professor
School of Urban and Environmental Studies, Northwestern University, China
Land mixed use: measurement, characteristics, and impacts

Xu Li, Professor
School of Architecture and Urban Planning, Chongqing University, China
Smart analysis of building a living environment adapting to regional climate: Taking the Chengdu-Chongqing region as an example

Chunping Miao, Professor
School of Architecture, Chang'an University, China
Air quality assessment of urban blocks: Methods, distribution, and drivers

Bin Cheng, Professor
School of Civil Engineering and Architecture, Southwest University of Science and Technology, China
Impact of urban microclimate environment on the thermal comfort of open spaces

Contents

Decarbonization and Carbon Neutral Solutions

Sustainable Land Use and Management

Urban Heat Mitigation and Adaptation

Urban Environmental Quality Assessment and Improvement

Sustainable Urban and Rural Planning

Resilience to Climate and Health Challenges

Editors and Contributors

About the Editors

Bao-Jie He is a Professor of Urban Climate and Sustainable Built Environment at Chongqing University. He is the leader of Centre for Climate-resilient and Low-Carbon Cities at Chongqing University and an Honorary Research Fellow of NERPS at Hiroshima University. Prior to these, He worked at UNSW Sydney as a PhD candidate, Research Associate, and Postdoc Research Fellow. His research is on heat-resilient urban planning and design and net-zero carbon built environment. He has published more than 150 peer-reviewed papers in high-ranking journals and delivered more than 30 invited talks in reputable conferences/seminars. He has been involved in several large research projects on urban climate and built environment in China and Australia. He has been invited to act as Associate Editor, Topic Editor-in-Chief, Leading Guest Editor, Editorial Board Member, Conference Chair, Sessional Chair, Scientific Committee by a variety of reputable international journals and conferences. He received the received the titles of the Most Cited Chinese Researchers in 2024, Highly Cited Researcher (Clarivate) in 2022 and 2023, the Sustainability Young Investigator Award in 2022, the Green Talents Award (Germany) in 2021, and National Scholarship for Outstanding Students (China) in 2019. Baojie is a World's top 2% Scientist (Stanford University) from 2020 onwards.

Scientia (Distinguished) Professor Deo Prasad AO FTSE is an international authority on the sustainable built environment, working at UNSW Sydney, Australia. Professor Prasad currently is the CEO of the NSW Decarbonisation Innovation Hub exploring next generation technologies, tools and systems for a more sustainable, regenerative, resilient and healthy cities. Professor Prasad previously was the CEO of the Co-operative Research Centre for Low Carbon Living based at the University of New South Wales in Sydney. Professor Prasad has in excess of 30 years of expertise and is a Fellow of the Australian Academy of Technological Sciences and Engineering. He is also a Fellow of the Royal Australian Institute of Architects, and one of the Ambassadors for Sydney (Business Events, NSW Govt) with special focus on promoting Sydney as a destination for high end environmental-scientific events. Professor Prasad has supervised more than 100 masters students and more than 80 higher degree doctoral students – his students' work places have varied from the 'White House' energy initiative in USA to the Energy Directorate in the European Union to many professors and other professionals in around 25 countries. Professor Prasad has published in excess of 400 refereed publications including ten books in the decarbonization and sustainable built environment.

Li Yan graduated from the Urban and Rural Planning Department of Chongqing University in 2021 with a PhD Degree. She is mainly engaged in research on urban design, climate and urban planning, health guidelines for public space, and evaluation and intelligent optimization of urban ventilation corridors. In the past five years, she has published one monograph, more than 10 academic papers in domestic and international academic journals, participated in the compilation of one high-quality planning textbook for the "14th Five-Year Plan", and obtained two software copyrights related to the research. At the same time, she focuses on natural ventilation in cities and applies research results to specific practical projects, achieving good results and receiving recognition from experts in the relevant research field. She has participated in six urban planning and design practices.

Ali Cheshmehzangi is a Full Professor and Head of the School of Architecture, Design and Planning at the University of Queensland. He has been a World's top 2% field leader since 2022, recognised by Stanford University. He is among the top 30 global scholars in urban sustainability research area. Ali is internationally known for his scholarly contribution to climate resilience and sustainable urbanism research. Prior to joining UQ, Ali held several strategic leadership and senior managerial roles, such as Vice-President for International Engagement and Global Partnership, Founding Director/Head of the Center for Innovation in Education and Research, Head of the Department of Architecture and Built Environment, Founding Director of the Urban Innovation Lab, Director of a university-wide Teaching and Learning platform, Director of International Research Network for Rural and Urban Development, Head of Research Group for Sustainable Built Environment, co-director of university-wide research priority areas, Director of Center for Sustainable Energy Technologies, and Interim Director of Digital Design Lab. So far, Ali is active in research and has published over 500 journal papers, articles, conference papers, book chapters, and reports. He also has 26 other academic books, three of which have received awards at the national, provincial, and municipal levels.

Gloria Pignatta is a Scientia Lecturer and City Futures Research Centre (CFRC) Fellow in the School of Built Environment, Faculty of Arts, Design and Architecture (ADA) at UNSW Sydney, focusing on building energy efficiency and sustainability in the built environment. She is the head of the Advanced Materials and Building Physics (AMBP) research group, one of the members of High Performance Architecture Research Cluster, and Investigator at the Material and Manufacturing Institute (MMFI) at UNSW. She is a Civil Engineer with a Ph.D. in Energy Engineering from the University of Perugia, Italy. In 2017, she worked as a Postdoctoral Associate and Building and Construction expert at the Singapore-MIT Alliance for Research and Technology (SMART), in Singapore. In 2016, she was a Postdoctoral Fellow at the Inter-University Research Center on Pollution and Environment "Mauro Felli" (CIRIAF), University of Perugia, Italy. Her research collaboration network touches different countries: Italy, Cyprus, Greece, Denmark, UK, France, Germany, Spain, Singapore, Sweden, USA, Canada, India, China, and Australia.

Contributors

Karam M. Al-Obaidi Department of the Natural and Built Environment, College of Social Sciences and Arts, Sheffield Hallam University, Sheffield, UK

Akram Ahmed Noman Alabsi College of Architecture and Urban Planning, Fujian University of Technology, Fuzhou, China

Shady Attia Sustainable Building Design Lab, Deptartment of UEE, Faculty of Applied Sciences, University of Liège, Liège, Belgium

Xianyun Cai College of Architecture and Urban Planning, Chongqing Jiaotong University, Chongqing, China

Jing Chen Southwest University of Science and Technology, Mianyang, China

Jinmin Chen College of Forestry and Landscape Architecture, South China Agricultural University, Guangzhou, China

Shuang Chen Institute of Fluid Physics, China Academy of Physics Engineering, Mianyang, China

Chen Cheng Chinese Academy of Meteorological Sciences, Beijing, China

Xiang Cheng Centre for Climate-Resilient and Low-Carbon Cities, School of Architecture and Urban Planning, Key Laboratory of New Technology for Construction of Cities in Mountain Area, Ministry of Education, Chongqing University, Chongqing, China

Ali Cheshmehzangi School of Architecture, Design and Planning, University of Queensland, Brisbane, QLD, Australia

Qing-Wen Deng College of Civil Engineering and Architecture, Southwest University of Science and Technology, Mianyang, China

Bart Julien Dewancker Faculty of Environmental Engineering, Kitakyushu University, Fukuoka, Japan

Liang Dong School of Architecture, Huaqiao University, Xiamen, China

Lili Dong School of Architecture and Urban Planning, Chongqing Jiaotong University, Chongqing, China

Wei Dong China Meteorological Administration, Liaoning Branch of Cadre Training College, Shenyang, China

Feng Du College of Architecture and Urban Planning, Fujian University of Technology, Fuzhou, China

Hu Du School of Civil Engineering and Built Environment, Liverpool John Moores University, Liverpool, UK

Xiaoge Du School of Civil Engineering and Architecture, Southwest University of Science and Technology, Mianyang, China

Xingzhou Fan Institute of Fluid Physics, China Academy of Physics Engineering, Mianyang, China

Xiaoyi Fang Chinese Academy of Meteorological Sciences, Beijing, China

Wenyan Feng School of Civil Engineering and Architecture, Southwest University of Science and Technology, Mianyang, China

Juanlin Fu School of Civil Engineering and Architecture, Southwest University of Science and Technology, Mianyang, China

Kar Kheng Gan TCL Studio, Adelaide, SA, Australia

Yuejing Gao School of Architecture and Civil Engineering, Xi'an University of Science and Technology, Xi'an, Shaanxi, China

Qiang Gong Shenyang Regional Climate Center, Shenyang, China

Yunyi Guo Institute of Fluid Physics, Academy of Physics Engineering, Mianyang, China

Yunyi Guo Institute of Fluid Physics, China Academy of Physics Engineering, Mianyang, China

Guoyang Hai School of Civil Engineering and Architecture, Southwest University of Science and Technology, Mianyang, China

Aprilia Nurul Hanissa Architecture Study Program, Universitas Pendidikan Indonesia, Bandung, Indonesia

Xuejun Hao School of Environmental and Energy Engineering, Beijing University of Civil Engineering and Architecture, Beijing, China

Bao-Jie He School of Architecture and Urban Planning, Key Laboratory of New Technology for Construction of Cities in Mountain Area, Ministry of Education, Chongqing University, Chongqing, China

Hongman He School of Management Science and Engineering, Guizhou University of Finance and Economics, Guiyang, China

Yanling He School of Civil Engineering and Architecture, Southwest University of Science and Technology, Mianyang, China

Mohataz Hossain Department of the Natural and Built Environment, College of Social Sciences and Arts, Sheffield Hallam University, Sheffield, UK

Mei Huang College of Architecture, Xi'an University of Architecture and Technology, Xi'an, China

Mengyuan Jia School of Architecture and Urban Planning, Beijing University of Civil Engineering and Architecture, Beijing, China

Zhengyang Jiang College of Architecture and Urban Planning, Chongqing Jiaotong University, Chongqing, China

Xiaoting Jing School of Architecture and Urban Planning, Chongqing University, Chongqing, China

Joni Jupesta Institute for the Advanced Study of Sustainability (UNU-IAS), United Nations University, Tokyo, Japan

Gon Kim Department of Architectural Engineering, Kyung Hee University, Gyeonggi, Republic of Korea

Jiaxin Li School of Civil Engineering and Architecture, Southwest University of Science and Technology, Mianyang, China

Xuan-Yan Li School of Civil Engineering and Architecture, Southwest University of Science and Technology, Mianyang, China

Xuanyan Li School of Civil Engineering and Architecture, Southwest University of Science and Technology, Mianyang, China

Yaling Li School of Civil Engineering and Architecture, Southwest University of Science and Technology, Mianyang, China

Yichen Li China Academy of Urban Planning & Design Western Branch, Chongqing, China

Qiaoyuan Lin School of Civil Engineering and Architecture, Zhejiang University of Science and Technology, Hangzhou, China

Yu Liu Xinjiang Jialian Urban Construction Planning and Design Institute Co., Ltd, Urumqi, China

Yuzhen Liu School of Civil Engineering and Architecture, Southwest University of Science and Technology, Mianyang, China

Yu Long School of Civil Engineering and Architecture, Southwest University of Science and Technology, Mianyang, China

Jie Luo School of Civil Engineering and Surveying, Southwest Petroleum University, Chengdu, China

Bingxia Ma Zitong Meteorological Bureau, Sichuan, China

Qian Ma Yuechi County Bureau of Natural Resources and Planning, Guang'an, China

Junbo Mu School of Civil Engineering and Surveying, Southwest Petroleum University, Chengdu, China

Jack Ngarambe Department of Architectural Engineering, Kyung Hee University, Gyeonggi, Republic of Korea

Muhammad Rabbani Nurlette Architecture Study Program, Universitas Pendidikan Indonesia, Bandung, Indonesia

Beta Paramita Low Carbon Building Material and Energy, University Center of Excellent, Universitas Pendidikan Indonesia, Bandung, Indonesia

Huiyun Peng School of Civil Engineering and Architecture, Southwest University of Science and Technology, Mianyang, China

Yu Peng Zhongnan Hospital of Wuhan University, Wuhan University, Wuhan, China

Indra Permana Siliwangi University, Tasikmalaya, Indonesia

Xiaolei Qiu College of Architecture and Planning, Fujian University of Technology, Fuzhou, China

Yuxin Qiu College of Forestry and Landscape Architecture, South China Agricultural University, Guangzhou, China

Yubin Rao School of Aerospace Engineering, Xiamen University, Xiamen, China

Chuan Ren Liaoning Meteorological Information Center, Shenyang, China

Hao Ren Kangping County Meteorological Bureau, Shenyang, China

Joseph Scibetta School of Built Environment, Engineering and Computing, Leeds Beckett University, Leeds, UK

Shaohang Shi School of Architecture, Tsinghua University, Beijing, China

Jiahao Tian Department of Geological Engineering, Qinghai University, Xining, China

Dachuan Wang School of Civil Engineering and Architecture, Southwest University of Science and Technology, Mianyang, China

Genhou Wang Department of Geological Engineering, Qinghai University, Xining, China

Jiawen Wang China Academy of Urban Planning and Design, Beijing, China

Jiayu Wang School of Architecture and Urban Planning, Beijing University of Civil Engineering and Architecture, Beijing, China

Jing Wang Department of Computing, College of Business, Technology and Engineering, Sheffield Hallam University, Sheffield, UK

Ting Wang College of Forestry and Landscape Architecture, South China Agricultural University, Guangzhou, China

Xiaolan Wang School of Civil Engineering and Architecture, Southwest University of Science and Technology, Mianyang, China

Yu Wang School of Civil Engineering and Architecture, Southwest University of Science and Technology, Mianyang, China

Yue Wang Shenyang Meteorological Bureau, Shenyang, Liaoning, China

Yue Wang Shenyang Meteorological Bureau, Shenyang, China

Zhengyang Wang College of Architecture and Urban Planning, Tongji University, Shanghai, China

Rihong Wen Institute of Atmospheric Environment, China Meteorological Administration, Shenyang, China

Jiang Wu College of Architecture and Urban Planning, Chongqing Jiaotong University, Chongqing, China

Jing Wu School of Architecture and Urban Planning, Chongqing Jiaotong University, Chongqing, China

Silin Wu School of Civil Engineering and Architecture, Southwest University of Science and Technology, Mianyang, China

Wangqiang Xiao School of Aerospace Engineering, Xiamen University, Xiamen, China

Sisi Xie Design Institute No. 8, China Southwest Architectural Design and Research Institute Corp. Ltd., Chengdu, China

Changrong Xiong Zhongnan Hospital of Wuhan University, Wuhan University, Wuhan, China

Yuan Xiong School of Architecture and Civil Engineering, Guizhou Minzu University, Guiyang, China

Qian-Ming Xue College of Architecture and Urban Planning, Lanzhou Jiaotong University, Lanzhou, China

Li Yan School of Civil Engineering and Architecture, Southwest University of Science and Technology, Mianyang, China

Rui Yang School of Urban Construction and Design, Guizhou Vocational College of Agriculture, Guiyang, China

Siliang Yang Mott MacDonald, Leeds, UK

Wenyue Yang College of Forestry and Landscape Architecture, South China Agricultural University, Guangzhou, China

Yifeng Yang School of Marxism, Southwest University of Science and Technology, Sichuan, China

Tong Yao School of Economics and Management, Sichuan Tourism University, Chengdu, China

Minghong Yu School of Civil Engineering and Architecture, Southwest University of Science and Technology, Mianyang, China

Ying Yu Chinese Academy of Meteorological Sciences, Beijing, China

Geun Young Yun Department of Architectural Engineering, Kyung Hee University, Gyeonggi, Republic of Korea

Mingying Zeng School of Civil Engineering and Architecture, Southwest University of Science and Technology, Mianyang, China

Linxin Zhan School of Architecture and Urban Planning, Beijing University of Civil Engineering and Architecture, Beijing, China

Yuan Zhan Zhongnan Hospital of Wuhan University, Wuhan University, Wuhan, China

Baoqing Zhang Institute of Fluid Physics, Academy of Physics Engineering, Mianyang, China

Baoqing Zhang Institute of Fluid Physics, China Academy of Physics Engineering, Mianyang, China

Haoran Zhang China Academy of Urban Planning and Design, Beijing, China

Hongliang Zhang School of Power and Mechanical Engineering, Wuhan University, Wuhan, China

Peng Zhang School of Civil Engineering and Surveying, Southwest Petroleum University, Chengdu, China

Quanping Zhang Institute of Applied Ecology, Chinese Academy of Sciences, Shenyang, China

Shuai Zhang Shenyang Meteorological Bureau, Shenyang, China

Shuo Zhang Chinese Academy of Meteorological Sciences, Beijing, China

Yuanlong Zhang Liaoning Meteorological Information Center, Shenyang, China

Yuanxu Zhang College of Architecture and Urban Planning, Chongqing Jiaotong University, Chongqing, China

Yue Zhang Mianyang Meteorological Bureau, Sichuan, China

Chunrong Zhao School of Civil Engineering and Architecture, Southwest University of Science and Technology, Mianyang, China

Fuyun Zhao School of Power and Mechanical Engineering, Wuhan University, Wuhan, China

Jingyuan Zhao School of Architecture, Chang'an University, Xi'an, Shaanxi, China

Jinpeng Zhao Sichuan Provincial Rural Economic Information Center, Chengdu, China

Xiang Zhao College of Civil Engineering and Architecture, Southwest University of Science and Technology, Mianyang, China

Ziqi Zhao Institute of Atmospheric Environment, China Meteorological Administration, Shenyang, China

Ziqi Zhao China Meteorological Administration, Institute of Atmospheric Environment, Shenyang, China

Xiang Zheng Science and Technology Service Center, Jiangxi College of Applied Technology, Ganzhou, China

Haizhu Zhou China Academy of Building Research, Beijing, China

Zhiqiang Zhou School of Architecture, Huaqiao University, Xiamen, China

Ye Zhu School of Environmental and Energy Engineering, Beijing University of Civil Engineering and Architecture, Beijing, China

Climate Predication and Impact Assessment

Impact of Meteorological Disasters and Climate Change on Agricultural Economic Growth: A Meta-analysis

Jingyuan Hu[1,2], Hongmin Ji[2], and Jinpeng Zhao[2(✉)]

[1] School of Atmospheric Sciences, Chengdu University of Information Technology, Chengdu 610225, China
[2] Sichuan Provincial Rural Economic Information Center, Chengdu 610072, China
scmb_hujy@139.com

Abstract. Existing research findings on how climate disasters and change affect agricultural economic growth are not entirely consistent. In order to explore the general law, more than 600 estimates are incorporated to analyze the sources of bias in the variability of the research results, and summarize the general pattern of climate disasters and change impacts on agricultural economic growth by using Meta-analysis. It is based on five aspects of data, including the caliber of economic statistics, meteorological variations, irrigation situations, adaptive measures adoptions, and regional development situations. The results show that: First, meteorological disasters and climate change have a significant impact on agricultural economy, and publication bias is not significant. Second, whether irrigation has a significant impact on agricultural economy, game theory should be used to plan irrigation facilities reasonably in the context of climate change. Third, extreme weather causes more significant agricultural economic losses than meteorological disasters, and future research can further refine extreme weather data indicators.

Keywords: Meteorological Disasters · Climate Change · Agricultural Economy · Meta-Analysis

1 Introduction

Meteorological disasters and climate change are issues of concern to many countries around the world and often have negative economic, social and environmental impacts [1, 2]. In recent years, the empirical literature on meteorological disasters and climate change affecting agricultural economic growth has increasing, but the research conclusions are divergent. The impact of meteorological disasters is represented by the loss at the time of the event and the loss caused by the subsequent production interruption. Liu and Yan [3] found that meteorological disasters had caused serious damage to the overall economy, meanwhile Fomby [4] and Shi [5] announced that meteorological disasters promoted or delay-promoted the agricultural economic growth. The impact of climate change refers to meteorological conditions such as temperature and precipitation affect the output benefits of farmland or the GDP of agricultural economy. Benhin [6] and

© The Author(s) 2025
B.-J. He et al. (Eds.): UCSUD 2023, LNCE 559, pp. 3–15, 2025.
https://doi.org/10.1007/978-981-97-8401-1_1

Hossain [7] consistently found a positive correlation between increased temperature and rainfall and net farmland income, while Huong [8] found the opposite conclusion when they studied the impact of climate change on the economic value of household income in northwest Vietnam.

Meta-analysis can examine the true effects of meteorological disasters and climate change on agricultural economic growth by collecting the results of empirical studies. Meta-analysis and literature review studies on the impact of meteorological disasters and climate change on the economy have been carried out [9, 10], but agricultural economy field has not been detailed discussed. Therefore, the paper conduct a Meta-analysis on impact of meteorological disasters and climate change on agricultural economic growth to discuss the true value and put forward corresponding measures to ensure the development of agricultural economy. In order to explore the impact of meteorological disasters and climate change on agricultural economy, determine the focus of future research on meteorological services for agriculture, and provide reference for the government to formulate corresponding meteorological disaster prevention and reduction policies, the paper adopts the Meta regression method, including 23 research literature and 604 parameter estimates. Based on data from five aspects: economic statistical caliber, meteorological impact, irrigation situation, adaptive measures taken, and regional development, the paper analyzes the empirical research on the impact of meteorological disasters and climate change on agricultural economy.

The rest of the paper is organized as follows: Sect. 2 reviews the specific work and the included data overview of the included empirical studies, Sect. 3 outlines the methods of significance test, asymmetry test, Sect. 4 contains the results of the meta-regression analysis, and Sect. 5 provides the conclusions and countermeasure suggestions of the literature analysis.

2   Literature Review

In empirical research, the main models used to study the impact of climate change and meteorological disasters on agricultural economy include the Cobb-Douglas production function, fixed effects model, and Ricardian model.

The C-D production function takes climate change or meteorological disasters as factors that affect economic output in conjunction with conventional economic factors such as capital, labor, and energy. The concepts of meteorological factors, economic elasticity elasticity of agricultural output to meteorological factors, and range rates of agricultural economic output to weather variability were introduced to characterize the magnitude of the impact of individual and comprehensive meteorological factors on agricultural economic output [3].

The fixed effects model is a regression analysis of panel datasets on the specific impacts of climate change and meteorological disasters on agricultural economy in different regions. This method adds regional economic control variables and lagged variables, which can simultaneously evaluate the impact of different regions and duration [2].

The Ricardian model is the most widely used model in the included studies, with 97% of the included literature using this model for regression empirical research. Since

Mendelsohn first applied the Ricardian model to empirical research on the impact of climate change on agricultural economy in 1994, to quantitatively evaluate the impact of climate environment on farmland value. Since then, a large number of empirical studies have used this model. Compared with the production function model, the Richard model does not predict the adverse effects of climate change as significantly. This is because many researches which uses the production function model does not take into account the adaptation of farmers to climate change. Even if the adaptation of farmers to irrigation, fertilization, and cultivation techniques is considered, adaptation measures such as changing crops and land use are ignored, which will overestimate the adverse effects of climate change on the agricultural economy. The basic assumption of the Ricardian model is that climate changes the production function of crops. The Ricardian climate agricultural economic model is given by

$$pL = \frac{[P_i Q_i - C_i(Q_i, R, E)]}{L_i} \tag{1}$$

where pL is the agricultural profit (farmland value or farmland net income), P_i is the price of i farmland output, Q_i is yield; C_i is the value function of all production inputs except land, R is the vector of the price of the price of production means, E is the external environmental input vector required for production, including temperature, precipitation, soil status, all factors that affect yield; L_i is the vector of labor carrier, including employment or cultivation, education, and family members.

3 Data and Method

3.1 Data

SCIE, EBSCO and CNKI were used as the main data source, and meteorological disaster, climate change, extreme weather, agricultural economy, gross agricultural product and agricultural output value were used as keywords, and the search deadline was May 1st, 2023. After eliminating duplicate and review articles, 489 articles were remained from 658 initial articles. Then filter available articles according to the following rules: First, reading the topics and abstracts of the remained articles, and picking out 93 articles related to the research question. Second, reading the full text and picking out 27 articles that the t-value or regression coefficient and standard error were counted. Third, inspecting the detailed data information and the research quality and the repeatability of the research objects and picking out 23 articles.

The t-values density distribution of all included studies is shown in Fig. 1. The figure shows that the included studies had a tendency to report positive t-values, which the positive t-values were approximately 56%. Considering the quantile range, about 4% of t-values could be considered as outliers, the average t-value of the sample before outlier removal was about 0.16, and the average t-value drops to 0.01 after outlier removal.

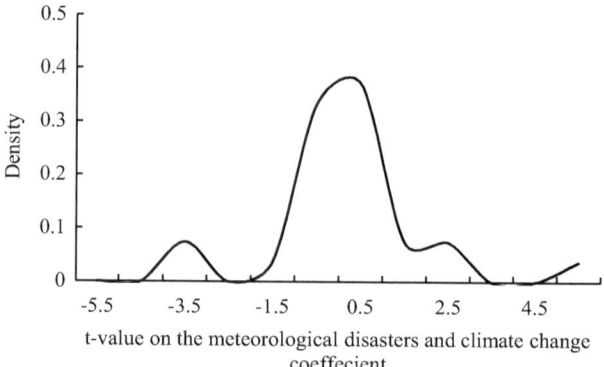

Fig.1. Density distribution of t-value on the meteorological disaster and climate change coefficient

3.2 Meta Regression Analysis Approach

The key research purposes in the study are whether exists a publication bias and whether a meaningful effect of meteorological disasters and climate change on agricultural economic growth remains. Meta Significance Test (MST) is used to evaluate the objective impact of meteorological disasters and climate change on agricultural economy as follows

$$ln\left|t_{ij}\right| = \alpha_0 + \alpha_1 lnObs_{ij} + \varepsilon_{ij} \tag{2}$$

where t_{ij} is the mean of t-values present in the j regression of i study, Obs_{ij} is the number of samples, and ε_{ij} is the meta-regression disturbance term.

The Funnel Asymmetry Test (FAT) is used to conduct further heterogeneity test, considering both the influence of publication effect and the existence of real effect as follows

$$t_{ij} = \beta_0 + \beta_1(\frac{1}{SE_{ij}}) + \varepsilon_{ij} \tag{3}$$

where t_{ij} represents the jth t-value present in study i, β_0 represents the publication bias, β_1 represents true value, and ε_{ij} is the meta-regression disturbance term.

The source of heterogeneity was analyzed using the t-value in the included literature as the explanatory variable, the equation as follows

$$t_{ij} = \gamma_0 + \gamma_1(\frac{1}{SE_{ij}}) + \gamma_k(\frac{Z_{ij}^k}{SE_{ij}}) + \varepsilon_{ij} \tag{4}$$

where the meaning of coefficients is similar to the Eq. (3), Z_{ij}^k is a vector of meta-independent variables, γ_k is the meta-regression coefficient.

3.3 Source of Heterogeneity Indicator Selection

In Meta-analysis, the observed subject was not a single study, but rather all estimates of the included studies. For example, many studies have included more than one regression when testing climate factor sensitivity. In addition, multiple meteorological variables and disaster variables were included in the same regression study. The causes of literature heterogeneity caused by subgroup analysis are used to explain the influencing factors and characteristics of each study model.

The 1th set of variables are economic indicators, including macro indicators (gross agricultural product) and micro indicators (farmland value, farmland net income and farmers' direct income). The 2nd group of variables is meteorological indicators. The impact time and results of meteorological disasters and climate change are not consistent. Meteorological indicators are divided into meteorological disasters (flood, drought, and extreme weather) and climate change (temperature, precipitation, and interaction term). The 3rd group of variables is the irrigation index, and whether the measured agricultural land is irrigated has different effects on the agricultural economy. Studies have shown that in areas without irrigation, the net income per mu compared with that without irrigation will have no significant impact on farmland value. Whether to increase irrigation measures to cope with climate change and meteorological disasters is the concern of this paper. The 4th set of variables are the indicators of meteorological disasters and climate adaptation measures, such as changing planting crops, sowing and harvesting time, and implementing disaster management. The 5th group of variables is the regional development index. The degree of economic development will affect the government subsidy policy and the education level of farmers, thus leading to the unequal change of farmers' income.

4   Results and Discussion

4.1   Results of MST and FAT

As shown in the MST results in Table 1, both the constant and the parameter are significant. In this method, the logarithm of degrees of freedom is regressed to the absolute t-value.The coefficient of the log number of observations is significant at 0.01 level, meaning that impact of meteorological disasters and climate change on agricultural economic growth is true value.

An alternative way to test is to use the FAT, which can test real effect and publication bias at the same time. The parameter of the inverse standard errors is negative but not significant, which indicates the impact of meteorological disasters and climate change on agricultural economic growth is negative but the validity is uncertain. The FAT results also showed a non-significant positive publication bias by the parameter of constant. To sum up the first result, there exists a negative impact of meteorological disasters and climate change on agricultural economic growth while the publication bias is not significant.

4.2   Results of Subgroup Analysis

In the first regression of Table 2 we control for the macro-economic and micro-economic statistical indicators used. The impact of meteorological disasters and climate change on agricultural economy is negative on the macro level (gross agricultural product) and positive at the micro level (farmland value, farmland net farm income and farmers' income),

Table 1. The result of Meta Significance Test and Funnel Asymmetry Test

| | MST | | FAT | |
| | $log|t_{ij}|$ | | t_{ij} | |
	Coefficient	z-value	Coefficient	z-value
Constant	-0.81***	-2.857	0.103	0.423
Log number of observations	0.232***	5.754		
Inverse standard erros			-0.001	-0.821
Number of observations	604		604	

Note: * means significant correlation at 0.1 level, ** means extremely significant correlation at 0.05 level, *** means extremely significant correlation at 0.01 level

which is consistent with the conclusion of Musolino [11]. When crops are impacted by meteorological disasters and climate change, farmers will transfer production losses by adjusting agricultural prices, which will lead to the growth of agricultural economy at the micro level. Therefore, the government should consider the transmission of disasters in the economic system when formulating subsidy policies, and the subsidies for agriculture should be distributed as a whole.

In meteorological indicators we control for the meteorological disasters and climate change. The impact of meteorological disasters and climate change on agricultural economic growth is negative, and the impact of meteorological disasters is greater than that of climate change. This is because the short-term impact of meteorological disasters will report more economic losses, while the negative effects of climate change is long-term and difficult to detect.

To control for the irrigation situations, we find that studies have shown that irrigation farms are more adaptable against meteorological disasters and climate change, and only rising irrigation areas have a clear positive impact on the agricultural economy [7, 12]. Analysis of the literature found greater climate vulnerability than non-irrigation farms, which is contrary to the findings above. The reason may be that irrigation farms with artificial water supply use local groundwater or surface water from other areas, and the use of water resources is not free. Meteorological disasters and climate change will increase the production cost of irrigation farms.

To control for the measures adopted or not, we include farms that tackled meteorological disasters and climate change will receive more benefits than those that did not, consistent with Ali [13] and Ngondjeb [14]. There is a complex relationship between the factors of taking measures to resist meteorological disasters and climate change and agricultural economy. The future research needs to further analyze the specific role of different measures in resisting meteorological disasters and climate change, and clarify its action mechanism on agricultural economy.

Furthermore, in regional development indicators, we control for the developed countries and developing countries. The agricultural economy of developing countries and developed countries has different responses to meteorological disasters and climate change, that is, it has a positive effect on promoting the agricultural economy of developed countries and a negative inhibitory effect on the developing world. The reason may be that the governments of developed countries provide high agricultural subsidies, and the government subsidies always have a promoting effect on the agricultural economy [15]. Therefore, government-led agricultural subsidies are indispensable in the process of dealing with meteorological disasters and climate change, and the government should make a specific analysis on how to maximize the subsidy benefits of the whole society from the perspective of game [16] (Table 2).

We further refines the meteorological data indicators to get more refined conclusions. Meteorological disasters were refined into flood, drought and extreme weather, and climate change was refined into temperature, precipitation and interaction terms, which mean that the variables in the study regression were calculated after interacting temperature and precipitation into a variable.

The specific analysis shows that: First, flood has a promoting effect on the development of agricultural economy, and the drought is harmful but not obvious. After the disaster, government subsidies and post-disaster reconstruction make farmers' income increase or remain stable rather than decrease [11]. However, extreme weather may not have caused obvious disaster reports, and most of the losses are borne by farmers themselves. Therefore, the government should pay attention to the damage caused by extreme weather in the follow-up subsidy policies and insurance design to ensure farmers' income. Second, The increase of temperature and precipitation increase both have a negative impact on agricultural economic growth, but when they rise at the same time, they will promote the economy. At present, various climate backgrounds are consistent in the prediction of future temperature rise, while the prediction of precipitation change varies greatly [17]. Therefore, under the condition of uncertain future precipitation, the reasonable construction of irrigation measures can promote the growth of agricultural economy under the general trend of climate warming (Table 3).

According to the results of the analysis and suggestions of the literature are as follows: first of all, when designing agricultural insurance and subsidy policies, the government should analyze the transmission of meteorological disasters and climate change in the economic system according to the specific local climate and economic background, and classify the impact of extreme weather into the scope of subsidies. Secondly, climate warming can increase the production capacity of irrigation farms, but it will not bring new economic income, and may even cause more economic losses than non-irrigation farms. The reasonable layout and construction of irrigation measures should be studied in detail. Finally, further research on the response mechanism of agricultural economic system to meteorological disasters and climate change measures can provide reference for the formulation of regional climate response strategies.

5   Conclusion

The paper collects relevant research as much as possible. Based on the analysis of the impact of meteorological disasters and climate change on agricultural economy, Meta regression analysis is used to explore the general laws of the impact of meteorological disasters and climate change on agricultural economy in existing literature from five

Table 2. Meta-analysis of the subgroups

	Economic statistical indicators		Meteorological indicators		Irrigation indicators		Response measures indicators		Regional development indicators	
	Coefficient	z-value	Coefficient	z-value	Coefficient	z-value	Coefficient	z-value	Coefficient	z-value
Constant	0.09	0.38	0.099	0.406	0.08	0.327	-0.066	-0.313	-0.298	-1.08
Macro-economy	-0.001	-0.852								
Micro-economy	0.001	0.103								
Meteorological disasters			-0.001	-0.812						
Climate change			-0.001	-0.188						
Irrigation farms					-0.006**	-2.219				
Non-irrigation farms					0.002	0.655				
Measures adopted							0.004	0.373		
Measures non-adopted							0.001	0.107		
Developed countries									0.017	0.758
Developing countries									-0.001	-0.856
Number of observations	604		604		604		195		604	

Note: * means significant correlation at 0.1 level, ** means extremely significant correlation at 0.05 level, *** means extremely significant correlation at 0.01 level

Table 3. Refined analysis of the meteorological indicators

	Meteorological disasters		Climate change	
	Coefficient	z-value	Coefficient	z-value
Constant	-0.876	-1.441	0.200	0.776
Flood	0.004	1.104		
Drought	0.000	-0.504		
Extreme weather	-0.010*	-1.932		
Temperature			-0.072	-1.383
Precipitation			-0.001	-0.166
Interaction term			0.179	1.132
Number of observations	42		562	

Note: * means significant correlation at 0.1 level, ** means extremely significant correlation at 0.05 level, *** means extremely significant correlation at 0.01 level

aspects: economic indicators, meteorological indicators, irrigation indicators, adaptation measures indicators, and regional development indicators. The main conclusions are as follows: First, the MST and FAT results indicate that meteorological disasters and climate change have a significant impact on agricultural economy, and publication bias is not significant. Second, whether irrigation has a significant impact on agricultural economy, game theory should be used to plan irrigation facilities reasonably in the context of climate change. Third, extreme weather causes greater agricultural economic losses than meteorological disasters, and future research can further refine extreme weather data indicators.

Acknowledgement. This paper is financially supported by the Second Tibetan Plateau Scientific Expedition and Research Program(2019QZKK0303).

Appendix

See Table A1.

Table A1. The studies included

Year published	Author	Number of observations	Number of estimates	Average t-value
1999	Mendelsohn and Nordhaus [18]	2938	32	-0.904
2003	Reinsborough [19]	267	48	0.476
2004	Liu et al. [20]	1275	74	2.534
2005	Schlenker et al. [21]	2938	8	1.674
2006	Schlenker et al. [22]	2398	30	-0.827
2007	Mendelsohn et al. [23]	1942	30	-0.002
2007	Mendelsohn and Reinsborough [24]	3195	100	0.624
2009	Molua [25]	721	16	0.617
2010	Ajetomobi et al. [12]	1200	43	0.155
2010	Wang et al. [26]	8405	64	0.433
2012	Liu et al. [27]	1079	11	-3.620
2013	Ngondjeb [14]	708	96	-0.367
2016	Yu et al. [28]	220	8	-3.261
2016	Arshad et al. [29]	360	6	-0.658
2018	Ge [30]	300	7	-0.599

(continued)

Table A1. (*continued*)

Year published	Author	Number of observations	Number of estimates	Average t-value
2018	Huong et al. [8]	1055	20	2.715
2019	Hossain [7]	396	16	0.396
2020	Ojo and Baiyegunhi [31]	360	8	0.075
2021	Ali [32]	635	20	0.039
2021	Ojo and Baiyegunhi [33]	529	10	0.007
2021	Weerasekara et al. [34]	162	9	-0.023
2023	Luo et al. [35]	528	12	1.643
2023	Jatuporn and Takeuchi [36]	1900	10	5.025

References

1. Xiao, F.J., Zhang, H.D., Wang, C.Y., et al.: Impact of climatic change on agriculture and its adaptation countermeasures in China. J. Nat. Disasters **15**(6), 327–331 (2006)
2. Weerasekara, S., Wilson, C., Lee, B., et al.: The impacts of climate induced disasters on the economy: Winners and losers in Sri Lanka. Environ. Sci. Pollut. Res. Int. **185**, 1–11 (2021)
3. Liu, T., Yan, T.C.: Main meteorological disasters in China and their economic losses. J. Nat. Disasters **20**(2), 90–95 (2011)
4. Fomby, T., Ikeda, Y., Loayza, N.V.: The growth aftermath of natural disasters. J. Appl. Economet. **28**, 412–434 (2013)
5. Shi, P.J., Ying, Z.R.: Impacts of meteorological disaster on economic growth in China. J. Beijing Normal Univ. (Nat. Sci.) **52**(6), 747–753 (2016)
6. Benhin, J.K.: South African crop farming and climate change: an economic assessment of impacts. Glob. Environ. Chang. **18**(4), 666–678 (2008)
7. Hossain, M.S., Arshad, M., Qian, L., et al.: Economic impact of climate change on crop farming in Bangladesh: an application of Ricardian method. Ecol. Econ. **164**, 106354 (2019)
8. Huong, N.T.L., Bo, Y.S., Fahad, S.: Economic impact of climate change on agriculture using Ricardian approach: a case of northwest Vietnam. J. Saudi Soc. Agric. Sci. **18**(4), 449–457 (2018)
9. Dell, M., Jones, F.B., Olken, A.B.: What do we learn from the weather? The new climate-economy literature. J. Econ. Lit. **52**(3), 740–798 (2014)
10. Tan, L., Wu, X.H., Li, L.S.: Impact of climatedisasters on economic development: a meta-analysis. Stud. Sci. Sci. **38**(2), 208–217 (2020)
11. Musolino, D.A., Antonino, M., Alessandro, M.: Does drought always cause economic losses in agriculture? An empirical investigation on the distributive effects of drought events in some areas of Southern Europe. Sci. Total. Environ. **633**, 1560–1570 (2018)
12. Ajetomobi, J.O., Abidun, A., Hassan, R.M.: Economic impact of climate change on irrigated rice agriculture in Nigeria. In: 2010 AAAE Third Conference/AEASA 48th Conference. Cape Town, South Africa (2010)
13. Ali, U., Wang, J., Ullah, A., et al.: The impact of climate change on the economic perspectives of crop farming in Pakistan: Using the Ricardian model. J. Clean. Prod. **308**, 127219 (2021)
14. Ngondjeb, Y.D.: Agriculture and climate change in Cameroon: an assessment of impacts and adaptation options. Afr. J. Sci. Technol. Innov. Dev. **5**(1), 85–94 (2013)
15. Mendelsohn, R., Reinsvorough, M.: A Ricardian analysis of US and Canadian farmland (Special Issue: Measuring climatic impacts with cross sectional analysis). Climatic Change **81**(1), 9–17 (2007)
16. Zhong, L., Nie, J.J., Yue, X.H., et al.: Optimal design of agricultural insurance subsidies under the risk of extreme weather. Int. J. Prod. Econ. **263**, 10892 (2023)
17. Huang, C., Li, N., Zhang, Z., et al.: Assessment of the economic cascading effect on future climate change in China: evidence from agricultural direct damage. J. Clean. Prod. **276**, 123951 (2020)
18. Mendelshon, R., Nordhaus, W.: The impact of global warming on agriculture: a Ricardian analysis: reply. Am. Econ. Rev. **89**(4), 1053–1055 (1999)
19. Reinsborough, M.: A Ricardian model of climate change in Canada. Can. J. Econ. **26**(1), 21–40 (2003)
20. Liu, H., Li, X.B., Fischer, G., et al.: Study on the impact of climate change on China's agriculture. Clim. Change **65**, 125–148 (2004)
21. Schlenker, W., Hanemann, W.M., Fisher, A.C.: Will U.S. Agriculture really benefit from global warming? Accounting for irrigation in the hedonic approach. Am. Econ. Rev. **95**(1), 396–406 (2005)

22. Schlenker, W., Hanemann, W.M., Fiser, A.C.: The impact of global warming on U.S. agriculture: an econometric analysis of optimal growing conditions. Rev. Econ. Stat. **88**(1), 113–125 (2006)
23. Mendelsohn, R., Basist, A., Kurukulasuriya, P., et al.: Climate and rural income. Clim. Change **81**, 101–118 (2007)
24. Mendelsohn, R., Reinsborough, M.: A Ricardian analysis of US and Canadian farmland. Clim. Change **81**, 9–17 (2007)
25. Molua, E.L.: An empirical assessment of the impact of climate change on smallholder agriculture in Cameroon. Global Planet. Change **67**, 205–208 (2009)
26. Wang, J.X., Mendelsohn, R., Dinar, A., et al.: The impact of climate change on China's agriculture. Agric. Econ. **40**, 323–337 (2009)
27. Liu, J., Xu, X.F., Luo, H.: An empirical research on the impacts of extreme weather and climate events on agricultural economic output in China. Sci. Sin. Terrae. **42**, 1076–1082 (2012)
28. Yu, Z.Y., Li, Z.Q., Gao, D.W., et al.: A method of evaluate the impacts of extreme weather on agricultural economic output quantitatively. Clim. Change Res. **12**(2), 147–153 (2016)
29. Arshad, M., Kachele, H., Krupnik, T.J., et al.: Climate variability, farmland value, and farmers' perceptions of climate change: implications for adaptation in rural Pakistan. Int J Sust Dev World **11**, 1–13 (2016)
30. Ge, J.Y.: Impact of Climate Change on Peasants' Agricultural Income in China's Ecologically Vulnerable Areas. Beijing Institute of Technology, China (2018)
31. Ojo, T.O., Baiyegunhi, L.J.S.: Determinants of climate change adaptation strategies and its impact on the net farm income of rice farmers in south-west Nigeria. Land Use Policy (2019)
32. Ali, U.: Farmers' choice of adaptation strategies and its economic impact in Pakistan. Northwest A&F University, China (2021)
33. Ojo, T.O., Baiyegunhi, L.J.S.: Climate change perception and its impact on net farm income of smallholder rice farmers in South-West, Nigeria. J. Cleaner Prod. **310**, 127373 (2021)
34. Weerasekara, S., Wilson, C., Lee, B., et al.: The impacts of climate induced disasters on the economy: Winners and losers in Sri Lanka. Ecol. Econ. **185**, 107043 (2021)
35. Luo, H., Liu, J., Wang, L., et al.: Impact assessment of climate change and extreme meteorological disasters on agricultural economic output in Ningxia. J. Catastrophol. **38**(2), 74–78 (2023)
36. Jatuporn, C., Takeuchi, K.: Assessing the impact of climate change on the agricultural economy in Thailand: an empirical study using panel data analysis. Environ. Sci. Pollut. Res. **30**, 8123–8132 (2023)

Impact of Climate Change and Human Activities on Ecosystem Health in the Poyang Lake City Group, China

Yaoyao Chen[1,2], Xiang Zheng[3], Linghua Duo[1,2(✉)], Yi Zeng[1,2], and Xiaofei Guo[1,2]

[1] Key Laboratory of Mine Environmental Monitoring and Improving Around Poyang Lake of Ministry of Natural Resources, East China University of Technology, Nanchang 330013, China
dlh_123@ecut.edu.cn
[2] School of Surveying and Geoinformation Engineering, East China University of Technology, Nanchang 330013, China
[3] Science and Technology Service Center, Jiangxi College of Applied Technology, Ganzhou 341000, China

Abstract. Ecosystem health refers to a state where the interactions and relationships among the internal components of an ecosystem and its external environment are in a balanced and stable condition. A healthy ecosystem can maintain its structure and functions, possessing the capacity for self-regulation, self-repair, and resilience to external disturbances. This study constructed an assessment framework for ecosystem health based on the PSR model. It evaluated the ecosystem health of the PLCG in 2010, 2015, and 2020, exploring the impacts of climate change and human activities. The results indicate a declining trend in the EHI from 2010 to 2020, with most regions falling into the medium level. Precipitation and land use were identified as dominant factors influencing ecosystem health, and the interactions between any two influencing factors enhanced the variability in ecosystem health. This research contributes to expanding scientific understanding of the Earth's ecosystems, providing a foundation for further developments in the field of ecology. By deeply understanding the influencing factors on ecosystem health, it can accurately identify potential ecological risks and promptly take measures to prevent or mitigate these risks, providing important scientific basis for formulating environmental policies and management strategies.

Keywords: Ecosystem health · PSR model · Geographical detector

1 Introduction

In today's world, climate change poses extremely serious and escalating challenges. The variability and instability of the climate not only impact the Earth's atmospheric and oceanic systems but also profoundly shape and threaten the overall health of ecosystems. Simultaneously, human activities exacerbate this challenge, particularly through excessive industrialization, overexploitation of natural resources, and the emission of greenhouse gases. This is not merely an environmental concern but a fundamental issue

B.-J. He et al. (Eds.): UCSUD 2023, LNCE 559, pp. 16–26, 2025.
https://doi.org/10.1007/978-981-97-8401-1_2

related to human survival. Greenhouse gas emissions from industrialization and large-scale agricultural activities not only contribute to global warming but also directly cause air and water pollution. These pollutants pose a severe threat to the ecological health of plants, animals, and microorganisms, disrupting the food chain and ecological balance. Extreme weather events resulting from climate change increase the frequency of natural disasters, directly affecting people's habitation, food supply, and water resource availability [1]. Studying the impacts on ecosystem health is of paramount importance for maintaining earth's ecological balance, ensuring sustainable resource utilization, and effectively addressing climate change.

Ecosystem health refers to the biotic and abiotic factors operate in a coordinated manner to maintain biodiversity, ecological processes, and environmental quality. By delving into the interactions between species, the functioning of ecological processes, and the mechanisms of resource cycling, the intrinsic complexity of ecosystems is revealed. By monitoring indicators such as species distribution, quantity, and genetic diversity, the level of biodiversity within the ecosystem is assessed to understand the health status of the ecosystem [2]. It also encompasses the study of human activities' impact on ecosystems. Anthropogenic activities, such as urbanization, industrialization, and agricultural expansion, may lead to land use changes, habitat loss, and overexploitation of resources [3]. Researchers can use models and scenario analyses to predict the potential effects of climate change, population growth, and other factors on ecosystems, providing insights for future conservation and management strategies [4]. Through the study of ecological characteristics, water resource features, environmental characteristics, and ecological pressures in lake and marine watersheds, the introduction of diversity indices to characterize nutrient levels and the distribution of planktonic organisms can reveal the characteristics of the health of aquatic ecosystems [5]. The research contributes to guiding environmental policies and facilitating more effective global responses to the threats ecosystems face.

Developing a thorough and efficient set of indicators for assessing ecological health is essential to ensure precise quantitative evaluations of ecosystem well-being. Several frameworks for ecosystem health assessment exist, such as SENCE model, PSR model, DSR model, DPSIR model. PSR model as a framework for assessing ecosystem health, offers advantages such as systemic thinking, multidimensional evaluation, policy guidance, wide applicability, and ease of understanding and communication. In contrast, other assessment systems like the DPSIR model may emphasize human-driven factors and ecosystem impacts, yet their complex structures and higher data requirements may increase evaluation complexity. While the VOR model considers ecosystem status comprehensively, it necessitates a more intricate framework and implementation process, along with higher data and information demands. Additionally, methods like ecological footprint and ecosystem services assessments provide intuitive measures of ecological pressure and human utilization, yet they may overlook some ecosystem dynamics and complexities. Overall, the PSR model stands out for its comprehensiveness, simplicity, and practicality, offering advantages in the field of ecosystem health assessment.

Research on ecosystem health has become relatively mature, with many scholars applying different evaluation models and indicator systems to assess the health of ecosystems in various regions. For example, using administrative divisions at the municipal level as evaluation units, researchers have constructed an indicator system for evaluating the health of coastal wetland ecosystems in Guangxi and Guangdong provinces using the PSR model, obtaining scores for each aspect and overall health status through weighted aggregation [6]. Additionally, for the river ecosystem of the Yangtze River Basin, researchers have developed an evaluation indicator system from the perspectives of integrity, stability, and sustainability, and employed scoring methods for evaluation [7]. Regarding the water ecosystem of Poyang Lake, a comprehensive indicator system has been utilized to assess its health status across multiple dimensions [8]. As urban clusters are densely populated, resource-intensive, and economically prosperous regions, the health of their ecosystems directly impacts residents' quality of life, economic development, and the sustainability of the ecological environment. A deeper understanding of the ecosystem health of urban clusters can reveal bottlenecks and vulnerabilities in ecosystem carrying capacity, prompting the formulation of corresponding protection and management strategies. Poyang Lake, as one of China's important wetland ecosystems, possesses significant ecological functions but faces numerous ecological pressures, including forest destruction, water pollution, and land use changes.

Given the escalating environmental challenges globally, the study of ecosystem health has assumed greater urgency. The expanding footprint of climate change and human activities has wrought widespread and profound repercussions on ecosystems worldwide. Focused on the Poyang Lake City Group (PLCG), this research aims to: (1) Develop a PSR model to gauge ecosystem health in the PLCG; (2) Analyze the changes in ecosystem health over space and time between 2010 and 2020; (3) Probe into the impacts of climate change and human activities on ecosystem health in the area. This study is dedicated to thoroughly comprehending and evaluating the present condition of ecosystem health, along with its underlying drivers. By furnishing scientific evidence and proposing solutions, it endeavors to foster sustainable development in the region and beyond.

2 Materials and Methods

2.1 Study Area

The PLCG is situated in the northern region of Jiangxi Province (Fig. 1). The PLCG holds distinction as one of the most advanced regions within Jiangxi Province. Within this urban landscape, the climate exhibits characteristics typical of a subtropical monsoon climate, characterized by lengthy winters and summers, brief springs and autumns, copious rainfall year-round, and generous sunlight. These climatic attributes contribute to the region's ecological dynamics and overall environmental conditions, shaping its socio-economic landscape and urban development trajectory. Geographically, the PLCG is predominantly hilly terrain, featuring extensive basins and valleys. With its rich natural resources and advantageous geographical location, this urban agglomeration has stood out, laying a solid foundation for the prosperity and sustainable development of the region [9]. However, with the acceleration of urbanization, the ecosystem has suffered severe damage in the PLCG. The water quality of the lake deteriorates, wetland ecosystems are destroyed, biodiversity is threatened. Additionally, urban discharges such as sewage and garbage directly or indirectly exacerbate environmental problems.

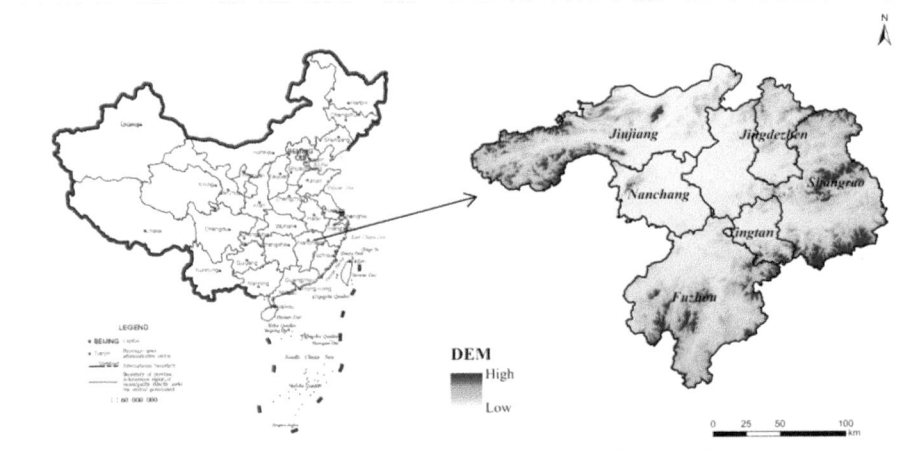

Fig.1. Study area.

2.2 Data Sources

The data used in this study can be found in Table 1. Carbon storage was calculated from the InVEST model [10]. A long-term night lights dataset for China was obtained by preprocessing, calibrating, and fusing DMSP/OLS Version 4 Non-radiance Calibrated Annual Average Nighttime Light Intensity data, NPP/VIIRS monthly data, and EVI data [11].

Table 1. Data sources.

Specific data	Data source
Administrative division	Resource and Environmental Science and Data Center of the Chinese Academy of Sciences (http://www.resdc.cn/)
Land use (LU)	
Digital Elevation Model	
Temperature (TEM)	
Precipitation (PRE)	
Evaporation (EVP)	
Population Density (POP)	
Gross Domestic Product (GDP)	
Normalized Difference Vegetation Index	
Air Quality Index	China National Environmental Monitoring Centre (http://www.cnemc.cn/)
Urbanization Rate	China Statistical Yearbook, Statistical yearbooks of different provinces
Scenic Spots	CnOpenData (https://www.cnopendata.com/)

2.3   Ecosystem Health Assessment Framework

The PSR model, highlighting the relationships between pressure, state and response, provides a holistic perspective on ecosystem health formation and evolution [12]. Reflecting the characteristics of the current PLCG ecosystem, a comprehensive evaluation index system was established, incorporating pressure, state, and response layers, totaling six indicators. Utilizing the Analytic Hierarchy Process (AHP), weights for each indicator were determined, as outlined in Table 2 [13].

Table 2. EHI Weightings.

target layer	Criterion layer	Indicator layer	Weights
Ecosystem health index (EHI)	Pressure	Urbanization Rate P1	0.178
		Night Lights P2	0.229
	State	Air Quality Index S1	0.153
		Scenic Spots S2	0.130
	Response	Normalized Difference Vegetation Index R1	0.163
		Carbon Storage R2	0.147

The formula for calculating EHI is as follows [14]:

$$EHI = \sqrt[3]{P \times S \times R} \tag{1}$$

In the equation, P, S, and R respectively stand for the baseline index, pressure index, state index, and response index. The formula is as follows:

$$P = \sum\nolimits_{P=1}^{n} W_i \times X_i \tag{2}$$

$$S = \sum\nolimits_{S=1}^{n} W_i \times X_i \tag{3}$$

$$R = \sum\nolimits_{R=1}^{n} W_i \times X_i \tag{4}$$

n represents the number of evaluation indicators, W_i and X_i respectively denote the weight coefficient and the normalized value of the i evaluation indicator.

2.4   Geographical Detector

Geographical Detector is utilized in this study to examine geographic phenomena and spatial data. It is particularly valuable for investigating driving forces and conducting factor analysis [15].

Factor detection helps unveil spatial heterogeneity, indicating that the effects of factors vary across different geographic locations. The q-statistic is employed to quantify the impact of explanatory variables, ranging from 0 to 1. The formula is expressed as follows [16]:

$$q = 1 - \frac{\sum_{n=1}^{F} N_n \sigma_n^2}{N \sigma^2} \tag{5}$$

In the equation: $n = 1, \ldots, F$ represents the strata for variable Y or factor X; N_n and N are the number of units in local; σ_n^2 and σ^2 are the variance of units.

Through interaction detection, it is capable of delving into the spatial interactions among different factors, thereby providing a more comprehensive understanding of the spatial distribution patterns of geographical phenomena [17].

3 Result

3.1 Spatiotemporal Changes of Ecosystem Health

This study used factors such as urbanization rate (UR), night lights (NL), air quality indexs (AQI), senic spots (SS), normalized difference vegetation index (NDVI), and carbon storage (CS) as variables to derive the EHI of the PLCG (Fig. 2). Overall, the EHI of the PLCG exhibited significant spatial distribution differences. In 2010, the EHI exhibited a pattern of elevated values in the southeast and diminished values in the northwest, indicating that the majority of areas boasted comparatively robust ecosystems, especially in the cities of Shangrao, Yingtan, and Fuzhou, where the ecosystem health was higher than in other regions. By 2015, the southern regions had a higher EHI compared to the northern regions, and Fuzhou became the main concentration area with a higher EHI. In 2020, the EHI in the eastern and western regions had shown relatively higher values. Specifically, the cities of Jiujiang, Shangrao, and Fuzhou still maintained higher EHI values than other areas.

This categorization was used to obtain the normalized results for the EH of the PLCG for the years 2010, 2015, and 2020. Overall, the EH of the PLCG is predominantly at medium and higher levels, but there are significant differences among different regions. During the period from 2010 to 2020, the high level EH areas in the northern region showed a shrinking trend, while those in the southern region experienced an initial expansion followed by a reduction. In the western region, the high-level EH areas continued to expand, whereas in the eastern region, they exhibited a trend of initial expansion followed by reduction. Overall, the EH of the PLCG shows a trend towards medium levels and indicates a declining trend. This suggests an imbalance in the ecosystem of this region between internal and external factors, necessitating further measures to maintain ecological balance.

By conducting an area-based analysis of the EHI of the PLCG (Fig. 3). In 2010, regions with a higher level constituted the largest proportion of the total area, amounting to 37.00%, followed by high-level areas at 29.13%. By 2015, areas with a medium level accounted for the highest proportion, reaching 28.22%, followed by regions with a higher level at 24.42%. In 2020, areas with higher level comprised the largest proportion at 32.14%, while medium level accounted for 28.61%. Throughout this period, the area of regions with low levels of EH continuously increased, accounting for 4.50%, 5.78%, and

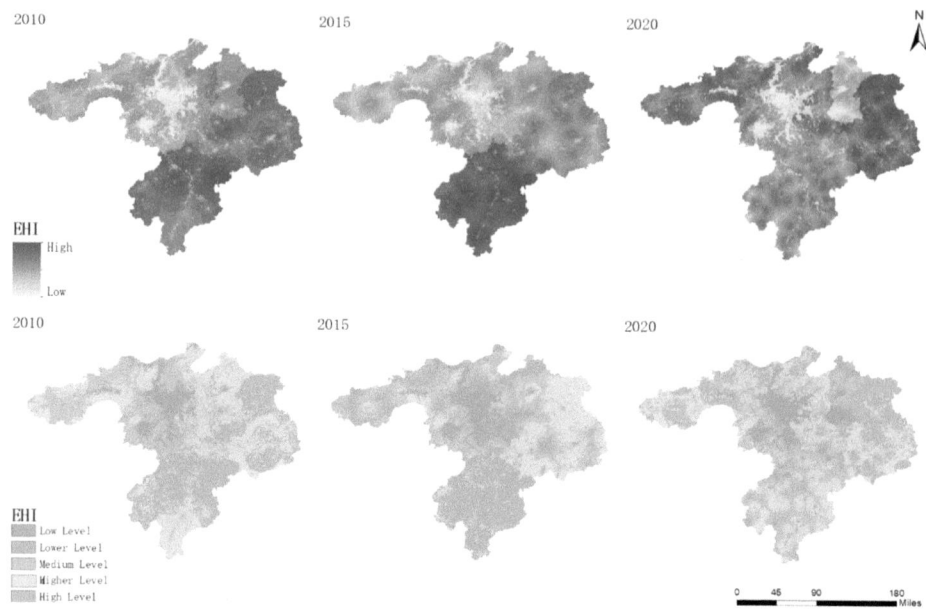

Fig. 2. Spatial distribution of EHI in the PLCG.

7.67%, respectively. In contrast, the area of regions with high levels of ecosystem health showed a decreasing trend, accounting for 29.13%, 21.37%, and 21.44%. Overall, the shifting dynamics of EH in the PLCG suggest a gradual decline in regions with higher and high levels, accompanied by a relative increase in areas with medium and low levels. It is essential to place greater emphasis on and implement appropriate ecological conservation measures to enhance sustainable resource utilization and alleviate the adverse impacts on the ecosystem.

Fig. 3. Changes of different ecosystem health levels.

3.2 Impacts of Climate Change and Human Activities on Ecosystem Health

By employing a factor detector to assess the influence of each indicator on PLCG ecosystem health, the findings revealed notable impacts, evidenced by p-values of 0 for all indicators (Table 3). In 2010, PRE and LU had a significant impact on EH, q-values is 0.46852 and 0.44894. By 2015, LU and POP emerged as the primary influencing factors, with q-values of 0.25291 and 0.18984. Meanwhile, in 2020, LU and TEM exhibited a more pronounced impact on EH, with q-values of 0.33595 and 0.17940, respectively. Additionally, variations in the impact of different environmental indicators across different periods underscore the importance of developing corresponding strategies for ecological protection and management based on temporal and spatial changes.

Table 3 q value statistics.

Indicators	2010			2015			2020		
	q	p	Order	q	p	Order	q	p	Order
TEM	0.06466	0	6	0.09171	0	4	0.17940	0	2
PRE	0.46852	0	1	0.11300	0	3	0.11436	0	3
EVP	0.07243	0	5	0.07628	0	5	0.03997	0	6
LU	0.44894	0	2	0.25291	0	1	0.33595	0	1
POP	0.12346	0	3	0.18984	0	2	0.08141	0	4
GDP	0.10593	0	4	0.05459	0	6	0.04618	0	5

By exploring interactions among the six factors, this study analyzes their collective impact on ecosystem health(Fig. 4). The findings unveiled dynamic patterns: in 2010, the PRE and LU had the most significant impact on the EH of the PLCG, with a q-value of 0.70810. By 2015, the PRE and POP became dominant, with a q-value of 0.45597. Meanwhile, in 2020, the TEM and LU had the most significant impact on EH, with a q-value of 0.43472.These results highlight varying interaction patterns among factors across different years, with a trend toward bilinear or nonlinear enhancement, underscoring that the interaction of any two factors augments explanatory power for EH.

In summary, the study indicates that from 2010 to 2020, the intensity of interaction between climate change and human activity factors surpassed internal interactions of each factor.

4 Discussion

4.1 Driving Factors of Ecosystem Health Change

Climate change serves as a fundamental influencing factor, indirectly affecting the ecosystem by altering natural environmental conditions. Meanwhile, human activities act as direct intervention factors, shaping the structure and function of the ecosystem. Climate variations alter habitats and species distributions, while human activities like

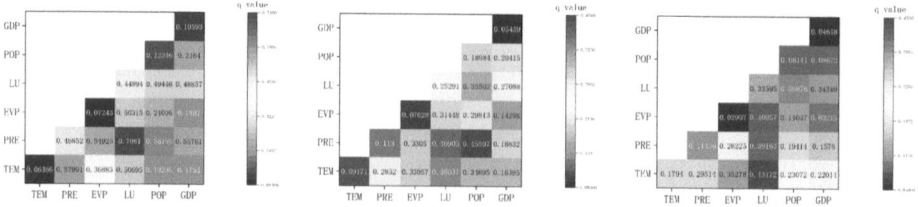

Fig.4. Interactive detector results.

deforestation and pollution further disrupt the delicate balance of ecosystems. These changes can lead to biodiversity loss, habitat degradation, and increased vulnerability to diseases and invasive species. These intertwined forces contribute to the complex dynamics observed in the PLCG's ecosystem health [18].

The research suggests that across different time spans, the cumulative influence of human activities shows a consistent upward trend, while the impact of climate change remains relatively limited. This suggests a growing significance of human activities in shaping and influencing ecosystem health within the PLCG. In 2010, PRE was considered the main factor causing changes in ecosystem health. In 2015 and 2020, LU were seen as the primary factor leading to a decrease in the EHI. This may be related to changes in LU caused by human activities such as urbanization, agricultural expansion, and industrial development [19]. As the economy develops, the urbanization process accelerates, and activities such as industry and agriculture expand, human activities' interference and impact on the environment continue to increase [20]. Climate change may have profound effects on the ecosystem over a longer time scale, but in the short term, the direct interference of human activities may be more significant [21]. LU is a major manifestation of human activities, directly affecting the ecosystem, making it more vulnerable and susceptible to external pressures [22].

4.2 Suggestions for Optimizing the Changes of Ecosystem Health

Drawing from the assessment findings of ecosystem health in the PLCG, the following recommendations are put forward to tackle the challenges posed by declining ecosystem health. Firstly, in response to the significant impact of LU on the ecosystem health, it is recommended to strengthen land use management, restrict unreasonable land development, and take measures to protect critical ecological areas to mitigate the deterioration trend of ecosystem health. Secondly, considering that PRE is a major factor affecting the health of the ecological environment to climate change, it is suggested to strengthen adaptation measures to climate change, improve water resources management, strengthen flood control measures, and enhance the ecological system's resilience to disasters. Additionally, recognizing the significant impact of interactions on the ecosystem health, it is advised to formulate more comprehensive and effective policy measures. Furthermore, to effectively respond to changes in ecosystem health, it is recommended to establish a sound monitoring and evaluation system, continuously track the changes in ecosystem health trends, and promptly identify problems and take measures. Finally, to achieve sustainable development of ecosystem health, it is essential to strengthen inter-departmental cooperation, form a joint force for ecological system protection and management, and ensure the long-term stability and sustainable development of ecosystem health.

5 Conclusions

This study utilizes the PSR model to evaluate the EHI of the PLCG for the years 2010, 2015, and 2020. Over the studied period, the low level health regions showed continuous expansion. Among human activities, LU emerges as the most significant contributor to this trend. Climate change, particularly PRE is identified as the primary factor. The study underscores the importance of interactions among various factors, highlighting their more pronounced impact compared to individual factors alone. The findings offer vital insights for policymakers, emphasizing the need for timely measures to address the deteriorating ecosystem health in the PLCG. Effective policies for the protection and management of the ecosystem are imperative, given the study's results.

References

1. Palmeiro-Silva, Y.K., Lescano, A.G., Elaine, C. et al.: Identifying gaps on health impacts, exposures, and vulnerabilities to climate change on human health and wellbeing in South America: a scoping review. Lancet Regional Health Am **26** (2023). https://doi.org/10.1016/j.lana.2023.100580
2. Lee, C.-C, Hsieh, C.-Y., Chen, C.S., Tien, C.-J.: Emergent contaminants in sediments and fishes from the Tamsui River (Taiwan): their spatial-temporal distribution and risk to aquatic ecosystems and human health. Environ. Pollut. **258** (2020). https://doi.org/10.1016/j.envpol.2019.113733
3. Shen, W., Li, Y., Qin, Y., Cheng, J.: Influencing mechanism of climate and human activities on ecosystem health in the middle reaches of the Yellow River of China. Ecol. Indic. **150**, 2023. https://doi.org/10.1016/j.ecolind.2023.110191
4. Pan, Z., He, J., Liu, D., Wang, J.: Predicting the joint effects of future climate and land use change on ecosystem health in the Middle Reaches of the Yangtze River Economic Belt, China. Appl. Geogr. **124** (2020). https://doi.org/10.1016/j.apgeog.2020.102293
5. Zhao, X., Huang, G.: Urban watershed ecosystem health assessment and ecological management zoning based on landscape pattern and SWMM simulation: a case study of Yangmei River Basin. Environ. Impact Assess. Rev. **95** (2022). https://doi.org/10.1016/j.eiar.2022.106794
6. Zhi,H., Zhijun, Z.: Health assessment of coastal wetland ecosystems in Guangxi and Guangdong. Sci. Technol. Eng. **23**(34), 14896–14904
7. Su, Y.F., Li, W.M., Ai, Z.Q., et al.: Establishment and application of the index system for health assessment of the middle and lower reaches of the Hanjiang River. Acta Ecol. Sin. **39**(11), 3895–3907 (2019)
8. Zhiyu, M., Xu, L., et al.: Assessment on ecosystem health of Lake Poyang based on a comprehensive index method. J. Lake Sci. **35**(3), 1022–1036 (2023)
9. Deng, Y., Shao, Z., Dang, C., Huang, X., Zhuang, Q.: The impact of policies on land cover and ecosystem services dynamics in the Poyang Lake Ecological Economic Zone, China. Ecol. Indic. **156** (2023). https://doi.org/10.1016/j.ecolind.2023.111169
10. Hu, J., Le, X., Wang, W., Xiong, Y., Tan, X.: Temporal and spatial evolution and prediction of ecosystem carbon storage in Jiangxi Province Based on PLUS-InVEST Model. Environ. Sci. (2023). https://doi.org/10.13227/j.hjkx.202305239

11. Zhong, X., Yan, Q., Li, G.: Development of time series of nighttime light dataset of China (2000–2020). J. Global Change Data Discov. **6** (2022). https://doi.org/10.3974/geodb.2022.06.01.V1

12. Zhao, H., Yan, X., Wang, F., Kang, P.: Assessment on ecosystem health of Sanmenxia Reservoir wetland based on PSR model. Water Resour. Prot. **36** (2020)

13. Mao, Z., Xu, L., et al.: Assessment on ecosystem health of Lake Poyang based on a comprehensive index method. Lake Sci. **35** (2023)

14. Li, W., Liu, C., Su, W., et al.: Spatiotemporal evaluation of alpine pastoral ecosystem health by using the Basic-Pressure-State-Response Framework: a case study of the Gannan region, Northwest China. Ecol. Indic. **26** (2021). https://doi.org/10.1016/j.ecolind.2021.108000

15. Wang, J., Li, X., et al.: Geographical detectors-based health risk assessment and its application in the neural tube defects study of the Heshun Region, China. Int. J Geogr. Inf. Sci. **24** (2010). https://doi.org/10.1080/13658810802443457

16. Peng, W., Fan, Z., et al.: Assessment of interactions between influencing factors on city shrinkage based on geographical detector: a case study in Kitakyushu, Japan. Cities. **131** (2022). https://doi.org/10.1016/j.cities.2022.103958

17. Wu, W., Zhang, J., Sun, Z., et al.: Attribution analysis of land degradation in Hainan Island based on geographical detector. Ecol. Indic. **141** (2022). https://doi.org/10.1016/j.ecolind.2022.109119

18. Chen, Y., Duo, L., Zhao, D., Zeng, Y., Guo, X.: The response of ecosystem vulnerability to climate change and human activities in the Poyang lake city group, China. Environ. Res. **233** (2023). https://doi.org/10.1016/j.envres.2023.116473

19. Guo, W., Jin, L., Li, W., Wang, W.: Assessing the vulnerability of grasslands in Gannan of China under the dual effects of climate change and human activities. Ecol. Indic. **148** (2023). https://doi.org/10.1016/j.ecolind.2023.110100

20. Zhang, S., Chen, Y., Zhou, X., Zhang, Y.: Climate and human impact together drive changes in ecosystem multifunctionality in the drylands of China. Appl. Soil Ecol. **193** (2024). https://doi.org/10.1016/j.apsoil.2023.105163

21. Weiskopf, S.R., Rubenstein, M.A., Crozier, L.G., et al.: Climate change effects on biodiversity, ecosystems, ecosystem services, and natural resource management in the United States. Sci. Tot. Environ. **733** (2020). https://doi.org/10.1016/j.scitotenv.2020.137782

22. Li, W., Wang, Y., Xie, S., Cheng, X.: Spatiotemporal evolution scenarios and the coupling analysis of ecosystem health with land use change in Southwest China. Ecol. Eng. **179** (2022). https://doi.org/10.1016/j.ecoleng.2022.106607

Application Method of Multi-model Fusion for Heavy Rainfall Forecast

Jing Liu[1,2,3], Chuan Ren[4(✉)], Ziqi Zhao[1], Yue Wang[3], Rihong Wen[1], Hao Ren[5], and Yuanlong Zhang[4]

[1] Institute of Atmospheric Environment, China Meteorological Administration, Shenyang 110166, China
[2] Key Opening Laboratory for Northeast China Cold Vortex Research, Shenyang 110166, China
[3] Shenyang Meteorological Bureau, Shenyang 110166, Liaoning, China
[4] Liaoning Meteorological Information Center, Shenyang 110166, China
49952706@qq.com
[5] Kangping County Meteorological Bureau, Shenyang 110166, China

Abstract. Using the hourly rainfall data in the Data as a Server, based on the North China Numerical Forecasting System (CMA-BJ), the Mesoscale Weather Numerical Forecasting System (CMA-MESO), and the Northeast Numerical Forecasting System (CMA-DB) of the China Meteorological Administration, a multi-model fusion rainfall forecasting method is proposed and tested in Laoning from 2022 to 2023. The results show that without considering the deviation within a radius of 40km, during the Northeast cold vortex rainfall process, the success rate of the fusion forecast in 2022 increased by an average of 13.08%, and the hit rate increased by an average of 0.57%. In 2023,these numbers are 25.7% and 17.5%.During the subtropical high-pressure rainfall process, the success rate of the fusion forecast in 2022 increased by an average of 16.8%, and the hit rate increased by an average of 1%.In 2023,the hit rate increased 289%. During the typhoon storm process, the success rate of the fusion forecast in 2022 increased by an average of 34.21%, and the hit rate increased by an average of 14.9%.In 2023, these numbers are 30.7% and 651%.When short-term heavy rainfall is caused by the influence of systems such as upper troughs, the success rate of the fusion forecast in 2022 increased by an average of 32.2%, and the hit rate increased by an average of 4.07%. Under the influence of systems such as the Northeast cold vortex and upper troughs, the success rate of the fusion forecast shows positive growth, while under the background of subtropical high pressure and typhoons, the fusion forecast is a negative skill compared to the original numerical model.

Keywords: Fusion forecast · Short-term heavy rainfall · Neighborhood method · Rainfall typing · Forecast verification

1 Introduction

The heavy rain in Liaoning Province is characterized by frequent occurrences, short duration, and severe disasters [1]. In 2022, the average rainfall in Liaoning was 420.6mm, 70% more than the same period in normal years, exceeding the total rainfall in normal summers, and the most in the same period in nearly thirty years [2]. Short-term

B.-J. He et al. (Eds.): UCSUD 2023, LNCE 559, pp. 27–46, 2025.
https://doi.org/10.1007/978-981-97-8401-1_3

heavy rain often occurs under the trigger of mesoscale systems [3]. Making full use of numerical models is the basis for good heavy rain forecasting. High-resolution models are better at forecasting heavy rain induced by mesoscale systems than global models. Verification is an effective way to improve the understanding and application ability of high-resolution models, and a reasonable assessment of the forecasting performance of numerical models for short-term heavy rain can help to issue red warning signals for heavy rain in advance. Verification is an effective way to improve the understanding and application ability of high-resolution models [4, 5], and a reasonable assessment of the forecasting performance of numerical models for short-term heavy rain can help to issue red warning signals for heavy rain in advance [6].

The storm system in Liaoning Province is primarily influenced by the subtropical high pressure, the Northeast cold vortex, and northward landing typhoons, among others. The Northeast cold vortex is a deep cold low-pressure system that occurs in the troposphere in the Northeast Asia region. Its abnormal activity often brings great uncertainty to summer precipitation forecasts. Under the large-scale situation where the shape, position, and strength of the subtropical high pressure are conducive to heavy precipitation in Liaoning, the abundant warm and humid air continuously transported by the low-level jet on the west side of the subtropical high pressure interacts with the dry and cold air in the upper atmosphere at the same location for a long time, providing a favorable environmental background for the occurrence and maintenance of heavy precipitation [7]. The water vapor carried by the northward typhoon itself and the water vapor transported on the south side of the vice high provide sufficient water vapor for the occurrence of extreme heavy rain; the unchanged structure during the weakening process of the northward typhoon, the convergence of the boundary layer northeast wind jet and southeast wind jet provide a strong dynamic mechanism for extreme precipitation.

In terms of fusion technology, after correcting the CMA-GD(R3) model through the optimal TS scoring correction method, it has good correction ability for precipitation of ≥ 1 mm/h and above [8]. By iteratively integrating the precipitation forecasts of CMA-GD, CMA-SH9, and ECMWF through the optimal area threshold selection scheme, the accuracy of high-resolution grid weather forecasts in Hainan Island can be effectively improved [9]. The graded correction method for rainfall based on the optimal neighborhood probability threshold effectively enhances the objective forecasting ability of precipitation, and the weather forecast is all positive skills compared to the model, with a TS score of 0.89 or above [10]. By comparing the extreme rainfall intensity, dispersion, and cumulative rainfall neighborhood score skill scores of CMA's different resolution ensemble forecasts, it is found that the extreme rainfall intensity and dispersion are closely related to the model's horizontal resolution. The higher the resolution, the higher the skill score of the extreme rainfall probability forecast, indicating that the convective scale ensemble forecast can better describe the uncertainty and extremeness of the extreme rainfall forecast [11]. The main role of improving the model's spatial resolution is to reduce the errors in the mid-low layer height field, temperature field, and horizontal wind field, improve the forecast of the intensity of the direct triggering system of precipitation in the lower troposphere, and thus improve the precipitation score of heavy to torrential rain [12].

In terms of verification technology, the current storm forecast verification uses a two-category event verification method, which has a serious "false report" and "missed report" double punishment, and does not consider the uneven distribution of storms in time and space and the comparability of forecast scores. Based on the analysis of forecasters' expected values for storm forecast scores, a new method and calculation model for storm forecast verification scores based on predictability have been designed, which can better parse TS 0 forecast scores [13]. The neighborhood ensemble probability method and the probability matching average method have much higher score skill scores for extreme precipitation than the traditional ensemble average, making up for the deficiency of the ensemble average's low ability to forecast extreme precipitation [14]. The precipitation neighborhood ensemble probability method has good application prospects, and the appropriate neighborhood probability method and neighborhood radius can obtain more reasonable precipitation probability forecast results [15].

Multi-directional evaluation can mine the added value of multi-regional models and provide forecasters with more comprehensive and objective reference information. This paper uses the hourly rainfall data in the meteorological big data cloud platform, based on the China Meteorological Administration's North China Numerical Forecasting System (CMA-BJ), the China Meteorological Administration's Mesoscale Weather Numerical Forecasting System (CMA-MESO), and the Ruitu Northeast Numerical Forecasting System (CMA-DB), proposes a multi-model fusion precipitation forecasting method, and conducts verification analysis during the main flood season in Liaoning in 2022–2023.

2   Data and Methods

2.1   Data Source

Observation Data. The data used comes from the meteorological data as a server platform "Tianqing", using hourly rainfall data from 1812 observation stations in Liaoning Province, removing outliers to ensure the continuity and accuracy of rainfall data.

Forecast Data. The forecast data comes from the operational CMA-BJ, CMA-MESO, and CMA-DB models, with the initial field of the three regional models being the NCEP-GFS model.

2.2   Research Method

Meteorological Typing. Sun Xin [1] analyzed 60 heavy rain events that occurred in Liaoning area from 1960 to 2013, pointing out that the influencing systems of Liaoning heavy rain mainly include Northeast cold vortex, rear part of the subtropical high, cyclone (including Hetao cyclone, Jianghuai cyclone, Mongolian cyclone, etc.) influence, typhoon northward, etc. This paper defines a short-term heavy rain process as one where the number of rainfall stations with a rainfall amount greater than 20mm per hour exceeds 20, and selects 50 heavy rain processes that occurred in Liaoning Province in 2022–2023, totaling 1200 times, for regional model 1-h heavy rainfall forecast verification analysis. The rainfall types are briefly referred to as Northeast cold vortex rainfall, subtropical high rainfall, typhoon rainfall, and other rainfall (upper trough, cyclone, shear line). Forecast Technology Method Competition "Inspection Scheme" issued by the China Meteorological Administration in June 2020 (hereinafter referred to as the "Inspection Scheme"), precipitation with a rainfall amount ≥ 20 mm per hour is defined as short-term heavy rainfall in the Liaoning region (Table 1).

Table 1. The number of different types of rainfall events

Num	Northeast Cold Vortex Rainfall	Subtropical High Pressure Rainfall	Typhoon Rainfall	Others (Upper Trough, Cyclone, Shear Line)
Days	15	15	9	11
Times	360	360	216	264

Inspection Validity. More detail about the modelling time period and the test time period can be seen in Table 2. In order to assess the stability of regional model forecasts, spatial inspections are conducted on 12 h to 36 h forecasts from different start times. For example, if the inspection period is from 08:00 on August 13, 2023, to 07:00 on August 14, the selected regional model times are the 12 h to 35 h forecasts starting at 20:00 on August 12, where the 12 h to 23 h forecast is referred to as the 12 h forecast, and the 24 h to 35 h forecast is referred to as the 24 h forecast.

Table 2. The modelling and test time period

	Time
Modelling time	08:00 AM; 20:00 PM
Test time	08:00 AM–07:00 AM (next day)
Test date	June 5, 2022–August 13, 2023

Fusion Method. Liu Jing and others, through evaluating the forecasting effect of regional models on heavy rainfall during the flood season in Liaoning Province [16], concluded three basic facts: There are differences in the forecasting performance of different regional models. Even for regional models with good inspection scores, there may be varying degrees of false alarms or missed reports compared to other regional models under the same (or different) weather systems.

Based on the basic facts of the inspection, a simple high-threshold fusion and low-threshold elimination are performed on the short-term heavy rainfall forecast of the regional model, resulting in a fused forecast product (hereinafter referred to as Forecast fusion, abbreviated as CMA-FP). According to the forecast success rate as screening criteria, taking CMA-MESO as the background field, when the precipitation forecast is greater than 20mm, the CMA-BJ and CMA-DB models are integrated, and when the precipitation forecast of any numerical forecast is less than 0.5 mm, the short-term heavy precipitation forecast of the precipitation at the station is eliminated.

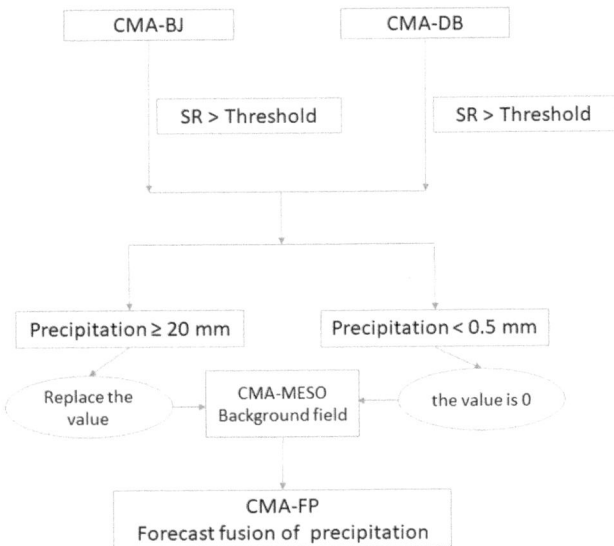

Fig. 1. The Roadmap of Fusion Method The threshold is determined by the numerical forecast evaluation results

Verification Method. The regional model grid forecast uses the nearest point interpolation method to interpolate to the verification site and compare with the corresponding actual verification site observation.

This paper uses the neighborhood method to perform spatial verification of short-term heavy rainfall forecasts. The spatiotemporal scale of short-term heavy rainfall is small and highly localized, and conventional ground observation stations are difficult to fully observe. The traditional "point-to-point" verification method is difficult to objectively and accurately reflect the quality of regional model forecasts. Short-term heavy rainfall verification often uses the American SPC's "point-to-face" spatial verification method, that is, whether a short-term heavy rainfall has occurred on the circular surface with a radius of 40 km centered on each regional site to judge whether the short-term heavy rainfall forecast is correct [17]. This paper refers to this method for short-term heavy rainfall verification.

$$POD = \frac{\sum_{i=1}^{n} A_i}{\sum_{i=1}^{n} A_i + C_i} \tag{1}$$

$$SR = \frac{\sum_{i=1}^{n} A_i}{\sum_{i=1}^{n} A_i + B_i} \tag{2}$$

The calculation methods of hit rate (*POD*) and success rate (*SR*) are as shown in formulas 1 and 2, where *n* is the number of samples participating in the average, and the definitions of A, B, and C are referred to in Table 3.

Table 3. Rainfall verification classification table

Observation	Forecast	
	Yes	No
Yes	A	C
No	B	D

3   Results Analysis

3.1   Evaluation of the Effect of the Integrated Forecast in Northeast Cold Vortex Rainfall

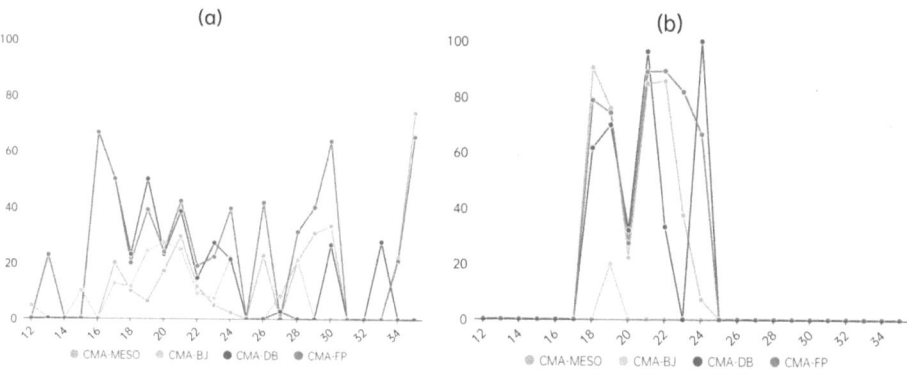

Fig. 2. Success rate of short-term heavy rainfall forecast during the Northeast cold vortex rainfall process 2022 (a), 2023 (b)

During the Northeast cold vortex rainfall process, based on the neighborhood method, the success rate index is used to compare the integrated forecast products of short-term heavy rainfall and high-resolution models. There are differences in the quantity and characteristics of rainfall in different years, and this paper separately evaluates 2022 and 2023. In order to analyze the differences in the performance of short-term heavy rainfall forecasts of multi-regional high-resolution models at different forecast times, by analyzing the success rate of short-term heavy rainfall forecasts within 12 h ~ 35 h (24 times a day) (Fig. 1), it can be seen that: within a radius of 40km without considering bias, the average success rate of the integrated forecast is 25.3%, with a maximum of 89%.

During the rainfall process in 2022, compared with the CMA-MESO model, the success rate of the integrated forecast increased by 11.67%. In terms of time, the 12 h integrated forecast (increased by 15.3%) was better than 24 h (increased by 8%). In terms of time, the success rate increased significantly after 5 ~ 8 times (12:00 ~ 15:00 in the afternoon) and 17 ~ 19 times (0:00 ~ 2:00 at midnight), which were 34.98% and 16.7% respectively. Compared with the CMA-BJ model, the success rate of the integrated forecast increased by 17.7%. In terms of time, the 12 h integrated forecast (increased by 14.5%) was lower than 24 h (increased by 20.9%). In terms of time, the success rate increased significantly after 5 ~ 6 times (12:00 ~ 13:00 in the afternoon), 18 ~ 19 times (1:00 ~ 2:00 at midnight), and 23 ~ 24 times (6:00 ~ 7:00 in the morning), which were 52.1%, 51.8%, and 43.2% respectively. Compared with the CMA-DB model, the success rate of the integrated forecast increased by 9.86%. In terms of time, the 12 h integrated forecast (increased by 1%) was lower than 24 h (increased by 18.7%). In terms of time, the success rate increased significantly after 17 ~ 19 times (0:00 ~ 2:00 at midnight) and 23 ~ 24 times (6:00 ~ 7:00 in the morning), which were 36.1% and 43.2% respectively.

The integrated forecast has improved the high-resolution model's forecast effect of short-term heavy rainfall at midnight and in the morning. Short-term heavy rainfall in Liaoning Province often occurs at midnight and in the morning [18], and the increase in the success rate of the integrated forecast has practical significance for targeted severe rainstorm forecast warnings.

In 2023, due to the influence of the cold vortex date and the number of rainfall events, from the data statistical analysis, the integrated forecast only has differences with the high-resolution model at 19 ~ 25 times, still showing positive forecast skills. The success rate increments of the integrated forecast for the CMA-MESO, CMA-BJ, and CMA-DB models are 14.8%, 69.6%, and 16.3% respectively, which are consistent with the trend in 2022, and the integrated forecast has the best correction effect on the CMA-BJ model.

The hit rate ignores the impact of false alarms, reflecting how many of the observed positive samples were forecasted. In the integrated forecast, if only high-threshold integration technology is used, the hit rate will definitely increase. This paper uses high-threshold integration and low-threshold elimination of false alarms to increase the hit rate of short-term heavy rainfall as much as possible while reducing the false alarm rate. By analyzing the hit rate of short-term heavy rainfall forecast within 12 h ~ 35 h (24 times a day) (Fig. 2), it can be seen that: without considering the deviation within a radius of 40km, the average hit rate of the integrated forecast is 1.11%, with a maximum of 16.67% (Fig. 3).

During the rainfall process in 2022, compared with the CMA-MESO model, the hit rate of the integrated forecast increased by 0.37%. In terms of validity, the 12 h integrated forecast (increased by 0.56%) was better than 24 h (increased by 0.17%). In terms of time, the hit rate increased significantly for the 8 ~ 13 times (15:00 ~ 20:00) after the start of the forecast, which was 1.27%.Compared with the CMA-BJ model, the hit rate of the integrated forecast increased by 0.67%. In terms of validity, the 12 h integrated forecast (increased by 0.51%) was lower than 24 h (increased by 0.83%). In terms of time, the hit rate increased significantly for the 10 ~ 12 times (10:00 ~ 12:00) and 23 ~ 24 times (6:00 ~ 7:00 in the morning) after the start of the forecast, which were 1.3% and 3.26% respectively. Compared with the CMA-DB model, the hit rate of the integrated forecast increased by 0.67%. In terms of validity, the 12 h integrated forecast (increased by 0.45%) was lower than 24 h (increased by 0.89%). In terms of time, the hit rate increased significantly for the 17 ~ 19 times (0:00 ~ 2:00 at midnight) and 23 ~ 24 times (6:00 ~ 7:00 in the morning) after the start of the forecast, which were 1.08% and 3.26% respectively. Combining the success rate and hit rate indicators, combined with the data analysis of hits and false alarms, the correction effect of the integrated forecast on CMA-BJ is the best. The 12 h integrated forecast effect of the CMA-MESO model is

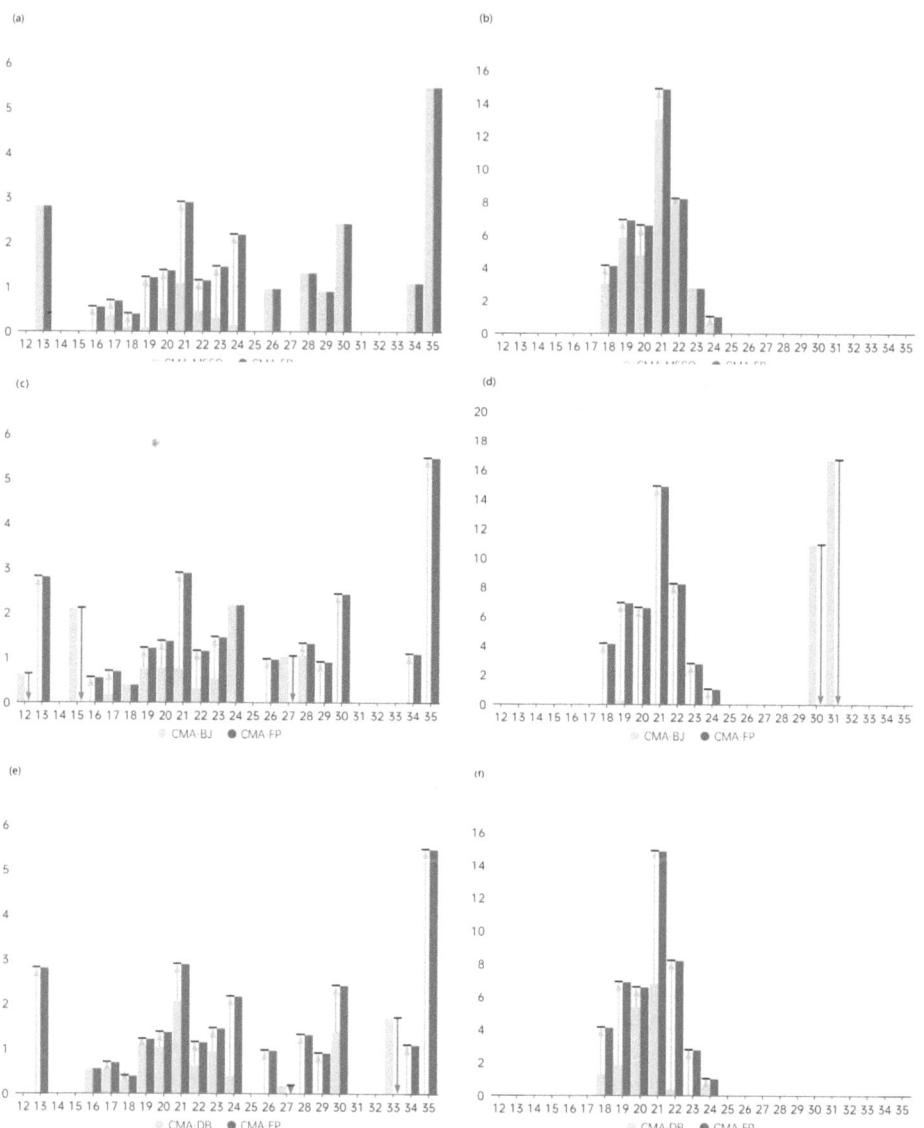

Fig. 3. Hit rate of short-term heavy rainfall forecast during the Northeast cold vortex rainfall process (unit: %) 2022 (a, c, e), 2023 (b, d, f) CMA-FP and CMA-MESO (a, b), CMA-FP and CMA-BJ (c, d), CMA-FP and CMA-DB (e, f)

better than 24 h, while the CMA-BJ and CMA-DB are the opposite. In terms of forecast time, the integrated forecast has a good correction effect in the middle of the night and early morning.

In 2023, the hit rate increments of the integrated forecast for the CMA-MESO, CMA-BJ, and CMA-DB models were 0.94%, 6.29%, and 4.03% respectively. The trend is consistent with the performance in 2022, and the average hit rate of the integrated forecast for the 19 ~ 25 times is 6.32%.

3.2 Evaluation of the Effect of the Integrated Forecast in Subtropical High-Pressure Rainfall

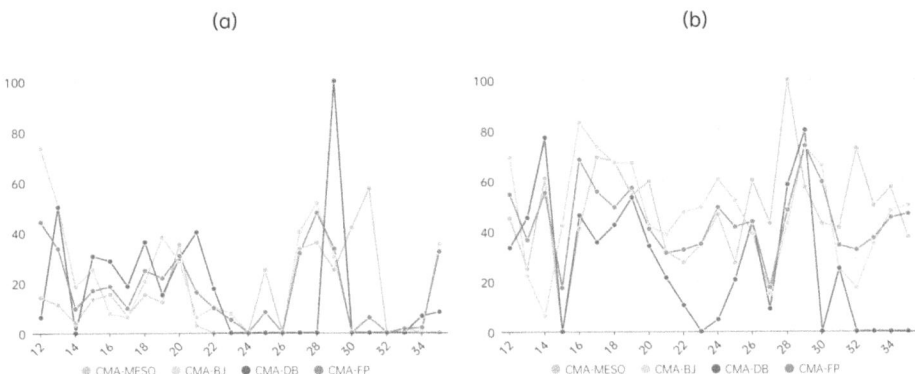

Fig. 4. Same as Fig. 1, but for subtropical high-pressure rainfall.

During the process of subtropical high-pressure rainfall, by analyzing the short-term heavy rainfall forecast success rate (Fig. 4) within 12 h ~ 35 h (24 times a day), it can be seen that: without considering the deviation within a radius of 40km, the average success rate of the integrated forecast is 30.5%, with a maximum of 73.8%.

Compared with the CMA-MESO model, in the rainfall process in 2022, the success rate of the integrated forecast increased by 6.93%. In terms of time, the 12 h integrated forecast (increased by 9.26%) was better than 24 h (increased by 4.6%). In terms of time, the success rate increased significantly after 12 ~ 13 times (8:00 ~ 9:00), 21 ~ 22 times (17:00 ~ 18:00), and 28 ~ 29 times (0:00 ~ 1:00), respectively 26.1%, 11.7%, and 10.1%. The increase in the success rate of the integrated forecast in 2023 was lower than in 2022, and the 12 h integrated forecast was not as good as 24 h. The increase was significant between 1:00 and 2:00 at midnight, with an average of 16.5%.

In 2023, the integrated forecast showed negative skills in short-term heavy rainfall forecasts for 12 times compared to the CMA-MESO model.

Compared with the CMA-BJ model, in the rainfall process in 2022, the success rate of the integrated forecast increased by a negative 1.95%. In terms of time, the 12 h integrated forecast (increased by a negative 4.26%) was lower than 24 h (increased by 0.37%). In terms of time, the success rate increased significantly after 16 ~ 18 times (12:00 ~ 14:00), 20 ~ 21 times (16:00 ~ 17:00), respectively 6.28% and 6.03%. The increase in the success rate of the integrated forecast in 2023 was lower than in 2022, and the 12 h integrated forecast was not as good as 24 h. The increase was significant between 9:00 and 10:00 in the morning, with an average of 31.5%. The integrated forecast showed negative skills in short-term heavy rainfall forecasts for an average of 12 times compared to the CMA-MESO model over the two years. Compared with the CMA-DB model, in the rainfall process in 2022, the success rate of the integrated forecast increased by 0.67%. In terms of time, the 12 h integrated forecast (increased by a negative 2.62%) was lower than 24 h (increased by 3.96%). In terms of time, the success rate increased significantly after 27 ~ 28 times (23:00 ~ 0:00 the next day), at 39.6%. The increase in the success rate of the integrated forecast in 2023 was higher than in 2022 (17.6%), and the 12 h integrated forecast was not as good as 24 h. The increase was significant between 2:00 ~ 7:00, 18:00 ~ 21:00, with averages of 38.3% and 30.6% respectively. Over the two years, the integrated forecast showed negative skills an average of 6.5 times. A comprehensive analysis of two years of data shows that the integrated forecast has improved the success rate of the CMA-MESO and CMA-BJ models in 2022, and the CMA-DB model in 2023. The integrated forecast has more instances of negative skill compared to CMA-MESO in 2023, more than 11 instances compared to CMA-BJ in both years, and more than 4 instances compared to CMA-DB in both years. Compared to CMA-MESO, the integrated forecast has better prediction results for short-term heavy rainfall at midnight (0:00–2:00). Compared to CMA-BJ, the integrated forecast has better correction results for short-term heavy rainfall from 8:00 to 9:00. Compared to CMA-DB, the integrated forecast has better correction results for short-term heavy rainfall from 2:00 to 7:00.

In summary, although the success rate of the integrated forecast is higher than the original high-resolution model in the process of subtropical high-pressure rainfall, there are more instances of negative skill, which may be related to the higher false alarm rate of the integrated forecast. Further analysis of the hit rate (Fig. 4) is used to find the advantageous features and error causes of the integrated forecast (Fig. 5).

Without considering the deviation within a radius of 40km, the average hit rate of the integrated forecast is 6.8%, with a maximum of 37.3%. Compared with the CMA-MESO, CMA-BJ, and CMA-DB models, the increase in the hit rate of the integrated forecast during the 2022 rainfall process is lower than in 2023, and it is generally better for a 12-h integrated forecast than a 24-h one. Only CMA-MESO had 3 instances in 2022 where the hit rate of the integrated forecast was negative.

In the process of subtropical high-pressure rainfall, the high-threshold fusion technology significantly improved the hit rate of short-term heavy rainfall forecasts, but it also increased the false alarm rate. Therefore, the increase in the success rate is not large, and there are more instances of negative success rate skills.

3.3 Evaluation of the Effect of Integrated Forecast in Typhoon Rainfall

During the typhoon rainfall process, by analyzing the short-term heavy rainfall forecast success rate (Fig. 6) within 12 h to 35 h (24 times a day), it can be seen that: without considering the deviation within a radius of 40km, the average success rate of the integrated forecast is 33.7%, with a maximum of 84.4%.

Compared with the CMA-MESO model, in the rainfall process in 2022, the success rate of the integrated forecast decreased by 2%. In terms of time, the 12 h integrated forecast (increased by 2%) was better than 24 h (decreased by 6%). In terms of time, the success rate increased significantly for the 34th to 35th times (5am to 7am) after the start of the forecast, at 23.6%. The increase in the success rate of the integrated forecast in 2023

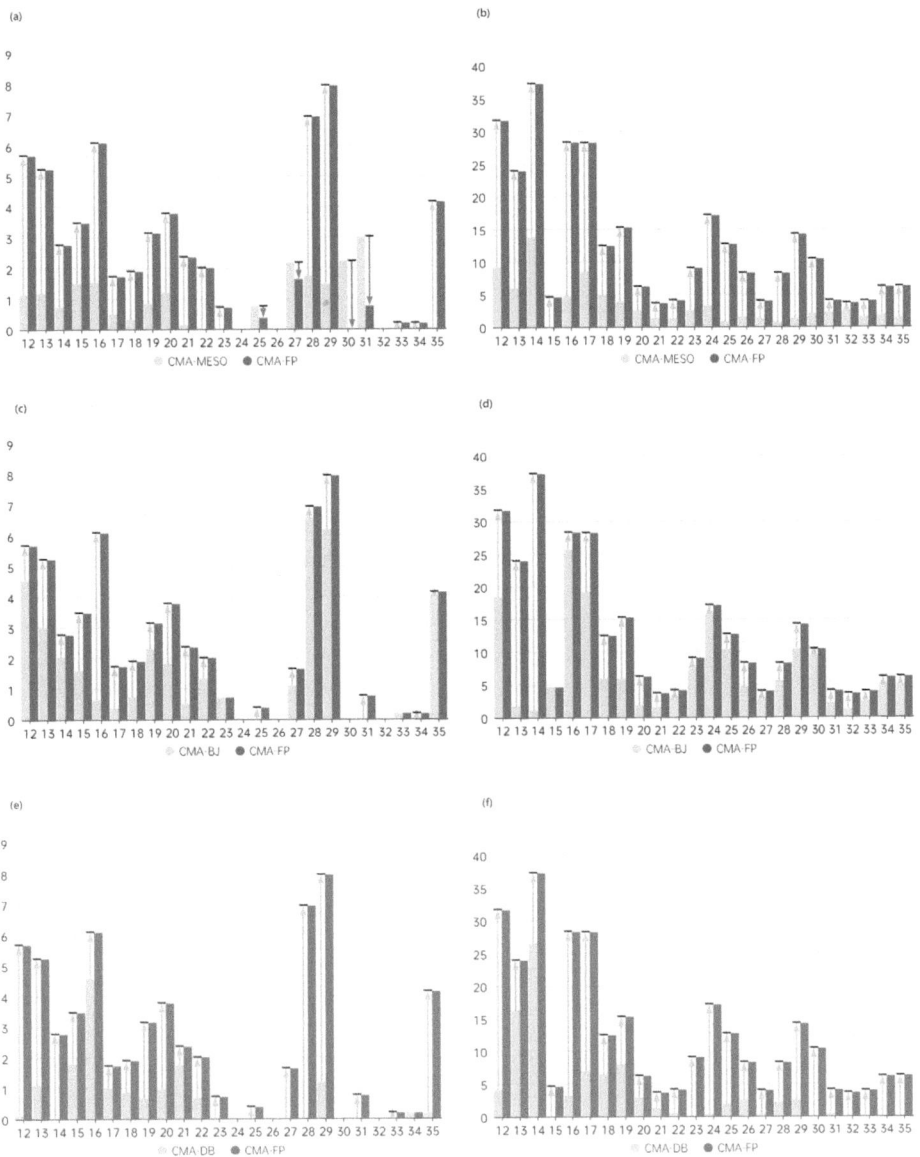

Fig. 5. Same as Fig. 2, but for subtropical high-pressure rainfall.

was significantly higher than in 2022, with an increase of 7.78%. The 12 h integrated forecast was better than 24 h in terms of time. The increase was larger between 13:00 and 14:00 in the afternoon, with an average of 41.7%. In 2022 (2023), the integrated forecast had 12 (6) times when the short-term heavy rainfall forecast showed negative skills compared to the CMA-MESO model. Compared with the CMA-BJ model, in the rainfall process in 2022, the success rate of the integrated forecast increased by 3%.

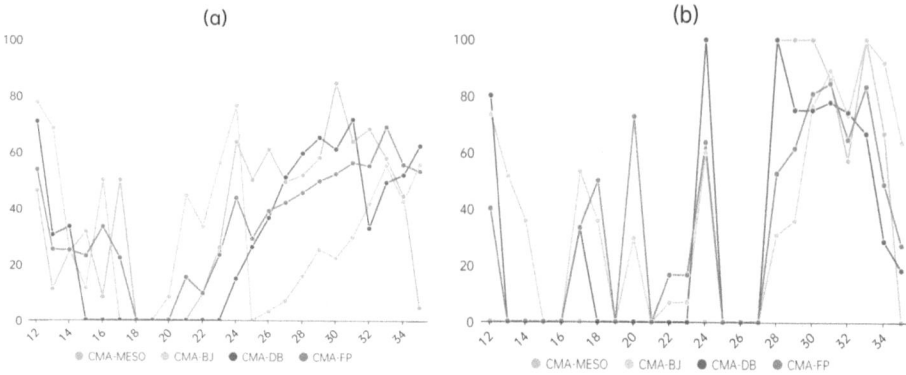

Fig. 6. Same as Fig. 1, but for typhoon rainfall.

In terms of time, the 12 h integrated forecast (decreased by 12%) was lower than 24 h (increased by 18%). In terms of time, the success rate increased significantly for the 25th to 31st times (21:00 to 03:00 the next day) after the start of the forecast, at 30.1%. The increase in the success rate of the integrated forecast in 2023 was lower than in 2022. The 12 h integrated forecast was not as good as 24 h in terms of time. The increase was larger between 18:00 and 20:00 in the evening, with an average of 7.6%. Both years had 9 times when the integrated forecast showed negative skills compared to the CMA-MESO model's short-term heavy rainfall forecast. Compared with the CMA-DB model, in the rainfall process in 2022, the success rate of the integrated forecast increased by 4%. In terms of time, the 12 h integrated forecast (increased by 8%) was better than 24 h (increased by 0.7%). In terms of time, the success rate increased significantly for the 15th to 17th times (11:00 to 13:00) after the start of the forecast, at 26.2%. The increase in the success rate of the integrated forecast in 2023 was lower than in 2022. The 12 h integrated forecast was better than 24 h in terms of time. The increase was larger between 18:00 and 19:00, at 16.7%. On average, there were 7 times when the integrated forecast showed negative skills over the two years.

Based on the analysis of two years of data, the integrated forecast improved the forecast success rate of the CMA-MESO model in 2023, the CMA-BJ model in 2022, and the CMA-DB model in both years. Compared to CMA-MESO, the integrated forecast had more times (12) showing negative skills in 2022, 9 times in both years compared to CMA-BJ, and more than 7 times in both years compared to CMA-DB. Compared to CMA-MESO, the integrated forecast had a better forecast effect for short-term heavy rainfall at noon (13:00 to 14:00). Compared to CMA-BJ, the integrated forecast had a better correction effect for short-term heavy rainfall from 21:00 to 03:00 the next day. Compared to CMA-DB, the integrated forecast had a better correction effect for short-term heavy rainfall from 11:00 to 13:00. During the typhoon rainfall process in 2022, the integrated forecast included the rainfall amounts of the CMA-BJ model on July 5, July 6, and August 1, and the CMA-DB model on July 5 and 6. Through the increase in the 12 h hit rate, it can be seen that the integrated forecast has a larger negative growth compared to the CMA-BJ model, indicating that there is a high false alarm rate in the integrated model products, which is not difficult to understand that this model is the CMA-DB model with the highest growth in success rate. Further analysis through the hit rate (Fig. 7) is used to find the advantageous features and error causes of the integrated forecast. Within a radius of 40km without considering deviation, the average hit rate of the integrated forecast is 11.53%, with a maximum of 48.3%. Compared with the CMA-MESO, CMA-BJ, and CMA-DB models, the increase in the hit rate of the integrated forecast during the rainfall process in 2022 is higher than that in 2023, and it is generally shown that the 12 h integrated forecast effect is better than 24 h. Only when compared with the CMA-BJ model, the hit rate of the integrated forecast has a negative skill at 3 time points.

During the typhoon rainfall process, the high-threshold fusion technology significantly improved the hit rate of short-term heavy rainfall forecasts, especially the 12 h hit rate significantly increased, but it also correspondingly increased the false alarm rate, so the increase in success rate is not large, and there are more times with negative success rate skills. In summary, during the typhoon rainfall process, the appearance of negative values in the success rate after the integrated forecast reflects the increase in the false alarm rate, that is, a larger hit rate is obtained at the expense of a smaller false alarm rate.

3.4 Evaluation of the Effect of Integrated Forecast in Other Rainfalls

Heavy rainfall caused by the influence of systems such as upper troughs, cyclones, and shear lines is referred to as other rainfall processes. Because there were fewer such heavy rainfalls in 2023, separate analysis may affect the rigor of the data, so this article only tests and integrates the forecasts of 111 days and 264 time points in 2022. Through the success rate indicator (Fig. 7), it can be seen that: within a radius of 40km without considering deviation, the average success rate of the integrated forecast is 32.3%, with a maximum of 58.3%.

Due to data quantity and quality, here we only analyzed the data from 2022. Compared with the CMA-MESO model, the success rate of the integrated forecast increased by 9%. In terms of time, the 12 h integrated forecast (increased by 7.8%) is lower than 24 h (increased by 10.2%). In terms of time points, the success rate increase is larger at 33–34 time points (5–6 h) after the start of the forecast, which is 31.2%. Compared with the CMA-BJ model, the success rate of the integrated forecast increased by 3.9%. In terms of time, the 12 h integrated forecast (increased by 8.3%) is better than 24 h (decreased by 0.4%). In terms of time points, the success rate increase is larger at 31–32 time points (2–3 h) after the start of the forecast, which is 21.4%. Compared with the CMA-DB model, the success rate of the fusion forecast has increased by 8.5%. In terms of time, the 12-h fusion forecast (an increase of 5.5%) is lower than the 24-h forecast (an increase of 11.5%). In terms of time, the success rate increases significantly for the 28–29 h after the start of the forecast (23:00 to 00:00 the next day), reaching 41.4%.

The fusion forecast has significantly improved the success rate of the three models. In terms of time, the CMA-MESO model and the CMA-DB model have better improvement effects at 24 h than at 12 h, increasing the success rate of forecasting heavy rainfall caused by systems such as upper troughs in the long term.

Further analysis through the hit rate (Fig. 8) shows that, without considering the deviation within a radius of 40km, the average hit rate of the fusion forecast is 4.07%,

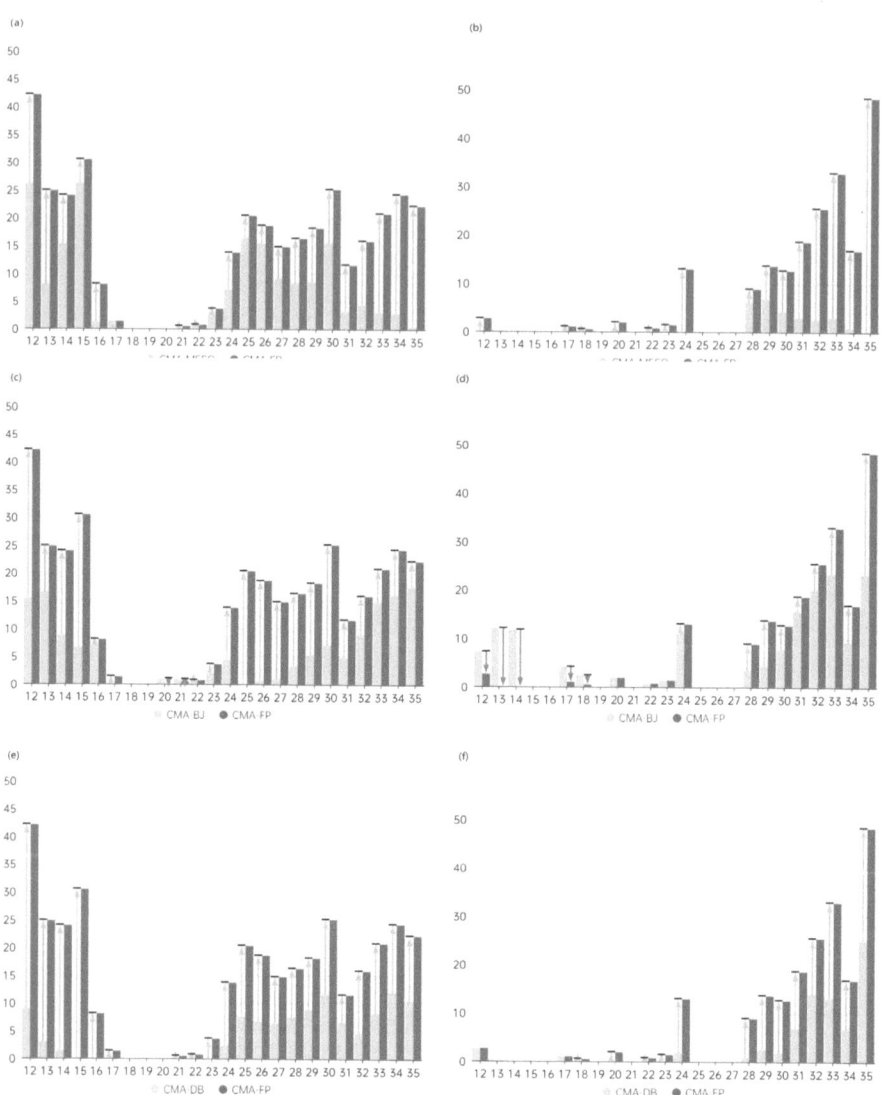

Fig. 7. Same as Fig. 2, but for typhoon rainfall.

with a maximum of 8.65%. In terms of time, the increase in the hit rate of the 12-h and 24-h fusion forecasts is basically the same, ranging from 2.11% to 2.96%.

3.5 Overall Evaluation of Fusion Forecast

Comparative Analysis of Fusion Forecast and High-Resolution Model Short-Term Heavy Rainfall Forecast. Figure 9 analyzes the success rate of short-term heavy rainfall forecasts of the fusion forecast and each original model in different weather systems. It can be seen that the fusion forecast shows a clear advantage in the Northeast cold vortex

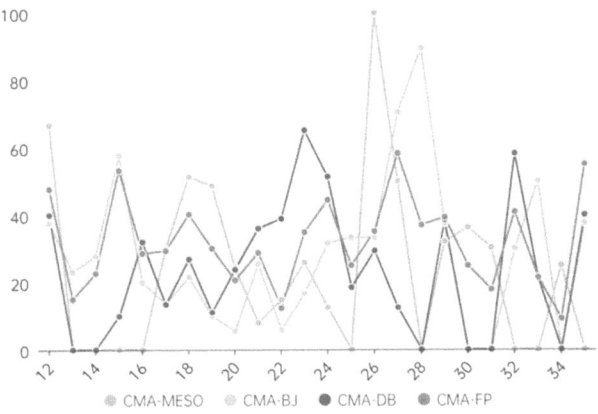

Fig. 8. Success rate of short-term heavy rainfall forecast during the Northeast cold vortex rainfall process in 2022

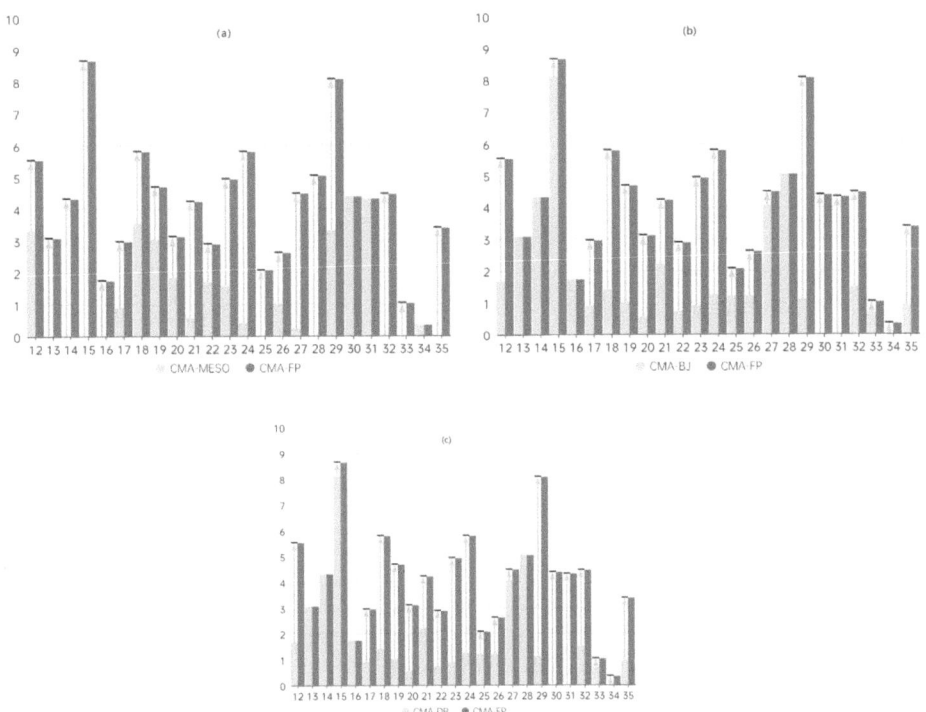

Fig. 9. Same as Fig. 2, but for other rainfall.

and other rainfall, with a success rate increase of 46% to 216% in the Northeast cold vortex rainfall, and both CMA-MESO and CMA-DB show that the success rate of 12 h is greater than 24 h; the success rate increase in other rainfall is 14.18 to 38.79%, and CMA-BJ has a higher success rate at 24 h (Fig. 10).

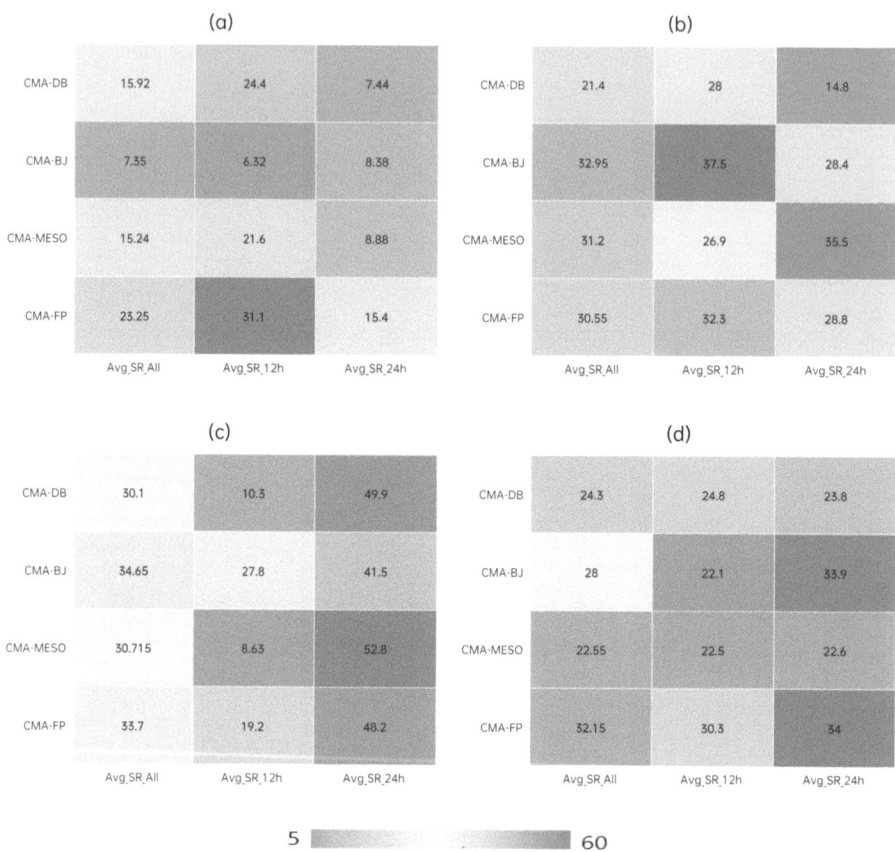

Fig. 10. Success rate of short-term heavy rainfall forecast of different numerical models (unit: %) (a): Rainfall in Northeast cold vortex; (b) Rainfall in subtropical high pressure; (c) Typhoon rainfall; (d) Other rainfall

In the subtropical high pressure and typhoon rainfall, the success rate of the combined forecast is not significantly higher than each original model. The possible reason is that the impact of the false alarm rate is relatively large, and while the hit rate is improved, the success rate also decreases. In the subtropical high pressure, the success rate of the combined forecast is only higher than the CMA-DB model, with a success rate increase of 42.6%. The biggest difference is the CMA-BJ model, with a success rate decrease of 7%. In the typhoon model, the success rate of the combined forecast is not as good as the CMA-BJ model, with a success rate decrease of 2.7%. However, the success rate of the combined forecast is 11.9% higher than the CMA-DB model. By comparing the two types of rainfall, it is not difficult to see that the forecast effect of the CMA-BJ model is better, the false alarm rate of the CMA-DB model is higher, therefore, after integrating the CMA-DB model, the success rate of the combined forecast is lower than the CMA-BJ model and higher than the CMA-DB model.

Comparative Analysis of Short-term Heavy Rainfall Forecast in Different Types of Rainfall by Combined Forecast. As can be seen in Fig. 11, We analyzes the TS score of short-term heavy rainfall forecast in different types of rainfall by the combined forecast. It can be seen that the success rate of the combined forecast in typhoon rainfall is the highest, reaching 5.26%, and it is only 0.51% in the Northeast cold vortex rainfall. From the perspective of time, the combined forecast in the Northeast cold vortex and subtropical high pressure both show that the 12 h forecast success rate is greater than 24 h, while it is the opposite in typhoons and other types of rainfall. Some studies have shown that numerical models are insufficient in forecasting typhoon landfall rainfall, especially the difference beyond 12 h is larger [19, 20]. In this study, the 24 h forecast TS score is higher than 12 h, which has practical significance for improving the accuracy of long-term typhoon rainfall forecast.

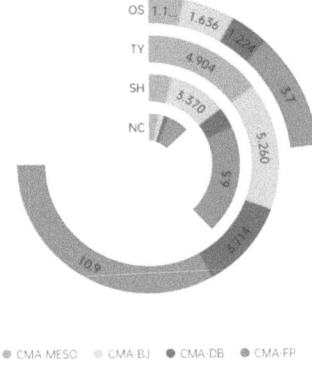

Fig. 11. Comparative Analysis of Short-term Heavy Rainfall Forecast in Different Types of Rainfall (unit: %). NC: Northeast cold vortex rainfall; SH: Subtropical high-pressure rainfall; TY: Typhoon rainfall; OS: other systems

4 Conclusions and Discussion

At present, there are few studies on precipitation fusion correction based on test results, and it is not common to analyze hourly rainfall tests under different weather systems. Liaoning Province has now included hourly rainfall in the warning signal release standards. This paper combines business reality and conducts a short-term heavy rainfall neighborhood test on the commonly used multi-regional models in the northern region, digging regional model additional information from different angles to provide reference for forecasters' decision-making.

(1) In the process of Northeast cold vortex rainfall, from the overall analysis of different forecast periods, the combined forecast has a better correction effect on CMA-BJ than the other two regional models. CMA-MESO's 12 h forecast is better than

24 h, and CMA-BJ and CMA-DB's 24 h forecast is better than 12 h. Without considering the deviation within a radius of 40km, compared with the three original high-resolution models, the success rate of the combined forecast in 2022 increased by an average of 13.08%, and the hit rate increased by an average of 0.57%, in 2023, these numbers are 25.7% and 17.5%. The combined forecast has a higher correction effect for midnight and morning than other times.

(2) In the process of subtropical high pressure rainfall, from the overall analysis of different forecast periods, the combined forecast has a better correction effect on CMA-MESO than the other two regional models. CMA-MESO's 12 h forecast is better than 24 h, and CMA-BJ and CMA-DB's 24 h forecast is better than 12 h. Without considering the deviation within a radius of 40km, compared with the three original high-resolution models, the success rate of the combined forecast in 2022 increased by an average of 16.8%, and the hit rate increased by an average of 1%.In 2023, the hit rate increased 289%. The combined forecast has a higher correction effect for 8–9 am in the morning than other times.

(3) In the process of typhoon rainfall, from the overall analysis of different forecast periods, the combined forecast has a better correction effect on CMA-DB than the other two regional models. CMA-MESO and CMA-DB's 12 The forecast for h is better than 24 h, and CMA-BJ's 24 h forecast is better than 12 h. Within a radius of 40km, without considering bias, compared to the three original high-resolution models, the success rate of the combined forecast in 2022 increased by an average of 34.21%, and the hit rate increased by an average of 14.9%. In 2023, these numbers are 30.7% and 651%. The combined forecast has a better correction effect for 18:00 to 19:00 in the evening than other times.

(4) When affected by systems such as upper troughs, surface cyclones, and low-level shear, overall analysis from different forecast durations shows that the combined forecast has a better correction effect on CMA-MESO than the other two regional models. All three models have a 12 h forecast better than 24 h. Within a radius of 40km, without considering bias, compared to the three original high-resolution models, the success rate of the combined forecast in 2022 increased by an average of 32.2%, and the hit rate increased by an average of 4.07%.

(5) The analysis of the performance results of the combined forecast under different rainfall systems shows that: under the influence of the Northeast cold vortex and other rainfall systems, the success rate of the combined forecast shows positive growth. In the context of subtropical high pressure and typhoons, the combined forecast and the original numerical model are negative skills. It is not difficult to understand that while improving the hit rate of the model, it also increases the false alarm rate of the model. The short-term heavy rainfall forecast success rate of the combined forecast in typhoon storms is the highest, at 33.69%.

This study aims at short-term near-warning services, and develops forecast products by combining commonly used multi-regional models, on the one hand, to provide some references for forecasters to use multi-regional models, and on the other hand, to provide a basis for model developers to discover model forecast bias tendencies from the phenomenon level and thus make targeted improvements to the model. Due to the limitations of the time scale and the number of rainfall dates, the test results cannot fully accurately describe the ability of regional model forecasts. Different regional models have large differences in forecast performance during the rainfall process in Liaoning, and the factors that lead to different forecast effects need further research.

Acknowledgement. This paper is jointly funded by the 2023 Open Fund Project of Shenyang Institute of Atmospheric Environment, China Meteorological Administration and Northeast Cold Vortex Research Key Open Laboratory (NO. 2023SYIAEKFMS17), Annual Fund Project of Liaoning Provincial Meteorological Bureau (2023) and Annual Scientific Research Project of Shenyang Meteorological Bureau (2023).

References

1. Xin, S., Chuanlei, C., Han, L., et al.: Synthetic analysis of severe storm environment field and physical quantity field in Liaoning. J. Meteorol. Environ. **32**(5), 40–46 (2016)
2. Ying, L., Dianxiu, Y., Ge, G., et al.: Climate characteristics and major weather/climate events in China in the summer of 2022. J. Atmos. Sci. **46**(1), 110–118 (2023)
3. Jing, L., Chuan, R., Wei, D., et al.: Diagnostic analysis of a heavy rainfall process in Liaoning. J. Meteorol. Environ. **32**(2), 18–27 (2016)
4. Gofa, F., Boucouvala, D., Louka, P., et al.: Spatial verification approaches as a tool to evaluate the performance of high resolution precipitation forecasts. Atmos. Res. **208**, 78–87 (2018)
5. Chakraborty, A.: The skill of ECMWF medium-range forecasts during the year of tropical convection 2008. Mon. Wea. Rev. **138**(10), 3787–3805 (2010)
6. Jing, L., Chuanlei, C., Jun, Y., et al.: Distribution and characteristic analysis of red warning signals for heavy rain in Liaoning Province from 2015 to 2019. J. Meteorol. Environ. **37**(1), 100–105 (2020)
7. Chuanlei, C., Zhaoyong, G., Saidi, W., et al.: Analysis of environmental characteristics of long-duration heavy rainfall in Liaoning. J. Atmos. Sci. **40**(3), 321–332 (2017)
8. Duanling, L., Chao, C.: Preliminary application and evaluation of the optimal TS scoring method in hourly precipitation correction. J. Trop. Meteorol. **38**(4), 611–620 (2022)
9. Feng Xiao, W., Yumei, L., et al.: Grid-based sunny and rainy forecast verification for Hainan Island based on grid point observations. J. Trop. Meteorol. **38**(1), 91–100 (2022)
10. Cong, L., Yang, S., Naigeng, W., et al.: Hourly precipitation correction method for GRAPES rapid update cycle assimilation forecast system based on optimal neighborhood probability. J. Tropic. Meteorol. **37**(4), 569–578 (2021)
11. Zhuoheng, W., Jing, C., Hanbin, Z., et al.: Experimental ensemble forecast of extreme rainfall intensity in the "7·20" Zhengzhou cloudburst. Atmospheric Science (2023)
12. Fei, Y., Liping, H., Liantang, D.: Analysis of the impact of different spatial resolutions of the GRAPES-MESO model on China's summer precipitation forecast. Atmos. Sci. **42**(5), 1146–1156 (2018)
13. Fajing, C., Jing, C., Qing, W., et al.: A new method for storm forecast scoring based on predictability II: Storm verification scoring model and evaluation experiment. J. Meteorol. **77**(1), 28–42 (2019)
14. Shenjia, M., Chaohui, C., Xiefei, Z., et al.: Evaluation and verification of convective scale ensemble forecast based on spatio-temporal uncertainty. J. Meteorol. **76**(4), 578–589 (2018)
15. Xueqing, L., Jing, C., Fajing, C., et al.: Scale sensitivity test of precipitation neighborhood ensemble probability method. Atmos. Sci. **44**(2), 282–296 (2020)
16. Jing, L., Chuan, R., Ziqi, Z., et al.: Verification analysis of high-resolution model strong precipitation forecast in multiple regions. Meteorology **48**(10), 1292–1302 (2022)

17. Wenyuan, T., Qingliang, Z., Xinhua, L., et al.: Verification analysis of national-level severe convective weather classification forecast. Meteorology **43**(1), 67–76 (2017)
18. Jing, L., Chuanlei, C., Jun, Y., et al.: Distribution and characteristics analysis of red warning signals for heavy rain in Liaoning Province from 2015 to 2019. J. Meteorol. Environ. **37**(1), 100–105 (2020)
19. Yu, G., Kan, D., Jun, X., et al.: Weather check characteristics of GRAPES-GFS model storm forecast. Meteorology **44**(9), 1148–1159 (2018)
20. Xue Wenbo, Y., Hui, T.S., et al.: Verification evaluation of near-surface wind speed forecast of Shanghai Rapid Update Assimilation Numerical Forecast System (SMS-WARR). Meteorology **46**(12), 1529–1542 (2020)

Evaluation of High-Resolution Model Heavy Rainfall Forecast Under Different Weather Systems

Jing Liu[1,2,3], Chuan Ren[4(✉)], Ziqi Zhao[1], Wei Dong[5], Yue Wang[3], Hao Ren[6], and Shuai Zhang[3]

[1] China Meteorological Administration, Institute of Atmospheric Environment, Shenyang 110166, China
[2] Key Opening Laboratory for Northeast China Cold Vortex Research, Shenyang 110166, China
[3] Shenyang Meteorological Bureau, Shenyang 110166, China
[4] Liaoning Meteorological Information Center, Shenyang 110166, China
49952706@qq.com
[5] China Meteorological Administration, Liaoning Branch of Cadre Training College, Shenyang 110166, China
[6] Kangping County Meteorological Bureau, Shenyang 110166, China

Abstract. Using hourly rainfall data from the meteorological big data cloud platform and based on the fuzzy verification neighborhood method, the forecast performance of the China Meteorological Administration's North China Numerical Forecast Model System (CMA-BJ), China Meteorological Administration's Mesoscale Weather Numerical Forecast System (CMA-MESO), and the Ruitu Northeast Numerical Forecast Model System (CMA-DB) during the rainfall process in the main flood season of Liaoning from 2022 to 2023 was evaluated. The results show that without considering the bias within a radius of 40 km, the average hit rate of the model forecast during the Northeast cold vortex rainfall process is below 0.8%, with a maximum of 3.94%. The success rate of the three models' 12-h forecasts is higher than 24 h, with an average of 14.5%. During the subtropical high rainfall process, the hit rate and TS score of all times of the three models are below 9%, with a maximum of 8.2%. The forecast effect during the typhoon rainstorm process is significantly better than the Northeast cold vortex and subtropical high rainstorm, with an average hit rate of 3.2%, and the false alarm rate of short-term heavy rainfall forecast at the near moment is high. When short-term heavy rainfall is caused by the influence of systems such as the upper trough, the average hit rate of the model forecast is only 1.54%. In all types of rainfall processes, a comprehensive analysis of indicators such as the success rate of the forecast shows that the CMA-MESO model forecast effect is better than the other two models.

Keywords: Multi-region model · short-term heavy rainfall · neighborhood method · rainfall typing · forecast verification

© The Author(s) 2025
B.-J. He et al. (Eds.): UCSUD 2023, LNCE 559, pp. 47–60, 2025.
https://doi.org/10.1007/978-981-97-8401-1_4

1 Introduction

The heavy rain in Liaoning Province is characterized by frequent occurrences, short duration, and severe disasters [1]. In 2022, the average rainfall in Liaoning was 420.6 mm, 70% more than the same period in normal years, exceeding the total summer rainfall in normal years, and the most in the same period in nearly thirty years [2]. Short-term heavy rain often occurs under the trigger of mesoscale systems [3], and making full use of numerical models is the basis for good heavy rain forecasting. The high-resolution model's forecast effect on heavy rain induced by mesoscale systems is better than the global model.

Testing is an effective way to enhance our understanding and application capabilities of high-resolution models [4]. A reasonable assessment of the numerical model's forecasting performance for short-term heavy rainfall can help issue early red warning signals for storms [5] [6] [7]. The storm impact system in Liaoning Province mainly includes subtropical high pressure, northeast cold vortex, and northward landing typhoons, etc. The northeast cold vortex is a deep cold low-pressure system that occurs in the troposphere in the Northeast Asia region. Its abnormal activity often brings great uncertainty to summer rainfall forecasts [8]. The water vapor carried by the northward typhoon itself and the water vapor transported by the southern side of the subtropical high provide sufficient water vapor for the occurrence of extreme storms; the unchanged structure during the weakening process of the northward typhoon, the convergence of the boundary layer northeast wind jet and southeast wind jet provide a strong dynamic mechanism for extreme precipitation [9]. In terms of testing technology, due to the influence of the spatial and temporal resolution of observation data, traditional classification testing methods are subject to a "double penalty" brought about by minor spatial and temporal differences in high-resolution precipitation testing [6]. Improving precipitation assessment capabilities requires the development of testing technology suitable for high-resolution ensemble precipitation forecasts [10]. Li et al. believe that the neighborhood method can provide more information about the forecast on different scales and evaluation strategies [11]. The neighborhood method usually uses the Fractional skill score (FSS) as an evaluation indicator. Wang et al. pointed out that the FSS score can better reflect the differences in the forecasting capabilities of different models in a quantitative way compared to the traditional TS score [12]. Liu et al. applied the neighborhood method to test the radar echo composite reflectivity factor in Liaoning area, and found that the FSS score under the 11 km neighborhood radius is higher [13]. In terms of test results, during the "21.7" extreme storm in Henan, the CMA-BJ model generally overestimated the rainfall intensity compared to the actual situation, while the CMA-MESO heavy rainfall forecast was more accurate [14]. The regional model's heavy rain forecast is mostly overestimated, with a high false alarm rate. The East China regional model's ability to forecast the rapid movement of typhoons is relatively weak [15]. The average hit rate of the CMA-MESO forecast is higher than that of CMA-DB and CMA-BJ. The three regional models score higher in short-term heavy rainfall during cyclonic precipitation than in post-high-pressure precipitation, but all three perform poorly in typhoon precipitation [16]. Liu et al. pointed out that high-resolution models such as CMA-MESO that start forecasting at 20:00 perform significantly better than those that start at 08:00 [17].

Therefore, this paper compares the precipitation forecasts of high-resolution models that start at 20:00. Multi-faceted evaluation can uncover the added value of multiple regional models, providing forecasters with more comprehensive and objective reference information. At present, there are not many studies on the verification of heavy rainfall forecasts under different weather systems for high-resolution models. To comprehensively evaluate the performance of high-resolution models in forecasting summer heavy rain in Liaoning, this paper conducts a neighborhood method short-term heavy rainfall forecast verification for 50 weather events (a total of 1200 times of heavy rain) that occurred in Liaoning Province from 2022 to 2023. The performances of the China Meteorological Administration's North China Numerical Forecast Model System (hereinafter referred to as CMA-BJ), the China Meteorological Administration's Mesoscale Weather Numerical Forecast System (hereinafter referred to as CMA-MESO), and the China Meteorological Administration's Ruitu Northeast Numerical Forecast Model System (hereinafter referred to as CMA-DB) are studied. The rainfall forecasts for post-high-pressure, Northeast cold vortex, typhoon, and other (upper trough, cyclone, shear, etc.) weather processes are evaluated, and the advantages and bias characteristics of different model forecasts are concluded.

2 Data and Methods

2.1 Data Sources

Observational Data. The data used comes from the meteorological platform "Data as a server (Daas)", using hourly rainfall data from 1812 observation stations in Liaoning Province, removing outliers to ensure the continuity and accuracy of rainfall data.

Forecast Data. The forecast data comes from the operational CMA-BJ, CMA-MESO, and CMA-DB models, the horizontal resolution of the models are all 3 km, and the initial field of the three regional models is the NCEP-GFS model.

2.2 Research Methods

Weather Typing. Sun Xin et al. (2016) analyzed 60 heavy rain events that occurred in Liaoning from 1960 to 2013, pointing out that the influencing systems of Liaoning heavy rain mainly include Northeast cold vortex, post-subtropical high pressure, cyclone (including Hetao cyclone, Jianghuai cyclone, Mongolia cyclone, etc.) influence, typhoon northward, etc. This paper identifies a short-term heavy rain event as one where the number of rainfall stations with a rainfall amount greater than 20 mm in one hour is more than 20. We selected 50 heavy rain events that occurred in Liaoning Province from 2022 to 2023, totaling 1200 times, for regional model 1-h heavy rainfall forecast verification analysis. The rainfall types are referred to as Northeast cold vortex rainfall, subtropical high-pressure rainfall, typhoon rainfall, and other rainfall (upper trough, cyclone, shear line). The specific typing dates are detailed in Table 1.

Different Rainfall Definitions. According to the "Intelligent Forecast Technology Method Competition Inspection Scheme" issued by the China Meteorological Administration in June 2020 (hereinafter referred to as the "Inspection Scheme"), short-term heavy rainfall in the Liaoning area is defined as rainfall ≥ 20 mm per hour.

Table 1 Dates of individual cases under different influencing weather systems.

Num	Northeast cold vortex	Subtropical anticyclone	Typhoon	Others (Upper trough, cyclone, shear line)
1	June 5, 2022	June 29, 2022	July 6, 2022	June 16, 2022
2	June 6, 2022	June 30, 2022	July 7, 2022	June 17, 2022
3	June 8, 2022	July 1, 2022	July 31, 2022	July 10, 2022
4	June 10, 2022	July 2, 2022	August 2, 2022	July 13, 2022
5	June 13, 2022	July 29, 2022	August 3, 2022	July 14, 2022
6	June 22, 2022	July 30, 2022	August 4, 2022	July 22, 2022
7	June 23, 2022	August 6, 2022	July 27, 2023	July 23, 2022
8	June 25, 2022	August 7, 2022	August 12, 2023	August 12, 2022
9	July 16, 2022	August 18, 2022	August 13, 2023	August 14, 2022
10	July 17, 2022	July 24, 2023		August 15, 2022
11	July 20, 2022	July 25, 2023		June 26, 2023
12	July 21, 2022	August 4, 2023		
13	June 11, 2023	August 16, 2023		
14	June 12, 2023	August 18, 2023		
15	July 7, 2023	August 21, 2023		
Total	15 days 360 forecast times	15 days 360 forecast times	9 days 216 times	11 days 264 times

Inspection Method. In order to assess the stability of the regional model's forecast, spatial inspections are conducted on 12 h to 36 h forecasts from different start times. For example, if the inspection period is from 08:00 on August 13, 2023 to 07:00 on August 14, the regional model times selected are 12 h to 36 h forecasts starting at 20:00 on August 12, where the 12 h to 24 h forecast is referred to as the 12 h forecast, and the 24 h to 36 h forecast is referred to as the 24 h forecast.

Inspection Method. The regional model grid forecast uses the nearest point interpolation method to interpolate to the inspection site and then compares it with the corresponding actual inspection site observation.

This paper uses the neighborhood method to conduct spatial inspections of short-term heavy rainfall forecasts. Short-term heavy rainfall has a small spatial and temporal scale and strong locality, and conventional ground observation stations are difficult to fully observe. Traditional "point-to-point" inspection methods are difficult to objectively and accurately reflect the quality of regional model forecasts. Spatial inspection methods are widely used in the inspection of severe convective weather forecasts [18].

The short-term heavy rainfall test often adopts the "point-to-face" spatial test method of the American SPC, that is, whether the short-term heavy rainfall forecast for each regional site is correct or not, is judged by whether there has been short-term heavy rainfall on the circular surface with the site as the center and a radius of 40 km [19]. This paper refers to this method for the short-term heavy rainfall test. The "Test Plan" stipulates that the short-term heavy rainfall test indicators are hit rate and false alarm rate. This paper uses these two indicators to analyze the forecast effect of multi-regional models, and also uses TS score and success rate as supplementary evaluation indicators. The calculation methods of probability of detection (POD), false alarm rate (FAR), TS score(threat score), and success rate (SR) are as shown in formulas 3 and 4, where n is the number of samples participating in the average, and the definitions of A, B, and C are shown in Table 2 (Images of formulas and tables).

$$POD = \frac{\sum_{i=1}^{n} A_i}{\sum_{i=1}^{n} A_i + C_i} \tag{1}$$

$$FAR = \frac{\sum_{i=1}^{n} B_i}{\sum_{i=1}^{n} A_i + B_i} \tag{2}$$

$$TS = \frac{\sum_{i=1}^{n} A_i}{\sum_{i=1}^{n} A_i + B_i + C_i} \tag{3}$$

$$SR = \frac{\sum_{i=1}^{n} A_i}{\sum_{i=1}^{n} A_i + B_i} \tag{4}$$

3   Results Analysis

3.1   Test Results of Multi-Regional Model's Forecast for Northeast Cold Vortex Rainfall

During the rainfall process in the Northeast Cold Vortex, based on the neighborhood method, a hit rate and false alarm rate scoring test was conducted for short-term heavy rainfall forecasts. In order to analyze the differences in the performance of short-term

Table 2 Classification for precipitation verifications

Observation	Forecast	
	Yes	No
Yes	A	C
No	B	D

heavy rainfall forecasts from multi-regional high-resolution models at different forecast times, through analyzing the hit rate(see Fig. 1), TS score, and success rate (see Fig. 2) within 12 h~35 h, it can be seen that: during the Northeast Cold Vortex rainfall process, the forecast hit rate and TS score of all times from the three models are all below 6%. Analysis of the average value (Avg_12) of the 12 h short-term heavy rainfall forecast test shows that, without considering the bias within a radius of 40 km, the average model forecast hit rate is below 0.8%, with a maximum of 3.94%. CMA-MESO and CMA-DB have higher forecast hit rates and TS scores within 12 h than 24 h, while CMA-BJ is the opposite.

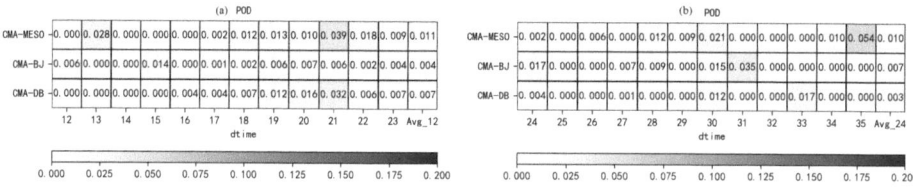

Fig. 1 Verification scores of short-term heavy rainfall caused by northeast cold vortex rainfall: (a) 12 h and (b) 24 h.

The success rate reflects the proportion of actual occurrences in the forecast positive samples. The forecast success rate of the three models within 12 h is higher than 24 h, with an average of 14.5% and a maximum of 73.9%. The above results show that during the Northeast Cold Vortex rainfall process within the study period of this paper, the forecast effect of CMA-MESO is better than the other two regional models (higher hit rate, TS score, and success rate, and lower false alarm rate). The possible reasons for the low hit rate and high false alarm rate of short-term heavy rainfall forecasts by multi-regional high-resolution models include: Northeast Cold Vortex rainfall mainly occurs in the mature stage of the cold vortex, the development of the cold vortex provides a suitable circulation background for rainfall [20], numerical model forecasts of the situation of low-level wind field convergence and high-level divergence, moisture transport conditions, etc., may all lead to a low hit rate; there is a certain error in the model itself; the spatial resolution determines that all three models have limited forecasting ability for small and medium-scale systems that produce short-term heavy rainfall. From the analysis at different times, the short-term heavy rainfall forecast of the regional model in 12 h and 24 h both show that the last three times are better.

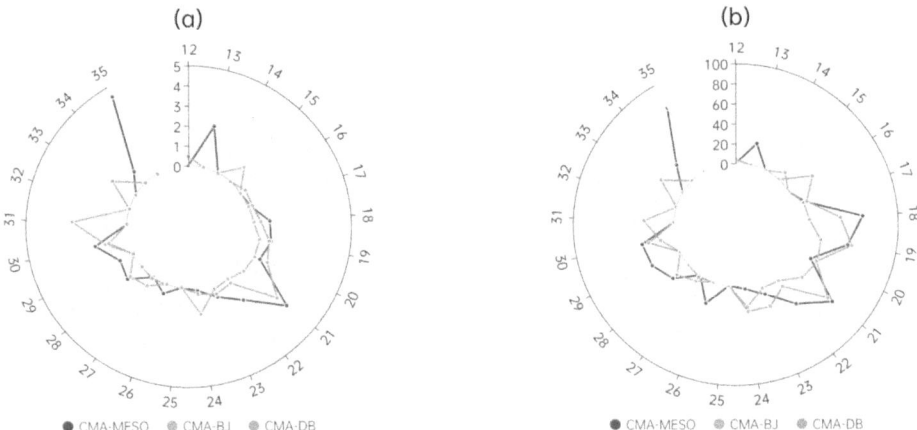

Fig. 2 Verification scores of short-term heavy rainfall caused by northeast cold vortex rainfall: (a) TS score and (b) success rate.

3.2 Results of the Verification of the Forecast of Subtropical High-Pressure Rainfall by Multi-Regional Models

During the process of subtropical high-pressure rainfall, in order to analyze the differences in short-term heavy rainfall forecasting performance of multi-regional high-resolution models at different forecast times, the hit rate and false alarm rate (see Fig. 3) of short-term heavy rainfall forecasts within 12 h~35 h, TS scores and success rates (see Fig. 4) were analyzed. The results show that during the rainfall process affected by subtropical high pressure, the forecast hit rate and TS score of all times of the three models are less than 9%, with a maximum value of 8.2%. Analysis of the average value (Avg_12) of the 12 h short-term heavy rainfall forecast verification shows that, without considering the bias within a radius of 40 km, the average hit rate of the model forecast is 2.4%. The forecast hit rate and TS score of CMA-MESO and CMA-DB within 12 h are higher than those within 24 h, while CMA-BJ is the opposite, which is consistent with the performance of regional models in the process of rainfall in the Northeast cold vortex.

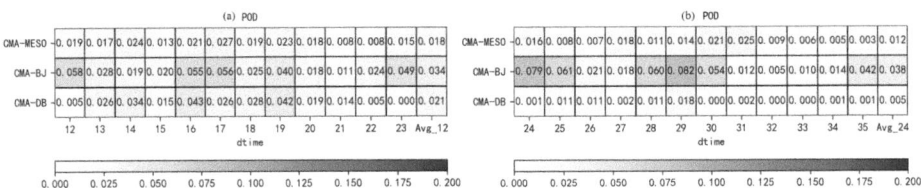

Fig. 3 Verification scores of short-term heavy rainfall forecasts caused by subtropical anticyclone: (a) 12 h and (b) 24 h.

The forecast success rate of the three models within 12 h is higher than that within 24 h, with an average value of 29.49%, and the highest reaching 85.71%. From the verification results of hit rate, TS score and false alarm rate, it can be seen that the forecast effect of CMA-BJ is better than the other two regional models (both hit rate and TS score are high, while the false alarm rate is low); however, from the analysis of the proportion of actual occurrences in the positive samples forecasted by the regional model (success rate), the forecast effect of CMA-MESO model is better than the other two models. This also objectively shows that although CMA-BJ has a lower miss rate and has achieved a higher hit rate and TS score, its false alarm rate is higher, so the success rate is lower. From the analysis of different forecast times, the short-term heavy rainfall forecast of the regional model performs better in the middle of the 12 h and 24 h forecast times than at the ends.

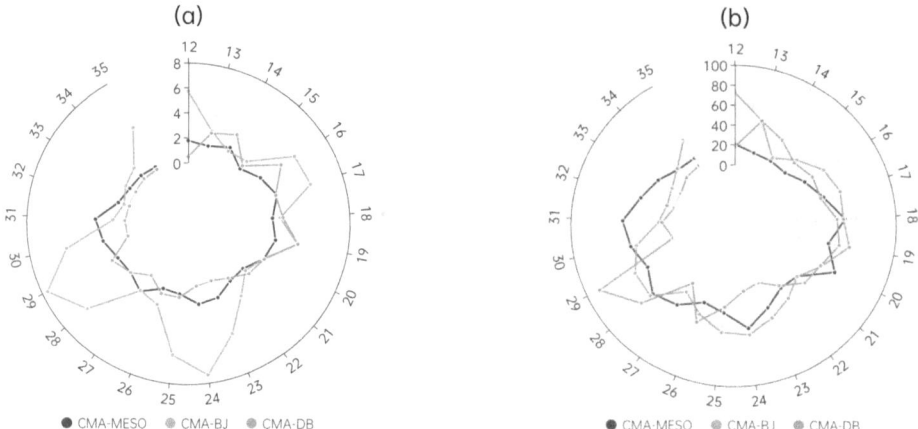

Fig. 4 Verification scores of short-term heavy rainfall caused by subtropical anticyclone: (a) TS score and (b) success rate.

3.3   Verification Results of Multi-Regional Model Forecasts for Typhoon Rainfall

Liaoning Province is located in the mid-latitude region far from the typhoon source. The heavy rain weather here is often caused by the interaction of typhoons and westerly belt systems, referred to as typhoon long-distance heavy rain [21]. According to numerical test results, the intensity of the typhoon and the westerly trough affects the size of the typhoon long-distance heavy rain. Strengthening (or weakening) the typhoon (westerly trough) intensity will correspondingly increase or decrease the intensity of the heavy rain [22]. This type of heavy rain caused by the long-distance influence of typhoons often lasts for more than 6 h, and the 24-h cumulative precipitation exceeds 100 mm [23]. In 2019, Shandong Province was affected by the long-distance super typhoon "Lekima" (1909), with more than 80 stations having hourly rainfall greater than 40 mm, and the number of stations with short-term heavy rainfall was even greater, and the distribution of heavy rainfall stations was uneven [24].

During the typhoon rainfall process, in order to analyze the differences in the short-term heavy rainfall forecast performance of multi-region high-resolution models at different forecast times (see Fig. 5), the hit rate,TS score, and success rate of short-term heavy rainfall forecasts within 12 h and 24 h were analyzed. The results show that: affected by the long-distance rainfall of the typhoon, the forecast hit rate and TS score of all times of the three models are less than 19%, with a maximum of 18.6%. Analysis of the average value of the 12 h short-term heavy rainfall forecast verification (Avg_12) shows that, without considering the bias within a radius of 40 km, the average hit rate of the model forecast is 3.2%. The forecast hit rate and TS score of the three regional models are higher within 24 h than within 12 h, and the hit rate and success rate of short-term heavy rainfall significantly decrease after adjusting for the near time. From the results of the hit rate tests (see Fig. 5), it can be seen that the hit rate of CMA-BJ is higher than the other two models, but its false alarm rate is also higher. Therefore, the TS score (see Fig. 6) does not show a clear advantage. This method of sacrificing false alarm rate for hit rate also leads to a lower success rate. From the analysis at different times, the short-term heavy rainfall forecast of the regional model performs better in the later periods within 12 h and 24 h than the earlier periods. The performance of the CMA-BJ model is the most obvious, with the highest hit rate for short-term heavy rainfall forecasts in the 33 to 35 periods.

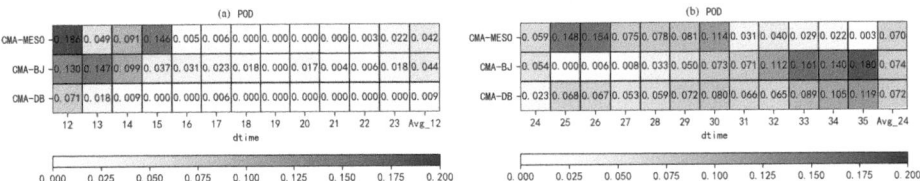

Fig. 5 Verification scores of short-term heavy rainfall forecasts caused by typhoon: (a) 12 h and (b) 24 h.

3.4 Results of Rainfall Forecast Test for the Influence of Other Systems by Multi-Regional Models

Summer heavy rain in Liaoning Province is mainly affected by the Northeast cold vortex, subtropical high pressure, typhoon system, and other systems such as shear line, upper trough, and surface cyclone (referred to as other systems). The hit rate, false alarm rate (see Fig. 7), TS score, and success rate (see Fig. 8) of short-term heavy rainfall forecasts under the influence of other systems by regional models can be seen: the hit rate and TS score of all forecasts of the three models are less than 8%, with a maximum value of 8.1%. The average value (Avg_12) of the short-term heavy rainfall forecast test at the near time (12 h) shows that the average hit rate of the model forecast is 1.54% without considering the deviation within a radius of 40 km. The hit rate and TS score of the CMA-BJ model within 12 h are higher than 24 h. The performance of the other two models after adjustment at the near time is not much different from before the adjustment.

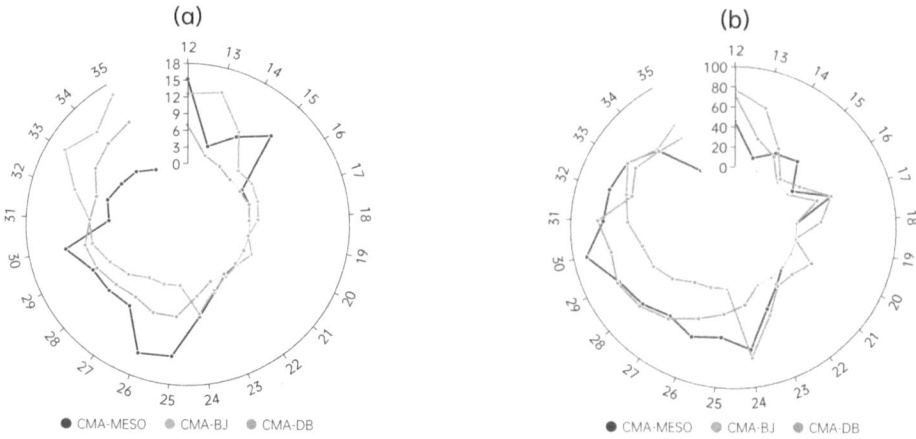

Fig. 6 Verification scores of short-term heavy rainfall caused by typhoon: (a) TS score and (b) success rate.

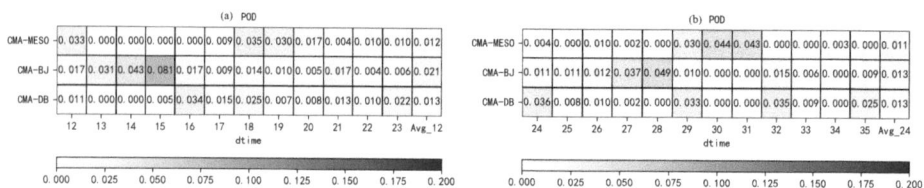

Fig. 7 Verification scores of short-term heavy rainfall forecasts caused by other systems: (a) 12 h and (b) 24 h.

From the results of the hit rate, TS score, and false alarm rate tests, it can be seen that 12 Within the 12-h period, the hit rate of CMA-BJ is higher than the other two models, but its false alarm rate is also higher, resulting in a lower success rate. Within the 24-h period, the hit rate of CMA-BJ is close to the other two models, but its false alarm rate is the lowest, therefore, its success rate is also the highest, at 31.5%. The level of the false alarm rate determines the size of the success rate. Analyzing the hit rate and TS score at different times, the short-term heavy rainfall forecast of the regional model performs better in the 12 h and 24 h periods, with the middle period within 12 h performing better than the two ends.

3.5 Main Evaluation Scores by All Models

Table 3 showcases that in the system of Northeast cold vortex, the CMA-MESO model performed good on short-term heavy precipitation where the TS score was the highest of three models (1.039%). However, in the system of subtropical high-pressure, typhoon and other systems, the CMA-BJ showed the best performance. The CMA-MESO model correctly predicted the number of rainfall stations, but the false alarm was also high, resulting in the low TS score.

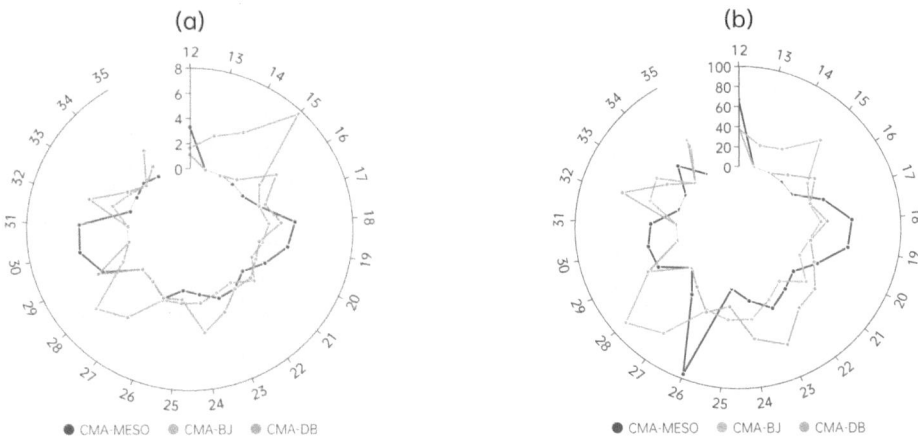

Fig. 8 Verification scores of short-term heavy rainfall caused by other systems: (a) TS score and (b) success rate.

Table 3 TS scores of all the models (unit: %)

Models	CMA-MESO	CMA-BJ	CMA-DB
Northeast cold vortex rainfall	1.039	0.511	0.506
Subtropical high-pressure rainfall	1.405	3.370	1.256
Typhoon rainfall	4.904	5.260	3.714
other systems	1.114	1.636	1.224

4 Conclusion and Discussion

Currently, there are many studies evaluating the effects of long-duration rainfall forecasts. For multi-regional high-resolution models, there are not many studies based on different weather systems and hourly rainfall checks. Liaoning Province has now included hourly rainfall in the warning signal release standards. This paper combines practical business and conducts a short-term heavy rainfall neighborhood test on the commonly used multi-regional model in the northern region, digging from different angles to provide reference for forecasters.

(1) In the process of Northeast cold vortex rainfall, overall analysis from different forecast periods shows that the CMA-MESO forecast effect is better than the other two regional models. CMA-BJ performs better in the 24 h forecast than in the 12 h, and CMA-DB performs better in the 12 h forecast than in the 24 h. The three models have an average hit rate of less than 0.8% within a radius of 40 km without considering bias. The low hit rate and high false alarm rate result in an average success rate of less than 15% for the three models, with CMA-MESO having the highest success rate within 24 h, with an average value of 20%.

(2) In the process of subtropical high-pressure rainfall, the CMA-BJ forecast effect is better than the other two regional models, and both CMA-MESO and CMA-DB show better results for nearby forecasts. In the 12 h period, the average hit rate of the three models is below 2.4%, the TS score average is 2%, and the hit rate (false alarm rate) is high (low), so the success rate of the three models is relatively high, with an average value of 31.25%, with CMA-BJ having the highest success rate, with an average value of 40%. The model's forecast performance in the process of subtropical high-pressure rainfall is better than that of Northeast cold vortex rainfall.

(3) In the process of typhoon rainfall, the CMA-MESO forecast effect is better than the other two regional models, and all three models show better results for the 24 h forecast. In the 24 h period, the average hit rate of the three models is 7.2%, the TS score is 6.6%, the forward forecast is less and the false alarm rate is low, so the success rate is relatively high, with an average value of 46%, with CMA-MESO having the highest success rate, with an average value of 56%. The regional model's forecast performance in the typhoon rainstorm process of 2022–2023 is significantly better than that of the Northeast cold vortex and subtropical high-pressure rainstorm.

(4) When affected by systems such as upper troughs, surface cyclones, and low-level shear, the CMA-MESO forecast effect is better than the other two regional models, and all three models show better results for the 12 h forecast. In the 12 h period, the average hit rate of the three models is 1.5%, the TS score is 1.5%, the success rate is an average of 22.7%, with CMA-DB having the highest success rate, with an average value of 24.3%. This study compares the test features of commonly used multi-regional models in business, providing a reference for forecasters to use multi-regional models, and providing a basis for model developers to discover model forecast bias tendencies from the phenomenon level and make targeted improvements.

The comprehensive test results believe that: in the process of Northeast cold vortex rainfall, the model's forecast effect on heavy rainfall before and after the evening is better; in the process of subtropical high-pressure rainfall, CMA-BJ has a higher hit rate for heavy rainfall after 20 o'clock; in the process of typhoon-type rainfall, attention should be paid to the 24 h forecast results, and the deviation is larger after adjusting the nearby 12 h forecast.

Here didn't use the reanalysis data as reference mainly for two reasons, it must contain errors induced by numerical models, assimilation schemes and changes in the observable system [25]. The values of reanalysis precipitation is generally higher than the observation [26]. Due to the limitations of the time scale and the number of rainfall dates, the test results cannot fully accurately describe the ability of regional model forecasts. Different regional models show significant differences in their forecasting performance during the rainfall process in Liaoning area, and the factors leading to different forecasting effects need further research.

Acknowledgements. This paper is jointly funded by the Open Fund Project of Shenyang Institute of Atmospheric Environment, China Meteorological Administration and Northeast Cold Vortex Research Key Open Laboratory (NO. 2023SYIAEKFMS17), and the annual fund project of Liaoning Provincial Meteorological Bureau (2023).

References

1. Sun, X., Chen, C., Liang, H., et al.: Synthetic analysis of severe storm environment and physical quantity field in Liaoning area. J. Meteorol. Environ. **32**(5), 40–46 (2016)
2. Li, Y., Ye, D., Gao, G., et al.: Climate characteristics and major weather/climate events in China in the summer of 2022. J. Atmos. Sci. **46**(1), 110–118 (2023)
3. Liu, J., Ren, C., Dong, W., et al.: Diagnostic analysis of a heavy rainfall process in Liaoning area. J. Meteorol. Environ. **32**(2), 18–27 (2016)
4. Qi, L., Xu, J.: Short-term forecast analysis and reflection of the "7·9" heavy rain in northern Henan. Meteorology **44**(1), 1–4 (2018)
5. Gofa, F., Boucouvala, D., Louka, P., et al.: Spatial verification approaches as a tool to evaluate the performance of high resolution precipitation forecasts. Atmos. Res. **208**, 78–87 (2018)
6. Chakraborty, A.: The skill of ECMWF medium-range forecasts during the year of tropical convection 2008. Mon. Wea. Rev. **138**(10), 3787–3805 (2010)
7. Liu, J., Chen, C., Yan, J., et al.: Distribution and characteristics analysis of red warning signals for heavy rain in Liaoning Province from 2015 to 2019. J. Meteorol. Environ. **37**(1), 100–105 (2020)
8. Hao, L., He, L., Ma, N.: Characteristics of the Northeast Cold Vortex Climate and Its Impact on Summer Precipitation in the Haihe River Basin. J. Meteorol. **81**(4), 559–568 (2023)
9. Zhang, J., He, L., Li, J., et al: Preliminary investigation of the characteristics and causes of the "23·7" extreme rainstorm in Hebei. J. Atmos. Sci. https://link.cnki.net/urlid/32.1803.p.20231007.1804.006
10. Ben Bouallegue, Z., Theis, S.E.: Spatial techniques applied to precipitation ensemble forecasts: Form verification result probabilistic. Meteorol. Appl. **21**(4), 922–929 (2014). https://doi.org/10.1002/met.1435
11. Li, B., Dai, J., Zhang, X., et al.: Fuzzy verification test and comparison of three types of severe convective weather nowcasts. Meteorology **42**(2), 129–143 (2016)
12. Wang, X., Li, H.: Spatial verification evaluation of multiple numerical models for typhoon rainstorm process forecast. Meteorology **46**(6), 753–764 (2020)
13. Liu, J., Cai, K., Tan, Z.: Analysis of the forecasting ability of high-resolution model radar echo in different types of rainfall. Meteorology **45**(12), 1710–1717 (2019)
14. Li, H., Wang, X., Zhu, F.: Comprehensive evaluation of the forecasting performance of multiple models for the "21·7" extreme rainstorm in Henan. J. Atmos. Sci. **45**(4), 573–590 (2022)
15. Xue, W., Yu, H., Tang, S., et al.: Verification evaluation of near-surface wind speed forecast of Shanghai Rapid Update Assimilation Numerical Forecast System (SMS-WARR). Meteorology **46**(12), 1529–1542 (2020)
16. Liu, J., Ren, C., Zhao, Z., et al.: Verification analysis of strong precipitation forecast of multi-region high-resolution model. Meteorology **48**(10), 1292–1302 (2022)
17. Su, X., Liu, M., Kang, Z., et al.: Verification of short-term rainstorm forecast in Jiangsu Main Flood season in 2020. Meteorology **48**(3), 357–371 (2022)
18. Zheng, Y., Zhou, K., Sheng, J., et al.: Progress in monitoring, forecasting and warning technology for severe convective weather. J. Appl. Meteorol. **26**(6), 641–657 (2015)
19. Tang, W., Zhou, Q., Liu, X., et al.: National-level severe convective weather classification forecast verification analysis. Meteorology **43**(1), 67–76 (2017)
20. Liu, D., Wang, L.: The structural characteristics and precipitation causes of a Northeast cold vortex in early spring. J. Atmos. Sci. **45**(3), 456–468 (2022)
21. Chen, L., Li, Y., Cheng, Z.: An overview of research and forecasting on rainfall associated with landfalling tropical cyclones. Adv. Atmos. Sci. **27**(5), 967–976 (2010)

22. Zhu, H., Chen, L., Xu, X.: The interaction of mid-latitude circulation systems and the simulation study of their storm characteristics. Atmos. Sci. **24**(5), 669–675 (2000)

23. Cote, M.R.: Predecessor rain events in advance of tropical cyclones. MS thesis, Department of Atmospheric and Environmental Sciences. University at Albany, State University of New York (2007)

24. He, L., Chen, S., Guo, Y.: Observational characteristics and causes of extreme heavy rainfall from Typhoon Lekima (1909). J. Appl. Meteorol. **31**(5), 513–526 (2020)

25. Zhao, T., Fu, C., Ke, Z., et al.: Global atmosphere reanalysis datasets: current status and recent advances. Adv. Earth Sci. **25**(3), 242–254 (2010)

26. Wang, C., Huang, A., Zheng, P., et al.: Applicability evaluation of China's First Generation of Global Land Surface Reanalysis(CRA40/Land)Air Temperature and Precipitation Products in China Mainland. Plateau Meteorol. **41**(5), 1325–1334 (2022)

Comparative Analysis of the Causes of Two Sudden Mountainous Rainstorms Occurred in Mianyang in 2022

Benhe Yuan[1,3], Qingyan Zhang[2,3(✉)], Zichuan Sun[2,3], Cong Wan[2,3], Xin Ouyang[2,3], and Meng Meng[2,3]

[1] Sichuan Provincial Meteorological Observatory, Chengdu 610072, China
[2] Mianyang Municipal Meteorological Bureau, Mianyang 621000, China
395247424@qq.com
[3] Key Laboratory of Plateau and Basin Rainstorm Drought and Flood Disaster, Chengdu 610072, China

Abstract. Compared to general rainfall events, sudden rainfall has always been a challenge and focus of weather forecasting and research. In mid-July 2022, sudden heavy rainfall occurred in the northwest of the Sichuan Basin, resulting in the severe disasters. To better understand the causes of such rainfall events, a comparative analysis of the physical environment and dynamic characteristics of these two sudden rainfall events is conducted based on observational data and ERA5 reanalysis data. The results showed that both temperature and moisture environments of the events exhibited sudden changes. Before the onset of rainfall event, energy, moisture, and instability are increased explosively in a short period, reaching extreme conditions. Intense upward motion occurred as mid-to-upper-level cyclonic vorticity developed downward to the lower levels. Rainfall events is occurred during the optimal period when vorticity and moisture were coupled, with a good correspondence between the maximum wet vorticity area and the rainfall area, indicating the diagnostic significance of wet vorticity for identifying rainfall areas.

Keywords: Slope Rainstorm · Comparative Analysis · Wet Helicity · Model Rainfall · Forecast Ability

1 Introduction

Rainfall, characterized by multiscale interactions of different spatial and temporal features, induces significant hazards in meteorology. The rainfall events are prone to occurred in Sichuan Basin, accompanied with the westward subtropical high (hereafter referred to as the high) in summer, resulting in convective weather processes in China [1–3]. Within the boundary of the high at 500hPa [4], and the widespread sinking airflow results in hot and dry conditions. Conversely, outside the boundary, due to significant moisture and dynamic thermal instability, convection, especially when the westerly trough approaches, is easily generated. Therefore, rainfall weather various

B.-J. He et al. (Eds.): UCSUD 2023, LNCE 559, pp. 61–71, 2025.
https://doi.org/10.1007/978-981-97-8401-1_5

uncertainties, especially sudden rainfall near its boundary, which poses challenges for forecasting. Sichuan is located in the east of the Qinghai-Tibet Plateau, where systematic and persistent rainfall events often occur in summer. These rainfall events are typically associated with the "east high, west low" pattern formed by the high and low-value systems on its northwest edge [5–8]. Wang et al. pointed out that convective activity is more likely to occur on the outer side of the dynamic boundary, near the moisture-thermal and convective instability boundary. From a mechanistic perspective, Chen et al. showed that Sichuan rainfall is often related to the coupling of upper and lower-layer precipitation weather systems, gravity waves near terrain, secondary vertical circulation, and enhanced upward motion, such as the dragging effect of terrain waves in the western mountainous areas on the warm and humid southeasterly airflow, which further strengthens horizontal convergence and upward motion in the western rainfall area [9]. These studies have enhanced our understanding of Sichuan rainfall.

In early July 2022, Sichuan experienced high temperatures and little rainfall during the day and sudden heavy rainfall at night. Among these events, on July 11th, from 23:00 to 4:00 on July 12th, short-term heavy rainfall occurred, leading to debris flows in Muzuo Township, Pingwu County, resulting in 4 deaths and 14 missing persons. In the early hours of July 16th, the heavy rainfall event causes flash floods broke out of the river channel, 8 deaths and 10 missing persons. The precipitation event appeared in Pingwu County, Mianyang City, from 0:00 to 6:00 on July 16th, with a maximum hourly rainfall intensity of 55.7 mm and a maximum cumulative rainfall of 192.1 mm over 6 h, exhibits significant characteristics of sudden heavy rainfall, such as rapid development, intense rainfall, and short duration. Model forecasting skill for such rainfall events is relatively low. This paper selects two events for comparative analysis aim to identify the evolution of relevant physical quantities and the causes of occurrence, improving the forecasting ability for such events and provide a basis for preventing disasters such as debris flows, flash floods, and urban waterlogging.

2 Data

Meteorological data is from the China Meteorological Administration's Meteorological Business Observation Network at 08:00 and 20:00 every day, including upper-air data; FY-2G satellite cloud top brightness temperature data from the China Meteorological Administration's Meteorological Satellite Center, with a horizontal resolution of $0.1° \times 0.1°$ and a temporal resolution of 1 h; Hourly encrypted observation precipitation data from Sichuan Province's surface; ECMWF's ERA5 reanalysis data, with a horizontal resolution of $0.25° \times 0.25°$ and a temporal resolution of 1 h.

3 Process Comparison

3.1 Background Field

Before the onset of the event, the Sichuan Basin experienced clear and hot weather with little rainfall, with energy gradually accumulating. Pseudo-equivalent potential temperature (θse) can effectively reflect the moist and warm conditions of the entire atmospheric column and the unstable state of the atmospheric stratification with changes in height. Moisture in the atmosphere is mainly concentrated below 700 hPa, and the sum of specific humidity (Q) at 700 hPa, 850 hPa, and 925 hPa can approximately represent the water vapor content in the atmospheric column. Analysis in Table 1 shows that within 12 h before the onset of the event, unstable energy, moisture, and unstable stratification increased rapidly, with surface maximum temperatures exceeding the convective temperature (Tg) by more than 2.0 °C. Various physical quantities exhibited explosive growth in a short period as the event approached, reaching relatively extreme levels.

Table 1. Characteristics of the element field at Wenjiang Station before and during the process

Time	CAPE (J/kg)	K (°C)	SI (°C)	Q 850+700+925 (g/kg)	$\theta_{se500-850}$ (°C)	θ_{se850} (°C)	Tg (°C)	Highest daytime temperature (°C)
071108	1892	43.9	−1.7	53	−11	93	32.1	34.2
071120	3271	43.9	−2.0	53.8	−27	100	34.8	
071508	2521	48.1	−4.6	56.0	−28	99	34.5	36.9
071520	4968	52.0	−8.6	59	−30	108	36.5	

Note: CAPE (Convective available potential energy), K (Kindex), SI (Showalter index), Q (specific humidity), θse (pseudo-equivalent temperature), Tg (convective temperature)

3.2 Upper Air Situation

At 500hPa, before the start of the 7.12 process, there is a broad trough over the northwest region, with many short-wave activities at the bottom of the trough. The sub-high ridge line is near (25° N,102° E) (Fig. 1(a)). The western part of the basin is always located on the northwest side of the 588 line, which is conducive to the occurrence of heavy rain in the western part of the basin [9] The situation before the 7.16 process is similar to the 7.12 process, but the sub-high is stronger, with the ridge line located near (30° N, 90° E) (Fig. 1(b)). Sichuan is controlled by the subtropical high pressure, with wind speed of 4–6 m/s at 08:00 on the 15th, decreasing to 2–4 m/s within the sub-high by 20:00, indicating a weakening trend of the sub-high. It reduces the suppression of the rising air flow. Moreover, the northern boundary of the sub- high falls southward, and the western part of the basin is on the northwest side of the sub-high. The influencing systems of both processes at 500 hPa are relatively weak, belonging to weak baroclinic processes, but the cold advection in the 7.16 process is significant.

At 700 hPa, before the 7.12 process, the flow from Yunnan and Guizhou to southern Shaanxi is southerly, there is a negative temperature change zone of 2–4.7 °C in the northern part of Gansu, and there is a cold shear from southern Gansu to the northwest of the basin. At 20:00 on the 11th, the wind in Mianyang is a southeast wind of 4–8 m/s, forming a dynamic convergence with the low-level northerly flow. At 01:00 on the 12th (Fig. 2(a)), the southerly wind in the eastern part of Mianyang increased to 8–12 m/s,

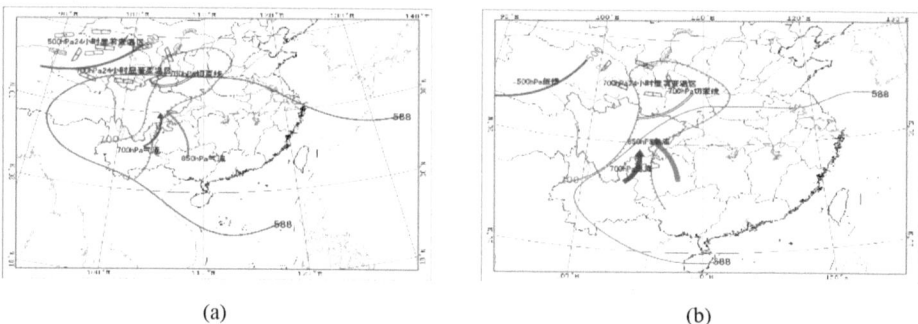

Fig. 1. (a) System configuration at 20:00 on July 11th; (b) System configuration at 20:00 on July 15th.

while the wind speed in the southwest is 4 m/s, forming a wind speed convergence. Due to the weak cold air force, strong convergence is occurred in Gansu, while heavy rainfall only occurred in the northwest of Mianyang; during the 7.16 process, at 20:00 on the 15th, the wind in the middle of Mianyang is a southwest wind of 2–4 m/s, while the east is a southeast flow of 4–6 m/s. There is a strong negative temperature change zone from Qinghai to Gansu, with a negative temperature change center of 7.2 °C in Xining, and a shear from southern Shaanxi to the north of our province. At 23:00 on the 15th, Mianyang city turned into a southeast flow of 2–6 m/s, and at 04:00 on the 16th, the wind speed in the eastern part of Mianyang increased to 6–8 m/s (Fig. 2(b)), and the easterly component significantly increased, while the wind speed in the middle remained at 4 m/s, and the west turned into a northeast wind of 4–8 m/s. Therefore, there is not only wind speed convergence but also wind direction convergence over the west of Mianyang,. The easterly component began to decrease at 5 o'clock, and the precipitation weakened.

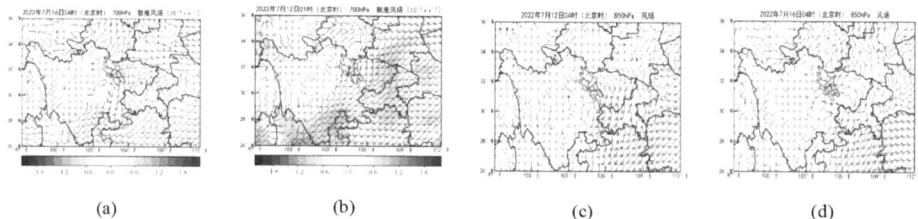

Fig. 2. 700 hPa divergence wind field (a) at 1:00 on the 12th and (b) at 4:00 on the 16th, and 850 hPa divergence wind field (c) at 4:00 on the 12th and (d) at 4:00 on the 16th (divergence unit:10^{-5} s^{-1}).

At 850 hPa, during both processes, the basin maintains an inverted trough, and the west changes from an easterly flow to a northeasterly flow. The difference is that the cold air force is weaker in the 7.12 process, and the northeasterly flow that moves south has a wind speed of only 2–4 m/s (Fig. 2(c)), while the northeasterly wind reaches the jet stream in the 7.16 process (Fig. 2(d)).

4   Coupling Diagnosis Analysis of Rainstorm Area

Based on the analysis of vorticity, vertical velocity, and moist potential vorticity (MPV) evolution in height-time cross-sections every three hours is conducted to reflect the dynamic-moisture coupling characteristics of the two events. The dynamical-moisture coupling characteristics of two processes were analyzed using ERA5 data. The center of the heavy rain on July 12th is located in Muzuo Township, Pingwu County (104.5° E, 32.6° N), therefore, the vorticity, vertical velocity, and moist helicity were analyzed based on this grid point, observing their 3-hourly height changes over time. As seen in Fig. 3(a), before the heavy rain at 9:00 on the 11th, the center of the cyclonic vorticity is located near 700 hPa, with a maximum value reaching $15 \times 10^{-5} \cdot \mathrm{s}^{-1}$. At 10:00, vertical upward motion is established, above -4×10^{-1} Pa·s^{-1}. During the heavy rain, the cyclonic vorticity developed in the lower layer, from 16:00 on the 11th to 7:00 on the 12th, the maximum upward speed reached below -10×10^{-1} Pa·s^{-1}. Near 200 hPa, the vertical upward motion rapidly increased. During the strongest period of the heavy rain from 22:00 on the 11th to 4:00 on the 12th, the lower layer cyclonic vorticity strengthened, with a central value of $-16 \times 10^{-5} \cdot \mathrm{s}^{-1}$, concentrated below 600 hPa. From the evolution of moist helicity and wind field (Fig. 3(b)), the large negative helicity area is located between 450 and 650 hPa, reaching below -45×10^{-10} kg·m^{-2}·s^{-3}, appearing during the strongest precipitation period from 22:00 on the 11th to 4:00 on the 12th. To further illustrate the connection between moist helicity and heavy rain, the 650hPa moist helicity is selected for analysis. As can be seen from Fig. 4, during the heavy rain, the negative moist helicity over Muzuo Township reached below -20×10^{-10} kg·m^{-2}·s^{-3}. Starting at 23:00 on the 11th, the negative moist helicity continued to increase, corresponding to an increase in precipitation, until 4:00 on the 12th, when the negative moist helicity began to decrease, and the precipitation weakened.

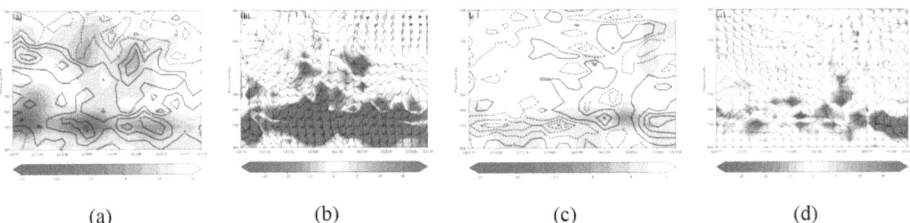

(a)   (b)   (c)   (d)

Fig. 3. Time-height cross-sections of (a) vorticity (contour lines, unit: $10^{-5} \cdot \mathrm{s}^{-1}$), (b) moist helicity (unit: 10^{-10} kg·m^{-2}·s^{-3}), (c) vertical velocity (color-filled, unit: 10^{-1} Pa·s^{-1}) and (d) horizontal wind (unit: m/s) at the centers of heavy rain of (104.5° E, 32.6° N) and (103.9° E, 32.05° N).

The center of the large negative moist helicity reached below -20×10^{-10} kg·m^{-2}·s^{-3}. Starting at 23:00 on the 11th, the negative moist helicity continued to increase, corresponding to an increase in precipitation, until 4:00 on the 12th, when the negative moist helicity began to decrease, and the precipitation weakened.

The rainstorm center on July 16th appeared in Qingpian Township, Beichuan County (103.9°E, 32.05°N). Before the rainstorm at 10:00 on the 15th, as shown in Fig. 3(c), the cyclonic vorticity center is near 700hPa, with a maximum value of $16 \times 10^{-5} \text{ s}^{-1}$, and the vertical upward movement is established at 16:00. During the rainstorm from 1:00 to 10:00 on the 16th, the maximum upward speed is below $-10 \times 10^{-1} \text{ Pa s}^{-1}$, the height is below 600 hPa, the vertical upward movement rapidly increased, and the cyclonic influence system is still concentrated in the lower layer, but the intensity did not increase. From the moisture vorticity and wind field evolution in Fig. 3(d), it can be seen that the negative vorticity high value area is located at 600–750 hPa, reaching below $-45 \times 10^{-10} \text{ kg·m}^{-2}\text{·s}^{-3}$, and it corresponds to the strongest precipitation period from 1:00 to 10:00 on the 16th; the 650hPa moisture vorticity is selected for analysis, as can be seen from Fig. 5, during the rainstorm, the sky above Qingpian Township is the negative moisture vorticity high value center reaching below $-20 \times 10^{-12} \text{ kg·m}^{-2}\text{·s}^{-3}$; starting at 0:00 on the 16th, the negative moisture vorticity continues to increase, corresponding to an increase in precipitation.

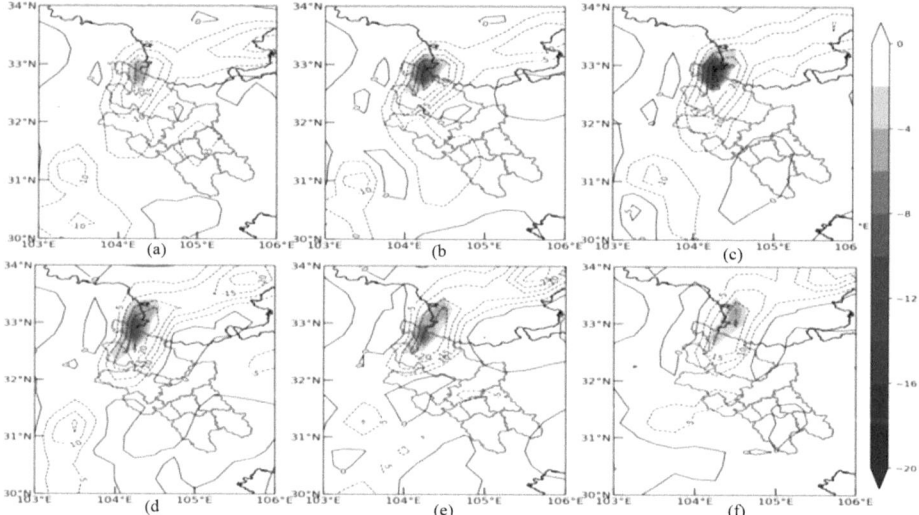

Fig. 4. 650h Pa moisture vorticity (dashed line, unit: $10^{-10} \text{ kg·m}^{-2}\text{·s}^{-3}$) and vertical velocity distribution (filled color, unit: $10^{-1} \text{ Pa·s}^{-1}$) (a-f represent the hourly changes from 23:00 on the 11th to 04:00 on the 12th, the blue box in a represents the Pingwu rainstorm area).

5   Satellite Cloud Images, Radar Echo Comparative Analysis

Studies have shown that rainfall in the Sichuan region is often associated with the activity of mesoscale convective systems (MCS), especially during short-duration rainfall events [10–15]. Both events exhibited significant convective activity. The development and evolution of MCS will be explored through analysis of satellite and radar data. Satellite cloud images, on July 11th, the cloud system on the periphery of the sub-high is active during the day, there is a northeast-southwest shear cloud system in the western part of the basin, and convection began to develop on the south side of the shear cloud system in

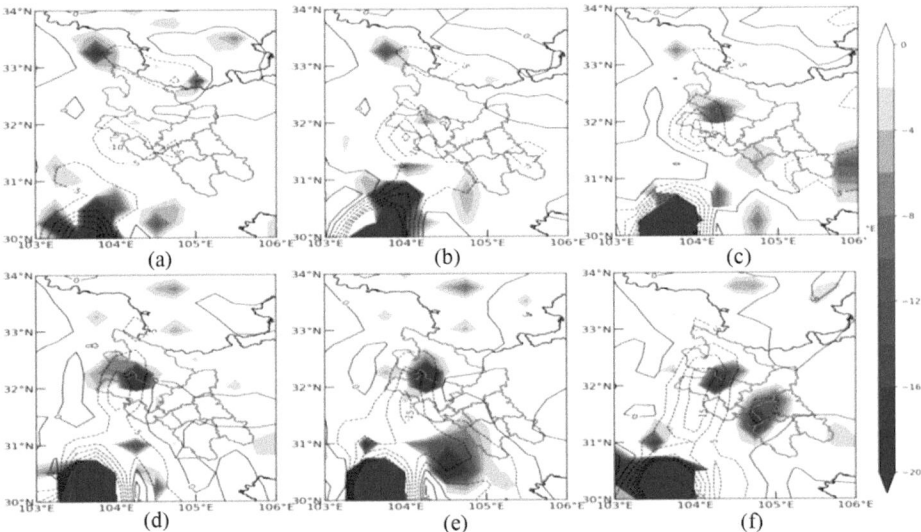

Fig. 5. 650 hPa moisture vorticity (dashed line, unit: 10^{-12} kg·m^{-2}·s^{-3}) and vertical velocity distribution (filled color, unit: 10^{-1} Pa·s^{-1}) (a-f represent the hourly changes from 0:00 to 5:00 on the 16th, the blue box in a represents the Beichuan rainstorm area).

the evening as shown in Fig. 6(a). Around 23:00, the convective cloud on the northwest side of the shear cloud system merged with the convective cloud group on the south side as shown in Fig. 6(b). At 2:00 on the 12th, a northward cloud group formed in the southwestern part of the basin merged with the convective cloud developing on the shear line in the west of Mianyang as shown in Fig. 6(c). At this time, the hourly rainfall in Guixi Yaowang Valley in Beichuan reached 42.6 mm; at 06:00, the cloud group structure is loose, and the rainfall intensity is significantly weakened.

The weather is fine during the day on the 15th, and convective cloud clusters formed in the slope area around 20:00. At 23:00, they intensified and merged into a band as shown in Fig. 6(d). At the same time, strong convective clouds developed in the southern part of Gansu, and heavy rain began on the western slope. At 1:00 on the 16th, the rain cloud cluster in the southern part of Gansu had moved into the western part of Mianyang City, merging with the cloud cluster that had formed and developed in the southwest of the basin and moved northward. This period is the strongest rainfall period in Mianyang City, with the maximum hourly rainfall in Anzhou from 2:00–3:00 reaching 80.7 mm. At 3:00, the western part of Mianyang City is covered by strong rain cloud clusters as shown in Fig. 6(e), and at 6:00, the rain cloud cluster pressed south to the southwest of the basin as shown in Fig. 6(f), and the heavy rainfall in Mianyang weakened.

Radar echo, the main rainfall period of the 7.12 process is analyzed from 2:00–5:00 on the 12th: the echo gradually developed from 2:00, and a northeast-southwest oriented band echo formed at 2:30, with convective cells developing within the echo band. The strongest echo in Muzuo reached 45dbz. Around 4 am, the echo in the central and southern parts weakened, but the strong echo in the northern section persisted, and convective cells continued to form on its southwest side and moved towards Muzuo as

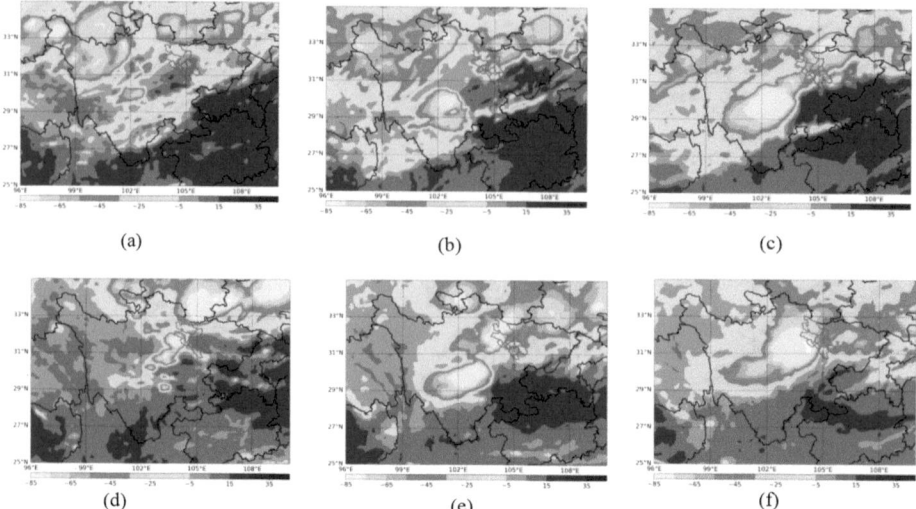

Fig. 6. Cloud evolution in July 2022: (a) 20:30 on 11th, (b) 23:15 on 11th, (c) 2:00 on 12th, (d) 23:15 on 15th, (e) 3:00 on 16th, and (f) 6:00 on 16th.

shown in Fig. 7(a)(b) within the red box area, forming a train effect. During this period, Muzuo and its northern Longnan area continued to rain. The vertical section shows that the centroid height is maintained at about 4km, with the maximum value of the centroid reaching 55 dbz. The horizontal distance of the high-value area is small, the precipitation efficiency is high, and it showed typical small-scale convective cell characteristics.

The process analysis of the strongest precipitation period from 1:00–4:00 on the 16th shows that at 2:00 (Fig. 8(c)), there is a strong echo in the northern part of Mianzhu to the eastern part of Mao County and the junction of Pingwu, Beichuan, and Songpan counties. The former gradually propagated in the northwest direction, and the area with an echo of 40 dbz or above gradually expanded and approached the latter. At the same time, the echo at the junction of the three counties developed southward. At 3:30, the echoes merged into a north-south band of echoes and propagated westward. Multiple single-body echoes were successively generated and strengthened at the boundary of the green piece on the west side of the band echo. At 4:00, the echo covered the green piece and Baishi Township, with the single body stable and less moving. At 4:39, it reached the strongest moment of the process with a combined reflectivity of 54 dbz and a horizontal scale of 14.2 km. The centroid of the reflectivity factor maintained a vertical height of 4 km, with a maximum value of 55 dbz or above, and the horizontal distance of the maximum value is 2.8 km.

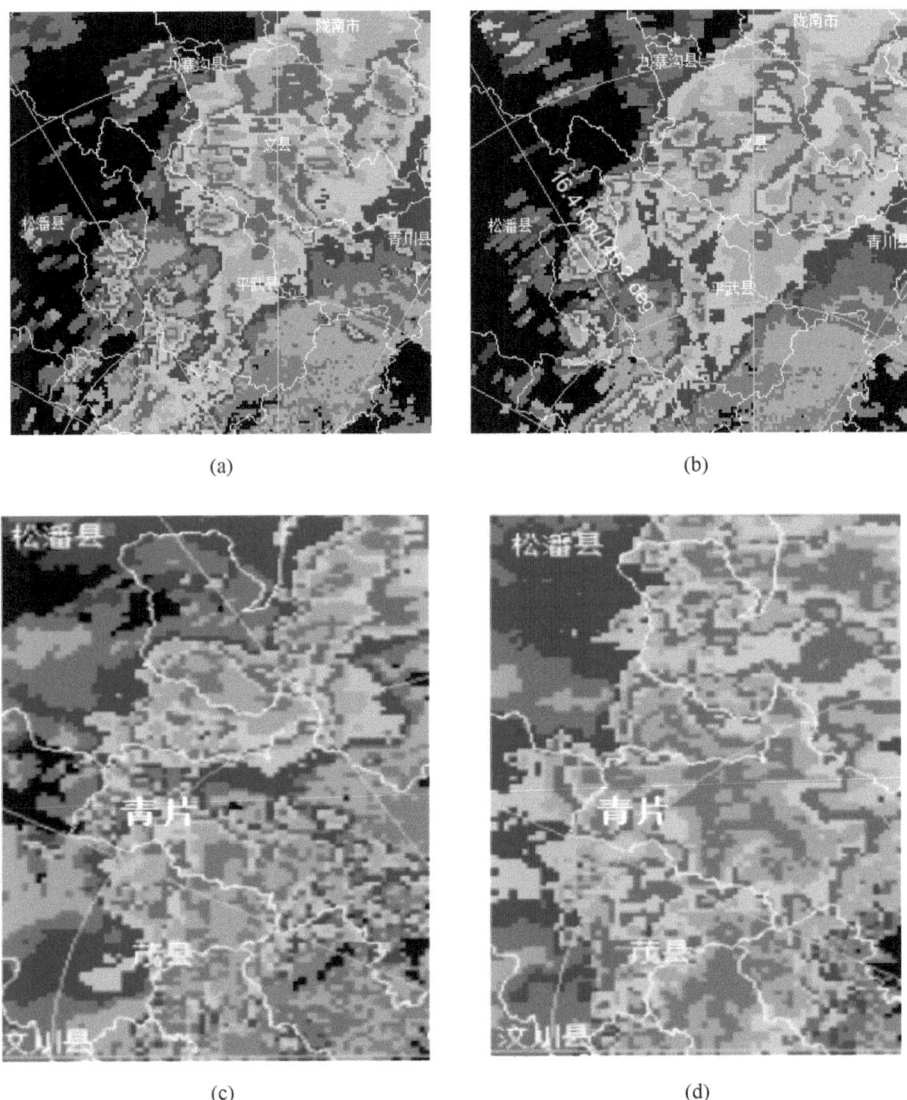

(a)   (b)

(c)   (d)

Fig. 7. Combined reflectivity at (a) 3:57 on 12th, (b) 4:48 on 12th (c) 2:01 on 16th, and (d) 3:31 on 16th.

6   Conclusion

This study conducts a comparative analysis of two sudden heavy rainfall events that occurred in Mianyang City in July 2022 by using encrypted data, Chinese FY meteorological satellite TBB data, and ERA5 reanalysis data. The main conclusions obtained are as follows:

1) Before the onset of the rainfall, there is an explosive growth of moisture, dynamics, and thermodynamics, reaching extreme values in a short period.
2) Satellite and radar data indicate that intense rainfall echoes began on the windward slope of the Longmen Mountains, and strong echoes moved northeastward along the mountain range, gradually intensifying and causing heavy rainfall.
3) Coupled diagnosis of rainfall areas revealed that cyclonic vorticity preceded the formation of strong upward motion. Vertical upward motion is established under the influence of cyclonic vorticity, with cyclonic vorticity developing at lower levels and vertical upward motion rapidly intensifying during the rainfall event. The rainfall occurred during the optimal period of coupling between vorticity and moisture. The maximum wet vorticity area corresponded well with the rainfall area. Compared to single physical quantities representing vertical upward motion, wet vorticity has more indicative significance for diagnosing rainfall areas and can be used as a diagnostic physical quantity for sudden heavy rainfall.

Acknowledgement. This paper is supported by project of "In-depth Review and Summary of Two Sudden Mountain Rainstorms in Mianyang in 2022" from China Meteorological Administration (No. FPZJ2023–113).

References

1. Zongmin Wang, Yihui Ding, Yingxin Zhang, et al.: Edge characteristics of the Western Pacific Subtropical High and the cause of the convective rain belt in its adjacent warm area. J. Meteorol. **72**(3), 1148–1157 (2014)
2. Yongren Chen, Yueqing Li, Dongmei Qi: The changes of the South Asian High and the Western Pacific Subtropical High and their relationship with precipitation. Plateau Meteorol. **30**(5), 1148–1157 (2011)
3. Chunguo Wang, Yongren Chen, et al.: Analysis of atypical storm process in Sichuan Basin. Journal of Yunnan University (Natural Science Edition) **33**(5), 540–547 (2011)
4. Yimin Liu, Guoxiong Wu: Review of subtropical high research and re-understanding of several basic problems. J. Meteorol. **58**(4), 500–512 (2000)
5. Yongren Chen, Chunguo Wang, Yueqing Li: Preliminary analysis of a rainstorm process in Sichuan Basin. Rainstorm Disaster **27**(4), 301–306 (2008)
6. Rui Shi, Yongren Chen, Hongru Xiao: Comparative analysis of the persistent heavy rainfall process in Sichuan Basin in 2013. Plateau Mt. Meteorol. Res. **34**(4), 11–15 (2014)
7. Chen, Y.R., Li, Y.Q., Qi, D.M.: An index reflecting mesoscale vortex-vortex interaction and its diagnostic applications for rainstorm area. Atmospheric Science Letters **20**(6), 1–10 (2019)
8. Yongren Chen, Yueqing Li: Analysis of the causes of the sudden rainstorm in the Panxi region's mianning "6.26". Plateau Mt. Meteorol. Res. **41**(4), 10 (2021). https://doi.org/10.3969/j.issn.1674-2184.2021.04.002
9. Sichuan Provincial Meteorological Bureau: Sichuan Province Forecaster's Handbook, pp. 1–200. Southwest Jiaotong University Press, Chengdu (2014)
10. Chen, Y.R., Li, Y.Q.: Convective characteristics and formation conditions in an extreme rainstorm on the eastern edge of the Tibetan Plateau. Atmosphere **12**(3), 381 (2021)
11. Tao Dai, Yongren Chen, Yueqing Li: Cause analysis of a heavy rainfall event in early summer 2014 in Luzhou. Plateau Mt. Meteorol. Res. **34**(3), 17–22 (2014)

12. Yongren Chen, Yueqing Li: Characteristics of MCS and its formation conditions in a non-typical heavy rainfall event. Plateau Mt. Meteorol. Res. **35**(3), 18–25 (2015)
13. Dixiang Xiao, Nini Tu, Shengxiu Qi: Diagnostic analysis and numerical experiment of heavy rainfall processes along the Longmen mountain. Plateau Meteorol. **34**(1), 113–123 (2015)
14. Jiaolan Fu, Xuekuan Ma, Tao Chen, et al.: Characteristics and meteorological causes analysis of the extreme heavy precipitation event on July 16 in North China. Meteorology **43**(5), 528–539 (2017)
15. Ying Wang, Shaowen Shou, Jun Zhou: Water Vapor Helicity and its application in the analysis of a heavy rainfall event in Jianghuai region. J. Atmos. Sci. **30**(1), 101–106 (2007). https://doi.org/10.3969/j.issn.1674-7097.2007.01.014

Characteristics of Climate Element Changes and Mango Planting Climate Suitability Changes in Panxi Region Under Global Warming

Xuemei Yin[1,2], Ke He[2], Dongdong Chen[3(⊠)], Xuan Li[2], Yuzhu Li[2], and Mingtian Wang[4(⊠)]

[1] Key Laboratory of Plateau and Basin Rainstorm, Drought and Flood Disasters of Sichuan Province, Chengdu 610072, China
[2] Panzhihua Meteorological Bureau, Panzhihua 617000, China
[3] Sichuan Provincial Agricultural Meteorological Center, Chengdu 610072, China
[4] Sichuan Provincial Meteorological Observatory, Chengdu 610072, China
274625239@qq.com

Abstract. Based on the observation data of 32 meteorological stations in Panxi and surrounding areas from 1961 to 2020 and Digital Elevation Model (DEM) data, the climate suitability zoning indicators for mango planting in Panxi region are comprehensively determined. Using linear trend analysis and regression methods, a grid calculation model for zoning indicators is established to analyze the changes in zoning indicators and the suitability of mango planting in Panxi region during the two periods of 1961–1990 and 1991–2020. The results show that the annual average temperature, $\geq 10°C$ accumulated temperature, and overwintering average temperature in Panxi region show a high in the south and low in the north characteristic, increasing or adding at a rate of 0.2°C/10a, 86°C•d/10a, 0.3°C/10a respectively; the sunshine duration is generally distributed in a zonal pattern, increasing at a rate of 16.3h/10a. The main mango planting areas in Panxi region are concentrated in the southwest, the suitable areas are mainly distributed in the central part of Panzhihua City, the sub-suitable areas are mainly located in the central and southern parts of Panzhihua City and Huili City, and the unsuitable areas are mainly concentrated in the central and northern parts. After 1990, the area of suitable climate for mango planting in Panxi region increased by 21.55×10^2 km^2, an increase of 3.2%; the area of sub-suitable area increased by 50.96×10^2 km^2, an increase of 7.5%; the area of unsuitable area decreased by 72.51×10^2 km^2, a decrease of 10.7%. The highest altitude of the suitable area increased by 161m, the highest altitude of the sub-suitable area increased by 221m, the suitable range moved to high-altitude areas, the planting boundary moved northward, and global warming is generally beneficial to mango planting in Panxi.

Keywords: Climate change · Panxi region · Mango · Planting · Climate suitability · Zoning

© The Author(s) 2025
B.-J. He et al. (Eds.): UCSUD 2023, LNCE 559, pp. 72–89, 2025.
https://doi.org/10.1007/978-981-97-8401-1_6

1 Introduction

According to the IPCC's sixth assessment report, the temperature from 2011 to 2020 has risen by 1.09°C compared to 1850–1900 [1, 2]. In recent decades, the trend of temperature rise in China is basically consistent with the trend of global warming, and even slightly higher than the global level [3–6]. Global warming has multiple impacts on human society and agricultural production, causing most of China's agricultural climate zones to shift northward [7–10], crop planting boundaries to move to higher latitudes and altitudes [7, 11], and the planting area for thermophilic crops to increase [12–15]. However, the degree of global warming and its impact on agriculture vary significantly by region [16]. The Panxi region is in the dry and hot valley of the Jinsha River, where the four seasons are not distinct, the climate is warm, the heat is abundant, and the sunlight is sufficient. The climatic resources are very favorable for the development of mango production. It is the world's highest latitude, highest altitude, and latest maturing mango production base.

Mango, also known as lemon fruit or stuffy fruit, is a perennial fruit tree of the Anacardiaceae family, originally from India. It is one of the world's five major fruits, along with citrus, bananas, grapes, and apples. Its pulp is delicate, with a unique flavor, and it is known as the "king of tropical fruits" [17–22]. Climate conditions are a significant factor affecting mango yield and quality. In recent years, there have been many studies on the relationship between mango safe production and climate. For example, Wu et al. [21] selected six climate factors such as annual average temperature and ≥ 10°C active accumulated temperature based on meteorological station data to complete the climate zoning of mango planting in Guizhou Province. Zhang et al. [23] conducted a study on the suitability of mango planting climate zoning nationwide based on national basic meteorological station data. Su and Li [24] used 300m × 300m small grid data to calculate zoning indicators and used GIS technology to conduct a study on the climate zoning of mango planting in the Baise area of Guangxi. Li et al. [25] used GIS technology to spatially interpolate climate elements (90m × 90m) based on ground meteorological station data, and completed the fine zoning of mango planting climate in Yuanyang County. Yang et al. [26] conducted a study on the climate risk zoning of mango planting in Fujian based on meteorological observation station data and cold damage process; Yu et al. [27] collected station climate statistical data and conducted a study on the characteristics of climate change in Eastern Fujian and its impact on mango growth based on the meteorological conditions required for mango growth. Previous studies on the suitability and risk zoning of mango climate in Guizhou, Guangxi, and Fujian are numerous, and most of the data used are based on sparsely distributed meteorological station data. Few uses fine grid data, and existing studies on the suitability and risk zoning of mango planting climate are static.

Under the backdrop of global warming, there have been no reports on the study of the suitability zoning of mango climate in the Panxi region and the impact of climate change on the change of mango planting climate zoning in the Panxi region. Moreover, there have been many studies on the climate zoning of banana, soybean, apple and other crops under the backdrop of global warming, and the changes in the area and boundaries of each suitable area [6–8, 11–15, 28–37], but there are no reports on the study of changes in the altitude suitable for planting. Therefore, using the meteorological observation data of Panxi and surrounding areas, we selected the annual average temperature, $\geq 10°C$ accumulated temperature, average temperature during the overwintering period, and sunshine duration as zoning indicators, established a calculation model for the spatial grid data of each climatic element of the zoning indicators, and studied the changes in the climatic suitability of mango planting in the Panxi region under the background of climate change. This is expected to provide a scientific basis for the full utilization of climate resources in the Panxi region, rational agricultural zoning, and adjustment of mango production development planning, and provide a reference for taking measures to cope with future climate change.

2 Materials and Methods

2.1 Overview of the Study Area

The Panxi region is in the southwest of Sichuan Province, at the southeastern edge of the Qinghai-Tibet Plateau and the Hengduan Mountains. Administratively, it includes Liangshan Prefecture and Panzhihua City, with latitudes between $26°03' \sim 29°27'N$ and longitudes between $100°15' \sim 103°53'E$ (Fig. 1). The terrain is high in the north and low in the south, with rugged terrain and mountains mostly running north south. The climate varies greatly between the north and south of the Panxi region. The southern part has abundant heat resources, sufficient sunlight, concentrated precipitation, and an annual average temperature of $\geq 15°C$ in most areas; the northern part has a relatively lower annual average temperature, more precipitation than the southern part, and more humid air; meteorological disasters include drought, lightning, heavy rain, high temperature, hail, strong wind, etc. [38].

2.2 Data Source and Processing

Meteorological data comes from the daily meteorological data such as average temperature and sunshine duration of 32 meteorological stations in Panxi and surrounding areas from 1961 to 2020, and the data of individual stations that have been relocated have been excluded. The accumulated temperature of $\geq 10°C$ is obtained by calculating the cumulative value of the daily average temperature of $\geq 10°C$. All meteorological data have undergone strict quality control. If the temperature data for one day is missing, it is replaced by interpolation. If the temperature data for more than one day is missing, it is replaced by the multi-year average value of the same day. Missing sunshine data is directly replaced by the multi-year average value of the daily value.

2.3 Research Method

Determination of Zoning Indicators. Temperature is an important factor for crop growth and development [39], it is a decisive factor for the survival of mangoes, and it is the most important climate indicator for planting, among which the weight coefficient of the annual average temperature is the largest [19], an annual average temperature of $\geq 20°C$ is suitable for mango growth, and mango growth stops when the annual average

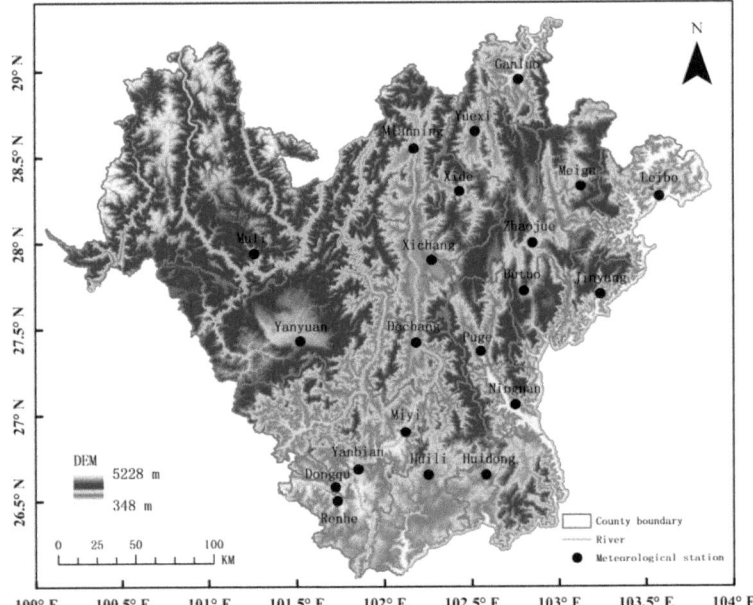

Fig. 1. The distribution of terrain and meteorological stations in the Panxi region.

temperature is $\leq 18°C$. The accumulated temperature of $\geq 10°C$ is one of the important heat conditions required for mango growth, and $\geq 6500°C$ can meet the heat required for mango growth in one year. The average temperature during the overwintering period is one of the important factors for the mango tree's cold resistance and whether it can safely overwinter. Mango trees like light, and sufficient light can satisfy the normal growth of the plant and help to ensure the quality of the mango.

Based on the above analysis, combined with the research results of predecessors [23–25, 40] Through consulting experts and conducting field surveys, the annual average temperature, accumulated temperature $\geq 10°C$, average temperature during the overwintering period (December to February of the following year), and sunshine hours are used as climate suitability zoning indicators for mango cultivation in the Panxi region (Table 1).

Spatial Grid Calculation Model and Zoning Method for Zoning Indicators. In order to analyze the spatial distribution evolution characteristics of the zoning indicators of mango planting climate suitability in the Panxi region from 1991 to 2020 compared to 1961–1990, a multiple linear regression equation is used. The annual average temperature, $\geq 10°C$ accumulated temperature, overwintering average temperature, and sunshine duration of the two periods before and after 1990 are related to the longitude, latitude, and altitude of the site to establish a spatial grid calculation model (Table 2).

The expression of the spatial grid calculation model is as follows:

$$X = f(x, y, z) + a \tag{1}$$

Table 1. Climate suitability zoning indicators for mango cultivation in the Panxi region.

Zoning index	Suitable region	Sub-suitable region	Unsuitable region
X_1: Annual average temperature (°C)	$X_1 \geq 19$	$15 \leq X_1 < 19$	$X_1 < 15$
X_2: ≥ 10°C accumulated temperature (°C•d)	$X_2 \geq 6500$	$5500 \leq X_2 < 6500$	$X_2 < 5500$
X_3: Average temperature during overwintering period (°C)	$X_3 \geq 12.0$	$8.5 \leq X_3 < 12.0$	$X_3 < 8.5$
X_4: Sunshine hours (h)	$X_4 \geq 1900$	$1200 \leq X_4 < 1900$	$X_4 < 1200$

where X represents the zoning indicators, x is the longitude (°), y is the latitude (°), z is the altitude (m), a is the residual term, which can be seen as the influence of the underlying surface and small terrain factors (slope direction, slope, etc.) on the climate.

As can be seen from Table 2, the R values of each model are between 0.966 and 0.990, all of which have passed the 0.01 significance test, indicating that the regression equation has a good regression effect. In order to facilitate superposition, assign 1, 2, and 3 to the suitable, sub-suitable, and unsuitable areas corresponding to each zoning indicator, respectively. By superimposing with the same weight, the comprehensive suitability zoning map of mango planting climate in the Panxi region can be obtained.

Table 2. A spatial grid calculation model for various regional indicators in the Panxi region.

Climatic Element	Time Slot	Calculation model	R
Annual Average Temperature	1961–1990	114.368-0.467x-1.419y-0.008z	0.990**
	1991–2020	112.807-0.432x-1.507y-0.007z	0.988**
≥ 10°C Accumulated Temperature	1961–1990	100 707.247-730.399x-623.168y-2.178z	0.968**
	1991–2020	90 180.353-636.476x-588.287y-2.100z	0.971**
Average temperature during overwintering period	1961–1990	258.15-1.768x-2.195y-0.006z	0.968**
	1991–2020	232.679-1.576x-1.984y-0.005z	0.974**
Sunlight hours	1961–1990	45163.439-336.415x-322.814y + 0.156z	0.966**
	1991–2020	43783.064-319.646x-339.418y + 0.205z	0.967**

Note ** represents significance < 0.01

Rate of climate change. When calculating the trend of each climate element, a linear equation is generally used to represent the trend change of the climate element [27], that is.

$$X_t = a_0 + a_1 t, t = 1, 2, 3, \cdots, n \tag{2}$$

where X_t is the fitted value of the climate element, a_0 represents the equation intercept, a_1 represents the equation slope, t is the sequence number of the research period, and the climate change rate is represented by 10 times the linear regression coefficient a_1.

2.4 Research Significance

The study on climate suitability regionalization of mango planting is related to the sustainable development of mango industry, which is helpful to adjust mango planting strategy in time, reduce the negative impact of climate warming on mango industry, optimize the allocation of agricultural resources, improve the yield and quality of mango, enhance the competitiveness of mango industry, promote the development of regional economy, and enhance the ability of mango planting to resist disasters. It has important theoretical and practical significance for promoting the healthy development of local mango industry.

3 Results and Analysis

According to previous studies, the change in climate elements, especially temperature, began to have a significant turning point in the 1990s [41–45]. This study takes 1990 as the dividing point and compares the changes in annual average temperature, $\geq 10°C$ accumulated temperature, average temperature during overwintering period, and sunlight hours in the Panxi region from 1961–1990 and 1991–2020.

3.1 Analysis of Changes in Mango Division Indicators in the Panxi Region

Characteristics of annual average temperature change. As can be seen from Fig. 2, the spatial distribution of the annual average temperature in the Panxi region shows a characteristic of being higher in the south and lower in the north. The annual average temperature was between $3 \sim 21°C$ from 1961–1990, and between $2.9 \sim 21°C$ from 1991–2020, with the latter period being 0.3°C higher than the former. The rate of climate change from 1961–2020 was between $0.0 \sim 0.6°C/10a$, with an average of 0.2°C/10a, all showing an upward trend, with the upward trend in Leibo being the most obvious. The $\geq 20°C$ region in the Panxi region has moved northward, increasing the area of this region by $17.27 \times 10^2 km^2$; the $15 \sim 20°C$ region has moved northward to varying degrees in Yanbian County, Renhe District, Xichang City, Huili City, and Huidong County, increasing the area of this region by $80.72 \times 10^2 km^2$, with the most noticeable northward movement in Huili City and Xichang City.

Characteristics of $\geq 10°C$ accumulated temperature change. As can be seen from Fig. 3, the spatial distribution pattern of the $\geq 10°C$ accumulated temperature in the Panxi region is like the annual average temperature. The $\geq 10°C$ accumulated temperature was between $3034.5 \sim 7619.1°C•d$ from 1961–1990, and between $3207.3 \sim 7643.2°C•d$ from 1991–2020, with the latter period increasing by 214.8°C•d compared to the former. The climate change rate from 1961–2020 was between $26.6 \sim 181.5°C•d/10a$, with an average of 86°C•d/10a, all showing an increasing trend, with the increasing trend in Muli being the most obvious. The $\geq 7000°C•d$ accumulated temperature zone in the

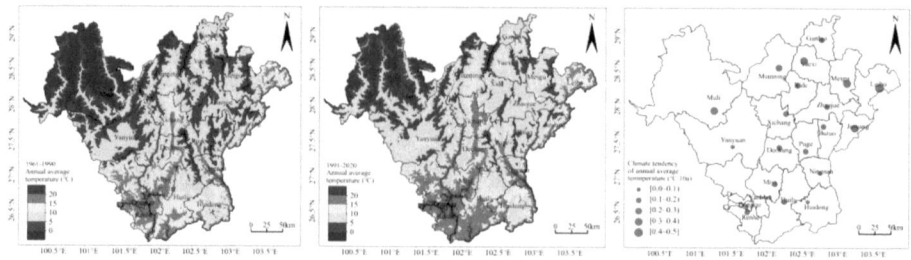

Fig. 2. Annual average temperature and its climate tendency rate in Panxi region.

Panxi region has moved northward in Panzhihua City and Huili City, increasing the area of this region by $3.34 \times 10^2 \text{km}^2$; the $5500 \sim 7000°C•d$ accumulated temperature zone has moved northward to varying degrees in Xichang City, Huili City, Huidong County, and Puge County, increasing the area of this region by $33.17 \times 10^2 \text{km}^2$, with the most noticeable northward movement in Xichang City and Huili City.

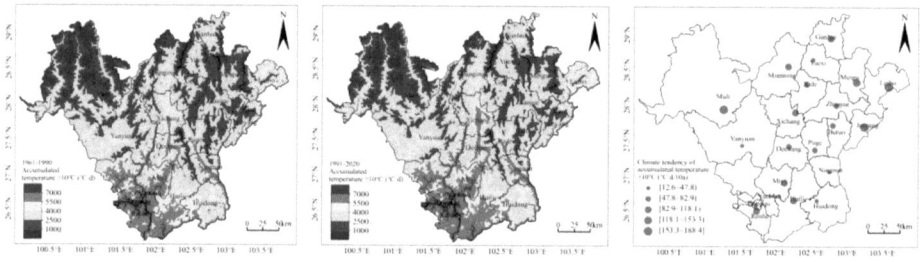

Fig. 3. Accumulated temperature $\geq 10°C$ and its climate tendency rate in Panxi region.

Characteristics of average winter temperature changes. As can be seen from Fig. 4, the spatial distribution of average winter temperature in the Panxi region shows a characteristic of being higher in the south and lower in the north. The average winter temperature from 1961–1990 was between $2.5 \sim 14.5°C$, and from 1991–2020 it was between $3.2 \sim 14.5°C$, the latter period was $0.6°C$ higher than the former. The climate change rate from 1961–2020 was between $0.08 \sim 0.6°C/10a$, with an average of $0.3°C/10a$, all showing an upward trend, with the most obvious upward trend in Leibo and Daocheng. The $\geq 10°C$ region in the Panxi region has moved northward in Yanbian County, Miyi County, Renhe District, Dechang County, Huili City, and Huidong County, increasing the area of this region by $79.86 \times 10^2 \text{km}^2$, with the most obvious northward movement in Dechang County and Huili City; the $5 \sim 10°C$ region has moved northward to varying degrees in Miyi County, Huili City, Huidong County, Yanyuan County, and Mianning County, increasing the area of this region by $199.87 \times 10^2 \text{km}^2$, with the most obvious northward movement in Mianning County and Yanyuan County.

Characteristics of Sunshine Duration Changes. As can be seen from Fig. 5, the spatial scale of sunshine duration in the Panxi region generally shows a zonal distribution. The sunshine duration was between $1204.4 \sim 2668.6h$ from 1961 to 1990, and between

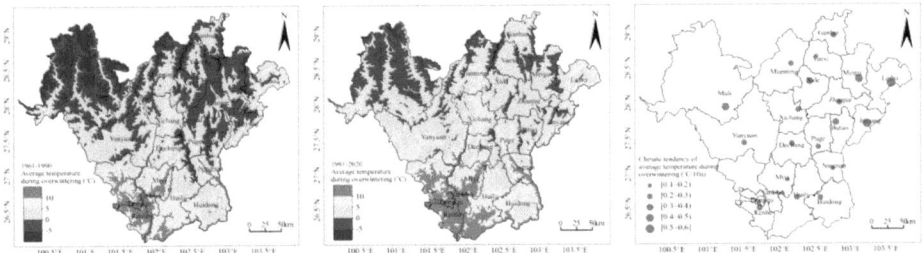

Fig. 4. Average temperature and climatic tendency during overwintering in Panxi region.

1032.7 ~ 2689.8h from 1991 to 2020. The climate change rate from 1961 to 2020 was between -57.1 ~ 60.5h/10a, with an average of 16.3h/10a. Except for Minning County, Yuexi County, Ganluo County, Leibo County, and Huidong County, where the climate change rate was negative, the rest of the sites were positive, with most areas showing an increasing trend, and the increasing trend in Muli was the most obvious. The area of the Panxi region ≥ 2500h increased by 5.75 × 10²km², the area of 2200 ~ 2500h increased by 4.84 × 10²km², and the area of 1800 ~ 2000h decreased by 11.75 × 10²km².

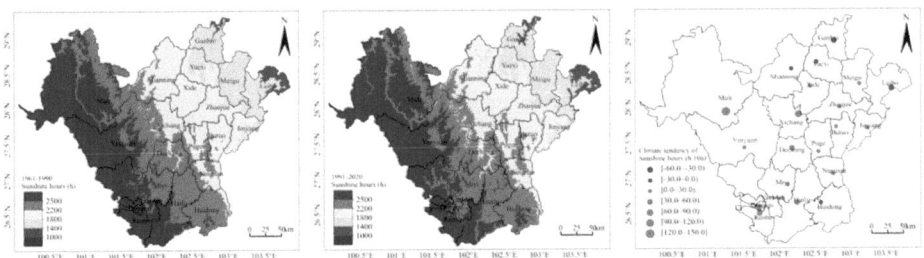

Fig. 5. Sunshine hours and climate tendency rate in Panxi region.

3.2 Changes in the Zoning of Climate Suitability for Mango Cultivation

Comprehensive zoning and changes. As can be seen from Fig. 6, the main mango planting areas in the Panxi region are concentrated in the southwest, with suitable areas located in the central part of Panzhihua City and a very small part of the southeast of Huili City; the moderately suitable areas are located in the central and southern parts of Panzhihua City and Huili City, and a small part of Huidong County, Ningnan County, Dechang County, Puge County, Yanyuan County, Leibo County, Butuo County, and the eastern part of Jinyang County; the unsuitable areas are mainly concentrated in the central and northern parts. After 1990, the area of suitable regions increased by 21.55 × 10²km², an increase of 3.2%; the area of moderately suitable regions increased by 50.96 × 10²km², an increase of 7.5%; the area of unsuitable regions decreased by 72.51 × 10²km², a decrease of 10.7% (Table 3). The increase in the area of suitable regions is most obvious in the central part of Panzhihua, with a very small part of Huili City and Ningnan County changing from moderately suitable to suitable, and the central and southern parts of Xichang City, the central and southern parts of Huili City, and a small part of Huidong County changing from unsuitable to moderately suitable. The suitable planting range in Huili City shows a trend of expanding to the north, west, and east. The suitable planting range generally shows a trend of expanding to the north, with an increase in the planting area. The maximum altitude of the suitable area increased by 161m, the moderately suitable area increased by 221m, and the climatically suitable range for mango planting moved to the altitude area (Table 3).

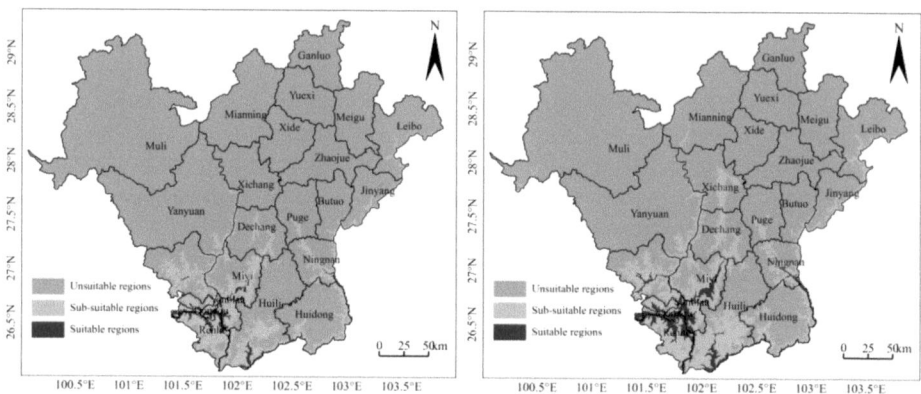

Fig. 6. Climatic zoning of mango planting in the Panxi region from 1961–1990 (left) and 1991–2020 (right).

Table 3. Area statistics and maximum altitude of climate suitability zoning for mango cultivation in the Panxi region during different climatic periods.

Zoning	1961–1990			1991–2020		
	Area (10^2 km^2)	Proportion (%)	Highest Altitude Maximum altitude (m)	Area (10^2 km^2)	Proportion (%)	Highest Altitude Maximum altitude (m)
Suitable area Suitable region	24.90	3.7	1442	46.45	6.9	1603
Sub-suitable area Sub-suitable region	117.40	17.4	1867	168.36	24.9	2088
Unsuitable area Unsuitable region	532.84	78.9		460.33	68.2	

Changes in planting boundaries. In Fig. 7, the blue line represents the planting boundaries of the suitable area (left of the figure) and the sub-suitable area (right of the figure) from 1961 to 1990, and the red line represents the planting boundaries of the suitable area and the sub-suitable area from 1991 to 2020. On the left of Fig. 7, after 1990, the planting boundaries of the suitable areas in Yanbian County, Miyi County, Huili City, Huidong County, and Ningnan County have all moved north to varying degrees, with the maximum northward movement reaching 17.76km; on the right of Fig. 7, the planting boundaries of the sub-suitable areas in Xichang City, Huili City, Huidong County, Muli County, Puge County, and Mianning County have all moved north to varying degrees, with the maximum northward movement reaching 22.2km; the changes in the boundaries of the suitable areas and sub-suitable areas in other regions are not obvious. The planting boundaries of the suitable areas and sub-suitable areas have generally moved north.

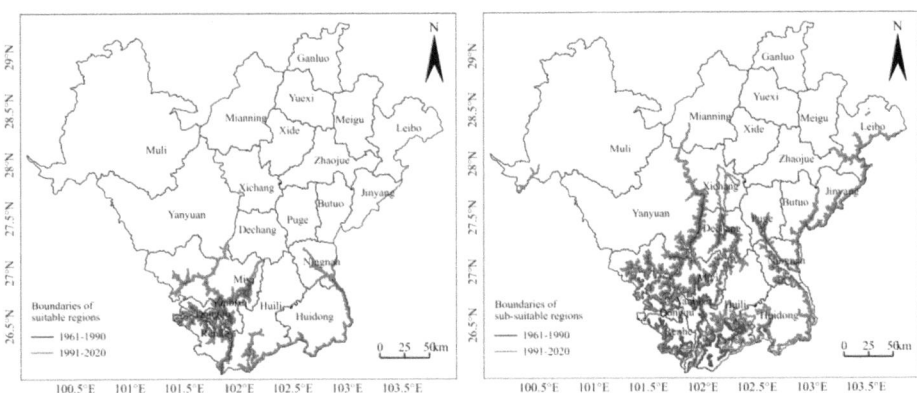

Fig. 7. Changes in the boundaries of suitable areas (left) and sub-suitable areas (right) before and after 1990.

Field Verification of Mango Planting Climate Suitability Zoning Results. To verify whether the zoning results are consistent with the actual situation of local mango planting, 14 mango planting points were selected in the climate suitable and sub-suitable areas for field investigation. The investigation targets were mango orchards with a tree age of more than 5 years. The investigation content included the latitude and longitude of the mango orchard, the altitude, the distribution of mango planting in the past 5 years, the growth, fruiting and other growth performance. According to the judgment standard of whether the deviation, general, and good growth of the fruit trees in the investigated mango orchard is consistent with the different climate zoning levels, a consistency analysis was conducted. The results show that the zoning results are basically consistent with the actual planting distribution of the mango orchard, with a consistency rate of about 79% (Table 4).

Table 4. Field validation of climate suitability zoning results for mango cultivation in the Panxi region.

County (District)	Verification Point	Latitude and Longitude (°), Altitude (m), Growth Momentum	Zoning Results	Consistency	Notes
Yanbian County	Yumen Town Rhino Village	101.55514,26.91109, 1253.7, Average	Sub-suitable	√	Rare cold damage during flowering period in 2022
	Tongzilin Town Jinhe Village	101.81584,26.72377, 1317.2, Good	Suitable	√	
	Guosheng Township Taoshui Village	101.54411,26.94974, 1363.9, Average	Moderately Suitable	√	Rare cold damage during flowering period in 2022
	Tongzilin Town Musala Village	101.87552,26.62569, 1286.9, Good	Suitable	√	
Renhe District	Futian Town Wuzitian Village	101.39553,26.59655, 1442.7, Average	Suitable	×	Agricultural meteorological observation point, poor management, distinct microclimate, rare cold damage during flowering period in 2022, yield reduction of about 30%

(continued)

4   Discussion

The increase in temperature and $\geq 10°C$ accumulated temperature in the Panxi region is conducive to the expansion of mango cultivation in the region, allowing the boundary of mango cultivation to move towards higher latitudes and altitudes. The increase in average winter temperature is beneficial for the safe overwintering of mangoes, making the overall trend of mango cultivation in Panxi favorable. However, the increase in temperature will intensify surface evaporation, making agricultural drought more severe

Table 4. (*continued*)

County (District)	Verification Point	Latitude and Longitude (°), Altitude (m), Growth Momentum	Zoning Results	Consistency	Notes
	Zhongba Township Xuefang Village	101.67217,26.42005, 1426.3, Average	Moderately Suitable	√	Agricultural meteorological observation point, average management, rare cold damage during flowering period in 2022
	Dalongtan Township Hunsala Village	101.85444,26.43606, 1160.9, Average	Suitable	×	Agricultural meteorological observation point, average management, rare cold damage during flowering period in 2022, yield reduction of 50%
	Bude Town Xinqiao Village	101.55418,26.66554, 1367.9, Average	Suitable	×	Low-lying area, distinct microclimate, suffered from rare cold damage during flowering period in 2022
	Dalongtan Township Xinjie Village	101.81185,26.40231, 1461.4, good	Suitable	√	

(*continued*)

and frequent [46], posing a threat to the production and development of mangoes in Panxi. In the context of global warming, efforts should be made to breed drought-resistant mango varieties, develop drought-resistant and water-saving cultivation techniques, and ensure the sustainable development of the mango industry in Panxi.

The suitable range for mango cultivation generally shows a trend of expanding northward, with the overall boundary moving northward and the cultivable area increasing,

Table 4. (*continued*)

County (District)	Verification Point	Latitude and Longitude (°), Altitude (m), Growth Momentum	Zoning Results	Consistency	Notes
	Ala Township Dazhu Village	101.68583,26.30307, 1691.9, average	Moderately suitable	✓	Agricultural meteorological observation point, well managed, affected by frost in 2021, 50% of fruit trees suffered from frost, yield decreased by 60%
	Dalongtan Township Lazha Village	101.89458,26.44459, 1013.7, good	Suitable	✓	
Miyi County	Wanqiu Township Reshui Village	102.16076,27.09452, 1333.0, average	Moderately suitable	✓	Well managed, affected by high temperature in 2023, yield decreased by 25%

(*continued*)

which is consistent with previous research results [7–15]. This is very beneficial for the vigorous development of the "mango economy" in the southwestern part of the Panxi region. It is recommended that relevant government departments adjust the layout of mango production in a timely manner, appropriately introduce and expand the cultivation of mangoes in suitable and sub-optimal areas, and further promote the sustainable and healthy development of the mango industry in Panxi.

Currently, the highest altitude for mango cultivation is around 2000m. Through consultations with mango experts and growers, field surveys, and other methods, it was found that when the altitude is above 1900m, the risk of frost and cold damage is higher. New orchards should fully consider whether it is suitable for cultivation based on factors such as topography, slope, aspect, and local microclimate. In already cultivated areas, measures should be taken to prevent frost and cold during the overwintering and flowering periods.

Overall, the zoning results of the field survey verification are in good consistency with the actual distribution of mango cultivation, indicating that the zoning indicators are

Table 4. (*continued*)

County (District)	Verification Point	Latitude and Longitude (°), Altitude (m), Growth Momentum	Zoning Results	Consistency	Notes
	Xinshan Township Zhongshan Village	102.15037,26.81149, 1695.9, average	Moderately suitable	✓	Well managed, suffered from rare cold damage during flowering period in 2022, yield decreased by 80%
	Binggu Town Bajiaoqing Village	102.11934,26.80557, 1287.7, average	Moderately suitable	✓	Average management, suffered from rare cold damage during flowering period in 2022, yield decreased by 60%

reasonably selected. The field survey found that although the mango orchards visited are all located in suitable or sub-suitable areas, due to the influence of local microclimate, terrain, soil, management level, etc., the growth of mango trees is uneven. The verification results are in good consistency with the actual distribution of mango cultivation in suitable areas; the consistency in sub-suitable areas is good. There is a 21% inconsistency in the results, mainly reflected in: first, although Xin Qiao Village in Bu De Town is in a climate suitable for mango cultivation, the growth is average, and it suffered from cold damage during the flowering period in 2022. The reason is that a large area of mango trees are planted in low-lying areas, and when the temperature drops sharply, cold air is prone to accumulate in the low-lying areas, causing the fruit trees to freeze or reduce production; second, although Wu Zi Tian Village in Fu Tian Town is zoned as a suitable area, the local microclimate is obvious, and the management is poor, so the growth of the fruit trees is average. The survey also found that the mango orchard in Re Shui Village, Wan Qiu Township is in the mountains, without irrigation facilities, and was severely affected by persistent high temperatures in 2023, causing severe drought and significant reduction in production. Therefore, in the future development of the mango industry in the Panxi region, attention should be paid to: first, new mango orchards should fully consider local microclimate, terrain, soil and other factors, avoid building mango orchards in low-lying areas prone to frost damage, and strengthen the defense against winter frost damage and cold damage during the flowering period in existing mango orchards; second, choose places with irrigation conditions to develop the mango industry, and strengthen the construction of irrigation facilities; third, actively carry out artificial hail prevention operations, and new orchards should avoid areas with frequent hail.

The planting area of mangoes is not only affected by heat conditions, but also related to planting techniques, soil, socio-economic factors, etc. Therefore, in the later stage, all aspects should be considered comprehensively to formulate a more realistic zoning and development plan for mango planting, which is one of the important research tasks for the future development of the mango industry in the Panxi region.

5   Conclusions

The annual average temperature, $\geq 10°C$ accumulated temperature, and average winter temperature in the Panxi region show a spatial distribution of being higher in the south and lower in the north. The rate of climate change is all positive, showing an upward or increasing trend. After 1990, the annual average temperature and average winter temperature increased by 0.3°C and 0.6°C respectively, and the $\geq 10°C$ accumulated temperature increased by 214.8°C•d. The spatial scale of sunshine hours generally shows a zonal distribution, with the average rate of climate change from 1961 to 2020 being 16.3h/10a; the area of regions with $\geq 2500h$ and $2200 \sim 2500h$ has increased, and the increase in strong sunshine hours plays a positive role in ensuring good mango quality.

After 1990, the area of suitable regions increased by $21.55 \times 10^2 km^2$, an increase of 3.2%; the area of sub-optimal regions increased by $50.96 \times 10^2 km^2$, an increase of 7.5%; the area of unsuitable regions decreased by $72.51 \times 10^2 km^2$, a decrease of 10.7%. The northern boundary of the climate suitable for cultivation moved up to 17.76km, and the sub-optimal area moved up to 22.2km. After 1990, the highest altitude of the suitable area for mango cultivation in the Panxi region was 1603m, an increase of 161m, and the highest altitude of the sub-optimal area was 2088m, an increase of 221m, indicating that the suitable range for mango cultivation has moved towards higher altitudes.

Acknowledgements. The Second Comprehensive Scientific Expedition to the Qinghai-Tibet Plateau (2019QZKK0303); Science and Technology Development Fund of the Key Laboratory of Plateau and Basin Storm, Drought and Flood Disaster of Sichuan Province (SCQXKJYJXMS202108); Mango Meteorological Service Innovation Team (PZHQXJ202011).

References

1. Climate Change, I.P.C.C.: The physical science basis, p. 2021. Cambridge University Press, Cambridge and New York (2021)
2. Xianli, Z., Cai, F., Kangkang, D., et al.: Trend of phenological changes of spring maize in Jinzhou area and its relationship with thermal and water conditions. J. Meteorol. Environ. **38**(5), 64–71 (2022)
3. IPCC. Climate Change 2014: impacts adaptation, and vulnerability. Cambridge University Press (2014)
4. Ying, L., Ge, G., Lianchun S.: New understanding of climate change risk and risk management in IPCC Fifth Assessment Report. Prog. Clim. Chang. Res. **10**(4), 260–267 (2014)

5. Dahe, Q.: Facts, impacts and countermeasures of climate change. J. Dipl. Acad. **2004**(77), 14–22 (2004)
6. Yueying, Z., Buchun, L., Zhijuan, L., et al.: Northward and westward expansion of suitable planting areas for apples in China under the background of climate change: zoning analysis based on high-resolution grid meteorological data. Chin. Agric. Meteorol. **40**(11), 678–691 (2019)
7. Xiaoguang, Y., Zhijuan, L., Fu, C.: Possible impact of Global Warming on China's Planting System I. Analysis of possible impact of Global Warming on the Northern Boundary of China's Planting System and Grain Yield. Chin. Agric. Sci. **43**(2), 329–336 (2010)
8. Liuhong, Z., Minyan, L., Yaodong, D.: The impact of climate change on the climate zoning of Shatian Pomelo Planting in Guangdong. Meteorol. Environ. Sci. **45**(3), 43–50 (2022)
9. Liuhong, Z., Yaodong, D., Jiaming, D.: The impact of climate change on the climatic suitability of Banana Planting in Guangdong. Ecol. J. **45**(3), 43–50 (2023)
10. Zhang Shanqing, P., Zongchao, L.J., et al.: Changes in the Cotton Planting Zone in Southern Xinjiang under the Background of Global Warming. Chin. Agric. Meteorol. **36**(5), 594–601 (2015)
11. Jin, Z., Xiaoguang, Y., Zhijuan, L. et al.: Possible impact of Global Warming on China's Planting System II. Characteristics of climate element changes in the Southern Region and possible impact on the boundary of the Planting System. Chin. Agric. Sci. **43**(9), 1860–1867 (2010)
12. Li Jinglin, P., Zongchao, Z.S., et al.: The impact of Climate Change in Northern Xinjiang over the past 52 years on the climate suitability zoning of cotton planting. Cotton Sci. **27**(1), 22–30 (2015)
13. Zongchao, P., Shanqing, Z., Chunrong, J., et al.: The Impact of climate change on the climate zoning of Hami Melon Planting in Xinjiang. Prog. Clim. Chang. Res. **11**(2), 115–122 (2015)
14. Shaojun, L., Guangsheng, Z., Shibao, F., et al.: The impact of future climate change on the climate suitability zone of natural rubber planting in China. J. Appl. Ecol. **26**(7), 2083–2090 (2015)
15. Xiaoyang, D., Liuhong, Z., Yujiao, D., et al.: The impact of climate change on the climate zoning of litchi planting in Guangdong Province. Ecol. J. **41**(10), 1998–2007 (2022)
16. Jun, S., Jun, G., Weiguo, S.: The impact of climate change from 1961 to 2010 on the heating period and summer air conditioning period in Tianjin. J. Meteorol. Environ. **29**(2), 61–67 (2013)
17. Yuping, L., Ye, L., Weihong, L., et al.: Current status and development strategies of Mango Industry Data Resources in China. Trop. Agric. Sci. **40**(8), 105–109 (2020)
18. Leilei, X.: 2015 Mango Industry Development Report and Situation. World Trop. Agric. Inf. **2016**(11), 14–26 (2016)
19. Meifang, Z., Shouzhi, C., Xi, C.: Statistical analysis of the climatic ecological adaptability of Mango Planting in Yunnan. Heilongjiang Agric. Sci. **2009**(1), 68–69 (2009)
20. Biao, D., Zhenfeng, M., Chengxun, D. et al.: Risk assessment of meteorological disaster factors for Mango Planting in Panzhihua. Plateau Mt. Meteorol. **38**(1), 91–95 (2018)
21. Xiaobo, W., Shuquan, H., Hai, H., et al.: Division of Mango Planting Climate Zones in Guizhou Province Based on GIS and RS. Jiangsu Agric. Sci. **45**(20), 268–271 (2017)
22. Lisheng, T., Hua, W., Fei, H. et al.: Damage symptoms and physiological responses of mango seedlings under low temperature stress. Ecol. J. **35**(10), 2627–2636 (2016)
23. Mingjie, Z., Jinghong, Z., Yajie, Z., et al.: Division of suitable climate zones for mango planting in China. Jiangsu Agric. Sci. **50**(2), 124–130 (2022)
24. Yongxiu, S., Zheng, L.: Agricultural climate zoning for mango planting supported by GIS. Guangxi Meteorol. **23**(1), 46–48 (2022)
25. Yanchun, L., Yinghai, X., Liyang, Z.: Climate suitability zoning for mango planting in Yuanyang County based on GIS. Green Sci. Technol. **2020**(6), 83–85 (2020)

26. Kai, Y., Binbin, C., Hui, C., et al.: Climate Risk Zoning for mango planting in Fujian based on cold damage process. Chin. Agric. Meteorol. **40**(11), 723–732 (2019)

27. Huikang, Y., Anfang, C., Cuibing, R., et al.: Analysis of climate change and mango growth in Eastern Fujian. China Agric. Resour. Zoning **35**(4), 63–68 (2014)

28. Zheng'e, S., Zhijuan, L., Wanrong, Y., et al.: Analysis of suitable climate zones for mechanical grain harvest of mid-late maturing spring maize in Northeast Region under future climate change scenarios. Chin. Agric. Meteorol. **44**(8), 649–663 (2023)

29. Hongqun, L., Xiaoli, L., Jianhua, W., et al.: Impact of future climate change on suitable planting areas for pickled mustard in Chongqing. J. Appl. Ecol. **29**(8), 2651–2657 (2018)

30. Liang, H., Liuxi, M.: Changes in climate suitability of soybean planting areas in Northeast China under climate change. Chin. J. Eco-Agric. **31**(5), 690−698 (2023)

31. Cao, Z., Jianyu, Y., Bo, H., et al.: Impact of climate change on corn planting zoning in rainfed agricultural areas of Gansu Province. Arid. Land Agric. Res. **39**(2), 211–219 (2021)

32. Zhang Shanqing, P., Zongchao, L.J., et al.: Impact of climate change on climate zoning for jujube planting in Xinjiang. Chin. J. Eco-Agric. **22**(6), 713–721 (2014)

33. Weikun, L., Meng, L., Xueqiong, H. et al.: Impact of climate change on potential rubber planting areas in Yunnan. J. Appl. Meteorol. **34**(3), 379−384 (2023)

34. Zhang Shanqing, P., Zongchao, J.C., et al.: Impact of climate change on climate zoning for wine grape planting in Xinjiang. China Agric. Resour. Zoning **37**(9), 125–134 (2016)

35. Huiqing, H., Jiaoyan, Z., Yingjia, Z., et al.: The impact of future climate change on the climatic suitability of kiwifruit cultivation in Guizhou. J. Southwest For. Univ. **38**(4), 46–52 (2018)

36. Zongchao, P., Shanqing, Z., Jianhua, B., et al.: The impact of global warming on the division of cotton planting areas in Wuchang, Xinjiang. Adv. Clim. Chang. Res. **8**(4), 257–264 (2012)

37. Pu Zongchao, Zhang Shanqing.: The impact of global warming on the climatic suitability of walnut cultivation in Xinjiang [J]. Chinese Agricultural Meteorology **39**(4), 267–279 (2018)

38. Jianming, C., Yunpeng, L., Ling, C., et al.: Spatial distribution characteristics of geological disasters in Panxi mining area based on kernel density analysis. World Nonferrous Metals **2022**(1), 152–154 (2022)

39. Chen, C., Yanchun, C., Zhiqiang, D., et al.: Study on cold damage warning indicators of typical thermophilic vegetables in greenhouses in Central Shandong. J. Meteorol. Environ. **39**(2), 92–99 (2023)

40. Qiren, L., Zhende, W.: The impact of meteorological factors on mango yield. Chin. Agric. Meteorol. **16**(1), 13−15 (1995)

41. Wei, J., Xingna, D., Li, L.: Analysis of the spatiotemporal evolution characteristics of temperature and precipitation in Sichuan Province. Anhui Agric. Sci. **51**(16), 205−211 (2023)

42. Chao, C., Yanmei, P., Xuebiao, P., et al.: Analysis of the trend of climate resource changes in Sichuan Province under the background of climate change. Resour. Sci. **33**(7), 1310–1316 (2011)

43. Li, W., Shujun, L., Weiwei, G., et al.: Spatiotemporal distribution and regional characteristics of strong cooling in Sichuan based on different standards. Plateau Mt. Meteorol. Res. **42**(1), 70–76 (2022)

44. Wenbo, G., Peng, H., Zhengyu, L., et al.: Study on the spatiotemporal variation characteristics of thermal resources in Sichuan Province under the background of climate change. J. Agric. Big Data **2**(1), 60–69 (2020)

45. Xiaoqiang, Y., Liqun, Z., Shuai, L., et al.: The impact of climate warming from 1980 to 2008 on soybean cultivation in Heilongjiang Province. J. Meteorol. Environ. **29**(2), 96–100 (2013)

46. Fenghua, S., Mingyan, L., Qingfei, Z.: Estimation and evaluation analysis of the impact of climate change on runoff in the Liaohe River Basin. J. Meteorol. Environ. **38**(3), 156−161 (2022)

Risk Analysis of Blueberry Gale Disaster in Liaodong Green Economic Zone Based on Machine Learning

Hai- tao Dong[1(✉)], Qing Sun[2], Lu- lu Shan[1], Xi Meng[1], Ru-nan Li[1], and Yi-he Fang[3]

[1] Dandong City Bureau of Meteorology, Dandong 118000, China
165175634@qq.com
[2] Chinese Academy of Meteorological Sciences, Beijing 100081, China
[3] Climate Center of Liaoning Province, Shenyang 110166, China

Abstract. Based on the long time series gale data of 9 national meteorological stations in Liaoning province and the data of blueberry growth period, the optimal machine learning model is chosen to extend the time series of maximum wind speed, using the maximum wind speed and disaster information data of the corresponding stations, considering the frequency and duration comprehensively, the meteorological index thresholds of different grades of high wind disaster risk during the ripening period of blueberry were determined, the risk grades of mild, moderate and severe gale disasters were established, and the spatial and temporal distribution characteristics of blueberry gale disaster risk were analyzed by the frequency of disasters and the ratio of stations. The results showed that the stochastic forest model had a high simulation precision and could extend the maximum wind speed time series, and the maximum wind disaster risk threshold was ≥ 13.9 ms^{-1} in the mature period of blueberry, the results were verified to be in accordance with the actual situation. During the whole mature period of blueberry in 30 years, the impact of Gale Disaster Risk tended to be mitigated, and the frequency of occurrence showed a non-significant decreasing trend, the Xiuyan area showed the most significant decrease, while the Kuandian Manchu Autonomous County area showed a significant increase trend. The probability of occurrence of gale hazard over the whole mature period of 30 years is 83.3%, and the probability of occurrence of gale hazard over two years ($\geq 50\%$), among which the Fushun region has the highest risk degree, the middle risk probability is 17.8%, the Qingyuan region has the highest risk degree, and the severe risk degree is 5.2%, mainly in Fushun and Benxi. In general, the northwest part of the Green Economic Zone is a high risk area for gale disasters, which are widespread, frequent and severe, mainly distributed in Fushun, Benxi, Qingyuan Manchu Autonomous County and Xifeng County.

Keywords: Blueberry · Maturity · Wind disaster · Machine learning · Risk probability

© The Author(s) 2025
B.-J. He et al. (Eds.): UCSUD 2023, LNCE 559, pp. 90–101, 2025.
https://doi.org/10.1007/978-981-97-8401-1_7

1 Introduction

With the aggravation of global climate change, extreme meteorological disasters such as high temperature and heavy precipitation occur frequently in northern China [1–4]. Liaodong Green Economic Zone is one of the main producing areas of blueberry in China, it is very affected by wind disaster in the ripening period of blueberry, results in fruit powder damage and fruit drop phenomenon, affect the quality and yield [6]. Therefore, it is of great practical significance to explore the change rule of blueberry weather and to evaluate the risk degree of wind disaster in the background of climate change in order to promote the development of blueberry industry. Based on the maximum and maximum wind speed data of 9 national weather stations in Liaodong Green Economic Zone from 1991 to 2020, and referring to the criteria and indexes of meteorological gale disasters, combined with the intensity and frequency of Gale, the paper divides the three grades of Gale disaster into light, medium and heavy. Based on the existing research results [7–9], through the comprehensive analysis of the different levels of Gale disaster risk, this paper discusses the characteristics of the temporal and regional changes of gale disaster risk from multiple angles, and applies the information diffusion theory [10–12], the risk probability and the law of reappearing period of different grades of gale disasters at different stations were obtained, which provided the reference basis for dealing with gale disasters in mature period of blueberry.

2 Data and Methods

2.1 Study Areas and Data Sources

The study areas were nine counties (cities) in eastern Liaoning province, including Qingyuan, Kuandian Manchu Autonomous County, Xiuyan Manchu Autonomous County, Fushun, Xinbin Manchu Autonomous County, Qingyuan Manchu Autonomous County, Benxi, Huanren Manchu Autonomous County and Xifeng County (Fig. 1). The data were obtained from the "Tianqing" big data platform (http://10.86.104.50:8088/cma daas/) of the Liaoning Meteorological Bureau. The data were the average wind speed, maximum wind speed and maximum wind speed of the nine weather stations from 1991 to 2020. Blueberry phenology data from the Dandong Centre for agricultural and rural development.

2.2 Division of the Growth Period of Blueberries

The growth period of blueberry in Liaodong Green economic zone can be divided into germination period, flowering period, fruit setting period, fruit growing period, fruit colouring period, harvest period, flower bud differentiation period and defoliation period. In order to facilitate the study, the fruit coloring period-ripening and harvesting period is collectively called ripening period. The average date of ripening period is from late June to early August. The mature period of blueberries varied with varieties (Table 1). Blueberries were classified into early, middle and late ripening varieties.

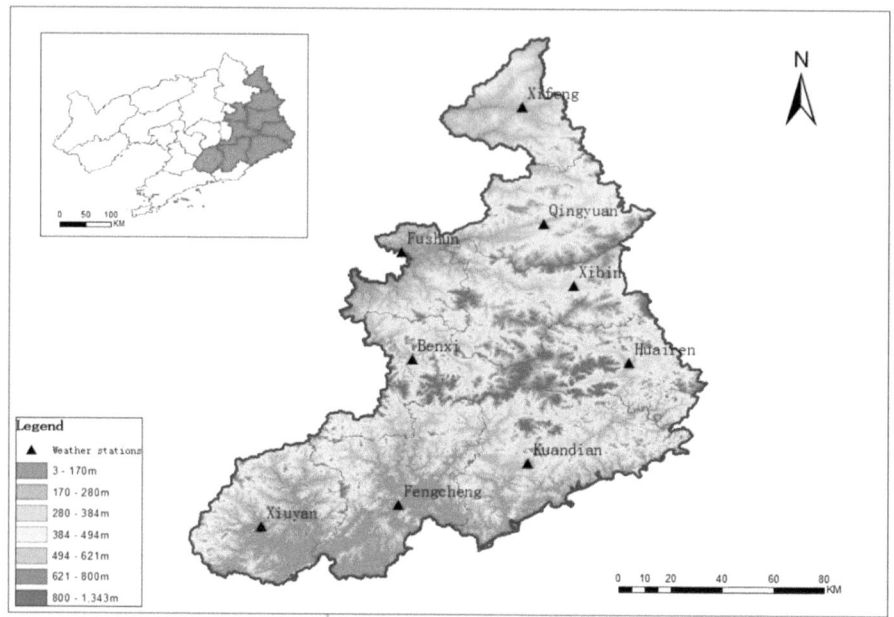

Fig. 1. Elevation and meteorological station distribution in Liaoning Green Economic Zone

Table 1. Average time of each mature stage of blueberry in Liaodong Green Economic Zone

Type	Species	Date period
Early maturity	Rekha, Duke, Duke	Late June to early July
Medium maturity	Lanfeng, Draper, Kitagawa	Early to mid-july
Late maturing	Freedom	Mid-july to early August

2.3   Research Methodology

3   Establish the Optimal Model of Prolonged Maximum Wind Speed Time Series

Since 2004, it is necessary to extend the time series data of maximum wind speed, and to evaluate and verify the historical gale. Four commonly used machine learning models were selected to simulate the maximum wind speed time series [13–18], namely multivariate linear regression (LR), random forest (RF), decision tree (DT) and K-nearest neighbor (KNN). The input data is the daily maximum wind speed of 9 weather stations in Liaodong Green Economic Zone. Using random sampling method, 75% of the data are used for training and 25% for prediction and verification. Correlation coefficient (r) and root mean square error (RMSE) were used to evaluate the results of the model simulation.

4   Frequency of Occurrence of Major Wind Disaster Risk

During the mature period of blueberry, it is sensitive to strong wind disaster, which directly affects the yield and quality of blueberry. Taking into account the results of studies by Hejinna et al. [19] and Zhou Yu et al. [20] and referring to the climate criteria for high winds in the regional blueberry growing season disaster survey and the agri-meteorological observation code [21], the maximum wind speed ≥ 13.9 m · S-1 or above during the ripening period of blueberries was taken as a process of occurrence of a strong wind disaster risk.

5   Risk Classification Standard for Major Wind Disasters

A comprehensive gale hazard risk index was used to grade the gale hazard risk of blueberry in ripening period. The maximum wind speed ≥ 13.9 ms^{-1} disaster risk index (DRI) was calculated for each station during the mature period from 1991 to 2020, and the actual disaster risk situation of blueberry growers was referred to, the risk indexes of light, moderate and severe strong wind disaster in the mature period of blueberry were classified. See Table 2 for details.

Table 2. Risk Classification of high wind disaster in mature period of blueberries in Liaodong Green economic zone (V is maximum wind speed, ms^{-1})

Disaster risk rating	Criteria	
	Wind speed index	Blueberry main performance
Mild	$13.9 \leq V<17.1$	The fruit powder was slightly damaged, which affected the fruit quality
Moderate	$17.2 \leq V<20.7$	Most of the powder were damaged obviously, which affected the fruit quality and yield
Heavy	$20.8 \leq V$	The fruit powder was seriously damaged and the fruit dropped, which affected the fruit yield

5.1   Probability Assessment of Gale Disaster Risk

In this study, information diffusion theory was used to evaluate the risk probability of blueberry in mild, moderate and severe gale. Information diffusion theory is a fuzzy mathematical approach to deal with the case of insufficient information in the sample [10]. The 30-year wind disaster risk index information from 1991 to 2020 at a certain station is diffused into a fuzzy data set, and the single-value sample point is changed into a set-value sample point. This method solves the problem of parameter estimation with limited sample size.

6   Results and Analysis

6.1   The Maximum Wind Speed Time Series is Extended

As can be seen from Fig. 2, the R of all models was above 0.9, and the test samples of both RF and DT models were distributed on both sides of the 1:1 line, with the highest R and all above 0.95, with the same RMSE, the results of KNN model were the worst, R the lowest, 0.9427, RMSE the highest, 1.06 ms^{-1}. In a word, RF model of random forest is the best, which can be used to extend the time series of historical maximum wind speed in Liaodong Green economic zone.

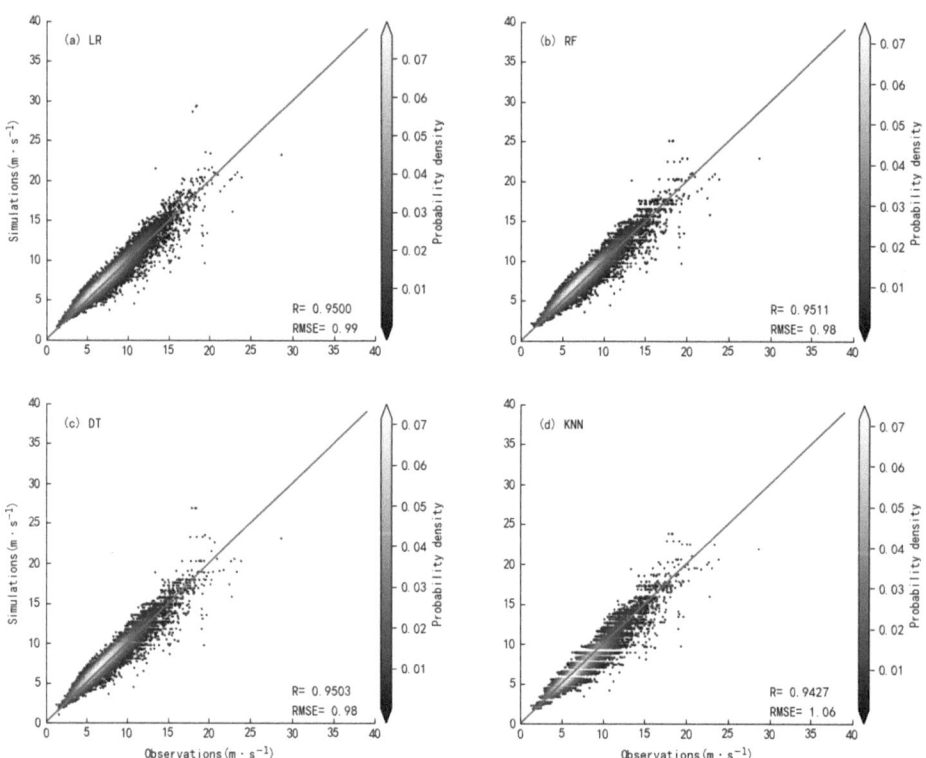

Fig. 2. Simulation results of time series of extended maximum wind speed by different methods

6.2   The Frequency of Occurrence of Gale Disaster Risk and Its Changing Characteristics

From Table 3, it can be seen that the average number of gale disasters occurred during blueberry ripening period in each station of Liaodong Green Economic Zone from 1991 to 2020 is 33.7, and the number of Gale Disasters during 2001 to 2010 is more. The frequency of strong wind disaster was the most in early-ripening period, the average frequency was 15.3, the next was 13.4 in middle-ripening period, which was 1.2

times different from that in late-ripening period.It shows that when the spring and summer alternate in the study area, the climate is complex and changeable, which is more conducive to the formation and development of strong winds.

Table 3. The average of the occurrence frequency of high temperature and strong wind disaster in the mature period of blueberry in Liaodong Green Economic Zone

Type	1991–2000	2001–2010	2011–2020	1991–2020
Maturity	11.3	11.5	10.9	33.7
Early maturity	5.3	5.8	4.2	15.3
Medium maturity	4.3	5.2	3.9	13.4
Late maturing	4.2	3.1	4.9	12.2

6.3 The Occurrence Range and Its Changing Characteristics of Gale Disaster Risk

The ratio of stations and the average ratio of stations during 1991–2020 are calculated. As can be seen from Fig. 3, there was a non-significant increase in maturity, with an average increase of 0.5 percentage points per 10a. From the early, middle and late maturity, the ratio of gale stations showed a non-significant reduction trend, especially in the early maturity, it is mainly related to the highest number of strong winds in blueberries in the early maturity period, and the change trend is also the most obvious. In general, the risk of wind disaster increased during the whole ripening period, but decreased during the early, middle and late ripening periods.

6.4 The Changing Characteristics of the Risk of Different Grades of Gale Disasters

1991–2020 are, the risk of high wind damage in different grades of blueberries during ripening stage varied significantly (Fig. 4). The average number of high wind damage in ripening stage was 1.25 times a year, and the risk of mild, moderate and severe high wind damage decreased in turn. In terms of each mature stage, the light disaster risk is the most serious in the early mature stage, and the moderate and severe disasters are in the late mature stage, while the moderate and severe disasters risk is relatively light in the middle mature stage, 0.07 and 0.03, respectively. In general, the risk of gale disasters in the early maturity period is higher than that in the middle and late maturity periods, which is consistent with the analysis results of 2.2 early maturity blueberry with the highest number of gales, which has a great impact on the yield and quality of early maturity blueberry.

6.5 Risk Probability Analysis of Strong Wind Disaster

According to the statistics of the risk grade of the gale disaster in the past years, based on the information diffusion theory, the risk probability $p(u_i)$ of the gale disaster risk in

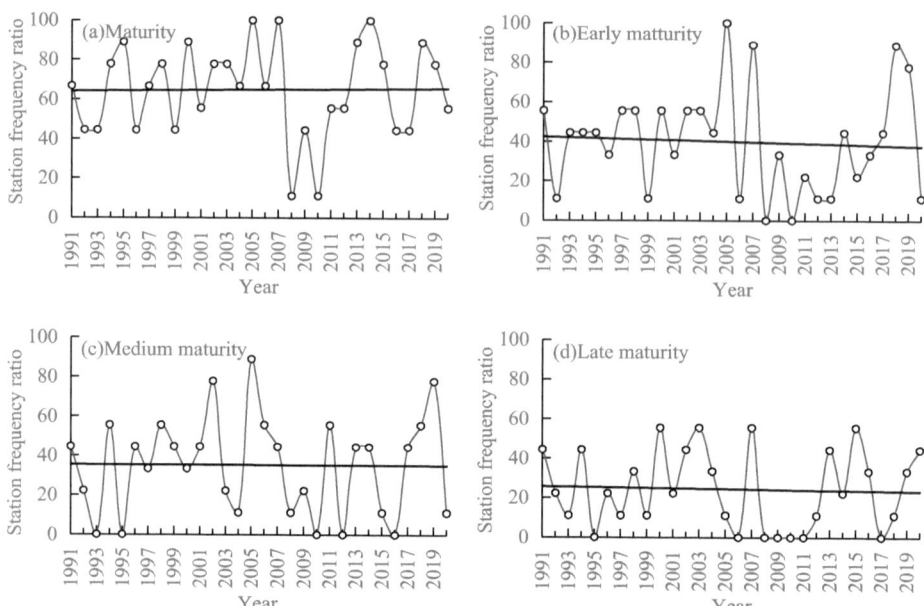

Fig. 3. Interannual variation characteristics of gale stations at mature stage (a), early mature stage (b), middle mature stage (c) and late mature stage (d) of blueberry in Liaodong Green Economic Zone from 1991 to 2020

each site is obtained. It can be seen from Fig. 5 that the probability of occurrence of light gale disaster in mature period is higher than that in southern region, and the probability range is 26.7% ~ 53.2%, which is more than once every five years (≥ 20%). In terms of the early, middle and late maturity stages, the risk probability of light gale disaster is generally high, which occurs more than once every five years (≥ 20%), and the risk probability of early maturity stage is 27.8%, which is the highest average value of each site, about 66.7% of the stations are at the frequency of once every three years, and the Fushun area is a high risk area.

From the probability distribution characteristics of the high wind disaster risk occurrence index Lu of each mature period in Fig. 6, the probability of high wind disaster risk occurring in the whole mature period of Liaodong Green Economic Zone is 61.1%, and the occurrence of disaster risk occurs more than once every two years, the probabilities of light, moderate and severe disasters were 38.1%, 17.8% and 5.2% respectively. In general, the blueberry disasters in the green economic zone of eastern Liaoning mainly occurred in the middle ripening stage, with the highest risk of mild and severe gale disasters occurring more than once in ten years (≥ 10%), fushun in particular.

7   Conclusion and Discussion

On the basis of meteorological data, according to the criteria of Meteorological Gale Disaster, the discriminant criterion is established by considering the intensity and frequency of Gale disaster, and the risk of Gale disaster in past years is classified, it can

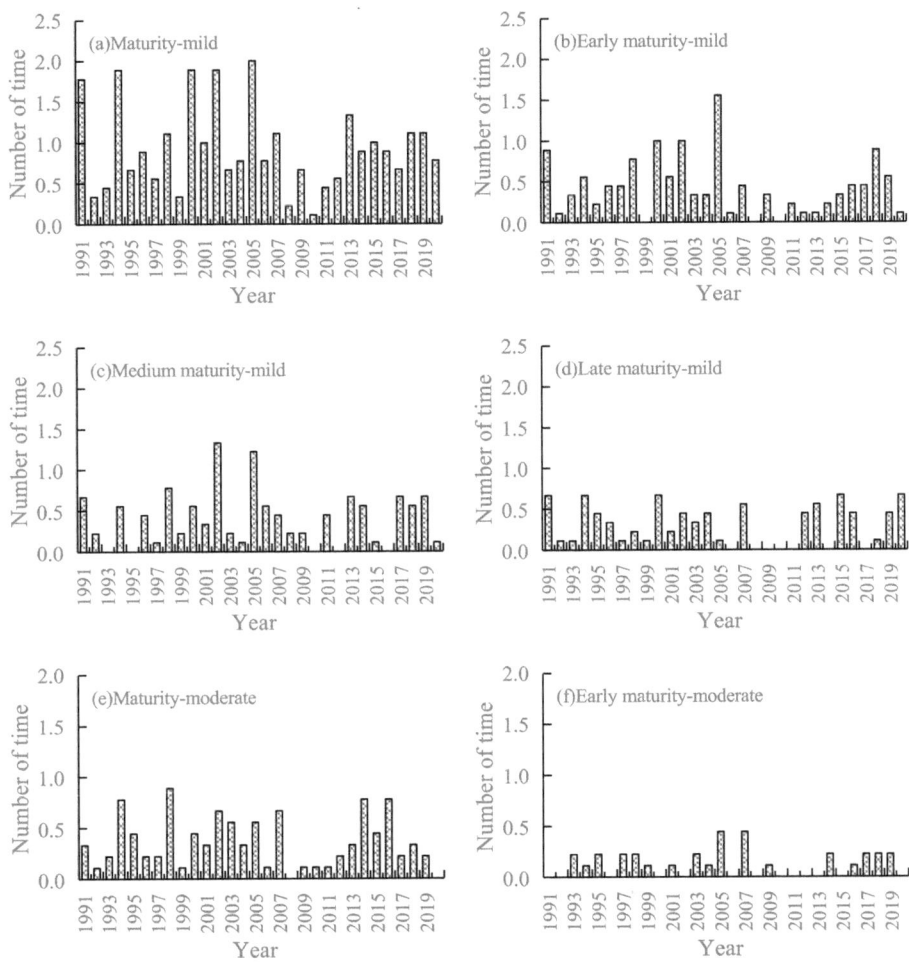

Fig. 4. 1991–2020 time series of cumulative number of gale of different grades during blueberry ripening period in Liaoning green economic zone

reflect the risk degree of Gale, judge the rule of Gale Disaster Risk and evaluate the risk scientifically. It has certain reference value for the development of blueberry industry. The results showed that during 1991–2020, the blueberry ripening period in Liaodong green economic zone was influenced by Gale disaster risk, and the occurrence frequency was not significantly reduced, especially in Xiuyan area, but the Kuandian area presents the remarkable increase trend. The probability of occurrence of gale hazard is 61.1%, which occurs more than once every two years (≥50%). The Fushun region has the highest risk level, and the Kuandian Manchu Autonomous County region has the lowest risk level. Based on the theory of information diffusion, the probability of Gale disaster risk at each station in Liaodong Green Economic Zone is evaluated, and the exceeding probability of Gale disaster risk at different degrees in each mature period is calculated, it can

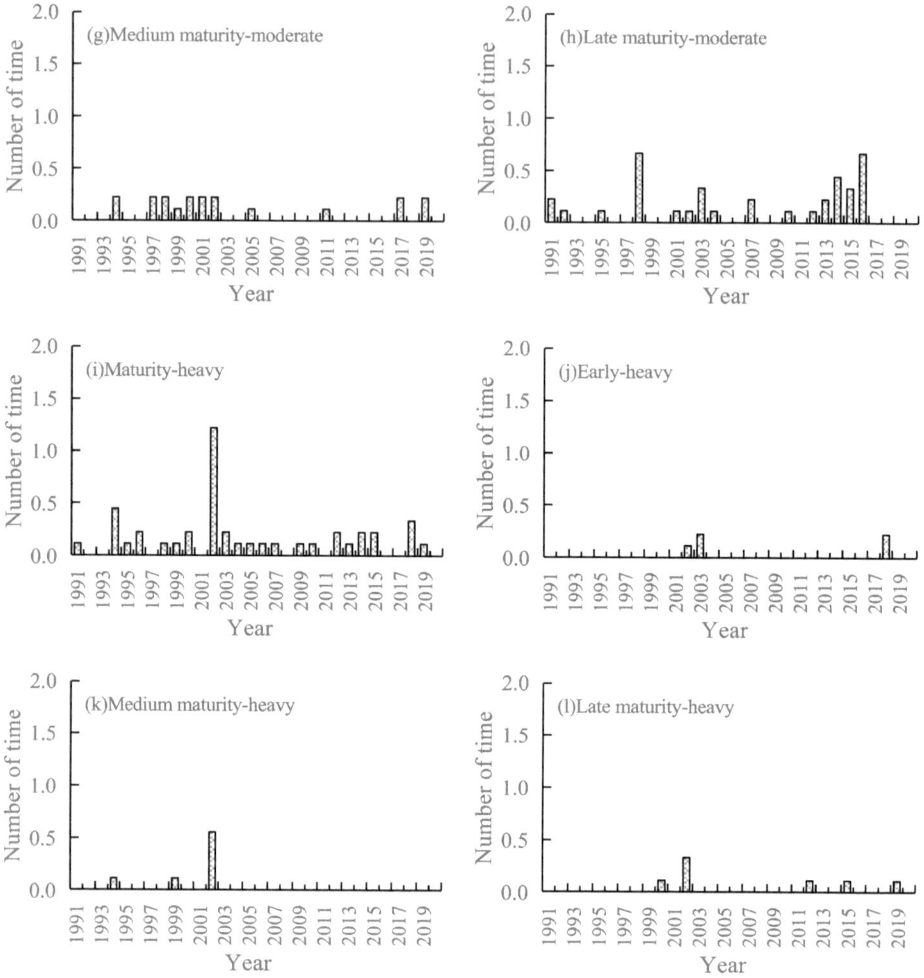

Fig. 4. (*continued*)

effectively reduce the error caused by the limited sample data, and reflect the high-risk area of the strong wind disaster.

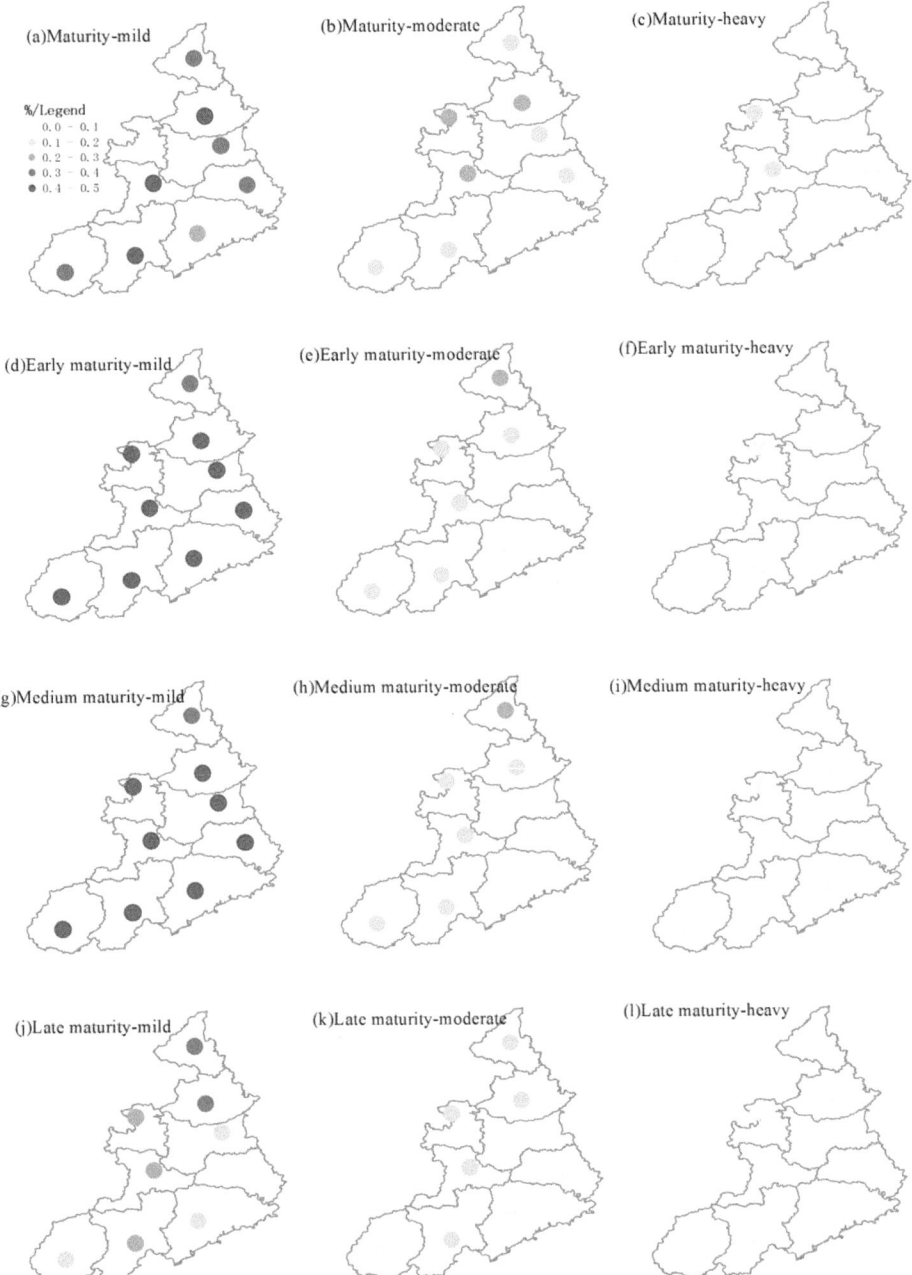

Fig. 5. 1991–2020 risk probability distribution of three levels of strong wind at each mature stage of blueberry in Liaodong Green Economic Zone

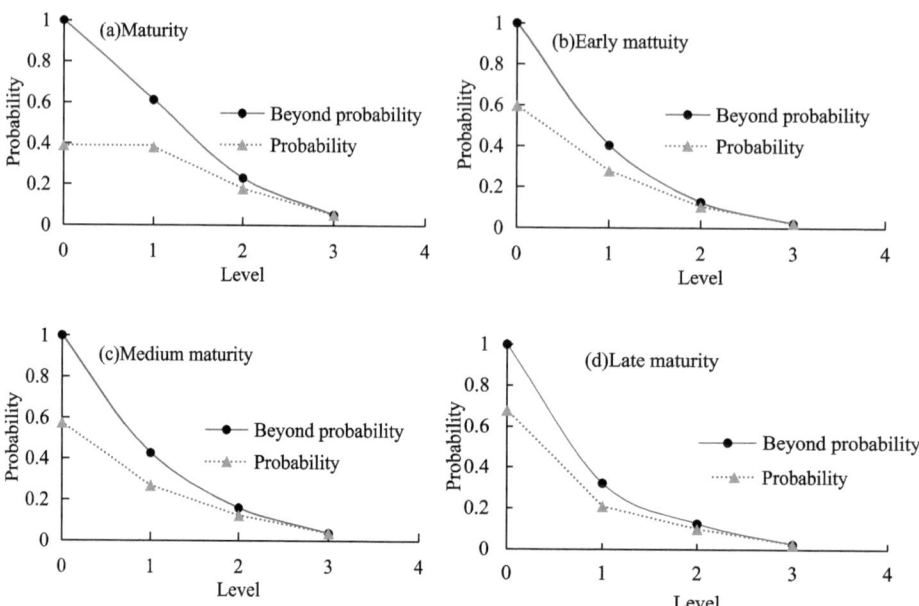

Fig. 6. 1991–2020 changes of probability and exceeding probability density of blueberry disaster risk in each mature period in Liaodong Green Economic Zone

References

1. Panmao, Z., Zhen, L., Yang, C.: A review of precipitation persistence and phase change under the background of climate warming[J]. J. Meteorol.Meteorol. **75**(4), 527–538 (2017)
2. Huizi, B., Dengpan, X., Jianfeng, L., et al.: Spatial and temporal patterns of extreme weather events and agrometeorological disasters in North China during the period of 965–2014[J]. Geogr. Geogr. Inf. Sci. **34**(05), 99–105 (2018)
3. Yuxi, L., Jingquan, R., Yue, S., et al.: Characteristics and impacts of agro-meteorological disaster losses in northeast China from 1971 to 2016[J]. Dry Weather. **38**(04), 647–654 (2020)
4. Edenhofer, O., Seyboth, K.: Intergovernmental panel on climate change(IPCC)[J]. Encycl. Energy Nat. Resour. & Environ. Econ. **26**(D14), 48–56 (2013)
5. Dong Haitao, Liu E, Li Runan, et al. Analysis of meteorological conditions in main planting counties of blueberry in Liaoning Province[J]. Liaoning Agric. Sci. (04), 58–62 (2023)
6. Yuan xuecuo, Wang Gang, Wu Jinxiang, et al. Meteorological factors analysis and service plan for blueberry production in fengyang county[J]. Anhui Agric. Bull. **29**(3), 68–72 (2023)
7. Zhang Li,Marjorie, Stephen, and so on.The effect of meteorological disasters on blueberry in southern Anhui province and its preventive measures[J].Modern Agricultural Technology,2013(02):245–247
8. Li Yadong, Sun Haiyue, Chen Li. Report on the development of our blueberry industry[J]. Chin. Fruit Trees (05), 1–10 (2016)
9. Lee Ah Tung, Pegbo, Chen Li, et al. China blueberry industry annual report 2020[J]. J. Jilin Agric. Univ. **43**(01), 1–8 (2021)
10. Xuelin, W., Qinqin, H., Jun, L.: Meteorological disaster risk assessment of double-cropping early rice in South China based on information diffusion theory[J]. China's Agrometeorol. **40**(11), 712–722 (2019)

11. Guan Yue, Liu Jiahong, he qijin, et al. Risk probability of high temperature damage at flowering stage of summer maize in North China Plain based on information diffusion theory[J]. China Agrometeorology **42**(7), 606–615 (2021)
12. Guilin, L., Yuanxin, Li., Xiaoxiao, Li., et al.: Typhoon disaster risk analysis in Guangdong province based on information diffusion technology[J]. J. Saf. Environ.Saf. Environ. **21**(04), 1684–1692 (2021)
13. Kuwata, K., Shibasaki, R.: Estimating corn yield in the United States with MODIS EVI and machine learning methods[J]. ISPRS Ann. Photogramm., Remote. Sens. Spat. Inf. Sci. **3**(8), 131–136 (2016)
14. Clifton, A., Kilcher, L., Lundquist, J. K. et al. Using machine learning to predict wind turbine power output[J]. Environmental Research Letters **8**(2), 024009 (2013)
15. Khosravi, A., Machado, L., Nunes, R.O.: Time-series prediction of wind speed using machine learning algorithms:a case study Osorio wind farm, Brazil[J]. Appl. Energy **224**, 550–566 (2018)
16. Ghorbani, M.A., Khatibi, R., FazeliFard, M.H., et al.: Short-term wind speed predictions with machine learning techniques [J]. Meteorol. Atmos. Phys.. Atmos. Phys. **128**(1), 57–72 (2016)
17. Ak, R., Fink, O., Zio, E.: Two machine learning approaches for short-term wind speed time-series prediction[J]. IEEE Trans. Neural Netw. Learn. Syst. **27**(8), 1734–1747 (2015)
18. Sallis, P.J., Claster, W., Hernández, S.: A machine-learning algorithm for wind gust prediction[J]. Comput. Geosci.. Geosci. **37**(9), 1337–1344 (2011)
19. He jinna, Liu Buchun, Liu Yuan, et al. Design of weather index insurance for table grape: taking the disaster of overcast rain in Wafangdian, the main producing area around Bohai Sea as an example[J]. China Agrometeorology **43**(10), 810–820 (2022)
20. Zhou Yu, Xie Liangmei, Dong Tong, et al. Analysis of meteorological conditions of blueberry cultivation in Dongxiang District of Fuzhou, Jiangxi[J]. Mod. Agric. Technol. (06), 160–163 (2023)
21. National Weather Service. Specification for Agrometeorological Observation (Vol. 1)[m]. Beijing: Meteorological Press (1993)

Spatial Distribution and Determinants of Convalescence Climate Tourist Attraions in Sichuan, China

Yanchuan Zhong[1,2], Yunmeng Peng[3(✉)], and Yiheng Cai[2]

[1] Institute of Plateau Meteorology, CMA, Chengdu 610072, China
[2] Sichuan Provincial Climate Center, Chengdu 610072, China
[3] Chongqing Meteorological Service Center, Chongqing 401147, China
`cctomato@qq.com`

Abstract. This study analyses the distribution pattern, directionality, and regional aggregation characteristics of W-CCR, S-CCR, and Y-CCR in Sichuan Province, using POI (Point of Interest) data from 538 Convalescent Climate Resorts (CCR) and daily data from 156 national meteorological stations from 1991 to 2020. The factors that influence the spatial distribution of CCR are discussed from the perspectives of climate, natural environment, and human society. The data suggests that CCR is concentrated along the southern edge of the basin, the eastern side of the Hengduan Mountains at the border of the basin and plateau, and the Anning River Valley in the Panxi region, following a northwest-southeast orientation. However, the degree and pattern of aggregation vary among different types of CCR. The CCRs that avoid cold have the highest centrality and are mainly located in the Panxi region, which is a rare convalescent climate resource in Sichuan. The CCRs that avoid heat and those for general convalescence have a wider distribution range and strong directionality. Their hotspots are distributed in the northeastern, southwestern, and southern parts of the basin, as well as most regions of the Panxi area. The development of CCR is influenced by various factors, including convalescent climate resources, urbanization, economic development levels, population size, and policy guidance. This study aims to explore local climate resources and promote the transformation of climate resource advantages into tourism economic development advantages.

Keywords: Convalescent climate resorts · Distribution pattern · Aggregation characteristics · Climate factors · Sichuan province

1 Introduction

The study of high-quality tourism climate resources can help to establish and promote distinctive convalescent climate brands, contribute to the growth of the tourism economy, and promote the development of climate tourism. Sichuan has unique convalescent tourism climate resources that span two climatic zones within the province, resulting in significant thermal differences. The Sichuan Basin in the east falls under the subtropical

B.-J. He et al. (Eds.): UCSUD 2023, LNCE 559, pp. 102–123, 2025.
https://doi.org/10.1007/978-981-97-8401-1_8

monsoon climate zone, while the Sichuan Plateau in the west is part of the alpine climate zone of the Tibetan Plateau. The region's large-scale topography and interlacing mountains create diverse vertical climate types. Optimizing the spatial structure of regional tourist attractions is an essential prerequisite for achieving a long-term mechanism for coordinated regional tourism industry development [1]. The spatial distribution patterns and influencing factors of convalescent climate tourist attractions can be scientifically analysed to provide references for addressing issues such as improving accessibility, enhancing tourism transportation, upgrading industries, increasing the environmental carrying capacity of scenic areas, preventing seasonal mismatches of resources, and reducing resource constraints and waste. Research into the spatial distribution and influencing factors of convalescent climate tourist attractions can aid in the optimal allocation of tourism resources in time and space. It can also provide insights into resource integration and possess practical significance for the planning and development of convalescent climate tourism.

In recent years, significant progress has been made in researching the spatial distribution of tourist attractions and their influencing factors. Scholars have approached this topic from various perspectives and methodologies. Research studies have analysed not only the grade and type of attractions [2, 3], but have also conducted in-depth research using multiple indicators such as geomorphological identification [4], institutional tourism materials, multimedia data [5], and geospatial data [6]. These studies have examined both domestic and international tourist landscapes, revealing the relationships between tourism development and population [7], economy, and society [8]. The distribution of tourist attractions can, to some extent, indicate the development status of regional tourism. However, research on the spatial distribution of tourist attractions is still limited to a few samples or small regions. Network analysis is commonly used to study this topic, with data from travel blogs, reviews, or social media platforms [9–11]. Regarding research methods, traditional approaches mainly rely on GIS spatial analysis and mathematical statistics. They often use POI (Point of Interest) data for spatial distribution analysis and various visualization methods to display hotspots, density, and trends. This way, they characterize the macro-distribution features of the research subjects [12]. However, in recent years, new methods have emerged, such as neural network prediction models [13], deep learning [14], gravity analysis [15], cluster analysis [16], spatial measurement [17], and social network analysis [18–20]. These methods aid in accurately revealing the characteristics of spatial distribution of tourist attractions and their influencing factors. The spatial distribution of tourist attractions is mainly concentrated on aggregation and imbalance phenomena [21]. Future research should continue to focus on these aspects to provide valuable insights for the development of the tourism industry.

The Sichuan region is an important part of China's major national development strategies and plays a leading role in the construction of the 'Belt and Road.' Tourism plays a significant role in regional economic development. Scholars in the Sichuan region have conducted preliminary analyses of convalescent climate resources and attractions [22, 23]. However, further research is needed to enrich the spatial distribution patterns and attribution analysis of convalescent climate attractions. Based on the latest cultural and tourism resource survey results, this study classifies and statistically analyzes Sichuan's convalescent climate tourist attractions using classic scenic area analysis methods. The language used is clear, concise, and objective, with a formal register and precise word choice. The text adheres to conventional structure and formatting features, including consistent citation and footnote style. The logical flow of information is maintained with causal connections between statements. The text is free from grammatical errors, spelling mistakes, and punctuation errors. No changes in content have been made. The study uses kernel density, standardized ellipse, and hotspot analysis, among other classic techniques, to examine the spatial distribution characteristics of various types of convalescent climate tourist attractions in Sichuan. It also explores the factors that influence them, providing scientific support for the development and optimization of the spatial structure of these attractions.

2 Data and Methods

2.1 Study Area

Sichuan Province is located in the southwest region of China, upstream of the Yangtze and Yellow Rivers. The terrain slopes from northwest to southeast, with significant variations in elevation. The western part of the province constitutes the southeastern edge of the Tibetan Plateau and the Hengduan Mountains, while the eastern part is surrounded by the Sichuan Basin, which is encircled by mountains. Sichuan's topography comprises the Sichuan Basin, the Western Sichuan Plateau, and the Southwestern Sichuan Mountains. The region's diverse climate conditions and unique natural environment offer a range of convalescent climate tourism resources, which provide ample material and support for the development of the convalescent tourism industry.

2.2 Data Sources

The "Convalescent Climate Tourist Attractions" (CCR) in this study refer to the point of interest (POI) data obtained from the 2021 Sichuan Province Cultural and Tourism Resources Census Database. These data were derived from the Sichuan Province Tourism Resources Classification, Survey, and Evaluation, which classifies tourism resources based on their characteristics. The study focuses on the main category of "Astronomical and Climatic Landscapes," specifically the subcategory of "Weather and Climate Phenomena," including "Summer Retreat Climate Zones" and "Cold-Avoidance Climate Zones," as well as "Convalescent Climate Landcapes." For the purpose of this research, these are collectively referred to as Convalescent Climate Resorts (CCR). After data cleaning, a total of 538 CCR POI data points were obtained.

Following the 'Interpretation of Sichuan Province Tourism Resources Classification, Survey, and Evaluation,' three categories of CCRs have been identified based on their resource characteristics: Winter Shelter Convalescent Climate (W-CCR), Summer Retreat Convalescent Climate Resort (S-CCR), and Year-round Convalescent Climate Resort (Y-CCR) (Table 1 and Fig. 1).The study calculates the 'Convalescent Climate Resources' based on meteorological data collected over many years. The data used for these calculations were sourced from the Sichuan Provincial Meteorological

Data Detection Center and include daily records from 156 national meteorological stations in Sichuan Province from 1991 to 2020. The tourism and accommodation data were obtained from the 2021 Sichuan Province Cultural and Tourism Resources Census Database. The topographic data and administrative boundary map of Sichuan Province were obtained from the China Meteorological Administration's Geographic Information Database. The data was extracted using the regional administrative boundary data of Sichuan Province.

Table 1. Classification of POI for CCR in Sichuan

Types of CCR	Quantity/pc	POI content
W-CCR	27	Attractions that provide shelter from the cold during the colder seasons and where the human body is more comfortable
S-CCR	274	Attractions suitable for summer holidays during the hot season and where the human body feels more comfortable
Y-CCR	247	Areas with favourable climatic conditions and ecological environment, suitable for recreational and therapeutic rest, and attractions with an all-year holiday climate index of suitable or very suitable for more than six months

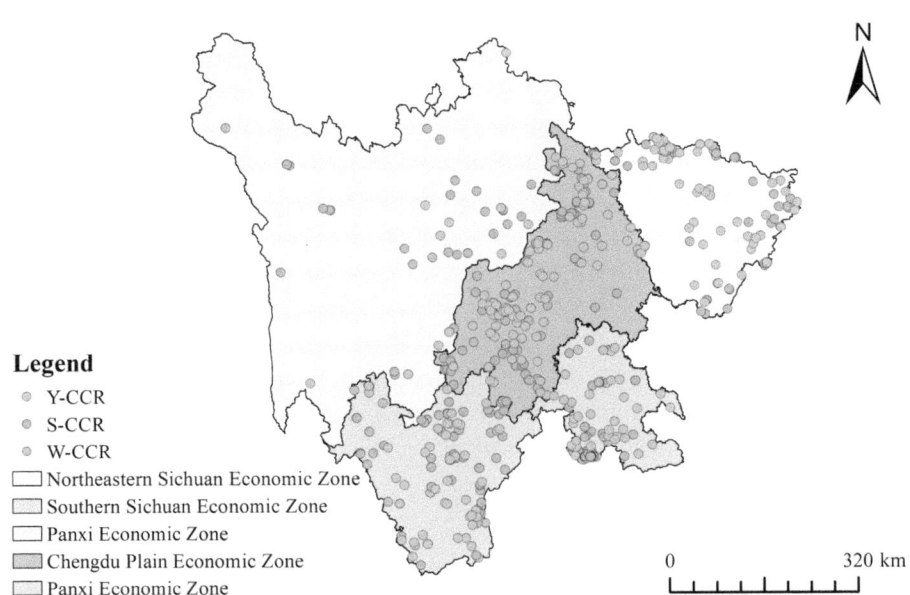

Fig. 1. Distribution of classified CCR in Sichuan Province

2.3 Research Method

2.3.1 Kernel Density Estimation

The POI data of convalescent climate attractions in Sichuan Province were analysed using the kernel density estimation method [24]. This method can describe the basic characteristics of POI point distribution in traditional regional analysis. By expressing the density calculation results in a two-dimensional grey scale, the aggregation or dispersion distribution characteristics of the POI point group can be obtained. The formula is as follows:

$$f_n(x) = \frac{1}{nh} \sum_{i=1}^{n} k\left(\frac{x - x_i}{h}\right) \tag{1}$$

K is the kernel function, x is the location of the POI within the study area, x is the centre of the circle, h is the radius, xi is the location of the POI within the area formed, and n is the number of samples.

2.3.2 Standard Deviation Ellipse Analysis

Calculating the standard deviation ellipse [25] allows for a clearer determination of trends. The standard deviation ellipse analysis method was used to analyze the overall directional distribution trend and core distribution range of the convalescent climate attractions in Sichuan Province after abstracting them into POI data. By creating a standard deviation ellipse, the spatial characteristics of geographic elements can be summarized, including central tendency, dispersion, and directional trends. By analyzing the centrality and aggregation of the distribution ellipse, important spatial structural characteristics of CCR efficiency can be reflected, and spatial dynamics interpretation can be provided. The standard deviation ellipse is defined as:

$$SDEx = \sqrt{\frac{\sum_{i=1}^{n} (x_i - \bar{x})^2}{n}} \tag{2}$$

$$SDE_y = \sqrt{\frac{\sum_{i=1}^{n} (y_i - \bar{Y})^2}{n}}$$

where x_i and y_i are the coordinates of element i, $\{\bar{X}, \bar{Y}\}$ denotes the mean centre of the element, and n is the total number of elements.

2.3.3 Hot Spot Analysis

The Getis-Ord Gi* hot spot analysis method was used to analyse the classification and aggregation of POI data for convalescent climate attractions in Sichuan Province. This method describes local autocorrelation, reflecting the density of high and low values of CCR points in the study area and expressing their second-order distribution properties [26]. Z score values are used to distinguish high-value areas. The expression is as follows:

$$G = \frac{\sum_{i=1}^{n} \sum_{j=1}^{n} \omega_{i,j} x_i x_j}{\sum_{i=1}^{n} \sum_{j=1}^{n} x_i x_j}, \forall_j \neq i \tag{3}$$

In the equation, x_i and x_j are the attribute values of features i and j, respectively, $\omega_{i,j}$ is the weight between spatial features i and j. n is the number of features in the dataset, and $\forall_j \neq i$ indicates that features i and j cannot be the same. After counting the number of three types of CCR points in each county, the hot spot analysis was performed to obtain the aggregation distribution of various CCRs in Sichuan Province and visualize the results.

2.3.4 Moran's I Index Analysis

Global autocorrelation analysis of CCR was conducted using the Global Moran's I index. The Moran's I index describes the degree of association between all spatial units and their surrounding areas across the entire region, transforming qualitative descriptions of spatial autocorrelation in geographic data into quantitative analysis. The global Moran's I index M is expressed as:

$$M = \frac{N \sum_{i=1}^{N} \sum_{j=1}^{N} W_{ij}(z_i - \overline{z})}{(\sum_{i=1}^{N} \sum_{j=1}^{N} W_{ij}) \sum_{j=1}^{N} (z_i - \overline{z})} \tag{4}$$

In the equation, $\overline{z}$ is the mean value of variable z; $\overline{z}$ and z_j are the values of the variable at spatial locations i and j, respectively (i ≠ j); W_{ij} is the spatial weight coefficient between z_j and z_i [27].

2.3.5 Spatial Distribution Influencing Factors

To investigate the factors affecting the spatial distribution of CCR, we utilized the expert scoring method and geo-detector to identify the relevant factors for a more targeted and scientific analysis. We also employed methods such as raster calculator and buffer analysis to overlay the identified factors and perform the analysis. [28].

3 Contents and Analysis

3.1 Analysis of Spatial Distribution Characteristics of CCR

3.1.1 Quantitative Analysis of Spatial Distribution of CCR

Table 2 shows that the spatial layout of Sichuan CCR is uneven, with a more significant difference between the eastern and western regions than between the southern and northern regions. The distribution of the three types of CCRs in different economic zones varies. S-CCR and Y-CCR are present in all economic zones, while W-CCR is mainly found outside the Chengdu Plain Economic Zone and the Southern Sichuan Economic Zone. This suggests that Sichuan CCRs are diverse in type but unevenly distributed geographically. The number of S-CCRs is the highest on a provincial scale, totaling 274, and ranking first among the total number of CCRs in each economic zone, particularly in the Chengdu Plain Economic Zone, where the count reaches 87. The economic zones are ordered by S-CCR quantity from high to low, starting with the Southern Sichuan Economic Zone, followed by the Panxi Economic Zone, the Northeastern Sichuan Economic Zone, and the Northwestern Sichuan Ecological Economic Zone. The number of

Y-CCRs is the second highest, with a total of 247, and the highest proportion is found in the Chengdu Plain Economic Zone, amounting to 84. The economic zones are ordered by Y-CCR quantity from high to low, with Northeastern Sichuan Economic Zone, Panxi Economic Zone, Southern Sichuan Economic Zone, and Northwestern Sichuan Ecological Economic Zone following this order. In contrast, the number of W-CCRs is the lowest, only 27, which is significantly different from the numbers of S-CCRs and Y-CCRs. Of these, 89% of W-CCRs are concentrated in the Panxi Economic Zone, with the remaining few distributed in the Northeastern Sichuan Economic Zone and the Southern Sichuan Economic Zone. In China, cold-avoiding climates are considered scarce holiday resources. The number of W-CCRs in Sichuan Province is far less than that of S-CCRs and Y-CCRs, which is similar to the national situation regarding the proportion of cold-avoiding and heat-avoiding resources [29]. In the economically underdeveloped Panxi region, the development and construction of CCRs depend primarily on the existing climate resource endowment. Additionally, the high proportion of S-CCRs and Y-CCRs in the Chengdu Plain Economic Zone indicates that the more economically developed Chengdu Plain Economic Zone has not only fully utilized the original climate resource endowment but also achieved a higher level of artificial development and innovation in CCR construction.

Table 2. Number of classified CCR in all economic regions of Sichuan Province

Type of economic zone	Quantity distribution		
	W-CCR	S-CCR	Y-CCR
Chengdu Plain Economic Zone	0	87	84
Northeastern Sichuan Economic Zone	1	44	68
Southern Sichuan Economic Zone	2	58	40
Northwestern Sichuan Ecological Economic Zone	0	31	12
Panxi Economic Zone	24	54	43
Total	27	274	247

3.1.2 Analysis of Spatial Patterns of CCR

(1) Differences in the scale and structure of spatial distribution.

   1) Average Nearest Neighbor Analysis. The results of the analysis show that CCRs in Sichuan Province are distributed in a distinct cluster, with an average observed distance of 9195.1140 m and an expected average distance of 18511.3321 m. The nearest neighbor ratio R is 0.4967, with a Z score of -22.3526 and a significance level $P < 0.01$. Upon further analysis, it has been discovered that the spatial aggregation of Y-CCRs and S-CCRs is particularly significant, with the likelihood of randomly generating such clustering patterns being less than 1%. However, there is no significant difference between the distribution of W-CCRs and random patterns.

2) Municipal (state) distribution differences. The concentration and dispersion characteristics of CCRs can be determined by the nearest neighbor index, while the density value represents the sparsity of CCRs. According to relevant literature [30], CCRs at the municipal scale are classified into eight types: dense clustered, general clustered, sparse clustered, dense uniform, general uniform, sparse uniform, general discrete, and sparse discrete (Table 3). From a sparsity perspective, the cities and states with a CCR density greater than 0.002 per 100 km^2 are Leshan, Meishan, Dazhou, Guangyuan, Ya'an, Mianyang, and Yibin. These areas are located in the Chengdu Plain, which is known for its dense population, developed economy, and mature tourism markets with great potential for leisure tourism. Cities and states with a population density lower than 0.0009 per 100 km^2 include Ganzi, Ziyang, Suining, Aba, Nanchong, Deyang, and Luzhou. Ganzi and Aba are located in high-altitude and high-latitude regions with small populations and low levels of economic development.The distribution of CCRs varies greatly among cities and states in terms of concentration and dispersion characteristics, with nearest neighbor index values ranging from 0.6 to 2.8. Thirty-eight percent of cities and states are of the clustered type, distributed along the western edge of the basin, as well as in the plateau region of Ganzi Prefecture and the Liangshan Prefecture in the Panxi region. Liangshan has the highest degree of concentration among the clustered cities and states, indicating that the development of CCRs is greatly influenced by resource constraints, leading to concentrated distribution in specific areas. Ya'an exhibits a lower degree of concentration and tends towards a balanced distribution, indicating that the spatial development of CCRs is relatively advanced.

3) Kernel Density Analysis

Kernel density analysis was used to reveal the distribution pattern of different types of CCRs in Sichuan Province. The results show that Sichuan CCRs exhibit a 'ring-basin cluster-band' spatial characteristic (Fig. 2a). This band-shaped area is created by connecting point elements, mainly concentrated in the southern part of the basin, along the eastern side of the Hengduan Mountains at the border of the basin and plateau, and along the north-south line of the Anning River Valley in the Panxi region. The region is influenced by complex terrain and crisscrossing mountains, resulting in a variety of vertical climate types. This provides rich climatic resources for the development of CCRs. Significant differences exist in the distribution of various types of CCRs. Y-CCRs are mainly concentrated in high-density areas located in the region east of the Hengduan Mountains. Compared to S-CCRs, Y-CCRs have a higher density in the central part of the basin but a lower density in the western and northern parts of the plateau. The distribution pattern of Y-CCRs is similar to the overall distribution characteristics of CCRs, presenting a 'ring-basin cluster-band' spatial characteristic (Fig. 2b). The distribution of W-CCRs is scattered, but high-density areas have formed in the Panxi region and the northern part of Muli County in southern Sichuan Province. A secondary density area is also present near Gong County in the southern part of the basin (Fig. 2c). The high-density areas of S-CCRs are mainly located along the mountainous perimeter of the basin and the main core area of the Panxi region, with sub-core areas in the plateau region near the basin (Fig. 2d). The Panxi region is a common high-value area for all three types of CCRs in terms of spatial distribution. The surrounding areas of the basin have a moderate climate, developed economy, and dense population, resulting in the formation

Table 3. CCR density and agglomeration characteristics

City (state)	Number of CCR	density(Units/100km^2)	Nearest neighbour index	Distribution type	City (state)	Number of CCR	density(Units/100km^2)	Nearest neighbour index	Distribution type
Chengdu	25	0.0017	0.6680	General Aggregate	Meishan	15	0.0021	0.9967	Dense and homogeneous
Zigong	8	0.0018	1.6228	General Discrete	Yibin	73	0.0055	0.7909	Densely aggregated
Panzhihua	12	0.0016	0.8979	General Uniform	Guang'an	8	0.0013	0.9226	Generally homogeneous
Luzhou	11	0.0009	1.4707	Sparse Discrete	Dazhou	35	0.0021	0.9144	Dense homogeneous
Deyang	4	0.0007	2.6084	Sparse Discrete	Ya'an	41	0.0027	0.8195	Densely aggregated
Mianyang	57	0.0028	0.6451	Dense aggregate	Bazhong	21	0.0017	0.6968	General aggregated
Guangyuan	44	0.0027	0.7770	Dense aggregate	Ziyang	1	0.0002	/	Sparse Discrete
suining	1	0.0002	/	Sparse Discrete	Aba Prefecture	23	0.0003	1.1129	Sparse Uniform
Neijiang	7	0.0013	1.6116	General Discrete	Ganzizhou	21	0.0001	0.6662	Sparsely aggregated
Leshan	26	0.0020	0.8868	Dense Uniform	Liangshan Prefecture	100	0.0017	0.6081	General aggregate
Nanchong	4	0.0003	2.5850	Sparse Discrete					

Note "/" indicates that there are too few elements for an average nearest neighbour analysis of numerous tourist attractions. The mountains' high altitude, river valley topography, and abundant forest resources provide favourable economic and population support, as well as superior climatic and natural environmental conditions for the development of CCR attractions in the basin's peripheral areas. The Panxi region is the most densely populated area for CCR types due to its warm winters, cool summers, and year-round suitability for recuperation and leisure activities. Additionally, the region boasts abundant sunshine and clean air. In contrast, the Western Sichuan Plateau has a lower density of CCRs due to its high altitude and cold temperatures, which result in poor tourism resource support conditions. However, during the summer months, this region is an ideal destination for leisure, summer retreats, and health recovery activities for those living in hot and humid basin areas.

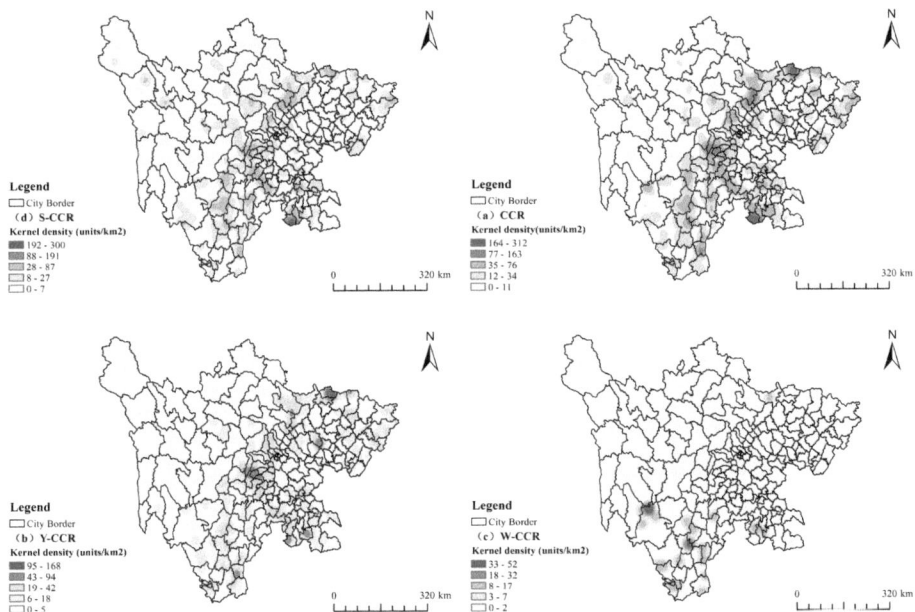

Fig. 2. Kernel density of classified CCR in Sichuan Province

(2) Directions and trends in distribution.

To examine the direction and trend of the distribution of the three types of CCRs, we conducted a standard deviation ellipse analysis on their spatial distribution. This resulted in the production of standard deviation ellipse distribution maps (Fig. 3) and parameter tables (Table 3) for various CCRs in Sichuan Province. The spatial distribution of all three types of CCRs exhibits a northwest-southeast orientation from the perspective of the azimuth angle. This orientation aligns with the direction of the Hengduan Mountains at the border of the Sichuan Basin and the Tibetan Plateau, revealing the significant impact of geographic location on the distribution of CCRs. When comparing the shape parameters, it becomes apparent that each type of CCR has its own unique characteristics. The X-axis lengths of S-CCR, Y-CCR, and W-CCR decrease in descending order, with

S-CCR being the most directional and W-CCR being the weakest, while Y-CCR falls in between. Additionally, the Y-axis length and ellipticity follow the ranking of Y-CCR, S-CCR, and W-CCR. This indicates that the distribution range and centripetal force of the three types of CCRs decrease in intensity, respectively. The long half-axis of the standard deviation ellipse represents the direction of data distribution, while the short half-axis reflects the range of data distribution. The centripetal force of the data increases as the short half-axis decreases. Additionally, the directionality of the data becomes more significant as the difference between the long and short half-axes, or ellipticity, increases. The analysis shows that of the three types of CCRs, W-CCR has the most concentrated distribution direction, the smallest distribution range, and the strongest centripetal force, but the least significant directionality. S-CCR has a more dispersed distribution direction, a moderate distribution range, and moderate centripetal force and directionality. Y-CCR's distribution direction is between the two, with the widest distribution range, the weakest centripetal force, but the most significant directionality (Table 4).

Fig. 3. The standard deviation ellipse diagram of classified CCR in Sichuan.

Table 4. The standard deviation ellipse parameters for each type of recreational climate tourism attraction in Sichuan.

Type of Resorts	X-axis length /km	Y-axis length /km	Divergence angle $\theta/(°)$	Flatness
W-CCR	1.5444	1.8720	55.6421	0.3276
S-CCR	2.1304	2.9196	51.8047	0.7892
Y-CCR	1.6062	3.2361	44.5454	1.6298

(3) Spatial Autocorrelation.

1) *Spatial correlation analysis. Spatial* Correlation Analysis. Table 5 shows the global Moran's I results for various CCRs. The global Moran's I value for total CCR is 0.775, indicating a positive spatial correlation for CCRs in the Sichuan region. The Z test was conducted at a significance level of 0.001 (P = 0.000), indicating that CCRs in Sichuan do not exhibit random distribution but rather significant clustering phenomena.

W-CCR, on the other hand, did not pass the significance test and showed a random spatial distribution. S-CCR exhibited insignificant clustering in space. In contrast, Y-CCR demonstrated the most significant spatial clustering.

Table 5. Global Moran's I of CCR in Sichuan

Type of Resorts	Expected index	Variance	Global Moran's I	Z	P	Mode
CCR	−0.001859	0.000023	0.028165	6.29678	0	Clustered
W-CCR	0.038462	0.000392	−0.0571	−0.9417	0.346371	Random
S-CCR	−0.00365	0.000153	0.017325	1.696	0.089886	Clustered
Y-CCR	−0.004032	0.000099	0.062912	6.72513	0	Clustered

2) *Distribution hotspot characteristics.* Additionally, the distribution hotspot characteristics were analysed. This study uses the local association index Getis-Ord Gi* to explore hotspot and coldspot areas within the region and reveal the local spatial distribution patterns of CCRs. The regional Z values are divided into three categories - hotspot areas, intermediate areas, and coldspot areas - using the natural breakpoint method. Figure 4 shows the hotspot characteristics of CCR type distribution in each county. The results of the analysis suggest that the W-CCR hotspot areas are significantly clustered, mainly concentrated in the Panxi region. In contrast, the coldspot areas are concentrated in most parts of the central and eastern basin, with the plateau area, the southern part of the basin, and the northern part of the basin belonging to the intermediate areas. The distribution of W-CCR hotspot areas extends from north to south along the Hengduan Mountains at the border of the basin and plateau. There are fewer hotspot counties in the north and more in the south, particularly in the Panxi region and the southern part of the plateau. Coldspot areas are mainly concentrated in most parts of the basin, with the majority of the plateau and the southern part of the Panxi region being intermediate areas. Y-CCR hotspot areas are primarily located in the northeastern and southwestern parts of the basin, a small area in the southern part of the basin, and most of the Panxi region. The areas along the Hengduan Mountains from north to south are intermediate areas.

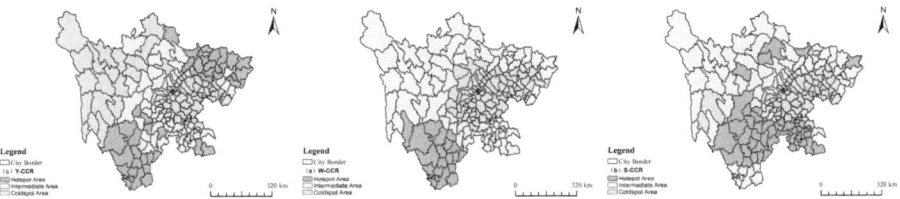

Fig. 4. Distribution of CCR cold spots and hot spots in the province of Sichuan

3.2 Analysis of Factors Affecting Spatial Distribution

3.2.1 Selection of Indicators for Influencing Factors

According to the evaluation criteria for convalescent, summer retreat, and winter retreat scenic spots in the 'Sichuan Province Tourism Resources Classification, Survey, and Evaluation Interpretation', and with reference to the climate zoning localization adjustment method in 'T/CMSA0008–2018, Health Climate Types' [31], Sichuan is classified into four convalescent climate zones: mild seasons, summer retreat, winter retreat, and pleasant sunshine. Based on the variety of climate zones, the factors that determine a restorative climate are identified. The selected influencing factor indicators for analysis are climate resource endowment (richness of convalescent climate types), natural geographical environment (altitude and forest cover), social development situation (population, GDP), and accommodation supporting conditions (hotels, guesthouses, homestays, agritainment). These indicators were chosen based on the research achievements of scholars such as Yang J [32], and by considering the principles of scientific nature, relevance, and availability.

3.2.2 Analysis of Factors Influencing CCR

This study evaluates Sichuan Province's CCR using the expert scoring method based on 12 indicators, including convalescent climate zone types, altitude, vegetation, different levels of accommodation conditions, economic development, and human resources. The evaluation is conducted from four aspects: climate resource endowment, natural geographical environment, social development situation, and accommodation supporting conditions. The evaluation results are classified into four categories using the natural breakpoint method. The geographic detector is used to calculate the impact of various factors on the spatial distribution of CCR across multiple dimensions (Table 5). In general, there are significant differences in the influence of factors such as climate resource endowment, natural geographical environment, social development situation, and accommodation supporting conditions on the spatial pattern of CCR. The largest influence among them is that of climate resource endowment, while the influence of natural geographical environment is relatively smaller (Table 6).

(1)*Climate Resource Endowment*. In terms of type, climate resource endowment has the greatest impact on the spatial distribution of CCRs (q = 0.2419), and the effect is significant (P < 0.001). Research by scholars such as LI L [33] has demonstrated that climate endowment resources significantly affect the distribution of scenic areas. The provision of resources for convalescent climates forms the foundation for the creation of CCRs and has a significant impact on the distribution of CCR agglomerations. Referring to the 'Classification of Health Climate Types' [31], this study divides climates suitable for convalescence into three types: winter retreat, summer retreat, and mild seasons. Figure 5 shows their distribution. The winter retreat climate zone is mainly concentrated in the southern and southeastern parts of Sichuan, including Liangshan Prefecture, Panzhihua City, Luzhou City, Yibin City, and Ya'an City. It has an average temperature above 8 °C in winter. The summer retreat climate zone, on the other hand, is mainly concentrated in the Western Sichuan Plateau and the Panxi region west of the basin. It has an average temperature below 25 °C in summer, and the number of days with

Table 6. Detected results of factors affecting of CCR in Sichuan

Factor	Indicator	Weight	Influence	q- value
14°C < T < 16°C; annual average number of days with "ideal" - "adequate" holiday index ≥ 200	Comfortable climate all year round	0.25	Climate Resources	0.2419**
The average temperature in June-August is between 15 °C and 25 °C; the number of days with the temperature and humidity index "hot and sticky" is less than five	Summer comfortable climate zone	0.25		
The average January temperature is over 8 °C	Winter comfortable climate zone	0.25		
Sunshine hours ≥ 1500 h, number of days above "ideal level of holiday index	Sunny comfortable climate	0.25		
Altitude	Elevation	0.6	Physical Geography	0.0740*
Forest Cover	Forest	0.4		
Hotels & Resorts	Level 1 accommodation	0.4655	Accommodation	0.0722*
Guest Houses	Level 2 accommodation	0.2465		
Homestay	Level 3 accommodation	0.2079		
Agritainment	Level 4 accommodation	0.0801		
GDP	Economic Development	0.6	Social Development Situation	0.1926**
Population	Human Resources	0.4		

Note * indicates P < 0.5 and ** indicates P < 0.001

muggy heat sensation is less than 15 days. The climate zone known as 'mild seasons' has an average annual temperature of approximately 15 °C. There are over 200 days with a suitable vacation index, primarily located along the border between the basin and the plateau from north to south. The mountainous regions of Southwestern Sichuan are ideal for convalescence tourism throughout the year. Particularly, the river valley areas of Panzhihua-Miyi are warm in winter and have a typical winter retreat climate. The Northwestern Sichuan Plateau, except for high-altitude areas that are unsuitable for convalescence tourism, is cool and comfortable in summer, typical of a summer retreat climate. When observing the aggregation effects of W-CCR, S-CCR, and Y-CCR within the three types of convalescence climates, it becomes apparent that there is an uneven distribution within the convalescence climate regions. This suggests that the development of CCR scenic spots primarily depends on the climate background resources, but is also influenced by other factors. In areas of Sichuan that are rich in convalescence climate resource endowments, there is still a significant amount of CCR development space.

Fig. 5. Distribution of healthy climate areas & CCR in Sichuan

(2) *Natural Geographical Environment.* The spatial distribution of CCRs is minimally affected by the natural geographical environment (q = 0.0740), and this result is statistically significant (P < 0.5). Sichuan Province has a diverse and complex terrain. It is bounded by the Longmen Mountains-Daliang Mountains line. The eastern part of the province consists of basins and hills, with altitudes ranging mostly between 500 and 2000 m. The western part comprises plateaus and mountains, with altitudes mostly above 3000 m. The terrain slopes from northwest to southeast, resulting in significant height differences. The climate in Sichuan is influenced by complex terrain and atmospheric circulation, resulting in noticeable zonality and vertical variation. Sichuan Province, situated in both mid-latitude and subtropical regions, has diverse climate types, with significant differences between the east and west. The western plateau has a cold-type climate, the basin area has a subtropical climate, and the surrounding mountainous areas exhibit mountain vertical climate types. Altitudes above 3000 m can significantly impact convalescence tourism activities. Short-term exposure may cause altitude sickness, while long-term living at such altitudes can have irreversible effects on the human body. Therefore, high altitudes restrict the distribution and development of CCRs. It is important to note that this information is objective and based on scientific evidence. However, the significant presence of vertical relief allows for the formation of diverse local climates. Additionally, the high forest coverage rate provides favourable natural underlying surface conditions for the development of CCRs.

(3) *Social Development Situation.* The social development situation has a significant impact on the spatial distribution of CCRs (q = 0.1926) (P < 0.001). The improvement of the socio-economic level provides a solid economic foundation for the development of the convalescence tourism industry [34], making it an important driving force for the construction of CCRs. The economic foundation is essential for developing product systems, enhancing supporting facilities, securing investment funds, and attracting talent and expertise [35]. Increased urbanization can have a ripple effect on the surrounding tourism industry, encouraging population growth, industrial coordination, and social governance in tourist towns, thus influencing the spatial distribution pattern of such towns. As urbanization levels increase, so does the public's demand for tourism and the

diversity of tourism services required. The population size supports the development of the convalescence tourism industry and provides sufficient human resources for the construction of CCRs. Furthermore, the population residing near CCRs constitutes a significant portion of the tourist market. The distribution and proximity of the tourist market are closely linked to the degree of attenuation of the tourism attraction of CCRs [36]. Additionally, policies and government emphasis play a crucial role in shaping and developing tourist attractions [37]. Sichuan Province prioritises the development of convalescence tourism, as evidenced by the 'General Plan for the Development of Traditional Chinese Medicine Health Convalescence Tourism in Sichuan Province' and the 'Sichuan Province Convalescence Tourism Development Plan (2015–2025)' among other plans. The province has implemented nearly a hundred convalescence tourism projects through the 'Double Hundred Billion' tourism investment project, hosting the China (Sichuan) International Tourism Investment Conference, and other activities. The language used is clear, objective, and value-neutral, with a formal register and precise word choice. The text adheres to conventional structure and format, with consistent citation and footnote style. The grammar, spelling, and punctuation are correct. The text is balanced and free from bias, and there are no changes in content. The development of CCRs is supported by good economic development, stable population support, and policy guidance. Based on the results of the survey on cultural and tourism resources in Sichuan Province, there has been a significant increase in newly discovered scenic spots, accounting for 46% of the total.

(4) *Accommodation Supporting Conditions.*The impact of accommodation supporting conditions on the spatial distribution of CCRs is relatively low (q = 0.0722), but still significant (P < 0.5). In Sichuan Province, areas with high densities of accommodation resources are mainly concentrated in the Chengdu Plain area in the central part of the basin, centred on Chengdu, gradually decreasing along the surrounding areas of Chengdu. In contrast, the accommodation resources in the Western Sichuan Plateau and Panxi region are relatively weak. However, near high-grade scenic spots, accommodation supporting resources have continuously improved and developed over the years.

(5) *Interactive Effects.* Revealed an interactive effect between various factors (3.2.5). The social development factor is enhanced by the natural geographical environment (0.2635), climate resource endowment (0.3740), and accommodation supporting conditions (0.2379). The natural geographical environment and climate resource endowment (0.2915) have a dual-factor enhancement effect, while their interaction with accommodation supporting conditions (0.1891) shows a nonlinear enhancement effect. The factor of climate resource endowment and accommodation supporting conditions (0.3037) also exhibits a dual-factor enhancement effect. In general, the enhancement effect between climate resource endowment and social development factors is relatively significant. For instance, the Chengdu Plain Economic Zone, despite not having the best climate resource endowment, is the most economically developed and densely populated area in Sichuan Province. This is due to its developed economy, convenient transportation, and strong tourism demand, resulting in the highest number of CCRs in the region. Additionally, it boasts higher-grade scenic spots and the most complete tourism supporting facilities. In contrast, the Panxi region, although possessing abundant convalescence

climate resources, has fewer CCRs and lower grades due to factors such as an underdeveloped economy and transportation, a smaller local population, and incomplete tourism
supporting facilities.

3.2.3 Countermeasures

The research findings suggest that the development of CCR scenic areas is significantly
influenced by the climate resource endowment. Therefore, it is crucial to take into account
the climate resource endowment and conduct detailed surveys of climate resources based
on local conditions to enhance the scientific assessment of regional convalescence climate resources. The task is to actively carry out the declaration work of brand climate
places in conjunction with meteorological departments. This involves creating distinctive convalescence climate brands that highlight the benefits of climate branding. To
achieve this, it is important to broaden marketing channels and introduce new media
such as streaming media. Additionally, improving the level of tourism informatization is
necessary to carry out precise and effective publicity and enhance the popularity of convalescence climate brands. In addition, the development of scenic areas is significantly
influenced by the socio-economic situation. Due to the varied distribution of economic
development in Sichuan Province, it is essential to adopt a province-wide perspective.
This will enable us to guide cities and prefectures to learn from each other's strengths and
weaknesses, and develop a unique and distinctive tourism path based on the provincial
convalescence climate tourism development plan. The development of convalescence
climate scenic areas requires the interactive and collaborative configuration of multiple
factors, such as climate resource endowment, natural geographical environment, socio-
economic development, and accommodation facilities. In regions with a large number
of convalescence climate resources and developed regional economies, it is possible to
enrich the types of such scenic areas and encourage localities to explore them deeply.
To meet the new needs of tourists, upgrade traditional business format attractions and
develop new business format convalescence climate scenic areas based on local climate
and tourism resource characteristics. Efforts should be intensified to support the development and construction of convalescence climate resources in areas that are lagging
behind. This can be achieved by improving tourism infrastructure, transportation network systems, and enhancing the accessibility of convalescence climate scenic areas. It
is important to optimize the last mile of transportation systems leading to scenic areas,
build traffic connections between scenic areas, and reduce tourists' travel time and economic costs. At the same time, it is important to improve the infrastructure and tourism
service facilities in the locations of convalescence climate tourist attractions. This will
help achieve regional shared facilities and services, and improve the continuity of service
areas. It is important to take full advantage of high-level superior tourist attractions to
drive lower-level attractions and create a clustering advantage from the perspective of all
regions in the province. To create a satisfying convalescence climate tourist attraction, it
is important to utilise tourism growth poles provided by superior scenic areas and make
reasonable use of big data to understand tourist needs. This can be achieved by upgrading
and enriching the number and types of convalescence climate scenic areas, improving
tourism services, and constructing a multi-dimensional support and supplementary convalescence tourism development axis. It is crucial to maintain a balanced and objective

tone throughout the text, avoiding biased or emotional language. Additionally, precise subject-specific vocabulary should be used when it conveys the meaning more precisely than a similar non-technical term. Finally, the text should be free from grammatical errors, spelling mistakes, and punctuation errors.

4 Conclusion and Discussion

4.1 Conclusion

Using classic geographic information system analysis methods, this study analyses the distribution characteristics and differences of three types of convalescent climate resources (S-CCR, Y-CCR, and W-CCR) in Sichuan Province, based on existing data of scenic areas for convalescent climate tourism and population economic distribution. Using classic geographic information system analysis methods, this study analyses the distribution characteristics and differences of three types of convalescent climate resources (S-CCR, Y-CCR, and W-CCR) in Sichuan Province, based on existing data of scenic areas for convalescent climate tourism and population economic distribution. The study concludes that:

(1) The distribution of CCRs in the study area is uneven, with a significant difference between the east and west, but not between the north and south. It is important to note that this is an objective evaluation based on the data collected. S-CCRs are the most numerous, concentrated in the Chengdu Plain Economic Zone, while W-CCRs are the least numerous and primarily located in the Panxi Economic Zone. Regarding the spatial distribution of high-value kernel density areas, the CCRs in Sichuan exhibit a spatial feature that resembles a ring basin band. The Y-CCRs are mainly located in the area to the east of the Hengduan Mountains, while the W-CCRs are primarily found in the Panxi area and the northern part of Muli County in southern Sichuan. The S-CCRs are mainly situated along the perimeter mountain range of the basin and in the Panxi area. In the study area, CCRs display obvious agglomeration, particularly Y-CCRs and S-CCRs, while the distribution of W-CCRs does not significantly differ from a random pattern. At the city scale, CCRs can be classified into eight types, with higher CCR densities in cities such as Leshan, Meishan, and Dazhou, and lower densities in areas like Ganzi and Ziyang. Spatial autocorrelation analysis shows that Sichuan's CCRs exhibit positive spatial correlation and significant aggregation, especially for Y-CCRs. The most significant clustering of W-CCR hotspots is concentrated in the Panxi area, while coldspots are primarily found in most areas of the central and eastern parts of the basin.

(2) The distribution characteristics and trends of CCRs in Sichuan's geographical space are as follows: The spatial distribution of the three types of CCRs (S-CCR, Y-CCR, and W-CCR) all exhibit a northwest-southeast orientation, which is consistent with the direction of the Hengduan Mountains at the junction of the Sichuan Basin and the Tibetan Plateau. This indicates that geographic location significantly impacts CCR distribution. The directionality of S-CCRs is the most pronounced, while W-CCRs are relatively weak, with Y-CCRs falling between the two. Additionally, Y-CCRs

have the widest distribution range, the weakest centripetal force, but the most significant directionality. CCRs in the Sichuan region exhibit positive spatial correlation and significant aggregation. Y-CCRs show the most significant spatial aggregation, while W-CCRs display a random distribution. The Panxi area exhibits the most significant clustering of W-CCR hotspots, while coldspots are primarily found in most areas of the central and eastern parts of the basin. Y-CCR hotspots are mainly located in the northeastern and southwestern parts of the basin.

(3) The development of CCRs is influenced not only by the endowment of convalescent climate resources but also by urbanization and economic development levels, population size, and policy guidance factors. The distribution of CCR is most significantly impacted by the climate resource endowment, while social development also plays an important role. The interaction between climate resource endowment and social development factors has a relatively large enhancing effect. In economically underdeveloped plateau and Panxi areas, the development of CCR is more influenced by natural resources. The dominant factor affecting the distribution and development of scenic areas is the endowment of climate resources. In economically developed areas close to the Chengdu Plain, the development of CCR is more influenced by socio-economic factors. Strong market demand, convenient transportation, and strong supporting construction are also important factors.

(4) Developing Sichuan's convalescent climate tourism scenic areas can be achieved through fully exploring available resources, actively collaborating with meteorological departments to make brand climate place declarations, creating distinctive convalescent climate brand attractions, planning convalescent climate tourism at the provincial level, guiding regions to learn from each other's strengths and weaknesses, and pursuing differentiated development. Developing CCRs in Sichuan requires increased investment in tourism infrastructure in areas with favourable climate conditions.

4.2 Discussion

(1) This research uses the latest cultural and tourism resource census data to analyze the spatial distribution characteristics of convalescent climate tourism scenic areas in Sichuan. The focus is on the spatial distribution pattern of convalescent climate resources, which is compared to existing domestic studies. The study conducts an analysis of impact factors using multi-year daily scale climate comfort data and comprehensive tourism resource data. The research offers a certain level of innovation in terms of research subjects, materials, and methods.

(2) This study presents a qualitative analysis of the factors influencing the spatial distribution of classified CCRs (Cold-Avoidance, Heat-Avoidance, and Convalescence) in Sichuan Province. It investigates the current status and reasons for distribution differences among various regions within the province and explores potential CCR scenic area development regions. The research direction is novel in the study of convalescent climate scenic resource potential development.

(3) This study has some limitations. The scenic area distribution analysis did not include weight judgments for CCR scenic area levels, resulting in errors between the analysis results and actual conditions. When applying the research results to real-world

tourism development, it is important to consider factors such as rivers, forests, transportation accessibility, and medical facilities, as they can impact CCR development. Additionally, this study solely analyses the spatial distribution characteristics of convalescent climate tourism scenic areas and their influencing factors. However, it does not comprehensively discuss future development trends and research directions of convalescent tourism. Future research should aim to enrich and perfect these aspects.

Acknowledgement. This study is financially supported by the Heavy Rain and Drought-Flood Disasters in Plateau and Basin Key Laboratory of Sichuan Province Science and Technology Development Fund Project of Sichuan Key Laboratory (No. SCQXKJQN2020025), Sichuan Local Standard Preparation and Revision Project (No. 202204/T008, No. 202337/T003) and Soft science research topics of the Provincial Meteorological Bureau in 2023 (No. 2023/11).

References

1. Zhang, K., Su, X.L., Su, K.H., Wang, Y.: Research on spatial differentiation of tourism resources in Beijing, Tianjin and Hebei based on POI big data. Areal Research and Development **40**(01), 103–108 (2021)
2. Liu, M., Hao, W.: Spatial distribution and its influencing factors of national Alevel tourist attractions inShanxi Province. Acta Geogr. Sin. **75**(4), 878–888 (2020)
3. Hu, W.X., Liang, X.T., Sang, Z.Y.: Analysis on the characteristics and causes of the spatial-temporal evolution of 3A and above tourist attractions in Shanxi Province. Journal of Arid Land Resourcesand Environment **34**(12), 187–194 (2020)
4. Dóniz-Páez, F.J., Becerra-Ramírez, R., Carballo-Hernández, M.: Proposal for an urban geo-tourism route in Garachico (Tenerife, Canary Islands, Spain). Investigaciones Geográficas **66**, 95–115 (2016)
5. Paül, I.A.D.: Territorial distribution of tourist attractions. Comparing projected and perceived image in Uruguay. Economía Sociedad Y Territorio, **18**(58): 735–762 (2018)
6. Dewi, S.M., H.: The spatial distribution of tourist attractions in Jakarta. IOP Conference Series: Earth and Environmental Science **338**, 012013 (2019)
7. Wang, F., Liu, Z., Shang, S., Qin, Y., Wu, B.: Vitality continuation or over-ommercialization Spatial structure characteristics of commercial services and population agglomeration in historic and cultural areas. Tour. Econ. **25**(8), 1302–1326 (2019)
8. Sun, Y., Duru, O.A., Razzaq, A., Dinca, M.S.: The asymmetric effect eco-innovation and tourism towards carbon neutrality target in Turkey. J. Environ. Manage. **299**, 113653 (2021)
9. Kim, G.S., Chun, J., Kim, Y., Kim, C.-K.: Coastal tourism spatial planning at the regional unit: identifying coastal tourism hotspots based on social media data. ISPRS Int. J. Geo Inf. **10**(3), 167 (2021)
10. Jiang, W., Xiong, Z., Su, Q., Long, Y., Song, X., Sun, P.: Using geotagged social media data to explore sentiment changes in tourist flow: a spatiotemporal analytical framework. ISPRS Int. J. Geo Inf. **10**(3), 135 (2021)
11. Shi, J., Xin, L., Liu, Y.: Simulation of tourists' spatiotemporal behaviour and result validation with social media data. Transp. Plan. Technol. **43**(7), 698–716 (2020)
12. Wu, R.L., Li, H.Y., Tian, F.J.: The spatial distribution and its influencing factors of China 's national study travel bases. Scientia Geographica Sinica **41**(7), 1139–1148 (2021)
13. Li, D.H., Zhang, X.Y., Lu, L., Zhang, X., Li, L.: Spatial distribution characteristics and influencing factors of high-level scenic spots in the Yellow River Basin. Econ. Geogr. **40**(05), 70–80 (2020)

14. Taecharungroj, V., Mathayomchan, B.: Analysing tripadvisor reviews of tourist attractions in Phuket, Thailand. Tour. Manage. **75**, 550–568 (2019)
15. Zhang, Y., Li, X., Robert, C., D. A., & Liu, Y.: Calculating theme parks' tourism demand and attractiveness energy: a reverse gravity model and particle swarm optimization. J. Travel Res. **61**(2), 314–330 (2022)
16. Duarte-Duarte, J.B., Talero-Sarmiento, L.H., Rodríguez-Padilla, D.C.: Methodological proposal for the identification of tourist routes in a particular region through clustering techniques. Heliyon **7**(4), e06655 (2021)
17. Kim, Y.R., Liu, A., Stienmetz, J., Chen, Y.: Visitor flow spillover effects on attraction demand: A spatial econometric model with multisource data. Tour. Manage. **88**, 104432 (2022)
18. Gan, C., Voda, M., Wang, K., Chen, L., Ye, J.: Spatial network structure of the tourism economy in urban agglomeration: A social network analysis. J. Hosp. Tour. Manag. **47**, 124–133 (2021)
19. Chung, M.G., Herzberger, A., Frank, K.A., Liu, J.: International tourism dynamics in a globalized world: a social network analysis approach. J. Travel Res. **59**(3), 387–403 (2020)
20. Bustamante, A., Sebastia, L., Onaindia, E.: Can tourist attractions boost other activities around A data analysis through social networks. Sensors **19**(11), 2612 (2019)
21. Zhang, C., Weng, S., Bao, J.: The changes in the geographical patterns of China's tourism in 1978–2018: Characteristics and underlying factors. J. Geog. Sci. **30**(3), 487–507 (2020)
22. REN, Xuanyu et al.: Study on convalescence climate in Panzhihua and its cause. J. Panzhihua Univ. 36.2: 1–5 (2019)
23. Zhong, Y., Guo, H., Cai, Y., Yuan, M., et al.: Matching degree analysis between Sichuan Healthy Climate Resources and Utilization. Plateau and Mountain Meteorology Research **43**(1), 146–150 (2023)
24. Wang, F., Wang, M.F.: Spatial aggregation characteristics and influencing factors of Taobao Village based on grid in China. Scientia Geographica Sinica **40**(2), 229–237 (2020)
25. Xiong, Y., Zhang, F.: Thermal environment effects of urban human settlements and influencing factors based on multi-source data: A case study of Changsha city. Acta Geogr. Sin. **75**(11), 2443–2458 (2020)
26. Yanping, G., Min, L.: Classification and spatial distribution characteristics of tourist attractions in Shanxi Province based on POI data. Scientia Geographica Sinica **41**(7), 1246–1255 (2021)
27. Zhang, K.K., He Jing, Zhong, Y. X. et al.: Identification of soil heavy metal sources around a copper-silver mining area in Ningxia based on GIS. Environ. **43**(11): 5192–5240 (2022)
28. Wang Jinfeng, X., Chengdong.: Geodetector: Principle and prospective. Acta Geogr. Sin. **72**(1), 116–134 (2017)
29. Deng, L.Z., Bao, J.G.: Spatial distribution of summer comfortable climate and winter comfortable climate in China and their differences. Geogr. **39**(01): 41–52 (2020)
30. Wang, Z.F., Shi, W.J.: Spatial distribution characteristics and influencing factors of China's beautiful leisure villages. Scientia Geographica Sinica **42**(1), 104–114 (2022)
31. T/CMSA0008–2018,Classification for health preservation climate
32. Yang, J., Zhang, Y.H., Xi, J.C.: The comprehensive evaluation of suitability of summer tourism base in China. Resour. **38**(12): 2210–2220 (2016)
33. Li, L., Tao, Z.M., Lu, L., et al.: Structural characteristics and influencing factors of summer tourism flow network in Guizhou province. Geogr. Res. **40**(11), 3208–3224 (2021)
34. Xie, Z.H., Wu, B.H.: Research on the tourism spatial structure of China's resource-based scenic spots. Scientia Geographica Sinica **28**(6), 748–753 (2008)
35. Xie, H., Li, Y.H., Wei, Y.Y.: Spatial structure characteristics and influencing factors of characteristic towns in Zhejiang Province. Scientia Geographica Sinica **38**(8), 1283–1291 (2018)

36. Sun, F., Wang, D.G.: Spatial distribution and development model of famousscenic towns and villages in China. J. Tour. **32**(5), 80–93 (2017)
37. Jin, C., Lu, Y.Q.: Evolution of Jiangsu province's economic spatial pattern based on county units. Acta Geogr. Sin. **64**(6), 713–724 (2009)

Rainstorm Waterlogging Simulation and Risk Assessment in Central Urban Area of Chengdu, China

Rui Sun[1(✉)], Jinxia Xu[1], and Liang Zhang[2]

[1] Sichuan Provincial Climate Center (Southwest Regional Climate Center), Sichuan 610072, China
619499692@qq.com

[2] Sichuan Provincial Meteorological Disaster Defense Technology Center, Sichuan 610072, China

Abstract. Urban rainstorm waterlogging disasters affect the sustainable urban development seriously. The distribution of waterlogging disaster risk is not only affected by rainstorm, but also closely related to urban terrain, population distribution and urban built environment. In this paper, based on the soil conservation service (SCS) runoff generation model and GIS tools, we selected Chengdu as the research area and proposed a simplified urban rainstorm waterlogging disaster model which can simulate the 3h waterlogging scenarios under the 10, 50, and100 year return periods of rainstorm. The proposed model overcome the drawbacks of time-consuming calculation, high data demands and low applicability of the previous hydrodynamic model. The results show that: (1) Under the rainstorm return period of 10a, 50a and 100a, the maximum inundation elevation in the central urban area can reach 0.67m, 0.72m and 0.79m, respectively. The flooded area accounts for 4.3%, 12.6% and 18.4% of the central urban area, respectively. (2) The ROC value of MaxENT waterlogging probability prediction model developed in this paper is above 0.8, and the distance from road traffic is the most contribution factor. (3) The risk level of rainstorm and waterlogging in the south of the Chengdu is significantly higher than that in the north, and gradually decreases from the urban center to the surrounding suburbs. This risk distribution is closely related to the precipitation climate distribution and the terrain.

Keywords: Chengdu · Rainstorm Waterlogging · SCS hydrologic model · GIS · Situational Simulation · MaxENT model · Risk assessment

1 Introduction

The new concept of urban development that is safe, ecological, and sustainable is an important issue in current urban development. Among them, urban meteorological disaster prevention is an important component of ensuring urban safety, and urban inundation, which arises with the process of urban modernization, is the most prominent type of urban meteorological disaster. Under the dual effects of rapid urbanization and climate

B.-J. He et al. (Eds.): UCSUD 2023, LNCE 559, pp. 124–137, 2025.
https://doi.org/10.1007/978-981-97-8401-1_9

warming, urban inland inundation disasters are becoming increasingly severe and the probability of occurrence is also increasing [1, 2]. According to a special survey conducted by the Ministry of Housing and Urban-Rural Development, from 2007 to 2015, 62% of cities nationwide experienced urban inland inundation disasters, and 137 cities suffered from inland inundation disasters more than three times. On July 20, 2021, the extreme rainstorm and waterlogging event in Zhengzhou City attracted national attention. Disaster risk assessment is one of the important ways of disaster prevention. It is not only the core content of disaster risk management, but also important basic research for human society to prevent natural disasters, control and reduce the risk of natural disasters [3, 4]. Chengdu is a new super large city in western China, with a high level of urbanization. Therefore, the risk assessment of rainstorm and waterlogging in Chengdu is of great practical significance for the prevention and management of waterlogging disaster risk. Urban rainstorm waterlogging risk assessment belongs to the cross field of meteorology hydrology risk management [5–10]. At present, the research on risk assessment of rainstorm and waterlogging in Chengdu is rare, and the existing studies still have the many defects, such as time-consuming calculation, high data requirements and low applicability [11–13]. Therefore, this paper selects Chengdu as the research area, and develop a simplify urban rainstorm waterlogging model in order to improve the efficiency of assessment. The simplify model overcome the disadvantages of preparation of redundant drainage pipe network data and the complex process of hydrological and hydrodynamic modeling. In addition, the risk distribution of urban rainstorm waterlogging is simulated based on the historical waterlogging disaster data, which can provide scientific support for urban rainstorm waterlogging early warning and urban planning and construction policies [14].

2 Data and Methods

2.1 Study Area

Chengdu has a high level of urbanization, especially in the central urban area, where the population is concentrated and the economy is developed. Therefore, this article selects the central urban area of Chengdu (within the Ring Expressway) as the study area (Fig. 1).The research data mainly includes environmental data and historical waterlogging point disaster data. The detailed information are as follows:

(1) Disaster causing factor data: the disaster causing factor of precipitation is calculated by the latest revised urban rainstorm intensity formula.
(2) Disaster-prone environment data:30-m resolution elevation data (DEM) from the Geospatial Data Cloud (https://www.gscloud.cn/); The land use data comes from the 2017 global 30 m dataset of Tsinghua University's Gong Peng, which is further reclassified into six land use types: arable land, forest land, grassland, water area, construction land, and unused land; The 1:1 million Chinese soil data comes from the Food and Agriculture Organization of the United Nations (FAO) (https://www.fao.org/soils-portal/en/), and reclassified the soil database of Chengdu, which was originally classified using the FAO-90 soil classification system, to obtain the hydrological soil classification results of Chengdu that meet the requirements of the SCS

model; The data of Chengdu's urban road network in 2019 is from Beijing Navinfo Co., Ltd.

(3) Disaster bearing body data: generalized as population and GDP distribution indicators, data is from the Data Center of Resources and Environmental Sciences, Chinese Academy of Sciences (https://www.resdc.cn/). (4) Historical flood disaster data: Through collecting news reports and literature related to urban flooding in recent years, historical flood prone points were vectorially collected, and quality control was performed on the same grid unit to avoid spatial autocorrelation caused by multiple distribution points. Ultimately, a total of 112 historical flood prone points in the central urban area of Chengdu were obtained. According to the research scale, all data were finally unified under the GCS_WGS_1984 coordinate system, and the spatial resolution of the raster data was processed to 90*90m.

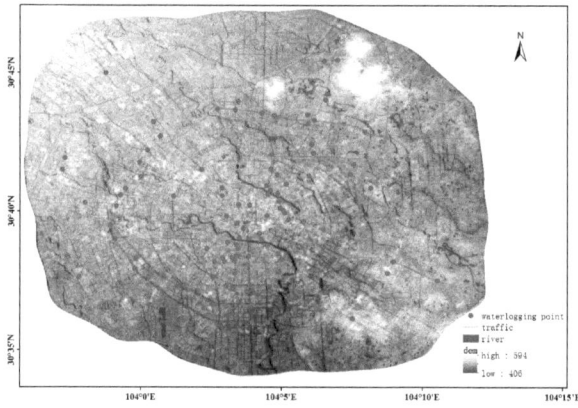

Fig. 1. Distribution of historical waterlogging points in the study area

2.2 Urban Rainstorm Waterlogging Scenario Simulation

(1) Urban Rainstorm Intensity Calculation: Using the newly revised empirical formula of rainstorm intensity in 2015 to calculate rainstorm intensity under different return periods in Chengdu, the formula is calculated by Sichuan meteorological bureau and recognized by Chengdu Municipal Government:

$$I = \frac{44.594(1 + 0.651 \lg P)}{(t + 27.346)^{0.953(lgP()^{-0.017})}} (mm/min()) \tag{1}$$

where I is the Rainfall Intensity (mm/min); t is the Rainfall Duration (min); P is the Recurrence Interval (a).

(2) Surface Runoff Calculation: The empirical hydrological model SCS (Soil Conservation Service) developed by the US Department of Agriculture in 1954 was used as the runoff model.

$$Q = \frac{(P - I_\alpha)^2}{(P - I_\alpha) + S} \tag{2}$$

$$I_\alpha = \lambda S \tag{3}$$

$$S = \frac{25400}{CN} - 254 \tag{4}$$

where Q is the runoff volume (mm); P is the rainfall (mm); I_a is the initial loss of rainfall in the catchment area before runoff generation, which is composed of plant interception, initial infiltration, and filling of depressions (mm); S is the possible maximum intercepted amount after runoff generation (mm); usually taking the empirical value of 0.2; CN reflects the characteristics of the watershed before rainfall, and its size is determined by three factors: land use type, soil type, and previous soil moisture conditions (AMC).According to the National Engineering Manual Chapter 4, a CN value lookup table can be used to calculate the CN value. Since the SCS-CN model does not consider the impact of slope on runoff and confluence, while in practice the slope factor has a significant impact on runoff and confluence, the parameter CN value obtained from the lookup table needs to be corrected by the slope value to modify the SCS-CN model.

$$CN_\alpha = CN \frac{322.79 + 15.63\alpha}{\alpha + 323.52} \tag{5}$$

where, CN_α is the CN value after slope correction, a dimensionless parameter, and α is the slope value of the polygon, expressed as a percentage (%).

(3) Local Isovolumetric Method: Urban rainstorm flooding can be classified into active flooding and passive flooding according to the cause of formation. Without considering the connectivity of the inundation zone and the surface runoff velocity, passive inundation includes all low-lying areas with an elevation below the given water level as part of the inundation zone. In this paper, it is assumed that the waterlogging in different areas of the city is in a relatively static state within a certain period of time, so the waterlogging in rainstorm simulated in this paper can be regarded as a passive inundation state. Furthermore, based on the principle that the total runoff volume is equal to the waterlogging volume, the DEM data is used to simulate and calculate the range and height of waterlogging in the region. The calculation method for urban waterlogging water volume W is as follows.

2.3 Probability Risk Prediction of Urban Rainstorm Waterlogging

The maximum entropy (MaxENT) model is a prediction model based on the principle of maximum entropy [15], which is commonly used to predict the potential distribution of species. It can infer or predict the suitable habitat area of species from incomplete known information under certain constraints. This article introduces it to the prediction of the probability of precipitation inundation events. Based on historical "inundation point" events and environmental variables that affect their occurrence, constraints are derived to explore the possible distribution of maximum entropy under these constraints, thus predicting the occurrence distribution of inundation events in the study area. The principle formula involved in the model is as follows.

$$SH(p) = -\sum_{x \in X} p(x) \ln p(x) \tag{6}$$

$$\sum_{x \in X} p(x) = 1 \tag{7}$$

where x represents a known single inland inundation event, $p(x)$ represents the probability risk of a single inland inundation event under the influence of certain disaster-causing, disaster-pregnant, and disaster-bearing environments, and $p(x) > 0$. X represents the known historical inundation point event dataset; $H(p)$ represents the probability distribution of inland inundation risk with the largest entropy. The MaxENT method has been widely used by many studies and their results show the method is very effective and robust [16, 17].

2.4 Comprehensive Risk Zoning of Urban Rainstorm Waterlogging

Risk matrix diagram, also known as risk matrix method, is a risk assessment analysis method that can visually comprehensively evaluate the probability of occurrence of hazards and the severity of injury (Table 1). Two-dimensional table is used to conduct semi-qualitative analysis of risks, which has the advantage of being simple and fast to operate. Using the risk matrix method, the rainstorm waterlogging disaster risk map and the waterlogging probability risk map are superimposed to form the urban rainstorm waterlogging disaster comprehensive risk zoning map, and then the location and scale of the urban risk area are identified.

Table 1. Comprehensive risk level of urban rainstorm and waterlogging disaster

Disaster risk level / Probabilistic risk level	low	medium	high
high			
medium			
low			

high	medium	low

3 Application Example

3.1 Simulation Results of Urban Rainstorm Waterlogging Scenarios

(1) Construction of simplified urban rainstorm waterlogging disaster model: According to the formation mechanism of urban rainstorm waterlogging disaster, urban underlying surface and terrain are the disaster environment of waterlogging; Rainfall is a disaster-causing factor. When the excess rainfall exceeds the drainage capacity of urban pipe networks, urban rainfall may become a disaster-causing factor of urban inland inundation. At the same time, considering that the hydrodynamic model has problems such as time-consuming calculation, high data requirements, and low applicability, this paper constructs a simplified urban rainstorm waterlogging disaster model based on GIS on the basis of comprehensive consideration of the disaster pregnant environment and disaster causing factors of waterlogging disaster, to simulate rainstorm waterlogging scenarios in Chengdu under different return periods of rainstorm (Fig. 2). The model consists of three parts: basic data module, mathematical calculation module, and GIS spatial analysis module. The basic data module mainly includes urban rainfall models, ground feature models, terrain models, runoff models, and drainage models, which provide data support for the calculation of urban waterlogging and waterlogging. The mathematical calculation module provides the key theory, algorithm and formula of urban rainstorm waterlogging disaster; The spatial analysis module provides spatial solutions and schemes for the spatial problems involved in the mathematical calculation module, including the simulation and display of urban rainstorm and waterlogging simulation process.

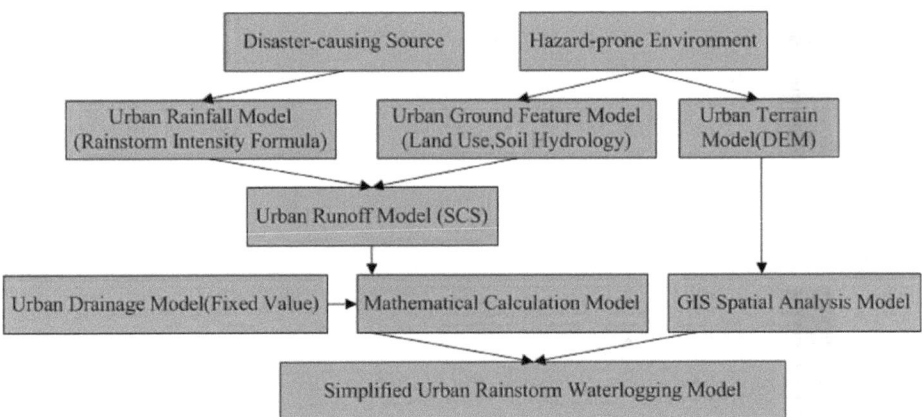

Fig. 2. Simplified Urban Rainstorm Waterlogging Model

(2) Simulation of urban waterlogging under different rainstorm return periods:
According to experience, rainstorm in Chengdu is generally concentrated in the first three hours, so the project has designed and calculated the 3h (180 min) rainfall scheme under three different rainstorm return period scenarios of 10a, 50a and 100a for rainstorm (Table 2).

Based on the modified SCS-CN runoff generation model, the hydrological distribution characteristics of surface runoff in the central urban area of Chengdu under the return period of 10a, 50a and 100a rainstorm scenarios are simulated and analyzed. The high value areas of runoff in Chengdu are highly consistent with the impervious surface of the city in space. With the increase of rainfall intensity, the high value areas of surface runoff expand in space. Under the 10a, 50a and 100a rainstorm scenarios, the average runoff depth in the central urban area is 4.5cm, 6.9cm and 8.0cm respectively (Fig. 3).

Table 2. Simulation Results of rainstorm Intensity in 10a, 50a and 100a Return Periods in Chengdu

Return Period (year)	Rainfall Duration (min)	Rainstorm Intensity (mm)
10	180	83.5
50	180	111.4
100	180	123.5

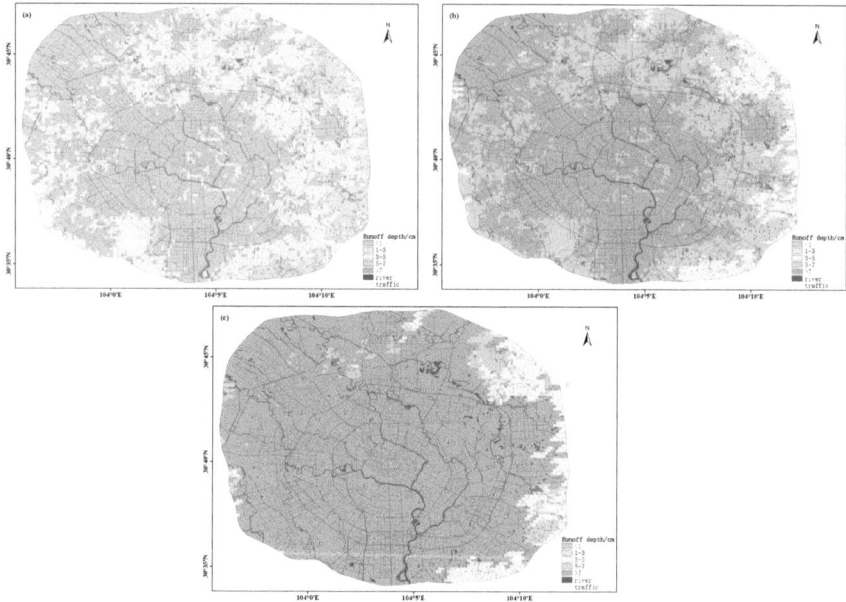

Fig. 3. Simulation diagram of surface runoff depth under different rainstorm return periods in Central Chengdu. (a) Once in 10 years, (b) Once in 50 years, (c) Once in 100 years

The GIS based simplified rainstorm waterlogging disaster model built above is used to simulate the distribution of waterlogging and inundation in the central urban area of Chengdu under the 3h rainstorm scenarios with 10a, 50a and 100a return periods (Fig. 4). The waterlogging areas are mainly distributed along the river and on both sides of the road in terms of spatial distribution. Under different rainstorm return periods, there are significant spatial-temporal differences in the inundation range and elevation in the central urban area of Chengdu, and the inundation range in the low altitude area in the south of the urban area is generally larger than that in the north; In addition, under different rainstorm return periods, the larger the return period is, the larger the scope and maximum elevation of waterlogging inundation are. Under the scenarios of 10a, 50a and 100a rainstorm, the maximum inundation elevation in the central urban area can reach 0.67m, 0.72m and 0.79m respectively; The area of the flooded area accounts for 4.3%, 12.6% and 18.4% of the central urban area, respectively.

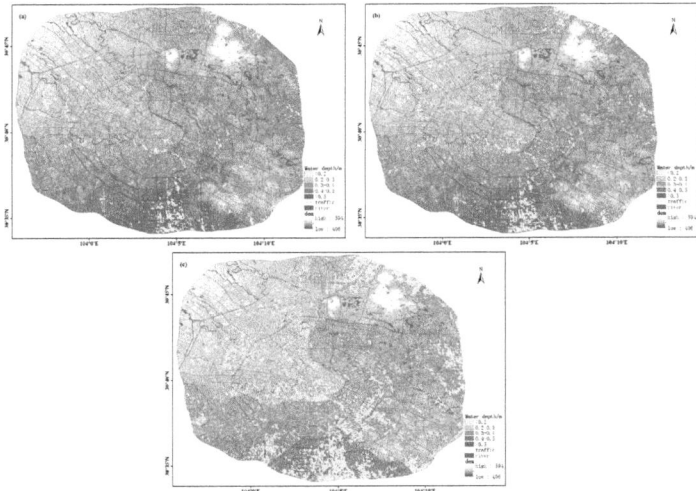

Fig. 4. Simulation Diagram of Waterlogging and Inundation under Different rainstorm Return Periods in Central Chengdu. (a) once in 10a, (b) once in 50a, (c) once in 100a

The model simulation results were verified and analyzed. The 24-h daily rainfall of Pujiang and Wenjiang representative stations during the "7.2" rainstorm in 2018 was 117.8mm and 110.3mm respectively, approaching the rainfall of 3h before the return period of 50a.Therefore, three historical waterlogging data (Fig. 5) caused by the "7.2" rainstorm in 2018 were collected through waterlogging related news reports and literature, and the simulation results of the validation model were compared with the simulated submerged elevation values in the 50a return period. The results show that although the error between the simulated values and the actual values of the three waterlogging points is generally small, it is within an acceptable range (Table 3).

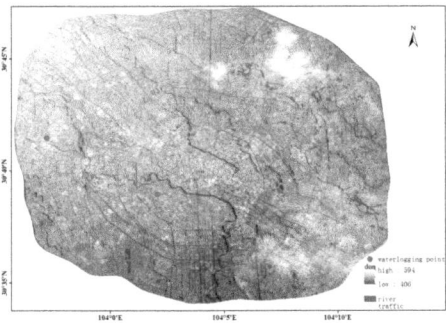

Fig. 5. Spatial distribution of historical inundation collection points for comparative verification

Table 3. Comparison and verification of model simulation errors and historical floods

Area	Longitude	Latitude	Real Situation (m)	Simulation Value(m)
Dashinan Road	104.035°	30.649°	0.15	0.11
Near Chengdu Aircraft Industry Hospital	103.954°	30.691°	0.4	0.32
Mingshu New Village	104.195°	30.654°	0.5	0.40

3.2 Probability Prediction of Urban Rainstorm Waterlogging

The eight environmental variable factors (Table 4) affecting the disaster pregnant environment and disaster bearing body environment of urban rainstorm waterlogging were selected to build a MaxENT probability prediction model based on the maximum entropy of urban rainstorm waterlogging based on the historical "waterlogging point", and the probability distribution of rainstorm waterlogging in the central city of Chengdu was simulated. The receiver operating characteristic (ROC) curve is a curve based on a series of different binary classification methods, with true positive rate as the vertical axis and false positive rate as the horizontal axis. As shown in Fig. 6, the receiver operating characteristic (ROC) curve is close to the upper left corner of the graph, and the ROC values of the training data and test data for the prediction model are both above 0.8, which indicate the model are very skillful. AUC is defined as the area enclosed below the ROC curve. The AUC value of the interpretation of each environmental variable on the prediction model reached 0.82, indicating that the model has good ability to distinguishes the positive and negative samples.

Table 4. Eight environmental variables for MaxENT prediction models

Name	Code
Digital Elevation Model	dem
Slope	slope
Landuse	landuse
Hydrological sensitivity	hyd-sen
Distance from the river	dis-riv
Distance from road traffic	dis-tra
Population distribution	pop
Gross Domestic Product	gdp

As shown in Figs. 7 and 8, the high-risk areas of urban waterlogging are mainly distributed in the dense urban road network, so the contribution rate of disaster-bearing factors is 38.7% based on the distance from road traffic. This is largely due to two reasons: First, the type of road land use is hard surface, with poor rainwater permeability. Second,

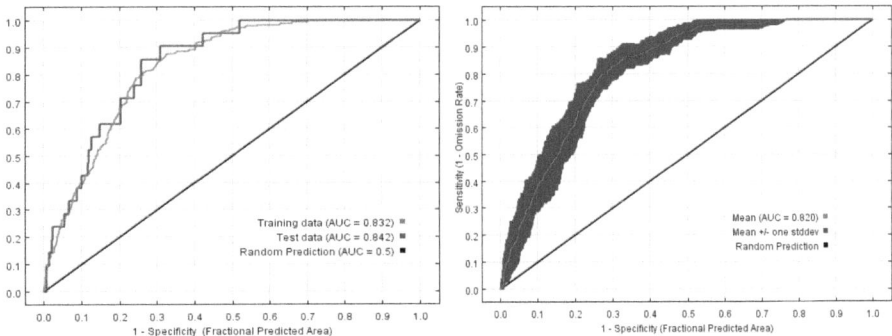

Fig. 6. ROC curves of model training data and testing data, as well as the overall ROC curve of the model

the road rainwater pipe network is the main part of the urban underground drainage system, and the design of the drainage pipe network is small. Besides, the blocked rainwater inlet, mixed rain and sewage and the ineffective maintenance, which also greatly reduced the drainage capacity. The southern part of the high-risk area of urban waterlogging is significantly more than the northern part, which is closely related to the northwest-southeast sloping terrain of Chengdu. With the continuous advancement of urbanization, population and wealth are continuously gathering towards the urban center, and the population and wealth exposed to the risk of waterlogging are correspondingly increasing. Therefore, the overall contribution rate of the two socioeconomic disaster-bearing environmental factors of population distribution and GDP distribution reaches 26.5%. The elevation factor contributes 10.5%, and the river network is the main drainage way of Chengdu's drainage system. When continuous heavy rainfall causes the water level of the river to rise rapidly, the pipe network system will be affected to some extent by the rise of the water level of the river. The discharge method of "gravity flow" makes the southern river network system more stressed than the northern part, and in severe cases, even reverse irrigation occurs, causing heavier disasters. Therefore, the contribution rate of the distance from the river factor reaches 13.8%. Most of the urban construction land is impervious surface, with relatively poor rainwater permeability compared to forest land and grassland land types. Therefore, the contribution rate of land use type factor reaches 7.3%. The slope and hydrological sensitivity factors contribute relatively little to the occurrence probability of waterlogging. Among them, the hydrological sensitivity factor is fixed due to the land use type at a small scale, and the regional variation of soil type is relatively small. In the early stage, when the soil AMC is set to be spatially uniform, its spatial distribution does not vary greatly.

3.3 Comprehensive Risk Zoning of Urban Rainstorm Waterlogging

The inundation height (>0.1m) of rainstorm waterlogging in the central urban area during the return period of 100a rainstorm is classified as high risk level, the inundation height (0–0.1m) is classified as medium risk level, and the rest of the areas (0) are classified as low risk level, which is regarded as the risk of precipitation disaster causing factors in the comprehensive risk zoning of rainstorm waterlogging. At the same time,

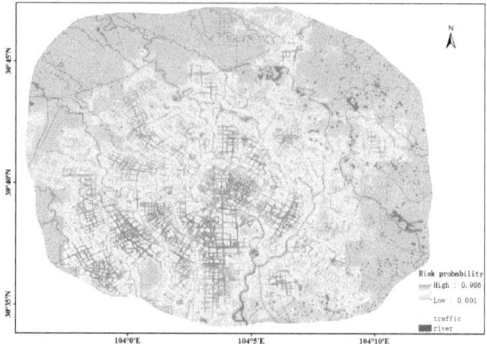

Fig. 7. Probability distribution diagram of rainstorm waterlogging in Central Chengdu City based on historical "waterlogging point"

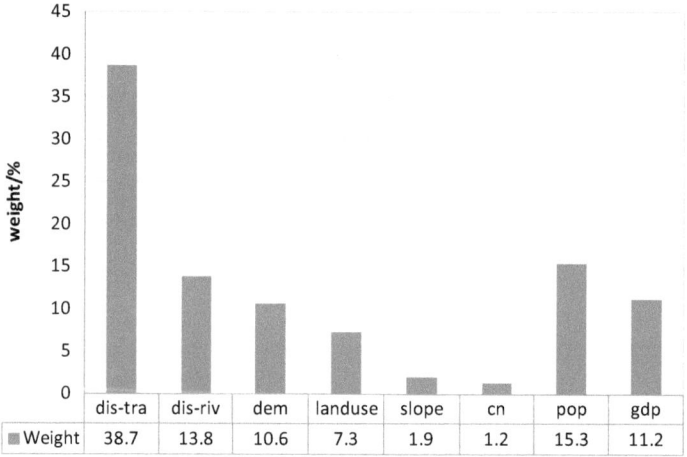

Fig. 8. Contribution rate of various environmental variable factors to probability risk of rainstorm waterlogging (%)

the probability of rainstorm waterlogging based on the "historical waterlogging point" and the simulation prediction of disaster pregnant and disaster bearing environmental factors is divided into three risk levels under the combination of disaster pregnant environment and disaster bearing environment according to low (<0.3), medium (0.3–0.7) and high (>0.7). Using the risk matrix method, the risk level map of urban rainstorm waterlogging disaster and the probability risk level map of rainstorm waterlogging are superimposed for comprehensive visual assessment, and the comprehensive risk zoning map of rainstorm waterlogging disaster in the central city of Chengdu is obtained (Fig. 9).

In terms of spatial distribution, the risk level of rainstorm and waterlogging in the south of the urban area is significantly higher than that in the north, and gradually decreases from the urban center to the surrounding suburbs, which is closely related to the climate distribution characteristics of "more rainfall in the south and less rainfall in the north" and the terrain of "high rainfall in the north and low rainfall in the south" in the flood season of Chengdu; Meanwhile, high-risk areas are mainly concentrated on both sides of roads and rivers. The percentage of low-risk, medium-risk and high-risk areas in the urban center of the city is 62.1%, 27.1% and 10.8% respectively.

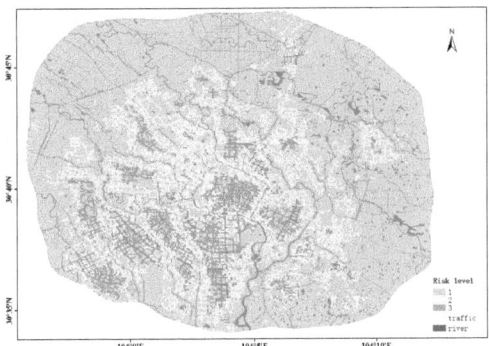

Fig. 9. Comprehensive risk zoning map of rainstorm and waterlogging disaster in central Chengdu (1–3 levels represent 1 low, 2 medium, and 4 high)

3.4 Conclusion and Discussion

In this paper, based on SCS runoff generation model, local isovolumetric method, Max-ENT model and historical waterlogging event records, we develop an simplified urban rainstorm waterlogging prediction model, and analyze the prediction results of rainstorm waterlogging occurrence probability of central Chengdu area. The results are as follows:

(1) Under the rainstorm return period of 10a, 50a and 100a, the maximum inundation elevation in the central urban area can reach 0.67 m, 0.72 m and 0.79 m, respectively. The flooded area accounts for 4.3%, 12.6% and 18.4% of the central urban area, respectively. In terms of spatial distribution, waterlogging areas are mainly distributed along rivers and on both sides of roads.
(2) The waterlogging probability prediction model developed in this paper is above 0.8. The distance from road traffic is the most important contribution factor, followed by two socioeconomic disaster-bearing environmental factors of population distribution and GDP distribution. The slope and hydrological sensitivity factors contribute little to the occurrence probability of waterlogging.
(3) As for comprehensive risk zoning of rainstorm and waterlogging in central Chengdu area, the risk level of rainstorm and waterlogging in the south of the Chengdu is significantly higher than that in the north, and gradually decreases from the urban center to the surrounding suburbs. The percentage of low-risk, medium risk, and high-risk areas in the central urban area is 62.1%, 27.1%, and 10.8%, respectively.
(4) Based on the above research conclusions, it is suggested that when carrying out urban planning and construction and sponge city reconstruction, reasonable zoning

discussions should be conducted and measures should be taken according to local conditions. For the suburbs near the third ring road, the original ecological landscapes such as forest land, rivers, and lakes should be maximally protected to maintain the natural hydrological characteristics before urban development. For the central urban areas with rapid urbanization, the proportion of impervious surface areas should be controlled, and the types of regional landscapes should be rationally allocated to minimize the damage of urban development and construction to the original ecological environment, so as to achieve low-impact development. In addition, urban planning and construction should try to avoid areas prone to floods, and reasonably guide the spatial distribution of population and scientific planning of industries.

Acknowledgments. This research was jointly funded by the Project of Meteorological Service of China Meteorological Administration (JCZX2023010); Key projects of research-based business of Heavy Rain and Drought-Flood Disasters in Plateau and Basin Key Laboratory of Sichuan Province(SCQXKJYJXZD202202), Innovation Team Fund of Southwest Regional Meteorological Center, China Meteorological Administration (XNQYCXTD202201).

References

1. Alexander, L.V., et al.: Global observed changes in daily climate extremes of temperature and precipitation. J. Geophys. Res. **111**(d5): D05109-1–D05109-22 (2006)
2. Amengual, A., Romero, R., Gómez, M., Martín, A., Alonso, S.: A hydrometeorological modeling study of flash-flood event over Catalonia, Spain. J. Hydrometeorol. **8**(3), 282–303 (2007)
3. Crichton, D.: The risk triangle in Ingleton. Nat. Disaster Manag.:102–103 (1999)
4. Da-ming, L., Hong-ping, Z., Bing-fei, L., Yi-yang, X., Pei-yan, L. and Su-qin, H.: Basic theory and mathematical modeling of urban rainstorm water logging. J. Hydrodyn. **16**(1), 17–27 (2004)
5. Carrara, A., Guzzetti, F., Cardinali, M., Reichenbach, P.: Use of GIS technology in the prediction and monitoring of landslide hazard. Nat. Hazards **20**(2–3), 117–135 (1999)
6. Chen, J., Hill, A.A., Urbano, L.D.: A GIS-based model for urban flood inundation. J. Hydrol. **373**(1–2), 184–192 (2009)
7. Li, Q., Zhou, J.Z., Liu, D.H., Jiang, X.: Research on flood risk analysis and evaluation method based on variable fuzzy sets and information diffusion. Saf. Sci. **50**, 1275–1283 (2012)
8. Zhang, D.D., Yan, D.H., Wang, Y.C., Lu, F., Liu, S.: Research progress on risk assessment and integrated strategies for urban pluvial flooding. J. Catastrophol. **29**(1): 144–149 (2014)
9. Zhan'e, Y.I.N., Shiyuan, X.U., Jie, Y.I.N., Jun, W.A.N.G.: Small-scale based scenario modeling and disaster risk assessment of urban rainstorm water-logging. Acta Geogr. Sin. **65**(5), 553–562 (2010)
10. Meixia, L.: Study on rainstorm waterlogging disaster base on the construction of sponge city: a case of Xiamen. Xi'an University of Science and Technology, Shaanxi Province (2018)
11. Benito, G., et al.: Use of systematic, palaeoflood and historical data for the improvement of flood risk estimation, review of scientific methods. Nat. Hazards **31**(3), 623–643 (2004)
12. Yang Sen, Zhu Gang, Lu Ke, Yi Xiao-nan: Introduction to drainage capacity improvement plan for Chengdu Central City area. China Water Wastew. **31**(03): 135–138 (2015)
13. Liu, Q., Wang, Y.K., Peng, P.H., Lu, Y.F., Chen, Y.F., Wang, S.: Characteristics of distribution and migration of species in Sichuan under the climate change. Mount. Res. **34**(6), 716–723 (2016)

14. White, I., Kingston, R., Barker, A.: Participatory geographic information systems and public engagement within flood risk management. J. Flood Risk Manag. **3**(4): 336–347 (2010)
15. Phillps, S.J., Anderson, R.P., Schapire, R.E.: Maximum entropy modeling of species geographic distributions. Ecol. Model. **90**: 231–259 (2006)
16. Li, H., Wang, Q., Li, M., Zang, X., Wang, Y.: Identification of urban waterlogging indicators and risk assessment based on MaxEnt model: a case study of Tianjin Downtown. Ecol. Ind. **158**, 111354 (2024)
17. Huang, Y., Lin, J., He, X., Lin, Z., Wu, Z., Zhang, X.: Assessing the scale effect of urban vertical patterns on urban waterlogging: an empirical study in Shenzhen. Environ. Impact Assess. Rev. **106**, 107486 (2024)

Decarbonization and Carbon Neutral Solutions

Study on Carbon Emission Data of Urban Buildings

Chen Yang[1] and Tong Yao[2(✉)]

[1] School of Civil Engineering, Jinjiang College, Sichuan University, Meishan 620860, China
[2] School of Economics and Management, Sichuan Tourism University, Chengdu 610100, China
Yaotong2011tony@163.com

Abstract. In pursuit of global carbon peak and neutrality, precise carbon emission measurement using advanced management tools and technologies is crucial for effective urban carbon reduction and sustainable growth. We propose classifying building emissions into three distinct stages—construction, operation, and maintenance—by utilizing mathematical methods and Building Information Modeling (BIM) for separate calculation. Moreover, we recommend implementing an annual carbon emission management system tailored to each building.

Keywords: Carbon emission · Building Information · BIM modeling · Urban buildings · Management system

1 Introduction

Climate change is one of the biggest global challenges, with man-made greenhouse gas emissions making a significant contribution to climate change. In order to address climate change, the international community has put forward the goal of carbon neutrality, that is, to achieve a balance between global greenhouse gas emissions and absorption by reducing greenhouse gas emissions and increasing the capacity to absorb carbon dioxide. Achieving carbon neutrality is one of the most important ways for the global response to climate change, and it is also the common responsibility of governments and enterprises. The Intergovernmental Panel on Climate Change (IPCC) released its Sixth Assessment Report, which states that limiting global warming to below 1.5 °C– 2 °C will be unattainable unless greenhouse gas emissions are reduced immediately, rapidly, and on a large scale (IPCC, 2021). In addition, according to the statistics of the World Energy Statistics Yearbook released by British Petroleum (BP) 2022: global energy carbon dioxide emissions in 2022 increased by 0.9% year-on-year to 34.4 billion tons, and fossil fuels accounted for as much as 82% of the global disposable energy consumption, of which China's carbon emissions accounted for 30.7% of the world (BP, 2022), ranking first in the world (Liu et al., 2023).

Meanwhile, the International Energy Agency (IEA) emphasized in its World Energy Outlook Special Report that buildings will occupy a pivotal position in the transition towards a clean energy future (IEA, 2022). The attribution of greenhouse gases stemming

B.-J. He et al. (Eds.): UCSUD 2023, LNCE 559, pp. 141–150, 2025.
https://doi.org/10.1007/978-981-97-8401-1_10

from the production of building materials and the operational functions of buildings comprises a substantial 30.5% of overall industrial emissions (Liu et al., 2023).

In summary, China accounts for the highest share of carbon emissions globally, and reducing carbon emissions in the construction industry is key to achieving carbon neutrality. Drawing upon the existing carbon emission database for buildings, the article authored by G. Verbeeck and his colleagues conducted an in-depth study on the carbon emissions of different types of buildings throughout their entire lifecycle, meticulously developing a comprehensive carbon emission calculation model (G. Verbeeck, 2010). Chinese scholar established a double regression prediction model of building carbon emissions to predict the carbon emissions of urban buildings in Huangpu District from 2012 to 2025 (Wang and Bao, 2021). However, European scholars primarily establish evaluation systems based on the characteristics of carbon emissions from buildings in their own countries, which are not applicable to Chinese buildings. In China, research still focuses primarily on carbon emission measurement, lacking a specific evaluation mechanism. Furthermore, all extant studies necessitate the prolonged tracking of intricate influencing factors to derive the evaluation outcomes of building carbon emissions. Such a methodology is evidently incompatible with the unique circumstances prevalent in China (Yang et al., 2020; Li et al., 2017; Liu et al., 2018). Therefore, in the context of carbon neutrality and peak carbon dioxide emissions, this paper will take urban buildings as the research object, and use mathematical models and BIM models to count and evaluate the "annual" carbon emissions of single buildings. In the subsequent phase, The utilization of Python to develop operable carbon emission factor codes will facilitate the automatic collection and evaluation of carbon emissions from urban buildings, thereby providing valuable insights for promoting carbon reduction and sustainable development in urban construction.

The life cycle assessment (LCA) of a building encompasses multiple distinct stages, encompassing project construction, the production and transportation of raw materials, the utilization of various construction equipment, the operational phase, and ultimately the disposal of building waste (Chen et al., 2011). As the majority of carbon emissions from buildings occur during the construction and operational stages (Blengin and Carlo, 2010), the focus of this paper is primarily on the calculation of carbon emissions during these two phases.

2   Methods and Materials

2.1   BIM Modeling

Because the carbon emissions of urban buildings mainly come from the two phases of building construction and building operation and maintenance, it is necessary to calculate and summarize these two phases separately.

This study focuses on a 9,100 m² office building in Chengdu, Sichuan Province, China. By establishing a BIM model, the carbon emissions during the construction phase and the material consumption of the building entity have been quantified. The BIM model is presented in Fig. 1.

The material usage in the model is then extracted as shown in Fig. 2.

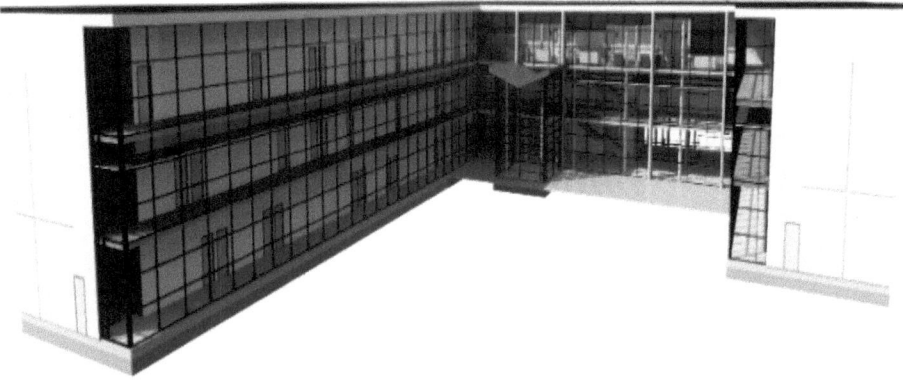

Fig. 1. BIM model

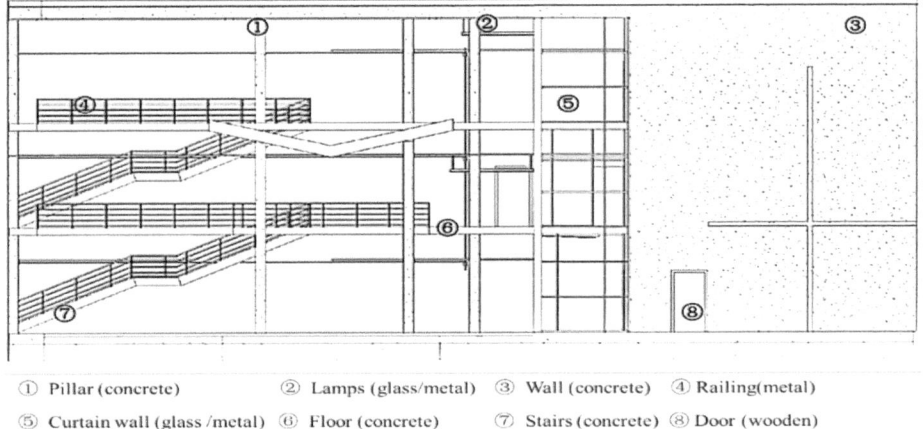

① Pillar (concrete)   ② Lamps (glass/metal)   ③ Wall (concrete)   ④ Railing(metal)
⑤ Curtain wall (glass /metal) ⑥ Floor (concrete)   ⑦ Stairs (concrete) ⑧ Door (wooden)

Fig. 2. Examples of building material information that can be extracted from a BIM model

2.2 Mathematical Modeling

The building material usage obtained from the BIM model can be solved by utilizing the Building Carbon Emission Calculation Standard (GB/T 51366-2019) (National Standard of the People's Republic of China, 2019), which allows for the calculation of whole-life-cycle building carbon emissions: the product of various building materials/energy usage and the corresponding carbon emission factor.

The carbon emissions of the two phases were calculated through mathematical modeling.

(1) Building construction course of events

$$C_{SC} = \sum_{i=1}^{n} M_i F_i$$

In formula (1), C_{SC} is Carbon emissions for building component materials ($kgCO_2/e$),

M_i is the consumption of the i major building material, F_i is the carbon emission factor of the i major building material ($kgCO_2$/ Quantity of building materials per unit).

(2) Building operations and maintenance activities

$$C_M = \frac{\left[\sum_{i=1}^{n}(E_i EF_i) - C_p\right] y}{E_i = \sum_{i=1}^{n}(E_{i,j} ER_{i,j})}$$

$$CCCC = \frac{x_{ij} - x_{min}}{x_{max} - x_{min}}$$

In formula (2)C_M is the carbon emissions of the building during the operation phase of the building ($kgCO_2/m^2$), E_i is the annual consumption of category i energy in buildings (units/a), EF_i is a carbon emission factor for Type i energy ($kgCO_2/kWh$ or $kgCO_2/kg$), A is floor area (m^2), C_p is the annual carbon reduction of the building green space carbon sink system ($kgCO_2/a$), y is the building design life (a). $E_{i,j}$ is the consumption of type j energy for type i systems (units/a), $ER_{i,j}$ is the amount of type j energy consumed by type i systems supplied by renewable energy systems (unit/a), i is the type of end-use energy consumed by the building, j is the type of building energy system.

Within the specified system of equations, the carbon emission factor can be accessed from the carbon accounting database. Once the quantities of building materials and energy are determined, the system becomes closed. Consequently, numerical calculations can be effectively utilized to solve the system.

2.3 Python Model Based on RBF Neural Network

The carbon emissions calculated (as outlined in Sect. 2.2) serve as the input data for the Python model. This model is based on the Radial Basis Function (RBF) neural network and derives an accounting model for the operation and maintenance phase of urban buildings. The specific steps of this model are highlighted in the detailed flowchart presented in Fig. 3.

2.4 Mechanism for "Years" Management of Carbon Emissions from Urban Building

To achieve the goals of Carbon Dioxide Peaking and Carbon Neutrality, it is imperative to conduct statistics on the carbon emissions of buildings. Consequently, a carbon emission accounting model can be established simply by deriving the carbon emissions from building materials based on the BIM model and calculating the carbon emissions during the operation and maintenance phase on a yearly basis using Eqs. 1 and 2. This accounting model enables the computation of the total carbon emissions from urban buildings.

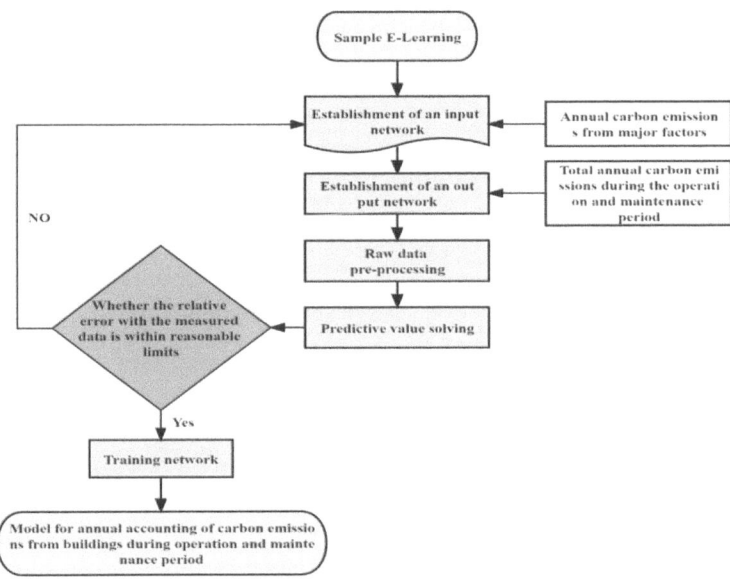

Fig. 3. RBF Neural Network Flowchart

3 Study of Carbon Emissions Data from Urban Building

3.1 Two-Stage Carbon Emissions Calculation

This paper takes an urban building with a floor area of 9100 m^2 as an example for research and analysis. The actual building drawings are used as a reference to establish the 3D model of the building using the Revit software, as shown in Fig. 3. The important material parameters are extracted by using ueBIM for the Revit plug-in, as shown in Table 1.

Table 1. Quantity of building materials used.

Title	Concrete (m^3)	Steel (t)	Wood Formwork (m^2)	Concrete blocks (m^3)	Mortar (m^3)	Doors and Windows (m^2)
material consumption	4096.51	458.9	28473.5	1561.45	642.95	2755.5

According to the Building Carbon Emission Calculation Standard (GB/T 51366-2019), the carbon emission factor of building materials is given in Table 2.

The carbon emissions of the main consumables can be calculated separately according to Eq. (1), as shown in Table 3.

The typical lifespan of a building is generally expected to be 70 years (Li et al., 2017), however, it is observed that the majority of buildings fail to reach this expected

Table 2. Carbon emission factors for related building materials (National Standard of the People's Republic of China, 2019)

Type of building material	Carbon emission factors for building materials
Ordinary silicate cement (market average)	735 kgCO$_2$e/t
C30 Concrete	295 kgCO$_2$e/m^3
Lime production (market average)	1190 kgCO$_2$e/t
Sand (f = 1.6~3.0)	2.51 kgCO$_2$e/t
Concrete Block	336 kgCO$_2$e/m^3
Molded Steel Window	121 kgCO$_2$e/m^2
Hot rolled carbon steel bars	2340 kgCO$_2$e/t
Water	0.168 kgCO$_2$e/t
Wooden Formwork (renovated production)	17.8 kgCO$_2$e/m^2

Table 3. Carbon emissions from major consumables

Title	Concrete	Steel	Wood Formwork	Concrete blocks	Mortar	Doors and Windows	Consumption
Carbon emissions (kiloton)	1.208	1.073	0.506	0.524	2.418	0.333	6.062

lifespan. Therefore, this study adopts a 50-year calculation cycle as the operational and maintenance phase for the analysis. The carbon emissions during the operation and maintenance period are mainly concentrated in the three aspects of electricity, heat, and gas. Consequently, this study has obtained the annual energy consumption data of the building from the property management department (Table 4).

Table 4. The energy consumption during the annual operation and maintenance period of buildings

Type	Power/(MW·h)	Diesel oil/t	Gas/10000 Nm^3
Consumption	4439.45	1.463	1.84

Utilize Python in conjunction with formula (2) to calculate and compile a comprehensive summation of the carbon emissions arising from the operation and maintenance phase. Finally, utilizing the data presented in Tables 3 and 4, the aggregate carbon emissions during the construction phase and the operation and maintenance phase can be accurately computed, as detailed in Table 5.

Table 5. Two-stage carbon emissions.

Stages	Carbon emission/kiloton	Annual carbon emissions per square meter/kg/ $(a \cdot m^2)$
Construction	6.062	13.32
Operate	167.55	368.25
LCA (50 years)	173.612	381.57

As highlighted in Table 5, the total carbon emission of the life cycle of buildings in the city is 173.613 kilotons, and the annual carbon emission per square meter is 381.57 kg/ $(a \cdot m^2)$. The carbon emission in the operation and maintenance stage is the most, 167.55 kilotons, accounting for 96.50% of the overall emission. In comparison, the carbon emission in the construction stage of the building is 6.062 kilotons, accounting for 3.50% of the total emission, and the annual carbon emission per square meter is 13.32kg/ $(a \cdot m^2)$.

3.2 Model for Evaluating Carbon Emissions from Buildings on a "Year" Basis

This paper adopts the entropy weight method to evaluate the carbon emissions of buildings.

Selection of evaluation indicators: Based on the data obtained from the comprehensive analysis of the relevant studies on carbon emissions from buildings, some representative indicators are selected, such as energy consumption of buildings, carbon emissions from building materials, the indoor environment of buildings, etc.

Data standardization: There may be differences in scale and order of magnitude for different indicators, so it is necessary to standardize the data. Specifically, for positive indicators, the following formula (3) can be used for standardization:

$$x'_{ij} = \frac{x_{ij} - x_{min}}{x_{max} - x_{min}}$$

For negative indicators, the following Eq. (4) can be used for normalization:

$$x'_{ij} = \frac{x_{max} - x_{ij}}{x_{max} - x_{min}}$$

where x_{ij} represents the raw data for the i building on the j indicator x_{max} and x_{min} denote the maximum and minimum values of the first metric over all buildings, respectively.

Calculating information entropy: For each indicator, its information entropy can be calculated. Specifically, it can be calculated using the following formula (5):

$$e_j = -k \sum_{i=1}^{n} p_{ij} ln p_{ij}$$

where e_j denotes the information entropy of the j metric, k is a constant, and p_{ij} denotes the standardized value of the i metric on the j building.

Calculation of weights: Based on the results of information entropy, the weights of each indicator can be calculated. Specifically, the following formula (6) can be used for calculation:

$$w_j = \frac{1 - e_j}{m - \sum_{i=1}^{m} e_i}$$

where w_j denotes the weight of the j indicator, and m is the number of evaluation indicators.

Analysis of evaluation results: Based on the overall score of each building, it can be judged whether its carbon emissions meet the requirements. Specifically, if the score is higher than a certain threshold, the building's carbon emissions are considered to meet the requirements; if the score is lower than a certain threshold, the building's carbon emissions are considered to be non-compliant. Since a large number of data samples are needed as support, this paper will not expand too much.

4 Conclusions and Suggestion

4.1 Conclusions

Employing mathematical techniques in conjunction with BIM models and Python software, this study conducted a thorough analysis of the dual-stage carbon emissions of urban buildings on an annual basis. A carbon emission evaluation model tailored to the Chinese context was developed, thereby laying the groundwork for the establishment of an effective carbon emission management framework for urban buildings. In terms of the proportion of carbon emissions throughout the LCA of buildings, it is evident that the carbon emissions generated during the operation and maintenance phase significantly exceed those produced during the materialization phase. This observation aligns with the outcomes reported by the National Institute of Standards and Technology (NIST) in collaboration with other organizations, which evaluated the carbon emissions associated with domestic architectural products through the utilization of the BEES computer evaluation system in tandem with the Life Cycle Analysis (LCA) methodology. Consequently, the primary focus of energy conservation and emission reduction measures for urban buildings ought to be on optimizing the operation and maintenance phase.

4.2 Suggestions

The pursuit of carbon neutrality stands as a critical endeavor for the global community in addressing climate change, representing a shared responsibility among nations and enterprises. To achieve this objective, a comprehensive approach encompassing various strategies is imperative. These strategies include emission reduction measures, new energy development, technological advancements, and policy adjustments. Emission reduction measures entail restructuring energy, industrial, and transportation systems, along with the utilization of carbon capture and utilization technologies. New energy development involves the research, promotion, and utilization of clean energy sources, such as solar, wind, hydroelectric, and nuclear power. Technological advancements necessitate the further development of smart grid technology, energy storage technology, and novel material technologies. Policy adjustments, on the other hand, require the formulation and refinement of tax policies, subsidy mechanisms, and regulatory frameworks.

However, research findings indicate that the operational phase of buildings constitutes the stage with the highest carbon emissions. Consequently, enhancing management measures during this phase is of utmost importance. Governments must strengthen quantitative carbon emission management for urban buildings, establish an annual carbon emission accounting system, and enforce the imposition of appropriate carbon emission fees on buildings exceeding prescribed emission quotas. These measures will effectively mitigate emissions during the operational phase of buildings, thereby contributing significantly to the attainment of carbon neutrality. Furthermore, it is imperative to consider the impact of the design phase of buildings on carbon emissions during the subsequent operational phase. To achieve this, it is advisable to integrate artificial intelligence (AI) into the Building Information Modeling (BIM) that is established during the design phase. This approach allows AI to be trained to understand the energy consumption distribution of the building, thereby enabling it to effectively manage energy allocation during the operational phase.

In conclusion, a concerted effort, encompassing various strategies and enhanced management measures, is essential in the pursuit of carbon neutrality. By focusing on the operational phase of buildings, governments can make significant progress in reducing carbon emissions and contributing to the global effort to address climate change.

Acknowledgement. This study is financially supported by the Chengdu Green Low Carbon Development Research Base Project (No. LD23YB24) and the Sichuan University Jinjiang College Young Teachers' Research Fund Project (No. QNJJ-2023-A01).

References

Blengini, G.A., Carlo, T.D.: The changing role of life cycle phases, subsystems and materials in the LCA of low energy buildings. Energy Build. **42**(6), 869–880 (2010)

British Petroleum (BP): bp World energy statistics yearbook. https://www.bp.com/content/dam/bp/countrysites/zh_cn/china/home/reports/statistical-review-of-world-energy/2022/2022sr book.pdf (2022)

Chen, G.Q., Chen, H., Chen, Z.M., et al.: Low-carbon building assessment and multi-scale input-output analysis. Commun. Nonlinear Sci. Numer. Simul. **16**(1), 583–595 (2011)

IEA: Perspectives for the clean energy transition, the critical role of buildings. https://www.iea.org/reprots/the-critical-role-of-buildings (2022)

Li, T.Y., Sun, J., Shi, C.Q., et al.: Carbon emission accounting and evaluation of the whole life cycle of large public buildings. Green Technol. **8**(16), 13–15 (2017)

Liu, N.X., Wang, J., Li, R.: Calculation method of emissions from urban settlements in China. J. Tsinghua Univ. (Nat. Sci. Ed.) **05**(49), 1433–1446 (2008)

Ma, Y.L.: Research on green building evaluation system based on carbon emission analysis of green buildings. Harbin Institute of Technology, Harbin (2021)

Shang, C.J., Zhang, Z.W.: Carbon emission accounting for building life cycle. J. Eng. Manage. **24**(1), 7–12 (2010)

United Nations Intergovernmental Panel on Climate Change IPCC: Report of Working Group I on the Sixth Report: Climate Change 2021: Fundamentals of physical science (2021–08–06). https://www.ipcc.ch/report/ar6/wg1/downloads/report/IPCC_AR6_WGI_Chapter_03.pdf

Verbeeck, G.: Life cycle inventory of buildings: a calculation method. Build. Environ. 3(45), 1037–1041 (2010)

Wan, Z., Guo, Y.H., Li, S.C.: Discussion on building energy-saving measures based on the whole life cycle. J. Jiaying College (Nat. Sci. Ed.) 27(6), 56–59 (2011)

Wang, J.Y.: Research on carbon emission measurement and benchmark management throughout the lifecycle of green buildings based on BIM technology. Hunan University of Technology, Hunan (2019)

Wang, S.Q., Bao, L.J.: Research on the current status of carbon emission in the operation phase of public buildings based on double regression prediction model. HVAC 8(30), 114–118 (2021)

Yang, S.H., Liu, J., et al.: Research on carbon emission accounting of public buildings in the process of carbon trading. Build. Sci. 36(11), 326–330 (2020)

Climate Adaptation in Carbon Neutral Cities: The Role of Clustered Buildings for Enhancing Energy Efficiency and Decarbonization in West Asian Cities

Akram Ahmed Noman Alabsi$^{(\boxtimes)}$ and Feng Du

College of Architecture and Urban Planning, Fujian University of Technology, Fuzhou 350118, China

AlabsiAkram@fjut.edu.cn

Abstract. This research focuses on the pivotal role of clustered building designs in enhancing energy efficiency and decarbonization strategies in West Asian cities, shedding light on an innovative approach to sustainable urban development. As climate adaptation measures gain momentum globally, carbon reduction and carbon neutrality stand as the primary objectives for major cities. West Asia confronts unique challenges that amplify the repercussions of climate change across various sectors. The literature reveals a lack of focus on climate adaptation and a limited emphasis on the integration of adaptation measures in carbon-neutral city strategies. This research addresses this gap through an inductive approach, employing case studies and a proposed mathematical model to evaluate the impact of climate adaptation strategies on carbon neutrality in West Asian cities. The results of our study underscore the immense potential of climate-adaptive clustered building designs in reducing building operating costs and carbon footprints. The synergy between these technologies presents a powerful avenue for achieving energy efficiency and emissions reduction while driving a green transformation in society's economic and social activities. Integrating climate adaptation techniques with renewable energy sources demonstrates significant improvements in energy efficiency and cost-effectiveness, bringing us closer to optimal carbon neutrality. Finally, our research paves the way for rethinking urban planning and design in West Asian cities and beyond. It emphasizes the critical role of clustered building arrangements and climate adaptation techniques in achieving carbon neutrality, offering a sustainable and cost-effective path toward a more resilient urban future.

Keywords: Climate Adaptation · Clustered Buildings · Energy Efficiency · Carbon-Neutral Cities

1 Introduction

The global phenomenon of climate change, marked by escalating temperatures, heightened frequency of extreme weather events, and rising sea levels, is significantly impacting cities worldwide [1]. Urban areas, with their dense populations, intricate infrastructure,

© The Author(s) 2025
B.-J. He et al. (Eds.): UCSUD 2023, LNCE 559, pp. 151–163, 2025.
https://doi.org/10.1007/978-981-97-8401-1_11

and robust economic activities, stand particularly vulnerable to these effects [2]. This vulnerability underscores the critical importance of urban resilience—the capacity of urban systems to withstand and recover from shocks and stresses. In response to these challenges, effective climate adaptation measures are increasingly recognized as vital components of urban resilience strategies. In West Asian countries, where rapid urbanization is underway due to population growth and infrastructure development, cities are experiencing climate changes induced by their own development. While these nations have a relatively modest impact on global greenhouse gas emissions, such emissions are key contributors to climate change within the region [3]. The West-Asian and Middle East regions are particularly susceptible to climate change effects, with projections indicating a substantial rise in average temperatures by 2060. Cities overlooking the Arabian Gulf are expected to experience a temperature increase of 7 degrees Celsius, while the rest of the region may see an increase ranging from 3.1 to 5.7 degrees Celsius. Under extreme emissions scenarios, the temperature rise from 1990 to 2100 is projected to be 5.6 °C, a figure that drops to less than 1.6 °C with significant emission reductions. Amid these environmental transformations and escalating climate challenges, there is a growing urgency for scientific research aimed at developing effective strategies to adapt to climate variations and enhance environmental sustainability [4]. This urgency is particularly crucial in regions highly vulnerable to climate impacts, such as West Asia. Globally, governments and institutions are actively working towards achieving carbon neutrality and promoting the transition to carbon-neutral urban living [5].

In the face of these challenges, the construction sector in West Asia stands at a pivotal juncture, poised to become a leader in decarbonization. Rapid urbanization, booming international investment, and the appeal of luxury projects define the landscape. The incorporation of cutting-edge sustainable technologies and design principles into projects not only meets the growing demand for luxurious, eco-friendly living spaces but also significantly reduces the carbon footprint associated with urban development [6]. Carbon-neutral buildings aim for a net-zero carbon footprint, addressing emissions across scopes through energy-efficient design, renewable energy, and sustainable practices [7]. The unique challenges faced by the West Asia region exacerbate the impacts of climate change across various sectors, highlighting the imperative for effective climate adaptation measures [8]. In response, carbon-neutral cities in this region must aim to harmonize environmental, social, and economic well-being, considering the cumulative benefits of urbanization in both developed and developing countries. However, despite the global emphasis on carbon neutrality, existing literature reveals a noticeable gap in addressing climate adaptation and a limited focus on integrating adaptation measures into strategies for achieving carbon-neutral cities. Enabling carbon neutrality in cities emerges as one of the most important pathways to a sustainable future. Modular buildings, a critical component contributing to energy efficiency and decarbonization goals, play a central role in this transition. Efficient mitigation methodologies at the urban scale, such as passive cooling techniques and strategic urban design, are paramount for reducing carbon emissions and enhancing climate adaptation, emphasizing the significance of clustered buildings over individual structures [9]. This study focuses on the role of clustered buildings as a key factor in mitigating the effects of climate change and promoting sustainability in cities [10]. The research explores how the integration of architectural designs and modern technologies can enhance energy efficiency and reduce the environmental impacts of buildings in urban architecture. The methodology of this research is based on a comprehensive analysis of scientific literature specialized in the field of climate adaptation and carbon neutrality strategies. It involves scrutinizing combined building design strategies and technologies used to improve energy efficiency and reduce emissions [11]. Practical case studies will be analyzed to evaluate the experiences of West Asian cities in integrating clustered buildings into their urban context. The research results highlight the importance of climate-resilient cluster building designs in reducing building operating costs and mitigating carbon footprints. The synergy between these technologies emerges as an effective means of achieving energy efficiency and reducing emissions, propelling the green transformation in various economic and social activities. Furthermore, the integration of climate adaptation technologies with renewable energy sources exhibits substantial improvements in energy efficiency and cost-effectiveness, steering us towards optimal carbon neutrality. The importance of urban planning is accentuated, rooted in local climate considerations and integrated with architectural and urban design elements along with renewable energy components. This holistic approach advocates for maximum shading, reduction in heat gain, energy consumption, and carbon emissions, fostering a paradigm of near-zero energy urban architecture through the application of photovoltaic envelope technologies. Clustered building arrangements, coupled with urban open spaces and the utilization of low-cost, environmentally friendly materials, assume a pivotal role in constructing an urban fabric that not only enhances social engagement and walkability but also diminishes the need for extensive vehicular networks.

1.1 Climate Change, Carbon Neutrality, and Energy Efficiency in West Asia

Over the past 15 years, the global energy landscape has undergone unprecedented changes marked by substantial growth in renewables, technological advancements, and increased diversification of energy sources. These developments have led to falling prices and a notable decoupling of economic growth from greenhouse gas (GHG) emissions. In the context of this global transformation, West Asia, with its oil-rich yet water-scarce profile and strategic geopolitical location, faces unique challenges [12]. The West Asia and North Africa region, particularly the Arab world, holds significant hydrocarbon reserves, constituting over 50% and 30% of the world's oil and gas reserves, respectively. Buildings emerge as crucial in energy consumption, according to the West Asia Climate Action Report. While west Asia has a relatively modest impact on global greenhouse gas emissions, such emissions are key contributors to climate change within the region (refer to Fig. 1). Buildings constitute the largest source of energy consumption and emissions globally, with around 75% of electricity in the US dedicated to building operations [3]. The building sector's energy consumption is expected to outpace that of industry and transportation, projecting a significant increase by 2030. These findings underscore the importance of addressing energy efficiency in the building sector to mitigate climate change and enhance sustainability in West Asia [13].

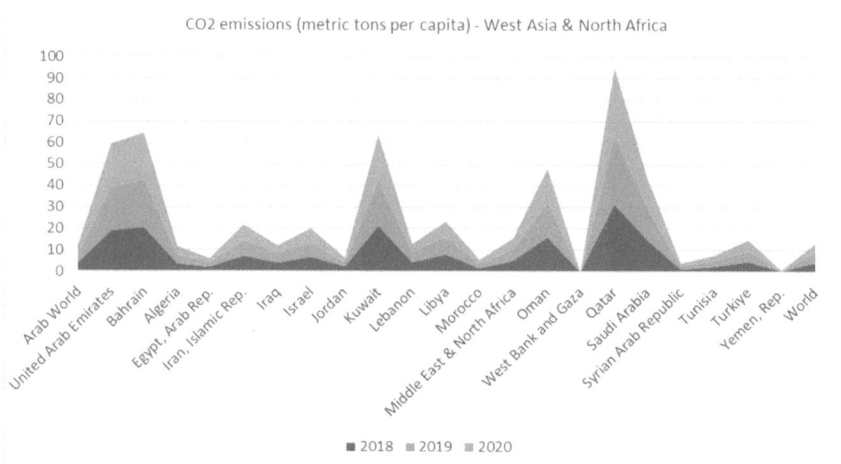

Fig. 1. National share of 2018, 2019, 2020 global carbon emissions across West Asia and North Africa. Source: by authors derived from World Bank data.

2   Methodology

This article employs an inductive approach, utilizing case studies and a proposed mathematical model to assess the impact of climate adaptation strategies on achieving carbon neutrality in cities across West Asia. The methodology involves extrapolating insights from prior research, incorporating global findings, and conducting a region-specific study to address challenges related to carbon neutrality strategies. A key emphasis is placed on the analysis of case studies and the evaluation of emissions calculations to formulate a mathematical model aimed at improving contemporary energy practices in modern architecture, ultimately striving for carbon neutrality.

3   Results and Discussion

3.1   The Role of Climate Adaptation in Cluster Building Planning

Climate significantly shapes the architectural landscape of West Asia and comparable climatic zones, influencing building material choices and design solutions to facilitate human adaptation to challenging conditions. While traditional architecture adeptly aligned with the region's realities, modern urban development, driven by global trends, often neglects the psychological and physical comfort of its residents. Climate emerges as a pivotal environmental factor in architectural formation, exerting a distinct impact on the urban fabric of Western Asia. Building styles in arid zones differ from those in rainy areas, implying a climatic influence on city planning and street layouts. Coastal cities, for instance, feature terraced buildings to harness sea breezes, while deep valleys optimize winter sun exposure, creating a humid climate [2]. These diverse environmental challenges pose complexities for planners and architects alike. Given the prevalence of desert conditions, local architecture in the region naturally aligns with the climate, prompting architects to devise adaptive solutions. Noteworthy among these is the concept of clustered buildings, characterized by several key features. These include the adoption of compact building forms, bringing houses together to minimize facade exposure to adverse weather conditions such as sunlight and wind. Drawing from Islamic architectural principles, this compact form, known as "one weft," mitigates temperature elevation within buildings. Additionally, the significant height of exterior walls provides shading for large roof areas, contributing to a cooler internal environment. In direct response to the challenges posed by the harsh desert climate in West Asia, the inhabitants developed unique planning and design strategies within their traditional architecture. These practices aimed at creating habitable outdoor and indoor spaces, ensuring natural thermal comfort in an era predating the widespread use of energy and electricity. A notable characteristic prevalent in the historical cities of West Asia is the clustering of buildings and the adoption of a compact and dense urban fabric. This distinctive urban planning is prominently evident in renowned cities such as Yazd in Iran, Kashan, the ancient city of Baghdad in Iraq (refer to Fig. 2), Sanaa, Shibam Hadhramaut in Yemen, and Damascus in Syria (refer to Fig. 3). Numerous other cities across diverse regions of West Asia showcase similar innovative approaches to address the environmental challenges posed by their surroundings. The horizontal projection of building blocks exceeds that of open spaces like streets and inner courtyards, maximizing shade provision. Narrow streets and buildings' reduced height contribute to minimal solar radiation exposure, further enhancing thermal comfort. In specific hotspots, streets are strategically oriented from north to south, aligning vertically with the sun's movement, ensuring continuous shading throughout the day. The deliberate narrow and winding design of streets serves to disrupt air currents, moderating temperatures in both winter and summer. This refined street layout enhances air circulation within homes, optimizing comfort by allowing only essential airflows in both seasons, with street direction determined by prevailing winds [12].

The Impact of Climate Adaptation Technologies on energy efficiency and decarbonization: West Asia stands prominently among regions significantly affected by climate change, posing formidable challenges to both energy sustainability and carbon emission reduction. Addressing these challenges is imperative, and the pivotal role of climate adaptation technologies in attaining carbon neutrality cannot be overstated. Rooted in the preceding dialogues, this text delves into the climatic milieu of West Asia, a driving force behind architectural design and construction technology. In the case study of the Alnakheel housing project in Riyadh (refer to Fig. 4), the environmental response is closely tied to strategically positioning buildings in relation to the sun and wind. This arrangement allows for effective shading between structures [14]. Factors such as space efficiency, minimized energy consumption, simplified maintenance, and the consideration of the availability, suitability, and durability of building materials all play pivotal roles in the thoughtful design of housing clusters and adhesion patterns. While traditional architectural practices demonstrate a harmonious alignment with the climate, the transition to modern urban development necessitates the integration of adaptive technologies to sustain energy efficiency and achieve carbon neutrality.

A noteworthy technique in this pursuit is the design of clustered buildings, consolidating residential units to mitigate environmental impact and enhance energy efficiency

Fig. 2. Compact planning and Adhesion and clusters of buildings, (1, 2) the city of Yazd in Iran, (3, 4) the Al-Adhamiya neighborhood, the old city of Baghdad in Iraq.

Fig. 3. Compact planning and clusters of buildings, (1, 2) the old city of Sanaa in Yemen, and (3, 4) the old city of Damascus in Syria.

[4]. Drawing inspiration from Islamic architecture principles, this design mitigates temperature elevation within buildings, fostering a cooler indoor environment. Integration with renewable energy and thermal insulation technologies further amplifies energy efficiency and reduces emissions. The discernible impact of climate adaptation techniques is evident in the marked improvement of building energy efficiency, notably through the

utilization of advanced energy generation and storage technologies, encompassing solar energy systems and thermal energy for heating and cooling.

Fig. 4. The residential cluster of the Alnakheel project, Riyadh, Saudi Arabia.

Traditional methods, design evolution, and natural energy Integration in cluster Buildings for achieving carbon neutrality: The utilization of cluster building planning offers several thermal advantages, contributing to a sustainable and comfortable environment. The design (refer to Fig. 5) we proposed and tested in this study serves as a model for a unit or fundamental building block within the context of cluster building-style planning. Firstly, the arrangement of clustered buildings facilitates solar shading, where walls mutually shade each other, casting shadows in the courtyard, thereby minimizing heat gain from solar radiation. Courtyards, often adorned with trees and plants, enhance this shading effect, fostering a cooler environment both on building surfaces and the ground. Additionally, this design creates a locally desirable climate by reducing outdoor temperatures and increasing relative humidity. Courtyard surfaces dissipate heat at night, providing a cool ambiance during the day. The enclosed courtyard space facilitates the retention of cold air, further prolonging the cooling effect. This design strategy indirectly reduces air conditioning costs as occupants can spend significant portions of the day outdoors in the shaded courtyard. Moreover, cluster building planning strategically employs wind shading on upstream walls, minimizing wind pressure distribution and reducing heat gains through infiltration. The length-to-height ratio of the courtyard space influences this, with lower values contributing to substantial reductions in heat gains.

This approach also decreases wind velocity, subsequently lowering the convective thermal coefficient and minimizing building envelope heat gain by convection. Furthermore, radiant cooling becomes an efficient means of dissipating absorbed heat, particularly from the building's roof, to the upper atmosphere. This process is particularly effective in desert climates. In traditional courtyard designs, such as those seen in Baghdad, the floor and ceiling expose a large surface area to the cold night sky, facilitating effective radiation loss. During the day, the ground and roof act as heat sinks, absorbing solar radiation, and creating thermal currents that enhance comfort. At night, the accumulated heat emits into the night sky, and the building's thermal mass retains cold, contributing to prolonged cooling effects the following day. Overall, cluster building planning exemplifies a comprehensive approach to thermal management, optimizing both passive and active strategies for enhanced comfort and energy efficiency.

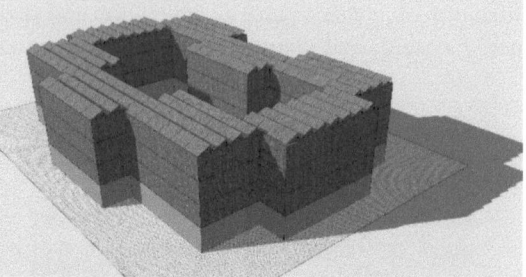

Fig. 5. The proposed design for clustered building types aiming at carbon neutrality entails minimizing exposed exterior surface areas to climatic conditions in a hot environment. This involves the utilization of efficient traditional materials and the comprehensive integration of solar panels on roofs to maximize renewable energy sources.

3.2 Methods for Evaluating and Calculating Carbon Neutrality (Proposed Model)

Carbon neutrality is a concept that signifies the equilibrium between carbon emissions released into the atmosphere and endeavors to diminish or counterbalance them. The objective is to attain carbon neutrality, wherein greenhouse gases released into the atmosphere are either eradicated or compensated for by alternative means. The mathematical model proposed by the study is grounded in the entire life cycle of the building. This encompasses scrutinizing the origins of materials and components, progressing through the planning and design stages, and then extending to implementation, operation, and maintenance. Additionally, it considers removal and recycling processes. All these phases are meticulously calculated, and the mathematical model operates on the principle of offsetting operational carbon emissions [10]. This is achieved by optimizing energy efficiency, reducing emissions through sustainable operational practices, and emphasizing the significance of negative carbon technology (refer to Fig. 6).

We employed the subsequent mathematical equation to monitor and compute emissions throughout the stages of implementation, operation, and removal, allowing us to approach a state of carbon neutrality:

$$E = \sum_{i=1}^{n} E_i - E_{KZS} - E_{TH}$$

where:

E——The total annual carbon dioxide emissions during the building operation phase, in tons of carbon dioxide equivalent (tCO2e);

i——the type of terminal energy consumed by the building, including electricity, gas, oil, municipal heat, etc.;

E_i——The total annual carbon dioxide emissions generated by the energy consumption of type i of the building, in tons of carbon dioxide equivalent (tCO2e);

E_{KZS}——the annual carbon dioxide emissions generated by the introduction of green electricity or self-produced green electricity in accounting unit i, in tons of carbon dioxide equivalent (tCO2e);

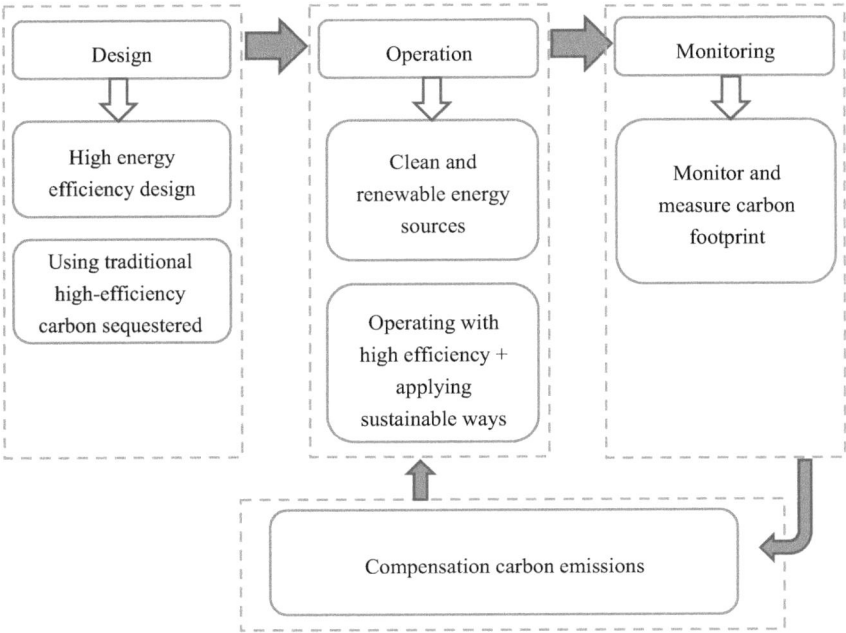

Fig. 6. The depiction and sequential delineation of procedures within the proposed framework to attain carbon neutrality.

E_{TH}———The annual carbon reduction amount of the building greening and vegetation carbon sink system, the unit is tons of carbon dioxide equivalent (tCO2e).

Carbon Neutral Design for Clustered Buildings: Carbon neutrality strives for equilibrium between carbon emissions and mitigation efforts. This concept entails offsetting or eliminating greenhouse gases released into the atmosphere. Achieving carbon neutrality involves reducing emissions through efficient technologies, utilizing renewable energy, and enhancing overall energy efficiency. Offsetting emissions includes supporting projects like tree planting and renewable energy initiatives. As in Table 1, we summarized our key findings. Carbon storage technologies aim to extract and store carbon dioxide in inaccessible locations, such as underground. Climate-adaption modular building designs are pivotal in minimizing operating costs and carbon footprints. The synergy of these technologies proves effective in enhancing energy efficiency and reducing emissions, catalyzing a green transformation in economic and social activities. Integration with renewable energy sources significantly improves efficiency and cost-effectiveness, advancing the journey toward optimal carbon neutrality. Urban planning assumes paramount importance, intertwining local climate considerations with architectural design and renewable energy components. This approach emphasizes maximum shading, diminished heat gain, reduced energy consumption, and lower carbon emissions, fostering near-zero energy urban architecture through photovoltaic envelope technologies. Clustered building arrangements, coupled with urban open spaces and cost-effective, eco-friendly materials, play a crucial role in crafting an urban fabric that enhances social engagement, walkability, and diminishes the reliance on extensive vehicular networks.

Table 1. Key design strategies for clustered buildings to achieve carbon neutrality.

Design Strategy	Description
Carbon Neutral Design	Aiming for equilibrium between carbon emissions and mitigation efforts, utilizing efficient technologies, renewable energy, and energy efficiency measures
Urban Planning	Aiming for equilibrium between carbon emissions and mitigation efforts, utilizing efficient technologies, renewable energy, and energy efficiency measures
Modular Building Designs	Intertwining local climate considerations with architectural design and renewable energy components to maximize shading, reduce energy consumption, and lower emissions
Renewable Energy Sources	Integration with renewable energy sources improves efficiency, cost-effectiveness, and progresses towards optimal carbon neutrality
Clustered Building Arrangements	Plays a crucial role in crafting an urban fabric fostering social engagement, walkability, and diminishing reliance on extensive vehicular networks
Eco-friendly Materials	Utilizing low-cost and eco-friendly materials contributes to sustainable urban development practices, aiding in achieving carbon neutrality
Photovoltaic Envelope Technologies	Foster near-zero energy urban constructions by harnessing solar energy and minimizing reliance on external power sources

4   Conclusions

In the face of rapid climate change, the imperative to mitigate its effects is paramount, particularly in regions highly susceptible to its impact, such as West Asia. Governments and institutions worldwide are striving to attain carbon neutrality and promote the transition to zero-carbon urban living. The escalating impacts of climate change coupled with rapid population growth in West Asia underscore the imperative to prioritize urban microclimate considerations and associated housing strategies. This study critically examined prior research investigating the influence of urban design and adaptation strategies on thermal performance and energy efficiency, emphasizing their role in emission reduction. Although limited research specifically evaluates building assembly within the urban context, analyses of traditional West Asian architecture underscore the significance of climate adaptation strategies. The design implications advocate for an optimal cluster urban fabric tailored to specific climate conditions, integrating selected plant species and strategically spaced tree ground cover. This housing environment vision aligns with decarbonization efforts, contributing to anticipated commitments to the United Nations Framework Convention on Climate Change (UNFCCC). Results highlight that enhancing building energy efficiency involves employing technologies such as heat pumps, integrating renewable sources like solar panels, and implementing energy storage systems. Thoughtful building design, incorporating materials that regulate temperatures, and the use of smart controls further contribute to wise energy use. Emerging technologies, including heat pumps, thermal energy storage, and grid-integrated building control, play pivotal roles in achieving net-zero and grid-responsive buildings, reducing operational costs and carbon footprints. Widespread adoption and integration of these technologies necessitate collaborative efforts to achieve broader building decarbonization. The direct correlation between building energy efficiency and the effective thermal and climatic performance of all components underscores the unique pathway of designing for climate adaptation to attain multiple benefits. This approach not only aligns positively with climate data and minimizes environmental impact but also seeks to reverse assumed environmental degradation through adaptation strategies, contributing to ecosystem enhancement and addressing environmental challenges. Climate adaptation, in this context, transcends mere compatibility with current conditions, evolving dynamically to achieve continuous thermal comfort amidst changing climates and fostering symbiotic interaction with the living environment. Our research underlines the significance of urban planning that prioritizes local climate considerations and the interconnectedness of architectural, urban design, and renewable energy components. This approach maximizes shading and minimizes heat gain, energy consumption, and carbon emissions, fostering near-zero energy urban constructions with photovoltaic (PV) envelope technologies. Clustered building arrangements, urban open spaces, low-cost, and eco-friendly materials play a pivotal role in creating an urban fabric that fosters social engagement and walkability while reducing the need for extensive vehicular networks. Finally, our research paves the way for rethinking urban planning and design in West Asian cities and beyond. It emphasizes the critical role of clustered building arrangements and climate adaptation techniques in achieving carbon neutrality, offering a sustainable and cost-effective path toward a more resilient urban future. Collaboration among stakeholders is essential to harness the full potential of these technologies and ensure their widespread adoption and dissemination, thereby advancing the goals of urban sustainability and climate resilience. The study's limitations, such as the exclusion of advanced field measurements and interdisciplinary integration, underscore the need for future research to incorporate advanced methodologies, including remote sensing and airborne techniques, to validate findings and contribute to sustainable urban development practices emblematic of adapting future societies to climate change's negative impacts.

References

1. Amini Toosi, H., Lavagna, M., Leonforte, F., Del Pero, C., Aste, N.: Building decarboniza-tion: assessing the potential of building-integrated photovoltaics and thermal energy storage systems. Energy Rep. **8**:574–581 (2022). https://doi.org/10.1016/j.egyr.2022.10.322
2. Rezvani, S.M.H.S., de Almeida, N.M., Falcão, M.J.: Climate adaptation measures for enhancing urban resilience. Buildings **13**(9) (2023) https://doi.org/10.3390/buildings130 92163
3. Alabsi, A.A.N., Song, D.: Sustainable adaptation climate and energy efficiency of traditional buildings technologies in the hot dry regions—case studies in West Asia, 1st ed., vol. 1. LAP LAMBERT Academic Publishing (2020). [Online]. Available: https://www.morebooks.shop/ shop-ui/shop/product/9786202525961. Accessed 5 Jan 2024
4. Fahmy, M., Mahmoud, S., Abdelkhalik, H., Abdelalim, M., Elshelfa, M.: Energy efficiency, carbon emissions, and thermal comfort comparisons between conventional and proposed clustered open courtyard housing using CEB blocks. In: IOP Conference Series: Earth and Environmental Science. Institute of Physics (2022). https://doi.org/10.1088/1755-1315/1056/ 1/012026
5. Fahmy, M., Elwy, I., Elshelfa, M., Abdelkhalik, H., Abdelalim, M., Mahmoud, S.: Energy efficiency and de-carbonization improvements using court-yarded clustered housing with Compressed Earth Blocks' envelope. Energy Rep. **8**, 365–371 (2022). https://doi.org/10.1016/ j.egyr.2022.01.051
6. Fahmy, M., Mahmoud, S., Elwy, I., Mahmoud, H.: A review and insights for eleven years of urban microclimate research towards a new Egyptian era of low carbon, comfortable and energy-efficient housing typologies. Atmosphere, no. 3. MDPI AG, March 1, 2020. https:// doi.org/10.3390/atmos11030236
7. Liang, R., Zheng, X., Liang, J., Hu, L.: Energy efficiency model construction of building carbon neutrality design. Sustainability (Switzerland) **15**(12) (2023). https://doi.org/10.3390/ su15129265
8. Shabb, K., McCormick, K.: Achieving 100 climate neutral cities in Europe: investigating climate city contracts in Sweden. npj Climate Action **2**(1) (2023). https://doi.org/10.1038/ s44168-023-00035-8
9. Fahmy, M., Sharples, S.: Urban form, thermal comfort and building CO_2 emissions—a numer-ical analysis in Cairo. Build. Serv. Eng. Res. Technol. **32**(1), 73–84 (2011). https://doi.org/ 10.1177/0143624410394536
10. Norouzi, M., Haddad, A.N., Jiménez, L., Hoseinzadeh, S., Boer, D.: Carbon footprint of low-energy buildings in the United Kingdom: Effects of mitigating technological pathways and decarbonization strategies. Sci. Total Environ. **882**(2023). https://doi.org/10.1016/j.scitotenv. 2023.163490
11. Ma, Z., et al.: An overview of emerging and sustainable technologies for increased energy efficiency and carbon emission mitigation in buildings. Buildings **13**(10). Multidisciplinary Digital Publishing Institute (MDPI) (2023). https://doi.org/10.3390/buildings13102658
12. Alabsi, A.A.N., Wu, Y., Koko, A.F., Alshareem, K.M., Hamed, R.: Towards climate adaptation in cities: indicators of the sustainable climate-adaptive urban fabric of traditional cities in West Asia. Appl. Sci. (Switzerland) **11**(21) (2021). https://doi.org/10.3390/app112110428
13. Alabsi, A.A.N., Song, D., Liu, Z.: Traditional solutions in climate adaptation and low energy buildings of hot-arid regions in west Asia. In 2016 International Conference on Green and Energy-Efficient Building At: Beijing-China, Bejing, pp. 1–7 (2016)
14. Abdelsalam, T.: Adapting the concept of courtyard in long-narrow attached houses as a sustainable approach: the Saudi experiment. Int. J. Contemp. Archit. **2**(2) (2015). https:// doi.org/10.14621/tna.20150303

Impact Mechanism of Residential Area Patterns Based on Carbon Emission Measurement on Residents' Travel: A Case Study of Mianyang, China

Guoyang Hai$^{(\boxtimes)}$, Chunrong Zhao, Li Yan, Minghong Yu, Huiyun Peng, and Jiaxin Li

School of Civil Engineering and Architecture, Southwest University of Science and Technology, Mianyang 621010, China
2670056124@qq.com

Abstract. Residents' over-reliance on motor vehicles has changed their travel modes, leading to many problems such as traffic and air pollution, and greatly affecting residents' low-carbon, healthy living. Reducing carbon emissions from residents' travel has become a social consensus. This paper studies the impact mechanism of residential patterns on residents' travel based on carbon emission measurement. An empirical study was carried out in eight neighborhoods in Mianyang City, first statistically analyzing residents' non-commuting trips, then calculating non-commuting carbon emissions by using the travel carbon emission measurement method, and then using multiple regression models to study the effects of each physical environment element in the settlement model on non-commuting carbon emissions. The aim is to explore the differences in the impact of different indicators, and the ultimate goal is to find out the factors that reduce residents' travel carbon emissions and are conducive to the construction of healthy residential areas.

The results show that there are differences in non-commuting travel and carbon emissions for different purposes in different residential areas, non-commuting travel is mainly by walking; carbon emissions from using service facilities are greater than carbon emissions from shopping and leisure travel. The impact of various elements of residential patterns is different, and the indicator elements with a greater degree of influence are floor area ratio, land use mix, and road network density.

Keywords: Residential pattern · Residents' travel · Carbon emissions · Impact mechanism · Mianyang City

1 Introduction

With the rapid development of urbanization and motorization, the number of cars in China has entered a period of rapid growth. Residents' over-reliance on motor vehicles has changed their travel modes, leading to many problems such as traffic and air pollution, and greatly affecting residents' low-carbon, healthy living. Therefore, guiding

© The Author(s) 2025
B.-J. He et al. (Eds.): UCSUD 2023, LNCE 559, pp. 164–182, 2025.
https://doi.org/10.1007/978-981-97-8401-1_12

residents to adopt green travel methods such as walking, cycling, and public transportation is particularly important, and reducing carbon emissions from residents' travel has become a social consensus. Residential areas, as microsystems of cities, provide a pleasant living environment for people on the one hand, and on the other hand, they are connected with other functional areas of the city through the arrangement of buildings, roads, and greening [1]. Therefore, in terms of building low-carbon cities and promoting green travel for residents, future residential planning is particularly important. The "Low Carbon Community Pilot Construction Guide" released in 2015 clearly states that new urban communities should be guided by low-carbon concepts and promote low-carbon lifestyles. People are gradually realizing that the key to achieving low-carbon residential construction lies in establishing a residential model suitable for residents' low-carbon travel.

In recent years, due to the rise of low-carbon cities, scholars have found that a simple qualitative description of residents' travel behavior cannot intuitively reflect the problems and differences caused to the environment, so carbon emissions are introduced into the research. In terms of road traffic facilities, street width, intersection density, and road network density have different degrees of impact on residents' travel carbon emissions. Among them, in related research, many scholars pointed out that small-scale, small-grid block models are conducive to reducing residents' travel carbon emissions. For example, scholars such as Huang, using Wuhan as an example, found that a greater number of bus routes and a higher density of intersections can help reduce carbon emissions from residents' travel when analyzing the relationship between the convenience of urban transportation facilities and residents' travel carbon emissions [2]. In the research of Yang and other scholars, it was found that the density of the road network in residential areas has a negative correlation with different types of travel carbon emissions, including dining, leisure, commuting, etc. That is, the greater the density of the road network, the smaller the carbon emissions from residents' travel [3]. In terms of land use, Man, Huang and other scholars have calculated the mixed use of land around residential areas and analyzed the impact of the mixed use of land around urban residential areas on travel carbon emissions [4, 5]. The results show that residential patterns with high mixed land use will promote low-carbon travel by residents.

However, the viewpoint that the residential pattern affects residents' travel has been widely recognized, and there has been some progress at multiple research levels, but its impact mechanism has not been given attention, and related research is less. Secondly, scholars mostly study residents' travel behavior from the perspective of transportation and urban economy, and less research considers the impact mechanism of residential pattern characteristics on residents' travel from the perspective of ecological economics, lacking mature quantitative analysis methods. Furthermore, although the research on residents' daily activities and travel is very mature, most of them focus on the research of residents' commuting behavior. In the existing research results, there is little in-depth research on non-commuting behavior, especially the refined classification research of non-commuting travel purposes.

Therefore, this article takes the carbon emission of residents' travel as the entry point to study the impact mechanism of the residential pattern on residents' non-commuting travel. An empirical study was carried out on 8 residential areas in Mianyang City, the non-commuting travel characteristics of residents in different residential areas were statistically analyzed, and the behavioral differences of residents in different residential areas were analyzed, and the impact relationship model between the index elements of the residential pattern and residents' non-commuting travel was constructed, aiming to explore the difference in the impact degree of different indicators, and the ultimate goal is to find out the factors that reduce residents' travel carbon emissions and are conducive to the construction of healthy residential areas.

2   Sample, Data, and Method

2.1   Regional Overview and Sample Selection

Overview of the study area. Mianyang City is located in the heart of the "Western Triangle" of Chengdu, Chongqing and Xi'an, and is an important node city of the Chengdu Plain Urban Agglomeration, which has become a sub-transportation hub city in northwestern Sichuan Province, and is also an important base of national defense, military and scientific research and production in China [6]. After the reform and opening up, the urban structural pattern gradually evolved, and the central city of Mianyang was centered on the old city at the confluence of the three rivers, namely, the Fujiang River, the Anchang River, and the Furong River, and developed to the periphery to form a cluster-type urban structure layout [7]. Under the promotion of rapid urbanization, the land use function, transportation system, and settlement construction of Mianyang city areas have been changed, and there are phenomena such as poor connection of clusters, imperfect transportation system, and unbalanced distribution of service facilities, etc. It is due to the uneven distribution that leads to the differences in residents' travel activities between different settlements, for example, residents of low-mixedness settlements have more motorized travel behaviors, which leads to more carbon emissions from residents' travel.

   Principles of sample plot selection ① All neighborhoods have occupancy rates greater than 80%. Higher occupancy rates can indirectly reflect the stability of the current conditions surrounding the neighborhoods. ② The selected neighborhoods should be representative of settlements with differences in the nature of housing, the degree of infrastructure, and the accessibility of road transportation, which have different residential populations and different travel need. ③ The sample neighborhoods have different site sizes, different planned land forms, and are located in different zones. ④ The term "neighborhood" in this paper refers not only to neighborhoods bounded by man-made walls, but also to some open living places. In the following specific measurement of some indicators, the scope of the survey is the center of the district as the center of the circle, with a radius of 800 m or 1,000 m covered by the circular range, the main reason is that 800 m or 1,000 m as the radius of the radius of the scope of radiation is equivalent to China's "Urban Residential Planning and Design Standards" (GB50180-2018) in the fifteen-minute living area of the scale of the living area, will be put into the urban environment to consider, not only within the district, helps to improve the study of the current residential living area. Put into the urban environment for consideration, not only limited to within the neighborhood, which is conducive to improving the applicability of this study to the current settlement planning. According to the questionnaire, walking is the more frequent way, so 1000 m is used as the radius of the neighborhood.

To reflect the situation of different community residential pattern characteristic elements and residents' travel, after careful visits, surveys, and comparative analysis, this article selected 8 communities in Fucheng District, Youxian District, and Jingkai District of Mianyang City as the research objects, as shown in Fig. 1 and Table 1.

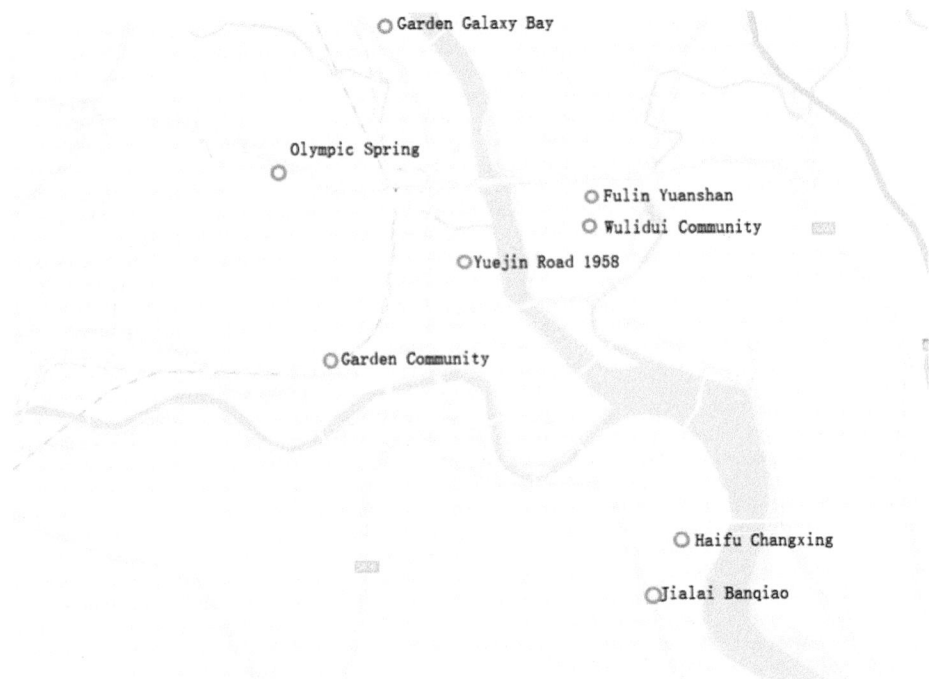

Fig. 1. Distribution of sample plots in Mianyang City

2.2 Survey and Data

In this study, because office workers are not easy to inquire on weekdays, the questionnaire survey will be conducted on both weekdays and non-working days to ensure the universality of the survey population. The data required for this study mainly includes three parts, as shown in Fig. 2.

Then, basic statistical analysis of the questionnaire data is carried out. This study filters out invalid questionnaires with special travel purposes, such as daily work for Didi travel, to ensure that each survey site distributes 40 questionnaires, a total of 320 questionnaires are distributed in 8 sample residential areas, 313 questionnaires are collected back, and the questionnaire effectiveness rate is 97.8%.

Table 1. Basic situation of 8 sample residential areas

Serial number	Survey community	Year of construction	Building density	Plot ratio	Land scale (hectares)	Urban district	Spatial form
1	Wulidui Community	2001	48%	1.7	16.66	Youxian District	Multi-layer high density
2	Jialai Banqiao	2006	42%	1.6	19.17	Jingkai District	Multi-layer high density
3	Garden Community	2001	30%	2.0	12.16	Fucheng District	Multi-layer medium density
4	Garden Galaxy Bay	2003	30%	1.5	16.04	Fucheng District	Multi-layer medium density
5	Olympic Spring	2009	20%	2.5	15.65	Science and Technology Park District	High-rise low density
6	Fulin Yuanshan	2011	20%	2.5	8.40	Youxian District	High-rise low density
7	Yuejin Road 1958	2016	25%	4.0	11.43	Fucheng District	High-rise low-density
8	Haifu Changxing	2017	25%	3.0	13.61	Economic Development Zone	High-rise low-density

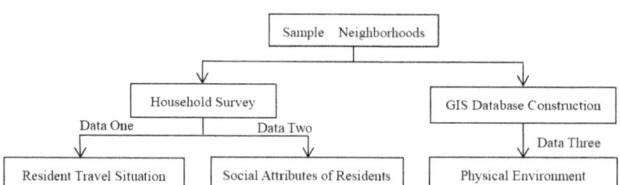

Fig. 2. Data acquisition and sources

2.3   Research Method

Linear regression model. Linear regression modeling is a more common statistical analysis method to study the influence relationship. In order to study the coupling relationship between the physical spatial environment of settlement patterns and the carbon emissions of residents' travel, this study mainly adopts the mathematical and statistical method of regression analysis on the basis of descriptive analysis statistics. When it is necessary to analyze the relationship between one independent variable and one dependent variable, a univariate linear regression model is usually constructed; when it is necessary to analyze the relationship between multiple independent variables and one dependent variable, a multiple linear regression model is usually constructed [8]. In the analysis of this study, since it contains multiple independent variables and the dependent variable is a continuous variable, the relationship between the settlement pattern factor variables and travel carbon emissions is analyzed by constructing a multiple regression model to obtain the results of the study. The model is as follows:

$$Y = b_0 + b_1X_1 + b_2X_2 + \cdots + b_kX_k + u \tag{1}$$

where Y - the dependent variable, carbon emissions from residential trips

X - independent variables, settlement pattern characterization variables.

b_0 - constant term (in math.)

b_k - Regression coefficients, the extent to which each factor affects carbon emissions from residential trips.

U - residual.

The degree of correlation of variables in regression analysis is usually expressed by the correlation coefficient B and significance, when $B > 0$, it means that there is a positive correlation between the two variables, and vice versa for negative correlation, the larger $|B|$, it means that the independent variable affects the dependent variable to a greater extent; significance < 0.05 means that the independent variable is correlated with the dependent variable to a certain extent (*), and when the significance is < 0.001, it means that the independent variable is correlated with the dependent variable with a high degree of correlation (**) [9].

Questionnaire survey. Select several typical residential areas in Mianyang City, conduct relevant interviews with community residents on issues such as transportation, to provide data support for carbon emission calculations in this study. The questionnaire includes residents' social attributes, travel characteristics, attitude preferences, etc.

Geographic Information System (GIS) spatial analysis method. This study needs to conduct a geographic information system spatial analysis of transportation facility elements, and analyze road traffic characteristics by counting road intersection density.

Travel carbon emission calculation model. Using a bottom-up approach (distance theory), carbon emissions are calculated based on the single-trip distance of different modes of transportation. The specific calculation method is to obtain detailed data such as different residents' travel methods, distances, travel frequencies, and types of vehicles used, calculate the carbon emissions of a single trip by residents, and finally calculate the total carbon emissions of different purpose trips. The carbon emission intensity of various modes of transportation is shown in Table 2:

Based on this, the carbon emissions of each resident's single trip are obtained through the calculation formula, specifically calculation formula 1, then multiplied by the number of times per week of the same travel method, and finally the total carbon emissions of travel are calculated using formula 2. Among them, travel distance, method, purpose, and frequency are all obtained from the travel activity records filled out by the residents.

$$CE = W_i + D_i \tag{2}$$

where CE—Carbon emissions of a single trip by residents (g); W_i—Carbon emission intensity of transportation mode i (g/person km); D_i Single trip distance using

Table 2. Carbon emission intensity of different modes of transportation (g/km)

Mode of transportation	Carbon emission intensity	Mode of transportation	Carbon emission intensity
Walking	0	Bus	30.096
Bicycle	5	Car	361.38
Electric vehicle	7.553	Taxi	259.89
Motorcycle	78.863		

transportation mode i (km).

$$E = n_1 CE_1 + n_2 CE_2 \cdots + n_n CE_n \qquad (3)$$

where E is carbon emissions from residents' travel (g), and n_i is the number of weekly trips using transportation mode i.

3   Characteristic Elements of Residential Patterns Factors Affecting Residents' Travel

3.1   Physical Environmental Factors Affecting Residents' Travel

This paper mainly studies the impact of residential patterns on residents' travel, so it is particularly important to determine the constituent elements of the residential pattern. This study will continue to use the built environment D variable concept proposed by American scholars Cervero and Ewing and will refer to the D variable and combine it with the actual situation, dividing the physical environmental factors into internal and external environments, with specific indicator elements as follows [10] (Table 3).

Table 3. Specific indicator elements of the physical environment

Physical environmental elements	Evaluation indicators	
Internal environment	Density category	Building density, plot ratio, entrance and exit density
External environment	Diversity	Land use mix, plot subdivision
	Connectivity	Intersection density, road network density

(1) Internal environmental elements

The internal environmental elements of the residential pattern mainly consider the housing aspect, and the three indicators of building density, plot ratio, and entrance and exit density are selected to quantify the internal environment (Table 4).

(2) External Environment Factors

Table 4. Measurement and formula of internal environmental assessment index

Serial number	Evaluation indicators	Indicator measurement	Indicator formula
1	Building density	Measures the proportion of the base area of the building occupying the ground area	$BD = ba/a$ ba—Residential building footprint (m^2) a—Gross floor area of residential buildings (m^2)
2	Plot ratio	Measures the proportion of the total building area within the plot occupying the ground area	$PR = ta/a$ ta—Gross area of all buildings in the settlement (m^2) a—Total land area of settlements (m^2)
3	Entrance density	Measures the convenience of travel in the study area	$ED = ne/a$ ne—Number of entrances and exits a—Total land area of settlements (m^2)

External environment factors mainly consider function, road, and space. Based on the analysis of the mechanism of action, the external space environment in the physical environment is summarized into diversity, connectivity indicator elements.

① Diversity Evaluation Indicators

The constituent elements of diversity attributes mainly include the content of land use functions. This study has drawn on the two indicators that often appear in related research in recent years, land use mix and plot subdivision, to evaluate the diversity attributes of residential patterns. The specific indicators are shown in Table 5.

② Connectivity Evaluation Index. The specific indicators are shown in Table 6.

3.2 Residential Pattern Variable Statistics

Based on the construction of the residential pattern variable model in the previous text, based on field investigations, the characteristics of the residential pattern are calculated to obtain 8 sample residential variables (Table 7).

4 Residents' Non-commuting Travel Behavior and Carbon Emissions

Although research on residents' daily activities and travel is well established, most of it focuses on the study of residents' commuting behavior, and there are few in-depth studies on non-commuting behavior in the existing research results, especially the

Table 5. Diversity evaluation indicator measurement and formula

Serial number	Evaluation indicator	Indicator measurement	Indicator formula
1	Land use mix [8]	Measures the degree of mixing and distribution uniformity of various types of functional land in the study area	$Mix\ Used = \frac{(\sum_{i=1}^{N} p_i ln p_i)}{ln N}$ p_i—Ratio of the area of category i of each land-use type to the area of the region in which it is located N—Types of different land use types
2	Plot subdivision [11]	Measure the degree of fragmentation of the land within the research area	$C = \Sigma n_i / F$ n_i—Number of land use boards in the study area in category i F—total area

Table 6. Connectivity evaluation index measurement and formula

Serial number	Evaluation index	Index measurement	Index formula
1	Road network density	Measure the total amount of street network	$ND = dl/F$ dl—the total length of the road F—the total area
2	Intersection density [12]	Measure the mechanical characteristics of the street network	$ID = ri/F$ ri—the number of road intersections F—the total area

study of the refined classification of the purpose of non-commuting travel. The present study addresses this point by exploring in depth the different factors that influence non-commuting trips.

Non-commuting behavior mainly refers to travel behavior other than going to work and school. This non-commuting trips are organized into three major categories, one is shopping and consumption trips, which mainly refers to trips with the trip purpose of going to restaurants, supermarkets, shopping malls and food markets; the second category is leisure and entertainment trips, which mainly refers to trips with the trip purpose of going to parks and plazas, movie theaters, pubs and bars, KTVs, gymnasiums, and libraries; and the third category is trips with the use of services, which means trips with trip purposes that include general hospitals, clinics, educational facilities, and banks, among other trip purposes.

Table 7. Statistics of various element variables in different residential areas

	Building density	Plot ratio	Entrance density (per/km^2)	Land use mix	Parcel subdivision	Intersection density (per/km^2)	Road network density (km/km^2)	Distance to the nearest bus stop (m)	Distance to green spaces and squares (m)	Bus stop density (per/km^2)
Wulidui Community	48%	1.7	168.07	0.55	7.96	26.11	7.84	138	50	14.01
Jialai Banqiao	42%	1.6	41.73	0.54	22.61	12.10	4.39	327	1175	10.83
Garden Community	32%	2.0	49.35	0.63	43.95	23.25	6.59	151	617	11.29
Garden Galaxy Bay	28%	1.5	24.93	0.51	22.93	5.10	3.45	175	1766	9.55
Olympic Spring	20%	2.5	38.33	0.77	32.48	16.24	6.02	166	684	11.78
Fulin Yuanshan	20%	2.5	35.87	0.47	8.28	25.16	8.24	105	945	15.92
Yuejin Road 1958	28%	4.0	43.74	0.70	40.76	11.15	4.77	478	1440	10.51
Haifu Changxing	25%	3.0	51.44	0.51	21.97	10.19	4.49	183	324	7.64

4.1 Sample Residential Area Non-commuting Travel Situation

(1) Different Residential Area Non-commuting Travel Modes

Before analyzing the impact of residential pattern elements on non-commuting travel carbon emissions, a descriptive analysis of non-commuting travel is first conducted. As seen from Fig. 3, unlike commuting modes, non-commuting travel is mainly by walking, accounting for 61.77% of total travel modes. In daily activities, whether for consumption, leisure, or using service facilities, residents prefer to walk. The next most common mode is car travel, with 21.78% of people choosing to travel by car for non-commuting activities. The least common mode is motorcycle travel, at only 0.31%.

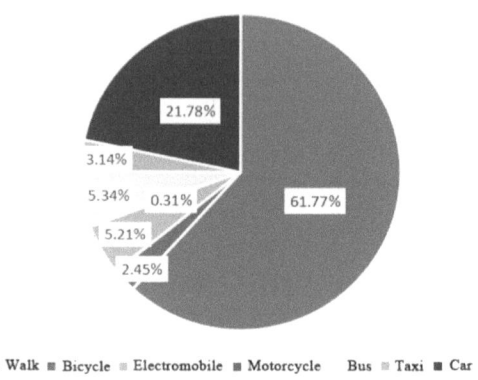

Fig. 3. Proportion of non-commuting travel modes in sample residential areas

Non-commuting travel is predominantly on foot (Fig. 4), except for the Wulidui community, where the second most common mode of travel in other residential areas is by car. Large shopping malls, comprehensive hospitals, and other facilities are concentrated in the city center, which is relatively far compared to other destinations, so residents going to these places use cars more frequently. The Wulidui community has a high density of roads around it, the roads are narrow, and there are more trips by electric bicycles. The least used mode of travel is motorcycles. Upon field investigation, it was found that the number of people buying electric bicycles is far greater than motorcycles for two reasons: one is that motorcycles require a driving license to be on the road, and now electric bicycles are low in price and easy to use, becoming a more popular choice for travel; second, Mianyang city has implemented restrictive traffic control measures for motorcycles in the city.

(2) Travel situation for different travel purposes

Among the 2402 travel records, there are 2089 non-commuting travel records, of which, there are 900 shopping consumption travel records, 504 leisure and entertainment travel records, and 685 service facility use travel records. In the proportion of travel modes for the three travel purposes (Table 8), it can be seen that walking is still the most chosen mode by residents, followed by car travel. Electric bicycle travel is more common in shopping consumption travel, while bus travel is more common in leisure and entertainment and service facility use travel.

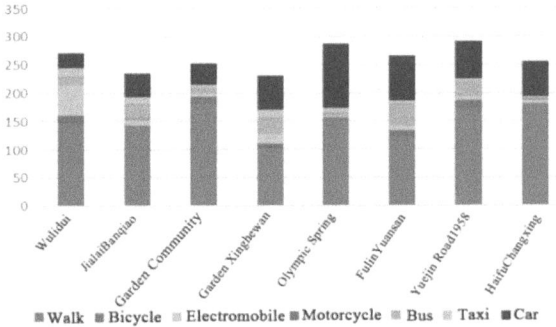

Fig. 4. Comparison of non-commuting travel modes among residents in different residential areas

Table 8. Proportion of travel for different travel purposes (%)

		Walking	Bicycle	Electric Bicycle	Motorcycle	Bus	Taxi	Car
Shopping consumption	Total	68.11	1.89	6.33	0.22	5.44	2.56	15.44
	Restaurant	17.00	0.33	0.89	0.00	0.89	0.56	4.33
	Supermarket	28.44	0.22	1.89	0.00	0.33	0.11	1.33
	Shopping mall	7.11	0.56	1.44	0.22	4.00	1.78	9.44
	Vegetable market	15.56	0.78	2.11	0.00	0.22	0.11	0.33
Leisure entertainment	Total	44.44	2.18	3.57	0.20	9.33	5.36	34.92
	Park square	29.96	1.19	0.79	0.00	2.98	1.19	9.52
	Cinema	11.51	0.20	1.59	0.20	2.78	3.17	11.71
	Bar	0.40	0.20	0.00	0.00	0.99	0.99	5.16
	Stadium	0.99	0.00	0.00	0.00	0.20	0.00	2.58
	Library	1.59	0.60	1.19	0.00	2.38	0.00	5.95
Use service facilities	Total	54.60	4.09	4.82	0.44	8.18	2.92	24.96
	General hospital	5.11	0.58	0.29	0.15	4.53	2.19	11.24
	Clinics, pharmacies	25.99	1.61	1.46	0.15	0.00	0.00	0.29
	Schools	6.86	1.46	1.75	0.00	1.46	0.15	8.91
	Banks	9.78	0.44	1.17	0.00	0.29	0.00	2.34

4.2 Characteristics of Carbon Emissions from Residential Non-commuting Travels

Based on the constructed travel carbon emission calculation model, the total non-commuting carbon emissions of residents, as well as the carbon emissions from shopping consumption travel, leisure and entertainment travel, and service facility use travel, are calculated separately, and the calculation results are statistically analyzed, as shown in Table 9.

Table 9. Non-commuting carbon emissions of residents in different residential

	Non-commuting carbon emissions	Shopping consumption travel carbon emissions	Leisure and entertainment travel carbon emissions	Service facility use travel carbon emissions
Wulidui Community	49050	10668	18625	19756
Jialai Banqiao	86466	36445	14871	35150
Garden Community	137213	21166	21998	94049
Garden Galaxy Bay	342376	177982	102120	62273
Olympic Spring	154806	16826	47066	90915
Fulin Yuanshan	331768	73901	82121	175747
Yuejin Road 1958	80840	17171	25081	38588
Haifu Changxing	143273	12654	18294	112325

Firstly, the overall non-commuting carbon emissions are statistically analyzed (Fig. 5). Overall, there are significant differences between the various residential areas. The highest non-commuting carbon emissions are from the Garden Galaxy Bay and Fulin Yuanshan, while the lowest are from Yuejin Road, Jialai Banqiao, and Wulidui Community. Haifu Changxing, Olympic Spring, and Garden Community have moderate emission levels. Notably, Wulidui Community and Fulin Yuanshan are very close in location, but there are significant differences in results.

Looking at the carbon emissions from shopping consumption, leisure and entertainment, and the use of service facilities (Fig. 6), the highest carbon emissions come from the use of service facilities, accounting for 47% of total carbon emissions. The proportions of shopping consumption and leisure and entertainment are not much different, accounting for 28% and 25% respectively. In the above statistics, it is found that the

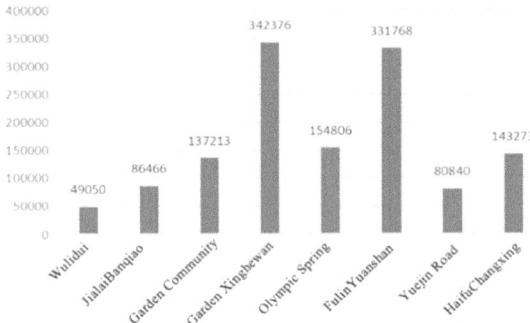

Fig. 5. Comparison of total non-commuting carbon emissions in sample residential areas

frequency of shopping consumption is much higher than that of leisure and entertainment, but the carbon emissions are not much different. The frequency of using service facilities is also less than that of shopping consumption, but the carbon emissions are the highest. This indicates that residents who travel for the purpose of using service facilities and leisure and entertainment prefer to use motor vehicles compared to shopping consumption.

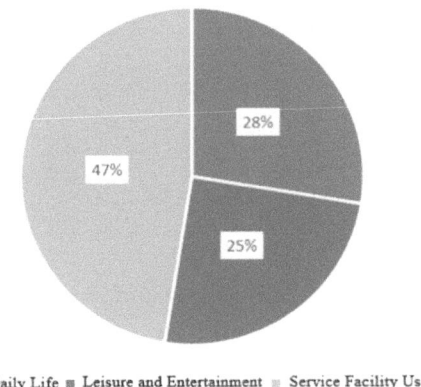

■ Daily Life ■ Leisure and Entertainment ▨ Service Facility Use

Fig. 6. Proportion of non-commuting travel carbon emissions

5 Analysis of the Impact of Residential Patterns on Residents' Non-commuting Behavior

5.1 The Impact of Density Indicators

In Table 10, the dependent variables of Model 1, 6, and 12 are shopping consumption, leisure and entertainment, and total carbon emissions from using service facilities, respectively. From the coefficient table, it is known that, first, the factor that has the greatest impact on shopping consumption and leisure and entertainment is the plot ratio. The higher the plot ratio, the lower the carbon emissions from trips to restaurants, shopping malls, cinemas, KTV bars, and libraries. The plot ratio to some extent reflects the nature of the residence and the attributes of the residents. Families with a high plot ratio are more likely to own cars. Secondly, the factor that has the greatest impact on the use of service facilities is the building density. The greater the building density, the smaller the carbon emissions from using service facilities. Finally, the factor most affected by the entrance and exit density is the carbon emissions from shopping consumption. The greater the entrance and exit density, the smaller the carbon emissions from shopping consumption.

Table 10. Impact of density indicators on non-commuting carbon emissions

Influencing factors	Plot ratio		Building density		Entrance and exit density	
	B	Sig.	B	Sig.	B	Sig.
Model 1 (Shopping consumption trip)	−43188.020	.000	−1351.478	.000	−702.729	.000
Model 2 (Restaurant)	−22571.805	.000	−2098.242	.000	215.241	.000
Model 3 (Supermarket)	1169.886	.000	2.669	.000	−3.212	.000
Model 4 (Mall)	−28683.114	.000	−3427.260	.000	434.068	.000
Model 5 (Vegetable market)	1637.854	.000	−87.105	.000	−29.693	.000
Model 6 (Leisure and entertainment travel)	−21911.076	.000	−2964.403	.000	−439.282	.000
Model 7 (Park)	5653.271	.000	−1061.924	.000	−221.747	.000
Model 8 (Cinema)	−24202.855	.000	−2054.038	.000	552.508	.000
Model 9 (KTV bar)	−6204.418	.000	−719.218	.000	94.421	.000
Model 10 (Stadium)	11242.943	.000	−364.206	.000	−253.366	.000
Model 11 (Library)	−1341.785	.000	−782.327	.000	−139.723	.000
Model 12 (Use of service facilities trip)	−2313.577	.319	−6564.607	.000	−544.103	.000
Model 13 (Hospital)	−9698.128	.000	−901.454	.000	−390.630	.000
Model 14 (Clinic)	−12.429	.706	−42.995	.000	−2.194	.000
Model 15 (School)	4777.941	.062	−4186.636	.000	−367.703	.000
Model 16 (Bank)	−193.376	.000	−70.753	.000	−9.263	.000

5.2 The Impact of Diversity Indicators

Looking at the impact results (Table 11), the carbon emissions from trips to markets and stadiums do not show a significant relationship with the mix of land use; apart from the carbon emissions from park squares and libraries being positively correlated with the mix of land use, the rest of the travel carbon emissions are negatively correlated with the mix of land use, i.e., the higher the mix of land use, the lower the carbon emissions from residents' travel in that residential area. In terms of plot subdivision, supermarkets, shopping malls, and KTV bars have no significant relationship with plot subdivision. Among the remaining travel carbon emissions, the carbon emissions from restaurants, cinemas, and schools increase with the increase of plot subdivision, while the carbon emissions from markets, park squares, stadiums, libraries, general hospitals, clinics, and banks decrease with the increase of plot subdivision.

Table 11. Impact of diversity indicators on non-commuting carbon emissions

Influencing Factors	Land use mix		Plot subdivision	
	B	Sig.	B	Sig.
Model 1 (Shopping consumption trip)	−324777.007	.000	954.789	.000
Model 2 (Restaurant)	−155859.265	.000	530.755	.000
Model 3 (Supermarket)	−9476.587	.000	2.896	.811
Model 4 (Mall)	156857.526	.000	465.776	.005
Model 5 (Vegetable market)	−2583.577	.025	−44.638	.000
Model 6 (Leisure and Entertainment travel)	−98271.042	.000	175.339	.350
Model 7 (Park)	39295.574	.000	−370.226	.000
Model 8 (Cinema)	−82713.973	.000	425.119	.000
Model 9 (KTV bar)	−18760.498	.000	73.051	.027
Model 10 (Stadium)	16014.732	.036	−318.207	.000
Model 11 (Library)	17955.358	.000	−144.525	.000
Model 12 (Use of service facilities trip)	142460.292	.000	103.507	.734
Model 13 (Hospital)	−98041.807	.000	−703.062	.000
Model 14 (Clinic)	−1168.760	.000	−14.384	.000
Model 15 (School)	236677.267	.000	837.671	.000
Model 16 (Bank)	−2208.250	.000	−18.601	.000

5.3 The Impact of Connectivity Indicators

From the coefficient table (Table 12), in Model 3, there is no significant relationship between supermarket trip carbon emissions and intersection density. In Model 9, there is no significant relationship between KTV, bar trip carbon emissions and connectivity indicators. The carbon emissions of all other types of trips decrease with the increase of intersection density. Except for the increase in carbon emissions from cinema trips with the increase in road network density, the carbon emissions from all other trips decrease with the increase in road network density. Among them, for shopping and consumption trips, leisure and entertainment trips, and the use of service facilities, the impact of road network density is greater than that of intersection density. Among these three types of trips, the impact on the use of service facilities is greater than the other two types of trips. The analysis found that the daily facilities used by residents are basically located along the external streets of the residential area. Increasing the road network density helps to increase the number of external facades, reduce their rents, thereby increasing the density of service facilities, reducing the frequency of motor vehicle trips, and increasing the proximity of residents' walking trips.

6 Conclusion and Discussion

This paper takes the quantification of carbon emissions from residents' travel as the entry point, studying the impact mechanism of residential patterns on residents' travel. Empirical research is carried out on 8 residential areas in Mianyang City. The specific conclusions can be summarized as follows:

(1) In shopping consumption travel, leisure and entertainment travel, and service facility use travel, walking is the main mode, followed by car travel.

(2) In terms of travel carbon emissions, the carbon emissions from service facility use travel are greater than those from shopping consumption and leisure and entertainment travel. Among the 8 residential areas, Garden Galaxy Bay and Fulin Yuanshan have larger non-commuting travel carbon emissions.

(3) Using a regression model to analyze the factors affecting residents' travel, the results show that the higher the plot ratio, building density, entrance and exit density, land use mix, intersection density, and road network density, the lower the non-commuting travel carbon emissions. The larger the plot subdivision, the higher the non-commuting travel carbon emissions. The elements with a larger impact are plot ratio, land use mix, road network density, and bus stop density. Therefore, by increasing the mix of land use, subdivision of land parcels, density of intersections and increasing the density of the road network around the settlements, carbon emissions from travel within the settlements can be effectively reduced.

Residential areas, as important units of the city, must actively embark on the low-carbon path. From the perspective of low-carbon development, this study introduces the carbon emission measurement tools in ecological economics, and explores the influence of the characteristic elements of the settlement pattern on residents' travel by analyzing the theoretical relationship between the settlement pattern and residents' travel as well as the constructed calculation model of travel carbon emission, which is of positive

Table 12. Impact of diversity indicators on non-commuting carbon emissions

Influencing Factors	Intersection density		Road network density	
	B	Sig.	B	Sig.
Model 1 (Shopping consumption trip)	−7566.028	.000	−21266.660	.000
Model 2 (Restaurant)	−2543.094	.000	−5356.448912	.095
Model 3 (Supermarket)	−261.060	.083	−831.145	.000
Model 4 (Mall)	−4330.549	.000	−12621.529	.000
Model 5 (Vegetable market)	−431.324	.000	−2457.535	.000
Model 6 (Leisure and entertainment travel)	−6004.605	.000	−26183.283	.000
Model 7	−3294.748	.000	−15945.149	.000
Model 8 (Cinema)	−850.786	.063	−143.281	.829
Model 9 (KTV bar)	−917.459	.000	−2258.683	.000
Model 10 (Stadium)	−2486.421	.000	−14943.511	.000
Model 11 (Library)	−1701.751	.000	−7437.940	.000
Model 12 (Use of service facilities trip)	−12939.141	.000	−69109.555	.000
Model 13 (Hospital)	−4119.816	.000	−19288.774	.000
Model 14 (Clinic)	−112.103	.000	−661.617	.000
Model 15 (School)	−8448.132	.000	−47886.864	.000
Model 16 (Bank)	−240.540	.000	−1204.044	.000

significance in enriching the theoretical research related to the planning pattern of urban settlements and residents' travel. Meanwhile, the research conclusions can provide an effective implementation path for future construction of healthy residential areas, and can also provide a reference basis for the standard design of urban residential spatial form, road traffic facilities, etc., making the residential planning pattern and residents' travel behavior more reasonable. This paper focuses on the explanation of the elements affecting residents' travel, and uses a regression analysis model. However, subsequent studies can consider using more complex models according to different focuses, such as considering the use of structural equation models to explore the mutual relationship between multiple endogenous variables.

Funding. This research was supported by Sichuan Science and Technology Program (Grant No. 2023NSFSC1051).

References

1. Ji Ze: Research on residential road system and its related indicators guided by low-carbon travel. Nanjing University of Technology (2018)
2. Huang Jingnan, Gao Haowu, Han Sunsheng: The impact of road traffic facility convenience on daily family transportation carbon emissions—taking Wuhan as an example. Int. Urban Plan. **30**(03), 97–105 (2015)
3. Yang Wenyue, Cao Xiaoshu: The mechanism of travel carbon emissions in Guangzhou from the perspective of residential self-selection. Geogr. J. **73**(02), 346–361 (2018)
4. Man Zhou, Zhao Rongqin, Yuan Yingchao, et al.: The impact of land mix around urban residential areas on residents' commuting carbon emissions—taking a typical residential area in Jiangning District, Nanjing as an example. Hum. Geogr. **33**(01), 70–75 (2018)
5. Huang Jingnan, Du Ningrui, Liu Pei, et al.: Research on the impact of land mix around the home on daily family transportation carbon emissions—taking Wuhan as an example. Int. Urban Plan. **28**(02), 25–30 (2013).
6. Xu Yang, Yao Yong, Zhang Zhaoqiang: Feasibility analysis study of construction industrialization in Mianyang City. Sichuan Arch. **35**(04), 107–109 (2015)
7. Zhang Hongyu, Zhou Bo, Wang Bo: Research on urban characteristics based on landscape patterns—taking Mianyang City as an example. Planers (04): 31–33 (2007)
8. Yao Yu: Study on the impact of built environment on urban residents' travel and carbon emissions. Harbin Institute of Technology (2015)
9. Wu Yan: The influence of urban built environment on residents' walking behavior. Jiangxi Normal University (2017)
10. Yang Yang: Quantitative study on the relationship between residential built environment and household travel energy consumption in Jinan City. Tsinghua University (2013)
11. Huang Yuangang: Study on the relationship between urban block land use and pedestrian travel. Chongqing University (2014)
12. Liu Chang: Study on the impact of land use characteristics in Chengdu City on residents' travel mode. Southwest Jiaotong University (2016)

Analysis of Carbon Emission Impact Factors and Trend Prediction Based on LMDI and ARIMA Models: A Case Study of Zhejiang Province

Peng Zhang, Junbo Mu$^{(\boxtimes)}$, and Jie Luo

School of Civil Engineering and Surveying, Southwest Petroleum University, Chengdu 610500, China

mjb_15183967214@163.com

Abstract. The present study proposed a method to examine the carbon emissions of various departments in Zhejiang Province from 2003 to 2020 using the IPCC sectoral method. The use of the LMDI model analyzed the factors that influence carbon emission change in Zhejiang Province. The ARIMA prediction model and grey prediction model are utilized to forecast carbon emissions of Zhejiang Province in the future. The proposed measures for carbon emission reduction in Zhejiang Province are given, and some reference basis is provided for similar provinces to carry out low-carbon transformation. The results demonstrated that: (1) The carbon emission of Zhejiang Province from 2003 to 2020 shows a linear increase trend, with a growth rate of 172% during the 18 years. (2) The energy structure of Zhejiang Province is developing towards energy cleanliness. (3) Energy intensity and industrial structure are inhibiting effects, economic output and population size are promoting effects, and energy structure has both inhibiting and promoting times. (4) ARIMA's prediction of carbon emissions in Zhejiang Province in the next few years is more accurate than that of the grey prediction model. The prediction results of ARIMA show that Zhejiang Province will usher in the carbon peak in 2025, while the grey prediction results show that it will not usher in the carbon peak before 2027.

Keywords: Carbon emissions · Energy and carbon emissions · ARIMA prediction model · LMDI model · Zhejiang Province

1 Introduction

CO_2 emissions from energy consumption constitute the primary origin of greenhouse gases [1]. The repercussions of the greenhouse effect lead to adverse weather conditions, impacting national economic development, human health, and causing sea level rise that inundates land, posing a threat to national and regional security. The Yangtze River Delta stands as one of China's most dynamic, innovative, and open regions, pivotal in the country's comprehensive opening-up strategy and modernization agenda. Zhejiang Province, being a key economic contributor in the Yangtze River Delta and ranking

© The Author(s) 2025
B.-J. He et al. (Eds.): UCSUD 2023, LNCE 559, pp. 183–198, 2025.
https://doi.org/10.1007/978-981-97-8401-1_13

fourth in GDP nationwide [2], holds significant economic weight and plays a crucial role in national economic growth. However, Zhejiang faces the dual challenge of fossil energy and scenic resource scarcity, with 100% foreign dependence on coal, oil, and natural gas, necessitating urgent low-carbon development initiatives.

Both domestic and international research on carbon emissions primarily focus on several aspects: carbon emission calculation and prediction, examination of key factors affecting carbon emissions, and examination of low-carbon emission reduction technologies and policies. The 2006 IPCC (Intergovernmental Panel on Climate Change) Guidelines for National Greenhouse Gas Inventories serves as a widely adopted international method for carbon emission accounting. Building on the IPCC framework, Li et al. [3] conducted research on carbon dioxide energy-related carbon emissions in Ningxia Province, revealing notable discrepancies between accounting results based on energy balance tables and total energy consumption. Other scholars recalculated CO_2 emissions for Chinese provinces from 2000 to 2012 using the apparent energy consumption method and revised emission factors, narrowing the gap between national and provincial CO_2 emission measurements [4]. Understanding and predicting carbon emissions are crucial for formulating effective low-carbon emission reduction strategies, requiring a comprehensive analysis of influencing factors. Scholars like Qi [5] and Xin et al. [6] utilized the EIO-LCA model to investigate CO_2 emission influencing factors, highlighting the role of import/export and inflow/outflow structures in emission reduction. The LMDI (Logarithmic Mean Divisia Index) decomposition model, Tapio decomposition model, and distorted Kaya identity are commonly employed to analyze driving factors affecting carbon emissions, suggesting that energy consumption intensity and structure inhibit carbon emissions, while energy and industrial structure and economic growth promote them [7–10]. Tong et al. [11] established a VAR model to study the driving role of influencing factors on carbon emissions in different industrialization stages across 34 countries, offering insights for China's current low-carbon development phase. Formulating a systematic emission reduction plan requires a thorough understanding of influencing factors and their contextual characteristics. Chang et al. [12] utilized the GM(1.1) grey prediction model and geographical detector to study agricultural carbon emissions in Henan Province, proposing a feasible carbon neutrality plan.

While much research on carbon emissions focuses on global, national, or metropolitan levels, there is a notable gap in localized studies. Given its rapid economic development, Zhejiang Province serves as a crucial area for studying carbon emissions. Filling this gap can enhance our understanding of China's carbon emissions and aid in formulating and implementing effective carbon reduction policies. The characteristics and drivers of carbon emissions vary across regions. As one of China's leading provinces in economy and technology, Zhejiang plays a vital role in economic and technological advancements. However, its high dependence on coal consumption and unique economic and industrial structure pose challenges to achieving the "peak in 2029 and striving for the peak in 2027" target set by Zhejiang Province. Conducting in-depth research on carbon emissions in Zhejiang Province can provide valuable insights into low-carbon development paths, policy formulation, and sustainable regional growth.

Drawing on existing research, this study employs the IPCC method to calculate and analyze carbon emissions across various sectors in Zhejiang Province from 2003 to 2020.

It utilizes the LMDI model to investigate the influencing factors of carbon emissions in the province. Furthermore, ARIMA (AutoRegressive Integrated Moving Average) and grey prediction models are employed to forecast carbon emissions in the coming years, offering valuable recommendations for Zhejiang Province's low-carbon transformation and serving as a reference for similar provinces' development.

2 Method and Materials

2.1 Calculation of Carbon Emissions

In this study, CO_2 emissions resulting from fossil fuel combustion in Zhejiang Province are computed by using the ultimate energy consumption data of various departments by category (from the China Energy Statistical Yearbook over the years) and the methods in the Revised Guidelines for the Compilation of Greenhouse Gas Inventories in Zhejiang Province. CO_2 emissions from fossil fuel combustion for energy activities can be calculated employing a meticulous technology-based sectoral methodology (i.e., the IPCC approach). This approach is grounded in different industry sectors, different types of fuel consumption multiplied by the corresponding emission coefficient, and cumulative to obtain carbon dioxide emissions. The calculation formula is derived from the provincial greenhouse gas inventory compilation guide:

$$CO_2\text{emissions} = \sum\sum\sum (EF_{i,j} \times Activity_{i,j}) \qquad (1)$$

where EF is the emission factor (t-c/TJ); Activity is the fuel consumption (TJ) expressed by calorific value, which needs to be obtained by multiplying the physical quantity data by the average low calorific value, i represent sectoral activities (Primary industry: agriculture, forestry, animal husbandry and fisheries. Secondary industry: industry, construction. Tertiary industry: transport, storage and post; wholesale and retail trades, hotels and catering services; other. Residential: urban, rural); j stands for energy type (raw coal, cleaned coal, other washed coal, coke, gasoline, kerosene, diesel oil, fuel oil, liquefied petroleum gas, natural gas, electricity). The average low calorific value, carbon content per calorific value and rate of carbon oxidation of various energy sources in different industrial sectors are all derived from the revised guide for the Preparation of Greenhouse Gas Inventories in Zhejiang Province.

2.2 Impact Factors of the Carbon Emission

The Kaya equation is a model developed by Japanese energy economist Yoichi Kaya to measure the degree of human impact on carbon dioxide emissions [13]. The specific expression is:

$$C = \frac{C}{E} \cdot \frac{E}{G} \cdot \frac{G}{P} \cdot P \qquad (2)$$

where C stands for CO_2 emissions, E for energy consumption, G for gross domestic product, and P for permanent population.

Given that the consumption of fossil fuels such as raw coal, diesel oil, gasoline, and liquefied petroleum gas significantly impacts carbon emissions, and considering Zhejiang Province's predominant reliance on the manufacturing industry, which heavily utilizes coal, we have opted to investigate energy structure factors. Additionally, Zhejiang Province benefits from its proximity to the sea and unique geographical advantages, fostering thriving processing and manufacturing, transportation, and tourism industries. Hence, we have chosen to examine industrial structure factors. Between 2003 and 2020, Zhejiang Province experienced rapid economic growth, with a GDP increasing by 588.55% and a permanent resident population growing by 33.17%. Consequently, we have included factors related to economic output and population size in our research. Building upon this foundation, we introduce the expanded Kaya identity to comprehensively analyze the factors influencing carbon emissions changes in Zhejiang Province, as expressed below:

$$C_k = \sum_i \sum_j \frac{C_{ijk}}{E_{ijk}} \cdot \frac{E_{ijk}}{E_{ik}} \cdot \frac{E_{ik}}{G_{ik}} \cdot \frac{G_{ik}}{G_k} \cdot \frac{G_k}{P_k} \cdot P_k \tag{3}$$

where i represents industry (industry), j represents energy type, and k represents year. For $\alpha_{ijk} = \frac{C_{ijk}}{E_{ijk}}$, $\beta_{ik} = \frac{E_{ijk}}{E_{ik}}$, $\gamma_{ik} = \frac{E_{ik}}{G_{ik}}$, $\delta_{ik} = \frac{G_{ik}}{G_k}$, $\varepsilon_k = \frac{G_k}{P_k}$, the representative meanings and the units of each variable are indicated in Table 1.

Table 1. Represents the meaning and units of each variable

Symbol	Meaning	Units
C_{ijk}	Carbon emissions from end-consumption of type j energy in type i industry in year k	104t
E_{ijk}	End consumption of fossil energy in type i industry and type j in year k	104tce
E_{ik}	Energy consumption of type i industry in year k	104tce
G_{ik}	Gross production of industry type i in year k	100 million yuan
G_k	Gross production in year k	100 million yuan
P_k	The total population in year k	Ten thousand people
α_{ijk}	Carbon emission coefficient of type j energy of type i industry in year k	
β_{ik}	Energy mix of industry type i in year k	
γ_{ik}	Energy intensity of type i industry in year k	Ton of standard coal/ten thousand yuan
δ_{ik}	Industrial structure in year k	
ε_k	GDP per capita, economic output, in year k	Ten thousand yuan/person

Decompose the Kaya identity into:

$$\Delta C = C^k - C^0 = \Delta C_\alpha + \Delta C_\beta + \Delta C_\gamma + \Delta C_\delta + \Delta C_\varepsilon + \Delta C_P \tag{4}$$

Year 0 is the previous year of year k.
Carbon emission factor effect:

$$\Delta C_\alpha = \sum_i \sum_j L \tag{5}$$

Because the carbon emission factor is constant, $\Delta C_\alpha = 0$.
Energy structure effect:

$$\Delta C_\beta = \sum_{ij} \frac{C_{ijk} - C_{ij0}}{\ln\left(C_{ijk}/C_{ij0}\right)} \ln\left(\frac{\beta_{ik}}{\beta_{i0}}\right) \tag{6}$$

Energy intensity effect:

$$\Delta C_\gamma = \sum_{ij} \frac{C_{ijk} - C_{ij0}}{\ln\left(C_{ijk}/C_{ij0}\right)} \ln\left(\frac{\gamma_{ik}}{\gamma_{i0}}\right) \tag{7}$$

Industrial structure effect:

$$\Delta C_\delta = \sum_{ij} \frac{C_{ijk} - C_{ij0}}{\ln\left(C_{ijk}/C_{ij0}\right)} \ln\left(\frac{\delta_{ik}}{\delta_{i0}}\right) \tag{8}$$

Economic output effect:

$$\Delta C_\varepsilon = \sum_{ij} \frac{C_{ijk} - C_{ij0}}{\ln\left(C_{ijk}/C_{ij0}\right)} \ln\left(\frac{\varepsilon_k}{\varepsilon_0}\right) \tag{9}$$

Population size effect:

$$\Delta C_P = \sum_{ij} \frac{C_{ijk} - C_{ij0}}{\ln\left(C_{ijk}/C_{ij0}\right)} \ln\left(\frac{P_k}{P_0}\right) \tag{10}$$

The contribution rate of each influencing factor to the variation in carbon emissions:

$$R_n = \frac{\Delta C_n}{\Delta C_\alpha + \Delta C_\beta + \Delta C_\gamma + \Delta C_\delta + \Delta C_\varepsilon + \Delta C_P} \times 100\% \tag{11}$$

where $Rn(n = \alpha, \beta, \gamma, \delta, \varepsilon)$ is the contribution rate of an influential factor, where the final consumption of energy "0" is replaced by a minimum value of 1×10–50, which is easy to calculate.

2.3　Carbon Emission Prediction

2.3.1　ARIMA Model

ARIMA (p, d, q), proposed by American statisticians Box and Jenkins, is called the differential autoregressive moving average model [14], which is a fitting and forecasting method for studying non-stationary time series [15]. ARIMA model includes $MA(q)$ moving average and $AR(p)$ autoregressive processes. The $ARMA(p, q)$ model, which is composed of the $AR(p)$ model and $MA(q)$ model, can only study stationary time series. To deal with non-stationary time series, differential transformation or cointegration relationships can be adopted to make data stable. The corresponding $ARIMA(p, d, q)$ model is formed.

Assumed original sequence $Xt = \{x_1^0, x_2^0, ..., x_m^0\}$, Prediction sequence $X_{t'} = \{x_1^1, x_2^1, ..., x_m^1\}$, the expression based on AR model is as follows:

$$x_{t'} = \varepsilon + \phi_1 x_{t-1} + \phi_2 x_{t-2} + ... + \phi_p x_{t-p} + \mu_t \tag{12}$$

The expression based on MA model is as follows:

$$x_{t'} = \mu + \mu_t + \delta_1 \mu_{t-1} + \delta_2 \mu_{t-2} + ... + \delta_q \mu_{t-q} \tag{13}$$

The following ARIMA model can be created:

$$\begin{aligned} x_t' &= \varepsilon + \phi_1 x_{t-1} + \phi_2 x_{t-2} + ... + \phi_p x_{t-p} + \mu_t \\ &+ \delta_1 \mu_{t-1} + \delta_2 \mu_{t-2} + ... + \delta_q \mu_{t-q} \end{aligned} \tag{14}$$

Finally, the final predicted value is computed using the following formula:

$$X_{t'} = (1 - B)^d x_t \tag{15}$$

$$B = \begin{bmatrix} -\frac{x_1^1 + x_2^1}{2} & 1 \\ \vdots & \vdots \\ -\frac{x_{m-1}^1 + x_m^1}{2} & 1 \end{bmatrix} \tag{16}$$

where d represents the constant, ϕ represents the white noise sequence in the regression model, ε is the residual sequence, p is the order of the AR model, q is the order of the MA model.

2.3.2　Grey Prediction Model

The white system refers to the information inside the system that is known, completely sufficient, that is, can be directly observed. Black system indicates that the information inside the system is not directly available, only by establishing contact with the outside world can be known. Grey system refers to that part of the information inside the system is known, while the rest is unknown, and various internal factors have uncertain relationships [16]. At present, GM(1.1) (grey prediction model) is widely used to forecast the carbon emission value. In this study, the GM(1.1) model is also used to forecast the CO_2

emission trend of Zhejiang Province in the next few years, and the results of ARIMA prediction model are compared. The GM(1.1) model operates as follows:

Assumed original sequence:

$$X^{(0)} = \left\{ x^{(0)}(1), x^{(0)}(2), ..., x^{(0)}(n) \right\} \tag{17}$$

Through one accumulation, one accumulation sequence is obtained:

$$X^{(1)} = \left\{ x^{(1)}(1), x^{(1)}(2), ..., x^{(1)}(n) \right\} \tag{18}$$

The following differential equation is established, and the parameters can be solved:

$$\frac{dx^{(1)}(t)}{dt} + ax^{(1)}(t) = \mu \tag{19}$$

The expression based on GM(1.1) grey prediction model is as follows:

$$X^{(1)}(t) = \left(X^{(0)}(1) - \frac{\mu}{a} \right) e^{-a(t-1)} + \frac{\mu}{a} \tag{20}$$

where α is the developmental grey number and μ is the endogenous control grey number.

In this study, two models are employed to forecast the carbon emission of Zhejiang Province in the next few years, and the accuracy of the two models is compared.

3   Results and Discussion

3.1   Carbon Emission Characteristics in Zhejiang Province

From a temporal perspective, as depicted in Fig. 1, the total carbon emissions of Zhejiang Province surged markedly from 2003 to 2020. Specifically, carbon emissions stood at 141.41 million tons of carbon dioxide equivalent (MT CO_2e) in 2003, escalating to 384.44 MT CO_2e by 2020, representing a staggering growth rate of 172%. This trajectory underscores an unfavorable trend in Zhejiang Province's carbon emissions, necessitating further enhancement in carbon emission management practices.

Driven by population growth, improved living standards, and the continual expansion of residential infrastructure, the number of privately owned vehicles has risen steadily, contributing to the annual rise in carbon emissions across Zhejiang Province. Notably, the construction industry, transportation, storage and postal sectors, wholesale and retail trade, along with accommodation and catering services, have witnessed a progressively larger share of carbon emissions attributable to urban and rural residents over the years.

Conversely, the proportion of carbon emissions originating from agriculture, forestry, animal husbandry, fisheries, and industrial activities has exhibited a declining trend annually. This phenomenon has been instrumental in driving down the overall energy intensity of the province, indicative of proactive efforts by both government entities and enterprises to foster the establishment of a low-carbon ecological agricultural industry and a green recycling industrial system.

From an industrial standpoint, CO_2 emissions stemming from the primary sector constitute a modest 2% to 5% of the total carbon emissions. In contrast, the secondary

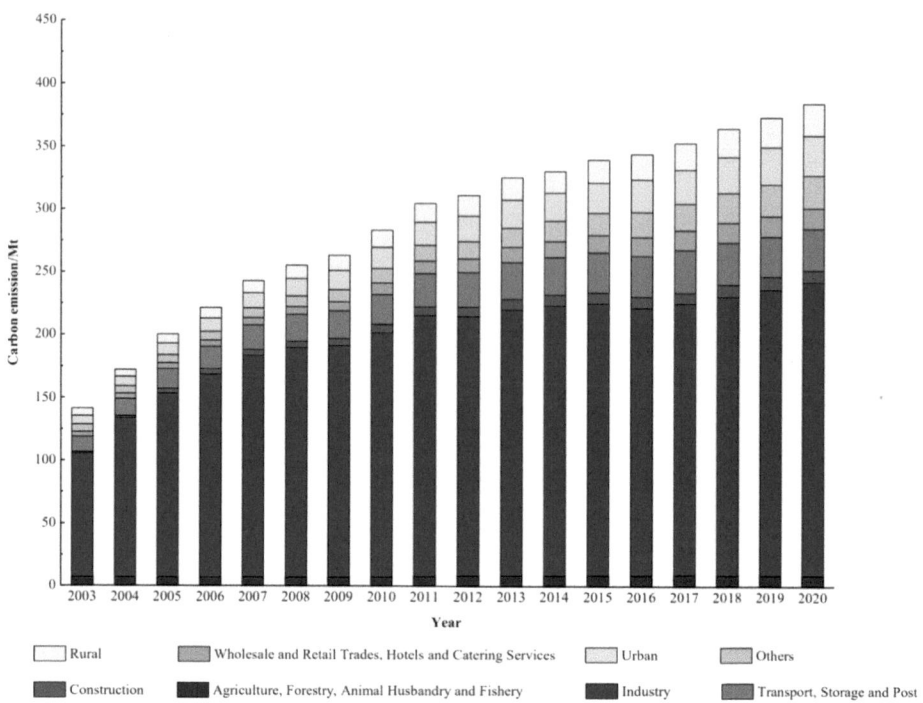

Fig. 1. Carbon emissions by sector in Zhejiang Province from 2003 to 2020

sector claims the lion's share of carbon emissions, ranging from 63% to 75% of the aggregate total, with industrial emissions constituting a significant portion, ranging from 60% to 74% of the overall figure. The tertiary sector, on the other hand, contributes 13% to 20% of the overall carbon emissions, underscoring the lower energy consumption associated with service-oriented economies compared to manufacturing-centric ones. Household consumption, comprising 7% to 14% of total carbon emissions, consistently registers higher carbon emissions among urban residents than rural counterparts.

In regions experiencing extreme climates, such as scorching summers or frigid winters, areas with higher living standards typically exhibit greater energy consumption for heating or cooling residential·and workplace environments. Conversely, regions with lower living standards tend to consume relatively less energy for these purposes. Furthermore, the presence of a substantial urban population base correlates with heightened energy consumption.

Turning to the energy consumption structure, as illustrated in Fig. 2 and Fig. 3, the utilization of raw coal and diesel oil has witnessed a declining trajectory. The proportion of raw coal usage plummeted from 38.84% in 2004 to 10.91% in 2020, while diesel consumption decreased from 17.08% in 2003 to 8.72% in 2020. Concurrently, the utilization of cleaner energy sources such as natural gas and electricity has exhibited an upward trend. Natural gas consumption surged from 0.07% in 2004 to 11.10% in 2020, while electricity consumption climbed from 30.13% in 2003 to 48.97% in 2020, with other energy consumption remaining relatively stable.

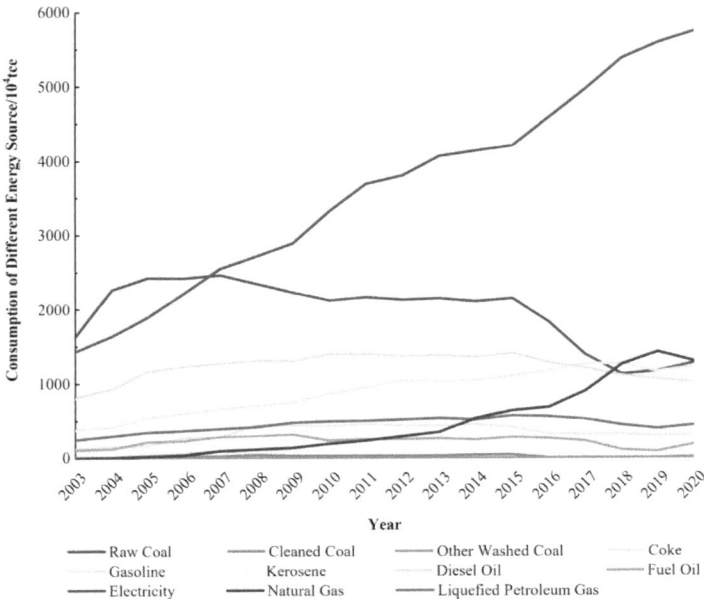

Fig. 2. Energy consumption by type in Zhejiang Province, 2003–2020

3.2 Influencing Factors of Carbon Emission Change

The effects of the five carbon emission factors are categorized into two groups: promoting effects and inhibiting effects. Energy intensity and industrial structure exert inhibiting effects, while economic output and population size demonstrate promoting effects. Energy structure exhibits both inhibiting and promoting effects.

As illustrated in Fig. 4, energy intensity factors: With the exception of 2005, the contribution value of energy intensity factors to carbon emissions in Zhejiang Province from 2004 to 2020 is negative. The average effect contribution value to carbon emissions from energy consumption in Zhejiang Province is $-1803.08 \times 104t$, indicating that this factor can suppress the expansion of carbon emissions in Zhejiang Province, consistent with the findings of Liu et al. [17].

Industrial structure factors: Except for 2009, the contribution value of industrial structure factors to carbon emissions in Zhejiang Province from 2004 to 2020 is negative. The average contribution value of industrial structure factors to carbon emissions from energy consumption in Zhejiang Province is $-1210.32 \times 104t$. Compared to energy intensity factors, industrial structure factors exhibit a slightly weaker inhibitory effect on carbon emissions growth in Zhejiang Province [18].

Economic output factors: From 2004 to 2020, the contribution value of economic output factors to carbon emissions in Zhejiang Province has consistently been positive, displaying fluctuations. The contribution value peaked in 2007, declined, rose to the second peak value in 2010, the third peak value in 2013, and the fourth peak value in 2018. The average effect contribution value of economic output factors to carbon emissions from energy consumption in Zhejiang Province is $2558.23 \times 104t$, significantly

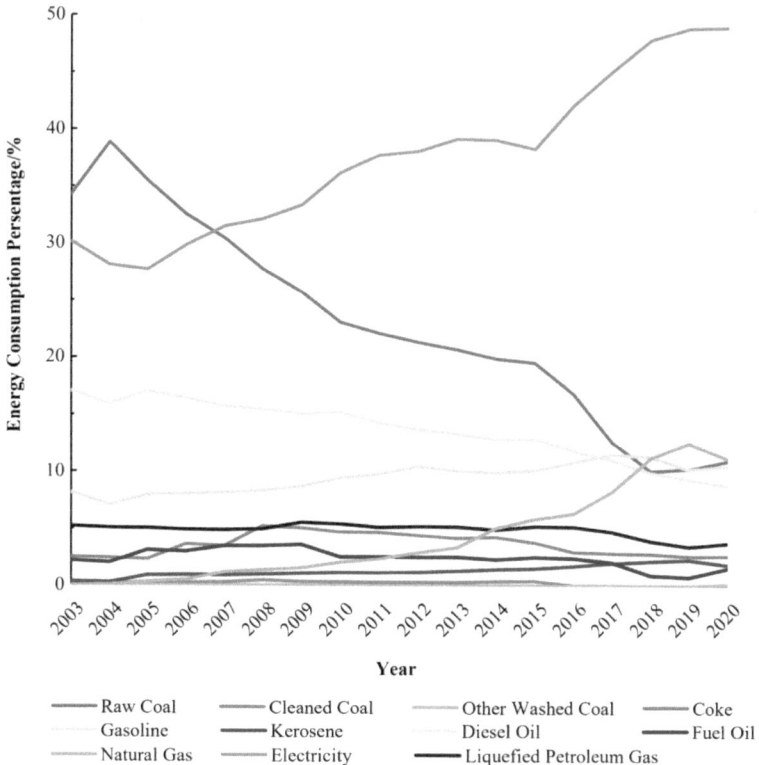

Fig. 3. Proportion of all kinds of energy consumption in Zhejiang Province from 2003 to 2020

facilitating the rise of carbon emissions in Zhejiang Province, corroborating the findings of Liu et al. [19].

Population size factors: The contribution value of population size factors to carbon emissions in Zhejiang Province from 2004 to 2020 has consistently been positive, albeit with a stable trend. The average contribution value to carbon emissions from energy consumption in Zhejiang Province is 493.99 × 104t. Compared to economic output factors, the promoting effect of population size factors on carbon emissions cannot be overlooked [20]. The trend showed an overall upward trajectory from 2004 to 2010, followed by a slow downward trend from 2010 to 2012, stabilizing thereafter.

Population size factors: The contribution value of population size factors to carbon emissions in Zhejiang Province from 2004 to 2020 has consistently been positive, albeit with a stable trend. The average contribution value to carbon emissions from energy consumption in Zhejiang Province is 493.99 × 104t. Compared to economic output factors, the promoting effect of population size factors on carbon emissions cannot be overlooked [21]. The trend showed an overall upward trajectory from 2004 to 2010, followed by a gradual decline from 2010 to 2012, stabilizing thereafter (Fig. 5).

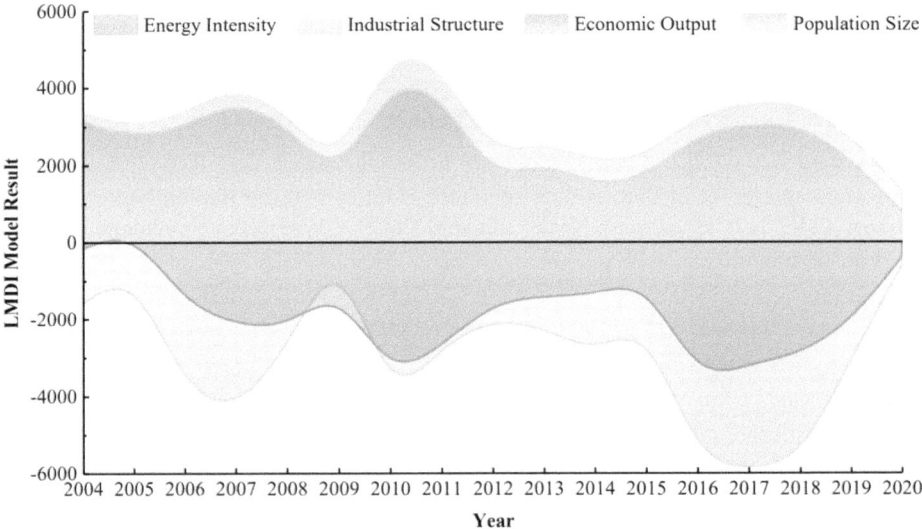

Fig. 4. Energy intensity, industrial structure, economic output, population size, 2003–2020 Contribution value affecting carbon emission change in Zhejiang Province

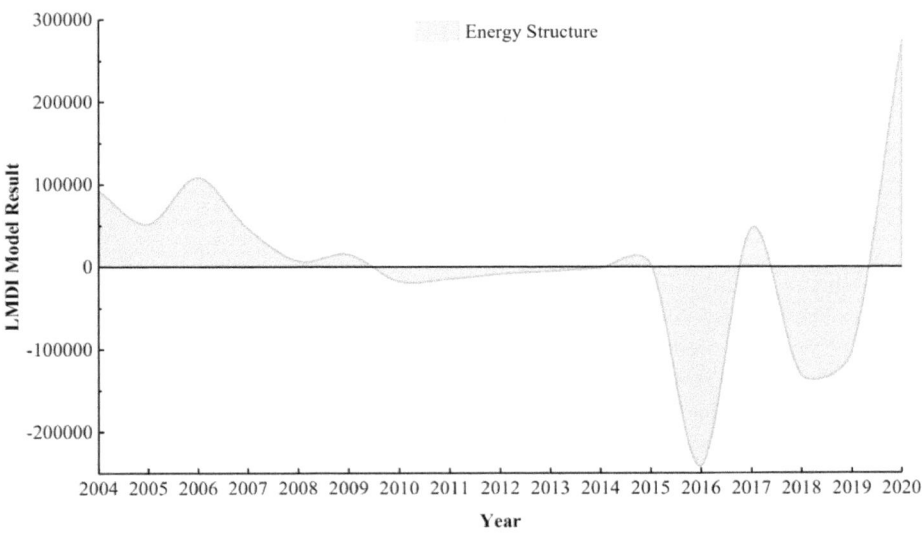

Fig. 5. Contribution value of energy structure to carbon emission change in Zhejiang Province from 2003 to 2020

Considering the contribution rate of each influencing factor to the change in carbon emissions in Zhejiang Province, energy structure emerges as the most influential factor, accounting for an average of 98.69%. Energy intensity follows as the second-most influential factor, with an average ratio of 3.35%. Economic output ranks third, with an average proportion of −2.04%. The impact of industrial structure and the population size is relatively small, with the average proportion of industrial structure being 1.33%, while the average proportion of population size is -1.33%.

3.3 Future Carbon Emission Trend in Zhejiang Province

As observed from Table 2, all p-values are more significant than 0.05, indicating no sequential correlation in the residual sequence used to test the model. Moreover, the value of T*R2 is 0.81, surpassing the threshold of 0.8, signifying an excellent fitting effect of the model.

Table 2. LM test of residual item of ARIMA prediction model

The F Statistic	0.2100	P value	0.8880
The T*R2 statistic	0.8110	P value	0.8470

Examining Table 3 reveals that the posterior difference ratio c falls within the range of $0.05 \leq 0.35$, indicating an excellent accuracy level of the model. Furthermore, the p-value of small error probability is 1.00, exceeding 0.95, which denotes a perfect accuracy level of the model.

Table 3. Grey prediction model construction results

Coefficient of development a	The grey action b	The posterior difference ratio c	Small probability of error p value
-0.0397	203.6953	0.0523	1

As illustrated in Fig. 6, the ARIMA model predicts a peak in carbon emissions in Zhejiang Province by 2025, reaching 404.43 million tons of CO_2. However, the upper limit of 95% of the predicted values exhibits a linear upward trend, with a growth rate of 19.00% from 2021 to 2027, suggesting a potential challenge in achieving carbon peak before 2027. The results from the grey prediction model project a continuous increase in carbon emissions for Zhejiang Province in the future. The growth rate of the predicted values from 2021 to 2027 is estimated at 26.89%. These findings underscore the considerable distance Zhejiang Province still needs to cover in its journey towards low-carbon transformation, urging the government, enterprises, and residents to bolster efforts in carbon control and emission reduction.

4 Conclusions and Suggestions

4.1 Conclusions

This paper utilizes annual energy terminal consumption data and social and economic data from Zhejiang Province spanning from 2003 to 2020 to examine the characteristics of carbon emissions across various industries and the province as a whole. Subsequently,

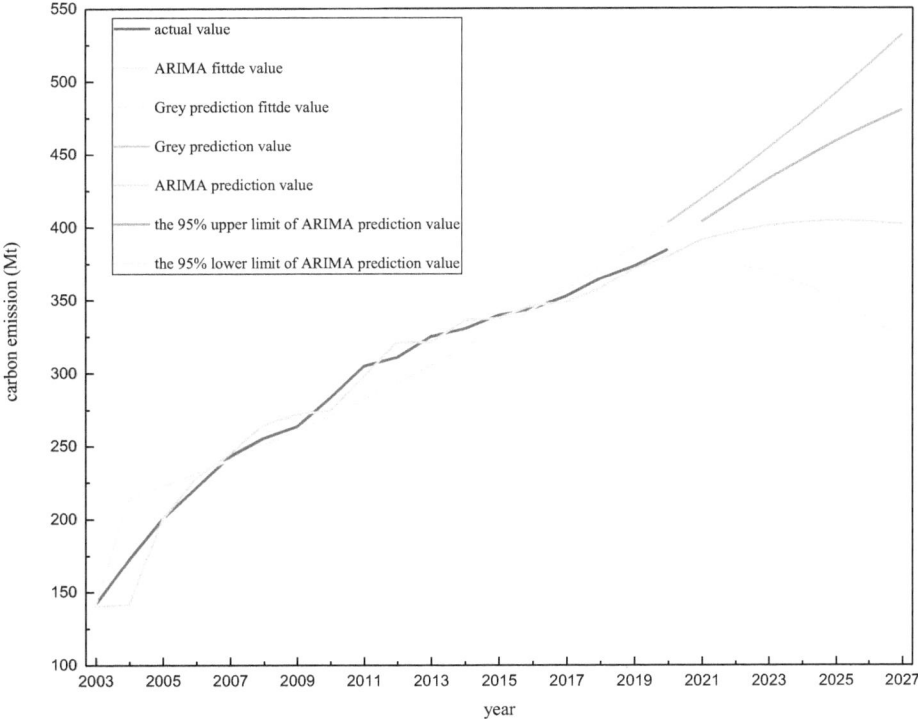

Fig. 6. Grey and ARIMA prediction value of total carbon emissions in Zhejiang Province

the LMDI model is employed to dissect the factors influencing carbon emissions in Zhejiang Province. Based on the findings presented in the figures and tables within the paper, the following key conclusions are drawn:

(1) Over the period from 2003 to 2020, carbon emissions in Zhejiang Province exhibited a consistent linear increase, with a notable growth rate of 172% over the past 18 years. Among the three industries, the secondary sector contributed the highest carbon emissions, averaging 69.22% of the total emissions in Zhejiang Province. Specifically, within the eight sectors analyzed, industry accounted for the most significant carbon emissions, comprising an average of 67.01% of the total emissions. Additionally, urban residents consistently exhibited higher carbon emissions compared to rural residents.

(2) Zhejiang Province is transitioning towards a cleaner energy structure, characterized by a decline in the consumption of raw coal and diesel, coupled with a steady increase in the utilization of natural gas and electricity each year.

(3) Economic output and population size factors have been factors driving the increase in carbon emissions in Zhejiang Province. Conversely, energy intensity and industrial structure factors have acted as inhibiting factors, constraining the increase in carbon emissions.

(4) Based on the current trajectory of carbon emissions, Zhejiang Province is projected to reach its peak in 2025. Without intensified efforts in carbon control and emission reduction, emissions are expected to continue growing, potentially delaying the peak until after 2027.

4.2 Suggestions

Drawing from the analysis of the research findings and the industrial characteristics of Zhejiang Province, the following sector-specific recommendations are proposed:

Agricultural Sector: Establish a low-carbon and efficient ecological agricultural system by supporting green and ecological development, discouraging crop straw burning, promoting comprehensive utilization of agricultural waste, and incentivizing the adoption of intelligent and eco-friendly machinery. Enhance forest conservation efforts and explore carbon sink resources such as forests and wetlands.

Industrial Sector: Transition to a green, low-carbon, and circular industrial system by phasing out backward production methods, investing in energy efficiency and carbon reduction technologies, and promoting the adoption of green and low-carbon equipment. Embrace emerging industries like clean energy and digital economy to drive high-quality industrial development.

Construction Sector: Construct energy-saving buildings and infrastructure, promote prefabricated construction methods, and integrate green and low-carbon principles into construction planning. Emphasize energy efficiency and resilience in building design and operation.

Transportation Sector: Establishing an environmentally friendly and low-carbon transportation system by investing in charging infrastructure, hydrogen refueling stations, and new energy vehicle technologies. Encourage the transition to new energy vehicles and discourage the use of high-emission vehicles.

Community Living: Encourage green and low-carbon lifestyles among residents by promoting public transportation, implementing tiered pricing for utilities, and raising public awareness about the benefits of sustainable living. Foster a culture of conservation and responsible consumption habits.

Energy Sector: Transition to a clean, low-carbon, and efficient energy system by reducing reliance on coal, promoting natural gas as a low-carbon alternative, and accelerating the development of renewable energy sources. Ensure energy security while prioritizing environmental sustainability.

In summary, Zhejiang Province, known for its commitment to environmental conservation, should continue on the path of green, low-carbon, and high-quality economic development. These efforts not only benefit the province itself but also serve as a model for other regions seeking to embark on a similar low-carbon transformation journey.

References

1. Wang, X., Wu, J., Wang, Z., Jia, X., Bai, B.: Analysis of urban CO_2 emission and its characteristics in China. Urban Environ. Res. **1**, 67–80 (2020)
2. Tian, Y.S., Mao, Q.H., Li, C., Qian, J.: Influencing factors of spatiotemporal evolution of ecological land in the Yangtze River Delta from the perspective of regional integration. Acta Ecol. Sin. **43**(13), 5406–5416 (2023)

3. Li, S., Cheng, Z., Wang, W., et al.: Accounting of CO_2 emission for Ningxia based on the energy balance. Environ. Eng. **33**(12), 130–133+137 (2015)
4. Yuli, S., Jianghua, L., Zhu, L., Xinwanghao, X., Shuai, S., Peng, W., Dabo, G.: New provincial CO_2 emission inventories in China based on apparent energy consumption data and updated emission factors. Appl. Energy **184**, 742–750 (2016)
5. Qi, M. Research on the Calculation of Carbon Dioxide Emission and Its Influencing Factors in Hebei Province. Hebei University (2016)
6. Xin, T., Fuli, B., Jinhu, J., Yang, L., Feng, S.: Realizing low-carbon development in a developing and industrializing region: Impacts of industrial structure change on CO_2 emissions in southwest China. J. Environ. Manage. **233**, 728–738 (2019)
7. Xi, C., Chenyang, S., Wu, Y., Yu, Z.: Analysis on the carbon emission peaks of China's industrial, building, transport, and agricultural sectors. Sci. Total Environ. **709**, 135768 (2020)
8. Ozturk, I., Majeed, M.T., Khan, S.: Decoupling and decomposition analysis of environmental impact from economic growth: a comparative analysis of Pakistan, India, and China. Environ. Ecol. Statistics (2021)(prepublish)
9. Wang, Q., Su, M.: Drivers of decoupling economic growth from carbon emission—an empirical analysis of 192 countries using decoupling model and decomposition method. Environ. Impact Assess. Rev. **81**, 106356 (2020)
10. Fu, Q., Gao, M., Wang, Y., Wang, T., Bi, X., Chen, J.: Spatiotemporal patterns and drivers of the carbon budget in the Yangtze River Delta Region, China. Land **11**(8), 1230 (2022)
11. Tong, X.H., Zhou, H.Y., Chen, W., et al.: Study on the measurement of carbon-driven effects from different development stages of industrialization. China Population Resour. Environ. **30**(5), 26–35 (2020)
12. Chang, Q., Cai, W., Gu, X., et al.: Spatial-temporal variation, influencing factors and trend prediction of agricultural carbon emissions in Henan Province. Bull. Soil Water Conserv. **43**(1), 367–377 (2023)
13. Kaya, Y.: Impact of carbon dioxide emission on GNP growth: interpretation of proposed scenarios[R]. In: Presentation to the energy and industry subgroup. Response Strategies WorkingGroup, IPCC (1989)
14. Sowell, F.,: Modeling long-run behavior with the fractional ARIMA model. J. Monetary Econ. **29**, 277–302 (1992)
15. Yu, H.: Prediction of China's carbon emissions based on ARIMA model [J]. China Economist **06**, 59–60 (2018)
16. Huang, X., Wu, J., Lin, W., et al.: Forecast of carbon emission of Jiangsu province based on GM(1, 1) model. Heilongjiang Sci. **13**(18), 26–28+32 (2022)
17. Liu, J., Feng, T., Yang, X.: The energy requirements and carbon dioxide emissions of tourism industry of Western China: a case of Chengdu city. **15**(6), 2887–2894 (2011)
18. Wang, F., Wang, C., Su, Y., Jin, L., Wang, Y., Zhang, X.: Decomposition analysis of carbon emission factors from energy consumption in Guangdong Province from 1990 to 2014. Sustainability **9**(2), 274 (2017)
19. Liu, M., Deng, X., Liu, S., et al.: Carbon emissions analysis of Tianjin City based on LMDI method and Tapio decoupling model. Environ. Pollut. Cont. **44**(10), 1397–1401 (2012)
20. Wang, C.J., Wang, F., Zhang, H.O.: The process of energy-related carbon emissions and influencing mechanism research in Xin jiang. Acta Ecol. Sin. **36**(8), 2151–2163 (2016)
21. Liu, J.: Research on the influencing factors of China's carbon emission and the countermeasures based on the LMDI model. China Journal of Commerce **20**, 146–148 (2022)

Technical Strategies for Low-Cost Adaptive Renovation of Traditional Dong Ethnic Group Residences in Southeast Guizhou

Sisi Xie[1(✉)], Mei Huang[2], Linxin Zhan[3], and Zhengyang Wang[4]

[1] Design Institute No. 8 , China Southwest Architectural Design and Research Institute Corp. Ltd., Chengdu 610041, China
mshomework365@163.com

[2] College of Architecture, Xi'an University of Architecture and Technology, Xi'an 710055, China

[3] School of Architecture and Urban Planning, Beijing University of Civil Engineering and Architecture, Beijing 100032, China

[4] College of Architecture and Urban Planning, Tongji University, Shanghai 200092, China

Abstract. The Dong ethnic residences in Southeast Guizhou constitute a significant part of China's traditional architectural cultural heritage. As times have progressed, these predominantly wooden structures are facing numerous issues related to the quality of living environments and housing safety performance. The flammability of the original fir wood used in construction, combined with the introduction of modern electrical appliances and non-standardized electrical usage, poses a significant fire hazard. Additionally, inadequate natural lighting, ventilation, sound insulation, and functional layout do not meet the demands of modern life. The emergence of disordered modifications and constructions that disrupt the traditional character of the villages represents a considerable challenge to the preservation of these traditional settlements and rural revitalization. This paper is based on the premise of limited funding for support and aims to address the issues of fire prevention, structural safety, lighting, ventilation, sound insulation, moisture proofing, and functional layout. It considers the common needs of villagers, local resources, and current market conditions to research and develop targeted technical measures for enhancement and renovation. It also summarizes and reflects on the designs implemented in selected demonstrative households to improve the living environment quality for villagers, preserve regional traditional culture, and contribute to the revitalization of rural culture.

Keywords: Wooden structure residences · Enhancement and renovation · Technical measures · Southeast Guizhou Dong villages

1 Introduction

The timber-framed traditional houses in Dong villages in southeastern Guizhou are the product of adapting to local climate and resource conditions for thousands of years. They represent the culmination of folk construction wisdom. However, these predominantly

B.-J. He et al. (Eds.): UCSUD 2023, LNCE 559, pp. 199–222, 2025.
https://doi.org/10.1007/978-981-97-8401-1_14

wood-based traditional houses possess inherent deficiencies in terms of housing safety performance and living environment quality, which impede their adaptation to modern living [1]. Consequently, the spontaneous and unregulated modifications, lacking design guidance, pose significant challenges to the preservation of traditional villages.

The preservation and inheritance of folk construction wisdom rely not only on the spontaneous efforts of villagers but also on their adaptation to their economic conditions while ensuring their integration into modern life. This study conducts in-depth investigations and research on the architecture of timber-framed houses in Dong villages in southeastern Guizhou, as well as the villagers' living habits and economic conditions. It addresses the existing issues related to fire prevention, structural safety, lighting, facility layout, ventilation, and moisture prevention. This endeavor aims to enhance the quality of the living environment while preserving local traditions, thereby contributing to the protection of traditional villages [2].

In 2014, the Ministry of Housing and Urban-Rural Development entrusted the Guizhou Provincial Department of Housing and Urban-Rural Development and Xi'an University of Architecture and Technology with a special research and demonstration project for the preservation and renovation of traditional houses, focusing on the Dong villages in southeastern Guizhou. The project received support and collaboration from the Wu Zhi Qiao (Bridge to China) Charitable Foundation in Hong Kong, Tamkang University in Taiwan, Chongqing University, Guizhou University, and local governments.

1.1 Context

Dong villages are mainly located in the hilly and valley basin areas of southeastern Guizhou, in the southern part of China at low to mid-latitudes. The region falls under a subtropical humid monsoon climate characterized by high air humidity, tall mountains, deep valleys, and abundant vegetation. Particularly, there is a significant presence of tall trees, such as Chinese fir, contributing to the unique architectural style of the Dong ethnic group, known as traditional Dong village houses, which are designed to adapt to the local climate and diverse topography. These traditional houses primarily feature wood-framed stilt architecture and its derivatives. They not only demonstrate the Dong people's remarkable ability to adapt to their environment but also showcase their superb construction techniques and wisdom in the past centuries. However, due to the economic conditions of the region and inherent issues related to housing safety performance and living environment quality in predominantly wood-framed traditional houses, there is a phenomenon of villagers spontaneously modifying their houses, which often leads to serious security issues and affects the living quality of interior spaces.

1.2 Research Aims

In response to the existing issues, and considering the economic conditions and usage requirements of the villagers, this study aims to develop renovation techniques that not only ensure the preservation of regional traditions but also integrate inheritance and contemporaneity. The proposed techniques seek to optimize the internal functional spaces, enhance the quality of the living environment, and reflect the regional characteristics.

Principle 1: economically affordable & local adaptable

Traditional villages in Dong regions of southeastern Guizhou are located in remote areas with relatively low levels of economic development. Considering the villagers' economic conditions and specific needs, this study proposes practical and feasible preservation and renovation techniques that are low-cost, high-value, and aligned with the local economic development level. In the renovation process, efforts are made to utilize local building materials such as Chinese fir, cobblestone, green tiles, and tree bark, while ensuring compliance with the local climate.

Principle 2: people-oriented & simply structured

Taking into account the local human resources characteristics and prioritizing the perspective of the villagers, this study emphasizes the adoption of low-tech, user-friendly, and easily operable technical measures that enable the residents to autonomously undertake the renovation process. It is preferable to engage and organize traditional craftsmen and villagers from the local community in the construction demonstration, thereby establishing a foundation for on-site showcases and technical training, which can facilitate the regional dissemination of these techniques.

2 Methodology

To achieve the objective of low-cost appropriate renovation, this study conducted surveys on the villagers' economic conditions and the current status of traditional local dwellings. It systematically analyzed and summarized the challenges encountered during the renovation process. Based on an understanding of the production and living needs of residents, as well as the availability of low-cost materials, research was conducted to develop renovation strategies, which were then applied to the demonstration households.

Various methods were employed to promote the renovation strategies, including a demonstration by the selected households and the distribution of renovation manuals. These measures were taken to ensure effective dissemination of the renovation strategies. The application of the renovation strategies and the effectiveness of the promotional efforts were evaluated through follow-up surveys [3, 4] (see Fig. 1).

3 Analysis of Current Situation and Existing Issues

3.1 Economic Status of the Local Population

The Qiandongnan Prefecture in southeastern Guizhou is home to at least nine ethnic minority groups, with the ethnic minority population accounting for nearly 80% of the total population. Furthermore, due to its mountainous and hilly terrain and restricted transportation conditions, the economic conditions and living patterns of the residents differ significantly from other regions and typical urban areas. Therefore, the economic status of the local villagers will serve as an essential criterion for determining the feasibility of low-cost and appropriate renovation measures.

Based on a sample survey conducted by the Statistical Bureau of Qiandongnan Prefecture, in 2013, the per capita net income of rural residents in the entire prefecture

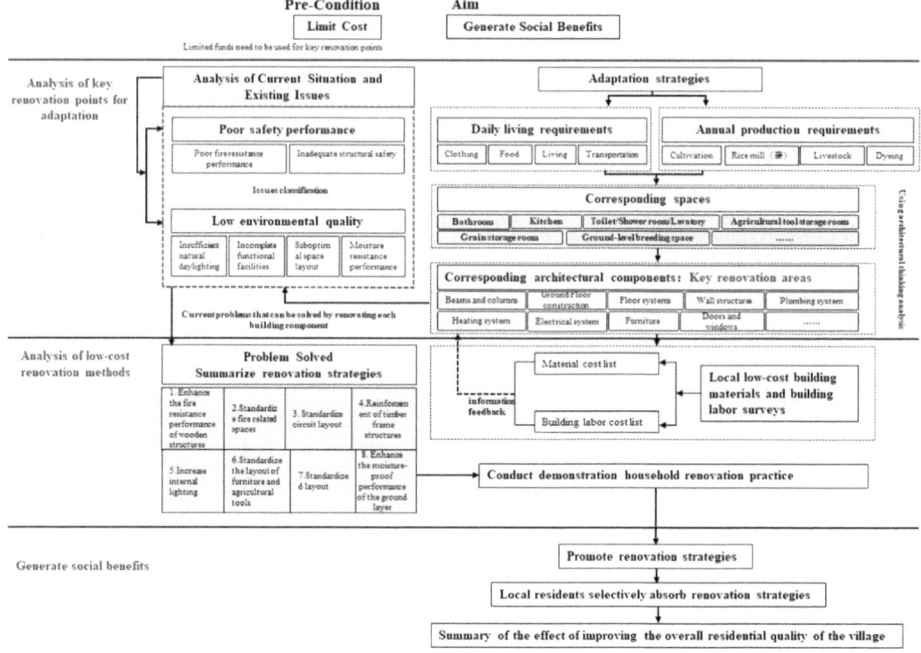

Fig. 1. Research mapping

was 5,345 yuan, while the per capita disposable income of urban residents was 19,640 yuan. The per capita living expenditure of rural residents was 5,013 yuan, whereas urban residents had an annual per capita consumption expenditure of 13,272 yuan. These figures indicate a significant income disparity between urban and rural residents in Qiandongnan Prefecture, highlighting the overall lower economic level of rural areas. After deducting living expenses, rural residents were left with a surplus of only slightly over 300 yuan per year. Such circumstances necessitate the implementation of effective measures to strictly control construction costs during the renovation of rural dwellings. This should be accompanied by the utilization of local construction tools and mutual assistance among village residents, as well as the utilization of available resources, such as privately owned lumber forests and abandoned bricks and tiles, to carry out the renovation works.

Additionally, looking at the proportion of housing expenditure in the consumption patterns of rural and urban residents in Guizhou Province, it was found that rural residents allocated 20.69% of their expenditure to housing, while urban residents allocated 10.92% of their total expenditure to housing during the same period. Considering the aforementioned per capita consumption expenditure in the entire prefecture, both rural and urban residents spent over a thousand yuan on housing. Furthermore, due to the substantial difference in total expenditure between urban and rural residents, the proportion of housing expenditure for rural residents far exceeds that of urban residents. This may be attributed to the fact that traditional building materials in the local area, such as wood (Chinese fir), constitute over 90% of the construction materials. These materials are prone to fire and decay, resulting in increased maintenance costs. Therefore, in the renovation process, it is advisable to use durable materials with lower maintenance costs, especially in areas prone to fire and water-related risks. (see Table 1.)

Table 1. Income and expenditure of residents

Measurement: YUAN per Year	Amount:	Annual surplus	Residential expenditures	Housing expenditure ratio
rural residents' disposable income	5345	332	1037	20.69%
rural residents' average spending	5013			
urban residents' disposable income	19640	6368	1449	10.92%
urban residents' average spending	13272			

Note: Data sourced from the Qiandongnan Prefecture Statistics Bureau of Guizhou Province

3.2   Present Condition of Local Traditional Dwellings

Substandard fire resistance performance

The architectural density in Dong ethnic villages is excessively high, and in the short term, it is challenging to organize villagers to reconstruct houses and create effective fire evacuation routes and fire isolation spaces. This "overcrowded" pattern has led to a trend of "fire spreading from house to house." Furthermore, the majority of residential buildings in southeastern Guizhou are constructed using fir wood, which contains oils. Once a fire occurs, the burning speed is astonishing. For instance, on July 7, 2014, the ZaiKeng Dong village, consisting of 28 contiguous households, turned into ruins within three hours due to a fire outbreak.

Issues originating from the ignition source

In the past, the majority of fire incidents were primarily associated with spaces involving open fires, such as hearths and kitchens, where activities like charcoal burning and cooking took place. Hearth, a multifunctional space used for heating, cooking, and gatherings, has commonly been identified as a prominent source of fires in rural villages. These hearths are often situated on the second floor of wooden structures and are fueled by logs and firewood, making them susceptible to fire outbreaks when exposed flames come into contact with wooden floors and partition walls. Furthermore, fires can occur when combustible materials are haphazardly piled or stored near the hearth, or when flammable items are hung above it. Additionally, fires can be caused by inattentiveness while drinking alcohol or mishandling fire in the kitchen.

However, with the renovation of hearths and kitchens, the primary causes of fires have shifted to electrical issues resulting from aging circuits and improper electrical usage. These issues include substandard wiring quality, exposed rubber wires, and haphazard wiring along wooden structures. Improper use of gas stoves, induction cookers, and electrical meter boxes also contribute to fire incidents. Furthermore, elderly individuals who are illiterate or do not understand the Chinese language often lack electrical knowledge when using modern household appliances, leading to incorrect usage and subsequent fire outbreaks.

The poor structural sustainability of the houses

The Dong ethnic group's timber-framed houses lack foundations, as the entire wooden structure is directly situated on compacted soil [5]. While these mortise and tenon structures possess inherent safety characteristics, the development of modern living has led to the storage of heavier objects indoors. As a result, the load-bearing capacity of the houses is exceeded, leading to severe tilting of the structures after a certain period [6].

Inadequate natural lighting

Excessive building density

Due to the compact layout of traditional Miao and Dong ethnic villages, the spacing between buildings is relatively small, resulting in insufficient natural light penetration into the interiors. As a result, the captured natural light inside the houses becomes particularly valuable.

The inner surface of the cladding structure exhibits low light reflectance

Over time, fir wood undergoes a color transformation from light yellow to dark brown. The inherent light reflectance of the wood itself is not high. Additionally, traditional cooking and heating methods in these houses involve the use of firewood without proper smoke extraction facilities. As a result, the inner surface of the wood cladding is darkened by the smoke. The reduced light reflectance of the darkened cladding structures contributes to dimness within the houses.

Low indoor space utilization and poor environmental quality

Inadequate spatial arrangement

The spatial layout of facilities is found to be impractical and inconvenient. For instance, many of the hearth used for cooking is still located on the second floor, resulting in difficulties in accessing water supply and drainage facilities. The placement of toilets is also deemed inappropriate, often situated outside the main residential building, and in some cases, lacking proper toilet facilities altogether, necessitating the use of makeshift outdoor latrines in distant wooded areas. Additionally, the lack of proper separation between livestock areas and human living spaces on the ground floor leads to inadequate segregation of animal and human spaces, resulting in poor living environment quality.

Inefficient space utilization

In a typical scenario, the ground floor of the residence is largely unutilized, except for livestock rearing, rendering the remaining space vast and primarily used for storing miscellaneous items. The second floor features a wide corridor area, which, if solely designated as a transit space, appears excessively vacant. Overall, the entire residential property lacks a systematic spatial utilization plan.

Decreased moisture mitigation performance

With the influence of modern lifestyles on traditional villages, villagers have started utilizing the elevated space on the ground floor, originally designed as a safeguard against wild animals and flooding. Consequently, the living spaces are directly in contact with the ground, resulting in a significant decline in ventilation and moisture resistance within the houses. This, in turn, leads to the corrosion of heavy modern agricultural machinery stored on the ground floor, food stored in large iron barrels, and modern vehicles such as motorcycles, due to the presence of moisture.

3.3 Issues Necessitating Resolution

The issues identified can be summarized as two major problems: poor residential safety performance (Poor fire resistance performance/ Inadequate structural safety) and low environmental quality (Insufficient natural daylighting/ Incomplete functional facilities/ Suboptimal space layout/ Decreased ventilation and moisture resistance performance), as shown in Fig. 2.

The poor fire resistance performance of local traditional wooden houses no longer meets the needs of residents. The firepit in traditional houses, located in the middle of rooms, easily ignite surrounding materials, leading to fires. Additionally, residents cooking on the second floor (while the first floor serves as a livestock shed) are at high risk of igniting the wooden floorboards (Fig. 2: line 1).

Local traditional cedar wood, due to the inherent characteristics of the material, can cause the main structural elements of buildings to tilt, impacting the use of architectural space and posing potential threats to the occupants' lives(Fig. 2: line 2).

The low environmental quality issues mainly manifest in the following four aspects: 1) local traditional houses suffer from poor natural lighting due to small windows, resulting in dimly lit interiors; 2) outdoor rudimentary dry toilets are inadequate to meet residents' living needs; 3) the interior functional space layout is unreasonable, with main function areas integrated into corridors; 4) the humid climate is unfavorable for grain storage (Fig. 2: line 3–6).

Therefore, when studying the renovation strategy, it is necessary to improve the quality of the residential environment and enhance the safety performance of the houses while maintaining the traditional style, adapting to the income of the villagers, and meeting their living needs.

4 Research on Renovation Strategies

4.1 Low-Cast Control Strategies

In light of the per capita net income of rural residents in the local area, and to control the cost of residential renovation, align with a low-cost renovation strategy, and consider transportation and construction costs, this renovation project will endeavor to utilize readily available local building materials. A survey of the target renovation village and its surroundings has been conducted to identify commonly used and easily obtainable construction materials, along with their corresponding prices, as presented in Table 2 below:

Current issues

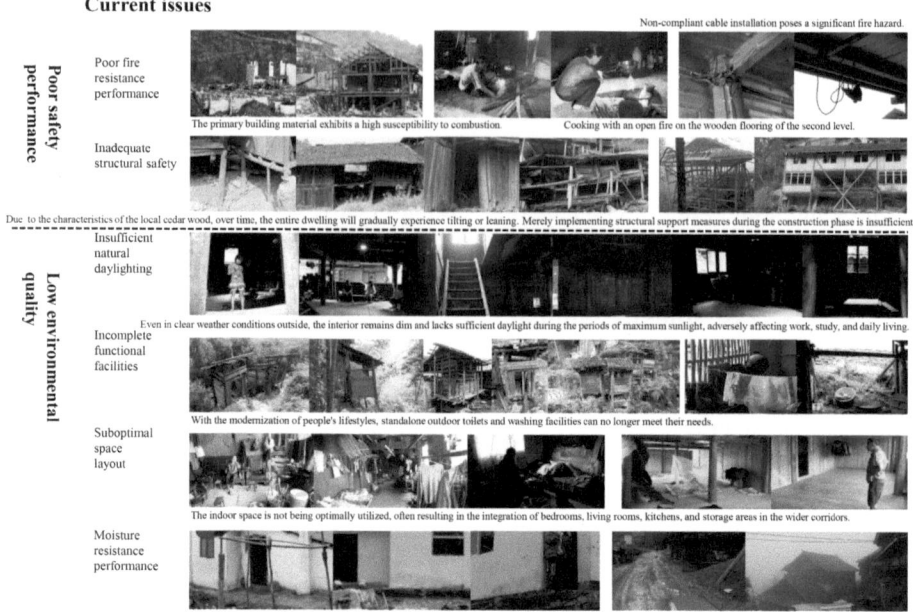

Fig. 2. Current issues

Table 2. Local building materials and prices

Material	Chinese fir	gypsum board	corrugated iron	copper wire
Price (YUAN)	0 (for private ownership by villagers)	4YUAN/m^2	2YUAN/kg	1-2YUAN/m
Material	burnt brick	cement mortar	river pebbles	Waterproof Coating
Price (YUAN)	0.5YUAN/piece	1YUAN/kg	0 (for private ownership by villagers)	6YUAN/Kg

Note: Data obtained from on-site market research

A portion of the materials required for the renovation is owned by the villagers themselves, which allows for significant cost savings in terms of material expenses. Furthermore, common home decoration materials found in urban areas, such as ceramic tiles, lightweight steel frame suspended ceilings, and latex paint, are difficult to obtain in the vicinity of the target renovation village. These materials can be considered relatively luxurious in comparison. Therefore, for this renovation project, the selection of such materials will be avoided.

4.2 Adaptation Strategies

In the field of architecture, suitability often refers to the alignment of a building's design, structure, and functionality with the needs of its users. Therefore, renovation strategies should not only address existing problems but also adapt to the essential living and production requirements of the villagers in order to achieve suitability goals. To ensure that the outcomes of the renovation are suitable and enhance the daily lives of the villagers while addressing the identified issues, we conducted observations of the daily activities of several groups of villagers. Based on these observations, we identified the activities they engage in and the corresponding residential spaces required for each activity. Subsequently, we developed renovation strategies specifically targeting the architectural components associated with these crucial spaces (see Table 3).

Daily living requirements

Annual production requirements

Based on the production behavior characteristics of Dong villagers in Southeastern Guizhou, the required functional spaces for production activities can be categorized into four main types: cultivation, sericulture, livestock rearing, and dyeing. Taking into account the existing problems and noteworthy aspects, corresponding strategies and measures are proposed to address these issues (see Table 4).

Conclusion

Based on the aforementioned investigation of the residents' living and production needs, an analysis was conducted to identify the architectural components corresponding to the required living spaces. Taking into account the utilization of low-cost materials, suitable renovation strategies were formulated based on the characteristics of these architectural components. The specific details are presented in the following Fig. 3.

4.3 Prominent Renovation Areas Necessitate the Implementation of Enhanced Renovation Techniques and Measures

Building upon the previous section's proposed renovation measures, specific renovation measures and practices are proposed based on the types of low-cost materials available and the local construction skills that can be utilized. The specific practices are outlined as follows (see Fig. 4):

Enhancement of housing safety performance

Enhancement of fire resistance performance in wood enclosure structural materials
It is recommended to enhance the fire resistance performance of timber enclosure structures through interior decoration. Specifically, this can be achieved by using fire-resistant board materials such as gypsum board or calcium silicate board to cover the wooden partitions on interior walls and ceilings. The surface of the interior walls can be left untreated according to the occupants' preferences or coated with white paint. By adopting this approach, not only can the fire resistance performance of the wooden walls be effectively improved, but it can also enhance the lighting effect.

Optimization of layout and facility retrofitting in fire-involved spaces

Table 3. Daily living requirements

Utilization requirements			Current issues	Corresponding spaces
Daily living requirements	Clothing	Storage	clothing, headgear, hosiery, and bedding	
			Clothing, bedding, and other items were haphazardly piled on the bedroom floor and bed or left exposed on bamboo poles	Bedroom
		shoes		
			The villagers' footwear is disorderly and piled up due to the absence of a designated area for shoe storage and changing upon entering or leaving	Parlor/Stairwell
		Cleaning	Washing clothes by the riverside is unhygienic	Toilet/Shower room/Lavatory
	Food	Storage	Burlap sacks and barrels containing a small quantity of grain, as well as various seasonings, kitchen utensils, tableware, and firewood, are haphazardly scattered throughout the interior	Kitchen

(continued)

Table 3. (*continued*)

Utilization requirements		Current issues	Corresponding spaces
	Cleaning	Water needs to be fetched for cleaning vegetables, fruits, and utensils	
	Cooking	The refrigerator, rice cooker, induction cooker, and other appliances are placed haphazardly on the floor of the main hall or kitchen without any order	Kitchen
Living	Excretion	There is no indoor sanitation facility, and the toilets are located far away in the fields. Alternatively, if there is an indoor toilet, the absence of a septic tank leads to water contamination in the village's small river (due to the absence of a water treatment plant, pollution poses difficulties in accessing clean drinking water)	

(*continued*)

Table 3. (*continued*)

Utilization requirements		Current issues	Corresponding spaces
	Personal hygiene	The washing and grooming activities are carried out in the kitchen, where personal hygiene items are mixed with seasonings and kitchen utensils. As for bathing, a makeshift shower area is created by surrounding the ground floor with plastic sheets, which is neither aesthetically pleasing nor hygienic	Toilet/ Shower room
	Recreation	The living room area, serving as the main gathering space, is spacious but lacks comfort. The second-floor wide corridor is used for social gatherings, yet when not in use, it occupies an excessive amount of space	Parlor/ Second-floor wide corridor
	Learning	The children resort to using benches as makeshift desks to do their homework on the dimly lit ground floor, as there is also a lack of dedicated writing space in the home	Second-floor wide corridor

(continued)

Table 3. (*continued*)

Utilization requirements		Current issues	Corresponding spaces
	Parking	The motorcycles are parked inside the ground floor, intermingled with storage and living room space	Ground floor
Transportation	Rain gear	The rain gear, such as umbrellas, is casually hung without proper arrangement. When placed in the kitchen, they can become contaminated with grease, resulting in an unhygienic condition	Entrance/Exit doorway
	Mud scraping	Before entering the living quarters, villagers often engage in mud scraping on the doorstep, which compromises indoor hygiene	

Table 4. Annual production requirements

Utilization requirements			Current issues	Corresponding spaces
Yearly production requirements	Cultivation	Storage		
		Compact farming tools	Agricultural tools can be seen everywhere inside the living space, without proper storage areas for orderly arrangement	Agricultural tool storage room \ Hanging racks in various locations
		A substantial quantity of grain	The large containers and barrels used for storing a substantial amount of grain are not easily movable to higher floors, so they are often placed in the main hall or wide corridors, lacking orderliness and cleanliness	Grain storage room
	Rice mill碾	Heavy-duty farming equipment storage	The storage of a series of large agricultural implements used for processing rice is required	Agricultural tool storage room
		Grain processing	The processing of rice, beans, and other crops requires a certain amount of operational and processing space	Processing room

(continued)

Table 4. (*continued*)

Utilization requirements			Current issues	Corresponding spaces
Livestock	nutrition	Swine	Currently, ground-level pig farming lacks concealment and standardization, which affects the cleanliness of the environment and is not adequately clean	Ground-level breeding space
		Cattle	Similar to the above-mentioned	
		Poultry	Similar to the above-mentioned	
Dyeing	Storage	Dyeing vat	Due to their relatively large size and the need to protect them from rain, dye vats require a spacious indoor area for storage	Dye storage room/Cabinet
		Dyed textile	While the required space is relatively minimal, it is still important to avoid haphazard stacking or storage	Storage cabinet
	Air drying		Similar to the aforementioned air drying of clothes	Second-floor wide corridor

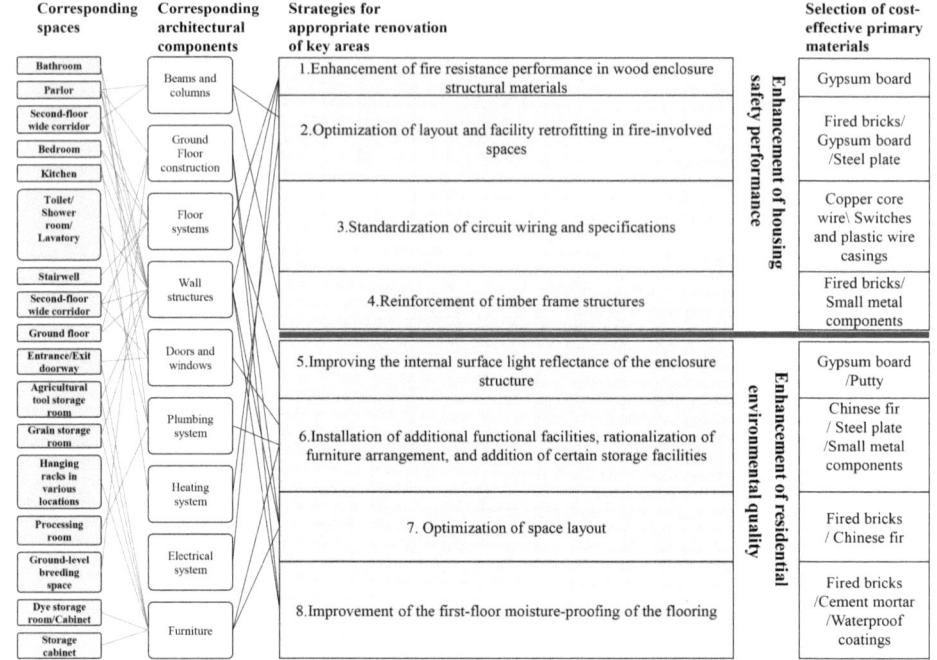

Fig. 3. Strategies for appropriate renovation of key areas

In conjunction with the adjustment and optimization of functional spatial layout in residential buildings, it is recommended to relocate fire-prone areas such as the kitchen and hearth to the relatively fire-resistant ground floor. This can be achieved by enclosing these areas with masonry or other materials with a higher fire resistance rating, creating segregated spaces that are isolated from other areas.

Standardization of circuit wiring and specifications

Firstly, it is important to standardize the wiring methods for electrical circuits. Subsequently, the specification and quality of electrical wires should be improved by utilizing fire-resistant BVVB sheathed cables that comply with fire safety standards (copper core wire with a cross-section ≤4 square millimeters, costing approximately 5 yuan/meter). Furthermore, the installation of air circuit breakers with short-circuit protection and alarm functions is recommended (32A circuit breaker, costing approximately 25 yuan), along with the use of standardized and qualified electrical appliances, sockets, and switches.

Reinforcement of timber frame structures

To address the periodic adjustment of the inclined support system, reinforcement of the safety of tenon and mortise joints, and replacement of decayed or insect-infested wooden components in the framework structure, considering the economic limitations, it is recommended that residents utilize small metal components such as lag screws and angle brackets to reinforce the horizontal connections of the wooden framework. This measure aims to prolong the lifespan of the structure.

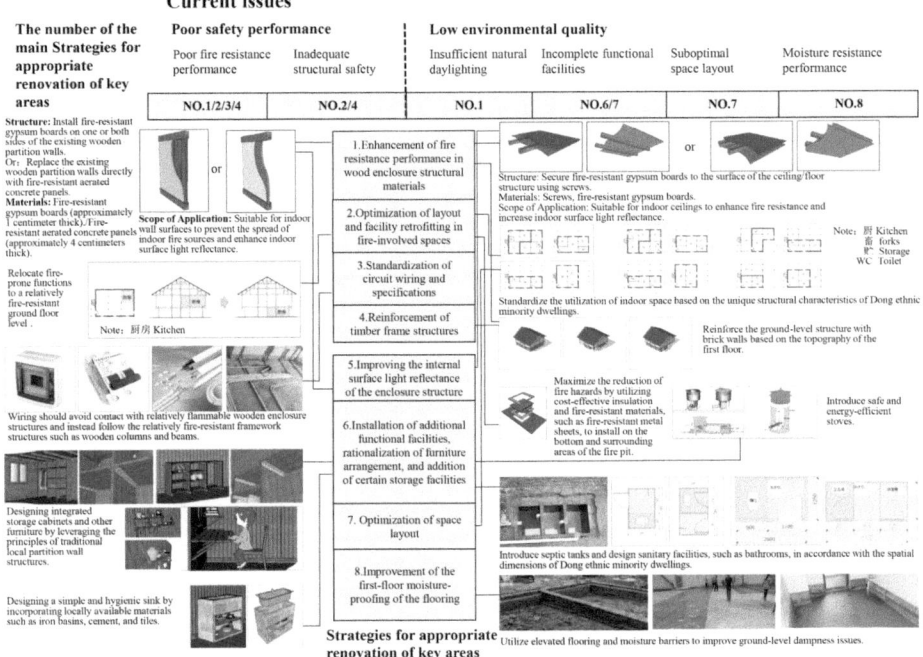

Fig. 4. Renovation technical measures and corresponding solutions to current issues

Furthermore, the reinforcement of the ground floor enclosure structure and the implementation of ceiling and wall construction techniques mentioned earlier contribute to enhancing the structural safety performance of timber-framed residences.

Enhancement of residential environmental quality

Improving the internal surface light reflectance of the enclosure structure

By incorporating improvements in interior fireproof materials, the use of fireproof boards such as plasterboards and calcium silicate boards on wooden enclosure structure walls and ceilings can effectively enhance the light reflectance of the internal surfaces. Applying these fireproof boards as finishes results in white walls, significantly improving the indoor daylighting effect and overall living environment quality within the enclosure structure.

Installation of additional functional facilities

By utilizing low-cost materials, it is advisable to employ simple methods to incorporate essential facilities such as septic tanks (for toilets) and energy-efficient stoves (for kitchens).

Utilizing local material resources, such as fir wood, is highly recommended for creating integrated wardrobes, shoe cabinets, and other household furniture, employing simple methods, such as using readily available boards, to meet functional requirements.

Optimization of space layout

When accommodating livestock on the ground floor, it is imperative to ensure proper segregation between human and animal spaces, achieving a clear separation. Additional facilities for daily living, such as toilets and shower rooms, should be incorporated. When livestock is not kept on the ground floor, the kitchen, toilet (combined with the shower room), storage rooms, and utility rooms should be relocated to the ground floor for the convenience of water supply, waste disposal, production processing, and storage.

Improvement of the first-floor moisture-proofing of the flooring

In the humid climate of Southeastern Guizhou, ventilation and dehumidification are the primary objectives of the raised ground floor in traditional dry-lattice-style dwellings. Therefore, when enclosing the ground floor, it is preferable to allocate spaces such as the kitchen, storage rooms, and toilets to this level, while avoiding the inclusion of bedrooms and living areas. If such functional spaces are included, enhanced ventilation and moisture prevention measures are necessary. The foundation of the walls and the bottom of the indoor floor should be compacted with a mixture of three parts lime and seven parts soil (known as "37 soil") with a minimum thickness of 300mm to ensure effective moisture prevention on the ground floor. The external enclosing walls should be equipped with an adequate number and size of doors, windows, or openings in different directions to ensure cross ventilation. Above these openings, provisions should be made for grain storage rooms, which effectively prevent rodent infestation.

Conclusion

By implementing the aforementioned measures, it is possible to assess the potential impact on the identified current issues faced by two local traditional dwellings during the problem analysis phase. Please refer to Fig. 5 for detailed information.

5   Practical Research

5.1   Current Status of the Target Dwelling

The selected household for renovation is a poverty-stricken household named "Shi," chosen by the village community. Both elderly members of the household have language barriers. Presently, the self-owned dwelling of the selected household has recently been completed, with the main structure being donated collectively by other local villagers. The overall construction is still unfinished (see Fig. 6).

5.2   Results of the Renovation

The renovation took place from late 2014 to 2015. Based on the established renovation objectives outlined in the aforementioned study, the following aspects were addressed: construction of a raised ground floor, installation of a septic tank for the kitchen and bathroom, improvement of doors and windows, enhancement of fireproofing and sound insulation for indoor floors and walls, implementation of fireproof wiring for electrical supply, and production of selected indoor furniture. The specific outcomes of the renovation are detailed in the Fig. 6.below:

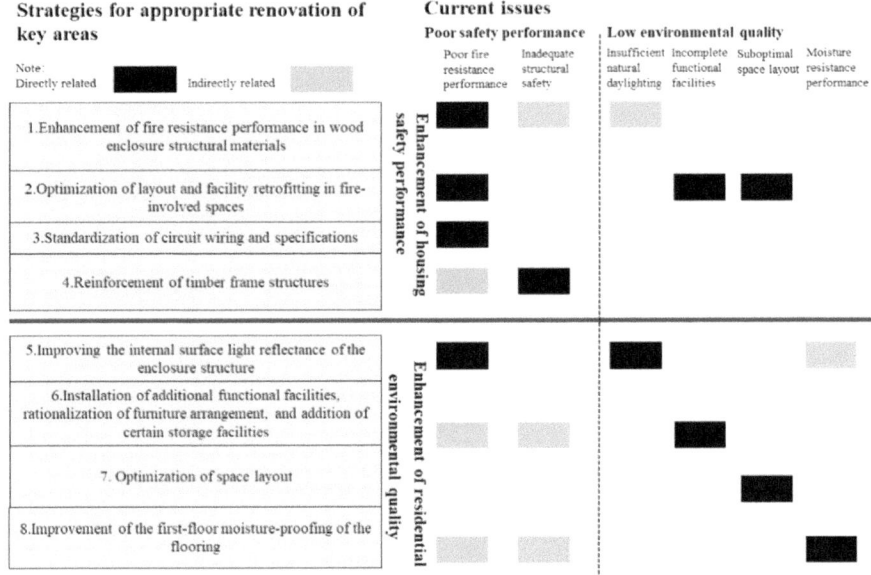

Fig. 5. The degree of relevance between the renovation strategies and the corresponding issues

6 Post-Renovation Evaluation

6.1 Evaluation of Cost Control in the Renovation

The renovation of the demonstration house in southeastern Guizhou, excluding the unfinished portions of the main structure, had a total cost of nearly 40,000 yuan. The total renovated area was approximately 120 square meters. The unit price for renovation was around 300–400 yuan per square meter, which accounted for only 20% of the construction cost of a new three-room Chinese fir wooden house in southeastern Guizhou in the previous year (15–20 thousand yuan in 2014, 20–30 thousand yuan in 2023). It constituted only 30% of the total project budget, indicating effective cost-control measures (see Fig. 7).

6.2 Evaluation of the Post-Renovation Effectiveness of the House

The renovation of the demonstration household was completed by the end of 2014. After the construction was finished, the team organized a detailed explanation campaign among the villagers regarding the renovation measures (see Fig. 8). This was done to facilitate the spontaneous adoption and application of the demonstration measures within the village. In 2023, a follow-up visit was conducted to the renovated household and surrounding villages. The purpose of the visit was to assess the maintenance status of the renovated areas in the demonstration household and to evaluate the effectiveness of the dissemination of the renovation measures. Please refer to Fig. 8 for specific details.

Based on the findings from the follow-up visits, the vast majority of rural households have voluntarily adhered to the team's formulated strategies to standardize the installation

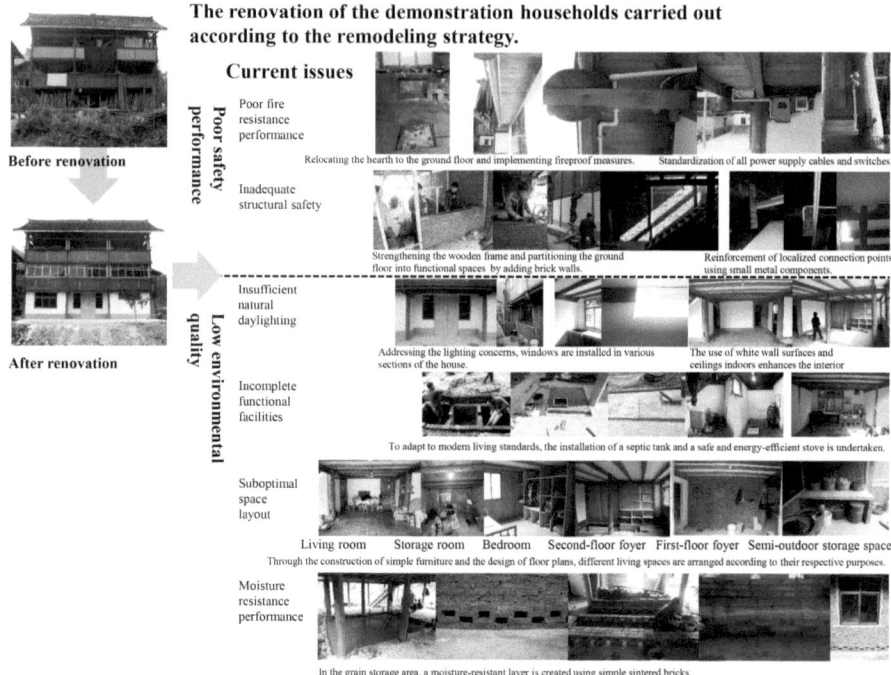

Fig. 6. The renovation of the demonstration households was carried out according to the remodeling strategy.

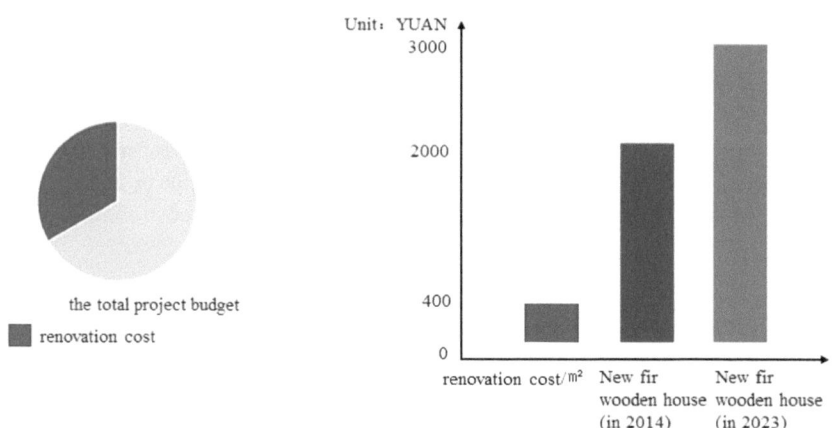

Fig. 7. The renovation cost

of electrical circuits, thereby enhancing safety performance. Notably, some villagers who have the means have used superior electrical materials compared to the demonstration households, indicating an increasing emphasis on fireproofing measures for the electrical components. Furthermore, a significant number of villagers have relocated fire-prone

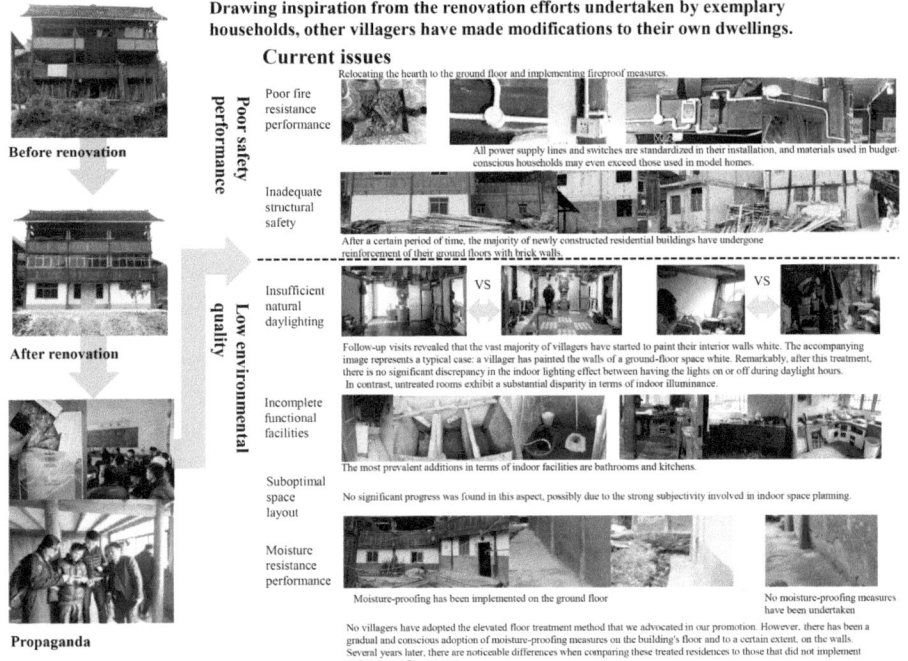

Fig. 8. Drawing inspiration from the renovation efforts undertaken by exemplary households, other villagers have made modifications to their dwellings.

equipment such as hearths and kitchen stoves to the ground floor and implemented fireproofing measures for the relevant facilities on that floor. Additionally, the majority of newly constructed rural residences have adopted brick wall maintenance structures on the ground floor, resulting in improved utilization and fire resistance of the ground floor space.

Regarding the measures implemented to enhance environmental quality, the majority of villagers, by the team's customized strategies, have utilized skim coating or installed gypsum boards to whiten the walls, thereby improving indoor lighting effects. Furthermore, nearly all villagers have installed bathrooms and energy-efficient stoves in their residences, addressing the functional deficiencies of some houses. Lastly, villagers have started to prioritize improvements in moisture resistance on the ground floor of their houses. Although the specific method of utilizing elevated floors, as proposed by the team, has not been employed, villagers have begun using waterproofing cloth to enhance moisture resistance of the ground floor walls and floors.

7   Conclusion

Under the influence of demonstration households and the promotional effect of transformation strategies, villagers commenced selectively absorbing and implementing self-housing renovation through the assimilation of said strategies. The final investigation revealed that, in addition to attaining the anticipated effects, the transformation strategies yielded unforeseen outcomes in the renovated dwellings due to the supplementary attributes of materials and structures. For instance, the gypsum board demonstrated fire-resistant properties and light diffusion capabilities, resulting in the attainment of not only the expected fire protection but also surpassing expectations by enhancing indoor lighting as a consequence of the gypsum board's light-colored surface layer. Moreover, this had an indirect protective effect on the wooden structure. Furthermore, the standardized layout of electrical wiring achieved fire safety objectives and, concurrently, the standardized arrangement of power outlets, switches, and other components regulated the positioning of electrical appliances, thereby indirectly influencing the spatial arrangement of functional areas within the interior. The reinforcement of the entire dwelling structure, including the establishment of masonry partitions, indirectly standardized the utilization of interior spaces. Additionally, proactive measures in spatial functional layout, such as the regulation of kitchen and bathroom positioning, facilitated fire protection, structural integrity, and moisture prevention for timber-framed houses by establishing norms for human activities in these designated spaces. See Fig. 9.

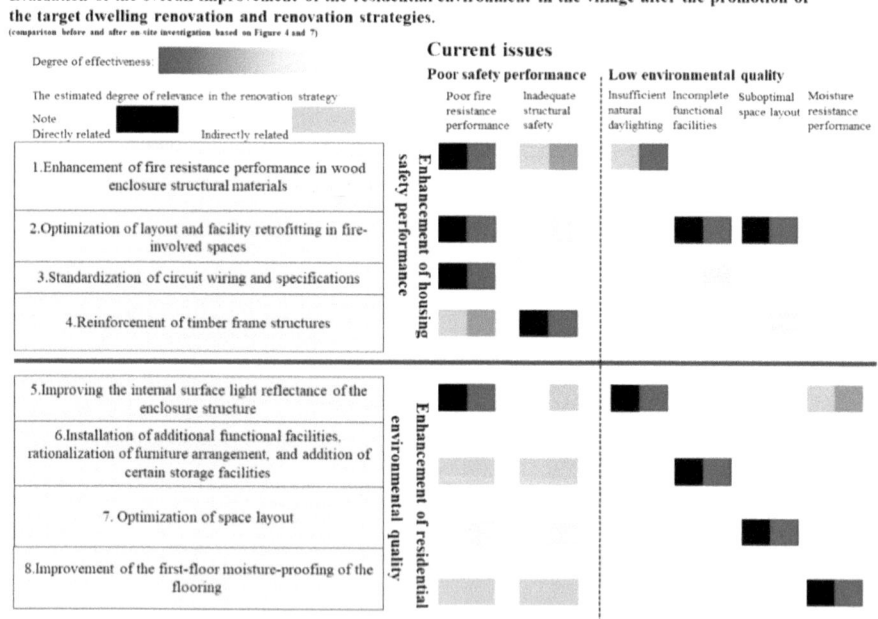

Fig. 9. Evaluation of the overall improvement of the residential environment in the village after the promotion of the target dwelling renovation and renovation strategies.

To address the existing challenges faced by timber-framed dwellings, this study combines local traditional building techniques with modern material processes to explore Low-Cost, adaptive, simple, and environmentally friendly measures for enhancing and renovating timber-framed dwellings. These measures are aligned with the local economic development level and the technical proficiency of residents, making them easily replicable. Furthermore, the research methodology and project operation system demonstrate a feasible and effective approach for upgrading traditional timber-framed dwellings in the Southeastern Guizhou region.

Acknowledgement. This research was supported by the National Key R&D Program of China under the 13th Five-Year Plan, focusing on the project "Key Technologies for the Adaptive Renovation and Utilization of Traditional Rural Architecture in Villages" (Project No. 2020YFC1522303). Note: The authors would like to express their sincere gratitude to Professor Mu Jun, Professor Duan Degang, Professor Zhou Tiegang from Xi'an University of Architecture and Technology, and Professor Bi Guangjian from Tamkang University (Taiwan) for their valuable guidance throughout the project.

References

1. Zhou, T.G., Laing, L.J.: Safety investigation of traditional wooden dwellings in Qiandong nan Dong villages. Constr. Sci. Technol. **5**, 21–23 (2017)
2. Mu, J., Zhou, T.G., Wang, L., et al.: Teaching fishing and local creation: a study on the post earthquake reconstruction of Ma'anqiao village in Liangshan, Sichuan Province. J. Archit. **12**, 10–15 (2013)
3. Wang, P., Zhang, Q., Sun, B.S., Zhou, Z.: Transformation of traditional wooden houses in rural areas of Southwest China. J. Anhuin Agri. Sci. **39**(21), 12979–12980, 13095 (2011)
4. Liang, L.: Investigation on the current situation and study on safety improvement technology of through type timber frame dwellings—Taking Bapa village of Qiandongnan as an example. Xi'an University of Architecture and Technology (2018)
5. Huang, M., Xie, S.S., Zhan, L.X.: Local construction—the analysis of the structural elements of traditional residence's wooden structure of Dong Villages in Southeastern. J. Hum. Settlements West China **34**(03), 110–116 (2019)
6. Cai, L.: Traditional Villages and Buildings in Dong Settlements. China Architecture & Building, Beijing (2007)

Energy Saving of City Road Tunnel Lighting Based on Optimal Luminaire Installation Method

Xianyun Cai[⊠], Zhengyang Jiang, Jiang Wu, and Yuanxu Zhang

College of Architecture and Urban Planning, Chongqing Jiaotong University, Chongqing 400074, China
kevinmayo@cqjtu.edu.cn

Abstract. Energy saving strategy of highway tunnel lighting mainly concentrated on shading facilities at tunnel entrances and exits, lighting intelligent control systems and lighting design methods. The contrast revealing coefficient q_c is widely recognized as a key index used to assess the quality of tunnel lighting. The optimal values of the contrast revealing coefficient are associated with the optimal visual effecicy in tunnel lighting. This study aimed to give a new energy saving strategy based on optimal luminaire installation methods. Light Emitting Diode lamps (LED) were selected in tunnel lighting simulation experiment. The optimal luminaire installation method was determined by considering both the lighting quality evaluation system and the optimal value of the contrast revealing coefficient. Then, the lighting conditions achieved through the optimal luminaire installation method can be assessed based on the detection distance of small targets. Results showed that lighting environment under optimal luminaire installation method can fulfill the criteria for safe driving and visual comfort. Compared with the original luminaire installation method, the optimal luminaire installation method can save 24.32% of electrical energy.

Keywords: Tunnel lighting · Energy saving · Luminaire installation method · Contrast revealing coefficient · Optimal value

1 Introduction

Regional European Electronic Toll Service (REETS) [1] prioritizes the energy consumption of tunnel lighting in the overall energy consumption of highway tunnel operation. Ministry of Transport of the People's Republic of China [2] recommended that researchers should strengthen studies on tunnel lighting, including energy conservation, low consumption, and low carbon. García and Martín [3] pointed out that pergolas can shift lighting threshold zone out of tunnels. Salam and Mezher [4] claimed that shading structures could reduce artificial lighting requirements in threshold zone and exit zone. Thus, shading structures saved up to 50% of electrical load. Wang et al. [5] developed a lighting intelligent control model which can achieve the goal of energy saving. Kimura et al. [6] proposed a high luminance uniformity method with LED. The result showed

© The Author(s) 2025
B.-J. He et al. (Eds.): UCSUD 2023, LNCE 559, pp. 223–233, 2025.
https://doi.org/10.1007/978-981-97-8401-1_15

that this method can save up to 20% of energy consumption. Liang et al. [7] explored the law of energy-saving in tunnel lighting. The result signified that the value of road surface luminance increased at least 10% using reflective and light-storing materials. Hirakawa et al. [8] proposed a light distribution method for highly efficient lighting. The result indicated that it can save approximate 20% of energy. The human visual science provided new horizons for energy saving strategy. For instance, as per the standards set by the International Commission on Illumination (CIE) [9], the contrast revealing coefficient q_c is delineated based on the perceived contrast method in tunnel lighting. This coefficient represents the ratio between the luminance of the road surface and the vertical illuminance of a small target situated at a specific location within the tunnel. Many specifications [9–12] recommended specific contrast revealing coefficient values for the threshold zone. Moreover, CIE [13] suggested that the contrast revealing coefficient can evaluate lighting system and lighting quality. The optimal installation approach for luminaires corresponds with achieving the best visual efficacy in tunnel lighting, as indicated by the optimal values of the contrast revealing coefficient (q_{cop}). The original evaluation index system for highway tunnel lighting quality consisted of average road surface luminance, average luminance of 2-m-high walls on both sides inside the tunnel, road surface luminance uniformity, glare control, and visual inducement. Cai and Weng [14–17] studied the contrast coefficient values systematically, including threshold values, range values and optimal values, and incorporated the contrast revealing coefficient into the evaluation index system for highway tunnel lighting quality. The optimal installation method of luminaires corresponds to achieving the best visual efficacy in tunnel lighting, as indicated by the optimal value of the contrast revealing coefficient (q_{cop}). The contrast revealing coefficient serves as a metric for assessing luminaire installation methods, with the optimal contrast revealing coefficient corresponding to the best luminaire installation approach.

This paper aimed to study energy saving strategy based on optimal luminaire installation method. The research ideas on energy saving strategy were explicit. 1) With the assumption of a symmetric lighting system, the optimal contrast revealing coefficient value under symmetric lighting was 0.2. LED was selected in tunnel lighting simulation experiment. 2) The optimal luminaire installation method was determined based on lighting quality evaluation system and optimal contrast revealing coefficient value. 3) Taking Dongyangguan highway tunnel in Shanxi Province, China as the case study object. The original luminaire installation method was optimized based on optimal value of contrast revealing coefficient. 4) The scientific rigor and practical viability of the optimal luminaire installation method were validated through field measurements within tunnels and the determination of small target detection distances. 5) Energy consumption of tunnel lighting under original luminaire installation method was compared with that under optimal luminaire installation method. Furthermore, the study demonstrates that the optimal luminaire installation method not only meets the standards of tunnel lighting quality evaluation systems but also significantly reduces energy consumption, providing valuable guidance for tunnel lighting design methods and actual lighting engineering projects.

2   Materials and Methods

In this study, the optimal luminaire installation method was determined with exhaustive method. Based on optimal values of contrast revealing coefficient and lighting quality evaluation system, the method for installing optimal luminaire can be determined through tunnel lighting simulation experiment. The contrast revealing coefficient can be obtained by Eq. (1).

$$q_{c0p} = \frac{\rho}{\pi} \cdot \left(1 - \frac{\Delta L_0}{L_b}\right)^{-1} \tag{1}$$

In Eq. (1), ρ is the reflection coefficient, the ratio of ρ to π is regular, ΔL_0 is minimum luminance threshold in cd/m^2, L_b is road surface luminance in cd/m^2. Moreover, the scientificity and feasibility of the optimal luminaire installation method were demonstrated with field measurement and the discovery distance of small target.

2.1   Optimal Luminaire Installation Method

Some researchers clarified the advantages and disadvantages of optical software DIALux and AGI 32 [18]. In this study, tunnel lighting simulation models were set up by DIALux and AGI 32 (Fig. 1). The calculation grids were set to 0.5 m × 0.5 m. The parameters were the same in the two lighting simulation models. To demonstrate the consistent of the models, the simulated values and relative errors were determined respectively (Table 1). The relative errors included road surface luminance, total uniformity of luminance and longitudinal uniformity of lane midline luminance. The relative errors varied from - 2.78% to -1.05%. The average value of relative errors varied from -1.42% to -1.88%. Thus, it is feasible to set up lighting simulation models by DIALux and AGI 32. In this study, small target visibility (STV) and glare threshold increment (TI) were calculated by AGI 32. Meanwhile, the rest of the lighting quality evaluation indices were obtained by DIALux. The advantages of LED included high luminous efficiency, good color rendering, long service life, easily control of light distribution and easily dimming of light spatial distribution. Moreover, compared with high pressure sodium lamps (HPS), LED saved more than 45% of energy consumption [19, 20]. Therefore, LED was selected in the tunnel lighting simulation models. The selected LED was C0820-YC-TL-30W.

Fig. 1. Tunnel lighting simulation model.

Fig. 2. Dongyangguan highway tunnel.

Table 1. Error analysis of simulation results using DIALux and AGI 32 software

Road surface luminance(cd/m^2)			Total uniformity of luminance			Luminance and longitudinal uniformity of lane midline luminance		
DALux	27.0	26.3	26.8	0.72	0.69	0.70	0.95	0.93	0.94
AGI 32	27.5	26.9	27.2	0.74	0.70	0.71	0.96	0.95	0.95
Error	-1.85 %	-2.28 %	-1.49 %	-2.78 %	-1.43 %	-1.43 %	-1.05 %	-2.15 %	-1.06 %
Average error	-1.87%			-1.88%			-1.42%		

According to optimal values of contrast revealing coefficient and lighting quality evaluation system, the optimal luminaire installation methods were determined using the exhaustive method. It is worth noting that the parameters of lighting simulation model included road surface luminance, lighting systems, light distribution curve, and among others [17, 21]. Determination steps for optimal luminaire installation method were clear. 1) Determine the threshold zone luminance through the tunnels environment [22, 23]. Thus, the road surface luminance in different regions can be obtained. 2) The range values of contrast revealing coefficient were created under symmetric lighting system in lighting simulation models. 3) Through visual efficacy experiments, the optimal value of contrast revealing coefficient can be obtained, which corresponds to the best visual efficacy and can be clearly determined as the optimal luminaire installation method. Luminaire installation method included basic and enhanced lighting conditions. The basic lighting conditions can be determined based on the optimal luminaire installation method in interior zone. The basic lighting conditions included lamp installation distance, lamp installation height, and among others. Thus, the enhanced lighting conditions can be obtained based on basic lighting conditions.

2.2 Case Study

Dongyangguan highway tunnel (Fig. 2) in Shanxi province, China was selected as a case study object. The details are clear under the original luminaire installation method. 1) Luminance outside tunnel portal was 5000 cd/m^2. Meanwhile, the ratio of sky area within 58.6 degrees of viewing field was 25% and the direction of the tunnel was north. 2) The designed driving speed was 80 km/h. 3) Unidirectional traffic flow was 392 veh/(h·ln). 4) The subjects have been driving for more than 5 years. They have not fallen victim to ophthalmic pathema. The visual acuity of the subjects is 5.0 or above. The luminaire types and quantity are shown in Table 2.

Table 2. Luminaire quantity under original luminaire installation method

Lighting zones	30 W LED	100 W HPS	150 W HPS	250 W HPS	400 W HPS
Threshold zone 1	11	——	18	——	37
Threshold zone 2	9	18	——	——	18
Transition zone 1	14	16	27	——	——
Transition zone 2	15	18	——	——	——
Interior zone	629	——	——	——	——
Exit zone	14	6	——	7	——

Optimization for original luminaire installation method

The reduction factor of threshold zone luminance k was 0.035 based on the surroundings of Dongyangguan highway tunnel [22, 23]. According to tunnel lighting specifications [9, 11, 12, 23], road surface luminance in other zones can be determined based on threshold zone luminance. The steps of optimization for original luminaire installation were explicit. 1) Tunnel lighting simulations can determine road surface luminance. 2) The parameters of lighting simulation model included light distribution curve, luminaire installation distance, and others. The contrast revealing coefficient under different luminaire installation methods was determined using the exhaustive method. 3) If the contrast revealing coefficient closely approaches the optimal values and lighting quality meets evaluation standards, the luminaire installation method is considered at the optimal condition.

If lighting quality matched the optimal value of contrast revealing coefficient and lighting quality evaluation system, the luminaire installation method was optimal. Moreover, driver's discovery distance of small targets was got with eye tracker and BeGaze software. If the discovery distance was not less than safety parking distance [17, 24], the lighting environment under optimal luminaire installation method can meet the requirements of safe driving and visual comfort.

Analysis of energy consumption under different luminaire installation methods

Energy consumption under different luminaire installation methods can be calculated based on luminaire power, luminaire quantity and lighting control time. If energy consumption under the optimal luminaire installation method was lower than that under the original luminaire installation method, the optimal luminaire installation method based on optimal values of contrast revealing coefficient was a good strategy for energy saving.

3 Results

3.1 Lighting Quality Under Original Luminaire Installation Method

The details about the original luminaire installation method are as follows. 1) Symmetric lighting system was used in Dongyangguan highway tunnel. 2) Luminaire installation height was 5.5 m. LED with 30 W was applied for basic lighting. 3) HPS lamps with different powers (400 W, 250 W, 150 W, 100 W) were used for enhanced lighting. Luminaire was arranged symmetrically on both sides of the wall.

According to arrangement of measuring points [22], lighting quality indices can be obtained under original luminaire installation method. 1) The measured value of road surface luminance was 29.6% higher than designed value of luminance. To ensure the efficacy of luminaire, maintenance value of road surface luminance was set to 70% of the initial luminance. 2) Overall uniformity of road surface luminance was greater than 0.4. The longitudinal uniformity of lane midline luminance was higher than 0.6. 3) The measured values of contrast revealing coefficient were lower than 0.2, which meant that lighting environment under original luminaire installation method cannot meet the optimal value of contrast revealing coefficient. The optimal contrast revealing coefficient values correlated with the highest visual efficacy in tunnel lighting. To summarize, lighting environment under original luminaire installation method was not in the best condition.

3.2 Optimal Luminaire Installation Method

The details about the optimal luminaire installation method are as follows. 1) The selected LED was C0820-YC-TL-30W. 2) Luminaire installation height was 5.0 m. 3) LED with 30 W was applied for basic lighting. Luminaire was staggered on both sides of the wall. 4) LED lamps with different powers (2×150 W, 120 W, 75 W, 60 W) were used for enhanced lighting. Luminaire was arranged symmetrically on both sides of the wall. Table 3 shows the luminaire types and quantity under optimal luminaire installation method.

Lighting quality and energy consumption under optimal luminaire installation method

The values of road surface luminance were higher than designed luminance values. The measured road surface luminance was 48.65% higher than designed luminance. Overall and longitudinal uniformity of road surface and lane midline luminance remain unchanged from original lighting conditions. Thus, the lighting quality under optimal luminaire installation method matched the lighting quality evaluation system. The measured contrast revealing coefficient is infinitely close to 0.2. In other words, lighting environment under optimal luminaire installation method meets the best visual efficacy. The measured contrast revealing coefficient and other results were affected by some

Table 3. Luminaire quantity under optimal luminaire installation method

Lighting zones	30 W LED	60 W LED	75 W LED	120 W LED	2 × 150 W LED
Threshold zone 1	10	——	16	——	33
Threshold zone 2	8	16	——	——	16
Transition zone 1	13	14	24	——	——
Transition zone 2	14	16	——	——	——
Interior zone	566	——	——	——	——
Exit zone	13	5	——	6	——

factors, including mutual reflection between tunnel interfaces, measurement errors and errors of processing data. Moreover, Table 4 shows the road surface luminance and average discovery distance of small target. The threshold zone luminance was not tested in accordance with the influence of natural light outside tunnels. Figure 3 shows discovery distance of small target of five drivers. The total number of measurements was 100 times, of which 95 measurements were valid. If driving speed is 80 km/h, the safety parking distance is 100 m. Five drivers' discovery distances are greater than safety parking distance. Moreover, driver's discovery distance gradually decreases with age. It is obvious that under optimal luminaire installation, the lighting environment meets visual comfort and safe driving requirements.

Table 4. The road surface luminance and average discovery distance of small target

Lighting zones	Road surface luminance (cd/m^2)	Average SD (m)
Threshold zone 2	91.4	106.3
Transition zone 2	7.67	111.9
Interior zone	2.41	113.7
Exit zone 1	8.21	103.1

According to lighting control time and luminaire quantity, energy consumption can be calculated under optimal luminaire installation method. The calculation processes of energy consumption are as follows. 1) Energy consumption of enhanced lighting is 294.88 kW within 24 h. 2) Energy consumption of basic lighting is 449.28 kW within 24 h. To summarize, energy consumption is 744.16 kW within 24 h under the optimal luminaire installation method.

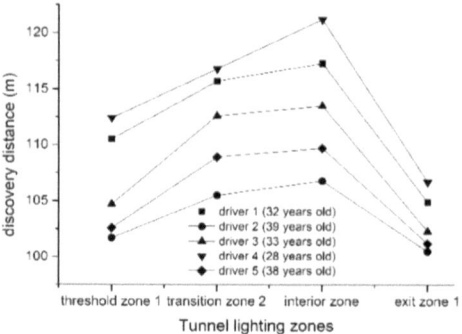

Fig. 3. Drivers' discovery distance of small target

4 Discussion

4.1 Comparison of Lighting Quality Under the Two Luminaire Installation Methods

Lighting quality was compared under different luminaire installation methods across eight aspects, including average luminance and overall uniformity. If the lighting quality index was higher under the optimal installation method than the original, the percentage difference was positive. Lighting quality indices were greater under the optimal method, ranging from 6.17% to 15.26% in difference, with an average of 9.04%. Overall luminance uniformity was 15.26% higher with the optimal method. Additionally, the average contrast revealing coefficient (qc) in the tunnel lighting evaluation system was 0.2. Under the original installation method, qc measured 0.1846, compared to 0.2012 under the optimal method. This suggests that lighting quality indices with the optimal installation method closely align with the evaluation system's standards.

4.2 Comparison of Energy Consumption Under the Two Luminaire Installation Methods

Based on lighting control duration and luminaire quantity, with interior zone lighting included as part of the basic lighting, energy consumption was calculated for both the original and optimal luminaire installation methods (Table 5). In summary, energy consumption within 24 h is 983.24 kW for the original method, whereas for the optimal method, tunnel lighting consumes 744.16 kW. This indicates a 24.32% reduction in electrical energy consumption for tunnel lighting with the optimal installation method in the Dongyangguan highway tunnel. Consequently, the optimal installation method can save 174,528.4 kW of electrical energy annually (365 days). Furthermore, tunnel lighting under the optimal installation method can save 25,779.68 kW/km.

4.3 Evaluation and Improvement of Research

The optimal luminaire installation method achieves lighting quality meeting the evaluation system and reduces energy consumption by 24.32%. This method ensures safe

Table 5. Comparison of energy consumption under the two luminaire installation methods (24 h)

Lighting zones	Threshold zone 1		Threshold zone 2	Transition zone 1	Transition zone 2	Exit zone	Basic lighting	Energy consumption
original installation method (kW)	245		118.8	72.7	18	30.5	498.25	983.24
optimal installation method (kW)	155.4	76.8		33.6	16	13.08	449.28	744.16

driving, visual comfort, and energy efficiency in tunnels, providing valuable guidance for lighting design and engineering projects.

A tunnel lighting system can be established for energy-saving. 1) Gathering information about the tunnel's surroundings, such as its orientation, original luminaire installation method, and lighting quality. 2) Road surface luminance can be determined based on luminance outside tunnel portals L20(S). 3) Optimizing the original luminaire installation method based on the optimal value of the contrast revealing coefficient, considering factors like the selected light source, light distribution curve, and lighting system. 4) By comparing the lighting environment under two luminaire installation methods, the system can provide detailed information on energy consumption.

It is important to acknowledge the limitations of this study. For example, the tunnel lighting simulation experiment and visual efficacy assessment only utilized LED light sources. However, various studies have indicated that different light sources exhibit varying visual efficacy [25], influencing factors such as perception probability and reaction time [26]. Liu et al. [27] demonstrated that under a constant background luminance, reaction times were shorter with LED lighting compared to high-pressure sodium (HPS) and metal-halide (MH) lamps. Therefore, future research should consider incorporating different light sources to broaden the applicability of findings.

5 Conclusions

According to optimal values of contrast revealing coefficient, this paper aimed to study energy-saving strategy based on optimal luminaire installation method. The energy consumption was also analyzed under optimal luminaire installation method. It included luminaire arrangement method, light distribution curve, among others.

(1) Based on the lighting quality evaluation system and the optimal values of the contrast revealing coefficient, the optimal luminaire installation method was determined through tunnel lighting simulation experiments. Subsequently, the original luminaire installation method can be refined based on the optimal value of the contrast revealing coefficient.

(2) The lighting quality achieved under the optimal luminaire installation method aligned with the tunnel lighting quality evaluation system. Moreover, lighting quality under optimal luminaire installation method was better than that under original luminaire installation method.

(3) Compared with the lighting environment under original luminaire installation method, the lighting environment under optimal luminaire installation method can reduce 24.32% of electrical energy. To summarize, the optimal luminaire installation method not only aligns with the lighting quality evaluation system but also reduces energy consumption in tunnel lighting.

Acknowledgement. This work was supported by the National Natural Science Foundation of China (52108071) and Project of science and technology research program of Chongqing Education Commission of China (KJQN202000715). The authors wanted to thank the project partners, Changzhi Expressway Co., Ltd for continuous support.

References

1. Peeling, J., Wayman, M., Mocanu, I., et al.: Energy efficient tunnel solutions. Trans. Res. Proc. **14**, 1472–1481 (2016)
2. Editorial Department of China Highway.: Focus on the "Thirteenth Five-Year Plan"— Overview of the core tasks of the 13th Five-Year Highway Traffic Development in various provinces (autonomous regions and municipalities), China Highway **5**, 38–62 (2016). https://doi.org/10.13468/j.cnki.chw.2016.05.005
3. Peña-García, A., Gil-Martín, L.M.: Study of pergolas for energy saving in road tunnels. Comparision with tension structures. Tunnelling Undergr. Space Technol. **35**, 172–177 (2013)
4. Salam, A.O.A., Mezher, K.A.: Energy saving in tunnel lighting using shading structures. In: 2014 International Renewable and Sustainable Energy Conference (IRSEC) (2015). https://doi.org/10.1109/IRSEC.2014.7059842
5. Wang, Y.Q., Zhang, Y.C., Xing, S.S., et al.: Research on intelligent energy saving control technology for entrance lighting of mountain road tunnel. Energy Conserv. Environ. Prot. Transp. **15**(69), 99–101 (2019)
6. Kimura, M., Hirakawa, S., Uchino, H., Motomura, H., Jinno, M.: Energy savings in tunnel lighting by improving the road surface luminance uniformity—a new approach to tunnel lighting. J. Light Visual Environ. **38**, 66–78 (2014)
7. Liang, B., Cui, L.L., Pan, G.B., et al.: Auxiliary tunnel lighting technology based on the light reflection and energy storage concept. Mod. Tunneling Technol. **51**(5), 15–22 (2014). https://doi.org/10.13807/j.cnki.mtt.2014.05.003
8. Hirakawa, S., Hayakawa, M., Takenouchi, M., et al.: Study of luminous distribution design of tunnel base lighting. J. Illum. Eng. Inst. Jpn. **100**(6), 217–223 (2016)
9. Commission Internationale de L'Eclairage.: CIE 88-2004 Guide for the lighting of road tunnels and underpasses. CIE Central Bureau, pp. 4–6 (2004)
10. Bommel, W.V.: Road Lighting—Fundamentals, Technology and Application, pp. 267–318. Springer International Publishing Switzerland (2015)
11. Standards Policy and Strategy Committee. BS 5489-2:2003+A1:2008 Code of practice for the design of road—Part 2: Lighting of tunnels. BSI Group. 17–18 (2003)
12. Comité européen de normalization. CR 14380-2003 Lighting applications—Tunnel lighting. CEN Brussels. 15 (2003)

13. Commission Internationale de L'Eclairage. CIE 189-2010 Calculation of tunnel lighting quality criteria. CIE Central Bureau, pp. 1–9 (2010)
14. Weng, J., Cai, X.Y., Du, F., et al.: The research of contrast revealing coefficient based on small target visibility in highway tunnel lighting. China Illum. Eng. J. **26**(6), 87–90 (2015). https://doi.org/10.3969/j.issn.1004-440X.2015.06.019
15. Weng, J., Cai, X.Y., Hu, Y.K., et al.: Analysis of threshold values of contrast revealing coefficient in highway tunnel lighting. Tunnelling Undergr. Space Technol. **92** (2019). https://doi.org/10.1016/j.tust.2019.103038
16. Cai, X.Y., Weng, J., Hu, Y.K., et al.: Research on optimal values of contrast revealing coefficient in road tunnel lighting. Sādhanā-Acad. Proc. Eng. Sci. **45**, 1–12 (2020)
17. Cai, X.Y., Guo, T.H.: Lighting Quality Evaluation of Highway Tunnels, pp. 85–162. China Architecture & Building Press (2022)
18. Yang, Q., Liang, B., Pan, G.B., et al.: Research on layout of measurement points in tunnel lighting experiment. Appl. Mech. Mater, 1024–1028 (2014)
19. Chen, T.L.: Application of LED in tunnel lighting under the new specification. Low Carbon World **2**, 156–157 (2017). https://doi.org/10.16844/j.cnki.cn10-1007/tk.2017.02.098
20. Karatekin, C.: Tunnel lighting design with high power led luminaires of an urban tunnel in Istanbul. Light Eng. **25**(4), 69–75 (2017)
21. Weng, J., Chen, X.W., Huang, K., et al.: Discussion on the tunnel lighting contrast revealing coefficient. Light & Lighting **37**(1), 4–7 (2013). https://doi.org/10.3969/j.issn.1008-5521.2013.01.002
22. Commission Internationale de L'Eclairage. CIE 88-1990 Guide for the lighting of road tunnels and underpasses. CIE Central Bureau (1990)
23. China Merchants Chongqing Communications Technology Research & Design Institute Co., Ltd. Guidelines for Design of Lighting of Highway Tunnels JTG/T D70/2-01-2014. China Communications Press Co., Ltd., pp. 14–18 (2014)
24. Ke, H.: Study on Perceived Contrast Method of Tunnel Lighting. PhD's thesis. Chongqing University, Chongqing, China (2016)
25. Liu, Y.Y., Weng, J., Chen, J.Z., et al.: Influence of light sources color on tunnel lighting. J. Civ. Environ. Eng. **35**(3), 162–166 (2013). https://doi.org/10.11835/j.issn.1674-4764.2013.03.026
26. Liang, B., He, S.Y., Tähkämö, L., et al.: Lighting for road tunnels: the influence of CCT of light sources on reaction time. Displays **61** (2020). https://doi.org/10.1016/j.displa.2019.101931
27. Liu, Y.Y., Chen, J.Z., Zhang, Q.W., et al.: Influence of light source color temperature on traffic safety at tunnel entrance based on reaction time. J. Highw. Transp. Res. Dev. **32**(2), 114–118+133 (2015). https://doi.org/10.3969/j.issn.1002-0268.2015.02.018

Enhancing Facade Design to Improve Energy Efficiency of Office Towers in the Hot and Humid Climate Region (Case Study: Bandung, Indonesia)

Beta Paramita[1]([✉]), Muhammad Rabbani Nurlette[2], and Aprilia Nurul Hanissa[2]

[1] Low Carbon Building Material and Energy, University Center of Excellent, Universitas Pendidikan Indonesia, Bandung, Indonesia
betaparamita@upi.edu
[2] Architecture Study Program, Universitas Pendidikan Indonesia, Bandung, Indonesia

Abstract. Performance Based Design (PBD) makes the design process easier to achieve optimal energy efficiency in office towers. Building performance is impacted by various factors, including decisions related to the shape of the building, its orientation, the building envelope, and the building façade. This study aims to determine the building energy consumption by applying the office building's secondary skin during the hottest season in the hot and climate region. A design solution was obtained by utilizing Sefaira with ASHRAE 90.1 – 2013-based parameters for the building performance analysis application. Building in the hot and humid climate region requires energy for cooling due to the long solar durations. The building facade then is the main architectural element that needs to design properly. Facade design, including the building height, orientation, shadowing, and wall window ratio, define the use of electric devices per square meter. The results showed the building's energy use before the special treatment was 126 kWh/m2/yr from the supposed 79 kWh/m2/yr. The design process that responds to the climate will then influence the building form and massing. It is the initial step for building envelope before requiring mechanical assistance to achieve thermal comfort. The design experiment later takes a 3-m-long balcony for all facades, reduce envelope area with compact form have proven to reduce the building energy use. Those experiments later found able to reduce energy use to 57 kWh/m2/yr or 69 kWh/m2/yr obtained.

Keywords: Tropical cilmate · Sefaira · Building envelope · Energy use

1 Introduction

Most office building facades are glass windows. These materials can bring both positive and negative experiences: access to views and sunlight, but also glare and thermal discomfort [1]. Building facades in tropical areas located on the equator face longer daily radiation than in other areas. High-rise structures located in hot and humid regions often face issues with overheating as a result of intense solar radiation [2]. The heat

B.-J. He et al. (Eds.): UCSUD 2023, LNCE 559, pp. 234–244, 2025.
https://doi.org/10.1007/978-981-97-8401-1_16

caused by this insolation induces heat that will affect the energy use for cooling. The heat loss through an envelope can occur due to differences in temperature outside and inside. The facade window-to-wall ratio is also mentioned to reduce the solar heat gain coefficient equal to 54% and 57% [3]. The building facades will achieve those performances if the layout and building groups respond to site conditions, especially on its microclimates, such as solar duration, daylight availability, and wind flow. In the low latitude, the variation of building group fabric (morphology) with high SVF (sky view factor) for high cooling and low solar load from the building height per distance. The characteristic of a building group with narrow H/W and increasing the SVF will provide good daylighting and radiant cooling, reducing the solar gain. The daylight factor 1,0 and the H/W with spacing angle Tg $\alpha = 1,7 - 2,0$ are the best suggestion for building groups in the hot and humid climate as in the low latitude [4]. The study on the comparison of five building groups in Bandung reveals that by facing the building orientation to North-South able to reduce air temperature. Meanwhile, to achieve the wind speed faster, the building group with the square plot is preferable [5]. The utilization of the Passive Performance Optimization Framework (PPOF) as an alternative approach involves enhancing the efficiency of daylighting, solar control, and natural ventilation. This method has been substantiated to effectively decrease the Energy Use Intensity (EUI) by a range of 4 to 17% [6]. A well-oriented building that incorporates passive design strategies can significantly reduce energy consumption over its lifespan [7].

Also, emphasize the importance of passive design strategies [8]; The selection of appropriate parameters for building orientation, shape, envelope system, passive heating, cooling mechanisms, shading, and glazing is crucial in order to minimize the energy consumption of residential buildings. On the other hand, the passive design for office tower is difficult to implement fully. It demands a more complex building performance due to achieving thermal comfort in large and multi-story building areas. Consequently, the implementation of mechanical support systems like automated motorized shading, smart properties, and adjustable electric lighting led to a significant decrease in energy consumption for cooling and lighting purposes [9, 10].

Several studies on the office building design in the hot and humid climate context show that energy for cooling is the most challenging. The investigation into the impact of various roof construction methods, glazing options, and sun-shading techniques on the energy usage of residential buildings in Indonesia has been conducted. The implementation of window shading has a profound effect on the overall energy consumption of a building, thereby enabling the attainment of energy efficiency [11]. Some design strategies are carried out to achieve proper energy consumption, such as adding the canopy with 3 m height and with the length of 2 m. The building orientation needs to be tilted 21^0 to the south orientation with 32% of the window-to-wall ratio's total façade [12]. The treatment of the building facades by providing canopies, double-skin facades, and material selection was found as important elements to reduce the building energy as much as 51.6% [13]. The use of electrical equipment and material selection also found as significant elements in the design process. Sefaira simulation tool is used to define the accurate calculation [14].

2 Method

Performance Based Design (PBD) makes the design process easier to achieve optimal energy efficiency in office towers. PBD is able to optimize the occupant experience, reduce resource use and minimize negative impacts on the environment [15]. The performance of a building is impacted by various factors including its shape, orientation, envelope, and facade. Throughout the design and construction phases, decisions are made that ultimately shape the building's performance. Choices made based on PDB data are particularly sought after as they are tailored to respond effectively to the local climate conditions. Figure 1 explains the office tower design process.

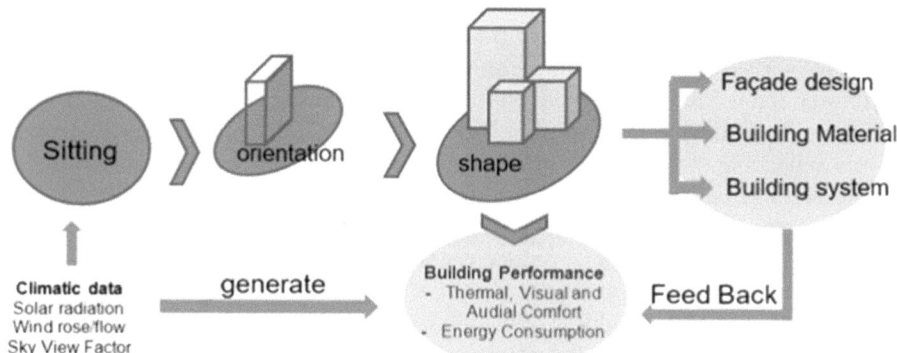

Fig. 1. Office tower design process based on PBD

2.1 Location

Climate data is an important part of the office tower design process. The existing data will influence the shape and mass of the building, such as solar radiation data (Fig. 2) and wind rose/flow data (Fig. 3).

The design location is set on the hot-humid climate region, with Bandung geographically sited on 6° 93′76.6 "S and 107° 70′51.5" E. The total area of the footprint is 5.890 m2 while the circumference is 304 m long. The diurnal temperature is in the range of 17 - 28°C, the hot and humid climate marked with high humidity at 60–80% daily and 0-3m/s wind speed [15]. The office building with 15 floors tower and flat rooftop, 2 floors as a podium. The existing condition in the northern part have some plants that beneficial to reduce the direct solar radiation.

The initial of building form and massing influenced by solar duration that hit to each facade, this is also affected by wind flow velocity and its direction. The initial building envelope later has significant contribution for passive design strategy to minimize the energy consumption. The initial process as seen at Fig. 2 shows that solar radiation hits the north façade longer than any other façade. Meanwhile the Fig. 3 shows that wind velocity that induced by this building envelope in the range of 2.6 to 4.6 m/s.

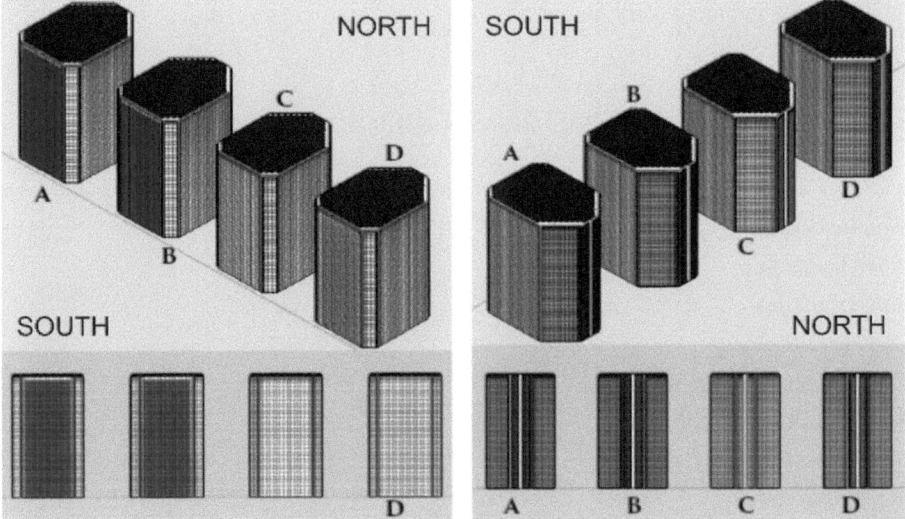

Fig. 2. Building form and massing based on solar duration

Trees are placed on the north side of the building to break the incoming wind, so that the incoming wind does not directly hit the building

Fig. 3. Wind velocity simulation

2.2 Sefaira

The Sefaira Plugin is an application designed for SketchUp/Revit modeling, which focuses on Performance-Based Design. It enables the analysis and simulation of a building's energy consumption by examining the impact of climate on both the building envelope and its shape. Sefaira's simulation to determine the annual energy use of office tower uses EnergyPlus and Radiance which are industry accredited with ASHRAE 90.1

as basic input and are widely known as accurate prognostic simulations [16, 17, 18, 19] (Table 1).

Table 1. The baseline for ASHRAE 90.1–2013.

ASHRAE climate zone	2
Wall Insulation	0,86 W/m2−k
Floor Insulation	0,61 W/m2−k
Roof Insulation	0,22 W/m2−k
Glazing U-Factor	2,27 W/m2−k
Visible Light Transmittance	0,42 W/m2−k
Solar Heat Gain Coefficient	0,25 SHGC
Infiltration Rate	7,2 m3/m2h
Ventilation Rate	15 L/S
Equipment	25 W/m2
Lighting	10 W/m2

2.3 Energy Performance

Next, to find out the energy performance produced in the initial building design, need to use Sefaira.

However, beforehand it is necessary to know the energy use intensity target for the office tower which must be achieved by 2030, which is 79 kWh/m2/yr (see Fig. 4).

Table 2. The analysis reveals that this initial structure has yielded an Energy Use Intensity (EUI) of 125 kWh/m2/year, as indicated by the results. Equipment here includes all the mechanical, electrical tools that are used to support the building operational. Most building owners do not realize that proper equipment can take up the largest portion of energy consumption. The operational schedule for some equipment also affects energy use. The second highest consumption from cooling marked that the heat transfer from the façade needs to address more. It includes inefficient energy due to sun exposure, thermal bridge leaks, and conduction to glass elements.

Various methods can be employed to lessen the strain on the cooling system, such as decreasing the quantity of glazing on the facade, implementing shading elements, or opting for high-performance glass materials. These methods can reduce the solar heat gain coefficient [18].

3   Result

Based on the preliminary examination, it is imperative to modify the architectural layout in accordance with the suggested design guidelines in order to minimize energy consumption.

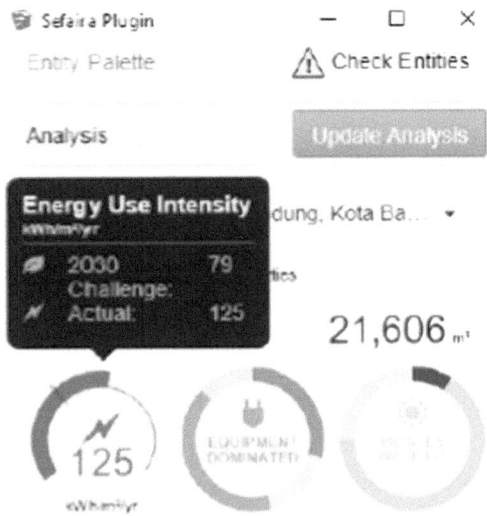

Fig. 4. Sefaira simulation results and target

Table 2. Energy Performance

Indicator	Initial Value
Total Area Floor	19,240 m^2
Energy Use Intensity (EUI)	125 kWh/m^2/yr
Equipment	712,964 kWh/yr
Dominated	399,461 kWh/yr
Cooling:	998,652 kWh/yr
Lighting:	308,259 kWh/yr
Equipment:	
Fans:	

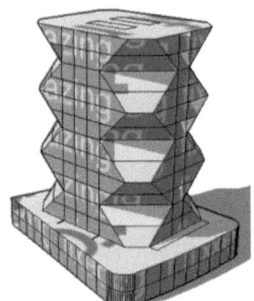

Fig. 5. The shape of the building before applying PBD

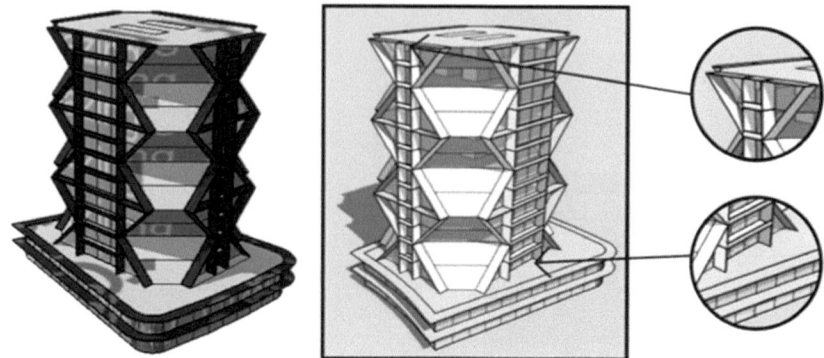

Fig. 6. The shape of the building after adopting the design strategy

Figure 6 shows the addition of shade on each side of the building where the previous building design (Fig. 5) resulted in large energy consumption. Shading is applied with a width of 3 m to reduce radiation exposure through windows and is adjusted on the sides of the building based on the duration of sunlight.

Next, re-analyze the use of energy consumption after changes in the shape and mass of the building. Apart from that, it is also necessary to select materials so that the desired energy efficiency is obtained.

Table 3 presents a comparison of the energy standard values utilized as a benchmark for assessing energy consumption in buildings. Notably, no alterations were noted in the HVAC Type, Base-line, and ASHRAE Climate Zone, as they have established themselves as crucial references for practical implementation. Nevertheless, modifications were implemented to yield outcomes that enable the reduction of energy consumption, emphasizing the utilization and characteristics of the materials intended for application in the constructions.

The EUI generated by the building was 69 kWh/m2/yr, and this is a significant reduction from the previous 126 kWh/m2/yr (Table 4).

Total area floor is later increasing due to building shade on each facade while the cooling experience reduced to 358,278 kWh/yr from the previous 712,964 kWh/yr. It means the application of shade provides several benefits to energy use in buildings by blocking direct sunlight and resisting the radiation that is supposed to go through the window. Later, the buildings' most lit areas increased to 59% from 50% while over lit reduced to 17% from 24% and underlit decreased to 24% from 26%. It means a higher percentage of light is now entering the building. The building's facade's energy load was successfully reduced by 50% based on the application of shade and changing the value of the indicator on the application. All these, therefore, caused the EUI of this building to be 69 kWh/m2/yr which is smaller compared to the energy use limit set at 79 kWh/m2/yr.

Table 3. Change in Energy Standards

Indicator	Energy standard	Change in Energy Standards
HVAC Type	VAV – Return Air Package (System 5/6)	VAV – Return Air Package (System 5/6)
Baseline	ASHRAE 90.1 – 2013	ASHRAE 90.1 – 2013
ASHRAE Climate Zone	2	2
Wall Insulation	0,86 W/m^2–k	0,10 W/m^2–k
Floor Insulation	0,61 W/m^2–k	0,10 W/m^2–k
Roof Insulation	0,22 W/m^2–k	0,10 W/m^2–k
Glazing U-Factor	2,27 W/m^2–k	0,10 W/m^2–k
Visible Light Transmittance	42%	100%
Solar Heat Gain Coefficient	0,25 SHGC	0,21 SGHC
Infiltration Rate	7,2 m3/m2h	1 m3/m2h
Ventilation Rate	15 L/S	0 L/S
Equipment	25 W/m2	15 W/m2
Lighting	10 W/m2	6.5 W/m2

Table 4. Sefaira Analysis Results (After Value Changes)

Indicator	Initial Value	Value Change	Information
Total Area Floor	19,2 40 m^2	22,851 m^2	–
Energy Use Intensity (EUI)	126 kWh/m^2/yr	69 kWh/m^2/yr	Decrease
Equipment Dominated			Increase
Cooling: **Lighting:** **Equipment:** **Fans:**	712964 kWh/yr 399461 kWh/yr 998652 kWh/yr 308259 kWh/yr	358278 kWh/yr 303627 kWh/yr 711626 kWh/yr 201311 kWh/yr	
Mostly Lit **Under Lit:** **Well Lit:** **Over Lit:**	26% 50% 24%	24% 59% 17%	Increase
East Solar	261,912 kWh/yr	109,088 kWh/yr	Decrease
West Solar	197,500 kWh/yr	76,017 kWh/yr	Decrease
North Solar	112,073 kWh/yr	57,824 kWh/yr	Decrease
South Solar	114,368 kWh/yr	43,924 kWh/yr	Decrease
Glazing **Conduction**	89,545 kWh/yr (Losses) 98,772 kWh/y (Gains)	121,364 kWh/yr (Losses) 97,153 kWh/yr (Gains)	Decrease
Wall Conduction	92,814 kWh/yr (Losses) 56,698 kWh/yr (Gains)	122,697 kWh/yr (Losses) 114,240 kWh/yr (Gains)	Increase
Roof Conduction	67,180 kWh/yr (Losses) 47,393 kWh/yr (Gains)	96,919 kWh/yr (Losses) 91,080 kWh/yr (Gains)	Decrease
Floor Conduction	1,659 kWh/yr (Losses) 71 kWh/yr (Gains)	6,059 kWh/yr (Losses) 146 kWh/yr (Gains)	Increase
Infiltration	27,470 kWh/yr (Losses) 91,831 kWh/yr (Gains)	1,826 kWh/yr (Losses) 23,559 kWh/yr (Gains)	Increase
HVAC Heating	2 kWh/yr (Losses) 334 kWh/yr (Gains)	72 kWh/yr (Losses) 228 kWh/yr (Gains)	–

4 Conclusion

The office tower facade in Bandung's hot and humid climate failed to meet the Sefaira 2030 energy use target of 79 kWh/m2/year with its initial design. Energy assessment was conducted utilizing the Sefaira plugin in conjunction with SkethUp modeling and ASHRAE 90.1 – 2013 as the foundational energy guideline. Initially, the energy consumption of the building amounted to 125 kWh/m2/yr. Subsequently, energy efficiency measures were implemented to curtail the building's energy consumption. Following the alteration of the building facade in alignment with the energy reduction design approach, the energy usage decreased to 69 kWh/m2/yr.

References

1. Aries, M.B.C., Veitch, J.A., Newsham, G.R.: Windows, view, and office characteristics predict physical and psychological discomfort. J. Environ. Psychol. **30**(4), 533–541 (Dec.2010). https://doi.org/10.1016/J.JENVP.2009.12.004
2. Ling, C.S., Ahmad, M.H., Ossen, D.R.: The effect of geometric shape and building orientation on minimising solar insolation on high-rise buildings in hot humid climate. J. Constr. Dev. Ctries. **12**(1), 27–38 (2007)
3. Talami, R., Jakubiec, J.A.: Early-design sensitivity of radiant cooled office buildings in the tropics for building performance. Ener. Build. **223**, 110177 (Sep.2020). https://doi.org/10.1016/J.ENBUILD.2020.110177
4. DeKay, M., Brown, G.Z.: Sun, Wind and Light, Second. John Wiley and Sons, Inc., Canada (2001)
5. Paramita, B., Fukuda, H., Khidmat, R. P., Matzarakis, A.: Building configuration of low-cost apartments in Bandung-its contribution to the microclimate and outdoor thermal comfort. Buildings (2018).https://doi.org/10.3390/buildings8090123
6. Konis, K., Gamas, A., Kensek, K.: Passive performance and building form: An optimization framework for early-stage design support. Sol. Energy **125**, 161–179 (Feb.2016). https://doi.org/10.1016/J.SOLENER.2015.12.020
7. Abanda, F.H., Byers, L.: An investigation of the impact of building orientation on energy consumption in a domestic building using emerging BIM (Building Information Modelling). Energy **97**, 517–527 (Feb.2016). https://doi.org/10.1016/J.ENERGY.2015.12.135
8. Pacheco, R., Ordóñez, J., Martínez, G.: Energy efficient design of building: A review. Renew. Sustain. Energy Rev. **16**(6), 3559–3573 (Aug.2012). https://doi.org/10.1016/J.RSER.2012.03.045
9. Tzempelikos, A., Athienitis, A.K.: The impact of shading design and control on building cooling and lighting demand. Sol. Energy **81**(3), 369–382 (Mar.2007). https://doi.org/10.1016/J.SOLENER.2006.06.015
10. Tzempelikos, A., Athienitis, A.K., Karava, P.: Simulation of façade and envelope design options for a new institutional building. Sol. Energy **81**(9), 1088–1103 (Sep.2007). https://doi.org/10.1016/J.SOLENER.2007.02.006
11. Setiawan, A.F., Huang, T.L., Tzeng, C.T., Lai, C.M.: The effects of envelope design alternatives on the energy consumption of residential houses in Indonesia. Energies (2015).https://doi.org/10.3390/en8042788
12. Khidmat, R.P., Ulum, M.S., Lestari, A.D.E.: Facade components optimization of naturally ventilated building in tropical climates through generative processes. case study: Sumatera Institute of Technology (ITERA), Lampung, Indonesia. IOP Conf. Ser. Earth Environ. Sci. **537**(1) (2020). https://doi.org/10.1088/1755-1315/537/1/012015

13. Amalia, M., Paramita, B., Minggra, R., Koerniawan, M.D.: Efficiency energy on office building in South Jakarta. IOP Conf. Ser. Earth Environ. Sci. **520**(1) (2020). https://doi.org/10.1088/1755-1315/520/1/012022

14. Nurlette, M.R., Paramita, B.: Optimization of energy usage of the building envelope material at the rental office buildings. IOP Conf. Ser. Earth Environ. Sci. **248**(1) (2019). https://doi.org/10.1088/1755-1315/248/1/012018

15. Meteoblue: Gedebage Weather (2018)

16. Kilkelly, M.: Five digital tools for architects to test building performance. Architech Magazine, Dec. 2015

17. Beidi, L.: Use of building energy simulation software in early-stage of design process. Dep. Civ. Eng. Archit. SE-100 44 Stock. Sweden (2017)

18. Sefaira: Sefaira. https://sefaira.com/. Accessed 30 Aug 2020

19. Paramita, B., Rabbani, B.A., Sari, D.C.P.: Energy optimization on preliminary design of the botani museum using sefaira®. Int. J. Eng. Adv. Technol. **8**(5), 2614–2618 (2019)

Thermal and Optical Performance of Semi-Transparent BIPV Windows in High-Rise Office Buildings: A Case Study in the UK

Joseph Scibetta[1], Shaohang Shi[2], Hu Du[3], Bao-Jie He[4,5,6], and Siliang Yang[7(✉)]

[1] School of Built Environment, Engineering and Computing, Leeds Beckett University, Leeds LS2 8AG, UK

[2] School of Architecture, Tsinghua University, Beijing 100080, China

[3] School of Civil Engineering and Built Environment, Liverpool John Moores University, Liverpool L3 3AF, UK

[4] School of Architecture and Urban Planning, Ministry of Education, Chongqing University, Chongqing 400045, China

[5] Key Laboratory of New Technology for Construction of Cities in Mountain Area, Ministry of Education, Chongqing University, Chongqing 400045, China

[6] CMA Key Open Laboratory of Transforming Climate Resources to Economy, Chongqing 401147, China

[7] Mott MacDonald, Leeds LS12 1BE, UK
siliang.yang@mottmac.com

Abstract. Semi-transparent building-integrated photovoltaic (BIPV) windows replace the external building glazing, allowing power to be generated as part of the existing building envelope. Due to the replacing of the BIPV windows, both thermal and optical performance of a building are worth exploring. This paper presents the results of a simulation study investigating the effects of applying BIPV windows to an existing office building within the UK, with a view to assessing the thermal performance, daylighting condition and BIPV electrical production capacity. It was considered on a typical floor of the office building, consisting of an open plan office space, two meeting rooms and a kitchen suite. A baseline scenario without BIPV window was modelled for comparison, giving recommendations to the viability of the BIPV window in terms of the overall performance. Basically, the use of the BIPV window resulted an annual cooling load reduction of 16.3% compared to the baseline. Heating loads were also reduced but only by 1.89%, while lighting energy usage was slightly increased due to a significantly worse performance in daylighting condition in the open plan office when applying the BIPV window. Specifically, daylight factor dropped over 50% in the office area, which, however, was already poor to begin with. The annual electricity generation of the BIPV window was little with only 902 kWh. This study can be a reference for future research on thermal and optical performance of high-rise office buildings using semi-transparent BIPV windows.

Keywords: BIPV Glazing · Energy Performance · Daylighting · Building Simulation · IES VE

© The Author(s) 2025

B.-J. He et al. (Eds.): UCSUD 2023, LNCE 559, pp. 245–262, 2025.

https://doi.org/10.1007/978-981-97-8401-1_17

1 Introduction

With the rapid development of urbanisation and the progressive implementation of energy saving and carbon emission reduction measures in countries around the world [1, 2], energy saving renovations for existing buildings are receiving increasing attention [3, 4]. The UK Green Building Council (UKGBC) [5] estimates that by 80% of buildings will exist in 2050 have already been built. This means that the majority of engineering work going forward will be targeting the existing building stock for retrofits and refurbishments, and a significant amount of decarbonisation work is required in order for most of the buildings to reach aspirational carbon targets in the UK [6]. A reasonably simple target to achieve is on-site energy generation, typically through the installation of photovoltaic (PV) systems [7]. Building-Integrated Photovoltaic (BIPV) is a novel technology that aims to integrate the generational aspects of PV modules into the elements of a building facade, for example in the roofing, cladding, shading or windows of a building envelope [8–12]. BIPV window glazing in particular, utilising the flexible amorphous silicon (a-Si) technology for transparency, could potentially offer an attractive alternative to traditional roof-mounted PV modules for those looking to retrofit an existing building, due to the reduced weight load on the existing structure [13–19].

When looking at the replacement of the external glazing, the thermal factors and daylighting performance of a building are worth exploring [20–22]. Traditional double glazing manages an average U-value of 3 $W/m^2 \cdot K$ approximately, with more novel modern technologies such as vacuum glazing being able to approach a U-value of 1.12 $W/m^2 \cdot K$ [23, 24]. Although BIPV glazing may be able to achieve the similar U-values owing to its complex multipart makeup, the visible light transmittance (VLT) and solar heat gain coefficient (SHGC) through BIPV glazing (such as the semi-transparent option) are likely to be considerably worse [25]. However, maintaining a comfortable indoor environment is important for the health of occupants and the energy efficiency of a building through the parameters of the building envelope design [26]. In particular, a combination of power generation, thermal regulation and daylight controlling may prove useful to the existing building stock, especially in translucent envelopes [27].

Numerous previous studies have analysed the comprehensive energy performance of translucent BIPV facades in typically warmer climates than the UK, for example in Australia [16, 28, 29] or the UAE [30], where the obvious differences in solar gains due to the climate and differing thermal performance of the facades are likely to have a more significant impact on building energy consumption and occupancy comfort. Conde and Shanks [30] found that for a given building in the UAE, the application of BIPV windows resulted in an increase on the building's peak cooling load by over 50%, and overall poor electrical performance when compared to the values quoted by the manufacturer, around 7% lower, due to the exceeding of the nominal operating cell temperature (NOCT). The results, from the study of Yang et al. [29], show that total annual energy reductions of 34.1%, 86% and 106% in three typical climate zones in Australia, although the technology of building-integrated photovoltaic/thermal double-skin façade (BIPV/T-DSF) is different, utilising double-skin facades (DSF), and so better thermal performance is expected due to the addition of an air gap within the facade.

Statistics show that there is a huge potential of solar energy sources in the UK, but the uptake of solar energy is still low especially for the BIPVs on building envelopes [31].

Roberts et al. [27] recently published a numerical study on an a-Si BIPV in double-skin façades (BIPV-DSF) under a typical UK climate. They reveal that the BIPV-DSF might not be a feasible solution help in the building performance improvement, due to the low visibility BIPV window caused excessive lighting energy demand, which could not be complemented by the PV electrical power production. Furthermore, some research has been conducted on polycrystalline type of BIPV glazing in the UK, for example cadmium telluride (CdTe), finding 73% lower solar heat gains through the windows, which, however, had relatively poor thermal performance with an averaged U-value of 2.7 W/m²·K [32]. Thus, it would not be a viable replacement for existing double glazing being frequently used in the UK. In terms of a comprehensive literature survey, Stephanos et al. [33] conclude that the CdTe based BIPV can potentially be used to achieve a net building energy saving up to 20% in comparison with the regular single glazing case in the UK context. By comparison, Khalifeeh et al. [34] found that a-Si based double-glazed BIPV has the best performance on energy saving for buildings within the UK climates, due to its nature of the low U-value and SHGC, although the daylighting performance was not reported. In addition, a novel BIPV vacuum glazing has been found to be able to maintain thermal comfort in buildings and hence improve the overall building performance [35]. Yu et al. [36] confirmed the potential for BIPV glazing to reduce energy consumption in hot climates, and highlighted the increasingly good thermal performance (that is, the U-value) of such technology to be around 1 W/m²·K for vacuum-glazed products; they also noted that the need for further study in cooler climate areas such as the UK, as most of the study and implementation so far has been carried out in hot climates.

In summary, little research has been conducted on the viability of BIPV windows as a retrofit technology for the existing building stock in temperate climates like the UK. More specifically, the determination of advantages of BIPV especially the BIPV windows is still in dispute. Therefore, this paper was aimed to present the results of a simulation-based study on the application of a-Si semi-transparent BIPV window to an existing 15-story office building in the UK. In the scenario of a BIPV retrofit, where the existing external plant arrangement and structural constraints ruled out a traditional roof-mounted PV installation, noting the effects on internal thermal comfort and daylighting when compared with the base case apart from the energy efficiency retrofit. The study was, however, intended to provide a data reference and research methodology for energy efficiency retrofitting of buildings in moderate areas. Specifically, the study looked at both thermal and lighting energy performance of the building when adopting the BIPV windows.

2   Materials and Methods

2.1   Case Study Model Descriptions

The case study office building model was acquired, which contained as-built information from a previous full building refurbishment, by completing with a modelling of off-site shading and adjacent buildings. The building modelled was a 15-story office block situated in Leeds, UK, and the modelling was conducted using IES VE, which is utilised as an international industry standard for modelling building thermal and energy performance [37]. Figure 1a and b show both south- and north-facing facades of the building, respectively.

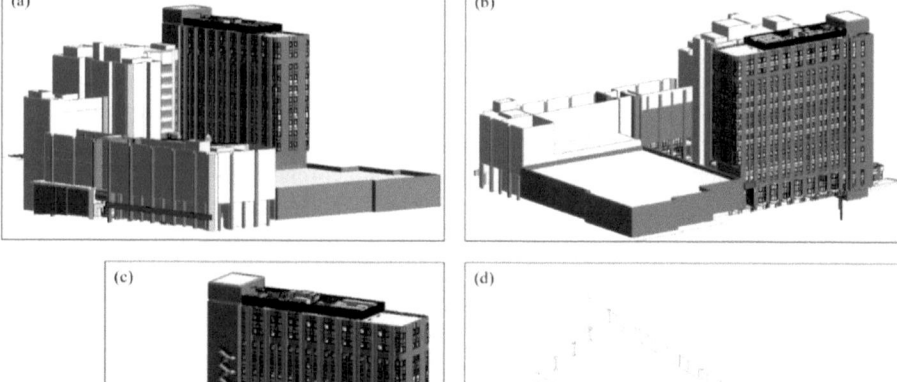

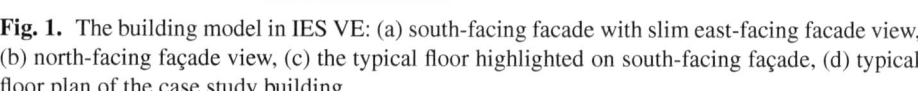

Fig. 1. The building model in IES VE: (a) south-facing facade with slim east-facing facade view, (b) north-facing façade view, (c) the typical floor highlighted on south-facing façade, (d) typical floor plan of the case study building.

It should be noted that the floor 7 was used for the simulation as it is the typical floor. Specifically, floor 7 contains an open plan office, two meeting rooms, and a kitchen suite. Figure 1c highlights the typical floor's south-facing façade, where the BIPV window was applied, with adjacent buildings removed for clarity only. Furthermore, the typical floor plan is shown in Fig. 1d. The typical floor's total area is about 535 m², consisting of a 323.13 m² open plan section, a 37.9 m² meeting room, a 53.8 m² meeting room, and a 120.02 m² kitchen. The PV glazing applied on the typical floor is 40.6 m², with the glazing type of semi-transparent GL-06 by Onyx Solar (the manufacturer). The baseline existing building fabric properties are listed in Table 1, with the design conditions and properties of replacement windows listed in Tables 2 and 3, respectively, following the existing schedule. Parameters for the BIPV windows were achieved within IES VE by creating a thermal construction element to represent the windows, where the a-Si semi-transparent GL-06 PV window was chosen.

Some design inputs of the building (such as the typical office latent and sensible heat gains) were checked against the NABERS UK criteria (a rating system for energy efficiency of office buildings in the UK) set out in the NABERS UK Guide to Design for Performance [38]. It should be noted that the NABERS is a sustainability rating scheme developed in Australia, becoming increasingly common across the world and

Table 1. Physical characteristics of existing facade elements.

Parameter	Value
External wall U-value (W/m^2·K)	0.21
Internal ceiling/floor U-value (W/m^2·K)	1.069
Internal partitions U-value (W/m^2·K)	1.857
External windows U-value (W/m^2·K)	1.68
External windows solar transmittance (%)	41
External windows visible light transmittance (%)	71

Table 2. Design conditions.

Parameter	Value
Weather file	Leeds_DSY1_2020High50_.epw
Cooling setpoint (°C)	25
Heating setpoint (°C)	18
People sensible gain (W/person)	75
People latent gain (W/person)	75
Lighting gain (W/m^2)	6
Miscellaneous equipment (W/m^2)	11
Occupancy density (m^2/person)	8
Infiltration rate (ACH)	0.5
Auxiliary ventilation rate (l/s/person)	10
Building surroundings	Urban
Design days	365

within the UK. There are no regulatory requirements for daylighting, but the BREEAM system set out a required average daylight factor of 2% and a minimum of 0.8% in a given occupied room, with a minimum uniformity of 0.3 U$_o$ [39], which was considered as the benchmark for lighting evaluation in this study.

As shown in Fig. 2, the building was simulated in the base condition for thermal effects (using Apache simulation in IES VE) and daylighting (using Radiance simulation in IES VE), the window constructions were then altered and simulated for both results.

2.2 Mathematical Models

The thermal effects (fabric thermal transmissions) within IES Apache were calculated using the CIBSE admittance method algorithm [40], the details of which can

Table 3. Physical characteristics of a-Si semi-transparent PV windows.

Parameter	Value
PV window U-value (W/m^2·K)	1.0
PV window solar transmittance (%)	15
PV window visible light transmittance (%)	20
Peak power per module area (W$_p$/m^2)	34
Installed azimuth from north (degree)	185
Peak power for studied floor (kW)	1.36
Installed inclination from horizontal (degree)	90
Module efficiency (calculated from peak power and size)	3.41%
PV window U-value (W/m^2·K)	1.0

be concluded as follows:

$$\bar{\theta}_{fa} = F_{cu} \sum A \bullet U \bullet (\bar{\theta}_{eo} - \bar{\theta}_c) \tag{1}$$

where F_{cu} is the mean fabric heat gain/loss to the air node, A is the area of the fabric, U is the thermal transmittance (U-value) of the fabric, $\bar{\theta}_{eo}$ is the mean outside temperature, $\bar{\theta}_c$ is the mean operative temperature.

Daylighting calculations were considered using the methods utilised by the Radiance within the IES VE, which uses the ray tracing method to enable lighting and daylighting simulations. Daylight Factor (DF) was a key parameter for the daylighting simulation, which was calculated within IES VE utilising the common DF equation, shown as below:

$$DF = \frac{E_i}{E_o} \bullet 100\% \tag{2}$$

where E_i is the daylight illuminance at a point on an internal working plane, E_o is the outdoor illuminance on a point from an unobstructed CIE overcast sky.

In addition, the E_i can be considered as three components, the sky component (SC), the externally reflected component (ERC), and the internal component (IRC), therefore:

$$E_i = SC + ERC + IRC \tag{3}$$

It should be noted that the horizontal working plane considered within all areas of the typical floor studied was set at a standard of 0.7 m [27].

3   Results and Discussion

3.1   Energy Performance of the Building with BIPV Windows

Apache dynamic simulations were carried out for both scenarios. The existing ventilation system was simulated during running of the scenarios (the existing window scenario and the PV window scenario) in order to determine true thermal loads (heating and cooling) for the areas given the expectation of a window retrofit.

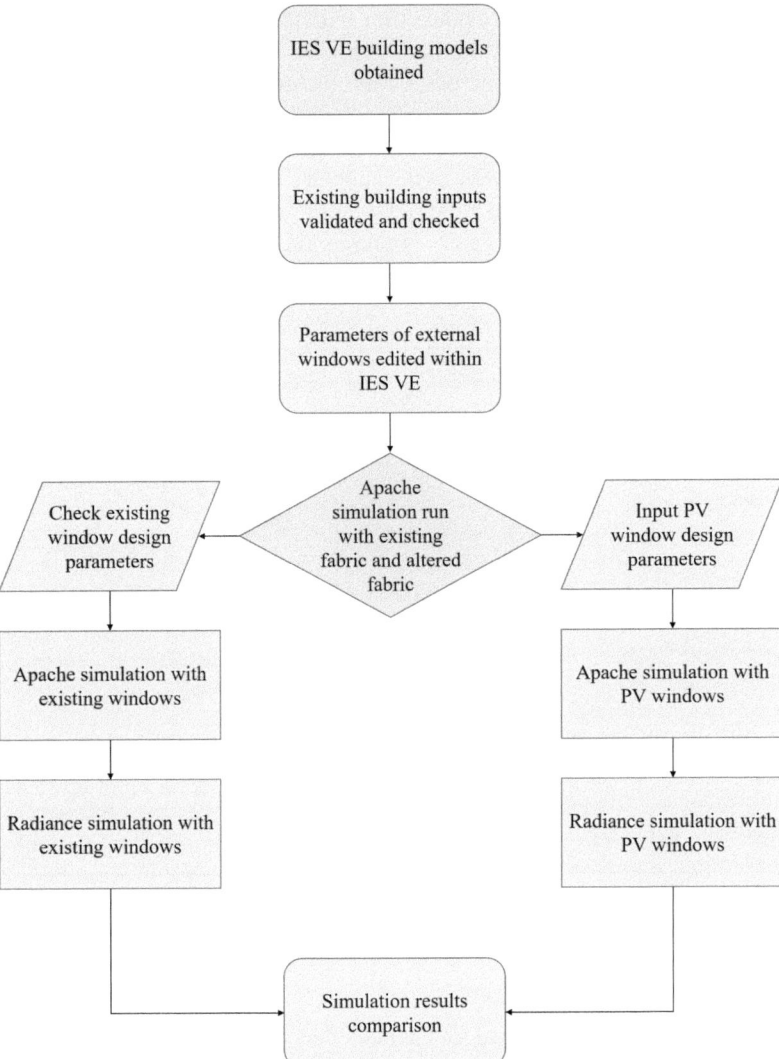

Fig. 2. Workflow of simulation.

As shown in Fig. 3, an overall reduction in heating and cooling loads was found as a result of applying the BIPV windows. Cooling loads specifically were reduced in all areas throughout the year, with the open plan office benefitting from a significant reduction. The peak monthly cooling load in the open plan office observed in June was reduced from 600 kWh to 460 kWh, representing a 23.3% decrease. All other areas besides the kitchen also received a notable reduction in cooling loads, though these were less significant, experiencing a reduction of the peak monthly load of 8.1% and 4.7% for the small and large meeting rooms respectively. It was expected that these reductions were lower, as the internal rooms would not benefit from the comparable reduction in solar gains experienced by the open plan office with the semi-transparent PV applied to its south facade. Heating loads, on the other hand, were increased slightly in some months for all areas; but in this case the increases were insignificant and cancelled out by the peak monthly heating load reduction, for example, which was observed in the open plan office during the month of December with a heating reduction of 7.5%. The kitchen has a single south-facing window, where was not applied PV glazing, and thus any effects on its internal heat situation would be gleaned from the adjacent rooms. Although cooling loads reduction in the kitchen area were observed, these were statistically minor when compared to the other areas on the same floor.

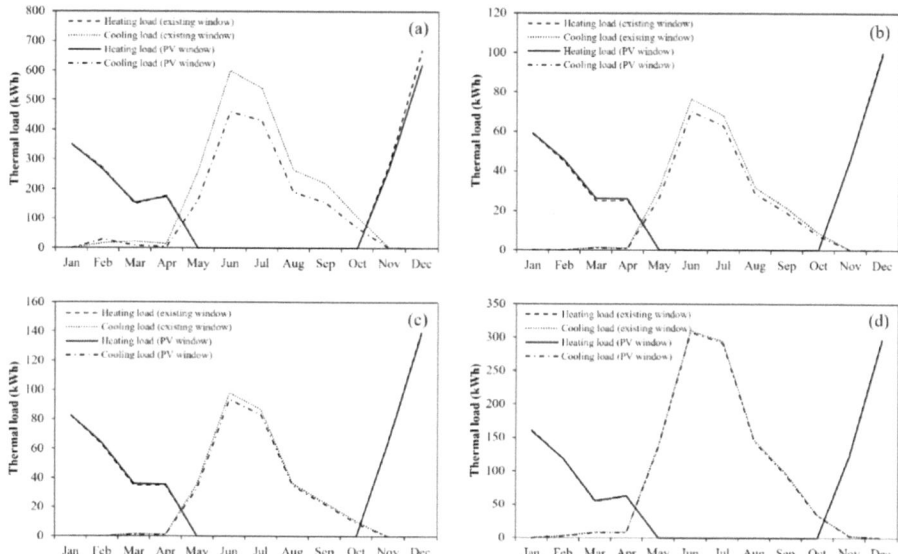

Fig. 3. Monthly thermal loads: (a) open plan office, (b) small meeting room, (c) large meeting room, (d) kitchen.

Furthermore, when taken in total for the modelled typical floor of the office building, the results of annual thermal loads can be shown in Fig. 4. It shows a reduction of 1.89% in heating loads and a 16.3% reduction in cooling loads throughout the year. Basically, the results are understood to be due to the significantly better U-value of the semi-transparent PV window (that is, 1.0 $W/m^2 \cdot K$) compared to the existing window (that is, 1.68 $W/m^2 \cdot K$), effectively offsetting the lack of solar gains experienced by the floor during winter.

3.2 Daylighting Performance of the Building with BIPV Windows

In addition to the changes to the heating and cooling loads of the building, the daylighting illuminance was expected to reduce significantly due to the low visible light transmittance of the semi-transparent PV windows of 20% versus the 71% of the existing windows. The baseline scenario calculation results are shown in Table 4, calculated on the equinox of

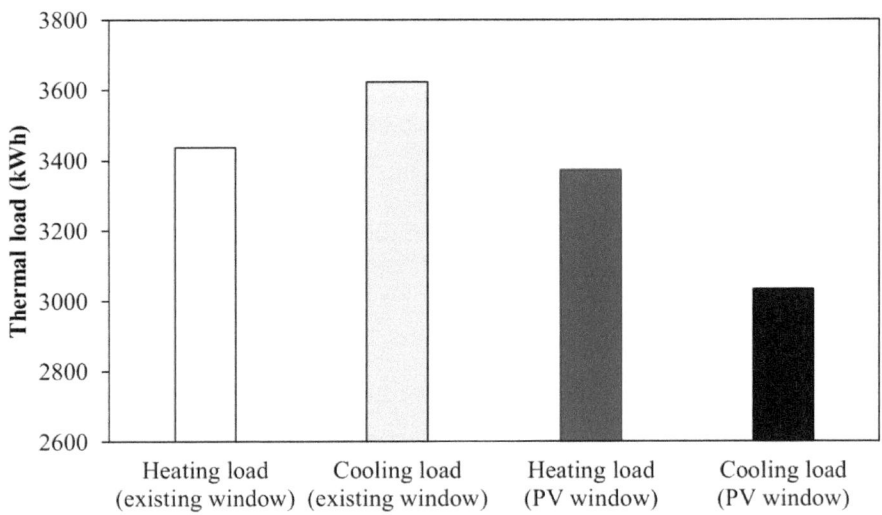

Fig. 4. Annual thermal loads of the entire floor under different scenarios.

September 23rd. The equinox was used as the hours of day and night are roughly equal, giving an average case scenario for daylight incidence when using the CIE overcast sky model.

Table 4. Baseline scenario daylight factors on the day of the equinox.

	Area (m^2)	Average DF	Minimum DF	Maximum DF	% floor >2% DF	U_o
Large meeting room	53.8	2.98	0.91	10.93	52	0.3
Small meeting room	37.96	2.94	0.98	10.6	55.5	0.33
Open plan office	323.13	2.52	0.68	11.94	39.32	0.27
Kitchen	120.02	7.69	1.96	41.37	99.76	0.255

As evidenced in the baseline scenario, all rooms except the open plan office could pass the BREEAM requirements for an average DF of 2% and a minimum of 0.8%, with a minimum uniformity of 0.3, although all areas perform poorly in terms of daylighting. Figure 5 shows the lux levels across the floor in the open plan office, where the issue with the existing design and its implications became clearer. The daylight lux levels dropped off significantly from either the north facade or south facade extending towards the centre of the room.

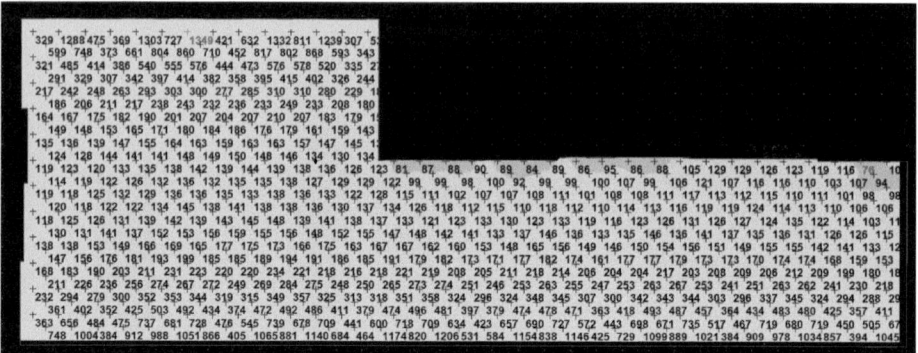

Fig. 5. Baseline scenario lux level for open plan office.

As can be seen in Fig. 6, the negligible levels of light from the south facade made it to the small meeting rooms, while a minimum of 107 lx was observed at the border of the large meeting room. Given this situation, it was expected that the application of the semi-transparent PV windows to the south facade of the open plan office would have little effect on the daylighting situations found in the north-facing rooms. Looking specifically at the open plan office, it is clear that significant artificial lighting is required in the existing design in order to meet the 500 lx requirement for the space, set out in BS EN 12464-1 [41]. Moreover, the kitchen was excluded as it shared no openings with any of the neighbouring rooms.

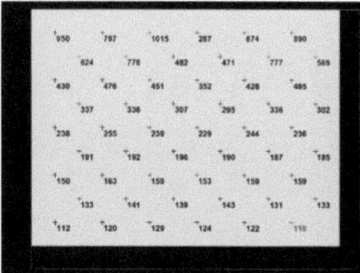 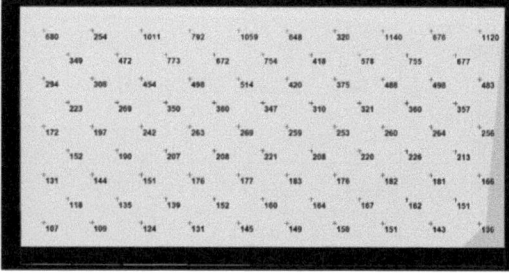

Fig. 6. Baseline scenario lux level: small meeting room (left), large meeting room (right).

Table 5 shows the daylight factor results with the semi-transparent BIPV windows applied to the south-facing facade (covering the open plan office). The results in the neighbouring rooms were basically unchanged as expected. The open plan office, however, would now be expected to fail the BREEAM requirements, with a poor average of 1.24 DF, U_o of 0.177, and an acceptable DF coverage of only 13.5% of the floor area. This represents a significant 50% decrease in DF for the open plan office adopting the

BIPV windows. While interpreting daylight factor results, a metric commonly used is the % floor >2% DF.

Table 5. BIPV window scenario daylight factors on the day of the equinox.

	Area (m^2)	Average DF	Minimum DF	Maximum DF	% floor >2% DF	U_0
Large meeting room	53.8	2.97	0.93	10.33	53	0.31
Small meeting room	37.96	2.9	0.92	10.61	52.78	0.317
Open plan office	323.13	1.24	0.22	11.89	13.5	0.177
Kitchen	120.02	7.69	2	41.42	99.76	0.26

Referring to the lux levels of the open plan office displayed in Fig. 7, the averages achieved in the daylight factors are skewed by the maximums at the north- and south-facing facades, demonstrated by the poor U_0 of 0.27 and 0.177 for both standard windows and BIPV windows, respectively. It follows therefore that the artificial lighting power required for the space to achieve the 500 lx at the working plane would be high for both scenarios. However, Fig. 8 shows an obvious difference in lighting power usage between both scenarios throughout the year; with the total annual lighting power usage being 4521 kWh for the BIPV window scenario, and 3284 kWh for the standard window scenario (baseline). It verifies that the BIPV window is not as good as the standard window in offering a better daylighting condition.

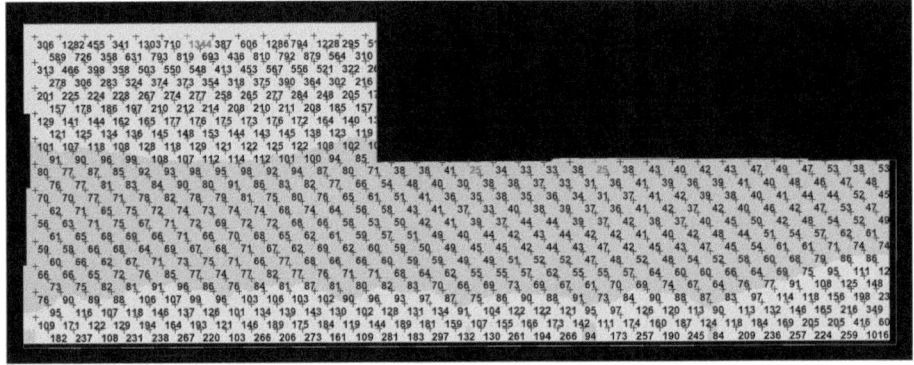

Fig. 7. BIPV window scenario lux level for open plan office.

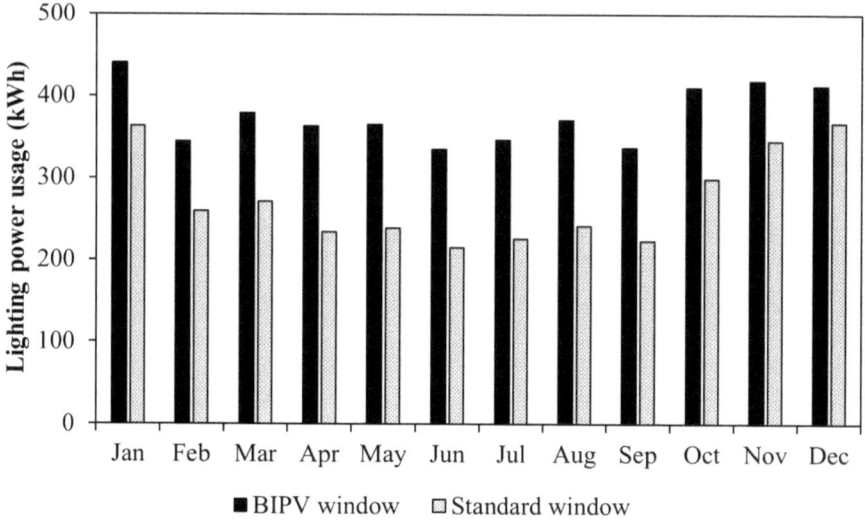

Fig. 8. Comparison of artificial lighting power usage.

3.3   Total Energy Performance of the Building with BIPV Windows

The additional power requirements for artificial lighting between the two scenarios is significant, with a difference of 1237 kWh over the year (as per Fig. 8). The useful surface area of the solar cells installed on the glazing was 40.6 m², at a 90° inclination from horizontal with a 185° azimuth clockwise from north. As shown in Fig. 9, the PV window's energy production would only cover about 73% of the increased demand from artificial lighting in the BIPV window scenario, with a total yearly expected output of 906 kWh. However, together with the better thermal performance achieved as a result of the windows (Fig. 10), the building overall with BIPV windows utilised less energy when considering the PV energy generated, although only by 3.1% (as per 10345 kWh for the existing window scenario and 10022 kWh for the BIPV window scenario).

4   Simulation Model Validation

In order to validate the simulation outputs with regard to the BIPV electric production, the following equation can be used to glean the energy generated per useful area:

$$E = A \bullet r \bullet H \bullet PR \qquad (4)$$

where E is the PV energy output (kWh), A is the area of BIPV installed (m²), r is the PV module yield (%), H is the in-plane solar irradiance (kWh/m²/year), PR is the PV performance ratio (the combined losses, assumed to be 0.9 here).

For the selected floor, the usable window area of PV module was 40.6 m² only. The utilised efficiency was 3.41% as per Table 3. The in-plane solar irradiance was determined using the solar resource map of the UK from the database – Solargis [42]. Therefore, the PV energy output can be calculated as 918 kWh under the standard

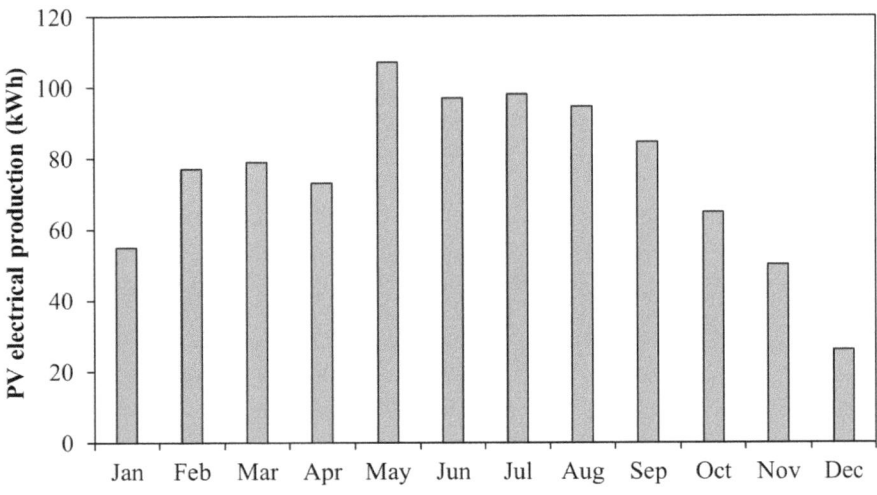

Fig. 9. Electrical production of the BIPV window.

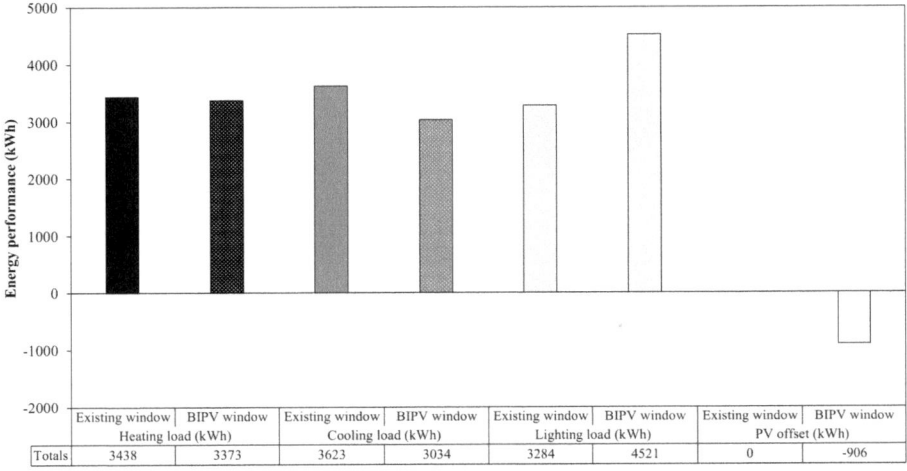

	Existing window	BIPV window	Existing window	BIPV window	Existing window	BIPV window	Existing window	BIPV window
	Heating load (kWh)		Cooling load (kWh)		Lighting load (kWh)		PV offset (kWh)	
Totals	3438	3373	3623	3034	3284	4521	0	-906

Fig. 10. Comparison of overall energy performance between the base and BIPV scenarios.

test condition (STC) in average irradiation conditions for Leeds. Compared to the IES gleaned results of 906 kWh for the year, it appears the simulation is representative of the likely real power output, which is validated within a good engineering tolerance (only 1.3% margin of error) [43]. Similarly, the simulated lighting results can be validated in order to deduce the accuracy of the simulations using the lumen method. The equation used for the validation is as follows:

$$F \bullet n = \frac{A \bullet E}{UF \bullet MF} \tag{5}$$

where F is the luminous flux, n is the number of lamps, A is the floor area, E is the required lux level on the working plane, UF is the utilisation factor, MF is the maintenance factor.

Using a typical mid-powered 600×600 LED lamp rated at 2300 lumens (20 W), while assuming a utilisation factor of 1 and a maintenance factor of 0.67 are used (based on the guidelines of a 'clean room, 3 year maintenance cycle' from BS EN 12464-1 [41]). In addition, the floor is to be lit to 500 lx as per a general office environment. Thus, 174 LED lamps are needed for the office floor (about 535 m^2 as per Table 5).

The building occupants confirmed that the dimming profile utilised within the office building aims to ramp up power usage as required to meet 500 lx at the typical working plane (0.7 m), aiming to use 50% of the illumination contribution from daylight, being switched on from 7 am – 9 pm on working days only (252 days in the UK). Therefore, the likely energy usage from the 174 LED lamps, under the assumption that zero daylight contribution is factored, is 12423.6 kWh. Looking at the simulated energy used in the artificial lighting of 4521 kWh (Fig. 10) and assuming that 50% daylight contributions were met, so the calculated usage of 12423.6 kWh can be divided by two to get 6211.8 kWh. The difference between both values is about 27.2%, which is still considered a fair accuracy [43]. Thus, the lighting part of simulation is deemed validated.

It should be noted that although the true power usage in the building will depend on actual usage patterns, the dimming profiles would likely be altered in the BIPV windows case to account for the significantly lower daylighting contributions.

5   Conclusions

This paper presents a case study of using semi-transparent BIPV glazing to retrofit a typical floor of a high-rise office building in the UK, which responds to the concerns of future research (as per the recently published studies [16, 27, 43]) on the novel BIPV technology applications in high-rise buildings in temperate climates. In terms of thermal energy and lighting performance, both baseline (existing standard window) and BIPV window scenarios for the typical floor were investigated comparably through simulations. The significant space on this floor was an open plan office, with additional meeting rooms (one large and one small) and kitchen area. The main research findings of the study can be summarised as follows:

1) The significant cooling load reductions were found for all the areas throughout the year when applying the BIPV windows bringing an overall reduction of 16.3% relative to the baseline, especially the peak monthly cooling load of the open plan office dropped significantly by 23.3%. The heating load performance remained similarly month to month, which achieved a reduction by 1.89% over the year.
2) In terms of daylighting condition, both scenarios were not desirable, meaning energy demand for artificial lighting would be increased. Although daylighting conditions of the meeting rooms and kitchen area were similar in both scenarios, the average daylight factor in the open plan office reduced over 50% from 2.52% to 1.24%. This would cause the spaces to fail the minimum requirements for daylighting (for example, the BREEAM).

3) Furthermore, the BIPV window produced electrical energy was little, with a total of 906 kWh for the entire office floor over a year. As such, the PV generation contribution only covered 73% of the increased lighting demand. Nevertheless, due to the significant reduction in cooling loads and slight reduction in heating loads, the electrical power generation from the BIPV windows appeared to be a slight improvement when compared to the baseline scenario.

However, this study was merely a desktop research based upon the computer simulation modelling. Future experiments and simulations should be carried out for different building types in the similar climate scenarios to further validate the findings in this paper.

References

1. Hepple, R., Du, H., Feng, H., Shan, S., Yang, S.: Sustainability and carbon neutrality in UK's district heating: a review and analysis. e-Prime. Adv. Electr. Eng. Electron. Energy **4** (2023). https://doi.org/10.1016/j.prime.2023.100133
2. Hancox, L., Yang, S., Hallam, P., White, M., Memon, S.: An assessment for the viability of recovering heat from a smoke extract system. Energy Built Environ. **4**, 458–466 (2023). https://doi.org/10.1016/j.enbenv.2022.03.003
3. Saretta, E., Caputo, P., Frontini, F.: A review study about energy renovation of building facades with BIPV in urban environment. Sustain. Cities Soc. **44**, 343–355 (2019). https://doi.org/10.1016/j.scs.2018.10.002
4. Turner, L., Yang, S., White, M.: Study of energy saving potential of solar shading devices in various climates. In: Proceedings of the International Conference on Building Energy and Environment, pp. 1181–1189 (2022)
5. UK Green Building Council. Net Zero Whole Life Carbon Roadmap: A Pathway for the UK Built Environment, London, UK (2021)
6. Alabid, J., Bennadji, A., Seddiki, M.: A review on the energy retrofit policies and improvements of the UK existing buildings, challenges and benefits. Renew. Sustain. Energy Rev. **159** (2022). https://doi.org/10.1016/j.rser.2022.112161
7. Shan, S., Yang, S., Becerra, V., Deng, J., Li, H.: A case study of existing peer-to-peer energy trading platforms: calling for integrated platform features. Sustainability **15** (2023). https://doi.org/10.3390/su152316284
8. Yang, S., Fiorito, F., Sproul, A., Prasad, D.: Study of building integrated photovoltaic/thermal double-skin facade for commercial buildings in Sydney, Australia. In: Proceedings of the Final Conference of COST TU1403 "Adaptive Facades Network", Lucerne, Switzerland (2018)
9. Yang, S., Fiorito, F., Sproul, A., Prasad, D.: Studies on optimal application of building-integrated photovoltaic/thermal facade for commercial buildings in Australia. In: Proceedings of the Proceedings of SWC2017/SHC2017, pp. 1–10 (2017)
10. Attia, S., Bertrand, S., Cuchet, M., Yang, S., Tabadkani, A.: Comparison of thermal energy saving potential and overheating risk of four adaptive façade technologies in office buildings. Sustainability **14** (2022). https://doi.org/10.3390/su14106106
11. Yang, S., Fiorito, F., Prasad, D., Sproul, A.: Numerical simulation modelling of building-integrated photovoltaic double-skin facades. In Bulnes, F., Hessling, J.P. (Eds) Recent advances in numerical simulations. IntechOpen, London, UK, pp. 61–75 (2021)
12. Gagliano, A., Tina, G.M., Aneli, S., Chemisana, D.: Analysis of the performances of a building-integrated PV/Thermal system. J. Cleaner Prod. **320** (2021). https://doi.org/10.1016/j.jclepro.2021.128876

13. Shukla, A.K., Sudhakar, K., Baredar, P.: Recent advancement in BIPV product technologies: a review. Energy Buildings **140**, 188–195 (2017). https://doi.org/10.1016/j.enbuild.2017.02.015
14. Li, X., Peng, J., Tan, Y., He, Y., Li, B., Ju, X., Ji, J., Zhang, S., Li, N., Chen, Y.: Optimal design of inhomogeneous semi-transparent photovoltaic windows based on daylight performance and visual characters. Energy Buildings **283** (2023). https://doi.org/10.1016/j.enbuild.2023.112808
15. Peng, J., Curcija, D.C., Lu, L., Selkowitz, S.E., Yang, H., Zhang, W.: Numerical investigation of the energy saving potential of a semi-transparent photovoltaic double-skin facade in a cool-summer Mediterranean climate. Appl. Energy **165**, 345–356 (2016). https://doi.org/10.1016/j.apenergy.2015.12.074
16. Yang, S., Fiorito, F., Sproul, A., Prasad, D.: Optimising design parameters of a building-integrated photovoltaic double-skin facade in different climate zones in Australia. Buildings **13** (2023). https://doi.org/10.3390/buildings13041096
17. Elhabodi, T.S., Yang, S., Parker, J., Khattak, S., He, B.-J., Attia, S.: A review on BIPV-induced temperature effects on urban heat islands. Urban Climate **50** (2023). https://doi.org/10.1016/j.uclim.2023.101592
18. Yang, S.: Studies on the Performances of Building Integrated Photovoltaic/Thermal Double-Skin Facade for Commercial Buildings in Australia. PhD thesis, University of New South Wales, Sydney (2020)
19. Chen, L., Yang, J., Li, P.: Modelling the effect of BIPV window in the built environment: Uncertainty and sensitivity. Building Environ. **208** (2022). https://doi.org/10.1016/j.buildenv.2021.108605
20. Ghosh, A.: Investigation of vacuum-integrated switchable polymer dispersed liquid crystal glazing for smart window application for less energy-hungry building. Energy **265** (2023). https://doi.org/10.1016/j.energy.2022.126396
21. Roberts, F., White, M., Memon, S., He, B.-J., Yang, S.: The application of human-centric lighting in response to working from home post-COVID-19. Buildings **13** (2023). https://doi.org/10.3390/buildings13102532
22. Moschella, A., Amato, D., Gagliano, A.: Lighting characterization of an Italian beginning twentieth-century school building. Renew. Energies Power Qual. **21**, 381–387 (2023). https://doi.org/10.24084/repqj21.330
23. Aguilar-Santana, J.L., Velasco-Carrasco, M., Riffat, S.: Thermal Transmittance (U-value) evaluation of innovative window technologies. Future Cities Environ. **6**, 12 (2020)
24. Memon, S., et al.: Modern Eminence and Concise Critique of Solar Thermal Energy and Vacuum Insulation Technologies for Sustainable Low-Carbon Infrastructure. Int. J. Solar Therm. Vac. Eng. **1**, 52–71 (2020). https://doi.org/10.37934/stve.1.1.5271
25. Yang, S., Fiorito, F., Prasad, D., Sproul, A., Cannavale, A.: A sensitivity analysis of design parameters of BIPV/T-DSF in relation to building energy and thermal comfort performances. J. Building Eng. **41** (2021). https://doi.org/10.1016/j.jobe.2021.102426
26. Mirrahimi, S., Mohamed, M.F., Haw, L.C., Ibrahim, N.L.N., Yusoff, W.F.M., Aflaki, A.: The effect of building envelope on the thermal comfort and energy saving for high-rise buildings in hot–humid climate. Renew. Sustain. Energy Rev. **53**, 1508–1519 (2016). https://doi.org/10.1016/j.rser.2015.09.055
27. Roberts, F., Yang, S., Du, H., Yang, R.: Effect of semi-transparent a-Si PV glazing within double-skin façades on visual and energy performances under the UK climate condition. Renew. Energy **207**, 601–610 (2023). https://doi.org/10.1016/j.renene.2023.03.023
28. Yang, S., Cannavale, A., Prasad, D., Sproul, A., Fiorito, F.: Numerical simulation study of BIPV/T double-skin facade for various climate zones in Australia: Effects on indoor thermal comfort. Build. Simul. **12**, 51–67 (2018). https://doi.org/10.1007/s12273-018-0489-x

29. Yang, S., Cannavale, A., Di Carlo, A., Prasad, D., Sproul, A., Fiorito, F.: Performance assessment of BIPV/T double-skin façade for various climate zones in Australia: effects on energy consumption. Sol. Energy **199**, 377–399 (2020). https://doi.org/10.1016/j.solener.2020.02.044

30. Conde, M.G.S., Shanks, K.: Evaluation of available Building Integrated Photovoltaic (BIPV) systems and their impact when used in commercial buildings in the United Arab Emirates. Int. J. Sustain. Energy Dev. (IJSED) **7** (2019)

31. Gholami, H., Røstvik, H.N.: Economic analysis of BIPV systems as a building envelope material for building skins in Europe. Energy **204**, 117931 (2020)

32. Alrashidi, H., Ghosh, A., Issa, W., Sellami, N., Mallick, T.K., Sundaram, S.: Thermal performance of semitransparent CdTe BIPV window at temperate climate. Sol. Energy **195**, 536–543 (2020). https://doi.org/10.1016/j.solener.2019.11.084

33. Constantinou, S., Al-naemi, F., Alrashidi, H., Mallick, T., Issa, W.: A review on technological and urban sustainability perspectives of advanced building-integrated photovoltaics. Energy Science & Engineering (2023)

34. Khalifeeh, R., Alrashidi, H., Sellami, N., Mallick, T., Issa, W.: State-of-the-art review on the energy performance of semi-transparent building integrated photovoltaic across a range of different climatic and environmental conditions. Energies **14** (2021). https://doi.org/10.3390/en14123412

35. Ghosh, A., Sarmah, N., Sundaram, S., Mallick, T.K.: Numerical studies of thermal comfort for semi-transparent building integrated photovoltaic (BIPV)-vacuum glazing system. Sol. Energy **190**, 608–616 (2019). https://doi.org/10.1016/j.solener.2019.08.049

36. Yu, G., Yang, H., Luo, D., Cheng, X., Ansah, M.K.: A review on developments and researches of building integrated photovoltaic (BIPV) windows and shading blinds. Renew. Sustain. Energy Rev. **149** (2021). https://doi.org/10.1016/j.rser.2021.111355

37. Futcher, J.A., Kershaw, T., Mills, G.: Urban form and function as building performance parameters. Build. Environ. **62**, 112–123 (2013)

38. NABERS UK. Guide to Design for Performance (2021)

39. The Society of Light and Lighting. The SLL Code for Lighting (2012)

40. Hepple, R., Yang, S., Khattak, S., Qian, Z., Prasad, D.: Comparative analysis of CIBSE admittance and ASHRAE radiant time series cooling load models. CivilEng **3**, 468–479 (2022). https://doi.org/10.3390/civileng3020028

41. British Standards Institution. BS EN 12464-1: Light and lighting. Lighting of work places-Indoor work places 2021

42. Solargis. Solar resource maps of United Kingdom. Available online: https://solargis.com/maps-and-gis-data/download/united-kingdom (accessed on 30 September)

43. Jhumka, H., Yang, S., Gorse, C., Wilkinson, S., Yang, R., He, B.-J., Prasad, D., Fiorito, F.: Assessing heat transfer characteristics of building envelope deployed BIPV and resultant building energy consumption in a tropical climate. Energy Buildings **298** (2023). https://doi.org/10.1016/j.enbuild.2023.113540

Application of BIM in Renovation Design of Existing Buildings

Yuan Zhan, Yu Peng, and Changrong Xiong[✉]

Zhongnan Hospital of Wuhan University, Wuhan University, Wuhan 430071, China
19947626742@163.com

Abstract. Based on the analysis of the reconstruction process and corresponding problems of the old building, the paper discusses the BIM modeling process based on three-dimensional laser scanning technology, namely, acquiring the point cloud data of the existing building through 3D laser scanning, importing the point cloud processing software for preprocessing, and Revit software, docking establishes the process of building a BIM model of an existing building. Through the application of BIM technology in the pre-reformation, design process, performance design and other aspects of the existing building renovation design, it is demonstrated that BIM has a positive effect on the design and management of the existing building renovation design stage.

Keywords: building information modeling (BIM) · existing buildings · renovation · design · concrete · three-dimensional scanning technology

1 Introduction

The challenges inherent in the renovation of existing buildings are multifaceted, stemming from both the characteristics of the buildings themselves [1], and the limitations of current computer-aided design (CAD) software [2]. Existing buildings pose unique challenges due to the loss and damage of information during construction and a lack of maintenance management during the operational period, leading to diminished accuracy and traceability of information. Simultaneously, the prevalent CAD software, designed to support traditional concepts and processes, falls short in achieving collaborative design, hindering the timely transmission of crucial information, and resulting in low utilization and susceptibility to distortion and loss. Moreover, the process of extracting information from drawing documents exhibits difficulties in querying, slow extraction speed, and low efficiency. Existing buildings in China lack effective operational maintenance management, and information traceability is poor [6]. In summary, compared to new construction projects, renovations of existing buildings face challenges such as high design difficulty, demanding quality and safety management, complex construction processes, poor site conditions, and dynamic project management requirements. The design phase of existing building renovation projects is time-consuming and inefficient. The technology roadmap is shown in Fig. 1.

Yuan Zhan and Yu Peng contributed equally to the work

© The Author(s) 2025
B.-J. He et al. (Eds.): UCSUD 2023, LNCE 559, pp. 263–271, 2025.
https://doi.org/10.1007/978-981-97-8401-1_18

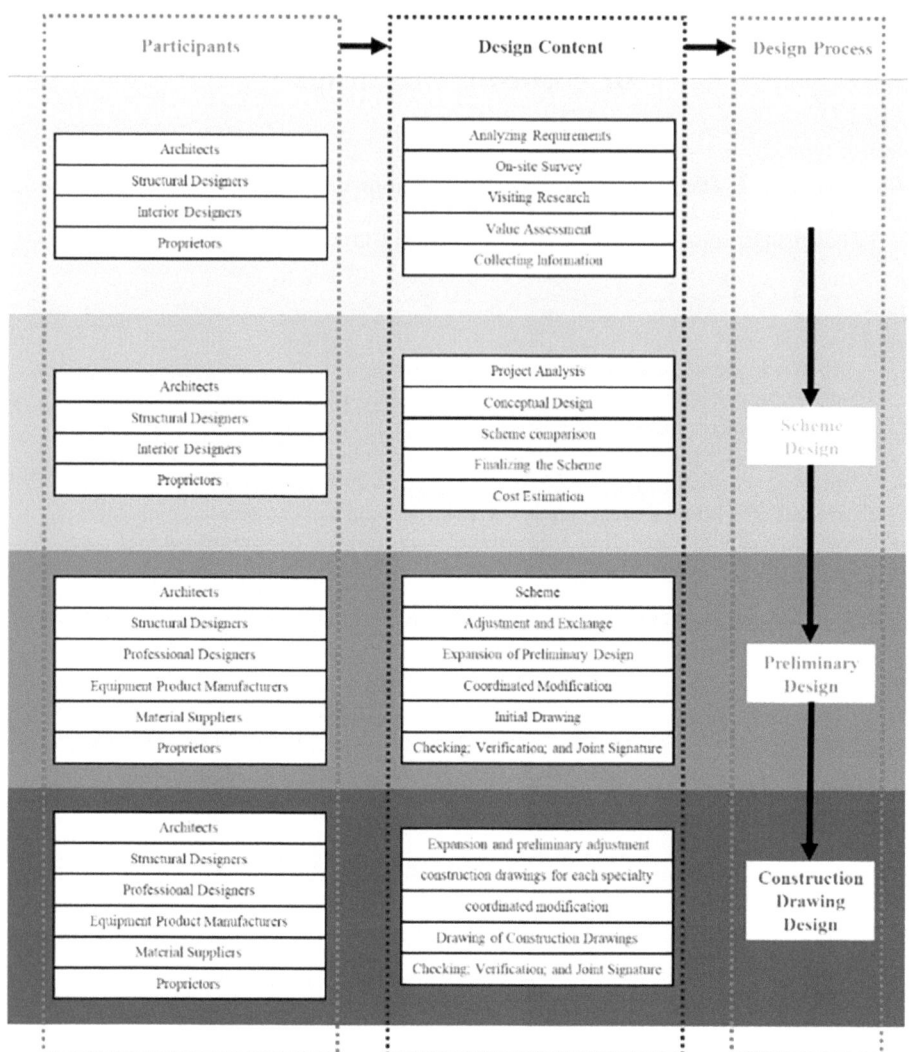

Fig. 1. Technology roadmap for existing building renovation design

This approach is suitable for relatively simple construction projects and can also enhance work efficiency. However, in the traditional process, progressive information transmission leads to delayed information, resulting in issues in professional design transformations that cannot be promptly identified and resolved. Consequently, the design cycle is prolonged, and efficiency is compromised [5].

Although the current design tools for architects and interior designers can improve work efficiency and drawing accuracy, they often cannot avoid the following three issues [5]: a) 2D drawing: using CAD as an example, can only present various 2D plans. It lacks the sense of hierarchy and construction information in 3D space, requiring high spatial imagination from designers and demanding high professional knowledge and experience from readers; b) 3D drawing: while it is possible to create a 3D geometric model of the original old building, the model information lacks the parameters of the components; c) project document management: due to numerous interim results and the presence of various types of documents, the integration and utilization of information pose significant challenges. In the design process of renovating existing buildings, design information is scattered, not easily integrated, and exhibits diverse forms. Therefore, in practical engineering, it is crucial to bolster the preservation and management of information pertaining to existing buildings, improve the efficiency and utilization of information transfer at various stages of construction. When considering the subsequent operation and utilization of the building, it is advisable to establish a standardized plan for daily maintenance management.

BIM technology is underpinned by cutting-edge scientific advancements and the principles of scientific digitization, enhancing traditional design concepts and processes while addressing the intricate challenges posed by conventional renovations of old buildings [6]. Enhanced by its distinctive features of visualization, associativity, coordination, optimization, and ability to generate graphics, BIM offers a groundbreaking, practical, convenient, and precise methodology for revitalizing existing structures. This not only positively influences the renovation design of historical buildings, but adeptly caters to the growing demands in the market for upgrading existing structures.

2   3D Laser Scanning Technology

3D laser scanning methods not only enhance safety and efficiency but yield data of superior accuracy, precision, density, and three-dimensionality. Furthermore, the obtained results can be seamlessly integrated into BIM software following necessary processing steps [7]. Moreover, 3D laser scanning technology utilizes a non-contact laser measurement method, eliminating the need for reflective prisms. This method captures the three-dimensional coordinates of visible target surfaces in the form of point clouds. In the initial phases of renovating old buildings, the utilization of 3D laser scanning technology for intricate or high-risk measurements, and achieves a high-precision, high-resolution architectural information model, is emerging as a prevailing trend in the engineering industry.

2.1   Measurement Principles

Operating on the principle of laser ranging, 3D laser scanners deploy multiple stations around the target object, executing comprehensive and dense point measurements at each station. This process entails emitting a series of laser beams toward the target object, subsequently reflected by it. Factoring in both the reflection distance and angle, the 3D coordinates, texture, color information, and other relevant data of the measured object at that station are acquired. This culminates in the creation of a three-dimensional digitized point cloud representing the target object. The alignment of point clouds from individual stations using specialized analysis software, followed by essential data editing, output procedures. Subsequent integration and synthesis of point clouds obtained from various stations, leads to the creation of a panoramic point cloud representing the target object. When determining the number and positioning of stations, it is crucial to select a setup plan that minimizes measurement blind spots, ensures reasonable perspectives, and facilitates efficient measurements.

2.2 3D Laser Scanning Technology Measurement Equipment

Measurement equipment requires the use of scanner, tripod, target, GPS locator and computer. In the process of measuring with a 3D laser scanner, it is necessary to establish multiple stations, and subsequently, integrate and stitch the information from each station to generate a three-dimensional model of the target object. The foundation and limitation for this stitching process rely on the overlapping information obtained from different stations. Consequently, when measuring between every two stations, it is imperative to set up three common measurement points. Targets are markers placed in the measurement scene before measuring at each station, ensuring that the positioning of the stitched common points is highly precise and accurate. This is an accessory device affixed to the scanner, and its usage can be opted for during measurements. GPS positioning allows for the measurement of ground coordinates for each station locator. Integrating the ground coordinates of the scanner at different stations with their respective measurement results enhances the accuracy in determining the positional coordinates of the target object. Throughout the scanning process and finished, it is essential to link scanner data to the computer, execute required processing, integrate data from various stations, and color assignment to point clouds.

2.3 Factors Influencing Measurement Accuracy and Control

The employment of laser ranging introduces factors such as the distance between the scanner and the measured object, laser beam intensity, and the reflectivity of the measured object, all of which collectively influence the accuracy of measurements. Point cloud data obtained through 3D laser scanning, in relation to the scanner, must undergo point cloud registration to be transformed into data represented in a consistent reference coordinate system. Point cloud registration methods encompass target registration and registration using corresponding point pairs. When employing registration with corresponding point pairs, the registration outcomes are contingent on factors including the data accuracy of the existing building point cloud, the distance between the scanning equipment and the measured object, and the reflectivity of the measured object. In a real-time environment, as the distance between the equipment and the object being measured increases, the laser spot area on the object also enlarges. The cumulative enlargement of the distance between adjacent measurement stations diminishes the accuracy and stitching quality of the measurement results.

3 Methodology

3.1 3D Scanning and Data Acquisition

By leveraging data from the existing building and conducting thorough on-site investigations, we determine the number and locations of each station, as well as the suitable target settings between different stations. For the assurance of measurement accuracy during station setup, it is imperative to maintain a distance between the equipment and the measured building of less than 50m when deciding the number and locations of stations. If this condition is not met, the requirement can be fulfilled by augmenting the number of stations. Additionally, the selection of suitable technology, in alignment with practical requirements, is crucial for determining the scanning main span and internal point cloud sampling interval. When determining the position of the target, ensuring measurement accuracy, and achieving an even distribution of different targets across various measurement locations, the specific characteristics of the scanning object should be the primary consideration. Additionally, efforts should be made to maximize the distance between the scanner and the corresponding target, keeping it within 50m. Conduct on-site measurements in accordance with the predetermined scanning plan to acquire point cloud data of existing buildings.

3.2 Data Processing

Point Cloud Pre-processing
Trimble Realworks, Autodesk Recap, Autodesk Revit, etc., to remove noise points and distorted elements from the original graphics. The point cloud data obtained from the scanner represent measurement values in the measurement coordinate system. To align with the project requirements, a coordinate transformation is necessary to convert these measurement coordinates into the spatial coordinates essential for the project. The directly acquired data contains datasets unrelated to the renovation design of existing buildings and the entire lifecycle.

Point Cloud Registration
To minimize point cloud registration errors, the choice of registration methods should align with the specific measurement conditions. Typically, the registration method is prioritized in the sequence of target stitching and same-name feature point stitching. Alternatively, a combination of these two stitching methods can be employed depending on the actual circumstances, aiming to attain superior registration results. Subsequently, the point cloud is consolidated based on the identified targets following the data processing from each station.

3.3 Model Construction

After conducting operations such as denoising, coordinate transformation, redundancy removal, and registration on the original point cloud, generate a data file in the *. Lass format. Import the point cloud data in *. Lass format into Revit software to create an RCP format through data indexing. Reinsert the point cloud data in RCP format into

Revit software. At this stage, the three-dimensional coordinate information of each point in the point cloud data is successfully imported into Revit software. To ascertain the three-dimensional positioning coordinates of components within the project site, utilize Revit software. Leveraging the visual representation and three-dimensional coordinate information provided by point cloud data on the computer, document the architectural information model. Generating new Revit families for the intricate elements of existing buildings. Subsequently, establish a connection to the outlined, identifiable three-dimensional positioning coordinate information. Upon delineating all components, eliminate the point cloud data to generate a BIM model. This enables the one-to-one mapping of building components on the construction site within the computer (see Fig. 2).

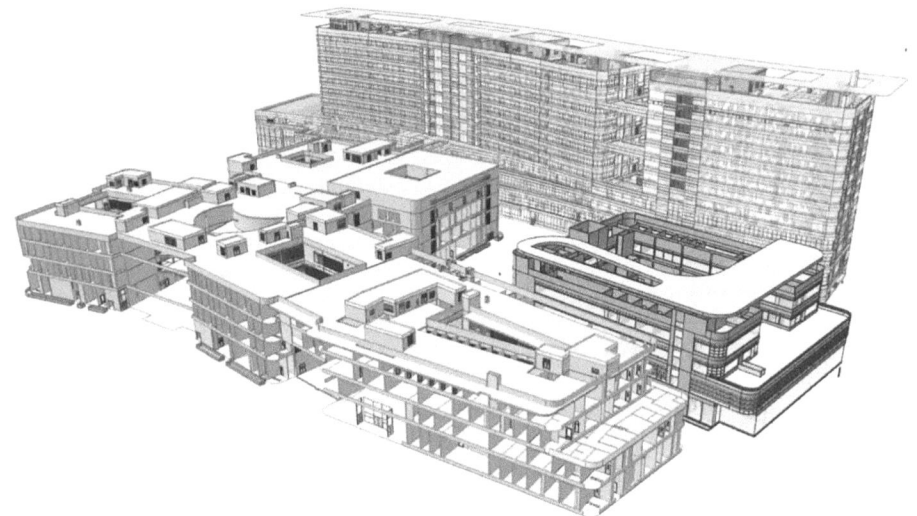

Fig. 2. Three-dimensional solid model of a building

4   The Application of BIM Technology in Various Stages of Renovation for Existing Buildings

4.1   Preliminary Planning for Renovation of Existing Buildings

In the initial phase of renovating existing buildings, the primary objective is to verify structural stability, safety, and conditions related to energy efficiency and material quality, necessitating substantial relevant data. However, the prolonged construction time of many existing buildings results in a lack of original data, complicating the tracing of material information, and obtaining structural details becomes relatively challenging. Moreover, the absence of energy analysis content in the original design underscores the need for a comprehensive analysis of existing buildings early in the renovation process. BIM technology emerges as a pivotal technical solution for energy analysis, intertwining with pertinent findings to guide the renovation design of old buildings' functional layout and facade [8]. Figure 3 (a) is the BIM-based solar analysis for buildings. Figure 3 (b) is the BIM-based simulation of the building's wind environment.

(a)   (b)

Fig. 3. Energy analysis based on BIM

Moreover, in the initial stages of construction, existing buildings often lack long-term considerations and limitations of paper document storage, mean that essential data for design primarily relies on on-site surveys. Modern methodologies, such as integrating three-dimensional scanning measurements with BIM software, enable the digitization of survey data, streamlining the rational design process. This not only enhances the efficiency and accuracy of data acquisition but also ensures the objectivity of survey results. In the preliminary survey of existing building renovation design, the application of BIM technology involves utilizing three-dimensional laser scanning technology to measure existing buildings. The acquired three-dimensional data and additional information are then imported into a computer, where BIM tools are employed for various performance analyses of existing buildings.

4.2 Optimization of Existing Building Renovation Design

The incorporation of BIM technology as a cornerstone in the early design phases, involving the input of necessary parameter information for the entire lifecycle of existing building renovations into BIM software, facilitates the establishment of an architectural information model. This model enables real-time, dynamic, interconnected, and consistent adjustments, serving as a foundational electronic document throughout the entire design process.

1). In the initial phases of design, it is imperative to refine and improve the model continually, striving to adjust the plan to the most suitable state. This proactive approach minimizes the necessity for repetitive modifications and mitigates additional workload later due to unmet specific requirements.
2). BIM systems establish uniform standards for the design process, govern the representation of BIM models, incorporating widely utilized IFC standards, and provide a unified design platform. In intricate designs for renovating existing buildings, diverse professionals can access essential information on the BIM design platform and

upload design outcomes from different stages. This ensures the real-time and effective transmission of information across various disciplines and stages, mitigating the risk of information loss and isolated islands.

3). Using models created in BIM software like Revit as an illustration, upon the completion of model construction, it can be seamlessly imported into clash detection software such as Navisworks. This offers a swift, precise, and intuitive method for pinpointing conflicts across various disciplines. Upon the completion of optimizing the information model for existing buildings, designers, at this stage, can maximize the graphical capabilities of BIM software to export necessary 2D drawings and essential architectural model images.

Consequently, during the initial planning and optimization design phases of old building renovation, the utilization of BIM tools facilitates real-time information exchange and communication. This ensures ongoing optimization of solutions during collaborative design, decreasing the design change workload, enhancing the depth of the design, ultimately resulting in an optimized model. It further contributes to the construction of an accurate information model for the entire lifecycle of existing building renovation.

4.3   Performance Analysis of Existing Building Renovation Design

In response to energy-saving requirements and other sustainable development needs, the current renovation of old buildings is primarily focused on enhancing various performance aspects. Within the BIM technology environment, building performance, covering aspects such as lighting, solar radiation, thermal conditions, wind exposure, and energy consumption, can be simulated and analyzed. Diverse design professionals can utilize BIM software platforms to construct models incorporating data for existing buildings. These models can be seamlessly integrated with pertinent energy analysis software to assess the performance of existing structures. This aids in identifying design solutions characterized by low energy consumption, high material utilization, and convenient construction, among other considerations, through a comprehensive evaluation process.

5   Conclusion

Through the analysis of the current situation of existing building renovations, we confirm that BIM software can achieve more rational collaborative optimization design. Through an exploration of the fundamental applications of BIM technology in the design phase, extensive research has delved into measurement principles, equipment, the scanning process, point cloud data acquisition and processing, and the construction of Revit models in 3D laser scanning technology. This offers an efficient approach for obtaining diverse building parameters and facilitating collaborative design in the renovation phase of existing buildings. Furthermore, it fully exemplifies the remarkable advantages of BIM technology. A detailed analysis of the exceptional application advantages of BIM technology, encompassing the acquisition of pre-renovation information, virtualization, and visualization of the design process, and intelligent performance design during the existing building renovation design phase, contributes to a clearer understanding of its remarkable benefits in this context. This paper only studies general buildings, such as residential buildings, like special building forms, such as medical buildings, which need to be further discussed in future studies.

References

1. Wang, Shengwei, Chengchu Yan, Fu Xiao: Quantitative energy performance assessment methods for existing buildings. Energy Build. **55**, 873–888 (2012)
2. Fernandes, Rui Pedro Lopes: Advantages and disadvantages of BIM platforms on construction site. Diss. Universidade do Porto, Portugal (2013)
3. Liu, Yang, Hongyu Chen, Xian-jia Wang: Research on green renovations of existing public buildings based on a cloud model–TOPSIS method. J. Build. Eng. **34**, 101930 (2021)
4. Wu, Xianguo, et al.: Intelligent optimization framework of near zero energy consumption building performance based on a hybrid machine learning algorithm. Renew. Sustain. Energy Rev. **167**, 112703 (2022)
5. Ham, Y., Golparvar-Fard, M.: An automated vision-based method for rapid 3D energy performance modeling of existing buildings using thermal and digital imagery. Adv. Eng. Inform. **27**(3), 395–409 (2013)
6. Volk, R., Stengel, J., Schultmann, F.: Building Information Modeling (BIM) for existing buildings—Literature review and future needs. Autom. Constr. **38**, 109–127 (2014)
7. Macher, H., Landes, T., Grussenmeyer, P.: From point clouds to building information models: 3D semi-automatic reconstruction of indoors of existing buildings. Appl. Sci. **7**(10), 1030 (2017)
8. Liu, Yang, et al.: Enhancing building energy efficiency using a random forest model: A hybrid prediction approach. Energy Reports **7**, 5003–5012 (2021)

Performance of a Mechanical Pump-Driven Two-Phase Cooling System for Aircraft Systems

Baoqing Zhang, Yunyi Guo, Xingzhou Fan, and Shuang Chen$^{(\boxtimes)}$

Institute of Fluid Physics, China Academy of Physics Engineering, Mianyang 621900, China
1321174@qq.com

Abstract. In order to solve the heat dissipation problem of high heat flux components of military electronic equipment, a mechanical pump-driven two-phase flow cooling system was designed in this paper. The thermal performance of the system was tested experimentally at room temperature, high and low temperature and vibration environment. By means of the experimental data and phenomenon analysis, the system had good heat transfer performance under room temperature, high and low temperature, vibration conditions. Under the condition of acceleration environment, the system liquid storage separator won't work normally, the internal liquid would be confused, gathering in a certain part and can't circulate in the system pipeline normally which would result the failure in system. In order to solve this problem, the system can keep working normally under the condition of multi-angle acceleration by increasing the amount of liquid filled in the system.

Keywords: two-phase flow · accelerated speed · vibration · active drive

1 Introduction

With the urgent need of miniaturization and high performance of airborne electronic equipment, the chip integration requirement was getting higher and higher, besides, the heat flux of the chip at work was also increasing. The heat generated by the power chip couldn't be taken away in time, and the long-term operation of the equipment at high temperature would affect the useful life of the chip, even caused the direct failure of the chip [1, 2]. High temperatures had become a major cause of electronic equipment failure. Data showed that in the operation failure of electronic equipment, the fault caused by temperature accounted for more than 55% [3]. With the increase of temperature, the reliability and mean time to failure (MTBF) of electronic equipment dropped sharply, and its failure probability increased exponentially [4].

Active driven two-phase flow cooling technology was a kind of high efficiency cooling technology, becasue of latent heat of vaporization of liquid working medium. This technology possessed high heat transfer efficiency and external force resistance, which can provide a new way to solve the problem of high heat flux of chip [5–15]. In this article, a mechanical pump-driven two-phase flow cooling system was designed to investigate the influence of high and low temperature, vibration and acceleration to active driven two-phase flow cooling system.

© The Author(s) 2025
B.-J. He et al. (Eds.): UCSUD 2023, LNCE 559, pp. 272–285, 2025.
https://doi.org/10.1007/978-981-97-8401-1_19

2 Principle Analysis

The active driven two-phase flow cooling system was mainly composed of a drive source, evaporator, condenser, liquid reservoir and separator, fluid pipeline and internal working medium, etc. Its working principle was shown in Fig. 1. The system taken the working medium as the heat carrier, absorbed the heat of the heating element in the evaporation section, so that the liquid working medium was heated and evaporated, flowed into the condenser in the form of gas, and become liquid under the cooling of the condenser, and the heat was transferred to the external heat sink through the wall of the condenser tube, so as to achieve the transfer of heat, which could reduce the working temperature of the chip. Under the driving force of the flow bump, the liquid working medium condensed in the condenser returned to the evaporator through the liquid pipeline, so that the liquid working medium could absorb the heat, evaporated and flowed in the evaporator constantly, and the heat in the evaporator was constantly transferred to the remote heat sink.

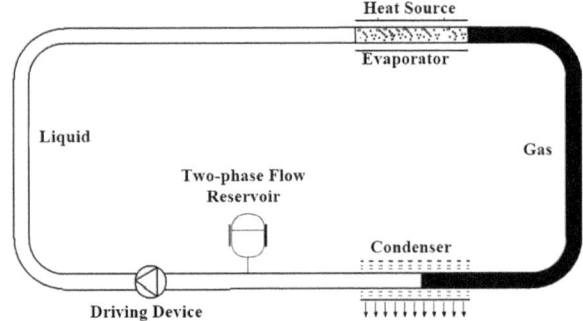

Fig. 1. Working principle of active driven two-phase flow cooling

3 Experimental Design

3.1 System Composition

The structure of the active driven two-phase flow cooling system was shown in Fig. 2. The evaporator, condenser and liquid storage separator were independently designed, and the driving source, temperature detection system, fan, pipeline and internal fluid were purchased parts. The driving source was a flow micro-pump, model GA-T23.DGF1.J, with environmental adaptability of -49–70°C, driving voltage of 24V, control voltage of 0-5V, size of about 80mm × 80mm × 93mm, maximum supply pressure of 21bar, maximum flow of 270ml/min. The flow of the pump could be changed by adjusting the control voltage value to study the thermal performance of the system under different flow rates. The pipe is made of copper with an outer diameter of Φ6 and an inner diameter of Φ4. The internal fluid used acetone with purity greater than 99.5% as the working medium, the specific heat capacity of acetone was 2.2×10^3J/kg·K, latent heat of vaporization was 52.46 × 104J/kg, and the density was 0.77g/cm³ (25°C ambient temperature). The temperature detection system was the intelligent temperature digital display instrument of Shanghai Chi Kong Automation Instrument Co., LTD. By integrating the thermocouple PT100 in the system loop, the intelligent temperature digital display instrument can display the temperature of the internal fluid in real time; The NX203100 fan was used for forced air cooling of the system.

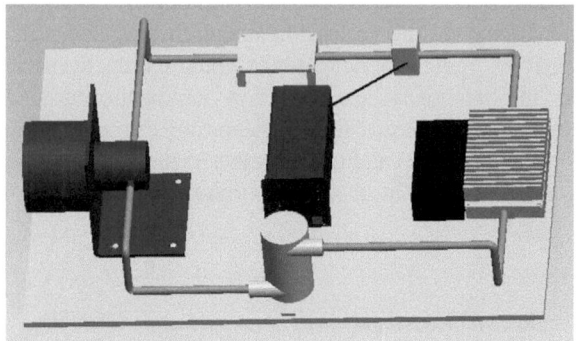

Fig. 2. Composition diagram of active driven two-phase flow system

3.2   Evaporator Structure Design

In order to meet the heat dissipation requirements of high heat flux heating devices, the structural design of the evaporator adopted the structural form of micro-channel, and the structure was shown in Fig. 3. A number of micro-channels (hydraulic radius 0.4mm) were made on an aluminum sheet with a thickness of 2mm to form a parallel flow channel evaporator. The liquid inlet of the evaporator micro-channel was provided with a shrink (0.2mm), when the liquid flows into the micro-groove through the shrink, due to the inlet effect, the liquid atomization was accelerated, and the heat become gas in the evaporator, which strengthened the heat transfer effect of the evaporator. Because there were a large number of grooves and fins in the micro-channel, and there was a large specific surface area, the working medium absorbed a lot of heat in the micro-channel, and the temperature of the evaporating end device was reduced.

3.3   Structure Design of Condenser

The cooling process of condenser consisted of two stages. The first stage was the condensation process when the gaseous working medium passed through the inner wall of condenser, transferring heat to the condenser matrix. The second was that the heat in the condensing matrix was transferred to the heat sink through the external fins of the condenser. In order to ensure the smooth progress of the first cooling process, the internal heat transfer performance was better, the internal condenser also adopted the structural form of micro-channel, because the condensing process of the condenser and

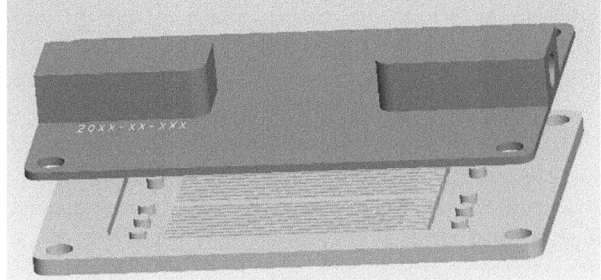

Fig. 3. Evaporator structure diagram

the evaporation process of the evaporator was actually the opposite phase transition process, its fluid flow state was consistent with that in the evaporator, so the size of the micro-tank in the condenser was consistent with the size of the evaporator micro-tank. A liquid collection area was also designed at the outlet of the condenser to reduce fluid flow resistance. The structure of the heat sink was shown in Fig. 4.

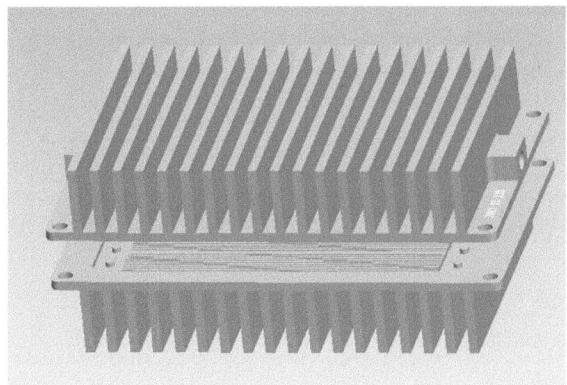

Fig. 4. Schematic diagram of internal structure of condenser

3.4 Structure Design of Liquid Storage Separator

The main role of the storage separator in the system was: first, the storage of excess liquid in the system to maintain system stability and avoid excessive internal saturation pressure. The second was to create gas-liquid two-phase separation, to avoid the incomplete cooling of the gas system into the pump to make the pump cavitation, while affecting the thermal performance of the evaporator. The outline design of the liquid storage separator was shown in Fig. 5. The gas-liquid two-phase separation of the liquid reservoir adopted the centrifugal principle. The gas-liquid mixture entered the pipeline and entered the gas-liquid separation chamber at a certain speed. Under the action of centrifugal force, the liquid gathered at the edge of the outlet and entered the liquid collecting chamber through the edge outlet, so as to ensure that the outflow from the outlet of the liquid reservoir was liquid. The gas was concentrated in the center of the gas-liquid separation chamber, so as to achieve gas-liquid separation. In order to achieve the above functions, the structure of the liquid reservoir was circular, the liquid inlet and the liquid outlet were arranged at both ends, and the liquid inlet should be set at the upper end, so that the direction of gravity was consistent with the direction of liquid flow. The pipeline should have a certain length in the axial direction, and the length was related to the diameter of the internal pipeline and the spiral spacing, so that the fluid fully formed a vortex flow state in the pipeline to achieve separation in the gas-liquid separation chamber.

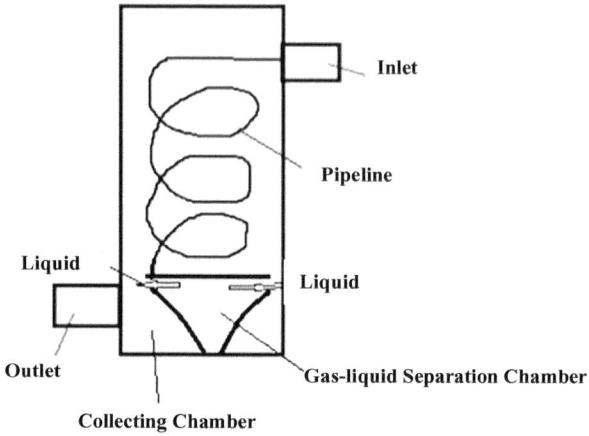

Fig. 5. Structure diagram of liquid reservoir

3.5 Test Content and Method

(1) Attach temperature sensors to the heat source, evaporator and condenser respectively. Heat source and evaporation of the cooling system were closely bonded by lamination. Heat source and the evaporator were required to be coated with thermal conductive grease or other interface materials to reduce the contact thermal resistance. Temperature heat transfer device was connected to the temperature acquisition system to test temperature of each component of the system in real time.

(2) High and low temperature experiments needed to be carried out at high and low temperature test chamber. Acceleration and vibration experiments needed to be carried out on the shaking table.

(3) Start the cooling fan at the condenser of the cooling system. Start the driven device so that the working medium inside the cooling system began to circulate and flow. Started the heat source load.

(4) After the measured temperature was stable, temperature of each component of the system was read from the temperature acquisition system, and temperature change curve of each component of the system was obtained from the temperature test system after the experiment was completed.

4   Experimental Result Analysis

4.1   Research on System Heat Dissipation Performance Under Normal Conditions

The heating sheet with diameter of Φ30 was used as the heat source, the input heat was 120W, the heating sheet was directly attached to the evaporator, the heat source and the evaporator were coated with thermal grease, the heat flux was 17W/cm², the pump worked under rated conditions, the flow rate of the pump was about 200ml/min.

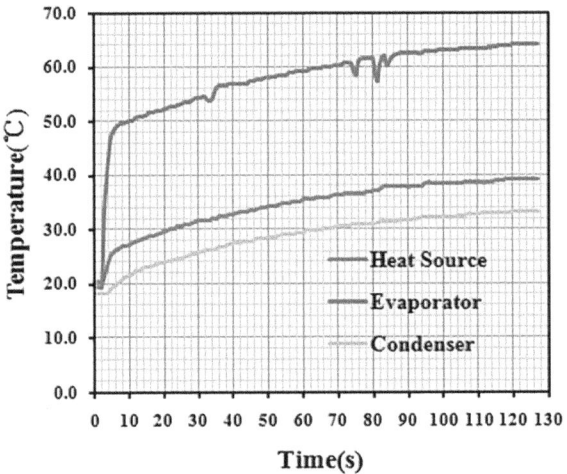

Fig. 6. System temperature and time curves

As shown in Fig. 6, the surface temperature of the heat source was 64.1°C, the temperature near the heat source on the evaporator was 39.3°C, and the temperature on the condenser was 33.3°C. Meanwhile, the temperature of the internal fluid (evaporator outlet) monitored by the built-in temperature sensor of the system was 37°C, and the temperature of the evaporator inlet was 32.8°C. The temperature difference between the surface temperature of the evaporator and the internal fluid was 2.3°C, and the temperature change of the fluid flow after passing through the evaporator was 4.2°C. The temperature difference between the condenser surface and the internal fluid of the system was 3.7°C. The temperature difference between the heat source and the evaporator surface was 24.8°C, and the temperature difference between the evaporator and the condenser was 6°C.

Since there were two forms of sensible heat and latent heat in liquid heat transfer, since the heat leakage of the system was small, without considering the heat leakage, the heat transfer of sensible heat and latent heat could be calculated by the following formula (1) and (2):

$$Q_s = CM \, \Delta T \tag{1}$$

$$Q_1 = Q - Q_s \tag{2}$$

where Q_s represented the sensible heat of the fluid, Q_l represented the latent heat of the fluid, and Q represented the total heat entering the system. C represented the specific heat capacity of the fluid in the system. M represented the mass flow of the fluid. ΔT represented the temperature difference between the input and output temperature of the evaporator.

From the test results of the above experiments, it could be seen that the system mainly conducted heat transfer through the latent heat of the internal fluid, and the internal micro-channel circulation heat dissipation effect was better. Because the contact thermal resistance between the heat source and the evaporator was large, the temperature difference between the heat source and the evaporator surface was large, and the maximum temperature difference in the circulation system was small, so the two-phase flow cooling circulation system itself had a good heat transfer performance.

4.2 Research on System Heat Dissipation Performance at High and Low Temperatures

The system heat input conditions at high and low temperatures were the same as the normal system heat input conditions. Figures 7 and 8 showed the temperature change curve of the two-phase flow cooling system working at high and low temperature.

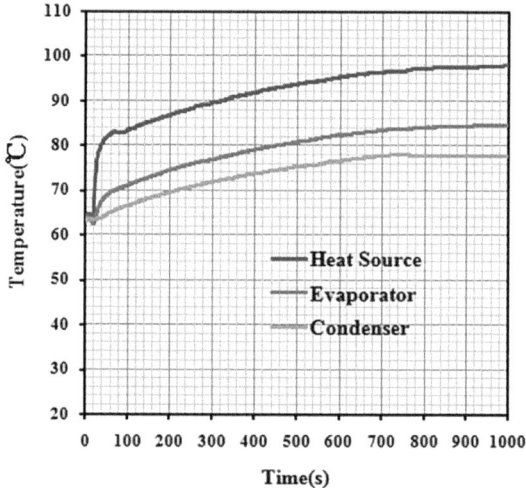

Fig. 7. Temperature change curve of the system under high temperature environment(+70°C)

The temperature difference between the evaporator and the condenser was about 6°C under the working environment of high temperature (70°C). In the low temperature (−45°C) working environment, the temperature difference between the evaporator and the condenser was about 10°C, the temperature difference between the evaporator and the condenser of the cooling system was small, and the heat dissipation performance was good, while the thermal performance of the low temperature environment system was not as good as that of the high temperature environment because the viscosity and surface

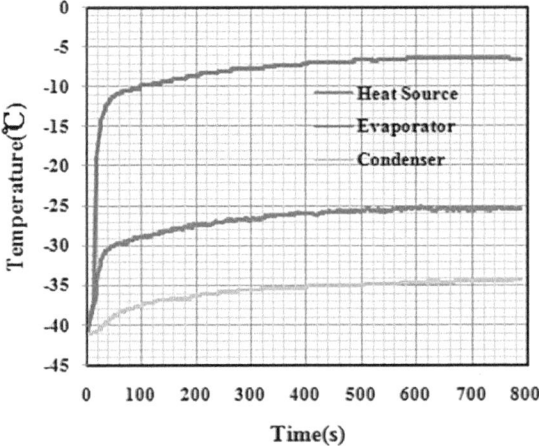

Fig. 8. Temperature change curve of the system at low temperature (−45°C)

tension coefficient of the fluid in the low temperature environment system increase, Resulting in reduced internal fluid flow performance.

4.3 Test Data and Conclusion of Thermal Performance of System Under Random Vibration Condition

The direction definition of the system was shown in Fig. 9. The random vibration method was adopted to study the anti-vibration performance of a typical airborne electronic device in the functional vibration environment. The vibration spectrum line was shown in Fig. 10. The thermal input conditions of the system in the vibration environment were the same as those in the normal system thermal test, and the vibration directions are X, Y and Z.

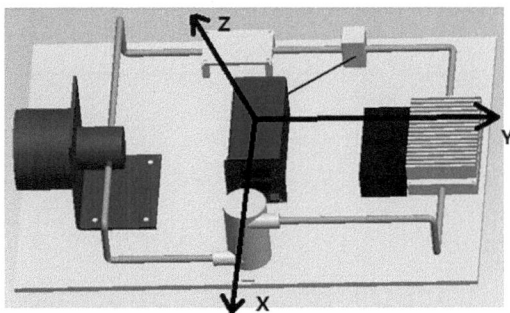

Fig. 9. Schematic diagram of system orientation

Figures 11, 12 and 13 showed the temperature variation curve of a two-phase flow cooling system operating in a vibration environment. The experimental results showed

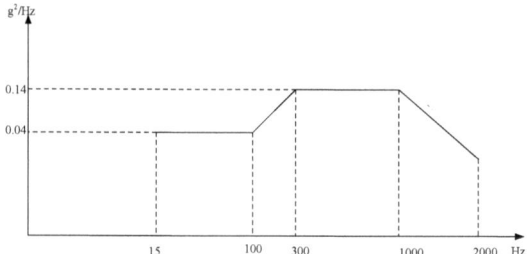

Fig. 10. Spectrum lines of functional vibration

that vibration had little effect on the thermal performance of the system. When the vibration was in X and Y direction, the temperature changed before and after the test was less than 2°C, and the influence was small. In the z-direction vibration, the temperature changed before and after the test was less than 5°C, which was slightly larger than that in the X and Y direction vibration. Mainly because the heat source was installed in the Z direction of the evaporator, the Z-direction vibration decreased the contact performance between the internal fluid of the system and the internal wall of the evaporator, and the heat transfer performance decreased.

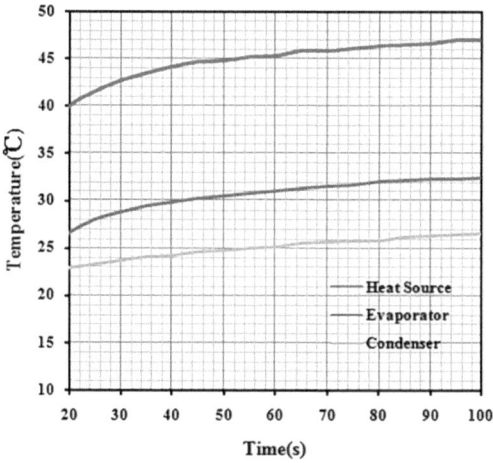

Fig. 11. Temperature change curve of the system under vibration environment (X direction)

4.4   Test Data and Conclusion of Thermal Performance of the System Under Acceleration Environment

In the acceleration environment, the system drive pump had a large driving force to overcome the acceleration resistance to drive the internal fluid flow in the pipeline. However, in the reservoir separator, the system relied on centrifugal force to realize the

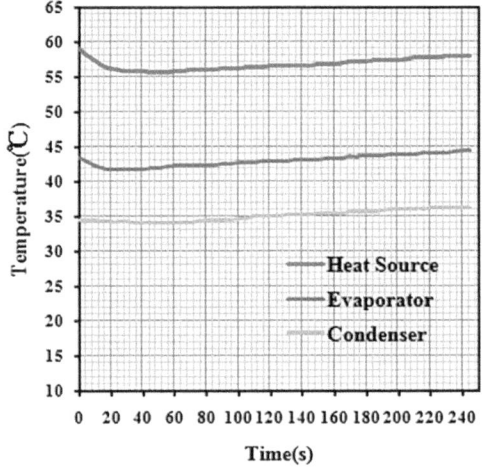

Fig. 12. Temperature change curve of the system under vibration environment (y-direction)

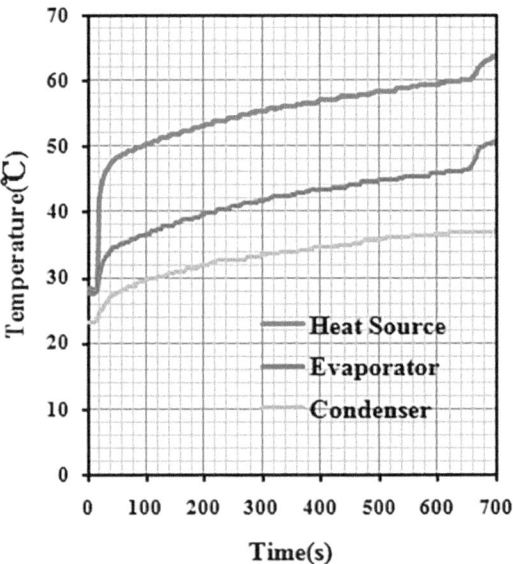

Fig. 13. Temperature change curve of the system under vibration environment (Z direction)

vapor liquid separation of the fluid, which would completely fail under the action of external acceleration, and the gas and liquid distribution inside the reservoir separator was chaotic. The fluid flowing out of the reservoir separator into the system pipeline circuit was a mixture of gas and liquid. It is found that the amount of liquid filled in the system had a great influence on the heat dissipation performance of the system under the acceleration state. In order to solve the influence of acceleration on the heat dissipation performance of the system, the influence of the amount of liquid filled in the system on the heat dissipation performance of the system under acceleration was discussed.

4.4.1 Performance of the System at 50% Liquid Filling

In this experiment, it was firstly studied that the liquid filling of the system was 50% and the acceleration direction was Y +. As shown in Fig. 14, under the action of Y + acceleration and gravity, the liquid in the reservoir mainly gathered at the bottom of the reservoir away from the outlet direction of the reservoir. Under the action of Y + acceleration, the driving current of the pump also decreased from 0.68A to 0.42A, and less liquid in the system could flow into the evaporator, resulting in a large temperature rise on the heat source and evaporator. At about 25°C, the system heat dissipation was greatly affected and the system failed.

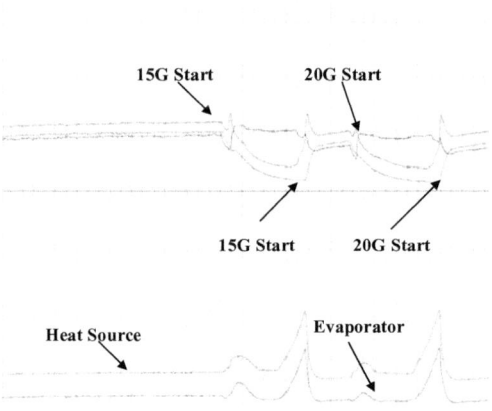

Fig. 14. Temperature change curve of the system under acceleration (Y +)

4.4.2 Performance of the System at 90% Liquid Filling

Increase the liquid filled amount of the system from 50 to 90%, and explored the influence of the acceleration of Y + on the heat dissipation capacity of the system.

As shown in Fig. 15, it can be seen from the test data that after changing the liquid storage capacity of the system, under the action of acceleration, the driving current of the pump fluctuated less within the normal range, and there was no sharp decrease, and the temperature of the heat source and evaporator did not rise sharply, that was, the system can work normally.

Under the acceleration environment, the amount of liquid filled in the system was less than the volume of the reservoir separator, thus the acceleration direction made the liquid in the system mainly gather in the reservoir separator, which leaded to less liquid flowed in the system circulation pipeline, resulting in system failure. While when the amount of liquid filled in the system was greater than the volume of the reservoir separator, even

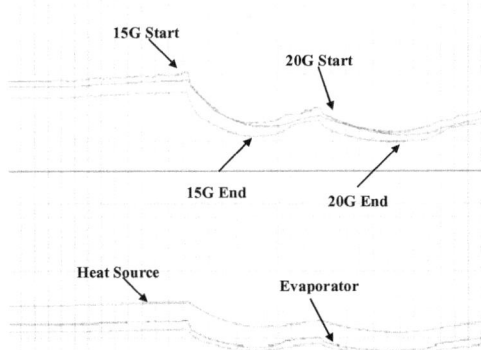

Fig. 15. Temperature change curve of the system under acceleration (Y +)

though the system was controlled by acceleration in the opposite direction, Because liquid gathered in the reservoir separator, the exceeded liquid was compelled to circulate in the pipeline under the action of the mechanical pump, so as to avoid system failure.

5   Conclusion

In this article, an integrated self-circulating pump-driven two-phase cooling device was designed. The refrigeration effect of the device under high and low temperature, random vibration and acceleration impact are discussed and studied. The experiment results showed that the device demonstrated a fine cooling ability under high and low temperature and random vibration, while the cooling effect under acceleration was related to the direction of acceleration and the amount of liquid filled in the circulated system.

1. The evaporator and condenser of the system adopt the structure form of microchannel, under the condition of applying 120W heat and 17W/cm^2 heat flux. The temperature difference between the evaporator surface temperature and the internal fluid was 2.3°C, the temperature difference between the condenser surface and the internal fluid was 3.7°C, and the temperature difference between the evaporator surface and the condenser surface was 6°C. The system had high heat transfer performance.
2. In the high and low temperature (high temperature 70°C, low temperature -45°C) environment, the system can work normally under the condition of 10.45G random vibration, and the system heat dissipation performance was good.
3. When the system was at horizontal amperage, the acceleration was Y + direction, the acceleration magnitude was 15G and 20G respectively, and the liquid filling amount of the system was increased from 50% to 90%, the heat dissipation performance of the system ws effectively improved.

In this paper, an active driven self-circulating pump driven two-phase heat dissipation device was investigated, which could effectively solve the heat dissipation problem of airborne high heat flow electronic chassis. However, there were still some shortcomings in this system, such as the impact of acceleration on the two-phase flow heat system, which needed more in-depth research and analysis. The size and volume of the system needed to be further reduced, especially when it came to the heat dissipation of airborne electronic chassis.

References

1. Gang, L., Yuting, W., Biao, L. et al.: Experimental research on micro-cooling System for Electronic Chip cooling. J. Refrig. **35**(6), 85–89 (2014)
2. Kai, Z., Huafeng, W., Jianhui, W., et al.: Experimental research and Numerical simulation of chip heat sink with enhanced Heat Transfer structure. J. Refrig. **36**(2), 46–51 (2015)
3. Janicki, M., Napieralski, A.: Modelling electronic circuit radiation cooling using analytical thermal model. Microelectron. J. **31**, 781–785 (2000)
4. Pedram, M., Nazarian, S.: Thermal modeling, analysis, and management in VLSI circuits: principles and methods. Proc. IEEE **94**(8), 1487–1518 (2006)
5. Qionghui, T., Jinliang, X., Yinhui, L. et al.: Experimental study on heat transfer performance of a new micro-heat pipe. Therm. Energy Power Eng. **21**(4), 350–354 (2006)
6. Jiaxuan, W., Xia, S., Tianyuan, G. et al.: Experimental study on two phase flow cooling System with high heat flux driven by avionics pump. **44**(1), 51–57 (2023)
7. Jianyun, T.: Research on thermal dynamic Characteristics of pump-driven Two-phase flow loop System. Southeast University, Nanjing (2020)
8. Jie, L., Nianqiang, P., Tingxun, L., et al.: Experimental study on overheating of two-phase cooling System of Mechanical Pump. J. Sun Yat-Sen Univ. (Natural Science Edition) **46**(6), 25–29 (2007)
9. Jie, L., Tingxun, L., Kaihua, G., et al.: Experimental study on characteristics of two-phase cooling system driven by mechanical pump of parallel evaporator. J. Refrig. **29**(6), 5–8 (2008)
10. Xuan, W., Guoyuan, M., Feng, Z.: Experimental study on pump driven two-phase cooling Unit and its energy saving effect. J. Refrig. **38**(2), 82–88 (2017)
11. Li, L., Qi, C., Rui, W.: Research on effect of charge volume of pump-driven two-phase flow cooling system. Electro-Mech. Eng. **37**(4), 26–28 (2021)
12. Thome, J.R.: State-of-the-art overview of boiling and two-phase flows in microchannels. Heat Transfer Eng. **27**(9), 4–19 (2006)
13. Cheung, M., Hoang, K., Ku, J. et al.: Thermal performance and operational characteristics of loop heat pipe. SAE-981813
14. Chiul, H.C., Jang, J.H., Yeh, H.W., et al.: The heat transfer characteristic of liquid cooling heat sink containing microchannels. Int. J. Heat Mass Transf. **54**(1), 34–42 (2011)
15. Kaya, T., Ku, J.: Experimental investigation of performance characteristics of small loop heat pipes. AIAA 1038–1042 (2003)

Organic Waste Management in Tasikmalaya City, Indonesia

Joni Jupesta[1](✉) and Indra Permana[2]

[1] Institute for the Advanced Study of Sustainability (UNU-IAS), United Nations University, Tokyo 1508925, Japan
jjupesta@yahoo.com

[2] Siliwangi University, Tasikmalaya 46115, Indonesia

Abstract. Waste sector remains the largest contributor to urban greenhouse gas (GHG) emissions after the energy sector in the city level. This study aims to assess the potential of circular economy by using Black Soldier Fly (BSF) as organic agent to processing the organic waste to the value-added products fish meal and organic fertilizer in the city. The study has been conducted at organic waste management site affiliated with the Tasikmalaya City. The harvesting process for the BSF larvae (BSFL) was conducted while the co-products BSFL frass and BSFL skin were analyzed in the laboratory to determine its chemical characteristics. The result of the study shows that in average 8.33 kg BSFL has been produced from 25 kg organic waste feedstock. The highest organic carbon content was shown in the mixture of BSFL frass and skin with a value of 15.6% and followed by the BSFL frass and BSFL skin samples independently with values of 12.8% and 10.5% respectively. All maggot residues showed similar pH ranging at 6.3–6.5. The mixture of BSFL frass and BSFL skin had a higher total N content compared to only BSFL frass or BSFL skin with respective values of 5.8%, 5.2% and 2.6%. The total P_2O_5 content in all treatments shown almost similar values range from 3.55%, 3,12%, and 3.55% for BSFL frass, BSFL skin and the mixture of BSFL frass and BSFL skin respectively. However, the total K_2O shown big discrepancy with value 1.42%, 0.53%, and 0.85% for BSFL frass, BSFL skin and mixture of BSFL frass and skin respectively. This study shown that the GHG emissions from waste in the Tasikmalaya city could be reduces by utilize organic agent such as BSF. This circular economy could create high economic value products such as BSF larvae (maggot) and co products BSFL frass and BSFL skin.

Keywords: GHG emissions · circular economy · waste management · Black Soldier Fly (BSF) · Indonesia

1 Introduction

Urban systems are critical for achieving deep emissions reductions and advancing climate resilient development [1]. Most future urban population growth will occur in developing countries including Indonesia, where per capita emissions are currently low but expected

B.-J. He et al. (Eds.): UCSUD 2023, LNCE 559, pp. 286–294, 2025.
https://doi.org/10.1007/978-981-97-8401-1_20

to increase with the construction and use of new infrastructure and the built environment, and changes in incomes and lifestyles [2]. Implementing waste management and wastewater recycling measures can provide additional sources of income for citizens and local authorities. Waste management and wastewater recycling is also a pathway for inclusion of the informal sector into the urban economy with high agreement and medium evidence. The waste sector is a significant source of GHG emissions, particularly methane (CH_4). Currently, the waste sector remains the largest contributor to urban emissions after the energy sector with 1,580 billion tonnes of CO_2e (CO_2 equivalent), equivalent to 3.2% of the total CO_2 emission worldwide [3]. In Indonesia, the waste sector is the third largest source of GHG emissions with a contribution of 7% of total emissions after forest and land conversion and energy with percentages of 50% and 34% respectively [4]. Since waste management systems are usually under the control of municipal authorities, they are a prime target for city-level mitigation efforts with co-benefits.

Home composting and compact urban form can also reduce waste transport emissions. Decentralised waste management can reinforce source-separation behaviour since the resulting benefits can be more visible. Fladeret *et. al.* Mentioned that decentralization of solid waste management requires stakeholder participation in a bottom-up process [5]. Community involvement is key to the success of CBSWM (community-based solid waste management) as it involves them in all stages: planning, technology, infrastructure, operations, development, and evaluation [5]. Integrated policymaking can increase the energy, material, and emissions benefits in the waste management sector. Organisational structure and programme administration poses demands for institutional capacity, governance, and cross-sectoral coordination for obtaining the maximum benefit. The informal sector plays a critical role in waste management, particularly but not exclusively in developing countries [6]. Integrating the formal and informal sectors through waste banks can increase waste management services in Yogyakarta City from 85% to 95.5%, Sleman Regency 30.7 to 31%, and Bantul Regency 7.49 to 7.7% [7]. Overall, the positive impacts of waste management on employment and economic growth can be increased when informality is transformed to stimulate employment opportunities for value-added products with an estimated 45 million jobs in the waste management sector by 2030.

Vickerson (2016) found that insects as potentially solving two major global problems—a lack of sustainable feed and wasted food [8]. Most of the complex nutrients in organic waste end up in landfill, compost or waste-to-energy facilities, or anaerobic digesters. Insect larvae, he realized, could become part of a closed loop, consuming recycled food and being harvested to create a renewable source of nutrients for livestock. Feeding waste to insects allows nutrients to be recovered and used as a valuable source of protein and fat, naturally bioconversion (see Fig. 1). Once the larvae are ready to be harvested, they are mechanically sifted to winnow out the 'frass', or manure. This is treated separately as a natural fertilizer certified for use in organic crop production. The larvae meal can be used in animal feed as a direct substitute for resource-intensive ingredients such as fish meal and soya-bean meal [8].

This study aims to assess the potential of circular economy by using BSF as organic agent to processing the organic waste to the value-added products fish meal and organic fertilizer in the Tasikmalaya city, Indonesia. This approach aligns with the broader goal of sustainable urban development and addresses the pressing issues of waste management and GHG emissions in urban areas. This research aligns with the government's proactive policy in addressing organic waste issues in Tasikmalaya City. The city government has established the Organic Waste Processing Group (GOSO), responsible for collecting organic waste and distributing it to maggot farmers. Therefore, the results of this research undoubtedly have a very positive impact on the local government's formulation of waste processing strategies using BSF larvae.

2   Literature Review

2.1   Tasikmalaya City Profile

Tasikmalaya city is in the West Jave province. It is on the southeast from Bandung, the capital of the West Java province. Tasikmalaya city has $183,85$ km^2 area which is consist of 10 districts and 69 villages. The area of Tasikmalaya Municipality based on Mayor Tasikmalaya Municipality Regulation No.29 in 2021 is 183.14 km^2 which is divided into 10 (ten) subdistricts, namely Kawalu Subdistricts, Tamansari Subdistricts, Cibeureum Subdistricts, Purbaratu Subdistricts, Tawang Subdistricts, Cihideung Subdistricts, Mangkubumi Subdistricts, Indihiang Subdistricts, Bungursari Subdistricts, and Cipedes Subdistricts. The territorial boundaries of Tasikmalaya Municipality in northern and eastern area bordered by Tasikmalaya Regency and Ciamis Regency, while in southern and western area bordered by Tasikmalaya Regency [9].

2.2   Waste Management System

According to BPS Statistic Agency of Tasikmalaya municipality, the total waste is116,475.95 ton in 2022 [10]. The waste composition is: 48.7% food waste, 5.8% paper/cartoon, 5.4% plastics, 6.3% glass, 1.9% clothes, 0.2% metals, and 31.6% others. The largest waste is organic waste from the food waste. This is mostly coming from the household waste from settlement. This organic waste needs to process properly through composting or waste collection due to the decomposition process which creates odor and leachate. Organic Waste processing in Tasikmalaya was conducted through two systems; landfill and community based. The waste collection site Ciangir in Tasikmalaya municipality located in Tamansari subdistrict with processing area 12 Ha and hoarding area 5 Ha. The waste collection site Ciangir starts into operation around 2000. Waste management using is transform from open dumping system; where the waste only piled and not covered into controlled landfill; the waste covered with land to avoid the odor [11].

Waste collection site Ciangir also do leachate management system. The leachate the was produced from the waste landfill will be flow to the leachate pond to avoid the leaching into surrounding. The methane emissions still not yet properly managed, only methane pipe to flow the methane to the open air. The other efforts are the monitoring the quality of river waste and ground water surrounding the waste collection site. Wate quality monitoring conducted frequently to assess the impact of waste collection site Ciangir to the ambient environment.

Waste procession by community still very low by do collection without waste reduction effort and waste separation. Community already conducts waste management in the few locations in Tasikmalaya municipality. Community based waste treatment conducted voluntarily by community as awareness to the high waste volumes in the municipality. There are eleven (11) community-based waste management has been reported. The community-based waste management are consisting of the non-organic waste processing and organic waste processing. Nonorganic waste which is usually plastic waste by separate and later sold to the buyers. Other activity is to recycle the plastic waste to become bag or table cover. The organic waste is through composting which later could be sold to the ornament plants sellers. The other organic waste that becomes focus of the attention is the BSF production which could create additional household revenue [10].

2.3   Black Soldier Fly (BSF)

The use of insects, including Black Solider Fly (BSF) or so called Hermetia illucens, is well known for playing a vital role in solving issues linked with high volumes of organic wastes distributed all over the world. It has progressively been employed in treating biological waste as it is seen as being an environmentally friendly and inexpensive process. BSF larvae (BSFL) was highlighted as potent recyclers of various types of wastes such as abattoir waste, food waste, fruits and vegetable waste, and human feces. BSF falls into the Diptera family from the order Stratiomyidae and inherently resides in temperate tropical areas. BSFL was noticed to have a remarkable ability (75%) of recycling the biological wastes in which 800 g of larval biomass would be produced from 4 kg of waste. To optimize its bioconversion efficacy, BSFL must be maintained under ideal environmental conditions, including parameters such as humidity, nutrient composition, physical properties, temperature, and oxygen level. Amongst rearing conditions, the temperature was recognized to play a significant role in the growth of BSFL in which the optimal temperature ranges between 25°C and 30°C. Like other living organisms, BSFL needs nutrients to support their growth. Therefore, for higher bioconversion performance, BSFL needs to feed on the organic wastes rich in digestible nutritive substances. In addition, it was stressed that BSFL can effectively decompose various types of organic waste if it contains an adequate amount of protein and carbohydrates. It is noteworthy that whole or processed BSF larvae or pupae can be incorporated into the diets of poultry, fish, pets, and pigs thereby serving as prospective alternatives of common feed ingredients namely soybean- and fish-based meal. Consequently, the usual feed ingredients, which insect products could replace, can be reserved for other uses including human consumption thus contributing to food security. Furthermore, the bioactive chemicals, such as antimicrobial peptides, present in BSFL could also add great benefits to animal diets [12]. While the most challenging from this bioconversion is the optimum BSF yield from organic waste, the next section will be discussed on the methodology for the BSF propagation from the organic waste and the characteristics of the BSF in terms of Carbon, Nitrogen, P_2O_5, K_2O_5 contents, and pH.

3 Methodology

3.1 Black Soldier Fly Larvae (BSFL) Production

The experiment was conducted at a designated organic waste management site affiliated with the Tasikmalaya City Organic Waste Group. Observations were carried out to see the potential of BSF larvae in decomposing waste that is often found in the city of Tasikmalaya, such as restaurant waste, market, and household waste. The research materials encompassed larvae eggs, market waste (vegetables and fruit), rice bran, household waste, and restaurant waste. An array of research tools was employed, including wooden containers, mesh nets, wood, scopes, basins, digital scales, mini drums, manual biopond, digital scales, analytical scales, brooms, and scopes. Every 3 g of eggs BSF was put into a biopond with dimensions of 60x80 cm. Biopond is a medium for rearing larvae made from wood or plastic tubs. The biopond functioned as both a cultivation and hatching medium for BSF maggot eggs, facilitating their transformation into baby maggots. The provided biopond was fashioned into a permanent shelf, with the closing finger opened to prevent the ingress of flies and ants that might damage the medium. Chalk scratches were strategically applied to the media walls to preclude ant entry.

In this study, the medium for hatching BSF maggot eggs comprised rice bran mixed with water, maintaining specific moisture conditions. After 6 days of hatching, BSF maggots were ready for placement into cultivation or grow-out media. During the growth phase, BSF larvae receive a staged allocation of food in the form of a mixture comprising household, market, and restaurant waste, with a total weight of 25 kg for each biopond. Harvesting of BSF larvae was systematically conducted at the 21-day mark, aligning with the life cycle of the BSF maggot, where the cessation of feeding and drinking occurs. Consequently, the harvesting process in this research was executed precisely at the culmination of the 21-day period. Parameters observed during harvesting included the weight of the produced BSFL and the quantity of remaining feed or BSFL frass.

3.2 Black Soldier Fly Larvae Frass (BSFLF) Characteristics

The residues generated during the organic waste decomposition process are collected and utilized as organic fertilizer due to their nutrient-rich content beneficial for soil and plants. The BSF larval residue is also referred to as Black Soldier Fly Larvae Frass (BSFLF). The produced frass is subsequently analyzed in the laboratory to determine its chemical characteristics such as pH value, organic carbon content, as well as the content of nitrogen, phosphorus, and potassium nutrients. The analysis is conducted at the soil laboratory of the Faculty of Agriculture at the Siliwangi University. In addition to frass production, the shells of the BSF larvae are also analyzed to understand their chemical content and their effects when mixed with the residues from organic waste decomposition. The residues generated during the organic waste decomposition process are collected and utilized as organic fertilizer due to their nutrient-rich content beneficial for soil and plants. The BSF larvae residue is also referred to as Black Soldier Fly Larvae Frass (BSFLF). The produced frass is subsequently analyzed in the laboratory to determine its chemical characteristics such as pH value, organic carbon content, as well as the content of nitrogen, phosphorus, and potassium nutrients. The analysis is conducted at the soil laboratory of the Faculty of Agriculture at the Siliwangi University. In addition to frass production, the shells of the BSF larvae are also analyzed to understand their chemical content and their effects when mixed with the residues from organic waste decomposition.

4 Results and Discussions

4.1 Black Soldier Flies (BSF) Larvae Production

Bioconversion is a natural process which consists of the transformation of nutrients via biodegradation, e.g., using insect larvae [13]. The transformation (conversion) rate can be expressed as the ratio between dry matter before and after the process. The amount of organic waste feedstock given during maggot cultivation for all organic waste until it can be harvested is 25 kg. Feeding 25 kg of mixture organic waste produces 8 to 9 kg of wet BSF larvae (maggot). Harvesting is carried out at 21 days after egg laying, when the larvae begin to darken slightly or enter the pre-pupal stage. The results are consistent with the study by Rachmawati et.al that showed an optimal time of 21 days [14]. Observed maggot production from three trials were shown in Fig. 1. In average 8.33 kg BSF larvae (BSFL) has been produced from 25 kg organic waste feedstock. A similar study was carried out by Fajri et al. (2021), which found that 0.50 g of BSF maggot eggs could produce 2.8 kg of adult maggots when fed with 7.3 kg of restaurant organic waste [15]. The culture time was shortened by maintaining important parameters such as temperature and humidity at the right setting. The growth and overall success of maggots are significantly impacted by the conditions and temperature of the growth media. Maggots thrive best in an ideal temperature range of $27^0C–30^0C$. It is crucial to note that maggots cannot survive temperatures exceeding 36^0C.

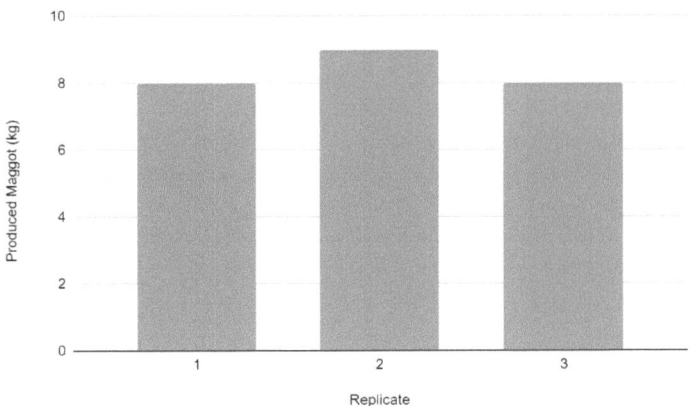

Fig. 1. The produced maggots (BSF Larvae) for each replication.

In overall the life cycle of H. *illucens* on mixture organic waste media based took 32–41 days from egg to imago.

4.2   Black Soldier Flies (BSF) Larvae Frass Characterization

The waste by-product of the BSF larvae (BSFL) digestion process is known as frass. Frass is a granular, nutrient-rich material that contains digested organic matter, exoskeletons, and microbial biomass. This frass serves as an excellent soil conditioner and fertilizer due to its high nutrient content. Analysis of frass properties was shown in Table 1. The highest organic carbon content was shown in the mixture of BSFL frass and skin with a value of 15.6% and followed by the BSFL frass and BSFL skin samples with values of 12.8% and 10.5% respectively. According to Basri et al., compost with a C/N ratio less than 20 is beneficial to plants because organic nitrogen is mineralized into inorganic nitrogen, which can then be absorbed by plants [16]. Organic fertilizers with a C/N ratio greater than 30 are more likely to fix nitrogen for plants to absorb. BSF larvae frass, skin and mix of both have relatively low C/N ratio (average 2.5–4.1) and are readily available for plant uptake.

Table 1. Compost analysis results

No	Parameter	Method	BSFL Frass	BSFL Skin	Mixture of BSFL frass and skin
1	Organic C	Walkley and Black	12.8%	10.5%	15.6%
2	C/N	–	2.5	4.1	2.7
3	pH	Potentiometry	6.3	6.5	6.5
4	Total N	Kjeldahl	5.2%	2.6%	5.8%
5	Total P_2O_5	Spectrophotometry UV-VIS	3.35%	3.12%	3.65%
6	Total K_2O	Atomic Absorption Spectrometry	1.42%	0.53%	0.85%

All maggot residues showed similar pH ranging at 6.3–6.5. This is in line with study from Song et al. that frass has a pH value of around 6–7 [17]. Elissen, van der Weide, & Gollenbeek mentioned that pH 6–8 is a good value for a mature compost for agronomic purposes [18]. The mixture of BSFL frass and BSFL skin had a higher total N content compared to only BSFL frass or BSFL skin with respective values of 5.8%, 5.2% and 2.6%. Beesigamukama et al. mentioned that BSFL frass has significantly higher N (20–130%) concentrations compared to frasses of other insects [19]. The total P_2O_5 content in all treatments shown almost similar values: 3.55%, 3,12%, and 3.55% for BSFL frass, BSFL skin and mixture of BSFL frass and skin respectively. However, the total K_2O shown big discrepancy with value 1.42%, 0.53%, and 0.85% for BSFL frass, BSFL skin and mixture of BSFL frass and skin respectively. The average from all three co-products could be characterized similar with compound NPK fertilizer with mass percentage 4.6% N, 3.4% P_2O_5, and 0.93% K_2O.

5   Conclusions

Urban population growth in developing countries like Indonesia will increase per capita emissions due to new infrastructure and lifestyle changes. Waste, particularly methane emissions, is a significant contributor to greenhouse gases (GHGs). Utilizing insects for waste processing, such as black soldier flies (BSF), can recover nutrients for valuable products like protein and fat. BSF larvae can replace resource-intensive ingredients in animal feed, contributing to a circular economy. This study in Tasikmalaya, Indonesia, demonstrated the production of 8.33 kg of BSF larvae from 25 kg of organic waste, with high organic carbon and nutrient content in BSF byproducts. Utilizing BSF can reduce GHG emissions from waste and create economically valuable products like BSF larvae and their byproducts.

Acknowledgement.   JJ received funding from Ministry of Environment Japan (MOEJ).

References

1. Lee, H.S., Romero, J.: Climate change 2023: synthesis report. contribution of working groups I, II and III to the sixth assessment report of the intergovernmental panel on climate Change Synthesis Report. IPCC, Geneva (2023)
2. Lwasa, S. et al.: Urban systems and other settlements. In: Climate change 2022: Mitigation of climate Change, pp. 861–952. IPCC, Geneva (2022)
3. Ritchie, H., Roser, M.: CO_2 and Greenhouse Gas Emission. Retrieved from: https://ourworldindata.org/co2-and-other-greenhouse-gas-emissions (2020)
4. Aprilia, A.: Waste management in Indonesia and Jakarta: challenges and way forward. 23rd ASEF Summer Univ [Internet]. 2021 (October):1–18. Available from: https://asef.org/wp-content/uploads/2022/01/ASEFSU23_Background-Paper_Waste-Management-in-Indonesia-and-Jakarta.pdf
5. Fladerer, F., Wahyudi, S., AlRasyid, H.H., Ismawati, Y.: ... Decentralized urban solid waste management in Indonesia: Final technical report. 2009 (103074):1–36. Available from: https://idl-bnc-idrc.dspacedirect.org/handle/10625/44814; https://idl-bnc-idrc.dspacedirect.org/handle/10625/44814
6. Wilson, D.C., Velis, C., Cheeseman, C.: Role of informal sector recycling in waste management in developing countries. Habitat Int. **30**(4), 797–808 (2006)
7. Purnama Putra, H., Damanhuri, E., Sembiring, E.: Integration of formal and informal sector (waste bank) in waste management system in Yogyakarta. Indonesia. MATEC Web Conf. **154**, 1–5 (2018)
8. Vickerson, A.: Transform waste into protein. Nature Comment **531**, 445–446 (2016)
9. BPS Statistics of Tasikmalaya Municipality: Regional Statistics of Tasikmalaya Municipality. BPS Statistics of Tasikmalaya Municipality, Tasikmalaya (2023)
10. Indonesian Ministry of Environment and Forestry Homepage, https://sipsn.menlhk.go.id/sipsn/public/data/komposisi, last accessed 2024/1/9
11. Environment Agency of Tasikmalaya Municipality: Laporan Kegiatan Bidang Kebersihan (in Bahasa), Environment Agency of Tasikmalaya Municipality, Tasikmalaya (2019)
12. Siddiqui, S.A., et al.: Black soldier fly larvae (BSFL) and their affinity for organic waste processing. Waste Manage. **140**(2022), 1–13 (2023)

13. Hem, S. et al.: Bioconversion of Palm Kernel Meal (PKM) as A business symbiosis with palm oil industry: a modern industrial ecologist approach. In: International Conference on Palm Oil (ICOPE), Bali, Indonesia (2007)
14. Rachmawati, R., Buchori, D., Hidayat, P., Hem, S., Fahmi, M.R.: Perkembangan dan Kandungan Nutrisi Larva *Hermetia illucens* (Linnaeus) (Diptera: Stratiomyidae) pada bungkil kelapa sawit (in Bahasa Indonesia). J. Entomol. Indones. **7**, 28–41 (2010)
15. Fajri, N.A., Hamid, A.: Production of BSF (Black Soldier Fliy) maggots as feed cultivated with different media. AGRIPTEK (Jurnal Agribisnis dan Peternakan). **1**(1), 12–17 (2021)
16. Basri, N.E.A., Azman, N.A., Ahmad, I.K., Suja, F., Jalil, N.A.A., Amrul, N.F.: Potential applications of frass derived from black soldier fly larvae treatment of food waste: a review. Foods **11**(17), 2664 (2022)
17. Song, S. et al.: Upcycling food waste using black soldier fly larvae: Effects of further composting on frass quality, fertilising effect and its global warming potential. J. Entomol. Indones. **288**, (125664) (2021)
18. Elissen, H., van der Weide, R., Gollenbeek, L.: Effects of black soldier fly frass on plant and soil characteristics- a literature overview. Wageningen University and Research, Wageningen (2023)
19. Beesigamukama, D., Subramanian, S., Tanga, C.M.: Nutrient quality and maturity status of frass fertilizer from nine edible insects. Sci. Rep. **12**(1), 7182 (2022)

Sustainable Development of Photovoltaic Market in Zhejiang, China

Gaochuan Zhang[1,2,3], Qiaoyuan Lin[1], and Bao-Jie He[4(✉)]

[1] School of Civil Engineering and Architecture, Zhejiang University of Science and Technology, Hangzhou 310023, China
[2] Zhejiang Southeast Architectural Design Group CO. LTD, Hangzhou 310023, China
[3] Joint Laboratory for Urban Renewal and Future City, Zhejiang 310023, China
[4] School of Architecture and Urban Planning, Key Laboratory of New Technology for Construction of Cities in Mountain Area, Ministry of Education, Chongqing University, Chongqing, China
baojie.unsw@gmail.com

Abstract. Greenhouse gases, such as carbon dioxide and methane, emitted from thermal power generation, pose significant threats to human survival and development, contributing to phenomena such as floods, droughts, and storms. The photovoltaic market, as an emerging source of clean energy in China, has experienced widespread adoption and rapid growth, emerging as a cornerstone industry driving local economic development. This study focuses on the photovoltaic (PV) market in Zhejiang Province, analyzing its fundamentals across four key dimensions: market conditions, economic benefits, technological advancements, and policy frameworks. Findings demonstrate that a PV power station operating over a 30-year lifespan could potentially generate up to 864.275 billion kWh of electricity, yielding tariff revenues of 358.93 billion yuan. Moreover, such installations could significantly mitigate environmental impact by reducing carbon dioxide emissions by 873.965 billion metric tons, nitrogen oxide emissions by 245,100 metric tons, sulfur dioxide emissions by 163,900 metric tons, and conserving 261 million metric tons of standard coal.

Keywords: PV markets · PV agriculture · PV buildings · power generation · emission reductions

1 Introduction

Amidst the global climate crisis, China is actively promoting the high-quality development of renewable energy with the goal of achieving carbon peak and carbon neutrality. Solar energy, among these renewable sources, stands out as a clean and sustainable option [1]. Photovoltaic (PV) power generation stands as one of the primary methods for harnessing solar energy resources [2], and it serves as a crucial tool in advancing the objectives of carbon peaking and carbon neutrality. Nevertheless, the construction of photovoltaic power plants necessitates significant land use, creating conflicts between land allocated for photovoltaics and other purposes, notably agricultural production [3].

© The Author(s) 2025
B.-J. He et al. (Eds.): UCSUD 2023, LNCE 559, pp. 295–307, 2025.
https://doi.org/10.1007/978-981-97-8401-1_21

To address the challenge of limited land resources in photovoltaic (PV) development, prior research has introduced two strategies: PV agriculture and PV integration with buildings, outlined in Table 1. PV agriculture entails integrating PV power generation with agricultural activities by installing photovoltaic systems in various agricultural settings such as greenhouses, farmland, vegetable plots, and pastures [4], aiming to optimize land utilization in three dimensions. In 1982, Goetzberger et al. proposed the concept of integrating solar energy conversion with crop cultivation, suggesting that synergistic development of photovoltaic power generation and agriculture could be achieved by raising the height of photovoltaic panels and optimizing inter-panel spacing [5]. This concept was implemented in 2011, when Dupraz et al. established the inaugural PV farm in Montpellier, France, and conducted controlled experiments to assess the impact of PV panels on crop growth. The findings indicated that erecting PV panels obstructed essential sunlight for crop growth, leading to reduced yields [6]. Subsequently, researchers conducted further investigations into the land utilization of agricultural photovoltaic systems by developing solar radiation interception models and crop models with varying PV panel densities. These studies revealed that agricultural photovoltaic systems could enhance land productivity by 60% to 70% overall, emphasizing the importance of striking a balance between photovoltaic power generation and crop cultivation. Subsequently, France, South Korea, and Japan conducted experiments on various crops such as root celery, grapes, and onions to assess the impact of PV panel installation on their growth. The findings demonstrate that through strategic placement of PV panel arrays and adjustments to crop planting cycles, it is feasible to maintain yields while enhancing land utilization rates. These studies demonstrate that the PV market model can facilitate synergistic development of photovoltaic power generation and agricultural production [7], offering a theoretical foundation for crop selection within this model. An alternative solution involves integrating PV power generation systems into buildings, thereby converting them into structures that not only serve traditional residential purposes but also generate clean electricity. The integration of photovoltaics with buildings was initially proposed in Germany in 1991 and has subsequently been adopted and promoted in numerous countries [8]. Subsequent research has progressed from optimizing PV panel distribution on individual building rooftops to identifying potential installation sites for PV systems on a city-wide scale, transitioning from micro-level to macro-city analysis [9–11].

Overall, photovoltaic (PV) technology has the potential to optimize land resource utilization in three dimensions by integrating PV power generation with agricultural and architectural practices. However, current research in the field of photovoltaics primarily concentrates on enhancing power generation efficiency and optimizing the layout of PV modules, with limited attention devoted to assessing the status of PV market development and associated emission reduction benefits. Consequently, this study aims to (1) investigate the current status and development prospects of the PV market in Zhejiang Province, (2) evaluate the electricity generation and environmental emission reduction benefits associated with PV power plant construction, and (3) identify novel approaches for enhancing the efficient development of the PV market in Zhejiang Province.

Table 1. Main classifications of the "PV market"

Photovoltaic Market Type		Contents	specificities
Photovoltaic agriculture	Photovoltaic cultivation	PV + Vegetables, PV + Flowers, PV + Saplings, PV + Tea Gardens, PV + Orchards, PV + Forestry, PV + Mushrooms, Chinese herbal medicine	High land use, resource efficiency, flexibility and economic efficiency
	Photovoltaic farming	PV + Livestock (cattle, sheep, pigs, chickens, birds, etc.), PV + Fisheries (freshwater aquaculture, mariculture)	Constant farming environment, reduced farming costs, reduced environmental pollution
Photovoltaic building	Single installation	PV + roof, PV + balcony, PV + window sill	Easy installation and power cost savings
	Building Integrated Photovoltaics	Integration of photovoltaic products into buildings	Overall design, beautiful appearance, long service life

2 Subjects and Methodology

This study focuses on the photovoltaic (PV) market in Zhejiang Province, with completed PV projects as the primary research subject. It encompasses various aspects including industrial scale, development mode, technological advancement, policy framework, market utilization, and other relevant factors. Data collection methods include literature review, field surveys, and interviews. This involves two approaches: firstly, gathering online official information from sources such as the China Knowledge Network, government WeChat public accounts, and news websites; secondly, conducting offline research and demonstrations by visiting representative PV enterprises and market users in Jiaxing City, Jinhua City, and Taizhou City.

3 Zhejiang Province PV Market Survey Results

3.1 Market Conditions:Macroscopic Scope and Potentiality

The photovoltaic market represents a new strategic industry prioritized for development by Zhejiang Province, with 77 companies having capital exceeding 1 billion yuan, and 1,374 companies having capital exceeding 100 million yuan, as depicted in Table 2. As of the end of 2022, Zhejiang Province's cumulative photovoltaic (PV) installed capacity reached 25.39 GW, ranking third nationwide. Notably, its distributed PV installed capacity stands at 19.26 GW, securing the second position in the country. In 2022, the export value of photovoltaic (PV) modules reached 84.31 billion yuan, representing approximately 28.9% of the nation's total PV module exports. In alignment with national strategic initiatives such as the "dual carbon target" and "energy transformation", Zhejiang Province is actively promoting the utilization of barren slopes, abandoned mines, and other resources for establishing centralized photovoltaic power stations. Additionally, efforts are underway to develop composite photovoltaic projects such as agro-photovoltaic and fungus-photovoltaic systems, along with water surface photovoltaic power generation projects.

3.2 Benefit Conditions: Microscopic Scope and Potentiality

In 2023, the first document issued by the central government emphasized the crucial deployment of "promoting the consolidation and upgrading of rural power grids and the development of rural renewable energy." In response, the Zhejiang government and enterprises collaborated actively to develop the rural market. In Longyou County, Quzhou City, Chunjin Village Zhixi Home, a comprehensive "village photovoltaic EPC" solution was implemented, encompassing the co-development, sales, survey and design, installation, and after-sales operation and maintenance of roof photovoltaic systems. This project, with a total installed capacity of 4MW, generates over 4 million kWh of electricity annually, resulting in an income exceeding 500,000 yuan per year for villagers. Similarly, in Deqing County, Huzhou City, New Town, Jucheng Village, and Songshi Village, the completion of a photovoltaic power station is projected to produce 99,000 kWh of electricity annually, yielding a collective income of 45,000 yuan for the two villages. These proceeds are utilized to support widows, orphans, and elderly individuals in the village, as well as to enhance village infrastructure and public service initiatives. Additionally, the "shrimp light complementary" project in Gulong Village, Changxing County, Huzhou City, results in a longer growth cycle for crayfish, yielding a net profit of over 20,000 yuan per acre annually from shrimp farming, in addition to an annual income exceeding 10 million yuan from electricity sales alone. Moreover, in the village of the end of the head, Wuyi County, Jinhua, the "fungus light complementary" power station, equipped with nearly 20,000 polycrystalline silicon solar modules, generates up to 5 million kWh of electricity annually, equivalent to an annual savings of 1,800 tonnes of standard coal and a reduction of 4,860 tonnes of carbon dioxide emissions. The increased income solely from electricity bills amounts to 4 million yuan annually. With the implementation of the "mushroom and light complementary" project, mushroom cultivation in Shangduantou Village has flourished, yielding over 100 tonnes of fresh mushrooms per mu and generating an income exceeding 1 million yuan per mu annually. This initiative has transformed the village into a new agricultural demonstration park for technology promotion, serving as a "green magic weapon" for mushroom farmers in Zhejiang to achieve prosperity.

3.3 Technological Conditions: Investment Scope and Potentiality

In 2022, the photovoltaic market achieved a regulatory output value of 90.2 billion yuan in Yiwu County, constituting over 30% of the photovoltaic market's output value in Zhejiang Province. The local production capacity of cells and modules has surged to 35 GW,

Table 2. Size and number of PV enterprises in Zhejiang Province (data as of 1 August 2023)

NO	Region	Capitalised at over $1 billion	Capitalised at over $100 million	Surviving, active	NO	Region	Capitalised at over $1 billion	Capitalised at over $100 million	Surviving, active
1	Hangzhou	21	384	8639	7	Taizhou	6	54	2630
2	Ningbo	14	237	8211	8	Huzhou	9	123	2160
3	Wenzhou	5	194	5352	9	Lishui	1	24	1300
4	JIAxing	15	207	4810	10	Quzhou	0	41	1244
5	Jinhua	0	15	3053	11	Zhoushan	3	33	490
6	Shaoxing	3	62	2910	12	Subtotal	77	1374	40799

representing nearly 20% of the global market share. Yiwu boasts 11 national high-tech enterprises within its photovoltaic market, along with the establishment of 4 academician workstations and 5 postdoctoral workstations. Moreover, the city has successfully attracted 11 academicians and gathered over 300 doctoral-level talents. Driven by market demand, several leading high-tech photovoltaic enterprises in the province invest approximately 200 million yuan annually in scientific research, with this investment steadily increasing over time. The annual allocation for research and development (R&D) and innovation ranges from 4% to 12% of sales revenue, significantly surpassing the minimum requirement of 3%. This substantial investment in product development serves as a robust guarantee for both technological advancement and economic prosperity.

3.4 Policies Conditions: Development Scope and Potentiality

For the PV manufacturing industry, as outlined in the "Zhejiang Province PV Market High-Quality Development Action Program (Draft)" released by the Department of Economics and Information Technology of Zhejiang Province, the target set for 2025 is to surpass a Zhejiang PV market output value of 250 billion yuan. The development objectives for the PV market in each city are detailed in Table 3, with projected PV cell and module production capacities exceeding 90 gigawatts (GW) and 110 GW, respectively. These supply and demand planning policies ensure the synchronization of PV production with market demand, fostering market growth. As of 2022, 31 cities and counties (districts) in Zhejiang have implemented PV subsidy policies, providing substantial impetus for the sustained development of PV market projects.

4 Assessment of PV Market Benefits in Zhejiang Province

According to the annual sunshine hours data released by the National Meteorological Bureau, the comprehensive benefits of the PV market in Zhejiang cities encompass two main aspects: the tariff benefits derived from PV power plants and the environmental benefits they offer.

Table 3. PV development targets of Zhejiang municipalities in 14th Five-Year Plan (Million kilowatts)

NO	Region	2020	2025	Added value	NO	Region	2020	2025	Added value
1	Hangzhou	130	230	100	7	Taizhou	119	219	100
2	Ningbo	274	474	200	8	Huzhou	154	304	150
3	Wenzhou	90	290	200	9	Lishui	117	317	200
4	JIAxing	177	277	100	10	Quzhou	16	116	100
5	Jinhua	270	370	100	11	Zhoushan	63	213	150
6	Shaoxing	107	207	100	12	Subtotal	1517	3017	1500

4.1 Electricity Benefit

Photovoltaic modules experience gradual efficiency attenuation with each passing year. As photovoltaic technology advances, the current market trend indicates an initial efficiency attenuation of approximately 1% in the first year, followed by a subsequent annual attenuation rate of 0.4%. Consequently, the lifespan of a photovoltaic power station typically spans 30 years [12]. Additionally, the efficiency of PV power stations may diminish due to factors such as weather conditions, temperature variations, and line losses. Research conducted on PV power stations in Zhejiang Province reveals that the typical system efficiency ranges from 75% to 83%, with an average system efficiency of 80%. Utilizing the Formula*Electricity Generation = Installed Capacity * Annual Peak Hours * Module Efficiency * System Efficiency * Operating Hours* (1), the annual power generation in Zhejiang Province is estimated at 864.275 billion degrees. The calculation of PV power generation in each city is detailed in Table 4. The power generation calculation formula is as follows:

$$Electricity\ Generation\ =\ Installed\ Capacity\ *\ Annual\ Peak\ Hours$$
$$*\ Module\ Efficiency\ *\ System\ Efficiency\ *\ Operating\ Hours$$
$$(1)$$

The revenue generated by PV power stations primarily depends on the unified acquisition of the national grid. In Zhejiang Province, the guideline price for acquiring electricity per degree under the FGD benchmark tariff is RMB 0.4153. By integrating Table 5 and Formula*Generation revenue = FGD benchmark tariff * annual generation * operating hours* (2), Zhejiang Province is estimated to accrue approximately RMB 358.93 billion in revenue over a 30-year period through PV power generation, in addition to the power generation revenues for each city outlined in Table 5.

The tariff revenue calculation formula is as follows:

$$Generation\ revenue\ =\ FGD\ benchmark\ tariff\ *\ annual\ generation\ *\ operating\ hours$$
$$(2)$$

4.2 Environmental Benefit

The emission reduction achieved by photovoltaic power plants primarily stems from the displacement of thermal power generation by photovoltaic power generation. Thermal power generation emits greenhouse gases and harmful gases such as carbon dioxide, dioxide, and nitrogen oxides. In contrast, photovoltaic power generation utilizes solar energy as a clean energy source. The increased power generation capacity of photovoltaic power plants leads to a decreased demand for thermal power generation, thereby reducing the emission of greenhouse gases and harmful gases. By calculating the power generation of PV power stations in Zhejiang Province over a 30-year period as shown in Table 4, we can deduce the energy savings and emission reductions achieved by these power stations.

Table 4. Calculation of photovoltaic power generation by cities in Zhejiang Province

NO	Region	Optimum inclination angle (°)	Annual peak hours (h)	Annual Electricity Generation (million kWh)	NO	Region	Optimum inclination angle (°)	Annual peak hours (h)	Annual Electricity Generation (million kWh)
1	Hangzhou	23	1271	6131914.1	7	Taizhou	23	1372	6302616.8
2	Ningbo	24	1380	13720821.1	8	Huzhou	23	1355	8640433.9
3	Wenzhou	22	1408	8564920.3	9	Lishui	23	1391	9249304.3
4	Jiaxing	23	1372	7971802.9	10	Quzhou	22	1404	3416235.3
5	Jinhua	23	1337	10376617.4	11	Zhoushan	23	1409	6295254.2
6	Shaoxing	21	1326	5757534.4	12	Subtotal			86427454.8

Table 5. Estimated 30-year PV plant power generation revenues by city in Zhejiang Province

NO	Region	Annual income (¥ million)	NO	Region	Annual income (¥ million)
1	Hangzhou	2546583.92	7	Taizhou	2617476.74
2	Ningbo	13720821.1	8	Huzhou	3588372.21
3	Wenzhou	8564920.3	9	Lishui	3841236.06
4	JIAxing	7971802.9	10	Quzhou	1418762.51
5	Jinhua	10376617.4	11	Zhoushan	2614419.07
6	Shaoxing	5757534.4	12	Subtotal	35893321.96

4.3 Calculation of Carbon Dioxide Emission Reductions

The average CO_2 emission factor for provincial electricity in Zhejiang Province, as released by China's Ministry of Environment and Ecology in 2021, is 0.5422 kgCO2/kWh. Utilizing Formula *CO_2 annual emission reduction = annual power generation of PV power plant * baseline emission factor of the regional grid to which it belongs Eq.* (3), it is determined that Taizhou City in Zhejiang Province achieves the least CO_2 emission reduction of 185.23 tonnes, while Ningbo demonstrates the highest reduction, amounting to 743.94 tonnes. The emission reductions for other cities are relatively evenly distributed, ranging from 300 to 500 tonnes, as depicted in Fig. 1. The calculation formula for CO_2 emission reduction is as follows:

$$CO_2 \ annual \ emission \ reduction \ = \ annual \ power \ generation \ of \ PV \ power \ plant \\ * \ baseline \ emission \ factor \ of \ the \ regional \ grid \ to \ which \ it \ belongs \ Equation \quad (3)$$

4.4 Calculation of Nitrogen Hydride Emission Reductions

In the latest "China Electric Power Industry Annual Development Report 2022" published by CEC in 2022, the national ammonia oxide emission per unit of thermal power generation is reported as 152 mg/kWh. By integrating Fig. 2 and Formula *Nitrogen hydride emission reduction = annual electricity generation * 152mg/KWh* (4), it is deduced that the emission reduction of ammonia oxides in Ningbo can reach up to 20,900 tonnes.

The calculation formula is as follows:

$$Nitrogen \ hydride \ emission \ reduction \ = \ annual \ electricity \ generation \ * \ 152mg/KWh \\ (4)$$

4.5 Calculation of Sulphur Dioxide Emission Reductions

According to the "China Power Industry Annual Development Report 2022" by CEC in 2022, the national sulfur dioxide (SO_2) emission per unit of thermal power generation is

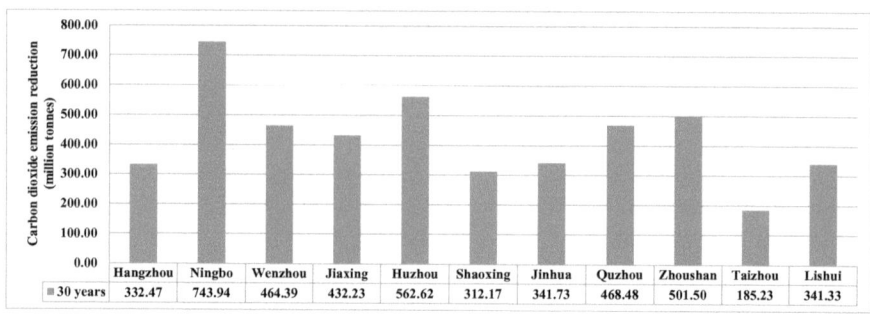

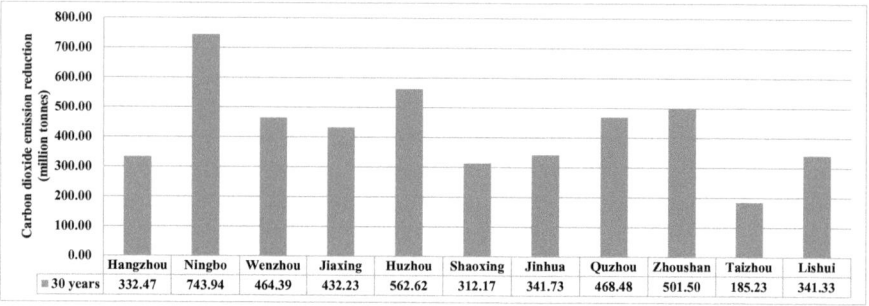

Fig. 1. CO2 reductions by cities in Zhejiang Province over 30 years

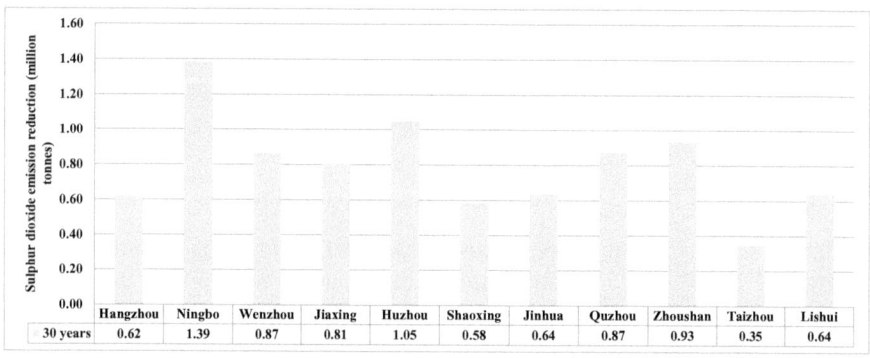

Fig. 2. 30-year nitrogen hydride emission reductions by cities in Zhejiang Province

documented as 101 mg/kWh. By integrating Formula *SO2 emission reduction = annual electricity generation * 101mg/KWh* (5) and Fig. 3, it is determined that Ningbo achieves the highest reduction in SO2 emissions, amounting to up to 13,900 tonnes.

The calculation formula is as follows:

$$SO2\ emission\ reduction\ =\ annual\ electricity\ generation\ *\ 101mg/KWh \qquad (5)$$

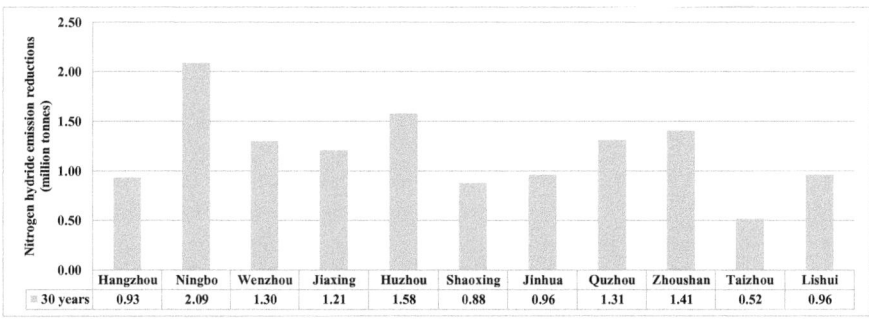

Fig. 3. Sulphur dioxide emission reductions by cities in Zhejiang Province over 30 years

4.6 Calculation of Standard Coal Savings

The calculation of standard coal saving is derived from the data presented in the latest "China Electric Power Industry Annual Development Report 2022" published by CEC in 2022. According to this report, the standard coal consumption for power supply from thermal power plants with capacities of 6,000 kW and above is recorded as 301.5 g/kWh nationwide. By incorporating Formula *Standard coal saving = annual power generation * standard coal consumption 301.5g/KWh* (6) and Fig. 4, it is established that Ningbo achieves the highest amount of standard coal saving, potentially reaching up to 41,368,000 tonnes.

The calculation formula is as follows:

$$Standard\ coal\ saving\ =\ annual\ power\ generation$$
$$*\ standard\ coal\ consumption\ 301.5g/KWh \qquad (6)$$

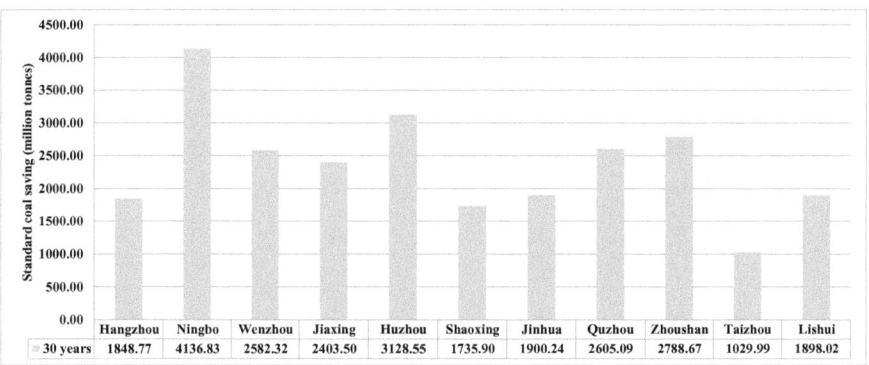

Fig. 4. Standard Coal Savings by Municipalities in Zhejiang Province over 30 Years

5 Conclusions

This paper presents a survey study on the development of the photovoltaic (PV) industry in China within the framework of the dual-carbon strategy. Utilizing Zhejiang Province as a case study, the findings (1) elucidate the PV market fundamentals from four key perspectives: market conditions, benefit conditions, technological conditions, and policy conditions; (2) based on the calculation of tariff benefits, project the future construction of 30.17 million kilowatts of PV power stations, estimating an income of 358.933 billion yuan over a span of 30 years; (3) through the assessment of environmental benefits, demonstrate that the operation of PV power plants for 30 years can effectively reduce carbon dioxide emissions by 873.965 million tonnes, nitrogen hydride emissions by 245,100 tonnes, sulphur dioxide emissions by 163,900 tonnes, and save 261 million tonnes of standard coal.

References

1. Tao, Z.: Interpretation of the peak carbon actionpProgramme to 2030. Ecol. Econ. **38**(01), 9–12 (2022)
2. Jie, J.: Developing and study of low-temperature solar thermal energy conversion applications. New Energy Progress **1**(01), 7–31 (2013)
3. Sacchelli, S., Garegnani, G., Geri, F., et al.: Trade-off between photovoltaic systems installation and agricultural practices on arable lands: An environmental and socio-economic impact analysis for Italy. Land Use Policy **56**, 90–99 (2016)
4. Chen, J., Wang, L.: Research review and outlook of agricultural light complementation. Jiangsu Agricultural Science **50**(05), 1–9 (2022)
5. Goetzberger, A., Zastrow, A.: On the coexistence of solar-energy conversion and plant cultivation. Int. J. Sustain. Energ. **1**(1), 55–69 (1982)
6. Dupraz, C., Marrou, H., Talbot, G., et al.: Combining solar photovoltaic panels and food crops for optimising land use: Towards new agrivoltaic schemes. Renewable Energy **36**(10), 2725–2732 (2011)
7. Zhang, H.: Research on the assessment of the utilisation potential of rooftop photovoltaic for urban buildings . Tianjin University (2018)
8. Cho, J., Park, S.M., Park, A.R., et al.:Application of photovoltaic systems for agriculture: A study on the relationship between power generation and farming for the improvement of photovoltaic applications in agriculture. Energies **13**(18): 4815 (2020)
9. Freitas, S., Serra, F., Brito, M.C.: PV layout optimization: String tiling using a multi-objective genetic algorithm. Sol. **118**, 562–574 (2015)
10. Zhong, Q., Tong, D.:Spatial layout optimization for solar photovoltaic (PV) panel installation. Renew. Energy **150**, 1–11 (2020)
11. Sun, T., Shan, M., Rong, X., et al.: Estimating the spatial distribution of solar photovoltaic power generation potential on different types of rural rooftops using a deep learning network applied to satellite images. Appl. Energy **315**, 119025 (2022)
12. Jinkosolar Homepage, https://www.jinkosolar.com/index.php?lan=cn. Last accessed 2024/04/27

Sustainable Land Use and Management

Analysis of Urban Planning and Construction for Regional Climate Change

Yue Zhang[1](✉), Yifeng Yang[2], and Bingxia Ma[3]

[1] Mianyang Meteorological Bureau, Sichuan, China
272810505@qq.com
[2] School of Marxism, Southwest University of Science and Technology, Sichuan, China
[3] Zitong Meteorological Bureau, Sichuan, China

Abstract. In recent years, a series of regional extreme weather events has emerged due to global climate change, resulting in various adverse impacts on regional ecological environments, human activities, and urban development. Addressing regional climate change has attracted widespread attention. Enhancing a city's ability to adapt to regional climate change is a pressing topic that requires both research and practical solutions. This study is grounded in the trends of regional climate change and takes a macroscopic perspective to explore the strategies and methods for addressing regional climate change in Mianyang City. Drawing upon the urban planning practices in Mianyang City, it presents measures and strategies for tackling regional climate change from the perspectives of urban spatial development, land use, transportation systems, and the construction of green spaces. These recommendations serve as a scientific foundation for creating a living environment well-suited to regional climate change.

Keywords: Region · Climate Change; · Mianyang City · Planning and Construction

1 Introduction

Climate change has increasingly significant impacts on human society, especially in a vast country like China. The country not only experiences the effects of global climate change but also faces a series of regional extreme weather events, such as seasonal heavy rains in the South, persistent drought in the Northwest, and frequent haze in the central and northern regions. These extreme weather events have had severe consequences on China's socioeconomic development and the living environment of its people. The frequency and intensity of these impacts are on the rise.

Cities are the largest interface between natural ecosystems and human social systems. Climatic conditions have changed, social conditions have changed with economic development, and many situations on the interface are changing, and cities should be a priority area for climate change adaptation. Compared to the agricultural sector, cities are more densely populated and highly exposed, and a large amount of social wealth is centrally exposed to climate change risks, making the task of adapting to climate

B.-J. He et al. (Eds.): UCSUD 2023, LNCE 559, pp. 311–320, 2025.
https://doi.org/10.1007/978-981-97-8401-1_22

change more urgent. With climate change, the frequent occurrence of extreme weather has brought more flooding, fog-haze and other problems to cities, making them more sensitive to climate, and the urban population has become even more environmentally and ecologically vulnerable.

Sichuan province has a complex and diverse natural environment with a 'sensitive body' that responds to global climate change. On January 15, 2024, *the Sichuan Provincial Climate Center* released the 2023 Sichuan Climate Bulletin, reviewing the major weather and climate events in Sichuan in 2023. The average temperature in the province in September was 21.6 °C, the was the highest for the same period in history. 2023 was the warmest year since 1961, and the annual average temperature in Sichuan province was 16.1 °C, 0.9 °C higher, ranking as the 1st highest in history. at the end of May 2023, the daily maximum temperature at 9 stations in Ningnan, Jinyang, Yanbian, Miyi, Pug, Xide, Coronation, Zhaojue, and Haili exceeded the historical maximum value.

Presently, China has conducted extensive research on urban planning in response to global climate change, with concepts like "low-carbon cities," "eco-cities," and "green cities" emerging as considerations for addressing global climate change. However, there remains a deficiency in the existence of planning mechanisms specifically customized to address regional climate change scenarios in China. Therefore, addressing regional climate change using appropriate technological measures is an important scientific problem that urgently requires exploration and implementation in China's urban planning research under the backdrop of regional climate change.

2 Urban Context and Climate Change

The topography of Mianyang City varies significantly from north to south, featuring diverse landscapes with varying elevations. The highest point, Xuebaoding in Pingwu County, stands at an elevation of 5,400 meters, while the lowest point is found in the Dujiangxia Valley in Santai County, with an elevation of 307.2 meters, resulting in an impressive elevation difference of 5,092.8 meters. Mianyang City falls within the North Subtropical Zone, characterized by a monsoonal climate with distinct seasonal variations [1].

Pingwu is located in the northern part of Mianyang (Fig. 1), and the average temperature is lower than the rest of the city, with the annual average temperature ranging from 13.88 to 15.65 °C, which is about 1 °C lower than the others area.. The unique topography and diverse landscapes contribute to the city's distinctive climate. The precipitation in Mianyang City is sufficient, and its inter-annual variation is large, although no significant overall trend has been observed. Notably, there is a substantial imbalance in rainfall distribution within Mianyang City, with the southern and northern regions experiencing lower rainfall while the central and western regions receive higher precipitation. With the global warming, meteorological disasters are getting more and more intense, and Pingwu County is affected by meteorological disasters that bring huge economic losses (Table 1). Drought is one of the main meteorological disasters in Pingwu County. Pingwu County has uneven distribution of precipitation during the year, and is prone to spring droughts and ambient droughts, especially spring and summer droughts. May ~ August is the season of high incidence of strong convective weather in Pingwu

County, and the strong convective weather is mostly accompanied by torrential rainfalls, gusty winds, and hailstones, etc. On July 12, 2022, under the influence of a short period of heavy rainfall in neighboring counties in the upper reaches of the county, flash floods broke out in Muzuo Tibetan Township of Pingwu County, Sichuan Province, resulting in casualties and economic losses. In addition, rainy weather occurs in Pingwu County every spring and fall.

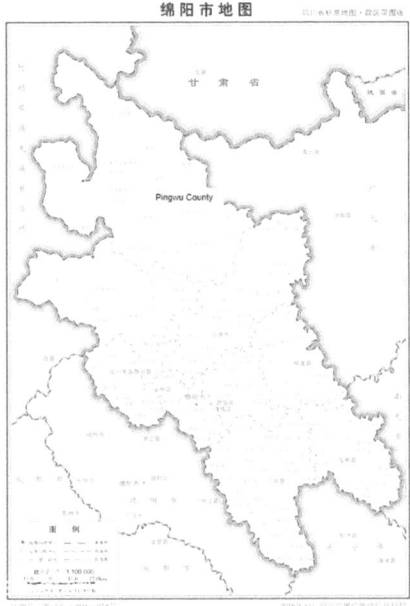

Fig. 1. Map of Mianyang city, divided by administrative districts.

Mianyang City, characterized by its extensive relief features, abundant rainfall, significant water resources, well-developed water systems, and dense river networks, exhibits a highly distinctive urban landscape. The city boasts over 3,000 small and large river channels, all part of the Jialing River system. The primary flood-prone season in the city occurs from June to September, particularly in the northern regions upstream of the Fujiang River, encompassing Beichuan County, Anzhou District, and Jiangyou City [2].

The city maintains a notably high vegetation coverage with various plant species, including both low-level and high-level vegetation. Additionally, there are 12 species of wild plants under state key protection and 38 species of wild plants under state secondary protection.

Table 1. Pingwu County has faced major climate change events in recent years

Date	Disaster type	Descriptions
2022–07–06	Drought	Precipitation was 71% lower than the same period in previous years, and drought conditions were observed in many places
2023–04–12	Hail	Intense hailstorms with thunder and lightning struck the area. The hail caused some damage to house roofs and crops
2022–7–12	Extremely heavy rainfall and mountain flash flood	As of 12:00 p.m. on the 14th, the disaster had resulted in 3 deaths and 15 missing persons
2018–6–26	Extremely heavy rainfall	The county's 25 townships and 248 villages were all affected by the disaster, the affected population amounted to 112,171 people, 8,209 people were relocated to emergency shelters, and the disaster losses amounted to 2.113 billion yuan

3 Regional Climate Change Technical Concepts and Planning Strategies

3.1 Technical Approach to Addressing Regional Climate Change

The interaction between natural conditions specific to different regions and global climate change has given rise to a series of extreme weather events, resulting in regional climate variations. Regional climate change is the outcome of the combined effects of regional natural conditions and global climate change, making its influencing factors more intricate. To effectively address regional climate change, it is essential to consider not only the regional climate change itself but also the response to regional natural climate conditions and regional extreme weather events.

3.2 Urban Planning Strategies for Climate Change

This section provides a summary of urban planning strategies to address climate change, framed as shown in Fig. 2.

Regional Climate Adaptation Strategies. China, with its vast expanse, encompasses five major climate zones characterized by features like "hot temperatures", "arid climates", "hot summers and cold winters", "cold climates", and "moderate temperatures" [4]. Each region exhibits distinct climate characteristics, necessitating the development of a comprehensive urban planning methodology tailored to the specific local

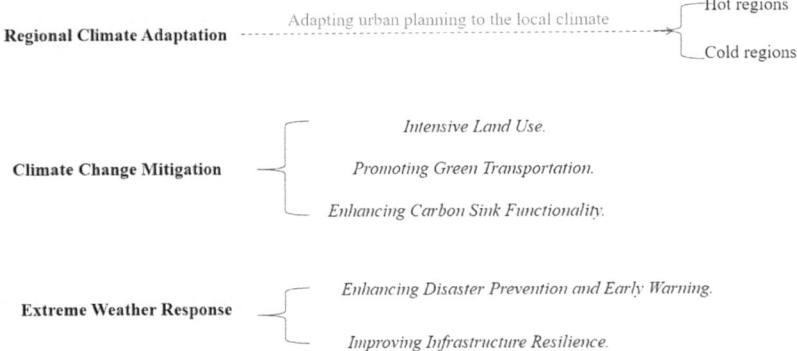

Fig. 2. Strategic framework for urban planning to address climate change.

climate features. Therefore, in the formulation of urban and rural planning, careful consideration should be given to the influence of regional climate characteristics on urban development and residents' lives. Specialized planning that aligns with the regional natural and climatic conditions should be incorporated to address these unique climatic features [3].

Adapting to regional climate change requires a holistic approach considering temperature, humidity, precipitation (snowfall), and wind. On a macroscopic scale, this involves the construction of urban spatial structures and road systems, the creation of urban ventilation environments, and the planning of green space systems. On a microscale, it encompasses considerations such as the spatial layout of buildings, building orientations, architectural forms, plant cultivation, and the design of indoor and outdoor spaces.

For hot regions, adapting urban planning to the local climate can involve the following measures:

1. Select locations with good ventilation and drainage to ensure the health and comfort of residents.
2. Utilize a low-density, wide-spacing, and open building layout, providing residents with more spacious living areas.
3. Plan the city layout according to the prevailing summer wind direction to maximize the use of natural airflow for cooling.
4. Incorporate green spaces and water bodies throughout the city as cooling sources, distributed in a scattered manner to provide natural cooling effects for surrounding buildings.
5. Prioritize ventilation and shading in public spaces to create a comfortable outdoor environment, allowing residents to enjoy fresh air and the benefits of shade.

For cold regions, adapting urban planning to the local climate can include the following measures:

1. Select suitable locations in sunlit areas and conduct detailed terrain and environmental surveys to ensure project feasibility and sustainability.

2. Place a strong emphasis on constructing urban protective forests in urban planning. Planting suitable trees and vegetation can effectively reduce air pollution, absorb noise, and improve the urban ecological environment.
3. Pay attention to lighting conditions when designing buildings. Through rational building layout and window design, buildings can maximize natural light utilization, enhance indoor lighting, and reduce energy consumption.
4. Ensure spacious streets in street planning to provide good sunlight conditions. This can enhance residents' quality of life and promote a healthy and vibrant urban development.
5. Consider wind protection measures in building design. By implementing appropriate building forms and facilities, you can reduce the impact of wind and provide a comfortable living environment.
6. Focus on the design of indoor public spaces in urban planning. Designing multifunctional indoor public spaces can meet various usage requirements in different seasons and weather conditions, offering more social and recreational areas.
7. Note the different usage patterns of outdoor spaces in the summer and winter seasons. Consider installing shading facilities, using suitable vegetation and materials, and providing seasonal landscapes to create a comfortable and pleasant outdoor environment.

Strategies for Mitigating Climate Change. Global climate change is a universal challenge, and as human society continues to experience rapid economic development, it profoundly affects our survival and development. The core objective of climate change mitigation strategies is to slow down the process of global warming.

Intensive Land Use. Intensive land use within a given land area can provide residents with more urban functions, enhance the city's service level, and increase production efficiency. It also reduces residents' travel time, leading to energy savings and emissions reduction. Studies comparing the urban densities of cities like Brasília and Curitiba have shown that higher-density cities have approximately 40% lower energy consumption. A well-planned mix of land uses is significant for lowering CO_2 emissions. Therefore, rational land resource utilization with a well-balanced allocation of various land functions can effectively control urban expansion, protect against carbon emissions, improve urban operational efficiency, reduce energy consumption, and minimize environmental pollution.

Promoting Green Transportation. Motor vehicles represent the primary source of greenhouse gas emissions in the atmosphere. Data indicates that the transport sector is responsible for around 15% of all greenhouse gas emissions, with even higher contributions in developed countries. Low-carbon transportation is the inevitable direction for future urban transportation development. In terms of travel modes, it is essential to develop green public transportation, and encourage cycling and walking. On a macro scale, connecting various functional zones with high-capacity Bus Rapid Transit (BRT) systems can improve accessibility and comfort, thus enticing more residents to choose public transportation [5].

Enhancing Carbon Sink Functionality. Increasing the carbon sequestration potential of ecosystems is a critical strategy for addressing climate change. Urban greening systems are essential in carbon sequestration, oxygen release, alleviating the urban heat island effect, and disaster risk reduction. In China, urban green spaces are often dispersed, and their resistance to natural disasters is weak, limiting their effectiveness in regulating urban climates. A network of interconnected green spaces is a more stable ecosystem, and the ecological network it forms is vital for maintaining the overall sustainability of the city. Additionally, a networked layout of urban green spaces can enhance the city's resilience to extreme weather events and allow for adjustments to the spatial form as the city develops, ensuring ecological stability. In urban planning, it is essential to integrate road green belts, natural water systems, expansion-resistant green belts, and ecological woodlands into a "point-line-plane" green space system, creating a combined system of "points" and "lines" in urban green spaces to enhance their carbon sink functionality and improve the region's response to climate change.

Strategies for Responding to Extreme Weather. By establishing efficient response mechanisms, cities need to respond to the increasingly frequent regional extreme weather events, including high temperatures, low temperatures, droughts, heavy rainfall, floods, typhoons, and dust storms.

Enhancing Disaster Prevention and Early Warning. In the context of frequent regional extreme weather events, cities should strengthen meteorological disaster warnings, enhance disaster prevention and mitigation for critical infrastructure, and provide a livable environment for residents. When formulating plans, various meteorological disaster risk assessments must be conducted to elevate the city's operational intelligence and enhance its ability to predict climate-related disasters. Additionally, it is essential to build a regional coordinated, unified disaster monitoring, prediction, early warning, information platform, and a comprehensive disaster response network. This network should include comprehensive disaster prevention and mitigation plans and reserve and supply emergency materials.

Improving Infrastructure Resilience. First and foremost, it is imperative to improve urban infrastructure to ensure rapid evacuation to nearby temporary shelters during sudden natural disasters and facilitate secure evacuation. Many disasters stemming from natural events pose a greater threat to public safety. Therefore, when upgrading urban infrastructure, it is essential to focus on enhancing the ability of the city's critical lifeline systems (transportation, energy, communication, water supply, and drainage) to cope with natural disasters and reduce secondary disasters caused by natural disasters. Upgrading and renovating urban infrastructure should be tailored to the natural disaster characteristics of different regions. For example, in areas with heavy rainfall, the focus should be on increasing drainage capacity, while in regions experiencing continuous drought, the emphasis should be on the deployment of water storage facilities.

4   Specific Planning Measures and Implementation

After the above analysis, combined with domestic and international experience, the development direction of combining urban land use and bus stations is proposed, and the Transit-Oriented Development (TOD) mode with residential and logistics land as the core can effectively prevent over-expansion, which is of great practical significance to alleviate the urban heat island effect. In addition, through the mixed land use can be avoided to a certain extent between the residential and work areas pendulum commuting to ease the pressure of urban transportation and reduce carbon emissions. Specifically, after analyzing the current situation in Mianyang, Sichuan Province, this paper proposes a strategy to reduce private vehicle emissions through the development of public transportation, especially high-speed rail transit, in order to reduce carbon emissions from urban transportation. Reducing inefficient travel while promoting public transportation is another key measure to achieve energy conservation and emission reduction. The details are shown below.

4.1 Urban Development Planning to Address Climate Change

Land Density with TOD Orientation. As China faces escalating energy crises and worsening air pollution, prioritizing public transportation has become necessary for urban development. Considering the current development status of Mianyang, Sichuan, and the ecological capacity of the region, a development direction combining urban land use with bus stations has been proposed. Drawing inspiration from foreign transit-oriented development models, the development process in Mianyang, Sichuan connects different urban functional groups using high-capacity public transit as a link. It revolves around bus stations, focusing on "nodes" and gradually reducing density from the "center" to the "periphery." This project centers around a rail transit hub, with commercial, office, and high-density residential areas as focal points. It forms TOD model with residential and logistics land at its core. This model effectively prevents excessive expansion and holds significant practical significance for mitigating the urban heat island effect.

Mixed Land Use. Combining land functions helps residents choose public transportation and fosters a vibrant urban living environment. The mixed land-use development with bus stations in Mianyang, Sichuan connects residential, commercial, public green spaces, and other functions within a convenient walking radius. Mixed land use can, to some extent, prevent pendulum-like commuting between residential and work areas. It also reduces unnecessary long-distance travel, positively alleviates urban traffic pressure, reducing energy consumption in transportation, and lowering carbon emissions.

4.2 Infrastructure Planning to Enhance Urban Resilience

Prioritizing Public Transportation. Developing public transportation is a significant avenue for reducing energy consumption and carbon dioxide emissions. After analyzing the current situation in Mianyang, Sichuan, a strategy for reducing emissions from private vehicles by developing public transportation, especially high-speed rail transit, has been proposed to lower urban transportation carbon emissions. It is crucial to effectively utilize public transit resources to encourage residents to voluntarily reduce the use of private cars to reduce carbon emissions. Firstly, at the overall planning level of the urban public transportation system, creating an excellent public transportation environment for urban residents to effectively channel them is essential. In the development of Mianyang, Sichuan, it is important to establish a comprehensive public transit network, connecting various functional groups through a high-capacity, high-speed transportation system. Integrating bus stations and service facilities and controlling the service radius of bus stations within 500 m facilitates seamless transfers between different public transit modes. Secondly, at the administrative level, affordable vehicle transfer parking facilities can be established in the suburbs, along with charges in the city center, to encourage citizens to use public transportation.

Reducing Inefficient Travel. While promoting public transportation, reducing inefficient travel is another critical measure to achieve energy savings and emissions reduction. Mianyang, Sichuan currently has an average per capita travel distance of 3.38 km, with a travel time of approximately 30 min, and overall, residents spend more time using various modes of transportation. Therefore, in the urban development process, it is vital to control the continuous increase in travel time and decrease the number of inefficient trips. In the land use layout of Mianyang, Sichuan Province, it is important to emphasize the balance between work and residence. This involves appropriately matching residential, office, and commercial land uses to avoid long-distance commutes and pendulum-like commuting patterns. Creating a comfortable and acceptable travel time for the public is important.

5   Conclusion

Taking Mianyang City in Sichuan Province as an example, this project focuses on guiding the organized spatial structure of the city, preserving urban ecological carbon sink spaces, and mitigating the impact of regional climate change on local conditions. It emphasizes mixed land development, optimizing urban function allocation, reducing inefficient travel, and decreasing carbon emissions. The promotion of public transportation has been emphasized to reduce regional carbon emissions.

While a preliminary framework for addressing regional climate change has been established, it is important to note that China's vast territory and diverse regional climates require more specific and targeted strategies. In future research, we will explore the development in greater depth and specificity by considering the unique characteristics of regional climate change.

References

1. Yu Xiangyu: Research on Optimizing Urban Design Practice Based on Regional Climate. South China University of Technology (2020)
2. Song Youliang, Wang Yingyi, Hong Liangping: Preliminary Study on the Pre evaluation Mechanism of Urban Planning Compilation in Response to Climate Change//China Urban Planning Society. Urban Rural Governance and Planning Reform - Proceedings of the 2014 China Urban Planning Annual Conference (01 Urban Safety and Disaster Prevention Planning), Vol 15 (2014). China Construction Industry Press
3. Liangping, H., Xiang, H., Zhilei, C.: Urban planning response to climate change. Urban Issues **07**, 18–25 (2013)
4. Zhou Jin: Research on the Techniques for Compiling Regulatory Detailed Plans to Address Climate Change. Huazhong University of Science and Technology (2014)
5. Cai Zhilei: Response to the Formulation of Urban Master Plans in Response to Climate Change. Huazhong University of Science and Technology (2013)

Open-Pit Image Detection Based on Improved Faster – RCNN

Rujin Huang[1,2(✉)], Genhou Wang[3], Jiahao Tian[3], and Quanping Zhang[4]

[1] School of Earth Sciences and Resources, China University of Geosciences (Beijing), Beijing 100083, China
3001230100@email.cugb.edu.cn

[2] Key Laboratory of Mine Environmental Monitoring and Improving Around Poyang Lake of Ministry of Natural Resources, East China University of Technology, Nanchang 330013, China

[3] Department of Geological Engineering, Qinghai University, Xining 810016, China

[4] Institute of Applied Ecology, Chinese Academy of Sciences, Shenyang 110016, China

Abstract. Remote sensing image open-pit mine monitoring is usually affected by speckle noise, multi-scale and other factors due to the limitations of landform and other conditions, and faces the problem of low availability of monitoring area effect. Therefore, this paper introduces an improved regional convolution neural network method. The network thinning process and the improved conditional random field are proposed respectively as a circular neural network, and the accurate classification and coordinate positioning of the target are completed by establishing the network thinning process in the output part to increase the classification and regression thinning of the target features; Remove the influence of color vector for the fully connected conditional random field and improve it and construct a recurrent neural network for the conditional random field. The experiment shows that the target detection accuracy of the improved Faster-RCNN network has achieved a breakthrough in the mine image detection details, with the overall recognition accuracy of 94.67% and the detection speed of 24.03 fps. Compared with Faster-RCNN, SPP-NET and YOLOV7, it effectively improves the accuracy and provides technical support for ecological restoration monitoring of open pit.

Keywords: Open pit · Ecological restoration · Faster-RCNN · feature extraction · Recurrent neural network · Full connection layer

1 Introduction

Due to the increasing industrialization of human society in recent centuries, fossil mineral energy has become the main driving force for social development in all countries of the world, but the ecological environmental protection and restoration after the development and utilization of mineral energy has been an important research topic that all countries of the world have been focusing their attention on during the late stage of the industrial process. With the continuous deepening of the development and utilization of China's mineral resources, the related ecological problems caused by mining have

© The Author(s) 2025
B.-J. He et al. (Eds.): UCSUD 2023, LNCE 559, pp. 321–335, 2025.
https://doi.org/10.1007/978-981-97-8401-1_23

become increasingly prominent. The effective monitoring of mine ecological environment has become an important link in mine production activities, and there is an urgent need to form the ecological problem monitoring technology in line with Chinese characteristics [1, 2]. The distribution of mineral resources in China is characterized by a wide range, multiple landforms and complex landforms. After mining, especially in open-pit mines, there will be topsoil stripping, soil water erosion, wind erosion, water pollution and other phenomena, which will seriously affect the safety of the mining area and regional ecological environment, and pose a threat to the safety of life and property of the people around the mining area [3]. In July 2016, the Ministry of Land and Resources jointly issued the Guiding Opinions on Strengthening the Mine Geological Environment and Comprehensive Treatment, emphasizing the establishment of a dynamic monitoring system by 2025 to comprehensively monitor the dynamic changes of the mine geological environment [4]. Therefore, it is an important part of realizing the integration of mine ecological environment and natural regional environment to provide targeted guidance on mine ecological management plan based on the effectively obtained data of mine ecological change.

The traditional mine monitoring mainly relies on historical data, actual adjustment, and remote sensing image manual calibration, but these methods cannot achieve long-term dynamic monitoring, and rely on manual conclusions with large errors [5–9]. People's requirements for image processing technology are also gradually increasing, and target detection, as an important part of the technology, also needs to be developed. Therefore, it is of great practical significance to study how to realize the rapid detection and information acquisition of mine pits. At the same time, mine pit detection is one of the basic conditions for realizing mine dynamic monitoring. The development of computer technology provides sufficient computing power for the development of remote sensing technology, making it possible to use depth learning technology for remote sensing image detection and information extraction. Traditional target detection methods first manually extract the features of the image, and then use common classification methods for classification, such as support vector machine, Bayesian classification [10, 11]. In short, it is the processing process from candidate region to manual feature extraction to classifier. This method is firstly that the manually extracted feature is less robust in the face of massive images. Secondly, because the region selection strategy based on sliding window is manually determined by strong subjective factors and the process is cumbersome and redundant, finally, the detection of open pit needs to be able to extract features more effectively, However, this shallow manual feature extraction method can no longer meet the needs of production. At present, the detection methods with good detection effect are divided into single-stage and double-stage. The single-stage means that the whole process only uses one network. Common models include SSD and YOLO series models, which are mainly based on regression for identification and classification [12]; Two-stage means that the whole process is divided into two steps. First, the region is recommended and then classified. Common algorithms include R-CNN, SPP-NET, FastR-CNN, and Faster-RCNN. They mainly use a large number of target candidate boxes to extract to help achieve detection. The Faster-RCNN model is proposed on the basis of Fast-RCNN. It completes the extraction of target candidate regions by replacing the selective search algorithm (ss) with the regional proposal networks (RPN), thus helping it achieve end-to-end [13, 14].

In recent years, many scholars have improved the algorithm of Faster-RCNN for target two-order detection, such as Huang Xiaohong [15], etc., by introducing the attention-guided context feature pyramid to optimize the feature extraction and use the cascade structure to improve the regression accuracy; Zhao [16] and others used ResNet as the network backbone to extract features, and obtained multi-scale feature maps through top-down feature fusion to enhance position information and classification information; Kang [17] and others used multi-layer fusion to build a CNN based on context region, in which the high-resolution RPN and the object detection network with context features are used; Cao Lei [18] et al. improved RPN by introducing VGG-16 to extract multi-layer features of the image, and carried out feature extraction and regularization on the three feature layers with the deepest convolution core to fuse their information. Dai [19] and others proposed R-FCN based on the problem of operation speed. The boundary layer without full connection, classification and regression becomes full convolution, which makes the classification position unchanged and the regression position variable to the maximum extent, and greatly improves the efficiency of the model. The above methods have achieved good results in the improvement of Faster-RCNN model, but there are still some problems in the practical application of open-pit detection: 1. The uneven distribution of image speckle noise and intensity; 2. The multi-scale performance of the mine pit is that the pixel value in the image is too small, resulting in insufficient feature extraction and small target information.

In order to solve the above problems, this paper proposes an improved open-pit mine detection model based on Faster-RCNN for detecting complex landforms, multi-forms and multi-scale. The network thinning process and the improved conditional random field are used as a recurrent neural network, and the accurate classification and coordinate positioning of the target are completed by establishing the network thinning process in the output part to increase the classification and regression thinning of the target features; The influence of color vector is removed from the fully connected conditional random field and the recursive neural network is constructed for the conditional random field.

The overall model still realizes end-to-end operation and controls the increase of calculation efficiency as much as possible while ensuring the detection accuracy.

2 Faster-RCNN Algorithm Foundation and Model Improvement

2.1 Faster - RCNN Network Infrastructure

Faster -RCNN is mainly composed of feature extraction network (convlayers), regional proposal network (RPN), interest pool (RoI pooling) and target classification and regression (Fig. 1). Its core is mainly composed of two subnetworks, including RPN and Fast-RCNN network [20]. Among them, Fast-RCNN is used to detect the region of interest (ROI) in the network. It mainly solves the problems existing in the original R-CNN, such as the need to calculate a large number of candidate boxes and the classifier training and position regression after feature extraction. The proposed feature blocks of uniform size are input into the final full-connection layer [21, 22]. RPN [23] is used to replace the

Selective Search in the RCNN model to generate higher quality candidate boxes. Its main purpose is to determine whether there are targets in the pre-selection box in all the previously set anchors, and at the same time correct the anchor to ensure the accuracy of the candidate box. RPN and Fast-RCNN will share these features into ROI pooling, and calculate the feature category and border regression function to accurately locate the location of the detection frame.

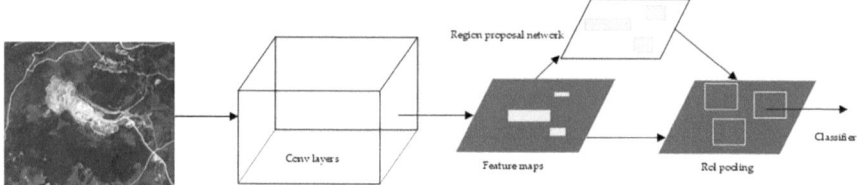

Fig. 1. Faster - RCNN Basic working principle.

2.2 RPN Structure

The core of the RPN structure is a full convolution neural network. Its core is to traverse all the input feature maps and generate corresponding anchor positions by taking each point extracted from the rough extraction as the center point, and map the anchor frame with different scales and proportions in the original map [24, 25]. Specifically, 40 × 60 Use 3 in the feature diagram × The convolution kernel of 3 constructs 9 different candidate frames at each center point, maps them to the rescale image, selects them, discards the parts beyond the boundary, and arranges them in order of size, selects the first 2000 for non-large value suppression, and then sorts them again according to size to output 300 Proposals to Faster-RCNN. As shown in Fig. 2, RPN performs sliding window convolution on the featuremap, where the convolution step is 1, and the padding = 2 is filled to obtain 256 features shared by the region, which are transferred to the classification layer and the regression layer, where the classification function outputs 2k scores for target classification, the border regression function outputs 4k vectors for target coordinate positioning, and then uses anchor boxes to correct the output target candidate box [26].

2.3 Model Improvement for Open-Pit Inspection

The characteristics of open pit under complex terrain and landform conditions, such as complex and changeable, disordered shape, vegetation change and image difference, are easy to affect the target results. Because the difference of mine pit size is generally larger than the actual situation when giving the proposal frame, so the network not only needs to process the target object, but also needs to calculate the changes in the shape of buildings around the mine pit, so as to reduce the detection effect. However, in the practical application of open pit, the uniqueness of the structure caused by the surrounding buildings of the pit is an important reference for feature extraction of the

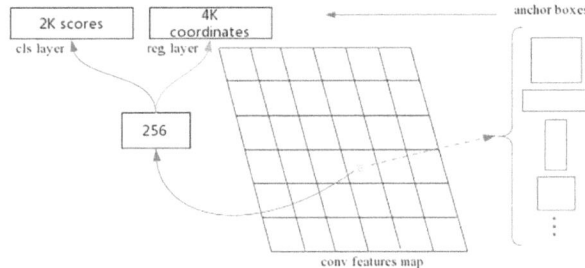

Fig. 2. RPN structure.

pit. Secondly, the Faster -RCNN network uses a fully connected network at the end. This part of the operation takes up most of the parameters of the whole operation, which makes the model operation difficult and data volume increased sharply. Therefore, the model optimization in this paper is mainly based on the above two points.

2.4 Network Refinement

In order to solve the problem that the model effect is reduced and the period becomes longer due to the large frame selection proposed by Faster -RCNN, a model accuracy improvement module is introduced after the Faster -RCNN structure. The final classification and regression are further refined, and a new full connection layer is built before the regression to help the whole Faster -RCNN network improve its accuracy. First, the output information is integrated with the new full connectivity layer, and then the convolution feature is combined with the ROI of the pre-model to carry out the classification and regression operation again. This can make the results more accurate than before, and also effectively remove the confusion effect of the background on the target. At the same time, the second regression operation further improves the position coordinates of the target object to ensure the combination and confirmation of the target object and multiple boundary boxes around it. It is divided into two steps to improve the overall accuracy of the model from the target object frame selection range and location information.

2.5 Conditional Random Field Is Improved to Make Cyclic Neural Network

The method constructed in this paper is to introduce the maximum posterior distribution of the mean field approximation on the basis of the introduction of the fully connected conditional random field (CRF) to solve the problem of the complexity and time-consuming of the algorithm in the process of minimizing Gibbs, and use the advantages of CRF to control the details to use the fully connected conditional random field as a circular neural network, and also realize that the improved Faster-RCNN can achieve end-to-end connection, While improving the detection accuracy, it also reduces the difficulty of model operation. The traditional conditional random field is mainly used to smooth the image noise. By coupling the adjacent nodes, it is helpful to allocate the same markers to the pixels that are relatively close in space. The commonly used convolutional neural network architecture score chart at this stage is usually very smooth, but when it is applied to the segmentation of open pit, there will be small structure omission, and it is impossible to realize the reproduction of local fine structure. Therefore, the introduction of fully connected conditional random fields can capture fine details to meet the requirements of the mine for details. All pixel labels are modeled as random variables, and Markov random fields are formed under global observation conditions. Assume that the image is I, xi is the label of pixel i, and take values from the label Li, X is the vector generated by random variables x1, x2,..., xN, and the relationship between I and x can be modeled as conditional random fields. The specific formula is shown in Formula (2):

$$P(X = x|I) = \frac{1}{Z(I)}\exp(-E(x|I)) \tag{1}$$

Thus, the Gibbs distribution can be defined as formula (2)

$$E(x) = \sum_i \psi_u(x_i) + \sum_{i<j} \psi_p(x_i, x_j) \tag{2}$$

$$\psi_p(x_i, x_j) = \mu(x_i, x_j), \sum_{m=1}^{M} \omega^{(m)} K_G^{(m)}(f_i, f_j) \tag{3}$$

$$Q_i^-(l) = \sum_{i \neq j} a_{i,j} Q_j(l) \tag{4}$$

$$H_1(t) = \begin{cases} softmax(U), t = 0 \\ H_2(t-1), 0 < t \leq T \end{cases} \tag{5}$$

$$H_2(t)\{f_0(U, H_1, I), 0 \leq t \leq T \tag{6}$$

$$Y(t) = f(x) = \begin{cases} 0, 0 \leq t \leq T \\ H_2(t), t = T \end{cases} \tag{7}$$

where: $\sum_i \psi_u(x_i)$ to measure the loss of pixel i taking label xi, it is obtained by deep convolution neural network; $\sum_{i<j} \psi_p(x_i, x_j)$ In order to measure the loss of pixel I and j taking labels xi and xj at the same time, depending on the image smoothing term, similar pixels are more likely to be labeled as the same label, As shown in (3), f_i, f_j is the feature vector of pixels i and j, usually two-dimensional coordinates and color vectors are selected; M is the number of Gaussian kernels, and the value is 1 or 2. Each $K_G^{(m)}$ is a Gaussian kernel acting on the eigenvector $\omega^{(m)}$; Linear combination for weight $\mu(x_i, x_j)$; Is a compatibility function.

The labeling result can be obtained at the minimum Gibbs distribution E(x), but the whole process is difficult and time-consuming, so the average approximate maximum posterior distribution is introduced for reasoning. Where Q(x) is an approximation of the conditional random field P(x), which is reconstructed as a recurrent neural network. This paper optimizes its message transmission process and weight adjustment. Specifically, the original Gaussian kernel considers the position and color vector of x and y, but in actual operation, the color vector has determined the prior probability of classification, so we can give up the vector distance of color vector in this process, and only consider the impact of position change, that is, the farther the distance is, the smaller the difference is. At the same time, The Gaussian kernel distance is replaced by the full-graph distance weight network and reset to the full-convolution network. The weight value is obtained from the training samples. This process is equivalent to a convolution operation. As shown in formula (4), a_i is the weight of the distance, l represents the category, and $Q_j(l)$ is the category probability. Secondly, as shown in Formula (5–7), the iterative mean field is taken as a circular network, and the results of the previous iteration are integrated iteratively by using the multi-layer mean field until this part meets the requirements, T Represents the average field iteration times, $H_1(t)$ is the result of the normalization of the deep convolution neural network, and $H_1(t)$ is a CRF operation.

2.6 Framework After Model Optimization

The improved algorithm flow through the above two parts is shown in Fig. 3 below. First, the result of image processing by Faster -RCNN is sent to the judgment box, and the output is carried out after the joint determination of the improved conditional random field's cyclic neural network and the network thinning module, and both confirm the fitting. In the network refinement part, add the new information obtained from the full connection layer integration to carry out the regression and classification operation again and send it to the judgment box; Secondly, the conditional random field is optimized and used as a recurrent neural network to iterate until the best fit. In terms of efficiency, the improved part is end-to-end connection, while the Faster -RCNN network itself belongs to the end-to-end connection network. In this way, the built model first ensures the integrity of the information transmitted, and then does not increase too much in the operation efficiency, thus maintaining the benign operation of the model.

3   Experimental Results and Analysis

3.1   Experimental Data Sources

The experiment detects open pit targets in remote sensing images under complex terrain conditions and analyzes the improved model in comparison with other advanced models. The experiment is carried out using a graphic workstation, the experimental framework is Anaconda3 + Tensorflow, and the relevant high-quality open pit dataset is constructed. The dataset sources mainly include the 2015 Guizhou Province pit vector released by Prof. Liu Shiliang's team at Beijing Normal University, as well as Gaofen-1, Gaofen-2, and Google earth images, and a total of 800 images are collected from Southwest China, including different types of mining pits and large quarries. A total of 800 images were collected from Southwest China, including pits of different mine types and large quarries, including 200 images with different cloudiness, seasons, topography, and resolutions. In order to solve the problem of large errors in the training process caused by the number of data sets, the image is expanded, and the original data is mirrored, rotated, scaled, and cropped. Finally, a total of 3200 images are obtained. Randomly select 60% as the training set, 30% as the verification set, and 10% as the test set. Secondly, through model training, it is found that the number of iterations of the convolutional neural network

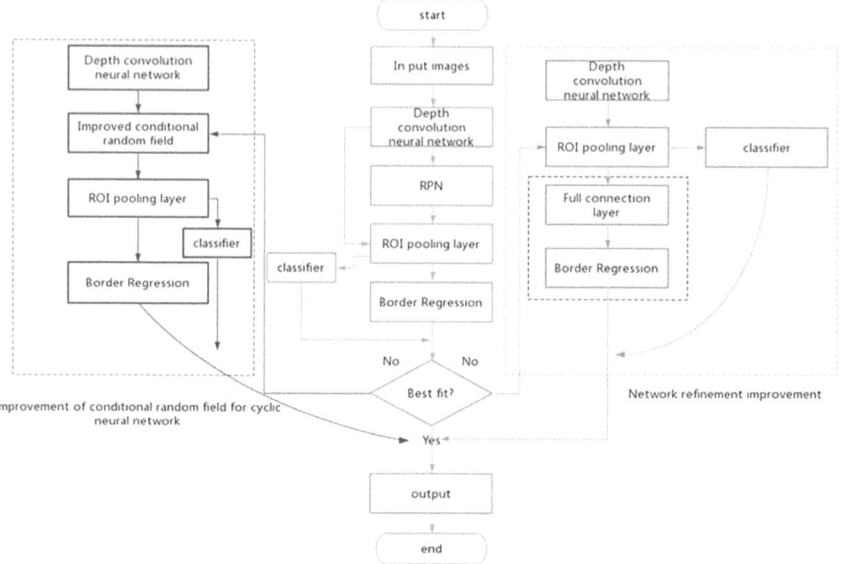

Fig. 3. Flow chart of improved model algorithm based on Faster -RCNN.

is designed to be 300, the batch size is 50, and the initial value of the learning rate is set to 0.001. In this process, in order to prevent the learning rate from being too large and causing shock in the process of converging to the optimal result, the learning rate should decline exponentially with the training change, Therefore, the weight attenuation is 0.0005 and the momentum factor is 0.9. The analysis of experimental results mainly includes the analysis of pit detection results, loss function, mAP and FPS.

3.2 Pit Detection Results

This experiment mainly compares and analyzes the open-pit detection under three conditions of thin cloud, hill and mountain, and compares the improved model with the basic model Faster-RCNN, SPP-NET and YOLOV7. The detection results are shown in Figure: under the condition of thin cloud in Fig. 4, all models have strong recognition ability and are not too affected by cloud cover. The overall results are above 0.90. Among them, the effect of the improved model in this paper is the most prominent, followed by the results of YOLOV7 and Faster-RCNN are 0.97 and 0.93, respectively, and the effect of SPP-NET is 0.90. The main reason is that it optimizes the transformation of the size of the feature vector, and does not fully grasp the characteristics of such a large single pit, so the detection effect is slightly lower than other models. Under the conditions of hills in Fig. 5 and in the face of large differences in detection targets, the detection effects gradually appear to be different. In addition to SPP-NET, the detection effects of all models in the main pit in the figure can be guaranteed, and the detection effects are respectively improved model 1.0, Faster-RCNN-0.90, YOLOV7–0.94. The main problems are concentrated in the small and medium-sized pits in the figure. The effect of the four models is not higher than 0.65. However, the improved model is better than other networks in feature extraction and detail processing of small pits under the support of the network refinement module. However, the improved model foundation Faster -RCNN in this paper has poor detection effect due to the lack of refinement steps. The effect of SPP-NET in the detection of small pits is slightly better than Faster-RCNN due to the addition of SPP layer. Figure 6 The detection effect of each model target under mountainous conditions increases the detection difficulty of each pit in the figure due to the variety of pit shape and the complex improvement of surrounding environment. First of all, the detection effect of the pit at the bottom right in the figure is relatively good, which shows that the model has strong detection ability for the pits with obvious characteristics and large area. In the face of a large number of mine production equipment, the detection accuracy decreases obviously because the complexity of ground objects is much higher than that of ordinary mines. However, the improved network model in this paper only retains the end-to-end connection between the position vector and the depth convolution neural network in the conditional random field to help it capture more detailed information, thus helping its detection accuracy to be better than other models. When the image contains many complex situations such as mountains, pits, mine buildings, etc., its accuracy is not ideal. YOLOV7 reaches 0.61, which is higher than other models. The improved model still has large room for improvement.

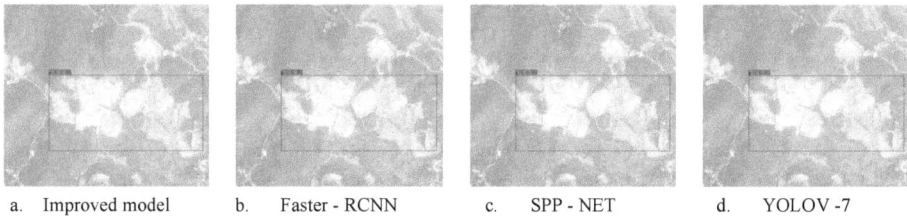

a.   Improved model   b.   Faster - RCNN   c.   SPP - NET   d.   YOLOV -7

Fig. 4. Comparison of detection effects under thin cloud conditions.

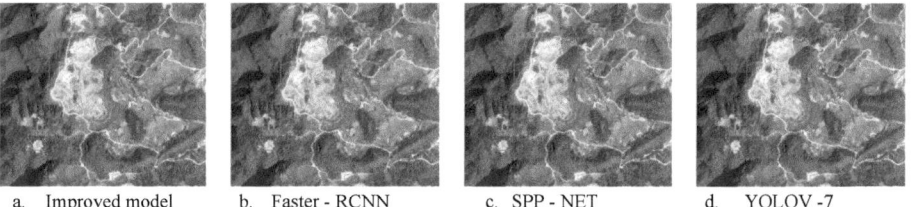

a.   Improved model   b.   Faster - RCNN   c.   SPP - NET   d.   YOLOV -7

Fig. 5. Comparison of detection effects under hilly conditions.

3.3   Loss Function

The loss function is a very important evaluation index in the process of deep learning target detection, which can intuitively show the advantages and disadvantages of the model [27]. This paper includes four parts: training sample loss function, test sample

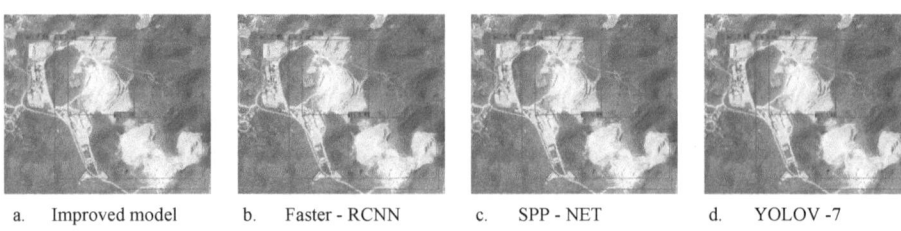

a.   Improved model   b.   Faster - RCNN   c.   SPP - NET   d.   YOLOV -7

Fig. 6. Comparison of detection effects under mountainous conditions.

loss function, training sample regression loss function, and test sample regression loss function, as shown in Fig. 7.

The loss of the four models compared in this paper has decreased rapidly, which shows that the overall initial conditions of the four models are good. Specifically, only YOLOV7 has a sharp fluctuation and a slight upward trend when it iterates to 15–50 times, which is mainly due to the fluctuation of YOLOV7 training with the increase of auxiliary head training cost. However, other models except the improved model have a function of around 1.5 when it iterates to 50 times, and the improved model has reached below 1.0, indicating that the learning effect of the improved model is better than the other three models. After that, the four models all entered the stage of rapid decline in loss function, and the loss function of the improved model was lower than 0.5 when the iteration reached 150 times, while the other models were still above. After that, the training sample loss function and regression function of the improved model continued to decline, and entered the stable stage after 210 times, and the final loss function remained stable at 0.23. The other three models, especially YOLOV7, still had slight fluctuations. Faster-RCNN and SPP-NET entered the stable stage after 220 times, respectively, but the loss functions of the three models were finally stable at about 0.5. On the whole, the performance of the improved model in training set and test set is better than other models.

3.4 Quantitative Analysis

In the experiment process of this paper, the mean average precision (mAP) and frames per second (FPS) are used as the evaluation indicators of open-pit target detection [28, 29]. First, mAP needs to be calculated according to the precision and recall, and the calculation formula is as follows: (8) (9). Second, the average precision (AP) is the calculation of the progressive average of the precision values of different recall levels. The higher the value is, the better the performance is. The calculation formula is as follows: (10). The calculation results are as follows:

$$Precision = \frac{TP}{TP + FP} \tag{8}$$

$$Recall = \frac{TP}{TP + FN} \tag{9}$$

where:

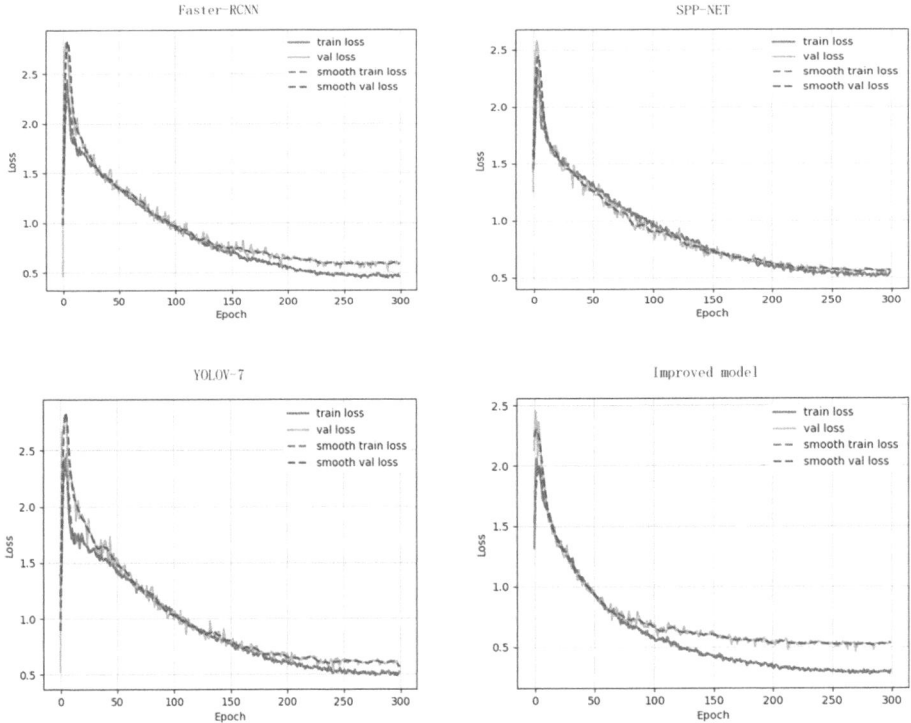

Fig. 7. Loss function curve of each model.

TP: An example of positive sample determined in the classifier and actually positive sample.

FN: An example of negative samples determined in the classifier but actually positive samples.

FP: An example of positive samples determined in the classifier but actually negative samples.

$$AP = \sum (R_n - R_{n-1})P_n \tag{10}$$

where:

Rn is the recall rate at the nth threshold.

Pn is the precision at the nth threshold.

Table 1 compares the average accuracy of the four models and the number of frames transmitted per second. It can be seen from the table that the detection speed of SPP-Net is 24.87 fps, which is the slowest of all models, and its detection results are at the rear of all models. This is because the feature extraction of SPP-Net needs to be written to the disk due to the existence of SVM, resulting in the reduction of the entire model speed. The detection accuracy and detection speed of YOLOV7 are relatively balanced among all models. In its training stage, the accuracy is improved by increasing the training cost, which does not affect the reasoning time. However, the accuracy is still far from the

Table 1. Quantitative analysis results of different models.

Method	mAP(%)	FPS/s
Faster-RCNN	89.95	23.36
SPP-Net	88.65	24.87
YOLOV7	90.17	21.76
Improved model	94.67	24.03

improved model in this paper, and the requirements for detection speed in the specific production process are not prominent factors. In the case of small difference in detection speed, the detection accuracy of the improved algorithm and its basic model Faster-RCNN has a large gap. The main reason is that the improved model has strengthened the extraction of target features and refined the network in the output process. The result is that the detection accuracy of Faster-RCNN is 89.95%, which is 4.72% behind the improved model. Compared with other networks, the improved model in this paper is slightly complex, but the detection accuracy of the model is significantly improved, which proves the effectiveness of the method and provides basic technical support for the monitoring of open-pit mines.

4 Discussion

Based on the current actual production development needs, this paper improves the detection model by combining the deep learning technology. The results show that the improved model can enhance the detection of open pit and meet the complex and changeable operation conditions of open pit. The improvement of secondary model will help to improve the monitoring capacity of ecological environment of open-pit mines and promote the ecological development of mines in the process of mining development, at the same time, this kind of targeted target detection model provides more application scenarios for target detection, such as coal mining area, ecological protection area and so on.

The basis of target detection comes from the evolving computer arithmetic and the continuous optimization progress of the model, but it also has certain limitations. For the application in the field of ecological restoration, the domain convolutional neural network model is a black-box model, so the construction of the model requires a lot of training and continuous attempts, which leads to a lack of theoretical explanations and mathematical reasoning, which ultimately affects the accuracy of the model, and due to the uniqueness of ecological restoration specialties and does not have the conditions to build a large-scale public dataset. At the same time, another model that can be trained on small datasets, such as U-Net, but the effect is limited by the amount of data, easy to encounter bottlenecks, and secondly, the ecological restoration problem is faced with the performance of different types of patches around the globe, with different landforms, different climates, and different latitudes, so that the transfer learning is a more desirable choice.

In the aspect of model sample database, although we strive to ensure the diversity of pit morphology to improve the detection and general ability of the improved model, it is not suitable for a wide range of applications due to the impact of the landform around the mine, so the pit database needs to be further expanded. Secondly, although the improved model has achieved good results in the detection, the pit detection under different types, different minerals and different geomorphic conditions is quite different in the process of progressive learning and training of samples. Therefore, it is necessary to further classify and itemize the pit detection data set and then conduct combined detection to achieve a balance. Finally, the improvement process of the model tends to pursue the detection accuracy and the learning ability of complex landform features. How to ensure the detection accuracy while taking into account the efficiency is the focus of future research.

This kind of target detection technology for ecological protection has become an important means in the field of ecological protection. From the practical application point of view, both in terms of accuracy and speed compared with the traditional methods have realized a leap-forward development. Compared with the same type of target detection model, the improved method proposed in this paper has higher accuracy and richer feature information, which is helpful for the further application of target detection technology in the field of ecological protection.

5 Conclusions

Deep learning provides an effective and fast feature learning method for large area and long time series open-pit detection. It can learn features from massive original data in a short time, and provides new ideas and methods for open-pit monitoring. In this paper, an improved open-pit detection model based on Faster-RCNN is proposed to solve the problems of low accuracy of open-pit image extraction, complex technical means and low utilization of results, the main conclusions are as follows:

1) The experimental results show that the improved model is better than other similar models in terms of accuracy, versatility, automation and efficiency. Meanwhile, through the experiments, the model has better applicability to different types of images, which expands the data source of open pit to a certain extent and further enhances the practicality and extensiveness of the model.
2) In this paper, the influence of background factors on the detection effect is effectively removed through network thinning for the model, and the influence of color vector is eliminated in the fully connected conditional random field. The improvement of position vector is strengthened and the conditional random field is constructed as a recurrent neural network. More details in the image are learned, which makes the model more capable of detecting complex mines. The detection effect of the improved model was significantly improved to 94.67%, and the detection time was 24.03 fps, and the detection accuracy under different conditions such as thin clouds, mountains and hills was higher than that of other models, which could meet the current ecological restoration monitoring needs of open pit.

References

1. Yue, H., Liu, Y., Zhu, R.: Monitoring ecological environment change based on remote sensing ecological index in Shendong Mining Area. Bullet. Soil Water Conserv. (2019)
2. Yu, L., Xia, J., Gu, J., Zhang, S., Zhou, Y.: Degradation mechanism of coal gangue concrete suffering from sulfate attack in the mine environment. Materials **16**, 1234 (2023). https://doi.org/10.3390/ma16031234
3. Wei, X.G., Liu, H.L., Li, G.H.: Influencing factors of eco-environmental safety of mines and their green development: a case study of Taoshan coal mine in Heilongjiang Province, China. Nat. Environ. Pollut. Technol 19(2), 831–838 (2020)
4. Min, W.: Vigorously promote the restoration and comprehensive treatment of mine geological environment. Chin. J. Geol. Disast. Prevent. 27(02), 3 (2016)
5. Steenkamp, N.C., Goosen, S.L., Bouwer, P.J.: Satellite applications in diamond exploration and mine monitoring. J. South. Afr. Inst. Mining Metal. 120(10), 575–580 (2020)
6. Tasionas G. et al.: UAV regular mapping focusing on surface mine monitoring. Eighth International Conference on Remote Sensing and Geoinformation of the Environment (RSCY2020), 11524 (2020)
7. Liu, H., et al.: Ecological environment changes of mining areas around Nansi lake with remote sensing monitoring. Environ. Sci. Pollut. Res. **28**(32), 1–13 (2021)
8. Deng, J.: Application of UAV oblique photogrammetry in mine ecological environment restoration. IOP Conf. Ser. Earth Environ. Sci. 719(4) (2021)
9. LUO Ming et al.: Technological model and benefit pre-evaluation of eco-environmental rehabilitation engineering of typical mines in the Nanling area of Northern Guangdong Province under the pilot framework of the eco-restoration of mountains-rivers-forests-farmlands-lakes-grasslands. Acta Ecol. Sin. 39(23) (2019)
10. Qiang Wenwen et al.: TSVM-M[formula omitted]: Twin support vector machine based on multi-order moment matching for large-scale multi-class classification. Appl. Soft Comput. J. 128 (2022)
11. Tarasova, O.A., et al.: Chemical named entity recognition in the texts of scientific publications using the naïve Bayes classifier approach. J. Cheminform. **14**(1), 55 (2022)
12. Yuhang, L., et al.: Spark plug defects detection based on improved Faster-RCNN algorithm. J. Xray Sci. Technol. **30**(4), 709–724 (2022)
13. Velumani, K., et al.: Estimates of maize plant density from UAV RGB images using faster-RCNN detection model: Impact of the spatial resolution. Plant Phenom. **2021**, 9824843 (2021)
14. Jules, R.K. et al.: Assets management on electrical grid using Faster-RCNN. Ann. Oper. Res. 308(1–2): 1–14 (2020)
15. Xiaohong, H., Ye, L., Rundong, Z., Shiqi, D.: Sshear crack detection of rock thermal infrared images based on improved Faster RCNN. Metal Mines 1–10
16. Jiakun, Z., Jun, S., Rui, H., etc. Based on the improved faster RCNN remote sensing image object detection. Comput. Appl. Soft. 39 (05): 192–196 + 290 (2022)
17. Kang, M.K. et al.: Contextual region-based convolutional neural network with multilayer fusion for SAR ship detection
18. Lei, C., Qiang, W., Runjia, S. et al.: Vehicle object detection method for Faster-RCNN network SAR images based on improved RPN. J. Southeast Univ. (Nat. Sci. Ed.), 51 (01): 87–91 (2021)
19. Dai, J., Li, Y., He, K., et al. (2016) R-FCN: Object detection via region-based fully convolutional networks. Curran Associates Inc. (2016)
20. Chen, K.B., Xuan, Y., Lin, A.J., et al.: Esophageal cancer detection based on classification of gastrointestinal CT images using improved Faster RCNN. Comput. Methods Progr. Biomed. (2021)

21. Jiang, D., Li, G., Tan, C., et al.: Semantic segmentation for multiscale target based on object recognition using the improved Faster-RCNN model. Fut. Gen. Comput. Syst. 123(1) (2021)
22. Razavian, A.S., Azizpour, H., Sullivan, J., et al.: CNN features off-the-shelf: an astounding baseline for recognition. 2014 IEEE conference on computer vision and pattern recognition workshops. IEEE (2014)
23. Peng, C., Zhao, K., Lovell, B.C.: Faster ILOD: incremental learning for object detectors based on faster RCNN. Pattern Recog. Lett. (2020)
24. Gu, J., Wang, et al.: Recent advances in convolutional neural networks. J. Pattern Recog. Soc. (2018)
25. Ren, S., He, K., Girshick, R., et al.: Faster R-CNN: towards real-time object detection with region proposal networks. IEEE Trans. Pattern Anal. Mach. Intell. **39**(6), 1137–1149 (2017)
26. Yewen, W., Mei, L., Yuanlin, X., et al.: Image detection of transmission line inspection based on improved Faster-RCNN. Electric Power Eng. Technol. **002**, 041 (2022)
27. Shi, J., Zhou, Y., Zhang, W.: Target detection based on improved mask RCNN in service robot. 2019 Chinese Control Conference (CCC) (2019)
28. Sriram, K.V., Havaldar, R.H. Analytical review and study on object detection techniques in the image. Int. J. Model. Simul. Sci. Comput 12(05) (2021)
29. Jiang, C. et al.: Object detection from UAV thermal infrared images and videos using YOLO models. Int. J. Appl. Earth Observ. Geoinform 112 (2022)

Measurement and Evaluation Efficiency of Ecological and Economic in the Upstream Region of the Yangtze River

Silin Wu and Dachuan Wang[✉]

School of Civil Engineering and Architecture, Southwest University of Science and Technology, Mianyang 621010, China
i@alwayswdc.com

Abstract. Ecological and economic efficiency level analysis is an effective means to evaluating and improving energy efficiency and resource utilization, and actively respond to climate change. This research gathered data from the three provinces and one directly-administered municipality in upstream region of the Yangtze River between 2011 and 2020. The data encompassing a total of 33 prefecture-level cities. Subsequently, an index system for evaluating ecological-economic efficiency was established, and the static and dynamic two-dimensional analysis of eco-economic efficiency was carried out by Super-efficiency DEA model and Malmquist index model. Two main conclusions have been found. Firstly, the eco-economic efficiency all below 1, level in the upstream region of the Yangtze River is relatively low, but the trend is increasing. Among them, Sichuan's ecological and economic efficiency is better than that of Chongqing, Yunnan, and Guizhou. Secondly, the average MI index in the upstream region of the Yangtze River is greater than 1, indicating a clear upward shift in ecological and economic efficiency. Technological progress (TC) is the dominant factor in promoting efficiency improvement. Tailoring strategies to local conditions, enhancing technological innovation, can effectively promote the enhancement of ecological economic efficiency, thereby steering development towards a green and low-carbon direction.

Keywords: Upper Yangtze River · Ecological and economic efficiency · Super-efficiency DEA model

1 Introduction

China, a major responsible nation dedicated to creating a global community with a shared future for all people, has promised to the world to fulfill "carbon peaking" before 2030 and "carbon neutrality" before 2060. To fulfill these commitments, it is imperative to depart from the previous extensive development model and shift towards a focus on improving ecological economic efficiency, which involves reducing inputs, increasing outputs, promoting resource conservation, and pay attention to the ecological environment health. Therefore, precise measurement and enhancement of ecological economic efficiency have become crucial aspects in promoting low-carbon development and actively addressing climate change.

© The Author(s) 2025
B.-J. He et al. (Eds.): UCSUD 2023, LNCE 559, pp. 336–349, 2025.
https://doi.org/10.1007/978-981-97-8401-1_24

The regions positioned in the upper divisions of the Yangtze River serve as a key support area for national strategies including China's "Western Development", "Belt and Road Initiative", "Yangtze River Economic Belt", and the "Chengdu-Chongqing Economic Circle". Currently, research on the ecological-economic efficiency in the upstream area is relatively scarce. Either due to the broad scope of studies or the inability to grasp the specific efficiency of individual cities in the upstream region, challenges arise in ecological-economic governance and enhancement efforts [1, 2]. This poses practical difficulties for ecological economic governance. By carefully choosing pertinent data from several regions situated in upper portion of the Yangtze River spanning the years 2011 to 2020 and constructing a rating index system using input-output indicators, this study employs the model named Super-efficiency DEA to statically analyze the ecological and economic performance. Subsequently, utilizing the Malmquist index to decompose the analysis, the study's central focus is to investigate the dynamic change's characteristics about ecological economic efficiency in cities in the upper sections of the Yangtze River. The primary objective is to provide valuable insights hat can guide the future progress of ecological economics in these regions of the Yangtze River, helping to shape policies and strategies for sustainable development in the area.

2 Existing Research and Research Significance

2.1 Existing Research

Currently, China exhibits low overall levels of energy and natural resource utilization, characterized by extensive and outdated utilization practices. Economic development in the country excessively relies on resource and energy investments, leading to significant waste and pollution [3]. This dependence has resulted in a poor state of ecological economic efficiency governance. Internationally, common evaluation methods for ecological economic efficiency include TOPSIS [4], index methods [5], and DEA [6, 7, 8, 9].

In terms of research subjects, assessments of ecological economic efficiency primarily focus on different industries [10, 11, 12, 13]. On a spatial scale, research is mainly concentrated at the regional, provincial, and municipal levels. It is evident that existing research on ecological economic efficiency lacks a significant focus on urban studies [14, 15].

Existing research on ecological-economic efficiency has relatively few studies focusing on cities. Additionally, there is a lack of comparative analysis between static and dynamic trends in research methodologies. In conclusion, further exploration is necessary to expand the study of ecological-economic efficiency in the upper Yangtze River region.

2.2 Research Significance

This research has a primary objective is to advocate for the principles of sustainable development and proactively address the challenges posed by climate change. It examines the ecological-economic efficiency across several regions situated in the upper portion of the Yangtze River, aiming to reveal innovative methods and practices to mitigate climate change. Additionally, the inclusion of several regions situated in the upper region of the Yangtze River into the scope of research provides a valuable supplement to the research on ecological-economic efficiency.

3   Research Methodology

3.1   The Super-Efficiency DEA Model

DEA (data envelopment analysis) is an analytical technique that examines the performance of units by examining the relationship between multiple input and output factors, because scholars continue to derive on top of their base model, resulting in the Super-efficient DEA model [16]. The efficiency measured by the Super-efficiency model can be greater than 1, making it possible to compare the efficiency of multiple DMUs with an efficiency of 1. The expression is:

$$\begin{cases} \min\left[\theta - \varepsilon\left(\sum_{i=1}^{p} s_i^- + \sum_{l=1}^{q} s_l^+\right)\right] \\ s.t. \sum_{j=1, j\neq k}^{m} \lambda x_{ij} + s_i^- \leq \theta x_o \\ \sum_{j=1, j\neq k}^{m} \lambda_j y_{lj} - s_l^+ \leq y_0 \\ \lambda_i \geq 0; s_i^- \geq 0; s_l^+ \geq 0 \\ i = 1, 2, ..., p; l = 1, 2..., q; j = 1, 2..., m \end{cases}$$

3.2   The Malmquist Index

The Malmquist index (MI) method combined with DEA can measure the ecological and economic efficiency across various time-frames and analyze it dynamically [16]. MI can be deconstructed into two distinct indices: EC, which measures technical efficiency, and TC, which gauges technical progress, such that MI equals EC multiplied by TC. An index surpassing 1 signifies a rising trajectory portrayed by the index; less than 1 indicates a decreasing trend.

3.3   Selection of Indicators

Through the literature combing method [16, 13, 17] Table 1. Outlines the set of indicators used to assess the efficiency of ecological and economic.

3.4   Research Area and Data Source

The geographical focus of the research is the upper Yangtze River region, which comprises the provinces of Sichuan, Yunnan, and Guizhou, along with the Chongqing municipality (Fig. 1). Due to data availability considerations, this study excludes Tibet. Additionally, there are several ethnic autonomous regions in the area where data collection

Table 1. Evaluating system for ecological and economic efficiency in the upstream regions of the Yangtze River

Type	Tier 1 indicators	Tier 2 indicators	unit
inputs	Labor factor	The number of employees per unit	Ten thousand people
	Capital factor	The fixed-asset investment	100 million yuan
outputs	Non-expected outputs	Industrial wastewater	10,000 tons
		Industrial exhaust	100 million standard cubic meters
		Industrial solid waste	10,000 tons
	Expected outputs	GDP	100 million yuan

is limited. To ensure data availability and reliability, the study focuses on 33 cities within the specified region. Data is sourced from the provincial and municipal Statistical Yearbooks as well as the Environmental Statistical Yearbooks of each city. When encountering missing values during data collection, the study fills these gaps by using the average of preceding and succeeding values or by extrapolating trends from previous years.

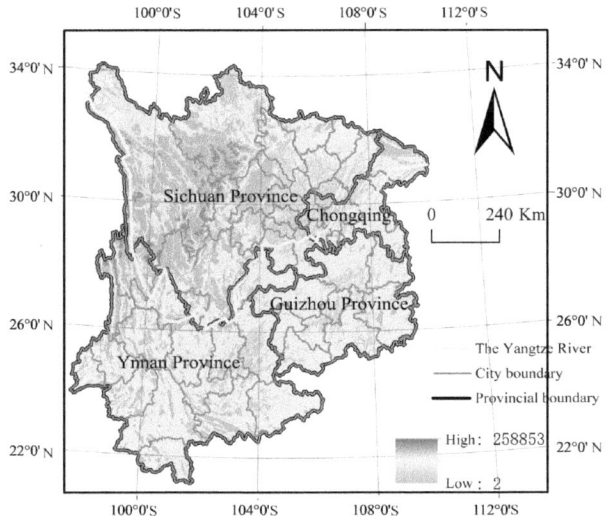

Fig. 1. Overview map of research area in the upper of the Yangtze River

4 Measurement and Evaluation

4.1 Static Empirical Analysis in Efficiency of Ecological and Economic in the Upstream Region of the Yangtze River 2011–2020

Based on the established indicator system, the eco-economic efficiency of 33 cities situated in the upstream portion of the Yangtze River was assessed using Maxdea for the period spanning from 2011 to 2020.

4.1.1 Overall Static Analysis of Ecological and Economic Efficiency

According to Fig. 2, the comprehensive progression of eco-economic efficiency in the upper Yangtze River region follows a "U-shape", and the average of the ecological and economic efficiency during the ten years from 2011 to 2020 is below 1, which indicates that the upper Yangtze River area demonstrates a comparatively poor environmental and economic effectiveness, pointing to a limited integration between financial progress and ecological sustainability.

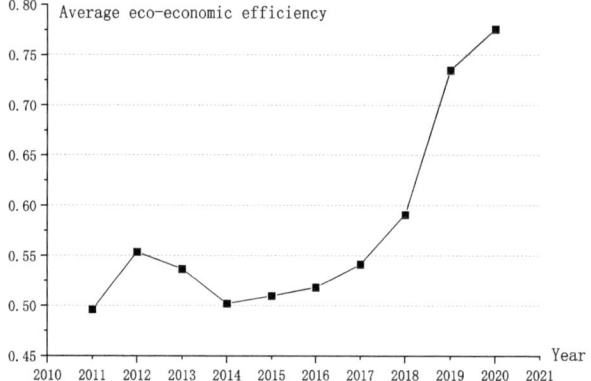

Fig. 2. Growth trend of eco-economic efficiency in the upstream of the Yangtze River between 2011 and 2020

The overall eco-economic efficiency can be split into two phases. The first phase spans from 2011 to 2014, during which cities positioned in the upper divisions of Yangtze River experienced an increase in ecological and economic efficiency followed by an annual decline. Specifically, the efficiency increased from 0.4959 in 2011 to 0.5537 in 2012, but then decreased by 9% to 0.5020 in 2014. The government has implemented expansionary policies to spur economic growth, increase resource and energy consumption, and increase industrial three-waste emissions during this time, which is still the financial crisis recovery period. As a result, ecological and economic efficiency have decreased. In the second stage, which runs from 2014 to 2020, there is an approximate 35% annual gain in ecological and economic efficiency. First, there is an economic restructuring component to this. For the first time since 2004, the projected objective for China's economic growth in 2012 was reduced by the 5th Plenary Meeting of the

National People's Congress in 2011, which set the goal at 7.5%. Reducing the GDP's reliance on natural resources can help to slow down environmental damage by lowering the growth target. Secondly, international collaboration for environmental preservation, like the signature of the Paris Agreement, encourages teamwork in the fight against climate change.

4.1.2 Static Analysis in the Efficiency of Ecological and Economic of Provinces and Cities

According to Table 2 and Fig. 3, the disparities in ecological and economic efficiency among the four major areas can be visualized.

Sichuan Province, known for its robust industrial sector and abundant raw material industries, excels in both economic development and ecological-economic efficiency compared to other regions. The Chengdu Plain Economic Zone stands out for its superior ecological-economic efficiency. Despite ranking lowest in per capita GDP in the province, Zigong City exhibits relatively high ecological-economic efficiency due to its significant emphasis on the primary sector, limited presence in the secondary sector, favorable ecological conditions, and minimal pollution from industrial waste.

Chongqing's ecological-economic efficiency fluctuated significantly from 2011 to 2020, showing notable ups and downs. It rose in 2011–2012, dropped to 0.3126 in 2014, stabilized for a while, and then started consistently increasing after 2016. The city's early heavy reliance on heavy industries caused severe pollution. After the 19th National Congress, Chongqing actively implemented Xi Jinping's directives, leading to an optimized economic structure and a marked boost in ecological-economic efficiency during the 13th Five-Year Plan.

Guizhou Province's ecological-economic efficiency saw a decline around the financial crisis, followed by a notable improvement after 2013. Initially focused on traditional agricultural methods, the province's economic development lagged, resulting in lower levels of ecological-economic efficiency. In recent years, Guizhou Province has thrived in industries like tourism, transportation, and finance, leveraging its rich natural and cultural resources to establish a strong industrial edge. This has led to a steady rise in ecological-economic efficiency.

Yunnan Province's ecological-economic efficiency has shown a fluctuating upward trend. Since 2017, the province has faced relatively lower ecological-economic efficiency due to two primary reasons. Firstly, its location in the southwestern region of China, surrounded by the Qinghai-Tibet Plateau, with high altitude and complex terrain, poses challenges in resource development and utilization due to natural constraints. Secondly, its relatively slower economic development and GDP lower than other provinces contribute to its lower ecological-economic efficiency compared to counterparts (Fig. 4).

4.2 Dynamic Analysis in Efficiency of Ecological and Economic in the Upstream Region of the Yangtze River Between 2011 and 2020

Draws on the Malmquist index to scrutinize the time-varying features of efficiency of ecological and economic about several regions situated in the upper divisions of the Yangtze River.

Table 2. The average of ecologic and economic efficiency from 2011 to 2020

		2011	2012	2012	2013	2014	2015	2016	2017	2018	2020	Average
Chongqing	Chongqing	0.4764	0.5273	0.3208	0.3126	0.3191	0.3209	0.5287	0.6140	0.6261	0.7365	0.4782
Average		0.4764	0.5273	0.3208	0.3126	0.3191	0.3209	0.5287	0.6140	0.6261	0.7365	0.4782
Sichuan	Chengdu	0.5397	0.5755	0.6053	0.4675	0.6545	0.5355	0.5451	0.6738	1.0107	1.1363	0.6744
	Zigong	0.7518	0.8140	0.8378	0.7832	0.7534	0.7955	0.8719	0.8842	1.3169	1.0868	0.8896
	Panzhihua	0.5637	0.6041	0.6396	0.5617	0.5895	0.5418	0.5782	0.6139	0.7168	0.5672	0.5976
	Luzhou	0.5289	0.6184	0.6235	0.5343	0.4836	0.4577	0.4613	0.5316	0.5543	0.6058	0.5399
	Deyang	0.6147	0.7005	0.7430	0.6774	0.7085	0.7284	0.7395	0.7798	0.9868	1.0679	0.7746
	Mianyang	0.4600	0.5361	0.5823	0.5454	0.5644	0.5795	0.5923	0.6489	0.7803	1.0244	0.6314
	Guangyuan	0.2950	0.3599	0.4289	0.4557	0.4959	0.5310	0.5675	0.6788	0.6415	0.6893	0.5144
	Suining	0.4515	0.5058	0.5031	0.4632	0.4752	0.5212	0.4444	0.7032	1.3543	0.5777	0.6000
	Neijiang	0.6558	0.7661	0.7992	0.7242	0.6417	0.6106	0.6185	0.7503	0.8087	0.7534	0.7129
	Leshan	0.4933	0.6026	0.6249	0.5862	0.6254	0.5782	0.6053	0.6057	0.7286	0.8102	0.6260
	Nanchong	0.5362	0.6419	0.5937	0.5140	0.5148	0.4889	0.5196	0.5502	0.8667	1.8423	0.7068
	Meishan	0.5396	0.6198	0.5949	0.5226	0.5159	0.5186	0.5453	0.6225	0.7229	0.7375	0.5940
	Yibin	0.5216	0.5874	0.6129	0.5873	0.5524	0.5478	0.5727	0.5987	0.6856	0.8273	0.6094
	Guangan	0.6655	0.6983	0.7031	0.6516	0.5910	0.5516	0.5690	0.4871	0.5324	0.4978	0.5947
	Dazhou	0.5474	0.6032	0.5899	0.5354	0.5275	0.5095	0.5053	0.4671	0.5486	0.6263	0.5460
	Yaan	0.3692	0.4609	0.4959	0.5194	0.4724	0.4802	0.5393	0.6595	0.7352	0.7503	0.5482

(continued)

Table 2. (*continued*)

		2011	2012	2012	2013	2014	2015	2016	2017	2018	2020	Average
	Bazhong	0.3754	0.4060	0.3602	0.2729	0.2532	0.2683	0.3324	0.4698	1.2711	0.7655	0.4775
	Ziyang	0.6690	0.8095	0.8130	0.6636	0.6605	0.6561	0.6892	0.8910	1.1880	0.8118	0.7852
Average		0.5321	0.6061	0.6195	0.5592	0.5600	0.5500	0.5720	0.6453	0.8583	0.8432	0.6346
Guizhou	Guizhou	0.3332	0.3475	0.3281	0.3299	0.3529	0.4460	0.4520	0.4493	0.4464	0.4734	0.3959
	Liupanshui	0.4926	0.4442	0.3333	0.4017	0.4183	0.5381	0.5389	0.5334	0.5184	0.4293	0.4648
	Zunyi	0.5669	0.5644	0.4794	0.4615	0.4093	0.6598	0.6925	0.8429	0.9004	1.0525	0.6630
	Anshun	0.5036	0.3889	0.3574	0.3203	0.3668	0.4891	0.5693	0.5903	1.1968	0.6395	0.5422
	Bijie	0.4794	0.4745	0.3856	0.4751	0.5023	0.5343	0.5307	0.5702	0.5718	0.6130	0.5137
	Tongren	0.4046	0.3534	0.3002	0.2890	0.3239	0.3482	0.5796	0.5338	0.6228	0.7014	0.4457
Average		0.4634	0.4288	0.3640	0.3796	0.3956	0.5026	0.5605	0.5867	0.7094	0.6515	0.5042
Yunnan	Kunming	0.3872	0.3680	0.4608	0.4446	0.5045	0.4908	0.4875	0.5429	0.5744	0.7809	0.5042
	Qujing	0.5399	0.6376	0.5888	0.5710	0.5507	0.5004	0.4566	0.4870	0.5188	0.8130	0.5664
	Yuxi	0.7604	1.2511	0.9269	0.8412	0.8071	0.7427	0.6327	0.6085	0.5847	1.2187	0.8374
	Baoshan	0.4490	0.4754	0.4910	0.4931	0.4925	0.4478	0.4010	0.3879	0.3742	0.5860	0.4598
	Zhaotong	0.4008	0.4330	0.4515	0.4503	0.4833	0.4706	0.4598	0.4403	0.4279	0.5343	0.4552
	Lijiang	0.2839	0.2986	0.3561	0.2965	0.3775	0.3807	0.3896	0.4469	0.4523	0.6276	0.3910
	Puer	0.3420	0.3692	0.4001	0.3935	0.4478	0.4613	0.4856	0.4611	0.6320	0.7785	0.4771
	Lincang	0.3673	0.4297	0.3770	0.4211	0.3848	0.3703	0.3670	0.3716	0.3593	0.4386	0.3887
Average		0.4413	0.5328	0.5065	0.4889	0.5060	0.4831	0.4600	0.4683	0.4905	0.7222	0.5100

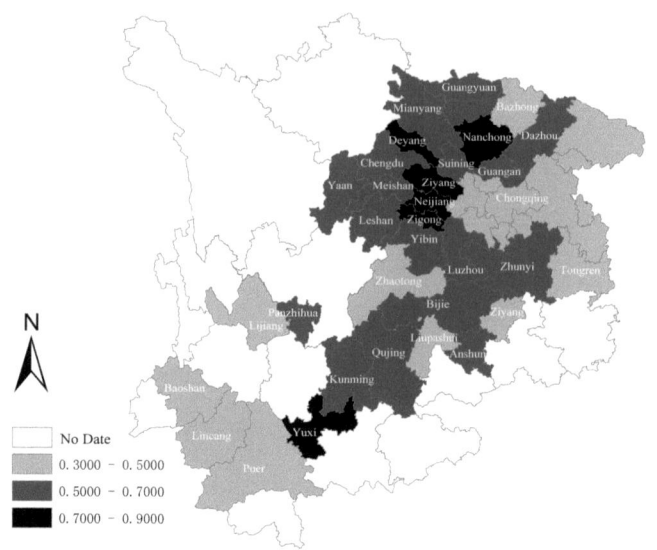

Fig. 3. The average of ecologic and economic efficiency from 2011 to 2020

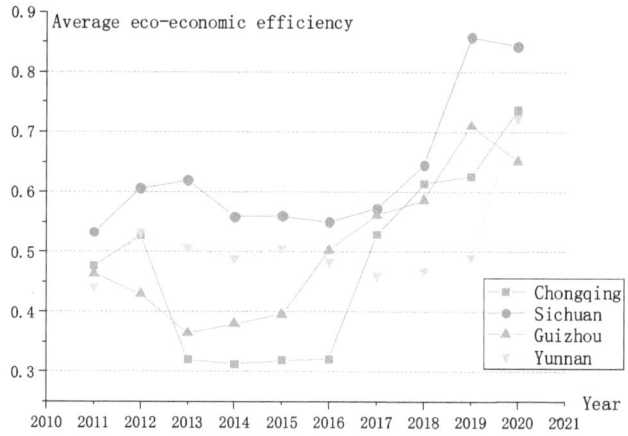

Fig. 4. Trends in ecologic and economic efficiency from 2011 to 2020

From 2011 to 2020, several regions situated in the upper divisions of the Yangtze River showcased a positive trend in their overall ecological and financial effectiveness, as reflected by an average MI index of 1.0649, greater than 1 (Table 3, Fig. 5).

Judging from the two decomposes of the MI index in 2011, The average TC index is 1.2781 is greater than 1, the average EC index is 0.9181 less than 1, and the technical efficiency shows a decreasing trend. In summary, technological progress (TC) serves as the primary impetus behind the elevation of ecological-economic efficiency within the study region. There remains significant room for advancement in technological efficiency (Table 4, Fig. 6).

Table 3. Total factor productivity and its decomposition from 2011 to 2020

	MI	EC	TC
2011–2012	1.0795	0.9181	1.2781
2012–2013	0.9794	1.1280	0.8781
2013–2014	0.9482	1.1100	0.8627
2014–2015	1.0278	1.0973	0.9733
2015–2016	1.0327	1.0125	1.0388
2016–2017	1.0616	1.0449	1.0621
2017–2018	1.0949	1.0229	1.1219
2018–2019	1.2346	1.0060	1.2726
2019–2020	1.1255	1.0311	1.1092
Average	1.0649	1.0412	1.0663

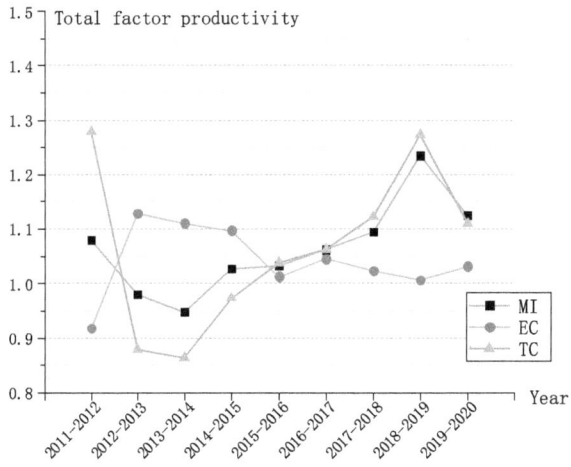

Fig. 5. Total factor productivity and its decomposition from 2011 to 2020

Figure 6 shows that the MI indices for Chongqing Municipality, Sichuan Province, Guizhou Province, and Yunnan Province stand at 1.0801, 1.0690, 1.0567, and 1.0601 respectively. Therefore, these four major provinces have all exhibited an upward trajectory in ecological-economic efficiency from 2011 to 2020. The values for technical efficiency (EC) and technological advancement (TC) also surpass unity.

Table 4. Dynamic decomposition of ecologic and economic efficiency from 2011 to 2020

Province	City	MI	EC	TC	Province	City	MI	EC	TC
Sichuan	Chengdu	1.1097	1.0676	1.0714	Guizhou	Guizhou	1.0431	1.0175	1.0411
	Ziyang	1.0421	1.0743	1.0918		Liupanshui	0.9974	0.9743	1.0425
	Bazhong	1.1860	1.1224	1.0961		Zhunyi	1.0900	1.0594	1.0363
	Zigong	1.0544	1.0509	1.0632		Anshun	1.0921	1.0962	1.1023
	Panzhihua	1.0072	0.9834	1.0375		Bijie	1.0331	1.0059	1.0361
	Luzhou	1.0202	0.9919	1.0401		Tongren	1.0844	1.1319	1.0812
	Deyang	1.0670	1.0463	1.0340	Average		1.0567	1.0475	1.0566
	Mianyang	1.0981	1.0962	1.0212	Yunnan	Kunming	1.0888	1.0556	1.0756
	Guangyuan	1.1020	1.2290	1.1824		Qujing	1.0621	1.0294	1.0376
	Suining	1.1051	1.0408	1.1114		Yuxi	1.1147	1.0138	1.0903
	Neijiang	1.0212	0.9914	1.0383		Baoshan	0.9481	0.9408	1.0538
	Leshan	1.0612	1.0294	1.0396		Zhaotong	1.0358	0.9682	1.1777
	Nanchong	1.1957	1.1656	1.1831		Lijiang	1.1032	1.0761	1.0378
	Meishan	1.0393	1.0024	1.0443		Puer	1.1022	1.0714	1.0421
	Yibin	1.0558	1.0249	1.0399		Lincang	1.0257	0.9952	1.0396
	Guangan	0.9710	0.9350	1.0695	Average		1.0601	1.0188	1.0693
	Dazhou	1.0190	0.9934	1.0376	Chongqing	Chongqing	1.0801	1.0329	1.0415
	Yaan	1.0866	1.0464	1.0515	Average		1.0801	1.0329	1.0415
Average		1.0690	1.0495	1.0696					

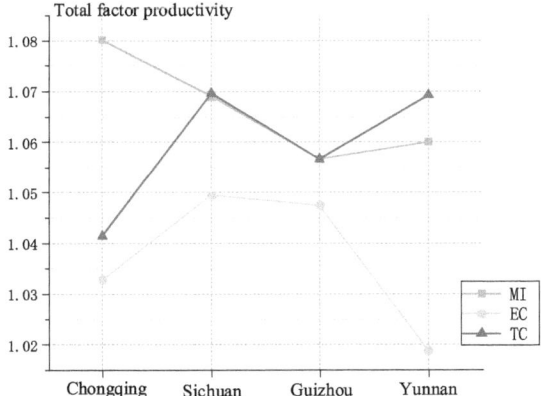

Fig. 6. Dynamic decomposition of ecological-economic efficiency of provinces in the research region of Yangtze River between 2011 and 2020

5 Conclusions

In this research, the Super-efficient DEA and Malmquist index models were employed to examine the efficiency of ecological and economic practices in several regions situated in the upper portion of the Yangtze River territory between 2011 and 2020, yielding the conclusions presented below.

1. The efficiency of ecological-economic in upper region of the Yangtze River region has been low but on the rise. Among the 33 cities studied, all efficiency values were below 1, suggesting a continued reliance on resource-intensive development models. Disparities persist in environmental concern and economic development levels between the upper reaches and middle-lower regions. Sichuan Province exhibits the highest efficiency, while Yunnan Province ranks lowest.
2. The average MI index in the region surpasses 1, signaling an upward trend in ecological-economic efficiency. Both EC and TC indices show positive inclines. Of the 33 cities, 30 have MI indices exceeding 1, with the remaining 3 cities displaying noteworthy patterns: EC below 1, TC above 1. Overall, technological progress emerges as a key driver of ecological-economic efficiency in the upper Yangtze region.

From the evaluation results, it is necessary to clearly outline some effective recommendations.

1. Tailored strategies. Cities like Lincang and Lijiang, with low ecological-economic efficiency, need to enhance their economic structure by emphasizing economic growth quality and ecological preservation. For cities like Chongqing, which exhibit high economic development but low ecological-economic efficiency, transitioning from secondary to tertiary industries is vital to align the industrial framework with goals of low pollution, low consumption, and low emissions.
2. Emphasis on technology and energy. Elevating technological prowess significantly boosts ecological-economic efficiency. Cities should amplify technology investments

and attract skilled tech professionals. Incorporating clean energy sources also aids in curtailing pollution emissions.
3. Regional cooperation. Substantial disparities in ecological-economic efficiency exist among cities in the upper Yangtze River region. Leveraging the Chengdu-Chongqing urban cluster as a center of growth and leveraging the Yangtze River's "Golden Waterway" can bolster overall ecological-economic efficiency through spill-over effects, thereby amplifying the region's ecological-economic advantages.

6   Discussion

In the context of China's transition to a new stage of development, enhancing the ecological-economic efficiency of regions is crucial to fulfilling the pledges of "carbon peaking" and "carbon neutrality" by specific timelines.

Existing studies [1, 13] have not focused specifically on the upper Yangtze River region in China, leading to challenges in governing cities' ecological economies effectively.

This study addresses the shortcomings of current evaluation methods by developing an index system that considers both positive and hindering factors for ecological-economic efficiency. The prefecture-level cities situated in the upper portion of Yangtze River were examined using avant-garde methods like the Super-efficiency DEA and the Malmquist index and provides insights into both static and dynamic trends. Empirical validation has demonstrated the efficacy and viability of this approach.

The methodology and deductions of this research can be used as valuable guides for forthcoming investigations on ecological-economic efficiency and in implementing policies to combat climate change, including the "dual carbon" initiatives. Moving forward, it is important to further explore the factors influencing the efficiency of ecological-economic in the upstream region of the Yangtze River.

References

1. Wang, Z., Liu, Q.: The spatio-temporal evolution of tourism eco-efficiency in the Yangtze River Economic Belt and its interactive response with tourism economy. J. Nat. Resour. **34**(9), 1945–1961 (2019)
2. Xie, Y., Ke, X., Min, Y., Guo, H., Wang, D.: Ecological efficiency evaluation and spatial evolution pattern analysis of the Three Gorges Reservoir area based on super efficiency DEA model. Chin. J. Environ. Manage. **12**(1), 113–120 (2020)
3. Miao, C., Sun, L., Yang, L.: A study of regional technical efficiency in China under the constraint of energy consumption and carbon emissions. Sci. Res. Manag. **37**(1), 1–8 (2016)
4. Sun, L., Miao, C., Yang, L.: Ecological-economic efficiency evaluation of green technology innovation in strategic emerging industries based on entropy weighted TOPSIS method. Ecol. Ind. **73**, 554–558 (2017)
5. Wang, Y., Zhang, Y.: Regional difference of ecological efficiency and its interactive spatial spillover effect with industrial structure upgrading. Scientia Geographica Sinica **40**(8), 1276–1284 (2020)
6. Ingaramo, A., Heluane, H., Colombo, M., Cesca, M.: Water and wastewater eco-efficiency indicators for the sugar cane industry. J. Clean. Prod. **17**(4), 487–495 (2009)

7. Kelly, J., Haider, W., Williams, P.W., Englund, K.: Stated preferences of tourists for eco-efficient destination planning options. Tour. Manage. **28**(2), 377–390 (2007)
8. Haibo, C., Ke, D., Fangfang, W., Ayamba, E.C.: The spatial effect of tourism economic development on regional ecological efficiency. Environ. Sci. Pollut. Res. **27**(30), 38241–38258 (2020). https://doi.org/10.1007/s11356-020-09004-8
9. Mocholi-Arce, M., Gómez, T., Molinos-Senante, M., Sala-Garrido, R., Caballero, R.: Evaluating the eco-efficiency of wastewater treatment plants: comparison of optimistic and pessimistic approaches. Sustainability **12**(24), 10580 (2020)
10. Kounetas, K., Stergiou, E.: Technology heterogeneity in European industries' energy efficiency performance. The role of climate, greenhouse gases, path dependence and energy mix. MPRA Paper (2019)
11. Munisamy, S., Arabi, B.: Eco-efficiency change in power plants: using a slacks-based measure for the meta-frontier Malmquist–Luenberger productivity index. J. Cleaner Prod. **105**(15), 218–232 (2015)
12. Xu, S., Ma, C., Zhang, S.: Measurement and influencing factors of eco-economic efficiency of China's construction industry. Environ. Pollut. Control. **44**(6), 833–840 (2022)
13. Peng, D., Zhang, X., Liu, S.: Research on the evaluation of eco-economic efficiency and the evolution of space and time in the Yangtze River Economic Belt. Ecol. Econ. **36**(6), 44–50 (2020)
14. Yang, K., Wang, W.: Spatial-temporal evolution of the coupling coordinated development among synergic innovation, industrial structure and ecological efficiency in Yangtze River Delta. Sci. Technol. Manage. Res. **40**(21), 80–87 (2020)
15. Hu, B., Sun, X.: The spatial-temporal evolution analysis of eco-economic efficiency in China's three urban agglomerations. J. Dalian Univ. Technol. (Social Sciences) **41**(1), 19–27 (2020)
16. Zhou, L., Che, L., Zhou, C.: Spatio-temporal evolution and influencing factors of urban green development efficiency in China. Acta Geogr. Sinica **74**(10), 2027–2044 (2019)
17. Li, C., Zhang, S., Zhang, W.: Spatial distribution characteristics and influencing factors of China's inter provincial industrial eco-efficiency. Sci. Geogr. Sinica **38**(12), 1970–1978 (2018)

Assessing Urban Tourism Resource Ecological Carrying Capacity: A "Desired-Undesired" Dual Perspective Approach

Rui Yang[1], Yuan Xiong[2], Hongman He[3(✉)], and Yu Liu[4]

[1] School of Urban Construction and Design, Guizhou Vocational College of Agriculture, Guiyang, China

[2] School of Architecture and Civil Engineering, Guizhou Minzu University, Guiyang, China

[3] School of Management Science and Engineering, Guizhou University of Finance and Economics, Guiyang, China

Hehongman2021@163.com

[4] Xinjiang Jialian Urban Construction Planning and Design Institute Co., Ltd, Urumqi, China

Abstract. This study introduces a new approach to evaluate the ecological carrying capacity of urban tourism resources (UTRECC), emphasizing the perspectives of "desired" and "undesired". Based on urban sustainability principles, this study categorizes the load of urban tourism resources into "desired load" and "undesired load". Using tourism system theory, the study examines the carriers of four key subsystems related to urban tourism resource carrying capacity, including tourism attraction carrier, atmospheric environmental support carrier, waterbody environmental support carrier, and land environmental support carrier. A calculation model for UTRECC is constructed based on 11 carrier indicators and 5 load indicators. This method elucidates the performance of ecological carriers in urban tourism, suggesting that, within certain parameters, a higher carrying capacity index is preferable. Empirical research was conducted in four case cities: Beijing, Shanghai, Chengdu, and Guangzhou. The results reveal significant variations in the tourism resource carrying capacities of these cities, with Beijing and Shanghai exhibiting relatively higher capacities compared to Chengdu and Guangzhou. The findings offer valuable insights for the planning, development, and management of urban tourism, providing a robust framework for future assessments and adjustments of urban resources.

Keywords: Urban tourism resources · Ecological carrying capacity · Desired and undesired effects · Sustainable development

1 Introduction

Evaluating the ecological carrying capacity of urban tourism resources (UTRECC) is crucial for the sector's sustainable growth. As urban tourism flourishes, its impacts on the ecological environment, particularly in the realms of air quality and climate, become increasingly pronounced. These effects can be broadly categorized into two dimensions: "undesired" and "desired."

B.-J. He et al. (Eds.): UCSUD 2023, LNCE 559, pp. 350–362, 2025.
https://doi.org/10.1007/978-981-97-8401-1_25

From the "undesired" perspective, the unchecked growth of tourism often precipitates a host of ecological challenges. Specifically, the surge in tourism transportation is frequently linked to exacerbated urban air pollution and the substantial influx of tourists can lead to elevated ground temperatures and an intensified urban heat island effect, further destabilizing the urban's climatic equilibrium. For instance, Jia et al. indicated that heightened traffic volumes during peak tourism seasons significantly increase concentrations of particulate matter (PM2.5) and carbon monoxide (CO) in urban atmospheres, negatively affecting local climates [1]. Furthermore, Yang et al. highlighted the substantial influx of tourists can lead to elevated ground temperatures and an intensified urban heat island effect, further destabilizing the urban's climatic equilibrium [2]. These undesired ecological ramifications not only pose a threat to the local environmental health but could also jeopardize the long-term sustainability of the tourism industry.

Conversely, when viewed from the "desired" angle, moderate tourism development can benefit the urban's ecological environment. For example, the promotion of ecotourism in various locales has encouraged the implementation of eco-protection measures, effectively mitigating pollutant emissions and thereby contributing positively to local air quality, as demonstrated by He et al. [3]. Additionally, the advancement of cultural heritage tourism has spurred the revival of traditional ecological practices, subsequently boosting the ecological resilience of cities, a point elaborated by Lobo and Moretti [4].

However, despite the extensive discussions on the impacts of urban tourism, there is a noticeable gap in the literature regarding a comprehensive assessment method that simultaneously considers both the negative and positive effects of tourism on urban ecology. Thus, this study aims to bridge this gap by proposing a novel assessment framework that integrates both UTRECC dimensions of "undesired" and "desired" outcomes concurrently. This assessment method that acknowledges the tourism sector's negative effects on urban ecology while also considering its positive contributions. By doing so, this study can ensure the genuine sustainability of urban tourism and provide a solid theoretical foundation for future decision-making.

2 Literature Review

Previous scholars have had many studies on the carrying capacity of urban tourism resources, which offers valuable insights for this study's assessment. Wang et al. explored the link between economic income maximization and sustainable development of tourism resources [5]. They applied economic principles of marginal benefits and costs, employing the fuzzy affiliation method to devise an urban tourism resource carrying capacity (UTRCC) evaluation model. Yuan et al. classified the UTRCC system in more detail, divided the urban tourism resource carrying capacity system into five systems, such as ecological, resource, economic, psychological and social [6]. They developed a comprehensive evaluation index system from these five subsystem perspectives. Zhang and Liu took Shandong Peninsula city cluster, constructed the measurement model of tourism resource carrying capacity (TRCC) of city cluster, combined with the existing index statistics, used the gray prediction method to forecast the dynamic data, and came out with the countermeasures which are beneficial to the development of Shandong Peninsula city cluster [7].

Wang divided the sustainable carrying capacity system of urban tourism resources into three subsystems: natural, economic and social, constructed seven state layers with three subsystems, and established an index system and measurement model, and finally studied five cities in southern Jiangsu Province as an example [8]. Navarro Jurado et al. introduced the concept of TRCC growth limit, using coastal urban tourism as a case study to develop and apply a method for assessing the growth limits of tourism destinations [9]. This method, aimed at aiding the management of urban tourism resources, the limit of growth is calculated based on a combination of indicators of two scenarios: weak and strong sustainability. Lobo et al. argue that UTRCC should focus on quantitative impacts consistent with a specific environment, and the process of studying urban tourism resources should take full account of stakeholder influence [10]. Xiong and Yang analyzed the connotation of the urban tourism resources spatial capacity, and constructed the measurement index system [11]. By studying urban mountainous tourism sites, they measured the instantaneous spatial capacity, daily capacity, and seasonal capacity during tourism off-seasons.

In the diverse field of urban tourism resources carrying capacity, a plethora of research has unfolded various dimensions and methodologies. Wang et al. have delved into the economic aspects, using innovative models like the fuzzy affiliation method, aiming for a balance between maximizing income and sustainable development [5]. Yuan et al., on the other hand, proposed a holistic classification, emphasizing the need to consider a gamut of factors including ecological, resource, and social impacts [6]. Such broad categorizations were further nuanced by studies like that of Zhang and Liu which emphasized regional specificities, as evidenced in their detailed evaluation for the Shandong Peninsula city cluster [7].

While these foundational studies have advanced our understanding, a common thread in recent scholarship has been a growing emphasis on the ecological aspects. Wang [8] and Navarro Jurado et al., while introducing subsystems and growth limit concepts, highlighted the interconnectedness of natural, economic, and social systems [9]. Lobo et al. pivotally recognized that while numerical impacts are vital, the unique environmental constraints of a city need paramount consideration [10]. This sentiment is echoed by Xiong and Yang who underscored the significance of spatial carrying capacity, especially concerning natural environments like city mountainous tourism sites [11].

Urban tourism's ecological carrying capacity has increasingly been spotlighted in recent research, accentuating the harmony between urban development and the environment. Hou et al. emphasized the ecological dimension by examining Shenzhen Urban's ecological protection line [12]. Their work underscored that balancing environmental quality, ecological function, and resource utilization is pivotal for sustainable urban tourism. Similarly, George and Kini proposed an optimization framework for tourism resource carrying capacity, suggesting that urban development guidelines can be instrumental in sustaining tourism growth [13].

The approach of Huamantinco Cisneros et al. is particularly noteworthy. By assessing carrying capacities and beach usage levels in coastal cities, they introduced a real-time data-based methodology that can significantly benefit tourism management in coastal cities [14]. Meanwhile, Cartenì et al. offered insights into the challenges of managing periodic overcrowding in Roman coastal cities, further indicating the urgency of understanding ecological carrying capacity in the context of urban tourism [15]. Such views reiterate the multifaceted nature of carrying capacities, encompassing not just environmental but also psychological and social aspects.

The work of Iliopoulou-Georgudaki et al. [16] and Qin [17] resonates with the afore-mentioned studies, with both emphasizing the need for sustainable tourism management and development, informed by comprehensive indices and evaluation systems. Recent studies, such as those by Zhang et al. [18] and Xian et al. [19], have employed advanced data analysis techniques, underscoring the shift towards more empirical and data-driven approaches in this domain. By analyzing spatial differentiation characteristics and drawing correlations between various facets like infrastructure and socio-economic indicators, these studies shed light on the intricate dynamics governing urban tourism's ecological carrying capacity.

Building upon the existing literature, it's evident that various dimensions of urban tourism carrying capacity have been explored, ranging from ecological considerations, psychological aspects, to comprehensive sustainable management approaches. Yet, there remains a notable gap in the literature: addressing the urban tourism resource carrying capacity from both "desired" and "undesired" perspectives. This study aims to bridge this void, offering a novel approach to understanding the nuances and complexities involved in urban tourism dynamics and its sustainable management. Only by balancing growth with ecological constraints can cities ensure a sustainable future for both residents and tourists.

3   UTRECC Evaluation Method

To establish an evaluation method for UTRECC from the "desired and undesired" view-point, the following steps will be undertaken: First, the UTRECC mechanism will be defined by the "load-carrier" theory. Then, evaluation indicators will be set based on this mechanism, with each selected indicator aimed at capturing the essential aspects of both the carrier's capacity and the load's impact on urban tourism resources. The selection of specific indicators under each element is grounded in their ability to reflect the critical dimensions of urban tourism's ecological and social effects comprehensively. Finally, the respective carrying capacity index will be calculated using the entropy weighting and linear weighted sum methods.

Shen et al. [20] and He et al. [21] propose that the evaluation of UTRECC should encompass both carrier and load perspectives. Drawing on the tourism system theory, this study categorizes urban tourism resource carrier system into four distinct types: tourism attraction carrier, atmospheric environmental support carrier, waterbody environmental support carrier, and land environmental support carrier. Furthermore, based on preceding discussions, the load system of urban tourism resources is bifurcated into two categories, namely desired load and undesired load.

Based on the index system of UTRECC in the article of He et al. [22], the newly added evaluation objects in the evaluation system of UTRECC based on "Desired and unde-sired" load are selected and supplemented from the indexes on the website of National Bureau of Statistics and China Tourism Statistical Yearbook to get the evaluation of UTRECC. The indicator system is shown in Table 1.

Table 1. The list of UTRECC indicators

Aspect	Element	Indicator	Code
Urban tourism resource carrier system	Tourism attraction carrier	Number of attractions at 5A level (pcs)	C_1
		Number of attractions at 4A level(pcs)	C_2
		Number of attractions at 3A level(pcs)	C_3
		Number of attractions at 2A level(pcs)	C_4
		Number of attractions at 1A level(pcs)	C_5
		Aggregation of scenic spots	C_6
	Atmospheric environmental support carrier	Forest coverage rate per million people (%)	C_7
		Natural protected area per million people (hectares)	C_8
	Waterbody environmental support carrier	Urban sewage daily processing capacity per million people (10,000 cubic meters)	C_9
	Land environmental support carrier	Urban domestic waste harmless treatment capacity per million people (tons/day)	C_{10}
Load system of urban tourism resources	Desired load	Domestic tourism income per tourist (yuan)	L_1
		Inbound tourism income per tourist (yuan)	L_2
	Undesired load	Carbon emission per tourist (tons of standard coal)	L_3
		Daily waste clearance per resident (tons)	L_4
		Total wastewater discharge per resident (tons)	L_5

The indicator system, as shown in Table 1, is designed to thoroughly evaluate the multifaceted interactions between urban tourism activities and the environment. Each indicator was chosen for its direct relevance to assessing either the supportive capacity of the urban environment (carrier system) or the impact of tourism activities (load system). For example, the number of attractions by level (C_1 to C_5) serves as a quantitative measure of the tourism attraction carrier, indicating the capacity to attract and accommodate tourists. The forest coverage rate (C_7) and natural protected area size (C_8) are critical for understanding the atmospheric environmental support carrier, reflecting the city's ability to maintain air quality and biodiversity, essential for sustainable tourism. Similarly, indicators like urban sewage daily processing capacity (C_9) and urban domestic waste treatment capacity (C_{10}) provide insights into the infrastructural support for managing the environmental impacts of increased tourism activity. On the load side, domestic and inbound tourism income per tourist (L_1 and L_2) reflect the economic benefits, while carbon emission per tourist (L_3) and waste clearance per resident (L_4) highlight environmental pressures. These indicators collectively capture the key aspects of urban tourism's ecological carrying capacity, offering a comprehensive framework for evaluating the sustainability of tourism activities within urban settings.

The variables in the indicator system differ significantly in terms of measurement scale, magnitude and data characteristics. In this study, the raw data of each variable were processed using the extreme difference standardization method. According to the indicator attributes, the indicators are divided into positive and negative indicators.

Positive indicators (desired indicators) are standardized using the formula:

$$x_{ij}^* = \frac{x_{ij} - \min x_{ij}}{\max x_{ij} - \min x_{ij}} \tag{1}$$
$$(i = 1, 2, \ldots, n)$$

Negative indicators (undesired indicators) are standardized using the formula:

$$x_{ij}^* = \frac{\max x_{ij} - x_{ij}}{\max x_{ij} - \min x_{ij}} \tag{2}$$
$$(i = 1, 2, \ldots, n)$$

where: x_{ij} represents its actual value of the i th indicator in year j.

x_{ij}^* denotes its standardized value of indicator i in year j.

$\max x_{ij}$ is the maximum value of the indicator i, and $\min x_{ij}$ is the minimum value of the indicator i in year j, and $0 \leq x_{ij}^* \leq 1$.

The rationale behind using extreme difference standardization is to normalize the indicators to a common scale, allowing for a comparative analysis despite the differences in measurement units and scales. This method ensures that both positive and negative indicators are equally represented in the evaluation, reflecting their respective impacts on the urban tourism resource carrying capacity. Whether it is the desired or undesired indicator, the larger the value is after the data standardization, the better.

Since the relative importance level of each indicator in assessing an urban area's tourism resources carrying capacity, it is necessary to assign appropriate weights within the index system prior to evaluation. To reduce subjective bias, this study adopts the method of entropy value method for weight allocation. This method is often adopted to measure the degree of variation of a certain index of the evaluation object. The entropy method of assigning weights has been relatively mature, and its calculation process is not explained in detail here, and the specific weight calculation can be carried out based on the study of He et al. [23].

The weights that can be calculated by the entropy weighting method are denoted by w_{ij}.

Therefore, the sub-index calculation formula can be calculated as follows, respectively.

$$\text{Carrier index} = \sum_{i=C1}^{C10} w_{ij} \cdot x_{ij}^* \tag{3}$$

$$\text{Desired load index} = \sum_{i=L1}^{L2} w_{ij} \cdot x_{ij}^* \tag{4}$$

$$\text{Undesired load index} = \sum_{i=L3}^{L5} w_{ij} \cdot x_{ij}^* \tag{5}$$

According to the "load-carrier" theory, the index of resource carrying capacity should be reflected from both the carrier and the load. Therefore, the measuring method of the UTRCC index can be proposed as follows:

$$
\begin{aligned}
\text{UTRECC index} &= \text{Desired Load index/Carrier index} \\
&\quad + \text{Undesired Load index/Carrier index} \\
&= \sum_{i=L1}^{L2} w_{ij} \cdot x_{ij}^* / \sum_{i=C1}^{C10} w_{ij} \cdot x_{ij}^* + \sum_{i=L3}^{L5} w_{ij} \cdot x_{ij}^* / \sum_{i=C1}^{C10} w_{ij} \cdot x_{ij}^*
\end{aligned}
\tag{6}
$$

4 Case Study

4.1 Data Collection

This study selected Beijing, Tianjin, Shanghai, and Chongqing as empirical research samples primarily because these cities not only represent significant economic and political hubs of China but also boast distinctive tourism resources and developmental contexts. Their tourism development rate and scale are among the forefront domestically, and each urban grapples with unique ecological carrying capacity challenges. The rationale behind choosing these four cities lies in their diverse ecological environments and the varying degrees of tourism development, which collectively provide a broad spectrum of scenarios for assessing the ecological carrying capacity. Beijing, being the capital, reflects the challenges and opportunities of managing historical and cultural tourism alongside rapid urbanization. Tianjin, with its unique coastal and urban blend, offers insights into balancing industrial development with marine and urban tourism. Shanghai showcases the dynamics of international tourism and its implications on urban ecosystems in a global city context. Lastly, Chongqing, with its mountainous terrain and river systems, presents a case of integrating ecological preservation with tourism growth in less urbanized, yet ecologically sensitive areas. Thus, examining these four cities provides a comprehensive perspective, shedding light on the complexities and diversities of urban tourism resource ecological carrying capacity, subsequently offering robust insights for pertinent policy formulation.

The data used in the case study was quantitative data sourced from official sources such as statistical yearbooks, government reports and official websites. To ensure the reliability and sufficiency of the data, this study selected the most recent publications and online resources that are regularly updated and maintained by authoritative government agencies. Specifically, the statistical yearbooks published by the National Bureau of Statistics of China and the local statistical bureaus of Beijing, Tianjin, Shanghai, and Chongqing provided comprehensive data on urban development, tourism statistics, and environmental indicators. Government reports, which include annual reports on environmental protection and tourism development plans, offered insights into policy directions and priorities. Furthermore, official websites served as supplementary sources for up-to-date information and specific data related to urban tourism and ecological indicators. The data collection process involved rigorous verification of the data sources to ensure accuracy and consistency across the different datasets.

4.2   Results

In this study, the corresponding index data of four case cities were collected, and then formulas (1)–(6) were applied for calculation, and the UTRECC results of the sample cities were obtained, as shown in Fig. 1.

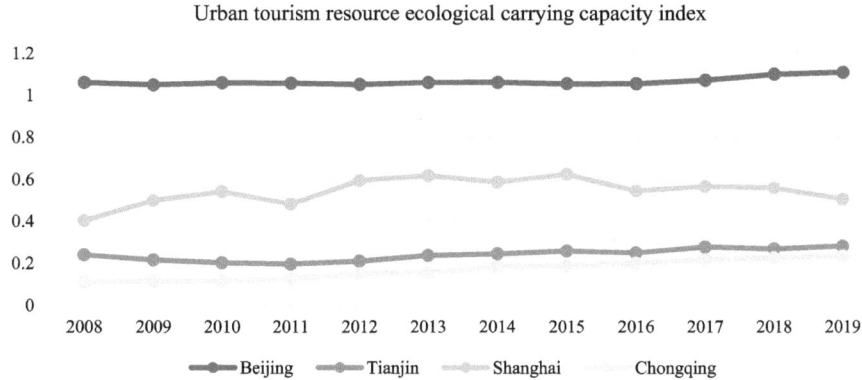

Fig.1. Result of UTRECC index

The graph illustrates the ecological carrying capacity of tourism resources across four major Chinese cities: Beijing, Tianjin, Shanghai, and Chongqing from 2008 to 2019.

Beijing consistently exhibits the highest ecological carrying capacity over the 12-year period. The capacity remains relatively stable, hovering around the 1.0–1.2 range, indicating that Beijing has maintained its ability to support tourism activities without significant ecological degradation. The consistently high carrying capacity of Beijing points towards effective sustainable tourism practices. This stability could be attributed to the city's rigorous management of its tourism resources, infrastructural developments, and successful implementation of environmental policies aimed at preserving its ecological integrity.

Shanghai witnessed a slight decline in its ecological carrying capacity from 2008 to 2014. However, post-2014, the capacity gradually increased, reaching nearly 0.6 by 2019. This slight decline could be attributed to rapid urbanization and industrial growth during this period, which potentially strained the city's ecological resources. This suggests that measures might have been taken in the latter half of the decade to enhance the Shanghai's capacity. The noticeable improvement in Shanghai post-2014 could be attributed to policy changes, infrastructural development, or increased ecological conservation efforts. Further research could shed light on the specific interventions leading to this positive trend.

Tianjin's ecological carrying capacity demonstrates a stable trend, remaining relatively unchanged throughout the period. Tianjin maintains its capacity around the 0.2 mark, indicating a lower carrying capacity compared to Beijing but consistent over the years. This stability might reflect the city's steady approach to managing its tourism and ecological resources, balancing development with environmental conservation efforts.

Chongqing's ecological carrying capacity has remained consistent across the years, hovering around the 0.2 range. This demonstrates Chongqing's ability to maintain a stable tourism support infrastructure despite facing challenges such as geographical constraints and the need for environmental conservation.

The unchanging nature of Tianjin's and Chongqing's ecological carrying capacities suggests that while they might not have witnessed significant growth in tourism infrastructure, they have also avoided degradation. The stable trend reflects a balance between tourism activities and ecological preservation.

The ecological carrying capacities of these cities provide valuable insights into the sustainability of tourism practices. While Beijing leads in maintaining a high capacity, Shanghai showcases potential with its upward trend. Tianjin and Chongqing, on the other hand, emphasize the importance of consistency and balance. For future tourism development, cities must monitor and manage their ecological carrying capacities to ensure both sustainable tourism and ecological preservation.

5   Discussion

Based on the empirical findings presented earlier, the discussion can be categorized into two main points.

5.1   Policy Recommendations for the Four Cities Based on Empirical Results

Given its consistently high ecological carrying capacity, Beijing's performance mirrors findings from similar studies [24], which highlight the city's robust environmental management practices. Policies could be designed to further harness this strength and make Beijing a model for other cities. Regular evaluations and impact assessments should be conducted to ensure the tourism activities do not exceed the ecological threshold.

The stability seen in Tianjin's ecological carrying capacity indicates a balance but also suggests there might be room for growth. Policymakers should consider initiatives that can enhance the urban's tourism infrastructure while ensuring that the ecological balance is not disrupted.

Shanghai was observed after 2014 productivity growth is promising. Policies should focus on understanding the key drivers of this positive change and reinforcing them. Given the urban's potential, investments in eco-tourism and green infrastructure can further boost its carrying capacity.

Consistency in Chongqing's ecological carrying capacity calls for an in-depth analysis to identify potential areas of growth. The urban could benefit from policies that encourage diversified tourism offerings, which spread the tourist load more evenly and reduce pressure on specific hotspots.

5.2 Validation of Methodological Effectiveness in Assessing UTRECC

The empirical results offer a testament to the effectiveness of the methodology employed in this study. By providing a detailed and differentiated overview of the UTRECC of four major cities over a decade, it not only highlights the performance trends but also suggests areas of intervention. This method can be efficiently used to assess the performance of UTRECC (Urban Tourism Resource Carrying Capacity) under different perspectives of desired and undesired loads. The clear distinction between the two allows cities to identify whether they are operating within sustainable limits or if there are risks of potential ecological degradation due to excessive tourism loads.

Comparing this study methodological approach with those in existing studies [25, 26], this study findings affirm the value of integrating both positive and negative indicators in assessing urban tourism ecological carrying capacities. This comprehensive approach enables a more nuanced understanding of urban tourism's impacts on ecological systems.

In conclusion, the methodological approach and empirical findings together provide valuable insights and directions for policymakers and urban planners focused on sustainable tourism development. The balance between desired and undesired loads is crucial for ensuring that cities can maximize their tourism potential without compromising their ecological integrity.

6 Conclusion

Building upon a review of literature concerning Urban Tourism Resources Carrying Capacity (UTRCC), this study revises its evaluation to consider both "desired" and "undesired" impacts, rooted in urban sustainability theory. It identifies two types of ecological loads for city tourism resources: Desired Load and Undesired Load. Utilizing tourism system theory, it examines the support carriers in four subsystems: tourism attraction carrier, atmospheric environmental support carrier, waterbody environmental support carrier, and land environmental support carrier. Ultimately, leveraging the "load-carrier" theory, a calculation model for UTRECC is established, reflected by 10 carrier indices and 5 load indices.

The proposed UTRECC evaluation method encapsulates the performance of the urban tourism resource ecological carrier under both "Desired Load" and "Undesired

Load". Concerning the carrying capacity indices corresponding to the desired and undesired loads, a larger carrying capacity index is generally favorable within a certain range.

This study makes a theoretical contribution by intricately weaving "desired" and "undesired" impacts into the fabric of urban sustainability theory, thereby offering a more comprehensive framework for understanding tourism carrying capacity. In practice, the UTRECC model as a tool that enabling stakeholders to pinpoint specific areas for sustainable growth and ecological preservation.

Nevertheless, this study reliance on currently available data may obscure the nuanced dynamics between urban tourism activities and ecological systems. This limitation underscores the need for future research to incorporate broader and more diverse data sets, potentially unveiling richer insights into these complex interactions. Future endeavors should focus on enhancing its predictive accuracy and applicability across diverse urban contexts. By integrating advanced data analytics and exploring nollvel indicators of ecological impact, researchers can further fine-tune the model, broadening its utility and relevance.

References

1. Jia, L.J., Wang, W.J., Yu, H., Cui, M.D.: Tourism development, traffic carrying capacity, and residents' income level. Transform. Bus. Econ. **18**(3C), 485–499 (2019)
2. Yang, J., Zhang, Z.C., Li, X.M., Xi, J.C., Feng, Z.X.: Spatial differentiation of China's summer tourist destinations based on climatic suitability using the Universal Thermal Climate Index. Theoret. Appl. Climatol. **134**(3–4), 859–874 (2018). https://doi.org/10.1007/s00704-017-2312-5
3. He, Y., Huang, P., Xu, H.: Simulation of a dynamical ecotourism system with low carbon activity: a case from western China. J. Environ. Manage. **206**, 1243–1252 (2018). https://doi.org/10.1016/j.jenvman.2017.09.008
4. Lobo, H.A.S., Moretti, E.C.: Tourism in caves and the conservation of the speleological heritage: The case of Serra Da Bodoquena (Mato Grosso Do Sul State, Brazil). Acta Carsologica **38**(2–3), 265–276 (2009)
5. Wang, H., Lin, J., Zhou, J.: Establishment and analysis of economic model of urban tourism environment carrying capacity. J. Dalian Marit. Univ. (03), 18–20+25 (2006). https://doi.org/10.16411/j.cnki.issn1006-7736.2006.03.005
6. Yuan, J., Yu, J., Yuan, H.: A preliminary study on the evaluation of urban tourism environment carrying capacity. Ind. Techno-Econ. (07), 130+134. (2006)
7. Zhang, G., Liu, J.: Regional differences and functional divisions of tourism environment carrying capacity in Shandong Peninsula City. Agglom. Reg. Res. Dev. (04), 77–80+85 (2008)
8. Wang, Y.: The carrying capacity of tourism environment in southern Jiangsu based on sustainable development. J. Liaoning Univ. Eng. Technol. (Nat. Sci. Ed.) **30**(05), 793–796 (2011)
9. Navarro Jurado, E., Tejada Tejada, M., Almeida Garcia, F., Cabello Gonzalez, J., Cortes Macias, R., Delgado Pena, J., et al.: Carrying capacity assessment for tourist destinations. Methodology for the creation of synthetic indicators applied in a coastal area. Tour. Manag. **33**(6), 1337–1346 (2012). https://doi.org/10.1016/j.tourman.2011.12.017
10. Lobo, H.A.S., Trajano, E., Marinho, M.D., Bichuette, M.E., Scaleante, J.A.B., Scaleante, O.A.F., et al.: Projection of tourist scenarios onto fragility maps: framework for determination of provisional tourist carrying capacity in a Brazilian show cave. Tour. Manag. **35**, 234–243 (2013). https://doi.org/10.1016/j.tourman.2012.07.008

11. Xiong, Y., Yang, X.: Analysis on the spatial carrying capacity of tourism resources in urban mountain-type tourist destinations—taking Yuelu Mountain scenic area as an example. China's Popul. Resour. Environ. **24**(S1), 301–304 (2014)

12. Hou, C., Han, Y., Li, D., Zhang, L., Deng, Y., Wang, X., et al.: Research on technical methods of ecological protection red line delineation in Dapeng New District, Shenzhen. J. Environ. Sci. **36**(03), 1106–1112 (2016). https://doi.org/10.13671/j.hjkxxb.2015.0581

13. George, R.M., Kini, M.K.J.P.T.: Formulating urban design guidelines for optimum carrying capacity of a place. Procedia Technol. **24**, 1742–1749 (2016)

14. Huamantinco Cisneros, M.A., Revollo Sarmiento, N.V., Delrieux, C.A., Cintia Piccolo, M., Perillo, G.M.E.: Beach carrying capacity assessment through image processing tools for coastal management. Ocean Coast. Manag. **130**, 138–147 (2016). https://doi.org/10.1016/j.ocecoaman.2016.06.010

15. Cartenì, A., Pariota, L., Henke, I.: Hedonic value of high-speed rail services: quantitative analysis of the students' domestic tourist attractiveness of the main Italian cities. Transp. Res. Part A: Policy Pract. **100**, 348–365 (2017). https://doi.org/10.1016/j.tra.2017.04.018

16. Iliopoulou-Georgudaki, J., Theodoropoulos, C., Konstantinopoulos, P., Georgoudaki, E.J.I.J.o.S.D.: Sustainable tourism development including the enhancement of cultural heritage in the city of Nafpaktos–Western Greece. Int. J. Sustain. Dev. World Ecol. **24**(3), 224–235 (2017)

17. Qin, C.: Research on the carrying capacity of urban tourism environment in Nanning. Travel Overv. (02), 137–139+142 (2018)

18. Zhang, G., Liu, J., Wang, L., Wan, R.: Comprehensive evaluation of the carrying capacity of tourism environment in Shandong Peninsula City cluster. Adv. Geosci. **27**(2) (2008)

19. Xian, W., Shang, G., Liu, Y., Liu, Q., Tang, L.: Spatial pattern and influencing factors of rural leisure tourism destination based on POI data: taking Miyun District, Beijing as an example. Jiangsu Agric. Sci. **49**(08), 15–22 (2021). https://doi.org/10.15889/j.issn.1002-1302.2021.08.003

20. Shen, L., Shu, T., Liao, X., Yang, N., Ren, Y., Zhu, M., et al.: A new method to evaluate urban resources environment carrying capacity from the load-and-carrier perspective. Resour. Conserv. Recycl. **154** (2020). https://doi.org/10.1016/j.resconrec.2019.104616

21. He, H., Shen, L., Wong, S.W., Cheng, G., Shu, T.A.: "Load-carrier" perspective approach for assessing tourism resource carrying capacity. Tour. Manag. **94** (2023). https://doi.org/10.1016/j.tourman.2022.104651

22. He, H., Luo, W., Cheng, G., Shen, L.: An Assessment model for city tourism resources carrying capacity from the "carrier-load" perspective. Paper presented at the International Symposium on Advancement of Construction Management and Real Estate (2019)

23. He, H., Shen, L., Du, X., Liu, Y.: Analysis of temporal and spatial evolution of tourism resource carrying capacity performance in China. Ecol. Indic. **147** (2023). https://doi.org/10.1016/j.ecolind.2023.109951

24. Fang, W., An, H., Li, H., Gao, X., Sun, X.: Urban economy development and ecological carrying capacity: taking Beijing city as the case. Energy Procedia **105**, 3493–3498 (2017)

25. Rahmani, A., Fakhraee, A., Karami, S., Kamari, Z.: A quantitative approach to estimating carrying capacity in determining the ecological capability of urban tourism areas (case study: Eram Boulevard of Hamadan city). Asia pacific journal of tourism research **20**(7), 807–821 (2015)

26. Adrianto, L., et al.: Assessing social-ecological system carrying capacity for urban small island tourism: the case of Tidung Islands, Jakarta Capital Province, Indonesia. Ocean Coast. Manag. **212**, 105844 (2021)

Provincial Land and Space Planning Meteorological Assessment Technology Research and Application for Climate Change Adaptation

Chen Cheng[1], Xiaoyi Fang[1(✉)], Jiawen Wang[2], Qiang Gong[3], Shuo Zhang[1], Jiyuan Hu[2], and Ying Yu[1]

[1] Chinese Academy of Meteorological Sciences, Beijing 100081, China
fangxy@cma.gov.cn
[2] China Academy of Urban Planning and Design, Beijing 100044, China
[3] Shenyang Regional Climate Center, Shenyang 110166, China

Abstract. Considering climate change adaptation in provincial land and space planning is of great significance for ensuring good ecology, agricultural safety, and urban livability. In conjunction with the compilation of the "Liaoning Provincial Land and Space Planning (2020–2035)", a land and space planning meteorological assessment technology suitable for climate has been established. From the three dimensions of climate resources, meteorological disaster risk, and wind-heat environment and meteorological diffusion capacity, 25 elements are selected to study the climate endowment and risk of land and space, and the spatial evaluation of agricultural and urban climate suitability and ecological climate importance is given; and using the climate change forecast results under different scenarios from 2020 to 2035, the spatiotemporal distribution of the province's future climate is analyzed. In summary, a comprehensive climate zoning of Liaoning Provincial Land and Space Planning is drawn, and climate adaptation strategy suggestions for each zone are proposed.

Keywords: Land and space planning · Meteorological assessment technology · Liaoning Province

1 Introduction

Respecting and adapting to climate and its changes in planning, optimizing land use and layout, so that cities, despite the continuous increase in vehicles and population, still maintain a stable level of urban heat islands and air pollution, and the climate environment is well protected is no longer an exception [1, 2]. According to the latest IPCC AR6, the degree and scope of climate change impacts have gradually increased [3], the process of global urbanization and climate change interact to exacerbate urban risks [4], and rational spatial planning is an important way to mitigate and adapt to climate change [5]. Considering climate adaptation before planning, combining historical, current and future climate information with planning issues, is of great significance for ensuring good ecology, agricultural safety and urban livability [6, 7].

B.-J. He et al. (Eds.): UCSUD 2023, LNCE 559, pp. 363–375, 2025.
https://doi.org/10.1007/978-981-97-8401-1_26

Previous studies have mostly focused on the analysis of wind, heat, precipitation, and meteorological disaster characteristics at the urban master planning scale [8], or on special planning such as the surface ventilation capacity in ventilation corridors, urban heat island intensity assessment [9], as well as the impact of building plot ratio, layout, and height on local wind and heat environment at the detailed planning scale [10]. For the weak provincial planning in our country's land and space planning, there are certain limitations in the current climate research: (1) More attention is paid to the impact of climate on ecology and agriculture, less on urban areas; (2) There are more studies on the impact of a single space, lacking comprehensive consideration of agriculture, ecology, and urban spaces, and insufficient research on comprehensive consideration of climate resources and risks; (3) Most are based on historical and current meteorological observation data for evaluation, lacking research on future climate change, making it difficult to support strategy formulation from a forward-looking development change time dimension.

The "Liaoning Provincial Land and Space Planning (2020–2035)", which began to be compiled in 2019, adhered to the path of ecological priority, integrated development and security, defined the overall pattern of provincial territorial space development and protection on the basis of the assessment of the carrying capacity of resources and the environment and the suitability of land and space development, coordinated efforts to draw red lines for ecological protection, permanent basic farmland, and urban development boundaries. Therefore, compiling the planning, a meteorological assessment technology for land and space planning adapted to the climate was established, realizing the comprehensive evaluation of climate resources [11], meteorological disaster risks [8], wind and heat environment and meteorological diffusion capacity [9, 12]. Based on this, evaluate climate suitability, coordinate the impact of meteorological environment on different territorial spaces, and predict future changes. Finally, provide climate zoning and develop planning recommendations for each region to support the construction of agricultural and urban spatial patterns, as well as the protection and restoration of ecological spaces.

2 Methods and Data

2.1 Methods and Technologies

Climate Resources and Risk Research. 25 elements were selected for the study of climate resourcefulness, meteorological disaster riskiness, wind and heat environment, and meteorological diffusion capacity, including:

Climate Resourcefulness Evaluation. The study of multi-year average climate elements (temperature, precipitation, humidity, wind) across the province, especially those related to crop growth such as active accumulated temperature (T_a), available precipitation (P_i), and climate production potential (T_{SPV}) and the spatialdistribution of climate resources. T_{SPV} is calculated by the Thornthwaite Memorial model [13]. P_i. is calculated by subtracting evaporation from precipitation. T_a. The calculation fmula is as follows:

$$T_a = \sum_{i=1}^{n} T_i, T_i \geq 10°C, or T_i = 0 \tag{1}$$

where, n is the total number of days in the initial and final periods, the initial and final dates of the active temperature calculated by the 5-day moving average method are used; T_i is the daily average temperature on the i day of the initial and final date period.

Meteorological Disaster Risk Assessment. Calculate the danger index of 9 types of meteorological disasters including heavy rain, drought [14], strong wind, hail, thunder [15], blizzard, fog, low temperature cold damage [16], high temperature, and then determine the types of disasters that affect agriculture, urban and ecological based on the number of occurrences, the number of affected people, crop and economic losses in historical disasters. Determine the impact weight of each disaster type through expert scoring method (see Table 1). Through the weighted comprehensive method, calculate the agricultural safety meteorological risk index (MRI_A), urban safety meteorological risk index (MRI_U) and the importance index of ecological response to meteorological risk (MRI_E), to carry out meteorological disaster risk analysis and evaluation.

Study on Wind-heat Environment and Meteorological Diffusion Capacity. Calculate the temperature and humidity index and wind effect index, evaluate the climate comfort of the living environment [17], calculate the annual average climate comfort days (CWD); remote sensing inversion of normalized vegetation index (NDVI), summer urban heat island index (UHI) and surface ventilation potential(SVP), evaluate the surface vegetation condition and heat environment [9]; calculate the A value of A–P Value Method of Atmospheric Environmental Capacity [18], Stable Weather Index (SWI) [19] and Evaluation on Meteorological Condition Index of $PM_{2.5}$ Pollution (EMI) [20], the weighted calculation of the Meteorological Pollution Index (MPI) is performed.

$$MPI = \frac{(SWI + EMI - A)}{3}.\tag{2}$$

Future Climate Change Assessment. Using the CMIP global climate model to drive the regional climate model RegCM [21], we estimate the changes in average temperature and precipitation, extreme temperature and precipitation relative to the 1986–2015 baseline period under different scenarios (RCP2.6, RCP4.5, RCP8.5) for 2020–2098, analyze future climate change trends and spatial changes during the 2021–2035 planning period, and provide support for the early prediction of future climate change impacts on territorial space.

Climate Suitability Evaluation. The indices of each climate resource and risk are normalized to 0–1, and the territorial space agricultural climate suitability index (CS_A), urban space climate suitability index (CS_U) and ecological climate importance index (CI_E) are calculated according to formula 3. The above 3 index are divided into 5 levels using natural breakpoints, and the climate suitability evaluations carried out.

$$CS_A = \frac{0.5 \times T_{SPV} + 0.5 \times T_a}{MRI_A}; CS_U = \frac{CWD}{0.5 \times MRI_U + 0.5 \times MPI}$$
$$; CI_E = \frac{NDVI}{UHI}\tag{3}$$

Suggestions for Land and Space Planning. Based on the comprehensive evaluation of climate resources and risks, future climate change, and climate suitability, we superimpose the spatial distribution results of $T_a, P_i. , T_{SPV}, MRI_A, MRI_U, MRI_E, CWD,$

NDVI, UHI, SVP, *MPI*, CS_A, CS_U, CI_E, DEM, as well as the changes in average temperature and precipitation, extreme temperature and precipitation by using ArcGIS, to draw a comprehensive climate zoning for the land and space of Liaoning, put forward planning suggestions for each zone, and assist in the formulation of plans.

2.2   Research Data

The daily average temperature, highest temperature, precipitation, number of windy days, hail days, foggy days, blizzard days, Meteorological Drought Comprehensive Monitoring Index (MCI), and lightning observation data of 61 national meteorological stations in Liaoning Province from 1961 to 2018 come from the Shenyang Regional Climate Center; the meteorological disaster data come from the "Chinese Disaster Dictionary Liaoning Volume" and the "Liaoning Province Historical Meteorological Disaster Database"; other data include the reanalysis data ERA-interim (0.125°) covering Liaoning Province from 2014 to 2018, the numerical simulation results of the China Meteorological Administration's Chemical Weather Forecast System (CUACE-EMI) from 2014 to 2019, the MODIS satellite's 8-day composite surface temperature (resolution 1km) in 2019, the land cover classification product (MCD12Q1) and 16-day composite NDVI (MOD13A2) in 2018.

3   Results Analysis

3.1   Climate Resources and Risk Distribution in Liaoning Province

The southwest of Liaoning Province is warmer than the northeast, the east is wetter than the west, the wind speed in Liaozhong Plain is relatively high, and the wind calm frequency in Liaodong and Liaoxi mountainous areas is high. Most areas of the province have an average T_a above 3000 °C, the plains of Liaozhong and the Bohai Sea rim are above 3600 °C, and the Liaodong mountainous area is less than 3200 °C. The annual average P_i in most areas of the province is 400–1000 mm, with more in the east and less in the west. The northwest of Liaoxi is generally less than 555 mm, and Liaodong is above 800 mm. The annual T_{SPV} is generally more in the east and less in the west, with most areas being 750–1550 g/m², the northwest of Liaoxi is less than 900 g/m², and the southeast of Liaodong is more than 1200 g/m². As for agricultural safety meteorological risks, the risk in Liaodong to Liaodong Peninsula is relatively low, the risk in Liaoxi to Liaoxi North is relatively high, and there is also risk in Liaozhong Plain. The urban safety meteorological risks in Liaozhong Plain, the west coast of Liaodong Peninsula, and Liaoxi towns are relatively low, while the eastern part of Liaodong Peninsula and the southwest of the Bohai Sea rim have relatively high meteorological risks. The importance of ecological response to meteorological risks is high in areas such as Liaoxi to the western part of Liaozhong Plain, the Bohai Sea rim, and Liaodong Peninsula. Most areas of the province have a CWD exceeding 100d, with more in the south and less in the north, and relatively more in Liaoxi South, Liaodong Peninsula, and Liaozhong Plain. The strong UHI above 3.0 °C in summer is mainly distributed in Liaoxi, Liaodong Peninsula, and Liaozhong urban areas, while Liaodong mountainous areas and coastal wetlands are

Table 1. Evaluation method of meteorological disaster risk.

Type	Evaluation Method	Charactestics	Weight
Heavy Rain	Calculate the cumulative rainfall R of all meteorological stations for at least 30 consecutive 1d, 2d…,10d heavy rain processes (at least one day's rainfall $\geq$ 50 mm), sort from small to large, and divide into 5 intensity levels using the percentile method. Calculate the total number of precipitation at each level of intensity, divide by the number of observation years of the station, and obtain the frequency of heavy rain at different levels. This risk index at the station is calculated using the weighted average method	High frequency, many affected and dead, significant agricultural and economic losses	A:0.25 U:0.3 R:0.4 (A: Agriculture; U: Urban; E: Ecology)
Drought	According to the MCI of each meteorological station for at least 30 years, 4 drought levels are divided [14], calculate the total number of days each drought level occurs, divide by the number of observation years, obtain the frequency of each drought level, and calculate the drought risk index of the station using the weighted average method	High frequency, many affected people, significant agricultural and economic losses	A: 0.4 U: 0.15 E: 0.2

(continued)

Table 1. (*continued*)

Type	Evaluation Method	Charactestics	Weight
High Wind	The total number of days of the corresponding weather phenomenon occurring at the statistical meteorological station for at least 30 years is divided by the observation years of the station to obtain the frequency of occurrence, which is used as the disaster risk index	High frequency, large number of victims and deaths, significant agricultural losses	A: 0.1 U: 0.05 E: 0.2
Hail	Same as High Wind	Same as High Wind	A: 0.1 U: 0.05
Lightning	The lightning risk index is calculated using lightning observation data as the disaster risk index [15]	High frequency and number of deaths	U: 0.15
Blizzard	Same as High Wind	Large number of victims, agricultural and economic losses	A: 0.05 U: 0.1
Heavy Fog	Same as High Wind	There are fatalities, significant impact on traffic	U: 0.15

(*continued*)

Table 1. (*continued*)

Type	Evaluation Method	Charactestics	Weight
Low Temperature Cold Damage	The occurrence of low temperature cold damage levels is determined by the latitude, altitude, and average temperature deviation from May to September during the growing season [16]. The total number of different levels of low temperature cold damage is then calculated, divided by the observation years of the station, to obtain the frequency of occurrence of different levels. The risk index is calculated using the weighted average method	Causes agricultural losses	A: 0.1
High Temperature	The extreme / average highest temperature, and duration of high temperature processes at the meteorological station are statistically analyzed to calculate the high temperature comprehensive intensity index (HTI). HTI is sorted from low to high and divided into 5 levels using the percentile method. The total number of occurrences of high temperature at different levels is counted, divided by the observation years of the station to obtain the frequency of each level, and the high temperature risk index is calculated using the weighted average method	Affects human health	U:0.05 E:0.2

cold islands. The MPI calculation results show that the overall meteorological pollution conditions in Liaodong mountainous area are relatively poor, while the western and coastal areas are better.

3.2 Future Climate Change Assessment

Under three RCP scenarios, the annual average temperature of 61 national meteorological stations in the province shows an upward trend from 2020 to 2098 (see Fig. 1a). From 2020 to 2035, the three scenarios will warm up by 1.2 °C, 1.3 °C, and 1.5 °C respectively. From 2046 to 2065, they will warm up by 1.3 °C, 2.5 °C, and 3.0 °C respectively. From 2081 to 2098, the warming will further increase to 1.4 °C, 3.4 °C, and 5.6 °C. The average annual precipitation of the provincial stations shows inter-annual fluctuations, the unevenness of annual precipitation distribution is enhanced, especially under the RCP8.5 scenario, the precipitation extremity is strong in some years after 2050 (see Fig. 1b). Under the RCP4.5 scenario with certain emission reduction measures, the warming range in the province from 2020 to 2035 is 1.2–1.5 °C, with relatively high warming in Liaodong North and Liaodong Peninsula, and locally up to 1.4°C, the precipitation in northern Liaoning, northwestern Liaoning, and the eastern coast of the Liaodong Peninsula will increase by 5–11%, while a decreasing trend will appear in southwestern Liaoning. The number of warm days and warm nights in the province has increased compared to the baseline period, and the intensity of high temperatures will increase. The number of warm days in the south and north has increased more, while the number of warm nights in Liaodong has increased relatively less. The number of heavy rain days will increase in most areas, with the most significant increase in northern Liaoning.

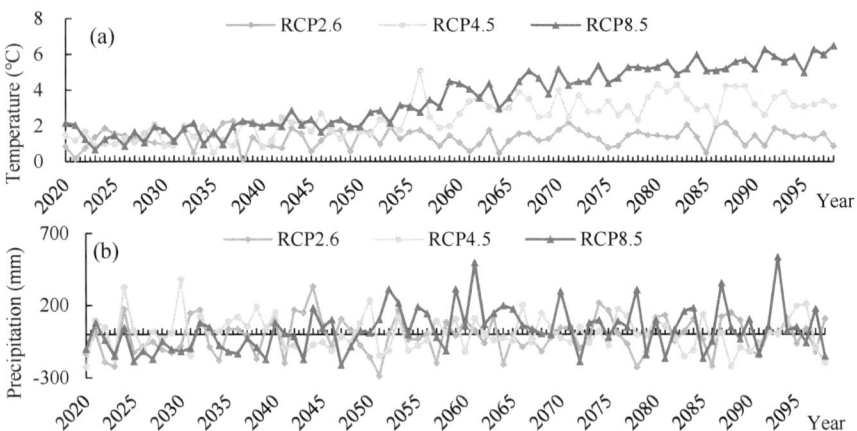

Fig. 1. Estimated changes in average temperature and precipitation at national meteorological stations in Liaoning Province from 2020 to 2098.

3.3 Results of Climate Suitability Evaluation

The overall agricultural climate suitability of the province is better in the southeast than in the northwest, and better in the plains than in the mountains. The Liaodong Peninsula is the most suitable for agriculture, followed by the plain from the Bohai Sea rim to central Liaoning, the Northern Liaoning is less suitable, and the mountainous area of western Liaoning is the worst (see Fig. 2a). The area of moderately suitable and suitable levels accounts for 58.2%, and the area of less suitable and unsuitable levels accounts for 22.2%. The climate suitability level of towns in southwestern Liaoning, central Liaoning to the Liaodong Peninsula is higher, while it is worse in northern Liaoning and the mountainous areas of Liaodong (see Fig. 2b). The area of towns with moderately suitable or above climate and less suitable or below climate accounts for 38% and 37.4% respectively. The ecological climate importance level is the highest in the mountainous areas of Liaodong and coastal wetlands, followed by the open spaces around the towns in the southwestern mountainous area of Liaoning and the central plain (see Fig. 2c). The above-mentioned areas have good vegetation cover and growth, or are covered by water bodies, which is beneficial to alleviate the heat island effect in summer and keep warm in winter. The area of important and very important levels accounts for 35.1%, and the area of less important and unimportant levels accounts for 28.2%.

3.4 Suggestions for Land and Space Planning

The comprehensive climate zoning of Liaoning Province's land and space planning is from Liaozhong to Liaobei Plain (A), Bohai Sea rim and Liaodong Peninsula (B), Liaodong Mountain Area (C) and Liaoxi Mountain Area (D). From the perspective of better utilizing climate resources, avoiding meteorological disasters, and improving the living environment, suggestions for agriculture, urban and ecological space protection and development are formulated for each climate zone, and the possible impacts of future climate change are considered in advance (see Fig. 3).

4 Conclusions and Discussions

4.1 Discussion

Based on the study, Liaoning is divided into temperate and warm temperate zones, humid and semi-humid zones, which support the spatial planning of agriculture and water resources in the plan. The results of climate suitability evaluation, meteorological risk evaluation respectively provide references for the "double evaluation" and "comprehensive disaster prevention" thematic research in the plan. Considering the uncertainty of climate change prediction, we can integrate the results of multiple climate models and scenarios, and scientifically discuss the support of climate change for land and space planning in the future. We can also further study the climate guidance area of ecological protection and restoration space, the agricultural space and the urban space, directly providing spatial information for the protection and development of land and space. In the urban master land and space planning, urban wind and ventilation potential, urban heat island, and meteorological disaster risk assessment may carry out to support the

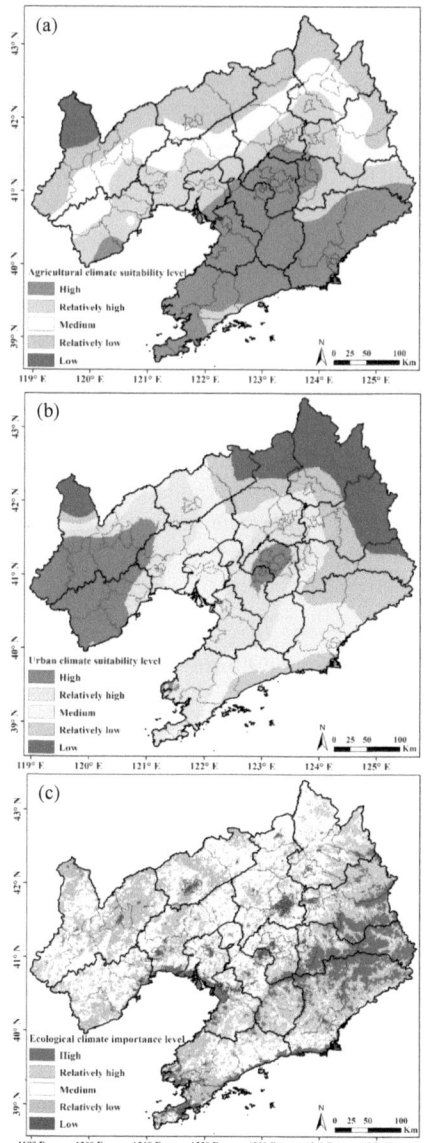

Fig. 2. The evaluation level of CS_A, CS_U and CI_E in Liaoning Province.

land use layout, urban form optimization, ventilation corridor planning, flood storage area planning, etc. At the scale of zoning planning, coupled with building height, density, floor area ratio, green space, water body, etc., ventilation, thermal comfort and urban waterlogging simulation shall be carried out to provide guidelines for ecological spatial planning, building height, density and layout.

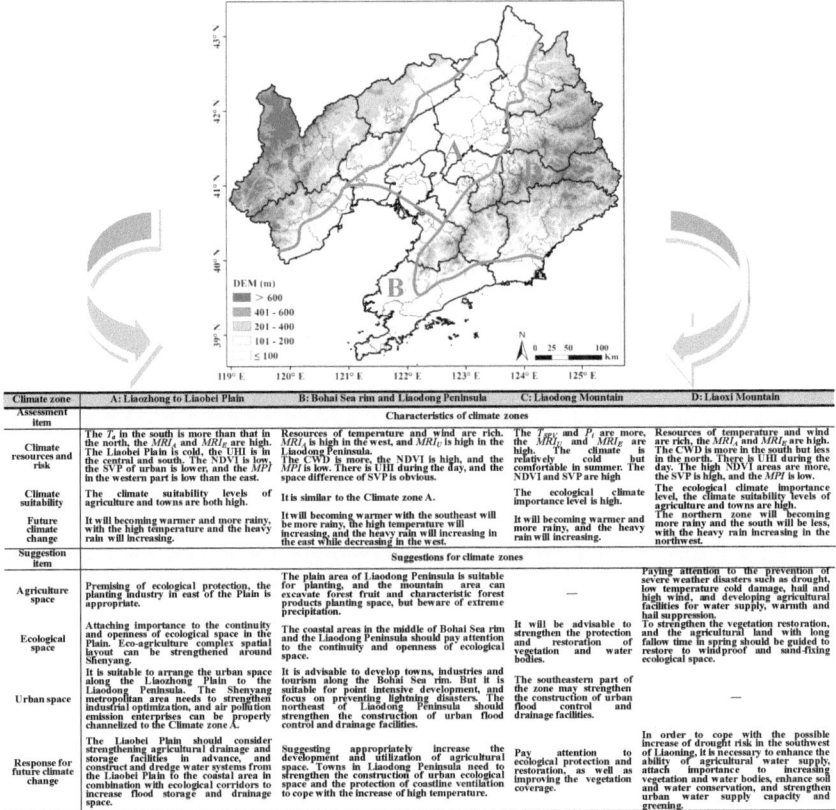

Climate zone	A: Liaozhong to Liaobei Plain	B: Bohai Sea rim and Liaodong Peninsula	C: Liaodong Mountain	D: Liaoxi Mountain
Assessment item	Characteristics of climate zones			
Climate resources and risk	The T_a in the south is more than that in the north, the MRI_{LF} and MRI_{IF} are high. The Liaobei Plain is cold, the UHI is in the central and south. The NDVI is low, the SVP of urban is lower, and the MPI in the western part is low than the east.	Resources of temperature and wind are rich, MRI_A is high in the west, and MRI_{IF} is high in the Liaodong Peninsula. The CWD is more, the NDVI is high, and the MPI is low. There is UHI during the day, and the space difference of SVP is obvious.	The T_{PEV} and P_i are more, the MRI_{LF} and MRI_{IF} are high. The climate is relatively cold but comfortable in summer. The NDVI and SVP are high	Resources of temperature and wind are rich, the MRI_A and MRI_{IF} are high. The CWD is more in the south but less in the north. There is UHI during the day. The high NDVI areas are more, the SVP is high, and the MPI are more.
Climate suitability	The climate suitability levels of agriculture and towns are both high.	It is similar to the Climate zone A.	The ecological climate importance level is high.	The ecological climate importance level, the climate suitability levels of agriculture and towns are high.
Future climate change	It will becoming warmer and more rainy, with the high temperature and the heavy rain will increasing.	It will becoming warmer with the southeast will be more rainy, the high temperature will increasing, and the heavy rain will increasing in the east while decreasing in the west.	It will becoming warmer and more rainy, and the heavy rain will increasing.	The northern zone will becoming warmer and the south will be less, with the heavy rain increasing in the northwest.
Suggestion item	Suggestions for climate zones			
Agriculture space	Premising of ecological protection, the planting industry in east of the Plain is appropriate.	The plain area of Liaodong Peninsula is suitable for planting, and the mountain area can excavate forest fruit and characteristic forest products planting space, but beware of extreme precipitation.	—	Paying attention to the prevention of severe weather disasters such as drought, low temperature cold damage, hail and high wind, and developing agricultural facilities for water supply, warmth and hail suppression.
Ecological space	Attaching importance to the continuity and openness of ecological space in the Plain. Eco-agriculture complex spatial layout can be strengthened around Shenyang.	The coastal areas in the middle of Bohai Sea rim and the Liaodong Peninsula should pay attention to the continuity and openness of ecological space.	It will be advisable to strengthen the protection and restoration of vegetation and water bodies.	To strengthen the vegetation restoration, the agricultural land with long fallow time in spring should be guided to restore to windproof and sand-fixing ecological space.
Urban space	It is suitable to arrange the urban space along the Liaozhong Plain to the Liaodong Peninsula. The Shenyang metropolitan area needs to strengthen industrial optimization, and air pollution emission enterprises can be properly channeled to the Climate zone A.	It is advisable to develop towns, industries and tourism along the Bohai Sea rim. But it is suitable for point intensive development, and focus on preventing lightning disasters. The northeast of Liaodong Peninsula need to strengthen the construction of urban flood control and drainage facilities.	The southeastern part of the zone may strengthen the construction of urban flood control and drainage facilities.	—
Response for future climate change	The Liaobei Plain should consider strengthening agricultural drainage and storage facilities in advance, and construct and dredge water systems from the Liaobei Plain to the coastal area in combination with ecological corridors to increase flood storage and drainage space.	Suggesting appropriately increase the development and utilization of agricultural space. Towns in Liaodong Peninsula need to strengthen the construction of urban ecological space and the protection of coastline ventilation to cope with the increase of high temperature.	Pay attention to ecological protection and restoration, as well as improving the vegetation coverage.	In order to cope with the possible increase of drought risk in the southwest of Liaoning, it is necessary to enhance the ability of agricultural water supply, attach importance to increasing vegetation and water bodies, enhance soil and water conservation, and strengthen urban water supply capacity and greening.

Fig. 3. Suggestion map of climate zoning for land and space planning in Liaoning Province.

4.2 Research Conclusions

In conjunction with the compilation of China's "Liaoning Provincial Land and Space Planning (2020–2035)", a land and space planning meteorological assessment technology adapted to the climate has been established and applied. The main conclusions are: (1) The spatial distribution characteristics of 25 elements for climate resources and risk, wind and heat environment, and meteorological diffusion capacity are mastered, as well as the spatial distribution of areas with high agricultural and urban meteorological risks and the important areas of ecological response to meteorological risks; (2) The planning period shows a trend of rising temperature, increasing precipitation, and strengthening of extreme warmth and extreme precipitation. Under the RCP4.5 scenario, the province will warm by 1.2–1.5 °C, precipitation in Liaobei, Liaoxi Northwest and the eastern coast of Liaodong Peninsula will increase by 5–11%, and precipitation in Liaoxi South will decrease. The number of warm days in the south and north of the province will increase more, warm nights in Liaodong will increase relatively less, and the number of heavy rain days in Liaobei will increase the most; (3) The agricultural climate suitability level in Liaoning is higher in the southeast than in the northwest, with nearly 60% of the area at a suitable or higher level. The urban climate suitability level in the central and southern parts of Liaozhong Plain and the Bohai Sea rim is relatively high, and the ecological climate importance level in Liaodong and Liaoxi South mountainous areas and coastal wetlands is relatively high; (4) The comprehensive climate zoning of land and space planning has been divided, and three types of land and space planning and climate change response suggestions have been formulated for each zone.

Acknowledgements. This project is financially supported by the National Key R&D Program of China (No. 2022YFC3090600); Natural Science Foundation of Hubei Province, China (No. 2023AFD106); Basic Research Fund of the Chinese Academy of Meteorological Sciences (No. 2023Z016).

References

1. Baumüller J., Hoffmann, U., Stuckenbrock, U.: Urban framework plan hillsides of Stuttgart. In: Paper presented at the 7th International Conference on Urban Climate, pp. 1–2. Tokyo Institute of Technology, Tokyo (2009)
2. Qian, Y., Chakraborty, T.C., Li, J.F., et al.: Urbanization impact on regional climate and extreme weather: Current understanding, uncertainties, and future research directions. Adv. Atmos. Sci. **39**(6), 819–860 (2022)
3. IPCC: Climate change 2022: impacts, adaptation, and vulnerability. Cambridge University Press, Cambridge (2022)
4. Wang, J.N., Qin, N.X., Jiang, T., et al.: Interpretation of IPCC AR6: impacts and adaptations of climate change on cities, settlements and key infrastructure. Adv. Clim. Change Res. **18**(4), 433–441 (2022)
5. Mi, Z.F., Zhang, H.R.: Interpretation of IPCC AR6 report: climate change mitigation of urban systems. Clim. Change Res. **19**(2), 139–150 (2023)
6. State Council of the PRC: Several opinions of the central Committee of the Communist Party of China and the state council on establishing a land and space planning system and supervising the implementation, p. 1. State Council of the PRC, Beijing (2019)
7. Ministry of Natural Resources of the PRC: Guidelines for compilation of provincial land and space planning (for trial implementation), p. 4. Ministry of Natural Resources of the PRC, Beijing (2020)
8. Fang, X.Y., Cheng, C., Du, W.P., et al.: Technical for climatic feasibility demonstration in master planning: GB/T37529–2019. Standards Press of China, Beijing (2019)
9. Du, W.P., Fang, X.Y., Cheng, C., et al.: Specifications for climatic feasibility demonstration—Urban ventilation corridor: QX/T 437–2018. China Meteorological Press, Beijing (2018)
10. Back, Y., Kumar, P., Bach, P.M., et al.: Integrating CFD-GIS modelling to refine urban heat and thermal comfort assessment. Sci. Total Environ. **858**(Part 1), 159729 (2023)
11. Ministry of Natural Resources of the PRC: Guidelines for evaluating the carrying capacity of resources and the environment and the suitability of territorial space development (for trial implementation), p. 18. Office of Ministry of Natural Resources of the PRC, Beijing (2020)
12. Wang, G.T., Jiao, J., Bao, Y.H., et al.: Technical Guide to Environmental Performance Assessment Guidelines for Urban Ecological Construction. China Building Industry Press, Beijing (2016)
13. Xu, Y.Q., Zhou, B.T., Yu, L., et al.: Climatic potential productivity and population carrying capacity in China from 1961 to 2010. J. Meteorol Environ. **35**(2), 84–991 (2019)

14. Zhang, C.J., Liu, H.B., Song, Y.L., et al.: Grades of Meteorological Drought: GB/T 20481–2017. Standards Press of China, Beijing (2017)
15. Cheng, X.Y., Tao, Y., Zou, J.J., et al.: Technical guidelines for lightning disaster risk zoning: QX/T405–2017. China Meteorological Press, Beijing (2018)
16. Wang, C.Y.: Research on chilling damage of crop in Northeast China. China Meteorological Press, Beijing (2008)
17. Feng, M., Mao, F., Wang, X.L., et al.: Climatic suitability evaluating on human settlement: GB/T 27963–2011. Standards Press of China, Beijing (2012)
18. Xu, D.H., Wang, Y.: Plume footprints analysis for determining the bearing capacity of atmospheric environment. Acta Sci. Circumstantiae **33**(6), 1734–1740 (2013)
19. Zang, H.D., Zang, B.H., Lv, M.Y., et al.: Development and application of stable weather index of Beijing in environmental meteorology. Meteorol. Monthly **43**(8), 998–1004 (2017)
20. Zhang, B.H., Liu, H.L., Zhang, D., et al.: Evaluation on meteorological condition index of PM 2.5 pollution: QX/T479–2019. China Meteorological Press, Beijing (2019)
21. Tong, Y., Gao, X., Han, Z., et al.: Bias correction of temperature and precipitation over China for RCM simulations using the QM and QDM methods. ClimDyn **57**, 1425–1443 (2021)

Spatial Impact of Surface-Water Bodies on Urban Expansion Using an Autologistic Regression Model

Mengyuan Jia[1](✉) and Haoran Zhang[2]

[1] School of Architecture and Urban Planning, Beijing University of Civil Engineering and Architecture, Beijing 100044, China
jiamengyuan@bucea.edu.cn
[2] China Academy of Urban Planning and Design, Beijing 100044, China

Abstract. In order to explore the differential impacts of various types of surface water bodies on urban expansion, Taking the data in 2000, 2010, and 2020 of Tianjin as research subjects, this paper establishes the Autologistic regression models to investigate the influence of factors such as proximity to different types of surface water bodies, presence within flood storage and detention areas, and their association with aquatic ecological corridors on urban expansion. The results show that 1) flood storage and detention areas are significant constraining factors for urban expansion and the primary rivers are a more pronounced driving effect compared to other surface water bodies; 2) over the period from 2000 to 2020, the driving effect of surface water bodies on urban expansion exhibits a declining trend; 3) The delineation of urban construction land boundaries and the establishment of ecological protection zones are identified as the principal external factors shaping urban expansion. The results further explained the relationship between cities and water, and can provide a theoretical basis for the harmonious coexistence of cities and water environment.

Keywords: Surface water bodies · urban expansion · Autologistic regression model · Tianjin

1 Introduction

Many cities are born and thrive because of water, and the characteristics of surface water environments such as river and lake systems are closely related to urban development. Traditional Chinese urban construction follows the philosophical concept of "following the laws of nature" [1], using the water environment as a medium for communication and connection between humans and nature, guiding the organization and evolution of urban space. Ancient cities' river and lake systems also carried important functions of transportation, forming a large number of docks and towns along the river. Today, although cities no longer rely on water transportation, the beautiful waterfront landscape and people's preference for living near water still influence the direction of urban development and construction, and many new urban areas and residential areas are built in waterfront

© The Author(s) 2025
B.-J. He et al. (Eds.): UCSUD 2023, LNCE 559, pp. 376–389, 2025.
https://doi.org/10.1007/978-981-97-8401-1_27

areas. However, the urban land expansion towards waterfront areas has interrupted the natural water cycle, affecting the healthy operation of the aquatic ecosystem, and weakening the ecological service functions of the natural water environment such as flood storage, water body purification, and microclimate regulation, causing problems such as eutrophication of water bodies, reduction of biological populations, and frequent flood disasters. Therefore, coordinating the relationship between urban development and water environment protection is an important link in promoting the harmonious coexistence of cities and the natural environment.

Urban land expansion is the result of the interaction of multiple driving forces. There are studies that combine historical data of land use changes to explore the impact of natural elements such as climate conditions, geographical location, topography and geomorphology, and engineering geology, socio-economic elements such as economic growth and industrial structure, and administrative elements such as urban planning and land policy on urban land expansion [2–5].

Among them, although some explanatory models of urban land expansion have found that the distance to lakes [4, 5] and the distance to rivers [6] are one of the driving factors of urban land expansion, the types of surface water environmental elements are rich and varied, and it is difficult to fully explain the impact of surface water environment with distance factors alone. The questions, such as the differences between big and small rivers, and the disparities between rivers, lakes, and wetlands in their impact on urban land expansion, are still unclear. In view of this, this paper hopes to further explore the differences in the impact of different types of surface water environmental factors on urban land expansion, to deepen the understanding of the relationship between cities and water, and provide a theoretical basis for the harmonious coexistence of cities and water environment.

2 Research Object and Method

2.1 Research Object

This paper takes the administrative jurisdiction of Tianjin as the research object (Fig. 1), with a total area of $11760km^2$. Tianjin's urban development has been closely related to water, from relying on river transport for trade development to building ports for sea transport. There are 19 primary rivers flowing through Tianjin, with a length of 1095.1 km, 79 secondary rivers with a total length of 1363.4 km, and several artificial rivers such as Ziya New River, Duliujian River, Chaobai New River, as well as long-distance water transfer projects such as Luan River into Tianjin and South-to-North Water Transfer. There are 14 large and medium-sized reservoirs in the city, including 3 large reservoirs, and many wetland resources such as Qilihai Ancient Lagoon Wetland, Beidagang Wetland, Tuanpowa Wetland, Dongli Lake and Guan Port Wetland. Therefore, the administrative jurisdiction of Tianjin covers a variety of types of water environmental elements such as rivers, lakes, wetlands, and coastlines, which is representative for research.

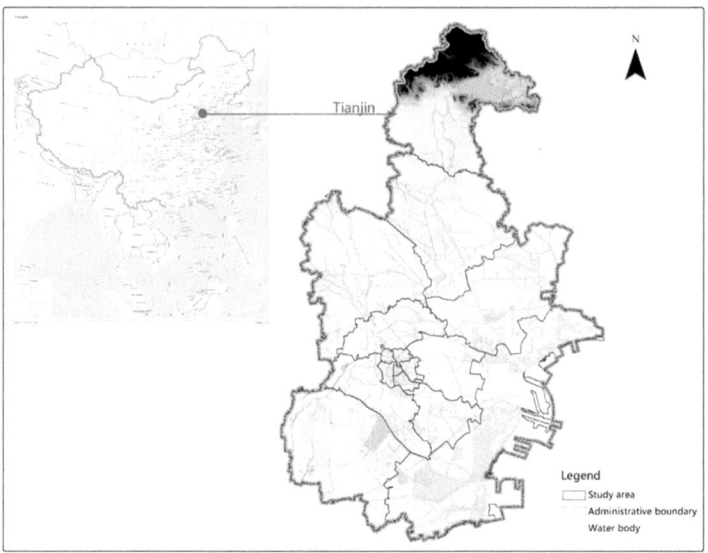

Fig. 1. The map of study area—Tianjin, China.

2.2 Research Method

There are currently various methods for identifying the driving factors of urban land expansion, such as spatial regression analysis [7], Logistic regression model [3, 8], Artificial Neural Network model method (ANN) [9], and Autologistic regression model which optimizes the spatial analysis capability of Logistic regression [10, 11]. Among them, studies have shown that the Autologistic regression model has the best fit, superior to the Logistic model and the Artificial Neural Network model [9]. The Autologistic model is a variant of the Logistic regression model after introducing spatial autocorrelation factors, and its advantage is that it can solve the impact of spatial autocorrelation in geographic information [10]. The Autologistic model was initially applied in the study of plant population competition [12, 13], and then extended to the analysis of all data with spatial information [14].

The structure of the Autologistic model used in this study is as follows:

$$\ln\left(\frac{P_i}{1 - P_i}\right) = \beta_0 + \beta_1 X_{1,i} + \beta_2 X_{2,i} + \cdots + \beta_n X_{n,i} + \gamma \, Autocov_i \quad (1)$$

where P_i is the probability of the cell i as urban built-up land, $X_{1,i}, X_{2,i}, ..., X_{n,i}$ are the value of driving factor n on cell i, $Autocov_i$ is the autocorrelation index of cell i; β_0 is a constant term, $\beta_1, \beta_2, ..., \beta_n$ are the regression coefficient of the driving factor n, γ is the regression coefficient of autocorrelation index.

The autocorrelation index is calculated as follows:

$$Autocov_i = \frac{\sum_{i \neq j} W_j y_j}{\sum_{i \neq j} W_j} \quad (2)$$

where, y_j represents the land use status of grid j, where urban construction land is assigned a value of 1, and non-urban construction land is assigned a value of 0; W_{ij} represents the spatial weight between grid i and j, which is determined by the inverse distance weighting method in this study. The specific calculation method is as follows:

$$W_{ij} = \begin{cases} \frac{1}{D_{ij}}, whenD_{ij} < d; \\ 0, D_{ij} \geq d, \end{cases} \tag{3}$$

where, D_{ij} is the Euclidean distance between grid i and j, the threshold d is determined by the spatial resolution of this study, taking d = 300m; that is, when $D_{ij} < 300m$, the spatial weight W_{ij} is the reciprocal of the distance between grid i and j, otherwise the spatial weight W_{ij} is 0.

For the Autologistic model, the β value reflects the correlation coefficient of the driving factors in the model, that is, when the driving factor $X_{n,i}$ changes by one unit, the corresponding variable's logarithmic advantage increases by β units. Exp(β) is the natural power exponent of the β coefficient with e as the base, indicating the change in the advantage ratio of urban construction land appearing for each additional unit of the driving factor. The algorithm for the advantage ratio of the appearance of construction land is $\frac{\mu}{(1-\mu)}$, where μ is the probability of the lattice value being 1. When Exp(β) > 1, it indicates that the odds ratio increases, and this factor has a positive driving effect on urban land expansion; when Exp(β) = 1, it indicates that the odds ratio remains unchanged; when Exp(β) < 1, it indicates that the odds ratio decreases, and this factor has a negative driving effect on urban land expansion. The regression model uses p = 0.01 as the significance level test threshold for explanatory variables.

This study selects land use data from the years 2000, 2010, and 2020 for analysis, and constructs Autologistic regression models respectively (Fig. 2). By comparing the results of the three regression models, the differences in the impact of water environmental factors on urban land expansion during the rapid urbanization process can be revealed.

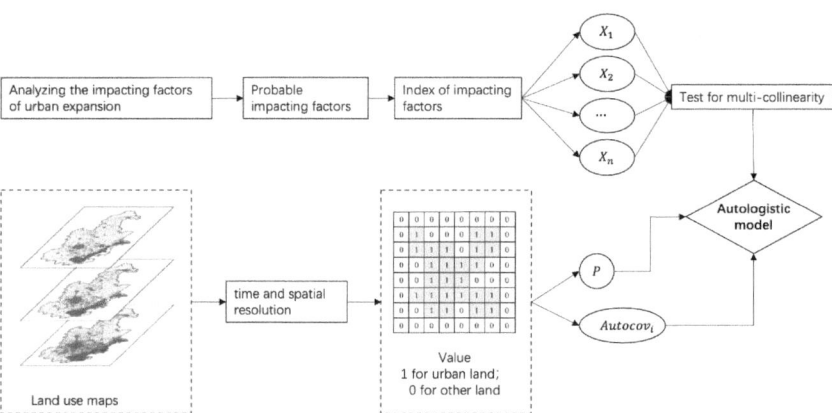

Fig. 2. Research framework for exploring the impact of water environmental factors using the Autologistic model

2.3 Data Sources and Processing

The study uses 30m resolution land use maps from 2000, 2010, and 2020, with land types including arable land, gardens, forests, urban construction land, water bodies, wetlands, unused land, etc. The data comes from the GlobalLand30 database (http://www.global landcover.com/GLC30Download). The data for the water environment driving factors comes from the "Tianjin Water System Special Plan (2005–2020)", "Tianjin Urban Master Plan (2005–2020)", "Tianjin Water Function Zone Division", "Tianjin Water Resources Bulletin", "Tianjin Yearbook", etc. Other required vector data includes administrative boundaries at the street level, main urban road networks, railway networks, rail transit stations, river system maps, 30m resolution DEM data, etc.

3 Research Results and Discussion

3.1 Extraction of Driving Factors

This study refers to the driving factor indicators of landscape pattern changes [15], land use/land cover changes [3], urban land expansion [5, 16, 17] and other studies, combined with the water environment characteristics of Tianjin, based on the principles of accessibility, representativeness, specificity, and diversity, extracts the driving factors of the Autologistic model from six aspects: water environment factors, natural terrain, socio-economics, location conditions, transportation infrastructure, and planning policies (Table 1).

Water Environment Factors

While many studies have used the distance to surface water bodies such as rivers and lakes as a general characterization of the driving role of the water environment, this study further refines the understanding of the driving role of the water environment on urban spatial growth. Based on different types of water bodies, six water environment driving factors ($X1$—$X6$) are set as explanatory variables in the Autologistic regression model.

Other Driving Factors

Natural terrain conditions are one of the important constraints on urban development, and the main factors that significantly affect urban land expansion are elevation [17], slope [15], topography [3], and precipitation [15] elements.

Socio-economic factors include population, industry, economy, and other aspects. Studies have found that urbanization rate, industrial activity intensity, GDP, and other factors are correlated with urban spatial growth [3, 15]. This study uses the total population ($X11$), population density ($X12$), and total GDP ($X13$) as explanatory variables for the changes in these three socio-economic factors from 2000 to 2020.

Location condition factors are mainly extracted through neighborhood analysis of the geographical location conditions of each grid. The city center gathers the main commercial service functions of the city, and the closer the grid is to the city center, the more obvious its development location advantage. According to the overall urban planning of Tianjin, this study extracts the main city center, secondary city center, and district-level center, conducts neighborhood analysis, and obtains the main center location factor

Table 1. The driving factors of the Autologistic model

Dimension	Driving Factors	Abbr	Data type
water environment factors	Distance to Hai River (m)	X1	Continues
	Distance to primary rivers (m)	X2	Continues
	Distance to secondary rivers (m)	X3	Continues
	Distance to lakes and reservoirs (m)	X4	Continues
	Located in flood storage area (0,1)	X5	Type
	Distance to aquatic ecological corridor (m)	X6	Continues
natural terrain	Elevation (m)	X7	Continues
	Slope (degree)	X8	Continues
	Terrain (0,1,2)	X9	Type
	Average Annual Rainfall (mm)	X10	Continues
socio-economics	Total Population Change (10 000 persons)	X11	Continues
	Population Density Change (person/km2)	X12	Continues
	Total GDP Change (10 000 yuan)	X13	Continues
location conditions	Distance to Main Center (m)	X14	Continues
	Distance to Sub-center (m)	X15	Continues
	Distance to District Center (m)	X16	Continues
transportation infrastructure	Distance to Main Traffic Road (m)	X17	Continues
	Distance to Train Station (m)	X18	Continues
	Distance to subway station (m)	X19	Continues
planning policies	Is it a planned construction land (0,1)	X20	Type
	Is it located within the planned ecological protection area (0,1)	X21	Type
	Is it located within the basic farmland protection area (0,1)	X22	Type

(X14), secondary center location factor (X15), and district-level center location factor (X16) as three explanatory variables.

Transportation facilities are closely related to the value and use of land. Convenient transportation facilities can promote the development and renewal of urban land. This study selects the distance to the main traffic arteries (X17), the distance to the train station (X18), and the distance to the subway station (X19) as evaluation factors for transportation infrastructure.

Planning policy is a key factor affecting urban land expansion [3, 16]. Only land designated as urban construction land in the overall plan can be legally developed, and land designated as ecological protection red lines and basic farmland protection areas will strictly limit development and construction. Therefore, this study selects whether it is planned urban construction land (X20), whether it is located within the planned ecological protection red line (X21), and whether it is located within the basic farmland protection area (X22) as three explanatory variables for the Autologistic regression. Each factor value is represented by 0 or 1, where 0 means not located within the planning area, and 1 means located within the planning area.

In summary, this study selects 22 driving factors as explanatory variables for the Autologistic regression. After spatial computation, the corresponding driving factor layers are obtained (see Fig. 3).

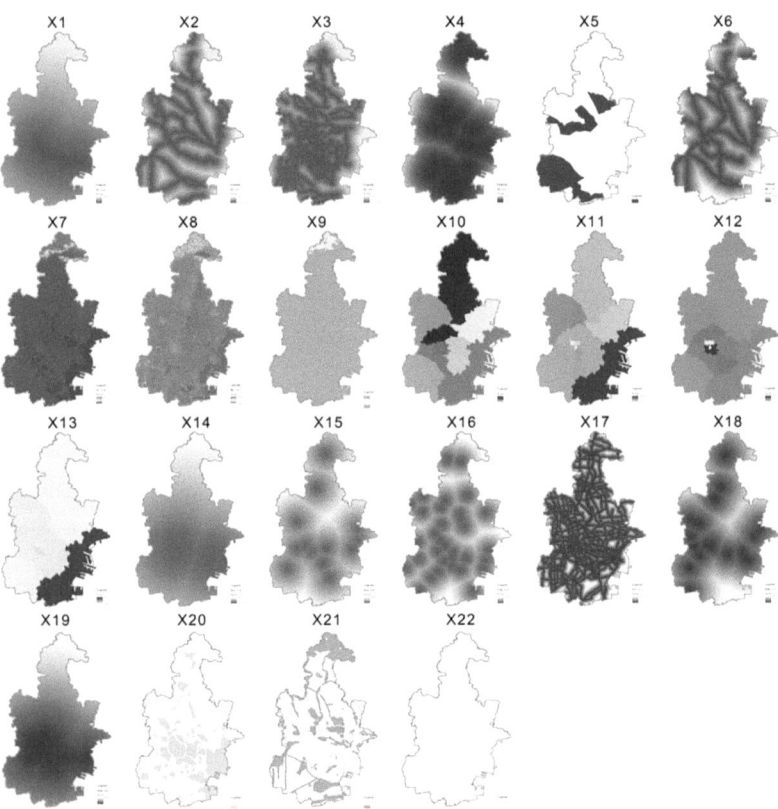

Fig. 3. Layers of driving factors for the 22-item Autologistic regression model

3.2 Model Construction and Verification

This study conducted a collinearity test on 22 driving factors, using the kappa coefficient and Variance Inflation Factor (VIF) to exclude factors with multicollinearity (Table 2).

The study found that 22 driving factors have strong multicollinearity issues with the distance from the Haihe River (X1), elevation (X7), distance from the main center (X14), and distance from subway stations (X19). Considering the preservation of the factor of distance from subway stations (X19), the other three factors were excluded. After factor screening, the kappa coefficient of the remaining 19 driving factors is 22.825, and the VIF is between 1.06 and 3.24, which can exclude the problem of multicollinearity and can all be used as explanatory variables for the Autologistic regression model.

Autologistic models were constructed for the years 2000, 2010, and 2020 respectively. The AUC values of the models are shown in Fig. 4. It can be seen that all model AUC values are above 0.9, indicating that the models have good explanatory power.

3.3   The Driving Role of the Water Environment

The regression model results of the driving factors of urban spatial growth in 2000, 2010, and 2020 are shown in Table 3. The results show that among the water environment factors, the range of the flood storage and detention area has the most significant impact on urban spatial growth and is an important restrictive factor for urban land expansion. The Exp(β) of the X5 factor in 2000 is 1.51377, indicating that under the condition that other driving factors remain unchanged, the probability of non-flood storage and detention area land being urban construction land is about 51.4% higher than that within the flood storage and detention area. Compared with other surface water environment factors, primary rivers have a more significant driving effect on urban spatial growth. The closer the distance to the primary river, the higher the advantage ratio of urban construction land appearing. Although the factors related to secondary rivers, lakes and reservoirs, and water ecological corridors are correlated with urban construction land, their β coefficients are smaller than those of road traffic factors and location condition factors, indicating that the driving effects of these water environment factors are less than those of road traffic and location conditions. Overall, the β coefficients of all water environment factors (X2—X6) in 2020 are smaller than those in 2000 and 2010, indicating a weakening trend in the driving role of water environment factors on urban spatial growth.

3.4   Other Driving Factors

Planning policy factors have the most significant impact on the spatial distribution of urban land. The planned ecological protection area and the basic farmland protection area are important restrictive conditions for urban spatial growth. Assuming other driving factors remain unchanged, according to the regression results, the probability of land outside the ecological protection area being urban construction land in 2000 was 39.3% higher than that within the ecological protection area, 62.4% higher in 2010, and 65.5% higher in 2020. This indicates that from 2000 to 2020, the restrictive effect of the ecological protection area on urban growth has been continuously strengthened. However, compared with other driving factors, the restrictive effect of the basic farmland protection area on urban spatial growth shows a weakening trend. In 2000, the probability of land outside the basic farmland protection area being urban construction land was 70.6%

Table 2. List of Pearson correlation coefficients and Variance Inflation Factors (VIF) for 22 driving factors

	X1	X2	X3	X4	X5	X6	X7	X8	X9	X10	X11	X12	X13	X14	X15	X16	X17	X18	X19	X20	X21	X22	VIF
X1	1																						148.7
X2	0.26	1																					2.25
X3	0.42	0.51	1																				2.52
X4	0.23	-0.07	0.02	1																			1.45
X5	0.08	0.12	0.19	0.04	1																		1.48
X6	0.08	0.53	0.19	0.03	-0.06	1																	2.24
X7	0.48	0.11	0.38	-0.08	0.09	0.04	1																11.15
X8	0.42	0.1	0.23	-0.05	0.07	0.01	0.71	1															2.19
X9	-0.4	-0.06	-0.34	0.08	-0.09	-0.03	-0.93	-0.63	1														8.3
X10	0.57	0.15	0.25	0.07	0.07	-0.07	0.28	0.26	-0.23	1													1.83
X11	-0.29	-0.01	0.08	-0.14	0.05	0.2	-0.1	-0.11	0.07	-0.08	1												2.65
X12	-0.18	0	-0.06	-0.05	0.05	0.02	-0.05	-0.04	0.04	-0.05	-0.35	1											1.63
X13	-0.36	0.05	0.16	-0.27	0.08	0.33	-0.15	-0.19	0.1	-0.12	0.63	0.1	1										3.67
X14	0.99	0.27	0.41	0.2	0.09	0.1	0.49	0.42	-0.4	0.57	-0.28	-0.17	-0.34	1									106.46
X15	0.28	0.15	0.11	0.19	-0.22	0.37	0	-0.03	0	0.04	-0.09	-0.11	-0.04	0.27	1								2.56
X16	0.4	0.21	0.46	-0.07	0.06	0.21	0.31	0.2	-0.28	0.17	0.07	-0.06	0.19	0.41	0.13	1							1.69
X17	0.08	0.17	0.26	0.04	0.01	0.24	0.09	0	-0.1	0.01	0.13	-0.02	0.25	0.08	0.23	0.12	1						1.29
X18	0.19	0.15	0.16	-0.1	-0.09	0.29	0.03	0	-0.06	-0.02	0.1	-0.02	0.27	0.23	0.63	0.25	0.31	1					2.57
X19	0.99	0.28	0.45	0.21	0.15	0.12	0.49	0.42	-0.41	0.55	-0.24	-0.17	-0.28	0.99	0.27	0.42	0.11	0.22	1				108.8
X20	-0.4	0.01	-0.08	-0.12	0.18	0	-0.12	-0.12	0.09	-0.1	0.22	0.13	0.24	-0.39	-0.33	-0.24	0	-0.21	-0.36	1			1.5
X21	-0.26	-0.04	-0.26	0.14	0	0.09	-0.25	-0.21	0.21	-0.14	0.04	0.04	0.02	-0.26	-0.05	-0.25	-0.13	-0.18	-0.26	0.3	1		1.33
X22	0.05	-0.04	-0.09	0.15	-0.12	0	-0.06	-0.04	0.06	-0.02	-0.1	-0.04	-0.15	0.05	0.1	-0.06	-0.04	-0.01	0.03	-0.12	0.06	1	1.07

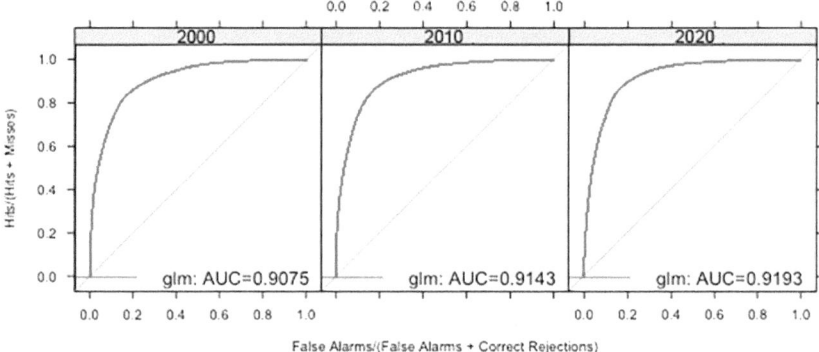

Fig. 4. ROC curve of the Autologistic regression model

higher than that within the protection area, but by 2020 this difference had decreased to 49.4%.

Among the remaining driving factors, natural terrain factors, location factors, and transportation infrastructure factors are significantly correlated with urban construction land. Specifically, the β coefficients of the distance from the main traffic arteries (X17), the distance from the sub-center (X15), and the distance from the district center (X16) indicate that the closer the distance to these factors, the higher the likelihood of urban construction land appearing. Natural terrain factors (X8, X9) show that lands with plain terrain and gentle slopes have a higher likelihood of urban construction land appearing. In socio-economic factors, the growth of the total population and the increase in population density are also positive factors driving urban spatial growth.

Table 3. Exponentiated β coefficients and Exp(β) of the Autologistic model

Driving Factors		2000		2010		2020	
		β	Exp(β)	β	Exp(β)	β	Exp(β)
(Intercept)		-7.19	0.00	-4.85	0.01	-6.37	0.00
Distance to primary rivers	X2	0.00	1.00	0.00	1.00	0.00	1.00
Distance to secondary rivers	X3	0.00	1.00	0.00	1.00	0.00	1.00
Distance to lakes and reservoirs	X4	0.00	1.00	0.00	1.00	0.00	1.00
Located in flood storage area	X5	0.41	1.51	0.46	1.58	0.23	1.26
Distance to aquatic ecological corridor	X6	0.00	1.00	0.00	1.00	0.00	1.00
Slope	X8	0.09	1.09	0.06	1.07	0.04	1.04
Terrain	X9	2.37	10.64	1.44	4.20	3.05	21.09
Average Annual Rainfall	X10	0.00	1.00	0.00	1.00	0.00	1.00
Total Population Change	X11	0.01	1.01	0.01	1.01	0.00	1.00
Population Density Change	X12	0.00	1.00	0.00	1.00	0.00	1.00
Total GDP Change	X13	0.00	1.00	0.00	1.00	0.00	1.00
Distance to Sub-center	X15	0.00	1.00	0.00	1.00	0.00	1.00
Distance to District Center	X16	0.00	1.00	0.00	1.00	0.00	1.00

(continued)

Table 3. (*continued*)

Driving Factors		2000		2010		2020	
Distance to Main Traffic Road	X17	0.00	1.00	0.00	1.00	0.00	1.00
Distance to Train Station	X18	0.00	1.00	0.00	1.00	0.00	1.00
Distance to subway station	X19	0.00	1.00	0.00	1.00	0.00	1.00
Is it a planned construction land	X20	1.96	7.10	2.01	7.47	1.94	6.92
Is it located within the planned ecological protection area	X21	-0.33	0.72	-0.49	0.62	-0.50	0.60
Is it located within the basic farmland protection area	X22	-0.53	0.59	-0.46	0.63	-0.40	0.67
Autocov		-0.05	0.95	-0.23	0.80	-0.37	0.69

4 Conclusion

This study takes Tianjin as an example, applying the Autologistic regression model to analyze the impact of water environment factors on urban land expansion. The results show that the flood storage area is an important limiting factor for urban spatial growth, and primary rivers, compared to other surface water environment factors, have a more significant driving effect on urban land expansion. Overall, comparing the Autologistic regression model analysis results of 2000, 2010, and 2020, it can be seen that the driving effect of water environment factors on urban land expansion has weakened from 2000 to 2020. In addition, the regression results show that planning policy factors are important factors affecting urban land expansion, among which the planned urban construction land range and planned ecological protection areas are the main external factors that play a constraining role.

The findings of this study suggest that to coordinate the relationship between urban expansion and water environment protection, we should be more careful with the planning policies that related to the delineation of flood storage area, ecological protection area and urban growth boundary. We also should pay more attention on controlling the urban development around the primary rivers as it has a more significant driving effect on land expansion. We expect these findings can provide a theoretical basis for the harmonious coexistence of cities and water environment.

Funding. 1. National Natural Science Foundation of China Youth Science Fund Project (52208040) "Study on the Associated Mechanism and Scenario Simulation Method of Land Use Layout under the Constraint of Water Resource Environment in the 'Basin-City'"

2. Ministry of Education Humanities and Social Sciences Research Youth Project (22YJCZH066) "Research on Coupled Water Resource Environment's Beijing-Tianjin-Hebei Urban Spatial Growth Simulation and Coordinated Planning Strategy".

References

1. Zhiqiang, W.: On urban planning and its ecological rationality in the new era. Urban Plan. J. **03**, 19–23 (2018)
2. Guoping, X.: Research on the Evolution of Urban Form in China Since the 1990s. Nanjing University, Nanjing (2005)
3. Liang, Z.: Research on Urban Growth Boundary Delimitation and Management Based on Ecological Security Pattern. Zhejiang University (2018)
4. Lei, Y., Flacke, J., Schwarz, N.: Does urban planning affect urban growth pattern? A case study of Shenzhen, China. Land Use Policy **101**, 105100 (2021)
5. Luo, J., Wei, Y.H.D.: Modeling spatial variations of urban growth patterns in Chinese cities: The case of Nanjing. Landsc. Urban Plan. **91**(2), 51–64 (2009)
6. Zhang, Y., Liu, Y., Wang, Y., et al.: Urban expansion simulation towards low-carbon development: A case study of Wuhan, China. Sustain. Cities Soc. **63**, 102455 (2020)
7. Fuzhuang, H.: Study on the Expansion and Spatial Impact Factors of the Central Urban Area of Guangzhou. Guangzhou University (2016)
8. Qiurong, X.: Construction and Application of Urban Development Boundary Model Based on Ecological Safety in Beijing. China University of Geosciences, Beijing (2019)

9. Hao, W., Zhixiong, M., Shiyun, L.: Dynamic simulation and analysis of land use change based on improved CLUE-S model—a case study of Zengcheng District, Guangzhou City. J. South China Normal Univ. (Nat. Sci. Ed.) **47**(06), 98–104 (2015)

10. Guiping, W., Yongnian, Z., Xuezh, F., et al.: Improvement of CLUE-S model and dynamic simulation of land use change—a case study of Yongding District, Zhangjiajie City. Geogr. Res.. Res. **29**(03), 460–470 (2010)

11. Peng, Z.: Measurement and Evaluation of Urban Residential Space Form. Wuhan University (2015)

12. Mead, R.: Models for interplant competition in irregularly distributed populations. Models for interplant competition in irregularly distributed populations. Int. Symp. Stat. Ecol. New Haven (1969)

13. Ord, K.: Estimation methods for models of spatial interaction. J. Am. Stat. Assoc. **70**(349), 120–126 (1975)

14. Dormann, C.F.: Assessing the validity of autologistic regression. Ecol. Model. **207**(2), 234–242 (2007)

15. Dinghua, O., Jianguo, X., Xingzhu, Y., et al.: Theories, Methods and Applications of Landscape Ecological Security Pattern Planning, p. 160. Science Press, Beijing (2019)

16. Linlin, Z.: Multi-scale Measurement, Inherent Mechanism and Control Research of Urban Sprawl in China during the Transition Period. Zhejiang University (2018)

17. Puertas, O.L., Henríquez, C., Meza, F.J.: Assessing spatial dynamics of urban growth using an integrated land use model. Application in Santiago Metropolitan Area, 2010–2045. Land Use Policy **38**, 415–425 (2014)

Impact of Residential Land Use Function Mix on Residents' Travel Behavior from the Perspective of Carbon Emissions: A Case Study of Gaoshui District in Mianyang, China

Chunrong Zhao[1(✉)], Li Yan[1], Juanlin Fu[1], Huiyun Peng[1], Yanling He[1],
Xiaolan Wang[1], Yanping Yi[1,2], Jiaxin Li[1], Guoyang Hai[1], and Wenyan Feng[1]

[1] School of Civil Engineering and Architecture, Southwest University of Science and Technology, Mianyang 621010, China
406986617@qq.com
[2] Economic Cooperation Bureau of Renshou County, Meishan 620000, China

Abstract. Residential areas, as the main places for urban population activities, are the basic spatial units for reducing carbon emissions. The mix of residential land use functions is closely related to residents' daily travel. This paper takes the Gaoshui district of Mianyang city as an example, uses carbon emissions as a measurement tool, constructs a family travel carbon emission quantitative estimation model and a land use function mix calculation model, compares and analyzes the differences in residents' travel behavior in each sample area, and uses a linear regression model to analyze the impact of function mix on residents' travel behavior. The results show that there are spatial differences in residents' shopping, entertainment, medical treatment, commuting and other travel behaviors in the 12 sample areas; the function mix has no significant impact on family commuting carbon emissions, but has a significant negative impact on non-commuting carbon emissions, and mixed development of residential areas has a positive significance for reducing carbon emissions.

Keywords: Residential function mix · Travel behavior · Household carbon emissions · Linear regression model

1 Introduction

CO_2 emissions have seriously affected the healthy development of the socio-economy [1], and the low-carbon urban development model has reached a global consensus [2]. According to statistics, carbon emissions from the household life of urban residents in the United States account for 40 per cent of total urban carbon emissions [3]; in most Japanese cities, household carbon emissions account for more than 15 per cent of total urban carbon emissions. As a major means of travelling, private cars has the highest carbon emission intensity compared to other means, and so it is particularly important to guide residents' travel behavior [4]. Residential area can achieve carbon emission reductions through spatially compact and land use mixed development [5]. Therefore, how to optimize the functional space of residential areas, change residents' travel modes, and reduce carbon emissions has become one of the hot topics for scholars at home and abroad.current research mainly focuses on the study of the impact of urban spatial morphology [6], urban spatial compactness [7, 8], and spatial structure on residents' travelling or transport carbon emissions [9], and some of the studies take into account the relationship between land use mixing and carbon emissions [10]. The mix of residential functions affects residents' travel distance, mode of transportation, frequency of travel, etc. Brown believes that single land use will significantly increase residents' commuting, schooling, shopping, entertainment and other daily travel distances [11]. Kim and Brownstone, through analyzing residents' travel behavior survey data, believe that high-density development of residential land is positively effective in reducing residents' daily travel [12]. Simma and Naess found that residents' travel behavior is inevitably linked to the density of service facilities around the residential area and the quality of service [12–15]Li Ming found that moderate land use mixing can significantly increase the proportion of slow travel [16]; Xu Siyang and others believe that mixed-function development can reduce long-distance travel by private cars and other modes of travel [17]; Zhai Qiang believes that mixed communities characterized by diverse land use functions have lower demand for car traffic and carbon emissions [18].

In conclusion, most of the current studies directly explore the relationship between urban spatial environment and residents' carbon emissions, and there are fewer studies on how urban spatial environment affects carbon emissions, and not enough attention has been paid to the interaction between land use mixed and carbon emissions. Taking the Gaoshui district of Mianyang city as an example, with carbon emission as a measurement tool and travel behaviour characteristics as an entry point, This paper studies the intrinsic relationship among residential behavior, carbon emission and functional mix, constructs a family travel carbon emission estimation model and a land use function mixing degree calculation model, compares and analyzes the differences in residents' travel behavior in each sample area, and uses Linear regression mode to analyze the impact of function mixing degree on residents' travel behavior.

2 Methods and Data

2.1 Research Object and Data

The research object is located in the Gaoshui district of Mianyang, China. The area as a whole presents characteristics such as diverse community types, large scale differences, uneven distribution of service facilities, and diverse modes of travel for residents. This study fully considers factors such as topography, road traffic, population concentration, community spatial features, etc., and divides the Gaoshui district into 12 sample areas as the basic research units(Fig. 1). Each research unit varies in size and community type. The northern communities of the research area have relatively backward supporting facilities; the central communities have built education, public activity spaces, commercial entertainment facilities, etc., with relatively complete supporting facilities; the south is adjacent to the city center, with convenient transportation.

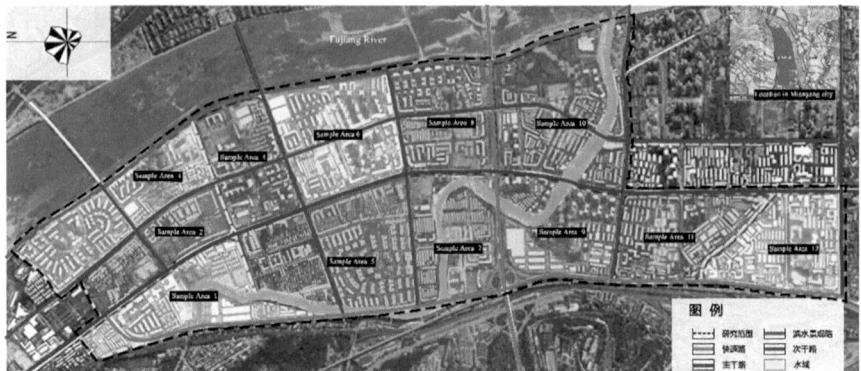

Fig. 1. Research Area Location and Sample Division

The data mainly includes the following three categories:

1) Travel behavior, divided into commuting and non-commuting travel behavior, including travel mode, travel time, travel distance, travel frequency or frequency, and other travel elements [19].
2) Household carbon emissions, referring to the total carbon emissions produced by residents' one-way travel from home to destination within a week, essentially the CO_2 produced by the energy consumption of transportation tools during use [20].
3) Function mixing degree, referring to the mixed state of no less than two functions within a certain spatial range, is a direct reflection of the high or low mixed state [21].

2.2   Research Method

2.2.1   Carbon Emission Calculation

The calculation of the total carbon emissions from the sample family's travel [22–24], see formula 1.

$$E = nW_i S_i. \tag{1}$$

where E —— Total carbon emissions from using this mode of transportation in a week (g/week)

n —— Total number of one-way trips in a week (times)

W_i —— Carbon emission intensity of transportation mode i (g/person km) [13].

S_i —— Distance traveled using transportation mode i (km)

The calculation of household travel carbon emissions, see formula 2:

$$U = E_1 + E_2 + E_3 + \ldots\ldots \tag{2}$$

In the formula, U represents the total carbon emissions from all family members' travel in a week (g/week).

E_1 represents the total carbon emissions from different travel purposes of family member one in a week (g/week).

E_2 represents the total carbon emissions from different travel purposes of family member two in a week (g/week).

E_3 represents the total carbon emissions from different travel purposes of family member three in a week (g/week).

2.2.2 Calculation of Function Mix Degree(FMD)

Two calculation methods are used: function mix entropy and function mix model. When there are many possible states of information, information entropy is usually used to judge the average amount of information [25]. Function mix entropy is a calculation method formed by referring to information entropy to judge the average distribution state of different land uses in a certain area, for specific calculation see formula 3 [26]:

$$W = -\sum_{i=1}^{n} \frac{k_i * \ln(k_i)}{\ln(n)} \tag{3}$$

In the formula, W represents the FMD of sample area i, n represents the types of building use functions in sample area i, and k_i represents the proportion of the building area of land use function i in the total building area of the sample area. The land use functions involved in this paper mainly include three categories: residential function, service function, and work function, and the service function covers types such as medical services, commercial services, entertainment services, and educational services.

The second more commonly used is the function mix model proposed by Dr. Hoek, and scholar Wang Haofeng further divided it based on this [27]. The degree of functional mix is determined quantitatively according to the proportion of different functional lands in the land unit, and the land use function mix status of the sample area is obtained by comparing the K values, where the calculation of K value is shown in formula 4:

$$K_i = \frac{S_i}{S_{total}} \tag{4}$$

In the formula, K_i represents the proportion of the i-th type of functional land in the sample area, i represents the type of land use function (such as residential function, service function, work function), S_i represents the building area of this land use function, and S_{total} represents the total building area in the sample area.

The first method is a quantitative judgment of the degree of functional mix; the second one of functional mix model research is the proportion of different functional buildings in all buildings, used for intuitive judgment of functional mix type. Combining them, this paper first determines the type of functional mix, and then uses the entropy method to calculate the degree of functional mix quantitatively.

3 Results and Discussion

3.1 Calculation Results

3.1.1 Carbon Emission Calculation

According to Sect. 2.2.1, the carbon emissions from household travel in each sample area are shown in Table 1.

Table 1. Summary of carbon emissions from household travel in each sample area

Sample area number	Commuting carbon emissions (g/week)	Non-commuting carbon emissions (g/week)	Total household travel carbon emissions (g/week)
1	537040.30	195253.80	732294.10
2	809913.11	236803.54	1046716.65
3	1289570.13	812049.73	2101619.86
4	909012.74	313112.79	1222125.53
5	1231914.29	121375.46	1353289.75
6	998980.19	147300.21	1146280.4
7	484871.16	101683.83	586554.99
8	494638.52	58384.13	553022.65
9	929043.18	165598.46	1094641.64
10	1667420.89	321560.13	1988981.02
11	1191931.85	189226.35	1381158.2
12	18422.68	60476.54	78899.22

Table 1 shows that the highest total carbon emissions from household travel are from sample 3, followed by samples 10, 5, and 11. Except for sample 4, the total carbon emissions from commuting in other sample areas are higher than those from non-commuting. The highest total carbon emissions from commuting are from sample 10, followed by samples 3, 5, 11. These four sample areas are closely related to residents choosing high-carbon travel tools and long-distance travel. The highest and lowest total carbon emissions from non-commuting travel are from samples 3 and 4, respectively.

3.1.2 Calculation of Function Mix

The closer the function mix is to 1, the more reasonable the distribution of various functional lands in the residential area (Table 2).

Table 3 shows that there are fewer samples with a functional mix above 0.6 (samples 2, 10, 12), among which sample 10 has the highest functional mix (0.784). In addition to the residential function, this sample area also has a large-scale park square land and some work function land. The functional mix between 0.4-0.6 belongs to the residential area with a medium functional mix (samples 4, 5, 7, 9, 11). Those with a functional mix less than 0.4 belong to the residential area with a low functional mix (samples 1, 3, 6, 8), and sample 3 has the smallest functional mix (0.160). This sample area is mainly residential, with fewer service and work function lands.

Table 2. Overview of function mix in each sample area

Sample Area	Function Mix	Level	Sample Area	Function Mix	Level
1	0.381	Low	7	0.461	Medium
2	**0.626**	High	8	0.289	Low
3	0.160	Low	9	0.448	Medium
4	0.460	Medium	10	**0.784**	High
5	0.441	Medium	11	0.574	Medium
6	0.295	Low	12	**0.659**	High

3.2 Differences in Residents' Travel Behavior from the Perspective of Carbon Emissions

After classifying the carbon emissions in Table 2, we obtained the proportion of carbon emissions from travel behavior in each sample area (Table 3).

Table 3. Proportion of carbon emissions from travel behavior in each sample area (%)

Sample Area Numbe	Shopping	Leisure and Entertainment	Recreation	Medical Treatment	Going to School or Taking Children to School	Going to Work
1	11.10	19.18	2.96	1.23	1.28	64.26
2	2.16	11.53	1.56	1.13	10.81	72.80
3	3.76	17.25	0.72	1.21	5.10	71.96
4	3.58	11.04	1.98	1.62	8.30	73.47
5	1.11	3.39	0.74	0.88	1.84	92.03
6	2.23	5.68	0.48	0.64	0.92	90.05
7	2.79	7.71	2.03	0.42	6.87	80.19
8	2.28	5.18	0.13	0.26	7.85	84.30
9	1.56	4.14	0.29	0.02	10.10	83.89
10	0.69	2.12	0.12	0.07	5.81	91.19
11	0.87	4.47	0.61	0.05	4.36	89.65
12	28.32	32.01	6.23	1.06	4.86	27.52

Table 3 shows that there are differences in the behaviors of residents in the 12 sample areas in terms of shopping, entertainment, medical treatment, and commuting. By comparison, it is found that: first, the carbon emissions from commuting to work by residents in the high-water area are high, indicating that the demand for working around the home has not been met; second, the 1st, 2nd, 3rd, and 4th sample areas in the northern part of the high-water area are far from the commercial center, and there is a lack of necessary commercial facilities inside, which leads to their shopping carbon emissions far higher than other sample areas; third, the carbon emissions from medical trips in the sample areas within 1 km from the hospital are very low, the survey shows that this is related to the medical level of the hospital; fourth, the carbon emissions from school trips in the sample areas with no educational facilities or only kindergartens are the highest.

3.3   The Impact of Functional Mix on Residents' Travel Behavior

3.3.1   Linear Regression Model

Regression analysis is a statistical analysis method to determine the quantitative relationship between two or more variables. According to the characteristics of the research variables (Table 4), a linear regression model is established to analyze the impact of functional mix on household travel carbon emissions (Formula 5). Because the data of each variable cannot be obtained completely with a random sample one, the model is also called sample regression model. The independent variable in this model is the functional mix, and the dependent variable is the carbon emissions from commuting and non-commuting travel. By observing the fit of this model and the significance of each variable, the relationship between functional mix and travel carbon emissions is obtained.

Table 4. Analysis of related variable characteristics

Variable name	Variable type	Variable definition and description
Functional mix	Continuous variable	The spatial mix of living, working, and service functions. The closer the value is to 1, the higher the mix; otherwise, the lower
Commuting carbon emissions	Continuous variable	
Non-commuting carbon emissions	Continuous variable	

$$y_i = a + bx_i + u \qquad (5)$$

In the formula, y_i. is the dependent variable, which refers to travel carbon emissions; x_i is the independent variable, that is, the functional mix; a is a constant term; u is an unobservable random variable.

3.3.2   Function Mix and Household Commuting Carbon Emissions

As shown in Table 5, the R-square of the established linear regression model is 0.15. The R-square is directly related to the number of independent variables and the total sample size. Since this model only considers the function mix, the model is relatively stable.

Table 5. Model Fit Test

Model	R	R-square	Adjusted R-square	Standard Estimation Error	Change Statistics				
					R square Change	Change in F-value	df1	df2	Significance F Change
1	.068a	.15	.13	29328.70556	.15	2.197	1	469	.139

Table 6. Model Variance Analysis

Model		Sum of Squares	df	Mean Square	F	Significance (Sig)
1	Regression	1889639023.570	1	1889639023.570	2.197	.139b
	Residual	403421122792.679	469	860172969.707		
	Total	405310761816.249	470			

a. Dependent variable: Carbon emissions from commuting b. Predicted value: (Constant), Function Mix

Tables 6 and 7 show that the independent variable, function mix, has no significant impact on the carbon emissions from household commuting. That is, the carbon emissions from household commuting do not change significantly with the change in the function mix. This may be because the places for work and school generally do not change easily, and there are fewer office areas in high-water areas, so they are not affected by the change in function mix.

Table 7. Model Estimation Results

Model		Non-standardized Coefficients		Standardized Coefficients	T	Significance (Sig)
		B	Standard Estimation Error	Beta		
1	(Constant)	16982.666	3959.038		4.290	.000
	Function Mix	11898.141	8027.542	.068	1.482	.139

3.3.3 Function Mix and Household Non-Commuting Carbon Emissions

As shown in Table 8, the R-square of the established linear regression model is 0.19, and the adjusted R-square is 0.17. Similarly, the R-square is directly related to the number of independent variables and the total sample size. Since this model only considers the function mix, the model is relatively stable.

Table 8. Model Fit Test

Model	R	R Square	Adusted R Square	Standard Estiated Error	Change Statistics data				
					R Square Change	Change of F value	df1	df2	Significance F Change
1	.137a	.19	.17	18552.0	.19	9.010	1	469	.003

As can be seen from the variance analysis of Model 9, the significance is 0.03, which is less than 0.05, therefore the established linear regression model is valid (Table 9).

Table 9. Variance Analysis of the Model

Model		Sum of Squares	df	Mean Square	F	Significance (Sig)
1	Regression	3100958254.651	1	3100958254.651	9.010	.003b
	Residual	161420509227.489	469	344180190.251		
	Total	164521467482.140	470			

From the estimation results of the Model 10 shown in Table 10, it can be seen that the independent variable, functional mix, has a significant impact on the carbon emissions of non-commuting family trips. The B value is negative, indicating that the functional mix has a significant negative impact on the carbon emissions of non-commuting family trips. That is, the higher the functional mix in high-water areas, the lower the carbon emissions from residents' non-commuting trips. This may be because in areas with a high functional mix, commercial, medical, educational and other service facilities are evenly distributed. Residents can meet their daily needs in the vicinity of their homes, thereby reducing the carbon emissions from daily activities such as shopping, entertainment, and dining.

Table 10. Estimation results of Model

Model		Unstandardized Coefficients		Standardized Coefficients	T	Significance (Sig)
		B	Standard Estimation Error	Beta		
1	(Constant)	14015.662	2504.319		5.597	.000
	Functional Mix	-15241.849	5077.882	-.137	-3.002	.003

4 Conclusion

Currently, there are studies analysing the factors influencing carbon emissions of households in settlements in terms of their internal attributes (e.g., education level of family members, household income, age group size, quality of life, etc.), but there is less literature analysing carbon emissions of urban households in terms of spatial dimensions, such as land use, neighbourhood scales, and land use functions. Therefore, exploring the effects and mechanisms of residential functional mixing on travel behaviour and carbon emissions is of positive significance in enriching the theoretical research on the spatial environment and carbon emissions. Meantime, comparing and analyzing the residents' travel behavior, household carbon emissions and functional mixing in the study area with a large number of old communities, it provides an effective way to realize the urban renewal and function repair.

This study shows that there are spatial differences in residents' shopping, entertainment, medical treatment, commuting and other travel behaviors in the 12 sample areas. The functional mix has no significant impact on the carbon emissions of family commuting trips, that is, the carbon emissions of residents' family commuting trips do not change significantly with the change of land use functional mix; it has a significant negative impact on non-commuting travel carbon emissions. Thus, it is positive to reduce carbon emissions from residents' shopping trips, recreational trips, medical trips, etc. through residential mixed development. Specific measures are as follows: Firstly, community life circle oriented mixed use development, which will rationally allocate commercial facilities, cultural facilities, medical facilities, educational facilities, health facilities and so on within the walking scale. Secondly, it is suggested that adjustment the settlement business through analysis of the current situation, which owing to different residential area of FMD differing in optimization programme, and it must also consider them specifically when FMD is not much different with same functional mixing type. Thirdly, creating a community commercial street that integrates family leisure shopping, cultural entertainment, catering and retailing, which not only meets the residents' desire to solve their shopping needs in the neighbourhood, but also reduces the frequency of going out to shop, shortens the distance of shopping trips and reduces the use of motor vehicles. Finally, the creation of shared community living spaces, such as community plazas, parks and green spaces, provides space for leisure and recreation, which addresses residents' needs for outdoor activities, and thus contributes to the mixed development of settlement functions.

In this study, the content involved is extensive and the impact relationship is complex. The shortcomings are as follows: (1) The travel data within a week is random, and different seasons or weather factors will affect the residents' travel mode or frequency. (2) When calculating the functional mix, some buildings have diverse uses, which are difficult to define, and this will affect the overall judgment of the functional mix.

Funding. This research was supported by Sichuan Science and Technology Program (Grant No. 2023NSFSC1051).

References

1. Kong, F., Wang, Y.F., Lv, L.L.: Progress and prospect of China's response to climate change under the background of global change. J. Anhui Agric. Sci. **46**(1), 18–23 (2018)
2. Gu, C.L., Tan, Z.B., Wan, L. et al.: Progress in research on climate change, carbon emissions and low carbon urban planning. Urban Plann. J. (3), 38–45 (2009)
3. Glaeser, E.L., Kahn, M.E.: The greenness of cities: carbondioxide emissions and urban development. J. Urban Econ. (3), 404–418 (2010)
4. Chen, F.: Research on Low Carbon Urban Development and Countermeasures-Empirical Analysis of Shanghai. China Architecture & Building Press, Beijing (2010)
5. Yu, P., Chen, X.Q., Ren, F.: Construction practice of low carbon residential areas abroad—a case study of the "city of tomorrow" demonstration residential area in Sweden. China Real Estate (4), 76–80 (2011)
6. Qiao, Y., Jiao, L.M.: Impacts of urban form on transportation carbon emissions from a global perspective. Geogr. Res. **42**(12), 3202–3218 (2023)
7. Guo, R., Wu, X.C., Rui, F.: Research on the impact of town space compactness on carbon emissions: taking Changxing county as an example. J. Hum. Settlements West China **38**(6), 122–128 (2023)
8. Yang, S.: The Correlation between Urban Spatial Compactness and Residents' Carbon Emissions Intensity: Based on the Analysis of 30 Large and Medium-sized Cities. Xi'an University of Architecture and Technology (2021)
9. Han, S.S., Miao, C.H., Li, Y.C.: Effects of urban polycentric spatial structure on carbon emissions in the Yellow River Basin. Geogr. Res. **42**(4), 936–954 (2023)
10. Xu, Z., Li, C., Niu, L.: Decoupling relationship between land mixed use and carbon emissions in Hohhot-Baotou-Ordos-Yulin urban agglomeration. Res. Environ. Sci. **35**(1), 299–308 (2022)
11. Brown, M.A., Southworth, F., Stovall, T.K.: Towards a Climate-Friendly Built Environment. Pew Center on Global Climate Change, Arlington, VA (2005)
12. Kim, J., Brownstone, D.: The impact of residential density on vehicle usage and fuel consumption. Economics 31–51 (2010)
13. Simma, A., Axhausen, K.W.: Interactions between travel behaviour, accessibility and personal characteristics: the case of the Upper Austria Region. Eur. J. Transp. Infrastruct. Res. 179–197 (2003)
14. Naess, P.: Accessibility, activity participation and location of activities: exploring the links between residential location and travel behavior. Urban Stud. 627–652 (2006)
15. Maat, K.: Influence of the residential and work environment on car use in dual-earner households. Transp. Res Part A 654–664 (2009)

16. Li, M.: Empirical analysis of the relationship between land use mix, block scale and traffic travel—case study of Zhangjiagang. China Urban Planning Society. Diversity and Inclusion—2012 China Urban Planning Annual Conference Proceedings (05. Urban Road and Traffic Planning). China Urban Planning Society: China Urban Planning Society, vol. 6 (2012)

17. Xu, S.Y., Chen, Z.G.: Mixed function development: towards a sustainable urban form of low carbon city. In: The 9th China Urban Housing Seminar, pp. 107–114. University of Hong Kong, Hong Kong (2011)

18. Zhai, Q.: Research on Urban Block Mixed Function Development Planning. Huazhong University of Science and Technology Wuhan (2010)

19. Yao, Y.: Research on the Impact of Built Environment on Urban Residents' Travel and Carbon Emissions. Harbin Institute of Technology, Harbin (2015)

20. Liu, Z.J.: Applied Research on the Impact of Internal Family Attributes and External Factors on Family Commuting Carbon Emissions. Chang'an University, Xi'an (2013)

21. Zhai, Q.: Research on Urban Block Mixed Function Development Planning. Huazhong University of Science and Technology, Wuhan (2010)

22. Chai, Y.W., Xiao, Z.P., Liu, Z.L.: Comparative analysis on CO_2 emission per household in daily travel based on spatial behavior constraints. Geogr. Sci. (7), 843–849 (2011)

23. Zheng, B.H.: Research on urban carbon emission intensity based on spatial compactness—a case study of Changsha. China Urban Planning Society, Nanjing Municipal Government. Transformation and Reconstruction—2011 China Urban Planning Annual Conference Proceedings. China Urban Planning Society, Nanjing Municipal Government: China Urban Planning Society, vol. 8 (2011)

24. Liu, P.: Research on the impact of urban spatial parameters at the community scale on daily family travel carbon emissions. Urban Planning Society of China. Diversity and Inclusion—Collection of Papers from the 2012 Urban Planning Annual Conference in China (06. Housing Construction and Community Planning) [C]. Urban Planning Society of China: Urban Planning Society of China, vol. 22 (2012)

25. Liu, C.X., Zhu, Q.: Research on the space-time differentiation of manufacturing structure in China based on Shannon Entropy. Urban Dev. Res. (4), 20–25 (2005)

26. Liu, C.: Study on the Impact of Land Use Characteristics in Chengdu on Residents' Travel Modes. Southwest Jiaotong University, Chengdu (2016)

27. Wang, H.F., Shi, S., Rao, X.J.: Street morphology and urban density. In: Proceedings of the 2013 Urban Planning Annual Conference in China, pp. 1–16 (2013)

Evaluation of Social Value in Urban Parks Through Implementation of SolVES Model in Mianyang People's Park

Mingying Zeng[1], Qian Ma[2], Kar Kheng Gan[3], and Zhanglei Chen[1,4](✉)

[1] School of Civil Engineering and Architecture, Southwest University of Science and Technology, Mianyang 621010, China
20181501001@cqu.edu.cn

[2] Yuechi County Bureau of Natural Resources and Planning, Guang'an 638300, China

[3] TCL Studio, 109 Grote Street, Adelaide, SA 5000, Australia

[4] School of Architecture and Urban Planning, Chongqing University, Chongqing 400044, China

Abstract. Urban parks bring multiple benefits in terms of ecology, economy, culture, and well-being. While evaluation of the ecosystem service value is conducive to integrating urban functions and green space, a lack of human-nature interactions leads to neglect in knowledge and assessment of the social value of urban parks. In this study, Visitor Employed Photography (VEP) and the Social Value of Ecosystem Services (SolVES) model were utilized to connect visitors' immediate perceptions with landscape elements, subjective will, and the geographical environment. This approach aimed to assess the social value of parks and quantify the public's preferences in landscape layout. The results compare the relative importance of nine social values and verify that different environmental variables affect types of value indices. It is observed that the public is particularly aware of the social value generated by good planting and landscaping, as well as by proximity to water. In addition, the type of environmental landscape also has a crucial impact on perceptions. Bodies of water, characteristic landscapes, and well-maintained convenient roads emerge as key elements in people's landscape preferences. The conclusions drawn from this study provide useful information for park management and landscape planning, particularly from the perspective of social value growth. Most importantly, they serve as a reference for future research.

Keywords: urban park · social value of ecosystem services · public preference and perception · VEP · SolVES · landscape optimization and management

1 Introduction

Ecosystems are life-support systems that create and sustain the environment necessary for human survival, forming the basis for human social progress, and providing services such as leisure, recreation, and aesthetic enjoyment [1]. The degradation of ecosystem service functions is a major environmental crisis facing the world today [2, 3], especially in the context of rapid urbanization. With an increasing number of urban dwellers having

© The Author(s) 2025
B.-J. He et al. (Eds.): UCSUD 2023, LNCE 559, pp. 402–423, 2025.
https://doi.org/10.1007/978-981-97-8401-1_29

limited daily interaction with urban ecosystem services, there exists a contradiction between people's aspirations for a better environment and the pursuit of sustainable development goals [4]. Addressing this contradiction requires a scientific approach in quantifying ecosystem services, as such, policymakers and the public then can better understand the significant services the urban ecosystem provide [5, 6].

The valuation of ecosystem services has emerged as a crucial tool for policymakers to reassess environmental priorities, offering a quantifiable means to raise awareness of ecological civilization through recognition of the value of natural resources [7, 8]. Software models such as i-Tree and InVEST have proven instrumental in assessing the regulating and supporting functional services provided by ecosystems [9–11]. Furthermore, research on the social value of parks has made notable progress, particularly in evaluating parks rich in natural, ecological, and cultural landscape resources, often employing monetization principles like conditional value and hedonic value [12–16]. Most recently, studies have recognized the necessity of incorporating quantitative and spatially explicit social value types into the valuation of ecosystem services, [17, 18], leading to the adoption of models like the Social Value of Ecosystem Services (SolVES) [19, 20].

Valuing ecosystem services in inner-city settings poses significant challenges due to variations in park types, users, and cultures, resulting in notable research gaps [21, 22]. While different types of park ecosystems may offer similar services, the diversity in use patterns, availability, accessibility, and quality of parks, as well as the involvement of various stakeholders, significantly impacts value assessment [23, 24]. Unlike forest and wetland parks, urban parks are heavily used by the public, closely connected to residents, and are in a symbiotic relationship with urban functions in city centers, making them ideal for generating social value [19, 25]. Nevertheless, research on urban parks is made difficult by the multitude of stakeholders involved [26, 27], leading to increased complexity in social and environmental data [21, 22]. To date, there has been relatively little quantitative study of public perceptions regarding landscape preferences in urban parks in China.

This study employs the Social Value of Ecosystem Services (SolVES) model, leveraging the public's visualization capabilities to enhance understanding of social value types and calculate environmental layer indicators. Unlike traditional economic relationship algorithms, the SolVES model offers a more comprehensive approach. Scholars have expanded the model's utility by integrating it with tools such as Maxent Maximum Entropy modelling software to explore the relationship between forest ecosystem service values. Moreover, scenario analysis and simulation using value transfer functions have been conducted to enhance the model's effectiveness and transplantability. Through mapping cold and hot spots of social value, some researchers have also evaluated and categorized 12 ecosystem service values in conjunction with public preferences [28]. As a result, the SolVES model has found international applications, particularly in established green space types, such as forests, national parks, coastal areas, marine park reserves, and agricultural ecosystems. Yet, its application in assessing the social value of urban parks in China remains undiscovered. To address this gap and ensure objective authenticity of data whilst minimizing public perception bias, this study combines Visitor Employed

Photography (VEP) with SolVES and appropriate underlying environmental variables to assess the social value of urban parks.

Hence, the aim of this article is to examine the linkages between urban park ecosystem services and park landscape environments, with the goal of understanding public perceptions and preferences relevant to landscape planning. This study addresses the following research questions: (1) How is the social value of urban parks expressed? (2) What is the relationship between park environmental variables and social value types? (3) How do plant landscape features affect value preferences?

2 Materials and Methods

2.1 Study Area

Mianyang People's Park, located in the oldest business district in the central area of Fucheng, Mianyang, Sichuan Province and was built in 1930. Recognized as the "No. 1 park in northwest Sichuan," it has undergone several construction projects over the years, resulting in the creation of five distinct functional zones. These includes spaces dedicated cultural education, ornamental tours, and children's play. The park's landscape is interconnected by prominent features such as the Cultural Square, Liberation Monument, and Artificial Lake, providing visitors with a cohesive and immersive experience. With such diversity of entertainment, leisure amenities, and lush greenery, the park is designed to enhance visitors' experience (Fig. 1).

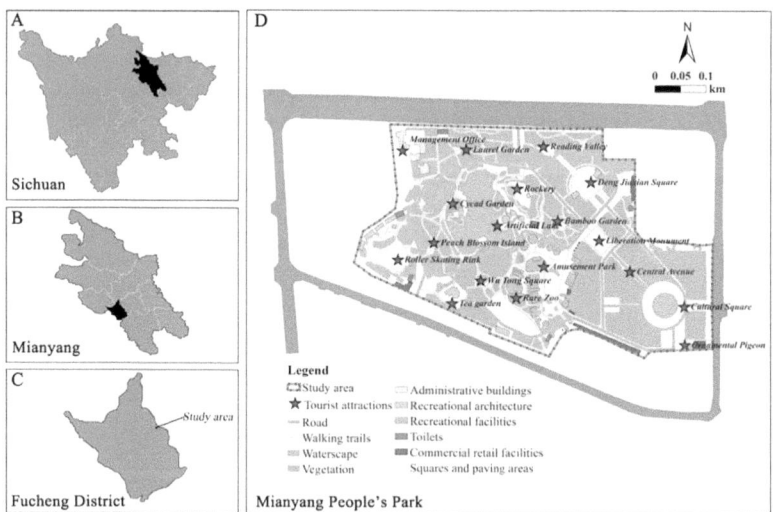

Fig. 1. Location analysis of the study area (A) Sichuan Province, (B) Mianyang City, (C) Fucheng District and the general plan of the study area (D) Mianyang People's Park.

2.2 Data Sources

VEP landscape preference data acquisition

Given that urban parks primarily serve public use, and considering the central role of visual representation in landscape perception [29], this study uses the VEP method proposed by [30] to acquire landscape preference data and identify research sites for assessing social value. VEP leverages public attitudes to understand their perceptions of resources and the environment. The results provide valuable information for understanding public preferences regarding park landscapes and social services [30, 31].

The participants in the experiments were primarily employed individuals using a product with the iOS mobile operating system (i.e., an iPhone or iPad), alongside with the proprietary Map Plus application to navigate the park and take photographs. This method made it possible to maximize the information gathered from the point in time at which the photographs were taken, drawing on the photographer's sensory perception and experience and capturing differences in the landscape preferences of visitors [32]. By minimizing the impact of temporal factors, this approach reduces the subjectivity and variability biases associated with non-locality [33, 34], thereby increasing the objectivity and authenticity of the data.

The experiment lasted for one week. To ensure that the sample size was appropriate for the implementation of VEP, 34 volunteers were selected [35–39]. The volunteers comprised residents of Mianyang aged 18–45 years, ensuring adequate gender parity and a diverse range of social characteristics, as well as backgrounds. Preference was given to those with iPhone devices. Before the final selection, potential participants were given a brief introduction to the filming software and a description of the nine social value types to ensure that they understood the definitions and were clear about the activities and tasks they would be required to perform in the experiment. Selected participants received compensation via online transfer for their participation.

At the end of the experiment, a total of 671 valid images were collected and then classified by landscape type according to the criteria outlined in Table 1. In cases where multiple images targeted the same object, the initial photograph taken was selected. In instances where a single image depicted multiple landscape types without a specified main landscape, grid-based analysis was employed to select the image with the widest coverage as the main landscape type. Finally, kernel density calculation was carried out on the photographed points to identify areas of preferred landscape hotspot, facilitating the integration of spatially distributed preferred landscapes for the selection of social value points.

SolVES model data acquisition

The SolVES model consists of three main submodules: social value, value mapping, and value conversion mapping [28]. The model uses questionnaires and interviews to obtain social value survey data on public attitudes and preferences, taking environmental factors into consideration. It incorporates the Maxent model to estimate the social value of ecosystem services in the study area. Through the process depicted in Fig. 2, the output assigns a non- monetized value index (ranging from 1 to 10) to each specific spatial location. In this study, the model is used to investigate the social value of Mianyang People's Park, focusing on the nine common value types, including aesthetic, historical

Table 1. Ten predefined landscape types. (adapted from [40])

Landscape	Description
Landform	Platforms, steps, lawns, squares, etc.
Vegetation	Trees, shrubs, herbal flowers, etc.
Animal	Birds, fishes, pigeons, etc.
Waterscape	Fountains, lakes, etc.
Architecture	Buildings, Pavilions, bridges, etc.
Art installation	Sculptures, stone scenes, etc.
Service facility	Rest seats, lamps, trash cans, etc.
Pedestrian system	Pedestrian trails, boardwalks, etc.
Athletic facility	Walking fitness, running field, roller skating, etc.
Recreational activity	Children's entertainment facilities, picnics, etc.

and cultural, and recreation and leisure, selected based on a review of relevant literature and the nature of the study area (Table 2).

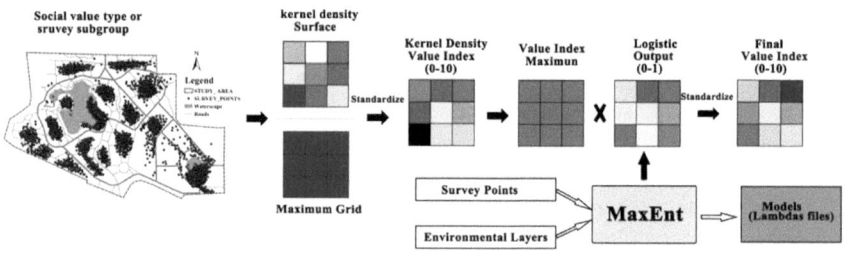

Fig. 2. Work flow of data processing in SolVES

Survey data

In accordance with requirements of the SolVES model, a questionnaire survey was conducted on the use of the park by visitors, taking into account the natural environment. A total of 205 questionnaires were distributed, and 200 valid questionnaires were returned, giving a valid return rate of 97.56%. The questionnaire consisted of three parts: (1) items on respondents' play characteristics and feelings; (2) items on respondents' allocation and marking of social value points; and (3) items on the basic characteristics of the respondents' circumstances.

Environmental GIS data

The SolVES model requires geospatial data and tabulated data (Table 3). First, the survey data were sorted, processed, and loaded into tables. Given that the park is open to the general public, we used standard demographic characteristics such as gender and level of education during the data analyzing process. Next, the geospatial data take the form of a Shapefile of the study area and a raster dataset. The Shapefile was obtained by

Table 2. Descriptions of the 9 types of social values employed in this study (adapted from [41–43])

Social value type	Description
Aesthetic	A place with a beautiful view
Biodiversity	A place with many kinds of animals and plants
Economic	Places with more small shops and rich retail
Historical and cultural	There are historical sites and places where you can learn about natural history
scientific research and educational	It can provide tourists and citizens with places to popularize science education and promote culture
Life sustaining	Places that can purify the air and adjust the climate
Recreation	A place for walking slowly
	A place for dancing, singing, and musical instrument sketching
	Children's play area
	A place for drinking tea and chatting, playing cards and chess, and amusing birds
	A place for meditation and reading
Spiritual	The surrounding environment is quiet and natural, a place that can make people sink their hearts, forget their worries, and purify their souls
Therapeutic	A place where you can exercise, soothe your body and mind, and release stress

using ArcGIS geographic information services to spatially align the park and carry out map vectorization operations to depict and digitize the social value points marked by the respondents, thereby obtaining a social value point layer. The raster dataset (i.e., the extraction of the park's environmental elements) was used as in previous comparable studies. Additionally, we calculated the Euclidean distances for the road, water, and toilet facilities, using DTR, DTW, DTT, DTCL. On top of that, the amount of infrastructure (including seats, dustbins, signs, and lighting) and greenery, as well as the type of parkland coverage (vegetation, paving, water, profit-making facilities, and infrastructure facilities) were added to the raster data.

2.3 Selection of Value Points

Processing of the VEP landscape preference data according to the criteria in Table 1 yielded the statistics for the main landscapes captured in the 671 photographs (Fig. 3A). We then carried out a kernel density analysis of this element using the coordinates of the photographs taken (Fig. 3B). And 15 areas were selected as the social value marker points for subsequent assessment.

Table 3. Feature class layer and environmental data layer of the social value model.

Types	Layer name	Formats	Description
Tables	ATTITUDE_TYPES	Table	Attitude types for use of PPMC
	USE_ATTITUDE		Allocation of attitude
	USE_TYPES		Use types of PPMC
	VALUE_ALLOCATION		Allocation of social-value types
	VALUE_TYPES		Social-value types of PPMC
Feature Classes	SURVEY_POINTS	Shapefile	Point layer of PPMC with linear units' projection
	STUDY_AREA		Polygon layer of PPMC with linear units' projection
Environ-mental Layers	DTR	Raster	Horizontal distance to nearest road
	DTT		Horizontal distance to nearest toilet
	DTW		Horizontal distance to nearest waterscape
	DTWT		Horizontal distance to nearest walking trail
	DTHL		Horizontal distance to nearest historic landscape
	NOSE		Number of seats
	NOTC		Number of trash cans
	NOSI		Number of signs
	NOL		Number of lamps
	GC		Green coverage rate of PPMC
	GR		Greening rate of PPMC
	LC		Land cover type of PPMC
	HPD		Hard pavement density of PPMC

3   Results

3.1   Survey Sample Statistics

Analysis of the sample characteristics shows that the proportion of male and female respondents is almost equal (male = 46.5%, female = 53.5%); that all age groups are represented, with 76.5% under 40 years old); and that the education level is generally

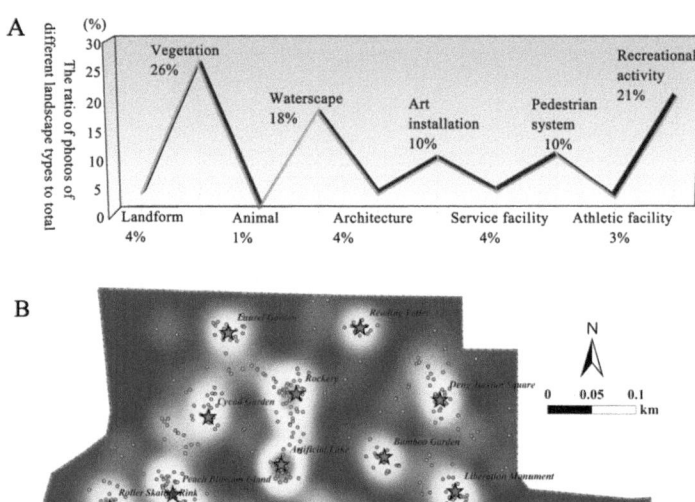

Fig. 3. Proportional of each landscape type of volunteer photos (A), Nuclear density analysis map of volunteer photo shooting in Mianyang People's Park (B)

high, with 90.5% of the respondents having a specialist or undergraduate degree or above. Thus, it is reasonable to assume that the respondents are generally young, capable of understanding the questionnaire, and likely to exhibit active thinking with a wide range of concerns. In addition, the monthly household income of the respondents is mainly concentrated in 3000–5000 (32.5%) and 5000–7000 (34%), and their occupations cover all kinds of jobs in the society, with the largest number of ordinary workers (32.5%). Finally, respondents are mostly local residents of Mianyang (68.5%), and usually choose non-working afternoons to enter and use the parks with their families, classmates, or coworkers.

The survey yielded a total of 3,493 social value points, and kernel density analysis showed that these points were concentrated in several areas, with the densest being Cultural Square, Rockery, and Artificial Lake, followed by Peach Blossom Island and Liberation Monument (Fig. 4). To identify meaningful value types for mapping and analysis, we used the SolVES model to generate mean nearest-neighbor statistics for the nine types of social value (Table 4). A larger M-VI indicates a stronger preference by the survey group, while an R-ratio of less than 1 and a smaller Z-score value reflects a point pattern of significant clustering identified through the mean nearest-neighbor statistic in a full spatial random test on the point data. This procedure focuses on users' attention to types of social value, identifying the most valuable hotspots [44]. Comparison of the three indicators led to the following ranking of the social value types, the higher-ranked value types (recreational, aesthetic, and historical and cultural) are selected for in-depth discussion.

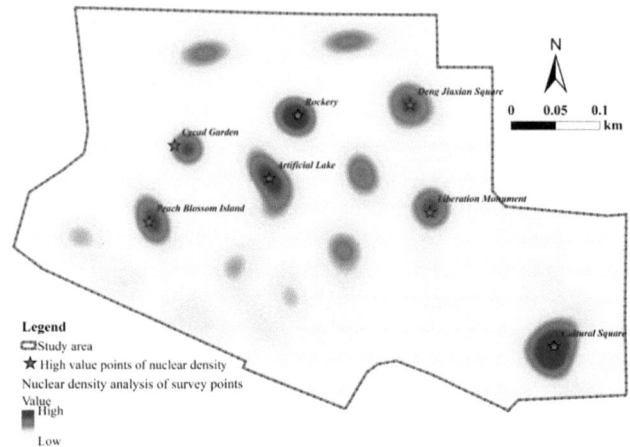

Fig. 4. Kernel density distribution of social value points in Mianyang Peoples's Park.

Table 4. The largest value index and spatial agglomeration table of the 9 social values of Mianyang People' s Park.

Social value type	Performance indicators			
	Sample size	R-ratio	Z-score	M-VI
Aesthetic	485	0.46	−22.69	7
Biodiversity	155	0.57	−10.24	2
Economic	189	0.36	−16.79	1
Historic and Cultural	315	0.33	−22.82	6
Research and Educational	128	0.50	−10.90	1
Life sustaining	134	0.56	−9.74	2
Recreation	1591	0.54	−35.21	8
Spiritual	161	0.50	−12.13	1
Therapeutic	335	0.51	−17.01	4

Note: R < 1 means clustering, R = 1 means random, and R > 1 means dispersion

3.2   Quantifying and Mapping the Social Value Space

Given the survey results for the social value distribution of this park (Fig. 5), the spatial distribution and perceived levels of different types of value can be shown by means of color distributions. The red areas in Fig. 5A are concentrated on Artificial Lake, Peach Blossom Island, Cycad Garden, and the fountain at Cultural Square, indicating that these areas generate high aesthetic benefits. Historical and cultural value is concentrated at Deng Jiaxian Square, Liberation Monument, and (slightly more weakly) Cultural Square. For recreation value, the high points are distributed in several areas of the park, namely Cultural Square, Deng Jiaxian Square, Liberation Monument, and Rockery.

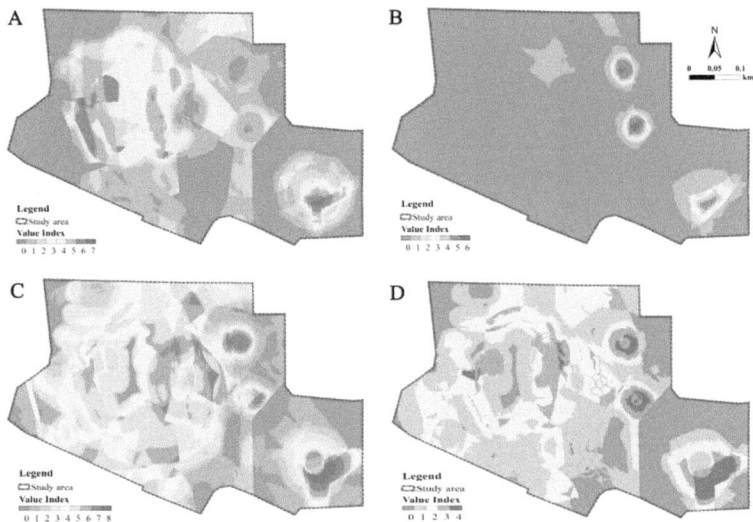

Fig. 5. Spatial distribution of four values of (A) Aesthetic, (B) Historic and Cultural, (C) Recreation, (D) Therapeutic based on all surveys.

3.3 Links Between Environmental Elements and Social Value

The response curve characterizing the relationship between social value types and environmental elements (Fig. 6) shows that recreational value decreases as DTR increases. In the range of 0–50 m, aesthetic value fluctuates, showing no significant overall effect, while the recreational and historical and cultural VIs also fluctuate, with an overall increase observed. At DTR > 50 m, the aesthetic and recreational VIs decrease significantly (Fig. 6A). With the exception of the fluctuating historical and cultural extremes, all types of value decrease significantly as DTW and DTWT increase (Fig. 6B, C). Aesthetic VI increases with increasing DTT, albeit with no major effect on the remaining value types (Fig. 6D). Aesthetic VI rises with increasing DTHL, while historical culture shows the opposite tendency, with no effect on the remaining two value types (Fig. 6E). Aesthetic VI tends to increase when the amount of litter bins, seats, and signage reaches a certain level, with the remaining value types fluctuating in the extreme range (Fig. 6F–H). Although the aesthetic VI decreases as the number of luminaires increases, the historical and cultural and recreational types of value do not fluctuate greatly. When the number of luminaires in the area is less than 20, the recreational VI decreases and then increases again as the number of luminaires increases (Fig. 6I). Aesthetic VI generally tends to decrease when the green space ratio and greenery coverage is less than 0.6, increasing significantly when levels greater than 0.6 are reached, a pattern that also applies to the historical and cultural value type, with no significant trend change in the remaining two value types (Fig. 6J, JK). With the exception of recreational and recreational value types, the remaining VIs decline as paving density increases (Fig. 6L). Aesthetic VI is mainly located in sites where there is vegetation and water cover; the recreational and historical and cultural VIs are mainly located where there are recreational profit-making facilities and hard paving (Fig. 6M).

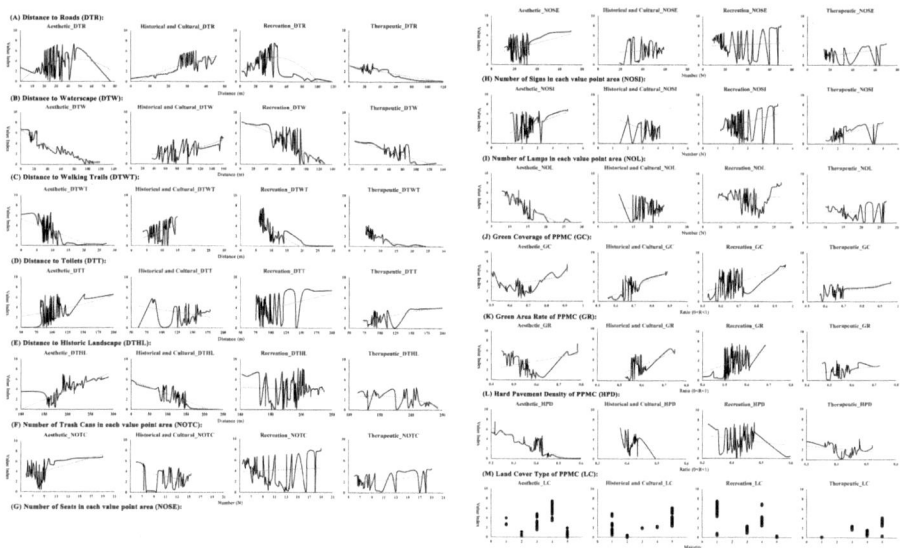

Fig. 6. Relationship between environmental indicator and value types. The majority category 1, 2, 3, 4, 5 in Fig. 6M represents the recreational and profit-making facilities, infrastructure facilities, vegetation, water and hard pavement.

3.4   Impact of Plant Landscape Features on Social Value

Correlation analysis was conducted between the assigned values and the landscape types labeled under each value (Table 5). The p-values offer initial insights into which landscape types visitors prefer and which could therefore be adjusted to generate higher social value. The correlations are ranked as follows (from highest to lowest): vegetation, waterscape, landform, athletic facility, architecture, pedestrian system, recreational activity, art installation, service facility, and animals. With the exception of animals, the values assigned are positively correlated with the number of marker points for the landscape type. The most significant positive correlation is vegetation and bodies of water, which indicates that bodies of water are perceived by the public as bringing higher social value in this environment. Diverse plant types and rich color combinations can be used to differentiate between open and closed spaces, as well as to delineate areas of movement from static zones. They can also directly mobilize people's emotional perceptions and generate different value preferences. In conjunction with the correlation results in Table 6, this study selected the vegetation landscape with the highest correlation as an influencing factor for social value; the more vegetation marker points in a certain category of value, the higher the value. Statistical analysis of social value types and landscape type marker points (Fig. 7A) indicates that the public perceives good planting and landscaping in the built landscape environment of this park as generating greater value in terms of biodiversity, life sustainability, and spirituality, either directly or indirectly, and most significantly in terms of aesthetics and recreation (Fig. 7B).

Analysis of the correlations between plant species, combinations of form, seasonal phases, and social value types (Table 6) shows that plant species have a significant positive correlation with the other value categories. There are, however, some exceptions; plant combination forms show a weak negative correlation with economic value, while the historical and cultural and scientific and educational value types show no significant effect ($p < 0.05$). The remaining value types are positively correlated, and the proportion of deciduous tree species shows a very significant positive correlation with the aesthetic, recreational, life sustainability, and spiritual value types (the exception being historical and cultural value).

Further field research was conducted to record and identify the plant species in the park. Plant species and combinations in each area were collated. The social value of each area was then calculated based on the correlation coefficient and plotted (Fig. 8). For aesthetic value, the ranking (from highest to lowest) runs from sightseeing area, leisure fitness area, entrance square, children's amusement area, to cultural educational area; for historical and cultural value, the ranking runs from cultural educational area, entrance square, children's amusement area, leisure fitness zone, to sightseeing area; for leisure and recreation value, the ranking runs from sightseeing area, leisure fitness zone, entrance square, children's amusement area, to cultural and educational area.

Finally, the AUC statistics generated during the SolVES analysis were used to check the credibility and applicability of the model. An AUC value less than or equal to 0.5 indicates that the model is at the stochastic prediction level or worse [45]. When the AUC value is greater than 0.7, the model assessment is considered valid, with a larger AUC value indicating higher credibility [46, 47]. In this study, the training AUC values for the social value types assessed are greater than 0.8 (except for recreational value at 0.784). The output for historical and cultural value even exceeds 0.9 (Table 6). These results indicate that the model assessment is credible and applicable to the study of the social value of Mianyang People's Park.

4 Discussion

4.1 Assessing the Social Value of Urban Parks

In response to the first research question, the results show that Mianyang People's Park is most successful in terms of the recreational and aesthetic value types, followed by the historical and cultural and recreational value types. The basic design specifications of the park are aligned with the needs of visitors, confirming that aesthetic value is

Table 5. Value Distribution and Correlation Analysis of Landscape Types.

Landscape types	Landform	Vegetation	Animal	Waterscape	Architecture	Art installation	Service facility	Pedestrian system	Athletic facility	Recreational activity	Value Allocation
Landform	1										
Vegetation	0.605	1									
Animal	0.376	0.267	1								
Waterscape	0.715*	0.924**	0.322	1							
Architecture	0.479	0.447	0.360	0.394	1						
Art installation	0.080	0.111	−0.225	0.021	0.506	1					
Service facility	−0.021	−0.104	−0.241	−0.193	0.462	0.958**	1				
Pedestrian system	0.550	0.037	−0.257	0.119	0.046	−0.167	−0.118	1			
Athletic facility	0.996**	0.542	0.332	0.665	0.443	0.074	−0.017	0.594	1		
Recreational activity	0.886**	0.605	0.580	0.766*	0.445	0.128	−0.002	0.144	0.866**	1	
Value Allocation	0.634	0.768*	−0.058	0.764*	0.522	0.325	0.192	0.461	0.598	0.462	1

Note: * and ** respectively indicate a significant correlation at the $P < 0.05$ level and a very significant correlation at the $P < 0.01$ level

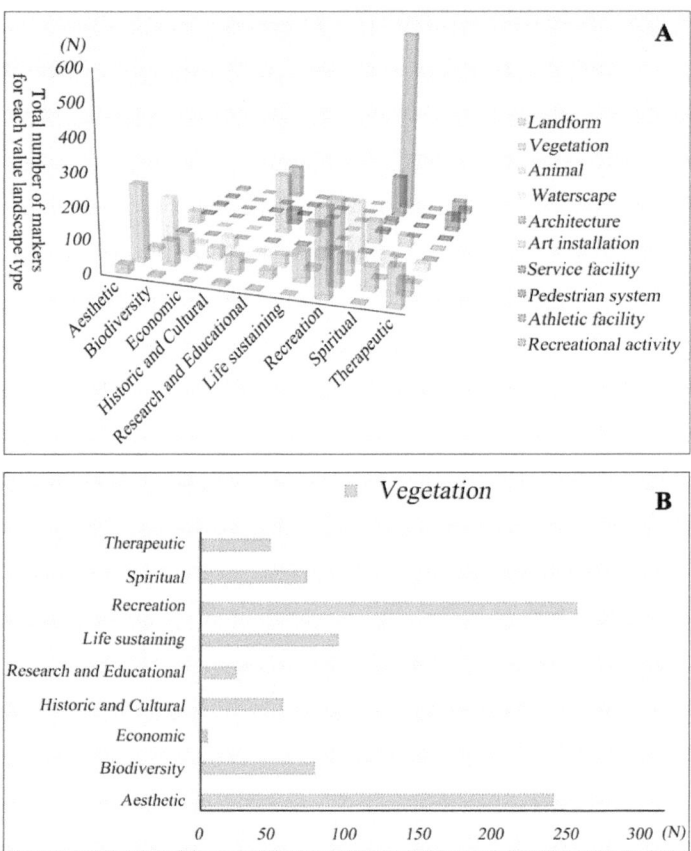

Fig. 7. Analysis of marking points of social values and landscape types (A for all landscapes, B for vegetation).

one of the most evident benefits of urban green space services and among the most frequently acknowledged ecosystem services [48–50]. Recreational value, on the other hand, produces the most obvious benefits in the social value assessment of urban parks. Nevertheless, Mianyang People's Park lacks an appropriate configuration of fitness facilities, which means that it does not fully meet the recreational needs of urban residents, providing relatively low value in this respect compared to aesthetics and recreation. The historical and cultural services that can make a valuable contribution to the development of urban green spaces and urban construction are not reflected in the actual layout and use of this park [51], with its single sculpture and limited historical attractions failing to provide the desired cultural atmosphere. Therefore, to avoid a disconnect between the park's unique cultural status and the public's perceptions of the services it provides, it is important to reinforce the promotion and education of history and culture there.

Regarding the second research question, there is a large influence of geography on the intensity and diversity of social value types [52]. Accordingly, as the DTR increases, the aesthetics, recreation, and recreation VIs decrease [28, 53, 54]. However, given the small

Table 6. Correlation analysis of plant species, combination modes and social values.

	Aesthetic	Biodiversity	Economic	Historic and Cultural	Research and Educational	Life sustaining	Recreation	Spiritual	Therapeutic
Floristics	0.826	0.928[*]	−0.105	−0.208	−0.247	0.898[*]	0.917[*]	0.899[*]	0.708
Combining form	0.426	0.808	−0.075	−0.037	0.006	0.294	0.673	0.382	0.308
Deciduous species	0.914[*]	0.790	−0.031	−0.444	−0.005	0.902[*]	0.911[*]	0.972[**]	0.917[*]

Note: * and ** respectively indicate a significant correlation at the P < 0.05 level and a very significant correlation at the P < 0.01 level

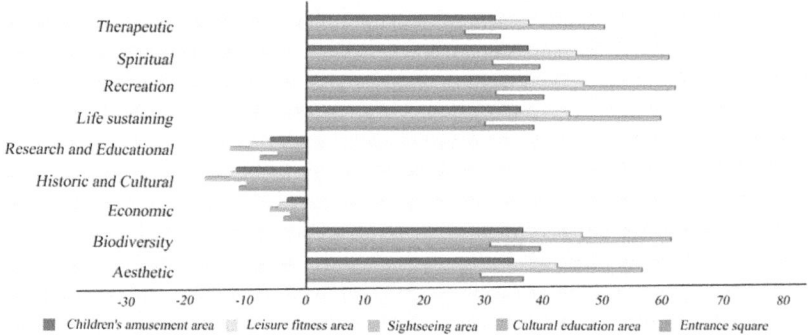

Fig. 8. Histogram of social value of each functional area based on plant characteristic ccoefficient.

size of the park and the density of the road network, studies suggest that the aesthetic, recreational, and recreational value types all decreased with increasing DTWT, which is in line with the findings of [40] and [55] at the Wusong Soak Terrace Wetland Park. There is also a clear preference for water landscapes, with areas near water being more popular with visitors than other areas, generating greater aesthetic and recreational value. This result is also consistent with the findings of [18, 56, 57].

In terms of the psychological and perceived factors of aesthetics, people generally do not consider the inclusion of amenities such as toilets and rubbish bins within the aesthetic comfort range, but recognize their necessity for park planning and services. It can be seen from the curve that each infrastructure has a degree of acceptability within which VI is not influenced by distance and quantity, but that beyond a certain point aesthetic VI increases with distance. Thus, a higher VI indicates that a district is likely to match people's landscape preferences and their preferences for where to stay, and so the amount of infrastructure needs to be increased to meet public demand [58]. In addition, as Fig. 6M indicates, aesthetic value is mainly distributed in areas of vegetation and bodies of water, whereas historical and cultural value is mainly located in areas where there are recreational and profit-making facilities. Thus, although the VI generally decreases as paving density increases, the value factor influence analysis suggests that aesthetic VI first decreases and then increases. In some areas of Mianyang People's Park, such as Cultural Square, although the area green area ratio and green coverage are high, visitors perceive the different elements and plant combinations at the main entrance of the park (an important landscape area and gateway that includes a large ornamental lawn area) as a single form [59, 60]. This unified perception leads to a weakening of aesthetic VI, which could be increased by better plant matching and by adding landscape features to other high-ratio areas.

Finally, in respect to the third research question, how plant landscape features affect value preferences, we find that people's landscape value preferences are closely related to the presence of vegetation, and that high-quality plant configuration landscaping contributes significantly to the enhancement of all types of value [61, 62]. Therefore, to increase the social benefits generated from urban parks, careful consideration must be given to planning the vegetation types and the structural layout of the green spaces. For example, the aesthetic, recreational, and recreational value of urban parks can be improved by planting more deciduous tree species. A carefully considered mix of evergreen and deciduous tree species with clear combination levels and seasonal staggering can create a strong visual effect that corresponds to public landscape preferences and is therefore conducive to the generation of social benefits.

4.2   Implications for Landscape Designers and Planners

This study shows that assessment of the social value of urban parks can be based on a comprehensive and detailed summary of the relationship between geography, landscape type, plant characteristics, and social services from the perspectives of relevant stakeholders. Unlike monetary assessment of ecosystem services, the spatialization of the social value assessment model enables the integration of people's preferences and the natural environment into decision-making processes [63]. Designers and planners can utilize social value assessment to propose strategies for optimizing landscape planning and management from a value growth perspective that fully incorporates public input.

4.3   Strengths and Limitations

By linking public perception to social value assessment, this study analyses the spatial distribution of social value and its influencing mechanisms in terms of perceptions of landscape type, environmental conditions, and plant characteristics, as well as enhancing understanding of the relationship between social value and landscape planning. The combination of VEP and SolVES model strengthens the link between social value and landscape [64] while facilitating public participation and ensuring accurate real-time data.

The limitations of the present study can be addressed in future research. Firstly, the utilization of the VEP approach relies on mobile phone software, potentially limiting the participation of older individuals and teenagers, thus leading to a missing sample that could introduce bias into the overall results. Secondly, although we selected several value types with high relevance based on actual park conditions, it is important to acknowledge that each park may encompass numerous potential values that need to be considered for a more comprehensive range of social value types.

5   Conclusions

This study, which identifies, quantifies, and maps the social value of ecosystem services in Mianyang People's Park, Mianyang City, Sichuan Province, allows three main conclusions to be drawn.

First, Mianyang People's Park provides well-developed social services and successfully generates high levels of aesthetic, recreational, and historical and cultural value, with recreational value being the most clearly perceived. In this respect, this urban park differs from most other ecological parks. However, our findings also indicate that several other types of value, not least economic, life sustainability, and spirituality value, tend to be overlooked by the public and should therefore be prioritized in future park development.

Second, the public's perceptions play an important role in the quantification and assessment of social value. Environmental landscapes have a significant impact on perception, with areas within close proximity to water, featuring characteristic landscapes, and benefit from well-maintained and convenient roads being favored among visitors. These areas are likely to be preferred for leisure activities, hence prolonged stays. Therefore, there will be a higher demand for environmental infrastructure in these locations, necessitating an appropriate increase in the provision of facilities and services.

Third, there is a clear link between plant characteristics and social value. Increasing the number of plant species, planting more deciduous species, and diversifying plant community structures, i.e., trees, shrubs, and grasses, are ways of increasing aesthetic, recreational, and biodiversity value. The social value assessment of parks thus provides useful information for park managers and park landscape architects seeking to improve the park environment and to enhance overall park performance. However, the results that incorporates public perceptions are subjective in nature, and evaluations are inevitably influenced by the cultural background [65], education level [66, 67], and age [68] of the relevant stakeholders, as well as their expertise [69] and familiarity with the environment [70]. Future research should include people from different social backgrounds and with specific demographic characteristics to identify similarities and differences in stakeholders' preferences and social perceptions in relation to ecosystem services.

References

1. Fu, B., Yu, D.: Ecosystem services tradeoffs and integration approaches. Resour. Sci. **38**, 1–9 (2016)
2. Liu, J., et al.: South–south cooperation for large-scale ecological restoration. Restor. Ecol. **25**, 27–32 (2017)
3. Mononen, L.: Monitoring ecosystem services and biodiversity: From biophysical metrics to spatial representations (Itä-Suomen yliopisto, 2017)
4. Zhang, J., Wang, S., Liu, Y., Fu, B.: Research progress on the interlinkages between the 17 Sustainable Development Goals and their implication for domestic study. Acta Ecologica Sinica **39** (2019)
5. Mustajoki, J., Saarikoski, H., Belton, V., Hjerppe, T., Marttunen, M.: Utilizing ecosystem service classifications in multi-criteria decision analysis – Experiences of peat extraction case in Finland. Ecosyst. Serv. **41**, 101049 (2020)
6. Everard, M., Longhurst, J., Pontin, J., Stephenson, W., Brooks, J.: Developed-developing world partnerships for sustainable development (2): An illustrative case for a payments for ecosystem services (PES) approach. Ecosyst. Serv. **24**, 253–260 (2017)
7. Wang, R., Ouyang, Z.: Social-economic-natural complex ecosystem and sustainability. Bull. Chinese Acad. Sci. **27**, 337–345, 403–404, 254 (2012)
8. Silori, C.: Ecosystem Services and Sustainable Development: Opportunities and Challenges (2015)
9. Rao, S., et al.: Study on the balance of grassland ecosystem services in Zhengqi county. Arid L. Resour. Environ. **29**, 81–86 (2015)
10. Chen-Xi, X. et al.: Application of ecosystem services value assessment in coastline protection and utilization planning of Qinzhou city

11. Ge, Y., Xin, B., Li, X.: Study on urban forest construction in warm temperate semi-humid area based on the improvement of ecosystem services. Beijing For. Univ. **42**, 127–141 (2020)
12. He, F., Dong, J., Xie, X., Xu, D., Wu, Z.: Evaluation of non-use value of constructed wetland ecosystem services in Beijing Olympic Forest Park. Resour. Environ. Yangtze River Basin **19**, 782–789 (2010)
13. Xie, X., Zhang, R., Zhang, J.: Analysis of recreational value of Lishui Baiyun national forest park based on CVM method. For. Ecol. Sci. **35**, 119–126 (2020)
14. Zhang, Y., Zhang, H., Yu, H.: Evaluation of the social value of ecosystem services based on tourists' perception: A case study of Qianjiangyuan national park. Tour. Sci. **34**, 66–85 (2020)
15. Deng, M., et al.: Improvement and application of the multi-destination travel cost sharing model in recreation value evaluation of Zoige county. Nat. Resour. **35**, 1090–1097 (2020)
16. Wang, F., Zhou, Z., Zheng, Z.: Evaluation on non-use values of typical lake wetlands in Wuhan. Acta Ecol. Sin. **30**, 3261–3269 (2010)
17. Szücs, L., Anders, U., Bürger-Arndt, R.: Assessment and illustration of cultural ecosystem services at the local scale – A retrospective trend analysis. Ecol. Ind. **50**, 120–134 (2015)
18. Plieninger, T., Dijks, S., Oteros-Rozas, E., Bieling, C.: Assessing, mapping, and quantifying cultural ecosystem services at community level. Land Use Policy **33**, 118–129 (2013)
19. van Riper, C.J., Kyle, G.T., Sherrouse, B.C., Bagstad, K.J., Sutton, S.G.: Toward an integrated understanding of perceived biodiversity values and environmental conditions in a national park. Ecol. Ind. **72**, 278–287 (2017)
20. Semmens, D.J., Sherrouse, B.C., Ancona, Z.H.: Using social-context matching to improve spatial function-transfer performance for cultural ecosystem service models. Ecosyst. Serv. **38** (2019)
21. Chan, K.M.A., Satterfield, T., Goldstein, J.: Rethinking ecosystem services to better address and navigate cultural values. Ecol. Econ. **74**, 8–18 (2012)
22. Gómez-Baggethun, E., Barton, D.N.: Classifying and valuing ecosystem services for urban planning. Ecol. Econ. **86**, 235–245 (2013)
23. Bertram, C., Rehdanz, K.: The role of urban green space for human well-being. Kiel Working Papers (2014)
24. Johnson, J.A., et al.: Mapping ecosystem services to human well-being: A toolkit to support integrated landscape management for the SDGs. Ecol. Appl. **29**, e01985 (2019)
25. McPhearson, T., Andersson, E., Elmqvist, T., Frantzeskaki, N.: Resilience of and through urban ecosystem services. Ecosyst. Serv. **12**, 152–156 (2015)
26. Markevych, I., et al.: Exploring pathways linking greenspace to health: Theoretical and methodological guidance. Environ. Res. **158**, 301–317 (2017)
27. Enssle, F., Kabisch, N.: Urban green spaces for the social interaction, health and well-being of older people— An integrated view of urban ecosystem services and socio-environmental justice. Environ. Sci. Policy **109**, 36–44 (2020)
28. Sherrouse, B.C., Semmens, D.J., Clement, J.M.: An application of Social Values for Ecosystem Services (SolVES) to three national forests in Colorado and Wyoming. Ecol. Ind. **36**, 68–79 (2014)
29. MacKay, K.J., Fesenmaier, D.R.: Pictorial element of destination in image formation. Ann. Tour. Res. **24**, 537–565 (1997)
30. McNally, C., Gold, A., Pollnac, R., Kiwango, H.: Stakeholder perceptions of ecosystem services of the Wami River and Estuary. Ecology and Society **21** (2016)
31. Rall, E., Bieling, C., Zytynska, S., Haase, D.: Exploring city-wide patterns of cultural ecosystem service perceptions and use. Ecol. Ind. **77**, 80–95 (2017)
32. Garrod, B.: Exploring place perception a photo-based analysis. Ann. Tour. Res. **35**, 381–401 (2008)

33. Costanza, R., Groot, R.D., Braat, L., Kubiszewski, I., Grasso, M.: Twenty years of ecosystem services: How far have we come and how far do we still need to go? Ecosyst. Serv. **28**, 1–16 (2017)
34. Enriquez-Acevedo, T., Botero, C.M., Cantero-Rodelo, R., Pertuz, A., Suarez, A.: Willingness to pay for Beach Ecosystem Services: The case study of three Colombian beaches. Ocean Coast. Manag. **161**, 96–104 (2018)
35. Dandy, N., Van Der Wal, R.: Shared appreciation of woodland landscapes by land management professionals and lay people: An exploration through field-based interactive photo-elicitation. Landsc. Urban Plan. **102**, 43–53 (2011)
36. Sugimoto, K.: Analysis of scenic perception and its spatial tendency: Using digital cameras, GPS loggers, and GIS. Procedia. Soc. Behav. Sci. **21**, 43–52 (2011)
37. Sugimoto, K.: Quantitative measurement of visitors' reactions to the settings in urban parks: Spatial and temporal analysis of photographs. Landsc. Urban Plan. **110**, 59–63 (2013)
38. Sugimoto, K.: Use of GIS-based analysis to explore the characteristics of preferred viewing spots indicated by the visual interest of visitors. Landsc. Res. **43**, 345–359 (2018)
39. Qiu, L., Lindberg, S., Nielsen, A.B.: Is biodiversity attractive?—On-site perception of recreational and biodiversity values in urban green space. Landsc. Urban Plan. **119**, 136–146 (2013)
40. Sun, F., Xiang, J., Tao, Y., Tong, C., Che, Y.: Mapping the social values for ecosystem services in urban green spaces: Integrating a visitor-employed photography method into SolVES. Urb. Fores. Urb. Green. **38**, 105–113 (2019)
41. Clement, J.M., Cheng, A.S.: Using analyses of public value orientations, attitudes and preferences to inform national forest planning in Colorado and Wyoming. Appl. Geogr. **31**(2), 393–400 (2011)
42. Sherrouse, B.C., Semmens, D.J.: Validating a method for transferring social values of ecosystem services between public lands in the Rocky Mountain region. Ecosyst. Serv. **8**, 166–177 (2014)
43. Daixin, S.*, Tongkai.: Using public participation geographic information system to evaluate cultural service of modern urban parks: Taking Shanghai Fuxing Park as Case Study. FJYL **26**, 95–100 (2019)
44. Brown, G.G., Reed, P., Harris, C.C.: Testing a place-based theory for environmental evaluation: An Alaska case study. Appl. Geogr. **22**, 49–76 (2002)
45. Phillips, S.B., Aneja, V.P., Kang, D., Arya, S.P.: Modelling and analysis of the atmospheric nitrogen deposition in North Carolina. Int. J. Glob. Environ. Iss. **6**, 231–252 (2006)
46. Swets, J.A.: Measuring the accuracy of diagnostic systems. Sci. **240**, 1285–1293 (1988)
47. Elith*, J. et al.: Novel methods improve prediction of species' distributions from occurrence data. Ecography **29**, 129–151 (2006)
48. Tyrväinen, L., Mäkinen, K., Schipperijn, J.: Tools for mapping social values of urban woodlands and other green areas. Landsc. Urban Plan. **79**, 5–19 (2007)
49. James, P., et al.: Towards an integrated understanding of green space in the European built environment. Urban Forest. Urban Green. **8**, 65–75 (2009)
50. Pietrzyk-Kaszyńska, A., Czepkiewicz, M., Kronenberg, J.: Eliciting non-monetary values of formal and informal urban green spaces using public participation GIS. Landsc. Urban Plan. **160**, 85–95 (2017)
51. Martín-López, B., et al.: Uncovering ecosystem service bundles through social preferences. PLoS ONE **7**, e38970 (2012)
52. Ives, C.D., et al.: Capturing residents' values for urban green space: Mapping, analysis and guidance for practice. Landsc. Urban Plan. **161**, 32–43 (2017)
53. Danyang, C., Mengting, L.I., Yangyang, D., Yue, C.: Assessment of the urban waterfront based on social values of ecosystem services: A case study of the Huangpu River Waterfront. Shanghai Urb. Plann. Rev. (2018)

54. 马桥. Assessment on Social Values of Wetland ecosystem service based on SolVES Model: A case of Chanba Wetland Park and HeChuan Wetland (西北大学, 2018)

55. Wang, Y., Fu Bi, T., Lyu, Y.P., Yang, K., Che, Y.: Assessment of the social values of ecosystem services based on SolVES model: A case study of Wusong Paotaiwan Wetland Forest Park, Shanghai, China. Ying yong sheng tai xue bao = J. Appl. Ecol. **27**, 1767–1774 (2016)

56. Sigao, 霍思高, Lu, 黄璐 , Lijiao, 严力蛟. Valuation of cultural ecosystem services based on SolVES: A case study of the South Ecological Park in Wuyi County, Zhejiang Province. Acta Ecologica Sinica **38**, 3682–3691 (2018)

57. Semmens, D.J., Sherrouse, B.C., Ancona, Z.H.: Using social-context matching to improve spatial function-transfer performance for cultural ecosystem service models. Ecosyst. Serv. **38**, 100945 (2019)

58. Tao, Y., Fu, B., Che, Y.: Spatial optimization of urban park environmental facilities with recreational preferences. Urb. Eniviron. Urb. Ecol. **29**, 21–26 (2016)

59. Liu, T.: Analysis of urban entrance landscape design —— Taking the entrance landscape design of Zhuhui Mountain Forest Park in Hengyang City as an example. Civ. Archit. **215**, 219 (2016). https://doi.org/10.3969/j.issn.1673-0232.2016.03.194

60. Wang, S., Wang, L.: Research on landscape design of urban park entrance based on environmental psychology. Chifeng Univ. Nat. Sci. **32**, 145–147 (2016)

61. Dallimer, M., et al.: Biodiversity and the feel-good factor: Understanding associations between self-reported human well-being and species richness. Bioscience **62**, 47–55 (2012)

62. Graça, M., et al.: Assessing how green space types affect ecosystem services delivery in Porto, Portugal. Landsc. Urban Plann. **170**, 195–208 (2018)

63. Ancona, Z.H., Semmens, D.J., Sherrouse, B.C.: Social-value maps for Arapaho, Roosevelt, Medicine Bow, Routt, and White River National Forests, Colorado and Wyoming.https://doi.org/10.3133/sir20165019

64. Swanwick, C.: Society's attitudes to and preferences for land and landscape. Land Use Policy **26**, S62–S75 (2009)

65. Nohl, W.: Sustainable landscape use and aesthetic perception–preliminary reflections on future landscape aesthetics. Landsc. Urban Plan. **54**, 223–237 (2001)

66. Lindemann-Matthies, P., Briegel, R., Schüpbach, B., Junge, X.: Aesthetic preference for a Swiss alpine landscape: The impact of different agricultural land-use with different biodiversity. Landsc. Urban Plan. **98**, 99–109 (2010)

67. Molnarova, K., et al.: Visual preferences for wind turbines: Location, numbers and respondent characteristics. Appl. Energy **92**, 269–278 (2012)

68. van den Bosch, M., Ode Sang, Å.: Urban natural environments as nature-based solutions for improved public health – A systematic review of reviews. Environ. Resea. **158**, 373–384 (2017)

69. Vouligny, É., Domon, G., Ruiz, J.: An assessment of ordinary landscapes by an expert and by its residents: Landscape values in areas of intensive agricultural use. Land Use Policy **26**, 890–900 (2009)

70. Howley, P., Donoghue, C.O., Hynes, S.: Exploring public preferences for traditional farming landscapes. Landsc. Urban Plan. **104**, 66–74 (2012)

Optimization of Spatial Layout of Community Elderly Care Facilities Based on POI Data in Fucheng District

Xiaoge Du$^{(\boxtimes)}$, Yu Wang, and Yaling Li

School of Civil Engineering and Architecture, Southwest University of Science and Technology, Mianyang 621010, China
18723215282@163.com

Abstract. Community elderly care facilities, as significant facilities for maintaining the health of the elderly and supplementing family care, are a vital component of the construction of urban elderly care service systems. Reasonable layout of elderly care facilities can improve the efficiency of elderly care resource utilization and is of great importance for accelerating high-quality progress of elderly care. Taking 7 administrative streets in Fucheng District of Mianyang City as the research area, through analyzing the current situation of community elderly care facilities, it is found that the overall number of elderly care facilities in Fucheng District is insufficient, and the distribution of high in the east and low in the periphery is uneven. Half of the residential areas and elderly people are not in the service area, and there are obvious loopholes in the coverage of community elderly care facilities in the study region. Combined with the actual residential points and spatial layout of functional facilities in the research area, the location allocation model was used to optimize the layout of community elderly care facilities, and finally 20 new community elderly care facilities are determined. The optimization results can meet the convenient travel of most elderly people, basically meet the requirements of the urban community "15-min elderly care service circle", and finally construct the community elderly care service facility network, in order to achieve a balanced and efficient layout of community elderly care facilities in Fucheng District.

Keywords: POI data · Community elderly care facilities · Spatial layout · Location allocation model

1 Introduction

China is stepping into a deeply aging society, and aging will continue to rise. According to the seventh national census, in 2020, there were 264.0187 million people over the age of 60 in mainland China, accounting for 18.7% of the Chinese population. From the development of cities, this phenomenon can also be seen. In the past 20 years, Mianyang City has entered a rapid aging stage due to a large base of middle-aged population, with an elderly population of 1.15 million, making it one of the cities with the most

© The Author(s) 2025
B.-J. He et al. (Eds.): UCSUD 2023, LNCE 559, pp. 424–437, 2025.
https://doi.org/10.1007/978-981-97-8401-1_30

serious aging in Sichuan Province [1]. Due to the historical characteristics of the old urban area, Fucheng District has a slower social composition update and is an area with deep aging and a concentrated elderly population. To prevent events such as "aging before preparation" from happening in the future, conducting research on the layout of elderly care facilities is one of the important tasks to handle the current population aging issue [2]. Community elderly care facilities, as an important facility for maintaining the health of the elderly and supplementing family care, are a significance component of the construction of urban elderly care service systems. The "Fourteenth Five-Year Plan for the Construction of Mianyang City's Elderly Care Service System" (hereinafter referred to as the Plan) issued by the People's Government of Mianyang City mentions that the layout of elderly care service facilities is more optimized, the home community elderly care service network is basically established, and the functional connection with other facilities is strengthened; by 2025, it is required that the coverage rate of community elderly care services in Mianyang City basically reaches 100%, and the urban community "15-min elderly care service circle" is basically established [3].

The spatial layout of community elderly care service facilities has always been a research hotspot for scholars at home and abroad. Community elderly care service facilities refer to various convenient conditions built in residential areas or cities to provide living and medical care for the elderly, in order to facilitate their independent living. With the continuous updating and iteration of technology, scholars are increasingly inclined to use GIS, modern communication and other technologies to conduct quantitative research on the spatial layout optimization of elderly care facilities [4]. The channels for data acquisition include online big data, open map platforms, etc. [5]. From the perspective of research content, research on optimizing the spatial layout of community elderly care service facilities mainly focuses on accessibility analysis or the needs of the elderly. Few scholars have explored the "15-min elderly care service circle" and have not constructed corresponding facility networks [6]. In addition, the proximity of elderly care facilities to medical facilities, leisure facilities, and public transportation facilities indicates that the more construction can meet the health, leisure, and travel needs of the elderly, the more reasonable the allocation of elderly care facilities [7]. Therefore, this article will use technologies such as GIS, Gaode Open Platform, and network big data to explore the spatial layout of community elderly care facilities in Fucheng District from the perspectives of facility proximity and location theory, in order to divide the "15 min elderly care service circle" in Fucheng District, build a network of community elderly care service facilities, and achieve a balanced and efficient layout of community elderly care facilities in Fucheng District.

2   Research Area and Method

2.1   Overview of the Research Area

Fucheng District is the central area of Mianyang, with a land area of 554.47km². The urbanization rate of Fucheng District is 86.76% by 2022. Fucheng district is the population gathering center of Mianyang City, and the permanent population has been increasing continuously in recent years, among which the elderly population is 203,300, accounting for 15.66% of the permanent population. The elderly population of Fucheng

District is densely distributed in the selected research area. The elderly population is 151,400, and 74.47% of the elderly in Fucheng district live in this area. As the root of urban development and an area where the elderly gather and distribute, there are a large number of old and new communities in this area, which gather a large number of leisure facilities, medical facilities and public transport facilities, etc. Their construction standards are different and the level of facilities is uneven. In order to provide a large number of elderly people with necessary old-age care services, It is necessary to combine with other functional facilities to improve the spatial layout of elderly care facilities. The study area of this paper is located within the scope of 7 administrative streets in Fucheng District, including Chengxiang Street, Gongqu Street, Puming Street, Chuangyeyuan Street, Shitang Street, Chengjiao Street, and Tangxun Street (Fig. 1).

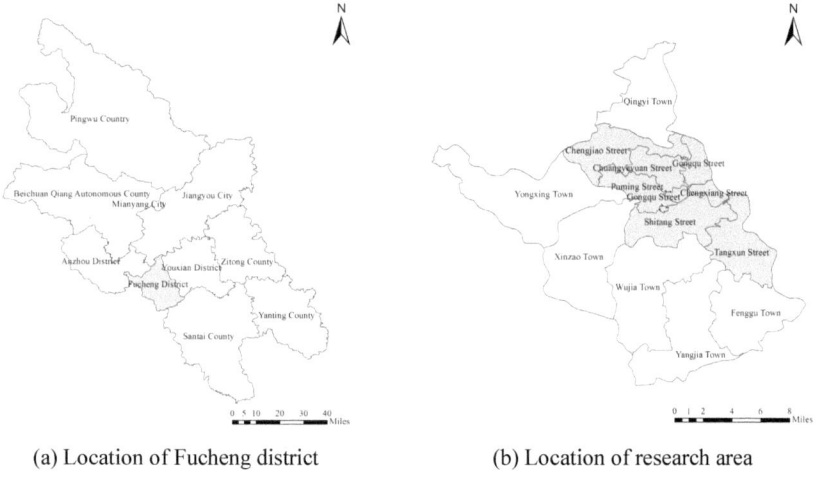

(a) Location of Fucheng district   (b) Location of research area

Fig. 1. Overview map of the study region

2.2 Data Source

The research data includes population data, administrative divisions, road networks, and POI data of various functional facilities (elderly care facilities, medical facilities, residential areas, leisure facilities, public transportation facilities, etc.) in Fucheng District. The population data comes from the "Seventh National Census Bulletin of Mianyang City"; the administrative divisions and road networks come from Bigemap GIS Office; according to the research of previous scholars, the POI data is obtained through the Amap Open Platform (http://lbs.amap.com), using python to crawl 2099 data, including 1219 residential areas, 26 community elderly care facilities, 523 public transportation facilities, 131 medical facilities, and 200 leisure facilities [8]. The community elderly care facilities in this paper include elderly daycare centers, nursing homes, and elderly service centers; medical facilities comprise comprehensive hospitals, specialty hospitals, and community hospitals; leisure facilities comprise parks, squares, libraries, attractions, cultural centers, and science and technology museums, etc.

2.3 Research Method

Service Area Analysis. Service area analysis is a method in GIS network analysis, which calculates the area that can be covered within the service radius of a certain type of public service facility with a service radius limit, based on the actual road network. According to the "Urban Residential Area Planning and Design Standards" (hereinafter referred to as the standard) published by the Ministry of Housing and Urban-Rural Development in 2018, the service range of community elderly care facilities should be combined with a fifteen-minute life circle, a ten-minute life circle, a five-minute life circle, and the service radius should not exceed 1500m, 1000m, 500m [9]. Related literature also shows that the activities of the elderly are highly concentrated within a range of 1500m, and the following will refer to this standard for service area analysis [10].

Location-Allocation Models

Maximum Coverage Model. The operating principle of this model is to cover the maximum number and range of demand points within the threshold, used to study the maximum coverage range under the condition that the facility point is determined [11]. The formula is as follows:

$$\min \sum_{i \in I} w_i x_i$$

$$s.t. \sum_{d_{ij} \leq D} y_j - x_i \geq 0, \forall_i \in I$$

$$\sum_{j \in I} y_i = p$$

$$y_i \in (0, 1), \forall_j \in J \tag{1}$$

In the formula: i represents the demand point, I represents the set of demand points, $I = \{i | i = 1, 2, \ldots, m\}$; J represents the set of candidate facility points, $J = \{j | j = 1, 2, \ldots, n\}$; d_{ij} represents the minimum distance between demand point i and candidate facility point j, $i \in I, j \in J$; D represents the maximum allowable distance between the demand point and the facility point; p represents the number of candidate facility points; $y_j = 1$ represents the demand point is selected at j, $y_j = 0$ represents the demand point is selected at other locations; y_j represents the service capacity of candidate facility point j; w_i represents the weight of demand point i.

Minimize Facility Point Model. The computation theory of this model is that each requirement spot is covered by at a minimum one facility spot, used to compute the minimum quantity of service facility spots demanded [12]. The formula is as follows:

$$\min \sum_{j \in J} y_j$$

$$s.t. \sum_{d_{ij} \leq D} x_{ij} \geq 1, \forall_i \in I$$

$$y_i \in (0, 1), \forall_i \in J \tag{2}$$

In the formula: i represents the demand point, I represents the set of demand points, $I = \{i | i = 1, 2, \ldots, m\}$; J represents the set of candidate facility points, $J = \{j | j =$

1,2,...,n}; d_{ij} represents the minimum distance between demand point i and candidate facility point j, $i \in I$, $j \in J$; D represents the maximum allowable distance between the demand point and the facility point; x_{ij} represents the demand point i being served by candidate facility point j; y_i represents the service capacity of candidate facility point j; $y_i = 1$ represents the demand point is selected at j, $y_i = 0$ represents the demand point is selected at other locations.

Accessibility Analysis. This study uses the closest facility spot method in GIS to calculate the accessibility of optimized community elderly care facilities. The nearest facility point analysis represents accessibility by calculating the walking distance from the requirement spot to the facility spot. This method assumes that the elderly always choose the nearest facility, which is applicable to the community elderly care facilities in this paper that choose the nearest travel mode.

3   Preliminary Analysis

3.1   Analysis of Community Elderly Care Facility Service Areas

So as to facilitate the analysis of the situation in the current service shortage areas within the research scope, referring to Perry's "neighborhood unit" theory, the research area is divided into 641 500m × 500m network units [13]. If there is a residential area within the network unit, it is considered that there is a demand for elderly care in this area; if there is no residential area, no mark is made. The residential areas are marked as the centroids of the network units, and there are a total of 259 residential points within the research scope, resulting in a residential point location map (Fig. 2(a)). Based on the range of elderly care life circles, the maximum service radius of elderly care facilities in the study region is install to 1500m, and different distance levels are defined on the basis of the convenience of reaching elderly care facilities. 0-500m is defined as a very convenient distance for elderly care services; 500-1000m is defined as a convenient distance; 1000–1500m is defined as a relatively convenient distance; areas more than 1500m away are considered unfavorable for elderly care services (Fig. 2(b)). Set the relevant parameters in ArcGIS, and statistically analyze data such as service area, service area coverage rate, service residential area rate, service elderly population, and service elderly population ratio (Table 1).

From a spatial distribution perspective, the service range of community eldercare facilities varies among the seven streets. In Gongqu Street and Chengxiang Street, where residential areas and roads are densely distributed, the covered service range expands in a circular pattern from the central area. The service areas of the remaining five streets are scattered, among which Chuangyeyuan Street and Puming Street have a large number of residential areas, but most are not within the service range, clearly failing to meet the full coverage needs of the "15-min eldercare service circle". Chengjiao Street, while not covered by the service area, also has a sparse distribution of residential areas, revealing a significant gap in the coverage of community eldercare facilities.

From the statistical table results, the area coverage rate of the "0-500m" service area of community eldercare facilities is only 5.12%. This range is the most comfortable walking distance for the elderly, with only 23.99 thousand elderly people within this service range, able to enjoy very convenient community eldercare services. The

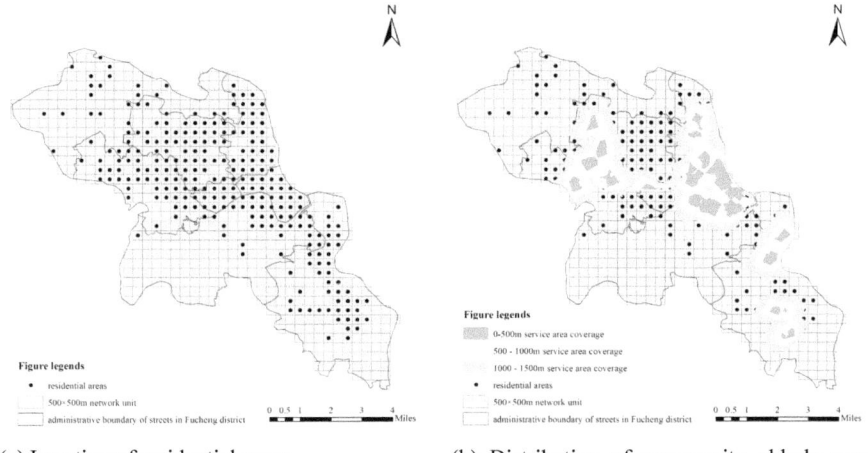

(a) Location of residential areas

(b) Distribution of community elderly care facilities service areas

Fig. 2. Location of residential areas and distribution of community elderly care facilities service areas

Table 1. Analysis of the current situation of services provided by community elderly care facilities

Convenience level	Service area (km^2)	Service area coverage rate (%)	Service residential area rate (%)	Serving the elderly population (thousand people)	The proportion of elderly people served (%)
Very convenient (0–500m)	8.18	5.12	12.36	23.99	15.84
Convenient (500–1000m)	20.44	12.80	23.55	32.16	21.23
Less convenient (1000–1500m)	25.45	15.94	17.37	25.48	16.83
Inconvenient (>1500m)	105.59	66.14	46.72	69.80	46.10

area coverage rate of the "500-1000m" service area of community eldercare facilities is 12.80%, and the proportion of residential areas served in this range is 23.55%. Community eldercare facilities can serve 33.86% of the land area within the "0-1500m" range, but nearly half of the residential areas and elderly population are not within this service range and cannot enjoy comfortable and convenient community eldercare services. The above analysis shows that there is considerable room for improvement in the spatial layout of community eldercare facilities in the study area, and the coverage of facility services needs to be improved.

3.2   Analysis of Service Areas for Buses, Medical and Leisure Facilities

While studying the current spatial layout and service area of community eldercare facilities, this paper also considers the impact of other functional facilities on its spatial layout. The following four figures mainly clarify the spatial relationship between residential areas, road networks, and various facilities (Fig. 3). According to scholars' research and planning standards, different walking thresholds are set for different types of facilities [8]. From the diagram, it can be seen that the service range of public transportation and medical facilities basically covers residential areas and road networks, while leisure facilities have certain coverage gaps in Puming Street and Chengjiao Street. Overall, the combined service areas of the three types of facilities have a high coverage rate, mainly distributed in Chengxiang Street and the industrial area. The optimization of community elderly care facilities will be based on the coverage range of the combined service areas of these three types of facilities for new site selection.

4   Research on Facility Layout Optimization

Based on the location allocation model research method, the optimization of elderly care facilities in the study region is carried out. According to the planned community elderly care service facility demand points and the service area range of various facilities mentioned above, the proposed new elderly care facility spots are based on the current spatial distribution of community elderly care facilities in the study region; then, with a 1000m walking service radius as the interruption distance, a maximized coverage range model formed by the road network is established; then the minimize facility spot model is established to obtain the least of facility spots needed for elderly care facilities within maximum coverage range; finally, the optimization results of the location allocation of community elderly care facilities in the research area are obtained by synthesizing the calculation consequences of two models, and the service area and the nearest facility point analysis optimization results are used, and a community elderly care facility network is established.

4.1   Determination of Proposed New Elderly Care Facility Points

The main problem with the current community facilities for the elderly is the uneven space pattern. According to the planning requirements, all existing facilities for the elderly in the research area are retained, and priority is given to adding new community elderly care service facilities within the analysis range of the three types of service areas during layout. Except for the saturation of elderly care facility points in Chengxiang Street, the other six streets all need to add multiple elderly care service facilities.

4.2   Maximizing Coverage Range Facility Point Layout

By establishing a road network dataset in the research area, a walking network between residential points and elderly care facility points is constructed, forming a path connection, configuring the problem type of maximizing coverage range. The result is that

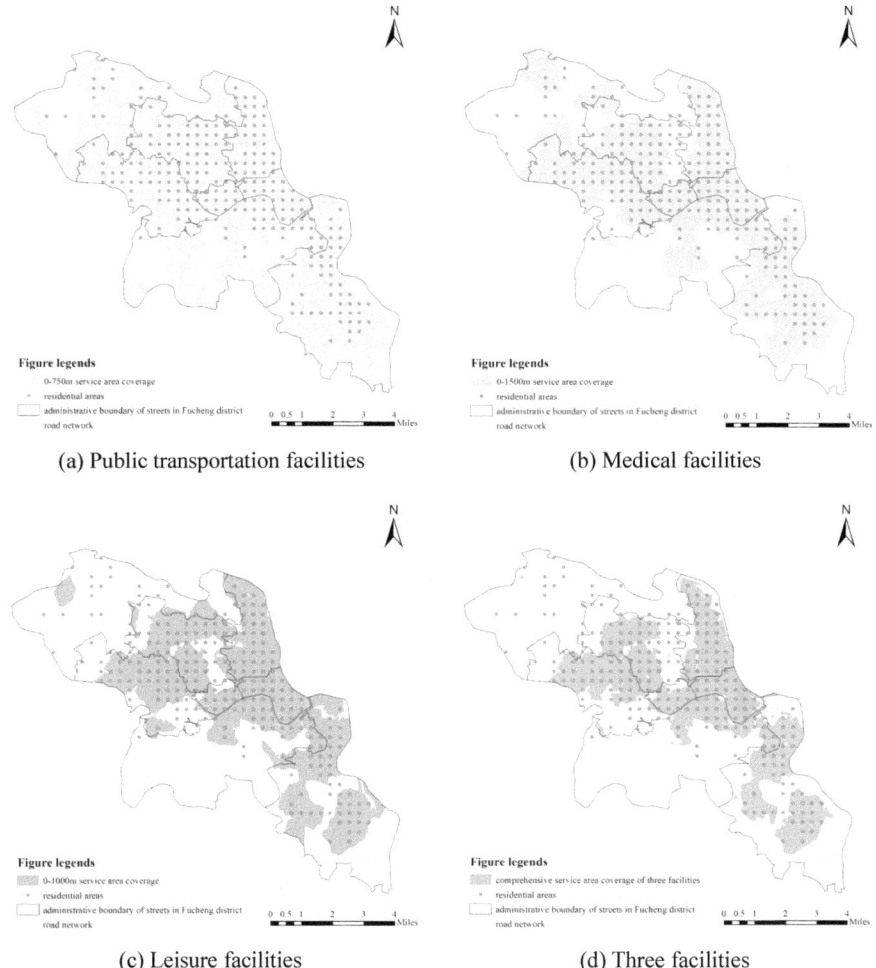

(a) Public transportation facilities

(b) Medical facilities

(c) Leisure facilities

(d) Three facilities

Fig. 3. Analysis of service areas for public transportation, medical and leisure facilities

28 new community elderly care facilities are proposed (Fig. 4), among which the new facilities for the elderly are primarily distributed in the Chuangyeyuan Street, Tangxun Street, and the industrial area. Due to the scattered residential points and imperfect public facilities in Chengjiao Street in the northwest of the research area, only one elderly care facility point has been added.

4.3 Minimizing Facility Point Layout Within Maximum Coverage

Based on the operation of the maximum coverage model, hiding the resident demand points that cannot be covered, and then running the minimum facility point model by setting the walking interruption distance to 1000m, we obtain the minimum number of facility points under non-full coverage. The result suggests that 20 new community elderly care facilities are to be added (Fig. 5).

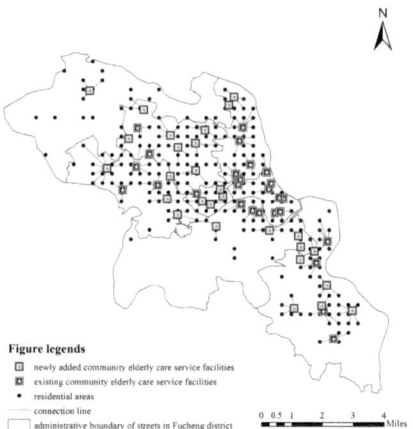

Fig. 4. Maximizing coverage model operation results

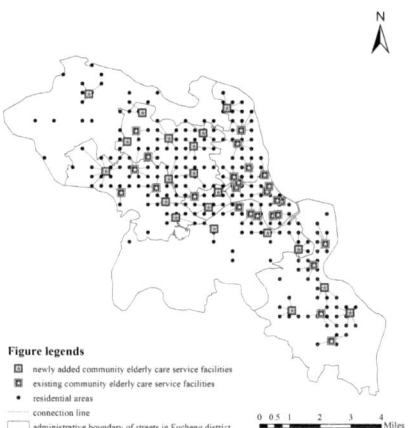

Fig. 5. Minimizing the number of facilities points model operation results

4.4 Optimization Layout Results and Analysis

According to the calculation results of the two problem-solving methods, considering the actual situation, the optimized layout of the minimum facility point model is finally adopted, determining the addition of 20 new community elderly care facilities. Among them, Gongqu Street, Shitang Street, and Tangxun Street each added 3, Chuangyeyuan Street added 6, Chengjiao Street added 1, and Puming Street added 4. The total number of community elderly care facilities after optimization reached 46. The optimized layout of community elderly care facilities can basically achieve full coverage within a 1500m walking range of elderly services.

Analysis of Service Area of Optimized Community Elderly Care Facilities. The service area analysis of the optimized community elderly care facility points (Fig. 6) shows that the coverage of the facility points has greatly increased, solving the serious shortage of community elderly services in Chuangyeyuan Street, Gongqu Street, and Tangxun Street. An elderly care facility was appropriately arranged in Chengjiao Street, and the service range includes the urban area of this street. Compared with the service area of the current facilities for the elderly, the service area after optimization increased by 29.91km^2, reaching 83.98km^2, and the coverage rate of the community elderly care service area increased by 18.74%. Using the nearest facility point to analyze the community elderly care facility points, judging the walking distance from the resident points of each street to the nearest community elderly care facility, most of the facilities for the elderly in the study region are within a 1500m walking range (Fig. 7), basically meeting the requirements of the urban community "15-min elderly service circle".

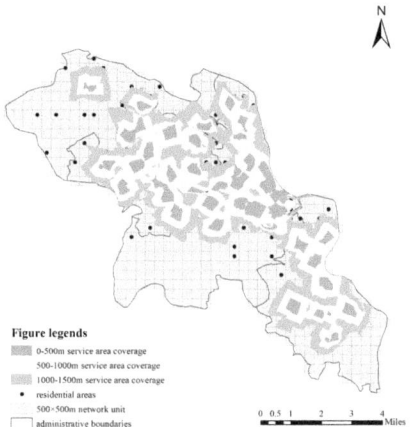

Fig. 6. Optimized distribution of community elderly care facilities service areas

Data Analysis of the Nearest Distance to Optimized Community facilities for the elderly. Based on the calculated consequences of the nearest distance, statistics such as the walking distance from the residential points of each street in the study area to the nearest community elderly care facility, the number and proportion of serviced residential points, and the number and proportion of serviced elderly people are compiled (Table 2). In the study area, the distance from most residential points to the nearest community elderly care facility is less than 1500m, and the coverage of residential points in 7 streets has significantly increased compared to before the optimization, all reaching over 80%. The average service distance of Chengxiang Street is the smallest, at 480m, which is within the very convenient travel range for the elderly; the average service distance of other streets is basically within the 1000m walking range, which can meet the convenient travel requirements of old people. The proportion of elderly people served by the optimized community elderly care facilities within a walking distance of

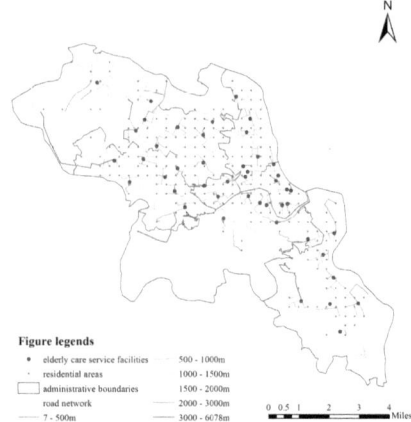

Fig. 7. Analysis of the nearest facilities of optimized community elderly care facilities

1500m has increased from 53.90% to 92.87%, and the total number of people covered by the service has reached 140.63 thousand, which can meet the community elderly care services of most elderly people in the study area.

Table 2. Analysis results of the nearest facilities of optimized community elderly care facilities

Service situation for walking distance <1500m

Street name	Serving the elderly population (thousand people)	The proportion of elderly people served (%)	Number of service residential area	Residential area coverage rate (%)	Average service distance (m)
Chengxiang	25.90	100.00	19	100.00	480
Gongqu	39.36	98.62	45	97.83	758
Shitang	15.78	84.53	27	81.82	639
Chengjiao	10.42	79.28	15	83.33	1072
Puming	20.07	89.42	41	93.18	762
Chuangyeyuan	12.88	93.20	48	88.89	747
Tangxun	16.22	92.48	40	91.00	754

Analysis of the Network distribution of Community facilities for the elderly. The existing squares, parks, attractions, universities for the elderly, cultural centers, libraries, science museums, medical facilities, etc. In the study region are marked, and the community facilities for the elderly are indicated on the map. With 500m as the service range for various facilities and 1000m as the walking range that the elderly can reach,

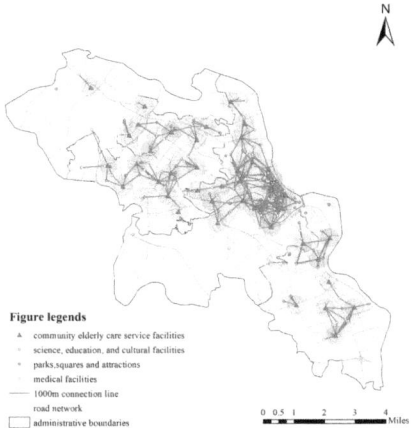

Figure legends
- ▲ community elderly care service facilities
- · science, education, and cultural facilities
- · parks,squares and attractions
- · medical facilities
- —— 1000m connection line
- road network
- ☐ administrative boundaries

0 0.5 1   2   3   4
                                    Miles

Fig. 8. Network distribution of community facilities for the elderly

a walking network analysis is conducted to connect nearby elderly care service facilities, forming a comprehensive and balanced network layout of community elderly care service facilities (Fig. 8). The research area has abundant green spaces and cultural and entertainment venues, which are closely connected with elderly care service facilities, and a home community elderly care service network has basically been formed.

5 Conclusion and Discussion

5.1 Conclusion

Based on POI data, this study starts from the perspective of location theory, using service area analysis, location configuration model, and nearest facility point analysis methods to analyze and optimize the spatial layout of community facilities for the elderly at the 7 streets of Fucheng District, Mianyang City, obtaining following conclusions:

The service area of the current community elderly care facilities mainly covers Gongqu Street and Chengxiang Street, and the other five streets are scattered. The service area of the current community facilities for the elderly can only cover 33.86% of the land area of the research area, and half of the residential points and elderly population are not within this range. There are obvious coverage gaps in community elderly care facilities in the research area, and the coverage of facility services needs to be improved.

After the calculations of two models, it is finally determined to add 20 new elderly care facilities, the general number of facilities for the elderly after optimization reaches 46. After optimization, the distance from most residential points to the nearest community elderly care facility is less than 1000m, which can meet the convenience of the elderly, and the proportion of the elderly served by community elderly care facilities within a walking distance of 1500m reaches 92.87%, basically meeting the requirements of the urban community "15-min elderly care service circle", forming a comprehensive and balanced network distribution of community facilities for the elderly.

5.2 Discussion

Optimizing the spatial layout of community elderly care facilities and building community elderly care living network and service circle are currently easy to realize community elderly care methods, and can also greatly improve the inclusiveness of the city to the elderly population. The integration of community elderly care service facilities with other public facilities such as green space, sports and sports, and medical care in the street can realize the comprehensiveness and diversification of community elderly care, and can also provide scientific site selection of elderly care facilities in the process of urban planning. However, the research method used in this paper still has some limitations, which cannot consider more influencing factors. For example, the actual needs of different elderly groups are not considered; The more advanced improved location allocation model was not used in the research. In an ideal state, the site selection of new elderly care facilities is carried out. Future studies can improve the location allocation model for facility layout, consider the subjective needs of the elderly population, and combine the overall urban planning of the study area, so as to make the optimization results more scientific and practical.

References

1. Xiaozhao, L.: Big data on aging in chinese cities: 149 cities deeply aged, concentrated in these provinces. First Financial Daily, 6 Sept 2021 (A06)
2. Jun, T., Weiwei, L.: The process and misunderstandings of aging development in China. J. Beijing Univ. Technol. (Social Sciences Edition) 18(04), 8–18 (2018)
3. Mianyang Municipal People's Government. The Fourteenth Five-Year Plan for the Construction of the Elderly Care Service System in Mianyang City [EB/OL] (23 Dec 2021). https://www.my.gov.cn/zwgk/zc/zfwj/27557541.html
4. Tsou, K.W., Hung, Y.T., Chang, Y.L.: An accessibility-based integrated measure of relative spatial equity in urban public facilities. Cities 6, 22 (2005)
5. Zenglin, H., Yuan, L., Tianbao, L., et al.: Spatial differentiation analysis of public service facility configuration in community living circles: a case study of Shahekou District, Dalian City. Prog. Geogr. Sci. 38(11), 1701–1711 (2019)
6. Fei, H., Renchao, L.: A study on the supply and demand matching of urban community home-based elderly care services based on accessibility measurement: a case study of Nanjing. Econ. Geogr. 40(09), 91–101 (2020)
7. Fangzhou, L.: Evaluation of spatial configuration of elderly care facilities in Jinan City. Shandong Jianzhu University (2018)
8. Lei, Q.: Research on the spatial layout optimization of community elderly care facilities in Yangzhou City. Yangzhou University (2023)
9. GB50180–2018. Urban Residential Area Planning and Design Standards
10. Kang, D.: Study on the evaluation and optimization of aging in community health service centers in Hefei City. Hefei University of Technology Architecture (2019)
11. Zhibiao, X., Yingying, X.: Research on the optimization of urban fire facilities site selection based on fire risk assessment - taking the central urban area of Gaozhou City, Guangdong Province as an example. Shanghai Urban Plann. 06, 117–123 (2020)
12. Shuo, Y., Taofang, Y.: The application progress of the "location-configuration model" in regional medical service facilities planning. Urban Planning Society of China, Dongguan Municipal People's Government. Sustainable Development Rational Planning - 2017 China

Urban Planning Annual Conference Proceedings (05 New Technology Application in Urban Planning). Tsinghua University Institute of Architecture and Urban Studies, Department of Urban Planning, 16 (2017)

13. Weitao, Z., Yingxia, Y.: The configuration and layout of shelters under the perspective of disaster risk overlapping differentiation - taking coastal port cities as an example. Urban Probl. **08**, 41–50 (2019)

Urban Heat Mitigation and Adaptation

Cooling Benefits of Urban Cooling Infrastructures: A Review

Yu Luo[1,2], Xiang Cheng[2], Bart Julien Dewancker[1], and Bao-Jie He[2,3(✉)]

[1] Faculty of Environmental Engineering, Kitakyushu University, Fukuoka 808-0135, Japan
[2] Centre for Climate-Resilient and Low-Carbon Cities, School of Architecture and Urban Planning, Key Laboratory of New Technology for Construction of Cities in Mountain Area, Ministry of Education, Chongqing University, Chongqing 400045, China
baojie.he@cqu.edu.cn
[3] Institute for Smart City of Chongqing University in Liyang, Chongqing University, Jiangsu 213300, China

Abstract. As climate change intensifies, a significant rise in urban temperatures is exacerbating the global menace of urban heat, increasingly impacting the human residential environment. This paper conducts a comprehensive analysis of existing literature and data to explore the multifaceted impacts of urban heat on global cities across social, environmental, health, and economic domains. The article elaborates on the definition, cooling mechanisms, and primary types of Urban Cooling Infrastructures (UCI), including blue infrastructure, green infrastructure, white and grey infrastructure, and urban design. It emphasizes the importance of UCI in improving the quality of the urban residential environment against the backdrop of global warming. By showcasing the comprehensive cooling benefits, the study contributes not only to enhancing overall urban resilience but also offers profound insights for sustainable urban development. The research aims extend beyond assisting policymakers and urban planners in formulating effective strategies for climate resilience and sustainability. It also seeks to guide future research directions in UCI within the context of widespread urban heatwaves.

Keywords: Urban heat · Human residential environment · Thermal comfort · Cooling infrastructure · Cooling benefit

1 Introduction

Overheating is considered "one of the most under-recognized hazards of climate change" [1, 2]. As greenhouse gases accumulate and more heat is trapped in the atmosphere [3], global temperatures have already risen by more than 1.2 °C compared to the industrial era [4]. In recent years, the rate has accelerated, and comparing to pre-industrial level, global average temperature is expected to rise by 2.7 °C by the end of the century [2]. Excessive heat stress and greenhouse gases are making cities overheat, affecting human habitats mainly in the form of urban heat islands, causing problems such as air pollution and resource depletion, which can lead to health problems and economic losses. Therefore, we need immediate and efficient actions to mitigate global urban heat.

© The Author(s) 2025
B.-J. He et al. (Eds.): UCSUD 2023, LNCE 559, pp. 441–455, 2025.
https://doi.org/10.1007/978-981-97-8401-1_31

Although traditional cooling measures such as air conditioning and electric fans provide urban residents with a certain level of thermal comfort, they are increasingly inadequate in addressing the escalating urban heat challenge and can exacerbate outdoor heat environmental issues. Consequently, governments, urban planners, and researchers have begun to seek more effective strategies for heat preparation and prevention, heat adaptation, and heat mitigation [5]. However, most of the current literature focuses only on specific types of certain prevention and mitigation strategies and lacks systematic exploration of their combined cooling effects. More importantly, UCI are considered one of the key measures to combat urban overheating. UCIs offer cooling solutions to cities through increasing urban green spaces, improving water body management, and other methods [6]. Yet, infrastructures to enhance climate resilience have been limited to addressing water hazards such as flooding and sea level rise, while UCI has not received much attention or research.

Therefore, through a literature review, the research proposes a framework of global urban heat hazards and cooling benefits of UCIs (Fig. 1), featuring three circles ded-icated to deeply exploring the following three core research questions:

1) What are the hazards of urban high temperatures on the human living environ-ment? It aims to comprehensively examine the impacts of urban over-heating on the environment, health, society, and economy, emphasizing the urgency of mitigating urban heat issues.
2) What are Urban Cooling Infrastructures (UCI), and what are their classifications and cooling mechanisms? By analyzing different types of UCIs and their cooling mechanisms, it assesses their effectiveness and applicability in practical applications.
3) What are the cooling benefits of Urban Cooling Infrastructures? This explores the direct and indirect contributions of UCI implementation to the environmental, health, social, and economic aspects.

This research introduces the concept of UCI, analyzes the interactions between the cooling benefits of UCI and global urban heat hazards, and emphasizes the environmental, health, social, and economic [6] significance of UCI for human settlements. And it aims to provide a comprehensive perspective to understand the potential of UCI to mit-igate the heat island effect in global cities, and to provide a scientific basis and practical guidelines for urban planning and policy making.

2   Methodology

This study employs a literature review methodology to systematically collect, review, and analyze existing research on the urban heat island effect and Urban Cooling Infras-tructures (UCI) to answer the three core research questions posed. The research method involves the following steps:

2.1   Data Resource

The study utilized a literature review approach. Multiple academic databases including Web of Science, Scopus, and PubMed were searched, combining themes of urban heat

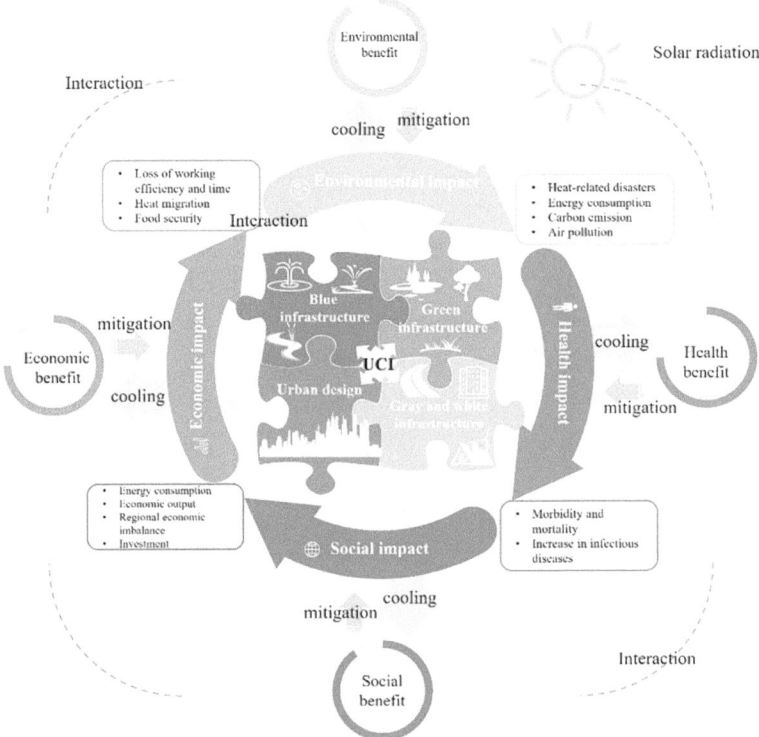

Fig. 1. The framework of global urban heat impacts and cooling benefits of UCIs

island effect and Urban Cooling Infrastructures with primary keywords such as urban heat island effect, urban cooling infrastructure, heat mitigation, blue infrastructure, green infrastructure, grey and white infrastructure, urban design, shading, ventilation, and their subsets and synonyms, to retrieve literature from the past decade (2013–2023).

2.2 Result

After rigorous screening to exclude irrelevant and duplicate entries, a total of 365,657 relevant publications were included, comprising 291,306 journal articles and 23,885 conference papers (as shown in Fig. 2). However, among these, only 18,668 publications involved 'infrastructure', with 13,083 being journal articles and 1,423 conference papers.

As shown in Fig. 2 and Fig. 3, the volume of published literature related to urban cooling has steadily increased from 19,956 articles in 2013 to 42,621 (Fig. 2(b)) articles in 2023, indicating a continuous rise in research interest in the field of heat mitigation as well as an intensification of scientific research activities. The literature related to Urban Cooling Infrastructures has grown from 2,023 articles in 2013 to 2,545 (1 (Fig. 3(b)) articles in 2023, indicating that the research field of cooling infrastructures is rapidly developing.

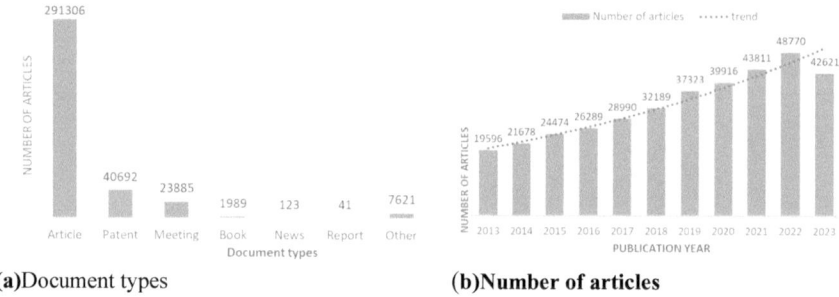

Fig. 2. Analysis of literature search results related to urban cooling

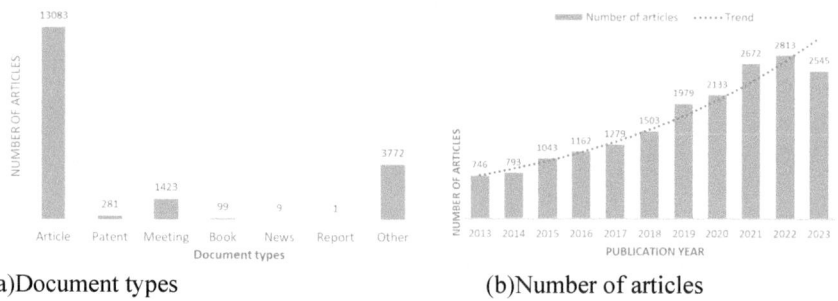

Fig. 3. Analysis of literature related to urban cooling infrastructure.

In summary, while the volume of literature on Urban Cooling Infrastructures (UCI) only constitutes a small portion of the total research on the urban heat island effect, its growing trend year by year highlights the importance of this field and the increasing attention from the scientific community. Having clarified the development trend and current status of the research literature, the following text will delve into the specific definitions of Urban Cooling Infrastructures and their cooling mechanisms. It will explore how these mechanisms function within the ever-changing urban environment and their potential positive impacts on the environment, society, health, and economy.

3   Urban Cooling Infrastructure

UCIs refer to integrated urban infrastructure (including fountains, wetlands, green spaces, and architectural or encircling structures employing cooling materials) comprising a series of preventive, adaptive, and mitigative strategies designed to withstand extreme heat conditions. These infrastructures are utilized to (directly or indirectly) reduce temperatures (indoor or outdoor). UCI strives to improve the microclimate in urban settings, reducing the negative effects of elevated temperatures on the environment, society, and health. Furthermore, they support social and economic growth, thus improving the quality of urban human residential environment overall. UCI includes blue infrastructure, green infrastructure, white and gray infrastructure. Urban planners and designers, through urban design, configure urban spaces in various forms to accommodate diverse urban open spaces and microclimates, regulating urban microclimates, and alleviating urban heat.

3.1 Blue Infrastructure

Blue infrastructure utilizes water and water-based facilities to enhance urban ecology and adapt to microclimates. Water, characterized by its high specific heat capacity, low thermal conductivity, and low emissivity [7, 8], absorbs latent heat through processes such as evaporation [9–11] and convection [12],significantly reducing environmental temperatures and contributing to the formation of cool islands (Fig. 4). Blue infrastructure typically comprises two main types: natural and artificial infrastructures, designed in various forms such as rivers, ponds and lakes, urban wetlands [6, 13], as well as water-based artificial facilities like fountains and spray systems [6, 14].

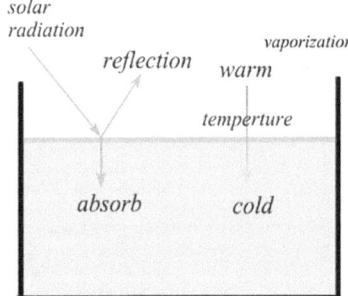

Fig. 4. Cooling mechanism of water

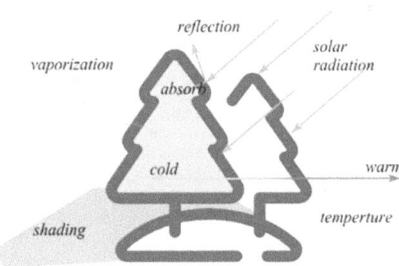

Fig. 5. Cooling mechanism of greenery

3.2 Green Infrastructure

Green infrastructure is typically described as a network of natural and semi-natural elements that are interconnected, which provides a broad range of ecosystem services, with a particular emphasis on microclimate regulation [15–19]. Green infrastructure achieves cooling by mitigating city's or buildings' surface temperatures through the absorption and reflection of solar radiation, evaporation [20], shading [21], and thermal insulation (Fig. 5) [6, 22, 23], with vegetation canopies intercepting a significant portion of the heat flux. Green infrastructures mainly include urban parks, forests, street greening, green roof, and vertical greening (green wall).

3.3 Grey and White Infrastructure

Grey and white infrastructure refer to the novel materials in the realm of built environment, characterized by lacking natural features and being entirely human-designed and engineered, primarily focusing on construction materials in this context. Typically, it involves altering the thermal features of urban surface to possess attributes including high reflectance, high permeability, high thermal conductivity, or high heat capacity, aimed at reducing solar radiation absorption, facilitating rapid heat dissipation, or storing heat (Figs. 6 and 7). This category encompasses reflective roofs, reflective pavements, permeable pavements [24–26] and thermochromic materials (phase-change materials, etc.) [5, 27, 28]. In recent years, the majority of research efforts have predominantly focused on cool pavements and cool roofs.

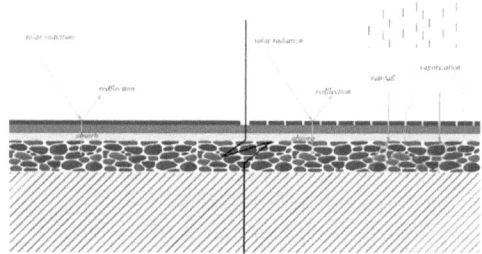

Fig. 6. Cooling mechanism of cool pavement

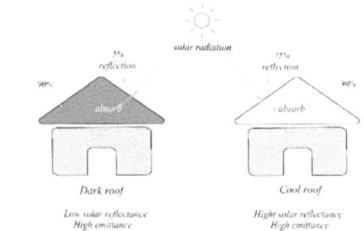

Fig. 7. Cooling mechanism of cool roof

3.4 Urban Design

Urban geometry modifies the urban thermal environment by altering the exchange of thermal radiation and convection in open spaces [29]. Urban design can achieve cooling objectives by strategically configuring buildings, streets, blocks, communities, and regions to enhance ventilation and to provide solar radiation obstruction (shading) [6, 30–32]. Urban design leverages and alters urban morphology to achieve optimal shading and ventilation. Shading devices and components can achieve shading [22], while urban ventilation is primarily addressed through the design of street orientation, width, building height, and density. These strategies contribute to the optimization of thermal environment, elevating the comfort and sustainability of cities.

4 Impacts of Global Urban Heat and Potential Cooling Benefits of UCIs

Urban regions are more prone to accumulating heat. Resulting in higher average temperatures in large cities than in rural settings. Such elevated temperatures have significant implications on human health and well-being, economic output, environmental conditions, and key infrastructure and services. The reciprocal interactions among them will exacerbate the impact of high temperatures on themselves, rendering the environment more adverse (Fig. 3). UCIs serve as a comprehensive strategy to address the challenges of urban heat, influencing urban ecology, social health, and the economy through direct temperature control and indirect benefits. UCIs aim to go beyond simple cooling, striving to create a more suitable living environment for humans and enhance the overall resilience of cities to heat impacts, becoming a key component of urban sustainable development and effectively tackling the challenges of urban thermodynamics.

4.1 Environmental Impacts and Benefits

Wildfires, floods, storms, and droughts, among other extreme weather events, pose serious challenges. The period between 2001 and 2004, as well as between 2016 and 2019, saw a marked rise in the number of days with a high risk of wildfires [4]. In 2019, vulnerable populations globally experienced an additional 475 million instances of heatwaves [4]. In China, elevated temperatures contribute to reduced precipitation during the summer and autumn seasons, leading to the phenomenon of "anti-drying during flood season" in the Yangtze River basin. Regionally and temporally, there is a distinct increase in both area-specific and episodic droughts. From September to early October 2022, the station-based occurrences of extreme high-temperature events nationwide reached 1.51, surpassing the annual average by 1.39 [33].

The urban overheating contributes to an increased demand for AC systems and cooling facilities in cities, leading to elevated energy consumption and carbon emissions. The escalating urban power demand affects electricity costs and system efficiency [34], consequently influencing the consumption of fossil fuels [35], which constitutes the largest carbon flux in the urban atmosphere [5]. High temperatures also amplify the dependence on water resources, resulting in excessive water use [36]. To alleviate the discomfort caused by high temperatures, individuals often resort to increased water intake, showering, and similar measures to enhance thermal comfort. Moreover, watering practices are employed to sustain private gardens, public green spaces, and urban forestry in response to the challenges posed by high temperatures, aridity, and drought [5].

Hot weather is typically accompanied by a decline in air quality. With the ongoing process of urbanization, the size of urban populations is steadily increasing, accompanied by a rising demand for energy. Anthropogenic activities, such as the combustion of fossil fuels, intensify the greenhouse effect (CO_2) and contribute to air pollution (SO_2, NO_2, CO, PM), as well as environmental pollution [37].

In this context, urban complexes, by introducing green infrastructure and enhancing vegetation cover, effectively mitigate the urban heat island effect and combat the challenges posed by extreme weather events. These measures not only improve the health of urban ecosystems but also provide cleaner air for city residents by reducing carbon emissions, improving air quality, and lowering pollutant concentrations [6]. This enhances the overall environmental quality, showcasing the significant role of urban complexes in promoting urban sustainable development and addressing climate change.

4.2 Health Impacts and Benefits

Extreme hot can impose significant health burdens [3, 38]. Prolonged exposure to overheating may result in heat stress and associated illnesses such as heatstroke, heat exhaustion, cardiovascular diseases, respiratory diseases, and an elevated incidence of morbidity and mortality. As temperature thresholds rise and heatwave durations extend, the risk of mortality correspondingly increases [39]. According to statistics, from 2016 to 2018, over one million people died annually due to air pollution caused by coal-fired power generation[4]. Vulnerable populations, including children [40], the old [41], pregnant women [42], and patients [43] are more susceptible to the impacts of high temperatures, influenced by differences in income, gender, age, and physical conditions [3, 36]. In the past two decades, individuals aged over 65-year-old experienced a 53.7% increase in mortality associated with high temperatures, with the death toll reaching 0.3 million in 2018 [3].

The continuously changing environmental conditions have exacerbated the transmission of various infectious diseases. In the last several decades,the regional increase in malaria and vibrio infections has been attributed to elevated temperatures. The intensity, frequency, and geographic range of West Nile virus outbreaks have also escalated. Additionally, Compared to 41 years ago, the blooming period of allergy-inducing trees like birch, olive, and oak has shifted earlier by 10 to 20 days, negatively affecting around 40% of Europeans who suffer from pollen allergies. [44].

UCIs effectively mitigate the adverse effects of high temperatures on the human body, reducing the incidence and mortality rates of heat-related stress and diseases [45]. Additionally, the implementation of UCIs has a regulatory effect on the reproduction of certain insects and the transmission of related diseases [46]. This not only helps alleviate the pressure on healthcare systems and reduce medical burdens but also provides residents with a healthier and safer living environment, ultimately increasing average life expectancy.

4.3 Social Impacts and Benefits

Loss of working hours and efficiency. The impacts of high temperatures on workers are evident in potential losses of working hours and efficiency [47]. Globally, in 2019, there may have been a loss of 3.02 trillion work hours, an increase of 1.03 trillion hours compared to the year 2000 [4]. Among the most affected are agricultural workers (the developed countries) and construction workers (the developing and poor countries). On the one hand, extreme heat, due to the human body's sensitivity to heat stress, can lead to reduced attention, thereby affecting work efficiency (reducing intensity and quality) and working time [36]. On the other hand, excessively high temperatures can cause equipment malfunctions, including failures in electrical connectors and power semiconductor devices [48, 49], which may lead to significant losses in productivity [36].

Heat migration. Extreme weather events (drought,,heatwave) are increasingly recognized as major drivers of population migration [2, 50–52]. It is projected that urban individuals will experience twice the exposure to heat compared to those in non-urban areas [53, 54]. Population migration can be categorized into permanent migration and temporary migration (comfort or seasonal migration) [55]. Temporary migration is likely to bring about irreversible changes to human habitats, ultimately evolving into permanent migration [2, 56–58].

Food Security. Food is a substance that globally satisfies the physiological and biochemical energy requirements of organisms. According to statistics from 1981 to 2019, due to rising temperatures, the potential yields of major staple crops (such as maize, wheat, soybeans, and rice) have consistently decreased by 5.6% on a global scale, relative to the baseline. By 2030, people suffering from malnutrition will increase to over 840 million due to food shortages [4, 59].

By improving the urban living environment and increasing residential comfort, UCIs help stabilize the urban resident population and reduce population outflow [55]. Simultaneously, the implementation of UCIs provides cities with more green spaces, enhancing the overall living environment and offering residents more livable living spaces. Furthermore, promoting UCIs also contributes to food security by fostering urban agriculture and the development of green spaces, providing residents with a richer food supply and improving the quality of urban life [59].

4.4   Economic Impacts and Benefits

Energy consumption. It is a critical aspect of urban dynamics. Urban not only generate more than 80% of global GDP, but also are responsible for consuming over 60% of energy. The challenge of mitigating urban heat to maintain thermal comfort places a significant energy demand on cities, thereby substantially affecting overall energy conumption. Urban heat Costs. The elevated costs incurred by humans due to high temperatures, including impacts on working hours and health, are associated with the economic consequences. In 2018, the financial cost associated with heat-related fatalities was roughly 1.2% of the regional Gross National Income. In 2014, Australia experienced an aggregate economic loss of 6.2 billion USD due to high temperatures [60], wherein productivity losses and work hour reductions may have resulted in average annual economic losses per person of 932 USD and 845 USD, respectively [36].

Infrastructure investment. High temperatures may induce infrastructure failures, such as power system overloads, road damage, and water resource shortages, thereby increasing urban maintenance costs. The life cycle costs of infrastructure depend on factors such as excessive temperatures, usage frequency, and equipment maintenance or replacement costs. As temperatures gradually rise, it forces refrigeration equipment to operate less efficiently, requiring higher energy loads [61]. Moreover, extremely hot environments can impede equipment functionality, reducing lifespan and operational efficiency, consequently raising replacement or maintenance costs [62].

Regional Economic Disparities. The seasonality and policy-induced migration associated with high temperatures contribute to imbalances in regional economic development. This phenomenon is particularly pronounced in small cities and low-income countries experiencing significant population outflows, resulting in diminished employment opportunities and reduced service efficiency. The migration patterns of high-income groups exhibit distinctive characteristics, fostering a predisposition towards the development of infrastructure in sectors such as science, education, culture, and healthcare. This inclination, in turn, stimulates a corresponding trend in the development of consumer markets [54].

UCIs yield significant long-term investment returns. Beyond direct economic benefits such as increased work efficiency, reduced energy costs, and savings in investment and maintenance expenses, the promotion of UCIs also contributes to regional economic development [63]. By creating a more livable urban environment, UCIs attract more businesses and talent, driving sustainable economic growth and providing robust support for the future development of cities.

UCIs function as a multifaceted solution, playing a constructive role in alleviating urban heat issues. Their positive impact extends beyond mitigating the immediate challenges posed by overheating, contributing not only to the enhancement of overall urban resilience but also providing profound implications for the sustainable development of cities.

5 Conclusion

The global urban heat issue has emerged as a formidable challenge, exacerbated by global warming. The myriad problems stemming from urban overheating are becoming increasingly pronounced, with projections indicating a further intensification in the future. This research provides a comprehensive analysis of the global impacts of urban heat, exploring the adverse effects of urban overheating on the environment, health, society, and economy. It underscores the profound threats posed by urban heat to the human residential environment. In addressing the challenge of urban heat, this paper proposes and underscores the significance of UCIs in temperature moderation, improvement of regional microclimates, enhancement of climate resilience, and elevation of the quality of the human residential environment. UCIs, as a comprehensive measure for mitigating urban heat, encompass various aspects such as blue infrastructures, green infrastructures, grey-white infrastructures, and urban design. Through a multidimensional and multilayered approach, UCIs aim to reduce urban temperatures and alleviate the adverse impacts of urban heat on cities.

This research provides a comprehensive analysis of global urban heat hazards and the cooling benefits of UCI, reveals the significant role of UCI in mitigating the urban heat island effect, improving the living conditions of urban residents, and promoting sustainable urban development. By systematically reviewing the current literature, this paper not only delves into the multifaceted impacts of urban heat hazards on the environment, health, society, and economy but also analyzes the classification and mechanisms of action of various types of UCI—including blue-green infrastructure, white-grey infrastructure, and urban design—and their effectiveness in mitigating these impacts. The research findings emphasize that UCI can significantly lower urban temperatures, reduce the incidence of heat-related diseases, enhance the quality of life for city dwellers, and foster economic sustainability. Moreover, the implementation of UCI has profound implications for improving the environmental quality of cities, enhancing urban ecological resilience, and addressing global climate change.

- From an environmental perspective, the implementation of UCIs contributes to the enhancement of urban green development, the mitigation of carbon emissions, improvement of air quality, reduction of pollutant concentrations, and the provision of fresher air to the urban environment, thereby reinforcing the overall environmental quality of the city.
- On the societal front, the introduction of UCIs stabilizes the urban resident population, enhances the quality of life for city dwellers, and provides a more livable human residential environment.
- Regarding health aspects, UCIs effectively alleviate the adverse impacts of high temperatures on the human body, reducing the incidence of heat-related stress and associated illnesses. This, in turn, furnishes residents with a healthier and safer living environment, contributing to an increase in average life expectancy.
- Economically, UCIs exhibit significant long-term investment returns. Beyond direct economic benefits such as enhanced work efficiency and reduced energy costs, they also facilitate regional economic development, providing robust support for the sustainable economic growth of the city.

However, this research merely elucidates the definition, cooling mechanisms, and cooling benefits of UCIs, providing readers with a comprehensive overview of the hazards posed by overheating and the efficacy of cooling measures. Numerous critical issues have not been addressed in this study, thereby delineating directions for future research endeavors.

- Individual facilities are also influenced by factors such as temporal and spatial conditions, climatic variations, land-use characteristics, and the physical properties of the elements themselves, exhibiting diverse cooling effects. Future efforts should focus on summarizing the influencing factors of UCIs measures and comprehensively understanding their cooling effects.
- Things do not exist in isolation, and it is imperative to consider whether each UCI measure within different combinations will yield positive or negative effects. Subsequent research needs to evaluate multiple combinations of UCI measures to determine the mutual influences on cooling effects, seeking optimal combinations for temperature reduction in specific regions.
- Urban cooling (heat mitigation) has implications for the environment, society, health, and the economy. How to quantify the impact of heat and cooling benefits? How to assess the cost-benefit ratios (economic benefit)? The research still requires further exploration and synthesis.

To sum up, UCIs, as a multifaceted solution, play a constructive role in mitigating urban heat-related challenges. Through the comprehensive manifestation of cooling benefits, UCIs contribute not only to the enhancement of overall urban resilience but also offer profound implications for sustainable urban development. In future urban planning and development endeavors, heightened emphasis and widespread adoption of UCIs should be prioritized to foster the creation of more habitable, healthy, and environmentally friendly cities.

References

1. Nature, Cities must protect people from extreme heat. Nature **595**(7867), 331–332 (2021)
2. Zander, K.K., Baggen, H.S., Garnett, S.T.: Topic modelling the mobility response to heat and drought. Climatic Change **176**(4) (2023)
3. Guo, Y., Li, S.: Heat exposure and human health in the context of climate change (2022)
4. Watts, N., et al.: The 2020 report of the lancet countdown on health and climate change: Responding to converging crises. Lancet **397**(10269), 129–170 (2021)
5. Wang, Z.-H.: Compound environmental impact of urban mitigation strategies: co-benefits, trade-offs, and unintended consequence. Sustain. Cities Soc. **75** (2021)
6. He, B.-J., et al.: A framework for addressing urban heat challenges and associated adaptive behavior by the public and the issue of willingness to pay for heat resilient infrastructure in Chongqing, China. Sustain. Cities Soc. **75** (2021)
7. Ghosh, S., Das, A.: Modelling urban cooling island impact of green space and water bodies on surface urban heat island in a continuously developing urban area. Model. Earth Syst. Environ. **4**(2), 501–515 (2018)
8. Wilson, J.S., et al.: Evaluating environmental influences of zoning in urban ecosystems with remote sensing. Remote Sens. Environ. **86**(3), 303–321 (2003)
9. Le Phuc, C.L., et al.: Cooling island effect of urban lakes in hot waves under foehn and climate change. Theoret. Appl. Climatol. **149**(1–2), 817–830 (2022)
10. Oke, T.: Boundary Layer Climates. Psychology Press (1992)
11. Amani-Beni, M., et al.: Impact of urban park's tree, grass and waterbody on microclimate in hot summer days: A case study of Olympic Park in Beijing, China. Urban For. & Urban Green. **32**, 1–6 (2018)
12. Spronken-Smith, R.A., Oke, T.R., Lowry, W.P.: Advection and the surface energy balance across an irrigated urban park. Int. J. Climatol. **20**(9), 1033–1047 (2000)
13. Teshnehdel, S., et al.: Improving outdoor thermal comfort in a Steppe climate: effect of water and trees in an Urban Park. Land **11**(3) (2022)
14. Ulpiani, G.: Water mist spray for outdoor cooling: A systematic review of technologies, methods and impacts. Applied Energy **254** (2019)
15. Bartesaghi Koc, C., Osmond, P., Peters, A.: Evaluating the cooling effects of green infrastructure: a systematic review of methods, indicators and data sources. Solar Energy **166**, 486–508 (2018)
16. Benedict, M.A., McMahon, E.T.: Green Infrastructure: Linking Landscapes and Communities. Island Press (2006)
17. EEA: Green infrastructure and territorial cohesion: The concept of green infrastructure and its integration into policies using monitoring systems. European Environment Agency: EEA Technical report, Vol. 18 (2011)
18. Mell, I.C.: Green infrastructure: concepts, perceptions and its use in spatial planning. Newcastle University (2010)

19. Roy, S., Byrne, J., Pickering, C.: A systematic quantitative review of urban tree benefits, costs, and assessment methods across cities in different climatic zones. Urban For. & Urban Green. **11**(4), 351–363 (2012)
20. Takebayashi, H., Moriyama, M.: Surface heat budget on green roof and high reflection roof for mitigation of urban heat island. Build. Environ. **42**(8), 2971–2979 (2007)
21. Hoyano, A.: Climatological uses of plants for solar control and the effects on the thermal environment of a building. Energy Build. **11**(1), 181–199 (1988)
22. Akbari, H., et al.: Local Climate Change and Urban Heat Island Mitigation Techniques – the State of the Art. J. Civ. Eng. Manag. **22**(1), 1–16 (2015)
23. Morakinyo, T.E., et al.: Right tree, right place (urban canyon): tree species selection approach for optimum urban heat mitigation - development and evaluation. Sci. Total. Environ. **719**, 137461 (2020)
24. Stempihar, J.J., et al.: Porous asphalt pavement temperature effects for urban heat island analysis. Transp. Res. Rec. **2293**(1), 123–130 (2012)
25. Ferrari, A., et al.: The use of permeable and reflective pavements as a potential strategy for urban heat island mitigation. Urban Climate **31** (2020)
26. Liu, Y., Li, T., Peng, H.: A new structure of permeable pavement for mitigating urban heat island. Sci. Total. Environ. **634**, 1119–1125 (2018)
27. Roman, K.K., et al.: Simulating the effects of cool roof and PCM (phase change materials) based roof to mitigate UHI (urban heat island) in prominent US cities. Energy **96**, 103–117 (2016)
28. Nagano, K., et al.: Thermal characteristics of magnesium nitrate hexahydrate and magnesium chloride hexahydrate mixture as a phase change material for effective utilization of urban waste heat. Appl. Therm. Eng. **24**(2–3), 221–232 (2004)
29. Lai, D., et al.: A review of mitigating strategies to improve the thermal environment and thermal comfort in urban outdoor spaces. Sci. Total. Environ. **661**, 337–353 (2019)
30. Oke, T.R.: Street design and urban canopy layer climate. Energy Build. **11**, 103–113 (1988)
31. Arnfield, A.J.: Street design and urban canyon solar access. Energy Build. **14**, 117–131 (1990)
32. Du, P., et al.: Understanding the seasonal variations of land surface temperature in Nanjing urban area based on local climate zone. Urban Climate **33** (2020)
33. Centre, N.C.: China Climate Bulletin (2022). N.C. Centre, Editor (2023)
34. Shickman, K., Rogers, M.: Capturing the true value of trees, cool roofs, and other urban heat island mitigation strategies for utilities. Energ. Effi. **13**(3), 407–418 (2019)
35. Ritchie, H., Rosado, P.: Electricity Mix. OurWorldInData.org (2020)
36. He, B.-J.: Green building: a comprehensive solution to urban heat. Energy Build. **271** (2022)
37. Tuomimaa, J., et al.: Developing adaptation outcome indicators to urban heat risks. Clim. Risk Manag. **41** (2023)
38. Moon, J.: The effect of the heatwave on the morbidity and mortality of diabetes patients; a meta-analysis for the era of the climate crisis. Environ. Res. **195**, 110762 (2021)
39. Kjellstrom, T., et al.: Heat, human performance, and occupational health: a key issue for the assessment of global climate change impacts. Annu. Rev. Public Health **37**(1), 97–112 (2016)
40. Mangus, C.W., Canares, T.L.: Heat-related illness in children in an era of extreme temperatures. Pediatr. Rev. **40**(3), 97–107 (2019)
41. Gronlund, C.J., et al.: Vulnerability to Renal, Heat and Respiratory Hospitalizations During Extreme Heat Among U.S. Elderly. Clim Change **136**(3), 631–645 (2016)
42. Ward, A., et al.: The impact of heat exposure on reduced gestational age in pregnant women in North Carolina, 2011–2015. Int. J. Biometeorol. **63**(12), 1611–1620 (2019)
43. Kenny, G.P., et al.: Heat stress in older individuals and patients with common chronic diseases. CMAJ **182**(10), 1053–1060 (2010)

44. van Daalen, K.R., et al.: The 2022 Europe report of the Lancet Countdown on health and climate change: towards a climate resilient future. Lancet Public Health **7**(11), e942–e965 (2022)

45. Orimoloye, I.R., et al.: Implications of climate variability and change on urban and human health: A review. Cities **91**, 213–223 (2019)

46. Yang, L., et al.: Can urban greening increase vector abundance in cities? The impact of mowing, local vegetation, and landscape composition on adult mosquito populations. Urban Ecosystems **22**(5), 827–839 (2019)

47. Kjellstrom, T., et al.: Estimating population heat exposure and impacts on working people in conjunction with climate change. Int. J. Biometeorol. **62**(3), 291–306 (2018)

48. Szendi, J.: Effects of the Heat Wave to the Object Security. Aerul si Apa. Componente ale Mediului, pp. 116–123 (2017)

49. Khazaka, R., et al.: Survey of high-temperature reliability of power electronics packaging components. IEEE **30**(5), 2456–2464 (2015)

50. Mueller, V., Gray, C., Kosec, K.: Heat stress increases long-term human migration in rural Pakistan. Nat. Clim. Chang. **4**(3), 182–185 (2014)

51. Mastrorillo, M., et al.: The influence of climate variability on internal migration flows in South Africa. Glob. Environ. Chang. **39**, 155–169 (2016)

52. Mueller, V., et al.: Do Social protection programs foster short-term and long-term migration adaptation strategies? Environ. Dev. Econ. **25**(2), 135–158 (2020)

53. Wouters, H., et al.: Heat stress increase under climate change twice as large in cities as in rural areas: A study for a densely populated midlatitude maritime region. Geophys. Res. Lett. **44**(17), 8997–9007 (2017)

54. Zander, K.K., Garnett, S.T.: The importance of climate to emigration intentions from a tropical city in Australia. Sustainable Cities and Society **63** (2020)

55. McLeman, R., Smit, B.: Migration as an adaptation to climate change. Clim. Change **76**(1–2), 31–53 (2006)

56. McLeman, R.: International migration and climate adaptation in an era of hardening borders. Nat. Clim. Chang. **9**(12), 911–918 (2019)

57. Ulus, T., Ellenblum, R.: How long and how strong must a climatic anomaly be in order to evoke a social transformation? Historical and contemporaneous case studies. Hum.Ities Soc. Sci. Commun. **8**(1), 252 (2021)

58. Zickgraf, C.: Climate change, slow onset events and human mobility: reviewing the evidence. Curr. Opin. Environ. Sustain. **50**, 21–30 (2021)

59. FAO, et al.: The State of Food Security and Nutrition in the World 2023. Rome, Italy: Urbanization, agrifood systems transformation and healthy diets across the rural–urban continuum. Rome, FAO (2023)

60. Zander, K.K., et al.: Heat stress causes substantial labour productivity loss in Australia. Nat. Clim. Chang. **5**(7), 647–651 (2015)

61. Otanicar, T.P., et al.: Impact of the urban heat island on light duty vehicle emissions for the Phoenix, AZ Area. Int. J. Sustain. Transp. **4**(1), 1–13 (2010)

62. Miner, M.J., et al.: Efficiency, economics, and the urban heat island. Environ. Urban. **29**(1), 183–194 (2016)

63. Elmqvist, T., et al.: Benefits of restoring ecosystem services in urban areas. Current Opinion in Environmental Sustainability **14**, 101–108 (2015)

Bibliometric-Based Analysis of Hotspots and Dynamics for the Urban Thermal Environment During 1990–2019

Yuejing Gao[1(✉)], Jingyuan Zhao[2], and Li Han[1,3]

[1] School of Architecture and Civil Engineering, Xi'an University of Science and Technology,
Xi'an 710054, Shaanxi, China
yuejinggao@xust.edu.cn
[2] School of Architecture, Chang'an University, Xi'an 710054, Shaanxi, China
[3] Geological Resources and Geological Engineering Postdoctoral Research Station, Xi'an
University of Science and Technology, Xi'an 710054, Shaanxi, China

Abstract. In the post-epidemic era and extreme climate context, the study of urban health development, especially the urban thermal environment, has increasingly received extensive attention. In this paper, using Web of Science Core Collection as data sources, we analyzed the relevant literature in the field of urban thermal environment research during 1990–2019 by using scientific knowledge mapping and bibliometric method. It is found that the research in this field tends to be diversified and integrated, and the observation methods are gradually transitioning from field measurements to numerical simulation combined with field measurements. Furthermore, the focus tends to be on the quantitative studies of the urban thermal environment combined with the corresponding response mechanism for the planning action, which is shifting from passive adaptation to active governance. In the future, the research on adaptation planning under the role of urban microclimate especially thermal environment, the quantitative indicators, and dynamic collaborative planning methods will be the focus of research in this field.

Keywords: Bibliometrics · urban thermal environment · research hotspots · visualization analysis · CiteSpace

1 Introduction

At the cost of high resource consumption and environmental damage, the rapid urbanization development pattern has made China's economy grow at a high speed of nearly 10% per year since the reform and opening up. This has led to a continuous improvement in people's living standards while also posing a threat to the healthy development of the city and the living environment of the inhabitants [1, 2]. The rapid urbanization process and the spillover spreading mode have led to disorganized urban space and heterogeneous spatial patterns [3]. A large number of natural surfaces within the city have been covered by artificial structures [4], with the surface roughness increasing, and the solar radiation,

© The Author(s) 2025
B.-J. He et al. (Eds.): UCSUD 2023, LNCE 559, pp. 456–466, 2025.
https://doi.org/10.1007/978-981-97-8401-1_32

temperature, humidity, and wind speed changing, exacerbating the urban heat island (UHI) effect in the city [5]. Coupled with global warming and urbanization, the frequent occurrence of extreme weather in recent years are increasing significantly [7], resulting in severe heat-induced mortality [8], energy consumption [9], economic impacts and social inequality [10, 11]. It is expected that the urban heat challenge will intensify and become a new normal in the coming decades, with greater impacts on urban built environment [12, 13]. As a response to the aforementioned impact, the research on urban climate adaptation represented by the urban thermal environment has received extensive attention from governmental organizations and academics. However, the current research on the urban thermal environment is often based on problem-oriented case studies and lacks a systematic review and summary from the perspective of literature analysis. This leads to difficulties in grasping the hot spots of urban thermal environment research and providing further support for the green and low-carbon transformation of urban and rural construction under the Carbon Neutral Targets in China. Thus, this paper mainly focuses on the urban thermal issues in built environment and adopts the bibliometric analysis method to analyze peer-reviewed academic papers. The data is from the Web of Science (WoS) Core Collection in the period of 1990–2019, to sort out the hotspots in the urban thermal environment, aiming to grasp the research trend and provide references for future academic research in the field of urban thermal environment.

2   Methods

Scientific knowledge mapping can quantitatively present the development, evolution, and research hotspots of various disciplines. CiteSpace, as the most widely used tool among many scientific mapping tools, can conveniently and intuitively discover the potential patterns in citation data through quantitative analysis. Therefore, this paper analyzes the research status of urban thermal environment and urban planning, especially for the built environment, by using the scientific network citation database as the data source using CiteSpace software. Through analyzing the domestic and international research hotspots and development trends of the urban thermal environment from 1990–2019, this paper clarifies the research progress, identifies the hotspots in this research field, and further provides support and basis for the subsequent research related to the urban thermal environment.

According to the requirements of CiteSpace for data sources, this paper uses WoS as the data collection platform. For the WoS database, the search terms are TS = (urban thermal environment* or urban heat island*) and TS = (urban planning* or building environment*), document type = (Article), and language = English. Cumulatively, 2532 literatures were obtained. Through the elimination of irrelevant literature, and merging of synonyms, 2369 literatures were screened. All the literature contains the following information: author, title, source publication, reference, etc.

To form a preliminary knowledge of the related research on urban thermal environment, this study combines bibliometric methods to analyze the keywords of the datasets using CiteSpace software in a quantitative visualization way to get the current research hotspots in the field of urban thermal environment. The specific parameters are set as follows: the time range is 1990–2019, and the time slice period is 3 years, the node type is set as keyword. The 50 keywords with the highest frequency of occurrence in each time slice are selected, and the rest are set as linear interpolation by system default. In addition, to compensate for the shortcomings of quantitative and visual analysis in mining detailed content, the highly cited literature is deeply profiled to deepen the understanding of urban thermal environment research.

3   Results

3.1   Overall Trend

As shown in Fig. 1, a time-series analysis of the above 2369 documents was conducted. It can be seen that research on the urban thermal environment from 1990 to 2000 was basically in a stagnant period. Although the first and second IPCC Climate Assessment Conferences as well as the 1997 Kyoto Protocol were held during this period, scholars had insufficient understanding and attention to climate change, especially urban microclimates. The period from 2001 to 2010 was revival, with a steady increase in the number of literatures. Major conferences held during this period, such as the Third IPCC Climate Assessment Conference in 2001, the Bali Roadmap in 2007, and the World Climate Conference in Copenhagen, have gradually made the research on climate change and urban planning a hot topic. Against this background, some meteorologists have begun to focus on the study of urban climate effects. Another group of urban planning scholars attempts to introduce the concepts of sustainable development and ecological cities into urban planning and construction, making beneficial attempts on urban climate adaptability. Since 2011, it has been a period of accelerated development, with a sharp increase in the number of literatures. The research focus has shifted from a single dimension to multiple dimensions. Adaptable concepts such as ecological cities, low-carbon cities, and healthy cities have received widespread attention, and research has entered an unprecedented period of prosperity.

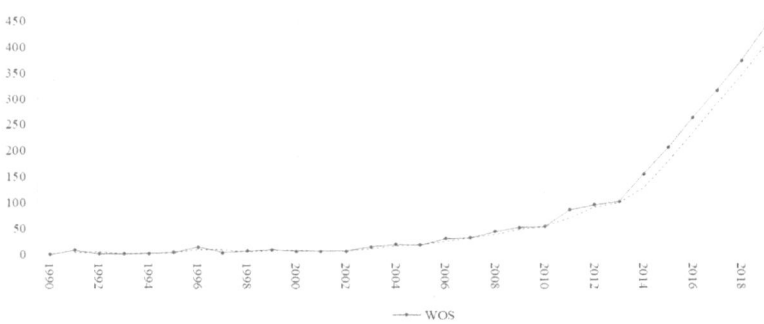

Fig. 1. Annual amount of publications for the thermal environment during 1990–2019

3.2   Keyword Analysis

Keywords and citation frequency can reveal the main research contents and hot spots [14]. Thus, this study, based on the above method, uses keywords as node types to highlight the main structural features. The keyword collinear map for the urban thermal environment is shown in Fig. 2. In addition to the UHI effect, city, and urban thermal environment, the results show that the core keywords with high centrality include thermal comfort, temperature, model, simulation, vegetation cover, built environment, etc. According to the frequency and centrality of keywords to further summarize the research content, research direction, and research methods of the urban thermal environment, the urban thermal environment can be roughly divided into two aspects of the measurement method and driving mechanism research (as shown in Table 1).

The measurement method contains two aspects of the conceptual definition and methodological data. In particular, the keywords of the conceptual definition mainly include land surface temperature (LST), temperature, heat island intensity, radiation temperature, remote sensing, surface temperature inversion, thermal comfort, model, simulation, pattern, CFD, etc. The driving mechanism mainly contains cover change, landscape pattern change, urban planning, etc. Specifically, the keywords for cover change include land use, and impervious surface. The keywords for landscape pattern change include NDVI, green space, green roof, etc. For the group of urban planning, the keywords mainly include built environment, urbanization, urban form, street canyon, air pollution, pollutant disperse, etc. It can be seen that the research related to the urban thermal environment mainly focuses on the impact of the thermal environment among different cities.

3.3 Burst Detection

Burst detection can statistically display the research hotspots that suddenly emerged in a certain period according to the change in word frequency [15]. Thus, this paper further analyzes the research in this field by using CiteSpace to detect the burst, obtains the time sequence chart of burst keywords, and indicates the burst intensity of the keywords with the start and end time, to further understand the research hotspots in this field.

From the keyword burst information presented in Table 2, it can be seen that the research on urban thermal environment presents a remarkable staging feature, with distinctive research themes at different stages. For the WoS data, the focus was still on climate change before 2000, while it gradually appeared the research on the mechanism of the thermal environment after 2000 with the emergence of keywords such as simulation, anthropogenic heat, green infrastructure, street valley, etc. At present, the keywords are mainly urban morphology, urban canopy model, CFD, urban design, etc. This indicates that the related research is gradually focusing on the relationship between urban thermal environment and urban planning from the quantitative perspective.

3.4 Highly-Cited Paper Analysis

The references constitute a knowledge database of cutting-edge hotspots in urban thermal environment research [16]. Thus, it is necessary to further summarize the references to find out the basic research literature that really promotes the development of urban thermal environment on the basis of the above analysis. Co-citation analysis was adopted in this section. Figure 3 displays the network diagram of literature co-citation for the

Table 1. Keyword list for the urban thermal environment during 1990–2019

Categorization		Keywords	Frequency	Centrality	Year
Measurement methods	Conceptual definition	Temperature	370	0.01	1996
		LST	218	0.03	1993
		Air temperature	126	0.05	1996
		Thermal comfort	576	0.01	1991
	Methodological data	Model	294	0.06	2003
		Simulation	280	0.06	1999
		Pattern	97	0.01	2011
		Remote sensing	80	0.02	2003
		CFD	101	0.03	2008
Driving mechanism	Land cover changes	Land use	121	0.04	2000
		Land cover	52	0.04	2008
		Impervious surface	41	0.04	2010
	Landscape pattern changes	Vegetation cover	227	0.01	1995
		Green space/area	81	0.01	1998
		Green/cool roof	49	0.01	2011
		NDVI	37	0.03	2006
	Urban planning	Built environment	224	0.05	2008
		Urbanization	203	0.01	2000
		Geometry/urban form	113	0.05	2002
		Street canyon	101	0.04	2006
		Air pollution/pollutant disperse	77	0.01	2008
		Wind path/environment	68	0.01	2005

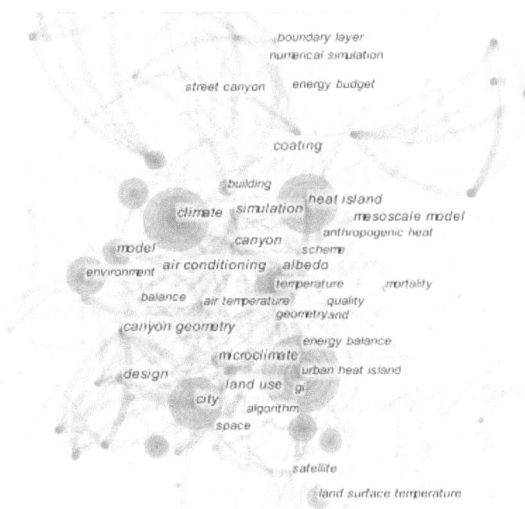

Fig. 2. Keyword collinear map for the urban thermal environment during 1990–2019

urban thermal environment. Furthermore, the five representative literatures with the high citation frequency were listed in Table 3.

As shown in Table 3, Steward ID, Oke TR [17] proposed a new local climate classification system. This zoning method comprehensively considers the interaction between spatial morphology patterns and UHI effect, skillfully integrating the local climate information into planning practice. It has been applied in the field of urban planning, and is also becoming a popular method for the coupling mechanism of the urban thermal environment and spatial morphology. Santanouris M [18] systematically reviewed the current literature related to mitigating the UHI effect by increasing urban albedo and increasing green roofs, and quantitatively analyzed its impact, which effectively promoted the prosperity of the research on the urban thermal environment. Li JX [19], taking Shanghai as an example, investigated the influence of the landscape pattern on the UHI, which promoted the dissemination and application of landscape ecology in the field of urban thermal environment. Bowier DE [20] systematically sorted out the cooling effect of green space on the countryside and quantitatively analyzed the research content and direction of this field, which is an important generalization and summary of the existing research on cooling of urban green space. Zhou WQ [21] pointed out the influence of urban surface cover on the land surface temperature from the perspective of landscape ecology, which to some extent provided ideological inspiration and technical support for later researchers [21].

Table 2. Sudden keywords for the urban thermal environment during 1990–2019

Data	Keywords	Strength	Begin	End	1990–2019
WOS	Urban climate	6.8951	1991	2013	
	Canyon geometry	5.185	1992	2011	
	Simulation	7.2329	1993	2012	
	Surface temperature	4.9564	1993	2010	
	Anthropogenic heat	4.5321	1999	2012	
	Air flow	4.5562	2001	2011	
	GI	12.7037	2002	2009	
	Thermal environment	5.9734	2003	2013	
	Urban area	5.9734	2005	2012	

(continued)

Table 2. (*continued*)

Data	Keywords	Strength	Begin	End	1990–2019
	Street canopy	5.8848	2006	2012	—
	Land use	4.6926	2008	2011	—
	CFD	4.422	2008	2013	—
	Urban thermal environment	3.479	2011	2016	—
	Urban canopy model	3.7907	2012	2016	—
	Urban design	4.3329	2014	2015	—

Fig. 3. Network diagram of literature co-citation during 1990–2019

Table 3. Highly cited central literature during 1990–2019

Ranking	Frequency	Year	Author	Article
1	178	2012	Steward ID, Oke TR	Local Climate Zones for urban temperature studies [17]
2	150	2014	Santanouris M	Cooling the cities–a review of reflective and green roof mitigation technologies to fight heat island and improve comfort in urban environments [18]
3	127	2011	Li JX	Impacts of landscape structure on surface urban heat islands: A case study of Shanghai, China [19]
4	113	2010	Bowier DE	Urban greening to cool towns and cities: A systematic review of the empirical evidence [20]
5	101	2011	Zhou WQ	Does spatial configuration matter? Understanding the effects of land cover pattern on land surface temperature in urban landscapes [21]

4   Conclusion

In this paper, the general situation and development trend of urban thermal environment from 1990 to 2019 are sorted out by using the knowledge graph visualization technology. It can be seen that the research about thermal environment mainly includes the following aspects: (1) urban thermal environment and global climate change, global temperature rise, sea-level rise, and biodiversity; (2) analytical methods, including scenarios simulation, remote sensing, numerical simulation, CFD, GIS technology, and urban vulnerability assessment, etc.; (3) the interaction mechanism between UHI effect and urban system, such as the influence of urbanization, land use, land cover, urban form, and landscape pattern on UHI effect; (4) the adaptive strategy and governance to cope with heat island effect.

In summary, the current research in this field tends to be in the direction of diversification and synthesis of the content system. The research methods are gradually from the field measurement to the combination of numerical simulation and field measurement. The research focus tends to be on the quantitative study of planning actions for the urban thermal environment and the construction of corresponding response mechanisms, shifting from passive adaptation to active management. Therefore, in the future, the microclimate adaptive planning especially the thermal environment, the construction of quantitative indicators and dynamic collaborative planning methods will certainly become the focus of research in this field, which will also have an important reference value for the research of urban planning in response to climate change.

Acknowledgements. This study was supported by the National Natural Science Foundation of China (No. 52400243, 52278087, 52300237); the Natural Science Basic Research Program of Shaanxi (No. 2023-JC-QN-0468); the Social Science Research Program of Shaanxi (No. 2023J325); the Social Science Research Program of Shaanxi Educational Department (No.22JK0118); Research on Major Theories and Realistic Problems of Philosophy and Social Sciences in Shaanxi Province (No.2022HZ1202).

References

1. Harmay, N.S.M., Choi, M.: The urban heat island and thermal heat stress correlate with climate dynamics and energy budget variations in multiple urban environments. Sustain. Cities Soc. (91), 104422 (2023)
2. Kotharkar, R., Ghosh, A.: Progress in extreme heat management and warning systems: a systematic review of heat-health action plans (1995–2020). Sustain. Cities Soc. **76**, 103487 (2022)
3. Gao, Y. et al.: Exploring the spatial heterogeneity of urban heat island effect and its relationship to block morphology with the geographically weighted regression model. Sustain. Cities Soc. **76**, 103431 (2022)
4. Chen, Y. et al.: Exploring the spatiotemporal patterns and correlates of urban vitality: temporal and spatial heterogeneity. Sustain. Cities Soc. (91), 104440 (2023)
5. Lan, T. et al.: The future of China's urban heat island effects: a machine learning based scenario analysis on climatic-socioeconomic policies. Urban Clim. (49), 101463 (2023)
6. Gao, Y. et al.: Quantifying the nonlinear relationship between block morphology and the surrounding thermal environment using random forest method. Sustain. Cities Soc. **91**, 104443 (2023)
7. He, B.-J. et al.: Progress, knowledge gap and future directions of urban heat mitigation and adaptation research through a bibliometric review of history and evolution. Energy Build. 112976 (2023)
8. Santamouris, M.: Recent progress on urban overheating and heat island research. Integrated assessment of the energy, environmental, vulnerability and health impact. Synergies with the global climate change. Energy Build. (207), 109482 (2020)

9. Santamouris, M., Kolokotsa, D.: Urban climate mitigation techniques. Routledge (2016)
10. Khosla, R. et al.: Cooling for sustainable development. Nat. Sustain. **4**(3), 201–208 (2021)
11. He, B.-J. et al.: A framework for addressing urban heat challenges and associated adaptive behavior by the public and the issue of willingness to pay for heat resilient infrastructure in Chongqing, China. Sustain. Cities Soc. **75**, 103361 (2021)
12. United Nations, Climate: This heatwave is the new normal, (July 20, 2022) (2022). https://unric.org/en/climate-this-heatwave-is-the-new-normal
13. Gao, Y. et al.: Effects of block morphology on the surface thermal environment and the corresponding planning strategy using the geographically weighted regression model. Build. Environ. **216**, 109037 (2022)
14. Hou, L., Jiang, Y., Shi, T., et al.: Overview and prospect of urban planning studies based on climate change. Urban Plan. **43**(03), 121–132 (2019)
15. Li, Z., Feng, L., Shen, R.: Scientific knowledge mapping of the evolution and frontier fields of space syntax research in foriengn countries. Planners **35**(08), 5–11 (2019)
16. Shi, B., Tian, Y., Zhao, J., et al.: Research progress and hotspots of ecological city: Based on bibliometrics and knowledge mapping analysis. J. Arid. Land Resour. Environ. **34**(03), 76–84 (2020)
17. Steward, I.D., Oke, T.R.: Local climate zones for urban temperature studies. Bull. Am. Meterological Soc. **93**(12), 1879–1900 (2012)
18. Santamouris, M.: Cooling the cities–a review of reflective and green roof mitigation technologies to fight heat island and improve comfort in urban environments. Sol. Energy **103**, 682–703 (2014)
19. Li, J., Song, C., Cao, L., et al.: Impacts of landscape structure on surface urban heat islands: a case study of Shanghai, China. Remote Sens. Environ. **115**(12), 3249–3263 (2011)
20. Bowler, D.E., Buyung-Ali, L., Knight, T.M., et al.: Urban greening to cool towns and cities: a systematic review of the empirical evidence. Landsc. Urban Plan. **97**(3), 147–155 (2010)
21. Zhou, W., Huang, G., Cadenasso, M.L.: Does spatial configuration matter? Understanding the effects of land cover pattern on land surface temperature in urban landscapes. Landsc. Urban Plan. **102**(1), 54–63 (2011)
22. Ng, E., Chen, L., Wang, Y., et al.: A study on the cooling effects of greening in a high-density city: an experience from Hong Kong. Build. Environ. **47**, 256–271 (2012)

Research Progress on Urban Street Canyon Heat Island Effect

Ye Zhu[1], Haizhu Zhou[2(✉)], and Xuejun Hao[1]

[1] School of Environmental and Energy Engineering, Beijing University of Civil Engineering and Architecture, Beijing 102600, China

[2] China Academy of Building Research, Beijing 100001, China
1620143510@qq.com

Abstract. With the continuous improvement of urbanization level and the increasing urban spatial density, the urban heat island effect is intensifying, resulting in the increasing range and intensity of extreme climate, which poses a threat to the physical and mental health of urban residents. Urban heat island effect has become a focus of multidisciplinary research. This paper reviews the technical development of the existing research methods on heat island effect, analyzes their advantages and disadvantages, and their applicability on different scales. This paper presents the street valley scale as an important method to study the heat island effect of compact city, and analyzes the influencing factors of heat island effect under this scale. Finally, according to the existing problems in the current research process, the research progress and development trend of urban heat island at street valley scale are proposed from the aspects of data and evaluation methods. The purpose of this paper is to provide a reference for the in-depth study of urban street valley heat island effect under the background of compact city development model, and to provide an effective way to mitigate urban heat island effect.

Keywords: Street canyon · Heat island effect · Technical methods · Influencing factors

1 Introduction

Under the influence of human factors, the local temperature, humidity, air convection and other factors of the urban surface change, resulting in urban microclimate change, and the phenomenon that the temperature of the urban center is higher than that of the suburban area, triggering the heat island effect. The heat island effect was first identified by British climatologist Lake Howard in 1833, [1] and documented in his book The Climate of London. On the one hand, urban heat island effect is greatly influenced by urbanization [2], including many factors such as changes in the physical properties of underlying surfaces, massive release of anthropogenic heat, reduction of green space, and urban population gathering [3]. On the other hand, the intensification of "greenhouse effect" [4, 5] leads to the frequent occurrence of high temperature weather in summer, which makes urban residents more dependent on the cooling system of buildings, and the urban heat island effect becomes more obvious, showing a positive causal relationship with building energy consumption [6, 7].

© The Author(s) 2025
B.-J. He et al. (Eds.): UCSUD 2023, LNCE 559, pp. 467–480, 2025.
https://doi.org/10.1007/978-981-97-8401-1_33

Serious urban heat island effect not only affects people's physical and mental health, but also becomes a restricting factor for the further improvement of people's life quality and the further development of cities [7]. Urban high temperature not only changes the urban microclimate [8, 9], but also increases the disease risk of urban residents, increases the incidence of heart disease and respiratory diseases [10, 11], and significantly increases the mortality and morbidity of urban residents [12, 13].

City street is an important part of the city. The street surface and the buildings on both sides form the street valley. The three-dimensional spatial form of the city changes the internal thermal environment of the city by affecting the surface energy balance process and air flow, and has a certain influence on the urban heat island effect. With the development of urbanization, the geometric characteristics such as high plot ratio and compact arrangement of buildings in the city center will lead to poor ventilation and large heat storage in the street valley, which intensifies the urban heat island effect. Street valley is an important part of the outdoor activity space of urban residents, and its good physical environment is an important prerequisite for improving the comfort of outdoor pedestrians and enhancing the vitality of the city [14]. Because there are too many factors affecting the heat island effect, it is very difficult to study the thermal environment of the whole urban area. Taking street valley as the research scale can narrow down the scope, so as to deeply explore its internal influence factors and interaction mechanism. Meanwhile, the integrated study of urban heat island effect can be carried out through the interconnection of multiple urban street valleys at the block scale. It is of great academic significance to understand and master the wind and thermal environment in urban streets and valleys, to create a good urban microclimate in compact cities, to adapt to urban development and to protect the health and safety of pedestrians.

2   Research Methods of Heat Island Effect

The research objects of street valley heat island effect are complex and different, such as the complexity of industrialization, climate background, development degree and regional space, so it is difficult to establish an accurate quantitative relationship model of street valley heat environment. Therefore, when conducting research on it, the research scale should be determined first, and then specific research methods or strategies should be determined [15].

2.1 Data Measurement Method

There are two main ways of data measurement method [16]: one is direct field measurement, and the other is remote thermal sensing method. The testing methods and advantages and disadvantages of the two methods are shown in Table 1.

2.2 Scale-Down Model Method

The scale-down model method is a method of conducting small-scale experimental research [37], mainly by creating a scaled-down model of actual city blocks, conducting wind tunnel or outdoor environmental experimental research [38, 39]. The scale-down

Table 1. Classification and Overview of Advantages and Disadvantages of Data Measurement Method

Main Research Methods		Method Overview	Advantages	Disadvantages	Scope of Application and Research Results
On-site Direct Measurement	Mobile Monitoring Equipment	Data is collected using mobile vehicles equipped with meteorological data sensors and data collectors	Accurate and fast data collection, small amount of equipment required, high flexibility [17, 18]	Data is limited by time synchronization, the instrument is easily affected by external environmental conditions, and is restricted by human and material conditions [19]	Basic understanding of the wind field characteristics and thermal environment around buildings [20–22]. Used to study the spatiotemporal variation of urban heat islands [19]
	Fixed Point/Meteorological Station Observation	Data is obtained from meteorological stations near the research area, or from portable meteorological stations, including horizontal and vertical observations [23]	The type and integrity of the detection data are high, and the measurement time step can be automatically set [23]	There are differences with the actual, the number of sites is limited, it is difficult to obtain a large range of data, and it is greatly affected by the surrounding environment [20]	

(continued)

Table 1. (*continued*)

Main Research Methods		Method Overview	Advantages	Disadvantages	Scope of Application and Research Results
Remote Thermal Sensing Method	Aerial Remote Sensing	Surface temperature is remotely sensed using satellites, airborne devices, or aircraft [24]	High resolution of data obtained, can avoid the impact of adverse weather [24]	High cost, cannot obtain static urban surface images	Obtain urban surface humidity, surface emissivity, surface reflectance, surface radiation, near-surface atmospheric motion and turbulent heat motion, etc. [25, 26]. Used to study the impact of factors such as water bodies [27], underlying surface [28], atmosphere [29] and vegetation [30] on the urban thermal environment; analyze the spatiotemporal pattern and evolution of heat islands [31–34]

(*continued*)

Table 1. (*continued*)

Main Research Methods		Method Overview	Advantages	Disadvantages	Scope of Application and Research Results
		Using micro unmanned aerial vehicles equipped with high-precision infrared thermal imaging equipment for ground fine temperature measurement [35]	Fine temperature measurement, low cost, high flexibility	High technical difficulty, complex operation, high cost	
	Ground remote sensing	Use infrared imaging equipment for ground temperature detection [36]	Low cost, simple operation, high precision, real-time dynamic information of ground temperature can be obtained [37]	Subject to time synchronization restrictions, difficult to obtain large-scale data	

model method has great significance for selecting key boundary condition parameters, can filter out meaningless dimensionless numbers, and can determine certain threshold conditions to reduce the impact of irrelevant criterion numbers. The wind tunnel simulation of the atmospheric boundary layer is a highly reliable prediction method. Wind tunnel experiments usually use hot wire or hot film technology [40], pulse hot wire technology [41], probe technology [42], Laser-Doppler Anemometry (LDA) [43], Particle Image Velocimetry (PIV) [44] and other measurement methods. In the study of urban heat islands, this method can usually only be used to study local areas of the city, that is, the mesoscale heat island effect. In the study of urban wind and heat environment, wind tunnel experiments have been determined to be applicable at the scale of urban blocks [45], and this method is not applicable for large-scale, complex architectural structure urban heat island spatiotemporal changes.

2.3 Numerical Simulation Method

With the improvement of mathematical modeling and numerical calculation technology, numerical simulation methods have gradually become the mainstream method of current academic research, the most commonly used models are the urban canopy model and the CFD model.

In the control body of the urban canopy model, the energy budget balance can be summarized as Equation (1). Basic equations governing the energy budget balance in the body:

$$Q^* + Q_F = Q_H + Q_E + \Delta Q_S + \Delta Q_A$$

In the equation, the definitions of each symbol are as follows, Q^* ——Net radiation heat flux, W/m 2; Q_F ——Anthropogenic heat production, W/m 2; Q_H ——Sensible heat flux, W/m 2; Q_E ——Latent heat flux, W/m 2; ΔQ_S ——Net storage heat flux, W/m 2; ΔQ_A ——Net advection heat flux, W/m 2. Net radiative heat flux Q^* can be simplified by a face-to-face radiation model.

This model considers the heat exchange process of various surfaces within the urban canopy and the change in air temperature. Generally speaking, early studies mainly considered the parameter changes in the height and width directions of the street canyon [1, 46–49], which are one-dimensional or two-dimensional models. In recent years, three-dimensional models have been increasingly adopted [50–52]. The characteristic of the urban canopy model is that the calculation speed is very fast, but the biggest shortcoming of the urban canopy model is the lack of simulation of the air velocity field [53]. Therefore, in various urban canopy models, logarithmic law or exponential law is often used to assume the distribution of air flow velocity, thereby improving the calculation results of the energy conservation equation.

Unlike the urban canopy model, the CFD model can simultaneously solve the coupling calculation problem of the velocity field and the temperature field. The conservation control equation in the CFD model includes multiple parameters such as mass, momentum, potential temperature, and air components [54–58]. Therefore, compared with the urban canopy model, the CFD model can often obtain more accurate heat island distribution information. In terms of computational technology, CFD models must be calculated through a large number of control bodies or nodes, and the scale of the research problem needs to be determined. CFD models can be roughly divided into mesoscale models and microscale models. Because the micro-scale model requires too much detailed data of urban building size, it is not suitable to extend to the calculation of heat island effect of the whole city.

In simulation studies, WRF (Weather Research and Forecasting) has also been applied to urban wind and heat environment research. Its simulation results can usually be used as the boundary conditions for CFD numerical simulation [59]. The advantage of the WRF model is that it can build models in combination with different land use type data, study the historical evolution and development trend of the heat island effect [60], intuitively display the impact of the spatial and temporal distribution of urban meteorological elements horizontally and vertically [61], and provide data support for subsequent quantitative analysis.

3   Street Canyon Heat Island Effect

The factors causing the formation of the heat island phenomenon are very complex, and the scales of these factors vary greatly [62]. The macro scale mainly focuses on the climate effects of the city and the processes of urban heat island circulation, focusing on the impact of the urban underlying surface on the atmosphere [63]; The meso scale research mainly discusses the air flow and heat transfer processes within the urban canopy [64]; The micro scale mainly studies the surface characteristics of individual buildings. The study of urban heat island phenomenon often needs to integrate these different scale factors. The study of street canyon scale can control the research scope, and at the same time, this scale includes most of the factors affecting the heat island effect in the city, which has a representative nature and is of great reference value for urban planning and improving the microclimate environment. In mesoscale studies, it is considered that geometric conditions such as road layout and building structure, as well as physical properties of underlying surface and building materials play an important role in the formation of the thermal environment inside urban canopy. Its parameters will be used as boundary conditions to simulate and analyze the air temperature, wind environment, building surface temperature, street surface temperature, etc. The study of street valley heat island effect mainly focuses on the spatial form of regular buildings, architectural planning and layout, urban ventilation, underlying surface properties, street plant configuration, etc. At the same time, the influence of individual building structure and building surface characteristics should also be considered.

3.1   Establishment of Street Canyon Model

When studying the heat island effect of street canyons, according to the parameter processing methods of energy and momentum, the model can be divided into single-layer (Single-Layer UCM) model, multi-layer (Multiple-Layer UCM) model, and surface layer (Surface-Layer UCM) model. Due to the limited use of the surface layer model, this article only introduces the single-layer and multi-layer urban canopy models. In the single-layer model, buildings will be treated in two dimensions such as symmetry and finite length, but radiation calculations will still use three dimensions. In the multi-layer model, buildings will be treated in three dimensions with symmetry, finite length, and height direction changes. The accuracy will be significantly improved. When the height difference of buildings in the area is not large, it is advisable to use a single-layer model, which can achieve higher accuracy and reduce computational difficulty.

3.2   Factors Affecting Street Canyon Heat Island Effect

Based on various research methods of urban heat island effect, the quantitative indicators of its influencing factors are shown in Table 2.

Table 2. Evaluation indicators and quantification results of influencing factors

Influencing factors	Evaluation indicators	Quantification results
Surface Characteristics	NDBI (Normalized Building Index)	Increase by 0.1, surface temperature increases by 2.85°C [65];
	NDVI (Normalized Vegetation Index)	An increase of 0.1 reduces the surface temperature by approximately 1.72°C [65];
	NDEI (Normalized Water Body Index)	An increase of 0.1 reduces the surface temperature by approximately 1.56°C [65]
	Surface Material Reflectivity	An increase of 10% can reduce the air temperature by 0.75°C [66]
Geographical Morphology	Surface Elevation Value	For every increase of 100 m, the surface temperature decreases by approximately 0.925°C (in the Yellow River area) [67];
Plant Configuration	Tree Rate	An increase of 10% can reduce the heat island intensity by 0.3°C [68];
	Closure Degree	An increase of 1% can reduce the air temperature by 0.14°C [69];
	Urban Green Coverage Rate	When it reaches 33%, it can reduce the air temperature by 1°C [70];

(*continued*)

Table 2. (*continued*)

Influencing factors	Evaluation indicators	Quantification results
	Shading Factor	D/PH (the ratio of canopy diameter to branch position height) increased from 0 to 8 and the average radiation temperature decreased by about 15 °C; When the leaf area density was 2.5 m^2/m^3, it had the best radiation shading effect [71]
Planning Layout	SVF (Sky View Factor)	Correlation with atmospheric temperature, correlation coefficient is 0.33 [72];
	H/W (Street Canyon Height-Width Ratio)	When the street height-width ratio is 1.0, the road (wall) reflectivity increases from 0.15 to 0.65, and the canyon reflectivity increases from 0.15 to 0.35 [72];
		An increase in the street canyon height-width ratio by one unit reduces the WBGT (Wet Bulb Globe Temperature Index) by approximately 1°C, and the heat island intensity decreases by about 1.3°C [73]
	Total Plot Overhang Rate	When the overhang rate of the first floor of a single building is 50%, it will increase the total overhang rate of the entire block by 22.5%, thereby causing the heat island intensity to decrease by 0.03°C [72];
	Windward Index	The daytime correlation coefficient with spatial surface temperature is 0.371, and at night it is 0.335, but this linear relationship may not be absolute due to the influence of other factors [74]
	Three-dimensional morphological indicators	Excessive building density and too many floors in the building complex will cause high temperatures [75]

4   Shortcomings and Prospects

The known theories and models at different scales still lack completeness in the analysis of urban heat island phenomenon. In the scope of the study, the influence factors around the street valley were not considered, and the external factors should be coupled with the internal factors in the follow-up study. In terms of research methods, we should explore an urban heat island research method that can integrate various scale factors in the future. Meanwhile, when selecting the boundary conditions of the model, it should be combined with the existing accurate monitoring system and prediction model. In the evaluation method, the discussion of outdoor thermal comfort should be strengthened. Based on the above shortcomings, three suggestions are put forward for the current research status in the field of urban street valley heat island:

4.1   Numerical Simulation Research Combined with Ground or Data Measurement

The simulation results of the numerical model are greatly affected by the input boundary conditions and the calculation range. In the future, the ground temperature data of different years, seasons and months can be obtained through remote sensing thermal infrared data of different phases at the same time, so as to form a complete time series and formulate an "urban environmental climate map" combined with the local climate characteristics. Surface temperature is determined by thermal radiation and thermodynamic properties of the surface, and is affected by thermal channels, ground humidity, surface reflectance, solar and atmospheric downgoing radiation, and near-surface air temperature. The air temperature is mainly affected by the heat flow from the surface, human activities and the background temperature of the surrounding landscape elements. At the same time, the discrimination and accuracy of various models are improved, and the influence mechanism of urban street valley heat island effect is analyzed quantitatively and accurately, so as to provide scientific theoretical basis and design guidance for urban street valley planning and design.

4.2   Expansion Research on Urban Heat Island Prediction Model

For the study of urban heat island effect at large scale in the whole urban area, the integrated study can be carried out through the interconnection of multiple urban streets and valleys at the block scale. Based on Kirchhoff's Law, the street valley network node method provides a solution for the urban canopy model to extend the simulation of a single street valley to multiple street valleys. Through the street valley network structure, block network structure and urban network structure, three urban heat island models of different scales are formed, and each other's boundary conditions are used to solve the expansion and nesting research problems of different scales.

4.3   Combination of Demand for Improvement of Outdoor Thermal Comfort and Form of Urban Micro-Renewal

As people's yearning for a better quality of life becomes more and more intense, human thermal comfort evaluation and analysis of street valley thermal environment can be carried out in the future. Correlation analysis and regression analysis can be used to screen relevant spatial form indicators that affect urban street valley thermal environment and human thermal comfort. Quantitative description of the relevant influencing factors and the relationship between them, combined with the field of sustainable development and human thermal comfort evaluation criteria, to explore the impact mechanism of street valley spatial form factors on the thermal environment of street valley, from the planning and application level to determine the optimization strategy for the spatial form of street valley in different regions.

References

1. OKE, T. R.: The energetic basis of the urban heat island [J]. Q. J. R. Meteorol. Soc. **108** (455) (1982)
2. Yuyu, R., Guoyu, R., Aiying, Z.: A review of the impact of urbanization on ground temperature change trends [J]. Prog. Geogr. **29**(11), 1301–1310 (2010)
3. Xinxin, L., Xingzhao, L.: Progress and hotspot analysis of urban heat island research based on CiteSpace [J]. Sichuan Arch. **41**(02), 20–23 (2021)
4. Pachauri, R.K., Climate change 2007: synthesis report. Contribution of working groups I, II and III to the fourth assessment report of the intergovernmental panel on climate change [J]. Speculum **77** (2), 586–588 (2007)
5. UK, P.S. Climate change 2014: mitigation of climate change, contribution of working group III to the fifth assessment report of the IPCC [J]. (2007)
6. Asimakopoulos, D. A., Santamouris, M., Farrou, I., et al. Modelling the energy demand projection of the building sector in Greece in the 21st century [J]. **49** (Jun.), 488–498 (2012)
7. Kapsomenakis, J., Kolokotsa, D., Nikolaou, T., et al. Forty years increase of the air ambient temperature in Greece: The impact on buildings [J]. Energy Convers. Manag. **74** (oct.), 353–365 (2013)
8. Cao, Lee, Xh, et al. Urban heat islands in China enhanced by haze pollution [J]. Nat Commun **1**(1) (2016)
9. Li, D., Bou-Zeid, E.: Synergistic interactions between urban heat islands and heat waves: the impact in cities is larger than the sum of its parts* [J]. J. Appl. Meteorol. Climatol. **52**(9), 2051–2064 (2013)
10. Curriero, F.C., Heiner, K.S., Samet, J.M. et al.: Temperature and mortality in 11 cities of the eastern United States. [J]. Am. J. Epidemiol. **155** (1) (2002)
11. Giuseppe, M., Ugo, F., Cristiana, V. et al.: Pattern and determinants of hospitalization during heat waves: an ecologic study [J]. BMC Public Health **7** (1) (2007)
12. L B A, A Z, J S.: The time course of weather-related deaths. [J]. Epidemiology (Cambridge, Mass), **12** (6) (2001)
13. Tan, Z., Lau, K. K-L, Ng, E. Urban tree design approaches for mitigating daytime urban heat island effects in a high-density urban environment [J]. Energy & Buildings **114** (1) (2016)
14. Shanglin, W., Yimin, S.: Study on microclimate simulation and improvement strategy of streets in Guangzhou area [J]. Urban Plan. Forum **1**(01), 56–62 (2016)
15. Mirzaei, P. A., Haghighat, F.: Approaches to study Urban Heat Island—Abilities and limitations [J]. Build. Environ. **45** (10) (2010)
16. Zhendong, X.: Study and analysis on the causes of urban heat Island effect [D]. Dalian Univ. Technol., Liaoning (2003)
17. Yanhong, L., Zhicai, L., Jinhong, Z., et al.: Study on Taiyuan urban heat Island based on automatic station data [J]. J. Arid. Land Resour. Environ. **27**(12), 173–179 (2013)

18. Yufeng, L., Zhihua, Y., Wei, K., et al.: Analysis of the trend and influencing factors of urban heat Island intensity in Xi'an urban area from 1993 to 2012 [J]. J. Nat. Resour. **30**(06), 974–985 (2015)
19. Dongfang, Z.: Research on effective wind energy utilization in building environment [D]. Shandong Jianzhu University, Shandong (2010)
20. Haihong, L.: Research on simulation analysis and optimization strategy of urban Mesoscale wind environment based on CFD [D]. Lanzhou University, Gansu (2021)
21. Stathopoulos, T., Storms, R.: Wind environmental conditions in passages between buildings [J]. J. Wind Eng. Ind. Aerodyn. **24**(1), 19–31 (1986)
22. Murakami, S., Iwasa, Y., Morikawa, Y.: Study on acceptable criteria for assessing wind environment at ground level based on residents' diaries [J]. J. Wind Eng. Ind. Aerodyn. **24**(1), 1–18 (1986)
23. Johansson, E.: Influence of urban geometry on outdoor thermal comfort in a hot dry climate: a study in Fez, Morocco [J]. Build. Environ. **41**(10), 1326–1338 (2006)
24. Weng, Q.: Thermal infrared remote sensing for urban climate and environmental studies: methods, applications, and trends [J]. Isprs J. Photogramm. Remote. Sens. **64**(4), 335–344 (2009)
25. Abahri, K., Belarbi, R., Trabelsi, A.: Contribution to analytical and numerical study of combined heat and moisture transfers in porous building materials [J]. Build. Environ. **46** (7) (2010)
26. Santamouris, Mat.: Heat Island Research in Europe: The State of the Art [J]. Adv. Build. Energy Res. **1** (1), 123–150 (2007)
27. Yang, X., Zhixiang, Z.: The impact of blue-green space landscape pattern on urban heat island [J]. China Garden **39**(01), 105–110 (2023)
28. Zhang, N., Yang, S., Fu, Yukai. et al.: Analysis of the spatiotemporal aggregation characteristics of color steel plate buildings in Lanzhou City and their impact on the urban heat island effect [J]. Geogr. Geogr. Inf. Sci. **38** (03): 43–49 (2022)
29. Meiyan, H., Xiaogang, F., Fengxia, L., et al.: Research progress on the interactive effects of urban heat island and aerosols [J]. Remote. Sens. Inf. **37**(04), 128–134 (2022)
30. Zhao, X.: Study on the thermal environment effect of Xi'an city and the mitigation role of green space [D]. Shaanxi; Northwest A&F University (2021)
31. Junzhi, Z., Yanan, L., Ji, W., et al.: Analysis of the spatiotemporal characteristics of the urban heat island effect in Beijing from 1981 to 2020 and its influencing factors [J]. J. Atmos. Sci. **1**(1), 1–15 (2023)
32. Qijiao, X.: Analysis of the characteristics and influencing factors of the urban heat island in Wuhan [J]. Resour. Environ. Yangtze Basin **25**(03), 462–469 (2016)
33. Tenglong, C., Jian, Z., Chen, L.: Study on the evolution mechanism and diffusion pattern of the thermal environment pattern in Wuhan City in the past 30 years [J]. Remote. Sens. Land & Resour. **29**(04), 197–204 (2017)
34. Xingwei, Y., Hongmei, Z., Meng, L.: Application of satellite data in the analysis of the thermal field in Pudong new area, Shanghai [J]. J. Appl. Meteorol. **1**(03), 369–373 (1994)
35. Yanyan, G.: Geothermal detection based on multi-source thermal infrared remote sensing technology [D]. China University of Mining and Technology, Jiangsu (2022)
36. Yang, L., Li, Y.: City ventilation of Hong Kong at no-wind conditions [J]. Atmos. Environ. **43**(19), 3111–3121 (2009)
37. Meroney, R., Lindley, D., Bowen, A. J.: Physical modelling of flow over complex terrain [J]. (1980)
38. Uehara, K., Murakami, S., Oikawa, S., et al.: Wind tunnel experiments on how thermal stratification affects flow in and above urban street canyons [J]. Atmos. Environ. **34** (10) (2000)

39. Flor, F.S.D.L., Dominguez, S.A.: Modelling microclimate in urban environments and assessing its influence on the performance of surrounding buildings [J]. Energy & Build. **36**(5), 403–413 (2004)
40. N. I, A.G. D.: Comparison of full-scale and wind tunnel wind speed measurements in the commerce court plaza [J]. J. Wind. Eng. Ind. Aerodyn. **1** (1) (1975)
41. Castro, I.P., Wiggs, G.F.S.: Pulsed-wire anemometry on rough surfaces, with application to desert sand dunes [J]. J. Wind. Eng. Ind. Aerodyn. **52**, 53–71 (1994)
42. Monteiro, J.P., Viegas, D.X.: On the use of Irwin and Preston wall shear stress probes in turbulent incompressible flows with pressure gradients [J]. J. Wind. Eng. & Ind. Aerodyn. **64** (1) (1996)
43. Blocken, B., Stathopoulos, T., van Beeck, J.P.A.J.: Pedestrian-level wind conditions around buildings: review of wind-tunnel and CFD techniques and their accuracy for wind comfort assessment [J]. Energy Sav. Build. **44**(5), 1 (2016)
44. Reyes, V.A, Sierra-Espinosa, F. Z., Moya, S. L. et al.: Flow field obtained by PIV technique for a scaled building-wind tower model in a wind tunnel [J]. Energy & Build. **107** (1) (2015)
45. White, B.R.: Analysis and wind-tunnel simulation of pedestrian-level winds in San Francisco [J]. J. Wind Eng. Ind. Aerodyn. **44**(1–3), 2353–2364 (1992)
46. Oke, T.R.: Boundary layer climates: second edition [J]. Geogr. Rev. **69** (4), 486 (1979)
47. Oke, T.R., Johnson, G.T., Steyn, D.G., et al.: Simulation of surface urban heat islands under 'ideal' conditions at night part 2: diagnosis of causation [J]. Bound.-Layer Meteorol. **56**(4), 339–358 (1991)
48. Johnson, G.T., Oke, T.R., Lyons, T.J., et al.: Simulation of surface urban heat islands under? IDEAL? conditions at night part 1: theory and tests against field data [J]. Bound.-Layer Meteorol. **56**(3), 275–294 (1991)
49. Mills, G.M.: Simulation of the energy budget of an urban canyon—I. Model structure and sensitivity test [J]. Atmos. Environ. Part B Urban Atmos. **1** (1) (1993)
50. Roulet, Y.A., Martilli, A., Rotach, M. W., et al.: Validation of an Urban Surface Exchange Parameterization for Mesoscale Models—1D Case in a Street Canyon [J]. Am. Meteorol. Soc. **1** (1) (2005)
51. Masson, V.: Urban surface modeling and the meso-scale impact of cities [J]. Theor. Appl. Climatol. **84** (1–3) (2006)
52. Kanda, M., Kawai, T., Kanega, M. et al.: A Simple Energy Balance Model for Regular Building Arrays [J]. Bound.-Layer Meteorol. **116** (3) (2005)
53. Tong, H., Walton, A., Sang, J., et al.: Numerical simulation of the urban boundary layer over the complex terrain of Hong Kong [J]. Atmos. Environ. **39**(19), 3549–3563 (2005)
54. Ashie, Y., Ca, V.T., Asaeda, T.: Building canopy model for the analysis of urban climate [J]. J. Wind. Eng. & Ind. Aerodyn. **81** (1) (1999)
55. Oleson, K.W., Bonan, G. B., Feddema, J., et al.: An urban parameterization for a global climate model. Part I: formulation and evaluation for two cities [J]. J. Appl. Meteorol. Climatol. **47** (4), 1038–1060 (2006)
56. Pielke, R.A., Cotton, W.R., Walko, R. L. et al.: A comprehensive meteorological modeling system RAMS [J]. Meteorol. Atmos. Phys. **49** (1) (1992)
57. Murakami, S.: Environmental design of outdoor climate based on CFD [J]. Fluid Dyn. Res. **38**(2–3), 108–126 (2006)
58. Mochida, A., Murakami, S., Ojima, T., et al.: CFD analysis of mesoscale climate in the Greater Tokyo area [J]. J. Wind Eng. Ind. Aerodyn. **67–68**(97), 459–477 (1997)
59. Li, J., Chao, L.: Shu Qian Analysis of the Impact of Urban Green Corridors on the Heat Island Effect Using the WRF-UCM Model - A Case Study of Shanghai [J]. Build. Energy Effic. **47**(10), 89–96 (2019)
60. Jin, L.: Simulation study on the impact of urbanization process and roof Albedo on the heat Island effect in Hangzhou [D]. Zhejiang University, Zhejiang (2020)

61. Xiaoyu, Z., Yongwei, W., Jihua, S., et al.: Numerical simulation study on the urban heat Island effect in Kunming [J]. Atmos. Sci. **46**(04), 921–935 (2022)
62. Li, Z.: Analysis of the changes in the three-dimensional urban heat Island effect [D]. Guilin University of Technology, Guangxi (2017)
63. Xiaojiao, W.: Study on the wind environment at pedestrian height of typical high-rise buildings under different incoming wind characteristics [D]. Hefei University of Technology, Anhui (2019)
64. Lilei, Z.: Study on urban street canyon wind environment and particle diffusion [D]. Shenyang University of Architecture, Liaoning (2017)
65. Zhang, C., Wan, Ji., Luo, H. et al.: Analysis of the characteristics of the urban heat island effect in changsha based on remote sensing technology [J]. Urban Survey (05), 90–95 (2022)
66. Alchapar, N.L., Correa, E.N.: The use of reflective materials as a strategy for urban cooling in an arid "OASIS" city [J]. Sustain. Cities Soc. **27**(1), 1–14 (2016)
67. Taibin, T., Bao, Z., Xiaomei, J., et al.: Study on the changes in surface temperature in the summer in the source area of the yellow river [J]. Arid. Zone Geogr. **1**(1), 1–14 (2023)
68. Du Xiaohan, Chen Dong, Wu Jie, et al.: The impact of street canyon geometry and greening on the summer thermal environment [J]. Build. Sci. **28** (12), 94–99 (2012)
69. Zhang Sihan, Xu Lihua.: Research progress on the impact of urban green space plants on thermal comfort [J]. Green Sci. Technol. **25** (07), 29–34+58 (2023)
70. Ng, E., Chen, L., Wang, Y., et al.: A study on the cooling effects of greening in a high-density city: an experience from Hong Kong [J]. Build. Environ. **47**(1), 256–271 (2012)
71. Zhao Xiaoyue, Zhou Yufei, Zhang Tailong, et al.: Research on the environmental influencing factors of vegetation on outdoor thermal radiation of buildings[J]. Build. Technol. **54** (11), 1375–1381 (2023)
72. Langtao, Z.: Study on the impact of urban geometric factors on the heat Island effect in Guangzhou [D]. Jinan University, Guangdong (2020)
73. Kanghao, T.: The mechanism of street canyon form and surface material affecting the urban heat Island effect [D]. Guangxi University, Guangxi (2017)
74. Zhangxian, F., Shijun, W., Shanhe, J., et al.: The impact of urban morphology and wind environment on surface temperature in Changchun [J]. Geogr. J. **74**(05), 902–911 (2019)
75. Weiqi, Z.H.O.U., Yunyu, T.I.A.N.: Research progress on thermal environment effects of urban three-dimensional spatial morphology[J]. Acta Ecol. Sin. Ecol. Sin. **40**(02), 416–427 (2020)

Construction of Block Microclimate Analysis Model and Strategy of Climate Adaptive City Construction Based on Grasshopper

Xianjun Zeng[1,2], Xiaolei Qiu[1], Jinmin Chen[3], Yuxin Qiu[3], Ting Wang[3], and Wenyue Yang[3(✉)]

[1] College of Architecture and Planning, Fujian University of Technology, Fuzhou 350118, China
[2] National Key Laboratory for Subtropical Building Sciences, School of Architecture, South China University of Technology, Guangzhou 510641, China
[3] College of Forestry and Landscape Architecture, South China Agricultural University, Guangzhou 510642, China
yangwenyue900780@163.com

Abstract. In the past, urban climate problems caused by rapid urbanization, mainly urban heat island effect, affected people's daily life; under the background of new urbanization, people pay more attention to climate perception on human scale, and explore green and comfortable urban environment construction path and economic and effective urban cooling strategy. At present, the study of urban planning pays attention to macroscopic land use layout or microscopic landscape design but lacks the discussion on the influence of architectural layout on microclimate construction. Based on Grasshopper software, this paper constructs a microclimate analysis model of urban block scale and simulates the influence of different architectural layouts on urban microclimate in urban blocks by using control variable method. Through analysis, it is concluded that the microclimate of urban blocks is affected by building height, density, overhead rate of the first floor and building layout. Finally, the index optimization of urban design is proposed to improve the microclimate of blocks in urban planning and promote the construction of climate-adapted cities.

Keywords: Microclimate · Thermal comfort · Building layout · Simulation · Grasshopper

1 Introduction

Under the background of rapid urbanization, the original natural environment has been replaced by many modern urban built-up areas, and the corresponding urban microclimate environment has also significantly changed [1, 2]. Traditional urban planning and design methods often ignore the influence of architectural layout and urban spatial form on microclimate. Spreading expansion and disorderly spatial layout will lead to overdraft of urban resources, which will restrict the livability and sustainable development of cities [3–5]. According to the data published by the WMO (World Meteorological

© The Author(s) 2025
B.-J. He et al. (Eds.): UCSUD 2023, LNCE 559, pp. 481–496, 2025.
https://doi.org/10.1007/978-981-97-8401-1_34

Organization), the average temperature near the surface all over the world has obviously increased and showed an accelerated development trend. At the beginning of the 20th century, the global average temperature rose by only 0.7 °C [6]. From the middle of the 20th century to the beginning of the 21st century, the average surface temperature in China rose by about 1.1 °C. From 2015 to 2019, it was the five years with the highest temperature in the global record data [7]. To cope with global climate change and promote urban green development, The World Bank launched the first pilot city of Sustainable Urban Cooling Project in China in September 2020. In addition, as of October 2020, 127 countries around the world have proposed to build "Cities of Carbon Neutrality" in 2050. In this context, it is urgency of the need for climate adaptability in urban planning. Strengthening the simulation of urban architectural layout and microclimate through planning auxiliary technologies is regarded as an important way and method to promote the planning and construction of low-carbon cities [8].

The research on the adaptation of urban planning to climate shows that the original urban planning mainly pays attention to macroscopic land use layout and microscopic landscape design, and the architectural layout often succumbs to control indexes such as FAR(floor area ratio), while the scientific discussion on the architectural layout of blocks created by microclimate under the overall network system such as urban air duct is still insufficient [9, 10]. It is difficult for the existing urban basic functional space to create a built environment that meets the demands of HTC (Human Thermal Comfort) [11]. The research of climate-adaptive planning needs to integrate multiple disciplinary systems. It is necessary to carry out microclimate assessment from the scales of regional planning, urban planning, block planning and architectural design and implement the core management and control contents in the planning at all levels [13, 14]. However, the current theoretical research is out of touch with practical operation, and the lack of quantitative and visual planning aids of diversified and effective simulation is also the key to restrict climate adaptation planning [15, 16].

Compared with CFD, Pheonics, Ecotect, WinAir, ENVI-met and other software used in urban microclimate research, Grasshopper has the advantage that it can realize multivariable performance leap-forward optimization through a series of plug-ins and integrate multidisciplinary theoretical knowledge and technical methods into one platform, which makes the planning scheme more scientific and diversified [17, 18]. Therefore, based on Grasshopper, this study builds a platform integrating "Parametric Modeling, Ecological Performance Analysis and Design Scheme Optimization" to explore the microclimate environment under different block layout scenarios. Analyze the wind and heat environment of the site by ecological performance simulation software such as Ladybug and Butterfly. And then optimize the design scheme by combining Galapagos genetic algorithm and TT Toolbox data analysis tools. To achieve the goal of climate-adaptive urban planning and obtain the approximate optimal scheme of the overall microclimate performance of the city.

2   Proposed Method

2.1   Grasshopper and Its Plug-In for Microclimate Simulation and Optimization

Grasshopper is a built-in plug-in of Rhino, which can transform practical problems into geometric models, realize parametric modeling and visual expression of site models, and achieve the effect of real-time linkage between control parameters and site models.

Install Butterfly and Ladybug in Grasshopper to simulate the wind and heat environment of the site, implement the quantitative analysis of the microclimate of the site, and provide a measure index for the ecological performance evaluation of the design scheme. And then, use the Galapagos algorithm arithmetic unit in the software to implement the site ecological performance optimization based on algorithm automatic optimization, and use the TT Toolbox data analysis tool to visualize the generated results. Finally achieve the design optimization.

2.2   Microclimate Evaluation Index

In the studies of urban microclimate, outdoor walking wind environment will have an impact on HTC [19], so it is necessary to construct the evaluation criteria of each influencing factor to quantitatively analyze the HTC characteristics [20]. In order to make the evaluation of site wind speed more intuitive, this study combines "Wind Effects on Structures: An Introduction to Wind Engineering" (Simiu Emil, Scanlan Robert H, 1992) with existing research, V<0.3m/s is classified as static wind, $0.3 \leq$ V<1.0m/s is classified as weak wind, 1.0m/s $\leq$ V $\leq$ 5.0m/s is classified as comfortable wind, and V>5.0m/s is classified as strong wind.

In addition, HTC evaluation indexes are introduced as the reference of urban thermal environment design. The commonly used HTC evaluation indexes in academic circles mainly include SET (Standard Effective Temperature), UTCI (Universal Thermal Climate Index) and PET (Physiological Equivalent Temperature), etc. Among them, UTCI can reflect the equivalent environmental temperature of human body in real environment based on artificially set reference environment. Compared with other indexes, UTCI has better HTC simulation effect in urban microclimate scale [**Error! Reference source not found.**][21]. The index is graded according to the physiological response of human body in the thermal environment [22].

2.3   Data

To reflect the influence of block space form on wind environment and HTC more accurately and intuitively, the object of this study is idealized model. However, due to relevant laws and regulations, economic benefits, and other reasons, some of the situations represented by the model will not appear in real life. Considering the time efficiency and the computing ability of computer, the buildings and their surroundings are simplified. In terms of land use size and scale, the road network is a chessboard layout, including 9 blocks, with a single block size of 100m × 100m and a road width of 20m.

This paper selects the meteorological data from 6:00 to 18:00 on August 26th in Guangzhou as the simulation experiment data, and the data comes from the general meteorological documents provided by EnergyPlus website (https://www.energyplus. net/). The solar radiation in this period is relatively strong, with an average radiation amount of 225.5W/m2, an average radiation temperature of 55 °C, and the dominant wind direction is southeast wind, which has typical summer sunny days such as high temperature (maximum temperature is 33.0 °C, minimum temperature is 25.1 °C, average temperature is 30.1 °C), high humidity (average relative humidity is 69.0%) and light wind (average wind speed is 2.4 m/s).

2.4 Route and Parameter

The route of the study mainly includes four steps: "Parametric Modeling, Ecological Performance Simulation, Design Scheme Optimization, Planning and Design Enlightenment" (**Fig. 1**). Parametric Modeling, using Rhino built-in plug-in Grasshopper to set up simulation environment and experimental groups. Ecological Performance Simulation, using Butterfly and Ladybug to simulate the wind and heat environment of the site and implement the quantitative analysis of the site microclimate. Design Scheme Optimization, using Galapagos algorithm optimization, data analysis manual screening and genetic algorithm automatic optimization, and then uses TT Toolbox to export data for subsequent analysis. Finally, based on the above operation, Summarize the enlightenment of planning and design.

3 Results

3.1 Influence of Height and Density of Building on Wind Environment and HTC

In Group A (**Table 1**), there are three building heights: 100m, 54m and 27m. The building density includes 33.33%, 21.33%, 12% and 5.33%. In terms of building type, the depth of each house is 12m, the bay is 8m, and every two houses are combined into one unit.

The simulation results of the influence of urban building height and density on wind environment and HTC are as follows (**Fig. 2**).

The solar radiation received by buildings and ground will decrease with the increase of building height or projected area, which reduces the overall heat island intensity of the site and optimizes the outdoor HTC of human body. Due to the influence of many factors, such as sunshine spacing, fire protection code, sky visibility, and height-width ratio of blocks, it is necessary to comprehensively consider the situation of blocks in actual construction. In the process of urban renewal and transformation, methods such as building facade shading and structure shading can be adopted to increase the effective shadow area of blocks and alleviate the urban heat island effect.

With the increase of building density, the difficulty of airflow entering the interior of the block increases obviously, and the proportion of comfortable wind area decreases first and then rises, while the HTC value shows a downward trend. When the building density decreases, the reduction of FAR is not conducive to the intensification of land resources, and then affects economic benefits; when the building density increases, the canyon zone formed by the reduction of building spacing is easy to produce "urban canyon effect" in the block with point layout, which leads to the strong wind. Because

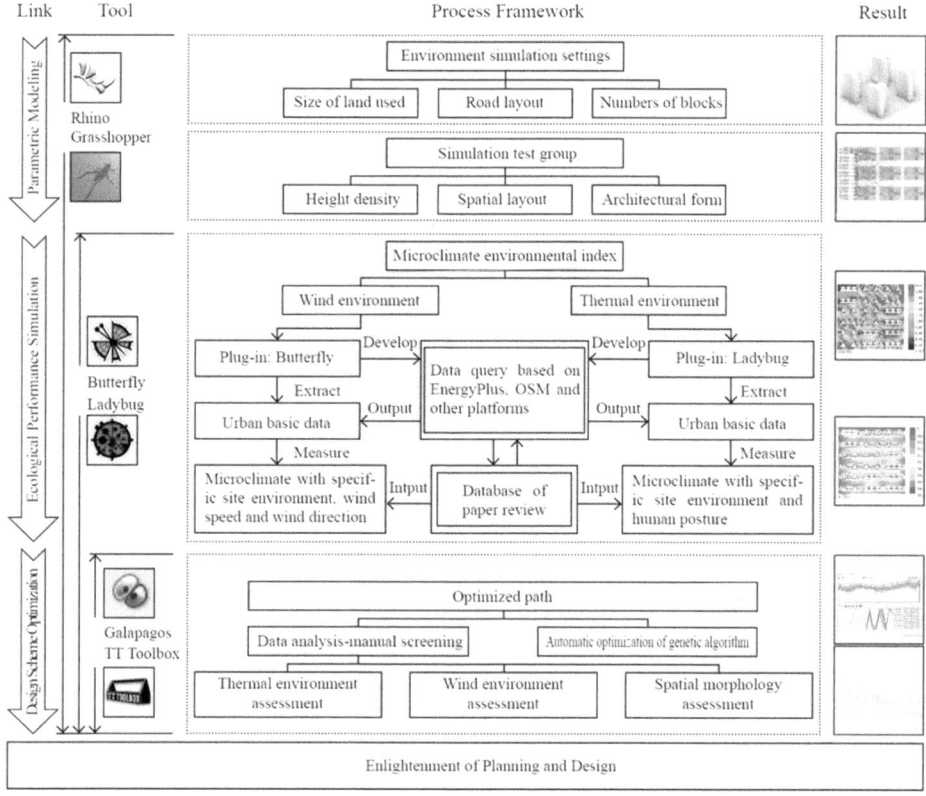

Fig. 1. The framework of this study.

Table 1. Group A: 6 models with different height and density of buildings.

Number of Group A	A1	A2	A3	A4	A5	A6
Height	100m	54m	27m	27m	27m	27m
Density	12%	12%	12%	5.33%	21.33%	33.33%
Number of buildings	3 × 3 × 9	3 × 3 × 9	3 × 3 × 9	2 × 2 × 9	4 × 4 × 9	5 × 5 × 9
Spacing (East-west)	26m	26m	26m	26m	12m	5m
Spacing (North-south)	32m	32m	32m	32m	17.34m	10m
Size of Building Base	16m × 12m					
Total Land Area	129600m^2					

the factors such as FAR, air circulation coefficient, sunshine spacing and fire protection code need to be considered in actual construction, although increasing building density is beneficial to improve the overall HTC of blocks while maintaining the same average height of blocks, there is a negative correlation between building height and building

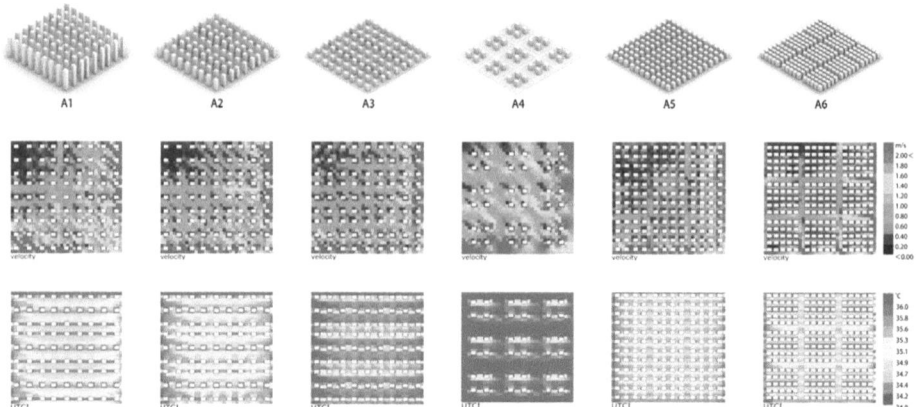

Fig. 2. Axonometric map, wind speed distribution map and HTC distribution map of Group A.

density while maintaining the same FAR. Excessive increase of building density will lead to marginal diminishing effect of HTC improvement, so that excessive occupation of public space cannot efficiently optimize urban thermal environment and reduce the quality of wind environment (**Table 2**).

Therefore, it is necessary to weigh the intensive degree of land resources and the quality of outdoor public environment on the premise of complying with laws and regulations. In the actual reconstruction, it is possible to increase the overhead space on the first floor while moderately adjusting the building density, thus optimizing the outdoor wind and thermal environment.

3.2   Influence of Spatial Layout of Buildings on Wind Environment and HTC

In simulation objects of Group B (**Table 3**), 6 groups of models each represent a typical block building arrangement form, and each group's "Continuity of Peripheral Interface" and "Integrity of Internal Space" are different from each other. The average building height is 27m, the building density is 18.67%, and the building type is the same as Group A.

The simulation results of the influence of urban building spatial layout on wind environment and HTC are as follows (**Fig. 3**).

It can be seen from above map and table of simulation results (Fig. 3 and **Table 4**): (1) In the 6 groups of models, the HTC is the best when the spatial layout is continuous in the east-west direction and scattered in the internal space; when the spatial layout of block buildings is continuous in east-west and north-south directions and the internal space is complete, the HTC is the worst. (2) Integrity or dispersion of the interior space has no obvious influence on the HTC, and the function of the interior space is mainly reflected in the visual environment rather than the wind environment. (3) When the interface of east-west and north-south direction is continuous, the environmental HTC is the worst; when the interface between east and west and north-south is discontinuous, the environmental HTC is in the middle; when the east-west interface is continuous, the environmental HTC is optimal. Therefore, the enclosing and point group layout of blocks

Table 2. Statistical table of simulation results of Group A.

	Classification	Static Wind	Weak Wind	Comfortable Wind	Strong Wind	Overall	HTC Value
A1	Area Proportion (%)	15.7	45.82	38.48	0	100	34.71°C
	Average Wind Speed (m/s)	0.18	0.63	1.53	--	0.91	
A2	Area Proportion (%)	17.53	48.9	33.57	0	100	34.92°C
	Average Wind Speed (m/s)	0.18	0.64	1.54	--	0.86	
A3	Area Proportion (%)	15.7	58.3	26	0	100	35.28°C
	Average Wind Speed (m/s)	0.21	0.63	1.54	--	0.8	
A4	Area Proportion (%)	6.14	41.49	52.37	0	100	35.76°C
	Average Wind Speed (m/s)	0.2	0.71	1.51	--	1.1	
A5	Area Proportion (%)	23.71	54.6	21.69	0	100	34.96°C
	Average Wind Speed (m/s)	0.21	0.59	1.59	--	0.72	
A6	Area Proportion (%)	13.93	55.84	30.23	0	100	34.62°C
	Average Wind Speed (m/s)	0.19	0.61	1.55	--	0.84	

is not conducive to steering flow, while the determinant architectural layout is conducive to the formation of ventilation corridors to guide the block airflow and improve the wind speed inside the block and form an outdoor open space with appropriate scale and improving the wind and heat environment.

3.3 Thermal Comfort Optimization of Urban Architectural Spatial Form Based on Genetic Algorithm

In simulation objects of Group C (**Table 5**), the architectural form adopts the 5 prototypes summarized above; the average building height is 0m, 9m, 27m, 54m and 80m respectively.

The simulation results of HTC optimization of urban architectural space form based on genetic algorithm are as follows (**Fig. 4** and **Table 6**).

Table 3. Group B: 6 models with different spatial layout.

Number of Group B	B1	B2	B3	B4	B5	B6
Continuity of Peripheral Interface	E-W and N-S continuous	E-W continuous	The E-W and N-S discontinuous	The E-W and N-S continuous	The E-W continuous	The E-W and N-S discontinuous
Integrity of Internal Space	Concentration	Concentration	Concentration	Disperse	Disperse	Disperse
Spacing (East-west)	12m	10m	12m	16m	12m	12m
Spacing (North-south)	14m.	32m	10m	22m	17m	17m
Building Size	16m × 12m					
Height	27m					
Density	18.67%					
Number of buildings	14 × 9					
Total Land Area	129600m^2					

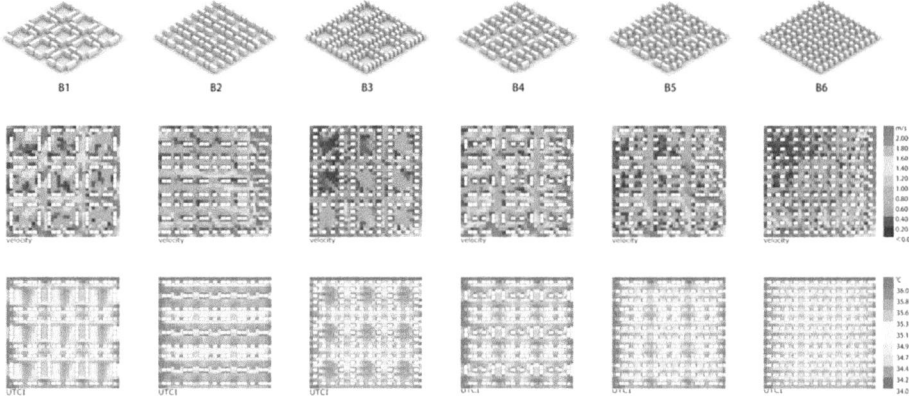

Fig. 3. Axonometric map, wind speed distribution map and HTC distribution map of Group B.

Sort out the data recorded and exported by the software, and then reordered according to the comfort index from good to bad, and various data are extracted to generate charts for analysis. From the system convergence and layout change trend of genetic algorithm interface (**Fig. 5**), and the initial generation, intermediate generation and optimal generation data extracted in the genetic process, the UTCI value decreases, which determines that Galapagos has an effective optimization effect on the target, and the system operation process is directional, and an ideal building layout can be obtained through system optimization. After continuous optimization process, Galapagos shows that the value approaches stability and then stops calculation and obtain 321 groups of valid data. All data are input into Excel and obtain the genetic line chart with UTCI value converging to the minimum (**Fig. 6**). In addition, from the trend line fitted by scatter plot, it can be seen that the change of comfort index shows a downward trend with the continuous operation of genetic algorithm (**Fig. 7**), from the highest 35.84 °C to 34.96 °C, and the temperature drop is as high as 0.88 °C, in which when n = 265 (Generation: 19, Solution: 15) is the ideal building layout under the set conditions.

Based on the above research process, this paper analyzes the characteristics of architectural space form optimization based on genetic algorithm from the following 4 aspects: (1) Architectural Form Combination: the three groups with the best comfort index contain the third and fourth architectural form prototypes, and the number of these two architectural form prototypes is almost equal; the three groups with the worst comfort index contain three or more architectural form prototypes, and the worse the comfort, the more types of architectural form prototypes. (2) Architectural Plane Layout: the three groups of outdoor open spaces with the best comfort index have moderate area and uniform layout in the site, which can form relatively obvious ventilation corridors; in the three groups with the worst comfort index, sporadic small-scale outdoor open space coexists with concentrated large-scale outdoor open space. (3) Building Height Layout: the three groups of building height layout with the best comfort index are unified and the building height changes smoothly; the three groups with the worst comfort index have chaotic building height layout. (4) Control Index of Planning: when the planning index of building density and FAR is the same, the change of plane and height layout

Table 4. Statistical table of simulation results of Group B.

	Classification	Static Wind	Weak Wind	Comfortable Wind	Strong Wind	Overall	HTC Value
B1	Area Proportion (%)	13.86	47.46	38.68	0	100	35.07°C
	Average Wind Speed (m/s)	0.21	0.64	1.53	--	0.92	
B2	Area Proportion (%)	9.85	53.31	36.84	0	100	35.01°C
	Average Wind Speed (m/s)	0.21	0.65	1.53	--	0.93	
B3	Area Proportion (%)	19.53	60.19	20.28	0	100	35.03°C
	Average Wind Speed (m/s)	0.21	0.6	1.59	--	0.72	
B4	Area Proportion (%)	9.99	48.65	41.36	0	100	35.04°C
	Average Wind Speed (m/s)	0.21	0.67	1.46	--	0.95	
B5	Area Proportion (%)	16.2	54.07	29.72	0	100	35.01°C
	Average Wind Speed (m/s)	0.2	0.62	1.54	--	0.83	
B6	Area Proportion (%)	20.79	55.4	23.81	0	100	35.04°C
	Average Wind Speed (m/s)	0.2	0.58	1.56	--	0.74	

of buildings in the block will lead to the difference of comfort; there are many kinds of architectural layout schemes corresponding to the same site comfort index, for example, there are 24 kinds of architectural layout schemes corresponding to the HTC value of 35.03 °C. From the perspective of microclimate, the above results enrich the existing research on the architectural layout of urban blocks and provide theoretical basis for urban planning technology.

4   Conclusion and Discussion

The contradiction between global climate change and urban sustainable development has become increasingly prominent, and the built-up environment of cities under climate change is closely related to HTC. It is increasingly urgent to promote climate-adaptive urban planning through computer simulation technology. Research on urban microclimate based on Grasshopper software is of great significance for climate-adaptive

Table 5. Group C: 5 models with different architectural form prototypes.

Number of Group C	Form 1	Form 2	Form 3	Form 4	Form 5
Prototype of building					
Area of Building	--	$200m^2$	$600m^2$	$400m^2$	$625m^2$
Number of buildings	--	9	6	4	2
Spacing(East-west)	--	15m	--	31m	--
Spacing(North-south)	--	30m	40m	48m	--
Height	--	18m	27m	54m	100m
Density	--	14.88%	29.75%	13.22%	10.33%
Total Land Area	$129600m^2$				

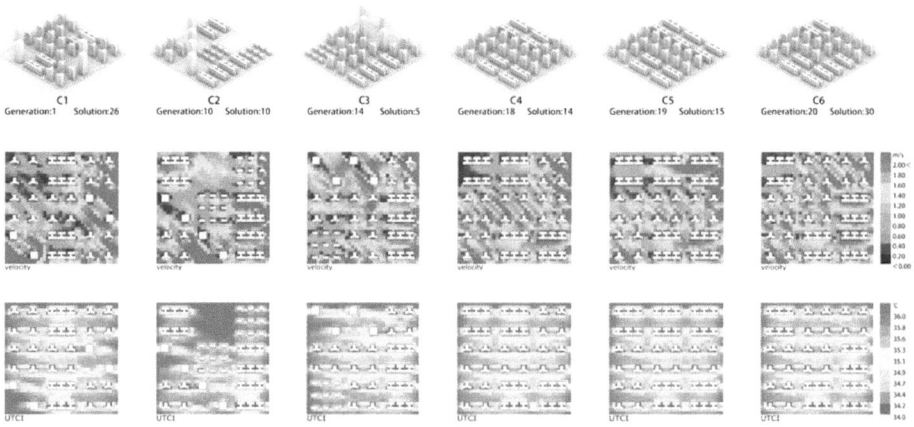

Fig. 4. Axonometric map, wind speed distribution map and HTC distribution map of Group C.

urban planning. Based on the above research results, the following three optimization suggestions are put forward:

(1) Increasing the building height of the block can increase the effective shadow area of the block, and moderately adjusting the building density and increasing the overhead space on the first floor of the building can optimize the wind and heat environment. In actual construction, it is necessary to comprehensively consider the factors such as sunshine spacing, fire protection regulations, sky visibility, and block height-width ratio. Under the condition of keeping the block FAR consistent, excessive increase of building density will lead to marginal diminishing effect of HTC improvement, which will lead to the decrease of wind environment quality. Therefore, in the process

Table 6. Statistical table of simulation results of Group C.

	Classification	Static Wind	Weak Wind	Comfortable Wind	Strong Wind	Overall	HTC Value
C1	Area Proportion (%)	13.86	47.46	38.68	0	100	35.07°C
	Average Wind Speed (m/s)	0.21	0.64	1.53	--	0.92	
C2	Area Proportion (%)	9.85	53.31	36.84	0	100	35.27°C
	Average Wind Speed (m/s)	0.21	0.65	1.53	--	0.93	
C3	Area Proportion (%)	19.53	60.19	20.28	0	100	35.05°C
	Average Wind Speed (m/s)	0.21	0.6	1.59	--	0.72	
C4	Area Proportion (%)	9.99	48.65	41.36	0	100	34.97°C
	Average Wind Speed (m/s)	0.21	0.67	1.46	--	0.95	
C5	Area Proportion (%)	16.2	54.07	29.72	0	100	34.96°C
	Average Wind Speed (m/s)	0.2	0.62	1.54	--	0.83	
C6	Area Proportion (%)	20.79	55.4	23.81	0	100	34.98°C
	Average Wind Speed (m/s)	0.2	0.58	1.56	--	0.74	

of urban renewal and transformation, the effective shadow area of the block can be increased by shading the facade of the building and shading the structure, and the air circulation of the block can be promoted by increasing the overhead space on the first floor of the building, thus reducing the unnecessary heat emission caused by the indoor refrigeration energy consumption of the building, thus effectively slowing down the heat island effect.

(2) The effect of block layout of determinant architecture on steering air flow and forming more pleasant outdoor open space thermal environment is more obvious than that of peripheral layout and point group layout. Therefore, the planning should scientifically and reasonably coordinate the combination relationship and mode between the architectural space layout and the public space such as roads, squares, and green spaces in the block. Sporadic small-scale outdoor open spaces and concentrated large-scale outdoor open spaces can be transformed into moderate-scale outdoor

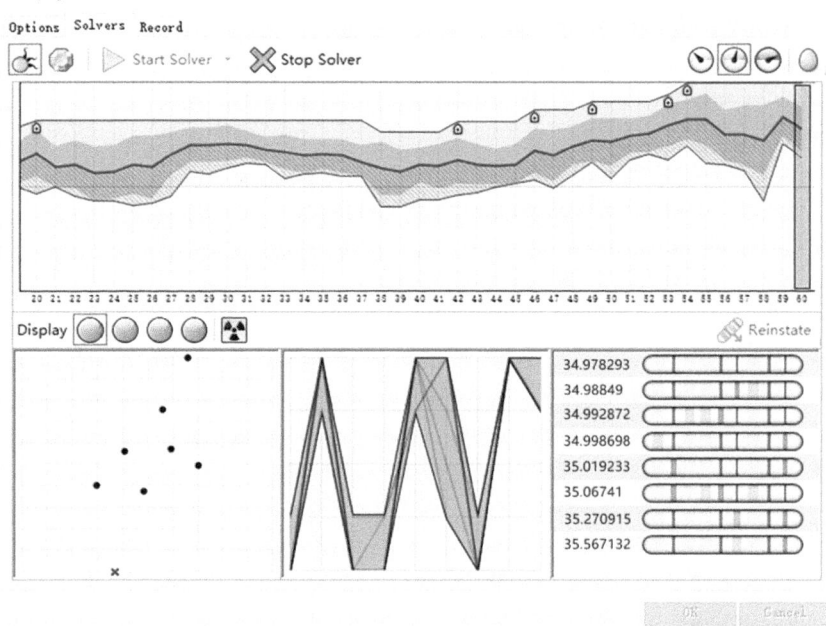

Fig. 5. Panel of genetic algorithm.

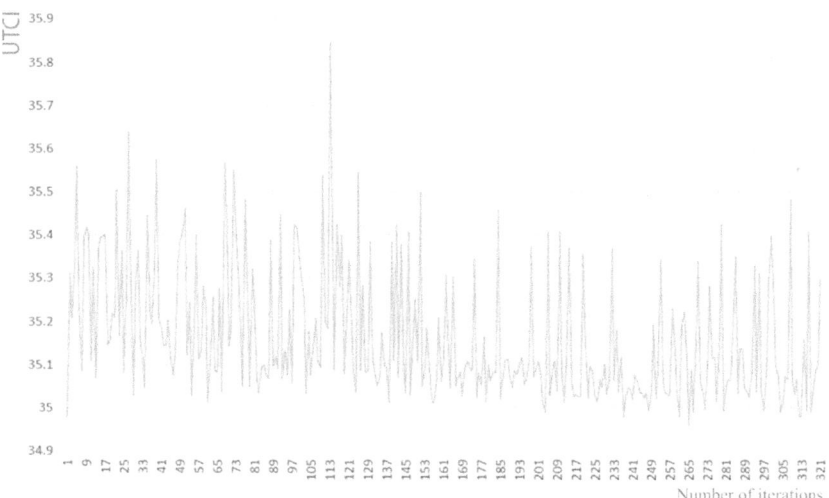

Fig. 6. Change trend line chart of comfort index in optimization process.

open spaces, and distribute in cities evenly, to create complete and continuous outdoor open spaces conforming to the dominant wind direction. By forming urban

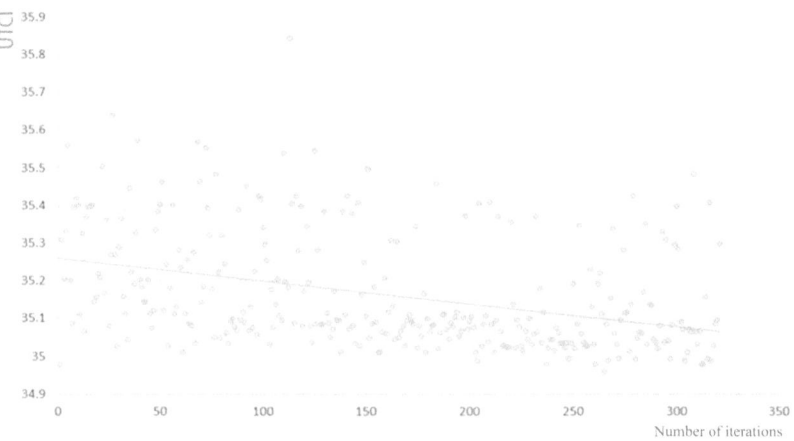

Fig. 7. Scatter diagram of comfort index in optimization process.

ventilation corridors, air circulation and surface heat dissipation in blocks are promoted. And finally, to achieve the objective of improving urban wind and heat environment.

(3) In the combination of block architectural forms, it is advisable to design single buildings and plan the layout of buildings based on the climate characteristics of the site, to solve the problems such as poor outdoor wind and heat environment of dense and low residential buildings or high-rise commercial office buildings. In addition, it is necessary to fully consider the ventilation, lighting, shading, and cooling effects of the site under various schemes to improve human comfort.

In the initial stage of urban planning, we mainly pay attention to the efficient layout of the overall spatial form of the site and realize the climate-adaptive urban planning by implementing planning control indicators, to achieve the best performance of the overall microclimate of the city. This study takes microclimate performance as the core driving force, draws lessons from multidisciplinary theories and methods, and constructs a theoretical model of site microclimate in urban blocks according to specific urban climate characteristics, which provides scientific basis for urban layout optimization. However, at present, this study only starts from the microscopic point of view and calculates based on the idealized mathematical model. The urban environment is complex and changeable, and macro conditions such as topography, climate and surrounding environment have not been fully considered in the model. Therefore, in the future, with the help of big data technology and large sample sampling, the research can be expanded from a single microclimate performance-driven design to an overall urban and rural planning performance-driven design.

Acknowledgement. This work was supported by the Guangdong Province Philosophy and Social Science Planning Project (GD24XSH11), the Central Government's Guidance on Local Science and Technology Development Special Fund (2023L3032), the National Natural Science Foundation of China (42471217), the Natural Science Foundation of Fujian Province, China (2021J011072), and the National Natural Science Foundation of China (41801160).

References

1. Estrada, F., Botzen, W., Tol, R.: A global economic assessment of city policies to reduce climate change impacts. Nat. Clim. Chang. **7**, 403–406 (2017)
2. Konstantinov, P., Varentsov, M., Esau, I.: A high-density urban temperature network deployed in several cities of Eurasian Arctic[J]. Environ. Res. Lett. **13**, 75007 (2018)
3. Fazia, A.-T., Helmut, M.: Numerical study on the effects of aspect ratio and orientation of an urban street canyon on outdoor thermal comfort in hot and dry climate. Build. Environ. **41**(2), 94–108 (2006)
4. Yingjun, Q., Shizhu, L.: Advances on impact of climate change on human health. Clim. Chang. Res. **6**(4), 241–247 (2010)
5. Shanglin, W., Yimin, S.: An analysis of the impact of Guangzhou's urban form on heat Island effect. Arch. J. **10**, 79–82 (2015)
6. WMO. World Climate in 2000. (2000)
7. WMO. State of the Global Climate in 2019. (2020)
8. Jingxiang, Z.: Reconsideration on the trend of low-carbon city development in China. Planners **26**(5), 5–8 (2010)
9. Xianhui, F.: Chu yanyan: The study of urban green space and local micro-climate effect based on air dynamics simulation. Chin. Landsc. Arch. **33**(4), 29–34 (2017)
10. Dai Fei, Chen Ming, et al: Effects of different green coverage in block scale on PM10 and PM2.5 Removal-A case study of the main city of Wuhan. Chin. Landsc. Arch. **34** (03), 105–110 (2018)
11. Deshun, Z., Zhen, W.: Micro-climate effect and human thermal comfort of square canopy in dense habitat-A case study of Shanghai knowledge and innovation community square. Chin. Landsc. Arch. **33**(04), 18–22 (2017)
12. Fei, G., Jun, Z., et al.: Multi-Model, multi-scale urban ventilation paths exploration and landscape strategy. Landsc. Arch. **27**(7), 79–86 (2020)
13. Chao, R., Chao, Y., et al.: A study of air path and its application in urban planning. Urban Plan. Forum **3**, 52–60 (2014)
14. Lili, W., Zhengbo, W., et al.: Spatial control of coastal area in response to climate change. Planners **37**(4), 11–16 (2021)
15. Zhiwei, C., Jie, Z., et al.: Process in application of parametric design and its inspiration to landscape design. Chin. Landsc. Arch. **28**(10), 40–45 (2012)
16. Xiaojian, B., Conghong, L.: Parametric design of energy efficiency through Ladybug + Honeybee a case study on the design of office complexes in cold regions in China. Arch. J. **2**, 44–49 (2018)
17. Capeluto, I.: G, Plotnikov B: A method for the generation of climate-based, context-dependent parametric solar envelopes. Archit. Sci. Rev. **60**, 395–407 (2017)
18. Amer, A.-J.: Jabi W: Spatial reasoning as a syntactic method for programming socio-spatial parametric grammar for vertical residential buildings. Archit. Sci. Rev. **63**, 135–153 (2020)
19. Kristina Kiesel, Kristina Orehounig, et al: Urban heat island phenomenon in Central Europe, The First International Conference on Architecture & Urban Design, 821–828 (2012)
20. Zhuolun, C., Lihua, Z., et al.: Field measurement and analysis of the microclimate in typical residential quarter of Guangzhou. Arch. J. **11**, 24–27 (2008)
21. Qian Zhang, et al.: Association between wind environment and spatial characteristics of high-rise residential buildings in cold regions through Field Measurements in Xi'an, Buildings 13.8(2023)
22. Peter, B., Dusan, F., et al.: Deriving the operational procedure for the Universal Thermal Climate Index (UTCI). Int. J. Biometeor. **56**(3), 481–494 (2012)

23. Steemers, K.: Sustainable urban design: issues, research and projects. World. Architecture **08**, 34–39 (2004)
24. Ministry of Housing and Urban-Rural Development of the People's Republic of China: Assessment Standard for Green Building, (2019)

Country-Wide Effects of Urban Heat Island on Cooling and Heating Energy Use—An Empirical Case of Office Buildings in South Korea

Jack Ngarambe, Gon Kim, and Geun Young Yun[✉]

Department of Architectural Engineering, Kyung Hee University, Gyeonggi 17104, Republic of Korea
gyyun@khu.ac.kr

Abstract. Urban Heat Islands (UHI) affect building energy use in many cities worldwide. The correlations between UHI and building energy use have mostly been studied via city-scale modeling simulations, making validation of the obtained results challenging. In this study, we use clustering and statistical methods to examine the relationship between various indicators of UHI and building energy use, utilizing archived empirical data on energy consumption in office buildings across the entire country of South Korea. Our findings reveal considerable differences in UHI behavior across provinces and cities in the country. These variations are driven by a complex interplay of factors related to geographic locations, urbanization levels, and the topography of the provinces. These results suggest that mitigative efforts for UHI in South Korea should consider targeted measures tailored to specific locales. We also identify strong positive correlations between various UHI indicators, particularly monthly average Urban Heat Island Intensity (UHII) and cooling energy consumption in office buildings. However, the relationship between UHII and heating energy consumption was largely non-existent. These findings offer an empirical foundation for the development of efficient and inclusive policies that promote livability in urbanized areas.

Keywords: urban heat islands · building energy use · spatial analysis

1 Introduction

The building industry plays a crucial role in the global energy system. In 2010, buildings worldwide consumed approximately 1.25 petawatt-hours (PWh) of energy solely for cooling purposes. Notably, 45% of this energy was attributed to the cooling of non-residential buildings, such as office buildings [1]. This high demand for space cooling energy is concerning, especially considering the expected increase in future temperature conditions and extreme heat events. Such concerns are more amplified for the urban building stock than for the rural building stock [2]. The large variations in energy use between urban and rural building stocks are primarily due to the energy budget differences between these environments. These differences are often induced by the nature

© The Author(s) 2025
B.-J. He et al. (Eds.): UCSUD 2023, LNCE 559, pp. 497–510, 2025.
https://doi.org/10.1007/978-981-97-8401-1_35

of the urban fabric, frequently resulting in urban heat islands (UHIs) – a phenomenon where urban areas register significantly warmer environments than rural and suburban areas [3].

The link between intensified Urban Heat Island (UHI) levels and building energy consumption is well-documented in scientific literature. Numerous studies have reported substantial influences of UHI on building energy use in major megacities [4]. The consensus from such studies is that UHI significantly increases cooling energy consumption while partially reducing heating energy consumption. For instance, studies have demonstrated that UHI increases the cooling energy consumption of buildings by 27.5% in Tokyo, Japan [5], 11.4% in Melbourne, Australia [6], and 11.28% in Beijing, China [7]. Li et al. [8] provide a succinct overview of the influence of urban heat islands on building cooling energy use across various cities.

However, the acquisition of real building energy consumption data presents significant challenges, leading most existing studies to rely on numerical modeling techniques. While these techniques are valuable, they are not without limitations. The accuracy of numerical models heavily depends on the input parameters and the expertise of the modeler, introducing potential uncertainties that can make validation of the results challenging. Consequently, this reliance on numerical modeling can yield varied and highly heterogeneous conclusions on the UHI effect and its impact on energy consumption. To the best of the authors' knowledge there is a single study at the moment that utilizes highly spatial data (i.e., city-wide) to explore the influences of synoptic weather conditions on building energy use [9]. While informative, the study considers buildings of diverse functions (i.e., mixed use buildings), making it difficult to estimate the contribution of space use behavior on the obtained results, although theoretically very small. This is critical as it could lead to biased estimations of the influence of local warming on building energy use. The aim of this study is to empirically explore the spatial effects of urban heat islands (i.e., considering the entire South Korea) on the energy consumption of office buildings. The concentration on office buildings alone ensures that the element of space use behavior is accounted for in the conducted analysis.

2   Methods

2.1   Building Property and Energy Use Data

Building energy data was obtained from an online repository run by the Korean government [10]. The repository contains monthly electricity and gas energy consumption data for the majority of building facilities in Korea. Electricity consumption data is collected via monitoring meters installed in most public and private facilities, whereas gas usage data is obtained from audit surveys conducted at district subdivisions. Additionally, a separate database with recorded information on the architectural characteristics of the facilities is provided. This database offers details related to building addresses, building function, number of floors, and other relevant information, and was used in the calculation of energy use intensities (EUI). It is important to note that the initial number of office buildings extracted from the public data repository was 6,648 office buildings. However, our analysis required segregating total energy use into specific end uses for heating and cooling. This necessitated the inclusion of only those buildings with comprehensive monthly energy data available for the entire year. Applying this criterion reduced the eligible building count to 2,299. Subsequent data processing to remove outliers and correct illogical entries, such as buildings with a recorded floor area of 0 square meters, further narrowed the dataset to 1,648 buildings. These office buildings are distributed across South Korea but are mainly situated in city centers which are often urbanized localities.

2.2 Cooling and Heating Energy Use

The energy data provided includes total monthly energy used for heating, cooling, and other purposes such as cooking, hot water systems, etc. However, the direct effects of UHI are predominantly related to space heating and cooling energy needs, and its influence on only these end-use energies was studied. The repository includes monthly electricity and gas energy consumption data for most building facilities in Korea. To segregate the specific monthly space cooling and heating energy from total energy use, we employed the disaggregation method. This method estimates space cooling and heating energy based on a short-term measurement method (STM) introduced by Robinson [11]. Furthermore, the building energy data repository mentioned above does not record information related to the equipment used for heating and cooling purposes in the buildings. Therefore, it was not possible to determine the specific energy source used for heating/cooling in each building. Consequently, based on our previous analysis [9] and other studies on South Korea end-use energy consumption [10, 12], we assumed that electricity is the dominant energy source for cooling, and gas is the primary energy source for heating.

2.3 Pairing Energy Use Data with Prevailing Weather Conditions

Hourly temperature data was obtained from automatic weather stations (AWSs) spread across South Korea. A total of 528 AWSs are present and are jointly run by the Korean Meteorological Agency and local metropolitan offices distributed throughout Korea.

 Automated Weather Stations (AWS) record many climatic elements using advanced and highly accurate methods. Temperature is measured using metallic sensors with thin films, capable of detecting temperatures in the range of -40 °C to 60 °C with an accuracy of $\pm$ 0.3%. Wind speed is gauged using ultrasonic sensors, measuring speeds from 0 m/s to 70 m/s with an accuracy of $\pm$ 0.5%. Relative humidity is determined using capacitive sensors, which can measure humidity levels ranging from 0% to 100% with an accuracy of $\pm$ 3.0% [13].

 After identifying all AWSs and their corresponding temperature recordings, we needed to pair each office building with the nearest AWS to capture the direct effects of local climatic conditions on the energy performance of each building. To do this, we first identified the geographical coordinates of each office building present in the governmental database described in Sect. 2.1 and, similarly, the geographical coordinates of the AWSs. We then employed the Haversine algorithm [14] via the Haversine library in Python to determine the closest AWS to each building. Figure 1 shows the distribution of the considered office buildings together with the paired AWSs.

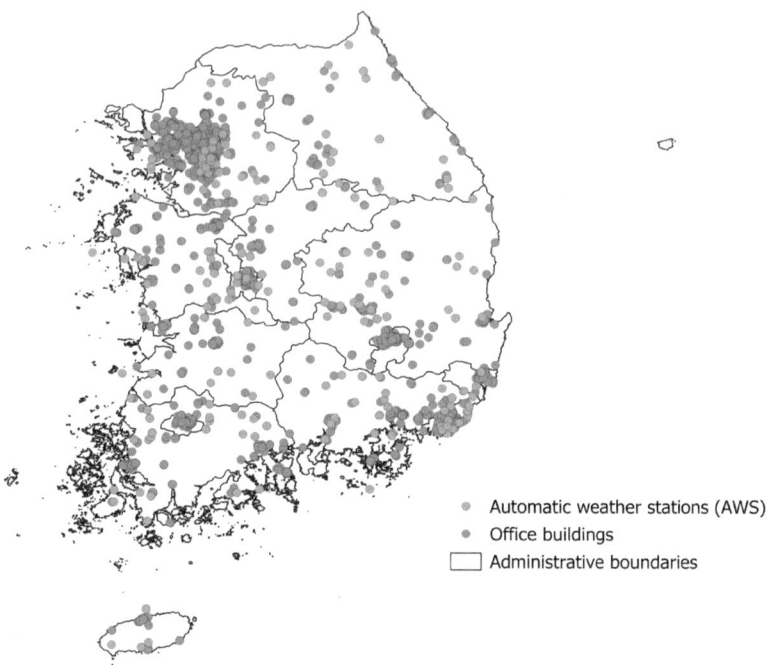

Fig. 1. Geographical Distribution of the Selected Office Buildings and the Proximity of Automatic Weather Stations (AWSs)

2.4 Quantifying Urban Heat Island

South Korea is divided into 17 upper administrative divisions (9 provinces and 8 special cities), which are further divided into multiple smaller divisions (i.e., counties, districts, towns, etc.). The lowest division is the neighborhood level (dong); therefore, a large province may consist of multiple dongs, each with multiple automatic weather stations (AWSs). As such, we needed to identify a reference rural station for each city. To determine which AWSs were located in relatively natural terrain and thereby resembled a rural area, we employed the normalized differentiated vegetation index (NDVI) as a criterion – NDVI is typically used to estimate the density of greenery within a bounded area. Consequently, we computed the mean NDVI within a 200 m radius buffer zone around each AWS. The AWS with the highest NDVI value within a city was then chosen as the representative reference rural area. One significant advantage of this approach is that it ensures the rural reference station utilized in Urban Heat Island (UHI) computations is in close proximity to, and likely shares similar geographical attributes with, the urban observatories of interest. This is crucial because it helps ensure that any observed climatic differences between the urban observatory and the reference rural stations are predominantly attributed to land use changes, thereby minimizing the influence of other confounding factors. However, a potential limitation of this method is the substantial time required to compute the NDVI values for each observatory and to undertake the iterative process of identifying an observatory situated in a more vegetated area compared to others. This aspect is less cumbersome in the conventional method, which involves selecting a single observatory, situated far from urban areas, and designating it as the reference rural station.

The urban heat island was quantified using urban heat island intensity (UHII) as an indicator; UHII is computed as the difference between the hourly atmospheric air temperature of one AWS and that of another AWS identified as belonging to a reference rural area (see Eq. 1 below). UHII was computed for each AWS, and since AWSs were paired with individual nearby buildings, the final dataset consisted of our initial office buildings spread across South Korea paired with monthly nearby AWSs and the corresponding hourly UHII values.

$$UHII = T_{AWS\_urban} - T_{AWS\_rural\_ref} \tag{1}$$

where UHII is the urban heat island intensity at a specific locality, T_{AWS_urban} is temperature recorded at urban automatic weather station and $T_{AWS\_rural\_ref}$ is temperature recorded at a designated reference rural automatic weather station.

2.5 Estimating Correlations Between Building Energy Use and UHII

To assess the potential correlations between UHI and building energy use, we organized our dataset to ensure that, for each province/special city, there was a corresponding data frame. This data frame included the individual energy use for each building paired with its respective UHII value, calculated as described in Sect. 2.4. Subsequently, within each data frame, we categorized the UHII data into bins, each differing by 2 °C intervals. We then compared the mean building energy consumption across the different UHII groups. To assess whether the observed energy variations among the UHII groups were statistically significant, we employed the Kruskal-Wallis H-test [15], as illustrated in the Eq. (2) below.

The threshold for statistical significance was established at a p-value of less than 0.05. A p-value below this threshold indicates that the observed differences in energy use across various Urban Heat Island Intensity (UHII) bin groups can likely be attributed to a genuine effect rather than random variation. Conversely, a p-value exceeding this threshold suggests that any observed differences may be due to chance.

$$H = \frac{12}{N(N+1)} + \sum_{i=1}^{k} \frac{R_i^2}{n_i} - 3(N+1) \tag{2}$$

where N = sum of the sample size for all samples, K = Number of samples, n = size of the ith sample and R_i = sum of the ranks in the ith sample.

3   Results and Discussion

3.1 Geographical, Social, and Weather Characteristics of the Administrative Divisions in South Korea

Table 1 shows the area, population (as of 2020), and typical monthly mean meteorological elements recorded in 2019 for each of the 17 top-tier administrative divisions in South Korea. The administrative divisions are divided into provinces (Gyeonggi, Gangwon, Chungcheongnam, Chungcheongbuk, Jeollanam, Jeollabuk, Gyeongsangnam, and

Gyeongsangbuk), metropolitan cities (Busan, Daegu, Incheon, Gwangju, Daejeon, and Ulsan), and special cities (Seoul and Sejong). In terms of population, Seoul is the largest city, followed by Busan, while Gyeonggi is the largest province. Urbanization in South Korea has increased rapidly over the years, bringing along substantial environmental modifications [16]. The summertime in Korea is characterized by high precipitation, partly from the East Asian monsoon and oceanic currents in coastal cities such as Busan and Incheon [17]. Conversely, the winters are quite dry, owing to high-pressure system regimes. Similarly, the humidity is often high during the summertime; for the reference year 2019, relative humidity greater than 65% was observed across the country, which slightly decreased during wintertime. As expected, the highest wind speeds, resulting from oceanic influences, are observed mainly in coastal cities/provinces (e.g., Busan, Incheon, Jeju), while the lowest are observed in inland cities and are relatively similar during both the summertime and wintertime. Cloud cover is highest in the summertime, largely as a result of the East Asian monsoon [18]. The information presented in this section deepens our understanding of Urban Heat Island (UHI) phenomena across each South Korean province and special city. Critical geographical factors, such as area size, as well as demographic aspects like population density, have been shown to significantly influence UHI formation. These factors, in turn, indirectly impact energy consumption patterns [19, 20].

3.2 Country-Wide Variations in UHII

Figure 2 demonstrates the variations in Urban Heat Island Intensity (UHII) across different provinces and cities in South Korea, presenting results for various monthly and daily UHII variants. This figure reveals substantial inter-city variations in UHII, suggesting distinct UHI characteristics across Korea, potentially influenced by factors such as topography, population density, and proximity to water bodies, among others.

For instance, daytime maximum UHII during summer (as depicted in Fig. 2d) shows the highest UHIIs in Gangwon (5.2 °C), followed by Gwangju (3.3 °C), Jeju (3 °C), and Seoul (2.8 °C). Gangwon is the largest province in Korea, spanning 20,569 km2. It comprises both relatively developed towns and typical rural areas with natural lands, leading to significant temperature variations over long distances and resulting potentially the reason for the observed pronounced UHI manifestations. In contrast, the high UHII observed in Gwangju could be attributed to the limited access to ventilative cooling characteristic of inland cities. At the same time, Seoul's UHI is predominantly a result of increased anthropogenic heat emissions, a high concentration of densely built facilities, and consequently restricted cooling mechanisms [20].

It is noteworthy that, in some areas during summertime, UHI is generally higher at night than during the day [see Fig. 2d and Fig. 2e]. Taking Seoul as an example, the difference between nocturnal UHII and daytime UHII reaches up to 2 °C. Such occurrences of elevated temperatures at night in city centers, relative to rural areas or those characterized by plain terrain, have been observed in various studies [21, 22] and are largely a result of elements that characterize the urban texture such as the sky view factors and the high thermal transmittance of urban materials [23, 24].

Comparing UHI manifestations, particularly daytime maximum UHII, between winter and summer reveals that UHI tended to be more intense during the winter than in

Table 1. Geographical, demographic and weather characteristics of the South Korean administrative divisions

Administrative division	Area (km²)	Population (millions)	Precipitation (mm)		Humidity (%)		Windspeed (m/s)		Cloud cover (Octas)		Temperature (°C)	
			Summer	Winter	Summer	Winter	Summer	Winter	Summer	Winter	Summer	Winter
Chungcheong Buk	7433	1,599,391	226.1	0.9	79	61	1.4	1.2	6.9	3.7	23.9	-0.3
Chungcheongnam	8204	2,122,455	141.1	6.4	76	62	1.2	1.0	6.6	3.5	23.8	0.6
Gangwon	20569	1,540,540	220.8	16.6	71	34	2.1	3.3	6.6	3.2	23.0	-0.1
Gyeonggi	10171	13,250,368	329.8	0.4	82	46	1.4	1.4	6.5	3.4	24.2	-0.7
Gyeongsangbuk	19030	2,662,508	159.2	13	82	49	2.1	2.7	7.3	3.2	23.7	1.9
Gyeongsangnam	10532	3,361,344	512.2	17.7	88	48	2.1	1.6	6.9	3.4	24.3	3.4
Jeollabuk	8043	1,817,302	142.9	15.4	85	71	2.6	2.6	6.9	5.1	23.9	1.4
Jeollanam	11858	1,864,712	247.4	10.1	81	60	2.1	2.6	7.1	4.4	23.6	3.5
Jeju	1849	6,707,49	191.3	17.4	95	69	2.5	8.1	6.9	6.6	24.2	8.0
Busan	769.89	3,411,819	358.9	12.7	85	39	3.7	3.1	7.2	3.1	24.1	6.0
Daegu	883.56	2,436,488	139.4	9.5	76	44	1.9	2.3	7.4	3	25.3	3.1
Incheon	1062.60	2,956,119	307.9	0	77	54	2.4	1.8	6.3	3.7	23.1	0.7
Gwangju	501.24	1,456,688	242.2	16.4	84	59	1.8	1.4	7.4	4.1	24.8	3.4
Daejeon	539.35	1,474,152	199.0	1.7	79	60	1.6	1.2	7	3.6	25.2	1.5
Sejong	465.23	342,328	274.4	12.7	77	39	1.3	2.3	6.3	3.7	24.4	2.0
Seoul	605.21	9,733,509	194.4	0	69	46	1.8	1.8	7.2	3.1	25.2	0.5
Ulsan	1061	1,147,037	206.5	15.5	84	40	2.1	2.5	7	3	24.0	4.7

the summer in all provinces/special cities except for Gwangju, Busan and Gyconggi where UHI was more intense in summertime than the wintertime. Specifically, considering the daytime maximum UHI in Seoul (see Fig. 2d), a difference of 0.7 °C was observed between the summer and winter periods. Similar differences are noted in other cities/provinces: Gwangju (0.98 °C), Ulsan (1.93 °C), Jeollanam (1.81 °C), Jeollabuk (1.97 °C), Incheon (1.80 °C), Gyeonggi (1.68 °C), Daegu (2.19 °C), Chungcheongnam (2.30 °C), and Chungcheongbuk (2.25 °C). These variations are primarily due to the monsoon rains characteristic of the Korean peninsula during summer, which often lead to increased evaporative cooling during this season.

3.3 Associations Between Building Energy Use and Urban Heat Island Intensity

3.3.1 Cooling Energy Use

Figure 3 illustrates the relationships between the cooling energy consumption of office buildings and six different UHII indicators: (a) monthly average UHII, (b) monthly maximum UHII, (c) daily UHII, (d) daytime maximum UHII, (e) nighttime average UHII, and (f) daytime average UHII. Our results indicate positive correlations between cooling energy consumption and various UHII indicators, with the exception of monthly maximum UHII. This correlation suggests that an increase in UHII generally corresponds to an increase in monthly cooling energy consumption.

For instance, examining the monthly average UHII, there is a notable positive correlation between monthly cooling energy use intensity and monthly average UHII, with an R^2 of 0.7 (see Fig. 3a). The monthly cooling energy consumption for a recorded UHII of -4 °C was 2.58 kWh/m^2, while it was 6.85 kWh/m^2 for a UHII of 6 °C. An increase in monthly average UHII of 0.5 °C corresponds to an increase in average monthly cooling energy consumption of 0.21 kWh/m^2. Similar patterns are observed when considering other variants of UHII. These observed differences are also statistically significant, as determined by a conducted Kruskal-Wallis test (see Table 2).

3.3.2 Heating Energy Use

Figure 4 illustrates the relationship between various UHII indicators and heating energy use. As depicted in the figure, the correlations between heating energy use intensity and UHII are generally weak for most indicators, with the notable exception of daytime average UHII, which shows an R2 of 0.28. For other indicators, the generated lines of fit are predominantly flat, indicating an almost non-existent linear relationship between heating energy use and UHI. The observed relationships were also statistically significant (see Table 3). This finding is somewhat contradictory to recent studies that indicate strong inverse relationships between UHI and building energy use in cold climates [25] and a comprehensive review reporting a median decrease of 18.7% in building heating energy consumption attributed to UHI [8]. Consequently, our results suggest that while UHI may confer benefits in terms of reduced heating energy requirements during winter, these benefits are not as direct or significant as the detrimental effects on cooling energy requirements observed during summertime. This observation could possibly be attributed to the insulation levels of buildings; office buildings in South Korea may be

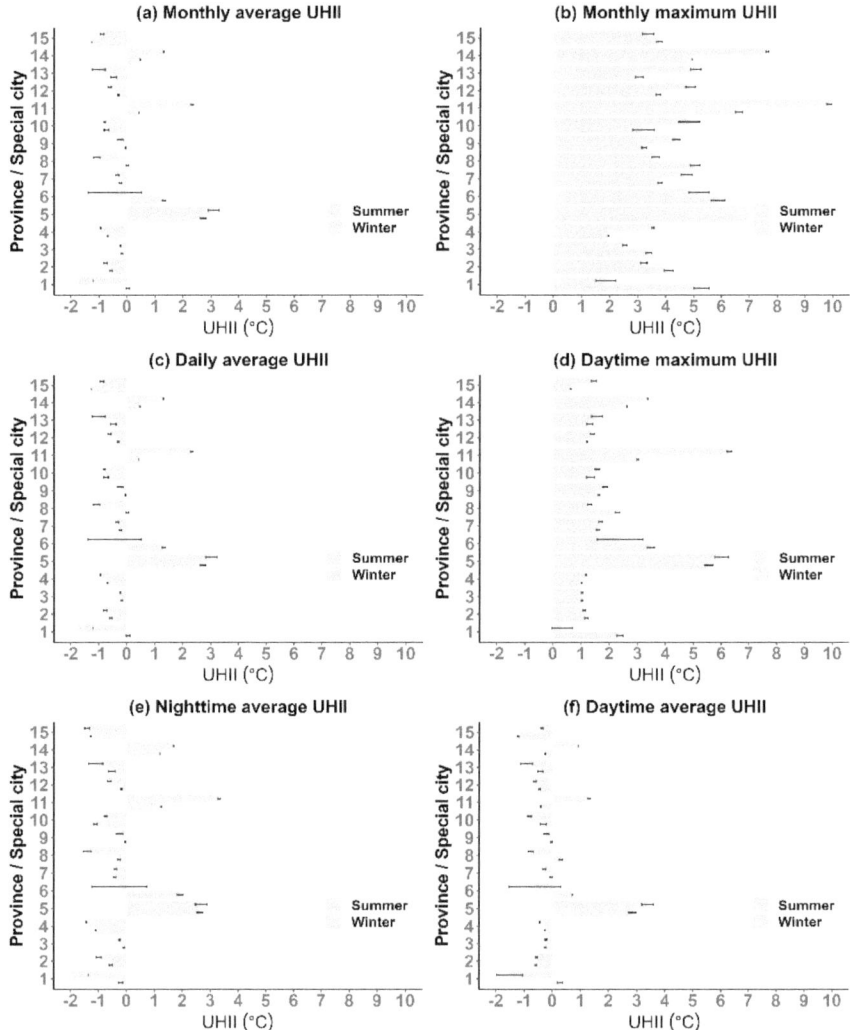

1. Busan 2. Chungcheongbuk 3. Chungcheongnam 4. Daegu 5. Gangwon 6. Gwangju
7. Gyeonggi 8. Gyeongsangbuk 9. Gyeongsangnam 10. Incheon 11. Jeju 12. Jeollobuk
13. Jeollonam 14. Seoul 15. Ulsan

Fig. 2. Inter-city variations in UHII indices during the summer

sufficiently well-insulated to require less heating regardless of UHI effects. This insula-tion efficacy could significantly contribute to the observed weaker correlation between UHI and heating energy use.

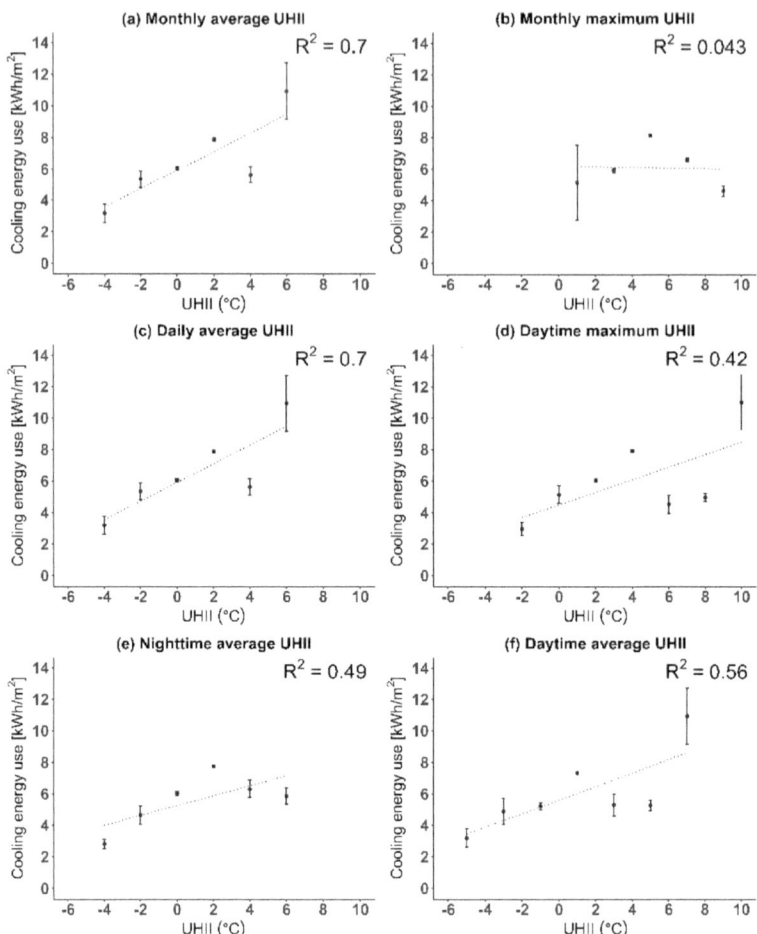

Fig. 3. Correlations between UHII and cooling energy use intensity

Table 2. Kruskal-Wallis results for UHII, cooling energy relationships

UHII indicator	X^2	df	P-value
Monthly average	593.21	20	< 0.005
Monthly maximum	887.23	25	< 0.005
Daily average	378.41	22	< 0.005
Daytime maximum	646.46	18	< 0.005
Nighttime average	728.74	22	< 0.005
Daytime average	730.2	22	< 0.005

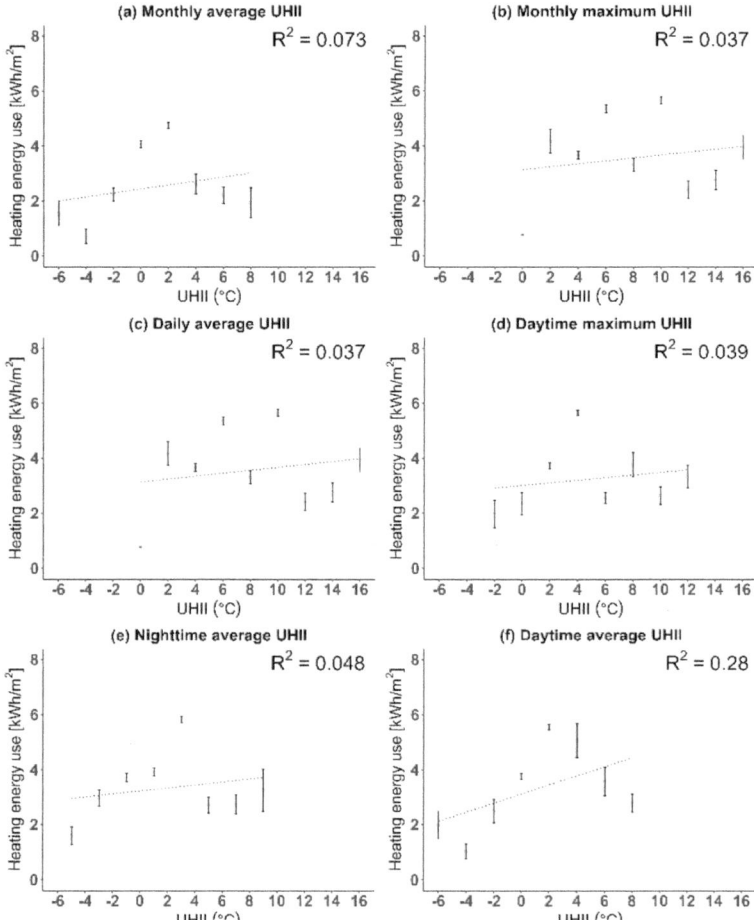

Fig. 4. Correlations between UHII and heating energy use intensity

Table 3. Kruskal-Wallis results for UHII, heating energy relationships

UHII indicator	X^2	df	P-value
Monthly average	60.20	17	< 0.005
Monthly maximum	62.83	25	< 0.005
Daily average	35.23	15	< 0.005
Daytime maximum	76.99	20	< 0.005
Nighttime average	58.48	18	< 0.005
Daytime average	60	17	< 0.005

4 Conclusions

We investigated the potential impact of Urban Heat Island (UHI) on the energy use of office buildings in South Korea using real energy data. Our findings reveal considerable differences in UHI behavior across South Korean provinces and cities, driven by a complex interplay of factors related to geographic locations, urbanization levels, and the topology of these provinces. These differences indicate that UHI mitigative efforts in South Korea should consider targeted measures tailored to specific locales. Additionally, we observe relatively higher UHII in Seoul compared to other localities, pinpointing urbanization as the primary driver of local climatic changes, given that Seoul is arguably the most urbanized area in South Korea. We also identify moderately strong positive correlations between UHII and cooling energy consumption in office buildings, but the relationship between UHII and heating energy consumption was largely non-existent.

These results provide empirical evidence on the impact of UHII on building energy use, particularly during the cooling period, further validating existing theoretical reports on the influence of local climates on building energy consumption and the associated socio-economic challenges, such as energy poverty, particularly among vulnerable demographics in metropolitan cities. Consequently, these findings offer an empirical foundation for the development of efficient and inclusive policies that promote livability in urbanized areas. Furthermore, the results highlight the critical need for local governments to invest in Urban Heat Island (UHI) mitigation strategies. Such strategies could include the utilization of high-albedo materials in urban infrastructure construction and the adoption of green roofs to reduce direct solar heat gains. These measures are essential in reducing building energy use and, indirectly, its global carbon footprint.

While useful, the current study has some limitations. For instance, the results obtained are only relevant to office buildings; outcomes may vary for buildings with different functions, such as residential buildings. Moreover, the end-use energies considered in this study are derived from a mathematical model, which may exhibit certain biases when compared to directly metered data. Consequently, future research should aim to further explore the impacts of urban warming on building energy consumption, utilizing directly metered end-use energy data. Additionally, it should encompass a wider range of building types beyond the office buildings considered in this study.

Moreover, it is crucial to acknowledge that during the process of pairing buildings with nearby Automated Weather Stations (AWSs), some AWSs may have been relatively far from the respective buildings. In our final dataset, the maximum distance observed between a building and its paired AWS was 45 km. Excluding such instances would have necessitated a scientifically substantiated definition of 'distance'—specifically, a threshold delineating the maximum horizontal distance at which the influence of microclimates is considered negligible. Defining this threshold is a complex matter that extends beyond the scope of this manuscript and presents a valuable avenue for future research.

It is also noteworthy that potential confounding factors, such as the efficiency of the employed Heating Ventilation and Air Conditioning systems that are likely to influence or mediate the relationship between Urban Heat Island (UHI) intensity and space cooling and heating energy use, were not considered in the current study. Incorporating these aspects into future research could yield a more comprehensive understanding of the intricate relationship between local climatic conditions and building energy consumption than what has been presented here.

Acknowledgements. This work was supported by the National Research Foundation of Korea (NRF) grant funded by the Korea government (MSIT) (No. 2020R1A2C1099611).

References

1. M. Santamouris, Cooling the buildings – past, present and future, Energy Build 128 (2016). https://doi.org/10.1016/j.enbuild.2016.07.034
2. G. Luber, M. McGeehin, Climate Change and Extreme Heat Events, Am J Prev Med 35 (2008). https://doi.org/10.1016/j.amepre.2008.08.021
3. T.R. Oke, The energetic basis of the urban heat island, Quarterly Journal of the Royal Meteorological Society 108 (1982). https://doi.org/10.1002/qj.49710845502
4. M. Santamouris, On the energy impact of urban heat island and global warming on buildings, Energy Build 82 (2014). https://doi.org/10.1016/j.enbuild.2014.07.022
5. Y. Hirano, T. Fujita, Evaluation of the impact of the urban heat island on residential and commercial energy consumption in Tokyo, Energy 37 (2012). https://doi.org/10.1016/j.energy.2011.11.018
6. Z. Ren, X. Wang, D. Chen, C. Wang, M. Thatcher, Constructing weather data for building simulation considering urban heat island, Building Services Engineering Research and Technology 35 (2014). https://doi.org/10.1177/0143624412467194
7. C. Li, J. Zhou, Y. Cao, J. Zhong, Y. Liu, C. Kang, Y. Tan, Interaction between urban microclimate and electric air-conditioning energy consumption during high temperature season, Appl Energy 117 (2014). https://doi.org/10.1016/j.apenergy.2013.11.057
8. X. Li, Y. Zhou, S. Yu, G. Jia, H. Li, W. Li, Urban heat island impacts on building energy consumption: A review of approaches and findings, Energy 174 (2019). https://doi.org/10.1016/j.energy.2019.02.183
9. M.A. Su, J. Ngarambe, M. Santamouris, G.Y. Yun, Empirical evidence on the impact of urban overheating on building cooling and heating energy consumption, IScience 24 (2021). https://doi.org/10.1016/j.isci.2021.102495
10. K.U. Ahn, H.S. Shin, C.S. Park, Energy analysis of 4625 office buildings in South Korea, Energies (Basel) 12 (2019). https://doi.org/10.3390/en12061114
11. D. Robinson, Pacific power: the use of short-term measurements to decompose commercial billing data into primary end uses, in: ACEEE 1992 Summer Study on Energy Efficiency in Buildings, n.d
12. IEA, https://www.iea.org/countries/korea/electricity, (n.d.)
13. Korea Meteorological Agency, Korea Meteorological Agency, (n.d.)
14. D.A. Prasetya, P.T. Nguyen, R. Faizullin, I. Iswanto, E.F. Armay, Resolving the shortest path problem using the haversine algorithm, Journal of Critical Reviews 7 (2020). https://doi.org/10.22159/jcr.07.01.11
15. MacFarland, T.W., Yates, J.M.: Kruskal-Wallis H-Test for Oneway Analysis of Variance (ANOVA) by Ranks. Introduction to Nonparametric Statistics for the Biological Sciences Using R (2016). https://doi.org/10.1007/978-3-319-30634-6_6
16. J.H. Kim, O.S. Kwon, J.H. Ra, Urban type classification and characteristic analysis through time-series environmental changes for land use management for 31 satellite cities around seoul, south korea, Land (Basel) 10 (2021). https://doi.org/10.3390/land10080799

17. Y.H. Kim, J.J. Baik, Daily maximum urban heat island intensity in large cities of Korea, Theor Appl Climatol 79 (2004). https://doi.org/10.1007/s00704-004-0070-7
18. D. Yihui, J.C.L. Chan, The East Asian summer monsoon: An overview, Meteorology and Atmospheric Physics 89 (2005). https://doi.org/10.1007/s00703-005-0125-z
19. T.R. Oke, City size and the urban heat island, Atmospheric Environment (1967) 7 (1973). https://doi.org/10.1016/0004-6981(73)90140-6
20. J. Ngarambe, J.W. Oh, M.A. Su, M. Santamouris, G.Y. Yun, Influences of wind speed, sky conditions, land use and land cover characteristics on the magnitude of the urban heat island in Seoul: An exploratory analysis, Sustain Cities Soc 71 (2021). https://doi.org/10.1016/j.scs.2021.102953
21. J. Nichol, Remote sensing of urban heat islands by day and night, Photogramm Eng Remote Sensing 71 (2005). https://doi.org/10.14358/PERS.71.5.613
22. S. Marina, C. Constantinos, Study of the urban heat island of Athens, Greece during daytime and night-time, in: Urban Remote Sensing Joint Event. URS **2007** (2007). https://doi.org/10.1109/URS.2007.371802
23. P.J.C. Schrijvers, H.J.J. Jonker, S. Kenjereš, S.R. de Roode, Breakdown of the night time urban heat island energy budget, Build Environ 83 (2015). https://doi.org/10.1016/j.buildenv.2014.08.012
24. J.M. Sobstyl, T. Emig, M.J.A. Qomi, F.J. Ulm, R.J.M. Pellenq, Role of City Texture in Urban Heat Islands at Nighttime, Phys Rev Lett 120 (2018). https://doi.org/10.1103/PhysRevLett.120.108701
25. Y. Fan, Z. Wang, Y. Li, K. Wang, Z. Sun, J. Ge, Urban heat island reduces annual building energy consumption and temperature related mortality in severe cold region of China, Urban Clim 45 (2022). https://doi.org/10.1016/j.uclim.2022.101262

Study on the Spatial Interaction of Habitat Quality Pattern and Thermal Environment Based on InVEST—A Case Study of the Guangdong-Hong Kong-Macao Greater Bay Area

Jiayu Wang[✉]

School of Architecture and Urban Planning, Beijing University of Civil Engineering and Architecture, Beijing 100044, China
wangjiayu@bucea.edu.cn

Abstract. With China's urbanization advancing into an advanced phase, urban clusters experiencing swift expansion of urban land have emerged as "disaster areas" where ecological environment problems are highly concentrated. Therefore, paying attention to the habitat quality of urban agglomerations and their ecological cooling effect is crucial for improving the quality of the living environment and fostering sustainable development in the region. This research uses the Guangdong-Hong Kong-Macao Greater Bay Area (GHM-GBA) as a case study, applies the InVEST platform, Geographic Information System (GIS), and Binary Moran's I index, and undertakes an investigation into the spatial interplay between the habitat quality pattern and the thermal environment within the urban clusters of the GHM-GBA between 2000 and 2020. Firstly, utilizing data on habitat threat density, the Habitat Quality Module on the InVEST platform is employed to measure the habitat quality and degradation extent of urban agglomerations. Secondly, based on MODIS satellite sensor data, the surface temperature is inverted and interpreted, and the spatial heterogeneity traits of the urban agglomeration's thermal environment are examined. Then, the Binary Moran's I index is used to clarify the spatial clustering and spatial dispersion relationship characteristics between habitat quality and thermal environment, including the network spatial distribution of regions with positive spatial correlation and negative spatial correlation. Ultimately, through the lens of urban design, this paper puts forward detailed planning recommendations to improve the quality of living spaces and thermal comfort within urban clusters. These suggestions serve as a reflection for the ecological sustainability, livability, and high-quality growth of the GHM-GBA.

Keywords: Habitat Quality · Thermal Environment · GHM-GBA · Surface Temperature · Spatial Interaction

© The Author(s) 2025
B.-J. He et al. (Eds.): UCSUD 2023, LNCE 559, pp. 511–522, 2025.
https://doi.org/10.1007/978-981-97-8401-1_36

1   Introduction

Climate change and globalization have altered the form and objectives of urban development. Urban agglomerations have become highly sensitive areas where the tensions between urban growth and ecological conservation are becoming more pronounced. High-density urban construction patterns erode ecological space with impermeable surfaces such as asphalt and cement, and the increase in anthropogenic heat emissions exacerbates the urban heat island effect. This not only changes the local ecological environment, such as reducing water and air quality and biodiversity, but also severely impacts human health, raising the likelihood of heatstroke and respiratory illnesses [1, 2]. As urban development transitions from traditional standalone cities to urban agglomerations, the development objectives of these agglomerations begin to pivot towards intensive, ecological, and sustainable growth. This approach is essential for realizing the objectives of building an ecological civilization and ensuring a high-quality living environment [3]. The Intergovernmental Panel on Climate Change (IPCC) pointed out that climate change has exacerbated the process of land degradation through high temperature stress and other means, and cities have been defined as major risk areas for climate change.

Since the Ming and Qing dynasties, the Guangdong-Hong Kong-Macao Greater Bay Area" (GHM-GBA) has been developing as a whole, serving as the core area of Cantonese culture. After more than twenty years of careful consideration, in 2016, China officially proposed to build the GHM-GBA into a world-class urban agglomeration. Thanks to the strong support of national policies, the GHM-GBA continues to optimize its industrial structure and ecological spatial pattern, gradually becoming an innovative experimental field for promoting the new pattern of urban clusters. However, when measured against other global bay areas, urban resilience and the quality of the living environment remain the most significant challenges for the GHM-GBA. On the one hand, the process of urbanization has led to rapid growth in energy consumption and carbon emissions. To accommodate the demands of increasing population and economic expansion, the GHM-GBA continues to develop and expand land. The result of this urban expansion is the destruction and deprivation of a large amount of natural habitat, leading to a reduction in biodiversity and degradation of ecosystem functions. Secondly, urbanization in the Bay Area has brought about serious thermal environmental problems. In the process of high-density urbanization, buildings and populations are highly concentrated, generating a large amount of heat island effect. The high-temperature heat island effect not only exposes urban residents to increasingly severe heat stress, but also affects the stability of urban ecosystems and the comfort of the urban living environment [4]. In 2021, the Bay Area experienced unprecedented temperatures since the start of weather recording, averaging 23.5°C. This was 1°C above the usual average of 22.5°C and 0.3°C higher than the previous record in 2020, which was 23.2°C, marking the highest since 1961. The area also recorded an average of 39.0 high-temperature days (where the daily maximum temperature was at least 35°C), which is 19.6 days more than the usual 19.4 days, thereby establishing a new historical high (see Fig. 1). On February 18, 2019, the Central Committee of the Communist Party of China proposing to build an ecological protection barrier and implement the strictest ecological environment protection system in GHM-GBA. Against the backdrop of widespread overdrawn ecological capital, the ecological environment and the quality of the living environment in the GHM-GBA face huge challenges.

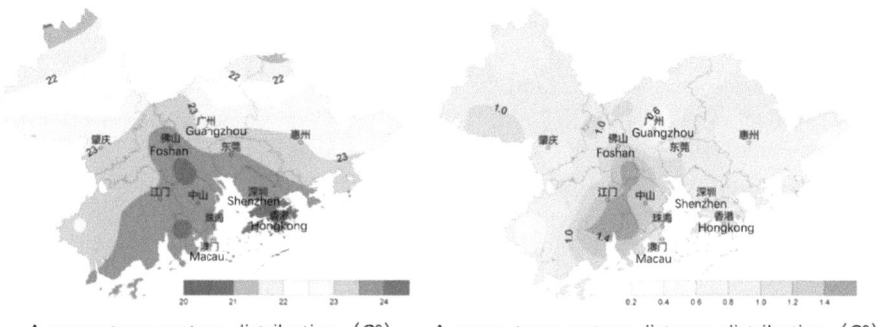

Average temperature distribution（C°）   Average temperature distance distribution（C°）

Fig. 1. Distribution of average temperature in the GHM- GBA (left) and distribution of average temperature anomalies (right). Source: Guangdong-Hong Kong-Macao Greater Bay Area Climate Monitoring Bulletin

In summary, in the face of these challenges, it becomes particularly important to study the spatial interaction between habitat quality and thermal environment. Habitat quality (HQ) refers to the quality and function of the ecosystem in providing materials and energy, and is a comprehensive indicator for assessing the health and sustainability of the ecosystem. The thermal environment (TE) is closely related to the city's temperature, climate, meteorology, etc., and directly affects the lives and health of urban residents. Numerous studies have demonstrated the significance of urban parks, ecological landscapes, urban water bodies and other ecological spaces have a significant effect on reducing surface temperature and alleviating urban heat island effect [5–7]. However, as further exploration into the cooling effects of ecological spaces continues, some researchers have found that the relationship between ecological spaces and the thermal environment is difficult to universally apply in urban planning practices. For instance, the cooling range of green spaces can vary significantly in the United States, Japan, Gothenburg, and Mexico, with their impact on the thermal environment differing based on the morphology, quality, and location of the ecological spaces[8–10]. Therefore, a comprehensive assessment of ecological spaces and their impact on the thermal environment is very important, the HQ possesses strong regional and spatial characteristics, while also focusing on the dynamism of ecological environmental flow processes. Considering this, the paper uses the GHM-GBA as a case study to concentrate on the mechanism of spatial interaction between HQ and TE, identifies the core areas with prominent contradictions and significant problems, and proposes planning optimization paths, to provide a scientific basis for improving the quality of life of residents in the GHM-GBA and addressing climate change. The research results have important practical value in guiding the ecological planning of urban clusters and promoting the goal of people-oriented new urbanization.

2 Research Scope and Methodology

2.1 Scope of Research

The GHM-GBA is a city cluster includes "9 + 2" cities in China's Pearl River Delta basin (see Fig. 2). This is a globally renowned bay area characterized by the harmonious interplay between land and sea, and the mutual dependence of mountains and waters [11, 12]. However, the GHM-GBA also faces many obstacles in its development process. Under the macro background of the global ecological deficit, our country is the only one among the countries where the world's four major bay areas are located that has an increasing per capita biological carrying capacity and the only one with an expanding per capita footprint, and the ecological deficit problem is severe. The GHM-GBA has a rich and diverse biodiversity, and the total area of about 57,000 km^2 presents a unique scene where the ecological space dominated by mangroves coexists with the high-density urban space characterized by high-rise buildings, and the situations in different cities are also different. The spatial pattern of the overall ecological carrying capacity of the GHM-GBA shows a "low near the sea, high far from the sea" distribution trend, with nearly 80% of the biological carrying capacity distributed in inland cities such as Zhaoqing, Guangzhou, and Jiangmen, while the total biological carrying capacity of Zhuhai, Shenzhen, Hong Kong, and Macao is less than 8%, and the efficiency of the coastal cities' ecosystems in producing natural resources is relatively low.

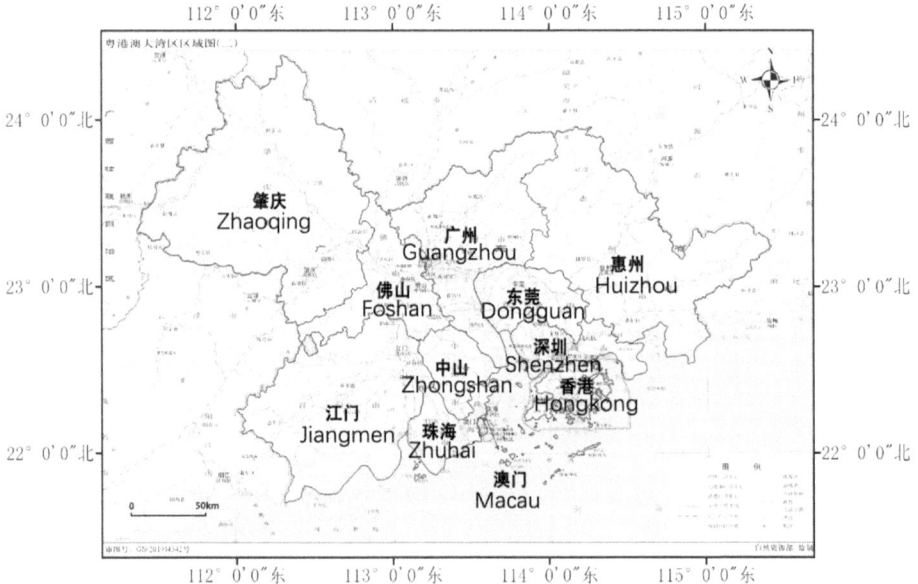

Fig. 2. Scope and administrative divisions of the GHM-GBA. Image source: The base map is from the standard map service website, and the review number is GS(2019)4342

2.2   Research Data

The research data includes the global 30m surface cover data for 2000 and 2020, and the MOD11A2 data from the MODIS (Moderate Resolution Imaging Spectroradiometer) satellite sensor for surface temperature inversion processing during the same period (https://ladsweb.modaps.eosdis.nasa.gov/search/). The administrative division data comes from the JSON data provided by the Aliyun Data Visualization Platform DataV.GeoAtlas. The analysis process is based on ArcMap, InVEST, and GeoDA platforms.

2.3   Research Method

2.3.1   Habitat Quality

HQ denotes the natural world's aptness for biological habitats, essentially the capacity of the ecological setting to furnish living conditions for groups of organisms. High-quality habitats possess comprehensive spatial shapes, structures, and functions, making the effects of land expansion and human activities on these habitats more pronounced [13]. D_{xj}^z is the disturbance index of grid x in land use type j.

Enhancing HQ is vital for the preservation of biodiversity and the maintenance of ecological safety, and enhancing human well-being. It is also a necessary condition for building a livable and sustainable GHM-GBA.

The HQ module from the InVEST platform was employed to assess the level of HQ in the GHM-GBA. The formula used for calculation is as follows:

$$Q_{xj} = H_j \left(1 - \frac{D_{xj}^z}{D_{xj}^z + k^z} \right) \tag{1}$$

In this formula, Q_{xj} denotes the index of HQ for grid x within land use type j, H_j signifies the suitability of habitat for land use type j, k is the half-saturation constant, often assumed to be the maximum of D_{xj}^z, z is the model's preset parameter, and D_{xj}^z indicates the extent of habitat degradation for grid x in land use type j.

Furthermore, referring to the InVEST user manual for the setting of threat sources and sensitivity parameters, specific parameters are set in combination with the actual situation of the GHM-GBA. Threat sources refer to elements that pose threats to the ecosystem. This study focuses on the factors that human interference has on the ecosystem, mainly considering the threats of farmland, construction land, forest land, and roads to habitats. Among them, the impact of construction land on HQ is the most significant, with a set weight of 1. The spatial decay type is set according to land use characteristics, with farmland and roads showing linear decay, and construction land and forest land showing exponential decay. Sensitivity parameters refer to the ability of the ecosystem to respond to threat sources under different scenarios, with values ranging from 0 to 1, where 0 is not suitable as a habitat, and 1 is completely suitable.

2.3.2   Surface Temperature

Surface temperature is one of the important indicators for studying the urban heat island effect, which helps to deepen the understanding of the impact of urban development on climate change, especially the study of surface temperature at the macro scale of cities and city clusters, which helps to visually compare the temperature rise in the city relative to the surrounding areas, thereby helping to formulate refined climate adaptation planning strategies. we obtain the surface temperature inversion grid map of the Greater Bay Area from MOD11A2-MODIS/Terra Land Surface Temperature/Emissivity Daily L3 Global 1km SIN Grid. Various calibration adjustments have been made to enhance the product, offering daily per-pixel land surface temperature and emissivity (LST&E).

2.3.3 Moran's Index

Moran's Index (Moran's I), serves as a measure of spatial autocorrelation, encompassing both global and local Moran's Index variants to quantify the level of spatial interrelation among binary variables. It can be used to analyze the similarity or difference of geographical phenomena in urban space [14]. Using the binary Moran's Index to evaluate the HQ and surface temperature of the city cluster in the GHM-GBA can reflect their spatial agglomeration, spatial heterogeneity, and spatial dependence. Spatial agglomeration explains whether there are more obvious spatial distribution characteristics of HQ and surface temperature, spatial heterogeneity reflects the spatial differences affected by factors such as land use, economy, human activities in different regions, and spatial dependence characterizes the relationship of mutual influence between adjacent areas in space.

3 Research Results and Discussion

3.1 Spatial Pattern of HQ

Changes in the Greater Bay Area HQ between 2000 and 2020 are illustrated in Fig. 3. Viewing from the standpoint of spatial distribution, the quality of the habitat demonstrates a pattern of being "higher in the center, low around the edges". In 2000, areas with no quality habitat accounted for only 7.75% of the total area, primarily concentrated in central cities of the GHM-GBA, such as Guangzhou and Hong Kong. With the expansion of construction land in urban agglomerations, habitats with no quality spread outwards along construction land and main traffic routes, and by 2020, the proportion of habitats with no quality had risen to 18.83%. This indicates that the relationship between human activities and the ecological environment is intense, and the quality of urban group habitats and the function of ecosystem services are under tremendous pressure [15]. The northern part of Foshan City (Sanshui District), although not a high-intensity urban construction area, has also seen habitats with no quality. This is due to the fact that some water bodies have been artificially developed into pits, beaches, and construction land, reducing habitat connectivity and increasing sensitivity. Upon examining the alterations in HQ between 2000 and 2020, it is evident that the HQ in every city has diminished in various extents, with the most noticeable change in the Macao Special Administrative Region. From 1912 to 2018, land reclamation activities were frequent, and its land area increased from 11.6 km^2 to 32.8 km^2, large-scale land reclamation projects pose a continuous threat to regional biodiversity, and environmental problems such as microclimate imbalance and heat island effects occur frequently. Considering the actual situation of GHM-GBA, the habitat quality is categorized into low, medium, and high for the ranges 0–0.25, 0.25–0.75, and 0.75–1, respectively (Table 1).

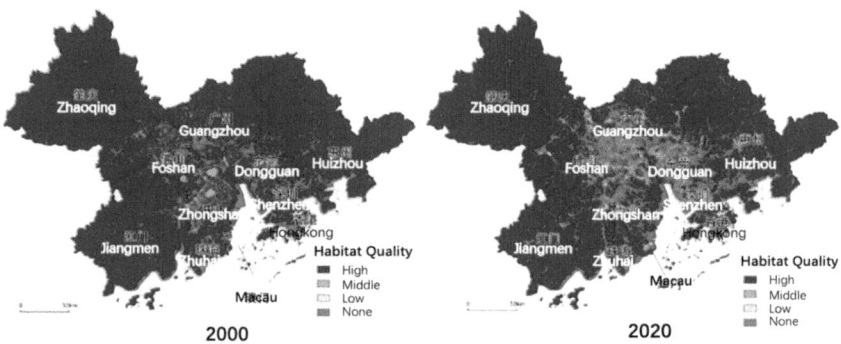

Fig. 3. Changes in HQ in the GHM-GBA between 2000 and 2020

Table 1. Proportion of different levels of HQ in the GHM-GBA between 2000 and 2020

Time	Proportion of Area with No HQ	Proportion of Area with Low HQ	Proportion of Area with Medium HQ	Proportion of Area with High HQ
2000	7.75%	0.02%	2.41%	89.82%
2020	18.83%	0.75%	3.67%	76.75%

3.2 Spatial Pattern of HQ

Referring to the surface temperature inversion grid map of the GHM-GBA city cluster, the TE level is delineated through ArcGIS and spatial data statistics and analysis are conducted (see Fig. 4). The spatial distribution pattern of surface temperature closely mirrors that of HQ, showing a "high in the middle, low around" characteristic. The TE in the Greater Bay Area has changed significantly, while also showing obvious spatial heterogeneity between 2000 and 2020. This is close to the carbon emission pattern of GHM-GBA analyzed by some scholars [11]. Areas of high temperature are predominantly located in the central regions of core cities such as Guangzhou, Shenzhen, and Foshan, where human activities are frequent and building density is high. Low-temperature areas are distributed on the outskirts of node cities such as Zhaoqing and Huizhou, as well as in the southern part of Jiangmen City, where vegetation coverage is high or near the coastal zone. Further statistics on the land use types in each temperature zone reveal that the proportion of construction land in high-temperature zones has increased from 38.79% to 65.53%, and the proportion of construction land in sub-high temperature zones has increased from 5.89% to 16.3%.

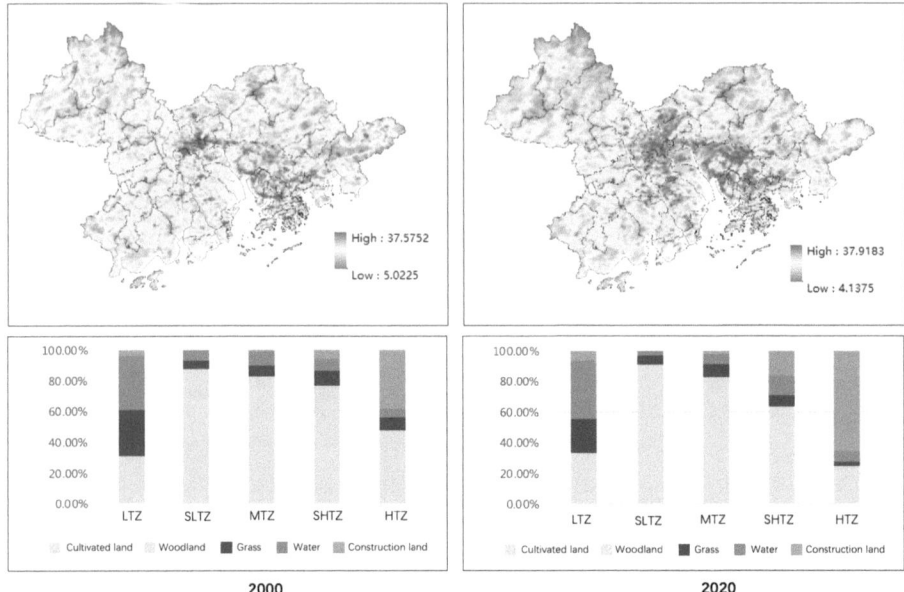

Fig. 4. Surface temperature and land use proportion in the GHM-GBA between 2000 and 2020 (Drawn by the author)

3.3 Analysis of the Spatial Interaction Mechanism Between HQ and TE a Subsection Sample

Based on the single-item evaluation and zoning results, the spatial agglomeration characteristics of HQ and TE in the GHM-GBA are obtained (see Fig. 5). In 2000, the LISA results showed that the Moran's Index was −0.375, indicating that HQ and TE had a relatively obvious local spatial negative correlation. In the agglomeration areas of each category tested at the significance level of $P \leq 0.05$, the "high-low" and "low-high" agglomeration areas dominated. High-quality habitats are mainly distributed in Huaiji County, Deqing County, Gaoyao District, and Sihui District of Zhaoqing City, where forest areas are concentrated, showing high biodiversity, crucial role in sustaining the services provided by the regional ecosystem. The "low-high" agglomeration areas are distributed in Liwan District, Yuexiu District, Tianhe District of Guangzhou, and Nanshan District, Baoan District of Shenzhen. This area had already shown a large-scale contiguous urban construction land spatial pattern in 2000, human activities significantly interfere with the flow of ecological elements, and it is far from large-scale ecological source areas, thus facing a series of severe urban problems brought about by the heat island effect, such as residents' health, water resource shortage, biodiversity reduction, and energy consumption increase.

In 2020, the LISA results showed that the Moran's Index was -0.273, indicating that HQ and TE had a relatively obvious local spatial negative correlation. Compared with 2000, the range of agglomeration areas tested at the significance level of $P \leq 0.05$ has significantly expanded, including "high-high", "high-low", "low-high", "low-low" four types of agglomeration areas, indicating that the spatial interaction between HQ and TE has a wider range and stronger effect. The "high-high" hotspots are mainly scattered in Longgang District of Shenzhen and Yau Tsim Mong District, Kowloon City District of Hong Kong. Although these high-density urban areas have high surface temperatures, they have high requirements for urban ecological green space construction. For example, the Hong Kong Special Administrative Region has gradually improved the urban country park system since the 1980s, carried out a series of park expansion projects, including Sai Kung Hoi, Lung Fu Shan, Lantau Island, etc., and designated special areas to protect areas with high ecological value, implemented strict legal systems and detailed planning strategies, and urban spatial expansion has not encroached on scarce urban ecological space [10]. The "low-high" area is distributed in the Nanhai District, Nansha District, Panyu District, Haizhu District of Guangzhou City, and Dongguan City. This area represents the neglect of high-quality habitat space in the process of urban land expansion, facing the severe challenge of urban development and ecological protection. It is a key area that needs to be optimized and regulated.

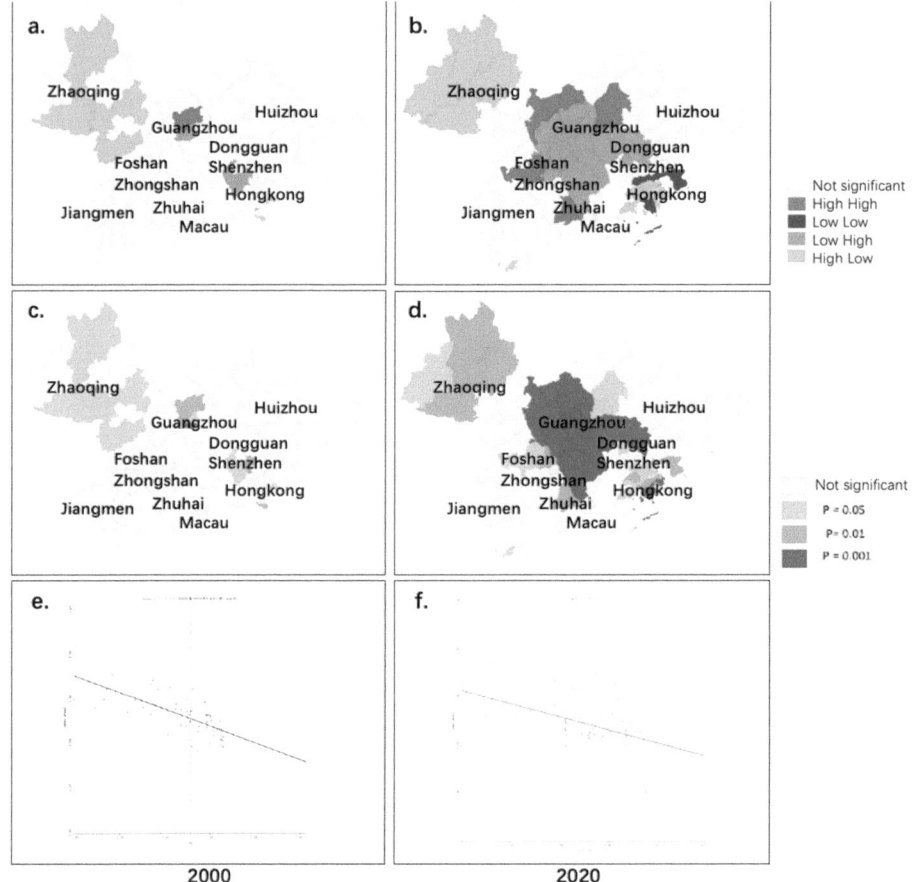

Fig. 5. Spatial clustering characteristics of HQ and TE in the GHM-GBA between 2000 and 2020 (drawn by the author)

4 Conclusion

This study uses the GHM-GBA as a case study, beginning with an analysis of the HQ pattern and TE, develops a research framework for the spatial dynamics of ecological conservation and living environment quality improvement in coastal city clusters. Based on land grid data and surface temperature inversion data, using ArcGIS, InVEST, GeoDa and other analysis platforms, the spatial pattern characteristics of HQ and TE in the GHM-GBA were evaluated respectively, and then the spatial interaction relationship between 2000 and 2020 was sorted out, the abnormal areas of HQ and TE interaction were identified, and the following main conclusions were drawn:

The HQ presents a "high in the middle, low around" distribution pattern. The HQ of each city has declined to varying degrees between 2000 and 2020. Human activities, urban expansion, and land reclamation have exerted a negative influence on the quality of habitats. The spatial pattern of surface temperature is similar to that of HQ. The TE has changed significantly, and at the same time, it shows obvious spatial heterogeneity between 2000 and 2020. According to the calculation of land area in each temperature zone, the area of construction land in the high temperature zone has significantly increased.

In 2000, the HQ and TE had already shown a fairly obvious local spatial negative correlation. The contrast between the spatial interaction of HQ and TE in core cities and node cities is strong. In 2020, the range of spatial interaction between HQ and TE has expanded, and the range and effect of spatial interaction between HQ and TE are wider and stronger. High surface temperature areas have gradually spread from the original southeast region inland and have intensified and clustered in the central and eastern wing regions along the river. The spatial interaction between HQ and TE has become more pronounced. In the development of future urban agglomerations, it is essential to control the expansion area and speed of construction land to reduce the loss of high-quality habitats. Moreover, given that wetlands and forests are vital sources of high-quality habitats, the GHM-GBA should protect these land cover types in urban planning to further mitigate the urban heat island effect.

The spatial interaction between HQ and TE effectively mirror the intricate interconnection between ecological conservation and human endeavors in the evolution of urban clusters. Limited by the difficulty of obtaining socio-economic data, this paper only abstractly characterizes human activities as urban TE, without the variances in how different population categories, like population density and building purpose, affect the TE. Uture studies may delve deeper into this field to efficiently mitigate the urban heat island phenomenon.

References

1. Xinyu, W., Jian, Z.: Research on disaster prevention planning strategies for urban high-temperature heat waves—Based on planning experience in Europe and America. Mod. Urban Res. **08**, 84–92 (2017)
2. Yaning, G., Xinliang, X., Jing, L. et al.: Research on the impact of the distribution of building density in Beijing on the heat island effect. J. Earth Inf. Sci. 18(12), 1698–1706 (2016)

3. Chaolin, G., Zongbo, T., Wan, L., et al.: Progress in research on climate change, carbon emissions and low-carbon urban planning. Urban Plan. J. 3, 38–45 (2009)
4. Tian, C., Junnan, L.: Resilience assessment and improvement strategy for high-temperature disasters in the Guangdong-Hong Kong-Macao Greater Bay Area. Shanghai Urban Plan. **01**, 9–17 (2023)
5. Wei, W., Long, Z., Yu, L.: Research on the layout and optimization of the park system to deal with the urban heat island effect—Taking the high-density city of Macau as an example. Fudan J. (Nat. Sci. Ed.), 62(02), 217–225. https://doi.org/10.15943/j.cnki.fdxb-jns.20230222. 001(2023)
6. Suiping, Z., Zhuo, S., Meifang, Z., et al.: The change of the buffering performance of urban water bodies to heat islands along the riverbank distance. Acta Ecol. Sin. 40(15), 5190–5202 (2020)
7. Tian, C., Chuanmiao, S., Gaoyuan, W.: Research on urban water environment resilience planning under the background of climate change—Taking Singapore as an example. Int. Urban Plan. 36(05), 52–60. https://doi.org/10.19830/j.upi.2021.430(2021)
8. Bowler, D.E., Buyung-Ali, L., Knight, T.M., et al.: Urban greening to cool towns and cities: A systematic review of the empirical evidence. Landsc. Urban Plan. **97**(3), 147–155 (2010)
9. Hamada, S., Ohta, T.: Seasonal variations in the cooling effect of urban green areas on surrounding urban areas. Urban Forest. Urban Green. **9**(1), 15–24 (2010)
10. Oliveira, S., Andrade, H., Vaz, T.: The cooling effect of green spaces as a contribution to the mitigation of urban heat: a case study in Lisbon. Build. Environ. **46**(11), 2186–2194 (2011)
11. Lin, B., Li, Z.: Spatial analysis of mainland cities' carbon emissions of and around Guangdong-Hong Kong-Macao Greater Bay area. Sustain. Cities Soc. **61**, 102299 (2020)
12. Hui, E.C.M., Li, X., Chen, T., et al.: Deciphering the spatial structure of China's megacity region: a new bay area—The Guangdong-Hong Kong-Macao Greater Bay Area in the making. Cities **105**, 102168 (2020)
13. Liu, Z., Wang, S., Fang, C.: Spatiotemporal evolution and influencing mechanism of ecosystem service value in the Guangdong-Hong Kong-Macao Greater Bay Area. J. Geog. Sci. **33**(6), 1226–1244 (2023)
14. Jiansheng, W., Jiaying, M., Qian, L. et al.: Research on urban growth boundary based on habitat quality—Taking the Yangtze River Delta region as an example. Geogr. Sci. 37(01), 28–36 (2017). https://doi.org/10.13249/j.cnki.sgs,01.004(2017)
15. Shifu, W., Xiaoling, Q., Zhaohua, D.: Research on the coordination degree of urban waterfront space vitality from the perspective of environmental behavior. Trop. Geogr. 41(05), 1009–1022. https://doi.org/10.13284/j.cnki.rddl.003393(2021)
16. Wang, J., Chen, T.: A multi-scenario land expansion simulation method from ecosystem services perspective of coastal urban agglomeration: a case study of GHM-GBA, China. 11(11), 1934 (2022)
17. Kai, Y., Qingji, S.: Research on the stage characteristics and control mechanism of Hong Kong Country Parks. Int. Urban Plan. **34**(03), 124–131 (2019)

Study on Vertical Greening Technology for Enhancing Environmental Performance of Existing Public Buildings in High-Density Mountainous Cities

Jing Wu[1], Xiang Cheng[1,2], and Lili Dong[1(✉)]

[1] School of Architecture and Urban Planning, Chongqing Jiaotong University,
Chongqing 400074, China
dongll@cqjtu.edu.cn

[2] Key Laboratory of the Ministry of Education of Mountainous City and Towns Construction
and New Technology, Chongqing University, Chongqing 400074, China

Abstract. To enhance the applicability of stereoscopic greening technology for the renovation of existing public buildings in high-density mountainous urban areas, this study conducted on-site investigations in a typical high-density mountainous urban area of Chongqing. This study collected data on the application of vertical greening technology in urban public buildings and analyzed the suitability of various vertical greening technologies for the renovation of public buildings. The results indicate that modular vertical greening is currently more suitable for the renovation of public buildings in such cities; however, further optimization is needed in terms of the overall compatibility, stability of modular units, and adaptability of modular plants to the local environment. This study can provide a theoretical basis for the feasibility of adopting vertical greening retrofitting for public buildings in high-density mountainous cities, and provide a reference for the planning and sustainable development of such cities.

Keywords: High-density mountainous urban areas · Existing public building renovation · Vertical greening technology

1 Introduction

Global urbanization is accelerating, and mountainous cities, characterized by complex terrain and limited land, face unique planning and development challenges [1, 2]. This urban form exacerbates environmental issues in the urbanization process, such as low spatial efficiency and ecosystem fragmentation [3]. A profound understanding of the challenges facing high-density mountainous urban areas is crucial for scientifically sound urban planning and redevelopment.

High-density mountainous urban environments have distinct characteristics, with scarce resources and a large and dense inventory of high-energy-consuming buildings, demanding an urgent improvement in urban quality [4–6]. As urban development enters

B.-J. He et al. (Eds.): UCSUD 2023, LNCE 559, pp. 523–535, 2025.
https://doi.org/10.1007/978-981-97-8401-1_37

a new phase emphasizing both new construction and renovation, public buildings, given their significant energy consumption, rapid growth, and high energy-saving difficulty compared to residential buildings, play a key role in urban sustainable development [7, 8].

With a growing focus on urban ecological and environmental issues, vertical greening and green roof have gradually gained attention[9]. Stereoscopic greening not only provides green coverage in urban spaces but also improves air quality, alleviates the urban heat island effect, and enhances ecosystem stability [10, 11]. In the context of sustainable urban development, stereoscopic greening is considered an effective solution for addressing environmental issues in high-density mountainous urban areas [12–14].

This study delves into the application of stereoscopic greening to the renovation of public buildings in high-density mountainous urban areas. Through on-site investigations and systematic research on the application of stereoscopic greening in public buildings in high-density mountainous urban areas, the feasibility and actual effects of this technology in specific urban environments were explored. Ultimately, by summarizing and analyzing the findings, this study provides an analysis of the suitability of various stereoscopic greening technologies for renovating public buildings. It aims to offer practical technical and experiential support for the sustainable development of high-density mountainous urban areas and serves as a valuable reference for the planning and redevelopment of such cities.

2 Materials and Methods

2.1 Study Area

Chongqing, a typical high-density mountainous city in the southwestern region, faces a series of challenges in the urbanization process owing to its unique geographical conditions and complex terrain. The main urban area has issues such as a large number of public buildings and a high energy consumption density. However, for a high-density city like Chongqing, its existing public buildings are often characterized by the following features: (1) Terrain adaptability: because Chongqing's terrain is predominantly mountainous and hilly, public buildings often need to adapt to complex and changing terrain conditions. This means that building design and structure must be flexible and diverse to accommodate uneven foundations and steep slopes. (2) High-density layout: In such cities, public buildings are usually concentrated in a limited space, resulting in small spacing between buildings. This high-density layout not only poses a challenge to urban planning, but also puts higher demands on the design and functionality of building retrofitting techniques. (3) Multi-functionality: Chongqing's public buildings often serve multiple functions, either as places to provide public services or as centers for cultural, educational or community activities. This requires that the building renovation design needs to be able to meet the diverse needs of the use. (4) Ecological considerations: Given the ecological sensitivity and green development goals of Chongqing and similar cities, the design and construction of public buildings is increasingly focused on eco-friendly and sustainable principles. This includes the use of green building materials, the adoption of energy-saving technologies, and the integration of elements such as urban greening. (5) Integration of cultural characteristics: Chongqing's public buildings are often designed to incorporate local historical and cultural elements, reflecting the city's unique identity and cultural heritage. This not only enhances the aesthetic value of the building, but also adds rich layers to the cultural landscape of the city. In summary, the characteristics of public buildings in Chongqing and other high-density cities reflect adaptability to complex terrain, efficient space utilization, multifunctional design, emphasis on ecology and sustainable development, and respect for and integration of local culture. Therefore, it is typical and informative to take Chongqing as a typical representative city for the research of existing public buildings. This study involved 50 public buildings or building clusters of various types, and the distribution of buildings in various districts of the main city is shown in Table 1 and Fig. 1.

Table 1. Summary of research building types by administrative area

Name of district	Commercial buildings	Research buildings	Educational buildings	Cultural buildings	Medical buildings	Sports buildings	Office buildings
Beibei			✔	✔			
Shapingba		✔	✔				
Jiulongpo			✔				✔
Dadukou					✔		
Banan					✔		
Nan'an	✔						
Yuzhong	✔				✔		✔
Jiangbei						✔	
Yubei	✔	✔	✔	✔	✔		✔

2.2   Method

The research adopted a typical sampling survey method to determine research projects. First, satellite maps were used to search for all types of important landmark public buildings in each district of Chongqing's main city, combined with the map of the street view function of the research project for initial screening, excluding the roof and façade, which are not green projects. The total number of projects after screening was 50, and the distribution of projects covered all urban areas in the main city and included eight types of typical public buildings. After the establishment of the research project on each project to carry out on-site field surveys, the survey content included all types of three-dimensional greening technology types, application of the project status quo, and so on.

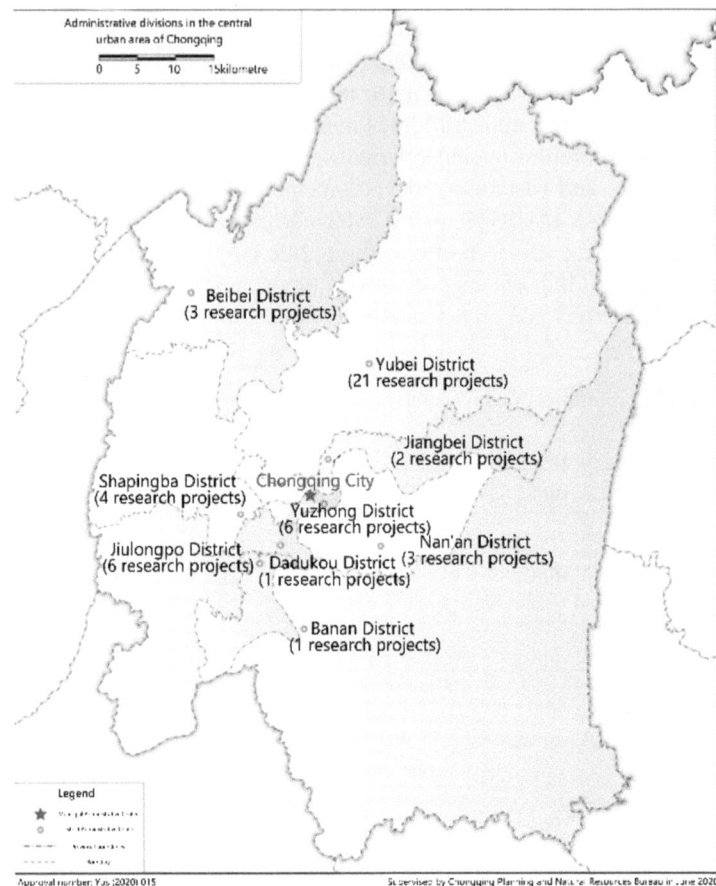

Fig. 1. Distribution of research projects in central business area of Chongqing

3 Results

Conduct a survey on the application types (Climbing Vertical Greening, Modular Vertical Greening, Paving Vertical Greening, Planting Trough Vertical Greening, as shown in Fig. 2) and conditions of stereoscopic greening technology in various existing public buildings in the Chongqing region, a comparative analysis of the four forms of wall greening is shown in Table 2. The survey covered eight categories of public buildings, including office buildings, sports facilities, and cultural buildings, as shown in Fig. 3.

The research findings reveal that there are relatively few existing public buildings in the main urban area of Chongqing that apply various stereoscopic greening technologies. Among the 29 surveyed projects, only three comprehensively applied four types of stereoscopic greening technologies. In terms of the distribution of technology applications, cost-effective climbing greening is widely used in various types of public buildings, particularly in offices and educational buildings. Modular and planter-based

Table 2. Horizontal comparison of four types of wall greening

Type	Maintenance frequency	Optional plant species	Time to develop green cover	Maintenance costs
Climbing Vertical Greening	low	fewer	long time	low
Modular Vertical Greening	high	wide range	short period	high
Paving Vertical Greening	medium	wide range	short period	medium
Planting Trough Vertical Greening	medium	wide range	short period	medium

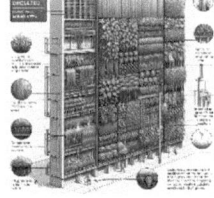

Climbing Vertical Greening   Modular Vertical Greening   Paving Vertical Greening   Planting Trough Vertical Greening

Fig. 2. Four types of vertical greening

greening are mainly applied in newer office buildings and large shopping malls, with planter-based greening mostly found in exhibitions and cultural buildings.

In terms of plant configuration, the plant species for climbing greening mainly include wisteria, honeysuckle, and bougainvillea, which fail to create high landscape value. The plant species for modular greening are not sufficiently diverse, and some projects are significantly affected by seasonal changes. Lush greenery in summer gives way to unique scenery, while withered plants clinging to the walls in winter affect the aesthetic appeal of the walls. Plant selection for planter-based greening is limited owing to the structural load-bearing capacity, resulting in a relatively monotonous landscape effect. Most planter-based greening allows users to choose their own plants, with limited application areas, a lack of scientific and systematic arrangement, and a poor match between greenery and energy efficiency. Improper selection of plant species in all four stereoscopic greening methods led to plant death or damage to buildings. In terms of later maintenance, modular and planter-based greening performs better, whereas climbing and paving greening are mostly poorly managed, leading to disorderly and chaotic greening effects, even affecting indoor lighting and normal plant growth.

Regarding the integration of stereoscopic greening with architecture, modular greening and climbing greening on the exterior of buildings have good integration with building facades. Paving greening is more commonly used internally; however, planter-based greening is applied haphazardly, lacking a close connection with the building, both

Building name	Building type	Climbing type Vertical greening	Modular type Vertical greening	Paved type Vertical greening	Planting trough type Vertical greening
Green Building Demonstration Building of Logistics Engineering College of the People's Liberation Army of China	Research building	✓	✓	✓	✓
Chongqing Western Information Technology Application Research and Development Headquarters	Research building		✓		✓
Teaching Building of Chongqing Yucai Middle School	Educational building			✓	✓
Jinmao Longyue Kindergarten Teaching Building	Educational building	✓		✓	
North Campus of Chongqing People's Primary School	Educational building	✓		✓	
Southwest University Administrative Building	Educational building	✓	✓		
Chongqing University Main Teaching Building	Educational building	✓	✓	✓	✓
Chongqing Yuelai Exhibition Park Exhibition Hall	Culture building	✓	✓	✓	✓
Southwest University Library	Culture building	✓		✓	
Chongqing Maternal and Child Health Hospital	Medical building	✓	✓	✓	✓
Chongqing First People's Hospital	Medical building			✓	
The Second People's Hospital of Dadukou District, Chongqing	Medical building	✓			✓
Chongqing Chengtou Rui'an Yishan Health Center	Medical building	✓		✓	
Shizishan Sports Park Comprehensive Hall	Sports building	✓		✓	
China Machinery Construction Technology Building	Office building	✓		✓	
Comprehensive Building of Chongqing Rail Transit Dazhulin Depot	Office building	✓	✓	✓	✓
Chongqing House Office Building	Office building	✓			✓
Chongqing Planning and Natural Resources Bureau	Office building	✓			✓
Chongqing Zhejiang Business Headquarters Economic Base	Office building	✓	✓		
The main building of the Guige Liangjiang headquarters	Office building	✓	✓		
Xie Xin Headquarters City Office Building	Office building	✓			✓
Liangjiang Smart Ecological City Yoshida Headquarters City	Office building			✓	✓
The former Municipal Bureau of Landscape Architecture	Office building	✓			✓
Jiulongpo District People's Congress Office Building	Office building	✓			
Xinguang Tiandi	Commercial building		✓	✓	✓
Chongqing Nanping Wanda Plaza	Commercial building			✓	
Yingli International Financial Cente	Commercial building	✓		✓	
Chongqing Convention and Exhibition Hall Phase II	Commercial building			✓	✓

Legend: Climbing type Vertical greening · Modular type Vertical greening · Paved type Vertical greening · Planting trough type Vertical greening

Fig. 3. Vertical greening application survey of existing public buildings in Chongqing

indoors and outdoors. Therefore, it can be concluded that modular stereoscopic greening technology is suitable for the existing public buildings in Chongqing. The reasons for the suboptimal application of stereoscopic greening in the surveyed projects can be summarized as follows:

1. Poor Implementation of Stereoscopic Greening Policies

Although the local government in Chongqing emphasizes the importance of stereoscopic greening and has introduced related policies, the benefits of stereoscopic greening have not been fully realized. There is a lack of specific regulations for the protection of stereoscopic greening and there are no well-established evaluation criteria for its utilization. Standard symbols for the norms, guidelines, and signage for greening have not been fully implemented in stereoscopic greening in Chongqing. The absence of educational institutions that provide training on stereoscopic greening restricts the development of specialized maintenance personnel for stereoscopic greening. At the same time, this also reflects that some existing stereoscopic greening technologies may not be suitable for the local soil and climate conditions in Chongqing, resulting in poor energy efficiency and a lack of enthusiasm in the implementation process by the entities responsible for the renovation.

2. Incomplete Stereoscopic Greening Technology System

In recent years, Chongqing has intensified efforts to renovate stereoscopic greening. However, owing to the incomplete renovation system, the quality of greening transformations is unsatisfactory. Currently, the direction and goals of stereoscopic greening in Chongqing are unclear, often resulting in time-consuming and inefficient greening processes. Additionally, the purpose of stereoscopic greening needs to be more precise, as most greening activities create independent areas along the exterior of buildings, with no functional connection to the interior. The existing stereoscopic greening in Chongqing primarily involves greening climbing plants on walls, mainly in the gaps between fences and walls, resulting in relatively small-scale greening. The greening of one or two open platforms or rooftops on buildings is often limited by conservative technical approaches, with limited applications of new technologies in stereoscopic greening construction. The forms of stereoscopic greening applied to building renovations lack innovation, and the incomplete system has resulted in a few systematic cases of stereoscopic greening transformations in existing public buildings in Chongqing.

3. Improper Maintenance and Management of Stereoscopic Greening

Stereoscopic greening not only requires the implementation of policies but also demands a sound management mechanism for regular supervision and maintenance. However, there is currently no established long-term management mechanism for stereoscopic greening in Chongqing. Moreover, the personnel responsible for stereoscopic greening are often not professionally trained, leading to a lack of expertise in maintenance techniques and suboptimal operational effects. The lack of professionalism in technology and lax supervision often results in the excessive growth of climbing plants in later stages, with accumulated dead branches causing high density and potential loss of the original functionality of stereoscopic greening [15]. Additionally, failure to implement supporting measures, such as fertilization, irrigation, and pest control, is also a contributing factor to the inefficient performance of stereoscopic greening.

4. Lack of Site-Specific Design in Stereoscopic Greening

Chongqing's sunlight conditions and suitable temperatures are conducive to the growth of most plants, and the direct application of existing stereoscopic greening technologies can effectively reduce air conditioning energy consumption in the summer for buildings in the Chongqing region. However, for the winter season with less sunlight in Chongqing, the plants on the facade walls provide some insulation but show limited energy-saving effects. In some cases, they may even increase lighting energy consumption and impact indoor human comfort on sunny days to some extent. Therefore, it can be observed that throughout the year, stereoscopic greening technology has a certain energy-saving effect on buildings in the Chongqing region, but lacks optimization based on local conditions. Chongqing has rich plant resources, and tremendous potential to develop and utilize these resources. However, the current diversity of plants used in stereoscopic greening in Chongqing is relatively low, and the configuration does not fully reflect the plant characteristics and landscape requirements. Additionally, the current pattern of plant configuration in stereoscopic greening in Chongqing is monotonous, with low artistic skills and techniques, and the configuration does not adequately reflect the characteristics of buildings, plants, and landscape requirements.

Although various inadequacies in vertical greening technologies have had some negative impacts on buildings in research cases, overall, the benefits outweigh the drawbacks. However, for existing public buildings with a massive existing stock, there are relatively few systematic vertical greening retrofit products, and some of the applicable technologies for retrofitting are not suitable (Table 3), resulting in a lack of suitable vertical greening technologies for many existing buildings during the retrofitting process. Improving the applicability of vertical greening technologies in retrofitting existing public buildings is necessary to ensure the success of the retrofitting process.

The analysis shows that not all vertical greening technologies are suitable for building renovations. From a comprehensive perspective, modular vertical greening has fewer limitations and is the most suitable for building renovations. Compared with other vertical greening technologies, modular vertical greening technology has more obvious advantages and can be summarized as follows:

1. Flexibility and customizability: Modular design makes the vertical greening system more flexible and can be customized to fit a specific space, building structure, or need. This design can be adapted to buildings of different shapes and sizes, thereby making the greening system more maneuverable.
2. Convenience of maintenance and management: Each module can be grown and managed independently, making maintenance easier. When pruning, replacement, or maintenance is required, only the corresponding module must be dealt with without affecting the stability of the overall system. Modular vertical greening is typically equipped with a better irrigation system and high maintenance convenience.
3. Climate Adaptability: The flexibility of modular greening systems allows them to adapt better to climate change. It is possible to adapt to different climatic conditions by replacing plant species in a particular module, which improves the resilience of the system.
4. Multiple ecological benefits: Modular greening systems can introduce more plants into the urban environment, helping absorb harmful substances in the air, improve air quality, and reduce the urban heat island effect. Modular vertical greening for noise and dust isolation is better than other types of greening, and it is easier to form large-scale greening, which in turn produces a more stable biological chain, and has significant ecological benefits in terms of improving the quality of the surrounding environment.
5. Fast forming: Compared with climbing greening, modular greening plants can be pre-cultivated and used to quickly form a complete green layer.
6. Landscape effect richness: Compared to other vertical greening technologies, modular vertical greening combinations are more flexible and can create a richer landscape effect.

Table 3. Applicability analysis of vertical greening technology to public building renovation

Technology type	Applicable types	Limitations of application in building renovation
Wall mounted climbing for vertical greening	New construction or building renovation	1. Plant growth is more difficult to control and difficult to manage scientifically 2. Some building facades are not suitable for climbing plants 3. Difficult to pre-cultivate, long growing period of plants 4. The overall style of some buildings is not suitable
Potted vertical greening	New construction or building renovation	1. The inability to create a complete green layer on the building façade 2. Need to occupy the internal plane space of the building 3. Difficult to irrigate on a large scale, with high dependence on labor for management and maintenance
Modular vertical greenery	New construction or building renovation	1. Most of the product modules are small in scale, and the effect is too fragmented for covering large surfaces 2. Poor stability in some building surface installation, there are security risks 3. Geographical suitability needs to be improved
Cloth bag type vertical greening	New construction or building renovation	1. Weak aging resistance, complicated local replacement, mainly used for indoor 2. Poor stability of installation on some building surfaces, with security risks

(continued)

Table 3. (*continued*)

Technology type	Applicable types	Limitations of application in building renovation
Hydroponic vertical greening	New construction	1. The high cost of most products leads to high remodeling costs 2. Most of the product unit scale is small, and the effect is too fragmented for large surface coverage 3. Poor stability of installation on some building surfaces, with safety risks 4. Heavy overall mass, which may cause excessive loading to the building structure 5. For poor waterproofing or old building facade application may have the risk of leakage
Frame traction vertical greening	New construction or building renovation	1. Harder to control plant growth and difficult to manage scientifically 2. Difficult to pre-cultivate, long growing period of plants 3. Unsuitable overall style of some buildings 4. Difficulty in installing part of the building façade
Integrated vertical greening	New construction	1. Need to reserve the location and pipe network at the design stage, not applicable to building renovation 2. High cost of construction and operation and maintenance

4 Discussion

In the unique environments of high-density mountainous cities, traditional horizontal greening methods face dual constraints from natural terrain and urban space, making it difficult to effectively expand green areas. Such high-density layouts not only lead to persistently high urban temperatures during summer but also further degrade the overall quality of urban spaces, affecting residents' comfort and health. Faced with this challenge, vertical greening technology demonstrates its unique advantages and strategies for response. By efficiently utilizing the facades and rooftops of buildings, vertical greening significantly increases the urban green rate without additional ground occupation, beautifying the urban environment and improving the urban microclimate through the natural functions of plants, thus combating the effects of high temperatures.

The implementation of vertical greening in high-density mountainous cities is not merely a quantitative increase in urban green spaces; it represents a qualitative leap. The abundance of public buildings in high-density cities, such as schools, hospitals, and government buildings, provides unparalleled spaces and potential for transformation. The facades and rooftops of these buildings become ideal sites for vertical greening, not only bringing more greenery to the city but also becoming key to enhancing urban aesthetics and residents' well-being. Moreover, vertical greening can effectively reduce building energy consumption through the shading and evaporative cooling effects of plants, reducing reliance on air conditioning, thereby lowering energy consumption and improving indoor environmental quality.

Vertical greening is not just a technique or method; it is a new concept in urban ecological design. In high-density mountainous cities, it prompts us to rethink the relationship between humans and nature and how to create livable and sustainable ecological environments within limited spaces. Through vertical greening, we can create vibrant green spaces in every corner of the city, injecting new vitality and hope into urban life. With the integration of more innovative technologies and design philosophies, vertical greening will play an increasingly important role in the sustainable development of high-density mountainous cities, becoming a key force in shaping more livable, green, and harmonious cities. Although vertical greening technology provides an innovative greening solution for high-density mountainous cities, making it possible to significantly increase green areas within limited spaces, applying this technology to the renovation of existing buildings also faces a series of challenges and difficulties.

Firstly, structural safety poses a significant challenge in implementing vertical greening. Existing buildings may not have been designed to bear additional weight, especially in rooftop greening, where the added weight of soil, plants, and irrigation systems must be considered to avoid exceeding the building's original load-bearing capacity. Therefore, conducting detailed assessments and possibly reinforcing the building structure are necessary steps to ensure the safety of vertical greening.

Secondly, water management requires special attention in vertical greening. Improper water management can lead to poor plant growth and even structural issues in buildings, such as water infiltration and wall damage. Designing effective drainage and irrigation systems is crucial to maintaining plant health and building structural integrity.

Moreover, maintenance cost and complexity represent another challenge faced by vertical greening. Compared to traditional ground-level greening, vertical greening requires more frequent maintenance, including watering, pruning, and fertilizing, which not only increases the long-term maintenance costs but also requires specialized knowledge and skills, especially in situations where access is difficult due to height or location.

Finally, vertical greening renovation projects must consider plant selection and local climate adaptability. Not all types of plants are suitable for growing in specific vertical greening environments, and selecting species that adapt to local climates and can thrive in limited soil conditions is key to successfully implementing vertical greening.

5   Conclusion

At present, although there are many applications of vertical greening technology in most high-density mountainous cities, most of them are applied in new public buildings, and there are negative effects in the actual use owing to less optimization of the region. For most existing public buildings with high energy consumption, there is no systematic vertical greening transformation. Part of the vertical greening transformation of public buildings is only in the roof to provide a simple green plant cover and the use of climbing plants in the façade, and did not systematically upgrade the vertical greening. In all types of vertical greening technology on the suitability of the transformation of public buildings, from the application of the comprehensive effect analysis, it can be concluded that modular vertical greening technology is the most appropriate. Modular vertical greening technology combination of flexibility determines its similar to the quality of such high-density mountainous cities as Chongqing existing public buildings to enhance the high suitability and development potential, but in the existing technology used in existing public buildings in the field of transformation, still need to do further optimization in the overall compatibility, stability of the modular unit, modular plants in the ground to improve.

Moreover, this study provides baseline data on existing vertical greening projects in the case city, including the types of technologies used, species of plants employed, and maintenance conditions. These data are crucial for evaluating the current implementation effects and sustainability of vertical greening technologies, serving as an initial reference point for future technological improvements.

Funding.   This paper was supported by the Chongqing city management academic project: Optimization of Green Space Plant Configuration Patterns in the Residential Area of Chongqing Central business area Based on the Measurement of Green Capacity Rate and Assignment Number: Urban-Management Kezi 2022·No.(32).

This paper was supported by the Chongqing city management academic project: Study on Optimization of Plant Configuration Patterns in Chongqing Residential Green Areas Based on High Carbon Sink Efficiency and Assignment Number: Urban-Management Kezi 2023·No.(33).

References

1. Wan., G., Zhang, Y.: Accelerating urbanization explained—The role of information. Economics (Quarterly) **20**(2), 465–492 (2021)
2. Shaochun, Y., et al.: Design and evaluation of sponge city reconstruction scheme for old building district in mountainous city based on InfoWorks_ICM model. Water Resour. Prot. **36**(5), 43–49, 70 (2020)
3. Ribiao, C., et al.: A study of the multi-scale wind environment assessment methods from the perspective of ventilation energy conservation: A case study of the central area of Houhai in Shenzhen. Southern Architecture. **2**, 77–87 (2023)

4. Liu, S.F., Tan, S.H.: Building a new framework for urban parking facilities research with quality improvement: the case of Chongqing, China. Int. J. Environ. Res. Public Health. **20**(1) (2023)
5. Le, D., Li, Y.S., Ren, F.: Does air quality improvement promote enterprise productivity increase? Based on the spatial spillover effect of 242 cities in China. Front. Pub. Health. **10** (2022)
6. Han, L.J., et al.: Challenges in continuous air quality improvement: An insight from the contribution of the recent clean air actions in China. Urban Climate. **46** (2022)
7. Polychroni, E., Androutsopoulos, A., Iop.: Innovative financial schemes for buildings' energy renovation. In: Conference on Sustainability in the Built Environment for Climate Change Mitigation (SBE). Thessaloniki, GREECE (2019)
8. Zhang, H., et al.: The ambient air quality standards, green innovation, and urban air quality: evidence from China. Sci. Rep. **13**(1), 19684 (2023)
9. Li, Y., et al.: Analysis on the incentive policy of vertical greening in China: a case study of Shenzhen City. In: 5th International Conference on Environmental Science and Civil Engineering (ESCE). Nanchang, PEOPLES R CHINA (2019)
10. Zhao, M., et al.: A study on vertical greening practice emphasizing ecological benefits in Xiamen City of China. Int. J. Sust. Dev. World Ecol. **22**(4), 368–374 (2015)
11. Peng, Y., et al.: Experimental investigation on the effect of vertical greening facade on the indoor thermal environment: a case study of Dujiangyan City, Sichuan Province. In: International Conference on Advances in Civil Engineering, Energy Resources and Environment Engineering (ACCESE). Jilin Jianzhu Univ, Coll Civil Engn, Changchun, PEOPLES R CHINA (2019)
12. Qi, Z.Y.: Urban forest construction and vertical greening development under climate change. Landscape Archit. Frontiers **11**(1), 58–64 (2023)
13. Dorozhkina, E., and I.O.P. Publishing: Architectural structures for the formation of vertical landscaping of buildings. In: International Conference on Construction, Architecture and Technosphere Safety (ICCATS). Sochi, RUSSIA (2020)
14. Guo, S.T., Yang, F., Jiang, Z.D.: Thermal environmental effects of vertical greening and building layout in open residential neighbourhood design: a case study in Shanghai. Archit. Sci. Rev. **65**(1), 72–88 (2022)
15. Zeng, M., et al.: Does vertical supervision promote regional green transformation? Evidence from Central Environmental Protection Inspection. J. Environ. Manage. **326** (2023)

Thermal Comfort Study of Human Activities at Different Levels in Outdoor Garden Spaces: Based on the Living Lab Method in Hot Summer and Warm Winter Regions

Zhiqiang Zhou and Liang Dong[✉]

School of Architecture, Huaqiao University, Xiamen 361021, China
19011085006@stu.hqu.edu.cn

Abstract. This study, leverages the Living Lab research platform to explore the variances in thermal comfort between dynamic and static activities in garden spaces during autonomous activity states. Volunteers participated by completing thermal comfort questionnaires in the Living Lab's garden area. Through monitoring and observation of participants' activity states during survey completion, autonomous activities were categorized into static and dynamic activities, and the results and behavioral data of the two groups were analyzed comparatively. Spanning nine months, the study analyzed the impact of activity levels on garden thermal comfort and seasonal thermal adaptability. The findings reveal: 1) The neutral temperature for participants in garden spaces is highest in summer and lowest in winter, with dynamic activities having a lower neutral temperature than static activities, the largest difference observed in winter; 2) Dynamic activities have lower thermal sensitivity compared to static activities; 3) In thermal neutral and warm environments, the comfort level of dynamic activities is similar to static activities, but in colder environments, dynamic activities are more likely to feel comfortable; 4) Attendance for static activities in the garden is more susceptible to changes in the thermal environment compared to dynamic activities.

Keywords: Garden · Outdoor thermal comfort · Dynamic activity · Static activity

1 Introduction

Environmental factors significantly influence outdoor activities. The evaluation criteria for the quality of public spaces should include elements such as the protectiveness, comfort, and pleasantness of the space. Among these, the outdoor thermal environment is an important factor affecting spatial comfort. In recent years, outdoor thermal comfort has gradually become a hot topic in academic research [1]. Past studies on outdoor thermal comfort have been conducted in cities of different climatic types, examining people's thermal perception and activity types in various outdoor spaces [2–19]. These studies demonstrate that the thermal environment of a space has a strong impact on its usage.

© The Author(s) 2025
B.-J. He et al. (Eds.): UCSUD 2023, LNCE 559, pp. 536–550, 2025.
https://doi.org/10.1007/978-981-97-8401-1_38

Existing research indicates that the state of activity significantly affects thermal comfort in outdoor activities. Lai's research delineates outdoor activities into categories of low and high intensity, noting an apex in participation across all activities within the neutral thermal sensation zone. Nonetheless, alterations in the thermal climate reveal that high-intensity activities exhibit heightened sensitivity to warmer conditions, whereas low-intensity activities are more adversely affected by cooler climates [20]. Huang et al. divided outdoor activities into low, medium, and high intensity and obtained the UTCI value thermal comfort range for each type of activity as 15.1 ~ 28.5 °C, 18.2 ~ 30.6 °C, and 9.0 ~ 23.9 °C [21], respectively. Leng et al. discovered that, in the winter park environment, the neutral temperature for static activities (PET = 21 °C) is higher than that for dynamic activities (PET = 18.1 °C) [22]. This body of work, primarily focused on high-traffic areas such as parks and squares, and the activities investigated include sports exercises, rest, sightseeing and other leisure activities, as well as necessary activities like waiting and commuting.

Existing research has two main limitations: 1) Lack of specificity in studies of autonomous activities. Jan Gehl categorized outdoor activities into autonomous, social, and necessary activities [23]. Sharifi's study [24] in Australia demonstrated that these three modes of activity are differently affected by the outdoor thermal environment. However, most studies on thermal comfort across various activity intensities primarily take place in open squares and parks, without differentiating the autonomy of the activities. 2) The majority of research relies on single-instance data collection regarding individuals' subjective thermal sensation and activity patterns, thus lacking longitudinal insights into the thermal experiences and behaviors of a fixed population. This leads to potential significant errors in comparative analyses of seasonal adaptations. For instance, July and August constitute the hot summer months, but due to being the summer vacation period for students in China, there is a marked increase in the number of young people and adolescents participating in thermal comfort field surveys during the summer.

This paper focuses on garden spaces as the subject of study to investigate the thermal comfort characteristics of outdoor autonomous activities, aiming to analyze the differences in thermal comfort for rest activities of varying intensities. The proximity of the garden to indoor spaces of buildings provides a relatively stable indoor thermal environment, ensuring that different participants have similar thermal experiences before entering the study area, effectively minimizing research errors caused by different metabolic rates, thermal experiences, and expectations. Additionally, the relative enclosure of the garden reduces the interference of necessary activities on the research. This study categorizes garden activities into static and dynamic activities. Static activities refer to wakeful states of sitting or lying down with low energy expenditure (1 ~ 1.5 METs) but higher than the resting metabolic rate, such as watching TV, using computers, and mainly include sitting rest and chatting in outdoor activities [25]. Outdoor dynamic activities refer to other activities with higher metabolic rates than static activities, primarily involving light to moderate metabolic levels (2.5 ~ 5METs) in garden activities, such as strolling, body stretching exercises, gardening activities, etc.[26]. The objectives of this study are:

(1) To compare the differences in neutral temperatures for dynamic and static activities in the garden across different seasons;

(2) To compare the differences in thermal comfort between the two states of activity;
(3) To analyze the patterns of change in the number of people engaging in dynamic and static activities across summer, autumn, and winter as the thermal environment varies.

2   Methods

2.1   Research Site

This research was carried out in Xiamen, Fujian Province (N24°43′, E118°10′), a locale identified within China's architectural climate zoning as a region experiencing hot summers and mild winters. The study started on June 1, 2020, and ended on February 28, 2021. Changes in air temperature and relative humidity in the Xiamen area during the study period are shown in Fig. 1. The study defined June to September as summer, October to November as autumn, and December to February of the following year as winter.

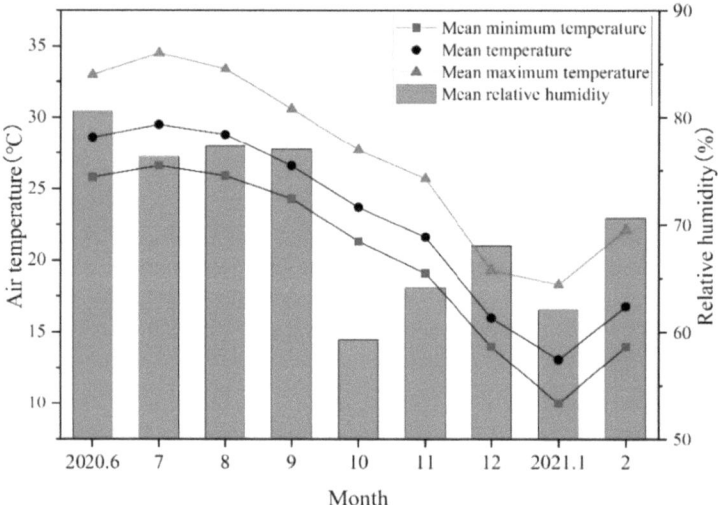

Fig. 1. Air temperature and relative humidity in Xiamen during the study period (Data source: China Meteorological Data Network)

The research site was an outdoor Living Lab, located on the fourth floor of the Architecture Discipline Laboratory Building at Huaqiao University. The foundation of the lab is a rectangular rooftop garden, covering an area of 230 m² (Fig. 2 (a)). The Living Lab is connected to the surrounding indoor work areas through a corridor (Fig. 2 (b)). The constructed garden features outdoor landscape elements such as water bodies, landscape architecture, trees and shrubs, and ground-cover plants. It is also outfitted with scientific research infrastructure such as meteorological monitoring, behavioral observation, and thermal environment control systems.

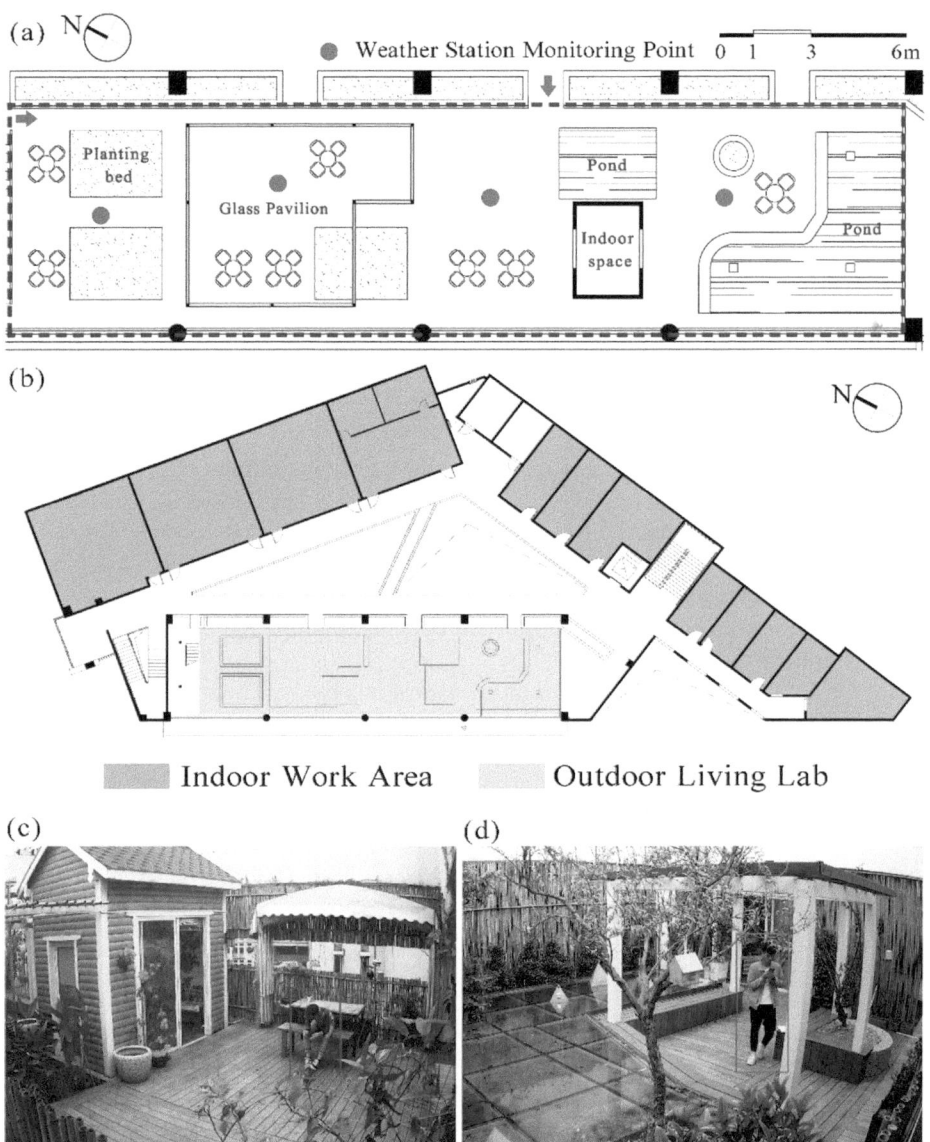

Fig. 2. (a) Floor plan schematic of the Living Lab. (b) Distribution of the Living Lab and surrounding work areas. (c) Static activities. (d) Dynamic activities

The Living Lab, as a garden, is open to the public on a daily basis, providing a resting place for nearby teachers and students. The main users of this site are graduate students from the Architecture College located near the Living Lab. In their free time, students often enter the Living Lab for various recreational activities or to organize gatherings.

2.2 Research Process

A total of 46 volunteers, all regular users of the Living Lab garden, participated in this study. They were allowed to freely enter and exit the Living Lab, engaging in various rest activities in its outdoor space as they wished (Fig. 2(c)(d)). During their visit, they had the opportunity to complete a thermal comfort questionnaire specifically designed for this research. The researchers evaluated the validity of the questionnaire responses.

The Living Lab is equipped with a behavioral observation system consisting of 6 sets of surveillance cameras, which use intelligent monitoring to observe activities on the site. Based on the state of activity, those filling out questionnaires and other active individuals were categorized into static and dynamic rest activities.

2.3 Questionnaire Design

This study distributed questionnaires to volunteers via mobile electronic forms. Researchers sent out questionnaires daily in a WeChat group for the volunteers. The questionnaire solicited participants' names and thermal sensation choices. Given that volunteers' age, height, and weight were documented beforehand, these details were excluded from the questionnaire. The area of activity and clothing of the volunteers at the time of filling out the questionnaire were obtained through the surveillance system of the laboratory.

In this study, the thermal perception questions involved a Thermal Sensation Vote (TSV), Thermal Comfort Vote (TCV), and acceptability vote. According to ISO 10551 (2019) standards, the TSV uses a 9-point scale, and the TCV uses a 5-point scale (Fig. 3).

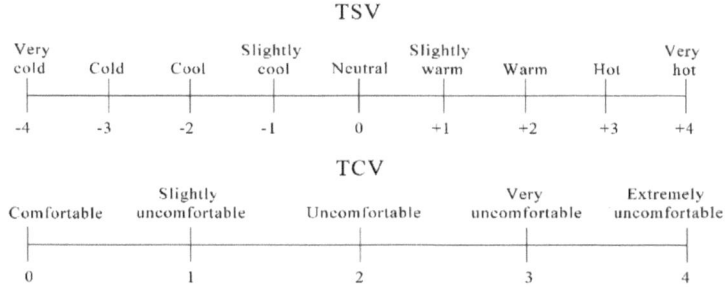

Fig. 3. Grading of Thermal Sensation Vote (TSV) and Thermal Comfort Vote (TCV)

2.4 Acquisition of Thermal Environment Parameters

Four meteorological stations were installed at the laboratory site. Each station is equipped with environmental temperature and humidity sensors, solar radiation sensors, and wind speed recorder (Table 1), placed in four different spaces within the site. All sensors were wired to an environmental weather monitor, which uploaded the data to a cloud platform for researcher access. These meteorological stations continuously monitored the site's thermal conditions round-the-clock, recording data at one-minute intervals. The stations are located in areas of high foot traffic and are arranged to minimize interference with the space's rest function, ensuring that the thermal environment data collected closely approximates the thermal environment experienced by humans in the Living Lab.

Table 1. Specifications of measurement devices

Equipment	Model	Measurement parameter	Range	Accuracy
Air temperature and humidity recorder	PTS-3	T_a (°C); RH (%)	−40–80 °C; 0–100%RH	±0. °C; ±2%RH
Global radiation sensor	TBQ-2	G (W/m^2)	0–2000W/m^2	<5%
Wind speed recorder	EC-9S	Ws (m/s)	0–70m/s	±0.3 m/s

This study selects the Physiologically Equivalent Temperature (PET) [27] as the objective thermal comfort indicator. PET is calculated using Rayman Pro software [6, 28]. Six parameters are required for the calculation (air temperature, air humidity, wind speed, solar radiation, clothing thermal resistance, metabolic rate). Among these, air temperature, air relative humidity, wind speed, and solar radiation are directly measured by the four groups of meteorological stations at the laboratory. Clothing thermal resistance and metabolic rate are obtained through video surveillance equipment. Metabolic rate is estimated based on ISO 8996 (2004), and clothing thermal resistance is estimated based on ISO 9920 (2007).

2.5 Selection of Valid Questionnaires

In the selection of behavioral data within the laboratory, several rules were followed to ensure the integrity of the research findings: Participants had to fill out the questionnaire at the Living Lab after engaging in static activities in an indoor environment and were required to have been in the lab for over 5 min for their responses to be valid. Activities primarily involved sitting without any adjustments for thermal comfort, such as drinking cold beverages or using fans. Data from questionnaires that did not match the thermal environment data due to participants being in blind spots of the meteorological instruments were deemed invalid. Moreover, participants' engagement in static or dynamic activities had to constitute more than two-thirds of their total time in the lab. To mitigate the effect of outliers, the number of valid questionnaires accepted from any single participant for a particular activity state was limited to 30; any excess in this number resulted in a random selection of 30 questionnaires for analysis.

3 Results and Discussion

The study amassed a total of 2,504 valid questionnaires. During the summer months (June to September), 997 questionnaires were collected, with 635 attributed to static activities and 362 to dynamic activities. In the autumn (October to November), 851 questionnaires were gathered, comprising 478 for static activities and 373 for dynamic activities. For the winter period (December to February of the following year), 656 questionnaires were collected, with 351 for static activities and 305 for dynamic activities.

3.1 Neutral Temperature

Figure 4 show the scatter distribution of the Thermal Sensation Vote (TSV) on PET for the two different activity states during summer, autumn, and winter. Linear regression was performed, and the PET where TSV = 0 was defined as the neutral temperature, while the PET within the range of TSV $(-1, 1)$ was considered the thermal neutral range. The fitting results are shown in Table 2.

The obtained neutral temperatures indicate that the highest neutral temperature is in summer, the lowest in winter, with autumn in between. This is consistent with the findings of Cheng [7], Pninit [9], Lai [11], Horizont [29], Li Kunming [13], and Chen [16]. Specifically, comparing different activity states, the neutral temperature for dynamic activities in all three seasons is lower than for static activities, with the differences being 1.1 °C, 1.3 °C, and 2.0 °C for summer, autumn, and winter, respectively. The smallest difference is in summer, and the largest in winter. The winter results are similar to Leng's findings in Harbin, China [22]. However, the winter neutral temperatures for static and dynamic activities in Leng's study are lower than in this research, due to different climatic conditions, but Leng did not compare the thermal comfort differences between the two types of activities in summer and autumn.

The analysis of the fitting line's slope across all three seasons reveals that the slope for static activities surpasses that for dynamic activities, and the thermal neutral range is wider for dynamic activities compared to static ones. This indicates that the thermal sensitivity of subjects in dynamic activities is lower than in static activities. The larger difference in slopes between the two in winter compared to summer suggests that the adaptability of dynamic activities in colder environments is more pronounced. This is not entirely in line with Lai's findings in Tianjin [20]. Lai's results indicate that intense activities are least sensitive to cold environments, while low-intensity activities are least sensitive to hot environments. However, in this study, static activities are more sensitive to both cold and hot environments than dynamic activities. The reason for this discrepancy might be that Lai's study of outdoor activities included high-intensity physical exercises and necessary activities, while the autonomous activities studied in this paper are of relatively lower intensity.

3.2 Thermal Comfort Vote (TCV)

The thermal environment corresponding to the questionnaire was grouped into intervals of 4 °C each, to analyze the distribution of the Thermal Comfort Vote (TCV) in the questionnaire results. Figure 5 show the distribution of TCV for static and dynamic activities in different thermal environments during summer, autumn, and winter. In summer and autumn, when the thermal environment is mainly hot or around neutral temperatures, the distribution of TCV for both static and dynamic activities increases with the rise in PET, with no significant overall difference between the two. However, in winter, when PET is less than 16 °C, the increase in TCV for static activities is faster than for dynamic

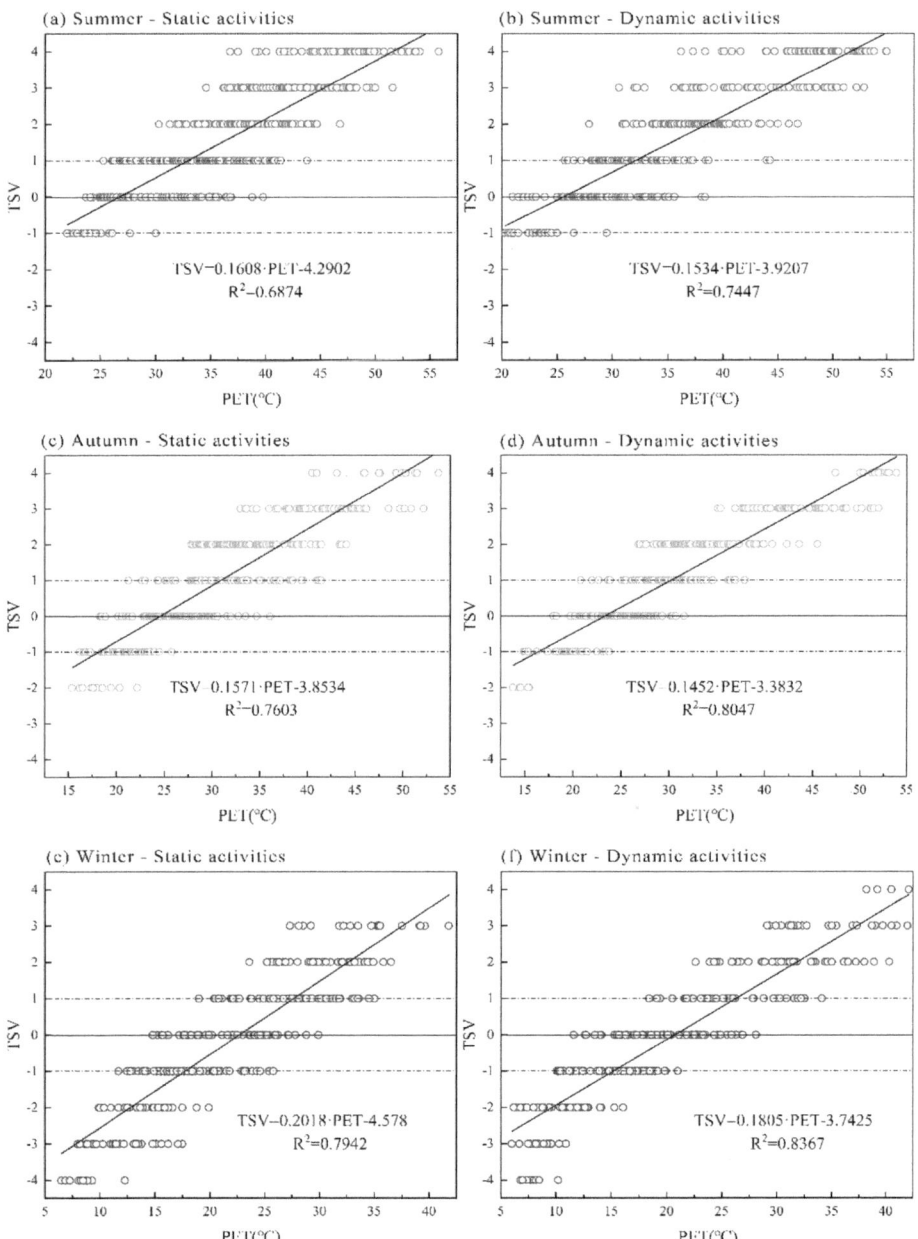

Fig. 4. Correlation between TSV and PET in different seasons

activities as PET decreases. In the thermal environment interval of PET $= 8\ °C$, the average TCV for static activities is 2.9, higher than the average TCV of 1.71 for dynamic activities.

Table 2. Neutral PET and neutral range

Seasons	Case	Neutral PET	Neutral range
Summer	Static activities	26.7	20.5–32.9
	Dynamic activities	25.6	19–32.1
Transition season	Static activities	24.5	18.2–30.9
	Dynamic activities	23.2	16.4–30.2
Winter	Static activities	22.7	17.7–27.6
	Dynamic activities	20.7	15.2–26.3

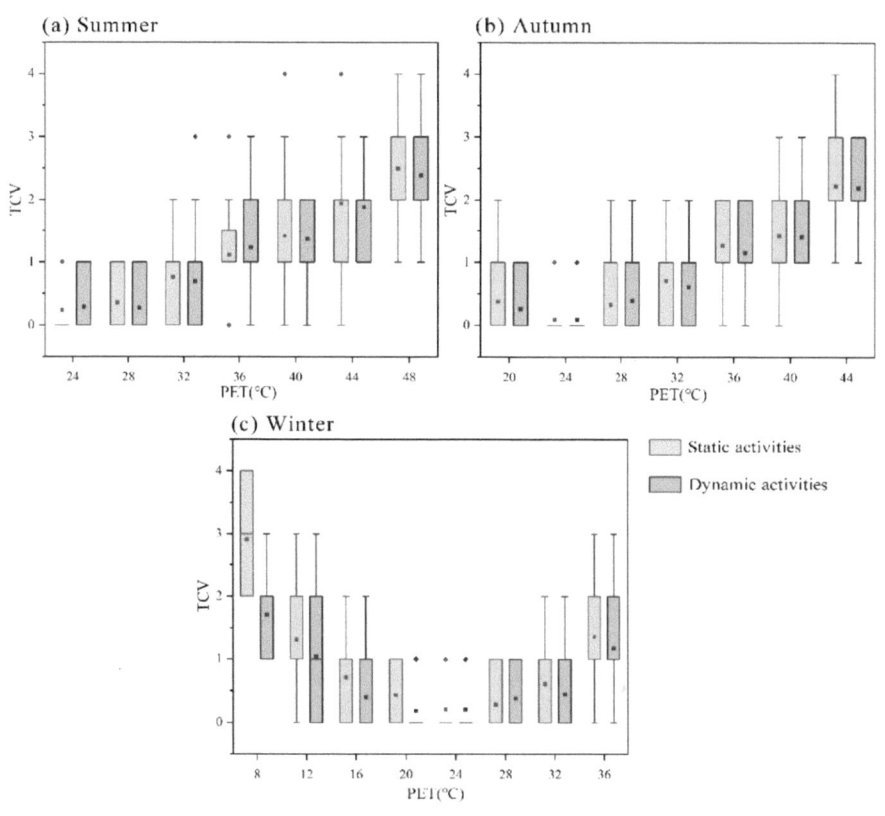

Fig. 5. TCV-PET

Observations indicate that in neutral to slightly warm outdoor thermal environments, the activity state has a negligible effect on individuals' thermal comfort. However, when the outdoor thermal environment is cold, people engaged in dynamic activities tend to feel more comfortable than those in static activities. This phenomenon may be related to human psychological activities; people in dynamic states have greater control over their environment and personal state than in static activities, thereby more easily achieving psychological comfort [30].

3.3 Attendance

Attendance reflects the popularity of a space in terms of its thermal environment. Data on activity attendance at the Living Lab were collected via its surveillance system. Each stay in the Living Lab lasting over 5 min was counted as one valid attendance; static activities lasting for two-thirds or more of the total stay time were counted as static activity attendances, and the same applied to dynamic activities.

The meteorological station in the Living Lab calculated the minute-by-minute PET during the observation period, and the data were grouped into 1 °C intervals. The range of PET values for activities in the Living Lab was from 4 °C to 58 °C, resulting in 55 groups of data. Specifically, the summer PET range was 20 ~ 58 °C, the autumn range was 14 ~ 56 °C, and the winter range was 4 ~ 46 °C. Given the variable duration of each PET interval in real-world scenarios, it was necessary to calculate the ratio k of the duration of each PET unit to the average duration, for the standardization of attendance data. The relative attendance per unit PET, which is the number of people present in the space per 1 °C PET for the same duration, was calculated as follows:

$$\text{Relative attendance} = \text{Attendance} \cdot \text{k} \tag{1}$$

where k is the ratio of the duration of a unit PET to the average duration of unit PET.

The relative attendance for dynamic and static activities in autumn and winter was calculated in the same way, using a Gaussian curve fitting. The thermal environment corresponding to the central axis of the Gaussian curve is defined as the most suitable thermal environment, and the thermal environment corresponding to 90% of the peak of the curve is defined as the comfort range. The results are shown in Fig. 6 and Table 3.

Previous research has indicated that in temperate and subtropical regions, the highest frequency of space utilization occurs when the thermal environment is near the neutral thermal sensation [20, 21]. In this study, during summer, the attendance for both static and dynamic activities diminish as the PET increases, with no discernible optimum temperature or comfort zone identified. In autumn and winter, the attendance for both static and dynamic activities first increases and then decreases with the increase in PET. The obtained optimal temperatures, i.e., the PET values corresponding to the peak of the curve, are close to the neutral temperatures obtained from the questionnaire. This is consistent with previous research findings [20].

In autumn, the optimal temperature (OT) for static activities (25.0 °C) is close to that for dynamic activities (25.5 °C); in winter, the optimal temperature for static activities (23.1 °C) is higher than that for dynamic activities (20.7 °C). Across all three seasons, the standard deviation σ of the Gaussian curve fitted for static activities is lower than that for dynamic activities, indicating that the distribution of static activity attendance over PET is more concentrated than dynamic activities, requiring a more stringent thermal environment. In all three seasons, the comfort range (CR) for dynamic activities is broader than that for static activities, similar to Huang's findings in Wuhan [21]. Nevertheless, unlike Huang's study, which did not differentiate between seasonal effects, this

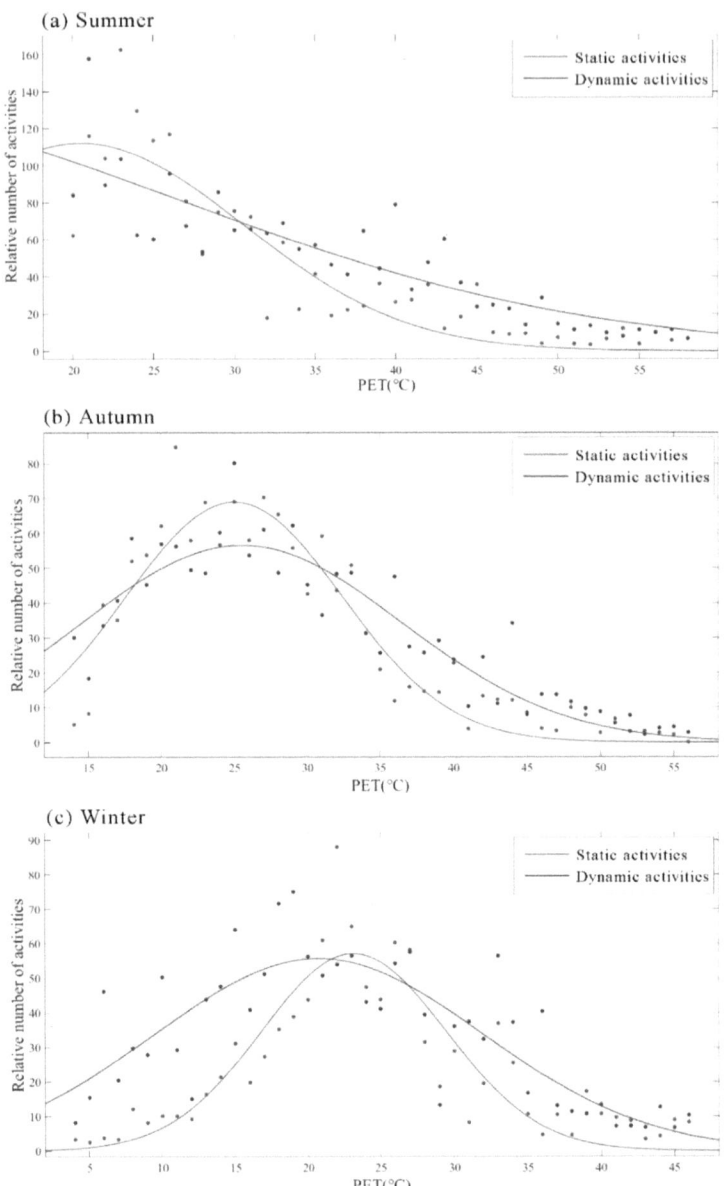

Fig.6. Relative attendance with thermal environment

investigation reveals that the disparities in thermal comfort between static and dynamic activities become more marked during the winter months.

Table. 3 Gaussian function fitting results

	Gaussian function fitting formula	R^2	σ	OT (°C)	CR (°C)
Summer					
Static activities	$Y = 112.1 \cdot e^{-\frac{(PET-20.5)^2}{203.35}}$	0.899	8.1	–	–
Dynamic activities	$Y = 134.5 \cdot e^{-\frac{(PET-1.15)^2}{1295.28}}$	0.742	25.4	–	–
Autumn					
Static activities	$Y = Y = \cdot e^{-\frac{(PET-24.99)^2}{109.62}}$	0.874	7.4	25.0	21.6 ~ 28.4
Dynamic activities	$Y = 56.53 \cdot e^{-\frac{(PET-25.52)^2}{241.49}}$	0.734	11.0	25.5	20.5 ~ 30.6
Winter					
Static activities	$Y = 57.06 \cdot e^{-\frac{(PET-23.12)^2}{80.1}}$	0.811	6.3	23.1	20.2 ~ 26
Dynamic activities	$Y = 55.63 \cdot e^{-\frac{(PET-20.73)^2}{253.13}}$	0.679	11.2	20.7	15.6 ~ 25.9

4 Limitations and Prospects

Due to the limitations of article length and research site, this study did not consider the impact of higher intensity sports on thermal comfort. The study chose university students as subjects, but physiological differences between different age groups may lead to varying impacts of activity state, which could be explored more deeply in future research.

5 Conclusion

This research adopted the Living Lab methodology to carry out a longitudinal survey of a consistent population within garden spaces located in a region characterized by hot summers and mild winters. It collected thermal sensation questionnaires across different seasons and observed and recorded people's behavior in the garden. The research examined the impact of activity levels within garden spaces on thermal comfort, as well as the differences in thermal adaptability across various seasons. The findings contribute to an understanding of the mechanisms of thermal comfort in small outdoor garden spaces, thereby providing a reference for optimizing the thermal environment of outdoor courtyards. The analysis yielded several key findings:

(1) The neutral temperature for static autonomous activities in the garden is slightly higher than for dynamic activities. The smallest difference is in summer (26.7 °C for static and 25.6 °C for dynamic), the largest in winter (22.7 °C for static and 20.7 °C for dynamic), and autumn shows an intermediate difference (24.5 °C for static and 23.2 °C for dynamic). In all three seasons, the sensitivity to thermal sensation is higher in static activities than in dynamic activities. The thermal neutral range for dynamic activities is broader than for static activities.

(2) In summer and autumn, the difference in Thermal Comfort Votes (TCV) between static and dynamic activities is minimal. Nonetheless, in winter, with PET below 20 °C, TCVs for static activities surpass those for dynamic ones, suggesting that individuals engaged in dynamic activities experience greater comfort in colder conditions.

(3) The impact of the thermal environment of the garden on activity attendance varies with the state of activity. In summer, attendance for both dynamic and static activities diminishes as temperatures rise, whereas in autumn and winter, attendance trends upwards then downwards. In autumn, the peak attendance temperature for static activities is similar to that for dynamic activities, but in winter, the optimal temperature for static activities is significantly higher than for dynamic activities. In all three seasons, the attendance for static activities is more significantly affected by the thermal environment. In thermal environments deviating from the comfort temperature, the attendance for dynamic activities is less affected.

Acknowledgments. The work was supported by the National Natural Science Foundation of China (No.51678253), the Scientific Research Funds of Huaqiao University (No.15BS302).

References

1. Lai, D., Lian, Z., Liu, W., Guo, C., Liu, W., et al.: A comprehensive review of thermal comfort studies in urban open spaces. Sci. Total Environ. **742**, 140092 (2020)
2. Zhou, Z., Jiao, R., Dong, L.: The influence of perceived control on outdoor thermal comfort: a case study in a hot summer and warm winter climate. Build. Environ. **245**, 110872 (2023)
3. Zhou, Z., Dong, L.: Experimental investigation of the effect of surgical masks on outdoor thermal comfort in Xiamen, China. Build. Environ. **229**, 109893 (2023)
4. Zacharias, J., Stathopoulos, T., Wu, H.: Microclimate and downtown open space activity. Build. Environ. **33**, 296–315 (2001)
5. Thorsson, S., Lindqvist, M., Lindqvist, S.: Thermal bioclimatic conditions and patterns of behaviour in an urban park in Göteborg, Sweden. Int. J. Biometeorol. **48**, 149–156 (2004)
6. Matzarakis, A., Rutz, F., Mayer, H.: Modelling radiation fluxes in simple and complex environments—application of the RayMan model. Int. J. Biometeorol. **51**, 323–334 (2007)
7. Cheng, V., Ng, E., Chan, C., Givoni, B.: Outdoor thermal comfort study in a sub-tropical climate: a longitudinal study based in Hong Kong. Int. J. Biometeorol. **56**, 43–56 (2012)
8. Lin, T., Tsai, K., Liao, C., Huang, Y.: Effects of thermal comfort and adaptation on park attendance regarding different shading levels and activity types. Build. Environ. **59**, 599–611 (2013)
9. Cohen, P., Potchter, O., Matzarakis, A.: Human thermal perception of Coastal Mediterranean outdoor urban environments. Appl. Geogr. **37**, 1–10 (2013)

10. Zhou, Z., Chen, H., Deng, Q., Mochida, A.: A field study of thermal comfort in outdoor and semi-outdoor environments in a humid subtropical climate city. J. Asian Arch. Build. Eng. **12**, 73–79 (2013)

11. Lai, D., Guo, D., Hou, Y., Lin, C., Chen, Q.: Studies of outdoor thermal comfort in northern China. Build. Environ. **77**, 110–118 (2014)

12. Song, G., Jeong, M.: Morphology of pedestrian roads and thermal responses during summer, in the urban area of Bucheon city, Korea. Int. J. Biometeorol. **60**, 999–1014 (2016)

13. Li, K., Zhang, Y., Zhao, L.: Outdoor thermal comfort and activities in the urban residential community in a humid subtropical area of China. Energy Build. **133**, 498–511 (2016)

14. Shooshtarian, S., Rajagopalan, P.: Study of thermal satisfaction in an Australian educational precinct. Build. Environ. **123**, 119–132 (2017)

15. Shih, W., Lin, T., Tan, N., Liu, M.: Long-term perceptions of outdoor thermal environments in an elementary school in a hot-humid climate. Int. J. Biometeorol. **61**, 1657–1666 (2017)

16. Chen, X., Xue, P., Liu, L., Gao, L., Liu, J.: Outdoor thermal comfort and adaptation in severe cold area: a longitudinal survey in Harbin, China. Build. Environ. **143**, 548–560 (2018)

17. Fang, Z., Liu, H., Li, B., Tan, M., Olaide, O.M.: Experimental investigation on thermal comfort model between local thermal sensation and overall thermal sensation. Energy Build. **158**, 1286–1295 (2018)

18. Cheng, B., Gou, Z., Zhang, F., Feng, Q., Huang, Z.: Thermal comfort in urban mountain parks in the hot summer and cold winter climate. Sustain. Cities Soc. **51**, 101756 (2019)

19. Yang, W., Wong, N.H., Zhang, G.: A comparative analysis of human thermal conditions in outdoor urban spaces in the summer season in Singapore and Changsha, China. Int. J. Biometeorol. **57**, 895–907 (2013)

20. Lai, D., Chen, B., Liu, K.: Quantification of the influence of thermal comfort and life patterns on outdoor space activities. Build. Simul. **13**, 113–125 (2020)

21. Huang, J., Zhou, C., Zhuo, Y., Xu, L., Jiang, Y.: Outdoor thermal environments and activities in open space: an experiment study in humid subtropical climates. Build. Environ. **103**, 238–249 (2016)

22. Leng, H., Liang, S., Yuan, Q.: Outdoor thermal comfort and adaptive behaviors in the residential public open spaces of winter cities during the marginal season. Int. J. Biometeorol. **64**, 217–229 (2020)

23. Gehl, J., Mortensen, L.: Livet mellem husene (2001)

24. Sharifi, E., Boland, J.: Passive activity observation (PAO) method to estimate outdoor thermal adaptation in public space: case studies in Australian cities. Int. J. Biometeorol. **64**, 231–242 (2020)

25. Biddle, S., Asare, M.: Physical activity and mental health in children and adolescents: a review of reviews. Br. J. Sports Med. **45**, 886–895 (2011)

26. Ainsworth, B.: 2011 Compendium of Physical Activities: a second update of codes and MET values. Med. Sci. Sport. Exerc. **43**, 1575–1581 (2011)

27. Matzarakis, A., Mayer, H.: Another kind of environmental stress: thermal stress. WHO Collab. Cent. Air Qual. Manag. Air Pollut. Control. **18**, 7–10 (1996)

28. Matzarakis, A., Rutz, F., Mayer, H.: Modelling radiation fluxes in simple and complex environments: basics of the RayMan model. Int. J. Biometeorol. **54**, 131–139 (2010)

29. Da Silveira Hirashima, S.Q., de Assis, E.S., Nikolopoulou, M.: Daytime thermal comfort in urban spaces: A field study in Brazil. Build. Environ. **107**, 245–253 (2016)

30. Nikolopoulou, M., Steemers, K.: Thermal comfort and psychological adaptation as a guide for designing urban spaces. Energy Build. **35**, 95–101 (2003)

Thermal Comfort in Pedestrian Spaces of Mountain Cities in Humid and Cold Environments

Ke Xiong[1,2(✉)], Shady Attia[3], and Bao-Jie He[1,2]

[1] Centre for Climate–Resilient and Low–Carbon Cities, School of Architecture and Urban Planning, Ministry of Education, Key Laboratory of New Technology for Construction of Cities in Mountain Area, Chongqing University, Chongqing 400045, China
kexiongarch@cqu.edu.cn

[2] Institute for Smart City, Chongqing University, Liyang 213300, Jiangsu, China

[3] Sustainable Building Design Lab, Deptartment of UEE, Faculty of Applied Sciences, University of Liège, 4000 Liège, Belgium

Abstract. This study aimed to explore the microclimate and outdoor thermal comfort characteristics of pedestrian spaces in mountain cities under humid and cold conditions. It focused on rainy and cloudy winter days in a typical mountain city (Chongqing), employing a combination of onsite thermal environment measurements and survey questionnaires. The research analyzed the Thermal Sensation Votes (TSV), thermal comfort evaluation indices (Universal Thermal Climate Index—UTCI, Physiological Equivalent Temperature—PET), and thermal environment parameters at representative sites. The findings revealed that firstly, outdoor thermal comfort and perception on cloudy winter days was minimally influenced by the microclimate. In contrast, on rainy winter days, it was significantly impacted by black globe temperature and wind speed. Secondly, the correlation between PET and Mean Thermal Sensation Vote (MTSV) was found to be higher than that between UTCI and MTSV, indicating that PET might be more aligned with the local climate and pedestrian activities. Lastly, the study determined the neutral PET range for different weather conditions in the area and compared it with existing research to identify discrepancies. This paper offers a reference for the neutral thermal comfort range in pedestrian spaces in regions with hot summers and cold winters under humid and cold winter climates, providing theoretical support for urban planning and design, with an emphasis on the results being presented in the past tense to reflect completed experiments.

Keywords: Microclimate · Outdoor thermal comfort · Mountain city · Walking space

1 Introduction

Against the backdrop of global warming, various extreme weather and climate events around the world have become more frequent, with their intensity, duration, and impact range all significantly increasing [1, 2]. However, people's willingness to engage in outdoor activities continues to rise. Outdoor activities are beneficial for increasing physical

© The Author(s) 2025

B.-J. He et al. (Eds.): UCSUD 2023, LNCE 559, pp. 551–562, 2025.

https://doi.org/10.1007/978-981-97-8401-1_39

activity, promoting interpersonal interactions, enhancing life satisfaction, and improving physical health [3]. Urban pedestrian spaces, due to their necessity, have always been a focus as the most commonly used spaces for outdoor activities [4, 5]. Yet, the increased frequency of extreme weather events due to climate warming can negatively impact people's outdoor activities by dampening their willingness to be active. Both extreme heat and extreme cold can lead to cardiovascular diseases and, in severe cases, threaten the health or even the lives of organisms, particularly humans. Therefore, thermal comfort plays an unparalleled role in assessing the quality of outdoor environments [6].

The attention of scholars to the microclimate of urban streets and their thermal comfort is gradually increasing, especially in terms of creating livable urban environments and improving local microclimates, which are receiving significant focus. Through interdisciplinary integration, mastering the correct principles of technical means and operational skills provides a reference for guiding climate-adaptive planning and design, conducive to forming design strategies and guidelines. However, thermal comfort indices based on human energy balance cannot fully reflect the complex ways people perceive their environment, change behavior, or gradually adjust their expectations to adapt to it. It requires integrating local comfort perceptions and adjusting the index ranges corresponding to thermal comfort. Yet, the current outdoor thermal comfort indices face great limitations due to regional climate and seasonal adaptability, leading to inconsistent neutral ranges for outdoor thermal comfort.

There are over 165 outdoor thermal comfort evaluation indices, with the most commonly used ones in outdoor thermal comfort research being UTCI and PET [7, 8]. However, due to individuals' adaptation to regional climates and seasons, there is significant variation in the evaluation results between these two indices [9, 10]. There is no definitive conclusion on which index provides a more accurate evaluation. Even when the same index is used for evaluation in the same region, inconsistencies in outdoor thermal comfort ranges may arise due to seasonal variations [11]. These limitations significantly constrain outdoor thermal comfort evaluation. Moreover, research on outdoor thermal comfort in hot summers and cold winters regions typically focuses on clear or cloudy conditions during summer or winter seasons [12, 13], neglecting the importance of the winter thermal environment in these regions, characterized by cold and humid conditions. Furthermore, most studies target pedestrians with light activity levels (1.1–1.9 Met) [7]. However, for outdoor environments in mountainous cities, where people typically engage in uphill walking as part of their daily activities due to the unique topographical features, the metabolic rate is higher, typically around 3.1 Met [14]. This discrepancy in metabolic rates may result in deviations in the outdoor thermal comfort evaluation indices for pedestrian spaces in mountainous cities. If standard range values are still applied, it may lead to misunderstandings in outdoor thermal comfort assessment in these regions and potentially misguide urban planning and design strategies for thermal environment optimization.

Therefore, the purpose of this study is to develop outdoor thermal comfort evaluation indices more suitable for winter conditions in mountainous cities and to determine the local thermal comfort neutral range. It focuses on the typical pedestrian spaces in the Yuzhong District of Chongqing, analyzing the spatial and temporal differences between microclimate and thermal comfort under the outdoor microclimate measured and questionnaire surveyed during the cold and humid winter weather. It clarifies the correlation between thermal perception, microclimate and thermal comfort. Through linear regression analysis of the correlation coefficients between UTCI and PET and MTSV, suitable outdoor thermal comfort evaluation indices for the local context are identified, and the neutral range of outdoor thermal comfort in pedestrian spaces in mountainous cities during cold and humid conditions is calculated. This study aims to provide theoretical basis for improving outdoor thermal comfort indices in mountainous cities and to offer technical support for strategies aimed at enhancing outdoor thermal environments in mountainous city.

2 Method

2.1 Study Sites

Chongqing, located in the southwestern part of China, features a subtropical monsoon humid climate, classifying it within the hot summers and cold winters region. The city's average annual temperature ranges between 17.5 to 20.0 °C, with the coldest month averaging temperatures of 4.0 to 8.0 °C. The average humidity often exceeds 70.0%, and over the past decade, the number of days with precipitation has reached more than 200 days a year, making it one of China's high-humidity areas. Even during the cold winter months (November to January of the following year), the number of days with precipitation can reach up to 19 days. In December 2020, the highest humidity reached 95.0%. Therefore, as one of the typical regions with hot summers and cold winters, the impact of Chongqing's cold and humid winter climate conditions on the urban outdoor thermal environment and human thermal comfort cannot be overlooked.

This study selected two pedestrian spaces with mountainous characteristics in Chongqing for its experimental sites: the First Mountain City Trail (Jianxing Ramp—JXR) and the Third Mountain City Trail (Shancheng Lane—SCL) (see Fig. 1). These trails integrate green corridors and urban balconies, serving as crucial pedestrian stairways connecting the upper and lower parts of Chongqing's main city, aiding in alleviating the inconvenience of vehicular traffic between these areas. Moreover, the buildings along these two streets are mostly traditional Bayu residences, showcasing typical mountain city spaces and traditional Bayu architectural styles, representing a microcosm of Chongqing's historical and cultural heritage. The primary users of these spaces are residents, with a small number of tourists also visiting.

2.2 Field Measurements

The field measurements were conducted on a rainy winter day and a cloudy day, specifically on January 10, 2021 (rainy day), and December 29, 2021 (cloudy day), from 8:00 to 17:00. The microclimate parameters measured included air temperature (Ta), relative humidity (RH), air velocity (va), and black globe temperature (Tg). These parameters are commonly used to analyze outdoor thermal environments and outdoor thermal comfort. The measurements were taken at 5-min intervals, with the average value for

Fig. 1. Study sites and measurement points in the sites

each hour being used for analysis. The sensors for measuring these parameters were positioned approximately 1.1 m above the ground, corresponding to the location of the human body's core temperature. Additionally, these devices were calibrated before the measurements to comply with the ISO 7726 standard [15].

Spatial heterogeneity can significantly impact the outdoor thermal environment [16]. Therefore, considering the high degree of heterogeneity displayed by different interfaces (buildings, mountains, water bodies, etc.), building heights, and vegetation coverage on either side of the pedestrian spaces, this study arranged 5 and 6 measurement points in characteristic locations along SCL and JXR, respectively, with the locations of these points shown in Fig. 1. These two streets differ in building heights, street orientation, and vegetation. The pedestrian space of SCL is very narrow, ranging from 2 to 5 m in width, while JXR is about 10 m wide [15]. The street direction of JXR runs from north to south, while that of SCL is northwest. Moreover, SCL is close to the Yangtze River, with one side of the street adjacent to the river and the other side against a mountain or buildings. The pedestrian spaces on either side of JXR are primarily composed of traditional commerce and residences.

2.3 Questionnaire Survey

During the measurements of the outdoor thermal environment in pedestrian spaces, a questionnaire survey was conducted simultaneously. Randomly selected respondents were asked to complete the questionnaire after staying at each measurement point for 3 to 5 min. The questionnaire consisted of two parts: respondent basic information and their thermal perception votes. The basic information section included gender, type of respondent (permanent residents, visitors from other places, etc.), age group, weight range, clothing condition, outdoor stay duration, and activity state before the survey. The thermal perception voting section involved an overall evaluation of the outdoor thermal environment and an assessment of individual thermal environment factors (air temperature, humidity, wind speed, and sunlight). The overall thermal sensation evaluation in the questionnaire was based on the 7-point thermal sensation vote (TSV) according to the ASHRAE 55–2013 standard, with the overall thermal comfort evaluation set according to a 5-point thermal comfort vote (TCV) [17]. The acceptability levels of microclimate factors (sunlight, temperature, humidity, and wind speed) were assessed using a 4-point voting index [9].

2.4 Outdoor Thermal Comfort Indices

This study employs PET and UTCI as indices for evaluating thermal comfort. It combines actual measurement data and questionnaire surveys to analyze which index is more suitable for evaluating the cold and humid winter conditions of mountain cities. During outdoor testing, the mean radiant temperature (MRT) can be approximately calculated from the black globe temperature (Tg) and air velocity (va) using a formula (see formula 1) [18]. PET and UTCI are calculated using the Rhino & Grasshopper platform, where the input microclimate parameters are obtained from actual measurements. In terms of individual human factors, except for the activity level, which is set based on the climbing slope (5°) value (3.1 Met), all other parameters are set according to the software's built-in winter settings. Additionally, the wind speed required for UTCI calculations is at the height of 10 m/s. This study approximates the calculation based on the formula provided by Bröde, P. et al. [19].

$$\text{MRT} = \left[\left(T_g + 273.15 \right)^4 + \frac{1.1 * 10^8 * v_a^{0.6}}{\varepsilon * D^{0.4}} \left(T_g - T_a \right) \right]^{0.25} - 273.15 \qquad (1)$$

ε: Emissivity of black bulb thermometer; D: Diameter of black bulb thermometer, mm; Diameter of black bulb in this study is 150 mm.

3 Results and Discussion

3.1 The Results of the Questionnaire Survey

The survey was conducted over two days, with 430 respondents interviewed. Almost all respondents were residents of Chongqing, accustomed to the local climate, and thus capable of accurately and objectively evaluating the thermal environment during the field measurements. The gender distribution among respondents was nearly equal, with a male to female ratio close to 1:1. The majority of respondents were aged between 18 and 40 years, with weights mostly ranging from 40 to 70 kg.

Based on the statistics related to different weather conditions, the outdoor activities of the respondents were analyzed (including outdoor stay duration, types of activities, and whether they had been in an air-conditioned room within 15 min prior to completing the questionnaire). The results revealed that respondents generally spent a long time outdoors. On cloudy winter days, over 45% of respondents stayed outdoors for 3–4 h, and even on rainy winter days, around 35% stayed outdoors for the same duration. Regarding the type of activities, walking was the most frequently mentioned activity. There was a higher proportion of people standing during the rainy days compared to cloudy days.

Nonetheless, over 50% of respondents reported walking as their activity on rainy winter days. In terms of thermal experience, the vast majority of respondents had not been in an air-conditioned room in the 15 min before completing the questionnaire.

However, the proportion of respondents who had been in such a room was higher on a cloudy day than on a rainy day.

The study analyzed the thermal sensation and thermal comfort voting results of respondents under different weather conditions, yielding the following findings. On rainy winter days, respondents tended to feel cold, with 47% feeling cold and 16.5% feeling cool. Only 14% of respondents felt neutral. On cloudy winter days, however, 38.7% of respondents felt neutral, and 9.5% felt moderately warm or warmer. Additionally, over 60% of respondents voted the thermal environment on rainy winter days as uncomfortable or slightly uncomfortable. In contrast, on cloudy winter days, more than 70% of respondents indicated they felt neutral, comfortable, or slightly comfortable.

The results regarding the acceptance of microclimatic elements by respondents in different winter weather conditions show clear differences in the acceptability of sunlight, temperature, humidity, and wind speed under different weather conditions. On a rainy day compared to a cloudy day, the acceptance of microclimatic elements was significantly lower. The largest gap was in the acceptance of sunlight, with only 37.5% of respondents on rainy days indicating they slightly accept or accept it. In comparison, more than 59.5% of respondents on cloudy days found it acceptable. The acceptance of temperature also showed that the rainy day had a lower acceptability compared to the cloudy day. On cloudy days, 59.1% of respondents indicated that they found the temperature acceptable, whereas on rainy days, only 39.5% of respondents reported finding it somewhat acceptable.

In summary, there are clear differences in outdoor activity preferences, thermal sensations, and thermal comfort evaluations among respondents under different weather conditions. These differences provide an important reference for environmental regulation and urban planning.

3.2   Heterogeneity of Microclimate and Thermal Comfort

This study conducted measurements and analysis of microclimatic elements (Ta, RH, va and Tg) in pedestrian spaces of a mountain city during the winter rainy and cloudy days, revealing the heterogeneity of its microclimatic elements and thermal comfort. The results showed that there were certain differences in microclimatic elements and thermal comfort on rainy and cloudy winter days, with RH showing the largest variance, followed by PET. Other elements and UTCI also demonstrated some degree of diversity. On the one hand, these differences are reflected in the spatial morphology, vegetation cover, and other distinctive features of the measurement points [14]. On the other hand, the variations are also evident across different times and weather conditions.

Table 1 presents the statistical results of microclimatic elements and thermal comfort evaluations, showing significant differences in the results of each microclimatic element and thermal comfort evaluation under different weather conditions. For example, on a rainy day, the average Ta was only 4.8 °C, with a standard deviation (SD) of 0.4; while on a cloudy day, the average Ta was 9.1 °C, with an SD reaching 1.1. Similarly, the Tg showed similar results. However, for va, the wind was stronger on a rainy day than on a cloudy day, with a larger SD. The maximum va on the rainy day could reach 2.0 m/s, while on the cloudy day, it was 1.3m/s. On a rainy day, the RH ranged from 69.4% to 99.9%, with an average RH of 87.3%; on a cloudy day, the RH ranged from 55.8% to

78.5%, with an average of 66.1%, more than 20% lower than on the rainy day. However, the SD of RH on both rainy and cloudy days, although relatively large, was close, at 7.6 and 6.8, respectively.

Regarding the comparison of thermal comfort indices under different weather conditions, it was found that the averages of PET and UTCI were significantly different. On the rainy day, the weather was coldest, with an average PET of only 2.6 °C and a minimum value of -1.9 °C. On a cloudy day, the average PET was 7.4 °C, nearly three times higher than on a rainy day. Similarly, the average UTCI on a cloudy day (10.6 °C) was more than twice that on a rainy day (5.2 °C), indicating significant variability in thermal comfort indices across different weather conditions.

Table 1. The statistics of microclimate parameters and thermal comfort indices

		Ta (°C)	RH(%)	Tg (°C)	va (m/s)	PET (°C)	UTCI (°C)
Rainy	Mean	4.8	87.3	4.8	0.5	2.6	5.2
	Max	6.3	99.9	6.0	2.0	6.3	7.2
	Min	4.0	69.4	4.1	0.0	-1.9	-1.2
	Mean ± SD	0.4	7.6	0.5	0.5	2.5	1.8
Cloudy	Mean	9.1	66.1	10.4	0.6	7.4	10.6
	Max	10.6	78.5	14.8	1.3	11.5	14.5
	Min	7.0	55.8	7.1	0.1	2.5	5.9
	Mean ± SD	1.1	6.8	1.8	0.3	2.0	1.8

3.3 Correlation Between Thermal Perception and Microclimate and Thermal Comfort

A single-sample Kolmogorov-Smirnov (K-S) test was performed on all variables to check for the normality of the samples. The results showed that the asymptotic significance (two-tailed) $p < 0.05$ for all variables, indicating that the samples of all variables are not normally distributed. Therefore, Spearman's correlation was used to analyze the mechanisms of how thermal perception is influenced by microclimate and thermal comfort. The results are presented in Fig. 2, where '*' indicates $p < 0.1$, '**' indicates $p < 0.05$, and '***' indicates $p < 0.01$.

Overall, the correlation between thermal perception and microclimate and thermal comfort on a rainy day is stronger. The specific findings are as follows: (1) On a rainy day, the effects of Tg and va on TSV and TCV are significant, with Tg having a positive impact on both TSV and TCV, and va having a negative impact. On a cloudy day, however, all microclimate parameters have no significant impact on TSV and TCV. This suggests that wind protection in urban design needs to be emphasized for rainy winter days. (2) On rainy days, both PET and UTCI show a positive correlation with TSV and TCV, indicating that these indices can accurately assess thermal perception to some extent during such weather. On a cloudy day, the impact of PET and UTCI on TSV and TCV is almost negligible, with only UTCI showing a positive correlation with TSV at $p < 0.1$. The result reveals that the applicability of PET and UTCI varies under different weather conditions, and even within the same weather conditions, their applicability can differ.

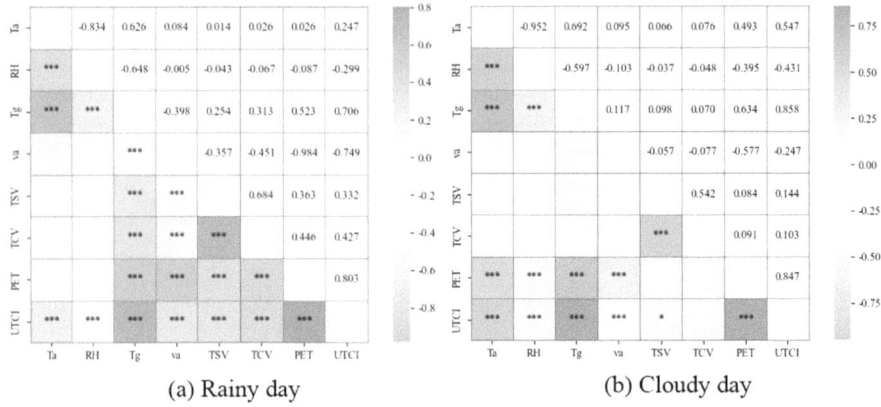

(a) Rainy day   (b) Cloudy day

Fig. 2. Correlation of thermal perception with microclimate and thermal comfort in different winter weather

3.4 Neutral PET Range

To further understand the applicability of different evaluation indices to human thermal sensation, this study used the temperature frequency method (Bin method) [20] to group PET and UTCI values from questionnaires under different weather conditions by intervals of 1 °C. It then calculated the average values of PET, UTCI, and thermal sensation votes (MTSV) for each interval, as well as the number of cases per interval. Using Python software, the study conducted statistical analysis on PET and UTCI against TSV, resulting in regression graphs of PET, UTCI, and MTSV under different weather conditions (see Fig. 3).

The analysis revealed the distinct applicability of thermal comfort indices under different weather conditions. On rainy winter days, the relationship between PET and MTSV was more pronounced, with a determination coefficient of 0.682. On cloudy winter days, the relationship between UTCI and MTSV was closer, with a determination coefficient of 0.547. Furthermore, on a rainy day, the slope of the regression equation between PET and MTSV was 0.162, which was higher than the slope on a cloudy day (0.085). This indicates that PET more influences MTSV on a rainy day than on a cloudy day. Combining the results from rainy and cloudy days, PET demonstrated stronger applicability than UTCI.

These findings suggest that for the cold and humid winter climate of the Chongqing area, using PET as an outdoor thermal comfort evaluation index is more in line with the local climate and pedestrian activities.

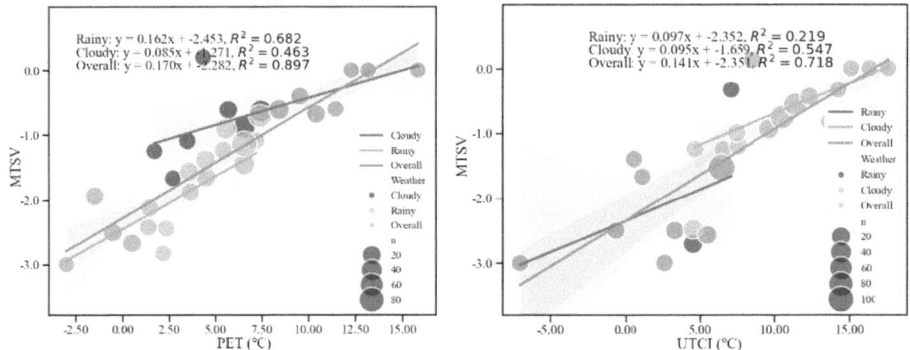

Fig. 3. The outdoor thermal perception benchmarks during the winter (left) based on the relationship between MTSV and PET, (right) based on the relationship between MTSV and UTCI

The findings suggest that using PET to predict thermal sensation is more accurate for this study. The study calculated the PET values corresponding to an MTSV between -0.5 and 0.5, which is defined as the Neutral PET Range (NPETR), to determine the neutral range of human thermal perception using PET. The NPETR varied significantly under different weather conditions (see Table 2). On the cloudy day, the neutral range was broader, from 9 to 21 °C, whereas on the rainy day, the NPETR was from 12 to 18 °C. Combining the results from rainy and cloudy days, the overall NPETR was determined to be from 10 to 16 °C.

Comparing this with the NPETR of other regions yields several insights (see Table 2). First, there is a considerable gap between the NPETR of different climatic zones and the initial NPETR. For instance, Chen, Hong et al.'s study on outdoor thermal comfort in Xi'an identified an NPETR of 7 to 16 °C [21], significantly lower than 18 to 23 °C. Secondly, even within the same climate zone, the NPETR can differ between cities. The results of this study are closer to He, Gao et al.'s findings on NPETR in Zhejiang [10] but diverge significantly from those for Shanghai [22], which also falls within the Cfa climate zone.

Overall, the neutral range for thermal comfort in outdoor studies varies. This variation can be attributed to differences in climatic zones, as well as the sample size, seasons, and weather conditions of the studies, which all influence the outcomes. Therefore, future research will need to include larger sample sizes to substantiate these findings further.

Table 2. Neutral PET range in different outdoor thermal comfort studies

Source	Climate zone	Season/ Weather	Site	NPETR(°C)	R^2
This study	Cfa	Winter/rainy	Chongqing, China	12–18	0.682
		Winter/cloudy		9–21	0.463
		Winter/overall		10–16	0.897
Matzarakis and Mayer(Initial) [23]	Cfb	Summer	Middle/western Europe	18–23	-
Yahia and Johansson [24]	BSk	Winter	Damascus, Syria	20–29	0.604
Chen, Wen et al. [22]	Cfa	Winter	Shanghai, China	15–29	0.74
Zhang, Wei et al. [25]	Cwa	Winter	Chengdu, China	11–21	0.356
Chen, Hong et al. [21]	Cwa to BSk	Winter	Xi'an, China	7–16	0.919
He, Gao et al. [10]	Cfa	Winter	Zhejiang, China	11–18	0.944

4 Conclusion

With the rapid changes in global climate due to significant greenhouse gas emissions, urban climate issues are becoming increasingly severe. Addressing how to improve urban climates and create comfortable pedestrian environments to meet people's aspirations for a healthy and livable living environment is an urgent issue. For regions with hot summers and cold winters, while the outdoor thermal environment in summer is important, thermal comfort outdoors during the cold, humid winter is equally critical. This study measured the microclimatic parameters of pedestrian spaces in a typical mountain city (Chongqing) during rainy and cloudy days in winter. At the same time, respondents were surveyed on their thermal perception, and the PET and UTCI thermal comfort indices were used to evaluate the outdoor thermal environment of the mountain city, leading to the following conclusions:

- The outdoor microclimate in winter is influenced by spatial heterogeneity, showing variations, especially in terms of humidity differences.
- On rainy winter days, thermal comfort is highly sensitive to black globe temperature and wind speed, which are the main environmental parameters affecting thermal comfort. In the future design of pedestrian spaces in mountainous urban areas, measures should be taken to prevent wind and minimize cold radiation. Additionally, it is advisable to increase the radiation heat sources appropriately, such as outdoor

vertical heaters, to improve outdoor thermal comfort during the winter's humid and cold seasons.

- Under the cold and humid conditions of Chongqing's winter, using PET as an outdoor thermal comfort index may be more aligned with the local climate and human activities. When assessing the neutral temperature for outdoor thermal comfort during winter, the NPETR is between 10 and 16 °C.

This study provides a foundation for research on outdoor thermal comfort during the cold and humid winter in regions with hot summers and cold winters. It also offers experience and reference for comfortable design in the construction of future urban environments that are livable and healthy.

Acknowledgments. The work was supported by the Graduate Scientific Research and Innovation Foundation of Chongqing, China (Grant No. CYB22035).

References

1. Meehl, G.A., Tebaldi, C.: More intense, more frequent, and longer lasting heat waves in the 21st century. Science **305**(5686), 994–997 (2004)
2. Mora, C., et al.: Global risk of deadly heat. Nat. Clim. Change **7**(7), 501 (2017)
3. Fuegen, K., Breitenbecher, K.H.: Walking and being outdoors in nature increase positive affect and energy. Ecopsychol. **10**(1), 14–25 (2018)
4. Sharmin, T., Steemers, K., Humphreys, M.: Outdoor thermal comfort and summer PET range: a field study in tropical city Dhaka. Energy and Buildings **198**, 149–159 (2019)
5. Karimi, A., et al.: Microclimatic analysis of outdoor thermal comfort of high-rise buildings with different configurations in Tehran: Insights from field surveys and thermal comfort indices. Build. Environ. **240** (2023)
6. Li, J., Liu, N.: The perception, optimization strategies and prospects of outdoor thermal comfort in China: a review. Build. Environ. **170**, 106614 (2020)
7. de Freitas, C.R., Grigorieva, E.A.: A comparison and appraisal of a comprehensive range of human thermal climate indices. Int. J. Biometeorol. **61**(3), 487–512 (2016). https://doi.org/10.1007/s00484-016-1228-6
8. Cheung, P.K., Jim, C.Y.: Determination and application of outdoor thermal benchmarks. Build. Environ. **123**, 333–350 (2017)
9. Wei, D., et al.: Variations in outdoor thermal comfort in an urban park in the hot-summer and cold-winter region of China. Sustain. Cities Soc. **77** (2022)
10. He, X., et al.: Study on outdoor thermal comfort of factory areas during winter in hot summer and cold winter zone of China. Build. Environ. **228** (2023)
11. Zhang, H., et al., Spatial differences in thermal comfort in summer in coastal areas: a study on Dalian, China. Front. Pub. Health **10** (2022)
12. Gu, H., et al.: Research on outdoor thermal comfort of children's activity space in high-density urban residential areas of chongqing in summer. Atmosphere **13**(12) (2022)
13. Lian, Z., Liu, B., Brown, R.D.: Exploring the suitable assessment method and best performance of human energy budget models for outdoor thermal comfort in hot and humid climate area. Sustain. Cities Soc. **63**,(2020)
14. Xiong, K., Yang, Z., He, B.-J.: Spatiotemporal heterogeneity of street thermal environments and development of an optimised method to improve field measurement accuracy. Urban Climate **42**, 101121 (2022)

15. Xiong, K., He, B.J.: Wintertime outdoor thermal sensations and comfort in cold-humid environments of Chongqing China. Sustain. Cities Soc. **87**,(2022)
16. Imam Syafii, N., et al.: Thermal environment assessment around bodies of water in urban canyons: a scale model study. Sustain. Cities Soc. **34**, 79–89 (2017)
17. Li, J., et al.: Exploration of applicability of UTCI and thermally comfortable sun and wind conditions outdoors in a subtropical city of Hong Kong. Sustain. Cities and Soc. **52**,(2020)
18. Thorsson, S., et al.: Different methods for estimating the mean radiant temperature in an outdoor urban setting. Int. J. Climatol. **27**(14), 1983–1993 (2007)
19. Bröde, P., et al.: Deriving the operational procedure for the Universal Thermal Climate Index (UTCI). Int. J. Biometeorol. **56**(3), 481–494 (2012)
20. Paxton, R.J., et al.: Associations of sociodemographic and community environmental variables to use of public parks and trails for physical activity. J. Inst. Health Educ. **43**(4), 108–116 (2005)
21. Chen, H., et al., *Effects of Acoustic Perception on Outdoor Thermal Comfort in Campus Open Spaces in China's Cold Region.* Buildings, 2022. **12**(10)
22. Chen, L., et al.: Studies of thermal comfort and space use in an urban park square in cool and cold seasons in Shanghai. Build. Environ. **94**, 644–653 (2015)
23. Matzarakis, A., Mayer, H.: Another kind of environmental stress: thermal stress. WHO Collaborating Centre for Air Quality Management and Air Pollution Control **18**, 7–10 (1996)
24. Yahia, M.W., Johansson, E.: Evaluating the behaviour of different thermal indices by investigating various outdoor urban environments in the hot dry city of Damascus. Syria. Int. J. Biometeorol. **57**(4), 615–630 (2013)
25. Zhang, L., et al.: Outdoor thermal comfort of urban park-A case study. Sustainability **12**(5), 1961 (2020)

Urban Environmental Quality Assessment and Improvement

Analyzing the Impact of Tall Building Geometries on Wind Environment in a Hypothetical Urban Context: A Typological and Parametric Study

Yihan Wu[1,2,3], Weifeng Li[1,2], Ningyi Zeng[1,2], and Xiaoxia Bai[1,2(✉)]

[1] School of Architecture and Urban Planning, Huazhong University of Science and Technology, Wuhan 430074, China
baixiaoxia@hust.edu.cn
[2] Hubei Engineering and Technology Research Center of Urbanization, Wuhan 430074, China
[3] Key Laboratory of Ecology and Energy Saving Study of Dense Habitat, Ministry of Education, Shanghai 200092, China

Abstract. Infill development has become a popular strategy for revitalizing old urban districts. However, the construction of new high-rise buildings can cause nuisance issues that are of great concern to indigenous residents. Our study seeks to address three key research questions by examining the impacts of 34 high-rise building typologies within a hypothetical urban context: 1) To what spatial extent does the construction of new high-rise buildings significantly affect their surroundings? 2) How do different tall building typologies modify their surrounding buildings' outdoor and indoor ventilation potentials? 3) What are the quantitative relations between building geometrical variables and wind performance indicators? The findings reveal that the correlation between building shape and wind behaviors is strongest within a radius of 0.7~1.3H of the target building. Moreover, the study observes that building width, building projected width, frontal area, and shape coefficient can substantially affect indoor and outdoor ventilation potentials of the surrounding spaces, but the influence of these parameters significantly varies depending on the incident wind direction. Certain building typologies can increase the outdoor ventilation potential of the neighborhood by 10–20%. Furthermore, the study employs an elastic net model to investigate the multivariate relations between building geometrical variables and wind performance indicators. The results indicate that the elastic net model outperforms ridge and lasso regression models in predicting wind performance indicators. These findings can enhance our understanding of how to design a tall building within an existing urban configuration.

Keywords: infill development · urban context · tall building · typology · parametric analysis

B.-J. He et al. (Eds.): UCSUD 2023, LNCE 559, pp. 565–577, 2025.
https://doi.org/10.1007/978-981-97-8401-1_40

1 Introduction

Urban infill development has gained significant momentum globally, particularly in cities characterized by aging infrastructure or space constraints due to rapid population growth [1]. Yet, the process of urban infill construction frequently involves the addition of tall or supertall buildings that permanently alter the wind environment in their vicinity, resulting in various positive or negative effects [2]. Beginning from the 1960s, many wind tunnel experiments have been carried out to investigate the effects of an individual building on its surrounding wind environment [3–5]. It was observed that the three-dimensional flow field around a building was complex and dynamic, characterized by the outward movement of incoming mean flows towards all front edges of the building, which leads to the development of cavity zones near the roof, leeward face, and side walls of the building [4, 6].

Following these qualitative observations, parametrization studies were developed to relate observed flow behaviors to building morphological quantities. For example, [7] reported that pedestrian-level wind (PLW) speed around a supertall building with a height of 200m and 400m was about 1.5 and 2.1 times higher than the incoming wind speed, respectively. [5] found that increased building width resulted in an elongated low-speed zone near the building's leeward face because the sheltering effect often became more remarkable. Although the acceleration of corner streams could only be slightly affected by increased building width, [8, 9] suggested that building width should be kept to less than 60 m to avoid wind blockage that might shelter a large downstream area. Moreover, several studies demonstrated that the mean wind speed surrounding the building remained relatively unaffected by minor changes in building depth [10, 11]. However, in cases where the increase in building depth was significant, the flow would re-attach to the side walls, resulting in negligible speed-up of corner streams [12]. [13, 14] found that the maximum value of the wind amplification factor increases with the building height-to-width ratio and building influence scale (i.e., $S = \left(B_L B_S^2\right)^{1/3}$, where B_L and B_S denote the larger and smaller values of building's windward façade dimensions. According to [10], the peak value of the wind amplification factor K_{corner} around building corners occurred at about $Y/S = 0.4$, where Y is the distance from the side wall of the building. And the peak value of the wind amplification factor in front of a building can be estimated using the equation $K_{front} = 0.3S^{1/2}$.

Moreover, tools for investigating urban wind environments have evolved significantly in recent decades. In addition to sensors and techniques such as Irwin probes, infrared thermography, and particle image velocimetry, which are particularly suited to detecting the dynamics of PLW behaviors [6], various computational fluid dynamics (CFD) models have been developed to uncover the underlying principles of complex urban wind phenomena. These models include Reynolds-averaged Navier-Stokes (RANS), large-eddy simulation (LES), and Direct Numerical Simulation (DNS) models, as described by [15].

Despite the significant research advancements that have been achieved to comprehend the complex flow behaviors around isolated buildings, there exist several research gaps that necessitate further investigations. First, it is essential to establish whether the relationship between buildings and wind is scale-dependent, and to what extent newly constructed buildings impact the wind dynamics in their surroundings. Second, while many studies have investigated the impact of buildings on pedestrian wind environment, it is crucial to examine how different tall building typologies influence the outdoor ventilation efficiency and indoor ventilation capacity of the surrounding structures. Last, it is imperative to determine the quantitative and multivariate relationships between building geometry and wind performance indicators, such that the impacts of new building designs can be predicted. Following these questions, we conducted CFD simulations for various common high-rise building designs, and examined the univariate and multivariate relationships between building typological factors and wind performance indicators by employing correlational and regression analyses.

2 Data and Methodology

2.1 Tall Building Typologies and Settings of Urban Context

In order to provide comprehensive insights into the impact of different building typologies on the wind environment, and to generate guidelines for urban infill development practices, we have collated a broad spectrum of typical building typologies from the studies of [16–18], as presented in Fig. 1. We believe that these studies have proposed a representative set of mainstream typological design concepts during the past two decades, from which we can derive valuable knowledge applicable to real-life practices. Nonetheless, in contrast to previous studies that have focused on buildings of 200m or 400m height, we have chosen to limit the building height to 100m. This is due to the regulations that many countries have implemented to cap building height, which serve to mitigate the energy consumption of skyscrapers, protect historic skylines, and ensure building structure stability [19–21].

In this study, our focus is not solely on analyzing wind patterns around individual buildings, but on examining the impact of a new building on the surrounding urban environment, which is a crucial aspect of designing urban infill development projects. To this end, we have selected the Case D model, made available by the Architectural Institute of Japan (AIJ), as the basis for our parametrization study [22]. This model not only provides a practical framework for examining how a tall building affects the wind environment of an existing neighborhood, but also furnishes detailed wind tunnel experimental data, which allows us to verify the accuracy of our CFD simulations. Specifically, the Case D model comprises a centrally located high-rise building of 100m in height, surrounded by 83 simplified low-rise building blocks, and the wind tunnel test was conducted at a scale of 1/400. We utilize this experimental setup for all building typologies under consideration.

2.2 Selection of Building Geometrical Parameters and Wind Performance Indicators

As the present study focuses on the complex interplay between the geometry of a high-rise building and various wind performance indicators, a set of building geometrical parameters pertinent to the building's aerodynamic features are selected and defined in Table 1, developed based on existing literature. While some conventional geometrical parameters such as building width, building aspect ratio, and building frontal area

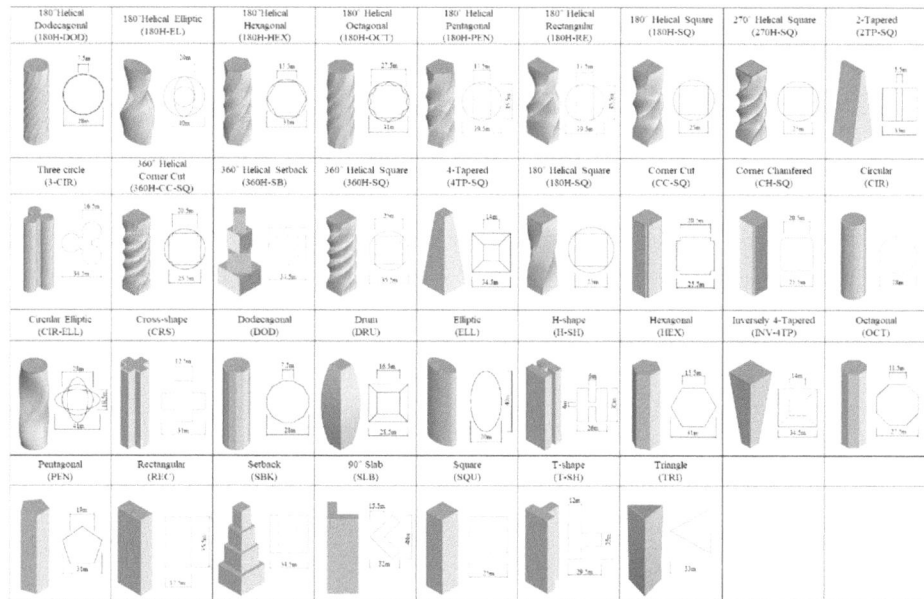

Fig. 1. Configurations of 34 tall building models (the heights of all buildings are 100m).

have been extensively studied, this research also investigates the impacts of a few new geometrical parameters, including the building's projected width at 1/3 and 2/3 building height, as previous studies assert that the projected building width at 0.3~0.6H are remarkably correlated with pedestrian-level high wind activities [7]. In addition, three compactness measurement variables, namely, φ_{ft}, φ_{sf}, φ_{vo}, are employed to quantify the degree to which a building typology resembles a circular cylindrical object, given the prior research indicating that the cylindrical structure can effectively diminish the high wind speed region near the building's base and generate a smaller wake area behind the building [7].

Two wind performance indicators are utilized, including indicators that evaluate the static surface pressure of buildings and air ventilation efficiency of the urban block. Specifically, the pressure coefficient (C_p) is adapted to measure low-rise buildings' indoor ventilation potentials surrounding the infilled tall building [23]. It can be obtained by:

$$C_p = \frac{P - P_{ref}}{\frac{1}{2}\rho U_{ref}^2} \tag{1}$$

where, P and P_{ref} represent the static pressure on the building surface and the static pressure at the reference point, respectively. ρ is the air density with a value of 1.225 kg/m^3 (at 101.325 kPa and 15 °C). The reference point is taken on the lateral boundary at the reference height and has the same distance from the inlet to the windward façade. And U_{ref} is the wind speed at the reference height. In this study, the mean C_p on the building façade is denoted as $\overline{C}_p$.

Moreover, we utilize the age of air to reflect the outdoor ventilation quality of the urban block, which is calculated by the following equation [24]:

$$\tau_p = \frac{C}{\dot{m}} \qquad (2)$$

The homogeneous emission source method is employed to calculate τ_p, meaning that the pollutant is assumed to release homogeneously with a rate of $\dot{m}$ (kg/m^{3}s) in the entire domain, where C denotes the local pollutant concentration (kg/m^3) of each cell within the domain. This research calculates the mean τ_p within the urban block, denoted as $\overline{\tau}_p$.

Table 1. Definitions of building geometrical parameters.

Geometrical Parameter	Definition
γ_{wd}	Building width
γ_{pw1}	The projected width at 1/3 building height
γ_{pw2}	The projected width at 2/3 building height
γ_{ar}	Building height to width ratio
γ_{is}	Building influence scale defined by Wilson: $S = \left(B_L B_S^2\right)^{1/3}$, where B_L, B_S refer to the larger and smaller values of building dimensions of the building's windward face
λ_{ft}	building footprint area
λ_{pj}	The building's projected area from the top view
λ_{fr}	The total areas of building facets facing the wind direction
λ_{ls}	The lateral surface area of the building, not including its top
λ_{sf}	The building's surface area
ω_{vo}	The volume of the building
ω_{sc}	The shape coefficient of the building, calculated by dividing building exterior area with building volume
φ_{ft}	An index that measures how the building's projected footprint resembles the footprint of its bounding cylinder
φ_{sf}	An index that measures the closeness of the building's lateral surface area to that of its bounding cylinder
φ_{vo}	An index that measures the closeness of the building's volume to that of its bounding cylinder

2.3 Machine Learning Models to Study the Building-Wind Relation

This study employs ridge, lasso, and elastic net regression models to fit the CFD simulation sampling dataset and compare their prediction errors to determine the best model for generalizing the "building-wind" relation. Ridge regression uses L2 regularization, lasso uses L1 regularization for covariate selection, and elastic net combines L1 and L2 regularization. Ridge regression minimizes a loss function with a shrinkage factor λ, while lasso limits the total absolute value of model coefficients. Elastic net's loss function includes parameters for balancing L1 and L2 regularization. Unlike ridge, lasso can force some coefficients to zero, excluding certain predictor variables. Whilst Elastic net offers a balanced trade-off between estimation bias and errors, outperforming lasso in predictive power while still enabling feature selection.

3 Parametric Studies of Scale-Dependent Building-Wind Relation

This study delves into the complex relationship between newly constructed buildings and wind behaviors in their vicinity in a hypothetical urban context, and examines the scale-dependent nature of this relationship. To this end, we established 27 buffer zones with radii ranging from 0.5H to 3.1H around a designed building (where H represents the height of the designed building), and calculated Pearson correlation coefficients between building geometrical factors and wind performance indicators to analyze the building-wind interaction as a function of spatial distance.

3.1 Effects of Tall Building Geometrical Parameters on Outdoor Ventilation

First, we explored the association between building geometrical parameters and outdoor ventilation efficiencies of the neighborhood. we analyzed their correlations across diverse spatial scales, by defining several "evaluation domains", as depicted in Fig. 2. Specifically, these 3D domains were delineated horizontally by varying buffering distances. This approach enabled us to uncover the extent to which a new building affects the air quality of its surrounding 3D space. The outcomes of our investigation are showcased in Fig. 3.

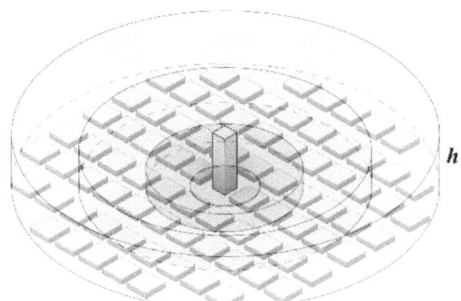

Fig. 2. 3D evaluation domains constructed to examine the scale-independent building-outdoor ventilation interactions (h represent 3D domains under building height).

The results indicate that the building had a significant impact on the air quality in its immediate surroundings (i.e., 0.7H~1.3H), as measured by $\bar{\tau}_p$. Furthermore, we

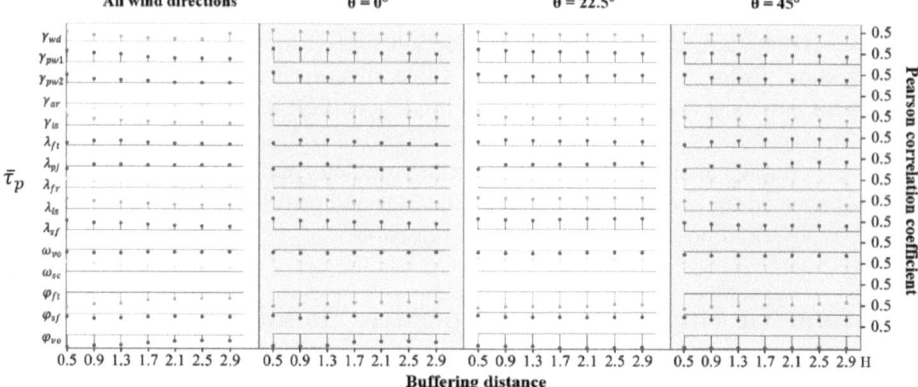

Fig. 3. Pearson correlation coefficients between building geometrical parameters and age of air under h at different buffering distances.

examined how incoming wind direction affected the air quality in the neighborhood. Our findings show that the relationship between building geometrical parameters and $\overline{\tau}_p$ was somewhat erratic when the wind approached perpendicularly to the building's front façade. In contrast, when oblique winds approached the experimental urban block, a more stabilized distance-based correlation pattern was observed.

Lastly, we tentatively explored the effects of individual building geometrical parameters on air quality. As presented in Fig. 3, several parameters including γ_{wd}, γ_{pw1}, γ_{pw2}, λ_{ls}, λ_{sf}, ω_{sc}, were positively and strongly associated with $\overline{\tau}_p$, whereas φ_{ft}, φ_{vo} were negatively and strongly correlated to $\overline{\tau}_p$. Notably, a few building geometrical parameters such as λ_{pj} were found to be highly nonlinearly correlated with $\overline{\tau}_p$ at various buffering distances. We suggest that building geometrical parameters such as γ_{wd}, ω_{sc}, φ_{ft}, φ_{vo}, etc. could be effectively employed to predict and evaluate the impact of a new building on the outdoor ventilation efficiency of its surrounding environment.

3.2 Effects of Tall Building Shape on Indoor Ventilation of Its Neighbors

The correlation coefficients between building geometric factors and $\overline{C}_p$ values at various buffering distances were plotted in Fig. 4. The results indicate that the majority of the building geometrical parameters were weakly or moderately correlated with $\overline{C}_p$ of the surrounding low-rise buildings. However, when the strength of the correlation was analyzed separately based on wind directions, strong correlations were observed between building geometrical parameters (e.g., λ_{fr}, λ_{ls}, ω_{sc}, φ_{ft}) and $\overline{C}_p$ values of low-rise buildings under oblique winds. In addition, the study found that the high-rise building had a more significant impact on the indoor ventilation potential of low-rise buildings in its immediate vicinity (within 2.1H) compared to those located beyond this range.

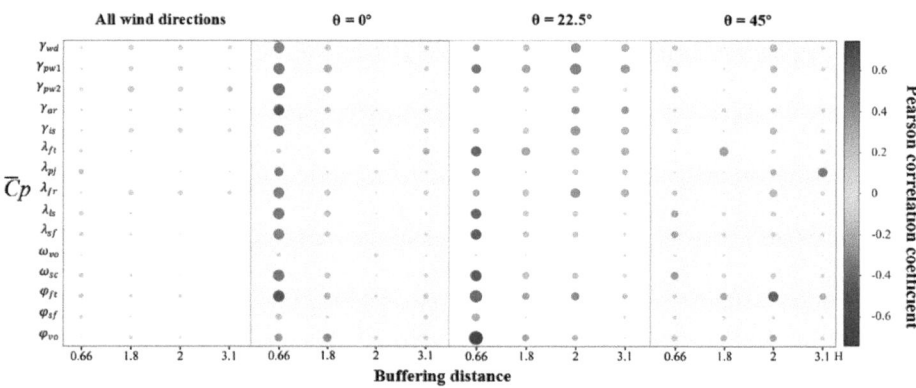

Fig. 4. Pearson correlation coefficients between building geometrical parameters and mean building pressure coefficient at different buffering distances.

3.3 Effects of Tall Building Typologies on Outdoor Ventilation

In parallel with the analysis in Sect. 3.1, we explored the effects of high-rise building typologies on outdoor ventilation efficiencies by examining the percentage changes of $\overline{\tau}_p$, which are illustrated in Fig. 5. The results show that the construction of high-rise buildings had led to a decrease in $\overline{\tau}_p$, thereby increasing the outdoor ventilation potential in the tall building's immediate vicinity. Specifically, typologies such as 180H-HEX, 360H-SQ, 360H-CC-SQ, CH-SQ, CIR, CIR-ELL dramatically reduced neighborhood $\overline{\tau}_p$ by 15~20%. These findings suggest that high-rise building typologies played a crucial role in shaping the outdoor ventilation potential of the surrounding areas.

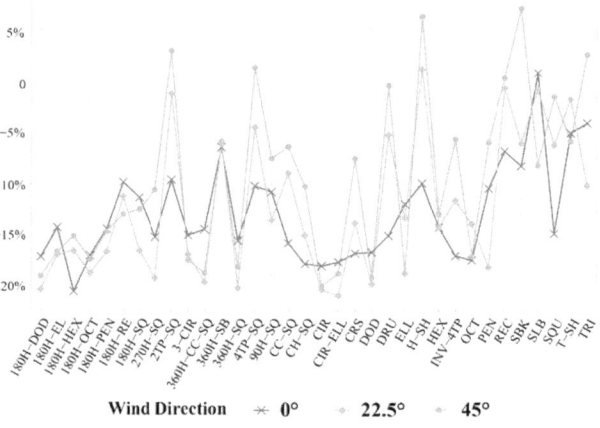

Fig. 5. Percentage changes of mean age of air when different building typologies are added to the urban block.

3.4 Effects of Building Typologies on Indoor Ventilation Potential of Its Neighbors

To measure the extent to which the newly built buildings influence the indoor ventilation capacity of their neighbors, the surface difference $\overline{C}_p$ of low-rise buildings within different buffering distances were plotted in Fig. 6. Figure 6 highlights that the 180H-HEX and CRS models notably improved the indoor ventilation potential of low-rise buildings under perpendicular winds. Under normal wind conditions, the indoor ventilation potential of low-rise buildings was reduced due to the blocking effect of the target building. Additionally, the surface difference $\overline{C}_p$ of low-rise buildings increased under oblique winds when 180H-OCT, 180H-SQ, 2TP-SQ,4TP-SQ, 90H-SQ, CC-SQ, CH-SQ, CIR-ELL, CRS, ELL, HEX, OCT, PEN, REC, SLB, SQU, T-SH were added to the site. However, the 270-SQ, 360H-SB, CIR, and SBK models reduced the surface difference $\overline{C}_p$ of the low-lying buildings.

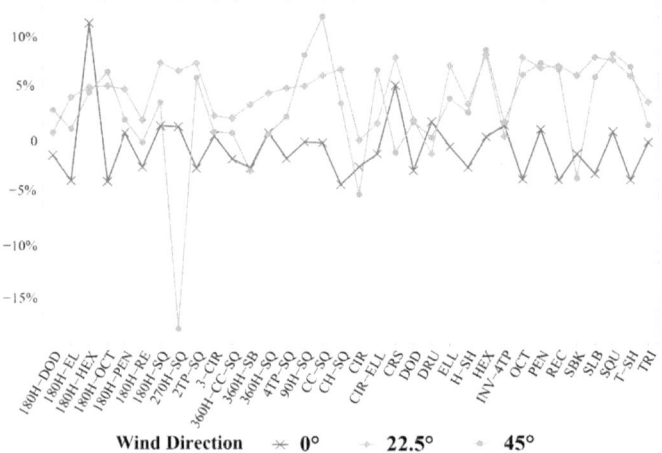

Fig. 6. Percentage changes of low-rise buildings' mean surface difference pressure coefficients when different building typologies are added to the urban block.

4   Multivariate Analysis by Ridge, Lasso and Elastic Net Models

In this section, we present an analysis of the synthetic effects of building geometrical parameters on wind behavior using ridge, lasso, and elastic net regression models. These models offer improved accuracy in coefficient estimates compared to multivariate linear regression models by mitigating multicollinearity effects through regularization.

[25] suggests that bootstrap can be used in conjunction with cross-validation to determine hyperparameters in ridge, lasso, and elastic net models. During the model training process, regression coefficients were estimated for each bootstrap sample by iterating over a set of hyperparameter values. Predictions on unchosen samples were then calculated and used to compute the RMSE (root mean square error) value for the

unchosen population. In the case of elastic net regression, a grid search strategy was employed to obtain the best-fit regression model.

Table 2 shows the mean RMSE, MAE (mean absolute error) and R^2 values of the models on the bootstrap resampling dataset when they were fitted to learn the building-wind relations. The results indicate that the elastic net model outperformed the ridge and lasso models in predicting most wind performance indicators. The relationships between building geometry and wind behavior are presented in Table 3, which demonstrates that wind direction variables significantly contributed to the explanation of the variances of $\overline{\tau}_p$ and $\overline{C}_p$. These empirical models can thus be further integrated into building design provisions to assist designers in formulating building geometries that facilitate better wind conditions.

Table 2. MAE, RMSE, R^2 of ridge, lasso, elastic net models.

Model	$\overline{\tau}_p$			$\overline{C}_p$		
	MAE	RMSE	R^2	MAE	RMSE	R^2
Ridge	0.007	0.009	0.724	0.009	0.011	0.533
Lasso	0.008	0.010	0.681	0.007	0.009	0.773
Elastic net	0.006	0.008	0.766	0.004	0.007	0.865

Table 3. Coefficient estimates of elastic net models of various ventilation indices.

	$\overline{\tau}_p$	$\overline{C}_p$
(Intercept)	0.285	-0.104
$cos\theta$	-	0.159
$sin\theta$	-0.027	0.079
γ_{wd}	-	-
γ_{pw1}	0.225	0.012
γ_{pw2}	-	0.041
λ_{ft}	1.023	-
ω_{vo}	-0.258	-
ω_{sc}	0.061	-
φ_{ft}	-0.016	-
φ_{sf}	-	-
φ_{vo}	-0.018	-

5   Conclusions

In this study, rather than focusing on the effects of tall building typology on pedestrian wind conditions, our primary interest lies in examining how building designs shape the quality of the three-dimensional wind environment in their surroundings. The results indicate that the range of 0.7 to 1.3 times the height of the tall buildings represents the critical zone where their geometric characteristics have the most notable impact on the surface pressures of adjacent low-rise buildings and the average age of air within the urban block. This observation leads to the recommendation that infill development analyses should explicitly consider buildings within a radius of 2H to better understand how new buildings affect wind behaviors in their vicinity. The construction of new high-rise buildings can enhance outdoor ventilation potential in their vicinity, reducing the mean age of air by 10–20% for all building typologies studied. This is primarily due to the tall building's windward façade capturing high winds, which are then introduced to the ground level. However, several models can decrease ventilation potential in their wake region, increasing the mean age of air by 20% when the 180H-DOD, 360H-CC-SQ, 360H-SQ, CIR, CIR-ELL, DOD, and PEN models are added to the site.

Buildings with helical and stacked structures can significantly increase the indoor ventilation potential of low-rise buildings under various wind directions. Besides, several building geometric parameters, including building width, building projected width, building surface area, building shape coefficient, are found to be positively and strongly associated with $\overline{\tau}_p$ (mean age of air). The findings suggest that strong correlations are observed between building geometrical parameters (e.g., building frontal area, building surface area, building shape coefficient) and $\overline{C}_p$ (surface pressure coefficient of low-rise buildings) under oblique winds. Furthermore, we find elastic net model can accurately fit multivariate building-wind relations. However, additional experiments may be necessary to determine the extent to which the predictive models derived from this study can be generalized to other types of buildings.

Acknowledgement. This work was supported by the open funding through Key Laboratory of Ecology and Energy Saving Study of Dense Habitat, Ministry of Education (No. 20230107), the project of the National Natural Science Foundation of China (No. 52378056, 52378019), the Key Technology Project of Central South Architectural Design Institute Co., Ltd (CSADI-2022-14, CSADI-2022-15).

References

1. Newton, P., Glackin, S.: Understanding infill: towards new policy and practice for urban regeneration in the established suburbs of Australia's cities. Urban Policy Res. **32**(2), 121–143 (2014)
2. Wu, Y., Zhan, Q., Quan, S.J., Fan, Y., Yang, Y.: A surrogate-assisted optimization framework for microclimate-sensitive urban design practice. Build. Environ. **195**, 107661 (2021)
3. Davenport, A.G.: Wind loading of structures. National Research Council of Canada, Division of Building Research, 1960, Technical Paper No, 88 (1960)
4. Peterka, J.A., Meroney, R.N., Kothari, K.M.: Wind flow patterns about buildings. J. Wind Eng. Ind. Aerodyn. **21**(1), 21–38 (1985)

5. Tsang, C.W., Kwok, K.C., Hitchcock, P.A.: Wind tunnel study of pedestrian level wind environment around tall buildings: Effects of building dimensions, separation and podium. Build. Environ. **49**, 167–181 (2012)

6. Blocken, B., Stathopoulos, T., Van Beeck, J.P.A.J.: Pedestrian-level wind conditions around buildings: review of wind-tunnel and CFD techniques and their accuracy for wind comfort assessment. Build. Environ. **100**, 50–81 (2016)

7. Xu, X., Yang, Q., Yoshida, A., Tamura, Y.: Characteristics of pedestrian-level wind around super-tall buildings with various configurations. J. Wind Eng. Ind. Aerodyn. **166**, 61–73 (2017)

8. Reiter, S.: Assessing wind comfort in urban planning. Environ. Plann. B. Plann. Des. **37**(5), 857–873 (2010)

9. You, W., Gao, Z., Chen, Z., Ding, W.: Improving residential wind environments by understanding the relationship between building arrangements and outdoor regional ventilation. Atmosphere **8**(6), 102 (2017)

10. Stathopoulos, T., Wu, H., Bédard, C.: Wind environment around buildings: a knowledge-based approach. J. Wind Eng. Ind. Aerodyn. **44**(1–3), 2377–2388 (1992)

11. Wu, Y., Zhan, Q., Quan, S.J.: A robust metamodel-based optimization design method for improving pedestrian wind comfort in an infill development project. Sustain. Cities Soc. **72**, 103018 (2021)

12. Oke, T.R., Mills, G., Christen, A., Voogt, J.A.: Urban Climates. Cambridge University Press (2017)

13. Kamei, I., Maruta, E.: Study on wind environmental problems caused around buildings in Japan. J. Wind Eng. Ind. Aerodyn. **4**(3–4), 307–331 (1979)

14. Wilson, D.J.: Airflow Around Buildings. ASHRAE Handbook of Fundamentals (1989)

15. Germano, M.: From RANS to DNS: towards a bridging model. In: Direct and Large-Eddy Simulation III, pp. 225–236. Springer, Dordrecht (1999)

16. Tanaka, H., Tamura, Y., Ohtake, K., Nakai, M., Kim, Y.C.: Experimental investigation of aerodynamic forces and wind pressures acting on tall buildings with various unconventional configurations. J. Wind Eng. Ind. Aerodyn. **107**, 179–191 (2012)

17. Tanaka, H., Tamura, Y., Ohtake, K., Nakai, M., Kim, Y.C., Bandi, E.K.: Aerodynamic and flow characteristics of tall buildings with various unconventional configurations. Int. J. High-Rise Build. **2**(3), 213–228 (2013)

18. Iqbal, Q.M.Z., Chan, A.L.S.: Pedestrian level wind environment assessment around group of high-rise cross-shaped buildings: effect of building shape, separation and orientation. Build. Environ. **101**, 45–63 (2016)

19. Bertaud, A., Brueckner, J.K.: Analyzing building-height restrictions: predicted impacts and welfare costs. Reg. Sci. Urban Econ. **35**(2), 109–125 (2005)

20. Brueckner, J.K., Fu, S., Gu, Y., Zhang, J.: Measuring the stringency of land use regulation: the case of China's building height limits. Rev. Econ. Stat. **99**(4), 663–677 (2017)

21. Mills, P.: The limited city: [building height regulations in the City of Melbourne, 1890–1955] (Doctoral dissertation, Monash University) (2022)

22. Mochida, A., et al.: AIJ Benchmarks for Validation of CFD Simulations Applied to Pedestrian Wind Environment around Buildings. Architectural Institute of Japan, Tokyo, Japan (2016)

23. Costola, D., Blocken, B., Hensen, J.L.M.: Overview of pressure coefficient data in building energy simulation and airflow network programs. Build. Environ. **44**(10), 2027–2036 (2009)

24. Peng, Y., Buccolieri, R., Gao, Z., Ding, W.: Indices employed for the assessment of "urban outdoor ventilation"—a review. Atmos. Environ. **223**, 117211 (2020)

25. Delaney, N.J., Chatterjee, S.: Use of the bootstrap and cross-validation in ridge regression. J. Bus. Econ. Stat. **4**(2), 255–262 (1986)

Impact of Building Type Selection on Environmental Ventilation Effects in Residential Communities

Li Yan[✉]

School of Civil Engineering and Architecture, Southwest University of Science and Technology, Mianyang 621020, China
anniey007@swust.edu.cn

Abstract. The selection of residential building types generally considers the economic benefits of land space utilization or the control of plot ratio by higher-level planning for current land utilization, with less consideration for the impact of buildings on the environment, especially the ventilation and heat dissipation of the site. In this study, three commonly used building forms, small-sized villa buildings, multi-story buildings commonly used in old communities, and high-rise buildings commonly selected in most residential areas, were chosen as research objects. The numerical simulation of wind field changes in residential areas was conducted using computational fluid dynamics (CFD) method. The conclusion drawn is that under the same climatic background, different building forms have a significant impact on the ventilation of the environmental space, and higher building heights have a significant strengthening effect on the surrounding wind environment.High-rise buildings can achieve an average wind speed increase compared to the original area Above 50%.On the sides parallel to the incoming wind direction, there will be obvious areas of increased wind speed, and with the increase of building height, the area of increased wind speed also increases.The maximum can be expanded to 3 times the original wind speed or more.On the leeward side of the building, there will be a more obvious area of air stagnation, and the lower the building height and the smaller the volume, the larger the area of air stagnation.

Keywords: ventilation · residential community · building type selection · CFD

1 Introduction

Current building type selection in residential areas focuses on the economic and efficient use of land space, the ecological beauty of living environments, and the rationality of facility allocation [1]. This sets corresponding economic and technical indicators such as scale, land use, and spatial layout. However, there is no unified conclusion on the standards for setting these indicators despite considering them in depth. Although the regulations mention the need to follow requirements such as sunlight, daylighting, ventilation, disaster prevention, and management, it is somewhat strained to effectively relate these principles to specific indicators [2, 3]. In setting these indicators, we tend to prioritize economic rationality of land use and functional suitability, with less consideration

B.-J. He et al. (Eds.): UCSUD 2023, LNCE 559, pp. 578–589, 2025.
https://doi.org/10.1007/978-981-97-8401-1_41

for ventilation effects. However, extensive research has shown that wind environmental factors play an increasingly important role in the layout of residential areas and even cities. Effective use of wind environment will improve the air quality of residential areas and reduce unnecessary energy consumption [4].

Therefore, this article intends to introduce a new perspective - wind efficiency, and introduce corresponding fluid computational simulation software to evaluate the impact of the differences in current residential building type selection on the environment (specifically referring to urban ventilation environment), in order to obtain more suitable building type selection guidelines and better serve the design practice of residential areas [5–7].

2 Methods

2.1 Numerical Calculation Method

In order to accurately obtain detailed parameters of the atmospheric wind environment in residential areas, this article uses computational fluid dynamics (CFD) method to numerically simulate the air flow in residential areas. In general, the wind speed in residential areas is much smaller than 0.3 times the speed of sound, so the air can be treated as an incompressible fluid for solution [8, 9]. This article uses the incompressible Navier-Stokes equations as the main governing equations and solves the problem of difficult convergence in numerical solution of incompressible Navier-Stokes equations through preprocessing methods [10–12]. The three-dimensional incompressible Navier-Stokes equations with preprocessing methods in a Cartesian coordinate system are as follows [13]:

$$\frac{\partial Q}{\partial t} + \frac{\partial}{\partial x}(F_1 + G_1) + \frac{\partial}{\partial y}(F_2 + G_2) + \frac{\partial}{\partial z}(F_3 + G_3) = 0 \tag{1}$$

where Q is the flow conservation variable, F1, F2, and F3 are the non-viscous fluxes, G1, G2, and G3 are the viscous fluxes, and the specific expressions for each term are:

$$Q = \begin{pmatrix} \rho \\ \rho u \\ \rho v \\ \rho w \\ e \end{pmatrix}, F_1 = \begin{pmatrix} \rho u \\ \rho u^2 + p \\ \rho v u \\ \rho w u \\ (e + p)u \end{pmatrix}, F_2 = \begin{pmatrix} \rho v \\ \rho u v \\ \rho v^2 + p \\ \rho w v \\ (e + p)v \end{pmatrix}, F_3 = \begin{pmatrix} \rho w \\ \rho u w \\ \rho v w \\ \rho w^2 + p \\ (e + p)w \end{pmatrix},$$

$$G_1 = \begin{pmatrix} 0 \\ -\tau_{xx} \\ -\tau_{xy} \\ -\tau_{xz} \\ \dot{q}_x - u\tau_{xx} - v\tau_{xy} - w\tau_{xz} \end{pmatrix}, G_2 = \begin{pmatrix} 0 \\ -\tau_{yx} \\ -\tau_{yy} \\ -\tau_{yz} \\ \dot{q}_y - u\tau_{yx} - v\tau_{yy} - w\tau_{yz} \end{pmatrix},$$

$$G_3 = \begin{pmatrix} 0 \\ -\tau_{zx} \\ -\tau_{zy} \\ -\tau_{zz} \\ \dot{q}_z - u\tau_{zx} - v\tau_{zy} - w\tau_{zz} \end{pmatrix}$$

Here, ρ is the density, u, v, and w are the velocity components in the x, y, and z directions respectively, p is the pressure, e is the total energy, τ is the shear stress, and α is the thermal conduction term. $\dot{q}$

At the same time, it satisfies the state equation:

$$e = \rho C_P T + \frac{1}{2}\rho\left(u^2 + v^2 + w^2\right) - \frac{p}{\rho} \tag{2}$$

In the atmospheric environment, due to the large scale of buildings, the airflow is mainly turbulent and there is almost no laminar flow. Therefore, the reasonable selection of engineering turbulence models has a significant impact on the numerical calculation results [14, 15]. This paper adopts the improved two-equation k-ε turbulence model from the literature, which is simple in form and easy to use, and has been widely applied to various engineering problems. The transport equations for turbulent kinetic energy k and turbulence dissipation rate ε are:

$$\frac{\partial(\rho k)}{\partial t} + \frac{\partial}{\partial x_j}(U_j\rho k) = \frac{\partial}{\partial x_j}\left[\left(\mu + \frac{\mu_t}{\sigma_k}\right)\frac{\partial k}{\partial x_j}\right] + P_k - \rho\varepsilon$$

$$\frac{\partial(\rho\varepsilon)}{\partial t} + \frac{\partial}{\partial x_j}(U_j\rho\varepsilon) = \frac{\partial}{\partial x_j}\left[\left(\mu + \frac{\mu_t}{\sigma_k}\right)\frac{\partial\varepsilon}{\partial x_j}\right] + (C_{\varepsilon 1}P_k - C_{\varepsilon 2}\rho\varepsilon + E)T_t^{-1} \tag{3}$$

The specific expressions of each term in Eq. (3) can be found in the literature.

Based on the finite volume method, the above equations are discretely iterated and solved. The spatial discretization adopts the second-order upwind format coupled with the TVD limiter, and the time discretization adopts the second-order implicit format. The velocity inflow boundary condition is used at the windward boundary of the computational domain, and the outflow boundary condition with zero back pressure is used at the outflow boundary [16]. The buildings and ground are treated with the no-slip adiabatic wall boundary condition.

2.2 Modification and Validation of Methods

The environment of residential areas is closely related to people's living conditions. Therefore, the height of the wind field evaluation in this paper is the pedestrian height, which is 1.5m. The entrance wind speed is selected at a height of 100m and is 5m/s. Based on the wind profile (as shown in Fig. 1) It can be seen that the wind speed will vary with height, after correction The simulated entrance wind speed in the paper is 1.5, meters high, only 2.6m/s at a height of 1.5m. The temperature of the boundary conditions is selected as 11°C, and the wind direction is southeast.

The computational grid used in the article adopts a structured grid, as shown in Fig. 2. This type of grid can better ensure the accuracy of numerical simulations.

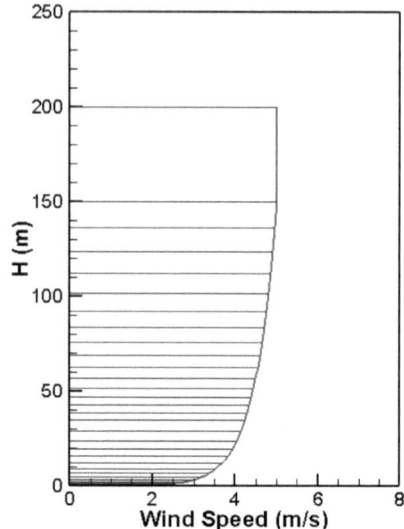

Fig. 1 velocity profile

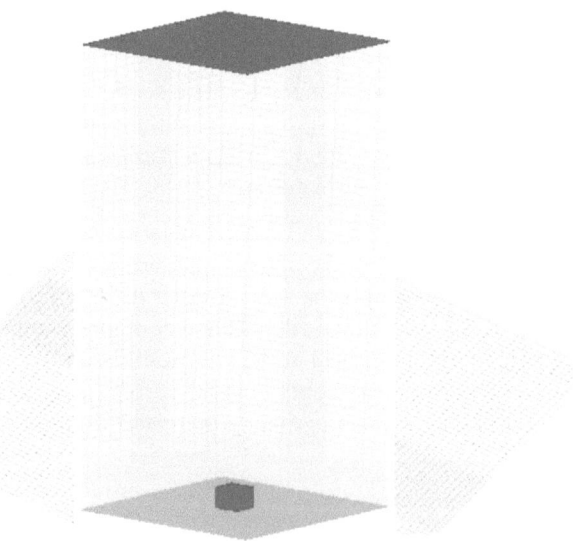

Fig. 2. Calculation Grid

2.3 Selection of Calculation Samples

Four common types of residential building forms were selected in the article: low-rise single buildings in villa areas, multi-story buildings in old communities, mid-rise buildings, and high-rise buildings with podiums. For the convenience of calculation, the geometric selection of buildings in the article was simplified by removing detailed geometric edges and trying to keep the volume as square as possible while maintaining the area [17]. To ensure the accuracy of the calculation results, the calculation domain, as shown in Fig. 3, was expanded by 10 times based on the original geometric area. Taking the single building villa as an example, the building size is 13m × 12m × 9m. The calculation domain range is 100m × 100m × 200m. The sizes of other calculation samples are shown in Table 1.

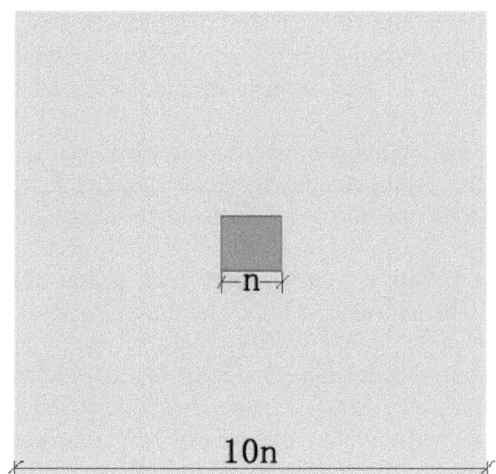

Fig. 3: Schematic of Calculation Domain

Table 1: Composition of Geometric Parameters of Sample Systems

Calculation Sample Group		Sample Parameter Settings		
		Dimensions (Length × Width × Height) in meters	Number of Floors	Floor Height (m)
Individual Building	Low-rise	13 × 12 × 9	3	9
	Multi-story	16 × 15 × 18 12 × 16 × 18 20 × 8 × 18	6	18
	Mid-rise	23 × 12 × 24 23 × 12 × 30	8 10	24 30
	High-rise	With Podium Layout	12 17	39 54
		Without Podium Tower Style	26	81

3   Results and Discussion

3.1   Calculation Results

The calculation results are shown in Table 2. For the four types of buildings in the selection, the highest wind speed for small-volume villa buildings is 5.53m/s, and the lowest wind speed is 0.0024m/s. The high wind speed area is only 110.86 m^2, and the low wind speed area is 104.69 m^2. The entire environmental wind field is not significantly affected by the buildings, with no obvious areas of wind speed enhancement, except for the formation of a large stagnant air area on the leeward side of the villa building.

The selected multi-story buildings have differences in planar geometry, with different proportions of width and depth. According to the calculation results, the multi-story buildings with a width-to-depth ratio of 1 have the highest maximum wind speed, which is 4.12m/s, and there are also more areas with high wind speeds, totaling 2063.17 m². The multi-story buildings with a width-to-depth ratio greater than 1 have the lowest maximum wind speed, which is only 3.83m/s. There are also fewer areas with high wind speeds, totaling 1379.71 m².

For low-rise buildings, the maximum wind speed of a 10-story building is greater than that of an 8-story building, and the wind speed impact on the environment is enhanced for the 10-story building, with a wind speed ratio of 1.02 compared to the original wind speed. The 8-story building has a wind speed ratio of 0.98, indicating a weakening effect.

High-rise buildings have a significant impact on the environmental wind field. The highest wind speeds for high-rise buildings with skirt houses (13 floors, 18 floors, and 27 floors) are 4.56m/s, 5.33m/s, and 6.77m/s, respectively. The ratio of average wind speed to inlet wind speed is 1, 1.08, and 1.19, respectively. The highest wind speeds for high-rise buildings with point houses (13 floors, 18 floors, and 27 floors) are 5.87m/s, 6.53m/s, and 7.39m/s, respectively. The ratio of average wind speed to inlet wind speed is 0.99, 1.05, and 1.56, respectively.

3.2   Conclusion and Discussion

Based on the above calculations, it can be seen that the differences in residential building types have a significant impact on the outdoor wind environment. Under the same climatic conditions, different building types show variations in the ventilation effects on the environment. The following conclusions can be drawn:

① Smaller villa buildings have a significant weakening effect on the wind speed in the environmental wind field, and a clear air stagnation zone is formed on the leeward side of the villa building.

② The impact of multi-story buildings and villa buildings on the environmental wind field is consistent, with no significant enhancement of ventilation effect and a significant weakening of local wind speed effect. The difference in span and depth of multi-story buildings has a prominent impact on wind speed in the wind field. When the span is significantly larger than the depth, a wind speed blocking zone is formed, resulting in a significant decrease in surrounding wind speed.

Table 2. The Impact of Different Building Type Selections on the Surrounding Environmental Wind Field

Sample		Maximum Wind Speed (m/s)	Minimum Wind Speed (m/s)	Area of High Wind Speed Generation (m²) Coordinate Area > 3m/s	Area of Calm Wind Zone (m²) <0.5m/s	Ratio to Original Wind Speed
Villa		5.53	0.0024	110.86	104.69	0.942
Multi-story		4.12	0.0172	2063.17	255.00	0.95
		4.01	0.0159	2058.78	189.67	0.95

(continued)

Table 2. (*continued*)

Sample		Maximum Wind Speed (m/s)	Minimum Wind Speed (m/s)	Area of High Wind Speed Generation (m²) Coordinate Area > 3m/s	Area of Calm Wind Zone (m²) <0.5m/s	Ratio to Original Wind Speed
		3.83	0.0047	1379.71	75.82	0.93
Mid-rise		4.41	0.0392	2705.60	164.18	0.98
		4.56	0.0586	2989.87	68.22	1.02

(*continued*)

Table 2. (*continued*)

Sample	Maximum Wind Speed (m/s)	Minimum Wind Speed (m/s)	Area of High Wind Speed Generation (m²) Coordinate Area > 3m/s	Area of Calm Wind Zone (m²) <0.5m/s	Ratio to Original Wind Speed
High-rise Building with Skirt House	5.53	0.0045	3176.47	258.16	1
	6.02	0.0179	3563.88	173.20	1.08
	6.77	0.0059	3907.98	164.04	1.19

(*continued*)

Table 2. (*continued*)

Sample	Maximum Wind Speed (m/s)	Minimum Wind Speed (m/s)	Area of High Wind Speed Generation (m²) Coordinate Area > 3m/s	Area of Calm Wind Zone (m²) <0.5m/s	Ratio to Original Wind Speed
	5.87	0.0136	2827.62	342.74	0.99
	6.53	0.0039	3272.72	427.04	1.05
	7.39	0.0279	3630.41	338.85	1.56

③ Compared to low-rise buildings, small high-rise buildings will have some more obvious wind speed enhancement zones, especially in the areas on both sides of the dominant wind direction.

④ As the height of high-rise buildings increases, there is a significant acceleration effect on the wind speed in the environmental wind field. The higher the height, the more pronounced this acceleration effect. Point-type high-rise buildings have a more obvious wind speed acceleration effect on the surrounding environment, resulting in more high-speed areas.

Based on the above results, when selecting building types in residential areas, it is advisable to choose high-rise buildings with point design in order to achieve good ventilation effects. This type of selection not only takes into account the spacing for building daylighting, but also creates a larger green space, thus providing a better environmental effect for the entire community. When windproofing and thermal insulation are needed, it is advisable to choose low-rise panel buildings and appropriately increase the ratio of span to depth, with the span being greater than the depth. Building type selection is only one of the factors that affect the ventilation environment in a community. To achieve optimal ventilation effects, other factors such as building layout and its relationship with the dominant wind direction should also be considered.

References

1. Sun, Y., Zhang, D., Lv, L., et al.: Analysis of landscape design in residential communities based on microclimate optimization: a case study of Yancheng journalist home. J. Central South Univ. of For. Technol. **32**(10), 7 (2012). CNKI: SUN: ZNLB.0.2012-10-038.
2. Zhang, L., Zhan, Q., Lan, Y.: Study on vegetation cooling and ventilation effect in Wuhan residential communities based on microclimate simulation. Chin. Landsc. Arch. **35**(3), 5 (2019). CNKI: SUN: ZGYL.0.2019-03-019.
3. Yue, Z., Zou, H., Ma, Y.: Analysis of the effect of residential community water bodies on microclimate. Arch. Budget (6), 3 (2020)
4. Yan, L., Hu, W., Gu, L., Li, L.: Analysis of the correlation between square air quality and spatial design elements: a case study of six design schemes in Urumqi diamond city square. Urban Plann. **44**(8), 61–70 (2020)
5. Yan, L., Hu, W., Yin, M.: Evaluation of urban square air quality and health place guidelines based on RANS. J. Appl. Ecol. **31**(11), 3786–3794 (2020)
6. Yan L., Yin, M., Yu, H., et al.: Public responses to urban heat and payment for heat-resilient infrastructure: implications for heat action plan formulation. Environ. Sci. Pollut. Res. **30**(57), 120387–120399 (2023)
7. Yan, L., Hu, W., Yin, M.-Q.: An investigation of the correlation between pollutant dispersion and wind environment: evaluation of static wind speed. Polish J. Environ. Stud. **30**(5), 4311–4323 (2021)
8. Yan, L.: Research on Ventilation Performance Under Urban Design. China Architecture & Building Press, Beijing (2023)
9. Gu, L., Hu, W., Xi, R., et al.: Research on the layout of residential areas under climate influence—taking eight residential communities in Xi'an as an example. In: 2017 (12th) Urban Development and Planning Conference. 6 Dec 2023
10. Mo, H., Cao, L., Qin, X.: Simulation study on the impact of residential building layout on outdoor thermal environment. Shanxi Arch. **48**(24), 34–37 (2022)

11. Hu, W., Xi, R., Yan, L., et al.: Research on Urban Design Optimization Strategies Based on the Influence of Wind Environment in Mountainous Waterfront Blocks: A Case Study of Chongqing Hualongqiao Area. China Urban Science Research Association. China Urban Science Research Association (2016)

12. Liu, Q., Yang, W.: Research on multi-objective optimization layout of outdoor thermal comfort in residential areas based on climate characteristics: a case study of Xi'an area. J. Xi'an Univ. Arch. Technol. Nat. Sci. Ed. **54**(1), 7 (2022)

13. Liu, Q., Yang, W.: Research on multi-objective optimization layout of outdoor thermal comfort in residential areas based on climate characteristics: a case study of Xi'an area. J. Xi'an Univ. Arch. Technol. Nat. Sci. Ed. **54**(1), 54–60 (2022)

14. Li, R.: Research on the Planning and Design of Mountainous Residential Areas in Coordination with the Thermal Environment: A Case Study of Fusheng Residential Area in Jiangbei District, Chongqing. Chongqing University (2020)

15. Li, Y., Hu, W.: Health Risk Assessment of Public Space Environment in Three Gorges Square Based on PM2.5. China Urban Planning Annual Conference. China Urban Planning Society (2019)

16. Hu, W., Lin, B.: Research on Wind Environment Optimization Strategies in Mountainous Waterfront Blocks: A Case Study of Shangqing Temple Area in Chongqing. China Urban Science Research Association. China Urban Science Research Association (2015)

Optimization Design of Tree Array Space in Pedestrian Street Based on PM$_{2.5}$ Distribution

Yichen Li[1]($\boxtimes$) and Xiaoting Jing[2]

[1] China Academy of Urban Planning & Design Western Branch, Chongqing 400045, China
arthen123@163.com

[2] School of Architecture and Urban Planning, Chongqing University, Chongqing 400045, China

Abstract. Wind, as one of the important elements of urban microclimate, plays a crucial role in the dispersion of air pollutants. Adequate improvement should be made to mitigate the impact of urban spaces on the wind environment. Green spaces along pedestrian streets are an essential component of the public activity space system, accommodating most outdoor communication activities for citizens. However, the design of existing green spaces along pedestrian streets is often based on past experience, lacking empirical research on the relationship between green spaces, wind environment, and pollutant dispersion. Therefore, the correlation research between tree array space and PM2.5 pollutant distribution in the air, and how to optimize the layout of tree array space by design means according to its distribution regularities, are the main research contents presented in this paper. Taking two important green spaces in Jiefangbei pedestrian street of Chongqing as the sample points for research, this paper adopts the methods of measurement and numerical simulation. Firstly, the pollutant concentration and wind environment in the green space are measured and analyzed, and then use the PHOENICS tool to simulate and compare the wind environment of the green space in the study area before and after optimization design. The results show that the concentration of pollutants is usually high in green space as the greening has aggregation of pollutants to itself; An obvious negative correlation is found between wind speed and the concentration of pollutants, good ventilation is an important means to alleviate the air with low quality; Reasonable design of tree array space based on microclimate can effectively improve the air quality in local space.

Keywords: Tree Array Space · Health · Pedestrian Street · Wind Environment

1 Introduction

Pedestrian street is the most dynamic place in the city where group activities are relatively concentrated, and the tree array space is a crucial space node where people are willing to stop and rest. Generally speaking, the tree array space of pedestrian street has the following important functions: Firstly, it could integrate greening into urban commercial pedestrian blocks with characteristics of mass hardened ground, which enriches the space function and adds landscape beauty. Besides, the tree array space itself plays a significant role in providing shadow, absorbing pollutants as well as releasing oxygen

B.-J. He et al. (Eds.): UCSUD 2023, LNCE 559, pp. 590–602, 2025.
https://doi.org/10.1007/978-981-97-8401-1_42

while absorbing carbon dioxide under the sunlight. Additionally, interspersing bustling commercial blocks with tree array space would bring people relaxed and comfortable psychological indication, which is considered as an effective method to create pleasant environment. On this basis, previous study reinforced the notation that the existence of tree array space is invariably beneficial to environment. However, whether there are negative environmental influence from tree array space and how to reduce negative effects through reasonable design is studied in this paper. Started from the adsorption effect on PM2.5 suspended particulate matter in the air, this research found the specific law of pollutant distribution among tree array space and air, then the optimization of environmental quality could be realized through reasonable design and layout of tree array space, which could provide support for building human health activity places.

Existing research on the correlation between greening and outdoor public space concerns many aspects, and two of them are frequently discussed nowadays: Firstly, the functional impact of green space on urban environment state, such as mitigation of urban heat islands [1], urban noise [2] as well as pollution [3], and its' important role of improving urban livability [4], which focus on the function of green space itself and clarify the influence pattern of greening upon urban environment. Secondly, these research regarded greening design [5, 6] (the types and size of trees, greening ratio, the shape and layout strategy of green space, etc.) as an important means to improve urban local environment. That is, greening design is not only considered as an absolutely necessary tool to beautify the environment, but also significant measure to adjust and improve urban microclimate. Relative studies focuses on greening as an effective means to regulate urban, and the optimization of urban microenvironment would be achieved through the change of greening design. However, existing research has paid more attention on the first aspect while the study relating to the second part, how to apply green design as kinetic energy to regulate urban microclimate, is still in the preliminary trial stage. This article attempts to integrate greening design with space design, looking for a desirable layout strategy of greening design to effectively improve local environmental air quality of the city.

2   Methodology

The research adopted a combined method of actual measurement and numerical simulation. Actual measurement was applied to obtain the association law between tree array spatial layout and pollutant distribution of the specific research site, and then the numerical simulation method was used to optimize the design of tree array space in pedestrian street.

2.1   Profile of the Research Object

Chongqing Jiefangbei Pedestrian Street Plaza was selected as the research object, Located in Yuzhong District, Jiefangbei Pedestrian Street Plaza is the most prosperous and densely populated area in Chongqing [7], and the selected tree array space among Jiefangbei pedestrian street is shown in Fig. 1. The research object was selected basing on the following reasons: Firstly, the representativeness of the space, Jiefangbei

Pedestrian Street is one of the most representative pedestrian streets in Chongqing, which has the characteristics of general pedestrian street with concentrated population, dense shops, and diverse functions. The research on it can represent most of the pedestrian street categories, thus the research could be both typical and universal; Secondly, the diversity of the space. Greening design is an important part of the pedestrian street space. It is of great significance to integrate healthy green space design into the overall pedestrian street for optimizing the outdoor public space; Thirdly, the integrity of the space. On the one hand, the two greening areas here are important and complete green areas of Jiefangbei Pedestrian Street. On the other hand, these areas are relatively complete independence, and are less affected by sudden pollution sources from the surrounding environment, which make it convenient to integrate it into the whole pedestrian street environment for research.

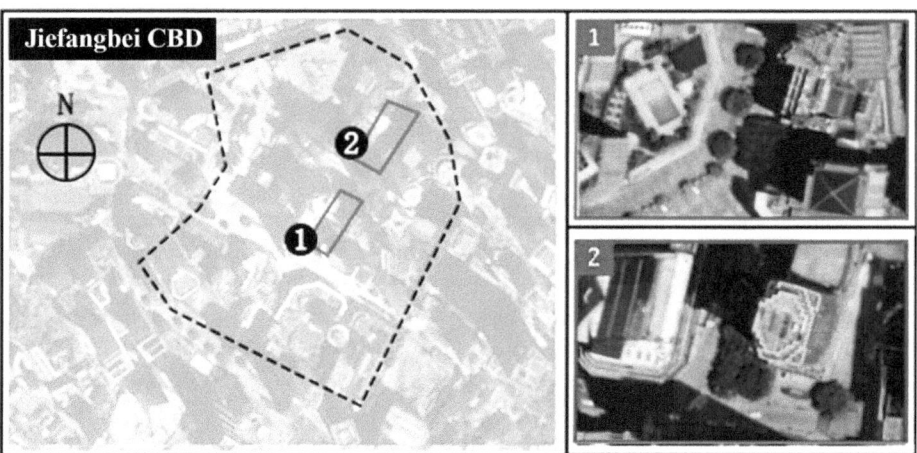

Fig. 1. Research Scope Street and Main Tree Array Space

2.2   Field Survey

The field measurement is conducted on December 15, 2019. During winter, the air quality is quite poor, thus the time selection for actual measurement during this period could better compare the removal and mitigation effects of wind speed on poor air quality.

Measured Setpoint. In this study, five points were selected in each green space. For the selection of points, on the one hand, the measurement workload was considered. Based on the scale of the study area, referring to the radius of the working point which has been measured before [8], these five points could roundly reflect the overall air quality of the study area. On the other hand, it is necessary to consider the relationship between important positions to balance the whole and the part. Combining these above consideration, the basic information of the final measurement point is shown in Fig. 2:

The distribution of the five measurement points in the two study areas was similar. Point 1 was set in the center of the street and inside the tree array space, Point 2 was

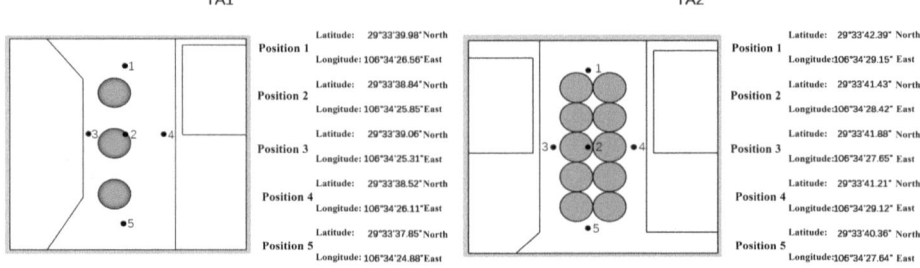

Fig. 2. Position of the Measurement Point

set outside the tree array and perpendicular to the street direction, and another two point were set outside the tree array and parallel to the street direction.

Measurement Tool. According to the elements which require measuring, professional tools were selected for measurement Tes-1341 hand-held anemometer was selected to collect data for the wind speed and direction. Braunton SMART-1265 air monitor With PM2.5 sensor equipped with laser, was applied to measure PM2.5 concentrations, has the advantage of more accurate measurement data and more sensitive response.

Measurement Method. In terms of selecting measurement dates, emphasis should be placed on seasons with high pollution levels. Air pollution in Chongqing is most severe in winter. This study conducted on-site measurements on Sunday, December 15, 2019. The weather on that day was overcast with weak solar radiation, temperatures ranging from 11 °C to 13 °C, humidity between 93–82%, an average PM2.5 concentration of $157\mu g/m^3$, wind direction shifting from easterly to southeasterly, and wind force of 1–2 on the Beaufort scale.

The research time span was set from 9:30 a.m. to 9:30 p.m. on December 15, 2019, synchronized with the main operating hours of the Liberation Monument Commercial District. This period represents the primary time range for pedestrian activities on the commercial street. Measurement intervals should be within 20–30 min. Wind speed, wind direction, and pollutant data were recorded every 30 min. To ensure data stability and reliability, three readings were recorded within 5 min upon initiating each recording session. These three readings should be the stabilized values after the instrument data had settled, excluding instantaneous fluctuations. During statistical analysis, the three data points were arranged in descending order, and the median value was taken as the data for the specific time on that day.

2.3 Numerical Simulation

After more than twenty years of development, the simulation technology of PHOENICS software has become highly mature and sophisticated, capable of effectively outputting relevant indicators for wind environment and thermal environment. For relatively complex tree array models, PHOENICS tool, compared to ENVI-met, can import complex models through data interfaces. The PARSOL technology in the LAIR architectural calculation module can also capture geometric features with reasonable accuracy. The

FOLIAGE plant module and pollutant dispersion calculation module of PHOENICS can effectively simulate and analyze the influence of vegetation on pollutant dispersion. Therefore, the PHOENICS tool has strong application value for the research content of this paper, and thus it was chosen for numerical simulation. The specific calculation module and parameter settings were as follows.

Environment Parameters Setting. Based on historical climate characteristics of Chongqing, the simulated boundary conditions follow the regional climate characteristics of Chongqing. Combined with the field measurement structure in winter, the main simulation wind direction was set as the northwest wind (45 °C from north to west), the same as the highest frequency wind direction of Chongqing [9]. The rest of the meteorological conditions all consider the situation of the measurement time, with an environmental temperature of 13 °C, relative humidity of 87%, and air pressure of 1013Pa.

Model Scale of Computational Domain. The size of the simulated computing domain still lacks unified standard. With reference to the results of the AIJ using wind tunnel experiments, the space size of the research model in the simulation should be less than one-third of the computational domain [10]. It was recommended that the length of both sides of the calculation domain was: 3W (W is the width of the area occupied by the building model), and the overall height of the calculation domain is: 3Hmax (Hmax is highest size of the building in the study area).

Mesh Generation. The total area of the research object in this experiment is 100m × 100m × 96m. According to the CFD simulation principle, the calculation range should be more than three times larger than the study area. Therefore, 320m × 320m × 320m was finally determined as the calculation boundary range for the computer simulation. Besides, the infinite boundary condition was set during the process to maximize the simulation of natural wind state, which could also reduce the errors caused by boundary conditions, and increase the accuracy and reliability of the calculation results.

Model Setup of the Plants. Large trees among the study area were selected, with crown diameter of 9m, trunk diameter of 0.45m, under branch height of about 3m. While the leaf area index was relatively small of about 0.25, and the coefficient of wind drag force was 1.5.

3   Results

3.1   Data Analysis

Statistics of Wind Environment. At the day of actual measurement, the TA1 (Tree array) space was dominated by the north and northwest wind, and the TA2 space was dominated by east wind. The average wind speed at the height of 1.5m of five measuring points in TA1 space and TA2 space was 0.9m/s and 0.79m/s respectively.

Statistics of Pollutants. The TA1 green space was composed of large trees in the layout pattern of dotted separation. Compare the average concentration of the five measuring points, the pollution concentration of point 2 was the highest. The concentration measured at point 1 near the pedestrian street was relatively small, and the concentration distributions of the other three points were similar; During the whole morning, the overall pollution concentration of TA1 street space steadily kept in a high level, and reach the highest value of 135μg/m3 at 12:30, and then decrease to the lowest value of 109μg/m3 at around 13:00. After that, the pollution concentration of the green space has always remained in a low level until 20:00, when a trend of increasing could be seen (Fig. 3). Compare and analyze the change of overall PM2.5 concentration with point concentration in the B1 area, It could be seen that when the pollutant concentration level of TA1 was low on that day, the pollution concentration inside the green space was also relatively low. However, during the morning when the overall pollution of TA1 street was relatively high, the pollution level of point 2, which was inside the green space, was significantly higher than other points.

The TA2 green space was also composed of large trees while in a different layout of intensive continuous pattern. Compare the average concentration of the five measurement points, the pollution concentration of point 5 was the highest, which was close to the pedestrian street center. The concentration of point 1 and point 5 was relatively small, and the overall pollution concentration of TA2 green space showed a clear and stable downward trend on that day from the highest value of 140μg/m3 at 10:00, then reached the lowest value of 119μg/m3 at around 21:30 in the evening, from which obvious pollutant purification and absorption effect could be seen. During 9:30–13:30 in the morning and 19:30–21:30 in the evening, the pollution level at point 5 was significantly higher than other place (Fig. 4).Compared with other points, the concentration of point 2 inside the green space was always at a lower level, which was less likely to accumulate pollutants.

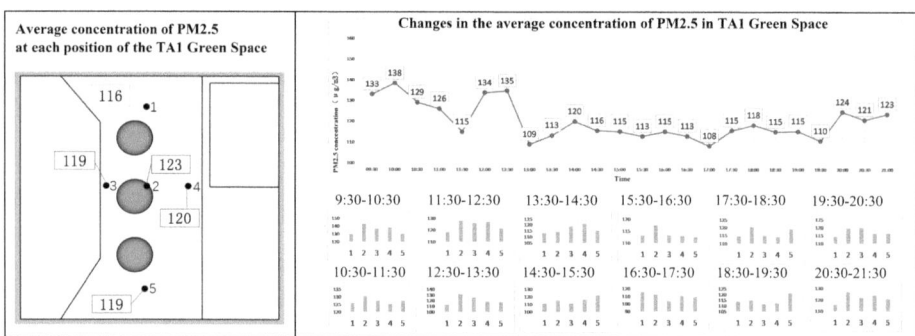

Fig. 3. Change of PM2.5 Concentration in TA1 Green Space

Pollutant Distribution Law among Tree Array Space. Based on the above measured data, it is not difficult to summarize the following rules concerns the relationship between tree array space and pollutant distribution:

(1) The pollutant concentration of a well-planted tree center were higher (point 2 of TA1 and TA2), which is contrary to conventional design wisdom that green plants release oxygen and absorb carbon dioxide when exposed to sunlight.

(2) The open space in the green tree array (point 4 of TA1, TA2) was found to be the area with lower pollutant concentration.

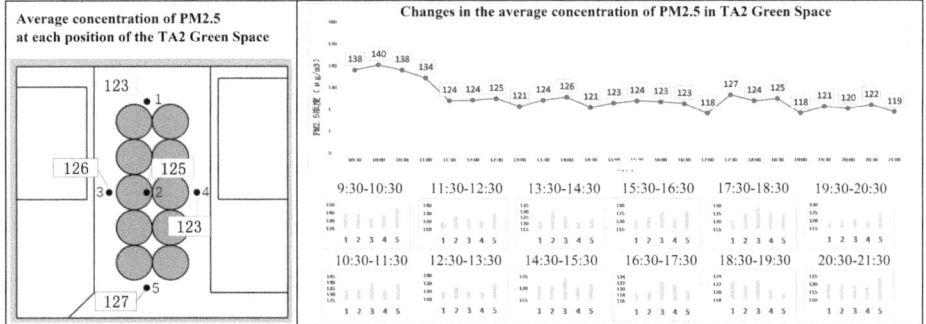

Fig. 4. Change of PM2.5 Concentration in TA2 Green Space

This is because the green space, especially the tree array space composed of large trees, has a strong adsorption effect on pollutants. When it locates in the air duct, air circulation would be blocked and the situation of pollutants accumulation would also be aggravated. Therefore, the following principles should be followed when design tree array layout composed of large trees: First, it is recommended to place the tree at the edge of the road to leave enough air duct space; second, try to leave gaps between tree arrays to avoid dense tree arrays layout which may form wind-break wall.

3.2   Space Design of Pedestrian Street Tree Array Based on Health

Site Analysis and Improved Greening Layout Design of TA1 Street Space. TA1 green space was located at the west side of Jiefangbei pedestrian street, the dominant wind direction of which was north. The tree array layout was unilateral uniform, consisting of large trees with large interval. Because the street where the TA1 tree array was located belongs to the traffic-oriented street space, except the windward part of the street may present a certain wind shadow area due to the influence of buildings, the wind environment of the main traveling space requires good ventilation. The wind speed in traffic space could be further promoted by refined position adjustment of the tree array.

The main optimization measures based on the actual measurement conclusions are: move large trees to the road edge and leave enough open space for people to walk through, and avoid pedestrians from staying in the negative space formed of pollutant absorption by trees; At the same time, sufficient and wide air duct should be left to improve ventilation in this area (Fig. 5). In order to verify the effectiveness of this optimization measure, numerical simulation on ventilation was carried out to compare the current situation and the optimized layout.

The simulation results showed that the optimization scheme could improve the wind environment of main travelling area. The tree array on the leeward side has little influence on the street airflow, and the maximum wind speed of the street increased from 1.45m/s to 1.57m/s, and the wind speed ratio increased from 0.79 to 0.86, indicating a significant improvement in the overall wind speed (Fig. 6).In key areas where a large number of people pass through, the static wind rate [11, 12] was reduced from 48.59% to 35.52%, and the regional ventilation environment has been improved to some extent.

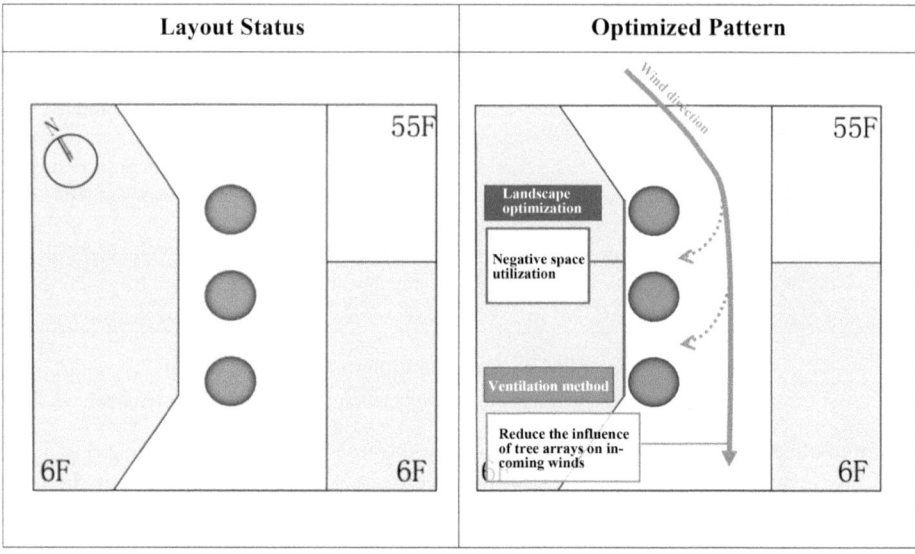

Fig. 5. TA1 Street Greening Space Layout Status and Optimized Pattern

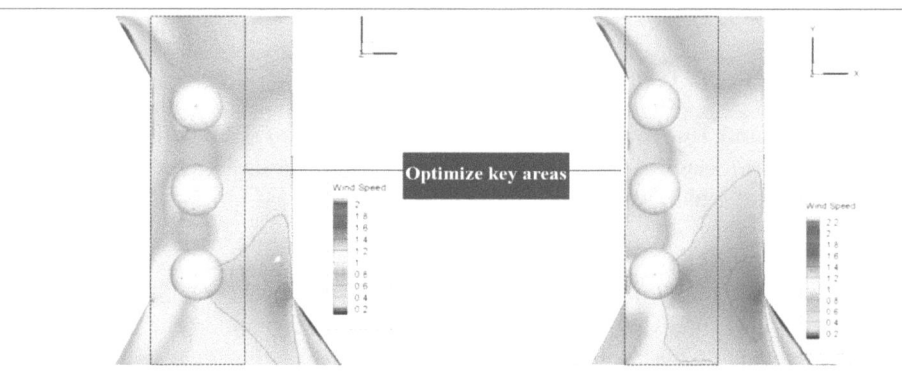

Current situation		**Optimized situation**	
Background wind speed (m/s)	1. 16	Background wind speed (m/s)	1. 16
Maximum wind speed (m/s)	1. 45	Maximum wind speed (m/s)	1. 57
Minimum wind speed (m/s)	0	Minimum wind speed (m/s)	0
Average wind speed (m/s)	0. 92	Average wind speed (m/s)	1
Static wind rate in key areas (%)	48. 59	Static wind rate in key areas (%)	35. 52

Fig. 6. Comparison of wind speed cloud map between TA1 and the optimized layout

Site Analysis and Improved Greening Layout Design of TA2 Street Space. TA2 green space was located at the north side of Jiefangbei pedestrian street, the dominant wind of which was east. The tree array was also consisted of large trees, while its' layout was evenly distributed and located at the central part with high spatial continuity, and crowd activities were mainly concentrated inside the green space. The simulation analysis of TA2 green space mainly highlighted the problem of tree array location. The tree array was close to the leeward side of the street, and the double-row layout further obstructed the smooth air flow into the green space interior, thus, a large area of calm wind appeared in the green space. Considering that the TA 2 belongs to recreational street space, so the wind environment inside the tree array should be optimized. On the premise of keeping the original layout unchanged with certain characteristics of rows and columns orderly arranged, the influence mechanism of tree array position as well as interval upon wind environment was utilized to optimize the original green space from the following two aspects (Fig. 7):

(1) Move the tree array to the windward side of the street. The shearing effect of the tree array on the incoming wind is exerted to introduce part of the air flow into the tree array.
(2) Slightly widen the intervals between tree arrays, open up the ventilation corridor inside the tree array to guarantee the air flow could run through the whole green space, and ultimately improve the overall wind speed among the tree array.

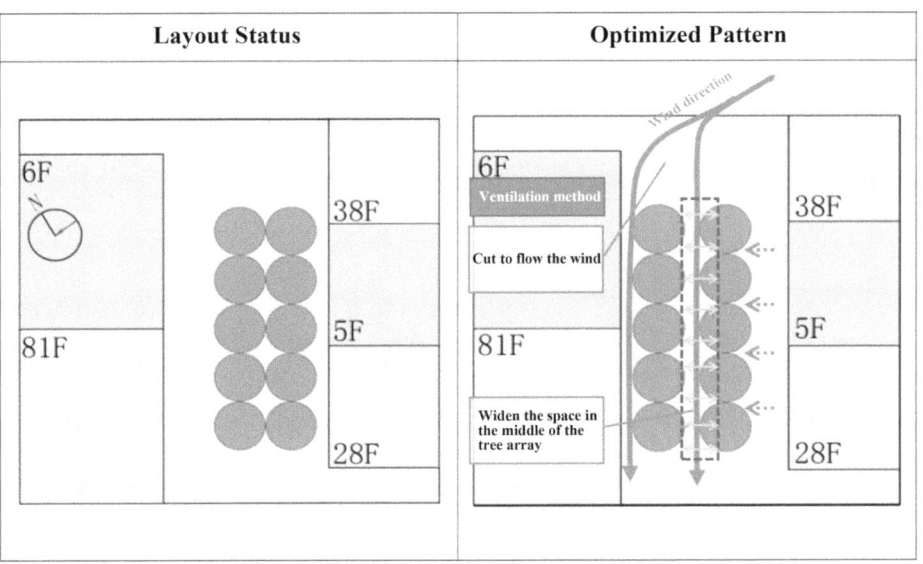

Fig. 7. TA2 Street Greening Space Layout Status and Optimized Pattern

Compare current layout with the optimized green space, the average street wind speed increased from 0.73m/s to 0.77m/s and the wind speed ratio increased from 0.61 to 0.64 on a holistic level. The wind environment inside the street greening space before and after optimization was obviously different (Fig. 8). The optimized scheme utilized the green space to shear the wind and introduce part of the air into the middle space of the tree array. The static wind rate of the original tree array was 93.3%, and that of the optimized tree array was 73.98%, revealing that the static wind area of the key area has been reduced by 19.32%.

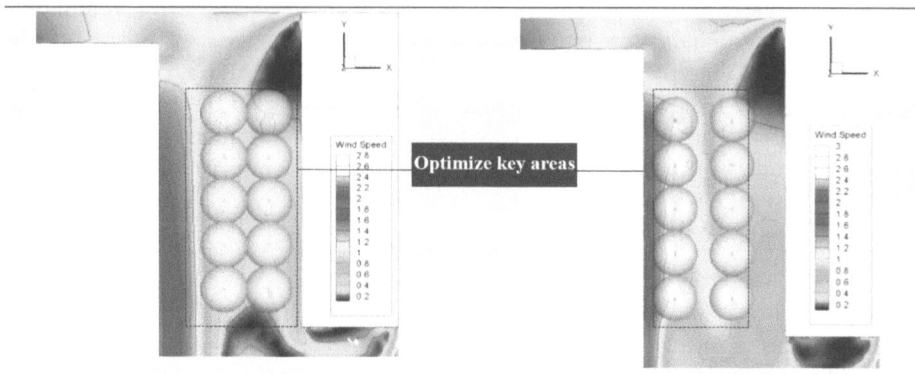

Current situation		Optimized situation	
Background wind speed (m/s)	1. 2	Background wind speed (m/s)	1. 2
Maximum wind speed (m/s)	1. 67	Maximum wind speed (m/s)	1. 76
Minimum wind speed (m/s)	0	Minimum wind speed (m/s)	0
Average wind speed (m/s)	0. 73	Average wind speed (m/s)	0. 77
Static wind rate in key areas (%)	83. 95	Static wind rate in key areas (%)	70. 33

Fig. 8. Comparison of Wind Speed Cloud Map between TA2 and the Optimized Layout

4   Disscussion

4.1   Analysis of the Relationship Between Wind Speed and Pollutant Concentration

Based on the measured data, it could be concluded that there was an obvious negative correlation between wind speed and pollutant concentration, and the higher the wind speed was, the lower the pollutant concentration would be. The study confirms the general conclusion that wind could carry and dilute pollutants, and effective ventilation is of great significance to improve air quality.

The measured overall wind direction of the pedestrian street greening space was consistent with the prevailing wind direction of Chongqing, that the northwest, north and west winds were the most frequent. Generally speaking, street space would exert disturbance to wind environment to some extent. The measured wind speed in street space was lower than the background measuring wind speed, indicating that the wind speed in street has a certain weakening phenomenon.

4.2 The Relationship Between Green Design and Air Quality

Most existing research verifies the relationship between green spaces, wind speed, and pollutants through software simulations or wind tunnel experiments [13, 14]. This study verified the influence patterns of street green spaces on pollutants through on-site measurements. In pedestrian blocks, the green space composed of large trees became a place where pollutants are easier to gather. In addition, the distribution of pollutant concentration in green space presented obvious spatial differentiation, that the concentration of pollutants in central green space was the highest, followed by the surrounding area. The farther away from the center, the pollutants were more easily taken away and diluted, and the lower the pollutant concentration would be.

The layout pattern of the tree array space strongly affected the concentration of pollutants in the air. The ventilation effect of tree array 1 with sparse single-column layout was much better than that of tree array 2 with dense double-column layout. Besides, the difference of these two type of layout between the static wind rate was more than doubled. That is, dense tree array had poor ventilation effect and higher pollutant concentration. Thus, in pedestrian blocks with crowds gathering, large trees should be arranged sparsely in a row along the edges, continuous pieces of flowers, shrubs and lawn are also dotted to form complete green space.

The green layout which is most conducive to the overall ventilation of the street is often located at the street edge, including single-row on the leeward side and double-row near the street side. For the green layout located at the street center, the most beneficial layout to reduce the quiet area of the green space includes single-row and double-row in the center. The number of tree array should not exceed three columns, otherwise it will have a great negative impact on street ventilation performance. At vertical intervals in the tree array, the green space with interval greater than 5m can effectively reduce the wind shadow area between tree arrays. As for the selection of tree types, small trees are most conducive to improving the overall ventilation of the street, while large trees can effectively reduce the static wind area under the tree.

4.3 Limitation of the Optimized Design

In this paper, the optimization strategy of greening design was just a tentative adjustment according to relevant conclusions obtained from actual measurement. The new green design scheme is considered to be favorable only when compared with the current situation, which may not be the optimal scheme, but it could provide directional guidance for future research. However, more accurate guidelines for healthy greening design still require more actual measurement and simulation.

5 Conclusion

This article has carried out a healthy greening design for the key green space of Jiefangbei Pedestrian Street. Based on the data analysis of actual measurement and simulation, the following conclusions can be drawn:

(1) The center of well-greened tree array was actually the place where pollutants accumulate, which was contrary to the traditional concept that greening could improve air quality. This is because the tree array reduced the efficiency of the wind field flow to a certain extent, thus increased the probability of air pollutants accumulation.

(2) Greening design contribute positively to improve local-regional outdoor air quality. Adjusting the layout of the tree array will increase air circulation efficiency in local areas, thereby optimize air quality. Such adjustment strategies include arranging greenery at the street edge and increasing the distance between double row plants.

(3) The integration of greening design and the functions of outdoor public spaces would become an effective means to solve the low-quality outdoor air condition. The green space owns the functions of releasing oxygen, absorbing carbon dioxide as well as pollutants, and guiding the flow of the wind field. Reasonable greening design is quite crucial for creating a both visually desirable and healthy outdoor public space.

References

1. Du, Y.X., Blocken, B., Pirker, S.: A novel approach to simulate pollutant dispersion in the built environment:Transport-based recurrence CFD. Build. Environ. **170**, 1 (2020)
2. Liu, B.Y., Wei, D.X.: Winds of change in Shanghai. Landscape Instit. 2017(Autumn), 45–51
3. Zhang, D.S., Wang, Z.: A study on microclimate effects and human thermal comfort of square canopies in high density areas: Taking Shanghai Chuangzhi Tiandi Square as an example. Class. Chinese Garden **33**(4), 18–22 (2017)
4. Liu, B.Y.: The trilogy of landscape planning and design: Seeking the basis for innovation in the development of landscape planning and design in China. New build. **5**, 1–3 (2001)
5. Lin, T.P.: Thermal perception, adaptation and attendance in a public square in hot and humid regions. Build. Environ. **44**(10), 2017–2026 (2009)
6. Jamei, E., Rajagopalan, P., Seyedmahmoudian, M., et al.: Review on the impact of urban geometry and pedestrian level greening on outdoor thermal comfort. Renew. Sustain. Energy Rev. **54**, 1002–1017 (2016)
7. Yang, C.Y., He, R., Chen, Z.L.: New design concept of modern urban commercial pedestrian street -- taking the reconstruction project of monument to the people's Liberation Commercial Pedestrian Street in Chongqing as an example. New Build. **4**, 23– (2006)
8. Shi, B., Wang, X., Zhao, D.: Numerical simulation of the impact of sky width on the thermal environment of urban residential communities. J. Appli. Mech. **34**(6), 1181–1186; 1227 (2017)
9. Cohen, P., Potchter, O., Matzarakis, A.: Human thermal perception of Coastal Mediterranean outdoor urban environments. Appl. Geogr. **37**, 1–10 (2013)
10. Badas, M.G., Ferrari, C., Garau, M., Querzoli, G.: On the effect of gable roof on natural ventilation in two-dimensional urban canyons. Wind. Eng. Ind. Aerodyn. **162**, 24 (2017)
11. Yan, L., Hu, W., Gu, L.: Correlation analysis between square air quality and spatial design elements: taking six design plans of Urumqi Diamond City Square as an example. City Plann. Rev. **08**, 61–70 2020
12. Yan, L., Hu, W., Yin, M.Q.: RANS based evaluation of air quality in urban squares and guidelines for healthy places. J. Appl. Ecol. **31**(11) (2020). https://doi.org/10.13287/j.1001-9332.202011.016
13. Wania, A., Bruse, M., Blond, N., Weber, C.: Analysing the influence of different street vegetation on traffic-induced particle dispersion using microscale simulations. J. Environ. Manage. **94**(1) (2012)

14. Zhou, S., Tang, R., Zhang, Y., Ma, K.: Simulation study on the influence of street canyon green belt setting on air flow field and pollution distribution. Acta Ecologica Sinica **38**(17), 6348–6357 (2018)

Applying Numerical Simulation to Identify the Suitable Block Scale for Improving Air Quality Inside Urban Streets

Wen Liu[1,2,3], Zhengdong Huang[1,2,3(✉)], Maopeng Sun[1,2,3,4], Hongliang Zhang[5], Fuyun Zhao[5], and Renzhong Guo[1,2,3]

[1] School of Architecture and Urban Planning, Research Institute for Smart Cities, Shenzhen University, Shenzhen 518060, China
zdhuang@szu.edu.cn
[2] State Key Laboratory of Subtropical Building and Urban Science, Shenzhen 518060, China
[3] Guangdong-Hong Kong-Macau Joint Laboratory for Smart Cities, Shenzhen 518060, China
[4] Shenzhen Urban Transport Planning Center Co., Ltd, Shenzhen 518000, China
[5] School of Power and Mechanical Engineering, Wuhan University, Wuhan 430072, China

Abstract. Rapid urbanization-induced poor ventilation makes it difficult for urban built-up environments to breathe fresh air and purify pollutants. Meanwhile, vehicle exhaust is emitted into urban streets and spread to nearby residential neighborhoods, severely threatening both the environmental quality of street space and the well-being of citizens residing adjacent to the road. The numerical simulation validated by wind tunnel measurements was adopted to examine the effects of the block scale (i.e. street length) on natural ventilation and exhaust pollutant dispersion within urban street canyons. The findings are as follows: (1) the airflow regime inside the street canyon is determined by the interaction of a canyon vortex at the middle and corner eddies at the ends, leading to notably elevated pollutant concentrations on its leeward and mid-section than on its windward and laterals; (2) the induced corner vortices at street ends can drive ambient pollutants toward the middle of the street canyon and subsequently be carried out by the canyon vortex, however there is a limit to this effect, which varies depending on the length of the street canyon with constant aspect ratio; (3) a proper block scale (street length) ranging from 200 m to 400 m is recommended, which could be used to strike a favorable balance between environmentally improving urban street air quality and morphologically creating a vibrant public space. Numerical simulation-based design strategies can be utilized to optimize existing urban design guidelines for creating a desirable scale in practical urban planning and construction.

Keywords: Street length · CFD Simulation · Ventilation · Pollutant Dilution

1 Introduction

Rapid urbanization has led to the expansion of urban boundaries and the occupation of natural surfaces by a vast array of buildings, particularly in developing countries. This phenomenon has made it challenging for urban built-up areas with limited ventilation capacity to breathe fresh air and purify pollutants [1]. Meanwhile, the continuous

© The Author(s) 2025
B.-J. He et al. (Eds.): UCSUD 2023, LNCE 559, pp. 603–616, 2025.
https://doi.org/10.1007/978-981-97-8401-1_43

emission of high traffic emissions in urban streets and their subsequent dispersion into adjacent neighborhoods have had a profound impact on air quality in street spaces, posing a serious threat to the physical and mental health of pedestrians and residents [2]. Streets are vital outdoor public spaces in cities, and it is imperative to strive for fresh air and comfortable living conditions, especially for the elderly and children who are particularly sensitive to the air environment [3].

Over the past three decades, natural ventilation and air pollution in urban streets have been extensively examined through field measurements, numerical simulations, and physical experiments [4, 5]. However, due to the significant labor and financial costs involved, on-site measurements and wind tunnel experiments are not easily feasible in urban planning and construction contexts. Conversely, numerical simulations based on Computational Fluid Dynamics (CFD) have gained popularity for investigating airflow and pollutant dispersion in urban street spaces because of their flexibility in implementation and cost-effectiveness [6]. Drawing upon the understanding of ventilation and pollutant dispersion mechanisms in urban street canyons, numerous urban design strategies have been proposed at the street scale to mitigate pollutant concentrations at the pedestrian level. These strategies aim to induce more fresh air into the street canyons, including low or non-uniform aspect ratios [7, 8], innovative building openings design [9], skewed street intersections [10], and vegetation configurations [11]. Nevertheless, it is noteworthy that in most of these studies, street canyons are often assumed to be infinitely long or fixed at a specific length, which may not fully capture the complexities and variabilities of real-world urban environments. Vardoulakis et al. [12] emphasized significant disparities in ventilation efficiency and pollutant dispersion behavior across street canyons of varying lengths. Chan et al. [13] revealed that canyon length profoundly impacts near-surface air pollution in short street canyons, influenced by airflow patterns at canyon ends. Li et al.[14]further demonstrated that lateral entrainment could significantly reduce pollutant concentrations by 78% in short (L/W = 1) and medium (L/W = 5) canyons. However, there remains a notable gap in understanding the impact of street length on ventilation and pollutant dispersion in street canyons, which could inform the development of practical urban and street design strategies aimed at improving air quality within these spaces.

Additionally, street length, defined as the distance between two intersections, is conventionally regarded as the block scale, serving as a metric for quantifying the dimensions of a planar urban block. The block scale or street length plays a pivotal role in urban road network planning [15], which is regarded as a guiding principle for land use planning in both the development of new urban areas and the revitalization of existing towns. It aims to mitigate various urban challenges, including enhancing street vitality, reducing traffic congestion, and improving spatial quality. The appropriate block size has been devoted to seeking for steering urban spatial planning and creating vibrant public spaces [15]. Moughtin and Shirley [16] proposed a sustainable urban block scale ranging from 70 to 100 m. Zang [17] argued that the basic block units typically span a range of 200 to 600 m, through an analysis of planning practices in modern eco-cities worldwide. Through meticulous measurements of block sizes in numerous prominent urban hubs globally, Huang and Sun [18] have observed a prevalent trend of blocks smaller than 200 m as illustrated in Fig. 1. When juxtaposed against the predominantly larger block dimensions prevalent in contemporary urban planning and development practices in China, a block size falling within the range of 150 to 200 m emerges as a judicious and viable option [18].

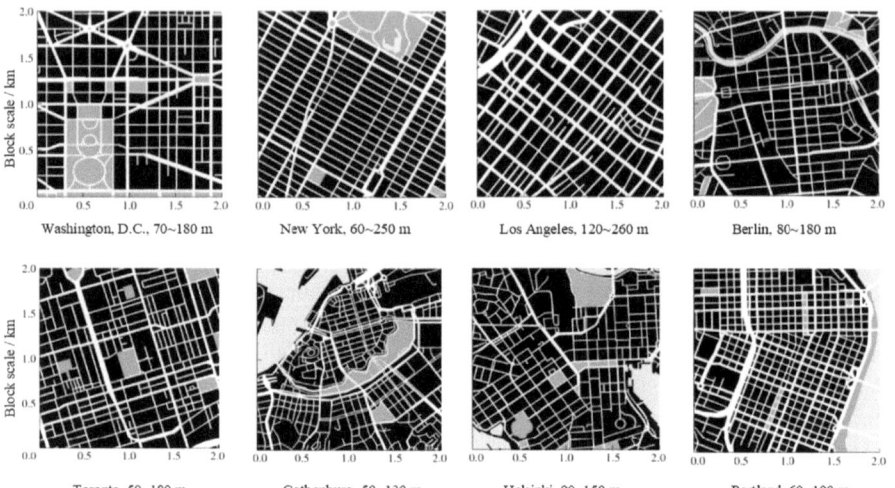

Fig. 1. Block scales in the central districts of some major cities (redrawn based on [18]).

Given the diverse geographic features and functional layouts of cities, a flexible block scale guidance (e.g., 100 m to 500 m street lengths) is believed to optimize urban spatial morphology, fostering lively social spaces [19]. However, these recommendations primarily focus on urban morphology and spatial psychology, neglecting the profound impact of block scale on microclimate elements. Notably, the microclimatic environment, particularly air quality, significantly influences urban health and livability. Consequently, the present study aims to provide a microclimatological perspective for enhancing the block-scale guidance scheme. To achieve this, a series of numerical simulations are conducted on street canyons with varying lengths, enabling a systematic investigation of the influence of block scale on air quality within urban streets.

Accordingly, the objectives of the present study are given as follows. It aims: (i) to qualitatively explore the influences of block dimension on the ventilation flow and pollutant concentration fields within urban street canyons; (ii) to quantitatively assess the effects of block scaling on the ventilation property and pollutant removal capability of the street canyons; and (iii) to identify the proper street scale for balancing the enhancement of spatial vitality in urban downtowns and the betterment of the air quality in street spaces.

2 Methodology

2.1 Domain and Boundary Conditions

In this study, three-dimensional regular street canyons were modeled to investigate the effects of block scale/street length on ventilation airflow and ambient pollutant fields. The buildings on two sides of the canyon were designed as residences with 6 floors, which means that the height (H) and depth (B) of them were fixed at 20 m and 15 m, respectively. The street canyon width was set as $W = 20$ m, implying that its aspect ratio (i.e. height-to-width or H/W) is 1.0, and the prevailing wind flows perpendicularly to the street axis. The traffic emission was assumed to be a surface source with two vehicle lanes at the ground level, that is, the width of the pollutant source was $We = 7$ m. The block scales were taken to be 40 m, 80 m, 120 m, 160 m, 200 m, 300 m, 400 m, 600 m, and 800 m, with the corresponding street length (L) falling in the interval of 2W to 40W. Additionally, a two-dimensional model with $H = W = 20$ m was also used as a base case, which was compared with three-dimensional cases to characterize the differences in street canyon airflow and pollutant concentration fields between the two-dimensional simulations.

The geometric dimensions of the computational domain are depicted in Fig. 2. The simulations followed well-known practice guidelines to establish the 3-D computational domain [20, 21]. Thus, in the CFD model, the distances from the outer boundaries of the physical street space model to three specific surfaces of the domain (i.e. inlet, outlet, and top) are 5H, 15H, and 6H, respectively. The Renormalization Group (RNG) k-ε turbulence scheme was chosen, based on the comparison of the validation results on three k-ε models in Sect. 2.4. Table 1 provides a comprehensive overview of the boundary condition settings. Drawing from prior studies, the symmetric boundary scheme was applied to the lateral boundaries of the computational domain to avoid the interference of the lateral flow, while the top of the domain was designated as a free-slip boundary [22]. The building walls and ground surface were configured with no-slip boundary conditions with standard wall functions. The emission rate of carbon monoxide (CO) as traffic exhaust was fixed as $Q_e = 0.003$ kg/s. It is noteworthy that background pollutant concentrations and variable meteorological factors were not incorporated into the numerical modeling.

2.2 Computational Girds

In all three-dimensional numerical simulation cases, the total mesh count ranged from 1.6 to 9.4 million, varying with the length of the street canyon. Non-uniform structured hexahedral meshes were generated, while the smallest meshes, cubic cells of 0.02 m (i.e., $0.0008H$) deployed near building walls and the ground surface. The meshes become progressively rougher from these solid surfaces outwards, while it expands by less than 1.08.

To ensure the numerical accuracy of our results, a grid sensitivity analysis was conducted on a canyon model characterized by an aspect ratio of H/W = 1 and a length of L = 200 m. Three computational grid schemes (coarse, basic, and refined) were constructed to simulate wind flow and pollutant concentration fields, respectively. Coarsening and

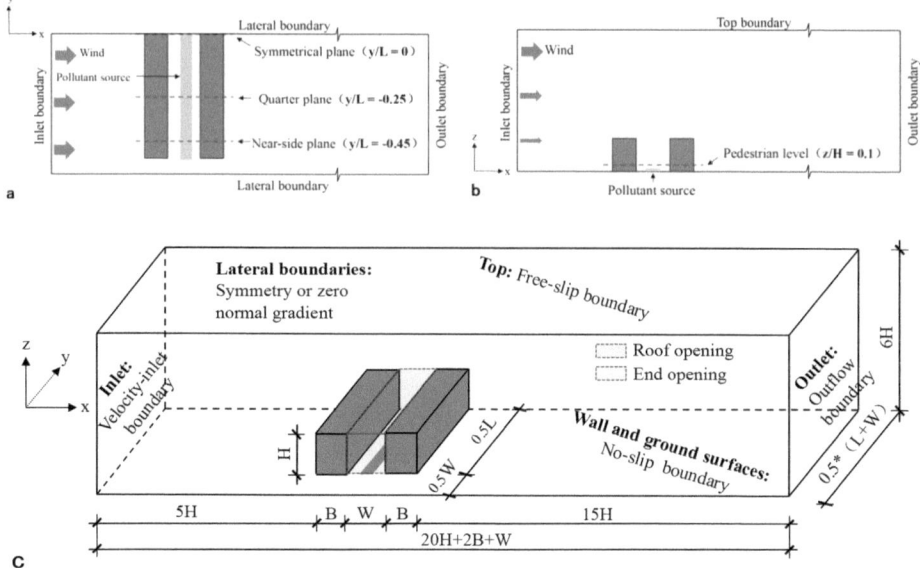

Fig. 2. An illustrative depiction of computational domain and boundary conditions in CFD simulations: (a) overhead, (b) lateral, and (c) isometric views, including boundary condition settings.

refining were performed by virtue of diverse grid sizes and stretch ratios of adjacent grids. The comparison of three grid systems based on dimensionless simulated values was implemented. Wind speed (u/U_{ref}) and turbulent kinetic energy (TKE) located at the center line ($z/H = 0.5$, along x-direction) in a regular street canyon of the three grid systems were compared, as shown in Fig. 3. Notably, the velocity (u/U_{ref}) and turbulent kinetic energy (TKE) fields exhibited minimal variation with the incremental increase in mesh density, from coarse to refined grids. This observation suggests that the basic grid system offers sufficient resolution for the purposes of the present study.

2.3 Solution Method

In this study, the Reynolds-Averaged Navier-Stokes (RANS) model is utilized to describe turbulence properties in CFD numerical simulations. Although large eddy simulation (LES) could accurately capture the structure and dynamic features of turbulence behaviors, it needs more computational resources and cost than the RANS-based model [23]. In a large number of recent investigations of urban street airflow and pollutant dispersion, the RANS-based CFD models have become a widely adopted numerical simulation technique, due to their ability of reasonably providing accurate results with high efficiency [24]. In this study, steady RANS simulations with high performance were conducted using the ANSYS Fluent 19.0 procedure. The RNG k-ε model is used to decipher the isothermal turbulence field, as it has been shown in early studies to reliably recharacterize the results of wind tunnel experiments and field measurements [25]. The air is assumed

Table 1. Boundary conditions setting in CFD simulations.

Boundary conditions	Specific settings
Inlet	Velocity-inflow boundary condition, according to the fitting results of the conducted wind tunnel experiment [23], $U(z)/U_{ref} = (z/H)^\alpha$, $k(z) = (U(z)I(z))^2, \varepsilon = \sqrt{C_\mu}k(z)\frac{dU(z)}{dz}$
Outlet	Outflow boundary condition
Top	Free-slip boundary condition
Lateral	Symmetric boundary conditions
Building walls and ground surface	No-slip boundary conditions
Pollutant source	$Q_e = 0.003$ kg/s, a surface source with two vehicle lanes at the ground level

Note $U(z)$ denotes the wind velocity at a specific height z (m), whereas U_{ref} represents the reference velocity observed at the building's height H. The power law exponent dependent on the terrain type is set as $\alpha = 0.26$. The inlet values of the turbulent kinetic energy $k(z)$ and dissipation rate ε profiles were precisely outlined following the above equations, respectively. $I(z)$ is the turbulent intensity, and $C_\mu = 0.0845$ is a constant. The predicted values of pollutant concentrations were calculated dimensionlessly for comparison among the simulation cases, following the equation $C^* = CU_{ref}HL/Q_e$.

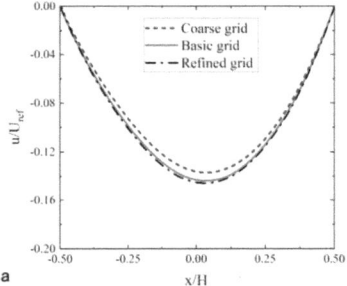 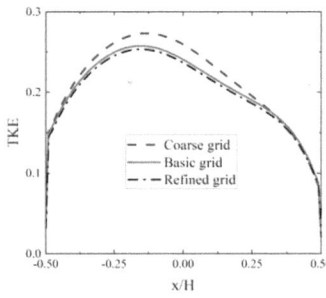

Fig. 3. A comparative analysis of the mean velocity (u/U_{ref}) (a) and turbulent kinetic energy (TKE) (b) at the center line ($z/H = 0.5$, along x-direction) in a regular street canyon.

to be an incompressible flow, i.e., the density of air and gaseous pollutants is considered to be unchanged. The second-order upwind discretization scheme is employed for the computation of momentum, turbulent kinetic energy, and turbulent dissipation rate, while the pressure-velocity coupling is handled through the application of the SIMPLE algorithm [26]. Until the residuals of continuity, momentum, and turbulence are less than 10^{-5} and simultaneously remain stable over hundreds of iterations, the convergence of the simulations is determined to be achieved.

2.4 Model Validation

The validation of turbulence models in RANS simulations is essential to ensure the accuracy of CFD simulations. Previously conducted wind tunnel experiments or field measurements have been routinely adopted as references for numerical simulations. In order to fulfill the objectives of this study, the obtained results of wind tunnel measurements completed at the Japanese Niigata Institute of Technology [23], were used for comparison with the flow and pollutant concentration fields in the numerical simulations. The model validation exercise has been detailed in our earlier investigations [3].

As a whole, the simulated concentrations using the RNG model agree relatively better with the wind tunnel experiment measurements, compared with those of adopting standard k-ε and realizable k-ε models. The simulation results of three k-ε models have some errors, which also represent an inherent flaw of the RANS models demonstrated in previous studies [23, 27]. Comprehensively, the suited RNG k-ε model was used to observe the varying patterns of urban street canyon ventilation and air quality with the block scale.

3 Results and Discussion

3.1 Airflow Structures

For street canyons with the perpendicular approaching wind, two distinguishable street canyon airflow characteristics were summarized: canyon vortex (rotating along the canyon axis, induced by the skimming flow through the roof opening) and corner vortex (rotating along the vertical axis, induced by the channelized flow through the end opening) [11]. Canyon eddies drive the transfer and accumulation of near-surface emissions released in the middle of the street canyon from the windward to the leeward side. This leads to a marked elevation in pollution levels on the leeward side when compared to the windward side. Meanwhile, corner vortices can repel pollutants near the street end toward the center, leading to noticeably worse air pollution conditions at the central section than that at the ends. Furthermore, within relatively long streets, both types of vortices were found to dramatically evolve into a complex spiral flow that fills the whole canyon space [28]. Figure 4 depicts the streamlines around buildings in three-dimensional short ($L/W = 2$) and long ($L/W = 10$) street canyons. For the short street (i.e. tiny neighborhood), two canyon vortex-dominated airflow sections can be observed, which are symmetrically aligned on both sides of the canyon midst. However, for the long street (i.e. large block), an unorthodox airflow structure is demonstrated that consists of canyon vortices near the two tail ends coupled with a corner vortex near the center. It is noteworthy that the spiraling portion mentioned above is observable at the waist ($y/L = \pm0.25$) inside the long street canyon, while the extent it develops will determine the maximum distance over which exhaust pollutants at the canyon ends are induced to be transported toward the canyon center.

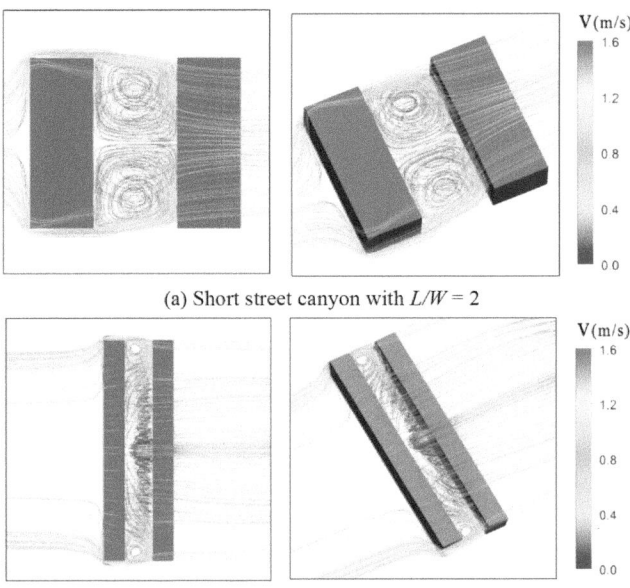

(a) Short street canyon with $L/W = 2$

(b) Long street canyon with $L/W = 10$

Fig. 4. 3-D streamlines within street canyons with $L/W = 2$ (a) and $L/W = 10$ (b).

3.2 Pollutant Concentration Fields

It is customary for residents to ensure a high-quality air environment by opening windows to introduce outdoor fresh air into the indoors [29]. Figure 5 shows the distribution of dimensionless pollutant concentrations in three-dimensional street canyons ($L/W = 2, 6$, 10, and 20) adjacent to the backwind and headwind walls. In general, the pollutant levels near the leeward facade are significantly higher than those near the windward facade, while the ambient pollutant concentrations gradually increase from the near-surface to the roof vicinity. Meanwhile, except for the short street canyon ($L/W = 2$), the air quality at the center of other long canyons is remarkably poorer than those at two street ends, but this dynamic differs between the leeward and windward facades. In the case of a street canyon with the block scale of $L/W = 20$, the concentration level on the leeward wall will increase along the y-direction from the end to the center and then remain nearly unchanged, while the concentration on the windward wall will reach the highest value and then drop slightly to a non-minimum condition. Obviously, this is strongly related to the airflow pattern and driving mechanism within the street canyon caused by the street length (i.e., block scale), especially the shear force intensity of the corner vortex induced by the passage flow at the street terminal.

3.3 Air Ventilation

Natural ventilation significantly drives atmospheric contaminants transmission and dispersion around buildings [30]. The average air exchange rate *AER* was applied to evaluate the air-changing performance of the street canyon at different vents [31]. The equations

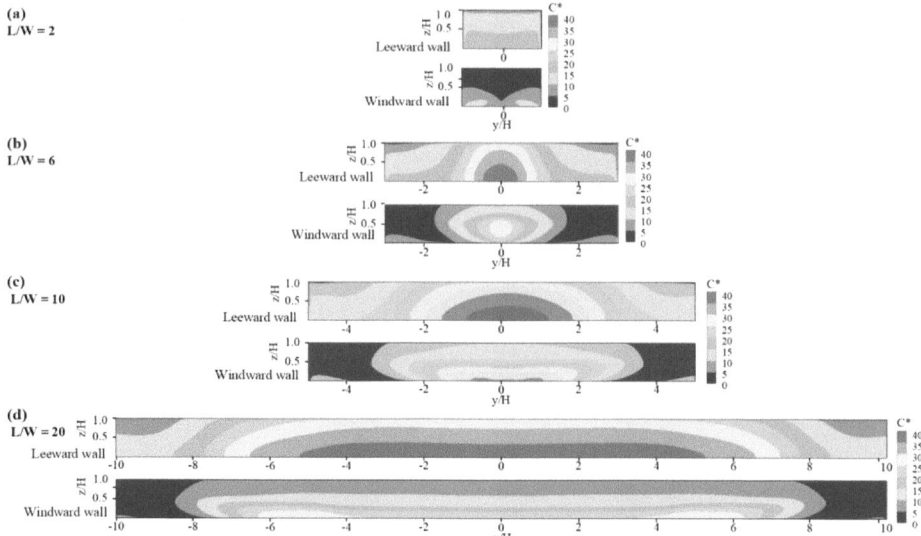

Fig. 5. Normalized pollutant concentration at y-z planes adjacent to the leeward and windward facades for three-dimensional street canyons.

for calculating the mean component $\overline{AER}$ and the turbulent component AER' of the air exchange rate at different openings have been detailed in our earlier work [3]. In this study, the total ventilation AER_c for the street space will be the sum of AER_{roof} and AER_{end}, following the formula $AER_c = AER_{roof} + AER_{end}$.

The air exchange rate as a function of length-to-width ratio (L/W) for the vents in the two-dimensional street canyon and the three-dimensional street canyon are plotted in Fig. 6 Overall, the air exchange rates of 3-D urban street canyons decrease with the length/width ratio (street length) until a constant level. This indicates that oversized block scales, i.e., excessively long streets, are not conducive to improving natural ventilation within the street canyon, which is consistent with the underlying idea of enhancing the city's breathability as put forward by some scholars [32, 33]. More specifically, the air exchange rate of the roof vents in three-dimensional urban street canyons generally increases progressively with the street canyon length (i.e., block scale) first up to a state of constancy, while the air exchange rate of the end openings behaves oppositely. This suggests that the air movement induced by the skimming flow over the rooftop gradually dominates as the length of the street canyon increases, while the driving force from street ends slowly weakens the improvement of ventilation within the canyon. Subsequently, the turbulence effect of air exchange is stronger than the role of its mean component for both roof and end vents, which agrees well with the findings in many previous studies [5, 30]. It is interesting to note that the air change capacity of the rooftop in the three-dimensional street canyon approaches very closely the constancy of a two-dimensional street canyon. This means that the two-dimensional model can only be used as a substitute for the infinitely elongated three-dimensional model in order to evaluate its natural ventilation performance (see Fig. 6a–b). The threshold length-to-width ratio range for the typical three-dimensional street canyon with $H/W = 1$ is identified as $L/W = 10 \sim 20$, implying that when the block scale ranges from 200 m to 400 m, the street end makes the greatest contribution to the domestic ventilation in urban street space.

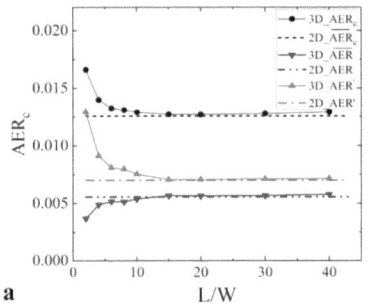

a

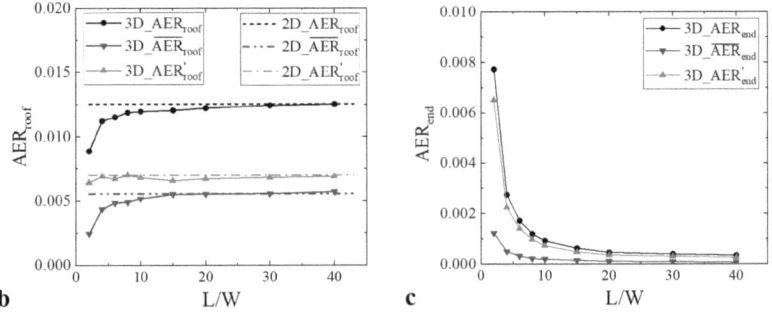

b   c

Fig. 6. Relationships between street length-to-width ratio (L/W) and various air exchange rate parameters, namely the overall air exchange rate (AER_C), mean exchange rate ($\overline{AER}$) and turbulent exchange rate (AER').

3.4 Pollutant Dispersion and Removal

The net escape velocity NEV, the average retention time ART and the mean pollutant concentration at pedestrian level C^*_{ped} will be employed to evaluate the pollutant removability of these simulated cases. The corresponding equations have also been elaborated in previous studies [3, 32].

Figure 7 presents a comprehensive illustration of the net escape velocity, average residence time, and mean pollutant concentration at the pedestrian level, all plotted as functions of the length-to-width ratio for both two-dimensional and three-dimensional street canyons. This study aims to find the optimal block scale to achieve a reasonable balance between the concept of "Narrow Roads and Dense Road Network" in the urban design guidelines and the goal of improving the air quality of streets in the environment governance. An effective block scaling (street length) guidance based on the dispersion mechanism of ambient pollutants can be utilized to create high-quality urban public spaces by enhancing the air environment. Similar to the description of the air exchange rate in the former section, the net escape velocity of pollutants inside three-dimensional street canyons drops stepwise to a steady condition with the value of L/W, while its average retention time and pedestrian-level mean pollutant concentration rises to a constant level accordingly. It indicates the effect of removing pollutants from the end vents and driving them toward the center of the street canyon would be limited, and the critical length-to-width ratio for effective pollutant removal is also considered to be $L/W = 10$ ~ 20 (i.e., a block size of 200 m to 400 m). Thus, the two processes of ventilation and pollutant dilution are proven to be synergistic [14]. Clearly, the block-scale interval oriented towards the improvement of atmospheric quality in urban street spaces has been identified, and that is when the role of street end openings in pollutant dispersion and clearance reaches a threshold. In addition, the maximum of the street end-driving forces eventually approaches the condition of the two-dimensional street canyon, also suggesting that the two-dimensional model can only be utilized to replace three-dimensional models with a specific length-to-width ratio (street length) to evaluate pollutant removal performance in the street canyon.

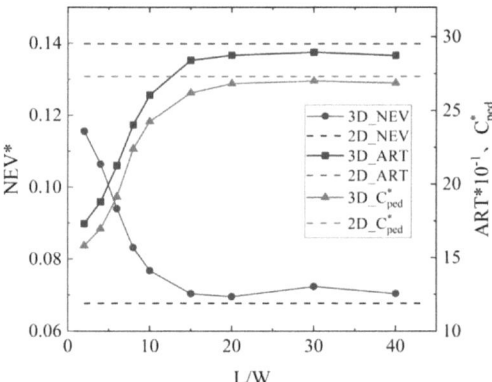

Fig. 7. The net escape velocity NEV^*, average retention time ART^* and pedestrian-level mean pollutant concentration C^*_{ped} against length-to-width ratio L/W.

4   Conclusions

The typical regular street canyon with the perpendicularly approaching wind is three-dimensionally modeled to simulate the processes of air flow and pollutant dispersion. Nine cases of fixed-width street canyons are investigated at block scales (street lengths) ranging from 40 m (L/W = 2) to 800 m (L/W = 40). The constant-rate source of pollutant emissions is assumed to be on the near-ground level in the middle of the street canyon, which contains two motor vehicle lanes. The corresponding two-dimensional model with H/W = 1.0 is employed for comparison with the three-dimensional models to identify the feasibility and applicability of substituting the real urban street canyon scenarios with the two-dimensional simulation. The CFD numerical simulation procedure is validated by comparing the accessible wind tunnel measurements, with the accompanying symmetric meshing approach.

The detrimental impact of block scale on ventilation flow and subsequent air quality within a conventional urban street canyon has been rigorously examined. It has been found that the airflow dynamics within the street canyon are predominantly influenced by a vertically rotating vortex at its midsection, alongside two horizontally rotating corner eddies situated at the extremities. Collectively, these vortices shape the overall airflow pattern within the street canyon. In alignment with the observations made by Gromke et al. [27] and Tsai et al. [28], both well-characterized vortices couple into a spiraling airflow, which causes the air pollution level of the canyon to be significantly higher on the leeward facade than on its windward one, and greater at the center than at the ends. Further, the corner eddies induced by the perpendicularly incoming channel flow can drive ambient pollutants towards the mid-street canyon, which are subsequently carried out of the street canyon space by the skimming flow-induced canyon vortex over the rooftop. These corner vortices are highly conducive to facilitating air circulation and pollutant dispersion in the short street canyon (small-scale street), whereas their repelling effect would be very weak in the long street canyon (large-scale block). Moreover, the driving force has been proven to be limited, which depends on the street length (i.e., block dimension) at a constant aspect ratio.

In the present study, it is crucial to emphasize that the recommended range for the street canyon's length-to-width ratio falls within $L/W = 10$ to 20, which corresponds to an acceptable street canyon length spanning from 200 m to 400 m. In other words, it is when the incoming airflow from the end openings almost expends all the energy to enhance the ventilation and air quality inside the urban street space. Notably, this interval falls within the range of diverse block scale (100 m ~ 500 m) recommended in many existing urban design guidelines. Also, the suggested street size of 200 m for urban central region planning and design practices fits in the proposed band. It can be seen that when developing a suitable scheme for city block and road network planning, the identified block dimension based on numerical simulation strikes a balance between improving the air quality of urban streets and creating vibrant and orderly public spaces. Still, it should be mentioned that a smaller block scale/street length actually enhances the permeability of urban natural ventilation for improving the diffusion and removal of traffic exhaust from city street public spaces. Hence, when implementing urban planning and design, the differential standards of block/neighborhood size should be further rationalized and stipulated, according to the functional orientation of urban blocks and air quality requirements.

Acknowledgments. This work was supported by the China Postdoctoral Science Foundation (Grant No. 2023M732353), Guangdong Basic and Applied Basic Research Foundation Guangdong-Shenzhen Joint Youth Fund (Grant No. 2023A1515110097) and Shenzhen Science and Technology Program (Grant No. 20220809120650001). The authors would like to appreciate the generous technical support offered by Dr. Huai-yu Zhong and Mr. Yi Jing of the School of Power and Mechanics Engineering, Wuhan University, throughout the CFD simulations.

References

1. Nyberg, F., Gustavsson, P., Jarup, L., Bellander, T., Berglind, N., Jakobsson, R., Pershagen, G.: Urban air pollution and lung cancer in Stockholm. Epidemiology **11**, 487 (2000)
2. Gallagher, J., Lago, C.: How parked cars affect pollutant dispersion at street level in an urban street canyon? A CFD modelling exercise assessing geometrical detailing and pollutant decay rates. Sci. Total Environ. **651**, 2410 (2019)
3. Liu, W., Zhan, Q., Shao, Z., Qiu, C., Wen, C.: CFD simulation of the influence of street interface density on natural ventilation and pollutants diffusion in urban streets. Geomatics Inf. Sci. Wuhan Univ. (2022). https://doi.org/10.13203/j.whugis20210711. (in Chinese, online)
4. Li, Z., Ming, T., Shi, T., Zhang, H., Wen, C., Lu, X., Dong, X., Wu, Y., de Richter, R., Li, W., Peng, C.: Review on pollutant dispersion in urban areas-part B: local mitigation strategies, optimization framework, and evaluation theory. Building Environ. **198**, 107890 (2021)
5. Li, Z., Ming, T., Liu, S., Peng, C., de Richter, R., Li, W., Zhang, H., Wen, C.: Review on pollutant dispersion in urban areas-part A: effects of mechanical factors and urban morphology. Building Environ. **190**, 107534 (2021)
6. Tominaga, Y., Stathopoulos, T.: CFD simulation of near-field pollutant dispersion in the urban environment: a review of current modeling techniques. Atmos. Environ. **79**, 716 (2013)
7. Zhang, K., Chen, G., Wang, X., Liu, S., Mak, C.M., Fan, Y., Hang, J.: Numerical evaluations of urban design technique to reduce vehicular personal intake fraction in deep street canyons. Sci. Total Environ. **653**, 968 (2019)
8. Gu, Z., Zhang, Y., Cheng, Y., Lee, S.: Effect of uneven building layout on air flow and pollutant dispersion in non-uniform street canyons. Building Environ. **46**, 2657 (2011)
9. Fan, M., Chau, C.K., Chan, E.H.W., Jia, J.: A decision support tool for evaluating the air quality and wind comfort induced by different opening configurations for buildings in canyons. Sci. Total Environ. **574**, 569 (2017)
10. Yassin, M.F., Kellnerová, R., Jaňour, Z.: Impact of street intersections on air quality in an urban environment. Atmos. Environ. **42**, 4948 (2008)
11. Gromke, C., Ruck, B.: Influence of trees on the dispersion of pollutants in an urban street canyon - Experimental investigation of the flow and concentration field. Atmos. Environ. **41**, 3287 (2007)
12. Kumar, P., Ketzel, M., Vardoulakis, S., Pirjola, L., Britter, R.: Dynamics and dispersion modelling of nanoparticles from road traffic in the urban atmospheric environment-A review. J. Aerosol Sci. **42**, 580 (2011)
13. Chan, A.T., So, E., Samad, S.C.: Strategic guidelines for street canyon geometry to achieve sustainable street air quality. Atmos. Environ. **35**, 4089 (2001)
14. Li, Z., Zhang, H., Wen, C., Yang, A., Juan, Y.: The effects of lateral entrainment on pollutant dispersion inside a street canyon and the corresponding optimal urban design strategies. Building Environ. **195**, 107740 (2021)
15. Eggimann, S.: The potential of implementing superblocks for multifunctional street use in cities. Nat. Sustain. **5**, 406 (2022)
16. Moughtin, C., Shirley, P.: Urban design: green dimensions. Architectural Press, Oxford (2003)
17. Zang, X.Y.: The construction of technical indicator system for Eco-Cities at blocks scale. Urban Plann. Forum. **04**, 81–87 (2013). (in Chinese)
18. Huang, Y.Q., Sun, Y.M.: Judgement characteristics and quantitative index of suitable block scale. J. South China Univ. Technol. (Natural Science Edition) **40**, 131 (2012). (in Chinese)
19. Zheng, X., Yang, J.: Urban road network design for alleviating residential exposure to traffic pollutants: super-block or Mini-block? Sustain. Cities Soc. **89**, 104327 (2023)
20. Tominaga, Y., Mochida, A., Yoshie, R., Kataoka, H., Nozu, T., Yoshikawa, M., Shirasawa, T.: AIJ guidelines for practical applications of CFD to pedestrian wind environment around buildings. J. Wind Eng. Ind. Aerodyn. **96**, 1749 (2008)

21. Franke, J.O.R., Hellsten, A., Nzen, H.S.U., Carissimo, B., Grawe, D., N I.A.N.G., Jaňour, Z., Karppinen, A.: Best practice guideline for the CFD simulation of flows in the urban environment: COST Action 732 quality assurance and improvement of microscale meteorological models (2007)
22. Hang, J., Li, Y., Sandberg, M., Buccolieri, R., Di Sabatino, S.: The influence of building height variability on pollutant dispersion and pedestrian ventilation in idealized high-rise urban areas. Building Environ. **56**, 346 (2012)
23. Tominaga, Y., Stathopoulos, T.: CFD modeling of pollution dispersion in a street canyon: Comparison between LES and RANS. J. Wind Eng. Ind. Aerodyn. **99**, 340 (2011)
24. Allegrini, J., Dorer, V., Carmeliet, J.: Buoyant flows in street canyons: validation of CFD simulations with wind tunnel measurements. Building Environ. **72**, 63 (2014)
25. Tominaga, Y., Stathopoulos, T.: Numerical simulation of dispersion around an isolated cubic building: model evaluation of RANS and LES. Building Environ. **45**, 2231 (2010)
26. Patankar, S.: Numerical heat transfer and fluid flow. CRC Press (1980)
27. Gromke, C., Buccolieri, R., Di Sabatino, S., Ruck, B.: Dispersion study in a street canyon with tree planting by means of wind tunnel and numerical investigations - Evaluation of CFD data with experimental data. Atmos. Environ. **42**, 8640 (2008)
28. Tsai, M.Y., Chen, K.S.: Measurements and three-dimensional modeling of air pollutant dispersion in an Urban Street Canyon. Atmos. Environ. **38**, 5911 (2004)
29. Yang, F., Kang, Y., Gao, Y., Zhong, K.: Numerical simulations of the effect of outdoor pollutants on indoor air quality of buildings next to a street canyon. Building Environ. **87**, 10 (2015)
30. Hang, J., Sandberg, M., Li, Y.: Age of air and air exchange efficiency in idealized city models. Building Environ. **44**, 1714 (2009)
31. Hang, J., Li, Y.: Wind conditions in idealized building clusters: macroscopic simulations using a porous turbulence model. Boundary-Layer Meteorol. **136**, 129 (2010)
32. Hang, J., Wang, Q., Chen, X., Sandberg, M., Zhu, W., Buccolieri, R., Di Sabatino, S.: City breathability in medium density urban-like geometries evaluated through the pollutant transport rate and the net escape velocity. Building Environ. **94**, 166 (2015)
33. Ren, C., Yang, R., Cheng, C., Xing, P., Fang, X., Zhang, S., Wang, H., Shi, Y., Zhang, X., Kwok, Y.T., Ng, E.: Creating breathing cities by adopting urban ventilation assessment and wind corridor plan – The implementation in Chinese cities. J. Wind Eng. Ind. Aerodyn. **182**, 170 (2018)

Health Risk Assessment of PM$_{2.5}$ in Mianyang City Commercial District

Xuan-Yan Li$^{(\boxtimes)}$, Juanlin Fu, Li Yan, Minghong Yu, Yu Long, and Yuzhen Liu

School of Civil Engineering and Architecture, Southwest University of Science and Technology, Mianyang 621010, China

623835189@qq.com

Abstract. PM$_{2.5}$ is one of the main pollutants of air pollution. Long-term exposure to high concentrations of PM$_{2.5}$ polluted environment can easily increase the risk of developing various diseases including cardiovascular disease and respiratory diseases. As the main place for people's leisure and entertainment, the exposure level and health risk of PM$_{2.5}$ in urban commercial districts directly relate to people's health. Based on the principle of exposure assessment, this paper constructs a method for assessing the health risk of urban public space environment and carries out empirical research on the commercial district of Mianyang by using on-site measurement methods. On the basis of collecting PM$_{2.5}$ concentration data, combined with the type of crowd activity and stay time, important public spaces in the commercial district are selected for health risk assessment and optimization strategies are proposed. Researches Indicate that the change of PM$_{2.5}$ concentration in commercial areas is affected by traffic flow and wind direction; there is a high health risk in public spaces where crowds gather and stay for a long time. The research results have important practical significance for improving the public disease prevention ability and health management level in cities.

Keywords: Healthy risk · PM$_{2.5}$ · Commercial district · Empirical research · Mianyang City

1 Introduction

With the acceleration of urbanization, the harm of air pollution is becoming increasingly prominent, gradually threatening people's health. PM$_{2.5}$ as a kind of inhalable fine particulate matter, enters the lungs and blood circulation through the respiratory tract, causing various diseases [1]. Chen et al. show that pathogens absorbed by fine particulate matter are easy to enter the human body through the respiratory tract, causing inflammatory reactions and directly inducing infectious diseases of the respiratory system [2]. LEE YL et al. proposed that fine particulate matter causes the production of free radicals in the lungs through organic components and excessive elements on the surface, leading to airway inflammatory response and lung function damage [3]. Yang et al. showed that the organic matter inside the fine particulate matter would reduce the tumor suppressor genes, increase the infectivity of the population to lung cancer, and

© The Author(s) 2025
B.-J. He et al. (Eds.): UCSUD 2023, LNCE 559, pp. 617–629, 2025.
https://doi.org/10.1007/978-981-97-8401-1_44

improve the probability of cancer change [4]. Pinault et al. found that Canada in 1998 and 2011 There is an association between population mortality between years and fine particulate matter, where the number of deaths from cardiovascular disease increases by 1.25% per 10 ug/m^3 increase in PM$_{2.5}$ concentration [5]. Chen et al., in 2013 to 2015, conducted a study of deaths in 272 major cities and daily population in China, and the results showed that for every 10μg/m^3 increase in PM$_{2.5}$, deaths from respiratory diseases increased by 0.29% [6].

With the improvement of people's life quality, the improvement of urban public space environment has become the research focus of many scholars. Especially in the aspects of urban public space environment and microclimate construction, most of the current literature research is based on physical environment elements. Liu et al. found that spatial elements are an important factor affecting the microclimate scale of landscape architecture, and thermal comfort is the main comfort research theme at present [7]. Liu et al. studied the thermal comfort of microenvironment of four commercial streets in Tianjin, and the results showed that the temperature and humidity have different differences, which can be improved by microenvironment design [8]. PM$_{2.5}$ is difficult for the human body to directly perceive and requires professional equipment to detect. Wang proposed that past studies focused on the direct correlation between spatial form and PM$_{2.5}$ distribution, evaluating the impact of spatial design on air quality, and less on the pollutant exposure concentration characteristics and risk assessment of pedestrians and the environment [9]. The key to the health risk study of PM$_{2.5}$ pollution is to assess the exposure of pollutants. PM$_{2.5}$ intake in the population is mainly determined by the following three factors: PM$_{2.5}$ exposure concentration, exposure time, and respiration rate of the population [10]. City business district as a person The spatial characteristics of the main places of group leisure and entertainment will have a certain influence on the concentration distribution of PM$_{2.5}$ in the block.

As mentioned above, this article chooses PM$_{2.5}$ as the monitored pollutant, selects the Jianguomen commercial district of Mianyang City as the research object, combines the method of on-site measurement, analyzes the exposure concentration distribution of PM$_{2.5}$ in the commercial district, identifies important public spaces within the district, conducts health risk assessments for the district and proposes relevant strategies.

2   Methodology

2.1   Study Area

Mianyang City is located in the northwest of the Sichuan Basin, in the middle and upper reaches of the Fujiang River, and is the second largest prefecture-level city in Sichuan Province. According to announced the national ambient air quality status in December and January to 2023, Mianyang's air quality comprehensive index ranked last in the 168 key cities [11].

Field investigations were conducted in Jianguomen commercial district in Mianyang (Fig. 1), The business district covers an area of about 20.3 hectares, with various large commercial complexes and commercial facilities (such as Xinglida Square, Fu'an Department Store, and Maoye Department Store), which leads to a high concentration of people in the district, making it of great significance and value to carry out health risk assessments in the district.

2.2 Field Surveys

Measure time. Select the weekend period with large traffic flow and no rainfall, and conducted measurements in the commercial district on December 10, 2023, recording the PM$_{2.5}$ density data and population activity information.

Measuring instruments. PM$_{2.5}$ concentration is measured by the Broad Link SMART-128S air monitor, which has a laser PM$_{2.5}$ sensor. The measurement data is accurate and responsive. Its measurement range: 0–999 u g/m^3, resolution:1 u g/m^3, accuracy: $\pm$ 15%read value or $\pm$ 20 ug/m^3. Wind direction is measured by a hand-held weather station, which is a simple operation, integrating multiple meteorological elements into a portable meteorological observation instrument. Its measurement range: 0–360°, resolution: 1, accuracy: $\pm$ 1.

Measuring process. The experimental instrument was uniformly calibrated one week before the measurement ($\pm$5%). The measured day, in order to analyze the spatial feature of the overall pollution distribution in the Jianguomen commercial district, the research team set up monitoring points evenly based on the spatial structure characteristics of the block, with Xinglida Square as the core, extending to the "cross-shaped" commercial pedestrian street, to ensure the balanced distribution of each monitoring point. Considering the average breathing height of the population, it was uniformly set at 1.5 m. As shown in Fig. 1, 27 monitoring points were set up. To ensure the completeness and regularity of the actual measurement data results, each point is measured 12 times per hour, and a total of 180 sets of data are obtained in 15 h a day. At the same time, the data of the monitoring stations near the business district were recorded, and the number and activities of the number of people in the business district in each time period were counted.

3 Result and Discussion

3.1 PM$_{2.5}$ Data Analysis

Overall trend of PM$_{2.5}$ concentration in Jianguomen commercial block. The daily trend of PM$_{2.5}$ concentration at each measurement point in the Jianguomen commercial block, as well as the comparison with the data source the nearest air quality monitoring station of the Municipal People's Congress, is shown in Fig. 2. It will be seen that First, the overall trend of PM$_{2.5}$ concentration shows a characteristic of high in the morning-lower in the afternoon-high in the evening. The reason is that the temperature is low at night, the atmospheric activity is weak, causing PM$_{2.5}$ concentration to accumulate and reach a higher level, which lasts until the next morning, then with the rise of temperature, the atmospheric boundary gradually expands, leading to a gradual decrease in the overall level of PM$_{2.5}$ concentration [12]. The influence of the surrounding dining environment leads to a rise in PM$_{2.5}$ concentration in the evening. Second, referring to the real-time data of Mianyang Environmental Monitoring Center, the average concentration of the Municipal People's Congress monitoring station on the day is slightly lower than that of Jianguomen commercial block, and PM$_{2.5}$ concentration difference between Jianguomen commercial block and the Municipal People's Congress is more obvious from 9:00 in the morning to 14:00 at noon, and the gap gradually narrows in the evening. This may be due to the dense population and complex activity types in the Jianguomen commercial block, while the Municipal People's Congress monitoring station is far from the city and is less affected by population activities [13], so Both of PM$_{2.5}$ concentrations is different.

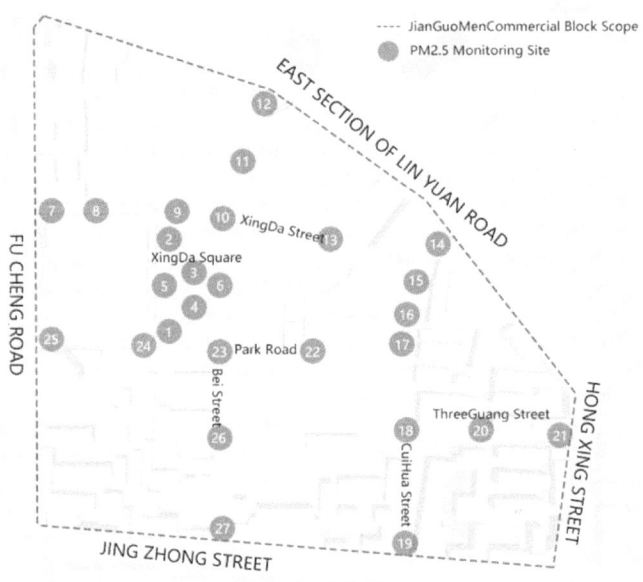

Fig. 1. Distribution map of PM$_{2.5}$ concentration monitoring points in Jianguomen commercial block

Spatial distribution characteristics of PM$_{2.5}$ concentration in Jianguomen commercial district. The distribution of PM$_{2.5}$ concentration in the district is related to traffic flow and wind direction (the dominant wind direction measured on the day was northwest). By comparing Fig. 3, the average PM$_{2.5}$ concentration at monitoring points such as No.7, No.21, and No.25 near the city's main roads is higher than that at monitoring points far from the main roads. This is because the closer to the roads with heavy traffic, the more severe the impact of car exhaust.2. The streets parallel to the wind direction are North Street and Cuihua Street, so the No.12 point on North Street is the wind inlet, the No.27 point is the wind outlet, the No.14 point on Cuihua Street is the wind inlet, and the No.19 point is the wind outlet. There is a certain difference in the PM$_{2.5}$ concentration values at the wind inlets and outlets of North Street and Cuihua Street, with the concentration values at the outlets slightly higher than those at the inlets. This is because

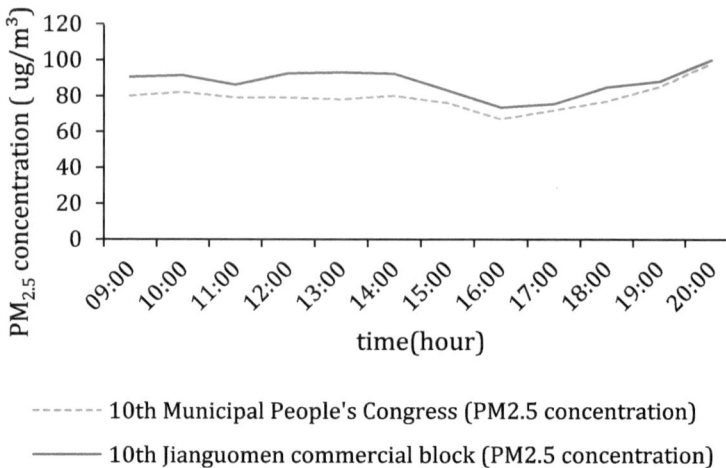

Fig. 2. 12-h trend of PM$_{2.5}$ concentration in Jianguomen commercial district

when the wind blows in a direction parallel to the street, it facilitates the diffusion of pollutants within the street canyon. PM$_{2.5}$ flows downstream along the street, resulting in a higher concentration of pollutants at the outlet than at the inlet [14]. As the height of the buildings in the street increases, the difference in concentration between the two decreases, with the maximum difference in concentration at the bottom of the street canyon.3. The PM$_{2.5}$ concentration at Xingli Plaza is "peak-shaped", with the average PM$_{2.5}$ concentration at point No.5 reaching the peak due to its proximity to the road and the impact of car exhaust. The spatial features of points No.3 and No.4 are similar, but the PM$_{2.5}$ concentration at point No.3 is significantly higher than that at point No.4. The reason for this phenomenon is that this measurement point has a wind channel from the northwest, but due to a man-made wall about 3 m high between the two points, which obstructs the flow of wind, the PM$_{2.5}$ concentration at point No.4 is significantly lower than that at point No.3.

3.2 Analysis of Crowd Activities

Summary of types of crowd activities. During the actual testing period on December 10,2023, namely 9:00–12:00, 13:00–17:00, 18:00–20:00, crowd behavior monitoring was carried out. The recreational behavior observation method was used to observe the crowd behavior (number of people, type of activity, duration of activity) at each monitoring point and record it hourly. Different public space characteristics have a great influence on crowd activities. Through the actual measurement, the present situation of crowd activity in different space nodes is analyzed, we can understand the interrelationship between crowd activities and different spatial forms. According to the recorded data, crowd activities can be roughly divided into two categories: the first is rest type:

Fig. 3. Average PM$_{2.5}$ concentration distribution map of each street and square in JianGuoMen commercial district

the most common type of activity in commercial blocks, mainly including rest, conversation, and card playing; the second is light exercise type: shop assistants use words and actions to promote products to attract customers, with the aim of selling products.

Spatial distribution characteristics of crowd activities. Based on a preliminary understanding of the regional distribution of the number of people in the block, the types of activities, and the distribution of time periods, combined with the team's statistics on the number of people at the measurement location, behavior records, and photos and other materials, using the GIS kernel density analysis tool, a spatial distribution map of the crowd at different time periods (Fig. 5) was drawn, and the types of crowd activities were distinguished. The following regular features can be found through comparison: The statistics of the crowd (Fig. 4) and spatial distribution (Fig. 5), it can be seen intuitively that the number of people at each monitoring point varies significantly. First, 1, 3, 6, 11, 19, 22, 24, 26 is higher than other areas throughout the day, among which there are many commercial shops around point 6, and the shop assistants promote products and ample rest facilities gather a large number of people, causing the total number of persons at point 6 to reach 486. From 9:00–12:00, the number of people at each monitoring point fluctuates slightly, and the total number of people does not exceed 100; Second, From 13:00–17:00, the number of people at each monitoring point is at a high level, and there is a significant downward trend after 17:00; Third, The rest facilities at points 1–4 and 6 are densely distributed, and the areas between each point are mainly paved with hard ground, the mobility of the crowd is large, and the gathering form presents a "face" feature; the crowd gathering at point 24 and the public space form are relatively matched, presenting a "point-line" combination; the crowd gathering at points 15 and 23 is at the intersection of paths The public node presents a "point" feature.

3.3   Health Risk Assessment of PM$_{2.5}$ in Important Public Spaces of Jianguomen Commercial Block

Screening of important public spaces. The walking space form in the **Jianguomen** commercial block is rich, the rest facilities are diverse, and the distribution of crowd activities is complex. Referring to the impact mechanism of PM2.5 on the human body,

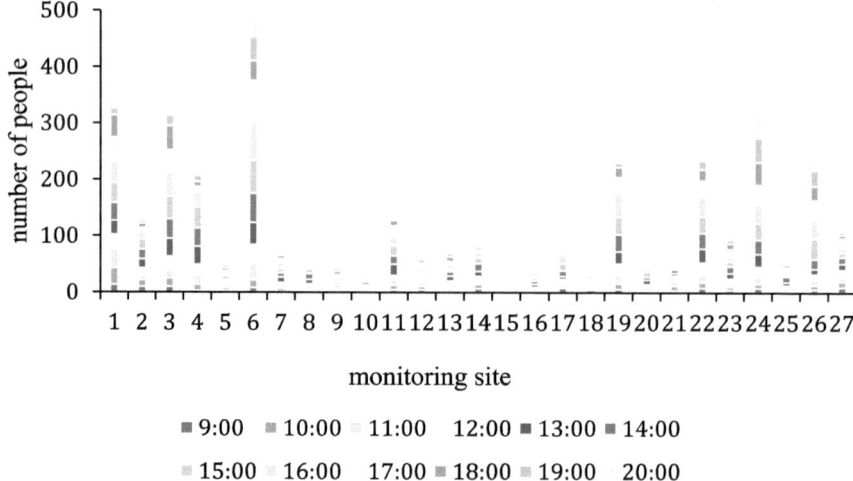

Fig. 4. Changes in the number of people at 27 monitoring points

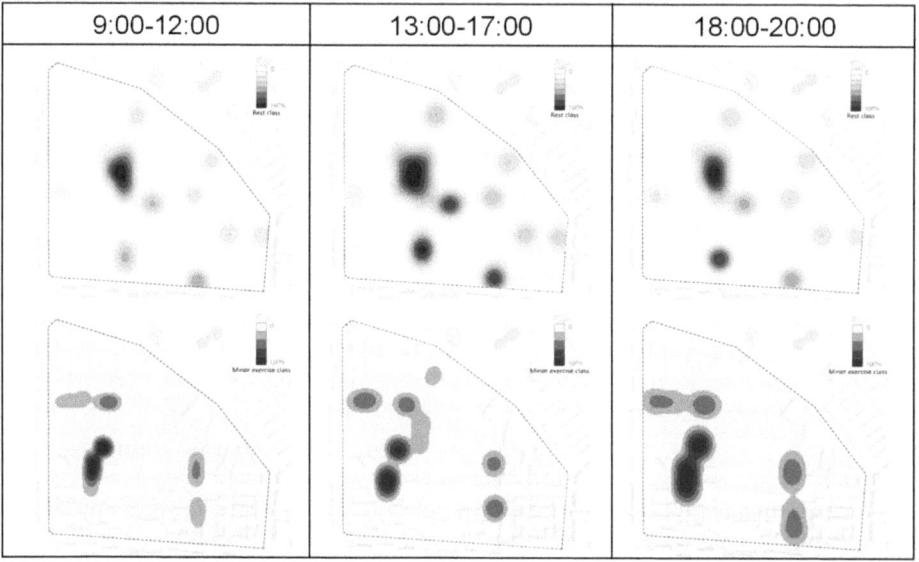

Fig. 5. Spatial-temporal distribution of crowd activities in Jianguomen commercial block at different time periods

targeted analysis is carried out on public spaces where there are more crowds and longer stay times, referring to PM$_{2.5}$ The mechanism of its impact on the human body should be specifically analyzed for spaces where there are many people gathered and the stay time is long.

Fig. 6. Important spatial node locations and real scene shooting conditions of JianGuoMen commercial block

Combined with the spatiotemporal distribution map of crowd activities (Fig. 5), the crowd activity situation of 27 measurement points in the Jianguomen commercial block is observed. According to the size of the crowd flow, important spatial nodes in the commercial block are screened, and finally 8 places are selected, including the fixed stall area of No.1, the north side seating area of No.3, the south side seating area of No.4, the tree array street of No.6, the entrance rest area of No.19, the tree array square area of No.22, the mobile vendor area of No.24, and the tree array rest area of No.26 (Fig. 6).

Application of exposure assessment in health risk assessment. Exposure assessment is an important part of environmental health risk assessment. To perform human $PM_{2.5}$ exposure studies, knowledge of possible sources of human exposure is needed. On the one hand, the amount of exposure to environmental pollutants through the respiratory tract, digestive tract and skin, on the other hand, the amount of a certain factor taken into the body from the external environment, the exposure dose [15]. Compared with exposure levels, potential doses can better reflect differences in exposure levels resulting from individual activity nature, activity duration, and physical function. The time-activity pattern, as a simplified calculation method of the potential dose of $PM_{2.5}$, According to the research of Guo et al. and other scholars, 8the daily exposure level of $PM_{2.5}$ in specific public spaces can be confirmed by the following formula [16]:

$$I = Cm \times BR \times T_m \tag{1}$$

I is the potential dose of $PM_{2.5}$ inhaled by the crowd in this environment (ug); Cm is the concentration of $PM_{2.5}$ in this environment (ug/m³), BR is the breathing rate of specific crowd (m³/min); Tm is the stay time in this environment (min).

C_m can be obtained from the actual measurement data. In this article, the daily average PM$_{2.5}$ concentration value of each measurement point is used as the C_m value, BR is mainly determined by factors such as age, health status, and type of activity [17]. Since it is difficult to obtain personal data such as age and health status on the spot, this article takes the type of activity of the crowd as the main factor affecting the respiratory rate. Combined with the research results of Liu et al., the respiratory volume of the crowd can be determined according to different activities [18]. Since the types of activities in commercial blocks are diverse, it is not comprehensive enough to calculate the PM$_{2.5}$ exposure dose of the crowd in the space with only one type of activity. Therefore, this article calculates the total PM$_{2.5}$ exposure dose of each measurement point according to the proportion of the types of activities in each measurement point. Use I x as the PM$_{2.5}$ exposure dose when performing various activities in space:

$$I_x = C_m \times BR \times T_m \times Q_x \tag{2}$$

where Q_X represents the proportion of various activities at each measurement point. By calculating and organizing, we can obtain the proportion of crowd activity types, breathing rate, and average stay time at each measurement point, as shown in Table 1:

Table 1. Proportion of crowd activities and their average activity time at important spatial nodes

	Point 1		Point 3	Point 4	Point 6	
Activity Type	Rest	Sales	Rest	Rest	Rest	Sales
Activity Proportion	0.2	0.8	1	1	0.9	0.1
Breathing Volume (L/min)	5.4	8.1	5.4	5.4	5.4	8.1
Activity Duration	Average 10min	Average 60min	Average 60min	Average 30min	Average 30min	Average 60min
	Point 19			Point 22	Point 24	Point 26
Activity Type	Rest		Sales	Rest	Sales	Rest
Activity Proportion	0.85		0.15	1	1	1
Breathing Volume (L/min)	5.4		8.1	5.4	8.1	5.4
Activity Time	Average 30min		Average 30min	Average 30min	Average 60min	Average 30min

Health risk Assessment of PM$_{2.5}$ in Crowds. As found in the previous sections, for the health risk assessment of PM2.5, First, the pollutant exposure needs to be assessed and then the type of crowd activity and stay time should be combined. as scholars like Zhu et al. have shown, an increase in daily concentration by $10\mu g/m3$will significantly increase the risk of respiratory diseases by 3.48% [19]. Referencing Liu et al. research, using 15.7L/d as the average daily breathing rate [18], we can calculate the short-term $PM_{2.5}$ If the intake increases by $157\mu g$ (daily average concentration increases by $10\mu g/m3$), the health risks of 8 important public spaces in the Jianguomen commercial district can be calculated:

$$H_X = \frac{(C_m - C_b) \times BR \times T_m \times Q_X}{157} \times 3.48\% \tag{3}$$

Through formula (3), (H_x is the health risk of the space (%), C_m is the average concentration at the specific location during that period (ug/m³), C_b is the national daily $PM_{2.5}$ concentration first-level standard 35ug/m³, BR is the breathing rate of performing a certain activity in that space, T_m is the duration of the crowd performing that activity, Q_X is the proportion of various activities at each measurement point), the health risks of 8 important public spaces can be calculated. The highest health risk in the 24nd mobile vendor area is 0.43%, followed by the 3rd north side seating area at 0.38%, the 1st fixed stall area at 0.36%, the 6th tree-lined street and the 19nd entrance rest area are 0.20%and 0.26%respectively, the 4th south side seating area and the 26nd tree-lined rest area are 0.16%, and the health risk of the 22nd tree-lined square is the smallest, at 0.15%. Through As shown in Fig. 7, the $PM_{2.5}$ concentration of points 1 and 24 is at a lower level, but due to the long stay of the crowd and high activity intensity, the health risk is at a higher level. The $PM_{2.5}$ concentration is high, so is the health risk. This is due to the low intensity of activities of the population, but the duration of stay is long. This indicates that the health risk assessment of $PM_{2.5}$ for the population not only considers the concentration of $PM_{2.5}$ but also combines the activity characteristics of the population.

3.4 Strategies to Reduce Health Risks in Important Public Spaces

There is a high health risk in public spaces where people gather and stay for a long time. For environmental problems in public spaces such as the 24th mobile stall area, the 1th fixed stall area, the 6th tree array street, and the 19th entrance rest area in the block, combined with the risk causes of different spaces, from the perspective of reducing $PM_{2.5}$The health risks of important public spaces can be reduced through three approaches: reducing $PM_{2.5}$ concentration, providing reasonable activity guidance, and reducing the dwell time of crowds.

The first is to reduce the concentration of $PM_{2.5}$. For areas with high health risks, relevant spatial transformations and greening design measures can be carried out to further reduce the concentration level of $PM_{2.5}$ in the space. Enhancing the openness and ventilation of public spaces, such as demolishing artificial walls in the seating area on the north side of No.3, and enhancing the spatial form response to the dominant wind direction can promote the air circulation within the public space; the green plant layout of the No.6 tree array street is changed to a dense interlaced layout of small crown diameter green plants, which will enhance the adsorption and sedimentation effect of green plants on $PM_{2.5}$ concentration.

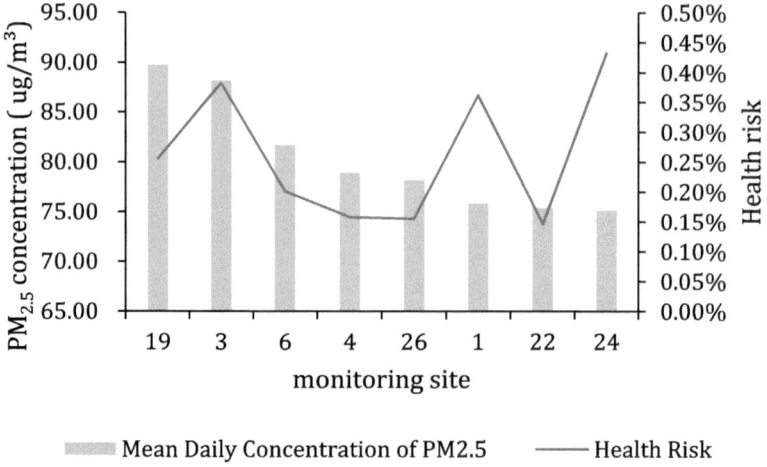

Fig. 7. Comparison of health risk assessment of important spatial nodes in Jianguomen commercial block

The second is reasonable activity guidance. For spaces with lower PM$_{2.5}$ concentration distribution in commercial blocks, certain landscape design means can be combined to enhance crowd participation, expand the function of such spaces, and enhance their spatial and environmental value. The No.26 tree array rest area can further expand the stay area, enhance crowd activities and exchanges through tree pools, flower pools, seat landscapes etc., and create a vibrant atmosphere. The No.4 south side seating rest area can design landscape sketches to enhance crowd interaction. Some spaces with higher PM$_{2.5}$ concentration distribution in commercial blocks should encourage crowds to pass quickly. The dwell time and activity intensity of the crowd in the No.1 fixed stall area and the No.24 circulation vendor area will affect their own health risks. Consider reducing the stall time of the stall owners.

The last is to reduce the dwell time of the crowd. For example, the rest facilities of the No.19 entrance square can be demolished to reduce the crowd stay, emphasize the passage function of the street entrance, and avoid arranging seats and other facilities for the crowd to play chess and cards in areas with higher health risks, thus avoiding the crowd's long stay in this space.

3.5 Limitations

This study has two limitations: 1) Limitations of the research perspective: the environmental elements of the business district are relatively complex, such as light, sound, microclimate and other elements, which have a direct impact on human physiological health. This study only selects fine particulate matter PM$_{2.5}$ as the main influence factor of health risk assessment, ignoring other physical environmental factors, so the health risks of various elements can be comprehensively discussed in the future; 2) Limitations of research samples: This paper selects the chronological commercial district, and the optimization space is limited in the later stage, which is difficult to significantly adjust the building space texture Therefore, in the future business district planning, health risk assessment can be incorporated into the consideration of the initial stage of spatial design.

4 Conclusions

Urban commercial district is the main place for people's daily activities, and the aggregation of $PM_{2.5}$ concentration is easy to cause virus transmission and cross-infection among individuals, causing negative effects on health. Assessing the impact of air pollution in urban commercial district has both theoretical and practical significance.

This study makes three contributions to $PM_{2.5}$ health risk assessment in urban commercial districts. First, we find that the concentration of $PM_{2.5}$ in the business district is affected by the traffic flow and the wind direction, so the future construction of the road inside the business district should consider the relationship between the wind direction and the street direction, because it determines the air flow pattern in the street, thus affecting the diffusion of $PM_{2.5}$. Second, studies have shown that there are high health risks in the public space with dense population and long stay. We can reduce the important public space by reducing the concentration of $PM_{2.5}$, reasonable activity guidance and reducing the residence time of the population in three ways Between the health risks. Third, our research system and operation methods are also applicable to the urban external environment with high crowd concentration, such as the outdoor public space of hospitals, and transportation hubs.

References

1. Liang Ruiming, Yin Peng, Zhou Maigeng: Progress in cohort studies on the health effects of long-term exposure to atmospheric PM2.5. J. Environ. Health **33**(2), 172–177 (2016)
2. Chen, F., Lin, Z., Chen, R., et al.: The effects of PM2.5 on asthmatic and allergic diseases or symptoms in preschool children of six Chinese cities, based on China, Children, Homes and Health (CCHH) project. Environ. Pollut. (2018)
3. Lee, Y.L., Shaw, C.K., Su, H.J., et al.: Climate, traffic-related air pollutants, and asthma prevalence in middle-school children in Taiwan. Environ. Health Perspect. **21**(6), 964–970 (2003)
4. Yang Yi jian, Song Hong. Toxic effects of fine particulate matter (PM2.5) on the respiratory system. J. Toxicol. **2**, 146–148 (2005)
5. Pinault, L.L., Weichenthal, S., Crouse, D.L., et al.: Associations between fine particulate matter and mortality in the 2001 Canadian Census Health and Environment Cohort. Environ. Res. **159**, 406–415 (2017)
6. Chen, R., Yin, P., Meng, X., et al.: Fine particulate air pollution and daily mortality. A nationwide analysis in 272 Chinese cities. Am. J. Respir. Crit. Care Med. **196**(1), 73–81 (2017)
7. Liu Binyi, Peng Xu Lu: Progress and enlightenment on microclimate comfort of urban streets. J. Chinese Gardens **35**(10), 57–62 (2019)

8. Liu Xiang, Zeng Jian, Ren Lanhong: Subjective study on the evaluation of thermal environment comfort of commercial pedestrian streets in Tianjin —— Take four commercial pedestrian streets in Tianjin as an example. Build. Energy Effic. **47**(5), 119–123 (2019)
9. Wang Jiwu, Zhang Chen, Feng Yujun: Review and Urban Planning response Framework. Res. Urban Dev. **19**(5), 82–87 (2012)
10. Yu Ping, Tan Haiping, Xiang Mingdeng, et al.: Research progress on PM2.5 exposure assessment methods. Environ. Occup. Med. **35**(9), 861–866 (2018)
11. The Ministry of Ecology and Environment announced the national ambient air quality status in December and January to December 2023, https://www.mee.gov.cn/ywdt/xwfb/202401/t20240125_1064784.shtml. Accessed 15 Dec. 2023
12. Zhang Yunwei, Wang Qingru, Chen Jia, et al.: Analysis of spatiotemporal variation and influencing factors of PM_(2.5) concentration in urban street canyons. China Environ. Sci. **36**(10), 2944–2949 (2016)
13. Zhang Hao: Health risk assessment of PM2.5 for people under the shade of public spaces - a case study of three gorges square; proceedings of the Chinese Landscape Architecture society 2020 ANNUAL Meeting, Chengdu, Sichuan, China, F (2020)
14. Zhang Hao: Research on spatiotemporal distribution of PM2.5 in street canyons of mountainous cities based on actual measurement-a case study of chongqing business district; proceedings of the 2020/2021 China Urban Planning Annual Meeting and 2021 China Urban Planning Academic Season, Chengdu, Sichuan, China, F, (2021)
15. Li Xiaodan: Assessment of the pollutant emission measurement and its health risk exposure in the urban public transport system. Chang'an University (2020)
16. Guo Shengli, Wang Xi, Huang Jun: Progress in research on human exposure to PM_(2.5) in various microenvironments based on time-activity pattern. Sci. Technol. Eng. **14**(27), 128–134 (2014)
17. Wang Zongshuang, Wu Ting, Duan Xiaoli, et al.: Research on exposure parameters of respiratory rate in environmental health risk assessment in China. Environ. Sci. Res. **22**(10), 1171–1175 (2009)
18. Liu Ping, Wang Beibei, Zhao Xiuge, et al.: Research on adult respiratory volume in China. J. Environ. Health **31**(11): 953–956 (2014)
19. Zhu Yu, Ma Zijian, Xiao Changchun: The relationship between PM2.5 pollution in hefei city and the number of hospitalizations for respiratory diseases. Mod. Prev. Med. **48**(6), 985–989 (2021)

Pollution Exposure Assessment of Pocket Park PM$_{2.5}$ Under the Influence of Urban Roads—A Case Study of Mianyang City

Jiaxin Li[✉], Chunrong Zhao, Li Yan, Minghong Yu, Huiyun Peng, Guoyang Hai, and Yuzhen Liu

School of Civil Engineering and Architecture, Southwest University of Science and Technology, Mianyang 621010, China
992210900@qq.com

Abstract. Pocket parks, as the main places for urban residents' daily recreation and communication, are usually arranged near urban roads for the convenience of residents, ignoring the risk of residents being exposed to traffic air pollution. Therefore, studying the potential pollution exposure of pocket parks is of great significance for reducing residents' exposure to air pollution and improving the health level of pocket park leisure activities. This paper carries out empirical research on the case of Qicai Pocket Park in Mianyang City from the perspective of PM$_{2.5}$, a traffic-derived air pollutant. Based on PM$_{2.5}$ monitoring and population activity statistics, a pollution exposure assessment is conducted on the public space nodes with dense population activities in Qicai Pocket Park, and the following conclusions are drawn: During the study, the average daily PM$_{2.5}$ concentration in Qicai Pocket Park is higher, and some spatial nodes with highly intensive population activities have higher pollution exposure; Meanwhile, the spatial and temporal distribution of PM$_{2.5}$ is significantly different, which is significantly correlated with urban roads, plant allocation and surrounding human activities. Based on the principle of population exposure assessment, this paper constructs a pollution exposure assessment method for pocket parks, in order to provide a scientific basis for improving the health quality of urban public space and optimizing the planning and design of pocket parks.

Keywords: Pocket Park · PM$_{2.5}$ · Pollution Exposure · Empirical Research · Mianyang City

1 Introduction

In recent years, air pollution has been a major global issue affecting the health of people around the world. According to the United Nations Environment Programme (UNEP), air pollution has led to over one million premature deaths and one million stillbirths globally each year [1]. Fine particulate matter (PM$_{2.5}$) is one of the major air pollutants in urban environments worldwide. PM$_{2.5}$ refers to particulate matter in the environment with an aerodynamic equivalent diameter less than or equal to 2.5μm, which is rich in

© The Author(s) 2025
B.-J. He et al. (Eds.): UCSUD 2023, LNCE 559, pp. 630–642, 2025.
https://doi.org/10.1007/978-981-97-8401-1_45

toxic and harmful substances and has a long stay time in the atmosphere and a long transport distance. Inhalation of particulate matter can cause a series of respiratory and cardiovascular diseases and increase the incidence of malignant tumors [2]. Studies in Europe, the United States, and many Asian cities have shown that the longer the exposure time and the higher the concentration of particulate matter, the greater the risk of disease and death [3, 4]. Therefore, the air pollution caused by PM$_{2.5}$ has attracted widespread attention from the public and the scientific community.

Urban green Spaces such as parks have long been seen as places of air quality improvement, playing an important role in reducing PM$_{2.5}$ in urban air [5]. However, as urbanization in China rapidly advances, major cities are confronted with the issue of inadequate green spaces in their central areas. The concept of pocket parks was initially introduced by American landscape architect Robert Zion [6]. It is generally believed by domestic scholars that pocket parks are publicly accessible, small-scale, and come in diverse shapes, serving as recreational green spaces ranging from 400 to 10,000 square meters [7]. Due to their compact size and flexible location, pocket parks are typically established as new additions within densely populated urban centers to meet the demand for green spaces among residents. Nevertheless, due to their limited size and proximity to urban roads, pocket parks may have a lesser impact on mitigating air pollution and could potentially pose health risks for park users.

Many studies have found that the trees, paths, seats, lights and entertainment facilities in a park usually affect the distribution of users in a park [8]. Meanwhile, the concentration of PM$_{2.5}$ in different activity Spaces in the park presents the characteristics of spatiotemporal uneven distribution [9]. This means that the layout of park spatial elements can change the diffusion of air pollutants and the distribution of park users. Therefore, the design of park space needs to be considered in more detail.

Therefore, this paper carries out empirical research based on the perspective of air pollutant PM$_{2.5}$, using Mianyang City's Pocket Park as a case study. Firstly, with the help of field observation methods, the study investigates the correlation between the spatiotemporal distribution of PM$_{2.5}$, the spatiotemporal distribution of human activities, and the spatial design features of the park, revealing the influencing mechanism of the spatiotemporal distribution of PM$_{2.5}$ in the Pocket Park. Meanwhile, based on the principle of population exposure assessment, the spatiotemporal distribution of PM$_{2.5}$ at important spatial nodes in the study area and the spatiotemporal distribution of human activities are overlaid for analysis, quantitatively assessing the pollution exposure distribution in important spaces in the Pocket Park. Finally, for different pollution exposure areas, based on the different design features of the Pocket Park and their impact on the spatiotemporal distribution of air pollutants and human activities, strategies for optimizing the spatial layout of the Pocket Park and guiding the population to reduce exposure to PM$_{2.5}$ are proposed.

2   Research Object and Method

2.1   Research Object

Mianyang City is located in the southwest of China. It is the second largest prefecture-level city in Sichuan Province, with geographical coordinates of 103°45′ ~ 105°43′E, 30°42′ ~ 33°03′N. The central urban area of Mianyang is a typical high-density development area with scarce land resources. The pocket parks are relatively small in size but have high utilization rates. Especially in the old city area in the center of Mianyang, which has a dense road network and large traffic flow, its pocket parks are more likely to be affected by traffic air pollution. Therefore, the study chose the best pocket park in Fucheng District of Mianyang City—the Qicai Pocket Park, as the field research object. This pocket park covers an area of about 19 acres and is currently the largest and most functional pocket park in the urban area. Meanwhile, this pocket park is located on both sides of an overpass (Fujiang Second Bridge) and is surrounded by city roads, which is more likely to have adverse effects on residents' health. Therefore, conducting experimental research on it has important significance and value (Fig. 1).

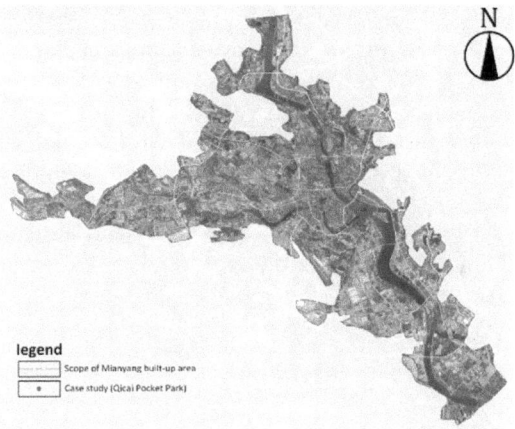

Fig. 1. Location map of Qicai Pocket Park in Mianyang City (*Photo source* Author's drawing)

2.2   Research Method

PM$_{2.5}$ Measurement Method. Mianyang city has severe pollution in winter, and the measurement was chosen to be conducted on December 9th and 11th, 2023, at the Qicai Pocket Park from 7:00 in the morning to 21:00 in the evening. The measurement used the Broad Air SMART-128S (Measuring range: 0-999ug/m^3, resolution: 1ppm, accuracy: ± 15% reading value or ± 45ppm) detector to measure the changes in PM$_{2.5}$ concentration. In terms of height, considering the average breathing height of the population, the measuring instrument was uniformly set at 1.5 meters. Meanwhile, in order to analyze the spatial characteristics of the overall pollution distribution in Qicai Pocket Park, this paper uses the public nodes of Pocket Park as the core, connected by walking paths, to evenly arrange various monitoring points. As shown in (Fig. 2), a total of 25 points were arranged. To ensure the completeness and regularity of the measurement data results, each point is measured an average of 12 times per hour. As Mianyang city is a city with a high frequency of quiet wind, and a handheld weather station instrument (SC-II-C) was used to monitor wind direction and speed during the actual measurement, it was found that the average wind speed was 0.1m/s for two consecutive days and there was no sustained wind direction, so the effect of wind on air pollutants was not mentioned.

Fig. 2. Distribution map of PM$_{2.5}$ concentration monitoring points in Qicai Pocket Park

Observation Method of Crowd Activities. This time, the observation of crowd activities was conducted using the recreational behavior observation method [10]. Firstly, random sampling interviews were conducted to obtain more detailed information about the crowd structure; secondly, while monitoring the PM$_{2.5}$data at each spatial node in Qicai Pocket Park, the number, type, duration, spatial distribution, and environmental characteristics of crowd activities were observed and recorded.

3   Research Results and Analysis

3.1 PM $_{2.5}$ Data Analysis

Overall trend of PM$_{2.5}$ concentration in Qicai Pocket Park. Based on the accuracy of the measurement results, combined with the changes in background concentration and the characteristics of crowd activities, the data for December 11, 2023, was selected for detailed processing. The overall trend of PM$_{2.5}$ concentration in Qicai Pocket Park (Fig. 3) was obtained. Through data comparison, it was found that: (1) The average daily PM$_{2.5}$ concentration in Qicai Pocket Park is relatively high. Referring to the real-time PM$_{2.5}$ data of the nearest national control monitoring station at Sanshui Plant from Qicai Pocket Park, it can be seen that the measured PM$_{2.5}$ concentration at each time point in Qicai Pocket Park is higher than the concentration at the Sanshui Plant monitoring station. (2) The PM$_{2.5}$ concentration in Qicai Pocket Park shows a "V" shaped upward trend throughout the day. From the field measurement, because the Qicai Pocket Park is adjacent to the city's main road, it has a great impact on the PM$_{2.5}$ concentration in the pocket park during the peak traffic hours of 8:00–9:00 and 18:00–19:00. The concentration of PM2.5 increased rapidly after 17:00, which may be affected by the dining environment around the pocket park.

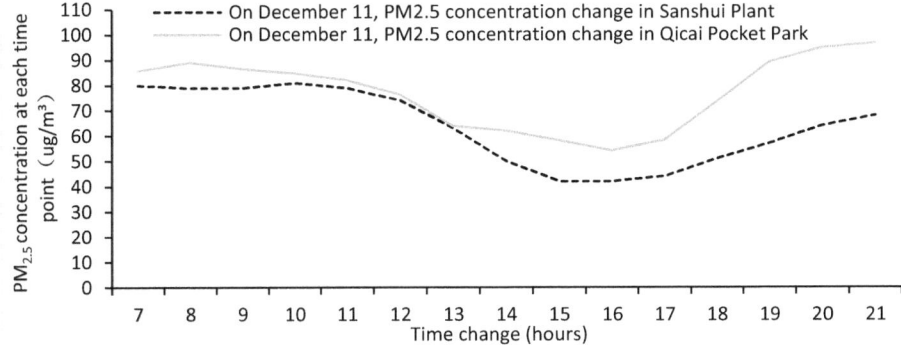

Fig. 3. 15-h trend of $PM_{2.5}$ concentration in Qicai Pocket Park and Sanshui Plant

Spatial distribution characteristics of $PM_{2.5}$ concentration in Qicai Pocket Park. By comparing the daily average concentration of $PM_{2.5}$ at 25 points, it can be found that the spatial distribution of pollutants presents a pattern (Fig. 4): (1) The distribution of $PM_{2.5}$ concentration in Qicai Pocket Park is significantly related to urban roads: the $PM_{2.5}$ concentration at points 1 and 18 near the city roads is higher, and the concentration at points 6 and 11 far from the road is relatively small, indicating that the distance from the pocket park to the city road and the $PM_{2.5}$ concentration show a significant negative correlation; meanwhile, the $PM_{2.5}$ concentration at point 1 is higher than that at point 18, indicating that the traffic flow of city roads and the $PM_{2.5}$ concentration show a significant positive correlation. (2) The distribution of $PM_{2.5}$ concentration in Qicai Pocket Park is significantly related to plant configuration: under the condition of similar traffic location, the concentration level of point 23 with tree type is higher than that of points 24 and 22 with lawn, and the concentration level of point 5 with shrub-grass type is higher than that of point 6 with lawn, indicating that in the public space with mild pollution ($75-115ug/m^3$), the resistance effect of plant canopy on wind environment is not conducive to the diffusion of pollutants, and the $PM_{2.5}$ concentration level under tree shade is generally high. (3) The distribution of $PM_{2.5}$ concentration in Qicai Pocket Park is significantly related to the surrounding nighttime dining activities (Fig. 5): After 19:00, the restaurants on the north side of Qicai Pocket Park start to operate, and the $PM_{2.5}$ concentration at points 4, 6, 7, 8, and 10 quickly rises overall, indicating that the $PM_{2.5}$ concentration is easily affected by surrounding activities. (4) A comparison of the daily average $PM_{2.5}$ concentrations at 25 points was conducted: The points ranked at the top are 13, 11, 7, 12, 14, 20, 22, where the daily average $PM_{2.5}$ concentration is not greater than the national secondary standard ($75ug/m^3$). Considering the spatial features of the Qicai Pocket Park, it can be seen that the points with better air quality have larger areas, flat and open spaces, and are relatively far from city roads. They are mainly distributed in the middle and east inner side of the Qicai Pocket Park. The points with poorer air quality are clearly on the east and west sides of the Qicai Pocket Park, these spaces are often near city roads and overpasses.

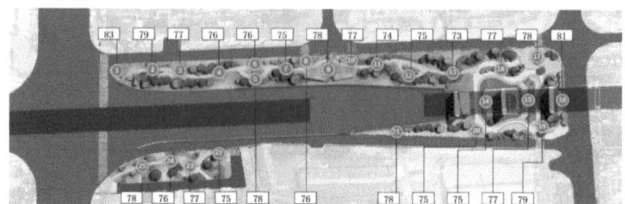

Fig. 4. Average PM$_{2.5}$ concentration distribution map of each point in Qicai Pocket Park

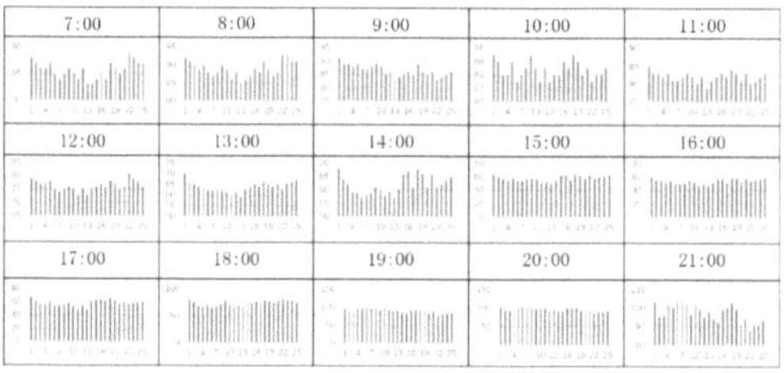

7:00	8:00	9:00	10:00	11:00
12:00	13:00	14:00	15:00	16:00
17:00	18:00	19:00	20:00	21:00

Fig. 5. PM$_{2.5}$ concentration changes at each measurement point in the Qicai Pocket Park

3.2 Analysis of Crowd Activities

Summary of Crowd Activity Types. During the actual measurement period on December 11, 2023, the behavior of the crowd at each measurement point was recorded hourly, and a total of 992 information samples were collected. According to the recorded data, the types of activities carried out in the pocket park can be summarized as rest, light exercise, and moderate exercise.

The impact of spatial features on crowd activities. A summary of the different spaces in the pocket park, the crowd using them, the content and type of activities, the duration of activities, and the time of activities. It can be found that the type of activities and the duration of activities in the crowd are closely related to the presentation of spatial features: (1) In smaller spaces, rest activities are dominant; in larger spaces, the types of activities are diverse, and the average duration of activities is higher than in other areas. (2) The degree of closure and the degree of line-of-sight blocking are both related to privacy. The higher the privacy, the more conducive it is to the development of rest activities. (3) The arrangement of facilities, comfort, and functionality in a space can affect the willingness of people to use the space. For example, in bridge and tunnel spaces with steps and platforms, there are more people and they are more inclined to carry out light and moderate exercises, while in the same bridge and tunnel spaces, there are fewer people and they are more inclined to rest activities. (4) As the buffer width of the space boundary and the accessibility of the site increase, the activities of the crowd will also increase accordingly.

The spatial distribution characteristics of crowd activities. According to the characteristics of the crowd activity in the Qicai Pocket Park, the kernel density analysis of the crowd activity was carried out by using GIS software, and the distribution map of the crowd activity in different time periods was obtained (Fig. 6). Through comparison, it is found that: (1) The spatial distribution of crowds in different areas has heterogeneity: The space area on the east side of the pocket park has a wide field of vision, entertainment and rest facilities are densely distributed, the mobility of the crowd is larger, and the gathering form presents a "face" feature; the crowd gathering and node space form of each green space node in the middle are relatively consistent, and are connected by paths, the gathering form presents a "point-line" combination; the environmental field of vision of each green space node is relatively high, and there are certain rest facilities, the crowd gathering form presents a "point" feature. (2) The crowd activities at different time periods have regular changes in spatial distribution: in the morning from 8:00–12:00, a small number of residents choose to go outdoors, mainly using bridge tunnel spaces and equipment fitness areas as places for rest and exercise. However, during school pick-up and drop-off times, the waiting area for picking up students (on the south side of the Qicai Pocket Park) has a large number of people engaging in tea-drinking, communication, and other rest activities. From 13:00–18:00, a large number of people choose to go outdoors for rest activities, mainly concentrated in various space nodes of the pocket park. From 19:00–21:00, the number of people starting to reduce outdoor activities, mainly using brightly lit, open-view space nodes as activity places. (3) Different types of activities have a degree of overlap in spatial distribution, such as places for light and moderate exercise activities often have people engaged in rest activities.

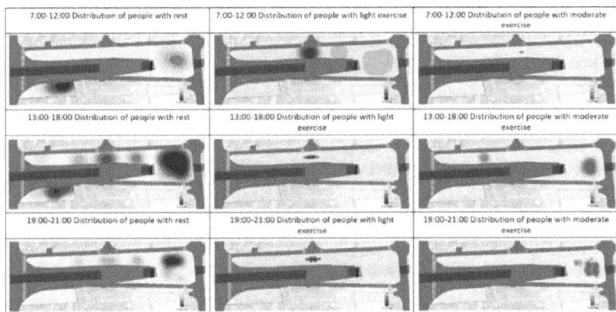

Fig. 6. Spatial-temporal distribution map of crowd activities at different time periods Pollution Exposure Assessment of Pocket Park

4   Pollution Exposure Assessment of Pocket Park

4.1   Important Space Node Screening

Referring to the impact mechanism of PM$_{2.5}$ on the human body, targeted analysis should be carried out on space nodes where there are more crowds and longer stay time. Combining the spatiotemporal distribution map of crowd activities (Fig. 5). According to the size of the crowd flow, important space nodes in the Qicai Pocket Park are screened, and finally 7 places are selected (Fig. 7).

Fig. 7. Location and real scene shooting of important space nodes in the Qicai Pocket Park

4.2 Time-Activity Pollution Exposure Assessment

The time-activity pattern is a common micro-scale human exposure assessment pattern. According to the research of Guo Shengli and other scholars [11], the daily exposure level of PM$_{2.5}$ in specific public spaces can be confirmed by the following formula:

$$I = C_m \times BR \times T_m \tag{1}$$

In it, I represents the exposure dose of microenvironment PM$_{2.5}$ (ug); C_m is the concentration of PM$_{2.5}$ in this environment (ug/m^3); BR is the breathing rate of a specific population(L/min); T_m is the duration of stay in this environment (min). C_m can be obtained from actual measurement data. BR is mainly based on the activity type of the population as the main influencing factor of the breathing rate. According to the research of Liu Ping on the respiratory volume of Chinese adults, the respiratory volume of the population is determined, that is, the respiratory volume of the resting population is 6.4 (L/min), the respiratory volume of the light exercise population is 8.1 (L/min), and the respiratory volume of the moderate exercise population is 21.5 (L/min) [12].

For microenvironments like pocket parks, due to the heterogeneity of spatial structure and function, the content and time of activities at different spatial nodes are significantly different. Therefore, this paper calculates the total PM$_{2.5}$ exposure dose at each measurement point based on the proportion of activity types within each measurement point:

$$I_x = C_m \times BR \times T_m \times Q_x \tag{2}$$

In formula 2, Q_x is the proportion of various activity types at each measurement point, and the comprehensive PM$_{2.5}$ exposure dose at each measurement point is the sum of the PM$_{2.5}$ exposure doses during various activities. Through calculation and sorting, seven important spatially related influencing factors are sorted out, as shown in the following table (Table 1).

Table 1. Proportion of activities and average activity time at important spatial nodes

Important spaces locations	No. 5		No. 9			No. 12	
Average daily $PM_{2.5}$ concentration	78 (ug/m³)		78 (ug/m³)			75(ug/m³)	
Activity type	Resting	Moderate exercise	Resting	Light exercise	Moderate exercise	Resting	Light exercise
Breathing rate (L/min)	6.4	21.5	6.4	8.1	21.5	6.4	8.1
Average daily duration of activity	20min	120min	30min	20min	120min	25min	50min
Activity proportion	0.91	0.09	0.47	0.46	0.07	0.94	0.06
Important spaces locations	No.15			No.16		No.22	No.23
Average daily $PM_{2.5}$ concentration	77(ug/m³)			77(ug/m³)		75 (ug/m³)	77 (ug/m³)
Activity type	Resting	Light exercise	Moderate exercise	Resting	Moderate exercise	Resting	Resting
Breathing rate (L/min)	6.4	8.1	21.5	6.4	21.5	6.4	6.4
Average daily duration of activity	35 min	30 min	80 min	50 min	10 min	15 min	135 min
Activity proportion	0.54	0.06	0.40	0.96	0.04	1.00	1.00

Studies both domestically and internationally have shown that long-term exposure to a $PM_{2.5}$ environment can increase respiratory inflammation, asthma, lung damage, and other respiratory diseases. For example, Prieto-Parra et al. stated that the smaller the particle size, the deeper the part that can reach the human body, and the unit toxic substances adsorbed by it are relatively more, so the harm of fine particles is greater [13]. Meanwhile, many scholars have proven a positive correlation between $PM_{2.5}$ concentration and the incidence and risk factor of respiratory diseases. For example, Cui et al. showed that from 2009 to 2011, when the average daily concentration of $PM_{2.5}$ in Harbin increased by 44, 35 and 60ug/m³, the number of patients with respiratory diseases increased by 11.6%, 18.9% and 35.8%, respectively [14]. Duan et al. showed that the average annual $PM_{2.5}$ concentration in Chengdu was positively correlated with the number of outpatient visits of respiratory system ($p < 0.05$), and it was predicted that the number of outpatient visits of respiratory diseases would increase by 0.58% when the average daily concentration of $PM_{2.5}$ increased by 10ug/m³ [15].

According to research by Dominici, an increase of 10ug/m³ in the daily average concentration of $PM_{2.5}$ increases the risk of various diseases, primarily respiratory, by 8% [16]. The long-term breathing rate of Chinese residents, 15.7 m³/d, is now taken as the daily average breathing rate [12], to calculate the short-term $PM_{2.5}$ Inhaling an additional

157ug greatly increases the risk of disease in the population. By comparing the pollution exposure in multiple public spaces within the research area, formula 2 can be evolved, using the average concentration of PM$_{2.5}$ at the San Shui Factory national control station as a reference for comparison, the basic calculation formula for the pollution exposure of Pocket Park is as follows:

$$H_X = \frac{(C_m - C_b) \times BR \times T_m \times Q_X}{157} \times 8\% \tag{3}$$

where H_X is the pollution exposure of performing a certain type of activity at a spatial node (%), C_b is the national daily average PM$_{2.5}$ concentration first-level standard of 35ug/m^3. By calculating and arranging formula 3, the pollution exposure of seven important public spaces are obtained. Meanwhile, the following regular features can be found by comparing (Fig. 8): (1) Daily average PM$_{2.5}$ Areas with higher concentrations pose a higher pollution exposure. The highest pollution exposure is for the crowd at the pick-up and drop-off area No.23, which is 1.849%. (2) The characteristics of crowd activities have a significant impact on pollution exposure. Although the PM$_{2.5}$ concentration and activity intensity at point No.23 are relatively low, the crowd stays for a longer time, so the pollution exposure is relatively high; the pollution exposure at points No.15 and No.9 also increase due to the high intensity of crowd activities.

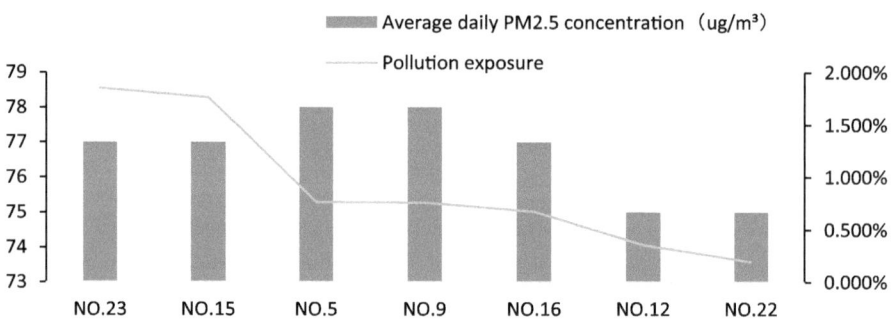

Fig. 8. Comparison of pollution exposure assessment of important spatial nodes

5 Conclusion and Discussion

5.1 Conclusion

This paper is based on field observations to study and analyze the changes in PM$_{2.5}$ concentration and crowd activity characteristics at various spatial nodes in the Qicai Pocket Park, and to evaluate the pollution exposure of important spaces in the pocket park. The main conclusions are: (1) Compared with the Sanshui Plant monitoring station, the PM$_{2.5}$ concentration in Qicai Pocket Park is always at a higher level, while some spatial nodes with highly intensive population activities have higher pollution exposure: Since the Qicai Pocket Park is adjacent to urban roads and viaducts, some spatial nodes in the pocket park have relatively poor pollution conditions, but their Spaces carry a lot of rich crowd activities. Using the time-activity pollution exposure formula, it is found that residents who engage in long-term activities in public Spaces with a high average daily concentration of $PM_{2.5}$. There is about 2.7–6.6 times more probability of respiratory related diseases than in a space with a lower average daily $PM_{2.5}$ concentration. In the long run, it will bring great health risks to the residents who frequent activities in these areas. (2) The spatial and temporal distribution of $PM_{2.5}$ in Qicai Pocket Park is significantly different, which is significantly correlated with urban roads, plant configurations and surrounding human activities: the farther away from urban roads and the smaller the traffic flow, the lower the concentration level of $PM_{2.5}$ at spatial nodes; In lightly polluted public Spaces, the resistance effect of plant canopy to wind environment is not conducive to the diffusion of pollutants, so the $PM_{2.5}$ concentration level is generally high in the shade space. Night catering activities around the Qicai Pocket Park have a greater impact on $PM_{2.5}$ concentration, and the closer the space to the restaurant, the higher the $PM_{2.5}$ concentration level.

In summary, for the spatial nodes with higher pollution exposure in the pocket park, human exposure assessment analysis can help to reduce pollution exposure in two ways: reducing $PM_{2.5}$ concentration and guiding activities reasonably: (1) Reducing $PM_{2.5}$ concentration: Due to the significant impact of urban roads on the $PM_{2.5}$ concentration in the Pocket Park, it is necessary to add multi-layered green spaces with high closure on the side adjacent to the urban roads, which can help reduce the impact of air pollutants such as $PM_{2.5}$ produced by urban roads on the pocket park. For example, at points 5 and 12, adding high-closure tree and shrub communities on the side near the overpass can effectively reduce the $PM_{2.5}$ concentration in the space. Meanwhile, for each green space node in the pocket park, the layout of the tree array can be changed, such as changing its direction, reducing the layout density, appropriately pruning branches and leaves to reduce the crown diameter, increasing the transparency of the leaf crown, etc., to reduce the barrier effect of green plants on the wind in the tree array, thereby reducing the aggregation of $PM_{2.5}$, achieving the effect of reducing the $PM_{2.5}$ concentration in the shaded space. (2) Reasonable activity guidance: On the one hand, for spaces in the Pocket Park with higher $PM_{2.5}$ concentrations, the number of people staying can be reduced by removing some of the rest facilities in the space, while sculptures can be arranged on flat and open spaces, or the form of paving bricks on the ground can be changed to guide people not to engage in square dancing and other entertainment activities in this space, thereby reducing the respiratory volume of the crowd in this space, for example, at point 15, the $PM_{2.5}$ concentration is higher at night due to the influence of surrounding catering, which can be reduced by reducing the duration of lighting, thus reducing the time people spend dancing. On the other hand, for spaces in the Pocket Park with lower $PM_{2.5}$ concentrations, the participation of the crowd should be enhanced, and certain landscape design methods can be combined to expand the functions of such spaces, enhance their spatial value and environmental value, for example, at point 14, the interaction and participation of the crowd can be enhanced by adding rest seats or step platforms, creating a vibrant atmosphere.

5.2   Discussion

Facing the increasingly prominent environmental issues and public health threats, it is urgent to explore the health impacts of public spaces on their users. As a pocket park adjacent to city roads, the pollution exposure assessment of the public space in the Qicai Pocket Park has strong reference value for other pocket parks near roads. As the experimental conditions are easily affected by a variety of uncertain factors, and there is no detailed study on the impact of vegetation on air pollutants, the measurement and evaluation methods need to be further improved in the future. It is believed that with the cross-integration and development of multi-disciplines, a perfect pollution exposure assessment system can be built, thus providing more scientific and reliable guidance and basis for the construction of healthy urban space.

Funding. This research was supported by Sichuan Science and Technology Program (Grant No. 2023NSFSC1051).

References

1. Wania, A., Bruse, M., Blond, N. et al.: Analysing the influence of different street vegetation on traffic-induced particle dispersion using microscale simulations. J. Environ. Manag. **94**(1), 91–101 (2012). https://doi.org/10.1016/j.jenvman.2011.06.036
2. Nowak, D.J. et al.: Air pollution removal by urban forests in Canada and its effect on air quality and human health. Urban For. Urban Green. **29**, 40–48 (2018)
3. Katsouyanni, K., Touloumi, G., Samoli, E., et al.: Confounding and effect modification in the short-term effects of ambient particles on total mortality: results from 29 European cities within the APHEA2 project. Epidemiology **12**(5), 521–531 (2001). https://doi.org/10.1097/00001648-200109000-00011
4. Schwartz, J.: Is there harvesting in the association of airborne particles with daily deaths and hospital admissions? Epidemiology **12**(1), 55–61 (2001). https://doi.org/10.1097/00001648-200101000-00010
5. Keming, M., Zhe, Y., Yuxin, Z.: Progress in the assessment of the dust retention effect and mechanism of green spaces. Acta Ecol. Sin. **38**(12), 4482–4491 (2018)
6. Haopeng, Z.: Interpretation and Enlightenment of NRPA Guidelines for Pocket Park Construction. China Urban For. **17**(6), 25–29 (2019)
7. Qiong, W., Zhigang, L., Min, W.: Research status and development trend of urban pocket parks. J. Earth Inf. Sci. **25**(12), 2439–2455 (2023)
8. Lingling, Z., Yuting, T., Xiaomiao, H. et al.: Study on the vitality of open boundary space in comprehensive parks in Zhengzhou. J. Henan Agric. Univ. 1–15 (28 December 2023). https://doi.org/10.16445/j.cnki.1000-2340.20230902.001
9. Huan, X., Hong, L., Taihao, J.: Study on the uneven distribution characteristics of PM$_{2.5}$ concentration in different activity spaces of urban parks. China Garden **34**(3), 117–122 (2018)
10. Zhongping, Z., Yali, W., Haoxuan, P.: Study on tourist recreational behavior based on SOPARC and KDE—a case study of Wuhan East Lake Greenway. Chin. Landsc. Arch. **35**(12), 58–62 (2019). https://doi.org/10.19775/j.cla.2019.12.0058
11. Shengli, G., Xi, W., Jun, H.: Progress in the study of human exposure to PM$_{2.5}$ in various microenvironments under the time-activity pattern. Sci. Technol. Eng. **14**(27), 128–134 (2014)
12. Ping, L., Beibei, W., Xiuge, Z., et al.: Study on adult respiratory volume in China. J. Environ. Health **31**(11), 953–956 (2014). https://doi.org/10.16241/j.cnki.1001-5914.2014.11.027

13. Prieto-Parra, L., Yohannessen, K., Brea, C. et al.: Air pollution, $PM_{2.5}$ composition, source factors, and respiratory symptoms in asthmatic and nonasthmatic children in Santiago, Chile. Environ. Int. **101**, 190–200 (April 2017)
14. Guoquan, C., Zhen, K., Song, L., Hongbing, L., Shaomei, L., Xiaobo, L., Lin, Z., Chao, Y., Hui, Y.: Impact of $PM_{2.5}$ pollution level on respiratory diseases in Harbin City. China Public Health **29**(7), 1046–1048 (2013)
15. Zhenhua, D., Xufang, G., Huilan, D., Fangkui, Q., Jun, C., Yong, J., Xia, W., Yan, H.: Study on the time series of PM2.5 concentration and respiratory disease outpatient visits in Chengdu. Mod. Prev. Med. **42**(4), 611–614 (2015)
16. Dominici, F., Peng, R.D. et al.: Fine particulate air pollution and hospital admission for cardiovascular and respiratory diseases. JAMA: J. Am. Med. Assoc. **295**(10), 1127–1134 (2006)

Health Risk Assessment of Park Environment Based on Particulate Matter (PM$_{2.5}$) - A Case Study of Mianyang City

Yu Long$^{(\boxtimes)}$, Juanlin Fu, Li Yan, Minghong Yu, Xuanyan Li, and Yuzhen Liu

School of Civil Engineering and Architecture, Southwest University of Science and Technology, Mianyang 621010, China
1586278269@qq.com

Abstract. Air pollution seriously harms human health, especially the fine particulate matter (PM$_{2.5}$) in pollutants is more harmful to the human body. Urban parks are important public places for residents' activities, and the air quality of the park's micro-environment directly affects residents' respiratory health. This article uses the method of field measurement, takes the People's Park of Mianyang City as the research object, conducts PM$_{2.5}$ monitoring and statistical analysis of crowd activities under light to moderate pollution weather, and according to the visit frequency, conducts health risk assessment of important spaces. The research results show: 1. The daily average PM$_{2.5}$ concentration in People's Park is high, and there are obvious differences in spatial and temporal distribution. 2. There is some overlap between the peak period of people flow in People's Park and the time period when the daily PM$_{2.5}$ concentration is high. 3. There is a high health risk in the space nodes where the activities of some people in the People's Park are highly concentrated. This article uses the principle of crowd exposure assessment to construct a health risk assessment method for parks, analyze the potential health risks of the spatial environment of People's Park, and provide new guidance for people to choose healthy activity places and urban public space renewal and optimization, which has certain practical significance.

Keywords: Urban park · PM $_{2.5}$ · Health risk · Field measurement study

1 Introduction

For a long time, urban parks are marked as the lungs of the city [1]. With the development of economy, air pollution has caused different degrees of negative effects. [2]. In particular, fine particulate matter PM2.5 has become the major factor affecting the quality of urban atmospheric environment and residents' health. PM$_{2.5}$ refers to particulate matter in the environment with an aerodynamic equivalent diameter less than or equal to 2.5μm, [3] When the logarithmic concentration of atmospheric PM$_{2.5}$ increases by one unit, the risk of respiratory symptoms increases by 1.79 times, especially for people who already have respiratory diseases [4]. Deaths caused by air pollution account for one-ninth of the total number of deaths worldwide, representing the greatest environmental health

B.-J. He et al. (Eds.): UCSUD 2023, LNCE 559, pp. 643–655, 2025.
https://doi.org/10.1007/978-981-97-8401-1_46

crisis we face. Traffic emissions are the main source of $PM_{2.5}$, affecting pedestrians and people working and living near roads [5].

Urban parks are the most frequently used public spaces by urban residents. Various elements of park design can change the diffusion of air pollutants and the distribution of people, the heterogeneity of exposure caused by the heterogeneity of pollutants and population distribution in the park [6]. At present, the research on $PM_{2.5}$ mainly focuses on source analysis, composition [7, 8], the impact of $PM_{2.5}$ on human health [9], plant dust retention capacity [10, 11], as well as the reduction [12, 13] and regulation [14, 15] of $PM_{2.5}$ in green space. At the same time, urban green belts can help improve the quality of the environment and thus improve human health [16]. Under light to moderate pollution weather, it is an important issue to study which activity space in the city park has a lower $PM_{2.5}$ concentration value. What is the most suitable activity in time and space has important research significance [17].

At present, there are few studies on the correlation between $PM_{2.5}$ and human respiratory health in parks in Mianyang area. This study starts from the perspective of respiratory health, identifies the pollution status of various public spaces in the park through actual measurement analysis. At the same time, the spatiotemporal distribution characteristics of crowd activities were evaluated by combining geographic information system spatial analysis. Then, based on the principle of crowd exposure assessment, the spatiotemporal distribution of $PM_{2.5}$ in the park is overlaid with the spatiotemporal distribution of crowd activities to quantitatively assess the health risk distribution of important spaces. Finally, for different health risk areas, spatial optimization and crowd guidance strategies that are conducive to reducing $PM_{2.5}$ exposure hazards are proposed. By improving the air quality of the park and stimulating the multi-functional role of urban parks, a win-win situation of social and ecological benefits can be achieved.

2   Research Method

2.1   Research Location

Mianyang City (latitude $30°42'–33°02'N$, longitude $103°45'–105°43'E$) is located in the northwest of Sichuan Province, China, with a subtropical humid monsoon climate. In winter, it is affected by the north wind, with a dry and cold climate. In summer, it is affected by the south wind, with a hot, rainy, and humid climate.

The study selected the representative urban park in Mianyang City - People's Park. People's Park is located in the center of Mianyang City, it is surrounded by major city arteries, with the City Traditional Chinese Medicine Hospital to the north., the east side is Xinglida Plaza, with a large flow of people. The main roads in the east and north of the park, namely Fuicheng Road and Linyuan Road, are the main sources of traffic pollution in the park. The People's Park has a rich landscape, providing a place for leisure and entertainment for the surrounding residents and tourists, and it is of great significance and value to carry out experimental research on it.

2.2   Research Method

$PM_{2.5}$ Measurement Method. The pollution in Mianyang City is severe in winter, and the measurement was chosen to be carried out in People's Park from 8:00 am to 9:00 pm on December 10, 2023, for a concentrated measurement of 14 h. The Broughton SMART-128S detector was used to measure the concentration changes of $PM_{2.5}$. Considering the average breathing height of the crowd, the measuring instrument was uniformly set at 1.5 m. To observe whether the particle concentration is the same in different micro-space environments, the study selected 34 monitoring points covering various space areas of the park. As shown in Fig. 1, each point is measured an average of 12 times per hour, with 168 sets of data measured in 14 h a day.

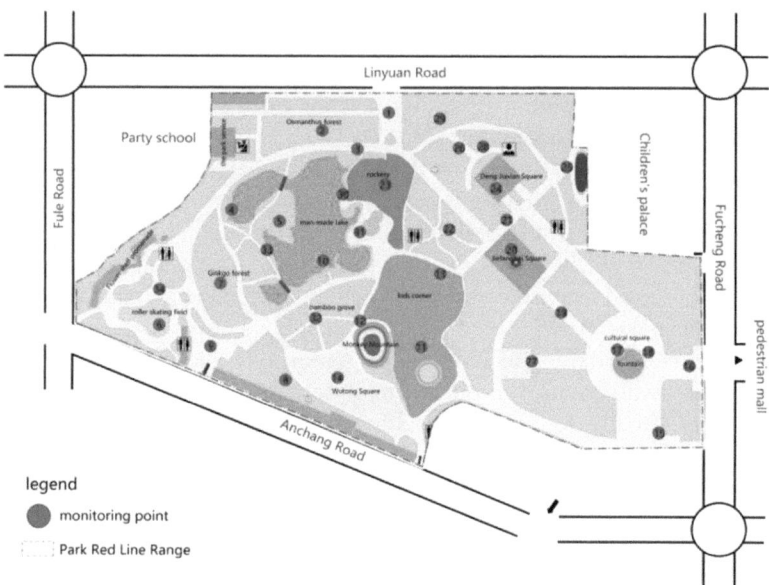

Fig. 1. Distribution map of PM2.5 concentration monitoring points in People's Park

Crowd Activity Observation Method. The recreational behavior observation method was used for this crowd activity measurement. While monitoring PM $_{2.5}$ data in People's Park, the number, type, duration, spatial distribution, and environmental characteristics of crowd activities were observed and recorded.

3 Measurement Data Analysis

3.1 $PM_{2.5}$ Data Analysis

Overall Trend of $PM_{2.5}$ Concentration. Based on the field measurement data, the overall trend of pollutant changes in People's Park was obtained. Through the comparison of charts, it can be found that: The average $PM_{2.5}$ concentration in the park is high, and the $PM_{2.5}$ concentration of the park measurement and the city's People's Congress monitoring station closest to the People's Park from 8:00–21:00 overall shows a trend of first decreasing and then increasing, and the measured concentration of the park at each time point is slightly higher than the monitoring station concentration. The concentration of the park changes throughout the day in a "V" trend, and the concentration at night is higher than the peak value during the day. (Fig. 2). Judging from the surrounding environment of the park, the city's main road adjacent to the park has an impact on the $PM_{2.5}$ concentration in the park during the peak traffic hours of 8:00–9:00 and 18:00–19:00.

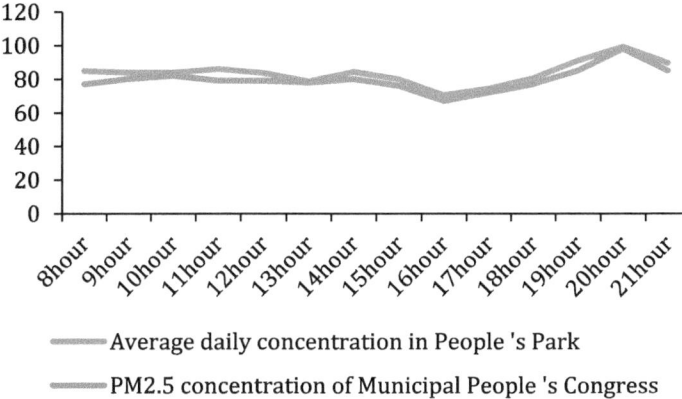

Fig. 2. 14-h variation trend of PM $_{2.5}$ concentration in People's Park and the Municipal People's Congress on December 10

Spatial distribution characteristics of $PM_{2.5}$ concentration. The park's overall concentration shows a trend of being high in the west and low in the east, and low in the middle. The areas with better air quality are larger, open and flat or near the lake, and are located in the center and east of People's Park; the areas with poorer air quality are clearly distributed in the northwest direction, these areas are rich in spatial forms, or close to the city's main roads (Fig. 3).

From 8:00 to 17:00, the pollution concentration at point 1 near Linyuan Road is higher, and the concentration at points 2 and 3 far from the road is relatively small, indicating that the closer to the road with large traffic flow, the more obvious the impact of car exhaust. During this period, the aggregation effect of tree space on $PM_{2.5}$ is greater than the impact of traffic source emissions on $PM_{2.5}$ distribution. The concentration of the measuring point is related to the traffic source and tree space.

The overall pollution level of the monitoring points around the artificial lake is relatively low, and the distribution of the concentration at each point shows a strong correlation with the openness of the building. The woodland area presents a situation of slight pollution during the day, and the internal green space has a significant reduction effect on the $PM_{2.5}$ of the city road to the south. The $PM_{2.5}$ concentration of the 7th is relatively low throughout the day, and the 8th tea garden and the 9th entrance space are at a higher level, indicating that the woodland green space has a purifying effect on $PM_{2.5}$. From 19:00 to 21:00, the background concentration slightly increased, and the pollutant concentration at points 16, 17, 18, and 20 increased less than other points, indicating that open space is conducive to the diffusion of pollutants.

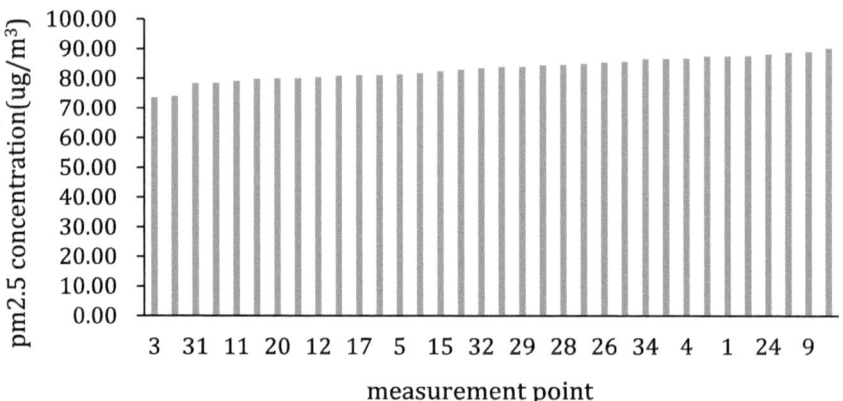

Fig. 3. Average PM$_{2.5}$ concentration distribution map of various points in People's Park on December 10

3.2   Analysis of Crowd Activities

Analysis of crowd activity types. While measuring the concentration of pollutants, the crowd behavior in each measuring point was recorded hourly. From the statistical results, it can be seen that the spatial characteristics of the People's Park have a significant impact on crowd activities. The author analyzes the current situation of crowd activities at different spatial nodes through field measurements, understanding the mutual connection between crowd activities and public space forms. According to the recorded data, the types of activities carried out in the park's public spaces can be summarized as rest, light exercise, and moderate exercise.

The impact of spatial characteristics on crowd activities. Summarize the users of different public spaces, the content and types of activities, the duration of activities, and the time of activities, it is found that they are closely related to spatial characteristics. The higher the privacy, the more conducive it is to the development of rest activities, and the duration of activities increases accordingly. The average rest time under the shaded trees is higher than that of the square space seats, and people are more inclined to carry out light and moderate exercises, such as the Square. The capacity, comfort, and entertainment activities of the venue will affect people's willingness to use the space, such as the Cultural Square is generally open, equipped with fountains, surrounded by seats, people can rest, talk, stall, feed pigeons, it is the public space with the most people in the park.

Spatial distribution characteristics of crowd activities. Through the statistics of the use of different public spaces in the previous text, the high-frequency time periods of the activity types corresponding to the crowd are summarized. Resting activities are from 8:00–18:00, light exercise is concentrated from 9:00–10:00, 15:00–16:00, morning light and moderate activities are concentrated from 9:00–11:00, the afternoon is mostly for rest and conversation, and square dancing is carried out from 19:00–21:00 at night. Combined with the actual measured activity number, behavior records, and photos, the GIS kernel density analysis tool is used to draw the spatial distribution map of the crowd at different time periods and different activity types (Fig. 4).

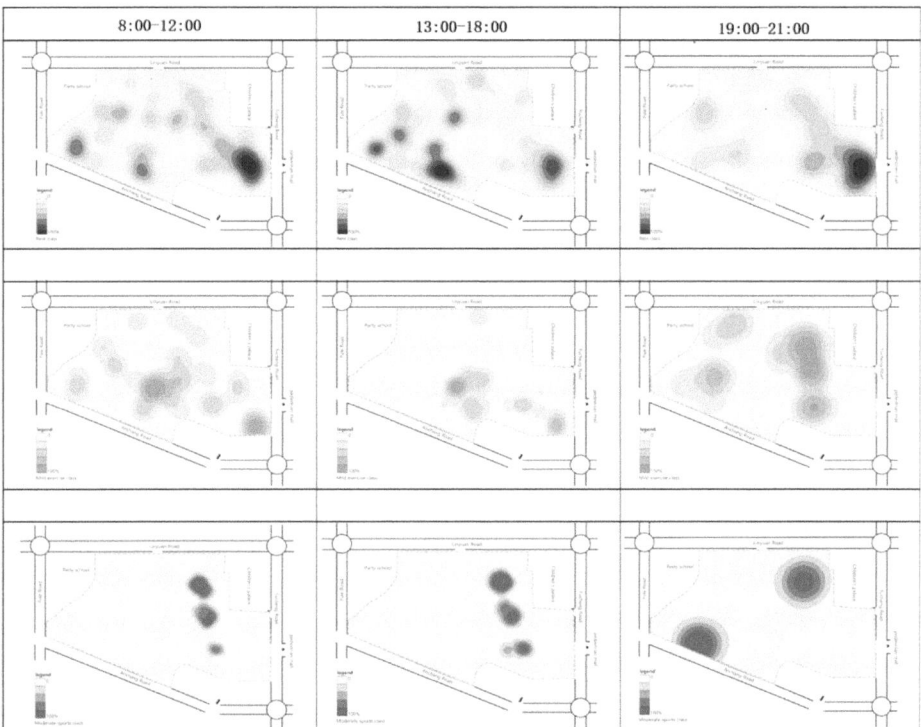

Fig. 4. Spatiotemporal distribution map of crowd activities in People's Park at different time periods on December 10

Through the spatial-temporal distribution, it can be intuitively seen that the number of people in different areas varies greatly. The number of people on the east side of the park remains at a high level throughout the day, while the southwest side and the middle fluctuate greatly, maintaining a high level from 8:00–18:00, and showing a significant downward trend after 19:00; the number of people on the north side is relatively small. (1) The spatial distribution of people in different areas is heterogeneous. The square space in the east and northeast of the park has a wide view, entertainment, and rest facilities are densely distributed, and the hard pavement between nodes is dominant, the mobility of the crowd is large; the green space and lakeside are more closed and private, with certain rest facilities, and are connected through various nodes and paths. (2) There are differences in activity types and spatial distribution at different times. From 8:00–12:00

in the morning and 13:00–18:00 in the afternoon, residents mainly use the square space and tea garden as places for rest and exercise. From 19:00–21:00 in the evening, they mainly use the brightly lit square space as activity places. (3) Different types of activities overlap in spatial distribution. Resting activities are distributed in various spaces of the park; light exercise activities are mainly distributed in node areas; moderate exercise activities are mainly distributed in the square space. At the same time, there are often people engaged in resting activities in places where light or moderate exercise activities are carried out.

4 Health Risk Assessment of People's Park

4.1 Selection of Important Spatial Nodes

Through the observation and summary of the spatial-temporal distribution and activity types of the crowd at various measurement points in People's Park, and referring to the impact mechanism of $PM_{2.5}$ on the human body, targeted analysis is carried out on public spaces where there are more crowds and longer stay times. Combining the spatial-temporal distribution map of crowd activities (Fig. 4), Finally, 10 important public spaces are selected, including the waterside area of the 4th water platform, the 6th roller skating rink, the 7th Peach Blossom Island, the 12th children's activity area, the 14th plane tree square, the 15th pigeon park, the 16th cultural square, the 20th liberation monument, the 22nd east side small square, the 24th Deng Jiaxian square, the 27th small square (Fig. 5).

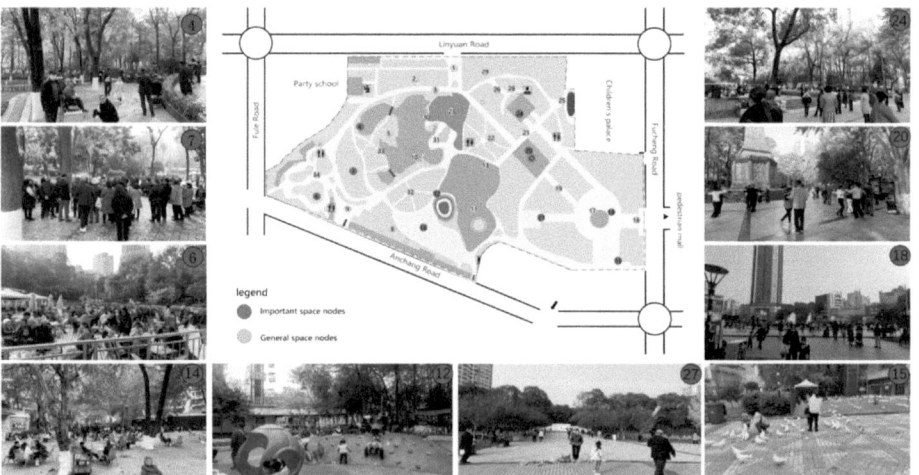

Fig. 5. Location and real scene shooting of important spatial nodes in People's Park

4.2 Health Risk Assessment of Time-Activity

The time-activity pattern is a common micro-scale human exposure assessment pattern [18], combined with micro-environment monitoring data, according to different individual activity characteristics and duration, thereby evaluating the corresponding health risks. According to the research of Guo Shengli and other scholars [19], the daily $PM_{2.5}$ exposure level of the population in specific public spaces can be confirmed by the following formula:

$$I = C_m \times BR \times T_m \tag{1}$$

where I is the $PM_{2.5}$ exposure dose in the microenvironment (ug); Cm is the $PM_{2.5}$ concentration in this environment (ug/m3); BR is the breathing rate of a specific population (L/min); Tm is the duration of stay in this environment (min). The formula shows that the mode and average duration of activities in public spaces determine the exposure level of $PM_{2.5}$. The daily average $PM_{2.5}$ concentration value at each measurement point is used as the Cm value. According to the research on the respiratory volume of Chinese adults by Liu Ping [18] and other scholars, combined with field observation data, the respiratory volume of the population is determined, and the type of activities of the population is the main factor affecting the breathing rate.

Due to the types and times of activities at different spatial nodes are significantly different. Therefore, this article calculates the total $PM_{2.5}$ exposure dose at each measurement point based on the proportion of activity types within each measurement point:

$$I_x = C_m \times BR \times T_m \times Q_x \tag{2}$$

In formula 2, Q_x is the proportion of various activity types at each measurement point, and the comprehensive exposure dose of $PM_{2.5}$ at each measurement point is the sum of the $PM_{2.5}$ exposure doses during various activities. 10 important spatial-related influencing factors are sorted out, as shown in the following table (Table 1):

Based on the time-activity exposure evaluation formula, this study uses the national $PM_{2.5}$ primary standard of $35\mu g/m3$ as a reference value; according to the research by Ma Zijian [20], for every $10\mu g/m3$ increase in $PM_{2.5}$, result in a 3.48% increase in total hospitalizations. An increase of $10\mu g/m3$ in daily concentration in the short term (absorbing an additional $99\mu g$ in the short term) is used as an effective indicator of disease probability, forming the health risk calculation method for this study. The basic calculation formula for the health risk of People's Park is as follows:

$$H_X = \frac{(C_m - C_b) \times BR \times T_m \times Q_X}{99} \times 3.48\% \tag{3}$$

where H_X is the health risk (%) of a certain type of activity at a spatial node, C_b is the national daily average $PM_{2.5}$ The primary standard for concentration is 35ug/m3. Through calculation and arrangement using formula 3, the health risks of 10 important public spaces are determined. At the same time, it can be found through comparison (Fig. 6) that squares often have the highest spatial health risks. The health risk of the crowd at Deng Jiaxian Square No. 24 is 3.33%, followed by the Liberation Monument Square No. 20 with 2.49%, Wutong Square No. 14 and Small Square No. 27 with 1.22%, and Cultural Square No. 18 has a relatively lower risk compared to other squares, at 1.12%. The health risks of Peach Blossom Island No. 4, No. 6, No. 15, No. 12, and No.

Table 1. Proportion of activities and average activity time at important spatial nodes

Important spatial locations	No. 4	No. 6		No. 7	No. 12	No. 14	
Daily PM$_{2.5}$ (ug/m^3)	87	89		80	80	90	
Activity Type	Resting	Resting	Mild Exercise	Resting	Mild Exercise	Resting	Mild Exercise
Breathing rate (L/min)	5.4	5	8	5.4	8	5.4	8
Average activity duration	60min	45min	30min	30min	30min	120min	30min
Activity proportion	1.00	0.07	0.93	1.00	1.00	0.96	0.04

Important spatial points	No. 15	No. 18	No. 20		No. 24		No. 27		
Daily PM$_{2.5}$ (ug/m^3)	83	84	80		88		81		
Activity type	Mild exercise	Resting	Resting	Moderate exercise	Resting	Moderate exercise	Resting	Mild exercise	Moderate exercise
Breathing volume (L/min)	8	5.4	5.4	21.4	5.4	21.4	5.4	8	21.4
Average activity duration	30min	120min	30min	90min	30min	90min	30min	60min	90min
Activity proportion	1.00	1.00	0.20	0.80	0.08	0.92	0.50	0.20	0.30

7 are relatively small, at 0.59%, 0.46%, 0.40%, 0.38%, and 0.26% respectively. At the same time, it can be seen that the PM$_{2.5}$ concentrations at No. 4, No. 6, No. 14, and No. 24 are at a higher level, but due to the longer stay and higher activity intensity of the crowd at No. 24, its health risk is greater than the other three places. No. 14 has a longer stay, and the daily average PM$_{2.5}$ concentration is slightly higher than that of No. 4 and No. 6. These phenomena prove that in addition to the concentration of PM$_{2.5}$, activity characteristics also have a significant impact on health risks.

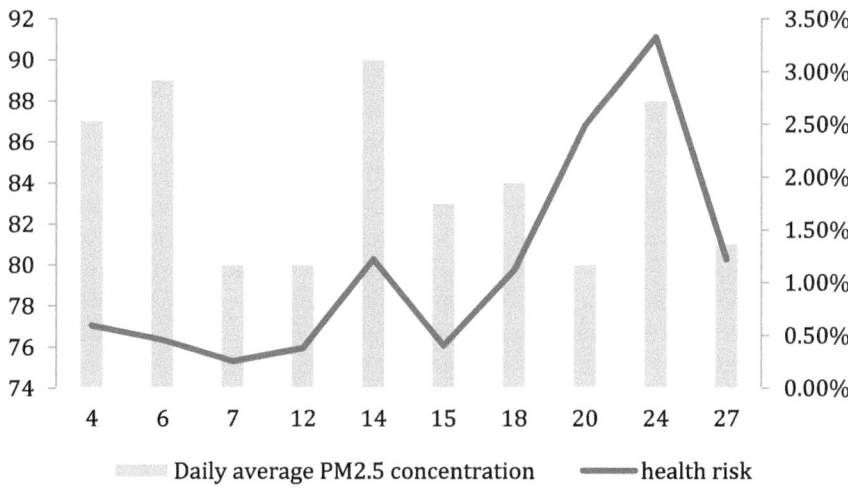

Fig. 6. Comparison of health risk assessment of important spatial nodes in People's Park

5   Conclusions and Prospects

5.1   Conclusion

This article analyzes the changes in PM$_{2.5}$ concentration and crowd activity characteristics in important public spaces in People's Park through field observations, The main conclusions are as follows: (1) The daily average PM$_{2.5}$ concentration in People's Park is high, and there are obvious differences in spatial and temporal distribution: compared with the city's People's Congress monitoring station, the PM$_{2.5}$ concentration in People's Park is always at a higher level. The larger and open square spaces tend to have lower PM$_{2.5}$ concentration levels. The farther away from the city road and the smaller the traffic flow, the lower the PM$_{2.5}$ concentration at its spatial nodes; Under the shade of trees, the denser the forest, the higher the PM$_{2.5}$ concentration. (2) The peak period of people flow overlaps with the period of high PM2.5 concentration: from the actual measurement, the overall change trend of the park's PM$_{2.5}$ concentration is "V"-shaped, with higher pollution concentrations in the morning and evening, and generally showing a downward trend in the afternoon, with a slight increase in the evening. The peak period of people flow is between 9:00–11:00 in the morning and 14:00–17:00 in the afternoon. The peak period of people flow lasts longer on weekends. (3) There are higher health risks in parts of People's Park where crowd activities are highly concentrated. According to actual crowd observations, a considerable number of people often stay in PM2. Residents carry out many sustained activities such as rest, communication, entertainment, and exercise in several public spaces with high $PM_{2.5}$ concentrations. Using the time-activity health risk formula, it was found that residents who carry out long-term activities in public spaces with high daily average $PM_{2.5}$ concentrations have a health risk about 2.65 times higher than those in spaces with lower daily average $PM_{2.5}$ concentrations. For residents who carry out long-term high-intensity activities in public spaces with high daily average $PM_{2.5}$ concentrations, their health risk is about 5.6 times higher than those who do light exercise or rest in spaces with similar $PM_{2.5}$ concentrations.

In summary, for the space nodes in the People's Park with higher health risks, two methods can be used: reducing $PM_{2.5}$ concentration and activity guidance. In terms of activity guidance, for spaces with high $PM_{2.5}$ concentrations, some rest facilities within the space can be dismantled to reduce the number of people staying, and sculptures can be arranged on flat and open spaces, or the form of paving bricks on the ground can be changed to guide people not to carry out entertainment exercises such as square dancing in this space. For example, only the passage and rest functions are retained in Deng Jiaxian Square at point 24, or space replacement is carried out, and moderate activities are carried out in other areas with good environment; Carry out space transformation to reduce $PM_{2.5}$ concentration: Enhance the openness and ventilation of public spaces, and promote the air circulation inside the public space; or increase the environmental humidity through the water body landscape to promote the settlement of $PM_{2.5}$, add fountains, water mist and other landscape features, enrich the water scene while also promoting the further settlement of $PM_{2.5}$.

This study is expected to provide guidance for park design and optimization. In the design of park space and landscape layout, priority should be given to the health of the population, and the accessibility and comfort of the active health space should be improved. Increase vegetation coverage as much as possible, introduce water, reduce large-area hard pavement, and select the underlying surface material that has the effect of regulating the microclimate of the park. For the area near the main road with large traffic flow, it can be improved by designing appropriate buffer distance, green belt and water body.

Through the study of air pollution in parks in Mianyang City, this paper fills the gap in such research in small and medium-sized cities, and enriches the understanding of the exposure risk pattern of population and $PM_{2.5}$ dynamic distribution in the park environment. This paper is compared with similar studies: Song et al.'s study on the pollution, concentration distribution characteristics, sources and health risks of polycyclic aromatic hydrocarbons (PAHs) in $PM_{2.5}$ in Beijing Olympic Park during heating and non-heating seasons shows that human respiratory exposure has potential carcinogenic risks in Beijing 's atmosphere [21]. Liu et al. found that the spatial structure of different green spaces in urban parks had significant differences in $PM_{2.5}$ reduction and diffusion. $PM_{2.5}$ concentration was affected by factors such as distance, air temperature and relative humidity, and was significantly negatively correlated with distance [22]. The results of this paper show that the concentration of $PM_{2.5}$ near the main road is higher, the health risk of the population is greater, and the daily average concentration and health risk of $PM_{2.5}$ under the forest are lower. Therefore, this study has certain rationality.

5.2 Outlook

Facing the increasingly prominent health problems, outdoor air pollution has also become a topic of concern to society and the people. Healthy outdoor spaces are more conducive to residents' health. As the public space with the highest usage rate by residents, parks have the characteristics of crowd gathering and continuity. Evaluating the health risks of the park environment can assist in optimizing the built environment and making rational and efficient use of public spaces. This article evaluates the health risks of People's Park, is of great significance for building healthy urban public spaces. Due to human resource constraints and other uncertain factors, this article only selected the winter period with the most severe $PM_{2.5}$ pollution for measurement. In future research, the $PM_{2.5}$ diffusion mechanism and crowd behavior in parks in other seasons can be explored.

References

1. Xing, Y., Brimblecombe, P.: Role of vegetation in deposition and dispersion of air pollution in urban parks. Atmos. Environ. **201**, 73–83 (2019)
2. Bao Hongguang et al.: Gradient distribution of PM2.5 concentration level in Haidian Park in summer. J. Environ. Eng. **11** (06), 3678–3684 (2017)
3. Johnson, P.R.S., Graham, J.J.: Fine particulate matter national ambient air quality standards: public health impact on populations in the Northeastern United States. Environ. Health Perspect.Perspect. **113**(9), 1140–1147 (2005)
4. Cui Guoquan et al.: The impact of PM 2.5 pollution level in Harbin on respiratory diseases in the population. Chin. Public Health. **29** (07), 1046–1048 (2013)
5. Li, L. et al.: Effects of green infrastructure on the dispersion of PM2.5 and human exposure on urban roads. Environ. Res. **223** (2023)
6. Xing, Y., Brimblecombe, P.: Urban park layout and exposure to traffic-derived air pollutants. Landsc. Urban Plan.. Urban Plan. **194**, 103682 (2020)
7. Gu Jinxia et al.: Pollution characteristics and source analysis of water-soluble inorganic ions in PM _ (2.5) in Tianjin. Environ. Monit. China **29** (03), 30–34 (2013)
8. Yao Zhenkun, Feng Man, Lyu Lin, et al.: Physicochemical characteristics and source apportionment of PM2.5 in Shanghai urban area and Lin 'an background station [J]. Environ. Sci. China **30** (03), 289–295 (2010)
9. Xu Jianjun et al.: Study on PM2.5 exposure levels and lung function in different working environments. J. Environ. Health **30** (01), 1–4 (2013)
10. Huixia, W., Hui, S., Yangyang, L.: Effects of leaf surface characteristics of urban greening plants on dust retention capacity. Appl. Ecol. **21**(12), 3077–3082 (2010)
11. Yu Linlin, Hu Haibo and Yu Wei.: Effects of urban green space types on atmospheric PM _ (2.5) concentration. J. Nanjing For. Univ. (Nat. Sci. Ed.) **44** (03), 179–184 (2020)
12. Wang Guoyu et al.: Analysis of green space design technology for reducing PM _ (2.5) and other particulate matter pollution in Beijing. Chin. Gard. **30** (07), 70–76 (2014)

13. Xiao et al.: The reduction effect of urban green space on atmospheric PM _ (2.5) in Beijing. Resour. Sci. **37** (06), 1149–1155 (2015)
14. Shi et al.: Study on the effect of PM _ (2.5) regulation in Wuhan garden green space. Anhui Agric. Sci. **50** (11), 101–105 (2022)
15. Sen, W., et al.: Comparison of air particulate matter concentrations and influencing factors in parks and squares. Gardens **39**(12), 129–134 (2022)
16. Islam, N.M., Rahman, K., Bahar, M.M., et al.: Pollution attenuation by roadside greenbelt in and around urban areas[J]. Urban For. Urban Green. **11** (4), 460–464 (2012)
17. Xu Huan, Li Hong, Jiang Taihao: Study on the uneven distribution characteristics of PM2.5 concentration in different activity spaces of urban parks. Chinese Gardens. **34** (03), 117–122 (2018)
18. Ping, L., et al.: Study on adult respiratory volume in China. J. Environ. Health **31**(11), 953–956 (2014)
19. Guo Shengli, Wang Xi, Huang Jun, Progress in the study of human PM2.5 exposure evaluation in various microenvironments based on time-activity pattern. Sci. Technol. Eng. **14** (27), 128–134 (2014)
20. Zhu Yu, Ma Zijian, Xiao Changchun. The relationship between atmospheric PM2.5 pollution and the number of hospitalizations for respiratory diseases in Hefei City[J]. Mod. Prev. Med. **48** (06), 985–989 (2021)
21. Song Guangwei, Hu Jian, Cui Meng, et al.: Characteristics, sources and health risk assessment of polycyclic aromatic hydrocarbons in PM _ (2.5) in Beijing Olympic Park during heating and non-heating seasons [J]. Ecol. J. **38** (11), 3400–3407 (2019). https://doi.org/10.13292/j.1000-4890.201911.037
22. Liu Qing, Liu Zhenmeng, Li Yaping, et al. Study on the reduction characteristics of PM _ (2.5) and PM _ (10) in urban parks based on DSM-Taking Nanchang People 's Park as an example [J]. J. Jiangxi Agric. Univ. **46** (01), 173–183 (2024)

Noise Reduction Performance of Metamaterials Sound Insulation Plate

Baoqing Zhang[1], Yubin Rao[2], Yunyi Guo[1], and Wangqiang Xiao[2(✉)]

[1] Institute of Fluid Physics, Academy of Physics Engineering, Mianyang 621999, China
[2] School of Aerospace Engineering, Xiamen University, Xiamen 361000, China
wqxiao@xmu.edu.cn

Abstract. In order to reduce the adverse effects of noise generated by urban vehicles on residents, a super-structured phononic crystals sound isolation plate is designed to enhance the low-frequency noise reduction capability of the plate by utilizing the special acoustic properties of the super-structure. The phononic crystals plate is modeled and analyzed using the finite element method to study its band gap. The energy band and transmission loss patterns of phononic crystals are obtained, and the noise reduction capabilities of traditional sound isolators and phononic crystals plate are compared. Through experimentation, the noise reduction effect of the phononic crystal sound insulation plate was found to be remarkable. This provides theoretical support for the use of phononic crystals sound insulation devices in managing urban noise.

Keywords: Metamaterials · Noise reduction · Phononic crystals

1 Introduction

Traffic noise is a common problem in urban, affecting people's quality of life and health. As an important part of the national economy and industry, urban traffic has a significant impact on people's well-being index. The braking and tire noise generated by large heavy-duty vehicles at night on the highway [1] seriously affects the lives of nearby residents. In order to effectively control traffic noise, people have taken a variety of measures. Rational planning and design of road network is one of the major control measures. Through rational planning and design of road networks, traffic flow can be divided rationally to reduce traffic congestion and noise generation [2]. The use of noise-reducing materials in pavement construction is also a common method of controlling traffic noise. Noise-reducing materials can effectively absorb and reduce the noise generated by vehicles, thus effectively reducing the level of traffic noise. By using these materials on major traffic arteries and in noise-sensitive areas, noise disturbance to nearby residents can be significantly reduced. Research into low-noise vehicles is also a meaningful measure. Traffic noise can be effectively reduced by improving engine and vehicle design, and reducing the noise generated during vehicle operation. In recent years, with the development and promotion of electric vehicles, their low-noise characteristics have become an effective means of reducing traffic noise. Rational greenbelt arrangement is also a

B.-J. He et al. (Eds.): UCSUD 2023, LNCE 559, pp. 656–666, 2025.
https://doi.org/10.1007/978-981-97-8401-1_47

common method to control traffic noise. Through the rational planting of vegetation on both sides of the road or around the noise source, it can absorb and isolate part of the noise, and play the effect of buffering and reducing traffic noise. At the same time, green belts can also improve the ecological environment of the city and provide air purification. Another widely used method of noise reduction is the installation of noise barriers. The advantages of noise barriers are their significant noise reduction effect, flexibility of installation, low cost and room saving. The installation of noise barriers can be tailored to specific needs to alleviate the impact of traffic noise on nearby residents and the environment. With innovations in technology and materials, the noise reduction effect of noise barriers is constantly improving, making them one of the most main means of urban traffic noise control in the world.

An effective sound insulation is needed to alleviate the impact of vehicle noise on nearby residents. In response to this need, the acoustic metamaterials based on phononic crystals acoustic insulator installed on both sides of the road is designed to reduce noise on urban highways through the special physical properties of acoustic metamaterials, such as bandgap properties and acoustic black holes. A large number of new acoustic metamaterials have been proposed: one of them is used for periodic open cellular structures (POCS) plate to absorb the noise. Liu [3, 4] and others proposed the local resonance type acoustic metamaterials, which can achieve the effect of blocking the low-frequency sound waves by using the flexible bag to wrap the small balls and then placing them into the periodic structure, but it is not effective for high-frequency sound waves. Yu and Lesieutre [5] designed honeycomb sandwich panels with high strength and full resin, which is a good choice for noise reduction, but its cost is high. Effectively solving the problem of poor structural static load adaptation, this structure is able to disperse and absorb energy when subjected to external pressure, thus improving the overall strength of the panel. In addition, all-resin honeycomb sandwich panels have better durability and corrosion resistance, and can be used for a long time in harsh environments. Phononic crystals [6–8], as a kind of periodically arranged composite acoustic metamaterials with unique propagation characteristics, can effectively play the role of sound insulation due to its elastic forbidden band property [9–13], which prevents the propagation of sound wave in the forbidden band region.

In this paper, an acoustic metamaterials phononic crystals sound barrier is designed to improve the low-frequency noise reduction capability of the sound barrier by utilizing the special acoustic properties of the structure. The finite element method is used to mathematically model the phononic crystals plate and to investigate the phononic crystals band gap. The energy band and transmission loss patterns of phononic crystals are obtained by COMSOL, and the noise reduction capabilities of traditional sound isolators and phononic crystals panels are compared. After that, the noise reduction effect of the phononic crystals sound insulation plate is obtained through experiments, which provides theoretical support for the phononic crystals sound insulation device in urban noise management.

2 Theoretical Analysis of Metamaterials Phononic Crystals Sound Insulation

According to the metamaterials bandgap theory and according to the definition of acoustic metamaterials, one or more kinds of elastic media are embedded in another elastic medium under the macro scale. Now we design the finite area type acoustic metamaterials, the material is aluminum profile, and the total thickness of the acoustic isolation plate is 12 mm, which consists of the upper and lower plates and the cylindrical scatterer embedded in them. The acoustic panel is 500 × 500 mm in length and width, the outer diameter of the cylinder is 34 mm, the inner diameter is 30 mm, which is very good to ensure the strength and quality of the acoustic metamaterials plate, and to ensure its antioxidant, fire-resistant and lightweight design, and due to the simple structure of the metamaterials, it is possible to simplify the processing process to reduce the cost of the processing, as shown in Fig. 1.

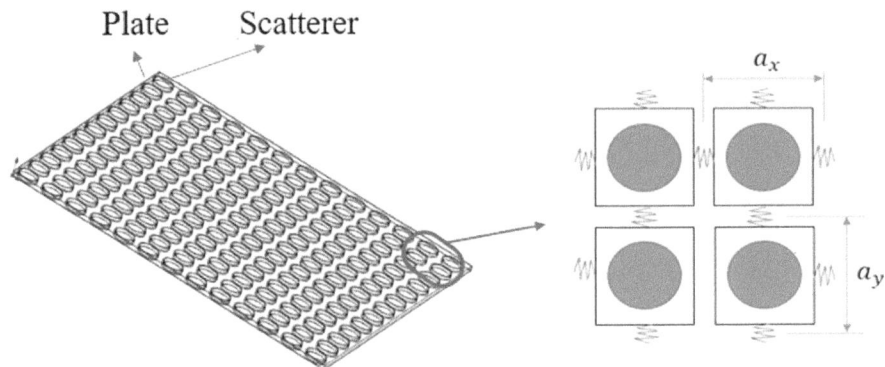

Fig. 1. Ultra-structural sound insulation panels and acoustic metamaterials model

The main focus of the metamaterials study is to determine whether there is a band gap and the frequency range in which it occurs. The core issue of the study is to understand the reason for the formation of the band gap. Figure 1 shows the acoustic metamaterials model.

From the energy band structure theory and the periodic Bloch's theorem, the Bloch function of the metamaterials is satisfied:

$$\psi_n(k, r) = e^{ik \times r} u_n(k, r) \tag{1}$$

where k is the Bloch wave vector taken in the first Bloch area and is satisfied for any lattice vector Rn:

$$u_n(k, r) = u_n(k, r) \tag{2}$$

Eq. $u_n(k,r)$ is shown to have the same periodicity as a lattice cell. As shown in the wave vector schematic in Fig. 1, typical of periodic structural cells, each cell is coupled and connected to the surrounding four cells. Assume that each lattice cell has n_x coupling coordinates on the left and right sides, whose coordinate arrays are denoted by $\{q_l\}$ and $\{q_r\}$, and at the same time, assume that each lattice cell has n_y coupling coordinates on the top and bottom sides, whose coordinate arrays are denoted by $\{q_t\}$ and $\{q_b\}$. Therefore, the force arrays acting on each lattice cell are $\{F_t\}$, $\{F_b\}$, and the n-order force and displacement arrays inside each lattice cell are $\{F_i\}$ and $\{q_i\}$.

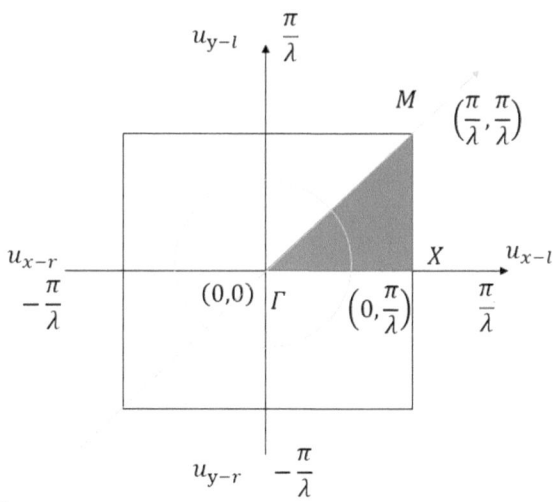

Fig. 2. Bloch area

When an elastic wave passes through a periodic cell, let the direction of elastic wave propagation make an angle of θ with the x-axis as shown in Fig. 2, then the projection along the direction of wave propagation is satisfied:

$$\vec{u_x} = ka_x cos\theta \tag{3}$$

$$\vec{u_y} = ka_y \sin\theta \tag{4}$$

where k is the wave loss satisfying $k = 2\pi/\lambda$, λ is the incident wave wavelength, and a_x and a_y are the unit lattice lengths. Therefore, $\vec{u_x}$ and $\vec{u_y}$ satisfy the relationship as follows:

$$\vec{u_y} = \vec{u_x}\frac{a_y}{a_x}tan\theta \tag{5}$$

When, $a_x = a_y$, $u_y = u_x$.

Therefore, based on the modal bandgap theory and the ideal metamaterials solution equation, we analyze the reasons for the formation of the bandgap of the ideal metamaterials by taking the metamaterials constructed by design as the research object.

When the sound source is far away from the obstruction, the waveform transmitted to the front of the demarcation can usually be equated to a plane wave. The sound wave incident from one medium to another medium on the acoustic boundary mainly consider two cases, oblique incidence and vertical incidence. Usually, the form of vertical incidence of sound waves is chosen as the main research object, so that the connection of sound waves in the transmission process can be obtained more simply. In a uniform medium sound field, $\rho_1 c_1$ and $\rho_2 c_2$ are assumed to be the characteristic impedances of background medium 1 and background medium 2, and $x = 0$ is considered to be the acoustic boundary coordinate point of the two mediums. When the plane acoustic wave in the background medium is vertically incident on the acoustic boundary, generally the first part of the acoustic wave will be reflected back by the boundary, and the other part of the acoustic wave enters into the background medium 2.

From the derived equation, the incident sound wave is p_i, where the amplitude of the incident sound wave is p_{ia}, then $p_{ia}e^{i(\omega t - k_1 x)}$ represents the sound wave in the forward direction, and the other term, $p_r = p_{ra}e^{j(\omega t + k_1 x)}$ is the reflected sound wave transmitted in the negative direction, so that it is obtained as:

$$p_1 = p_i + p_r = p_{ia}e^{i(\omega t - k_1 x)} + p_{ra}e^{j(\omega t + k_1 x)} \tag{6}$$

Assuming that there are no remaining obstacles behind medium 2, there are only transmitted sound waves in its sound field, denoted as p_t:

$$p_t = p_2 = p_{ra}e^{j(\omega t - k_2 x)} \tag{7}$$

And from the sound wave and mass velocity equations, the mass velocities v_1, v_2 in medium 1 and medium 2 are obtained:

$$\begin{cases} v_1 = v_{ia}e^{j(\omega t - k_1 x)} + v_{ra}e^{j(\omega t + k_1 x)} \\ v_2 = v_{ta}e^{j(\omega t - k_2 x)} \end{cases} \tag{8}$$

$$v_{ra} = \frac{p_{ia}}{\rho_1 c_1}, v_{ra} = -\frac{p_{ra}}{\rho_1 c_1}, v_{ta} = \frac{p_{ta}}{\rho_1 c_1} \tag{9}$$

Sound pressure and mass velocity are available at the boundary:

$$\begin{cases} p_{ia} + p_{ra} = p_{ta} \\ v_{ia} + v_{ra} = v_{ta} \end{cases} \tag{10}$$

$$\begin{cases} r_p = \frac{p_{ra}}{p_{ia}} = \frac{R_2 - R_1}{R_2 + R_1} = \frac{R_{12} - 1}{R_{12} + 1} \\ t_p = \frac{p_{ta}}{p_{ia}} = \frac{2R_2}{R_2 + R_1} = \frac{2R_{12}}{R_{12} + 1} \\ r_v = \frac{v_{ra}}{v_{ia}} = \frac{R_1 - R_2}{R_2 + R_1} = \frac{1 - R_{12}}{R_{12} + 1} \\ t_v = \frac{v_{ta}}{v_{ia}} = \frac{2R_1}{R_2 + R_1} = \frac{2}{R_{12} + 1} \end{cases} \tag{11}$$

The ratio of the mass velocity of the reflected wave to the mass velocity of the incident wave is represented by r_v, while tv represents the ratio of the mass velocity of the transmitted acoustic wave to the mass velocity of the reflected acoustic wave. Additionally, r_p and t_p represent the ratio of the reflected acoustic pressure to the incident acoustic pressure and the transmitted acoustic pressure to the incident acoustic pressure, respectively.

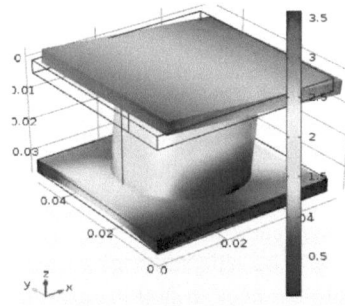

Fig. 3. Simulation model diagram of super-structured single cell

As when the excitation frequency converges to the complete band gap, the main vibration mode of the scatterer is excited, as shown in Fig. 3, the scatterer's Z vibration and the base plane vibration produced by the coupling, and inhibit the base vibration elastic wave is inhibited by the complete band gap formation. In turn, the transmission of sound waves can be suppressed, thereby reducing the noise to play a role in sound insulation. Next, the use of COMSOL software for acoustic super-structural panels for sound insulation and sound absorption simulation, so the simulation of the use of noise sources for a highway sound pressure level histogram (Fig. 4).

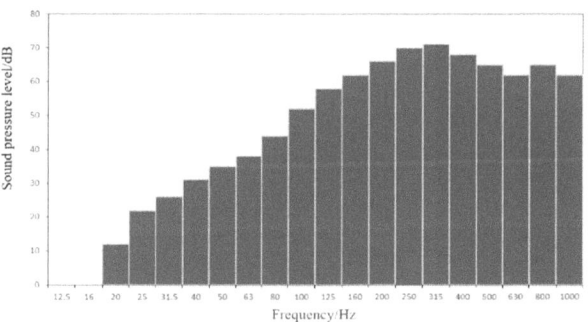

Fig. 4. Histogram of sound pressure level of a highway

The material is set to aluminum profile and its parameters are built-in parameters, the background pressure field and cavity in COMSOL are set to air and the periodic conditions are set around the unit body, Fig. 5 shows the sound isolation curve of the acoustic metamaterials plate:

As can be seen in Fig. 5, in the 50-500Hz range, the superstructure of the complete bandgap and the direction of the bandgap cause excitation of the scattering body in the main mode of vibration due to the acoustic wave. The scattering body of the Z vibration and the substrate plane vibration produced by the coupling are suppressed by forming a complete bandgap. This suppresses the substrate oscillation of the elastic wave and reduces the noise path to a large extent in the sound insulation board. This achieves

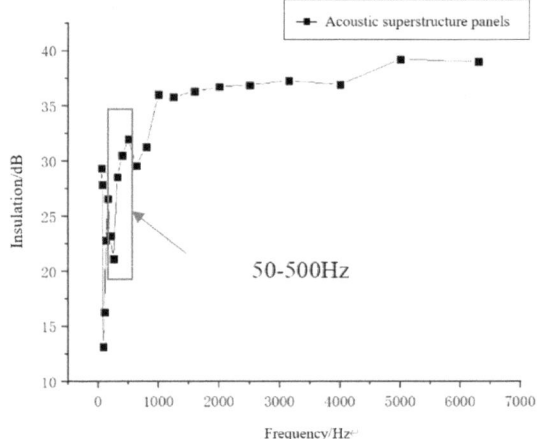

Fig. 5. Acoustic metamaterials plate sound insulation curve

better acoustic isolation, resulting in a weighted sound insulation of 32dB. Due to the phononic crystals bandgap, acoustic isolation panels have good acoustic isolation at specific frequencies, but the effect is not significant at certain non-bandgap frequencies.

To investigate the sound absorption and insulation effect of sound absorption and insulation structures with different slant plate angles. The structure of the metamaterial is 1mm aluminum alloy plate, 35mm metamaterial structure, 1mm aluminum alloy plate, and the total thickness is 37mm.

Four types of sound-absorbing structures are set up: the length of the straight plate is taken as 3500mm, the length of the inclined plate is taken as 693m, and the angles of the straight plate and the inclined plate are taken as 0°, 30°, 60° and 90° respectively. The slant direction of the inclined slab is toward the side of the highway (Fig. 6).

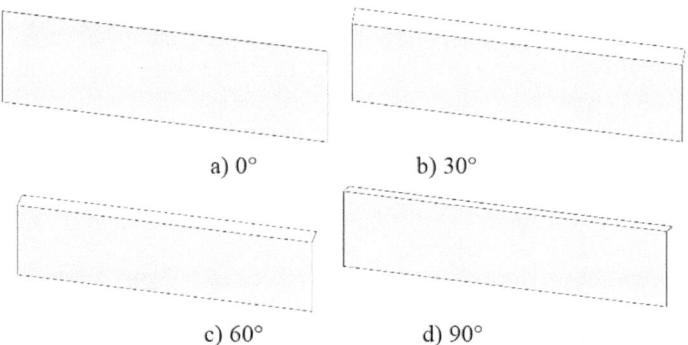

Fig. 6. Structural model of acoustic isolation with different sloped plate angles

After obtaining the three-dimensional model of the four types of sound absorption structures, mesh division is performed to obtain the finite element model, and then simulate the four types of sound-absorbing structures based on the method of analysis of sound-absorbing structures in turn, and then superpose the sound pressures measured by sensors based on the method of sound-pressure level superposition, to obtain the sound pressure level of the sound-absorbing structures with different angles of sloped panels in the same conditions at the monitoring points. The spectrograms of the sound pressure levels under the influence of the four types of sound-absorbing structures were plotted and then compared with the sound pressure levels monitored in the residents' homes (Fig. 7).

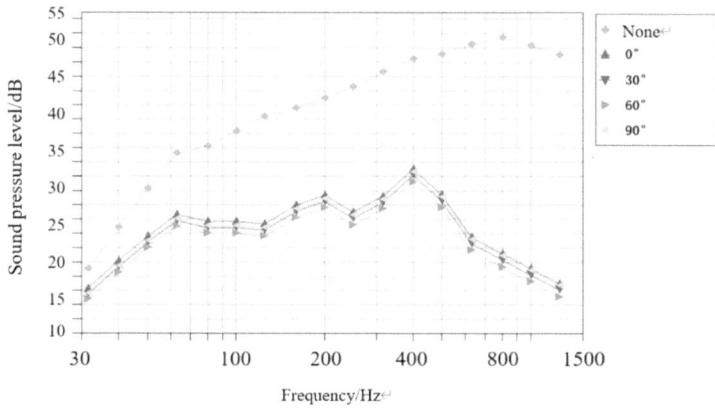

Fig. 7. Spectrum of sound pressure level of sound-absorbing structures with different angles of sloped panels

Under the influence of the acoustic metamaterials based on sound insulating structures, the sound pressure levels at the monitoring points of the residents' homes were reduced to below 40 dB. In addition, the acoustic isolation of the structure with a bevel angle of 60° is the largest, about 21.26 dB; the acoustic isolation of the structure with a bevel angle of 30° is the second largest, about 20.36 dB; and the acoustic isolation of the structure with a bevel angle of 90° is about 19.94 dB. However, the bevel angle of the acoustic isolation structure does not have a significant effect on the overall acoustic isolation. Therefore, based on the conditions of the same material and the same amount of material, the acoustic insulating structure with a ramp angle of 60° is the best, and the acoustic insulating structure with a ramp angle of 60° is selected for further analysis.

3   Noise Reduction Tests on Superstructural Sound Insulation Panels

The design of the experimental setup for the sound absorption and insulation test is shown in Fig. 8. Firstly, a window of 1000 × 1000 mm is opened in the partition wall in two rooms A and B for placing the sound insulation samples. A spherical sound source was arranged in room A at a distance of 1 m from the partition wall, and the noise data collected was used as the input noise excitation of the spherical sound source to simulate the noise generated by the public transportation, and a sound pressure transducer was arranged in room B at a distance of 1 m from the partition wall to simulate the measurement of the sound absorption data outside the room B after the acoustic isolation panel was installed. Firstly, the sound pressure was tested without the acoustic panels:

Fig. 8. Sound pressure test under empty working conditions

Table 1. Test data under empty working conditions

Condition	Total sound pressure level(dB)
None	87.6
	87.7
	87.6
Average	87.7

The average value of sound pressure measured by the test is 87.7 dB when in the null condition as shown in Table 1. To verify the accuracy of the simulation data, the test will evaluate the sound insulation capacity of the acoustic metamaterials plate.

Figure 9 shows the acoustic metamaterials plate in kind, with the size of 1m × 1m × 12mm, and the internal single cell is a scatterer with the outer diameter of 60mm and inner diameter of 56mm, and the material is aluminum profile.

Afterwards, the acoustic metamaterials panels were embedded in the wall and the sealing of the acoustic metamaterials panels was ensured, and then the acoustic insulation performance was tested.

The sound insulation effect of acoustic superstructural panels is evident from Table 2, with a reduction in the total extreme value of sound pressure by 31.5 dB.

Fig. 9. Acoustic metamaterials plate

Table 2. Acoustic metamaterials sound insulation test data

Condition	Total sound pressure level(dB)
Acoustic metamaterials plate	56.2
	56.1
	56.0
Average	56.1

4 Conclusion

In order to reduce the adverse effects of noise generated by urban vehicles on residents, metamaterial phononic crystals sound plate was designed. The following conclusions were obtained by simulating and testing it:

(1) The causes of the bandgap generated by the metamaterials are analyzed, and metamaterials phononic crystals sound isolation plate is designed, and the final noise reduction effect reaches 31.5 dB by COMSOL simulation of the metamaterials sound isolation plate.
(2) The laboratory test results of the phononic crystals plate are consistent with the simulation results, which proves that the super-structured band gap has an excellent blocking effect on the transmission of sound waves.
(3) The noise reduction ability of the phononic crystals plate, especially the vibration damping at specific frequencies, is consistent with the simulation results, which also confirms the feasibility of applying the super-structured phononic crystals plate to urban noise management.

References

1. Zhang, P.F., Yao, C.: Analyses of environmental pollution of traffic noise along expressway and urban road. Urban Environ. Urban Ecol. **12**(3), 29–31(1999)
2. Feng, Z.N.: Highway Traffic Noise Analysis and Prevention[J]. Communications Science and Technology Heilongjiang **07**, 92–94 (2007)
3. Liu, Z., Zhang, X., Mao, Y.: Locally resonant sonic materials. Science 289, 1734–1736 (2000)
4. Chen, L.F., You, S.H., Zhao, X.Y.: Study on sound insulation characteristics of two-dimensional phononic crystal thin plate. Materials Reports **34**(S1), 90–93 (2020)
5. Yu, T., Lesieutre, G.A.: Damping of sandwich panels via three-dimensional manufactured multimode metamaterial core. AIAA J. **55**, 1440–1449 (2017)
6. He, C., Gao, H.F., Xu, H.D., Li, Z.Q.: Dynamic characteristics of phononic crystals with locally resonant structure based on non-smooth system. J. Vib. Shock. **40**(22), 28–34 (2021)
7. Li, L.Y.: Research on the vibration reduction effect of track system based on phononic crystal theory. J. Railw. Eng. Soc. **37**(12), 64–69 (2020)
8. Zhao, W.J., Wang, Y.T., Zhu, R., Hu, G.K., Hu, H.Y.: Isolating low-frequency vibration via lightweight embedded metastructures. Scientia Sinica (Physica, Mechanica & Astronomica) **50**(09),162–175 (2020)
9. Zhou, X., Wang, L.: Opening complete band gaps in two dimensional locally resonant phononic crystals. J. Phys. Chem. Solids **116**, 174–179 (2018)
10. Chang, I.L., Liang, Z.X., Kao, H.W., et al.: The wave attenuation mechanism of the periodic local resonant metamaterial. J. Sound Vib. **412**, 349–359 (2018)
11. Yuan, B., Chen, Y., Jiang, M., et al.: On the interaction of resonance and bragg scattering effects for the locally resonant phononic crystal with alternating elastic and fluid matrices. Arch. Acoust. **42**(4), 725–733 (2017)
12. Lu, J., Qiu, C., Ye, L., et al.: Observation of topological valley transport of sound in sonic crystals. Nat. Phys. **13**(4), 369–374 (2017)
13. Krushynska, A., Miniaci, M., Bosia, F., et al.: Coupling local resonance with bragg band gaps in single-phase mechanical metamaterials. Extrem. Mech. Lett. **12**, 30–36 (2017)

Features of the Regional Layout of City Parks and Its Natural and Social Foundations: A Case Study of Mianyang, China

Qing-Wen Deng[1], Xiang Zhao[1(✉)], and Qian-Ming Xue[2]

[1] College of Civil Engineering and Architecture, Southwest University of Science and Technology, Mianyang 621000, China
1159799480@qq.com
[2] College of Architecture and Urban Planning, Lanzhou Jiaotong University, Lanzhou 730070, China

Abstract. Serving as a vital ecological space and public center for social activities, city play a key role in people's urban life. Rational and efficient layout is essential to promote citizens' participating level, which is the aim of building kinds of city parks. By investigating current situation of some typical city parks and inducing the data analysis method of GIS, POI and spatial autocorrection analysis, this research reveals the main feature of "high agglomeration and group" of city parks in major city centers in Mianyang and its positive effects. The research also shows that the number and spatial distribution density of city parks are strongly correlated with natural factors such as elevation, slope, and topography, as well as social factors including population, transportation, and economic growth, which causes an imbalanced layout geographically. Taking Mianyang as the research object, the aim of this paper is to visualize and quantitatively study the spatial distribution characteristics and its influencing factors, and provide basic analysis for the development of urban parks in Mianyang, and promote the implementation of Park City and Healthy City strategy in Mianyang.

Keywords: city park · regional layout · Mianyang City

1 Introduction

Politically, Mianyang, the only science and technology city of China, is also the second-largest city of Sichuan Province and a sub-center city in the Chengdu-Chongqing Dual-City Economic Circle, which grows faster than other cities (see Fig. 1). Economically, Mianyang City ranks the second in GDP, only lower than the capital city Chengdu in Sichuan Province, which is also the largest economy in West China. And gross domestic product of Mianyang reached 362.9 billion yuan in 2022, which equals 74322 yuan per capita and 6.9% per capita rate. Rapid growing inhabitant income causes appealing of public leisure spaces such as city parks, which number and regional layout consist of the main evaluating factors of a city's development [1]. Geographically, Mianyang City locates in the northwest of Sichuan Basin, which is surrounded by large mountains on

B.-J. He et al. (Eds.): UCSUD 2023, LNCE 559, pp. 667–679, 2025.
https://doi.org/10.1007/978-981-97-8401-1_48

both north and west sides. The climate of Mianyang is quite mild which means it is not too hot in summer and not too cold in winter. In 2022, the average temperature in the urban area was 18.5 °C, with 329 days of good air quality and an annual average precipitation of 1061 mm. Therefore, citizens here are used to outdoors life activities and they go to city parks to enjoy leisure times. Overall, the development of urban parks in Mianyang should be concerned.

Reviewing the literature on the spatial characteristics of urban parks, the research primarily focuses on aspects such as efficiency, equity, and supply, and accessibility aiming to optimize the spatial layout of urban parks from a humanistic perspective [2–4]. In terms of research areas, scholars have conducted specialized studies on different types of urban parks at national and provincial levels. Lately, with the development of park cities, the majority of researchers have shifted their focus towards the municipal level, and mainly focused on urban parks in provincial capitals and municipalities [5–8]. Regarding research methodologies, traditional methods primarily include GIS spatial analysis, econometric geographical models, and statistical analysis, with the advancement of digitalization, new technological such as big data have also begun to be applied in urban park research [9–12]. Given the research progress, China has accumulated certain achievements in the study of the spatial characteristics of urban parks, however, there is still room for further expansion in terms of research scale, perspective, and methodology. Hereby, this paper addresses the aforementioned weaknesses by taking the city of Mian yang as a case study, attempting to analyze the current characteristics of urban park layout and spatial structure in Mian yang and at the county level from a human geography perspective, and to deeply explore the geographical and human factors affecting their distribution, with the hope of providing theoretical strategies and practical suggestions for the high-level development of municipal and county urban parks, as well as the construction of park cities.

2   Data and Methods

2.1   Data Sources

The main source of data in this article is from the points of interest of park of the Amap platform, and based on the "Mianyang City Park Census Situation Table" and on-site investigations, tourist attractions, temple gardens, and comprehensive squares, etc. that meet the corresponding functions of parks were screened, a total of 171 park POI data of the park were obtained, including 54 in Fucheng District, 28 in Beichuan Qiang Autonomous County, 22 in Youxian District, 13 in Anzhou District, 21 in Jiangyou City, 11 in Zitong County, 10 in Santai County, and 6 in Pingwu County and Yanting County; Remote sensing image data came from the Landsat 8 satellite remote sensing image on August 30, 2023 from the Geospatial Data Cloud (http://www.gscloud.cn/); Social population and economic development-related data came from the "Mianyang City Statistical Yearbook".

2.2   Research Methods

In this article, relevant data of urban parks is introduced into ArcGIS 10.8 software and comprehensively applies spatial autocorrelation analysis methods to study the region

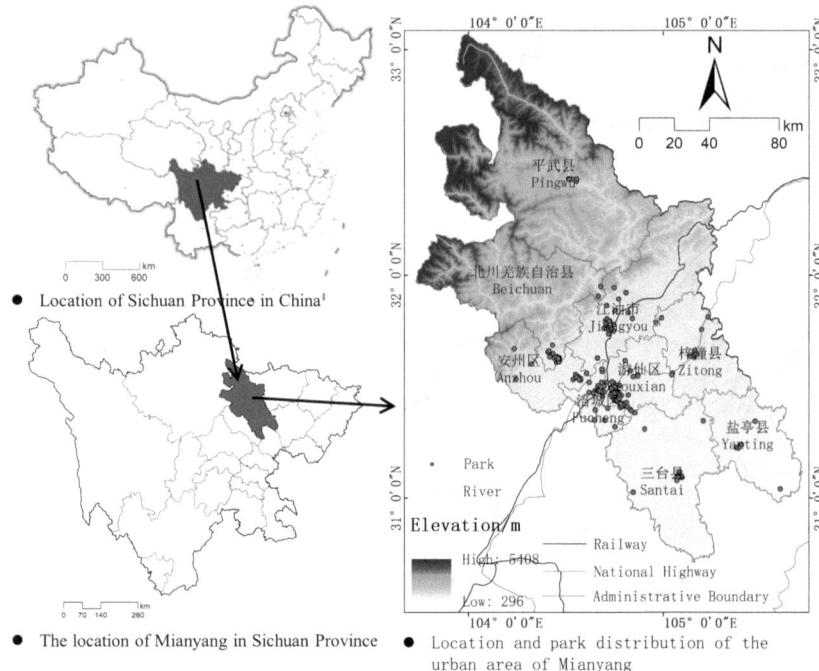

Location of Sichuan Province in China

The location of Mianyang in Sichuan Province

Location and park distribution of the urban area of Mianyang

Fig. 1. The diagram of local plan of Mianyang.

layout characteristics of parks in Mianyang and their influencing factors. These methods include the Nearest Neighbor Index to determine layout patterns, the Standard Deviation Ellipse method to assess centrality and directionality of the park layout, the Kernel Density Estimation method to analyze the park layout density of different districts and counties and the core area of park agglomeration, the Moran's I and Getis-Ord General G methods to explore spatial autocorrelation and heterogeneity and the Local Indicators of Spatial Association method to identify the presence of hotspots and coldspots in the park region layout of Mianyang districts and counties (see Fig. 2). Additionally, the study also explores the relationship between park region layout and factors such as elevation, transportation, and urbanization rate in Mianyang.

3   Results Analysis

3.1   The Region Layout Characteristics of the Park

For one thing, the average nearest neighbor index was -15.25, indicating that the region layout of urban parks in Mianyang shows significant agglomeration characteristics at the 0.01 level (see Fig. 3a). For another thing, the diagram of the Standard Deviation Ellipse (see Fig. 6a) reveals that the average center is located in the Shiban Street of Youxian District, suggesting a more significant agglomeration characteristics of parks on the Fucheng District side compared to the Youxian District side. Moreover, the main

Research on features of the regional Layout of city parks and its natural and social foundations: A Case Study of Mianyang City

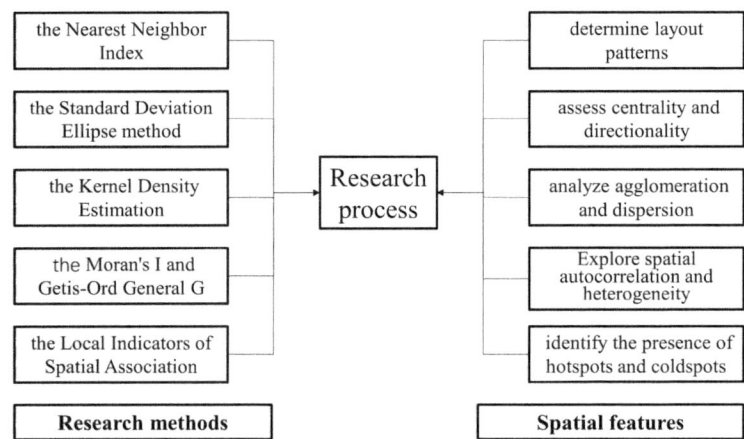

Fig. 2. The diagram of research process.

trend of the park layout is the northwest-southeast direction, with parks concentrated along this direction. Besides, the flatness rate of the standard deviation ellipse is 0.21, indicating that the region layout of Mianyang parks has a strong directionality and trend.

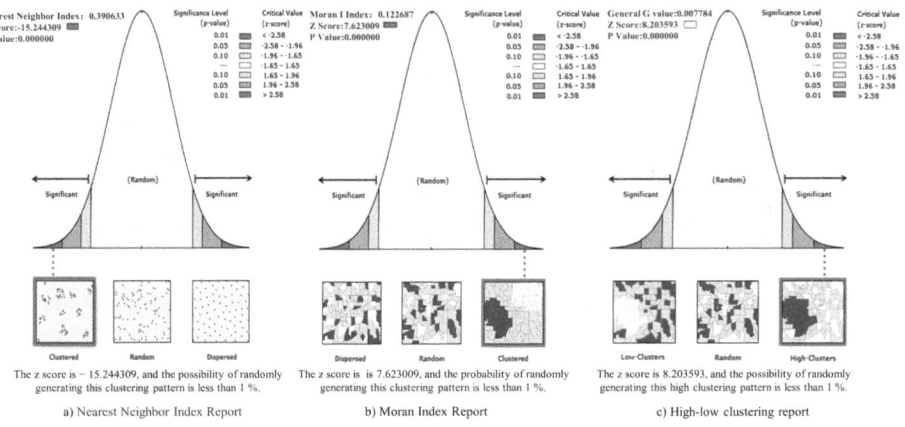

Fig. 3. The diagram of spatial analysis related reports.

3.2   The Region Layout Structure of the Park

Considering the characteristics of layout structure of park at different research scales, a kernel density analysis diagram of park regional layout was drawn with search radii of 500m, 1000m, 2000m and 3000m as the standard respectively (see Fig. 4) [13]. The resulting Kernel Density Estimation diagram of park region layout is divided into five levels of density. When the search radius is set to 500–1000 m, the region layout structure of the park shows a "multi-core, group distribution" feature, with high-density core areas forming in each county, and a relatively balanced park layout. However, as the search radius expand to 2000–3000 m, the park layout structure transitions to a "one core, two rings, single group" feature, with the high-density core area in the central urban area of Mianyang significantly expanded, and the surrounding areas, due to low park layout density, forming a ring-shaped cluster distribution around the central core area. The park in Pingwu County, due to its far distance from the central core area, still is distributed independently in a single group. Additionally, from the perspective of kernel density contour lines, the contour lines in the central urban area and its surrounding regions are denser, and the park aggregation characteristics are obvious, with dense park distribution. In comparison, the kernel density contour lines in the marginal area of Mianyang are sparse, the park group spacing is large, and the layout is scattered. In summary, the "high aggregation and group" layout structure characteristics of Mianyang parks are obvious. The three major urban areas consisting of Fucheng District, Youxian District, and Anzhou District within Mianyang City have the densest park distribution, forming a high-density core area, while other areas have low density, indicating an imbalance in the spatial distribution of parks within Mianyang.

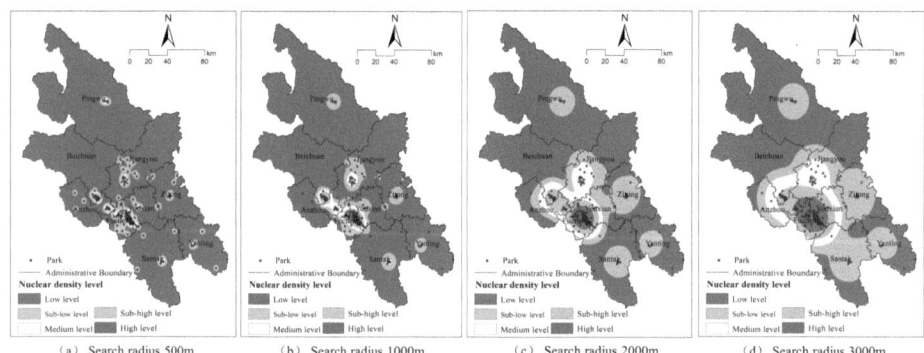

(a) Search radius 500m   (b) Search radius 1000m   (c) Search radius 2000m   (d) Search radius 3000m

Fig. 4. The diagram of Kernel Density Estimation of different search radius.

3.3   The Correlation and Heterogeneity of the Park Region Layout

Firstly, the Global Moran's I was 0.12 (see Fig. 3b), indicating that the region layout of Mianyang parks has a positive spatial autocorrelation at a significance level of 0.01. Next, according to the of Local Moran's I analysis diagram (see Fig. 5b), the Table 1 is drawn. Then, from the county level analysis, most of Fucheng District, the southwestern part of Youxian District, the northern and eastern parts of Anzhou District, the southeastern part of Beichuan County, and the southwestern part of Jiangyou City belong to the high-value aggregation area. The parks in these areas have a strong positive spatial influence on each other, promoting the coordinated development of urban parks in different counties.

Among them, although Anzhou District and Youxian District overall present aggregation characteristics, due to the relatively far distance of some parks from the high-value aggregation area, and park groups have not yet formed, so the spatial aggregation effect is relatively weak. Besides, the park layout in Santai County, Yanting County, and Pingwu County is mostly in a single group format, and the spatial aggregation effect of park layout is relatively weak. From the street level analysis, Fucheng District mainly has high-value aggregation type streets, relatively balanced park layout, and strong spatial aggregation effect. Youxian District has the larger area and a larger number of streets, but only 27% of the high-value agglomeration streets, which makes the park layout relatively uneven, with a weak spatial aggregation effect compared to Fucheng District. Although the number of streets in Anzhou District is fewer compared to the other two main urban districts of Mianyang, only 22% of the streets belong to high-high aggregation type, indicating that the spatial effect of park layout is weak in this urban district. The number of parks in Beichuan County is second only to Fucheng District, but only 4 high-high aggregation type streets, indicating that the balance of park layout in this county is relatively poor. The high-high aggregation type streets in Jiangyou City are concentrated in the southwest with fewer parks in the northeast, indicating an imbalanced park regional layout in this city. Pingwu County, Zitong County, Yanting County, and Santai County all have only high-low aggregation type streets, indicating that the parks in these areas are mostly independently distributed in different streets, with each park group being relatively far apart, weak spatial aggregation effect, and imbalanced layout. In addition, based on the high and low aggregation class report, further analysis was conducted on the main aggregation types of park regional layout. The observed value of 0.008 indicates that the park layout in Mianyang is mainly high value aggregation type (see Fig. 3c).

Table 1. The number of streets of different clustering types.

District/county	Number of H-H type streets	Number of H-L type streets	Number of L-H type streets	Number of L-L type streets	Number of streets per district/county
Fucheng	13(0.52)	—	12(0.48)	—	25
Youxian	8(0.27)	—	16(0.53)	—	30
Anzhou	4(0.22)	1(0.056)	10(0.56)	—	18
Jiangyou	9(0.2)	—	13(0.3)	7(0.16)	44
Zitong	—	5(0.16)	—	5(0.16)	23
Santai	—	5(0.08)	2(0.03)	16(0.25)	63
Pingwu	—	1(0.04)	—	1(0.04)	25
Yanting	—	3(0.08)	—	33(0.92)	36
Beichuan	4(0.17)	—	3(0.13)	5(0.16)	23

Note: () is the ratio of the total number of streets in the cluster to the total number of streets in the county;"–" is that there is no street of this type

3.4   The Hotspot of the Park Region Layout

Based on the hotspot map (see Fig. 5c and 5d) of Mianyang City park region layout and the Table 2, from the county level analysis, the hotspots include Fucheng District, Youxian District, Jiangyou City and Beichuan County. These regions are close to the average center and high-density core area, with high park distribution heat. Among them, Beichuan County due to its new planning, has a large number of pocket parks and belt parks and is a sub-hot spot area. Jiangyou City, because it does not fully utilize the aggregation effect of being close to the core area to develop the park group in its southwest area, and is also only a sub-hot spot area. Anzhou District, because its urbanization level is lower than Fucheng District and Youxian District, has a small number of park construction, and belongs to the only transitional area. The coldspots include Zitong County and Santai County, Yanting County and Pingwu County. The former is a secondary coldspots, and the latter is the high coldspots. The main differences between the two are: first, the economy of the former is more developed than the latter; second, the former is closer to the core area than the latter in terms of geographical location. In addition, Santai County, due to its failure to form park groups using the spatial aggregation effect of the core area, although it has a higher level of economic development and is closer to the core area compared to Zitong County, is still classified as a coldspots.

From the street level analysis, the highest proportion of hotspot streets in Fucheng District indicates that the park aggregation level in this area is the highest. And the lowest proportion of coldspot streets in all counties indicates that the park regional layout in this district is relatively balanced. Youxian District has the second-highest number of hotspot streets after Fucheng District, and its park regional layout is also relatively balanced. Anzhou District, Jiangyou City, Zitong County, and Santai County have a large number of secondary hotspot streets, but the park groups formed on adjacent streets are few, resulting in a higher proportion of coldspot streets compared to hotspot streets in these areas, and the park region layout is more scattered. Although the number of hotspot streets in Beichuan County is relatively high, the proportion of hotspot streets is lower than that of coldspot streets, indicating that the parks in this area are concentrated on a few streets, with strong spatial aggregation. However, it also indicates that the park regional layout in this county is imbalanced. Although the proportion of hotspot streets in Yanting County is higher than that in Pingwu County, it is the only county without high-hotspot streets, indicating that the park layout aggregation level of Yanting County is weaker than that of Pingwu County, with a scattered park layout. In addition, Yanting County has only 3 secondary hotspot streets, and the remaining streets are all coldspots, with a coldspot ratio of 0.92, second only to Pingwu County. This indicates that the park layout heat in this county is relatively low, and the attractiveness to park development is weak. Among all the counties and districts in Mianyang City, the lowest ratio of hotspot streets to total streets is in Pingwu County, at 0.04, indicating that the heat of parks in Pingwu County is the lowest and the attractiveness is the weakest. The ratio of coldspot streets to total streets in Pingwu County is 0.96, the highest among all counties, due to the small number of parks and their concentrated layout in Long 'an Street. The park layout heat on most streets is low and the park regon layout is imbalanced.

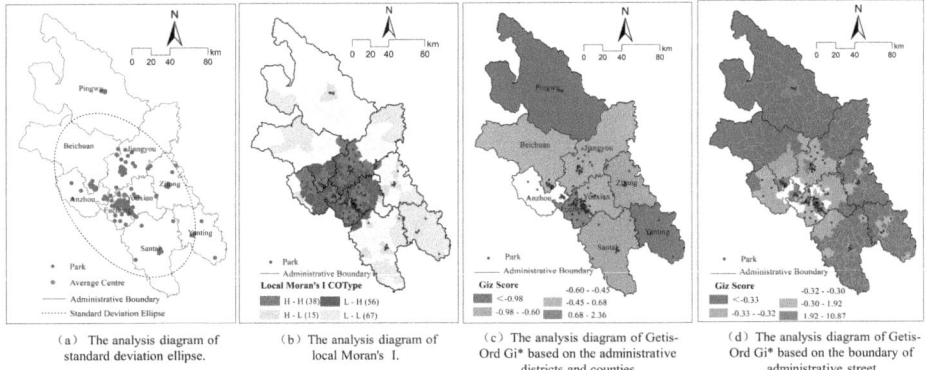

(a) The analysis diagram of standard deviation ellipse.

(b) The analysis diagram of local Moran's I.

(c) The analysis diagram of Getis-Ord Gi* based on the administrative districts and counties

(d) The analysis diagram of Getis-Ord Gi* based on the boundary of administrative street

Fig. 5. The diagram of spatial autocorrelation analysis of park layout.

Table 2. Statistical table of the number of streets with different heat

District/county	Hot spot zone	Sub-hot spot	Transition region	Sub-cold spot	Cold spot zone	Number of streets per district/county
Fucheng	4(0.16)	14(0.56)	5(0.2)	2(0.08)	—	25
Youxian	2(0.07)	8(0.27)	7(0.24)	8(0.27)	5(0.17)	30
Anzhou	1(0.06)	4(0.22)	3(0.17)	9(0.5)	1(0.06)	18
Jiangyou	1(0.02)	9(0.2)	2(0.05)	11(0.25)	21(0.48)	44
Zitong	1(0.03)	5(0.16)	—	—	26(0.81)	32
Santai	1(0.02)	5(0.08)	—	6(0.1)	51(0.81)	63
Pingwu	1(0.04)	—	—	—	24(0.96)	25
Yanting	—	3(0.08)	—	—	33(0.92)	36
Beichuan	2(0.09)	2(0.09)	—	4(0.17)	15(0.65)	23

Note: () is the ratio of the total number of streets in the cluster to the total number of streets in the county;"-" is that there is no street of this type

4 Impact Factors Analysis

4.1 Terrain Factors Analysis

Terrain is a main factor affecting the park region layout and it mainly include the flatness and relief of the terrain which are primarily determined by elevation and slope factors. Therefore, based on the 30-m resolution DEM data of Mianyang, elevation and slope analysis charts (see Fig. 6a and 6b) were drawn. Although the terrain of Mianyang is mainly flat plains and basins, the terrain relief is large. The geography is low in the southeast and high in the northwest. As a result, the parks are mainly concentrated in the southeast of Mianyang or gently terrain areas in each district and county. For instance, the parks in Beichuan County are mainly distributed in the southeast plain area, while the parks in Pingwu County are distributed in the central area with flat terrain. It can be seen from the Table 3 that areas with an elevation less than 875 m and a slope between 9.5° and 30° are more suitable for developing parks. Areas with an elevation of 875 ~ 1525m and a slope of 9.5° ~ 30° have fewer parks, mainly consisting of forest parks and sports parks. The results show that the number of parks decreases with the increase of elevation and slope, indicating a negative correlation between elevation and slope and the number of parks.

Table 3. Number of parks within different elevation and gradient.

Factor	Class	Number of parks/num	Ratio of the number of parks owned to the total number%
Elevation/m	296–875	164	96
	875–1525	4	2
	1525–5408	0	0
Gradient/(°)	0–9.5	133	78
	9.5–30	36	15
	30–78	2	1.2

4.2 River and Traffic Factors Analysis

Water is one of the main landscape elements in parks, and it has a good spatial relationship with natural water bodies such as rivers and lakes. And the use of parks by citizens is also affected by convenient transportation. Based on the main rivers and road data of Mianyang, the analysis charts of Mianyang rivers and traffic buffers (see Fig. 6c and 6d) were drawn. As shown in the figure and Table 4, the areas within 3000 m of rivers and within 500 m of roads are more suitable for developing parks. The number of parks with river distances between 3000 and 7000 m and road distances between 500 and 3000 m is relatively small, and these parks are mainly leisure parks such as sports parks and pocket parks that do not rely on river development, or forest parks and natural parks located in the suburbs with inconvenient transportation. It can be seen from the above that as the distance from the rivers and roads increases, the number of parks decreases, indicating that the distance from the rivers and roads has a significant impact on the park regional layout, that is, the distance from the rivers and roads and the number of park layouts have a negative correlation relationship.

4.3 Social Demographic and Economic Development Factors Analysis

Parks are the main public spaces serving the dwellers, and their layout is closely related to the number and density of the population, and social development level [13]. Therefore, this article uses population density, economic development level, and urbanization level as indicators to evaluate social factors. From the data in Table 5, although the population

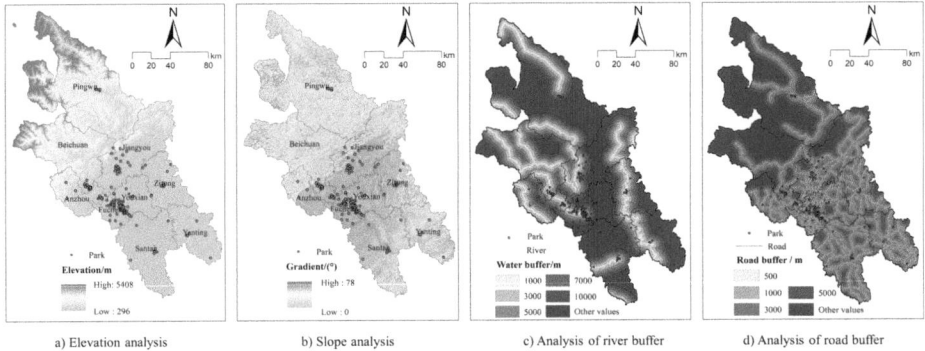

a) Elevation analysis   b) Slope analysis   c) Analysis of river buffer   d) Analysis of road buffer

Fig. 6. The analysis diagram of the influence of different factors on the park layout.

Table 4. Number of parks within different distances from rivers and traffic.

Factor	Buffer distance/m	Number of parks/num	Buffer area/km^2	Ratio of the number of parks owned to the total number/%
River	0–3000	95	6342.61	55
	3000–5000	23	3757.24	13
	5000–7000	36	5362.04	21
Road	0–500	132	4413.08	77
	500–1000	26	3321.01	15
	1000–3000	13	9841.21	7

density of Anzhou District is higher than that of Jiangyou City, due to the lower economic and urbanization levels of Anzhou District, the number of parks in Anzhou District is fewer than that of Jiangyou City. Although the population density of Pingwu County is the lowest, it has abundant forest resources, and the number of parks is comparable to that of Yanting County. Beichuan County, due to the natural disaster in 2008 that destroyed the old city, has a low population density and social development level, but has built a large number of pocket parks and linear parks in its new town planning. This makes its park number second only to Fucheng District. The above results show that the social factors such as population and economic development in Mianyang have a positive correlation with the number of parks distributed.

5 Conclusion and Discussion

5.1 Conclusion

(1) The spatial distribution of urban parks in Mianyang is primarily characterized by aggregation, presenting a "high agglomeration and group" feature in spatial structure, with the three major urban areas composed of Fucheng District, Youxian District, and

Table 5. Social population and economic development indicators.

District/county	Population density people/km^2	GDP/billion yuan	Urbanization rate/%	Number of parks/num
Fucheng	2363	1199.14	82.26	54
Youxian	557	446.74	61.23	22
Anzhou	384	218.26	44.90	13
Jiangyou	269	528.27	54.58	21
Zitong	355	450.10	30.03	10
Santai	193	164.82	33.72	11
Pingwu	222	191.32	38.69	6
Yanting	21	63.54	28.83	6
Beichuan	58	88.11	37.44	28

Anzhou District forming high-density core areas, and a strong spatial agglomeration effect. Parks in other areas also show a clustered distribution, but the distance between each group is large, the spatial agglomeration effect is weak, and the layout is uneven.

(2) The spatial correlation of urban parks in Mianyang is mainly positively correlated. Although there are different types of aggregation in different districts and counties, high-value aggregation is dominant overall.

(3) The distribution of urban parks in Mianyang forms a hot zone centered around Fucheng District, while the more distant counties form a cold zone, indicating the presence of spatial distribution imbalance.

(4) The number of parks in Mianyang is negatively correlated with elevation, slope, distance from rivers and roads, and positively correlated with social population and economic development levels.

5.2 Discussion

Considering the current state of Mianyang's park layout and combining relevant influencing factors, we propose the following suggestions to optimize the layout of Mianyang's parks from a holistic perspective and to promote the development of Mianyang's urban parks from a macroscopic viewpoint:

(1) Relevant departments should strengthen the utilization of park agglomeration effects of central urban areas in the urban planning strategy, appropriately strengthen the construction of parks in the periphery of the core area, the direction of sub-trends, and coldspots areas, thereby improving the agglomeration effect and balance of park regional layout in Mianyang as a whole.

(2) We should construct parks in accordance with local conditions. For one thing, it is necessary to strengthen the park construction in suitable areas with flat terrain or gentle terrain, close to the river, and convenient transportation. For other thing,

make full use of factors such as rivers, traffic, and terrain to develop different types of parks, such as riverside parks and forest parks.

(3) Each district and county should focus on economic construction and social development, providing more powerful economic and technical support for park construction. At the same time, people's culture and recreational needs should also be fully considered, and pocket parks and community parks should be built in densely populated and economically developed areas. we can make full use of the public areas on both sides of the road to establish belt parks to improve the utilization of urban public resources. Besides, in urban construction, adhere to the planning concept of "planting green first, planning later", improving the quality of people's living environment, enhancing urban health levels, and promoting the healthy and sustainable development of Mianyang.

In short, the value of this study lies in visualizing and quantifying the regional layout characteristics and influencing factors of urban parks, providing basic data analysis for the development of urban parks in Mianyang. Of course, the above research is still shallow and rough, and needs to be further explored. Otherwise, due to the limitations of the Amap platform, the POI data of parks is not complete, and the differences in park layout characteristics and influencing factors at different time sequences have not been explored. At the same time, the summary of influencing factors is also incomplete. In the future, we will conduct more in-depth research on these shortcomings to provide more scientific basis and data support for the development of parks in Mianyang.

References

1. Wang, L., Jiang, X., Ye, D.: Research Hotspots and Progress on Healthy City Planning in China: a Bibliometric Analysis Based on Citespace. Urban Development Studies (2020)
2. Zhang, J., Erbin, Xu, E.: Investigating the spatial distribution of urban parks from the perspective of equity-efficiency: Evidence from Chengdu, China. Urban Forestry and Urban Greening (2023)
3. Zhang, J.: Assessment of Spatial Equity of Urban Park Distribution from the Perspective of Supply-Demand Interactions. Urban Forestry and Urban Greening (2023)
4. Wang, Y.: Research on Optimization of Urban Park Green Space Layout Based on Accessibility Evaluation. Tianjin University (2018)
5. Zeng, H., Zuo, Y.: Spatial distribution pattern and its influencing factors of sports parks in China. J. Wuhan Inst. Phys. Educ. (2022)
6. Wang, C., Hong, X., Wang, Z.: Analysis of spatial distribution characteristics and influencing factors of Sichuan Geopark. J. Sichuan Geol. 691–696 (2021)
7. Zhou, J.: Study on the spatial distribution characteristics of park green space in the central urban area of Xi'an. Reform Open. Up **20**, 19–25 (2021)
8. Guo, J.: Research on the evaluation and optimization of the spatial characteristics of social services in urban parks in the functional core area of the capital based on POI data. Beijing University of Civil Engineering and Architecture (2022)
9. Yin, J., Liu, M., Liu, W.: Analysis of spatiotemporal heterogeneity characteristics and influencing factors of China National Geoparks. J. Nat. Sci. Hunan Norm. Univ. (2023)
10. Liu, H., Liu, Y., Song, J.: Spatial distribution and influencing factors of wetland parks in China. Wetland Science and Management, 63–68 (2020)

11. Li, C., Li, J.: Research on the distribution of streetside strip parks and the influencing factors of planning. Low Temperature Building Technology, 14–19 (2023)
12. Chen, Y., Li, D., Wang, Y.: Evaluation and layout optimization of urban parks and green space in Hefei central urban area based on POI data. J. China Agric. Univ. **28**(10), 87–97 (2023)
13. Ren, Q.: Research on Optimization of Spatial Pattern of Urban Parks. University of South China (2022)

Evaluation and Optimization Strategy of Community Environment Livability Based on New Urbanism: A Case Study of Mianyang, China

Jing Chen[✉]

Southwest University of Science and Technology, Mianyang 621000, China

Abstract. With the rapid development of China's social economy and the continuous improvement of urbanization, a range of issues impacting the quality of community spaces have emerged, including inadequate transportation, environmental degradation, insufficient amenities, housing and employment imbalances, as well as limited recreational areas. Consequently, there is an inherent public desire to establish healthy and livable communities that offer comfort, convenience, safety, and aesthetic appeal. This study examines the correlation between new urbanism and the concept of community environment livability by constructing an evaluation index system and model to quantitatively analyze Mianyang old city's overall level of community environment livability along with its influencing factors. Additionally, it summarizes various types of communities in terms of their spatial distribution and characteristics while proposing optimization strategies tailored to each type. The results of this study show that: (1) the livability of Mianyang Old city community decreases gradually from south to north, from east to west, and there is a big difference between the south and north. The areas with high livability evaluation value are concentrated in the central area of the old city, while the "poor" area is distributed in the western fringe area; (2) The service facilities and living environment of the old city community are highly valued, while the employment and public leisure areas need to be improved; (3) The 28 communities can be divided into five types: "environment improvement type, quality optimization type, basic improvement type, housing security type and function repair type", and the optimization strategy is proposed.

Keywords: New urbanism · Community environment · Livability evaluation · Old city · Mianyang City

1 Introduction

China first put forward the concept of urban livable construction in the Master Plan of Beijing City (2004–2020) in 2005 [1] Since then, the concept of building a livable city with beautiful environment and harmonious integration of man and nature has gradually become the goal of urban construction. In recent years, the outline of the 14th Five-Year Plan has clearly put forward the overall goal of building a harmonious and livable city.

© The Author(s) 2025
B.-J. He et al. (Eds.): UCSUD 2023, LNCE 559, pp. 680–692, 2025.
https://doi.org/10.1007/978-981-97-8401-1_49

In the Master Plan of Mianyang City, it is stated to develop Mianyang City into a modern livable city integrating diverse culture, landscape and natural ecology [2]. As an integral part of the city, the basic unit of residents' life, the foundation of livable city construction, and the internal need to enhance the attractiveness and competitiveness of the city.

In the early 1990s, the new urbanism movement was launched in the United States in response to the suburban sprawl. Peter Calthorpe proposed the theory of public transport-oriented development (TOD), which takes public transportation sites as the core for land planning and development. In Good Community: Theory and Practice of New Urbanism, Jill grant analyzed examples of new urbanism construction in many countries, and the results showed that new urbanism has certain guidance for future city and community construction [3]. Andrew Stanislava, by comparing new urbanist development zones with original urban suburbs and evaluating the livability of sustainable communities, concluded the importance of sustainable neighborhood design elements in new urbanist communities [4]. In China, the concept of new urbanism was initially applied to the real estate industry, and then in-depth research on the concept of new urbanism was carried out. Hu Sixiao introduced the idea of building traditional neighborhood communities and urban design based on public transportation stations [5]. Ma Yongjun, He Ping et al. analyzed the main planning and design ideas and connotations of new urbanism, and pointed out that the planning and construction of residential areas in China should be guided by natural ecological design, shared design, public participation, dynamic planning and diversified mixed design [6]. Ding Wenjing briefly described the background and concept of new urbanism and pointed out the guiding significance of humanism in the concept to China's environmental construction [7]. Li Xingju divided the types of suburban residential communities in China, analyzed the main characteristics and existing problems of suburban residential communities, discussed the relationship between new urbanism and community construction in China, and pointed out the reference significance of new urbanism for community construction in China [8].

In general, foreign research mainly focuses on the background, connotation and development prospects of new urbanism, while domestic research on new urbanism theories mostly focuses on theoretical interpretation and analysis, and the tracking of theories and practices remains to be studied. Although some scholars have carried out research on the combination of new urbanism and domestic cases. However, there are shortcomings that are not closely related to the national conditions of our country, and the combination and practice of the construction of livable communities in our country are still lacking, not comprehensive enough, and need in-depth analysis. Therefore, under the guidance of new urbanism, it is of great significance to carry out the research of community livable evaluation and optimization strategy.

2 The Evaluation Index System of Community Environment Livability Based on New Urbanism is Established

2.1 Analysis on the Relationship Between New Urbanism and Community Livability

New urbanism has certain practical exploration and guiding significance to guide the overall planning of city and community and the rectification of related problems, including community residential environment, service facilities, walking space and architectural design. The concept connotation of new urbanism is closely related to the requirements of livable community construction, and also coincides with many hot concepts and construction ideas at present. Therefore, new urbanism has guiding significance [9, 10].

2.2 Construction of Evaluation Index System of Community Livability

This paper analyzes the internal relationship between the concept of new urbanism and community livability, and according to the guiding concept and basic principles of indicators, refer to the representative livability evaluation index system and the research results of scholars for analysis [11, 12] and extracts relevant indicators as the content of this index system. From the objective point of view of living environment, work and employment, service facilities, transportation, public leisure and resource bearing 6 levels, subdivide 15 secondary indicators and 27 and tertiary indicators to establish a community livability evaluation index system (Table 1).

2.3 Index Calculation Method and Weight

This paper refers to "Urban Residential area Planning and Design Standards" (GB50180–2018), Shanghai "15-min Living Circle Planning Guidelines (Trial)" and relevant research results as index standards, and chooses AHP for weight calculation. According to the calculation of AHP (Table 2).

3 Empirical Research on Community Livability Evaluation

3.1 Research Area

The old city of Fucheng district in Mianyang city, Sichuan Province was selected for this study. Mianyang is a science and technology city of China, the second economy of Sichuan and the central city of Chengdu-Chongqing urban agglomeration. It is in the middle and upper reaches of the Fujiang River in northwest Sichuan. It covers an area of 20,200 square kilometers and has a population of 5,285 million. The main urban area has a built-up area of 167.58 square kilometers and a permanent population of 1.419,700 million.

Table 1. Contents of evaluation index system of community environment livability

Dimension level	Element level	Specific index
Living environment	Residential environment	Per capita land use index
		Age of housing construction
		Residential mixing degree
		Residential noise
	Accessory environment	Ratio of green space
Job employment	Local employment	Employment land proportion
	Job-residence relationship	Job-housing ratio
Service facility	Commercial facility	Coverage of convenient service facilities
		Comprehensive supermarket coverage
	Cultural facilities	Coverage of cultural facilities
	Educational facilities	Kindergarten coverage
		Elementary school degree coverage
		Secondary coverage
	Medical facility	Clinic coverage
		Coverage of community health service stations
		Coverage of hospital facilities

(continued)

Table 1. (*continued*)

Dimension level	Element level	Specific index
	Elderly care facilities	Coverage of elderly care facilities
Transportation trip	Road network	Density of road network
	Public transport	Bus station allocation rate
		Allocation rate of public transport facilities
Public leisure	Street space	Walking network connectivity
	Park Square	Park square coverage
	Leisure activity	Active center coverage
		Coverage of sports facilities
Resource bearing	Development intensity	Plot ratio
		Building density

Table 2. Analytic hierarchy process weight value

Dimension level	Dimension level weight	Element level	Relative weight of factor layer	Absolute weight of element layer	Concrete index layer	Relative weights of specific indicator layers
Living environment	0.2556	Residential environment	0.6158	0.1574	Per capita land use index	0.4087
					Age of housing construction	0.1218
					Residential mixing degree	0.1605
					Residential noise	0.3089
		Accessory environment	0.3842	0.0982	Ratio of green space	0.3842
Job employment	0.0947	Local employment	0.4849	0.0459	Employment land proportion	0.4849
		Job-residence relationship	0.5151	0.0488	Job-housing ratio	0.5151
Service facility	0.2519	Commercial facility	0.1853	0.0467	Coverage of convenient service facilities	0.6967
					Comprehensive supermarket coverage	0.3033
		Cultural facilities	0.0707	0.0178	Coverage of cultural facilities	0.0807
		Educational facilities	0.3262	0.0822	Kindergarten coverage	0.389
					Primary school coverage	0.4339
					Secondary coverage	0.1872
		Medical facility	0.2766	0.0697	Clinic coverage	0.198

(*continued*)

Table 2. (*continued*)

Dimension level	Dimension level weight	Element level	Relative weight of factor layer	Absolute weight of element layer	Concrete index layer	Relative weights of specific indicator layers
					Coverage of community health service stations	0.493
					Coverage of hospital facilities	0.3092
		Elderly care facilities	0.1411	0.0355	Coverage of elderly care facilities	0.1411
Transportation trip	0.1652	Road network	0.2987	0.0493	Density of road network	0.2987
		Public transport	0.7013	0.1158	Bus stop coverage	0.6889
					Allocation rate of public transport facilities	0.3111
Public leisure	0.0865	Street space	0.2928	0.0253	Walking network connectivity	0.2928
		Park Square	0.3492	0.0302	Park square coverage	0.3492
		Leisure activity	0.3581	0.031	Active center coverage	0.4167
					Coverage of sports facilities	0.5833
Resource bearing	0.1462	Development intensity	0.3804	0.0556	Plot ratio	0.3804
			0.6196	0.0906	Building density	0.6196

3.2 Data Sources and Evaluation Method

The research data of this paper comes from Mianyang district street resident demographic data, Baidu poi point data, housing network data, Baidu heat map data and so on. The data elements are divided into point elements, line elements and plane elements according to their spatial forms. GIS software is used to analyze the data, and the comprehensive livability score is calculated by combining the evaluation model.

The idea of livability evaluation in this paper is as follows (Fig. 1): Using analytic hierarchy process to measure the weight of indicators at each level; Then, according to the relevant standards and norms, the index is quantified by using the grading assignment method to obtain the grade index of the corresponding indicators in different communities. Finally, the comprehensive score of the community is calculated by the comprehensive index method, and the data is normalized. Finally, according to the cluster analysis method, the main internal problems of each dimension cluster are analyzed in detail, and then the community types are divided.

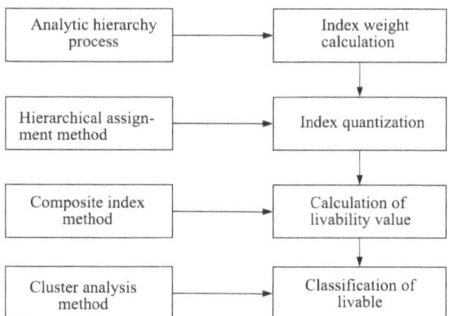

Fig. 1. Evaluation index model of community livability (self-drawn)

The data analysis of this paper is to use SPSS22.0, yahhp and other software to conduct statistical analysis of the collected data, and obtain the livability level of Mianyang community, which provides data support for the subsequent optimization strategy analysis. The main methods are as follows.

(1) Analytic hierarchy Process. This method is used to calculate the weight of livability evaluation index, and the weight values of different dimensions are obtained.

(2) Hierarchical assignment method. Using the method of grading assignment and combining with the relevant standards of the existing evaluation indicators, the index is graded to achieve the quantification of the index data.

(3) Comprehensive index method. The comprehensive index method is used to calculate the final community livability evaluation value of each dimension of the index by combining grading assignment and weight value.

(4) Cluster analysis. The software SPSS22.0 was used for cluster analysis, and the comprehensive index method was used to evaluate the value of each dimension, and the community was classified again, which is conducive to the next step to propose optimization strategies for the livable environment of the community.

(5) ArcGIS spatial analysis. First, kernel density calculation, service area analysis and spatial statistics based on ArcGIS are used to measure and make statistics on indicators; The second is to use ArcGIS to spatialize the results. The third is to use ArcGIS natural breakpoint method to obtain the community livable level.

3.3   Community Evaluation Results Analysis

The classification of livability was carried out by ArcGIS natural segment point method. According to the evaluation value, it was divided into five grades from small to large: "very poor", "poor", "average", "good" and "very good". Therefore, the evaluation value, livable grade, spatial distribution characteristics and trend of Mianyang city community are obtained (Fig. 2). Iron Cow Street is 3.6982, Jianguomen is 3.6926, Bell and Drum Tower is3.6054, Bell street is 3.5518, Flyingstone is 3.5138, Cheng Mian Road is3.3561, Yuebei is 3.3173, Deyue building is 3.3041, Garden South Street neighborhood is 3.2719, Jinjiang is 3.2356, Lingtong is 3.2284, Wenchang is 3.2120, Mianxing Road is 3.1998, Linyuan Road east section is 3.1931,Huafeng is 3.1917, Linyuan Road west section is 3.1676, Yingbin is 3.1660, Anchang is 3.1201, Mianzhou is 3.1166, Jinju is 3.1029, Xishan is 3.0949, Binjiang is 3.0518, Binfu is 2.9764, Jiuzhou is 2.9725, Huixian is 2.9616, Ziyunting is 2.3641, Chaoyang is 2.3574.1

From the perspective of overall spatial distribution, as shown in Fig. 2, the livability of Mianyang's central urban communities gradually deteriorated from south to north and from east to west, with a large difference between the south and the north. The communities with high livability are concentrated in the central area of the old city, Jianguomen Community and Zhongdrum Tower community; The "poor" communities are concentrated in the northern and western edges, namely Chaoyang community and Ziyunting community; Feishi community around the railway station and the west section of Linyuan Road are also relatively good.

From the perspective of the number and area ratio of communities at each level of livability, the number of communities with "good" livability is 35% of the entire old city, accounting for about 21% of the area, and the number of overall "poor" communities accounts for 21%, the area accounts for 33%. The "general" communities in the old city accounted for the proportion of about 42%, and the best livable communities in the central city were Jianguomen, Zhongdrum Tower and Jiefang Street.

3.4   Analysis on the Division and Optimization Strategy of Community Livable Types

In this paper, the evaluation value of six aspects of the index is clustered. According to the calculation results of the clustering center, 28 communities are divided into five categories: I, II, III, IV and V. The overall spatial distribution is shown in Fig. 3. And according to the characteristics of the community and the main influencing factors of livability analysis, the community is divided into five categories from the planning level and the optimization strategy is proposed, the main content is shown in Table 3.

4   Conclusion and Discussion

Based on the connotation and characteristics of new urbanism theory, combined with the theory of human settlement environment, sustainable development theory and the theory of community habitability, this paper takes the old city of Mianyang city as an example to evaluate the livability of community environment and analyze the optimization strategy.

Fig. 2. Livability grade and spatial distribution diagram

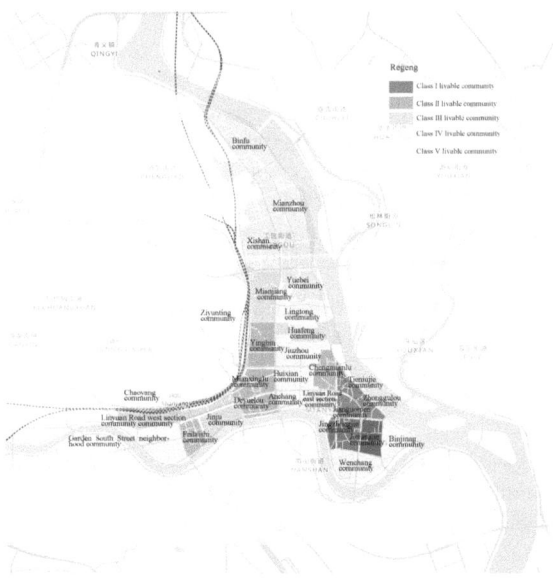

Fig. 3. Spatial distribution diagram of livable community type.

Based on multi-source big data data, this study quantitatively analyzes the livability of urban community environment in Mianyang city, evaluates the status quo of community construction in Mianyang City old city, finds the key factors of community livability, and explores the strategy of community renewal and optimization.

Table 3. Each type of community optimization strategy

Community classification	Contains Community	Type	Primary optimization strategy
Class I livable community	Tieniu Street community, Jianguomen Community, Bell and Drum Tower community, Bell Street community and Jiefang Street community	Environment Enhancement type	Improving the quality of living environment
Class II livable communities	Yingbin community, Feishi community, Mianxing Road community, Deyuelou community and Chengmian Road community	Quality optimization	Intensive land use, darning leisure space
Class III livable community	Binfu Community, Mianzhou community, Xishan community, Yuebei community, Mianjiang community, Huafeng community, west section of Linyuan Road community and Jinju community	Perfect Foundation type	Optimize the functional layout of service facilities and improve the road transportation system
Class IV livable community	Lingtong Community, Jiuzhou community, Huayuan South Street community, Huixian community and Anchang community	Housing security type	Living environment security, reduce community development, increase employment space
Category V livable communities	Chaoyang Community and Ziyunting Community	Functional repair type	Increase employment space, improve service facilities, clear the transportation system, and reshape leisure space

The livability of community environment involves all aspects of the community. Although this paper has done some exploration on the construction of the evaluation index system of community environment livability and the optimization strategy of community livability, it is not deep enough, and the paper still has the following shortcomings, which need to be further studied.

(1) The selection of indicators is not comprehensive enough, the lag of evaluation criteria is not well considered, and there is a lack of authoritative value range. Therefore, the selection of indicators and evaluation criteria need to be further improved in the future.
(2) The basic multi-source data of the objective evaluation research in this paper comes from the relevant big data system on the Internet. Due to the limitation of the availability and timeliness of the data, there are some differences between some data and the actual situation, so the conclusions obtained are somewhat different from the actual situation of the city. In future studies, more samples can be selected for investigation to broaden data sources and increase the reliability of data.
(3) Due to the subjective level of hierarchical analysis, the weight measurement cannot be completely accurate. In the future, multiple methods can be selected and different types of analysis can be carried out to analyze the inner problems of Mianyang Old city community in a more comprehensive way.

References

1. Overall Urban Planning of Beijing (2004–2020). Beijing Planning and Construction, 2005(2):5–51
2. China Academy of Urban Planning and Design. Master Planning of Mianyang City (2010–2020). Mianyang,2010.5
3. Quan, L.I.U.: Critical Thinking on "New Urbanism." Architect **03**, 50–53 (2006)
4. Stanislav A, Chin J T, Evaluating Livability and Perceived Values of Sustainable Neighborhood Design:New Urbanism and Original Urban Suburbs,Sustainable Cities and Society, https://doi.org/10.1016/j.scs.2019.101517
5. Sixiao, H.: Duany, Platerzyberk and "New Urbanism." Architectural Journal **01**, 67–72 (1999)
6. Yongjun, M., He Ping, H., Xijun, Z.Y.: New urbanism and modern residential area planning. Urban Problems **08**, 31–33 (2006)
7. Wenjing, D.: The humanistic concept of new urbanism and its application in China. Urban Problems **03**, 89–93 (2006)
8. Li Xingju. The development and application of new Urbanism in the construction of suburban residential communities. Hunan University,2006
9. Zhao, X.: Research on spatial allocation strategies of urban community convenience facilities from the perspective of new urbanism. Building of Beijing university (2021).
10. Lu Quan. Discussion on the adaptability and design Practice of "New Urbanism Community" in China. Tsinghua University,2014
11. China Society for Urban Studies. Livable city science evaluation standard [EB/OL.2017–04–08][2021–01–10]. https:/max.book118.com/html/2017/040799193500.shtm
12. Shanghai Municipal Administration of Planning and Land Resources. Guideline for 15-Minute Community Life Circle Planning in Shanghai (2016)

A Framework of Urban Renewal for Leisure Space in Neighborhood Streets: A Case Study of Shazheng Street in Chongqing, China

Yiqun Li[1,2], Ali Cheshmehzangi[3], and Bao-Jie He[1,2,4(✉)]

[1] Centre for Climate-Resilient and Low-Carbon Cities, School of Architecture and Urban Planning, Key Laboratory of New Technology for Construction of Cities in Mountain Area, Ministry of Education, Chongqing University, Chongqing 400045, China
`baojie.unsw@gmail.com`
[2] Institute for Smart City, Chongqing University, Liyang 213300, Jiangsu, China
[3] School of Architecture, Design and Planning, University of Queensland, Brisbane, QLD 4067, Australia
[4] CMA Key Open Laboratory of Transforming Climate Resources to Economy, Chongqing 401147, China

Abstract. Urban street spaces serve as a platform for fulfilling people's travel requirements and facilitating complicated and diverse social interactions. In this context, leisure spaces play a crucial role in urban streets. Nevertheless, with the acceleration of urbanization, the current design of street leisure space generally has problems of low quantity, inferior quality, and lack of humanism, which can not meet the needs of people's rest and communication in the street space. This study applied environmental psychology and typology theories, selected Shazheng Street in Chongqing as the research object, explored the link between the urban street environment and leisure behavior, and proposed a practical and effective design framework for the renewal of neighborhood street leisure spaces. The design framework first analyzes the current status of street leisure space in terms of the general street layout, the building interface along the street, and the structure of sidewalk cross-sections, and then discusses where leisure spaces can be designed for different structures of sidewalk cross-sections, as well as the types and characteristics of existing rest facilities. Then, we creatively combined the Baidu heat map data analysis, which investigates the vitality of a wide range of streets, with behavioral mapping and questionnaires, which accurately investigate people's needs, to investigate people's leisure needs accurately through the mutual validation of the results. Finally, we proposed zoning design strategies based on an analysis of the current situation and leisure needs. The findings of this study can serve as a reference for future renewal and design projects.

Keywords: Urban street · Leisure space · Urban regeneration

B.-J. He et al. (Eds.): UCSUD 2023, LNCE 559, pp. 693–705, 2025.
https://doi.org/10.1007/978-981-97-8401-1_50

1   Introduction

The first development of streets focused primarily on optimizing efficiency and convenience, with particular emphasis on facilitating the smooth flow of automobile traffic. Although the extensive urban transportation system facilitated rapid urban growth, it simultaneously reduced the available pedestrian space on the streets and disrupted the urban natural landscapes. Urban streets ceased functioning as public areas, resulting in a decline in inhabitants' social interactions and a growing disconnection between humans and nature [1, 2]. Jacobs emphasizes the significance of considering streets' varied social network characteristics in response to issues that arise from their fast development. When street infrastructure and facilities allocate sufficient space for pedestrians to engage in social activities, the street's social structure is reinforced, resulting in enhanced liveliness and economic growth [3]. There has been a recent surge in attention towards constructing public spaces on streets to facilitate people's daily interactions and activities [4, 5]. Specifically, the streets surrounding neighborhoods serve as primary locations for urban public activities, fulfilling the everyday demands of community residents [6]. Street leisure space is the area in the street environment that fulfills people's demand for rest. Being directly associated with people's fundamental necessities, a location in the street is an essential component of street public spaces [7]. Street leisure spaces alleviate individuals' exhaustion, facilitate social interaction, and augment the liveliness of streets [8]. Nevertheless, the present design of street leisure spaces is commonly marked by unreasonable design, inefficient use of space, absence of humanism, and numerous other issues, leading to unfulfilled requirements for relaxation and social interaction among pedestrians [9]. This study intends to examine the issues that arise in the development of neighborhood street leisure spaces, thoroughly investigate the correlation between the environment of street leisure spaces and leisure behavior and psychology, and propose a framework for the renewal design of street leisure spaces. Additionally, it utilizes Shazheng Street in Chongqing as a case study to demonstrate the efficacy of the renewal design framework in offering guidance for developing public spaces in neighborhood streets.

2   A Renewal Design Framework of Leisure Space in Neighborhood Streets

This study presents a design framework for the renewal of street leisure spaces in urban communities, consisting of three components: problem identification, problem analysis, and problem resolution (Fig. 1). Initially, the study examined the present conditions and challenges of street leisure spaces by analyzing the general street layout, the building interface along the street, and the structure of street cross-sections. Using humanism as a guiding concept, we delve further into people's demands for street leisure spaces. This study employs rigorous Baidu heat map data analytics, behavioral mapping, and questionnaire surveys to investigate the correlation between people's leisure behavior and urban street space.

Subsequently, this study examined a design approach for the renewal of street leisure spaces. To address issues related to neighborhood street leisure spaces, it is necessary to reevaluate the general arrangement of these spaces, considering the complex nature of the street environment and the unique characteristics of each road section. Next, it is necessary to conduct a comprehensive renewal design that considers various elements, including the layout of the street cross-section for each road segment [10], parking requirements [11], characteristics of nearby buildings [12], and needs of the primary user groups. The overall design of street leisure spaces prioritizes the principles of wholeness, diversity, openness, culture, and humanism [6]. This is done to ensure that the space is well-coordinated and reflects regional characteristics, while enhancing residents' cultural identity and sense of belonging. Sectional rehabilitation includes temporary leisure space, interactive leisure space, and street pocket parks, based on each road section's varying degrees of rest demand. A temporary leisure area, designed for pedestrians to briefly pause and relax, is typically furnished with basic seating facilities. The seat spacing varied between route portions with general and concentrated rest demands. In road sections with a high demand for rest, it is advisable to provide more chairs to cater to the resting needs of pedestrians. Interactive leisure space is suitable for areas with a high demand for rest, diverse crowd types, and spatial patterns on the street. It caters not only to the fundamental rest requirements but also facilitates diverse and affluent communicative behaviors. Street pocket parks are suitable for road sections with a significant demand for leisure spaces, many pedestrians, and ample street greenery and amenities. They serve as more engaging street leisure spaces. In the renewal design of neighborhood street leisure spaces, it is crucial to carefully assess the current issues with leisure spaces, the need for leisure spaces across different sections of the street, and sidewalk cross-sectional forms. The goal is to design the most appropriate leisure space for each road section.

Fig. 1. The renewal design framework of leisure space in neighborhood streets.

3   Analysis of Leisure Space Forms and Issues in Shazheng Street

3.1   Analysis of Street Leisure Space Forms

3.1.1   The General Street Layout

This study focused on Shazheng Street (SZS) in the Shapingba District in Chongqing. SZS spans 1320 m in length and, runs in a north-south direction. It has an average width of 16 m, which makes it a compact, diverse, and densely populated street. SZS, a typical community street, has many amenities, such as commercial malls, office buildings, residential areas, and schools. Furthermore, the street is surrounded by over ten bus stops and underground stations, ensuring convenient traffic flow and attracting a substantial influx of pedestrians. This abundance of foot traffic also provides many samples for research and study on urban street leisure space design.

3.1.2   The Building Interface Along the Street

The transparency, openness, and geometry of the street building interface on the ground floor significantly influences people's leisure behavior in street leisure spaces [12]. In mountainous cities, such as Chongqing, the building interface along the street is characterized by significant height differences and many steps. The study analyzes the situation of building interfaces in SZS and finds that building interfaces with steps is more likely to lead people to "sitting" because the presence of steps provides a place for this behavior. The form of the steps impacts people's resting behaviors, with upward steps leading to resting behaviors because of the sheltered nature of the back against the solid view in front of them for people to observe. Moreover, as the number of steps increases, the public area increases, producing more resting behaviors, such as chatting and sunbathing.

3.1.3   The Structure of Sidewalk Cross-Sections

A street is a three-dimensional space composed of many complex elements. The cross-section of the sidewalk consists of four parts: frontage zone, clear path, street furniture zone, and buffer zone [6]. The study researched all the major streets in Shapingba District, Chongqing, and categorized and coded the types of street cross-sections that affect the design of street leisure spaces, labeling the areas where leisure spaces can be installed. There are four main types of streets. Type 1 is a street with a clear path, generally narrower, and only suitable for pedestrians passing through. Type 2 is a street with a clear path and furniture zone. When the building interface is closed, the boundary effect can be reasonably utilized to provide seating and other resting facilities next to the closed interface. Type 3 is a street with a clear path, street furniture zone, and frontage zone (closed building interface). The rest of the facilities on this type of street are often designed together with a green landscape. Type 4 is a street with a clear path, street furniture zone, and frontage zone (open building interface). This type of street is generally broader, with leisure spaces located in the frontage zone, and has a high flow of people, making it an essential space for vitality and interaction in urban areas.

This study analyzed cross-sectional types of SZS. The width of street types I-1 and II-3 is only 2 m, which is only suitable for pedestrians to pass through and unsuitable for designing leisure spaces. Type II-3 has many road sections, and the rest of the facilities are in the street furniture zone. Type II-6 has a small road width, and the rest of the facilities are designed with the building interface. Type III-2 streets are campus entrance spaces with high foot traffic. People in the plaza often sit at the edge of the green belt to rest, gossip, and wait for others. This area is rich in behavioral types, and interactive leisure spaces can be designed according to the needs of the crowd. Streets of Types IV-1 and IV-2 are broader and can be considered street pocket parks (Fig. 2).

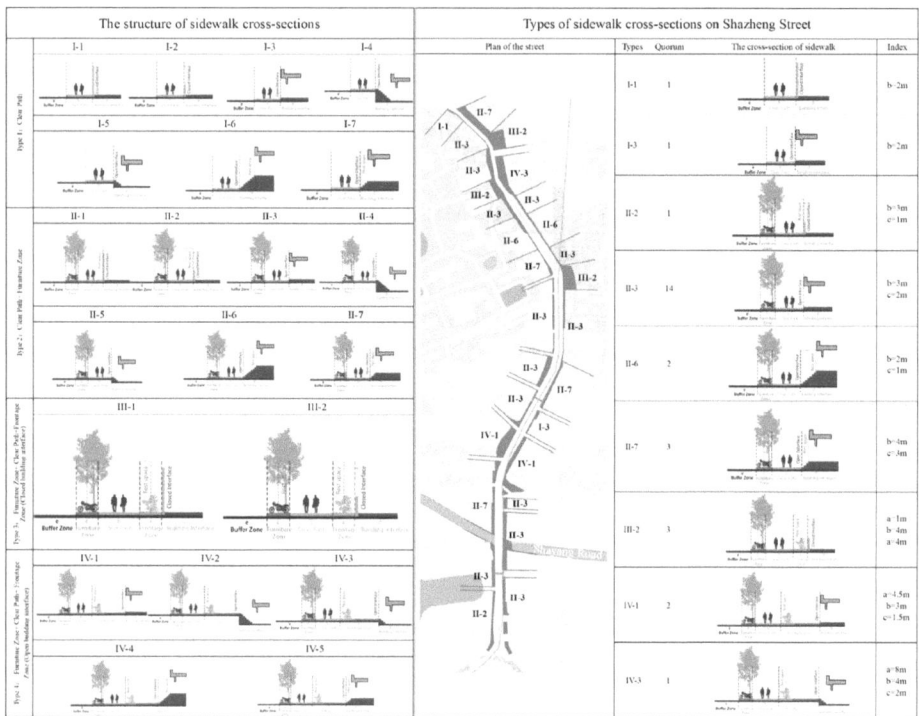

Fig. 2. Types of sidewalk cross-sections on SZS.

3.1.4 Types of Street Leisure Spaces

Street leisure space refers to any area on the street that can accommodate individuals to engage in leisure activities. This present study classifies street leisure spaces into subsidiary, functional, and temporary categories. A subsidiary leisure space on the street serves a distinct purpose while providing amenities for pedestrians to rest briefly. Examples include bus shelters, public exercise areas for residents, and seating areas outside stores. Functional leisure space refers to the space on the street specifically for people to provide leisure space, generally showing a point, line, surface, and other kinds of space form; the most common is in the walkway every distance set up by the street seating facilities. A temporary leisure space develops when insufficient leisure space is available on the street, prompting individuals to spontaneously utilize existing environmental features such as front stairs, kerb stones, traffic bollards, and fences as makeshift leisure spaces. When choosing a sitting area, people usually choose a space located at the edge, with good safety, without blocking pedestrian traffic, a comfortable environment, and ease of sitting down.

The study showed that subsidiary leisure spaces in SZS are primarily situated at bus shelters and shop entrances. These areas hinder pedestrian flow, offer limited comfort, and fail to facilitate interactions (Fig. 3). Functional leisure spaces have seating and green belt edges but lack leisure plazas. The rest of the seats are monolithic and wraparound, satisfying the fundamental purpose of providing rest; however, they do not promote social interaction among individuals. Additionally, untrimmed green shrubs hinder utilization of the rest facilities. Most temporary resting facilities are situated at the corners of the green belt. Because this is where spontaneous resting behavior of pedestrians occurs, it can serve as a focal point for designing street spaces that cater to people's resting requirements. Studying the resting behavior and psychology of people is necessary to create leisure spaces that are both comfortable and convenient.

3.2 Demand Analysis for Leisure Spaces in SZS

3.2.1 Heat Map Data

Studies on the liveliness of street crowds have mainly utilized methodologies such as heatmap, GPS tracking, social media, and cellphone signaling [13–15]. The primary purpose of the Baidu Heat Map is to visually represent the geographic and temporal patterns of population distribution [16, 17]. The system computes population density and mobility patterns using location data collected from mobile phone users who access the Baidu application products. The Baidu Heat Map shows real-time data with varying colors that indicate distinct population densities and distribution levels [18]. Recent research has confirmed that Baidu heat map data can serve as a reliable indicator for evaluating the liveliness of streets [19].

The study collected Baidu heat map data from the Baidu map platform to analyze the spatial distribution pattern of crowd activities on the street. The data included a period of 6:00–24:00 for seven consecutive days inside the SZS area. Mapping software was used to visualize the cumulative crowd density graphics for each day of the week (Fig. 4). The study analysis showed a more noticeable difference between weekdays and rest days in the crowd-gathering area. On workdays, in addition to road intersections, people gathered at the entrances of universities and high schools, particularly at the entrance of Chongqing No. 7 High School. On weekdays, people also gathered at road intersections and Xinyang Plaza, but the flow of people at school entrances was significantly lower. Therefore, for the five daily crowd gathering and stopping points on the SZS, where the intensity of vitality is higher, consideration can be given to adding interactive leisure spaces.

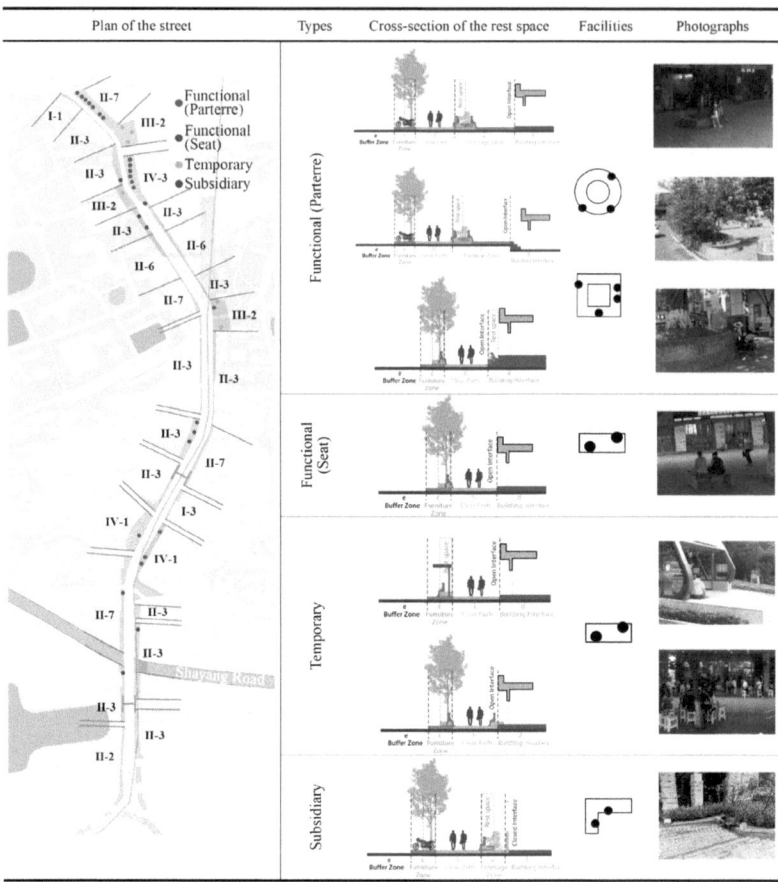

Fig. 3. Types of sidewalk leisure spaces on SZS.

3.2.2 Behavioral Mapping

Behavioral mapping is a methodology employed to systematically observe and record behaviors occurring in a specific location at a given moment [20]. Behavioral mapping methods can be used to study population needs directly and efficiently. The higher the frequency of spontaneous resting behavior in the street space, the greater demand for rest in that area [21]. The SZS behavioral mapping project spanned one week, commencing on October 23, 2022, and concluding on October 29. The researchers conducted multiple walks down the route, documenting spontaneous pedestrian resting behaviors by marking maps and capturing images as evidence. The study revealed that pedestrians tended to relax primarily in alleyways, street crossings, and university entrance spaces. Additionally, some people rested on the steps of closed banks. In addition, there was an age difference in the population's resting behavior at different locations (Fig. 5). Older adults on the street are generally residents of the neighboring community who prefer to chat and sunbathe alleyway where they live and are more in need of comfortable leisure spaces that meet their social needs. Younger people usually choose the more territorial

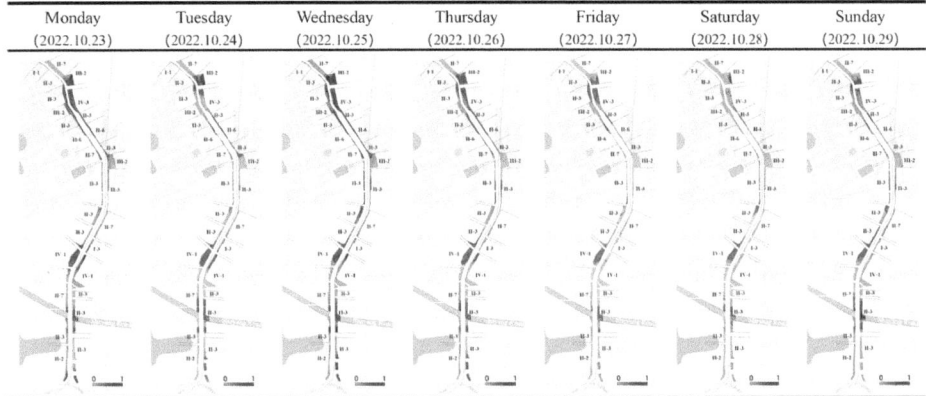

Fig. 4. Distribution of street vitality.

and private space of steps in front of the building for rest, with little need for social inter-action. Furthermore, porters tend to linger around crossings, with a substantial influx of individuals to secure employment. The design of street leisure spaces should fully consider the complex crowd structure around the street, and according to the needs of different types of people, design a leisure space with targeted functions in the appropriate location to effectively meet the needs of the crowd and improve the efficiency of the use of resting facilities.

3.2.3 Questionnaire Surveys

Conducting questionnaire interviews with pedestrians can be helpful for comprehending pedestrian satisfaction with street leisure spaces. The questionnaire had nine questions, including the respondents' general perception of the resting environment, their require-ments for resting in streets, and their recommendations on the current resting facilities. In total, 100 valid questionnaires were received during the experiment. Most respondents were 19–29 years old; 44% were students by occupation, and most came to SZS for entertainment and shopping. The demand survey showed that 56% of the respondents were not satisfied with the overall leisure space environment of the street; 64% of the respondents had unmet demand for leisure space; 72% of the respondents thought that the number of existing leisure space facilities was insufficient; 78% of the respondents wished for more leisure space while waiting for buses; and 72% of the respondents thought that there should be more street benches. In addition, the study also investigated people's suggestions on the interval of street rest facilities in the SZS, and most people thought that the interval distance of street rest facilities should be 50-100m (Fig. 6).

In conclusion, there are many problems with street leisure space on SZS, and people's satisfaction is low. Thus, the number of street seats must be increased. Researchers have found that spontaneous resting behavior occurs in many street sections. The inferior quality of the rest facilities and failure to adequately analyze the resting behavior of pedestrians in the street space resulted in the under-utilization of existing rest facilities. There is also a lack of green landscaping on the streets, located only at the entrances of

Plan of the street	Types of population		Spaces	Quorum	Actions	Demands	Photographs
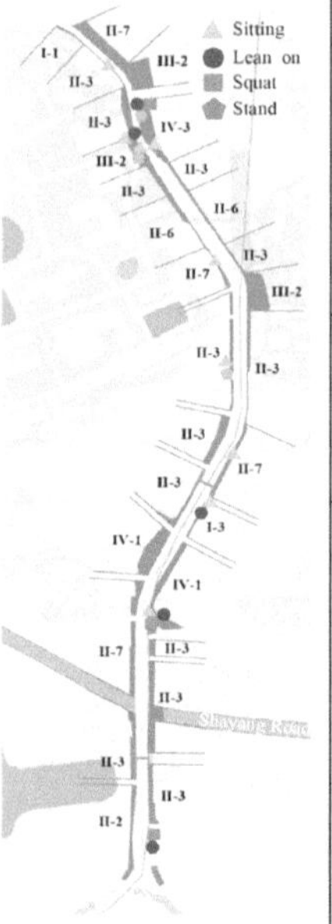	Senior citizen	Surrounding residents	Alley entrance	1-2	Sit	Sunbathing Relaxing Chatting	
			Near the wall	4-5	Sit	Playing mahjong Communication	
		Non-residents	Steps	1-2	Sit lean on	Resting Waiting	
	Children	surrounding residents	Campus entrance	1-2	Sit lie	Climbing Playing	
	Middle-aged and young peoples	Student	Campus and shop entrance	1-3	Stand sit	Resting Waiting	
		Staff	Steps	2-4	Sit lean on	Chatting Resting Waiting	
			Metro entrance	1-2	Sit	Resting Waiting	
		Service staff	Road intersection	3-4	Stand Squat lean on	Looking for jobs Chatting	

Fig. 5. The SZS behavioral mapping.

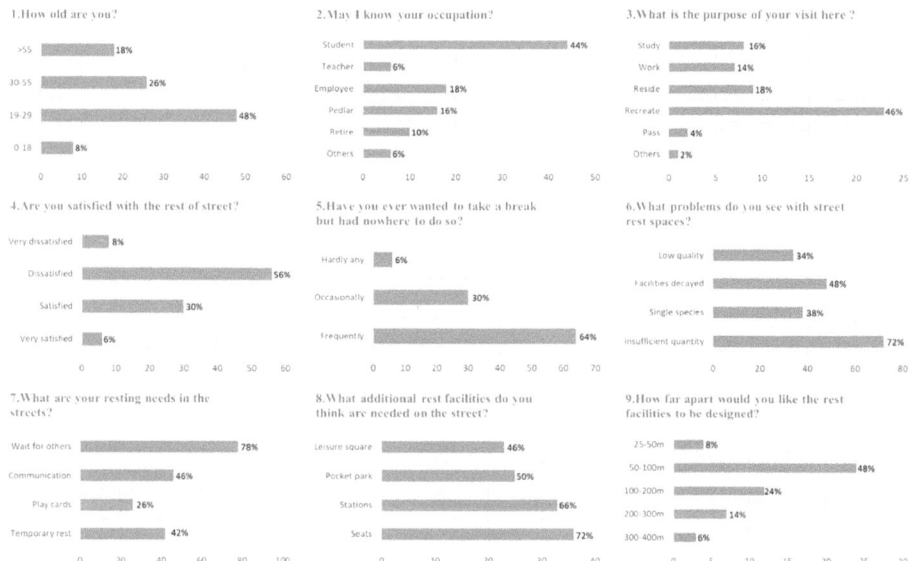

Fig. 6. Results of the questionnaire survey.

some schools, with a single form of green landscape and a lack of hierarchical changes. In addition, leisure space on the street lacks a unified design. It is yet to form its unique cultural characteristics, so residents do not have a sense of belonging to the neighborhood.

4   Renewal Design of Street Leisure Space on Shazheng Street

4.1   Overall Design

According to the analysis of the current situation and demand degree of leisure space on SZS, the overall design and zoning design of the street was carried out (Fig. 7). First, the rest of the street facilities were designed in an integrated manner, and the unique historical and cultural symbols of SZS were extracted and integrated into the renewal design of the leisure space to create a unique cultural memory of the street. While strengthening the continuity of street leisure spaces, it also preserves the unique history and culture of the street. It enhances residents' sense of identity and belonging to the neighborhood culture.

4.2   Zoning Design

Based on the current situation of street leisure space and the results of the demand analysis, street leisure space was modified by area. Streets were divided into high and low-demand sections. The rest of the facilities will be designed at 50 m intervals for high-demand sections and 100m for low-demand sections. The interactive leisure space is designed in three high-crowd-density road sections as the focus of the renovation design: Chongqing Seventh Middle School, Xinyang Plaza, and the intersection of SZS and Shayang Road. In addition, after comprehensively analyzing the resting needs of pedestrians, the type of pavement cross-section, and the location of bus stops, two street pocket parks were designed so that the leisure space in the street meets the needs of pedestrians.

5   Conclusion

Street leisure spaces are integral components of urban public spaces that directly impact people's everyday activities. While contemporary street construction increasingly focuses on the design of walkways and prioritizes pedestrians, more attention needs to be paid to addressing the design of leisure space renovations in existing streets. This study examined the forms of walkways in Chongqing, a city with many mountains. It uses a typological approach to summarize the placement of leisure spaces and common forms of rest facilities in different cross-sectional forms of pavements. People's resting behavior and their perception of leisure spaces were analyzed based on environmental psychology. This analysis also explored the relationship between street leisure spaces and people's resting behavior. The aim was to establish a foundational set of theories for the design of street leisure spaces.

Furthermore, this study establishes a design framework for the renewal of neighborhood street leisure spaces, using the SZS in Chongqing as a case study. First, the design

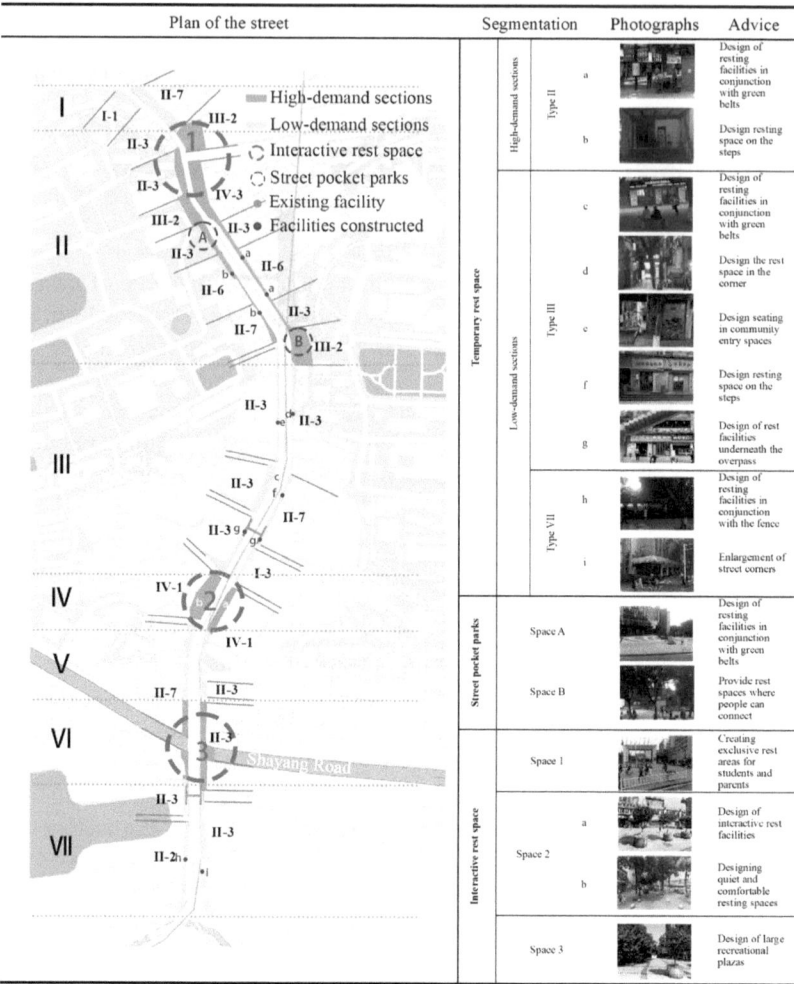

Fig. 7. Renewal design of leisure space on SZS.

employs Baidu heat map, behavioral maps, and questionnaires to examine the existing configuration of street leisure space and the demand for street leisure space amenities. Subsequently, the various cross-sectional configurations of different segments of the street and the requirements for people's rest are combined and examined to develop an appropriate design strategy for street renovation. This design framework is expected to serve as a reference for future renewal designs of leisure spaces in neighborhood streets.

There is room for improvement in this study. In addition to spatial factors, the design of street leisure spaces should consider the natural environment, climatic conditions, and sociocultural and economic factors of the neighborhood. In the future, several streets will be selected for comparative studies to analyze the impact of various factors on the design of street leisure spaces and to construct a more comprehensive and systematic design system.

References

1. Zavestoski, S., Agyeman, J.: Incomplete streets: processes, practices, and possibilities, 1st edn. Routledge (2015)
2. Karndacharuk, A., Wilson, D.J., Dunn, R.: A review of the evolution of shared (street) space concepts in urban environments. Transp. Rev. 34:190–220(2014)
3. Fuller, M., Moore, R.: The death and life of Great American Cities, 1st edn. (2017)
4. Levin, I.: The street: a quintessential social public space. 52:173–174 (2015)
5. El Khateeb, S., Shawket, I.M.: A new perception; generating well-being urban public spaces after the era of pandemics. Dev. Built Environ. 9, 100065 (2022)
6. Global Designing Cities Initiative: Global street design guide. Island Press, Washington, DC (2016)
7. Mehta, V.: The street: a quintessential social public space, 1st edn. Routledge, London (2013)
8. Jiang, Y., Han, Y., Liu, M., Ye, Y.: Street vitality and built environment features: a data-informed approach from fourteen Chinese cities. Sustain. Cities Soc. **79**, 103724 (2022)
9. Gerike, R., Koszowski, C., Schröter, B., Buehler, R., Schepers, P., Weber, J., Wittwer, R., Jones, P.: Built environment determinants of pedestrian activities and their consideration in urban street design. Sustainability (Switzerland) 1316 (2021)
10. Harvey, C., Aultman-Hall, L.: Measuring urban streetscapes for livability: a review of approaches. Prof. Geogr. **681**, 149–158 (2016)
11. Biswas, S., Chandra, S., Ghosh, I.: Effects of on-street parking in urban context: a critical review. Transp Developing Economies **31**, 10 (2017)
12. Li, Y., Yabuki, N., Fukuda, T.: Exploring the association between street built environment and street vitality using deep learning methods. Sustain. Cities Soc. **79**, 103656 (2022)
13. Wu, C., Ye, X., Ren, F., Du, Q.: Check-in behaviour and spatio-temporal vibrancy: an exploratory analysis in Shenzhen, China. Cities **77**, 104–116 (2018)
14. Wu, J., Ta, N., Song, Y., Lin, J., Chai, Y.: Urban form breeds neighborhood vibrancy: a case study using a GPS-based activity survey in suburban Beijing. Cities **74**, 100–108 (2018)
15. Kim, Y.-L.: Seoul's Wi-Fi hotspots: Wi-Fi access points as an indicator of urban vitality. Comput. Environ. Urban Syst. **72**, 13–24 (2018)
16. Lv, G., Zheng, S., Hu, W.J.G.: Natural hazards, risk: exploring the relationship between the built environment and block vitality based on multi-source big data: an analysis in Shenzhen, China. Geomatics Nat. Hazards Risk. **13**, 1593–1613 (2022)
17. Wang, F., Liu, Z., Shang, S., Qin, Y., Wu, B.J.T.E.: Vitality continuation or over-commercialization? Spatial structure characteristics of commercial services and population agglomeration in historic and cultural areas. Tourism Econ. **25**, 1302–1326 (2019)
18. Fan, Z., Luo, M., Zhan, H., Liu, M., Peng, W.: How did built environment affect urban vitality in urban waterfronts? A case study in Nanjing reach of Yangtze River. ISPRS Int. J. Geo Inf. **109**, 611 (2021)
19. Huang, J., Hu, X., Wang, J., Lu, A.: How diversity and accessibility affect street vitality in historic districts? Land **12**, 219 (2023)
20. Zhang, X., Cheng, Z., Tang, L., Xi, J.: Research and application of space-time behavior maps: a review. J. Asian Architect. Build. Eng. **20**, 581–595 (2020)
21. Vidal, D.G., et al.: Patterns of human behaviour in public urban green spaces: on the influence of users' profiles, surrounding environment, and space design. Urban For. Urban Greening **74**, 127668 (2022)

Sustainable Urban and Rural Planning

Data-Driven Urbanism: Image Processing Techniques for Urban Analytics

Karam M. Al-Obaidi[1(✉)], Jing Wang[2], and Mohataz Hossain[1]

[1] Department of the Natural and Built Environment, College of Social Sciences and Arts, Sheffield Hallam University, Sheffield S1 1WB, UK
`k.al-obaidi@shu.ac.uk`
[2] Department of Computing, College of Business, Technology and Engineering, Sheffield Hallam University, Sheffield S1 1WB, UK

Abstract. Geographic databases provided by open and public sources lack a high degree of accuracy. Although these sources were developed by collecting data from surveys, tracing from aerial imagery and freely licensed geodata sources, their reliability is questionable in testing new concepts for urban analytics and developing solutions for City Information Modelling (CIM). This study aims to examine a method using digital image processing to deliver precise information and accurate data for urban analytics. The research applied algorithmic solutions using content-based image segmentation, which accurately segments roof regions of buildings from aerial images. The study utilised an open access dataset annotated using 72 images grouped into 6 larger tiles from a joint project between Humans in the Loop with the Mohammed Bin Rashid Space Centre in Dubai, the UAE. The results show the efficiency of extracting buildings and their detailed features in an urban context. Finally, the study demonstrates the reliability of using the Base UNet model and the ResNet-based UNet, in analyzing urban aerial images.

Keywords: Urban Analytics · Image Processing · Segmentation · Geodata · Aerial Imagery · Buildings · Roofs

1 Introduction

Climate change and global warming are projected to intensify environmental risks in cities where new approaches to support climate resilience are needed [1]. Urban analytics is the practice of understanding urban and city processes and contexts, currently emerging as an interdisciplinary field, which is gaining popularity in exploring new advances in using Geodata and computational methods [2, 3]. Geographic databases have been widely used and practised in built environment studies to examine and assess urban scenarios [4]. These databases have played a role in defining new challenges in urban studies, such as population, infrastructures, environments, and policies [5].

The development of information technology has offered new possibilities to introduce advanced solutions for urban and city contexts. City Information Modelling (CIM) is a new movement that extends from the development of Building Information Modelling (BIM), a well-established process to provide smart visual models. CIM uses spatial

© The Author(s) 2025
B.-J. He et al. (Eds.): UCSUD 2023, LNCE 559, pp. 709–720, 2025.
https://doi.org/10.1007/978-981-97-8401-1_51

data from remote sensing and a geographic information system (GIS) by exchanging and sharing information associated with images and textures based on different levels of detail [6].

GIS and remote sensing practices demonstrated limitations in examining specific physical parameters in urban areas [7]. These include the accuracy of identifying, classifying, and quantifying physical characteristics, such as boundaries, materials, textures, and colours. In addition, the reliability of open geographic databases is questionable in testing for urban analytics and developing optimum solutions for CIM.

The integration of computer vision and image processing technologies provides a promising avenue to enhance the capabilities of CIM in gathering valuable data for urban analysis and decision-making. One possible solution is to use Artificial Intelligence (AI) to identify qualifiable information techniques such as image segmentation. Such an approach helps to examine aerial and satellite images and obtain relevant information. By applying these algorithms, it becomes feasible to identify and categorise physical elements such as roof areas and boundaries of buildings. Employing AI helps to address issues related to Urban Heat Island (UHI), energy consumption, and carbon emissions in urban contexts [8]. As a result, this research aims to explore an approach using digital image processing to accurately segment roof regions in aerial images to generate reliable information and accurate data for urban analytics that could help to develop efficient solutions for CIM.

2 Literature Review

Content-based image segmentation involves partitioning an image into meaningful regions based on the content or visual characteristics of the pixels. It has various applications in architecture and engineering, particularly in aerial and satellite image processing. It has the potential to extract buildings, roads, and other infrastructure elements from images, enabling detailed analysis and monitoring of urban areas. Studies in this area have focused on enhancing the accuracy of content-based image segmentation algorithms. Machine learning techniques, such as deep neural networks, have been used to enhance the segmentation results [9]. Additionally, the integration of other data sources, such as LiDAR or GIS data, has been explored to improve the accuracy of segmentation and information about the urban environment [10].

In the last decade, AI and machine learning techniques have revolutionised the area of satellite image analysis. The integration of deep learning algorithms, incorporating convolutional neural networks (CNNs) [11, 12] and recurrent neural networks (RNNs) [13], has brought about significant advancements in the accuracy and efficiency of analysing satellite imagery. These algorithms autonomously acquire knowledge by identifying patterns, features, and objects within satellite images. As a result, deep learning in satellite image analysis has effectively enhanced tasks such as classification, segmentation, and detection, leading to more precise and reliable results. However, this application has limitations when accurately segmenting complex and irregular shapes, such as building edges and road networks from satellite imagery and aerial photography. To overcome these limitations, researchers have turned to deep learning architectures like UNet [14].

In satellite image processing, UNet has shown promising results in segmenting buildings, roads, and other urban features [15]. By training the network on a large dataset of labelled satellite images, it can learn to identify and segment different objects of interest with high accuracy. However, training UNet requires a substantial amount of labelled data and computational resources. The network needs to be trained on a diverse set of satellite images to generalise well to different urban environments. Additionally, hyper-parameter tuning and data augmentation techniques may be necessary to optimise the segmentation performance. In addition, UNet segmentation has limitations when dealing with details such as the roofs of each building and smaller image objects. Due to its architecture and receptive field size, UNet may struggle to accurately segment these fine-grained elements. Additional techniques, such as post-processing or incorporating higher-resolution imagery, may be required to address this challenge.

3 Methodology

The study applied a quantitative approach to collect and analyse numerical data using different neural network frameworks and semantic segmentation of aerial imagery. In this research, a novel and innovative image segmentation algorithm called "Local Object Enhanced-UNet" has been developed. The main objective of this algorithm is to extract the regions of interest from individual building roofs. The Local Object Enhanced-UNet algorithm takes full advantage of the advanced convolution structure of the UNet while incorporating additional post-processing techniques based on low-level visual object features such as colours, textures, and edges. By leveraging these features, the algorithm achieves remarkable accuracy in identifying detailed roof regions. This breakthrough in image segmentation technology opens new possibilities for precise roof material identification and analysis.

Figure 1 illustrates the system pipeline of the proposed "Local Object Enhanced-UNet". In this figure, the input images are first pre-processed to remove any noise and ensure their suitability for the network's input structures. The UNet performs encoding and decoding operations, generating content-based image segmentation outputs. Each output pixel is assigned an index number that represents the semantic content of the pixel, which enables the identification and extraction of large areas of buildings. By extracting the area, we can specifically focus on the roof regions. The final outputs of this process consist of connected roof regions from each building, which are then ready for automatic roof material analysis.

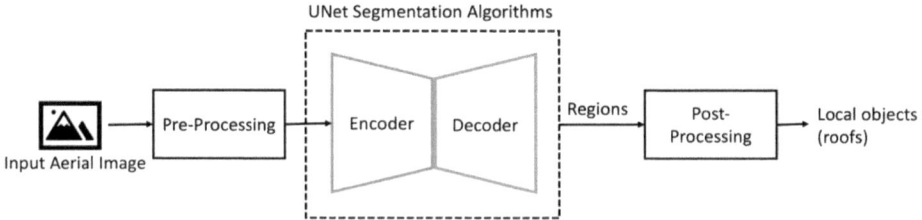

Fig. 1. Local object enhanced UNet system pipeline

Two different neural network frameworks were used to demonstrate the versatility of the proposed algorithm. These frameworks include the Base CNN, which serves as a solid foundation for the algorithm, and the Pre-trained ResNet 50, a well-established and widely used CNN model, which further enhances the performance and robustness of the algorithm. By employing these two distinct frameworks, we showcase the practicality and effectiveness of our proposed algorithm. In the following sections, each component from the system pipeline is introduced.

3.1   The Machine Learning Dataset and Data Pre-processing

The prototype is based on "Semantic segmentation of aerial imagery" [16], which has been published as an open access dataset annotated for a joint project between Humans in the Loop and the Mohammed Bin Rashid Space Centre in Dubai, the UAE. The dataset covers eight different areas and contains five main content categories, namely "building", "land", "road", "vegetation", and "water". Additionally, any other content that does not fall into these categories are labelled as "unlabelled". Figure 2 showcases a selection of samples from the dataset, displaying both the original aerial image and the manually annotated regions used for machine learning purposes.

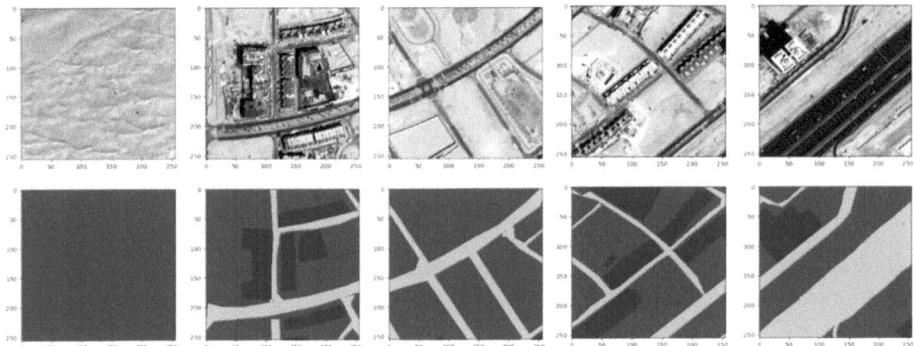

Fig. 2. Samples (top: original image; bottom: ground truth manual annotations) from "Semantic segmentation of aerial imagery"

The original aerial images and their annotation images are all cut into smaller tiles (255 pixels by 255 pixels). Generating smaller tiles is beneficial for several reasons. Firstly, it allows for an increased number of training samples, which allows the accuracy and robustness of machine learning models. Moreover, utilising a smaller tile size helps to reduce the overall data load required for machine learning, particularly during batch training, thereby streamlining and expediting the process. Furthermore, resizing the image into smaller tiles enables it to conform to the input data format of CNN and ResNet, facilitating efficient data training.

3.2   CNN-Based UNet

UNet is an improved CNN deep learning architecture introduced by Ronneberger et al. in 2015 [14], which is regularly utilised for image segmentation tasks. The UNet architecture contains an encoder-decoder structure. The role of an encoder is to capture context and extract attributes from the input image. The UNet encoders comprise convolutional layers with down-sampling procedures to lower the spatial dimensions of the feature maps while increasing the number of channels. The decoder part seeks to reconstruct the segmented image from the extracted features. It consists of up-sampling operations, such as transposed convolutions or interpolation, to progressively expand the spatial dimensions of the feature maps.

3.3 Training Approach

In a CNN structure, all the weighted filters in each feature map are ready to be adjusted through the training process. This adjustment allows the filters to learn and adapt to the specific features and patterns in the input data. To train the UNet model, a commonly used approach is to use a loss function to determine the variation between the predicted segmentation output and the ground truth (i.e., manual annotations of the segmentation), such as binary cross-entropy [17]. Optimization algorithms, such as backpropagation, play a crucial role in this training process. Backpropagation allows the model to iteratively update and adjust the weights in each filter based on the observed errors or discrepancies between the predicted segmentation output and the ground truth.

For instance, Fig. 3(a) demonstrates the learning process of our proposed CNN-based UNet model training from scratch until converged. The loss function gives less loss through the training process and the accuracy of the model showed improvement. In this study, we focused on measuring the segmentation accuracy of the system. This was done using the Intersection over Union (IoU) rate, a metric that is generally applied in image segmentation tasks. The IoU rate measures the overlap between the predicted segmentation mask and the ground truth mask. By calculating the IoU rate for each segmented object or region, we were able to assess the accuracy of our system by identifying and delineating different objects in the aerial images. This evaluation metric helped us understand the effectiveness of our segmentation algorithm and guided us in optimizing it for better accuracy and performance. Figure 3(b) demonstrates the progress of the average IoU rate increases throughout the training process over 100 training iterations (epochs).

3.4 Transfer Learning and ResNet 50-Based UNet

The key behind transfer learning is to transfer the learned features or representations from the pre-trained neural network model to a new model. In practice, transfer learning involves taking a pre-trained model, normally trained on a large dataset such as ImageNet [18], while removing the last few layers that are task specific. The pre-trained layers operate as a feature extractor, capturing common patterns and features that are useful for a wide range of tasks.

In this study, we adapted and adopted the original CNN structure through transfer learning. We selected the ResNet 50 model [19] which has been extensively used for diverse tasks. ResNet 50, short for Residual Network 50, is a deep convolutional neural network architecture, including image classification and object detection. We used a ResNet 50 model that has been pre-trained on the ImageNet dataset.

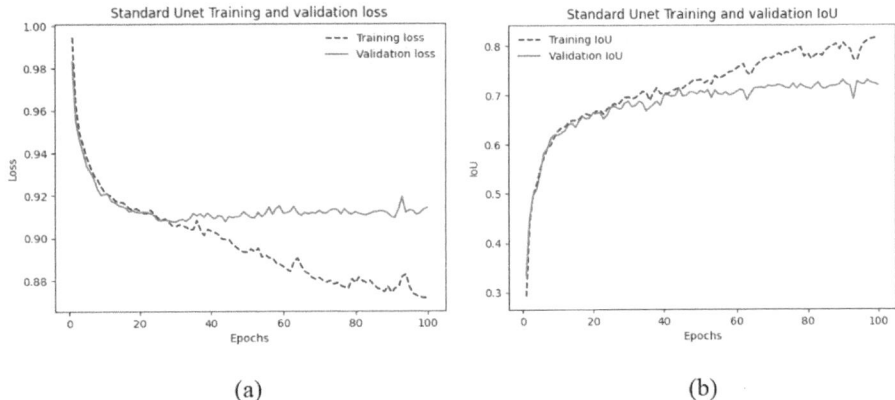

(a)   (b)

Fig. 3. (a) CNN-Based UNet model training and validation loss over 100 training iterations; (b) CNN-Based UNet model training and validation performance gain (IoU score) over 100 training iterations

3.5 Local Object Feature Extraction

One of the disadvantages of using a CNN structure for image segmentation is that it can sometimes overlook small details. This can happen due to the nature of the convolutional neural network architecture, which is primarily designed to capture and process larger patterns and features in an image. It is important to note that in aerial and satellite image segmentation tasks, the lack of detailed information can lead to under-segmented results. While the main region may be recognizable, it becomes challenging to distinguish individual objects within the region, such as each house in a residential area.

To effectively address the problem, we strategically implemented the localised, low-level image edge features. By doing so, we were able to significantly augment and amplify the previously overlooked intricacies and nuances, ultimately resulting in a more comprehensive and refined solution.

As illustrated in Fig. 4, the edge features in this study are comparable to the Canny edge detection algorithms [20], which are based on the gradient of the image. In addition, we used the Mean Shift clustering algorithms to cluster similar colours and pixel locations. This approach may result in the over-segmentation of regions, but it effectively preserved all the essential details, including the precise locations of the regions of interest for roof materials analysis. By incorporating these advanced techniques, we were able to obtain comprehensive and detailed insights into the analysis of roof materials. Since the Mean Shift clusters are unsupervised, one notable advantage is that the local shape has been extracted without requiring manual annotation. As shown in Fig. 4, the detailed building boundaries were not included in the annotation process due to the significant manual workload involved. However, the final output can automatically detect those details and improve the segmentation results.

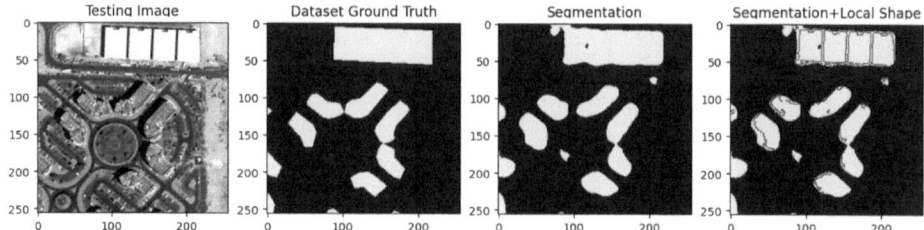

Fig. 4. Sample from output segmentation and local object shape feature extraction

4 Results and Analysis

The section presents a comprehensive evaluation to demonstrate a thorough examination of the segmentation performance using the reliability assessment metric, *i.e.*, Intersection over Union (IoU) for all the classes involved, namely "building", "land", "road", "vegetation", "water", and "unlabelled".

The IoU, recognized as the Jaccard index, is a metric utilised to quantify the accuracy of an image segmentation model. It assesses the overlap between the predicted segmentation and the ground truth. This is done by dividing the area of overlap by the area of union of both sets. The IoU score ranges from 0 to 1, where a higher score shows a better match between the predicted and actual segmentations. In this context, we use the IoU metric because it provides a reliable measure of how well our models are performing in terms of accurately identifying and classifying the different classes within our dataset.

The findings revealed that the average IoU score for the base CNN model stands at 58.4%, while the ResNet pre-trained model exhibited a higher average IoU score of 59.7%. These results are particularly noteworthy considering the complexity of the task, involving the accurate identification and classification of six distinct classes within a dataset comprising over 700 diverse data points.

The performance has been further compared as shown in Fig. 5. In this analysis, we separated the training and validation dataset results from the two proposed methods. After 100 training epochs, it is observed that all the modules have converged. It is worth mentioning that due to the scope of the models, the Base CNN UNet algorithms can reach their top performance earlier compared to ResNet, primarily because of their simpler structure. However, when considering the overall performance on both the training and validation dataset, the pre-trained ResNet outperforms the Base CNN UNet algorithms.

On the other hand, the learning process of the ResNet has some significant jumps during the training. These jumps are primarily caused by using a large learning rate, which can be adjusted and optimised through the iterative machine learning engineering process. By carefully analysing and monitoring the training progress, it is possible to identify these jumps and implement corrective measures to ensure the smooth and consistent learning of the ResNet model.

In addition to the experiments, we conducted further investigations that specifically examined the overall performance of the system by integrating both segmentation and local shape features. It is important to note that the annotation from the public dataset lacked any information regarding the local shape, therefore, to evaluate and discuss the

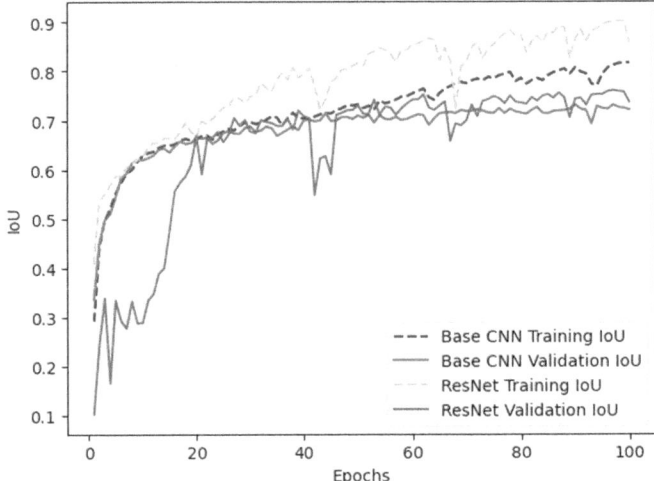

Fig. 5. IoU rate of two different neural networks across training and validation datasets

results obtained, we relied solely on subjective visual assessment using the following examples.

From Fig. 6, it is observed that each sample in the dataset is structured with two rows and four columns. The first rows represent the segmentation algorithms implemented using the Base UNet model, while the second rows correspond to the ResNet-based UNet. The first two columns of each sample contain the original tiles extracted from the dataset and the manual segmentation ground truth, respectively. Moving on to the subsequent columns, they showcase the segmentations obtained through various techniques, with the final columns specifically highlighting the enhanced version of the local shape features.

ResNet is known for its outstanding accuracy and precision in capturing the intricate details of various buildings. It excels in accurately preserving the sizes, proportions, and overall structure of the surrounding regions, demonstrating its exceptional capabilities in the field of image recognition and analysis. Additionally, ResNet has been proven to be highly proficient in maintaining the shapes and intricate details of objects compared to the base CNN version. This attribute makes ResNet a popular choice for tasks that demand a high level of preservation of shapes and fine details.

Focusing on the local shape feature enhancement, by using techniques such as Mean shift and gradient-based shape features, we were able to achieve the fine details from buildings in the local regions. The integration of these methods into our system significantly enhanced its performance in accurately capturing and preserving intricate architectural elements. This allowed us to focus specifically on the buildings and extract their detailed features with precision.

It is also worth mentioning that both algorithms demonstrate a certain level of robustness in dealing with shadows. This is an important characteristic as shadows can often pose challenges in satellite image processing tasks. Shadows can obscure important details and affect the accuracy of segmentation and shape analysis. Therefore, the ability of the algorithms to handle shadows effectively enhances their overall performance

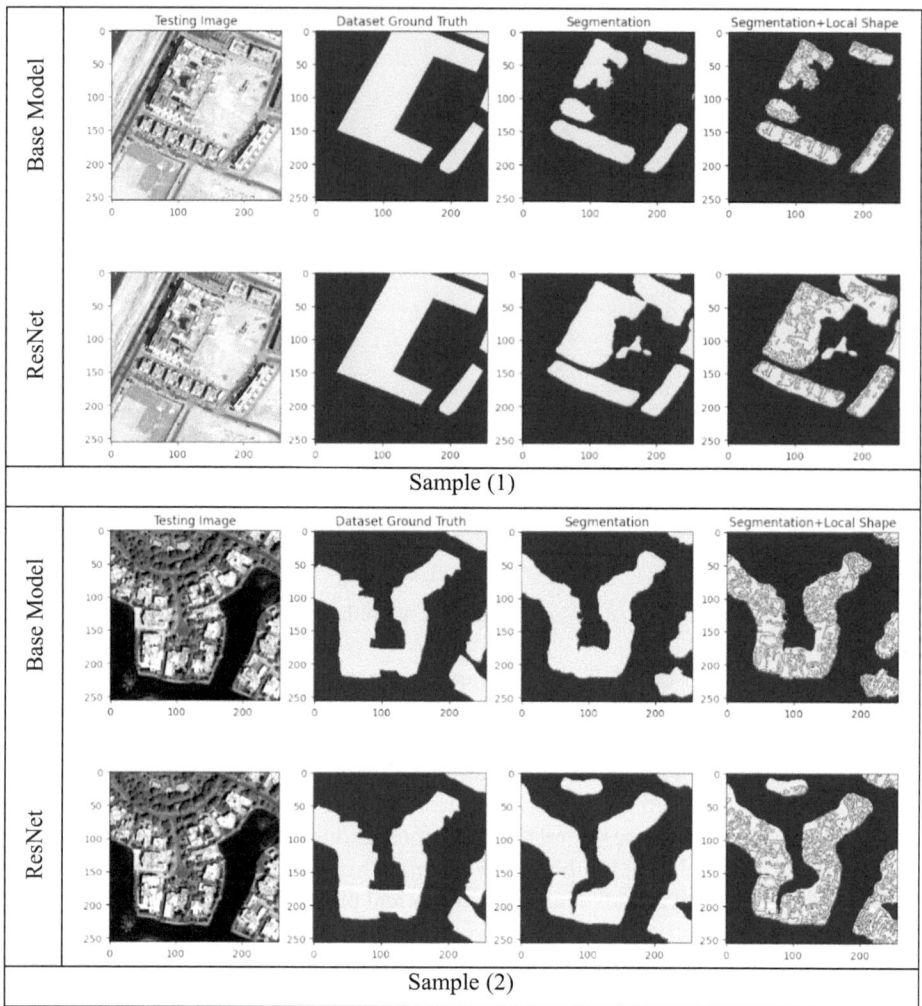

Fig. 6. Outputs (1–2) and side-by-side comparison between two algorithms (Base CNN model and ResNet model)

and reliability. By being robust in shadow handling, the algorithms can provide more accurate and reliable results, leading to improved image analysis and interpretation. This robustness ensures that the algorithms can effectively handle various lighting conditions and produce consistent and accurate outputs, regardless of the presence of shadows.

By integrating these methods into our system, we were able to specifically target the buildings in the images and extract their detailed features. This is crucial because, in urban areas, buildings are often the most prominent and important objects of interest. By enhancing the local shape features, we can effectively isolate the buildings from the surrounding environment and focus solely on extracting their architectural details.

This allows us to acquire a more thorough understanding of the urban landscape and facilitates further analysis and interpretation of aerial photography and satellite images.

The applied methods demonstrate the potential to produce accurate data from aerial imagery. The outputs provide insight into acquiring information on physical parameters and a deeper understanding of generating geo-located building geometry. The study identifies several implications of the applied methods to support initiatives for climate resilience. First, obtaining finer-scale details for improving urban climate simulation as many existing urban climate models lack a high level of detail. Second, providing a reliable tool to enhance climate-adaptation observation in terms of revealing data on how cities align with the climate goals. Third, testing the reliability of aerial imagery and satellite-based observations, the changing behaviour of the same environmental components and land-use activities. Fourth, identifying sources of heat stress, such as the density of built-up areas, percentages of greenery spaces, and aspects of biodiversity. Fifth, opening the possibility of enriching urban analytics with correct data to guide the development of environmental strategies for smart and sustainable cities. Finally, supporting green investments and establishing databases for future scenarios to support city regulators and urban planners.

5 Conclusion

The study demonstrated the efficiency of segmentation algorithms, particularly the Base UNet model and the ResNet-based UNet, in analysing aerial images of urban areas. UNet-based image segmentation was successful in extracting detailed features from buildings with high precision and has proven its effectiveness. The integration of UNet-based segmentation algorithms, including the CNN UNet model and the ResNet-based UNet, has significantly enhanced the image analysis process. In addition, by incorporating techniques such as Mean shift and gradient-based shape features, the study has success-fully enhanced the local shape features, enabling the precise capture and preservation of intricate architectural elements. This fine-grained analysis provides a comprehensive understanding of the urban landscape and facilitates further interpretation of the aerial images.

These methods can be seen as extended tools in the field of urban analytics, particu-larly in urban computing, a promising field that needs further research and development. These methods have specific implementation scenarios in the practice: (1) enabling active technologies such as City Information Modelling (CIM) to develop models for urban digital twins. These methods are capable of presenting correct digital replicas of physical systems or environments to guide in performing real-time monitoring, simulation, and analytics of ecosystems. (2) offering the possibility to provide computational resources to train models in machine learning and AI. (3) guiding the use of computer vision cou-pled with deep learning to value big data at urban levels. (4) eliminating predictions and advancing towards decision-making to support policymakers to optimise policy.

References

1. Hincks, S., Carter, J., Connelly, A.: A new typology of climate change risk for European cities and regions: principles and applications. Glob. Environ. Chang. **83**, 102767 (2023)
2. Singleton, A.D., Spielman, S., Folch, D.: Urban Analytics. Sage (2017)
3. Omrany, H., Al-Obaidi, K.M., Hossain, M., Alduais, N.A., Al-Duais, H.S., Ghaffarianhoseini, A.: IoT-enabled smart cities: a hybrid systematic analysis of key research areas, challenges, and recommendations for future direction. Discov. Cities **12**(1), 2 (2024)
4. Wang, G., Meng, X.: Evaluation and calculation of urban carbon emission reduction potential based on spatiotemporal geographic model. Int. J. Thermofluids **20**, 100478 (2023)
5. The Alan Turing Institute. https://www.turing.ac.uk/research/research-programmes/urban-analytics, last accessed 2023/12/29
6. Shi, J., Pan, Z., Jiang, L., Zhai, X.: An ontology-based methodology to establish city information model of digital twin city by merging BIM, GIS and IoT. Adv. Eng. Inform. **57**, 102114 (2023)
7. Mashala, M.J., Dube, T., Mudereri, B.T., Ayisi, K.K., Ramudzuli, M.R.: A systematic review on advancements in remote sensing for assessing and monitoring land use and land cover changes impacts on surface water resources in semi-arid tropical environments. Remote. Sens. **15**(16), 3926 (2023)
8. Srivastava, A., Maity, R.: Assessing the potential of AI–ML in urban climate change adaptation and sustainable development. Sustainability **15**(23), 16461 (2023)
9. Guo, Y., Liu, Y., Georgiou, T., Lew, M.S.: A review of semantic segmentation using deep neural networks. Int. J. Multimed. Inf. Retr. **7**, 87–93 (2018)
10. Ghaseminik, F., Aghamohammadi, H., Azadbakht, M.: Land cover mapping of urban environments using multispectral LiDAR data under data imbalance. Remote. Sens. Appl.: Soc. Environ. **21**, 100449 (2021)
11. Sultana, F., Sufian, A., Dutta, P.: Evolution of image segmentation using deep convolutional neural network: a survey. Knowl.-Based Syst. **201**, 106062 (2020)
12. Khryashchev, V., Ivanovsky, L., Pavlov, V., Ostrovskaya, A., Rubtsov, A.: Comparison of different convolutional neural network architectures for satellite image segmentation. In: 2018 23rd Conference of Open Innovations Association (FRUCT), pp. 172–179. IEEE (2018, November)
13. Marri, V.D., P, V.N.R.: RNN-based multispectral satellite image processing for remote sensing applications. Int. J. Pervasive Comput. Commun. **17**(5), 583–595 (2021)
14. Ronneberger, O., Fischer, P., Brox, T.: U-net: convolutional networks for biomedical image segmentation. In: Medical Image Computing and Computer-Assisted Intervention–MICCAI 2015: 18th International Conference, Munich, Germany, October 5–9, 2015, Proceedings, Part III 18, pp. 234–241. Springer International Publishing (2015)
15. Singh, N.J., Nongmeikapam, K.: Semantic segmentation of satellite images using deep-UNet. Arab. J. Sci. Eng. **48**(2), 1193–1205 (2023)
16. Roia Foundation, Semantic segmentation of aerial imagery. https://www.kaggle.com/datasets/humansintheloop/semantic-segmentation-of-aerial-imagery, last accessed 2023/12/29

17. Good, I.J.: Rational decisions. J. Roy. Stat. Soc.: Ser. B (Methodol.) **14**(1), 107–114 (1952)
18. Deng, J., Dong, W., Socher, R., Li, L.J., Li, K., Fei-Fei, L.: Imagenet: a large-scale hierarchical image database. In: 2009 IEEE Conference on Computer Vision and Pattern Recognition, pp. 248–255. IEEE (2009, June)
19. He, K., Zhang, X., Ren, S., Sun, J.: Deep residual learning for image recognition. In: Proceedings of the IEEE Conference on Computer Vision and Pattern Recognition, pp. 770–778 (2016)
20. Canny, J.: A computational approach to edge detection. IEEE Trans. Pattern Anal. Mach. Intell. **6**, 679–698 (1986)

Construction of Resilient City Based on the High-Quality Development of Fujiang River Basin City Cluster

Yuguo Liu[1], Yingwei Xiong[2,3](✉), Qian Li[3], Yanming Zhang[2], Xinghao Cui[3](✉), and Juan Chen[2,4]

[1] Institute of Industrial Economics, Chinese Academy of Social Sciences, Beijing 100006, China
[2] School of Urban and Rural Planning and Construction, Mianyang Teachers' College, Mianyang 621000, China
375459038@qq.com
[3] Technology Research Institute of Ecology, Energy-Saving, Environmental Protection and Low-Carbon Mianyang Science and Technology City New Area, Mianyang 621010, China
cuixinghao1997@163.com
[4] Engineering Research Center of Chuanxibei RHS Construction at Mianyang Teachers' College of Sichuan Province, Mianyang 621000, China

Abstract. The Fujiang River Basin, marked by its varied topography and frequent natural calamities, necessitates the establishment of resilient urban areas. This is essential for augmenting the region's capacity to manage disaster risks and for the enhanced development of its towns. This research based on the kernel density analysis method, analyzes and summarizes the distribution of geological hazards in the Fujiang River Basin, and then identifies the urgency of improving the resilience of urban agglomerations in the Fujiang River Basin, and proposes corresponding planning strategies. The preliminary conclusions are as follows: in the Fujiang River basin, the landforms prone to landslide, debris flow, collapse and slope instability are mainly distributed along the topographic inflection point. These geological disasters are mainly concentrated in Hechuan, Pingwu, Jiangyou, Santai, Tongnan and other places. In contrast, the rest of the tributary areas are less prone to such disasters due to their broad, basin-like topography.

Keywords: Fujiang River basin · Urban cluster · High quality development · Resilient city

1 Introduction

The formation of urban systems is randomly influenced by the variable internal and external urban environments, a result of the nonlinear interaction among diverse factors such as economic and social evolution, cultural adaptation, and resource integration [1]. A pressing issue in urban development is how to sustain the core structure and essential functions amidst a variety of unpredictable disturbances and pressures [2, 3], enhance the urban system's adaptability in everyday operations, its resilience against

B.-J. He et al. (Eds.): UCSUD 2023, LNCE 559, pp. 721–731, 2025.
https://doi.org/10.1007/978-981-97-8401-1_52

imminent threats, and its capacity for recovery after disasters, while concurrently mitigating the negative impacts of such disturbances [4]. Researchers such as the Comprehensive Resilience Alliance [5] and Yang Sun [6] define a resilient city as the ability of a city to resist and recover as a complex adaptive system in the face of environmental change. This capacity is essential for the sustainable and healthy development of cities. At present, some progress has been made in the field of resilient cities in the world [7–9] relatively speaking, the domestic research on resilient cities is still in its infancy, and the number of related studies is small, and the research on urban resilience focuses on the physical level, and pays less attention to the social and ecological aspects, and most researchers mainly focus on a certain natural disaster as the starting point, Du Jinying [10] took cities in the Pearl River Delta region as an example to evaluate and study the resilience of cities under the impact of tropical cyclone disasters.

These studies not only help us to better understand the connotation of resilient cities, but also provide us with valuable experience and enlightenment, pointing out the direction for the construction of resilient cities in the future.

2 Study Area

2.1 Study Area Profile

The Fujiang River Basin, nestled in the eastern part of the Sichuan Basin, spans between 102°20' to 106°48' east longitude and 30°42' to 32°30' north latitude. Its source lies at Xue Bao Ding, the highest peak of the Minshan Mountain, located between Songpan and Pingwu counties in Sichuan Province. The river courses southward, traversing Pingwu County, Jiangyou City, Santai County, Shehong County, Suining City, and the Tongnan and Tongliang Districts of Chongqing before joining the Jialing River in Hechuan District of Chongqing. Stretching approximately 670 km, the Fujiang River Basin encompasses a drainage area of 36400 square kilometers and maintains an average annual runoff of 572 cubic meters per second. It serves as a vital water source and a cornerstone for economic development and agricultural production in the region. The topography of the Fujiang River Basin is primarily characterized by basin plain landforms, with relatively flat terrain and minor elevation changes, conducive to agricultural activities and human settlement. The landscape transitions from northwest to southeast, beginning at the southeastern edge of the Zoige Plateau and descending into the basin. The upper reaches of the river feature steep terrain, narrow valleys, and pronounced valley slopes, often presenting as "V" shaped or box valleys. Overall, the Fujiang River Basin displays a diverse topography, with undulating mountains and narrow valleys in the upper reaches, gradually flattening slopes and widening valleys in the middle, and a predominantly flat, stable terrain in the lower reaches, with large terrace height difference between the two sides (Fig. 1). The research area mainly includes Pingwu County, Santai County, Fucheng District, Beichuan Qiang Autonomous County, Youxian District, Jiangyou City, Chuanshan District, Shehong County, Daying County, Pengxi County, Tongnan District and Tongliang District.

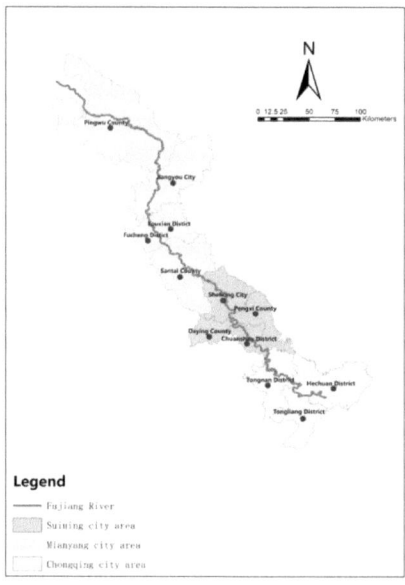

Fig. 1. Map of the Fujiang River Basin range

2.2 Study Area Profile of Natural Hazards

The Fujiang River Basin is notably marked by a fragile ecological environment and limited resistance to disturbances, making it susceptible to various geological disasters. This vulnerability also leads to ecological issues such as regional water level decline, soil erosion, accelerated land desertification, and ecosystem degradation. A review of disaster points from previous years reveals that the primary geological disasters in the Fujiang River basin include landslides, collapses, debris flows, and unstable slopes. In total, 2239 geological disasters have been identified in the area. The distribution of these hidden geological hazards in the Fujiang River basin is uneven, with certain areas exhibiting a higher concentration. Among the 2239 identified hazards (as depicted in Fig. 3), landslides are the most prevalent, with a total of 1165 incidents. This includes 735 small landslides, 343 medium landslides, 57 large landslides, and 7 super-large landslides, constituting 52.0% of the total hazards. Debris flows account for 2.7% of the total, with 60 incidents recorded, including 16 small, 37 medium, 6 large, and 1 super-large debris flow. Collapses are also significant, totaling 742 incidents, which comprise 469 small collapses, 209 medium, 25 large, and 3 super-large collapses, representing 33.1% of the total geological hazards. Unstable slopes constitute 11.7% of the hazards, with a total of 263 incidents including 1 super-large, 15 large, 93 medium, and 154 small unstable slopes. Additionally, the Fujiang River basin has experienced 3 small ground cracks and 6 small ground collapses (Fig. 2).

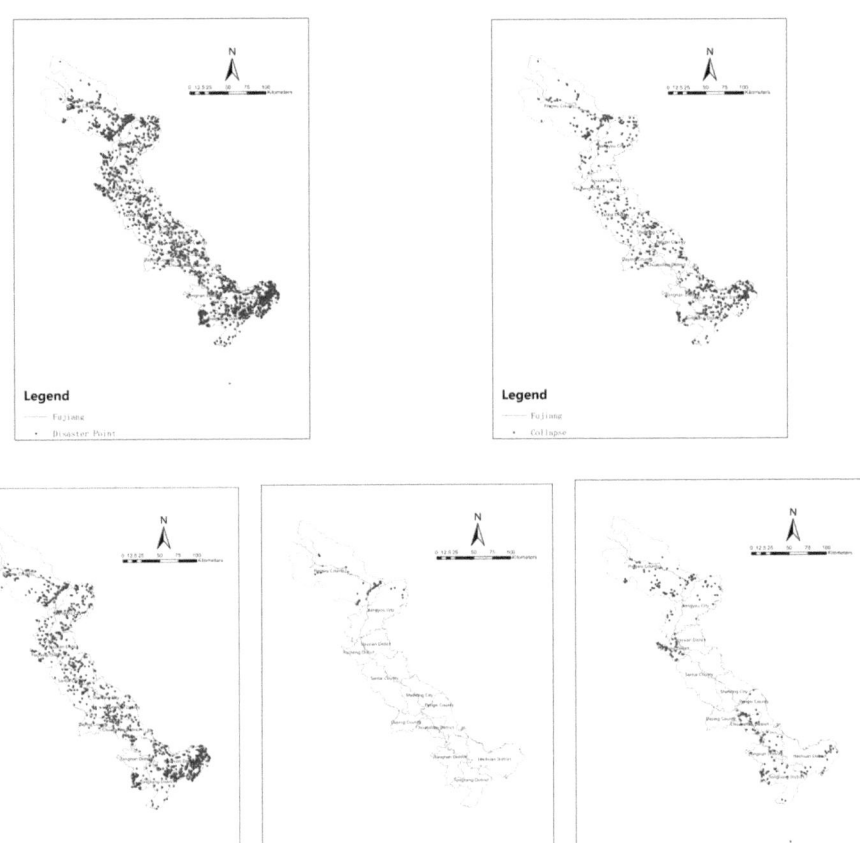

Fig. 2. Distribution map of geological disaster sites in the Fujiang River basin

3   Data Source and Research Method

3.1   Data Source

The geological hazard data was collected from the Resources and Environmental Science and Data Center of the Chinese Academy of Sciences.

3.2   Research Method

The research concept of this paper is defined as follows: It employs the kernel density method to analyze the spatial distribution of natural disasters like collapses, landslides, and debris flows within the region. This methodology is instrumental in summarizing the layout characteristics of these disasters in the study area. Identifying the core issues, the paper subsequently proposes planning suggestions, all within the ambit of building resilient cities, effectively addressing the identified challenges.

In the field of general geography, the predominant methodologies for point model analysis encompass the kernel density method. This approach is grounded in spatially representing the density of a region based on the frequency of aggregation, which is determined by the spatial interactions between a point and other locations. This method aligns with the distance decay effect as postulated in the first law of geography. According to this law, the radiative influence emitted by any point within a given area extends to its surrounding region, but the intensity of this influence diminishes as the distance increases, eventually reaching zero in the outermost regions. The kernel density analysis operates by computing a density grid based on this concept, where the average value of each grid cell equates to the sum of the mean number of all points within the region of each grid. As such, kernel density analysis not only adeptly captures the continuous variation in the effects of Points of Interest (POI) but also significantly mitigates noise disruption. The specific formula for this calculation is detailed in references [11, 12]:

$$f(s) = \sum_{i=1}^{n} \frac{1}{h^2} \varphi\left(\frac{s - c_i}{h}\right) \tag{1}$$

In Eq. (1), $f(s)$ denotes the kernel density computation function; h symbolizes the bandwidth of the continuous influence range; n represents the count of points within this influence range; c_i indicates the position of the i-element point; and φ is the kernel function. This kernel function is predicated on the quartic kernel function, which is computed as follows:

$$\varphi\left(\frac{s - c_i}{h}\right) = \frac{3}{4}\left[1 - \frac{(s - c_i)^2}{h^2}\right] \tag{2}$$

The selection of the maximum bandwidth h is a crucial step in kernel density calculations. This bandwidth is intimately linked to the dispersion of the input points; hence, a larger bandwidth is more appropriate for signals exhibiting high dispersion. For the analysis of clustered areas of various medical and health institutions in the Fujiang River basin, an adaptive bandwidth kernel density method was employed. This approach involves estimating the density range surrounding each point based on the density value of Points of Interest (POI). Subsequently, an optimal search radius is determined to accurately capture the density variations in the spatial linkage of the research subjects. It is important to note that a higher kernel density corresponds to a denser distribution of point elements.

4   Result Analysis

In accordance with the methodology of kernel density analysis, the resulting diagram depicting the distribution of geological disaster sites is illustrated in the figure. Typically, areas of high density are predominantly situated in both the upper and lower reaches of the Fujiang River basin, with isolated pockets of smaller clusters in other regions. This distribution manifests a spatial pattern characterized as "one primary nucleus plus multiple secondary nuclei." The principal nucleus is located at the confluence of Hechuan

District, Pingwu County, and Jiangyou City within the Fujiang River Basin, which harbors the highest concentration of disaster sites. Conversely, the number of disaster sites around Xuebaoding, the highest peak of Minshan Mountain and the origin of the Fujiang River basin, is minimal. Other districts exhibit localized concentrations, each forming their own distinct centers of disaster occurrence.

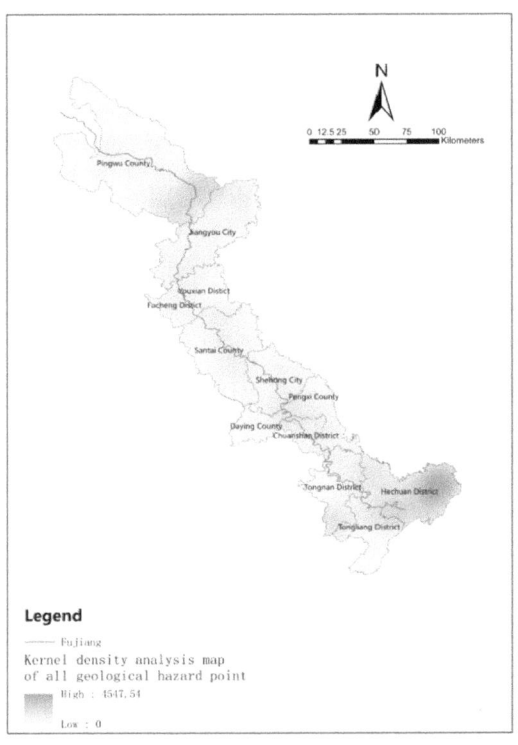

Fig. 3. The kernel density chart of geological disaster sites in Fujiang River basin

The analysis of geological disasters in the Fujiang River Basin, utilizing kernel density analysis and based on data of collapses, landslides, debris flows, and unstable slopes, is comprehensively illustrated in Fig. 4. From a categorical perspective, there are marked differences in the distribution of geological disaster points. Collapses and landslides constitute the majority, displaying a relatively uniform overall distribution. Their spatial distribution is characterized by a tendency toward clustering, with a significant concentration in Hechuan District of Chongqing and more dispersed occurrences across other areas. In contrast, the number of debris flow disaster points is the smallest. Spatially, these points exhibit a trend of unipolar concentration, particularly at the intersection of Pingwu County and Jiangyou City. The count of unstable slope disaster points falls in the middle of the spectrum. Their spatial distribution is identified by a multi-core pattern, with notable hotspots in Fucheng District, Chuanshan District, Pengxi County, and Tongnan District. The remainder of the region shows a more scattered distribution, with the least number of unstable slope disaster points occurring in Santai County and

Shehong City. In total, the distribution of geological disaster sites within the Fujiang River Basin spans 21 districts and counties, with Hechuan District recording the highest number at 683. Following Hechuan District, the districts of Pingwu, Jiangyou, Santai, and Tongnan each have over 160 geological disaster points. Conversely, there are seven other districts and counties where the count of geological disaster points is less than 120. In terms of distribution density across the region, geological disaster sites are dispersed at a rate of 6.25 per 100 square kilometers. Breaking this down by district, Hechuan District has a notably higher density of 32.5 per 100 square kilometers. Pingwu's density stands at 6.7, Jiangyou at 9.0, Santai at 6.8, and Tongnan at 10.2 geological hazards per 100 square kilometers. Analyzing the distribution maps and statistics by district and county reveals a distinct pattern: the landforms prone to landslides, debris flows, collapses, and unstable slopes are primarily found along the topographical turning points of the Fujiang River basin and adjacent to highways. These geological disasters are predominantly concentrated in Hechuan, Pingwu, Jiangyou, Santai, and Tongnan. In contrast, the remaining tributary areas, characterized by basin-like, broad terrain, show less development of such disasters (Fig. 4).

5 Planning Strategies and Programs

The Fujiang River Basin town group is an agglomeration of several towns within the Fujiang River Basin. In certain parts of this basin, the natural environment is notably harsh and the ecosystem fragile, making these areas particularly susceptible to natural disasters such as collapses and debris flows. A stable supply of water resources and a safe ecological environment are essential preconditions for the high-quality development of the town group in the Fujiang River Basin. Geological disasters in this region are characterized by their sudden onset and intense impact, coupled with considerable destructive power. These events pose a direct threat to the safety of residents' lives and property, as well as to the surrounding ecological environment. Therefore, analyzing and studying prevention and control measures for geological disasters, and achieving a scientific response, is of paramount importance. Such efforts are critical for reducing or avoiding disaster losses and ensuring the high-quality development of the urban clusters in the Fujiang River Basin.

5.1 Risk Assessment and Management Implementation Plan

Disaster data collection and analysis:Collect disaster data in the Fujiang River Basin in the past ten years, including disaster types, frequency and scope of impact, and use GIS system for spatial analysis to identify high-risk areas.

Establishment of risk assessment model: Combining geological, climatic, socio-economic and other data, a disaster risk assessment model is established to provide a scientific basis for formulating risk management plans.

Risk management plan formulation: According to the risk assessment results, formulate risk management plans for different types of disasters, including the construction of early warning systems, the allocation of emergency resources, and the construction of rescue teams.

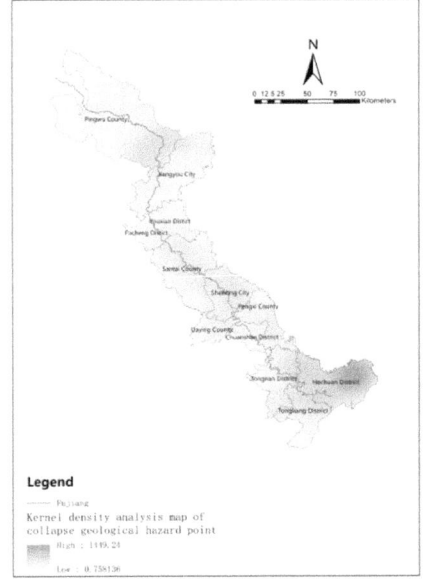
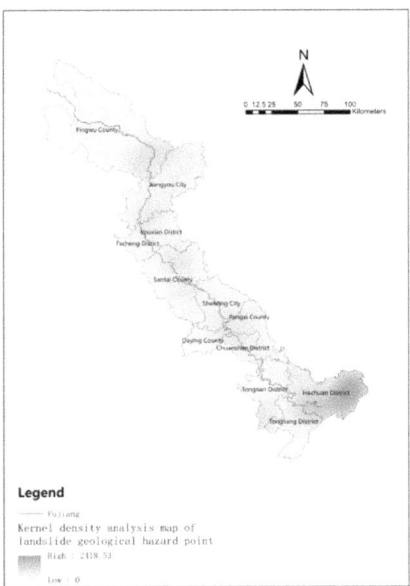
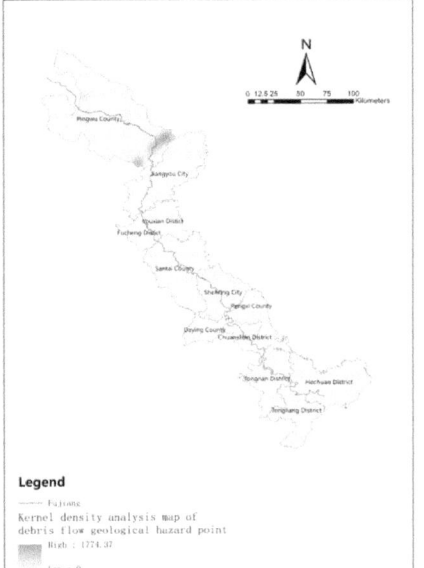
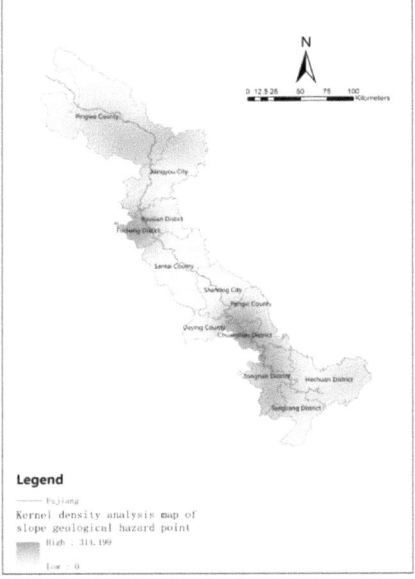

Fig. 4. The kernel density chart of various geological disaster sites in Fujiang River basin

5.2 Infrastructure Enhancement Implementation Plan

Flood control levee reinforcement: Conduct a comprehensive inspection of the existing flood control levees, and reinforce the embankment sections with hidden dangers to improve the flood control capacity.

Drainage system optimization: Renovation and upgrading of the drainage pipe network to increase rainwater collection outlets and drainage capacity to reduce the risk of urban waterlogging.

Promotion of earthquake-resistant buildings: In urban planning, it is clearly required that new buildings must meet certain seismic standards, and old buildings must be seismically reinforced.

5.3 Implementation Plan for Ecological Protection and Restoration

Wetland protection and restoration: Establish wetland reserves to carry out ecological restoration of damaged wetlands and restore their ecological functions.

Forest vegetation restoration: Implement the project of returning farmland to forest and grassland in mountainous areas to increase forest coverage and improve the stability of ecosystems.

Ecological corridor construction: Construct ecological corridors between cities and nature reserves to protect the migration corridors of wild animals and maintain ecological balance.

5.4 Community Participation and Education Implementation Plan

Disaster Risk Education Seminars: Regularly hold disaster risk education lectures, invite experts to explain disaster prevention knowledge, and improve residents' risk awareness.

Emergency drills: Organize residents to participate in emergency drills, simulate the scene when a disaster occurs, and let residents understand the methods and skills of emergency escape.

Community volunteer team: Establish a community volunteer team to participate in the construction of resilient cities, and form a good atmosphere for the participation of the whole people.

5.5 Planning and Policy Support Implementation Plan

Formulation of resilient city development plan: Based on the actual situation of the Fujiang River Basin, formulate a resilient city development plan, and clarify the development goals and tasks.

Policy guidance and support: Relevant policies have been introduced to reward and support units and individuals participating in the construction of resilient cities, so as to stimulate the enthusiasm of all aspects of society.

International cooperation and exchanges: Strengthen exchanges and cooperation with international organizations and other countries, learn from advanced experience and technology, and promote the development of resilient cities to a higher level.

5.6 Implementation Plan for Emergency Response and Recovery Plan

Rescue team construction: Strengthen the training and management of rescue teams, and improve the professional quality and emergency response capabilities of rescue personnel.

Rescue equipment configuration: Equipped with advanced rescue equipment and technical means to ensure that rescue can be carried out quickly and effectively in the event of a disaster.

Formulation and implementation of recovery plans: Formulate detailed recovery plans, including post-disaster reconstruction, production and life recovery, etc., to ensure that the city can resume normal operations as soon as possible.

6 Conclusion

Based on the kernel density analysis method, this paper analyzes and summarizes the distribution of geological hazards in the Fujiang River Basin, and then identifies the urgency of improving the resilience of urban agglomerations in the Fujiang River Basin, and proposes corresponding planning strategies. The preliminary conclusions of the study are as follows: landforms prone to landslides, debris flows, collapses and unstable slopes are mainly distributed near the topographic inflection points of the Fujiang River Basin. These geological disasters are mainly concentrated in Hechuan, Pingwu, Jiangyou, Santai, Tongnan and other places. In contrast, the rest of the tributary areas have less such disasters due to their basin-like topography and wide terrain. This paper has certain reference value for the construction of resilient cities in urban agglomerations in the Fujiang River Basin under the influence of natural disasters, but there are still shortcomings.

This paper only analyzes the improvement of urban agglomeration resilience from the perspective of natural disasters, but does not consider the economic and social impacts on the construction of urban resilience, and due to the limited data collection methods, the types of natural disasters analyzed do not fully represent the urban natural disaster resilience analysis system, which is the content that needs to be paid attention to in future research.

Acknowledgement. This study is financially supported by the Research on the construction and collaborative innovation of the whole element system of ecological environmental protection in Mianyang (NO. 2022ZYDF091).

References

1. Shi, Y., Zhai, G., Lihua, X., et al.: Assessment methods of urban system resilience: from the perspective of complex adaptive system theory. Cities **112**, 103141 (2021)
2. Meerow, S., Newell, J.P., Stults, M.: Defining urban resilience: a review. Landsc. Urban Plan. **147**, 38–49 (2016)
3. Zang, X., Wang, Q.: The evolution of the urban resilience concept, and its research contents and development trend. Sci. Technol. Rev. **37**(22), 94–104 (2019)
4. Chen, C., Lili, X., Zhao, D., et al.: A new model for describing the urban resilience considering adaptability, resistance and recovery. Saf. Sci. **128**, 104756 (2020)
5. Wilbanks, T., Sathaye, J.: Integrating mitigation and adaptation as responses to climate change: a aynthesis. Mitig. Adapt. Strat. Glob. Change **12**(5), 957–962 (2007)

6. Sun, Y., Zhang, L., Yao, S.: Evaluation of the resilience of prefectural cities in the Yangtze River Delta from the of social ecosystem. Population, Resour. Environ. China **27**(8), 151–158 (2017)
7. Li, T.: New progress in study on resilient cities. Urban Plann. Int. **32**(5), 15–25 (2017)
8. Prashar, S.K., Shaw, R.: Urbanization and hydro-meteorological disaster resilience: the case of Delhi. Int. Disaster Resilience Built Environ. **3**(1), 7–19 (2012)
9. Honghu, S., Xianfu, C., Mengqin, D.: Regional flood disaster resilience evaluation based on analytic network process: a case Chaohu lake basin, Anhui Province. China. Nat. Hazards **82**(1), 39–58 (2016)
10. Jinying, D., Xiaochun, T., Jiangang, X.: Study on urgency assessment of urban resilience promotion—a case study of typhoon disasters in the Pearl River Delta region. J. Nat. Disasters **29**(5), 88–98 (2020)
11. Wenhao, Y., Tinghua, A., Pengcheng, L., et al.: Network kernel density estimation for the analysis of facility POl Hotspots. Acta Geodaetica et Cartographica Sinica **44**(12), 1378–1383 (2015)
12. Xiaomeng, W., Wang, J., Qing, Z.: Identification of hollowing phenomenon in commercial space of six district of beiing based on checking-in data. Urban Dev. Stud. **25**(2), 77–84 (2018)

Theoretical Model for Urban Sustainability Planning Based on Metabolic Concepts

Juntao Ma[1], Ziyi Han[2,4], Yu Liu[2], Halike Sairjiang[2], Li Yan[3(⊠)], Hongjiang Wang[4], Jiekun Jiang[2], Xiaoyan Li[2], and Chunmin Luan[4]

[1] Faculty of Architecture and Urban Planning, Chongqing University, Chongqing 400045, China
[2] College of Civil Engineering and Architecture, Xinjiang University, Urumuqi 830017, China
[3] School of Architecture and Civil Engineering, Southwest University of Science and Technology, Mianyang 621000, China
anniey007@163.com
[4] Xinjiang Cultural Tourism Investment Group Co., Ltd, Urumuqi 830017, China

Abstract. In response to the core issue of how urban land planning can promote the level of sustainability, this paper takes the perspective of urban land change and draws on the requirements of the metabolic model in sociology. It takes the indicators of urban land planning and allocation as the core of regulation and constructs a theoretical model for urban sustainability planning in three dimensions: urban social, economic, and ecological environment. In the urban life support system, nine resource subsystems including land, water, atmosphere, population, transportation, energy, cultural tourism, basic supporting facilities, and public service facilities are constructed. At the same time, an empirical technical concept composed of a interdisciplinary quantitative model group is provided, aiming to provide ideas for the theoretical and empirical technology of urban sustainability planning.

Keywords: Planning theoretical model · sustainability · metabolism · urban life support system

1 Introduction

With the rapid progress of urbanisation and globalisation, the rapid expansion of urban land is restricting the sustainable development of global cities in terms of society, economy and environment. The concept of urban sustainability and rational land allocation based on system dynamics simulations or scenario projections applies to the terms land use planning and sustainability, which is also attributed to the fact that the primary means of urban planning is the power to own and allocate land and its internal subdivisions. For example, Saguin K. K. C. et al. [1] studied the environmental degradation caused by urban development and the general decline in the welfare of urban dwellers in the Philippines. They stressed that land use planning and policies are important interventions to achieve sustainable urban development and that social equity and justice are key components of sustainable development.

B.-J. He et al. (Eds.): UCSUD 2023, LNCE 559, pp. 732–744, 2025.
https://doi.org/10.1007/978-981-97-8401-1_53

The main task of urban management is to monitor the level of urban environmental sustainability and to regulate it according to land-use planning (allocation) indicators. Sustainable cities and communities—Indicators for city services and quality of life proposes that sustainability indicators should be used as the basis and criteria for judgement, and *Sustainable cities and communities Indicators for smart cities* provides a set of indicators and indicator systems for measuring urban sustainability to help cities assess their level of sustainability and develop measures for improvement. *Sustainable cities and communities - Indicators for resilience cities* include urban infrastructure and services, urban environment and resource management, social inclusion and equity, economic prosperity and innovation, etc., representing a coupling and collaborative study of multiple resource subsystems in determining urban sustainability. These tools need to be adapted to complex, non-linear and dynamic urban management issues, and they need to be based on the specific realities of the case cities, allowing management to constantly review sustainable land use and its effectiveness in order to develop differentiated regulatory policies. For example, Alexandre Repetti et al. [2] carefully designed a set of core indicators based on the most relevant elements of land use from three aspects: intrinsic quality of the indicators, position in the set and relationship with other indicators, and proposed a systematic relationship indicator model (RIM) based on the correlation of strategy, space, aggregation dimension and thematic dimension. Iannillo, Alessia et al. [3] based on the four challenges and case studies of urban green space, mobility and energy in three regions of Salerno University in Italy, concluded that proper planning and management of urban land can improve the sustainability and efficiency of residential areas. Although the concepts and ideas of urban land-use planning are constantly changing around the world, there are still some problems.

This paper mainly studies the theoretical model of urban planning, which provides a theoretical model line from theory to planning action, which is used to solve the quantitative control problem of land use and spatial layout planning. Taking the land use system and its optimal configuration parameters as the core of regulation and control, strengthening the ability of scientific regulation and control of urban sustainability construction, providing a consideration for the theory, technology and practice of urban sustainability planning.

2   Concepts Related to Current Urban Sustainability Planning Theory and Land Change

2.1   Sustainability

The concept of sustainability has long existed in the field of ecology. T. Kohlman [4] suggests that it was first used in German in 1713. In sociology, as a policy concept, it originated in Brundtland's 1987 report to the United Nations: "A global development process that minimizes environmental resources and reduces the impact on environmental resource carriers, while enhancing the economy and quality of life in the process. In the field of economics, it was first defined by the British Hicks J.R in 1946. Initially defined by British economist Hicks J.R. in 1946, this concept aims to enhance both the economy and overall well-being. As an extended concept from ecology, sustainability has gradually replaced the study of sustainable development and become an interdisciplinary integrated research field. As Starke. Points out, as a guide to the harmonious coexistence of nature and human beings, urban sustainability has relatively higher reliability and validity than sustainable development research. Moreover, sustainable development is only one of the development paths of a region or city towards sustainability, not all of them. Hence, this document focuses on identifying the sustainable condition as the primary goal of urban planning, to prevent conflicting with the value goals of sustainable development and the varied trends in implementation.

In short, this paper defines sustainability as: to judge whether the social, economic and ecological environment of a certain region or city is sustainable, sustainability indicators should be taken as the basis and criteria for judgment, and then strong sustainability and weak sustainability assessment or evaluation in the field of social, economic and ecological environment. The ideal state of urban sustainability is the maximum and most coordinated development of urban economy, society and ecological environment. However, for special cities, the requirement to achieve desirable results in all three areas of social, economic and ecological environment is a difficult criterion to implement. It is a practical problem to give priority to the practice order of the three. Therefore, this paper defines the state of urban sustainability as: under the premise of the improvement of one of the three aspects of ecological environment, society and economy, the other two do not regress, or the value goal judgment of one does not regress and one advances, the city also belongs to the ecological sustainable development mode.

2.2 Urban Support System

Currently, the urban life support system is defined as the urban lifeline system that is essential for maintaining the lives and production activities of urban residents, including transportation, energy, communication, water supply and drainage, and other urban infrastructure. For example, Wei, Lijun, and others developed a web-based prototype system by considering various infrastructure assets, triggering factors, and potential consequences as a holistic system concept. They demonstrated an intelligent approach to integrating and processing multi-source data, which has opened up a new method to assist complex decision-making processes with high social impact.

This article extends the urban material environment from the traditional single dimension to nine dimensions of the economic, social, and ecological environment, and determines nine resource and environmental systems from an interdisciplinary perspective. It covers the life systems that the urban material environment needs to maintain, including human resources, transportation, cultural tourism, water, atmosphere, public supporting service facilities, infrastructure services, energy, land, and other resources.

2.3 Ecological Environment

The ecological environment is a intricate system that is interconnected with the long-term growth of cities and economies. It is also a natural environmental system composed of certain ecological relationships. The safety of the ecological environment has become an important criterion for global urban or regional sustainable development. Land resources are divided into urban construction land and non-urban construction land, such as forests and grasslands.

This definition makes the research on social, economic, and ecological environment aspects in current urban planning theoretical models ambiguous and cannot differentiate between urban construction land and non-urban construction land. To address this issue, this paper subdivides the land resources in the ecological environment, separates urban construction land from land resources, and includes water resources, oil and gas resources, climate resources, and remaining land resources (non-urban construction land) in the concept of ecological environment. This concept definition is widely adopted in urban geography, economics, and ecology.

2.4 Urban Land

The current urban land classification system uses a single scale and classifies it based on land use. Under this system, the differences in land use intensity and land function characteristics are ignored. This system fundamentally limits the study of urban land changes and sustainability to a single discipline or an extension of the same discipline. Firstly, most of the current sustainability studies focus more on non-urban construction land such as forests, farmland, grassland, and minerals, while urban planning mainly focuses on the regulation of urban land at the scale of cities and blocks. This phenomenon neglects the main supporting role of urban areas in the sustainable development of multiple urban ecosystems. At the same time, urban areas are holistic units, and their reactions to similar influences are usually consistent regardless of size. This simple conceptual adjustment is caused by the use of classification maps and land classification schemes. Secondly, in urban land classification, there is both a binary structure of urban and rural areas and a ternary structure of cities, towns, and villages. Most of the current research results have the ambiguity of concepts such as urban and rural land or urban and rural land. Furthermore, in the research of urban land subdivision, grassland, forest land, and other urban land (non-urban area land) should be studied as a whole, while urban land should be studied separately, and urban land should be defined independently of the scope of urban ecological environment definition. This is a necessary setting for the independence of the ecological environment, social, and economic systems in the construction of urban planning theoretical models. Finally, urban storage land should be modified from secondary industry land to tertiary industry land. This makes storage logistics belong to the tertiary economic category in the national economic classification, and then establishes the correlation between urban storage land and its economic (GDP) indicators.

2.5 Urban Metabolism

Research on urban metabolism usually adopts explanatory and retrospective methods to study the driving factors of urban flow and stock. As shown in the research results of Wang, Xinjing, et al. [5], urban metabolism has evolved from an ecological discipline to a multidisciplinary field with three key knowledge bases: characterising urban ecosystems, developing urban metabolism theory, and creating methodological frameworks. The focus of research has shifted from quantifying environmental impacts to analysing the internal processes of urban systems. Gradually, simple socio-economic systems were

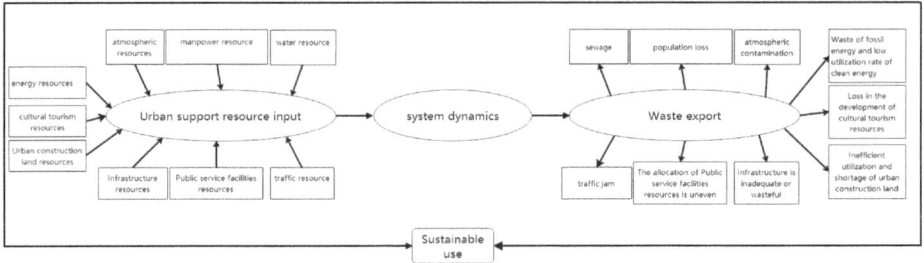

Fig. 1. Urban metabolism model

replaced by complex systems that involve interactions between nature, society, and the economy.

Figure 1 illustrates that this paper is based on the 'closed loop' process of destruction, utilization, restoration, and reuse between cities and natural materials. The "closed loop" of supporting nine types of resources in cities is regarded as the guarantee and support of urban life systems. The urban metabolism model is the physical process of transforming resources into public service products and generating waste. The amount of waste generated is determined by the demand for resources, and sustainability emphasizes the recycling of waste output, thereby forming a dynamic balance of regional resource input, waste output, and recycling. The integration of urban resource input and waste output must be closely maintained at a certain level of stability to achieve sustainability. It follows the natural laws of the ecological system for the nine types of resources in urban social, economic, and ecological environments.

3   Construction of Urban Sustainability Planning Theoretical Model

3.1   Design Route

Research Starting Point. We believe that the construction of theoretical models of urban sustainability planning is based on the concept of urban metabolism. The city is regarded as a self-correcting and self-improving material organic self-organising system. The theoretical and technical synergistic model for achieving the goal of sustainable development by continuously optimising the urban land allocation, scale and proportion of land use, and rationally regulating the links between various resource and environmental subsystems. The urban material environment is the result of a series of coupled and coordinated processes of social, economic, and ecological environments. Its functioning process is the core of sustainable development research, determining the rate at which the city tends towards "sustainability" or "extinction". Therefore, to establish a theoretical model for urban planning, it is necessary to first understand the urban support system required for human spatial needs, and to regard the urban material environment as a resource and environmental system that provides life support system services for the city. In the urban material environment, urban construction land dominates the circulation of new metabolic substances in the city, and it is the main means to implement urban sustainability planning in combination with urban and rural planning. By converting farmland into urban construction land, it affects the feedback in social, economic, and ecological environments, making the urban material environment tend towards the process of "sustainability" or "extinction". The city can be viewed as a material organism that constantly optimises the scale and proportion of land use, regulates various resource and environmental sub-systems, and achieves development goals through the linkage of multiple sub-systems.

Practice Route. The theoretical model of urban planning needs to understand the essence of the city through the spatial needs of urban residents, and establish a theoretical model of element network that meets the basic requirements of urban residents. Various element networks, together with urban planning methods, constitute an intermediate component of urban resource demand and environmental outcomes, making the public aware of the degree of environmental impact of planning methods. At the same time, AtKisson [6] and Ameen et al. [7] pointed out that there are few achievements in the research on the coupling operation mechanism of the large-scale metabolic system of urban agglomerations and their internal differentiated economic, industrial and social characteristics with the differences in urban metabolic efficiency.

Then, as shown in Fig. 2, aiming at the shortcomings of existing research, a research framework from theory to action on urban sustainability is explored and established. In 1990, Japanese scholar Mayumi [8] proposed a multi-scale comprehensive analysis of social metabolism, which was based on the available resources during human activities, including "external energy flow" resources and "value-added flow". However, in their analysis of Chinese societies and ecosystems, Geng et al. conclude that a multiscale approach to social metabolism is insufficient to analyze the specific metabolic factors that drive urban macroeconomic activities, such as industry and land use. Therefore, this paper constructs a theoretical model of urban sustainable development, improves the theoretical basis, and increases the metabolic mechanism of industry and land, as well as the impact mechanism of economic development.

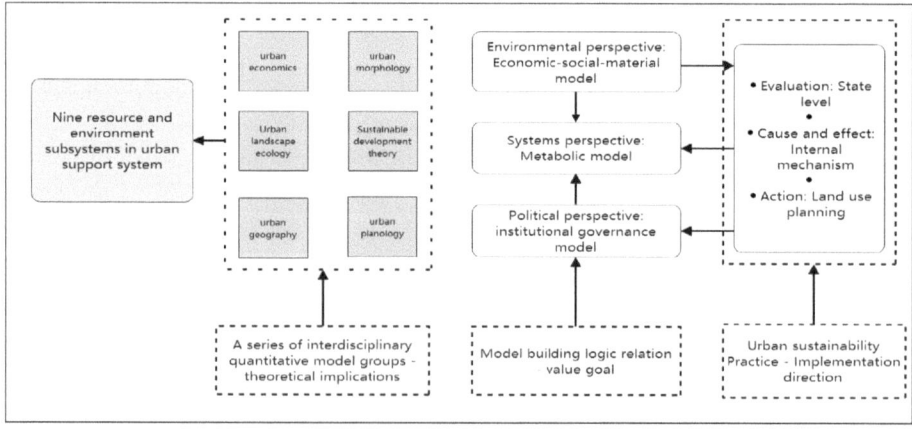

Fig. 2. Urban sustainability: from theory to planning action

3.2 Composition of Theoretical Model

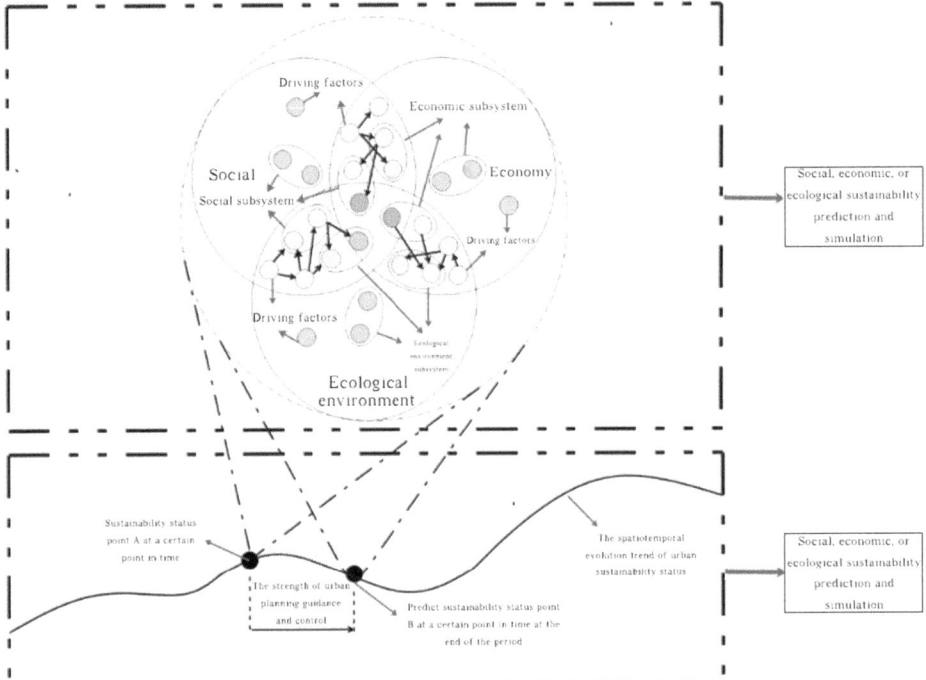

Fig. 3. Overall framework of the theoretical model for urban sustainability planning

Overall Framework of Theoretical Model. Cities are complex ecosystems comprising social, economic and ecological environments. The three subsystems of social, economic, and ecological environments have their own characteristics, structures, functions, and development principles, but their existence and development are limited by the structures and functions of other subsystems.

As shown in Fig. 3, the following is the general structure of the theoretical model for urban sustainability planning: in a future time period, the city is placed in a spatial-temporal framework composed of the three systems of "social, economic, and ecological environments" in a region. The coupling and coordination relationships between the elements of the nine resource subsystems in the three dimensions of social, economic, and ecological environments are conducted. Based on the parameters or indicators obtained from urban development prediction and simulation, as well as the evaluation of the results, the theoretical model uses urban planning methods to regulate urban construction. In other words, the framework includes all the driving factors of the systems or subsystems in the set, and the coupling and influencing relationships between the factors occur differently in different dimensions. The factors or categories of factors in "one-to-one," "one-to-many," and "many-to-many" relationships do not overlap and can clearly distinguish each factor or category of factors. That is, the classification of different systems or subsystems within the social, economic, and ecological dimensions in the model must meet the requirement of non-overlapping between systems, so that there is a coupling relationship between different systems or subsystems in the city, but with clear differentiation and non-overlapping classification.

Composition of the Theoretical Model

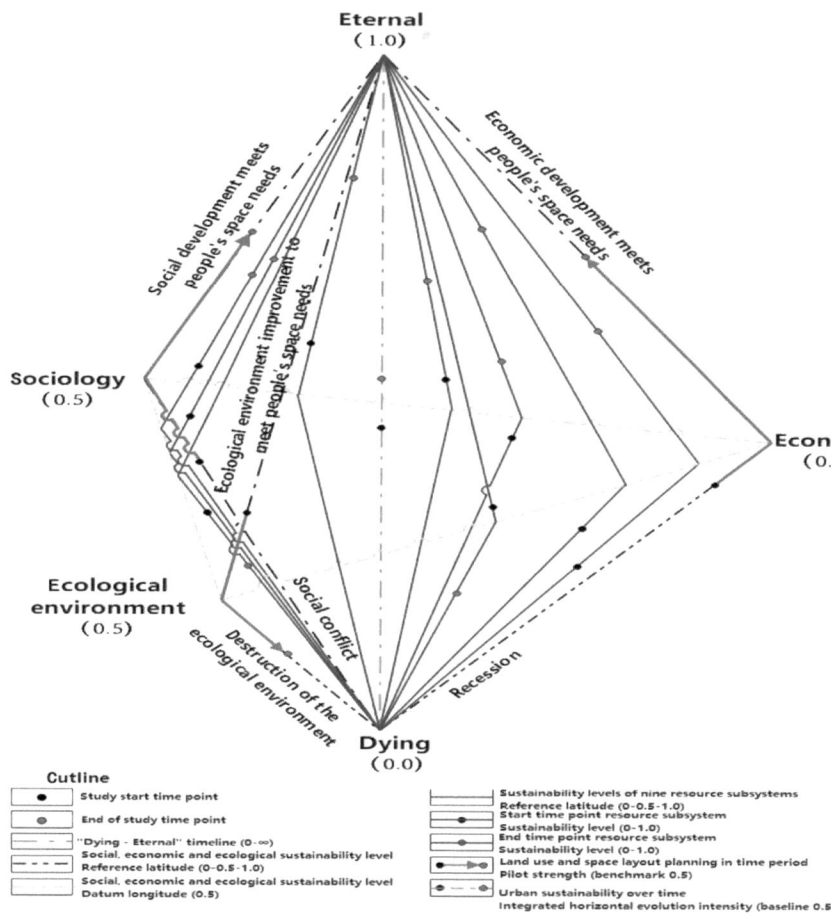

Fig. 4. Urban sustainability planning theoretical model

As a result of the above theoretical model, the theoretical model for urban sustainability planning is defined as follows: combining land use and spatial layout planning techniques to regulate the urban economy, social aspects, and ecological environment at different levels, different time stages, and different degrees to continuously meet the spatial needs of urban residents as they grow. It gradually tests the spatiotemporal evolution model of cities moving towards sustainability or decline between dying cities and sustainable cities, taking into account the specific conditions of China's regions or cities.

The research focuses on the temporal dependence relationship between the spatial needs of urban residents and the decline or sustainability of cities, the limiting relationship between the spatial needs of urban residents and the ecological environment, the temporal distribution relationship between the spatial needs of urban residents and society, and the temporal evolution relationship between the spatial needs of urban residents and the economy. Figure 4 provides a visual representation of these relationships.

Firstly, a city is a whole that includes the triple attributes of ecological environment, society, and economy, and uses urban land use and spatial layout methods to regulate the sustainable state of urban economy, society, and ecological environment in time and space, while maintaining coupling and coordination with "human spatial needs".

3.3   Theoretical Model Evaluation

Solving the Value Judgment Problem of Urban Sustainable Development Status. In the theoretical model of urban planning sustainability, three dimensions of sustainable development are taken into account: economic, social, and ecological. The ideal state of the three is to maximize and coordinate the development of economy, society, and ecological environment. However, in the face of uneven and unbalanced development in China, and relatively low efficiency in resource utilization and fund allocation, it is particularly difficult to achieve the maximization of economy, society, and ecological environment. Therefore, prioritizing the sequence of development among the three is a practical problem that must be faced [9].

Therefore, the practice standard is to meet the "urban people's spatial needs" in one aspect of ecology, society, and economy, without regressing in the other two aspects, or to judge the implementation direction of not regressing in one aspect and meeting the "urban people's spatial needs" in the other aspect. This provides direction and standards for subsequent quantitative evaluation research of theoretical models. Therefore, the theoretical model for urban sustainability planning includes three different implementation directions: urban ecological sustainability planning theoretical model, urban social sustainability planning theoretical model, and urban economic sustainability planning theoretical model.

The time axis is one of the core transformational elements of the theoretical model of urban sustainability planning, reflecting the coupling between the level of sustainable urban development and the "spatial desires of urban residents" in the time course. The timeline is not only a key element in urban planning research, but also a fundamental element in the development of urban sustainability, and it is widely used as a common independent variable in related disciplines. Therefore, the logical relationship between the timeline and other dimensions and resource subsystems in the theoretical model of urban sustainability planning is as follows:

Secondly, the logical relationship between the timeline and urban planning theory or practice: investigating past issues, setting current expected goals, and controlling future vision construction are the core or main work content of urban planning theory and practice. At the same time, the timeline, as an important coordinate of urban planning theory and practice, can be compared and determined with research results from related disciplines.

Third, the logical relationship between the timeline and ecological sustainability: the timeline is a basic element of the city, together with social, economic, and ecological environment, and it is also a common element of sustainable research in multiple ecological disciplines. The timeline is also a witness to the process of urban life (sustainability) and death (extinction).

Technical Empirical Aspects of Theoretical Models. First, the meridian line in Fig. 4 includes the input and waste output processes of the nine resource subsystems in Fig. 1. According to the current research, it can be determined adaptively which systems participate in quantitative analysis, and the uncertain resource subsystems in the future do not disturb the overall framework of the theoretical model, which is significantly different from the current ecological city theoretical model.

Thirdly, the latitude and longitude lines set in Fig. 4 meet the requirements of quantifying the ecological sustainability level of the nine resource subsystems and the "social, economic, and ecological environment" three dimensions in the urban metabolic model and urban sustainability theory into planning actions. It also conforms to the "white box" standard in the urban metabolic model, which means it is intuitive, clear, and can be understood and verified by the public, laying a good foundation for public participation.

3.4   Technical Validation of Theoretical Models

Based on the theoretical model of urban sustainability planning in China and taking city of Xinjiang as a sample, a technical methodology has been constructed from planning theory to land use planning action under multi-source data, multi-scale, multi-model, multi-index, and multi-software. The technical route followed is 'Expansion Motivation - Current Situation Evaluation - Vision Estimation - Planning Action'. The language used is clear, objective, and value-neutral, with a formal register and precise word choice. The text adheres to conventional structure and format, with consistent citation and footnote style. The structure is clear and logical, with causal connections between statements. The text is free from grammatical errors, spelling mistakes, and punctuation errors. No changes in content have been made. It has innovatively constructed a technical methodology of "from planning theory to land use planning action" for urban sustainability under multi-source data, multi-scale, multi-target, multi-model, multi-index, and multi-software. The study examines six indicators for controlling land use planning in industrial areas, including intensity, proportion, total scale, and annual increment. These indicators are considered as 'actions' at three different scales: provincial, urban, and neighbourhood. The study uses multi-source data, including satellite digitised maps, remote sensing data, statistical data, and historical documents. To analyse the data, the study employs ArcGIS 10.2, GeoDetector, Excel, SPSS, and AnyLogic 8. Literature visual analysis, Note Express data analysis, and other software were used for theoretical research and empirical analysis. Quantitative and qualitative analyses, social surveys, expert interviews, horizontal and vertical comparisons, and multiple indices and model analyses were employed to create a technical empirical model of urban sustainability planning theory in Xinjiang for the period 2017–2035. The language used is clear, objective, and value-neutral, with consistent technical terms and common sentence structure. The text is free from grammatical errors, spelling mistakes, and punctuation errors. The content of the improved text is as close as possible to the source text, with no additional aspects added (Fig. 5).

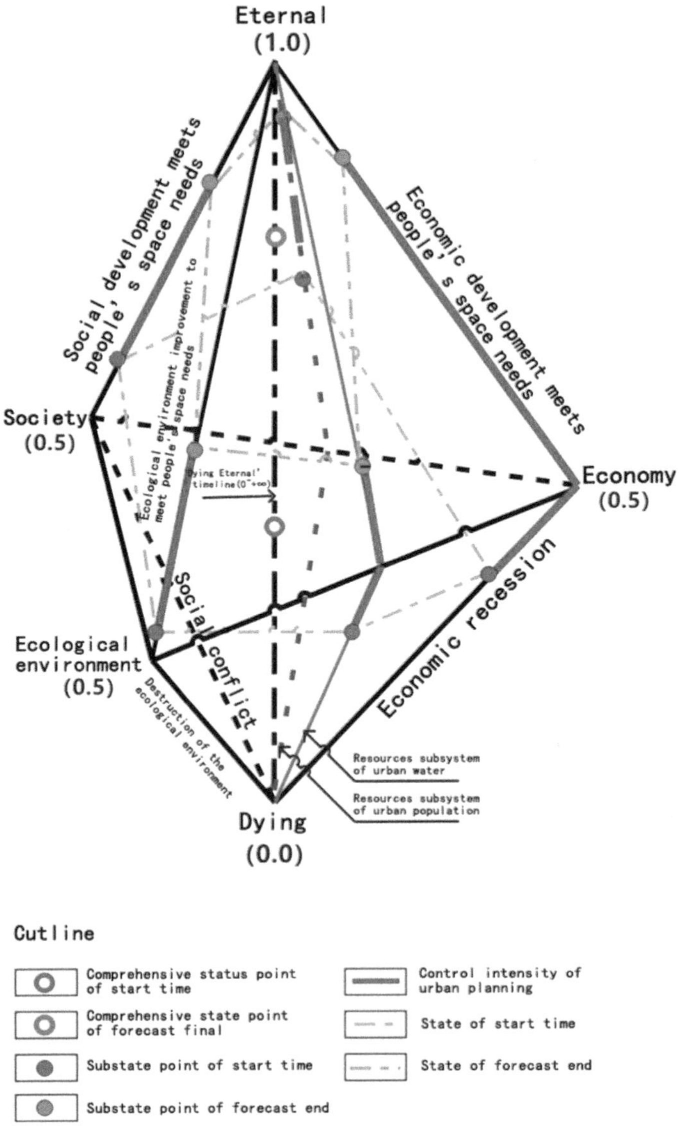

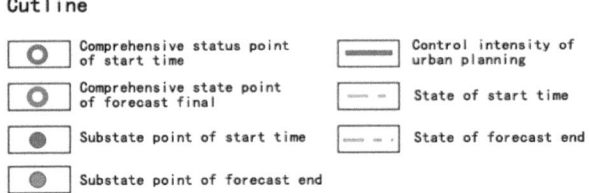

Fig. 5. Evaluation model of urban sustainability level in Xinjiang from 2017 to 2035

4 Conclusion and Prospect

Urban planning management is more about shifting, adjusting, and controlling land use development indicators. This model is particularly suitable for complex and dynamic resource system coordination problems. It can conduct real-time audits of current construction status and future sustainability in order to flexibly regulate the ideal urban sustainability status and formulate differentiated policies. These indicators indirectly assess urban land use by showing the state of urban sustainability. In particular, the indicators for each subsystem can be defined flexibly according to the specific situation of each city, thus clarifying the particular causal relationships between subsystems in particular cities.

The model established in this paper starts from urban land allocation and emphasizes the connection between the goals of managers and the urban support system in terms of sustainability. The urban sustainability theory model is still a prototype, demonstrating the original conceptual principles, but it faces limitations in practice due to the different weights of subsystems in exceptional city support systems. This can be achieved by regularly updating the subsystems and their indicators to determine certain evolutionary trends. More importantly, the causal relationships presented in each system can be simulated, which means that mathematical equations are needed to model them. The technical basis for the demonstration of this model has been verified in my doctoral thesis. Therefore, it provides a construction idea for translating urban sustainability values expressed through land allocation indicators into concrete actions.

References

1. Saguin, K.K.C., Chanco, C.J., Tan, A.I.S., Ortega, A.A.C.: Reclaiming social equity in land use planning for sustainable cities. Public Policy J. **18**, 99–126 (2017)
2. Repetti, A., Desthieux, G.: A relational indicatorset model for urban land-use planning and management: methodological approach and application in two case studies. Landsc. Urban Plan. **77**(1), 196–215 (2016)
3. Yujie, Jia, et al.: Land use performance evaluation based on attribute weight optimization algorithm and analysis of land use performance differentiation. J. Zhejiang Univ. (Science Edition) (2019)
4. Tom, K., John, F.: What is sustainability? Sustainability **2**, 11 (2010)
5. Wang, X., Zhang, Y., Zhang, J., et al.: Progress in urban metabolism research and hotspot analysis based on citespace analysis. J. Clean. Prod. **281**, 125224 (2021)
6. Atkisson, A.: Developing indicators of sustainable community: lessons from sustainable Seattle. Environ. Impact Assess. Rev. **16**(4), 337–350 (1996)
7. Ameen, R.F.M., Mourshed, M.: Urban sustainability assessment framework development: The ranking and weighting of sustainability indicators using analytic hierarchy process. Sustain. Cities Soc. **44**, 356–366 (2019)
8. Giampietro, M., Mayumi, K., Ramos-Martin, J.: Multi-scale integrated analysis of societal and ecosystem metabolism (MuSIASEM): theoretical concepts and basic rationale. Energy **34**(3), 313–322 (2009)
9. Jin, T., Tang, M.: Sustainable development: three models from goal to practice. Urban Plan. J. **01**, 86–89 (2005)

Investigating Urban Resilience Differences and Barriers to Regional Coordinated Development

Weimin Deng[1,3]([✉]), Wenjun Zhang[2,3], and Li Yan[1]

[1] School of Civil Engineering and Architecture, Southwest University of Science and Technology, Mianyang 621010, China
dengweimin@mails.swust.edu.cn
[2] School of Environment and Resource, Southwest University of Science and Technology, Mianyang 621010, China
[3] Tianfu Institute of Research and Innovation, Southwest University of Science and Technology, Chengdu 610299, China

Abstract. The article evaluates and analyses urban resilience and constraining factors in 30 provinces in China derived from panel data from 2006–2021 using entropy weighting, Gini coefficient and handicap model. The study shows that the current level of resilience in most cities is relatively low, with an overall upward trend, and the spatial distribution shows a decreasing 'high east, low west' pattern, While the majority of the nation's low-resilient communities are found in the west and center, and high resilient towns are mainly found in the east. Concerning the decomposition of regional differences, the eastern region has the largest intra-regional difference and the smallest increase, the central region has a smaller difference and the largest increase, the northern difference shows a decreasing trend and the western difference is increasing year by year; in terms of inter-regional averages, the east-west difference is the largest and the north-central difference is the smallest. When analysing the level of barriers, the level of barriers to business resilience is decreased and the degree of barriers to infrastructural resilience is enhanced.; the number of fixed internet users, revenue from local budgets and local education expenditure emerge as important factors in increasing urban resilience, but the level of barriers varies across regions. Using the conclusions of this paper, the article proposes policy advice for enhancing urban resilience in terms of strengthening the connectivity of regional cities, optimising the allocation of resources in cities, and adhering to multi-system coordination of cities.

Keywords: Resilience of urban areas · development over time and space · regional differences · barrier factors

1 Introduction

With China's accelerating globalisation and opening to the outside world, cities are growing more susceptible to various uncertainties, such as abrupt public health problems and warming temperatures. The development of urban resilience reflects the capacity

B.-J. He et al. (Eds.): UCSUD 2023, LNCE 559, pp. 745–760, 2025.
https://doi.org/10.1007/978-981-97-8401-1_54

of cities to weather and recover shocks, and is crucial to urban development. In the Recommendations for the 14th Five-Year Plan and the Vision to 2035, China clearly emphasises the building of safe, healthy and resilient cities. [1].

The Latin word "resilio" (which means "to return to the primitive state") is where the word "resilience" first appeared. [2]. In 1973, Holling introduced resilience to the ecological arena, suggesting 'ecological resilience', which is of great significance. [3]; Haixing Meng believes that urban resilience is a kind of defensive ability of the city to adapt and cope with change, and it is a kind of attribute inherent to the city [4]. Yiyan Duan et al. believe that urban resilience includes both the ability of cities to keep their basic state after change or disturbance, and the capacity of city systems for learning, adaptation and self-organisation [5]; The characteristics and determinants of China's regional economic resilience have been analysed by Tan Juntao and Hongbo Zhao et al. [6];Gang Chen and Han Zhang et al. assessed the ecological resilience of the Hancang River [7]; and Jiayi Ding and Lin Wang et al. applied the theory of urban resilience to public health construction and put forward the idea of public health safety construction [8]. Hong Liu and Lingli Jia applied urban resilience theory to the field of urban planning [9], and Jiayi Zhang and Yan Zhou et al. applied urban resilience theory to the field of landscape [10].Xicheng Pan used Terrell's index and natural fracture method to analyse urban resilience variation and evolution in Hubei Province [11];Spatial and temporal variation in urban resilience assessment by constructing assessment indicators under dual climate objectives by Lingna Liu [12]; Zhen Ma's study, which takes an urban resilience stance, examines the connection between smart city development and high-quality development in China. [13]; Fei Chen 's study on the resilience of Yangtze Delta cities through three main elements of density-distance-segmentation [14]; Shiju Zhang's evaluation of Wuhan's urban resilience construction level using the topsis entropy weight method [15].

According to the research on city resilience by scholars at home and abroad, most current research areas are small, and nationwide study on urban regions is lacking; the majority of studies assess the level of city resilience, and few of them involve the search of regional differences and influencing factors. Therefore, this paper takes the four major regions in China as the focus, sets up the scoring model, performs the calculation, analyses the regional differences, dynamic development and obstacle factors, and provides scientific decision-making reference for the growth of urban resilience.

2 Study Area, Data and Methodology

2.1 Study Area

The survey region is the centre of each province, autonomous region as well as municipalities directly subordinated to the central government.; the central city is the best equipped representative for economic, social, infrastructural and environmental resilience of each administrative region, and at the same time the 30 cities are divided into four major regions for the study, namely East, Central, West (excluding Tibet Autonomous Region) and North. (Fig. 1).

2.2 Data Acquisition

The China Urban Statistical Yearbook, the China city Construction Statistical Yearbook, the National Economic and Socio-Development Statistical Report, as well as the statistical yearbooks and reports of the provinces and municipalities, provided the statistical data, which were based on data from 2006 to 2021. Interpolation was used to complete and correct individual missing and erroneous data.

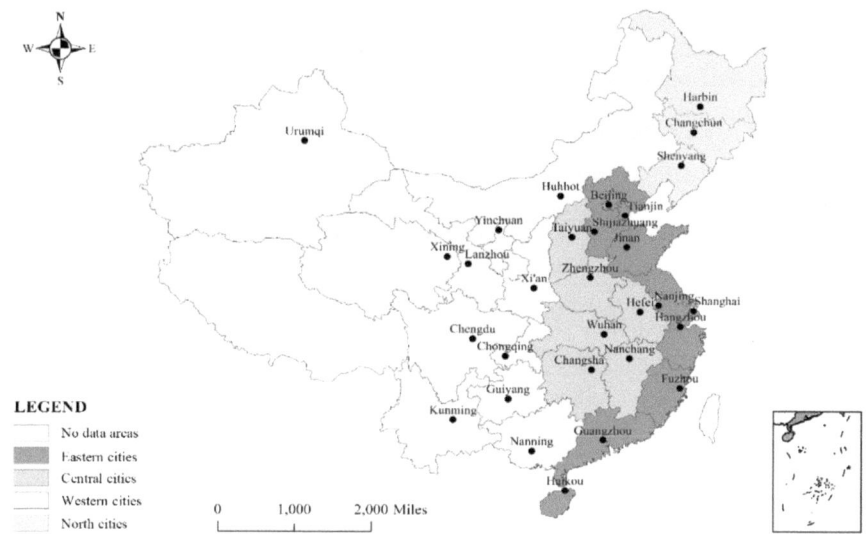

Fig. 1. Study area

2.3 Setting Up the System of Indicators and Research Methodology

In order to construct the urban resilience indicator system in terms of urban resilience features and constitutive urban resilience, this paper chooses four aspects of economic resilience, social resilience, ecological resilience, and infrastructural resilience based on a review of the literature and a reference to the nationally published Safety Resilience City Evaluation Guide (Table 1). Twenty assessment indicators are chosen as an organizing system for assessing urban resilience based on indicator selection concepts such as representativeness, scientificity, and systematicity.

2.3.1 The Method of Entropy Weight

Entropy power technique, as a kind of objective allocation method, can avoid the interference of subjectivity to a certain extent, in addition to different years, so that the urban resilience assessment is more accurate and comprehensive, the specific allocation is as follows:

Table 1. System for the evaluation of indexes

Target Plane	Normative Layer	Indicator Level	Meaning of indicator (Indicator type)	References
Urban resilience	Business resilience	(D1) GDP per head(CNY)	Levels of economic development (+)	(Xun X, Yuan Y[16];Wang B [17]; Xiong Y [18]; Shi C, Zhu X [19]; Li D, Yang W, Huang R. [20]; Han S [21])
		(D2) The tertiary sector's gross value added as a percentage of GDP (%)	Level of economic structural soundness (+)	
		(D3) Total investment in fixed assets (CNY)	The investment intensity of the national economy (+)	
		(D4) GDP growth rate (%)	Level of growth of the economy (+)	
		(D5) Revenue from the general budget of the local government (CNY)	Financial independence (+)	
	social resilience	(D6) Local government expenditure on education and training (CNY)	Intensity to invest in education (+)	
		(D7) Population density of cities (Person/km^2)	Size of urban population (-)	
		(D8) Number of participants in the urban basic health insurance scheme (person)	Ability to improve social security (+)	

(continued)

Table 1. (*continued*)

Target Plane	Normative Layer	Indicator Level	Meaning of indicator (Indicator type)	References
	ecological resilience	(D9) Number of students enrolled in higher education (person)	Socio-educational level (+)	
		(D10) Hospital beds per 10,000 population (person)	Capacity for socio-medical security (+)	
		(D11) Green area coverage in housing estates (%)	Level of urban greening (+)	
		(D12) Park area per capita (m²/Person)	Urban environmental protection level (+)	
		(D13) SO_2 emissions from industry (t)	Urban emissions pressure (-)	
		(D14) Rate of waste water treatment (%)	Capacity to re-develop urban (+)	
		(D15) Non-hazardous municipal waste treated (%)	Capacity for municipal waste (+)	
	Infrastructure resilience	(D16) Daily domestic water consumption (L)	Levels of urban water consumption (+)	
		(D17) Fixed internet users in the city (households)	Capacity for city outreach (+)	
		(D18) Penetration rate of urban gas (%)	City gas supply capability (+)	
		(D19) Urban roads per capita (m²/ Person)	Urban traffic capacity (+)	
		(D20) Density of sewers in	Municipal wastewater capacity	

(1) Construct the initial matrix. Construct a matrix with m objects and indicators: $X = [x_{ij}]_{m \times n}$

(2) Standardisation of indicator data.

Positive indicator:

$$x_{ij}' = \frac{x_{ij} - min(x_{ij})}{max(x_{ij}) - min(x_{ij})} \tag{1}$$

Negative Indicator:

$$x_{ij}' = \frac{max(x_{ij}) - x_{ij}}{max(x_{ij}) - min(x_{ij})} \tag{2}$$

(3) Weight of indicators:

$$P_{ij} = X_{ij}' / \sum_{i=1}^{n} X_{ij}' \tag{3}$$

(4) Indicator entropy e_j:

$$e_j = -k \left[\sum_{i=1}^{n} P_{ij} \ln(P_{ij}) \right] \tag{4}$$

(4) Determining the weighting of evaluation indicators:

$$w_j = (1 - e_j) / \sum_{j=1}^{m} (1 - e_j) \tag{5}$$

(5) Calculate your overall score S_i

$$S_i = \sum_{j=1}^{n} w_j \cdot P_{ij} (i = 1, 2, \cdots, m) \tag{6}$$

Following the construction and data collection for the Urban Resilience Index, an entropy weighing procedure is used to generate the weights for the Urban Resilience Assessment Indicators (Table 2).

2.3.2 Gini Coefficient

The causes of the spatial imbalance problem can be looked at applying the Gini coefficient breakdown approach. By calculating the overall Gini coefficient G, and decomposing it into three parts, G_w, G_b, G_t, this paper examines the differences in the contribution rates of the above three categories, exposing the makeup and causes of variations in a region's degree of urban resilience. The formula is as follows:

$$G = \frac{\sum_{j=1}^{k} \sum_{h=1}^{k} \sum_{i=1}^{n_j} \sum_{r=1}^{n_h} |y_{ji} - y_{hr}|}{2n^2 \bar{y}} \tag{7}$$

$$G_w = \sum_{j=1}^{k} G_{jj} p_j s_j \tag{8}$$

Table 2. Weighting of indicators to assess urban resilience

Target Plane	Normative Level	weight	Indicator Layer	weight
Urban resilience	economic resilience	0.2881	GDP per person (CNY)	0.0432
			The tertiary sector's gross value added as a percentage of GDP (%)	0.0305
			Total investment in fixed assets (CNY)	0.0867
			GDP growth rate (%)	0.0059
			Revenue from the general budget of the local government (CNY)	0.1218
	Social resilience	0.3332	Local government expenditure on education and training (CNY)	0.1106
			Population density of cities (Person/km^2)	0.0141
			Number of participants in the urban basic health insurance scheme (person)	0.1164
			Number of students enrolled in higher education (person)	0.0355
			beds in hospitals for every 10,000 people (person)	0.0566
	ecological resilience	0.0569	Coverage of green space in residential areas (%)	0.0084
			Park area per capita (m^2/Person)	0.0296
			SO$_2$ emissions from industry (t)	0.0047
			Rate of waste water treatment (%)	0.0083
			Non-hazardous municipal waste treated (%)	0.0059
	Infrastructure resilience	0.3217	Non-hazardous municipal waste treated (L)	0.0099
			Fixed internet users in the city (households)	0.245

(*continued*)

$$G_{nb} = \sum_{j=2}^{k} \sum_{h=2}^{j-1} G_{jh}\left(p_j s_h + p_h s_j\right) D_{jh} \tag{9}$$

Table 2. (*continued*)

Target Plane	Normative Level	weight	Indicator Layer	weight
			Penetration rate of urban gas (%)	0.0061
			Urban roads per capita (m²/Person)	0.0279
			Density of sewers in built-up areas (km/km²)	0.0328

$$G_t = \sum_{j=2}^{k} \sum_{h=1}^{j-1} G_{jh}\left(p_j s_h + p_h s_j\right)\left(1 - D_{jh}\right) \tag{10}$$

where y is a measure of city resilience; n is the overall count of cities ($n = 30$) and n_j is what proportion are cities in region j.; $p_j = \frac{n_j}{n}$, $s_j = \frac{n_j \overline{y_i}}{n \overline{y}}$, $j = 1, 2, \cdots, k$; D_{jh} is the relative influence between regions j and h.

2.3.3 Handicap Model

The hurdle factors influencing the enhancement of urban resilience are identified through the construction of the barrier degree model. The maximum barriers factors are identified by sorting the hurdle factors in order to determine the degree to which each obstacle factor has an effect on the growth of urban resilience.

$$O_j = \frac{\omega_j \left(1 - R_{\alpha ij}\right)}{\sum_{1 \omega j}^{m} \left(1 - R_{\alpha ij}\right)} \tag{11}$$

$$U = \sum O_j \tag{12}$$

where O_j is the obstacle degree for a given indicator j; ω_j is indicator weight; $R_{\alpha ij}$ is the standardised scalar value; and U is the criterion layer value.

3 Results

3.1 Urban Resilience Level Measurement Results

3.1.1 Urban Resilience Levels

The present research chooses the comprehensive score and ranking of city resilience exhibited at four time points: 2006, 2011, 2016, and 2021. It does so by using the entropy weighting method to determine the comprehensive score of urban resilience degree.

Using the interval classification method, this paper classifies the urban resilience level score (0, 1) in five levels, less than 0.2 for low resilience level, [0.2–0.4] for general resilience level, [0.4–0.6] for medium resilience level, [0.6–0.8] for higher resilience level, and [0.8–1] for high resilience level. Following the classification of city resilience stages, this research explores the features of the evolution of urban resilience over time and space.

Table 3. Urban Resilience Composite Score Scale

Region	City	2006		2011		2016		2021	
		Figures	Rank	Figures	Rank	Figurs	Rank	Figurs	Rank
North	Shenyang	0.279	8	0.325	8	0.208	20	0.193	17
	Changchun	0.222	15	0.264	18	0.216	18	0.179	24
	Harbin	0.207	18	0.269	16	0.221	17	0.167	26
East	Beijing	0.647	2	0.675	1	0.725	2	0.630	2
	Tianjin	0.397	3	0.486	4	0.485	4	0.328	7
	Shijiazhuang	0.209	17	0.285	13	0.221	16	0.188	20
	Shanghai	0.744	1	0.664	2	0.734	1	0.689	1
	Nanjing	0.303	7	0.318	10	0.348	8	0.308	8
	Hangzhou	0.305	6	0.339	6	0.373	6	0.370	5
	Fuzhou	0.192	21	0.240	21	0.208	19	0.210	14
	Jinan	0.256	10	0.301	11	0.270	11	0.237	13
	Guangzhou	0.359	4	0.427	5	0.442	5	0.588	3
	Haikou	0.198	19	0.234	23	0.202	21	0.181	23
Centre	Taiyuan	0.182	24	0.242	19	0.198	23	0.185	21
	Hefei	0.227	14	0.240	20	0.233	15	0.241	12
	Nanchang	0.194	20	0.205	28	0.173	26	0.191	18
	Zhengzhou	0.227	13	0.267	17	0.269	12	0.247	11
	Wuhan	0.256	11	0.321	9	0.362	7	0.299	9
	Changsha	0.228	12	0.278	14	0.283	10	0.258	10
West	Huhehot	0.217	16	0.273	15	0.244	14	0.162	27
	Nanning	0.184	23	0.205	27	0.158	28	0.203	16
	Chongqing	0.318	5	0.579	3	0.532	3	0.420	4
	Chengdu	0.259	9	0.339	7	0.331	9	0.358	6
	Guiyang	0.175	26	0.226	25	0.190	25	0.182	22
	Kunming	0.157	28	0.239	22	0.199	22	0.190	19
	Xian	0.189	22	0.285	12	0.254	13	0.208	15
	Lanzhou	0.127	30	0.195	29	0.170	27	0.135	30
	Xining	0.181	25	0.161	30	0.120	30	0.157	28
	Yinchuan	0.145	29	0.232	24	0.156	29	0.147	29
	Urumqi	0.163	27	0.214	26	0.197	24	0.167	25

3.1.2 Characterising Urban Resilience in Space and Time

According to (Table 3), the development level of urban resilience in the research area has continuously increased, with the growth law decreasing from east to west, with the high level mainly in the east. With the development strategy focusing on the west, cities with high levels of resilience have also emerged in the west. In terms of the timeline, the water resilience of each city increased from year to year in 2006, 2011 and 2016, but the resilience levels for the remaining cities fell by 2021, with the exception of a few cities that maintained a slight upward trend. This is because the 2021 New Crown Epidemic will have a significant impact on all aspects relating to economic and social growth in cities across the country, disrupting the functioning in integrated urban systems and affecting the development of resilient levels of urbanisation. Tianjin, Beijing and Shijiazhuang are the most affected, with declines of 32.37%, 13.10%t and 14.93% respectively in comparison with 2016, while Chongqing in the west also fell by 21.05%. This paper analyses the top 15 cities. The majority of cities are located in the east, followed by the centre, then the west and finally the north, with a largely stable ranking pattern. As a result, there are important differences to a certain extent urban resilience between regions, with the regional standard of resilience development showing a trend of East > Central > West > North.

Influenced by geographical location, degree of economic and social development and other factors, urban resilience is highly heterogeneous in its spatial distribution. This paper, based on ArcGIS 10.2, spatially characterises urban resilience levels for four time periods (Fig. 2). As can be seen from the figure, in 2006, the higher levels were only found in the eastern part of the country, in Beijing and Shanghai, While the lower levels were concentrated to the west, the central and northern sections were at general level.; in 2011, the resilience of cities across the country improved, with the exception of Xining, Lanzhou and Xi'an, of which Tianjin, Guangzhou and Chongqing improved to a medium level. In 2016, the resilience level of cities increased significantly, with all cities in the north, east and centre rising above the medium level, and the remaining cities in the west, except Xining, rising. In 2021, except for a few cities that maintained their original resilience level rating, all other cities declined, with the most severe decline in the west and north. This indicates that when cities are affected by disasters, their prevention and resilience performance is seriously out of line with their resilience development level.

3.2 Results of Result of the Gini Coefficient

The above study shows that spatial heterogeneity exists in the urban resilience level. To further analyse the local disparities and causes of city resilience in four selected time nodes, the Gini coefficient decomposition method is applied to measure the intra-zone, interval and contribution rate of the four time nodes. As shown in the result (Table 4).

See (Table 4), the overall difference coefficient of city resilience levels shows a "falling-rising-falling" trend, indicating that regional differences in urban resilience levels are changing in a wave-like manner, and that although the coefficient is slightly higher in 2021 than in 2006, the difference is getting larger and larger according to the overall coefficient development trend. According to the Gini coefficient and the decomposition method, the total variance with respect to the strength of urban resilience is

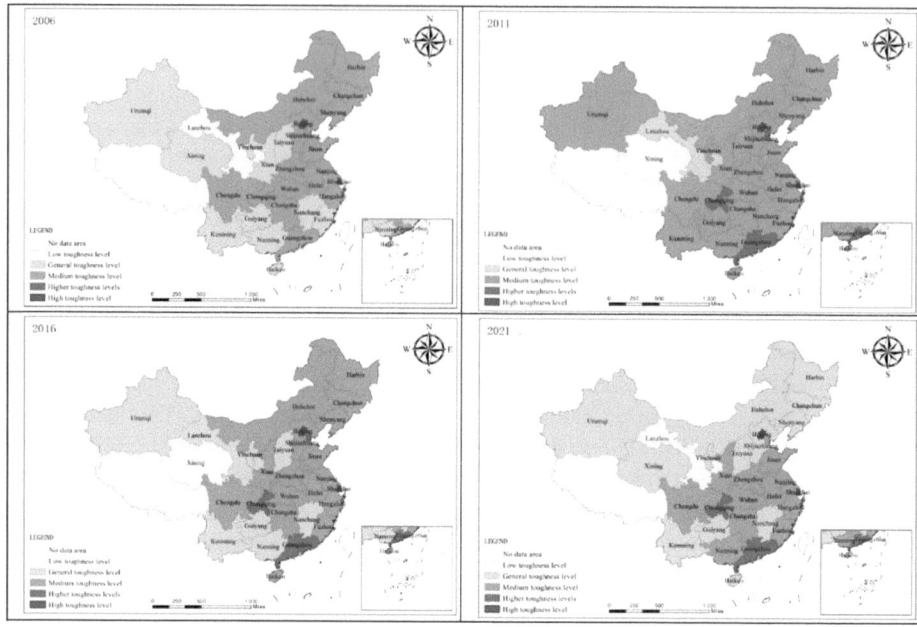

Fig. 2. Levels of resilience in urban areas

Table 4. Decomposition and sources of regional disparities in resilience

Year		2006	2011	2016	2021
Umbrella		0.229	0.200	0.259	0.256
Decomposition/ Contribution rate	G_w	0.057	0.055	0.069	0.066
	G_w (%)	24.90%	27.19%	26.46%	25.73%
	G_b	0.148	0.098	0.136	0.147
	G_b (%)	64.39%	48.98%	52.58%	57.48%
	G_t	0.025	0.048	0.054	0.043
	G_t (%)	10.71%	23.83%	20.96%	16.79%
G_w	North	0.068	0.047	0.013	0.032
	East	0.259	0.210	0.256	0.267
	Centre	0.060	0.077	0.136	0.091
	West	0.142	0.187	0.228	0.197

mainly derived originating in three components of the G_w, G_b and G_t variances. Over the sample period, G_w increased slightly, G_b remained essentially unchanged and G_t increased. In terms of contribution rate, the average contribution rate of $G_w(\%)$, $G_b(\%)$ and $G_t(\%)$ variances in 2021 is 25.73%, 57.48% and 16.79% respectively. As a result, the total sources of variance are, in order, $G_b(\%)$, $G_w(\%)$ and $G_t(\%)$. Therefore, reducing the spatial variance in urban resilience must start with reducing the inter-regional, regional and inter-regional cross-over variance together.

According to (Table 4), In the North, it starts to fall in 2006 and tends to bottom out in 2016, rising to 0.032 in 2021; The east had the smallest difference in 2011 and then increased each year, while the centre and west continued to grow, reaching their highest levels in 2016; In addition, the intra-regional differences are greater in the East than in other regions, suggesting that city resilience is more uneven in the East compared to other regions, with increasing differences and greater unevenness in the West.

According to (Table 5), there are large differences in the evolutionary trends of the coefficients between the regions, with the exception of the North-West, which will decrease by 3.97% between 2006 and 2021, and increasing differences between the regions, with the fastest growth rates in the North-East and West-Central, with growth rates of 41.37% and 40. 46% respectively; the difference between the North and the Centre increased by a factor of 1.8, and although the coefficient was the smallest, the difference increased the fastest; according to the mean value of the interregional coefficients, East-West is the largest with a measure of 0.315 and North-Central the bottom with an assessment of 0.104, indicating that East-West has the largest difference and North-Central the smallest. The focus of reducing regional disparities should therefore be on advancing the resilience of western towns and closing the resilience gap between regions.

In a word, the growing differences in the evolution of city resilience levels are mainly influenced by the contribution of intergroups, with the gap within each region showing a widening trend, except in the north, where the gap is narrowing; the interregional gap is widening fastest, although it is smallest in the north-central part of the country and largest in the east-west part of the country, where the gap is basically unchanged.

Table 5. Decomposition factor of the G_b

Year	2006	2011	2016	2021	Average
North-East	0.249	0.198	0.308	0.352	0.277
North-Central	0.076	0.077	0.123	0.140	0.104
North-West	0.151	0.152	0.166	0.145	0.154
Eastern-Central	0.269	0.230	0.279	0.274	0.263
East-West	0.327	0.262	0.335	0.335	0.315
West-Central	0.131	0.145	0.201	0.184	0.165

3.3 Identification of Barriers to Urban Resilience Factors

The impact of evaluation indexes on regional urban resilience is assessed applying the barrier degree model, and the top five hindrance factors influencing the development of urban resilience level are selected for analysis in each year and region. It can be observed from (Table 6) that the barrier to the improvement of urban resilience from 2006 to 2021 is constant change. In 2006, local financial general revenue (D5) and total fixed asset investment (D3) were the main obstacle factors, followed by local education expenditure (D6), GDP per person (D1), and the proportion of tertiary industry added value to GDP (D2). There are four indicators in the economic resilience criterion layer, indicating that economic resilience has an important effect on the development of urban resilience. In 2011, local financial general revenue (D5) and local education expenditure (D6) became the main obstacle factors, followed by the number of urban basic medical insurance enrollees (D8), the number of 10,000 college students (D9), the number of 10,000 hospital beds (D10), the social resilience of 3, and the economic resilience of 2, indicating that the impact level of economic resilience declined. The impact of social resilience has increased. In 2016, at the level of economic resilience, local financial general revenue (D5), total investment in fixed assets (D3), local education expenditure at the level of social resilience (D6), the number of urban basic medical insurance participants (D8), and the number of fixed Internet users in the city at the level of infrastructure resilience (D17).It shows that the factors affecting the level of urban resilience are diversified, and economic resilience and social resilience are the main obstacle factors at present. In 2021, the number of fixed Internet users (D17) and local education expenditure (D6) in the city ranked the top two, followed by rainage pipe density in built-up area (D20), the number of urban basic medical insurance participants (D8), and the general revenue of local finance (D5), indicating that infrastructure and education have become the main influencing factors, and the economic impact has decreased year by year. As a result, when urban resilience increases, the effect of the economy on resilience starts to decline, the influence of society and infrastructure on resilience increases, and the impact of ecology on resilience diminishes. There is basically no change in the barrier criterion level between regions, only the change of barrier factors, indicating that the process affecting urban resilience development is the same between regions.

Table 6. Key barriers to urban resilience ranked

Area	2006	2011	2016	2021
North	D5, D3, D6, D1, D2	D5, D6, D8, D10, D2	D5, D6, D8, D17, D3	D17, D6, D20, D8, D5
East	D5, D3, D4, D6, D2	D5, D6, D8, D9, D3	D5, D6, D8, D3, D10	D17, D6, D20, D8, D5
West	D5, D3, D1, D6, D2	D5, D6, D8, D10, D9	D5, D6, D8, D1, D3	D17, D6, D8, D20, D5
Centre	D5, D3, D6, D1, D2	D5, D6, D8, D9, D2	D5, D6, D8, D17, D3	D17, D6, D8, D20, D5

4   Conclude and Recommend

4.1   Conclude

The research quantifies the quality of urban resilience, looks at the spatio-temporal growth characteristics and regional disparities of urban resilience, and looks into the factors that inhibit the creation of urban resilience and their spatio-temporal heterogeneity. It relies on survey data which was collected from 30 Chinese cities between 2006 and 2021. The research indicates that urban resilience has significantly improved with the continuous advancement of economic and social development. It has a trend of progression from low to high resilience, while It has a pattern of increasing resilience from low to high decreases from east to west, exhibiting notable spatial disparities. The gap in regional development levels is narrowing, leading to a reduction in spatial differences and developmental imbalances in urban resilience. Although variations exist among regions, the disparity trend is diminishing. Among the factors hindering urban resilience, there is a shift from economic resilience to infrastructure resilience as the primary obstacle layer, with the main inhibiting factor transitioning from local budget revenue to the number of fixed Internet users within cities. Furthermore, local education spending has an important effect on urban resilience. The influence of different obstacle factors varies across regions due to distinct geographical locations and levels of economic and social development. Strengthening regional coordinated development by actively leveraging government leadership, prioritizing infrastructure construction planning, and enhancing comprehensive coordination capabilities are effective approaches for improving urban resilience.

4.2   Suggested

Drawing from the studies discussed, this paper proposes the following strategies for the urban sustainability: (1) Promote resource and spatial complementarity between regional cities. Cities have formed spatial complementarity and resource complementarity with each other according to their own resource endowment and characteristics, and in the meantime strengthened the radiating role of high-level urban resilience to form spatial complementarity, and secondly strengthened the construction of urban economic circle, where different regions and cities have different resources and exchange resources with each other to form resource complementarity. (2) Strengthen regional cooperation and interlinked development, reduce regional development disparities and promote common development. It insists on complementing each other's strengths, developing special and advantageous industries, perfecting the mechanism of cooperation in supporting each other, constantly improving the level of regional cooperation, promoting interregional coordination and interaction in infrastructure, industrial development, ecological and environmental protection, etc., and using its own strengths to strengthen cooperation and promote common development. (3) Leverage the government's leadership role in formulating city growth strategies that are tailored to regional conditions. The development levels of city resilience discriminate markedly across different regions, and the effects of obstacle factors are heterogeneous in their impact on these areas. Thus, each city should fully leverage the leading role of its government to accurately assess its own level of resilience and to develop planning strategies that are tailored to the specific focuses of different cities. (4) Focus on infrastructure development planning and strengthen overall coordination capacity. Infrastructure represents the overall strength of a region, and improving infrastructure construction capacity helps cities cope with risky disasters, while fully playing the role of the government and society, mobilising social resources to cooperate with the government, and making concerted efforts to build a resilient and efficient urban infrastructure system to promote urban resilience.

References

1. Bo, L.: Building a new pattern of land space development and protection. Decis. Inf. **12**, 30–31 (2020)
2. Chen, X., Wang, P.: Progress of resilient city research. World Earthq. Eng. **34**(3), 78–84 (2018)
3. Holling, CS.: Resilience and stability of ecological systems. Annu. Rev. Ecol. Syst. (4), 1–23 (1973)
4. Meng, H., Shen, Q.: The concept, characteristics and optimization of the resilience of megacities. Urban Dev. Res. **28**(7), 75–83 (2021)
5. Duan, Y., Zhai, G., Li, W.: Progress in international research on urban resilience measurement. Int. Urban Plann. **25**(18), 1–10 (2021)
6. Tan, J., Zhao, H., Liu, W., et al.: Analysis of the characteristics and influencing factors of China's regional economic resilience. Geoscience **40**(2), 173–181 (2020)
7. Chen, G., Zhang, H., Wang, L.: Evaluation of the suitability of restoring water ecological resilience in the Hancang River Basin. Environ. Prot. Sci. **47**(1), 71–75 (2021)
8. Ding, J., Wang, L., Yu, Q., et al.: Reflections on public health and safety construction in old urban communities based on resilience theory. Urban Arch. **17**(12), 14–15+29 (2020)
9. Liu, H., Jia, L.: Reflections on urban green space system planning based on resilience theory. Urban Arch. **18**(20), 167–169 (2021)
10. Zhang, J., Zhou, Y., Fan, L.: Research on landscape design strategy of Yanbian residential area from the perspective of resilience theory. J. Agron. Yanbian Univ. **44**(1), 89–96 (2022)
11. Pan, Y.C., Tian, J.H.: Spatiotemporal variations of urban resilience and their influencing factors: an empirical study in Hubei Province. Stat. Decis. **39**(23), 57–62 (2023). https://doi.org/10.13546/j.cnki.tjyjc.2023.23.010
12. Liu, L.N., Lei, Y.L., Zhang, W.Y., et al.: Resilience development of resource-based cities under the "dual carbon" goals. Geol. Bull. China, 1–15 (20 March 2024). http://kns.cnki.net/kcms/detail/11.4648.P.20230510.1945.004.html
13. Ma, Z.: The impact of smart city construction on high-quality economic development: an analysis from the perspective of urban resilience. East China Econ. Manage. 1–11 (20 March 2024). https://doi.org/10.19629/j.cnki.34-1014/f.230909015
14. Chen, F., Ma, X.Q., Li, Y.H.: The logical construction and empirical analysis of urban resilience in the Yangtze River Delta Region from the perspectives of density, distance, and division. Scientia Geographica Sinica **44**(2), 248–257 (2024). https://doi.org/10.13249/j.cnki.sgs.20221557
15. Zhang, S.J., Li, Q., Yu, S.W.: Evaluation and enhancement strategies for urban resilience construction in Wuhan City. Saf. Environ. Eng. 1–11 (20 March 2024). https://doi.org/10.13578/j.cnki.issn.1671-1556.20230402
16. Xun, X., Yuan, Y.: Research on the urban resilience evaluation with hybrid multiple attribute TOPSIS method: an example in China. Nat. Hazards **103**(1), 557–577 (2020)

17. Wang, B., Han, S., Ao, Y., et al.: Evaluation and factor analysis for urban resilience: a case study of Chengdu-Chongqing urban agglomeration. Buildings **12**(7), 962 (2022)
18. Xiong, Y., Li, C., Zou, M., et al.: Investigating into the coupling and coordination relationship between urban resilience and urbanization: a case study of Hunan Province, China. Sustainability **14**(10), 5889 (2022)
19. Shi, C., Zhu, X., Wu, H., et al.: Assessment of urban ecological resilience and its influencing factors: a case study of the Beijing-Tianjin-Hebei urban agglomeration of China. Land **11**(6), 921 (2022)
20. Li, D., Yang, W., Huang, R.: The multidimensional differences and driving forces of ecological environment resilience in China. Environ. Impact Assess. Rev. **98**, 106954 (2023)
21. Han, S., Wang, B., Ao, Y., et al.: The coupling and coordination degree of urban resilience system: a case study of the Chengdu-Chongqing urban agglomeration. Environ. Impact Assess. Rev. (2023)

Allocation of Public Service Facilities Based on Community Life Circles: A Case Study in Mianyang, China

Juanlin Fu[1(✉)], Li Yan[1], Chunrong Zhao[1], Minghong Yu[1], Yabin Zhang[2], Jiayue Yin[1], and Wenyan Feng[1]

[1] School of Civil Engineering and Architecture, Southwest University of Science and Technology, Mianyang 621000, China
fujuanlin1227@swust.edu.cn
[2] Chongxin County Urban Management Comprehensive Law Enforcement Bureau, Pingliang 744200, China

Abstract. The allocation of public service facilities based on community life circle is the main measure of urban social governance at present. Although there are many researches on the allocation of public service facilities, the evaluation on different levels of community life circle is still insufficient. Taking Mianyang city of Sichuan Province as an example, based on big data such as city blocks, roads and POI, the different levels of community life circles are delineated and their public service facilities are evaluated. The results show the following: (1) Combined 15, 10 and 5 min walking time with roads and administrative boundaries, three levels of community life circle is defined comprehensively. (2) The public service facility coverage rate and community life circle standard rate have gradually decreased from a 15-min community life circle(15min-CLC) to a 10-min community life circle(10min-CLC) and then a 5-min community life circle(5min-CLC). The unevenly distributed basic public service facilities, such as secondary schools and hospitals, mainly accounts for the low rates. (3) All categories of public service facilities present remarkable spatial auto-correlation, which shows on a characteristic transform from "core-periphery" to "one-core multi-center" distribution pattern with the decrease of the community life circle level. The study revealed the spatial distribution characteristics of urban public service facilities under different community life circle levels, and provided theoretical basis for optimizing the allocation of public service.

Keywords: Community Life Circle · Public Service Facilities · Spatial Distribution · POI · Mianyang City

1 Introduction

The configuration level of urban public service facilities, as an important part of residents' daily life, directly affects the life quality of residents. A well-configured public facility promotes social equity; improves the quality of urban construction; promotes residents' happiness, sense of access, and satisfaction. There is a long history of studying public service facilities. The existing studies mainly focus on accessibility through urban population distribution [1, 2],how people travel [3], the service radius of the facilities [4], and the difficulty of reaching the facilities [5, 6]; location issues for urban schools [7], medical and health facilities [8], parking lots [9], public sports facilities [10], and others [11]; and the equity and equalization of medical [12, 13], educational [2], sports [14], and other facilities [15]. With the creation of a 15min-CLC proposed in Shanghai's master plan in 2014, it has become important to investigate the configuration of public service facilities using the community life circle as the basic unit. The community life circle is a living unit that meets the material and living and cultural needs of residents within a suitable daily walking range.It advocates a convenient, healthy and low-carbon lifestyle for residents [16–18]. Upon analyzing the configuration [16, 19], coverage [20], and level of services [17], researchers found that there are issues such as single facility configuration, spatial agglomeration and imbalance in the community life circles.

Currently, most studies on community life circles or public service facilities focus on the megacities in the developed eastern regions of China [18, 19]. However, fewer studies have been conducted on the smaller cities in the west [6, 17].This study evaluates the allocation level of public service facilities in the community life circle of Mianyang, a smaller city in the west, using the the coverage rate, standard rate and spatial autocorrelation methods. The results will help to form a spatial pattern of balanced development of public service facilities in Mianyang city and improve the quality of life of citizens. At the same time, the study reveals the spatial distribution characteristics of urban public service facilities under different community living circle levels, which can provide theoretical basis for more cities to optimize the allocation of public services.

2 Study Area, Data and Methods

2.1 Study Area

Mianyang is the second largest city in Sichuan Province and the center of science and technology innovation in Chengdu-Chongqing Economic Zone. Taking the built-up area of Mianyang City as the research scope, the area is mapped, according to the satellite map of 2020, with a total area of about 140 square kilometers (see Fig. 1).

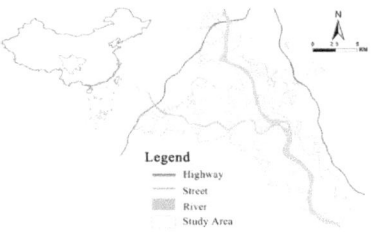

Fig. 1. Study area

2.2 Data Sources

This research mainly used two types of data, which are the basic geographic information and the Point of interest (POI) data of public service facilities. The basic geographic information is collected from Geospatial Data Cloud (https://www.gscloud.cn/) and the POI are collected from the Rivermap website (http://www.rivermap.cn/). After cleaning repeated and meaningless POI data points, we obtained 60,600 pieces of data. We divided the POI data into 3 major categories, 7 medium categories, and 22 subcategories. The specific types and numbers are shown in Table 1. All the data are organized through the WGS1984 coordinate system on ArcGIS10.7 software.

Table 1. Type and number of point of interest (POI) in public service facilities.

Major category (type/number)	Medium category (type/number)	Subcategory (type/number)
Basic public service facilities/8058	Education and research facilities/3775	Secondary schools/47, Elementary schools/62, Kindergartens/325, Education and training facilities/3341
	Medical facilities/3691	Hospitals/362, Basic medical facilities/3329
	Leisure facilities/592	Sports and leisure facilities/561, Cultural and leisure facilities/31
Transportation facilities/4785	Transportation facilities/4785	Parking lots/3989, Bus stops/796
Business facilities/47757	Dining facilities/21996	Chinese restaurants/11401, Other dining facilities/10595
	Shopping facilities/8975	Supermarkets/1127, Convenience stores/1542, Food markets/1137, Other shopping facilities/5169
	Other facilities/16786	Banks/1607, Beauty salon/9574, Laundries/716, Lottery shops/849, Public toilets/753, Logistics facilities/3287

2.3 Methods

Delineating the Community Life Circle.

Identifying the Center of the Life Circle. We choose urban feeder roads and restaurant POI points as the bases for community life circle delineation. Considering the width of the red line of the branch road and the distance and depth of the public buildings facing the street, a 50-m buffer zone is made for the branch road network.

Considering that community life circle centers should have a certain scale size, the present study [21] established that 15 min-CLC, 10 min-CLC, and 5 min-CLC centers should contain 10, 5, and 3 dining spots, respectively. The area of the community life circle center is calculated according to the minimum value of the indicator per 1,000 people for commercial service facilities and scale of the residential population in different community life circle levels as per the Urban Residential Area Planning and Design Standards (GB50180–2018). Subsequently, the area of the center of a 15min-CLC and a 10min-CLC is 1.60 hectares and 0.48 hectares, respectively; there is no mandatory requirement for a 5min-CLC.

Within the 50-m buffer zone of the urban branch road, patches with a POI number of dining facilities greater than 10 and an area greater than 1.60 hectares, greater than 5 and an area greater than 0.48 hectares, and greater than 3 were screened as 15min-CLC, 10min-CLC, and 5min-CLC centers, respectively.

*Determining the Community Life Circle Boundary.*Using the Euclidean distribution function in ArcGIS software to determine the community life circle boundary, we used the center of the community life circle as the starting point and carried out a buffer zone analysis considering radii of 1,000 m, 500 m, and 300 m. Upon erasing the area covering the water system, we ultimately obtained the initial ranges of 15min-CLC, 10min-CLC, and 5min-CLC.

Based on the data of the street administrative boundary and main road network, the community life circle was further amended.Finally, the standard shapes of 15min-CLC, 10min-CLC and 5min-CLC were obtained.

Determining the Coverage and Standard Rate. The coverage rate refers to the ratio of the total area served by public service facilities to the total area of the study area. The standard rate primarily measures the completeness of the facilities in the community life circle. This study calculates the coverage rate of public service facilities and standard rate of facilities in the community life circle by investigating whether a facility exists in the community life circle. The calculation formula is shown below:

$$C_{i,j} = \begin{Bmatrix} 1, \exists F_j \in C_i \\ 0, others \end{Bmatrix} \tag{1}$$

$$CF_j = \frac{\sum_{i=1}^{m} C_{i,j}}{m} \tag{2}$$

$$CR_i = \frac{\sum_{j=1}^{n} C_{i,j}}{n} \tag{3}$$

In the formula, C_i indicates community life circle i, and $C_{i,j}$ indicates whether public service facilities F_j exist within the community life circle i, if it exists it means the facility is covered by the community life circle. CF_j denotes the coverage of public service facility F_j, m is the number of community life circles, CR_i denotes the facility standard rate of community life circle i, and n denotes the number of essential facilities in community life circle i [22].

Spatial Differentiation. Moran's I is used to measure the spatial distribution characteristics of public service facilities. The Global Moran's I can reflect the overall spatial agglomeration characteristics of public service facilities, and the formula is detailed in the literature [23]. The Local Moran index Moran index is unable to show the heterogeneity of Local areas, so the Local Moran's I is used to test whether there is a phenomenon of similar or different attribute value aggregation in local areas. The formula is described in reference [24].

There are four spatial patterns in the Local Moran's I. The high cluster area (HH) and low cluster area (LL) represent the high or low number of public service facilities in the life circle and the surrounding life circles. The high-low agglomeration area (HL) and low-high agglomeration area (LH) represent that the number of public service facilities in the life circle is higher or lower than that in the surrounding living circle.

3 Results and Discussions

3.1 Division of the Community Life Circle

Combining the street administrative boundary and city main road network, the corrected community life circle area is shown to be close to the theoretical value. The values are listed in Table 2.

Table 2. Number and area of the initial and modified community life circles.

The level of community life circles	Initial community life circle		Modified community life circle		Theoretical area (km^2)
	Number (pcs)	Average area (km^2)	Number (pcs)	Average area (km^2)	
15min-CLC	157	0.78	39	3.15	3.14
10min-CLC	249	0.47	145	0.81	0.78
5min-CLC	531	0.21	394	0.28	0.28

The analysis shows that the proportions of the 15min-CLC, 10 min-CLC, and 5 min-CLC center areas to the community life circle area are 22.43%, 25.03%, and 25.21%, respectively. However, the ratio of the community life circle center to the number of various facilities covered by the community life circle is close to or more than 40%; sometimes, it even exceeds 50%, which means that it has a high degree of aggregation. Therefore, the constructed community life circle center and community life circle are considered reasonable.

3.2 Coverage Rate

The coverage rate of public service facilities is a basic indicator of the urban service system construction level. We calculated the coverage rate at all levels of the community life circle. The results are shown in Fig. 2.

As can be seen from Fig. 2, the coverage rate of various public services gradually decreases from 15 min-CLCand 10 min-CLC to 5 min-CLC. The coverage rate in 12 out of 20 facility categories is 100% in the 15min-CLC. The coverage rate in five facility

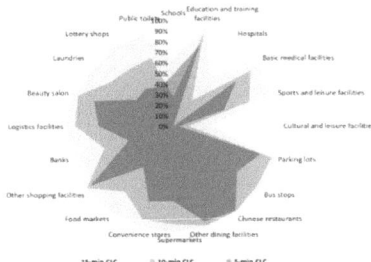

Fig. 2. Coverage rate of public service facilities in the community life circle at all levels.

categories exceeds 90%: sports and leisure facilities, banks, laundries, lottery shops, and public toilets. This result shows the high coverage rate of public service facilities in the 15 min-CLC at Mianyang. However, the coverage rates of secondary schools, hospitals, and cultural and leisure facilities are lower at 64%, 59%, and 36%, respectively. The coverage rate in 11 out of 20 categories exceeds 90% in the 10min-CLC. The coverage rate six facility categories, such as sports and leisure facilities, food markets, banks, and public toilets, is higher than 60%. The coverage rate in the 10min-CLC is relatively high, but the coverage rates of cultural and leisure facilities and elementary schools are extremely low at 14% and 28%, respectively. Among the 19 types of facilities in the 5min-CLC, only 3 have a coverage rate higher than 90%. The coverage rate of public service facilities in the 5min-CLC in Mianyang is poor; the coverage rate of cultural and leisure facilities is the lowest at 5%.

3.3 Standard Rate

The standard rate mainly measures the completeness of community life circle facilities, based on the requirements of different levels of community life circle facilities in the Urban Residential Area Planning and Design Standards to establish the necessary facilities in the community life circle at all levels (see Table 3). Although dining and shopping facilities are essential at all levels of the community life circle, they have essentially achieved full coverage as per the aforementioned coverage analysis; therefore, these facilities will not be analyzed for the standard rate.

The number of essential facilities that should be included in the 15min-CLC, 10min-CLC, and 5min-CLC, respectively, is 7, 5, and 3. The respective rates of the 15-, 10-, and 5min-CLC that fully meet the standard are 28%, 24%, and 24%; that lack one essential facility are 46%, 40%, and 37%; and that lack more than one essential facility are 25%, 36%, and 39%. The main facilities affecting the standard rate of the community life circle at all levels are secondary schools, hospitals, cultural and leisure facilities, food markets, elementary school, sports and leisure facilities, public toilets, kindergartens, and basic medical facilities.

3.4 Spatial Differentiation

The Moran's I is used to study the spatial distribution pattern of public service facilities. From the Global Moran's I of basic public service, commercial, transportation, and all

Table 3. Necessary facilities for each level of community life circle and the number of community life circles that they cover.

The level of community life circles	Number of community life circles	Number of required facilities	Types of facilities/Number of community life circles covered
15min-CLC	39	7	Secondary schools/25, Hospitals/23, Cultural and leisure facilities/14, Sports and leisure facilities/37, Bus stops/39, Banks/38, Logistics facilities/39
10min-CLC	145	5	Elementary schools/40, Basic medical facilities/131, Bus stops/140, Sports and leisure facilities/109, Food markets/112
5min-CLC	394	3	Kindergartens/190, Basic medical facilities/287, Public toilets/149

facilities in Table 4, it can be seen that all demonstrate significant spatial autocorrelation and clustered distributions, except transportation facilities, which demonstrate a random distribution pattern in the 5min-CLC. Global Moran's I for each type of facility increases sequentially from 15 min to 10 min and 5 min. In particular, the index value of basic public service facilities in the 5min-CLC reaches 0.76, showing a high degree of spatial autocorrelation, indicating a high degree of clustering. This explains the gradual decrease in the coverage and standard rates of facilities from 15 min to 10 min and then to 5 min as well as the low coverage and standard rates of educational facilities, hospitals, and leisure facilities.

The Local Moran's I is used to study the spatial variation characteristics of basic public service, transportation, and all facilities. (The total number of business facilities is 47,757, accounting for 78.8% of the total 60,600 facilities; Table 4 also shows that the index values of commercial facilities are close to those of all facilities, so we only study all facilities in this study.)

The local spatial autocorrelation analysis of basic public services (Fig. 3) shows that the area of HH clusters of 15min-CLC, 10min-CLC, and 5min-CLCs in the city center decreases. The 10min-CLC has only one HL outlier on the periphery of the city center, whereas the 5min-CLC forms two HH clusters and six HL outliers in the periphery of

Table 4. Spatial distribution pattern of public service facilities.

Facility types	15min-CLC/10min-CLC/5min-CLC		
	Global Moran's I	Z-score	Spatial distribution pattern
Basic public service facilities	$0.14^{***}/0.25^{***}/0.76^{***}$	3.28/8.75/4.15	Clustered/Clustered/Clustered
Business facilities	$0.18^{***}/0.19^{***}/0.44^{***}$	4.16/6.83/2.41	Clustered/Clustered/Clustered
Transportation facilities	$0.21^{***}/0.27^{***}/0.30$	4.74/9.48/1.62	Clustered/Clustered/Random
All facilities	$0.18^{***}/0.20^{***}/0.47^{***}$	4.17/7.30/2.58	Clustered/Clustered/Clustered

***indicates $P < 0.01$.

the city center. HH clusters and HL outliers in the periphery of the city center are formed mainly around schools.

Transportation facilities in the 15min-CLC, HH cluster are located in the city center. In the 10min-CLC outside the city center, one HH cluster is located in the west of the city and One HL outlier is located near the innovation center of Mianyang (see Fig. 4).

From the local spatial autocorrelation analysis of all public service facilities (see Fig. 5), it can be seen that the 15min-CLC, 10min-CLC, and 5min-CLC form a HH cluster area in the city center. The 10min-CLC forms a HL cluster area. The 5min-CLC forms two HH clusters and seven HL outliers in the periphery of the city center. The locations of the HH clusters and HL outliers in the 5min-CLC are generally consistent with the basic public service facilities.

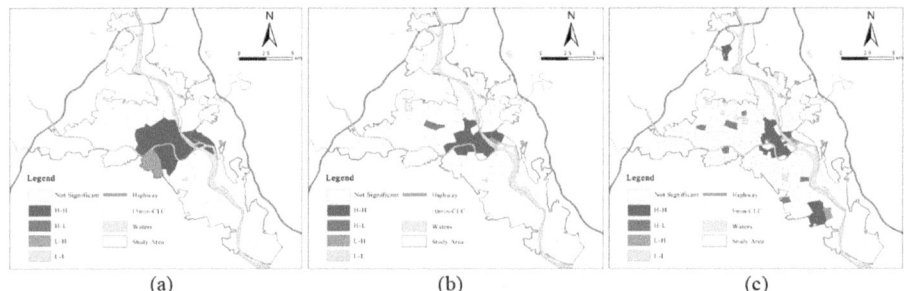

(a)   (b)   (c)

Fig. 3. Spatial differentiation characteristics for basic public service facilities at different levels of community life circles. (a) 15min-CLC; (b) 10min-CLC; (c) 5min-CLC

Generally, with the decrease of the living circle level, spatial distribution pattern of public service facilities shows on a characteristic transform from "core–periphery" to "one-core multi-center".The public service facilities demonstrate a high degree of clustering, with dense facilities in the city center, thereby forming a high clustering large-scale core, and fewer facilities in the periphery. Some small centers have been

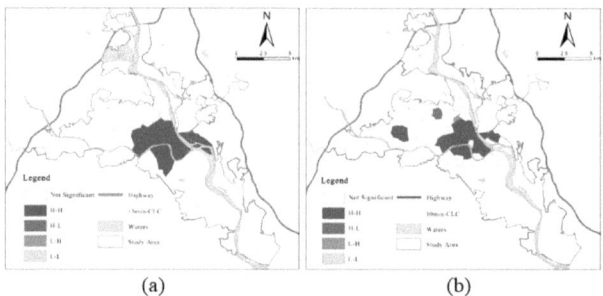

Fig. 4. Spatial differentiation characteristics of transportation facilities in different classes of community life circles. (a) 15min-CLC; (b) 10min-CLC

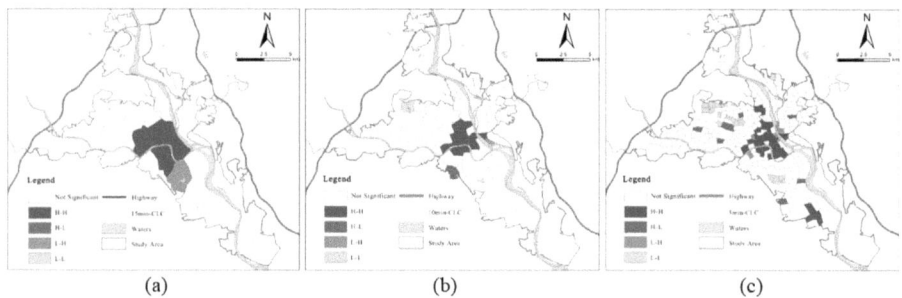

Fig. 5. Spatial differentiation characteristics of all public service facilities at different levels of community life circles. (a) 15min-CLC; (b) 10min-CLC; (c) 5min-CLC

formed, which is consistent with the development and construction history of Mianyang city.

4   Conclusion

The equalized construction of public service facilities is critical for ensuring people's happiness and promoting sustainable urban development. This study focuses on the spatial configuration characteristics of public service facilities in the built-up area of Mianyang City using POI of public service facilities and urban road network data with the help of the ArcGIS10.7 software. Based on the construction of community life circles, the conclusions are as follows.

1. From the 15min-CLC to 10min-CLC and then to 5min-CLCs, the coverage rate of various public service facilities gradually decreases, as does the standard rate of community life circles. The coverage rate of basic public service facilities, such as secondary and elementary schools, kindergartens, hospitals, sports and cultural facilities, and public toilets is low. These facilities mainly account for the low standard rate of the community life circle. Therefore, upgrading the above basic public service facilities is the focus of current urban planning. This is consistent with [25].

2. The spatial distribution of various public service facilities shows significant spatial autocorrelation. As a whole, spatial distribution pattern of public service facilities shows on a characteristic transform from "core–periphery" to "one-core multi-center" with the decrease of the living circle level. Dense facilities in the city center form a large-scale core with high clustering. However, only a small number of smaller centers lie in the periphery. Mianyang has a spatial structure of "one main center and many sub-centers" and its sub-centers are gradually being formed, but the existing scale is small. Therefore, it is necessary to strengthen the planning and construction of the public service facilities of the urban sub-center in order to balance its distribution in the whole city and facilitate the life of citizens. This is consistent with [16].

The study only considers the grade and service radius, and lacks the scale capacity, total population and population structure of the service. In addition, any community life circle will have different degrees of influence on the surrounding or further areas. This study focuses more on the "self-sufficiency" of the internal structure of community life circle, and lacks the "sharing" of the external association, that is, the common use of the same facility by multiple community residents. These shortcomings affect the improvement of community life circle facilities and decisions regarding the location, layout, and scale of their construction, which should be addressed in subsequent studies.

References

1. Falchetta, G., Hammad, A.T., Shayegh, S.: Planning universal accessibility to public health care in sub-Saharan Africa. Proc. Natl. Acad. Sci. **117**(50), 31760–31769 (2020)
2. Romanillos, G., Garcia-Palomares, J.C.: Accessibility to schools: Spatial and social imbalances and the impact of population density in four european cities. J. Urban Plan **144**(4), 04018044 (2018)
3. Hao, F.L., Zhang, H.R., Wang, S.J.: Spatial accessibility of urban green space in central area of Changchun: An analysis based on the multi-trip model. Scientia Geographica Sinica **41**(04), 695–704 (2021)
4. Wu, T.Y., Xia, Q.T., Chen, Z.: Spatial accessibility of elderly healthcare facilities based on an improved potential model: A case study on Fengxian District in Shanghai. J. East China Normal Univ. (Nat. Sci.) **1**, 85–96 (2022)
5. Rimmer, J.H., Riley, B., Wang, E., Rauworth, A.: Accessibility of health clubs for people with mobility disabilities and visual impairments. Am. J. Pub. Heal. **95**, 2022–2028 (2005)
6. Huo, Q.L., Tang, X.M., Wang, H.Y., Zhai, H.R: Research on the spatial distribution and accessibility of medical facilities in Liupanshan area. Sci. Survey. Map. **46**(07), 189–195 (2021)
7. Barbara, M., Rey, D.: Akbarnezhad, A.: Optimizing location of new public schools in town planning considering supply and demand. J. Urb. Plann. Develop. **147**(04), 04021057 (2021)
8. Flores, L.J.Y., Tonato, R.R., dela Paz, G.A., Ulep, V.G.: Optimizing health facility location for universal health care: A case study from the Philippines. PLOS One **16**(9), 1–15 (2021)
9. Aydin, N.: Decision-dependent multiobjective multiperiod stochastic model for parking location analysis in sustainable cities: Evidence from a real case. J. Urb. Plann. Develop. **148**(01), 05021052 (2022)
10. Jae-yeong, Y., Ko, J.-s.: A study on location suitability of public sport facilities by using the under - space of overpass in Seoul-Focused on GIS analysis. J. Urba. Des. Insit. Korea **17**(4), 91–106 (2016)

11. Jiang, Y.C., Sun, L.J., Xiao, K., Wang, T.T., Xu, F.: Optimization for location and layout of urban fire station. Sci. Survey. Mapp. **46**(09), 207–217 (2021)
12. Xu, F., Zhang, Q., Niu, J.Q., Liu, Z.B., Huang, R.C.: Research on spatial equality of medical resources in zhengzhou city based on accessibility. Area. Res. Develop. **40**(05), 37–43 (2021)
13. Zhang, X.Y., An, R., Liu, Y.F.: First aid resources accessibility analysis and site location optimization in Wuhan. J. Nanjing Norm. Univer. (Natural Science) **45**(01), 49–54 (2022)
14. Miok, K., Yu, J., Ko, J.-s.: Expansion of public sports facilities through improvement of legal system related to urban planning. J. Sport. Entertain. Law **21**(21), 87–112 (2018)
15. Zhang, Z.B., Chen, L., Da, X.J., Dong, J.H.: Spatial justice of public service facilities based on social stratification: A case study of the Central Urban Area of Lanzhou. City Plann. Rev. **45**(12), 48–58 (2021)
16. Zhao, P.J., Luo, J., Hu, H.Y.: Spatial match between residents' daily life circle and public service facilities using big data analytics: A case of Beijing. Progress Geograp. **40**(4), 541–553 (2021)
17. Du, X.J., Wang, W.J., Jin, G.: Studies on countermeasures for the public service planning in small and medium-sized cities from the perspective of life circle. J. Hum. Sett. West China **36**(06), 66–73 (2021)
18. Wu, H.Y., Wang, L.X., Zhang, Z.H., Gao, J.: Analysis and optimization of 15-minute community life circle based on supply and demand matching: A case study of Shanghai. PLOS One **16**(8), 0256904 (2021)
19. Jiang, M.Q., Wei, X.Y.: Research on the layout of Chain convenience stores in big cities from the perspective of life circle: Taking Beijing as an example. Urb. Stud. **28**(07), 91–98+123 (2021)
20. Xia, Y.J., Deng, S.Y., Wang, Y.: The rationality of spatial allocation of pension service facilities in Hefei City based on life cycle theory. Trop. Geograp. **41**(04), 769–777 (2021)
21. Sheng, Y.Y.: Study on the division of living circle and living atmosphere in downtown Shanghai. In: Proceedings of China urban planning annual conference, pp.1–11. China Building Industry Press, Beijing (2019)
22. Zhao, Y.Y., Zhang, B., Zhou, F.: Spatial measurement of Beijing 15 minutes community life circle based on POI. World Surv. Res. **5**, 17–24 (2018)
23. Zeng, D.H., Chen, C.J.: Spatial-temporal characteristics and influence factors of PM2.5 concentrations in Dhengdu-Chongqing Urban Agglomeration. Res. Environ. Sci. **32**(11), 1834–1843 (2019)
24. Getis, A.: A history of the concept of spatial autocorrelation: A geographer's perspective. Geograph. Analy. **40**(3), 297–309 (2008)
25. Zhu, P.J., Huang, Q.J., Wan, Y.L., Zou, Z.J.: Spatial differentiation and influencing factors of the allocation level of urban children's public service facilities in China. Econ. Geograp. **43**(11), 55–67 (2023)

Developing a Strategy for Child-Friendly Urban Planning: A Case Study of Feng County, Xuzhou City

Jia Tang$^{(\boxtimes)}$ and Guo Tong

Chongqing Planning Affairs Center, Chongqing 401120, China
tj8742@163.com

Abstract. In 2021, China issued the Guiding Opinions on Promoting the Construction of Child-Friendly Cities, and the construction of child-friendly cities was formally written into the national development plan, with cities such as Changsha and Shenzhen following suit to explore the construction of child-friendly cities. How to coordinate and plan the construction of child-friendly cities with top-level design will be the key content of the research on strategic planning of child-friendly cities. In this paper, from the spatial environment, policy system, service guarantee three aspects of friendly, and action manual to build a child-friendly construction and management of the development of a strategic framework, the formation of "pre-strategy" - "after the implementation of the" scientific closed-loop, building and sharing have been effective. Finally, it is practiced in the old urban area of Feng County, Xuzhou City, with a view to providing insights for a more in-depth development exploration of the subsequent construction of child-friendly cities.

Keywords: Child-friendly · Strategic Planning · Urban Renewal

1 Introduction

In 1996, UNICEF and UN-Habitat launched the Child Friendly Cities Initiative, which proposes that local governments and all sectors of society work together to mobilize efforts to create cities and communities suitable for children's needs [1]. Since then, many countries and cities have begun to pay attention to and promote the construction of child-friendly cities.

Nowadays, as China enters a period of high-quality development, the focus of urban development is shifting from quantity to quality, so that urban planning and city construction will pay more attention to the needs and feelings of the people [2]. Against this background, in 2021, China issued the Guiding Opinions on Promoting the Construction of Child-Friendly Cities, marking that the construction of child-friendly cities in China has become a wind vane under the goal of high-quality development and high-quality life in cities, and has become a beautiful trust for the people's aspirations for a better life and their expectations for children's healthy growth, and will surely open up a brand-new practice of China's new era of city construction [3]. The construction of child-friendly cities is a complex system involving multiple elements in the entire region, and a

B.-J. He et al. (Eds.): UCSUD 2023, LNCE 559, pp. 773–786, 2025.
https://doi.org/10.1007/978-981-97-8401-1_56

good top-level design must be made before the construction is implemented. Therefore, starting from the contemporary significance of the construction of child-friendly cities in China, this paper tries to construct a framework for the strategic planning system of child-friendly cities, and to clarify the focus and path that guides the construction of child-friendly cities, so as to promote the orderly and efficient construction of child-friendly cities in China.

2 Research and Synthesis on Child-Friendly Planning

The theoretical research on the construction of child-friendly cities in China started late, with the translation of foreign literature as the main focus in the early stage, and after the 21st century, children's games were used as the entry point to construct the association between "children" and "space", and gradually expanded to other types of urban space [4]. In 2014, research on child-friendly space design at the community level began to emerge in China, such as a series of discussions on community space planning, architecture and environment with children as the main body of community space design [5]. A series of community space planning, architectural and environmental explorations using children as the subject of community space design. Planning explorations of child-friendly space construction at the city level began to appear in 2017, such as the exploration of planning strategies and implementation paths for building a child-friendly city in Changsha [6], and the exploration of system design and paths for child-friendly planning in Beijing [7].

The practical history of the construction of child-friendly spaces in China can be roughly divided into two stages. The first is the stage of introducing construction concepts and values and exploring them first in a few cities. Since 2006, UNICEF has been promoting the construction of child-friendly cities in China. Since 2015, a few cities in China have begun to explore the construction of child-friendly cities, with Changsha and Shenzhen at the forefront of the country, and Shanghai, Beijing, Chengdu, Wuhan, Nanjing and other cities following one after another. Secondly, it is the stage of comprehensively launching practical actions for the construction of child-friendly cities.2021 In March, China's planning program explicitly proposed to comprehensively promote the concept of child-friendly in-depth urban planning, and included 100 pilot projects for the construction of child-friendly cities in the major projects.

Subsequently, according to the requirements of the document, many cities such as Chongqing, Hangzhou, Suzhou and so on have successively released the implementation program or action plan for child-friendly city construction, and the future practical actions in various places need to be further promoted.

China's existing theoretical research has the following two obvious shortcomings: first, most of it focuses on the construction and renovation and upgrading of individual categories of child-friendly spaces, and lacks a comprehensive and systematic system for the construction of child-friendly urban spaces. Secondly, there is a lack of standards for the construction of various types of child-friendly spaces and facilities in cities, and only the relevant standards for child-friendly spaces are sorted out, and it is suggested to build a new standard system with standards, guidelines and guidelines supporting each other.

The construction practices of Chinese cities are different in content, mostly focusing on individual categories of child-friendly space construction and renovation and upgrading research, lacking comprehensive and systematic system construction of child-friendly urban space construction [2]. From the perspective of overseas implementation paths, they all go through the complete process of "preparing strategic plans and action plans - putting them into practice in urban spaces and social areas - promoting children's in-depth participation - monitoring and evaluating dynamically - and improving the relevant laws, funds and institutional guarantees" [8].Therefore, China's child-friendliness is still in its infancy in terms of the construction of a top-level framework, building, recognition and evaluation, etc. The construction of a strategic planning system focusing on child-friendliness includes: firstly, answering the question of what is to be structured, and what common elements are to be organized around; and secondly, answering the question of how to be built, and how to carry out construction actions in a more orderly and effective manner.

3 Orientation and Focus of Strategic Planning for Child-Friendly Cities

A great deal of research has been carried out in the pre-planning stage, including the recognition of children's rights at home and abroad, the connotation of CFC, the framework system, the global experience, as well as the characteristics of domestic children's needs and the demands of urban development. On this basis, this paper focuses on the healthy development and real needs of child-friendly, and based on the common needs, puts forward four strategic elements: space-friendly, system-friendly, service-friendly and action manual, to constitute a comprehensive, systematic and effective planning system, so as to build a child-friendly city that is safe, fair, in line with the needs of children's healthy development, adapting to the future of the city, and characterized by a global humanistic approach (see Fig. 1).

3.1 Space-Friendly Planning Orientation and Landing Points

Practicing the concept of "95 cm to see the world", the plan puts children first, focuses on the trajectory of children's daily life, in order to increase children's social cognition and enhance children's social-emotional experience and shape good pro-social behaviors, and to satisfy children's needs for parent-child interactions, interpersonal exchanges, and natural contacts in different scales of space [9], and puts forward safety, knowledge, interesting, sports, imagination, and aesthetics as the principles of the design for building a child-friendly city.

First, the construction of a child-friendly spatial scene. Focusing on the psychological characteristics of children's individualized development and exploring the projection relationship between the mechanism of children's psychological development and the provision of urban space, we have systematically constructed a child-friendly spatial system and activity circles. In terms of shaping spatial quality, five major child-friendly spatial scenarios for health, culture and education, transportation, housing and recreation have been created.

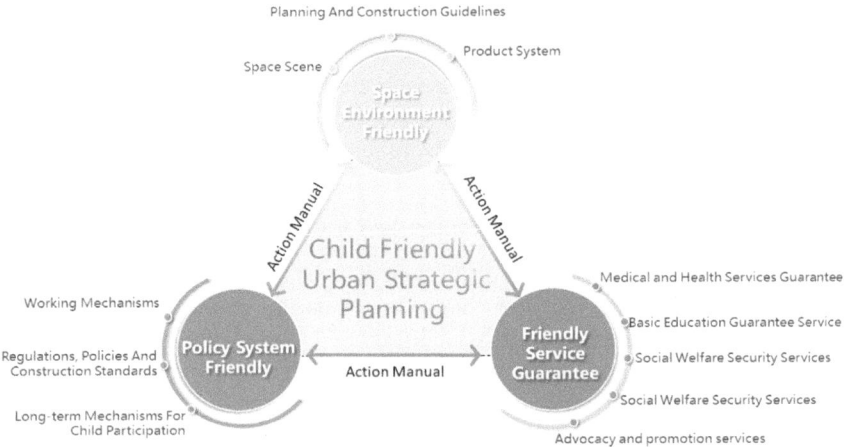

Fig. 1. Framework for a strategic system of child-friendly cities

Child-Friendly Health Space. This is a public healthcare facility to ensure the healthy development of children's bodies and minds by adding new children's health resources such as general hospitals, maternity and child healthcare centers, community hospitals, vaccination sites, and children's clinics, as well as new children's healthcare spaces full of children's interests.

Child-Friendly Cultural and Educational Spaces. This is to utilize green space plazas, build additional campuses with child-friendly outdoor activity spaces, promote the construction of public cultural facilities such as children's museums and science and technology centers, or utilize natural resources to create ecologically interesting nature study bases, so as to realize sufficiently diversified, affordable and friendly children's service facilities.

Friendly Transportation Travel Space. It refers to regulating motor vehicle traffic, providing a safe traffic environment for children, rationally laying out transportation stops for school trips, delineating safe, walkable and continuous walking and non-motorized space, and improving and optimizing the waiting areas for parents in kindergartens, primary and secondary schools in order to achieve the goal of building safe walking environments and interesting street space.

Child-Friendly Living Space. With children as the center and families as the basis, we continue to create a "happy paradise at the doorstep", supplementing and perfecting exclusive spaces full of children's innocence and interest such as community planting gardens, children's libraries, and the 1M Kingdom Appeals Hall, so that children can have what they want to learn, gain, and have fun in the community.

Child-Friendly Open Spaces. Create safe, fun and unique indoor and outdoor open spaces by creating back-to-nature, nature-inspired park systems, vibrant and friendly neighborhoods and street spaces, or large urban thematic playgrounds.

Second, formulating guidelines for the construction of child-friendly spaces. For child-friendly space scenarios, categorize and focus on common child-friendly space elements, form planning and construction guidelines, and guide subsequent design units to comply with important matters of strategic planning, so that child-friendly city strategies can be transformed from the virtual to the real, which is conducive to guaranteeing child-friendly space. At the same time, it is suggested that the combination of the city's regional characteristics and child-friendly space scenarios is conducive to highlighting the city's image and child-friendly recognizability.

3.2 Policy-Friendly Planning Orientation and Landing Points

In terms of the main bodies of implementation, the construction of child-friendly cities in China involves a number of main bodies, such as the government, social organizations, enterprises and individuals, with localities adopting differentiated organizational approaches in the process of exploration. As China comprehensively promotes the construction of child-friendly cities, the government will become the main investor and implementer of many projects in the future, and will comprehensively establish an appropriate and inclusive social security system for children.

First, improve the working mechanism. The Party has strengthened its leadership and organizational safeguards, and clarified the leading group and division of responsibilities for the construction of child-friendly cities. The leading department gives full play to its organizational advantages, strengthens coordination, research and supervision, establishes mechanisms for information exchange, incentives and research and supervision, and regularly reports upwards and downwards. Each department has implemented its main responsibility, and each department is responsible for its own duties, organically combining child-friendliness with its own functions, integrating it into all work, vigorously promoting its implementation, and regularly submitting progress reports to the overall responsibility and the lead unit. In terms of investment protection, all districts and departments have included funds related to the construction of child-friendly cities in their annual budgets, and have implemented targeted funding inputs to ensure the orderly implementation of various projects.

Secondly, a long-term mechanism for children's participation should be established. Respecting children's needs and safeguarding their right to participate, we have incorporated children's perspectives into the decision-making system of urban governance by fostering children's deliberative organizations, opening up channels for the expression of children's views, organizing children's participation in actions related to urban governance, and carrying out publicity and education on the protection of children's rights, so as to lead to the normalization and standardization of children's participation in an orderly manner. Pilot work on children's participation in friendly construction has been carried out using the improvement of the environment around schools and the transformation of communities as a vehicle.

Third, improving regulations, policies and construction standards. Strengthening the planning and construction of monitoring systems, preventing bullying among primary and secondary school students and safety around schools, and improving the supervision mechanism for the protection of minors have effectively safeguarded children's safety and built a support system for child safety protection regulations and policies. Researching and gradually compiling a series of construction standards and construction guidelines suitable for the evaluation index system of child-friendly city construction, policy standard research, and child-friendly spatial environment design, and incorporating the mandatory contents of building child-friendly cities into detailed land space planning, special planning, and so on.

3.3 Service-Friendly Planning Orientation and Landing Points

The 15-min community living circle residential area is the basic unit for control and transmission, and the promotion of the equalization of basic public education services, the high quality of medical service levels, and the high level of protection of women's and children's welfare services has been strengthened.

First, improving basic education coverage services. It has improved the coverage of basic education services such as primary and secondary schools, raised the level of supply of high-quality primary and secondary school places, further improved the quality of teaching, and promoted the balanced development of basic education resources. It has guided children to have more contact with nature, strengthened nature study education for children, and reinforced the educational and training functions of natural spaces.

Secondly, strengthening medical and health services. The focus has been on improving the supply of medical resources for children, improving general hospitals at all levels, maternal and child health-care institutions, township health-care centers and community health-care centers, supported by paediatrics and child health-care units, and upgrading the concept of development of children's welfare, with a view to transforming it from one of social assistance to one of social protection and social participation.

Thirdly, social welfare and security services have been improved. It has established and improved information accounts on children in difficulty, implemented regular home visits, and strengthened precise assistance for children from families with special difficulties. It has deepened the program of medical assistance and schooling for orphans, and has taken care of orphans, children with disabilities, and de facto unsupported children. It has strengthened the construction of community-based family development service centers and women's and children's homes, and has continued to promote the Women's Federation's thematic activities and caring actions for children.

Fourthly, it is exploring a child-friendly big data wisdom service system, constructing a panoramic wisdom governance model of "one network for unified management, one network for unified prevention, and one network for unified management", and realizing a full-process, full-closed-loop, intelligent service management for the maintenance of children's personal rights and interests.

Fifthly, city exchanges and publicity and promotion should be strengthened. It has continued to organize "child-friendly city" warm-up activities, providing children with platforms and spaces for learning, practice, participation, exchange and display. Publicize and practice the concept of child-friendly city construction in an all-round and multi-form manner, raise the awareness and participation of the whole society, and create a good social atmosphere for the construction of "child-friendly cities".

3.4 Implementation Orientation and Landing Points of the Action Manual

All public construction and resource allocation in the city is centered around different administrative levels (cities, urban areas, towns and neighborhoods), each of which has its own financial powers and responsibilities, and builds a public service system at different levels. In the future, the creation of child-friendly cities will combine the different levels of implementation, forming a multi-level paradigm unit of child-friendly cities, child-friendly neighborhoods, child-friendly communities, and so on. Although the construction contents of different levels have top-down penetration, the management focus and depth are different, so different construction points are formed by combining different administrative levels, building a list of core urban projects, a list of optional projects for neighborhoods, and a list of standard projects for communities, forming a standard system matching the characteristics of children's activities in different age groups, so that the main body of the construction of each level can have a basis to rely on.

The construction of the system is a long-term systematic consideration, and the greatest challenge will be how to maximize the effectiveness of child-friendly space construction in the face of limited local financial expenditures. Therefore, in conjunction with all kinds of urban construction and renewal activities, such as new community construction, renovation of old districts and micro-renewal, the concept of child-friendliness should be incorporated into them, so as to promote the child-friendly transformation of all kinds of spaces.

4 Feng County Development Overview

4.1 History and Current Features

Feng County is located in the northwestern part of Jiangsu Province, at the junction of four provinces of Jiangsu, Lu, Henan and Anhui, and in the center of Huaihai Economic Zone. It is an advanced county for plain greening, one of the top ten counties for fruit production in the country, a national ecological demonstration area, one of the top 100 counties for small and medium-sized cities with the best investment potential, and one of the counties with the best value for investment and business operation in the country (see Fig. 2).

Feng County has a long history of cultural heritage. It is the hometown of Liu Bang, the founder of Han Dynasty and Zhang Daoling, the originator of Taoism, and has the reputation of "Ancient Dragon Flying Ground, a generation of emperors and kings". Feng County is rich in natural resources, known as "the hometown of China's red Fuji", the ecological environment of forests and clear water, the forest coverage rate of 36% or more, ranking first in Jiangsu Province, the United Nations ecological demonstration area.

4.2 Planning and Updating Overview

In 2021, the Feng County government conducted a study on the construction control of key sectors in the central city, and put forward an action project to improve the spatial quality of the central city, focusing on building a livable city and carrying out an all-age care project that is friendly to women, children and the elderly. In the face of the lack of vitality in urban development, the loss of young adults, and the plight of the elderly and

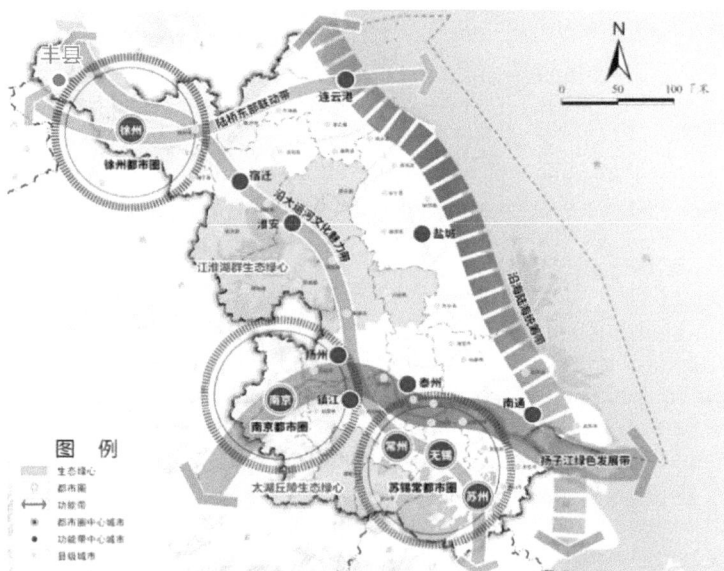

Fig. 2. Location plan.

children left behind in Feng County, it is proposed to magnify the advantages of local characteristics and resources, to build a life attraction of "small city with big happiness", and to take the construction of a child-friendly city as the core plan of action.

Therefore, the government of Feng County has seized the new opportunity of urban construction, taking child-friendliness as the starting point, taking high-quality renewal and development of the space in the center of the urban area as the cornerstone, and taking "children's participation, space creation, policies and systems, and social security" as the entry point, and comprehensively promoting urban construction by means of the strategic plan for child-friendly cities, boosting the return of young people, strengthening the rooting power of cities, and enhancing the happiness and sense of belonging of the city. It will also promote the return of young people, strengthen the rooting power of the city, and enhance the city's sense of happiness and sense of belonging.

5 Strategic Planning Strategies for Child-Friendly Cities in Feng County

Feng County has put forward the "3 + 1" strategy system in combination with the renewal of the central city, which is based on the spatial environment-friendly, policy system-friendly, service-security-friendly and child-friendly action manuals for the integrated planning of the construction of child-friendly cities, forming a scientific closed-loop of "pre-strategy"-"post-implementation". "After the implementation of the strategy, a scientific closed loop was formed.

5.1   Space Environment Friendly Strategy

Within the 38 square kilometers of the central city, through a combination of data from field surveys, WeChat questionnaires, government talks, and crowd interviews, five major child-friendly spaces will be formed, providing children's recreational spaces to return to nature and release their nature; building safe walking environments and interesting transportation and travel spaces; creating vibrant and friendly community interaction spaces; and realizing sufficient, diverse, affordable, child-friendly The children's cultural, educational and health space with adequate, diversified, affordable and child-friendly public service facilities.

Common elements are extracted for different spaces for planning guidance. For example, in terms of traffic safety, the four major control elements of "path-friendly, traffic stabilization, waiting area for parents, and traffic stops" were extracted. For example, to make the paths friendly, it is required to set up continuous and safe non-motorized lanes on the main roads for children going to and coming from school within the 15-min community, and pave them with special materials and colors to mark the right of way exclusively for children, as well as to require barrier-free design. Traffic calming requires speed limits for vehicular traffic at major school entrances and exits to prevent children from playing and disturbing them, which can easily cause traffic accidents. Extend the time of the green light in the traffic light, for children crossing the street to reserve enough time to cross the street safely. At the same time set up parents waiting area, at this stage of elementary school grade 1 and 2, not reserved class parents waiting area, resulting in a school time, parents and class students cause confusion, while congestion sidewalks, not conducive to safety.

5.2   Policy System Friendly Strategies

The leading group for the construction of child-friendly cities in Feng County has been established to carry out the pilot work of child participation in friendly construction by improving the environment around campuses and transforming communities, etc. It is required to incorporate the strategic planning of child-friendly cities into the overall planning of territorial space, the special planning of various fields of economic and social development, and the conditions of land transfer for land control plots, etc. to realize the simultaneous planning and implementation of children's development and economic and social development. It is required to incorporate the strategic planning of "child-friendly" cities into the overall planning of land and space, the special planning of education, health, culture, sports, food, medicine, social welfare and other fields of economic and social development, and the conditions of land transfer of controlled sites, so as to realize the synchronous planning and implementation of children's development and economic and social development, and to guarantee the construction and implementation of child-friendly cities. At the same time, we work according to the content of the clear division of responsibility, to the deputy governor in charge of the overall responsibility, to the county women's federation detailed development of the lead unit and participation in the unit, to ensure that the content can be implemented in an orderly manner to the grass-roots level, from the top to the practice of effective delivery of layers, so that the strategic plan is no longer just a thin piece of paper.

5.3 Service Assurance Friendly Strategy

Combining the urban area preparation unit of Feng County's spatial master plan and the 15-min community living circle of residential planning as the basic principle, the urban area will be divided into 12 community units (see Fig. 3), and the control and conduction will be carried out on the basis of the community units, and each community unit will improve the layout of basic education such as elementary school and middle schools, further improve the quality of teaching, and improve the children's medical security system. At the same time, Fengxian County can organize "City Youth Walk", "Good City Workshop", "Children's City Creation Conference" and other thematic warm-up activities to provide children with a platform and space for learning, participation, exchanges and demonstrations, and invite children to go deeper into the city. Children are invited to go deep into the city, have first-hand experience, think about natural open space, transportation safety, business formats, energy conservation and environmental protection, and make suggestions for urban development, so as to contribute to the construction of a child-friendly city.

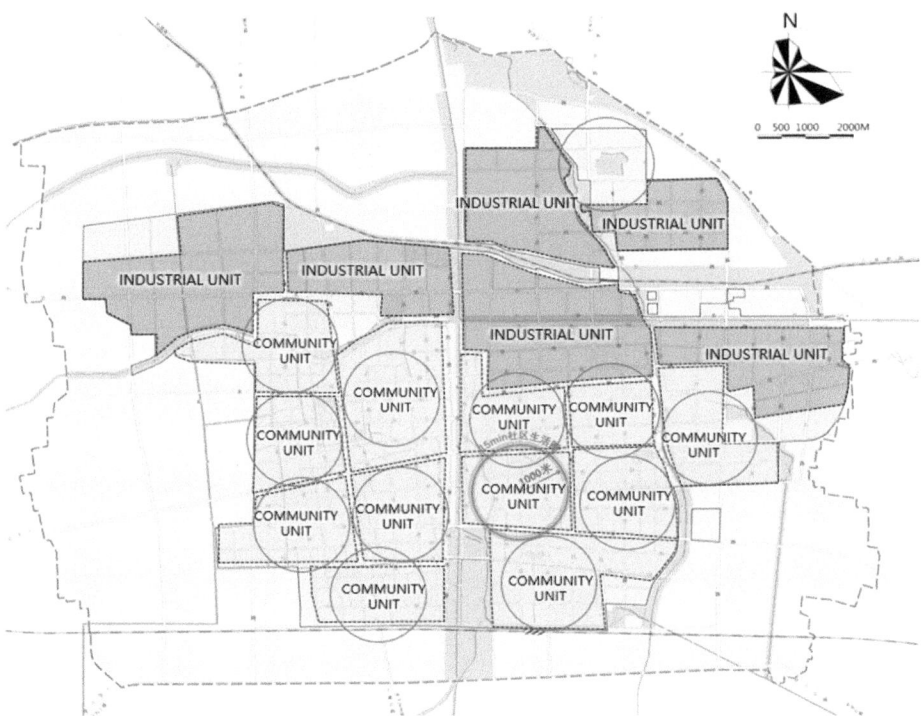

Fig. 3. The central city is divided into 12 units

5.4   Action Manual Implementation Strategies

The three strategic systems of child-friendly cities in Feng County further form an action manual, combining the standardization of land space planning, urban renewal and international child-friendly cities to form a county-level、community-level、optional product system of child-friendly cities (see Table 1), guaranteeing that the strategies of Feng County's child-friendly cities are implemented according to the needs. According to the creation strategy of top-level planning, pilot creation, standardization and gradual promotion, the child-friendly practice point will drive the whole situation and promote the construction of child-friendly cities in Feng County.

Table 1. Feng County Child Friendly City Product System.

	City-level project	Community-level project	Optional item
1	Children's hospital	Children's talk room	Youth extreme sports park
2	Child friendly model school	Room for mother and infant	Parent-child farming experience park
3	Super playground for kids	Children's recreation area	Children's science and technology museum
4	Youth activity centre	Children's library	Kid-friendly town
5	International Child Friendly Center	Children home bus stop	Pet park
6	Children's theme park	Home for women and children	Wow Farm
7	Children's sports park	—	Community children's art store
8	3KM Child Friendly Image Avenue	—	Community children's playground
9	6KM Child-friendly demonstration street	—	Children's rafting library
10	See Learning and Teaching Practice Base	—	Friendly way to school and home
11	—	—	Children's consultation room
12	—	—	Child care center

One is to build 10 urban core projects from the county level, to build a child-friendly urban skeleton, to effectively enhance urban rooting power, and to provide beneficial spatial scenarios for children (see Fig. 4). For example, the child-friendly demonstration elementary school, combined with the construction of the Wenbo Elementary School to create, require developers to build the Wenbo Elementary School, need to comply with the children's cultural and educational construction guidelines. Secondly, combined with the 12 community units standardized project, to build child-friendly community clusters by community unit. Each community unit implanted mother and baby rooms, children's open space and other projects on the basis of public service supporting products to create an ideal model of a child-friendly community (see Fig. 5).. Thirdly, it supplements the 12 optional projects with high-quality, demanding and effective products as personalized optional projects according to the future development trend of the community, so as to provide a complete product package for units with greater development potential.

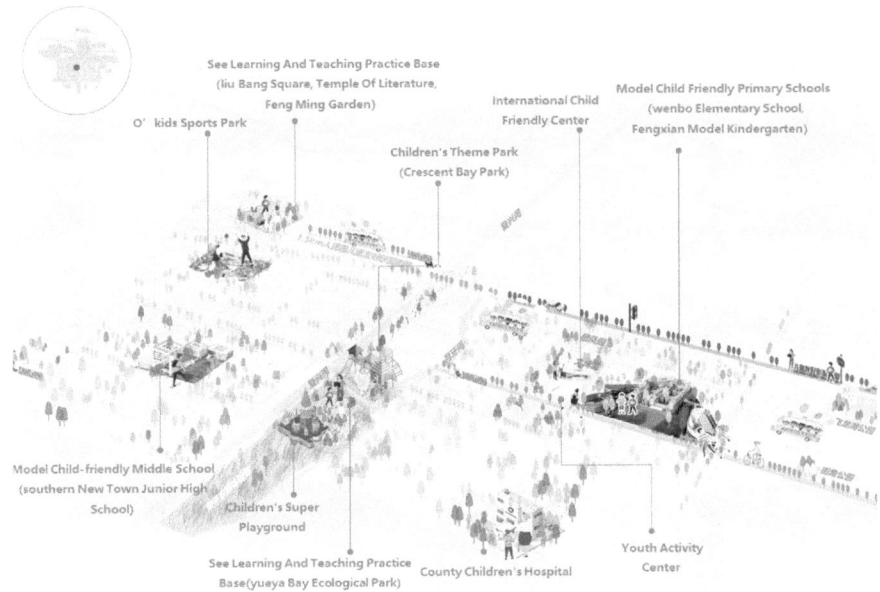

Fig. 4. Distribution map of 10 major urban core projects in Feng County.

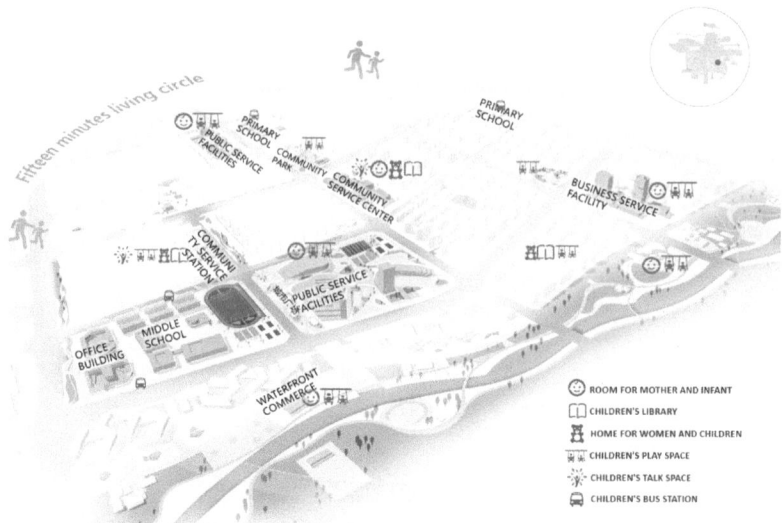

Fig. 5. Ideal Model for Child-Friendly Communities in Feng County.

6 Conclusion

Child-friendly cities are positive attempts by modern cities to address the immediate needs of children. As stated in China's Initiative for Child-Friendly Cities, "The spatial environmental form we create for children today, all the elements of its composition, and the perpetual values behind the form are the guarantee we give to the perpetual development of tomorrow's cities" [10]. A child-friendly city is perhaps one of the most long-term and thoughtful gifts we can give to our children. Through the preparation of the overall level of child-friendly city strategic planning, with the two lines of top-level planning and action plan in parallel, comprehensively and systematically constructing the framework of child-friendly city strategic planning system, we can better reach a consensus on the construction of child-friendly city at the city level, realize the coordinated use of various resources and space, clarify the phased goals in time sequence, the focus of the strategy in space, and the implementation of the path of action to build the construction of Safe, fair, healthy, fun and child-friendly city with local and global characteristics.

At the same time, as China is a vast country with varying levels of economic development and facility construction, the actual operation of the construction content system and implementation paths need to be deepened and adjusted according to local conditions. Looking ahead, it is hoped that further research can be carried out on the content of the system to expand the guideline standards and evaluation system for strategic planning of child-friendly cities.

References

1. UNICEF, UN-Habitat. Child friendly cities initiative (1996)
2. Wang, K., Yan, Y., Zhu, B.: Planning of the national strategic areas with new concepts. Urban Plann. Forum. **1**, 79–84 (2020)
3. Wang, J., Ye, Y., Dong, K.: From efficiency-priority to human-centrality: the value orientation of the territorial spatial planning based on the homo urbanicus theory. Urban Plann. Forum. **6**, 19–26 (2020)
4. Shi, W., Huang, C.: Review on research and practice of childfriendly space in China. Shanghai Urban Plann. Rev. **5**, 129–136 (2021)
5. Liu, Z.: Astudy of children friendly community space design. Southwest Jiaotong University (2014)
6. Gao, Y., Wang, H.: Exploration on the planning strategy and implementation path of building a childfriendly city in Changsha. Beijing Plann. Rev. **3**, 54–57 (2020)
7. Hong, Q.I.U., Lin, G.A.N., Jie, Z.H.U.: The path of international childfriendly city construction and its enlightenment to Beijing. Urban Dev. Stud. **29**(1), 1–5 (2022)
8. Zhang, J.: Scientifically shape the people oriented attractive space of urban and rural areas[N/OL]. Planning China (2021)
9. Jianhua, X.U., Yunyue, H.A.N.: Evelution of community planning practice guided by improving residents happiness: a case study of Shanghai New Jiangwan community. Urban Plann. Forum. **7**, 158–167 (2019)
10. Dongnan, F.U.: Build a new standard system that serves children's growth space-friendly. China Constr. **11**, 22–25 (2021)

Effects of the Built Environment on Physical Activity in the Elderly: A Comparative Study Based on Machine Learning and Logistic Regression

Junbo Mu[✉], Peng Zhang, and Xiaoping Wang

School of Civil Engineering and Surveying, Southwest Petroleum University, Chengdu 610500, China

mjb_15183967214@163.com

Abstract. In China, a conspicuous disparity exists among the built environments within urban and rural areas and the lifestyle requirements of the elderly, presenting a notable social issue. Throughout the urbanization process, the built environments in both urban and rural regions may not sufficiently address the preferences and requirements of the elderly demographic. This deficiency may encompass shortcomings in public transportation infrastructure, community services, and other aspects, consequently constraining the social engagement and activity scope of the elderly population. Despite the built environment at the community level being conducive to social interaction and physical activity among the elderly, the reality often deviates from their needs. Challenges such as insufficient pedestrian pathways, parks, and fitness facilities may impede the eagerness and capability of the elderly to actively participate in community life. To address these challenges, this study employed the Adaboost iterative algorithm and traditional logistic regression analysis to comprehensively grasp the influence of the constructed environment on the well-being of older adults. Through an examination of factors such as physical activity levels, dietary intake, and social interaction among the elderly, it was discerned that the built environment significantly influences the physical activity of older individuals, particularly in factors such as population density and residential areas. Additionally, a close correlation was observed between the built environment and the dietary intake of the elderly, indicating the potential influence of the built environment on their lifestyle. However, it is noteworthy that besides built environment factors, social demographic factors also play a pivotal role in determining social interaction among the elderly. This suggests that when considering the living environment of the elderly, attention should not solely be directed towards enhancing physical surroundings but also towards comprehending the influence of social environments on their lives. Therefore, future research endeavors should further propose strategic recommendations for creating "age-friendly" and "aging in one's original place" environments in urban and rural areas of China, aiming to better address the diverse needs of the elderly and provide a more scientifically grounded basis for relevant policy-making.

Keywords: Built environment · Physical Activity · Ageing · Machine learning

© The Author(s) 2025
B.-J. He et al. (Eds.): UCSUD 2023, LNCE 559, pp. 787–804, 2025.
https://doi.org/10.1007/978-981-97-8401-1_57

1 Introduction

Based on the findings from the seventh national population census in 2021, the total population of China is 1.4178 billion, the population aged 60 and above has reached 264.02 million, constituting 18.70% of the total population (with 190.64 million aged 65 and above, making up 13.50%). [1]. Compared to 2010, the percentage of individuals aged 60 and above has i risen by 5.44% [2]. Consequently, China's population aging is deepening, necessitating urgent solutions to address associated challenges.

However, challenges arise due to the urban-rural economic development divide, the distinct economic and social structures, and regional environmental disparities, complicating efforts for "aging in one's original place." As outlined in the 14th Five-Year Plan, China has made strides in ecological progress, notably enhancing living conditions in urban and rural areas. The plan underscores the importance of advancing human-centric urbanization, coordinating urban planning, construction, and management, determining urban scale, population density, and spatial structure, and fostering the harmonized growth of major, intermediate, and minor urban centers and settlements [3]. Additionally, the 14th Five-Year Plan emphasizes the execution of the national strategy for tackling population aging, promoting the coordinated development of pension services and industries, enhancing the basic pension service system, and fostering inclusive and mutual pension services [4, 5]. This policy directive poses challenges to urban and rural planners and policymakers in designing community-built environments [6]. The built environment has an important impact on people's behavior and emotions [7]. Therefore, to properly design the built environment, it is necessary to study the impact of the built environment on human behavior and emotions in advance. Frank et al. pointed out that the built environment is closely related to the physical activity of the elderly, which they divided into three aspects: physical activity, dietary intake and social interaction [8, 9]. Many foreign studies have demonstrated the relationship between the built environment and physical activity [7–10]. However, domestic research on the elderly group is still limited, and there is a scarcity of studies utilizing cross-sectional data to compare the impacts of the built environment on physical activity, dietary intake, and social interaction. Therefore, it is difficult to accurately assess the impact of the built environment on physical activity in older adults.

Adaboost excels in handling imbalanced datasets, a crucial factor when examining how the built environment affects behaviors that can greatly differ among individuals, such as physical activity, social interaction and dietary intake. Therefore, in this study, we employed the Adaboost iterative algorithm to examine and discuss the impact of the built environment on physical activity, dietary intake, and social interaction of the elderly population. By utilizing data from the CLHLS (Chinese Longitudinal Healthy Longevity Survey) 2018, we aimed to conduct a thorough analysis to elucidate the aforementioned issues. Our findings indicate a significant relationship between the built environment and the physical activity levels of older individuals, with particular emphasis on elements such as population density and residential area characteristics. This research endeavors to contribute to the academic discourse surrounding the design of "local pension" environments by offering empirical evidence and insights derived from rigorous analysis techniques.

Simultaneously, the logistic regression method was used for comparative analysis, and the results showed that the built environment exerted a notable influence on the physical activity of older people. In particularly, elements such as population density and area of residence. This study is anticipated to offer valuable insights and enlightenment for relevant research on the environment of "local pension".

2   Literature Review and Materials

2.1   Built Environment

The comprehensive evaluation of the built environment is pointed out in the study of Barnett et al.[7] in accordance with built environment of urban communities, which includes six elements: density, mixed use of land, street connectivity, block scale, aesthetic perception, and regional structure. This concept comprises the design of many physical components within the urban landscape, such as greenery, houses, streets, and so on. These are often designed to be more convenient for people's daily activities, such as parks to facilitate leisure walks. Brownson et al. pointed out in their study in 2009 that the influence of a community-built environment on physical activity should be evaluated mainly from four aspects: safety, aesthetics, transportation, and land utilization [11]. Frank et al. [12] put forward the view that community design is closely related to the moderate physical activity of young people. By objectively measuring samples and collecting indicators about walking and movement, they concluded that individuals residing in regions characterized by a higher connectivity and complex mixed land use were more active.

2.2   Research on the Relationship Between the Built Environment and Physical Activity

At present, some theories in the built environment and physical activity research have explored the relationship between the environment and the physical activity of older people. Mobile behavior, ecological environment, environmental stress, social learning, and social perception are typical [13]. Among them, mobility behavior predicts the relationship between transportation facilities and the travel mode of older people [14], and environmental stress implies the impact of the external environment of the elderly on mental well-being [15]. Lawrence et al. [16] established a theoretical framework for the impacts of the built environment on health, in which the mediating variables are behavior and direct exposure to the environment. Based on this, the author extracted the theoretical framework of this paper by referring to the framework of the built environment and physical activity, as shown in Fig. 1.

Just as transportation facilities and road conditions will affect the travel behavior of older people, different attributes of the built environment will affect the different physical activities of the elderly group [17]. Studies have found that restrictions such as distance or transportation to a supermarket or grocery store can impact dietary intake in older adults. The degree of mixed land usage, population density, and the number and scale of restaurants will also affect dietary intake [18]. Open community public Spaces, such

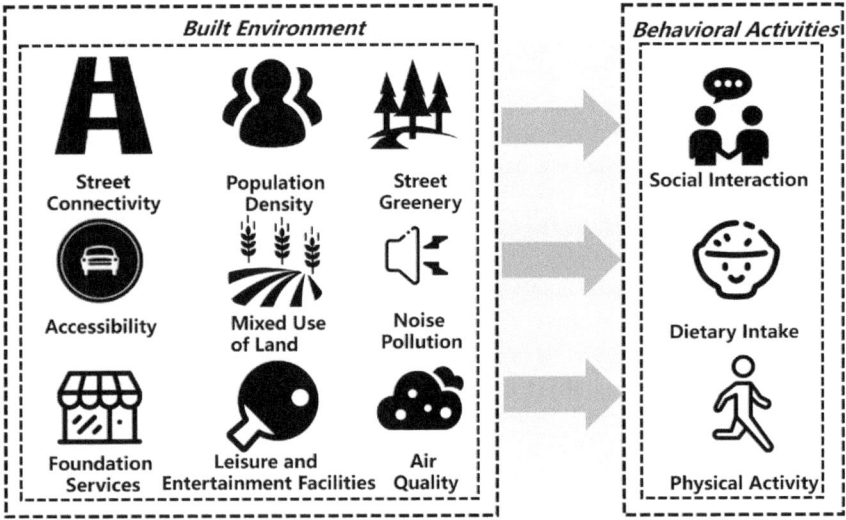

Fig. 1. Theoretical framework.

as parks and activity rooms, promote social interaction and enhance social connectivity and closeness.

For physical activity, many built environmental factors have an impact. Cheng et al. [19]studied the relationship between road safety, bicycle facilities, and street lighting and the walking tendency of older people; Kaczynski et al. [20] verified the impact of public parks, leisure facilities, and destination accessibility on physical activity of older people; Long et al. [21] mentioned that four community environmental factors could affect the physical activity of older people, namely, environmental safety, environmental leisure and entertainment, environmental walkability and environmental aesthetics. Wang et al. [22] studied the influence of spatial distribution density, actual travel distance, area scale, and type scale on the spontaneous or social outdoor activities of older people.

In the early stage of the investigations into the impact of the built environment on physical activity, most researchers used the traditional regression or literature review method. For example, Yan et al. [23] explored the influence of accessibility, quality, and safety of public space on improving community interaction by mining keywords and cited data. Procter et al. [24] used the logistic regression model to verify the influence of hospital and bus accessibility on the travel habits of older adults over the old. However, the results of traditional regression and literature review methods may be affected by sample limitations or literature selection bias, which limits the generalization and generalization ability of the results. Therefore, in recent years, machine learning technology has been gradually applied to this field. For instance, Yang et al. employed random forest to study the impact of street greening on the walking tendency of the elderly [25].

3 Methods and Data

3.1 Method

Adaboost's iterative algorithm is the abbreviation of "Adaptive boosting" in English, which Yoav Freund and Robert Schapire proposed in 1995 [26]. Enhanced learning algorithms represented by AdaBoost have succeeded dramatically in the past decade and developed into a machine learning technology with sound theory [27]. The fundamental concept is to train various classifiers (weak classifiers) on the same training set and subsequently combine these weak classifiers to construct a more resilient final classifier (strong classifier). Different from the Boosting algorithm, the Adaboost algorithm can improve the weak classifier with the general classification effect to the robust classifier with an ideal classification effect, and it does not necessitate prior knowledge of the upper limit of the error rate of the weak classifier so that many algorithms with unstable classification effect can be applied as the weak classifier in Adaboost algorithm [28]. The advantages of the Adaboost iterative algorithm are as follows: a. It makes effective utilization of weak classifiers to cascade; b. Different classification algorithms can serve as weak classifiers. AdaBoost demonstrates high accuracy. c. AdaBoost fully considers each classifier's weight relative to the bagging algorithm and Random Forest algorithm. In this paper, the traditional method was combined with the Adaboost iterative algorithm to rank the importance of the selected variables and explore and verify the main built environment factors that influence the physical activity levels of elderly individuals.

3.2 Data

Chinese Longitudinal Healthy Longevity Survey
CLHLS is China's the oldest and most enduring social science survey database. In this paper, the data of CLHLS 2018 (from the Research Center for Healthy Aging and Development of Peking University) are used, and the data of CLHLS 2014 are used to replace the missing data of built environment.The China Geriatric Health Survey 2018 follow-up data covered 7,193 older adults in 23 provinces, and all the surveyed older adults were over 65 years old. The questionnaire is supported by a personal questionnaire and a community questionnaire, among which the personal questionnaire is divided into the living elderly questionnaire and the dead elderly questionnaire. The survey covers the most basic personal information, physical health status, family situation, daily activities, social and economic support, community environment, medical services, elderly care services, and other aspects of older people. According to the questionnaire results, the authors divided the respondents into two categories: those who had participating in routine physical activity in the past week and those who had not participating in routine physical activity in the past week.

Descriptive statistics of variables
Considering the influence of various external factors, the article, combined with a substantial amount of domestic and international references, ultimately selected two significant categories of variables for exploration. One category is sociodemographic variables, and the other is built environment variables. Sociodemographic variables, determined

based on the literature reviewed, include age, gender, family size, family annual income, and educational level. The built environment variables selected include population density, residential area (where towns and cities are collectively referred to as urban areas), green area, destination accessibility (measured by the proximity of the residence to the urban core), land flatness, and environmental noise compliance rate (total community area divided by the area meeting national noise standards). Statistical summary for the variables are outlined in Table 1. The correlation heatmap for each variable is illustrated in Fig. 2, indicating weak, very weak, or no correlation among the selected variables, thus validating the rationality of our variable selection.

Table 1. Descriptive statistics for each variable

Variable	Average/Quantity	Error value
Sociodemographic variables		
Age	85.31	10.77
Gender		
Man	3 316	
Woman	3 876	
Number of households	2.66	5.73
Annual household income (yuan)	33 547	31 959.01
School age (level of education)	2.52	4.68
Built environment variables		
Population density (Unit: persons/km2)	634.72	474.953
Area of residence		
Town	3 222	
Countryside	3 980	
Green area	720.27	1 721.72
Destination accessibility (Unit: km)	170.57	126.72
Land leveling rate	49.97	35.72
Environmental noise compliance rate	59.19	69.41
Dependent variable		
Physical activity		
Exercise over the past week (days)	1 841	
No exercise in the past week (days)	5 051	
Dietary intake (50g)	5.58	5.78
Social interactivity		
Attend frequently	1 200	
Not often attended	5 829	

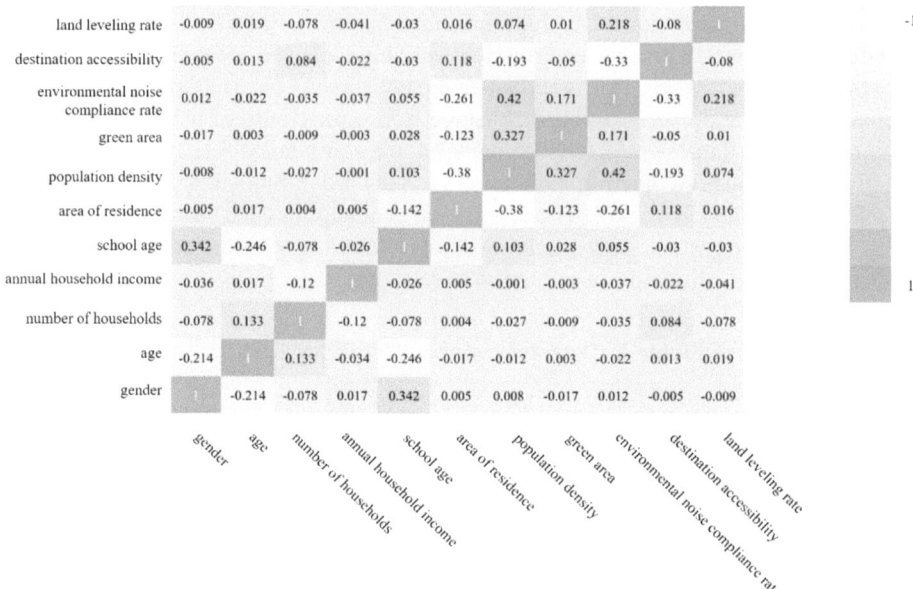

Fig. 2. Correlation between constructed environments and sociodemographic variables

4 Result

4.1 Impacts of the Built Environment on Physical Activity

Figure 3 and Table 2 show the ranking of the relative importance of sociodemographic and built environment variables that affect physical activity in the older age groups in the sample. Figure 4 indicates that the built environment variables are more able to affect the physical activity of elderly individuals. In the relationship with the physical activity of elderly individuals, the relative importance of sociodemographic variables accounted for 38% of the total, and the relative importance of built environment variables accounted for 62%. The green area has the most significant influence (27.95%), which is also in line with the findings of Wen et al. [29], who found that the walking tendency of local elderly groups is more significant in areas with better street greening. Other selected built environment variables, such as population density (16 percent) and destination accessibility (6.95 percent), were also more critical than most sociodemographic variables, such as school age (6 percent), family size (3 percent), and gender (0.1 percent). This could be attributed to the fact that regions with high population density often exhibit greater street connectivity, which can promote older people's recreational and physical activities,

implying that closer street connectivity offers additional transportation options for community residents to promote physical activity [30]. The outcomes of this study align with those of Sa et al. [31], who determined that street connectivity promoted recreational physical activities among older people. Shops, sidewalks, cycling facilities, as well as free or affordable recreational amenities are common in areas with high population density, which promote a higher likelihood of physical activity [32]. AO et al. [33] proposed that when accessibility is good, elderly residents are more active. Lin et al. [34] suggested that older people who are satisfied with the built environment are more likely to engage in physical activity. This indicates that aspects of the built environment, including population density and green area, exerts a considerable influence on the physical activity of older people. In contrast, sociodemographic variables, such as family size and gender, have minimal impact on physical activity. Nevertheless, age and annual household income on physical activity should not be undervalued, with age being the third most crucial variable and yearly household income the fifth most important variable. This parallels the findings obtained by Yang et al. [25], who used the random forest algorithm to explore the impact of greening on walking tendency. In their study, age ranked first.

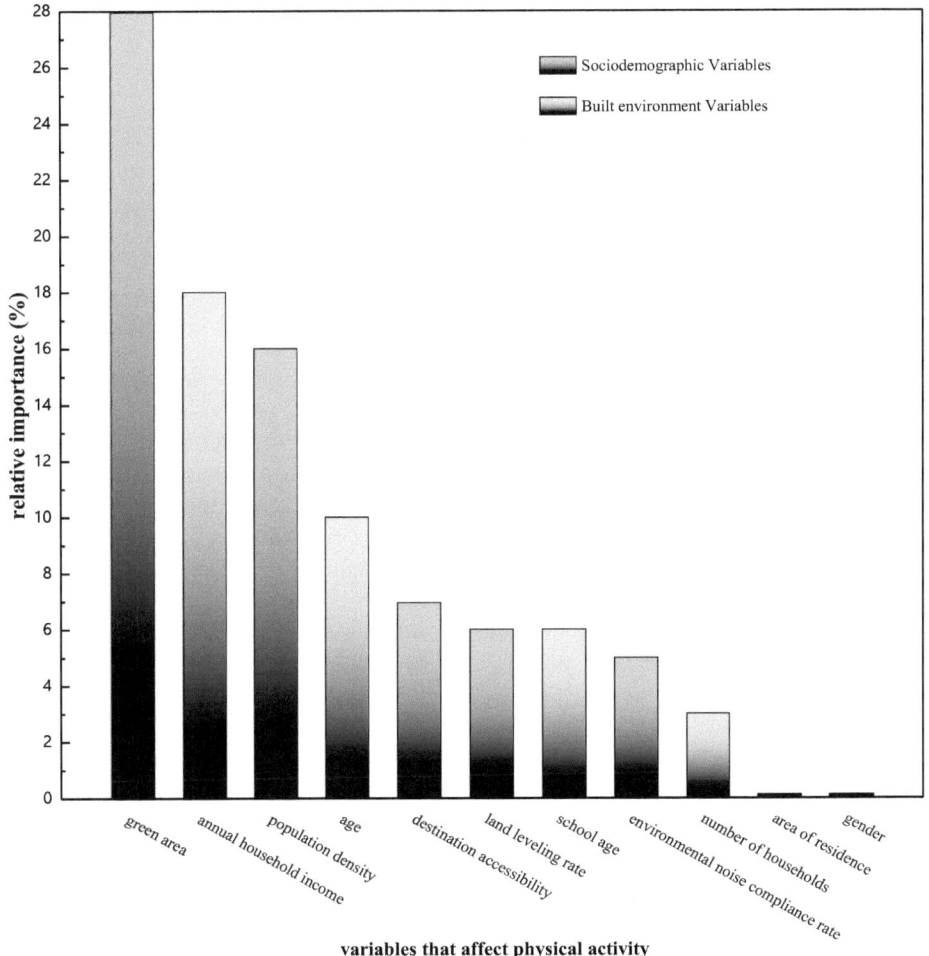

Fig. 3. Ranking of the relative importance of built environmental and sociodemographic factors on physical activity

Table 2. Results of the Adaboost iteration model of the built environment and physical activity

Variable	Relative importance (%)	Rank	Total (%)
Sociodemographic variables			38.00
Age	10.00	4	
Gender	0.10	9	
Number of households	3.00	8	
Annual household income	18.00	2	
School age	6.00	6	
Built environment variables			62.00
Population density (unit: persons/km2)	16.00	3	
Area of residence	0.10	9	
Green area	27.95	1	
Destination accessibility (unit: km)	6.95	5	
Land leveling rate	6.00	6	
Environmental noise compliance rate	5.00	7	

4.2 Effects of the Built Environment on Dietary Intake, Social Interaction

The relative importance of sociodemographic and built environment variables that affect the dietary intake of the sample elderly group is shown in Table 3, with the relative significance of sociodemographic variables accounting for 39% and built environment variables accounting for 61%. Household annual income (22%) was the most influential social demographic variable, and green area (20%) was the most influential built environment factor. The study showed that the built environment factor may support or limit dietary intake. For example, the accessibility and availability of grocery stores, farmers' markets, and fast-food restaurants can influence dietary habits [35]. As shown in Table 4, in the correlation between the built environment and sociodemographic factors affecting the social interaction of older people, the relative importance of sociodemographic variables accounts for 86.8% of the total, and the relative importance of the built environment variables accounts for 13.1% of the overall. The most influential sociodemographic variable was age (48.5 percent). The model shows that other selected socio-demographic variables, such as school age (19.9%), gender (9.1%), etc., are weighted more than the built environment variables of destination accessibility (3.5%), land level rate (1.8%), area of residence (1.7%), environmental noise compliance rate (1.4%), and green area (0.5%). This shows that sociodemographic variables have a more significant influence on the social interaction of older people. In contrast, most of the built environment variables, such as population density, destination accessibility, land level rate, living area, environmental noise compliance rate, and green area, have no significant impact on the social interaction of older people. Li et al. [36] believed that factors related to partial social demography can make older people have a better sense of belonging, thus

Table 3. Results of the Adaboost iteration model of the built environment and dietary intake

Variable	Relative importance (%)	Rank	Total (%)
Sociodemographic Variables			39.00
Age	11.00	5	
Gender	1.00	10	
Number of households	2.00	9	
Annual household income	22.00	1	
School age	3.00	8	
Built environment Variables			61.00
Population density (unit: persons/km2)	14.00	3	
Area of residence	1.00	10	
Green area	20.00	2	
Destination accessibility (unit: km)	12.00	4	
Land leveling rate	6.00	7	
Environmental noise compliance rate	8.00	6	

improving their social interaction. The results prove that the built environment's effect on older people's dietary intake must be addressed, which may be the effect of physical activity leading to dietary intake. In contrast, the impact of the inbuilt environment on older people's social interaction is minimal.

4.3 Compare the Effects of the Built Environment on Physical Activity, Dietary Intake, and Social Interaction

Through the comparison of Adaboost's iterative model results in Fig. 4, it can be seen that the built environment has the most significant influence on the physical activity behavior of older people (62%), followed by dietary intake behavior (61%) and social interaction (13.1%). Table 5 and Fig. 5 show the logistic regression analysis results of the built environment's impact on older people's physical activity. In Table 5, the closer the p-value is to 0, the more significant the linear correlation between the two variables is. The closer the points in scatter Fig. 5 are to the line $P = 0$, the more influential the built environment's impact on physical activity is. It can be inferred that no matter how the built environment's impact on social interaction changes, its effects on physical activity are significant. While dietary intake is closely related to physical activity, the built environment also affects dietary intake by affecting physical activity. Therefore, improving the built environment is expected to control the health burden caused by insufficient physical activity and unhealthy eating behaviors [37, 38]. This result parallels the study of Mikko et al. [39], who pointed out that built environment factors such as urban design and destination accessibility significantly impact physical activity and diet. Therefore, the authors suggest that urban and rural planners and policymakers should pay attention to the impact of the local built environment on the lives of older people,

Table 4. Results of the Adaboost iteration model of the interaction between the built environment and society

Variable	Relative importance (%)	Rank	Total (%)
Sociodemographic Variables			86.80
Age	48.50	1	
Gender	9.10	3	
Number of households	8.00	4	
Annual household income	1.30	10	
School age	19.90	2	
Built environment Variables			13.10
Population density (unit: persons/km2)	4.20	5	
Area of residence	1.70	8	
Green area	0.50	11	
Destination accessibility (unit: km)	3.50	6	
Land leveling rate	1.80	7	
Environmental noise compliance rate	1.40	9	

especially the effect of population density, accessibility to land, green areas, and land smoothness on the physical activity and diet of the elderly, and build a more suitable built environment for older people to fulfill the diverse requirements of the elderly.

The ROC curve of the logistic regression model is shown in Fig. 6, and closer the ROC curve is positioned towards the upper left, the higher the prediction accuracy. The area below the ROC curve is AUC (Area Under Cure). The closer the AUC is to 1.0, the greater the authenticity of the detection method; when it equals 0.5, the accuracy is the lowest and lacks practical value. The logistic regression model AUC = 0.703 has medium sensitivity, which shows that the results of logistic regression have moderate sensitivity, low false positive rate, and good performance of diagnostic methods. The Adaboost iteration involved in the study intersects with the logistic regression verification setting of 20% off, and the comparative outcomes are illustrated in Table 6, which indicates that the Adaboost iteration model has a higher accuracy and accuracy rate than the results of the logistic regression model.

5   Conclusions and Suggestions

The final research results reveal that both sociological factors and built environment factors have a significant impact on the physical activity of the elderly, with the built environment alone accounting for up to 62% of this impact. Specifically, among the various built environment factors affecting physical activity, social interaction and dietary intake among the elderly, the significance level stands at 1%. Notably, the physical activity of the elderly is closely linked to residential area, population density, destination accessibility, and other built environmental factors, which also exhibit certain correlations with

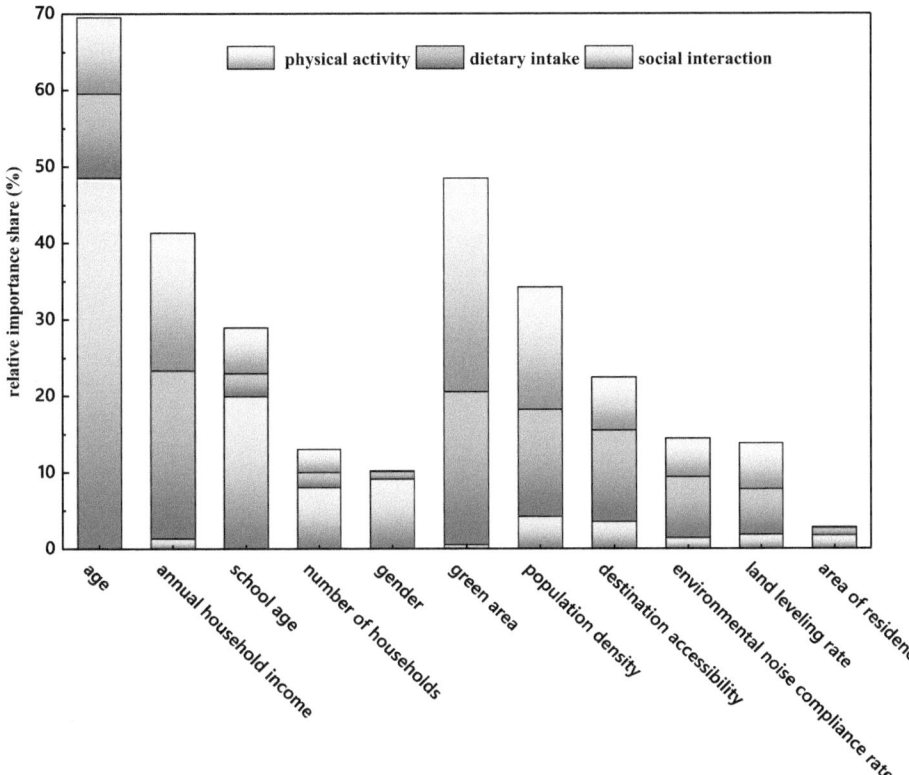

Fig. 4. Proportion of the relative importance of built-up environmental and sociodemographic variables on the physical activity of the elderly

dietary intake and social interaction behavior among the elderly. These findings offer valuable insights for urban and rural planners as well as policymakers when designing or enhancing residential environments for the elderly.

In the development of new villages and towns, it is imperative to consider reducing the distance between residential areas and daily destinations and transportation facilities. Optimal residential area layout should be pursued, alongside the provision of more accessible walking and cycling paths to encourage increased outdoor activity among the elderly. In rural settings, the density of construction should be appropriately increased to facilitate easier access to surrounding facilities for the elderly.

In urban locales, prudent control of population density is essential to ensure the tranquility and safety of elderly living areas, thereby fostering their enthusiasm for physical activity. Urban construction or renovation efforts should consider reducing community closures and enhancing destination accessibility through the establishment of interval roads. Moreover, it is crucial to ensure that the elderly have convenient access to public service facilities, commercial areas, and social venues, thereby stimulating their participation in social interactions. Urban green spaces should be adequately maintained to offer a more pleasant outdoor environment for the elderly. Concurrently, ongoing

Table 5. Logistic regression analysis of the effects of built environment on physical activity

Variable	Standardization coefficients	t test value	P-value
Population density			
Physical activity	0.153	12.945	0.000***
Dietary intake	−0.006	−0.486	0.627
Social interaction	0.047	3.975	0.000***
Area of residence			
Physical activity	−0.250	−21.642	0.000***
Dietary intake	−0.036	−3.128	0.002***
Social interaction	−0.027	−6.276	0.000***
Green area			
Physical activity	0.07	5.858	0.000***
Dietary intake	−0.012	−1.021	0.307
Social interaction	0.009	0.761	0.447
Environmental noise compliance rate			
Physical activity	0.119	10.085	0.000***
Dietary intake	0.044	3.738	0.000***
Social interaction	0.063	5.318	0.003***
Destination accessibility			
Physical activity	2.798	−0.115	0.000***
Dietary intake	4.895	−0.023	0.05**
Social interaction	3.246	−0.053	0.000***
Land leveling rate			
Physical activity	−0.049	−4.122	0.000***
Dietary intake	0.02	1.665	0.096*
Social interaction	0.024	1.975	0.048**

Note: ***, **, and * indicate the significance levels of 1%, 5%, and 10%, respectively

improvements in urban and town sanitation conditions can promote increased dietary intake and social interaction frequency among the elderly.

These recommendations are aimed at optimizing the living environment to provide more convenient, comfortable, and secure conditions for the elderly, thereby encouraging their active participation in physical activities, improving dietary habits, enhancing social interaction levels, and offering comprehensive support for elderly health.

As research into the alignment between human and environmental factors in aging environments progresses from a focus solely on indoor environments to encompass surrounding built environments, the research perspective becomes increasingly dynamic and multi-dimensional. This study utilized machine learning technology, specifically

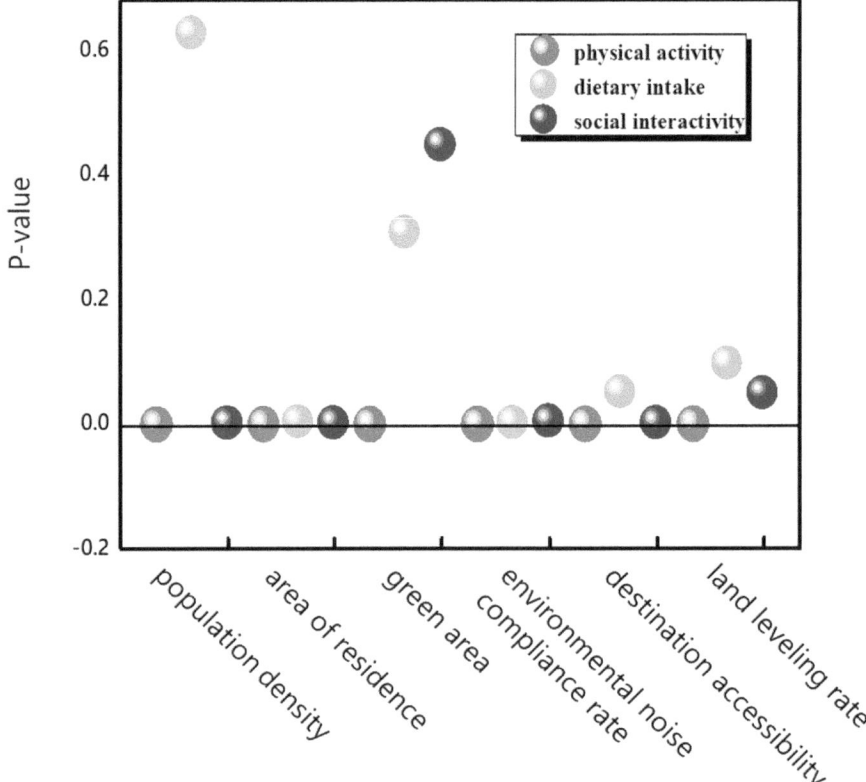

Fig. 5. Significance of the built environment associated with physical activity in the elderly

the Adaboost iterative algorithm, to investigate the influence of the built environment on the physical activity of the elderly. Compared to traditional regression methods, this approach offers higher precision and accuracy, facilitating a clearer understanding of the relationship between the built environment and physical activity while delineating the impact pathways of the built environment on elderly physical activity. These discoveries will offer valuable insights for the construction of urban community built environments in our country in the new era.

The results underscore the imperative of adopting a people-oriented design in the development of urban and rural public spaces, aligning elderly preferences with the built environment to meet the diverse needs of the elderly population and enable aging in place. Nonetheless, it is essential to recognize that studies on the correlation between the environment and physical activity among the elderly has been extensively verified in European and American countries.

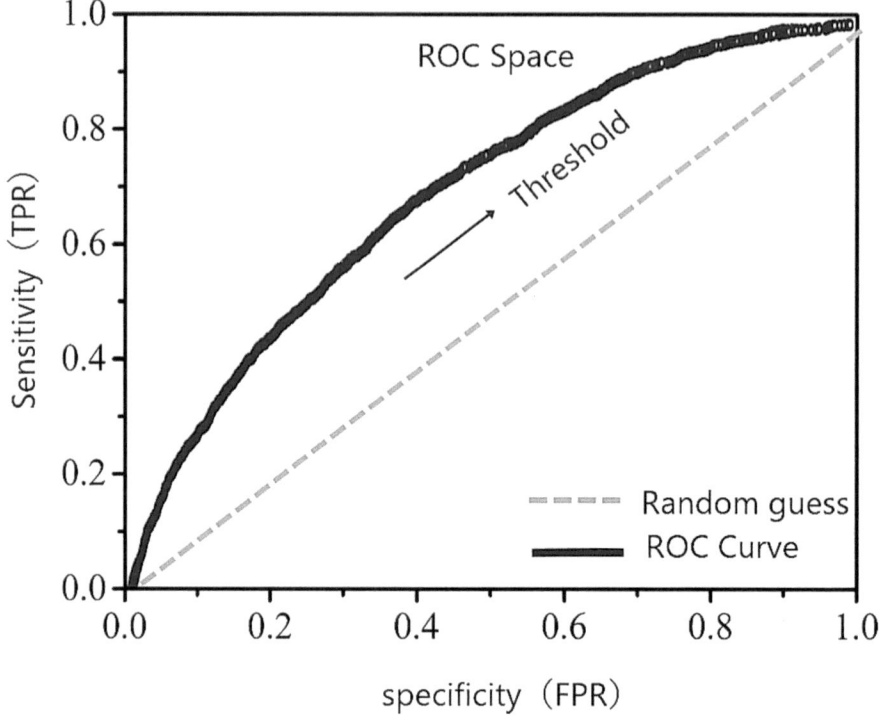

Fig. 6. ROC graph of logistic regression results

Table 6. Comparison results of Adaboost iteration model and logistic regression model

Model	Precision	Accuracy
Adaboost iterative algorithm model	0.846	0.857
Logistic Regression model	0.755	0.720

References

1. National Bureau of Statistics: Bulletin of the Seventh National Population Census (No. 5). Beijing, China (2021)
2. Jiehua, L.U., Qin, L.I.U.: The characteristics, influence and coping strategies of the new form of aging society in China: interpretation based on the data of the seventh census. Popul. Econ. **05**, 13–24 (2021)
3. Xinhua News Agency: Outline of the 14th five-year plan for national economic and social development of the People's Republic of China and the long-term Goals for 2035.] Beijing, China (2021)
4. Zhu, H., Lu, J.H.: The concept breakthrough of the national strategy of actively responding to aging, the evolution of context and the restructuring of system. On Social. Chin. Charact. **3**(2), 12–18 (2021)

5. Ye, J., He, C., Liu, J., Wang, W., Chen, S.: Left-behind elderly: shouldering a disproportionate share of production and reproduction in supporting China's industrial development. J. Peasant. Stud. **44**(5), 971–999 (2017)

6. Van Hoof, J., Kazak, J.K., Perek-Białas, J.M., Peek, S.: The challenges of urban ageing: Making cities age-friendly in Europe. Int. J. Environ. Res. Public Health **15**(11), 24–73 (2018)

7. Barnett, D.W., Barnett A., Nathan, A., Van Cauwenberg, J., Cerin, E., Grp, C.O.A.W.: Built environmental correlates of older adults' total physical activity and walking: a systematic review and meta-analysis.Int. J. Behav. Nutr. Phys. Act. 14 (2017)

8. Lawrence, D., Engelke, F.P.O.: The built environment and human activity patterns: exploring the impacts of urban form on public health. J. Plan. Lit. **16**(2) (2001)

9. Lagiewka, K.: European innovation partnership on active and healthy ageing: triggers of setting the headline target of 2 additional healthy life years at birth at EU average by 2020. Arch. Public Health **70**(1), 1–8 (2012)

10. Soltani, A., Hoseini, S.H.: An analysis of the connection between built environment, physical activity and health: comparing three urban neighbourhoods from Shiraz, Iran. Int. J. Urban Sci. **18**(1), 19–30 (2014)

11. Brownson, R.C., Hoehner, C.M., Day, K., Forsyth, A., Sallis, J.F.: Measuring the built environment for physical activity state of the science. Am. J. Prev. Med. **36**(4), S99–S123 (2009)

12. Frank, L.D., Schmid, T.L., Sallis, J.F., Chapman, J., Saelens, B.E.: Linking objectively measured physical activity with objectively measured urban form: findings from SMARTRAQ. Am. J. Prev. Med. **28**(2), 117–125 (2005)

13. Cunningham, G.O., Michael, Y.L.: Concepts guiding the study of the impact of the built environment on physical activity for older adults: a review of the literature. Am. J. Health Promot. **18**(6), 435–443 (2004)

14. Cervero, R., Seskin, S.: An evaluation of the relationships between transit and urban form[J]. TCRP Research Results Digest 7 (1995)

15. Lawton, M.P., Windley, P.G., Byerts, T.O.: Aging and the Environment: Theoretical Approaches, Vol. 7. Springer Publishing Company (1982)

16. Franka, L.D., Iroz-Elardob, N., MacLeodc, K.E., Hongd, A.: Pathways from built environment to health: a conceptual framework linking behavior and exposure-based impacts. J. Transp. & Health, 319–335 (2019)

17. Dixon, B.N., Ugwoaba, U.A., Brockmann, A.N., Ross, K.M.: Associations between the built environment and dietary intake, physical activity, and obesity: A scoping review of reviews. Obes. Rev. **22**(4), e13171 (2021)

18. Jennings, V., Bamkole, O.: The relationship between social cohesion and urban green space: an avenue for health promotion. Int. J. Environ. Res. Public Health **16**(3), 452 (2019)

19. Cheng, L., De Vos, J., Zhao, P., Yang, M., Witlox, F.: Examining non-linear built environment effects on elderly's walking: a random forest approach. Transp. Res. Part D: Transp. Environ. **88**, 102552 (2020)

20. Kaczynski, A.T., Henderson, K.A.: Environmental correlates of physical activity: a review of evidence about parks and recreation. Leis. Sci. **29**(4), 315–354 (2007)

21. Long Duoqi, A., Hang, M., Biao, Y., et al.: Review of research on the impact of community environment on elderly health abroad. Mod. Urban Res. **1**, 45–51+61 (2022)

22. Wang Xiaoyue: Research on the optimization of green space built environment for promoting physical activity of the elderly. Dalian University of Technology (2020)

23. Yan, S.Y., Zheng, X.: Promoting physical and mental health: research on the impact of community public space on social cohesion. Urban Development Research **28**(02), 117–124 (2021)

24. Procter-Gray, E.: Comparison of dietary quality assessment using food frequency questionnaire and 24-hour-recalls in older men and women. AIMS Public Health **4**(4), 326–346 (2017)

25. Yang, L., Ao, Y., Ke, J., Lu, Y., Liang, Y.: To walk or not to walk? Examining non-linear effects of streetscape greenery on walking propensity of older adults. J. Transp. Geogr. **94**, 103099 (2021)

26. Breiman, L.: Random forests. Mach. Learn. **45**(1), 5–32 (2001)

27. Cao Ying, MiAO Qiguang, LIU Jiachen, et al.: Research progress and prospect of AdaBoost Algorithm. Acta Automatica Sinica, **39**(6), 745–758 (2013)

28. Fang Kuang-nan, WU Jian-bin, ZHU Jian-ping, XIE Bang-chang: Review of adaboost iterative method. Statistics and Information Forum **26**(3), 32–38 (2011)

29. Wen, C., Albert, C., Von Haaren, C.: Equality in access to urban green spaces: A case study in Hannover, Germany, with a focus on the elderly population[J]. Urban Forestry & Urban Greening **55**, 126820 (2020)

30. Wu, Z., Song, Y., Wang, H., et al.: Influence of the built environment of Nanjing's Urban Community on the leisure physical activity of the elderly: An empirical study. BMC Public Health **19**(1), 1–11 (2019)

31. Sa, E., Ardern, C.I.: Associations between the built environment, total, recreational, and transit-related physical activity. BMC Public Health **14**(1), 1–8 (2014)

32. Ding, D., Adams, M.A., Sallis, J.F., et al.: Perceived neighborhood environment and physical activity in 11 countries: do associations differ by country? Int. J. Behav. Nutr. Phys. Act. **10**(1), 1–11 (2013)

33. Ao, Y., Zhang, Y., Wang, Y., et al.: Influences of rural built environment on travel mode choice of rural residents: the case of rural Sichuan. J. Transp. Geogr. **85**, 102708 (2020)

34. Lin Lin, FAN Yixin, Yang Ying, et al.: Threshold and influencing factors of walking distance among the elderly in Guangzhou from the perspective of health. Modern Urban Research **2**, 1–9 (2022)

35. Hino, A.A.F., Reis, R.S., Sarmiento, O.L., et al.: The built environment and recreational physical activity among adults in Curitiba, Brazil. Prev. Med. **52**(6), 419–422 (2011)

36. Jingwei, L., Li, T., Wei, O.: Study on the impact of community environmental safety on the health of the elderly – a case study of Beijing. Modern Urban Research **02**, 17–23 (2022)

37. Duan Yin-juan, LI Li-ming, LyU Jun: Association between community built environment and physical activity and dietary behavior of residents. Chin. J. Epidemiol. **40**(4), 475–480 (2019)

38. Sun Bindong, Yan Hong, Zhang Tinglin: The impact of community built environment on health: an empirical study based on overweight residents. Acta Geographica Sinica **71**(10), 10 (2016)

39. Kärmeniemi, M., Lankila, T., Ikäheimo, T., et al.: The built environment as a determinant of physical activity: a systematic review of longitudinal studies and natural experiments. Ann. Behav. Med. **52**(3), 239–251 (2018)

Impact of Community Park Recreational Space on the Physical Leisure Activities of the Elderly Based on the SEM Model

Lu Wang[✉], Yu Wang, and Yuxin Lu

School of Civil Engineering and Architecture, Southwest University of Science and Technology, Mianyang 621000, China
951188426@qq.com

Abstract. The situation of aging in China is becoming increasingly severe. As one of the main daily activity spaces for the elderly, community parks carry the function of meeting the activity needs of the elderly, affecting their choices of different physical leisure activities, and thus affecting their physical health and emotional perception. Faced with the problems of low-quality community park environment, lack of recreational facilities, and unreasonable layout of recreational spaces, how to build a livable and healthy community park environment for the elderly needs further detailed exploration. Under the guidance of the healthy aging strategy, multiple evaluation indicators were selected based on domestic and foreign literature. Three typical community parks in Chengdu were taken as research samples, and 218 scale questionnaires were collected as data basis. The SEM model was constructed by combining SPSS and Amos software for analysis. The results show that the built environment of community park recreation space has a positive impact on the emotional perception of the elderly through leisure interaction activities and sports and fitness activities, and all factors are interrelated. The three latent variables of service environment, accessibility and equipment are the most important influencing factors for the elderly to engage in leisure physical activities, among which the service environment has the most significant influence, which proves that the elderly pay more attention to the comfort of seats, the location distribution of activity squares, fitness facilities, shading facilities, seats, public toilets, pavilions and garbage cans when engaging in leisure physical activities. Promoting the construction of community parks suitable for the elderly is practically important. Finally, the paper puts forward some strategic suggestions for urban planning and design.

Keywords: Community park · Built environment · SEM model · Physical leisure activities

1 Introduction

In recent years, China has seen a noticeable increase in its aging population. Data from the seventh national census shows that there are 264 million people aged 60 and above, comprising 18.7% of the population, and 190 million people aged 65 and above, accounting for 13.5%. The proportion of elderly individuals continues to rise annually, reflecting

B.-J. He et al. (Eds.): UCSUD 2023, LNCE 559, pp. 805–826, 2025.
https://doi.org/10.1007/978-981-97-8401-1_58

the ongoing trend of population aging in China. Given the significant concentration of elderly residents in urban areas, community parks serve as crucial venues for their daily activities and social interactions. These parks not only cater to the activity needs of the elderly but also influence their choices of physical leisure pursuits. However, most of the existing community parks can only meet the basic needs of the elderly for leisure and social interaction, and the attraction of outdoor physical activity spaces for the elderly is relatively low [1].The existing community parks generally over-pursue visual design sense and material landscape effects, neglecting the care of society and users carried by the park [2], resulting in low use and high waste of some spaces in the park [3], and the number of community parks with "caring", "aging-friendly", and "practical" characteristics is decreasing year by year. In the planning and construction of urban community parks in China, the needs of the elderly for recreational space and leisure experience have not received due attention [4]. Secondly, the landscape function of the activity space in the community park is single, and it cannot fully meet the activity needs of the elderly. Existing designers often overlook the psychology of the elderly [5], physiological, behavioral characteristics, with more emphasis on the aesthetic function of the landscape.

Faced with the low-quality environment of community park recreational spaces, the lack of and outdated recreational facilities, and the unreasonable layout of recreational spaces [6].This study focuses on the leisure physical activity spaces for the elderly in Chengdu community parks, aiming to understand their leisure physical activity patterns, identify environmental factors influencing their activities, and pinpoint features in community park spaces that can enhance daily physical activities for the elderly. The research delves into the current shortcomings in the built environment of community park spaces and explores strategies to optimize and enhance these environments within the context of China's national conditions. By enriching activity spaces for the elderly and fostering a greater sense of active engagement in recreational physical activities, this study seeks to improve the overall well-being of elderly individuals. Truly create a good physical and mental health, comfortable, safe and high-quality urban activity space for the elderly group, but also make the elderly group truly realize the "old have some support" and "old have fun".

From the 1920s, when Perry proposed the theory that there should be recreational green spaces in the center of residential areas, to Jan Gehl's in-depth study of public spaces in residential areas in "Life Between Buildings", and then to Marcus's "Human Places—A Guide to Designing Urban Open Spaces" which provides operational guidelines for community park design [7]. The theoretical research of community parks has matured. Foreign theoretical research mainly studies the psychological feelings of residents at different spatial scales and their reactions to different activities and facilities from a humanistic perspective, in order to better provide spaces rich in humanity [8]. At present, domestic research on the built environment of community parks is still in its infancy. The research content is mostly reviews, mainly analyzing, summarizing, and learning from foreign research methods or results. He [9], Qiu [10],Wang [11] and other scholars have also proved through research that the high accessibility of the inner space of community parks can effectively promote the occurrence of physical activities. After summarizing the recent literature on community parks and physical activity in the United

States, Luo found that the safety of community parks was positively correlated with the occurrence of physical activity [12]. Zhang n [13], Wei [14] and Li [15] found that the space requirements of community park users are mainly reflected in the safety of fitness facilities, site safety and night lighting. The higher the safety of community park, the more likely physical activity is to occur; Tan found in his research that the comfort level of atmosphere (health degree), the safety of service facilities and the maintenance quality of site facilities are positively correlated with the spontaneity of physical activity [16]. Zheng found in his research that the perfection of supporting facilities in community parks, especially the recreation facilities, is positively correlated with the occurrence of physical activity [17]. The effects of landscape perception generated by the interaction between landscape perception elements and perception types in the physical environment of parks can meet the various needs of the elderly for activities [18]. Landscape comfort can enhance the elderly's contact with and love for outdoor activities [19]. Zhao [20] proposed that landscape participation can attract the elderly to carry out outdoor activities. Liu [21] believes that the creation of landscape diversity can relieve physical pressure and has a significant impact on the physical and mental health of the elderly. The allocation of public facilities is particularly closely related to the daily activities of the elderly. Song et al. showed that the type of facilities and the combination of auxiliary facilities and activity space can guide the outdoor activities of the elderly. Physical activity [22] (PA) was first proposed in medical exploration to refer to "any physical movement caused by skeletal muscle contraction that results in energy expenditure" and to indicate that physical activity affects public health. Leisure time physical activity, also known as leisure physical activity, mainly includes leisure communication and sports and fitness. Outdoor activities such as sports fitness equipment, walking, jogging, etc., are usually measured by frequency, duration, intensity, or other factors [23].

This paper proposes an identification method based on SEM model for the influence of built environment factors of community park recreation space on recreational physical activity of the elderly. Taking three community parks in the old city of Chengdu as examples, the influence factors and results of built environment factors of community park recreation space on recreational physical activity are explored and analyzed and summarized by comprehensively considering potential variables and observed variables. In order to provide a basis for determining the planning strategy of the built environment of community park recreation space.

2   Research Sample Overview

2.1   Selection of Research Samples

The community parks defined in this article refer to those built earlier near the old residential areas in the old city of Chengdu, with a large number of elderly people in the radiated residential areas, diverse forms, smaller areas, and uneven functions. For the elderly, community parks in the old city serve as vital outdoor activity spaces and concrete means to pursue a high-quality living environment. At present, there are great differences in the aging population in Chengdu, and the differences between the supply and demand of community, public space and resource allocation are also obvious. In fact, many community parks, due to unreasonable public space planning, low degree of greening, unscientific plant allocation, and weak sense of belonging to space and place, have seriously reduced the daily use and healthy activity needs of the elderly. Therefore, the author conducted a field survey on the aged fitness of the public space in which middle-aged and elderly people engaged in leisure physical activities in Chengdu community parks, in order to optimize the environmental elements of community parks. The research community park follows the following principles: (1) representativeness. This paper focuses on the community park in the old city as the research object. Through field investigation, Shahe City Park in Chenghua District, Jinsha Binhe Park in Qingyang District and Huangzhong Park in Jinniu District are selected as research samples. As shown in Fig. 1, these three community parks are all old city community parks, in which a large number of elderly people often gather for leisure physical activities. (2) There are more elderly people in community parks. The object group of this study is the elderly, so a large number of elderly people should be selected in the selection of community parks, which can meet the requirements of the number of samples. (3) Community parks have high vitality. The elderly have high frequency and rich types of recreational physical activities, which can almost cover most types of recreational physical activities. The sample data are more accurate, which makes the results of this study credible. This study seeks to investigate the relationship between factors influencing the leisure physical activities of the elderly and different built environment elements in community park recreational spaces, analyzing the impact of various variables. The research samples, representing natural landscapes, infrastructure, and elderly population density, have garnered positive feedback and serve as primary activity hubs for urban residents in their respective regions.

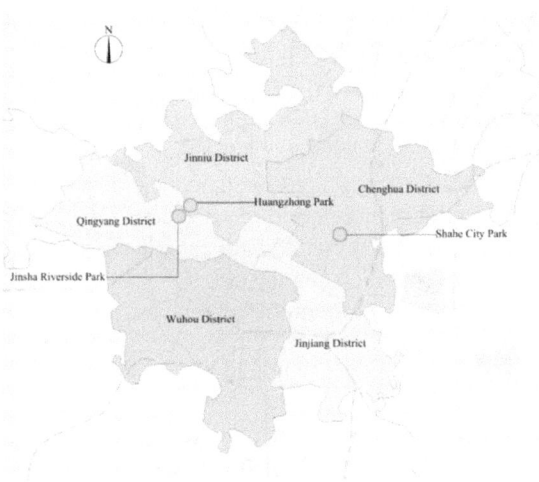

Fig. 1. Study sample distribution map

2.2 Research Scope

From a research perspective, the built environment can be categorized into objective indicators and subjective perception variables. The objective built environment includes facility diversity, and land use rate, etc. The subjective perceived built environment includes accessibility, safety, equipment completeness, landscape detail, management environment, service environment, etc. Based on existing research, studies on the built environment of community parks related to elderly activities mainly focus on behavior characteristic analysis and influencing factor identification, and emphasize the impact of the objective built environment on the physical activities of the elderly, lacking the elderly's subjective environmental perception. This study divides leisure physical activities into leisure social activities and sports fitness activities. In terms of element selection, this study starts from the perceptible factors of the built environment of the recreational space of three community parks in the old city area of Chengdu, divided into accessibility, safety, equipment completeness, landscape detail, management environment, service environment, and this study investigates how the built environment of the recreational space in the old city community park influences the leisure physical activities of the elderly from a multidimensional perception standpoint.

3 Research Design

3.1 Model Indicator Selection

Various elements of the built environment restrict and impact the physical activities of the elderly. The physical activity characteristics of the elderly group serve as dependent variables, while the built environment elements act as independent variables (Table 1). The study reviews and analyzes relevant literature from both domestic and international sources to categorize the impact of the built environment in community park recreational spaces. It then identifies evaluation indicators for characteristic variables. The article categorizes the perceptible elements of the built environment into accessibility, safety, equipment completeness, landscape detail, management environment, and service environment. Multiple secondary indicators are selected under each primary indicator as observation variables to investigate how the built environment of the recreational space in the old city community park affects the leisure physical activities of the elderly. The questionnaire is developed based on literature review and field research.

3.2 Model Construction

This paper aims to explore how different built environment factors impact the leisure physical activities of the elderly in community parks within the old city of Chengdu. The study focuses on the subjective experiences of the elderly to assess the influence of each variable, establishing causal relationships between these variables. To ensure the scientific and effective research on the impact of the built environment in community park recreational spaces on the physical activity of the elderly, it is essential to choose a quantitative method suitable for subjective assessment. This paper suggests an initial structural equation model to assess the impact of built environment factors in community

Table 1. Evaluation indicators of built environment elements of Old City Community Park

Latent variable	Observable variable
Accessibility	A1 Distance from home
	A2 Convenience of transportation to the park
	A3 Visibility within the park
	A4 Accessibility of public facilities
	A5 Accessibility of green landscape
	A6 Pedestrian path connectivity
	A7 Number of entrances and exits
Safety	B1 light level
	B2 Road surface smoothness
	B3 pedestrian system separated from vehicle system
	B4 Fitness equipment safety level
	B5 Adequate monitoring facilities and safety signs
	B6 Public security management
Equipment completeness	C1 Fitness equipment
	C2 Leisure facilities
	C3 Shade and rain shelter
	C4 dustbin
	C5 Rest room
Landscape detail	D1 Plant richness
	D2 Sculpture landscape
	D3 Walkway quality
	D4 Ground paving landscape
	D5 Water landscape
	D6 Color saturation
Management environment	F1 Sanitary environment
	F2 Security environment
	F3 Facility maintenance
	F4 Identification system
	F5 Popularity situation
Service environment	G1 Seat comfort
	G2 Seat position and distribution
	G3 Location and distribution of public toilets

(*continued*)

Table 1. (*continued*)

Latent variable	Observable variable
	G4 Location and distribution of pavilions
	G5 Location distribution of shading facilities
	G6 Activity square location distribution
	G7 Fitness facilities location distribution
	G8 Location and distribution of garbage cans
Leisure activities	U1 Chess and cards
	U2chat
	U3 onlookers
	U4 Sitting in meditation
	U5 Enjoying the scenery
	U6 Photography
	U7 Practice calligraphy
Sports fitness activities	P1 Sports fitness equipment
	P2 Walking
	P3 Square dance, fitness dance
	P4 Singing
	P5 Martial arts, Tai Chi
	P6 Jogging
	P7 Ball sports
Emotional perception	Q1 Leisure physical activities make me feel relaxed
	Q2 Leisure physical activities help me relieve stress
	Q3 Leisure physical activities fulfill my expectation of strengthening my body
	Q4 Leisure physical activities give me a sense of belonging

park recreational spaces in old urban areas on the physical activity of elderly individuals, depicted in Fig. 2. According to the literature review and the research focus of this paper, the built environment elements of community parks are divided into six aspects: accessibility, safety, equipment completeness, landscape meticulousness, management environment and service environment to explore the influence of leisure physical activity on the elderly. The structural equation model is essential for determining causal relationships between variables influencing the leisure physical activity of the elderly, ensuring the reliability and validity of research findings.

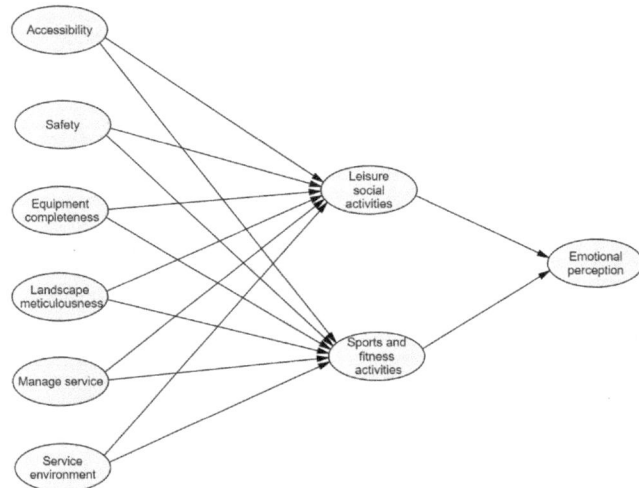

Fig. 2. Initial theoretical model

3.3 Questionnaire Design and Collection

In this study, the survey data mainly came from the elderly people in the parks of 3 sample communities. By engaging in discussions with elderly individuals and administering questionnaires, the questionnaire design comprises two parts. The first part focuses on collecting demographic information of the elderly respondents, such as age, gender, and frequency of travel. The second part is the evaluation of the built environment elements of the recreational spaces in three sample community parks in the old city of Chengdu and the satisfaction of the elderly with leisure physical activities. To obtain their cognitive evaluation of the surrounding environment is the item setting of latent variables and corresponding observed variables such as accessibility, safety, equipment completeness, landscape detail, management environment and service environment. It includes 6 items and a total of 37 questions, corresponding to the latent variable indicators and observed variable indicators in the above table; the second part is a scale question, using a Likert 5-point scale. Conduct on-site questionnaire interviews to directly communicate with the elderly, ensuring the questionnaire's validity. During October to November 2023, on-site questionnaire interviews were carried out in three selected community parks. The aim is to gather 80 questionnaires from each park, filtering out incomplete, repetitive, and contradictory responses. The number of valid questionnaires is 218. This study utilizes on-site questionnaires at the community park in the old city of Chengdu. However, it is important to note that the survey's findings may have limitations and may not fully represent the characteristics of elderly individuals in community parks in other areas.

4   Results Analysis

4.1   Population Characteristics Analysis

The recycled questionnaire results were statistically analyzed based on factors like age, gender, health status, and activity frequency (refer to Table 2, Fig. 3). The respondents exhibit a balanced gender ratio, with a majority of elderly individuals falling within the 60–69 age group. Most of the elderly at this stage are retired and their physical functions have not yet shown obvious degradation. They have increased energy for outdoor activities. Statistics show that 86.66% of the elderly frequently visit community parks for leisure activities like entertainment and fitness. Among these activities, the highest proportions are using sports fitness equipment (86.24%) and walking (83.49%).

4.2   Reliability and Validity Test

Before conducting confirmatory factor analysis using structural equation modeling, it is essential to test the reliability and validity of the questionnaire scale. This study employs Cronbach's Alpha coefficient to assess the credibility of the questionnaire scale. The overall Cronbach's Alpha coefficient is 0.921.The reliability test results of each indicator in the questionnaire scale exceed the standard of 0.7, indicating strong internal consistency in the collected scale data for further analysis. Additionally, after deleting item U7, the Cronbach's Alpha coefficient improved to 0.922, surpassing the overall Cronbach's Alpha coefficient of 0.921. Therefore, the reliability of this item is low, and this item will be considered for deletion in subsequent research.

Following the reliability analysis, item U7 was excluded for the validity analysis. Validity analysis used principal component analysis with maximum variance rotation to extract common factors from each measurement item. Following the KMO test and Bartlett's test of sphericity on the variables, the KMO value was 0.886, surpassing the benchmark of 0.7, and the significance level met the standard of less than 0.05, indicating good model validity. Factor loadings from the KMO test and Bartlett's test of sphericity (Table 3) were used to assess structural validity. As shown in Table 4, all measurement item factor loadings exceeded the extraction standard of 0.5, suggesting strong structural validity in the model, which was further confirmed.

4.3   Confirmatory Factor Analysis of Built Environment Elements

Amos software was used to conduct a confirmatory factor analysis of the dimensions of the built environment elements. The standardized factor loads of all measurement indicators of accessibility, safety, equipment completeness, landscape detail, management environment, and service environment are all above 0.6. The CR of each variable ranges from 0.842 to 0.932, all exceeding 0.7, indicating good internal consistency of the measurement results. The AVE for each variable is above 0.5, demonstrating good convergence validity (Table 5).

Discriminant validity assesses the distinction and variation among different potential variables in the model, typically evaluated using the square root of Average Variance Extracted (AVE). Good discriminant validity is confirmed if the square root of AVE for

Table 2. Basic characteristics of the elderly

Indicator	Category	Frequency	Percentage (%)
Gender	Male	106	48.62
	Female	112	51.38
Age	60–69 years old	145	66.51
	70–79 years old	52	23.85
	80–85 years old	16	7.34
	Over85 years old	5	2.29
Health status	Very Healthy	52	23.85
	Relatively Healthy	115	52.75
	Average	29	13.3
	Not Very Healthy	14	6.42
	Unhealthy	8	3.67
Importance of leisure physical activity	Very Important	82	37.61
	Somewhat Important	96	44.04
	Average	24	11.01
	Not Very Important	10	4.59
	Not Important	6	2.75
Frequency of visiting community park	Multiple Times a Day	36	16.51
	Once a Day	57	26.15
	Several Times a Week	96	44.04
	Once a Week	21	9.63
	Rarely	8	3.67
Duration of leisure physical activity	< 10 min	14	6.42
	10–30 min	36	16.51
	30–60 min	130	59.63
	Over 60 min	38	17.43
Time of leisure physical activity	6:00–9:00	183	83.94
	9:00–12:00	70	32.11
	12:00–15:00	74	33.94
	15:00–18:00	47	21.56

(*continued*)

Table 2. (*continued*)

Indicator	Category	Frequency	Percentage (%)
	18:00–21:00	94	43.12
Types of leisure social activities	Chess and Card Games	164	75.23
	Chatting	170	77.98
	Spectating	74	33.94
	Sitting and Contemplating	147	67.43
	Viewing Scenery	167	76.61
	Photography	76	34.86
	Practicing Calligraphy	36	16.51
	Others	12	5.5
Types of physical fitness activities	Sports Fitness Equipment	188	86.24
	Walking	182	83.49
	Square Dance, Fitness Dance	156	71.56
	Singing	52	23.85
	Martial Arts, Tai Chi	110	50.46
	Jogging	48	22.02
	Ball Sports	150	68.81
	Others	8	3.67
Main reasons for participating in leisure physical activities	Strengthening Physical Health	182	83.49
	Treating Diseases	65	29.82
	Killing time	177	81.19
	Making friends	65	29.82
	Pleasing mood	190	87.16
	Pursuing challenges	68	31.19
	Others	6	2.75

each variable is higher than its correlation coefficient with other variables. The correlation coefficient matrix of potential variables of each influencing factor and performance level is shown in Table 6. The values arranged diagonally in the table are the AVE square root of the latent variable, and the values under the diagonal are the correlation coefficients of the latent variable with other latent variables. By observing the data in the table, the first value of each column on the diagonal line is greater than all the values below, indicating that the discrimination validity of the measurement model is good.

4.4 Regression Analysis of Built Environment Elements

Using the Amos structural equation model, the study examines the causal relationship between the built environment elements of recreational spaces in parks across three sample communities in the old city of Chengdu and the satisfaction of elderly individuals with leisure physical activities. Among them, the overall fit index is used to test the fit between the hypothetical model and the sample data. As can be seen from the table below (Table 7), CMIN/DF is 1.423, which is less than the standard of 3; PGFI and PNFI are 0.686 and 0.748 respectively, meeting the standard of more than 0.5; IFI, TLI, CFI all reach the standard of more than 0.9, RMSEA is 0.044 which is less than 0.05, all the fit indicators meet the rationality standard, this model is considered to have a good fit.

After a series of tests on the data and models above, a research model on the influence of built environment of recreation space in three community parks in the old city of Chengdu on leisure physical activity of the elderly was finally determined. Data were imported into AMOS 23.0, calculated, and estimated using the maximum likelihood method. The results (Fig. 4) indicate that all types of built environment factors directly influence the activity level of the elderly, with interactions among the factors. The built environment influences the emotional perception of the elderly through leisure interaction activities and sports and fitness activities.

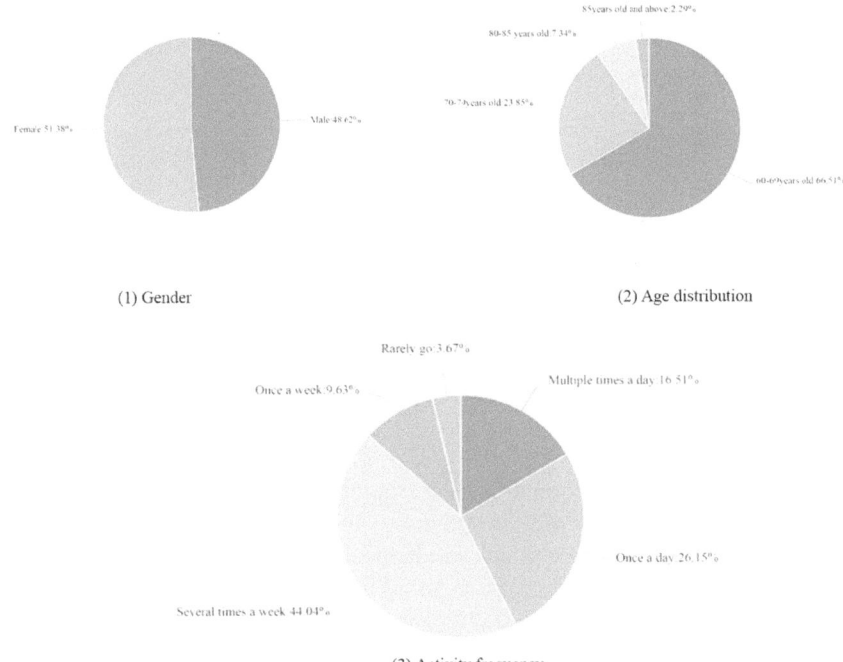

(1) Gender   (2) Age distribution

(3) Activity frequency

Fig. 3. Analysis of population characteristics

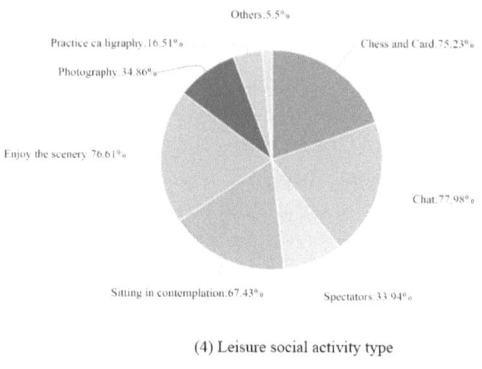

(4) Leisure social activity type

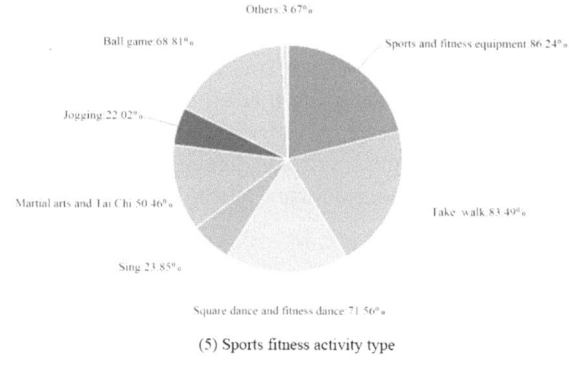

(5) Sports fitness activity type

Fig. 3. (*continued*)

Table 3. KMO and Bartlett's Test

KMO sampling adequacy		0.886
Bartlett's Test of Sphericity	Approximate Chi-Square	7777.918
	df	1485
	Sig	0.000

(1) Six dimensions of the built environment significantly and positively impact leisure social activities and sports fitness activities.. However, due to the excessive number of model indicators, the safety and landscape detail shown in the figure have negative path coefficients for leisure social activities and sports fitness activities. After separately analyzing the structural equation model of these built environment factors for leisure social activities and sports fitness activities, safety significantly and positively influences leisure social activities and sports fitness activities, with path coefficients of 0.20 and 0.21, respectively. Landscape detail has a significant positive direct impact on leisure social activities (path coefficient = 0.24), while

Table 4. Exploratory factor analysis

Component	Item	Factor loading
Accessibility	A1	0.818
	A7	0.812
	A3	0.767
	A2	0.755
	A6	0.750
	A4	0.714
	A5	0.707
Safety	B6	0.790
	B5	0.790
	B4	0.782
	B3	0.773
	B1	0.752
	B2	0.718
Equipment completeness	C5	0.791
	C3	0.760
	C2	0.746
	C4	0.734
	C1	0.718
Landscape detail	D3	0.815
	D5	0.787
	D4	0.781
	D6	0.769
	D2	.0766
	D1	0.753
Management environment	F1	0.811
	F2	0.800
	F5	0.800
	F4	0.759
	F3	0.756
Service environment	G1	0.787
	G7	0.783
	G3	0.761

(*continued*)

Table 4. (*continued*)

Component	Item	Factor loading
	G4	0.757
	G8	0.756
	G2	0.743
	G5	0.736
	G6	0.735
Leisure activities	U1	0.920
	U2	0.856
	U3	0.791
	U4	0.788
	U5	0.770
	U6	0.761
Sports fitness activities	P7	0.856
	P5	0.821
	P6	0.790
	P4	0.763
	P3	0.757
	P2	0.754
	P1	0,727
Emotional perception	Q3	0.769
	Q1	0.767
	Q4	0.730
	Q2	0.718

the management environment significantly influences sports fitness activities (path coefficient = 0.35).

(2) The degree of association of various elements of the built environment in community park recreational spaces with leisure social activities is service environment > accessibility > equipment completeness > management environment > safety > landscape detail. It can be seen that the elderly have higher requirements for the service environment of community parks, and the factor load value is 0.35, which proves that the comfort of seats in community parks, the location and distribution of built environments such as seats, public toilets, pavilions, shading facilities, activity squares, fitness facilities are extremely important for the elderly with declining physical functions. This shows that the elderly have higher requirements for the service environment of community parks. The comfort of the seats, the location and distribution of seats, activity squares, fitness facilities, etc., are very important for older people.

Table 5. Confirmatory factor analysis results

Latent variable	Item	Factor load	CR	AVE
Accessibility	A1	0.837	0.928	0.649
	A2	0.778		
	A3	0.748		
	A4	0.807		
	A5	0.831		
	A6	0.816		
	A7	0.817		
Safety	B1	0.796	0.899	0.599
	B2	0.846		
	B3	0.790		
	B4	0.789		
	B5	0.725		
	B6	0.685		
Equipment completeness	C1	0.750	0.879	0.593
	C2	0.774		
	C3	0.710		
	C4	0.829		
	C5	0.782		
Landscape detail	D1	0.761	0.888	0.570
	D2	0.787		
	D3	0.771		
	D4	0.756		
	D5	0.774		
	D6	0.675		
Management environment	F1	0.837	0.898	0.639
	F2	0.814		
	F3	0.783		
	F4	0.769		
	F5	0.792		
Service environment	G1	0.822	0.921	0.593
	G2	0.756		
	G3	0.820		

(*continued*)

Table 5. (*continued*)

Latent variable	Item	Factor load	CR	AVE
	G4	0.730		
	G5	0.801		
	G6	0.769		
	G7	0.722		
	G8	0.732		
Leisure activities	U1	0.858	0.932	0.695
	U2	0.809		
	U3	0.869		
	U4	0.795		
	U5	0.855		
	U6	0.814		
Sports fitness activities	P1	0.805	0.915	0.607
	P2	0.791		
	P3	0.804		
	P4	0.817		
	P5	0.737		
	P6	0.730		
	P7	0.763		
Emotional perception	Q1	0.763	0.842	0.572
	Q2	0.727		
	Q3	0.765		
	Q4	0.769		

(3) The degree of association of various elements of the built environment in community park recreational spaces with sports fitness activities is service environment > accessibility > equipment completeness > landscape detail > management environment > safety. Among them, the service environment has the most significant positive impact on the exercise and fitness activities of the elderly, and its factor load value is 0.30, which shows that the elderly also attach the most importance to the community park service environment when they carry out sports and fitness activities, such as walking, square dancing, singing, martial arts, jogging and ball games. Therefore, community parks should improve the comfort of seats, reasonably distribute seats, pavilions, shade facilities, activity squares, fitness facilities, etc., to enhance the emotional perception of the elderly and promote physical health.

(4) It can be seen from the figure that the factor load of leisure interaction activities on emotional perception of the elderly is 0.18, and that of sports and fitness activities

Table 6. Results of discriminative validity analysis

	AVE	Emotional perception	Sports fitness activities	Leisure activities	Service environment	Management environment	Landscape detail	Equipment completeness	Safety	Accessibility
Emotional perception	0.572	0.756								
Sports fitness activities	0.607	0.326	0.779							
Leisure activities	0.695	0.278	0.382	0.834						
Leisure activities	0.593	0.657	0.496	0.522	0.77					
Management environment	0.639	0.484	0.325	0.398	0.616	0.799				
Management environment	0.570	0.534	0.395	0.241	0.592	0.347	0.755			
Equipment completeness	0.593	0.455	0.409	0.398	0.535	0.438	0.461	0.77		
Equipment completeness	0.599	0.41	0.189	0.199	0.473	0.386	0.387	0.248	0.774	
Accessibility	0.649	0.328	0.411	0.484	0.519	0.323	0.324	0.362	0.304	0.806

Table 7. Model fit test results

Model fitting indicators	Optimal standard value	Statistic value	Fitting situation
CMIN/DF	< 3	1.423	Good Fit
PGFI	> 0.5	0.686	Good Fit
PNFI	> 0.5	0.748	Good Fit
IFI	> 0.9	0.925	Good Fit
TLI	> 0.9	0.921	Good Fit
CFI	> 0.9	0.925	Good Fit
RMSEA	< 0.05	0.044	Good Fit

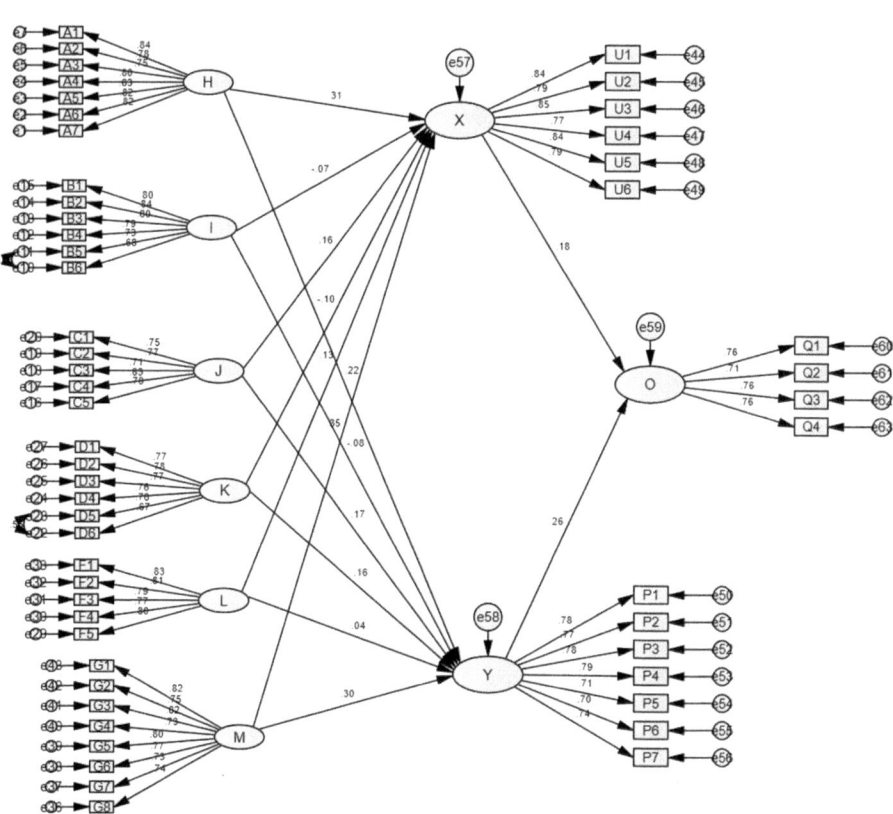

Fig. 4. Standardized path coefficient diagram of the model (*Note* In the latent variables, H represents accessibility, I represents safety, J represents equipment completeness, K represents landscape detail, L represents management environment, M represents service environment, X represents leisure social activities, Y represents sports fitness activities, O represents emotional perception)

on emotional perception of the elderly is 0.26. The elderly can have more emotional perception of relaxation, stress relief, physical fitness and belonging when they engage in sports and fitness activities such as exercise with sports and fitness equipment, walking, square dancing, fitness dancing, singing, martial arts, Tai chi, jogging and ball games. Therefore, community parks should improve the built environment such as sports fitness equipment and activity squares as much as possible, so as to strengthen sports and fitness activities for the elderly and enhance their emotional perception.

4.5 Suggestions for Urban Planning and Design

This study takes the community park with the highest frequency of the elderly as the starting point to study its built environment of recreation space. The study shows that when planning the service environment of community park recreation space, it is necessary to pay attention to the location distribution of seat comfort, activity square, fitness facilities, garbage cans, shading facilities, etc. Given the crucial role of built environment factors in the leisure physical activities of the elderly, it is essential to enhance the safety aspects by improving the public security environment in community parks, strengthening supporting facilities, enhancing lighting facilities, installing sufficient monitoring facilities, and safety signs. Additionally, focusing on accessibility, efforts should be made to enhance the transportation system and connect the road network. In terms of equipment completeness, we will increase fitness equipment, rest facilities, sunshade and rain shelter facilities. In terms of landscape detail, increase plant richness, improve walking quality, increase color saturation and so on. In terms of management environment, we conscientiously implement health environment, safety environment, facility maintenance, identification system, etc. This approach can help identify prevalent negative built environment issues affecting recreational physical activities of the elderly in community park spaces. Furthermore, it can address research gaps related to the elderly's recreational physical activities within community park environments in our country and serve as a guide for similar community parks in the region.

5 Conclusion

The aging trend in China is unstoppable, the community park environment is a prerequisite for building an elderly-friendly healthy community park. Creating a diverse, safe, well-serviced environment in parks that evoke a sense of place and belonging can effectively encourage outdoor activities among the elderly, leading to improved physical and mental well-being.

This study identified 37 evaluation indicators related to built environment elements and activities of the elderly through a literature review. Three community parks in the old city of Chengdu were selected as research samples, and 218 survey questionnaires were used as the data basis to explore and analyze the degree of association, impact pathways, and relationships between the elderly's emotional perception and various built environment elements. This study offers recommendations for enhancing the built environment of recreational spaces in community parks within the old city, with the goal of providing scientific research support for creating elderly-friendly community parks in the area.

The study establishes that the built environment elements of community parks in old urban areas influence the extent of leisure physical activity among the elderly. Firstly, government and urban planning departments should prioritize the top-level design of community parks in old urban areas, and build a safe, comfortable, and convenient community park environment. Secondly, the public service facilities around the community park should be reasonably arranged. Finally, creating a safe and comfortable service environment within the community park is of great importance for improving the leisure physical activity of the elderly and promoting physical and mental health.

This study acknowledges limitations in the selection of observed variables for constructing the system of recreational physical activity for the elderly in community park spaces within the old city of Chengdu. By synthesizing existing literature and employing methods such as field investigations and in-depth interviews, the paper develops an influence mechanism of the built environment in community park spaces on the leisure physical activities of the elderly. While the indicators are meticulously and comprehensively selected, factors such as time, region, and individual characteristics introduce uncertainties. Consequently, the construction of the evaluation system and the selection of influencing factors may not be entirely precise, potentially leading to some deviations. The conclusion of this paper is that the optimization strategy proposed for the current situation of community parks in old urban areas of Chengdu has certain limitations, and it cannot be ruled out that it cannot be applied to the built environment of community parks and recreation space in other areas.

References

1. Centre for Liveable Cities (CLC) and Civil Service College Singapore (CSC): Liveable & sustainable cities: a framework. Singapore: Centre for Liveable Cities (CLC) (2014)
2. Douglass, M.: From global intercity competition to cooperation for livable cities and economic resilience in Pacific Asia. Environ. Urban. **14**(1), 53–68 (2016)
3. Fujue, T.: Research on the optimization design of elderly physical activity space based on POE method. Jiangxi Agricultural University (2023). https://doi.org/10.27177/d.cnki.gjxnu.2022.000440
4. Yangshan, H.: Preliminary exploration of recreation. J. Guilin Tour. Coll.E **2**, 10–12 (2000)
5. Yulian, L., Zhengfan, H.: Environmental psychology. Beijing: China Architecture & Building Press (2000)
6. Qi, L.: The value of recreational space to people. Res. Dialectics Nat. **03**(22), 102–104 (2006)
7. Gehl, J.: Life between buildings. Translated by He Renke. Beijing: China Architecture & Building Press (1992)
8. Miaoru, C.: On the current situation and development of community parks in Shenzhen. Guangdong Landsc. Arch. **02**, 32–35 (2010)
9. Yu, H., Xiaomin, T.: Update the old community park activities from the perspective Evaluation of site suitability: A case study of Caoyang Park in Shanghai. 2021 Science and Technology Annual Meeting of Chinese Society of Environmental Science --Environmental Engineering Technology Innovation and Application Conference Proceedings (3)[C], 670–676 (2021)
10. Bing, Q., Fan, Z., Zhi, W.: Analysis of the main problems in domestic community park research. Mod. Urban Res. **3**, 35–41 (2019)

11. Xiaoyue, W., Dongfeng, Y.: How the built environment affects the frequency of green space use by the elderly: based on the dual perspectives of accessibility and attractiveness. Chin. Gard. **36**(11), 62–66 (2019)

12. Tianqing, L., Weiyun, F.: Case study of activity space and recreation facilities allocation in community parks under the background of population aging in Shanghai. Chin. J. Landsc. Arch. (4), 96–101 (2016)

13. Lin, Z., Suyan, L., Jia, H., et al.: Landscape design of urban community park based on public health needs. Chin. Urban For. **18**(6), 49–54 (2019)

14. Yuelu, W., Kesan, L.: Landscape design of community park based on children's health needs. Mod. Hortic. **45**(1), 172–174 (2002)

15. Man, L., Shaohua, T.: Study on community park environment based on health needs of the elderly. Arch. Cult. **7**, 193–194 (2017)

16. Shaohua, T., Lu, Z., Jize, S.: Study on the impact of community park Environment on the physical activity quality of the elderly -- Based on the perspective of individual activity and Social Interaction. 2020 Architecture of China Proceedings of the Annual Conference of the Society [C], 117–125 (2020)

17. Haohuai, Z., Feng, Q.: Discussion on the development and construction of community sports facilities abroad. Arch. J. **1**, 41–45 (2008)

18. Xiaofang, L., Tao, L., Yu, Z. et al.: Elements of urban park landscape and their effects on park activity patterns of different groups of people. J. Ecol. **40**(22), 8176–8190 (2020)

19. Chang, L., Ning, X., Jingda, S., et al.: Visitors' thermal comfort and space selection in urban forest parks. J. Ecol. **37**(10), 3561–3569 (2017)

20. Wanmin, Z., Changdong, L., Jiayao, Y.: Clu stering study on factors affecting the environment of urban parks suitable for the elderly. Chinese Landscape Architecture **37**(5), 50–55 (2021)

21. Boxin, L., Yue, X.: Research on the effects of different garden landscape types on the physical and mental health of the elderly. Landsc. Arch. (7), 113–120 (2016)

22. US Department of Health and Human Services. Physical Activity and Health:A Report of the Surgeon General. Atlanta, GA: U.S. Department of Health and Human Services & Centers for Disease Control and Prevention (1996)

23. Katzmarzyk, P.T., Tremblay, M.S.. Limitations of Canada's Physical activity data: implications for monitoring trends. Can. J. Public Health **98**(Supplement 2), 185–194 (2007)

Health-Oriented Walkability Measurement and Enhancement Strategies for Life Circle

Mingfeng Xie[✉] and Yu Wang

School of Civil Engineering and Architecture, Southwest University of Science and Technology, Mianyang 621010, China
a2430229739@163.com

Abstract. Walkability is an important characterization of the degree of pedestrian friendliness in cities, and is also an important element of healthy urban planning. However, due to the heat island effect and climate warming, the poor thermal comfort of walking in summer has seriously hindered residents' willingness to walk. In this paper, we take Nanhe Community 15-min living circle in Mianyang City, Sichuan Province as an example, simulate the thermal environment in the living circle in summer through ENVI-met software, and calculate the walkability based on the walk score, and simulate the identification of the thermal environment and walkability condition of the living circle. The results show that the overall thermal environment in the study area is poor, especially in the residential and commercial areas where buildings are concentrated, which affects the residents' willingness to travel, while the thermal environment at the edge of the study area is relatively good due to the factors such as the river. By combining the thermal environment simulation and walkability simulation results, the high walkability-low thermal comfort spaces and low walkability-high thermal comfort spaces are identified. Optimization strategies are proposed for these two types of spaces, in order to provide guidance for the improvement of walkability in the living area.

Keywords: Walkability · Walk score · Life circle · Thermal environment simulation

1 Introduction

In recent years, due to the development of urbanization, the rapid expansion of urban space, walking, cycling and other green travel modes have been replaced by motor vehicle travel, resulting in a large reduction in travel activities of residents, increasing the risk of health disease induced, and the health risk of residents is becoming increasingly serious [1]. At the same time, due to the increasingly serious urban heat island effect, the risk of urban heat exposure has also increased dramatically, in August 2020, the World Health Organization (WHO) data show that the growth rate of casualties triggered by high-temperature heat waves is much higher than that of other extreme weather. Hot environments not only increase the frequency of heat stroke and other thermal safety events and reduce human comfort, but prolonged exposure to hot weather significantly increases

© The Author(s) 2025
B.-J. He et al. (Eds.): UCSUD 2023, LNCE 559, pp. 827–840, 2025.
https://doi.org/10.1007/978-981-97-8401-1_59

the incidence of respiratory, cardiovascular, and other diseases [2]. Urban residents in the summer travel risk is too large, travel health benefits are low, so that residents in the summer more reluctant to choose walking mode travel. Therefore, when analyzing the built environment walkability, it is important to focus on the status of walkability under the influence of high-temperature heatwave climate as well as enhancement strategies. As the basic unit of China's future urban communities, the planning and construction of living areas should analyze the thermal environment and walkability in summer, and accurately identify the friendly spaces with high walking and low thermal environments, as well as the health-exposed spaces with low walking and high thermal environments, so as to encourage residents to choose walking as a travel mode more often in the summer, and to reduce the health risks of the residents.

The concept of walkability originated from the theory of sustainable urban design, and was initially proposed in transportation research in the United States, and is often referred to as "walkability" in related studies [3]. Because of the close connection between a good walking environment and the health of urban residents, economic development, social equity, etc., the research dimension of walkability has also expanded from urban transportation to sports, health, public health, urban planning and other fields, and it has been widely applied. Walkability score is an international method to quantitatively measure walkability, which has been well used in the ranking of urban walkability, spatial layout of daily service facilities [4–6], evaluation of urban housing prices [7], and research on the health level of urban community residents, etc., and has been widely used in the United States, Canada, Australia, the United Kingdom, New Zealand and other countries. This year, walkability has also been heavily utilized in the construction of healthy cities, for example, Leng Hong [8] and other studies have found that a good walking environment has a positive impact on human health. Chen Xi [9] and others explored the health effects of built environment based on the measurements of walkability and traffic pollutants. Xiao Yang et al. [10] explored the key points of healthy urban planning through the built environment from a health perspective.

Scholars have also carried out a large number of studies on the urban thermal environment, mainly focusing on the measurement of spatial and temporal changes in the heat island effect, the response to land use changes, and the excavation of factors affecting the thermal environment. In addition, the health effect of urban thermal environment is also a key area of concern for scholars. Huanchun Huang [11] and others found that the degree of harm and the spatial area of influence of thermal environment on the health of residents showed an increasing trend, which was mainly concentrated in the urban center area. Jingdi Nie et al. [12] found that the heat island effect will increase air pollution indirectly affecting human health. At the level of human thermal comfort, thermal comfort research has been conducted for a hundred years since the development of the first thermal comfort evaluation "effective temperature (ET)" [13], but due to the fact that the early thermal comfort evaluation were derived from the indoor thermal comfort theory, they were mostly based on the human body's ability to stay in contact with the environment for a long period of time and eventually reach the thermal comfort level. However, since the early thermal comfort evaluation followed the indoor thermal comfort theory and were based on the assumption that the human body is in contact with the environment for a long period of time and eventually reaches thermal equilibrium (i.e., the steady state situation), but in the actual outdoor environment the human body's thermal loads are in a constant state of flux, the evaluation based on the steady state heat transfer model are not considered to be able to effectively explain the dynamic aspects of the human thermal acclimatization process [14]. In this regard, Fiala et al. proposed a universal thermal climate score (UTCI) [15] based on a multi-node model of thermoregulation, aiming to more accurately simulate and predict human thermal sensation in non-steady state or transient conditions. And Wang Zuoxing et al. [16] demonstrated that UTCI has a good effect on the prediction of thermal environments in various climatic zones by combing the practical applications of UTCI evaluation models. Although a large number of studies have been carried out in the fields of walkability and thermal environment respectively, walkability measurements are still based on the pedestrian built environment such as service facilities and pedestrian streets, and there is still a lack of research related to integrated walkability and thermal environment.

In view of this, this study takes Nanhe Community 1 15-min living circle in Mianyang City, Sichuan Province as the study area, simulates the thermal environment in summer through ENVI-met software, and measures the walkability based on the walking score, and then identifies two types of spaces that need to be rectified, namely, high walkability-low thermal comfort and low walkability-high thermal comfort, and then proposes the strategy of improving walkability of the living circle based on it to provide guidance for the construction of a healthy living circle. Based on this, a walkability improvement strategy is proposed to provide guidance for the construction of a healthy living area.

2 Study Area, Data Sources and Methods

2.1 Study Area and Data Sources

Mianyang City, abbreviated as "Mian", also known as Mianzhou and Fucheng, a prefecture-level city under the jurisdiction of Sichuan Province, a provincial sub-center city, and also China's only science and technology city, Mianyang City, Nanhe Community(Fig. 1) is located in the main urban area of Mianyang City, the terrain is flat, there are many residential districts and residents, and the age of the different buildings spans over a large period of time, and the urban walking built environment is diversified, so the Nanhe Community of Mianyang City has a research value. Nanhe community in Mianyang city has a 15-min community living circle with research value, so it is selected as the research object. Meteorological data were obtained from the website of China Meteorological Administration, and the site and building height data of the study area were obtained from the sky map of China Geographic Information Public Service Platform, and the data were processed by Arcgis to emulate the thermal environment of the study area. The POI (point of interest) data were obtained from the Gaode map, with the time node of 2023, and each POI data contains name, category, latitude and longitude, etc. The road network traffic data comes from the mapping data of Mianyang city in 2023 and the field research data, which are combined to measure the walk source.

Fig. 1. Study area

2.2 Methods

2.2.1 Thermal Environment Simulation

In the study, based on the meteorological data of China Meteorological Administration, the meteorological data of Fucheng District, Mianyang City, in the summer time of 2023 were counted. The statistical results show that the highest temperature of the summer season in Mianyang in 2023 is the hot month of August, and the day in the hottest month when the daily mean values of temperature, daily difference, humidity, and solar radiation illuminance are the closest to the mean value of the month is August 19th, so the meteorological data of August 19th in 2023 is selected as the meteorological parameter used in ENVI-met simulation of the present study. The specific parameter values are shown in Table 1.

Table 1. ENVI-met simulated meteorological parameters.

Data category	Specific parameters	Data category	Specific parameters
Longitude and latitude	Longitude 31°31′ East, Latitude 104°40′ North	Minimum temperature	26.8 °C (07:00)
Simulation time	18:00	Highest temperature	36 °C (17:00)
Direction of the wind	245°	Minimum humidity	61% (17:00)
Wind speed	1.9m/s	Maximum humidity	96% (12:00)

Before the thermal environment simulation experiment, the author went to the Nanhe community in Mianyang City, Sichuan Province, to conduct on-site research, through the distribution of questionnaires to collect the use of public service facilities and summer travel frequency of residents in the study area, according to the results of the questionnaire, found that the residents of the study area walk during the daytime travel flow of the peak hours of 08:00–09:00 and 18:00–19:00, comparing the peak flow of these two Comparing these two peak traffic periods, the worst time for outdoor thermal environment is 18:00–19:00, so this time is chosen as the representative time for the more prominent thermal environment problems.

After completing the data collection, the study area was modeled in the ENVI-met software, in which the building height was obtained by processing the projection of the building satellite map through the Arcgis software, and corrected according to the field research to ensure that the building modeling parameters were compatible with the actual situation in the case area. Plant configuration According to the field research, it was found that the plants in the study area were mainly street trees and bushes, so the simple plants in the plant database in the ENVI-met system were modified and updated, and two kinds of plant models that meet the characteristics of the plants in the case area were reconstructed for modeling (e.g., Table 2).

Table 2. Plant modeling parameters table.

Plant height (m)	Crown (m)	Branch point (m)	Leaf albedo	Leaf transmittance
12	7	2	0.18	0.3
9	4	2	0.12	0.1

The setting of urban underlayment can be divided into man-made underlayment and natural underlayment. Through the field investigation of the case area, it is found that the man-made underlayment in the case area mainly includes asphalt pavement, soil, concrete pavement, permeable brick pavement and marble paving and other types of underlayment; the natural underlayment is mainly soil. With reference to the general parameter values of the sub-surface, the parameters of each type of sub-surface in the case area are assigned values, and the attributes of the sub-surface in the specific case area are shown in Table 3.

After obtaining the basic data, the model was built in ENVI-met software, and after setting the simulation duration to 18:00–19:00, the thermal environmental conditions in the study area were simulated.

2.2.2 Walkability Calculations

The calculation of walking score can be divided into single-point score calculation and facet score calculation. The calculation of single-point walking score can be divided into the following three steps: preparation of the facility classification table; calculation of the base walking score; and correction of the walking score. The surface area score is obtained by spatial interpolation of the single-point walking score. Referring to the

Table 3. Parameter table for modeling of the subsurface

Material	Intensity	Specific heat capacity	Thermal capacity	Thermal conduction	Thermal diffusion	Thermal absorption
Loam	1600	890	1420	0.25	0.18	600
Water	1000	4180	4180	0.57	0.14	1545
Pitch	560	800	1940	0.75	0.38	1205
Brick	1970	800	1370	0.83	0.61	1065
Concrete	2300	650	2110	1.51	0.72	1785

walking score calculation method. This study first classifies facilities based on Point of Interest data in the study area, covering the types of stores, restaurants, parks, schools, etc. The study then basically assigns different weights to various types of facilities in accordance with the indicators of the Walking score Company and adjusts them in combination with Chinese scholars [17] for the study of urban facility classification in the Walking score to obtain the final facility classification weight table (e.g., Table 4).

Table 4. Facility classification weight table.

Category 1 (i.e. class A)	Category II	Weights
Teach	Primary, secondary, kindergarten	1
Health care	Hospitals, pharmacies	1
For public use	Bus stops, subway stations park (for public recreation)	23
Financial Post and Telecommunications	Banks, ATMs webmail	11
Business Services	Supermarkets, stores Catering, restaurants Leisure and entertainment venues (movie theaters, etc.)	211

At the same time, the study area was divided into circles, and the corresponding attenuation coefficients were multiplied according to the weights within different circles. In order to facilitate the calculation, this study adopts a segmented function to set the attenuation coefficient for different distances: there is no attenuation of facility services within 400 m, the attenuation coefficient of 400 ~ 800 m is 0.9, the attenuation coefficient of 800 ~ 1200 m is 0.55, and the attenuation coefficient of 1,200 ~ 1,600 m is 0.25 (e.g., Table 5), and since there are no destinations exceeding 1,600 m within the 15-min circle walking range. Therefore, the interval of 1600 ~ 2400 m is not considered.

Finally, according to the circle range of 400 m, 800 m, 1200 m and 1600 m, the total score of the weight of the POI types around the street is counted to calculate the basic walking score of the study area, which is calculated as follows:

Table 5. Distance attenuation factor.

Timing	Reach	Distance decay law
5min	400m	not have
10min	400 to 800 m	0.9
15min	800 to 1200 m	0.55
20min	1 200 to 1 600 m	0.25

After obtaining the basic walking score, since only a simple distance attenuation calculation is carried out, while the real travel is along the road travel, the more road intersections, the shorter the block, the more varied the route choices for walking travel, it is necessary to further consider the impact of the actual walking environment by setting the density of road intersections in the area and the attenuation rate of the length of the block (e.g., Table 6), and using this as the basis for a corrective formula for its Correction, the formula is as follows:

$$W = W_n \times (1 - y) \times (1 - r) \tag{1}$$

where W is the single-point walking score, the is the single-point base walking score, y is the street intersection density decay rate, and r is the block length decay rate.s

Table 6. Attenuation factor of intersections and block lengths table.

Density of intersections/(pcs/km^2)	Attenuation rate/%	Block length/100m	Attenuation rate/%
≥ 77	0	≤ 120	0
58 ~ 77	1	120–150	1
47 ~ 58	2	150–165	2
35 ~ 47	3	165–180	3
23 ~ 35	4	180–195	4
< 23	5	>195	5

After obtaining the corrected single-point walking score, the surface area walking score of the study area is obtained by ordinary kriging interpolation, and finally the results are normalized to 0–100, which enables the visualization of the walking score of the study area.

3   Experimental Results and Analysis

3.1   Overall Thermal Environment in the Study Area

According to the simulation map of the thermal environment in the study area(Fig. 2), we can see that the overall thermal environment in the study area is relatively severe, with a higher risk of thermal exposure, and the high-temperature zones in the study area show a tendency to be concentrated in the center of the area, whereas the periphery of the study area, especially the northeastern area, has a relatively lesser problem of thermal exposure. This situation is caused by two main factors: first, the central location of the study area is farther away from the river zone than the peripheral area, and there is a lack of corresponding artificial lakes or rivers in the interior, which leads to higher temperatures compared with other areas; second, there are many high-rise buildings in the peripheral area of the study area, resulting in the absence of good ventilation corridors within the study area, which does not achieve the effect of heat dissipation, and the presence of many high-rise buildings in the peripheral area results in the absence of good ventilation corridors, which does not achieve the effect of heat dissipation. At the same time, due to the existence of many high-rise buildings in the surrounding area, the heat island effect is more serious in the center of the study area, which ultimately leads to a poor thermal environment in the area, and the outdoor walking thermal comfort of people is correspondingly lower.

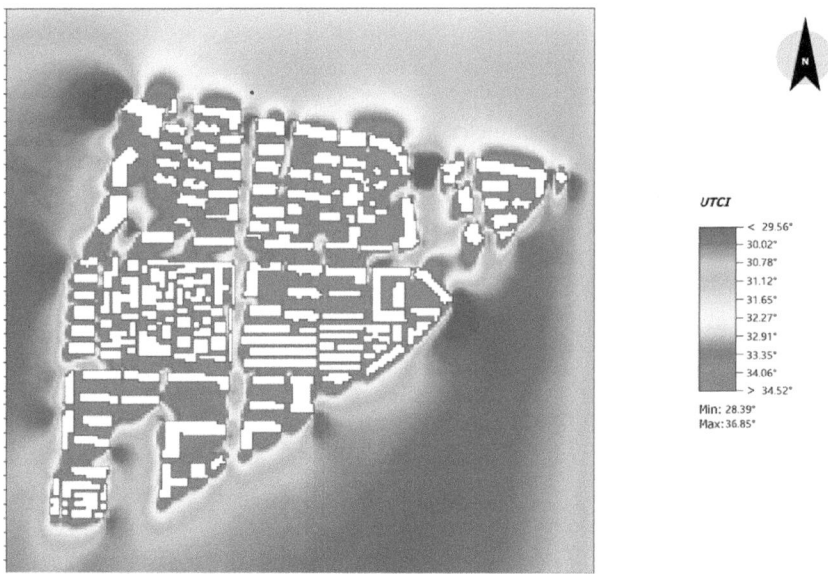

Fig. 2. Thermal environment simulation diagram

3.2 Zoning Status of the Thermal Environment in the Study Area

Based on the previously obtained thermal environment simulation of the study area, the study area was divided into six areas and their thermal environment conditions were analyzed separately(Figs. 3, 4, 5).

The overall ambient temperature in study area I shows a high trend, but the surrounding temperature is low compared to the temperature inside the building complex, and the thermal comfort is higher when people walk, and the greening inside the building complex is more concentrated in the area with a better thermal environment, which has less impact on the walking of people, but this area is not large and only accounts for a small part of the whole building complex, which is located in between the multi-storey building complexes.

The area of the western complex in study area II has a friendlier thermal environment and higher human walking thermal comfort due to its river-front location, but the interior of the eastern complex has a poorer wind corridor effect due to its enclosed location with more buildings, which results in a less human-friendly thermal environment in this area.

Study Area III has a better internal thermal environment due to its larger river frontage and the number and location of greening arrangements in the study area, thus ensuring that the walking thermal comfort is maintained at a high level during summer pedestrian trips.

The better thermal environment areas in Study Area IV are mainly located at the edges of the study area and are large enough to provide better thermal comfort for human walking, but there are still areas of high temperatures in the center, which to a certain extent affect the summer walking comfort of some residents.

Study Area V has a poor thermal environment due to dense buildings and outdated planning, especially in the central part of the complex, where the high temperature and heat wave conditions greatly affect the motivation and frequency of residents to travel, and only the part along the river has a relatively good thermal environment from the perspective of the whole area.

Study AreaVI has a better overall thermal environment due to its fewer buildings and better layout, thus reducing the impact on the thermal environment at the planning level. Coupled with the fact that the overall area is large enough to be adjacent to the river, the thermal comfort of people when walking will be further enhanced.

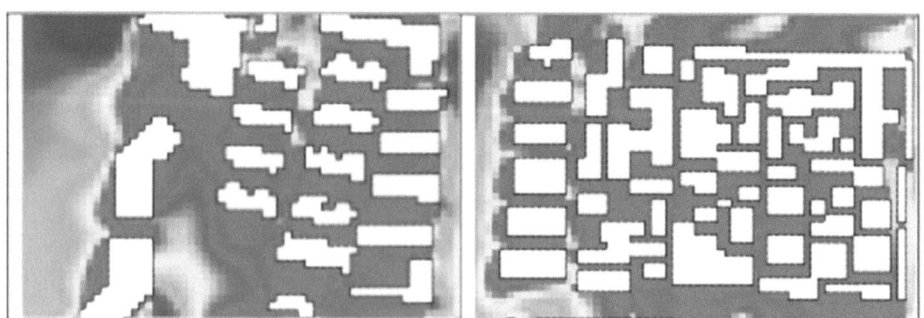

Fig. 3. Study area I and II

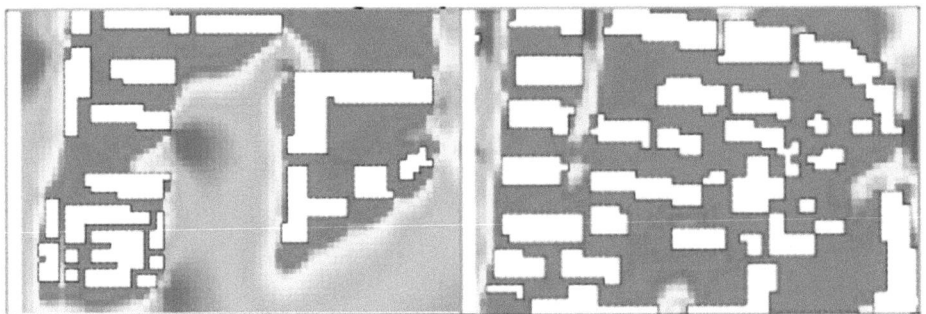

Fig. 4. Study area III and IV

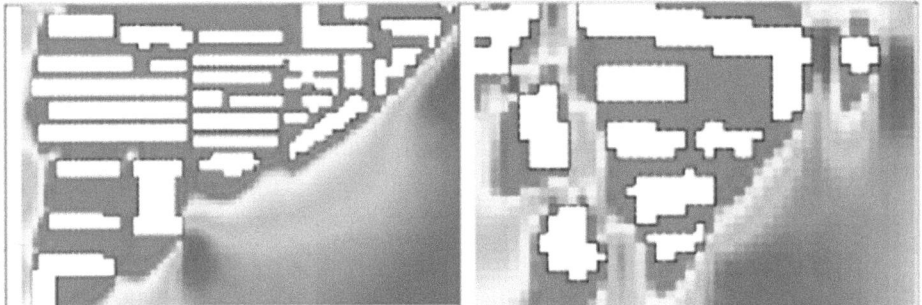

Fig. 5. Study area V and VI

3.3 Characterization of the Spatial Distribution of Walkability

After calculating the faceted walking score of the study area based on the corrected single-point walking score of the study area, referring to the relevant literature, in order to analyze the overall characteristics of the walking score and the internal differences of the walking score of the study area the walking score is divided into five levels: 0–0.26, poor walkability; 0.27–0.48, poor walkability; 0.49–0.69, walkability at a medium level; 0.81–0.89, high walkability, and 0.9–1, high walkability. The overall walkability of the study area was assessed based on the five levels of categorization criteria as shown in the Fig. 6. The total walkability of the study area is characterized by high in the middle and low on both sides, with the high walkability area mainly concentrated in the central area, which is distributed in a 45° band; from the high walkability area to the outer extension, the walkability is getting worse and worse, and the further away from the central area, the worse the walkability will be. According to the six small areas divided in the previous section, the analysis of their walkability shows that Study Area 1, Study Area 4 and Study Area 5 are more walkable, of which Study Area 5 has the best walkability, while Study Area 2, Study Area 3 and Study Area 6 are less walkable, of which the southwest direction of Study Area 3 and Study Area 6 have the worst walkability.

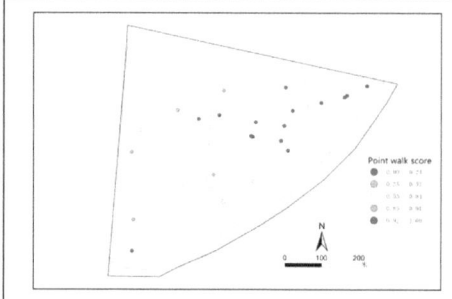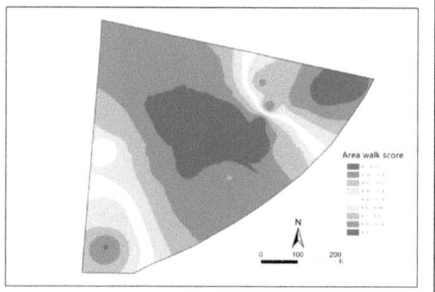

Fig. 6. Point walk score and area walk score

3.4   Walkability Enhancement Strategies for the Life Circl

By identifying the thermal environment and walkability of the study area, it is known that the high walkability-low thermal comfort space in the study area is study area V, while the low walkability-high thermal comfort space is study area III and study area VI, which need to be optimized and upgraded.

3.4.1   High Walking-Low Heat Comfort Space Enhancement Strategies

This type of area is highly walkable, but the poor summer thermal environment leads to a low willingness to travel and a decrease in the number of trips made by residents. The poor thermal environment also leads to residents being more susceptible to high-temperature heatwave weather while traveling, which just increases the number of thermal safety incidents, so an enhancement strategy is needed for this area, with the main goal of improving the summer thermal environment conditions in this area and enhancing the thermal comfort of residents traveling in the summer.

By understanding the basic overview of the study area V, we can know that the number of buildings in the area is large and the planning is relatively chaotic, and it is difficult to build a good ventilation corridor inside the study area. Since the area is mostly old buildings, it is necessary to dismantle the regional buildings to a certain extent when improving the area, to open up the channel between the area and the neighboring rivers, to promote a smoother flow of the wind environment inside the area, and to alleviate the thermal environment situation inside the area.

At the same time, through the field survey and satellite map, it can be learned that the area is relatively lacking in greening, and the greening types are single. Therefore, it is possible to introduce appropriate greening types and increase the greening area, in addition to planting street trees, it is also possible to focus on greening in several building enclosures that occupy a large area. In addition to the introduction of greening, we can also consider the introduction of artificial lakes in the region, through the heat-retarding function of the water body to further regulate the thermal environment of the region.

According to previous literature, facilities such as vignettes and seating can alleviate human thermal discomfort, so consideration can be given to providing facilities such as seating at intervals of distance in this area to provide a temporary rest stop in the middle of the walk, thereby improving the thermal comfort while walking.

3.4.2 Low Walking-High Thermal Comfort Space Enhancement Strategies

Study Area III Enhancement Strategy Through fieldwork and simulation studies, it was found that the low walkability of this area is mainly due to the fact that there are fewer daily service facilities in the vicinity, and only one middle school is a more important daily service facility. Therefore, in order to improve the walkability of this study area, it is mainly necessary to increase the types of daily service facilities in the area, especially stores, bus stops, and other daily life use of high-frequency facilities, so as to ensure that residents do not need to spend too much time to get to the facilities they need in order to promote the willingness of residents to go out on foot. It is important to increase the variety of daily facilities in the area, especially shops and bus stops, to ensure that residents do not need to spend too much time to get to the facilities they need in their daily life when they walk out, thus promoting their willingness to travel. Meanwhile, since the southern part of the area is a large area facing the river, it is recommended that a park or a riverfront embankment be installed in the southern part of the area in order to enhance the recreational function of the area and attract residents to travel.

Study Area VI Enhancement Strategies Although this area has some daily service facilities, its walkability is still low for two main reasons, firstly, it is far away from the rest of the study area and passes through more intersections, which affects the residents' travel, and secondly, although there are a certain number of daily facilities in the area, it lacks the facilities that are most frequently and heavily used by the residents, such as shopping malls, hospitals, and so on.

   To solve these two problems, the following measures are proposed: firstly, an embankment or trestle can be built in the riverfront area of Study Area 6 along with the whole study area, which is calculated to be a shorter walking distance to other areas than crossing the center of the area, and can reduce the number of intersections when walking. Second, for the lack of important daily service facilities, according to the actual situation of Study Area 6, small important daily service facilities, such as stores, can be set up inside the area, and set up a small park in the riverfront area below, and for hospitals, large CBD and other facilities with a large footprint and high frequency of use, can be set up in the Study Area 4, Study Area 4, which is not only closer to the Study Area 6, can better promote the development of Study Area 6, and can better improve the quality of the area. Study Area 4 is not only closer to Study Area 6, which can better enhance the walkability of Study Area 6, but the addition of important daily service facilities can also enhance the walkability of Study Area 4, which can further enhance the walkability of the entire study area. Study Area 5 is not considered for the addition of important daily service facilities because its walkability is already very high and it is farther away from Study Area 6 as compared to Study Area 4.

4 Conclusion and Discussion

4.1 Discussion

Increasing the walkability of built environments to promote physical activity and thus public health has become a hot topic, but a one-sided emphasis on increasing walkability while ignoring the risks of urban thermal environments is not conducive to improving public health. This adverse effect is especially significant in urban areas where the heat island effect is severe in summer. Therefore, when building a living area, the impact of the built environment on the health effects of residents needs to be considered comprehensively. Judging from both the possibility of walking activities and the risk of thermal environmental exposure, friendly spaces and warning spaces of the built environment can be effectively identified, so as to provide guiding suggestions for the formulation of scientific and reasonable policies for planning and management of living circles. Based on this, this paper combines thermal environment and walkability, seeks a method to identify thermal environment and walkability friendly spaces and warning spaces in the built environment, and proposes an optimization strategy for the warning spaces.

4.2 Conclusion

In this paper, the 15-min life circle of Nanhe community in Mianyang city, Sichuan Province, is taken as the research object, and the results within the research scope is analyzed. The research results show that:

(1) The overall thermal environment of the study area shows a trend of high in the middle and low on both sides, and the poor thermal environment is mainly concentrated in the area with more buildings, while the thermal environment of the edge area is better. (2) The high walkability area in the study area is distributed in a 45° band from the center, and spreads outward from the high walkability area, and the walkability of the study area shows a positive downward trend. (3) The low walkability zones in the study area are mainly concentrated in the southwest and northeast of the study area, which are caused by the small number of facilities, uneven distribution of facilities, and the influence of the built environment.

Although this study proposed a walking enhancement strategy for the living area by combining walkability and thermal environmental exposure, there are still many deficiencies, and the different spatial and temporal characteristics of users within the living area should be taken into account when selecting walkability indicators. Secondly, this article used road network data instead of the traditional Euclidean distance, but ignored the actual transportation road conditions. Finally, the social homogenization of walkability is not considered in the application of the distance decay law, and comparative studies should be conducted on the walkability of different types of people, and further in-depth research will be carried out in the future.

References

1. Cui Haiying,Yin Chaohui,Lu Xinhai et al.: Identification of urban thermal environmental exposure risk zones based on walkability and their health effects--a case study of Wuhan main city[J]. Geogr. Res. Dev. 42 (01), 75–80+107 (2023)
2. Jiansheng, W.U., Wei, Q.I.N., Jian, P.E.N.G., et al.: Rationality assessment of urban daily life facilities allocation based on walking score–taking Futian District of Shenzhen as an example[J]. Urban Dev. Res. 21(10), 49–56 (2014)
3. Cervero, R., Kockelman, K.: Travel demand and the 3Ds: density, diversity, and design [J]. Transp. Res. Part D 2(3), 199–219 (1997)

4. HUANG Jianzhong, HU Gangyu, LI Min.: Study on the suitability of community service facility layout from the perspective of the elderly: based on the method of walking score[.J]. J. Urban Plan. (6), 45–53 (2016)
5. Jiansheng, W., Nan, S.: Research on the social service function of park green space in Futian District, Shenzhen based on the walking score[J]. J. Ecol. **37**(22), 7483–7492 (2017)
6. Yue, F.: Waterfront slow walking space design based on walkability experience: an example of Sentosa crossing walkway in Singapore [J]. Decoration **8**, 130–131 (2017)
7. Gilderbloom, Ji., Riggs, W.W., Meares, W.L., et al.: The importance of walkability: the impact of walkability on house prices, foreclosure rates, and crime rates[J]. J. Urban Plan. (4), 121 (2015)
8. Leng Hong, Zheng Chunyu, Lu Yuwen.: A study on walkability of urban public spaces in cold areas under the perspective of fitness and mobility of the aging population [J]. Int. Urban Plan. **34** (5), 27–32 (2019)
9. Xi, C., Jianxi, F.: Health effects of built environment based on the comparison of walkability and spatial pattern of pollutant exposure: a case study of Nanjing[J]. Adv. Geosci. **38**(2), 296–304 (2019)
10. Xiao Yang, Sarkar, Webster.: The link between the built environment and health: an exploratory study from the Center for High Density Healthy Cities Research at the University of Hong Kong[J]. Times Arch. (5), 29–33 (2017)
11. Huang Huanchun, Yang Hailin, Deng Xin, et al.: Spatial evolution process of urban heat island impact on residents' health[J]. Remote. Sens. Inf. **36** (4), 38–46 (2021)
12. Nie Jingdi, Zhang Junhua, Huang Bo.: A review of research on the impact of urban heat island effect on human health[J]. Ecol. Sci. **40** (1), 200–208 (2021)
13. Aynsley, R., Marcus, S.: Thermal comfort models for outdoorthermal comfort in warm humid climates and probabilities of lowwind speeds [J]. Eng. & Ind. Aerodyn. **36**(1), 481–488 (1990)
14. Jendritzky, G., Maarouf, A., Fiala, D. et al.: An update on the development of a universal thermal climate score [C]// 15th Conf.Biomet.Aerobiol. and 16th ICB02 Aerobiol. and 16th ICB02, 129–133 (2002)
15. Shih, W.M., Lin, T.P., Tan, N.X., Liu, M.H.: Long-term perceptions of outdoor thermalenvironments in an elementary school in a hot-humid climate. Int J Biometeorol - 88, **61** (9), 1657–1666 (2017)
16. Wang Zuoxing, Guo Fei, Guo Ruonan.: Research on outdoor human thermal sensation in different climate zones based on UTCI score[J]. Low Temp. Build. Technol. **42** (12), 6–10, 18 (2020)
17. Yintao, L.U., De, W.A.N.G.: Progress of walkability measurement research in the united states and its implications[J]. Int. Urban Plan. **27**(1), 10–15 (2012)

The Scene Construction of Modern Community: A Case of the Residential Area of Wulidian Street in Chongqing, China

Lei Zhang[1], Jinyu Zeng[2], and Liming Dai[1,1(✉)]

[1] Center for Science and Innovation, Chongqing Design Group Urban Construction Strategy Research Institute, Chongqing 400072, China
csjsclyjy001@163.com

[2] Studio for Architectural Strategy Expert, Chongqing Design Group Urban Construction Strategy Research Institute, Chongqing 400072, China

Abstract. The community, as the fundamental unit of residents' lives, serves as the primary carrier of people's aspirations for a better life and also marks an important indicator of a city's level of modernization. Chongqing boasts abundant industrial heritage resources, and under policy support, the renewal projects focusing on outdated factories and their associated residential areas are a key direction for future urban renewal in Chongqing. This study takes the old community of Yinghua Lane in Wulidian Street, Jiangbei District, as a case study and is guided by the Modern Community Scenario Theory. Through the exploration of four characteristic scenarios this research provides solutions to urban governance issues within the area and offers solid theoretical support for the construction of modern communities in Jiangbei District, Chongqing, China.

Keywords: Modern community · Urban governance · Industrial culture

1 Foundations of Modern Community Theory Research

In recent years, concepts such as ecological communities, healthy communities, zero-carbon communities, smart communities, shared communities, resilient communities, and green communities have been proposed domestically and internationally. Notably, Zhejiang Province has introduced the concept of "Future Communities", which has had a profound influence nationwide. The Future Community concept features a "139" top-level design: one core—centered around the aspirations for a better life for the people; three value orientations—human-centric, ecological, and digital; and nine scenarios—neighborhood, education, health, entrepreneurship, architecture, transportation, low-carbon, services, and governance. The construction of Future Communities in Zhejiang provides a new paradigm for emerging first-tier and second- and third-tier cities (including megacities, Type I and Type II large cities, and medium-sized cities). This also offers fresh perspectives for urban renewal in Chongqing.

Future communities refer to the transformation of existing communities to better adapt to future needs, representing an innovation built upon existing urban structures.

B.-J. He et al. (Eds.): UCSUD 2023, LNCE 559, pp. 841–853, 2025.
https://doi.org/10.1007/978-981-97-8401-1_60

The focus of such transformations is to address the pervasive urban issues within communities—once these persistent problems are resolved, the community achieves a degree of future-readiness. Modern communities can be considered as a transitional phase towards future communities, representing a localized exploration of future community theory in conjunction with the current economic situation and urban renewal status in Chongqing.

In 2022, Chongqing's GDP per capita reached 90,663 Yuan, aligning with the World Bank's threshold for high-income economies(12,536$). Consequently, the demands of Chongqing residents have shifted from basic to more advanced developmental needs, giving rise to new lifestyles. However, issues of imbalance and inadequacy persist in urban and community construction and governance, which are closely linked to the daily lives of the people. Chongqing is at a critical historical juncture in initiating high-level modernization efforts. Against the backdrop of deepening and actualizing the construction of the Chengdu-Chongqing economic circle, the challenge lies in how to integrate the developmental strategy of "new era, new journey, new Chongqing" with the unique characteristics of Chongqing's communities. By focusing on the transformation of old urban residential areas and urban renewal, and enhancing the areas surrounding railway stations, the research on "modern communities" in Chongqing not only reflects the responsibility and mission of urban developers but also brings new business opportunities and prospects to the survey and design industry.

2   The Application of Modern Community Theory in Local Adaptation

2.1   Urban Development Foundation

Industrial development has played a pivotal role in the history of urban construction in Chongqing. With historical changes, a considerable amount of industrial heritage and cultural resources has been preserved in the city. As one of China's six major old industrial bases, Chongqing has experienced comprehensive development through various stages, including the port-opening period, the wartime period during the resistance against Japanese aggression, and after the establishment of the People's Republic of China. The transformation of industrial factories and their accompanying residential areas has witnessed the process of industrialization and urbanization in Chongqing.

2.2   Project Site Foundation

The project is located in Wulidian Street, Jiangbei District, directly above the Liyuchi subway station (Fig. 1). To the east, it adjoins the Jiangbei Central Business District, a core area of Jiangbei District with high accessibility to transportation. Within a radius of 1 km from the site, there is not only a wealth of cultural resources and multiple tourist attractions, but it also faces the municipal government across the river and is merely a street away from the headquarters of Chang'an Group (Fig. 2), embodying a profound cultural heritage.

Jiangbei has a historical foundation and a promising future. As the city's demonstration zone for "Two Highs" (high-tech industry and high-quality urban development),

Jiangbei District is dedicated to creating a city business card for a modern international metropolis with mountains and water. The urban renewal and transformation project in the former Chang'an Factory area has concentrated on the Chongqing's industrial culture, witnessing the old and new changes in Chongqing's modernization process. It is the root and soul of the older generation of Jiangbei people, and even all Chongqing people. Nowadays, with the development of the times and the relocation of the factory, the identity of the proud old Chang'an residents in this area has become the residents of old communities who urgently need to improve their living environment.

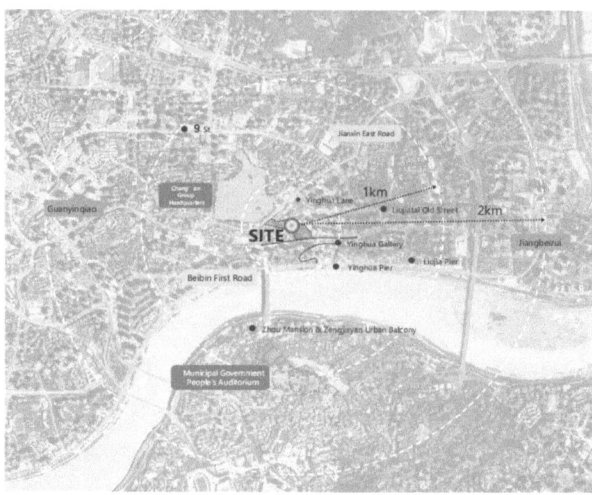

Fig. 1. The location of the project site and its connection with surroundings.

The Wulidian area bears the root and soul of Chongqing's industrial culture, witnessing the entire history of Chang'an from the construction of factories to the relocation. However, surrounded by high-rise buildings, the original residents feel the loneliness of being abandoned by The Times. The total area of the case transformation in this study is approximately 140,000 square meters, with nearly a hundred buildings. The surrounding geographical conditions are complex, with both new and old communities coexisting. The renovation and transformation need to consider not only the improvement of internal functions and quality, but also ensure the integration with surrounding buildings, transportation, and environment. Furthermore, it is necessary to deeply explore its regional context and historical fabric. Reshaping the neighborhood scene and awakening the memory of Chang'an is not only a local formation of modern community, but also a search for the source of Jiangbei regional identity.

How to focus on the "Modern Community Scene Theory" highlight the three main themes of organic renewal, cultural heritage protection, and functional improvement, and create a benchmark for Jiangbei and a demonstration for Chongqing in the construction of a "modern community" is a pioneering exploration for further improving the transformation of old urban communities.

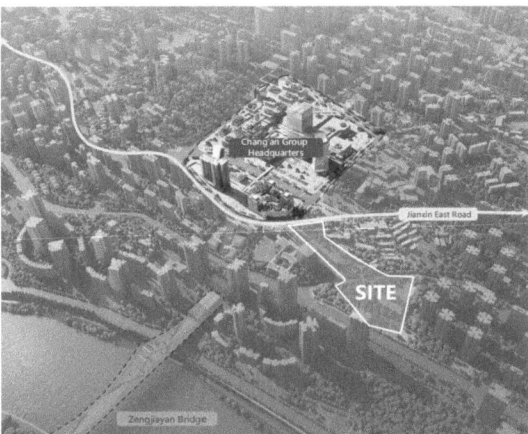

Fig. 2. The spatial relationship between project site and Chang'an Group Headquarters.

3 "1 + 4"——Modern Community Characteristic Scene Path

Based on the "Implementation Plan for Advancing Urban Modernization in Jiangbei District (Trial)" by the Jiangbei District Commission of Housing and Urban-Rural Development, the district adheres to the planning and deployment of "advancing urban modernization in a leading manner" as stated in the 13th Party Congress of Jiangbei District. It promotes the theory of modern community "one unity, three transformations, nine scenes" and the concept of culture-driven city, and proposes a modern community construction strategy guided by the "1 + 4 distinctive paths" in research (Fig. 3). This strategy includes one "exclusive" positioning approach and four distinctive scenes of modern community focusing on governance, entrepreneurship, transportation, and neighborhood. The integration of the "1 + 4 distinctive paths" aims to explore new paths for modern communities to demonstrate cultural heritage, continue historical context, strengthen digital applications, empower diversified entrepreneurship, and connect human emotions.

3.1 Digital "Governance" Scene

Through the innovative path of "breaking boundaries, crossing boundaries, and integrating boundaries," digital technology, digital thinking, and digital cognition are comprehensively utilized to promote the transformation of Jiangbei's urban renewal from "traditional management" to "holistic smart governance" and explore a digital governance model for modern community construction with multi-party participation.

Building a social intelligence service platform, realize the docking of the community with the urban brain, the Internet of Things perception network and other platforms, and integrate internal and external data. Make full use of the technology accumulation and data resources of the community information model platform in spatial governance to realize the digital application of modern community. The third party shared parking operation enterprises in the community are introduced, and the application community

1 "exclusive" positioning approach	4 distinctive scenes of modern community	Specific content
The concept of creating a modern community in Jiangbei with a rich industrial cultural heritage	Digital "governance" scene	Party leadership
		Digital system
		Hongyan Pioneer property services
		Planners, architects, and engineers enter the community
	Diversified "entrepreneurial" scene	Employment services
		Innovative industry incubation
		Talent security mechanisms
	Integrated "traffic" scene	Convenient transportation
		Mountain city trails
		Accessible facilities
		Smart shared parking
		Energy supply assurance and interface reservation
	Humanistic "neighborhood" scene	Neighbor culture
		Neighbor sharing mechanisms
		Neighbor mutual aid model

Fig. 3. The content of "1 + 4" for modern community pathway.

smart shared parking service platform is opened, providing online service functions such as information sharing, parking guidance, non-inductive parking, automatic settlement, and reverse car searching, and increasing the proportion of 5-min pickup and parking times.

Based on AI technology, refine community work tasks and develop an APP for community residents' clients through business deduplication and process reengineering. Covering scenarios such as community government services, public services, and commercial services, promote convenient and efficient community services for residents. Optimize the government service model of "Internet + government services", build a comprehensive community information database with "data sharing and coordinated linkage", realize real-time collection of information such as population, housing, enterprises, community components, events, public sentiment, disputes, and security incidents, and provide data support for community management service platforms.

3.2 Diversified "Entrepreneurial" Scene

In response to the requirements of urban innovation and entrepreneurship driven by "mass entrepreneurship and innovation," diversified entrepreneurial scenes are further created. Leveraging shared spatial resources, refined entrepreneurial services, and multi-level and multi-category talent introduction, the extension and development of "double innovation" in modern communities are realized to adapt to the integration trend of people's livelihood and employment.

First of all, we should adopt the operation management idea of "professional institution leading + government support, professional operation + market-oriented mechanism", plan supporting high-quality entrepreneurship space, introduce high-quality professional operators, integrate various innovative industry elements such as government, talent, capital, industry, etc., and establish a complete operation service system and platform for entrepreneurship and innovation space.

Build a digital platform for modern community entrepreneurship services, provide business segments such as entrepreneurial public services, industrial alliances, and innovation docking, and realize resource aggregation and exchange and information sharing.Provide shared office rental, property management, security and cleaning, conference services and other basic services to solve the operation of enterprises. It can also provide talent recruitment, resource links, entrepreneurship guidance, enterprise training, government support, remote collaboration, investment and financing services and other special services.

3.3 Integrated "Traffic" Scene

A future transportation scene system is built with "people-centered" as the core, focusing on the three key points of pedestrian smoothness, vehicle flow, and road network connectivity. It revolves around eight tasks such as barrier-free pedestrian networks, public transportation transfer connections, intelligent travel services, fine transportation organization, shared mobility services, comprehensive parking management, road network layout optimization, and tapping the potential of existing road resources, to achieve the goal of "transportation serving the city and the city promoting employment.

Optimize the way of traffic organization, open the roads, squares, green Spaces and other Spaces inside the community, and promote the formation of a block layout model with a high density of branch roads, composite development of land functions, and supporting and integrating public service facilities. Encourage eligible communities to plan and build in a fully open mode, and open roads between houses to slow and vehicular traffic.

Optimize the layout design of community branch roads, promote the organization of one-way traffic, turning restrictions and other traffic circulation organization patterns, optimize the community traffic flow line, and form a traffic circulation system within the community. For the community with the combination of demolition and renovation, the connection between the main entrance and exit of slow traffic and the nearest rail transit or bus station should be optimized.

Encourage the opening up of setback spaces in public buildings to create integrated pedestrian street spaces with diverse functions, considering different street types and spatial characteristics. Classify pedestrian spaces according to street nature, building features, and spatial scale, integrate land use, prioritize pedestrian spaces, and enhance the quality of walking spaces through facility arrangement and front area design. Integrate ground, aerial, and underground spaces to form a three-dimensional pedestrian system network, facilitating convenient connections between buildings, between buildings and rail stations, and within street spaces.

3.4 Humanistic Neighborhood Scene

With the expansion of the modernization process, the rampant growth of cities has led to a focus on residential areas rather than supporting facilities and speed over humanistic aspects in community construction. By creating neighborhood scenes with a focus on human-centeredness, building humanistic bonds that connect community emotions, and mobilizing the enthusiasm of community residents to participate in community governance, a virtuous cycle of long-term community operation is ultimately formed.

Explore the unique cultural characteristics of the neighborhood. Arrange centralized or decentralized community spaces for showcasing and expressing local cultural features. This includes community history displays, community cultural art exhibitions, as well as spaces for leisure, recreation, fitness, social gatherings, special festivals, and cross-community neighborhood activities. Extract elements from the local regional culture of the community and promote excellent traditional culture. Build distinctive cultural symbols that residents enjoy, guided by regional context and the value concept of "historical sustainability", to inherit collective memory.

Create shared neighborhood spaces, considering the core pain points of residents' lives and their aspirations for a better quality of life. Build different types of neighborhood interaction spaces, such as reading and learning spaces, children's play spaces, fitness and sports spaces, spaces for elderly activities, and public social spaces. Establish neighborhood community organizations based on the principle of diversity, tailored to the age, occupation, and family composition of the homeowners, including but not limited to interest groups, public service groups, life skill groups, and sports and fitness clubs. Ensure the availability of specific but non-exclusive activity facilities and venues, encourage resource sharing, and promote the sharing of venues and facilities. Develop neighborhood agreements that align with core social values, local cultural history, and the theme of community cultural characteristics, covering various scenarios of community life, such as neighborly respect and unity.

4   The Path of Modern Community Building

4.1   Digital "Governance" Scene

Current issue 1: The community infrastructure is old, and there are many difficulties in updating. The demands from the public are diverse (Fig. 4).

Remedial measures 1: Establish an online + offline model, create an investment, construction, and operation resource database online. Offline, set up a Jiangbei District Modern Community Construction Expert Workstation", and use it as a starting point to build a cooperative bridge and gather diverse forces. Firstly, establish a regular work meeting system, combine responsibilities and coordinate to solve key, difficult, and bottleneck issues in a timely manner. Secondly, collect, organize, submit, and process construction information through online data cloud. Thirdly, widely promote Jiangbei Modern Community Construction through various digital channels, build interactive platforms, lead public participation, and create a model of governance.

Current issue 2: The proportion of school-age children and elderly people over 60 years old in the pilot area is about 58%, and the usage rate of digital terminals is low.

Remedial measures 2: Firstly, build a community micro-brain. By integrating community data, intelligent perception, and business linkage, create a closed loop to construct a community smart service platform that combines government governance, community security, and homeowner services, and deepen the refinement of community governance (Fig. 5). Secondly, implement the generalist social worker + grid walking service model,

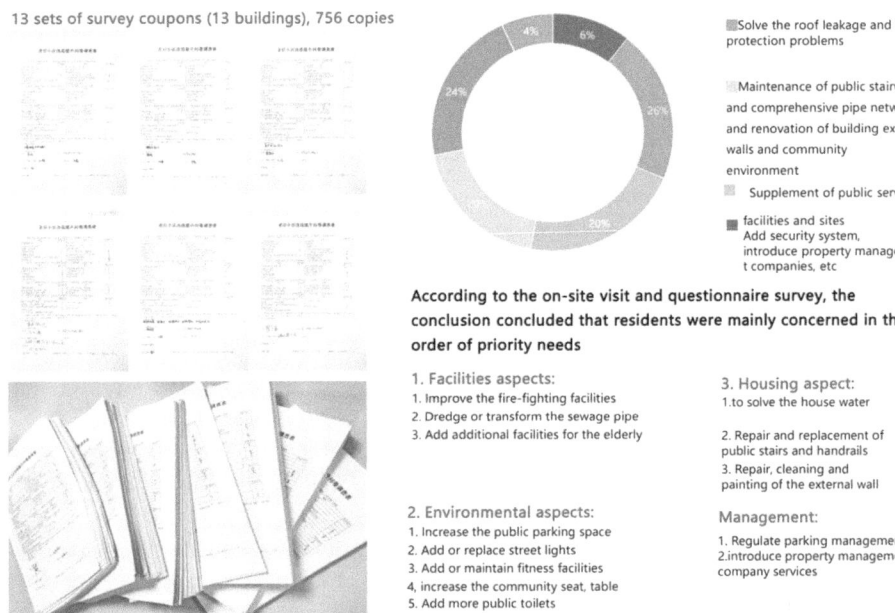

13 sets of survey coupons (13 buildings), 756 copies

- Solve the roof leakage and fire protection problems
- Maintenance of public stairwell and comprehensive pipe network and renovation of building exterior walls and community environment
- Supplement of public service facilities and sites
- Add security system, introduce property management companies, etc

According to the on-site visit and questionnaire survey, the conclusion concluded that residents were mainly concerned in the order of priority needs

1. Facilities aspects:
1. Improve the fire-fighting facilities
2. Dredge or transform the sewage pipe
3. Add additional facilities for the elderly

2. Environmental aspects:
1. Increase the public parking space
2. Add or replace street lights
3. Add or maintain fitness facilities
4. increase the community seat, table
5. Add more public toilets

3. Housing aspect:
1. to solve the house water
2. Repair and replacement of public stairs and handrails
3. Repair, cleaning and painting of the external wall

Management:
1. Regulate parking management
2. introduce property management company services

Fig. 4. The results of the questionnaire survey in four aspects.

rely on standardized community grid construction and social work teams, increase guidance and promotion for digital services to community residents, so that residents can truly experience the convenience brought by digital services.

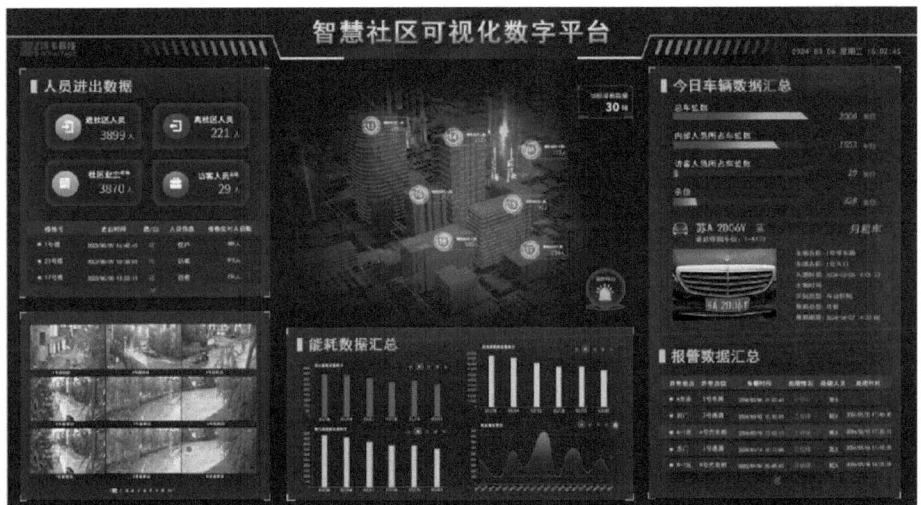

Fig. 5. The demonstration of the digital platform for smart community.

4.2 Diversified "Entrepreneurial" Scene

Current Issue 1: The positioning of urban entrepreneurship in the pilot area is not clear, and the format is relatively single.

Remedial measures 1: Create a diverse entrepreneurial business card and incubate a youth entrepreneurship community. Young talents have a strong desire to start business after the pandemic period and pursue personalized entrepreneurial environments. The "Workstation" will integrate various industrial resources in the pilot area, fully tap into the culture of Chang'an, and create a historical entrepreneurial village. Promote the introduction of relevant talent matching mechanisms and policies, and stimulate the unique entrepreneurial vitality of the pilot area. Emphasize the social role of young talents and attract "young people returning home" through talent placement subsidies and the establishment of youth talent apartments, injecting "modern elements" into the modern community.

Current Issue 2: The proportion of tenants in the pilot area (Fig. 6) is about 54%, most of whom are migrant workers with high mobility and low income, which does not match the existing community entrepreneurship scene.

Remedial measures 2: The Wulidian Street, in collaboration with the "Workstation", will create a "livable and business-friendly" entrepreneurial market, providing one-stop entrepreneurship and employment services. First, based on the characteristics of the population, provide "matchmaking" services for flexible workers in the community, such as registration and filing of flexible workers and posting employment information. Second, organize entrepreneurship training, such as home economics, elderly care, and childcare, to provide low-level professional entrepreneurship skills training, achieving employment opportunities, entrepreneurial support, and convenient services.

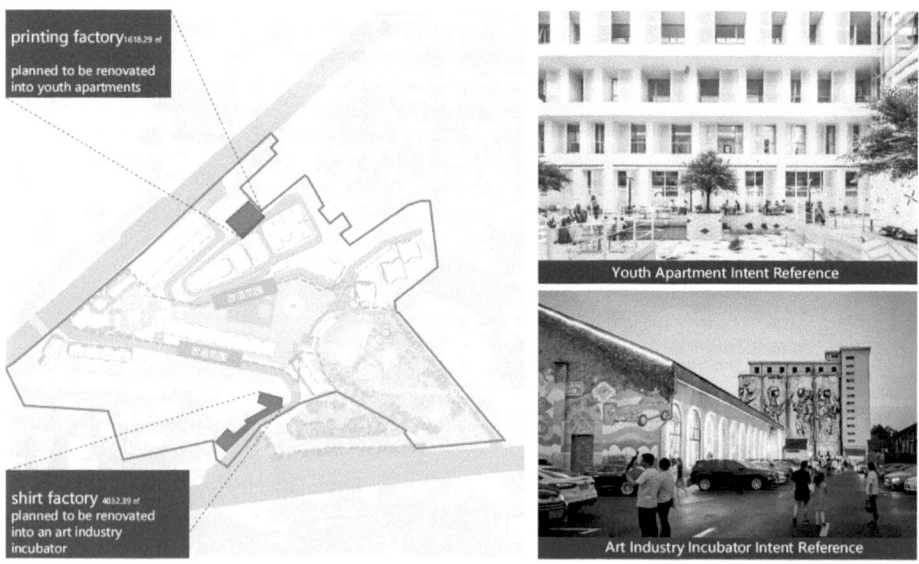

Fig. 6. The renovation of factory site to youth apartment and art industry incubator.

4.3 Integrated "Traffic" Scene

Current Issue 1: The construction standards of some secondary roads in the pilot area are low, with insufficient sidewalk width, steep slopes, and sharp turns, posing safety hazards and requiring traffic organization optimization.

Remedial measures 1: Implement standardized transformation projects for non-compliant roads such as Wanfeng Road, optimize road alignment, expand pedestrian space, and ensure safe passage for pedestrians and vehicles. Through one-way traffic and connecting internal roads, facilitate "microcirculation" and implement "lane slimming" on local roads to reduce the number of lanes and maximize the expansion of pedestrian space.

Current Issue 2: There is a lack of a regional-level slow traffic corridor, and important nodes such as scenic spots, schools, businesses, and residential areas lack efficient connections for slow traffic. The overall pedestrian network system in the pilot area has not yet formed, and considerations for the elderly and children are insufficient.

Remedial measures 2: Construct a skeleton pedestrian path in the pilot area, connecting community, schools, scenic spots, and commercial areas, extending the reach of the area. At the same time, upgrade and improve accessible transportation facilities for the elderly, such as installing escalators and other automated lifting facilities to meet the safe travel needs of the elderly and children.

Current Issue 3: The public transportation coverage in the pilot area is insufficient, and the connection between slow traffic and public transportation is not sufficient, with a lack of effective guidance for important nodes.

Remedial measures 3: Jointly with building renovations and entrance layout, add ground bus stops, transform linear bus stops into bays through the introduction of shuttle buses, plan shared transportation stops, and establish the capillary for the last kilometer. Construct a hierarchical signage system to achieve efficient information guidance.

Current Issue 4: Parking difficulties are prominent, and illegal roadside parking is common, with a low level of intelligent parking.

Remedial measures 4: Strengthen demand-side regulation of parking, implement comprehensive parking management, cautiously increase on-road parking spaces and mechanical parking buildings, prioritize the construction of underground parking lots, and reasonably allocate the number of charging stations. Increase law enforcement efforts against vehicles occupying roads, mitigate traffic quality by regulating zombie vehicles, and create a car-free and tranquil community.

4.4 Humanistic Neighborhood Scene

Current Issue 1: Important nodes lack guidance, local culture is missing, contextual relationships are disrupted, and the sense of identity in indigenous areas is generally low.

Remedial measures 1: With the core concept of "preserving the roots of industrial culture and retaining the soul of community development", we will deeply explore the history of Chang'an Factory and the street, reshape the industrial scene, and awaken memories of Chang'an. By creating community facilities with "Chang'an characteristic IPs" such as the "Memorial Park," "Red Pomegranate Community Garden," "No. 37

Mailbox," "State-owned 456 Story Store," and "Dustpan Stone Factory Trail," we will showcase interesting, cultural, and vibrant community elements to enhance residents' cultural confidence and cohesion (Fig. 7).

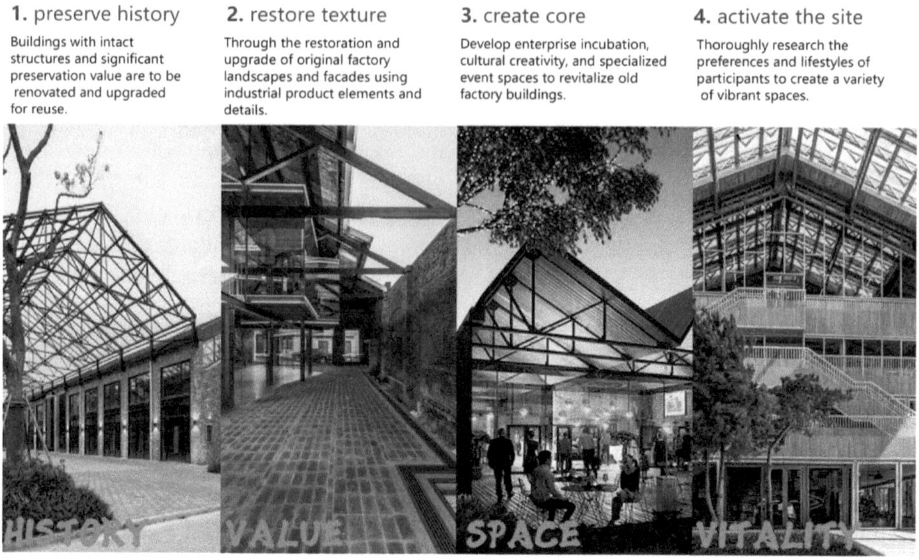

1. preserve history

Buildings with intact structures and significant preservation value are to be renovated and upgraded for reuse.

2. restore texture

Through the restoration and upgrade of original factory landscapes and facades using industrial product elements and details.

3. create core

Develop enterprise incubation, cultural creativity, and specialized event spaces to revitalize old factory buildings.

4. activate the site

Thoroughly research the preferences and lifestyles of participants to create a variety of vibrant spaces.

Fig. 7. The renovation philosophy of Chang'an Factory in terms of history, value, space, and vitality.

Current Issue 2: The complex property rights issue and the chaotic use of public spaces by different stakeholders have resulted in poor environmental quality.

Remedial measures 2: We will sort out and restructure the social relationships behind the spaces and integrate idle space resources through property transactions and rights transfers. The District Housing and Construction Commission will make full use of inefficient land, idle slopes, and passive spaces such as small and scattered "corner spaces" to plan and reshape functional land, achieving a balance between community characteristics and land use in the limited space of the area.

Current Issue 3: The functions of various parties in the community are fragmented, and the pilot area lacks connection, empowerment, and activation.

Remedial measures 3: Firstly, we will build a "neighborhood gathering" between "old and new villagers" in the pilot area, such as regularly organizing public community activities such as performances, party education, and festive events. Secondly, we will strengthen functional support and the co-construction and sharing of public spaces, implementing pilot area transformation and enhancement from a city perspective of "connecting points into lines and lines into areas," and scale up the development of an international landscape urban modern community cluster. Thirdly, we will combine the "Three Teachers Entering the Community" program, relying on the "Workstation" to regularly hold public opinion interviews, modern community construction salon forums, and sharing of neighborhood governance experiences, constructing a "people-oriented neighborhood" co-construction mechanism.

Current Issue 4: Community infrastructure is old, and the urban interface is dirty and messy.

Remedial measures 4: We will promote refined management of window areas, main and secondary roads, and back alleys in the community, empowering community management with digital, intelligent, and networked technologies. We will standardize "urban furniture," clean up advertisements and shop signs, promote "multiple poles in one" and "multiple boxes in one" and underground pipelines, promote the standard and tidy management of municipal facilities such as pipelines and road barriers, showcasing a good urban image. We will put meticulous efforts into community management and fine governance of various municipal "dirty and messy" issues.

5 Conclusions

This article explores the potential of combining the renovation of old residential areas with the historical culture of the Chang'an Factory. By excavating "Chang'an Stories," reshaping "Chang'an Memory," and perpetuating the "Chang'an Spirit," it introduces unique "governance scenes," "entrepreneurial scenes," "transport scenes," and "neighborhood scenes." Following the concept of "connecting dots into lines, and lines into areas," it organically links the Yinghua Lane Phase I and the Wanfeng Industrial School area renovation projects to stimulate community vitality. In the future, the area will deeply implement modern community construction concepts and evolve into a leading example of a "modern community" in Chongqing demonstration, creating the urban brand of a "modern international metropolis with beautiful landscapes."

References

1. Qin Mengdi Tongming: Community renewal involving multiple property rights entities - A case study of Huafu Community Public Lane. J. Ideal Space **91**, 54–57 (2023)
2. Urban China: Future Communities: A Global Concept of Urban Renewal with Six Samples, Zhejiang University Press, Chapter 2, pp. 48–116. (Year 2021)
3. Zhejiang Provincial Development and Reform Commission, Zhejiang Provincial Development Planning and Research Institute: Future Community - Theoretical and Temporal Exploration of Zhejiang. Chapter 7, pp. 94–135. Zhejiang University Press (1999)
4. Zou Jing, Shen Feiwei: Scene construction, resident integration and digital governance of future urban communities. e-government,https://kns.cnki.net/kcms/detail//11.5181.TP.202 30119.1004.009html
5. Lewis Murnford, The City in History[M]: The City in History (2005)
6. Gail, author; He Renke translate. Interaction and Space. Beijing: China Architecture & Building Press (2002)

Analysis of Market Demand for Medium and Long Distance High Speed Maglev Railway Passenger Transport

Jingjing Bao[✉], Xiao Yang, Yu Yang, and Weiwei Gong[✉]

Research Institute for Transportation & Economics, China Academy of Railway Sciences Corporation Limited, Beijing, China
baojingjing@rails.cn

Abstract. To promote the high-quality development of passenger transportation and fully meet the diversified transportation needs of passengers, it is necessary to enrich passenger transportation products. As the wheeled rail mode of transportation has basically reached a speed bottleneck, the study proposes that the high-speed magnetic levitation railway with a speed of 600 km per hour be involved in passenger transportation as a new transportation product. Comparing the advantages of high-speed maglev trains in terms of speed, safety, and economy with those of high-speed railway, airlines, and other modes of transportation, the study calculates the sharing rate of high-speed maglev trains in the transportation market, researches the induced incremental capacity generated by the new mode of transportation, and obtains the capacity of maglev railway on different routes. Research results show that 600 km/h Maglev train is the most competitive Maglev speed, with an advantageous distance between [500, 3000] kilometers. To realize medium and long-distance passenger transportation between the core cities of China's 4-pole urban agglomeration, Beijing-Shanghai, Beijing-Guangzhou and other trunk line corridors and point-to-point high-speed magnetic levitation trains will become the priority development direction.

Keywords: Maglev · Railway Passenger Volume · High-Quality Passenger Transport · Volume Prediction

1 Introduction

With the rapid development of social and economic, people on cross-regional, long-distance, large flow, high-density transportation put forward higher speed requirements, ultra-high-speed pipeline magnetic levitation transportation technology in recent years has become a key area of rail transportation research. The Central Committee of the Communist Party of China and the State Council officially issued the "Outline for the Construction of a Stronger Transportation State", which explicitly states that "the development of high-speed magnetic levitation system with a speed of 600 km per hour" [1]. China in the field of high-speed magnetic levitation technology has a certain amount of technical reserves and research foundations, and has the advantages of strong national support, superior R & D mode and so on [2].

© The Author(s) 2025
B.-J. He et al. (Eds.): UCSUD 2023, LNCE 559, pp. 854–863, 2025.
https://doi.org/10.1007/978-981-97-8401-1_61

In the 1960s, U.S. researchers put forward the idea of constructing vacuum pipe magnetic levitation lines, and Daryl Oster of the U.S. obtained a patent for the invention of vacuum pipe transportation system in 1999; in 2013, ElonMusk of the U.S. put forward a program of ultra-high-speed vacuum pipe transportation system; in 2016, the U.S. Super High Speed Rail Corporation for the first time carried out a propulsion system in the pipeline transport Public test, the test speed reached 186km/h; In 2017 months, HyperloopOne conducted a newest real test, the highest speed of pipeline super high-speed railway reached 320km/h. With the United States, represented by the ultra-high-speed electromagnetic levitation train technology has become mature [3–5].

In recent years, many places in China have carried out 600km/h and above ultra-high-speed maglev research. In 2018, the speed of 600kmh high-speed maglev transportation system technology program through, high-speed maglev transportation system key technology to achieve stage results [7]; in 2020, high-speed maglev test prototype car in Tongji University maglev test line successfully test run [8]; in 2021, the use of a five-section formation group of Train system in Qingdao off the line, the same year the ultra-high-speed maglev electromagnetic propulsion test [9]. In recent years, many places in China have successively carried out 600 km/h and above ultrahigh-speed maglev research, and have successively achieved a series of scientific research results [10–18].

Therefore, no matter from the viewpoint of national development strategy, or from the viewpoint of technological reserves, transportation improvement and innovative development, ultra-high-speed maglev in long-distance transportation has a huge innovative development space and broad development prospects [19]. High-speed maglev line (600 km/h speed class) as a kind of transportation mode to make up for the gap between high-speed railway (400 km/h speed class) and aviation (800 km/h speed class), in the middle and long-distance transportation has obvious advantages, and it can realize the rapid and direct point-to-point transportation between urban clusters [20]. It can fill the speed gap between aviation and high-speed railway and is expected to build multiple commercial operation lines between large cities, which can provide rail transportation support for the urbanization strategy, solve the vibration and noise problems of traditional rail transportation, and have a very broad development prospect in the future [21].

The current magnetic levitation research is mainly operating environment, vehicle structure and other technical aspects of the study more, on the magnetic levitation of the market demand, adapt to the opening range, the service object of the study less, on the magnetic levitation of the market operation of the necessity of the study less, this article can fill the gaps in this part of the research.

Therefore, this paper takes high-speed magnetic levitation railway as the research object, compares its advantages in speed, safety and economy with railways, aviation and other modes of transportation, calculates the sharing rate of high-speed magnetic levitation trains in the transportation market (Fig. 1), and at the same time researches the induced incremental capacity produced by the new mode of transportation, and obtains the capacity of different lines of magnetic levitation railways with a view to providing high-quality services for travelers, and realizing win-win situation between travelers and transportation enterprises.

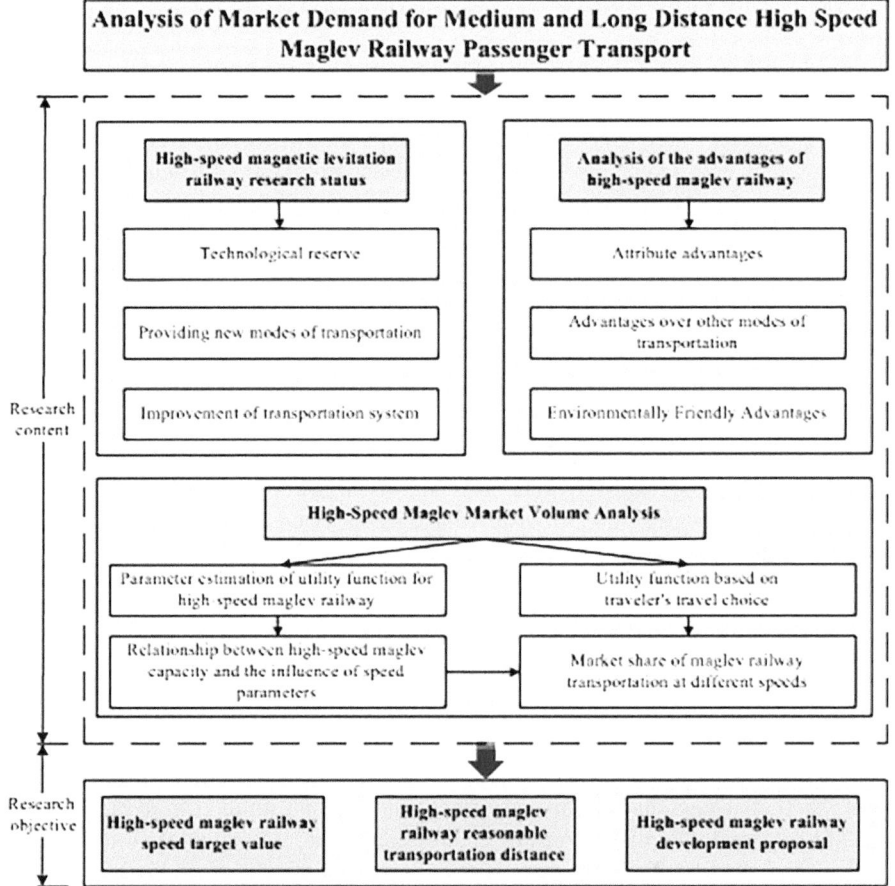

Fig. 1. Research ideas and technical roadmap.

2   Analysis of the Demand for High-Speed Maglev Railway

First, the operation of high-speed magnetic levitation railway helps to shorten the time and space distance between metropolitan areas, form a super metropolitan area, and stimulate the economic vitality of cities along the route. Second, the new high-speed maglev lines are conducive to making full use of underground space and upgrading existing ones to better implement the concept of station-centered urban development and renewal. Third, high-speed magnetic levitation railway can be used as a new mode of transportation, the development of differentiated development strategy, more conducive to China's comprehensive three-dimensional transportation network components and improve.

3   Analysis of High-Speed Maglev Railway Advantages

High Operating Speed and Short Travel Time. The development of the high-speed maglev train market benefits from multiple driving factors. First, high-speed maglev trains have high operating speeds and short travel times, which can meet people's needs for fast, efficient, and convenient transportation. The current research of ultra-high-speed maglev railway speed of at least 500 km per hour, and maglev railway acceleration and braking time is shorter, so that the train in the travel time and the short and medium-range aircraft is about the same.

Environmental Protection. High-speed maglev trains have better environmental performance, reducing dependence on traditional energy sources and helping to reduce air pollution and traffic congestion. In 25 m away from the line at a speed of 400 km per hour through, the maglev train noise bit 79dB, while the ICE3 wheeled train for 91dB, the noise is lower. And its energy consumption is not only lower than the high-speed train, more greatly lower than cars and airplanes, drive the same power respectively when cars and airplanes consume one-third and one-fourth of the energy.

Economy. High-speed maglev trains have better economics. Although the high-speed maglev train speed, but because of no wheel and rail friction, so energy consumption than ordinary high-speed train less than one-third, its cost and ordinary high-speed railway is almost equal, because there is no wheel and rail friction, so the maintenance cost is also low. Moreover, the magnetic levitation railway is not affected by the terrain and topography, and it can run on the line with a slope of 10%, and the curve radius is smaller than that of the ordinary high-speed railway, so it saves more land than the ordinary railway.

High Operating Safety and Comfort. High-speed maglev trains have high operational safety and passenger comfort, attracting more and more consumers to choose them. Maglev train structure to ensure no derailment, propulsion method to ensure no collision. The use of segmented power supply, no risk of tailgating, the train running power from the fixed track on both sides of the electromagnetic flow, the electromagnetic flow in the same area of the same strength, there will be no several trains at different speeds and the phenomenon of line, better guarantee the safety of transportation.

4   Ultra-High-Speed Maglev Market Volume Analysis

4.1   Characteristics and Positioning of Ultra-High-Speed Maglev Market

Ultra-high-speed magnetic levitation can fill the speed gap between high-speed rail and air transportation, forming a multidimensional transportation architecture that includes aviation, high-speed rail, high-speed magnetic levitation, and urban transportation with a more reasonable, efficient, flexible, and convenient speed gradient. With the gradual expansion of China's high-speed railway coverage and network operation, high-speed maglev lines are mainly applied to backbone transportation corridors with large transportation demand and high service quality requirements. Giving full play to the speed advantage of high-speed magnetic levitation railway, it is positioned to undertake long-distance passenger transportation. Mainly set up stations in mega-cities and big cities, the stations are mainly connected to the main hubs of city centers, and the services for peripheral urban clusters and surrounding small and medium-sized cities are mainly reached with the help of other modes of transportation in the hubs.

4.2 Ultra-High-Speed Magnetic Levitation Market Share Study

The advantages of ultra-high-speed magnetic levitation railway are studied from the dimensions of speed target value, fare and train departure frequency, and the market share of ultra-high-speed magnetic levitation in the passenger transportation market is studied and analyzed by using discrete mathematical model.

Calculation of market share of ultra-high-speed magnetic levitation railway

Using the study area's calendar year passenger volume data as a base, the natural growth rate of passenger volume was extrapolated by analyzing the study area's economic growth rate, and a discrete mathematical model was used to extrapolate the market share for each mode of transportation and to extrapolate the amount of passenger volume that would be shifted from other modes to the Maglev Railway on this baseline. Finally, by analyzing the induced incremental capacity brought about by the advantages of the Maglev Railway, the capacity of the Maglev Railway is predicted.

$$B_i = \frac{\exp(U_i)}{\sum_{n=1}^{N} \exp(U_n)} \tag{1}$$

$$U_i = \sum_j a_j X_{ij} \tag{2}$$

where B_i denotes the ratio of passenger volume to total passenger volume for travel mode i in the target year; U_i denotes the utility function for travel mode i; N denotes the total number of travel modes available; X_{ij} denotes the jth influencing factor for travel mode i; a_j denotes the coefficient to be determined.

Determination of speed target value of ultra-high-speed magnetic levitation railway

Under different speed conditions (Table 1), the market share of ultra-high-speed magnetic levitation railway is studied separately, and the reasonable speed target value of ultra-high-speed magnetic levitation railway is proposed.

Table 1. Maglev market share under different speed conditions.

Maglev speed	500 km/h	600 km/h	700 km/h	800 km/h
Maglev	55.05%	55.15%	55.03%	55.12%
High Speed Railway	28.14%	28.09%	28.16%	27.84%
Civil Aviation	16.81%	16.76%	16.81%	17.04%

Note: The original data are from the Yearbook of Civil Aviation and Railway Statistics 2012–2019

By calculating the market share rate of maglev trains at different speed levels, we can easily find that with the speed of maglev trains increasing, the market share rate of maglev trains basically shows a growing trend. Maglev 500 km/h, 700 km/h, 800 km/h passenger volume than under 600km/h were reduced by 3%, 4%, 2%, 600 km/h maglev train attracts more passenger flow, the market share rate reached 55.15% which are shown in Figs. 2 and 3.

The 600 km/h maglev train is between the two modes of transportation of high-speed rail (300 km/h) and aviation (800 km/h), effectively patching up the gap of high-speed transportation market in the transportation market, and is the most competitive maglev speed (Table 2).

Taking the intersection point of the sharing rate between the magnetic levitation transportation mode and the other two transportation modes as the boundary of the advantageous distance range, the study proposes that the advantageous distance range of the magnetic levitation transportation mode with a speed of 600 km per hour is between [500,3000] kilometers.

Table 2. The comparison of the Magnetic levitation (train), high-speed railway, and aviation.

	Safety	Speed	Fare Rate	Station to Departure (Arrival) Time
Magnetic levitation (train)	0.99	600 km/h	0.99 Yuan/km	60 min
High-speed railway	0.99	350 km/h	0.52 Yuan/km	60 min
Aviation	0.99	800 km/h	1.65 Yuan/km	120 min

Note: The original data are from the Yearbook of Civil Aviation and Railway Statistics 2012–2019

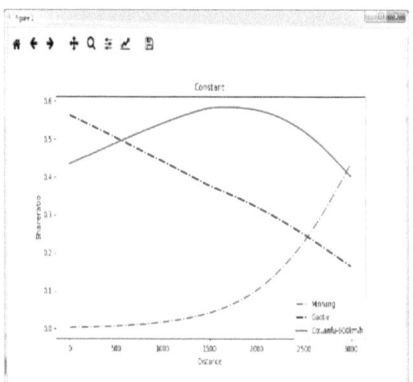 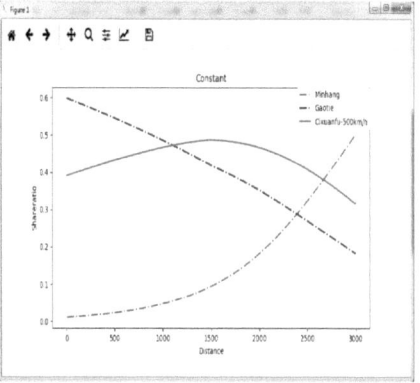

Fig. 2. Competitive Relationships in the Maglev Market (500 km/h left, 600 km/h right). Note: The original data are from the Yearbook of Civil Aviation and Railway Statistics 2012–2019

Beijing-Shanghai corridor ultra-high-speed magnetic levitation capacity study

Passenger flow of Beijing-Shanghai high-speed railway is characterized by large capacity, obvious regularity of spatiotemporal distribution and fluctuation of demand, diversified structure, and high proportion of cross-line passenger flow, etc. With the improvement of high-speed railway network, these characteristics will be increasingly

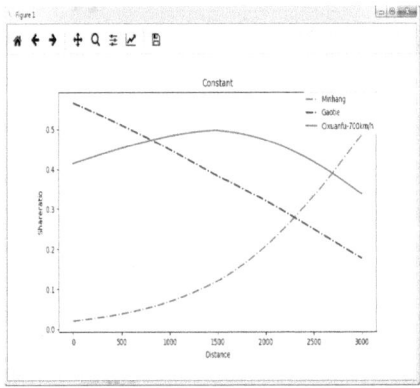 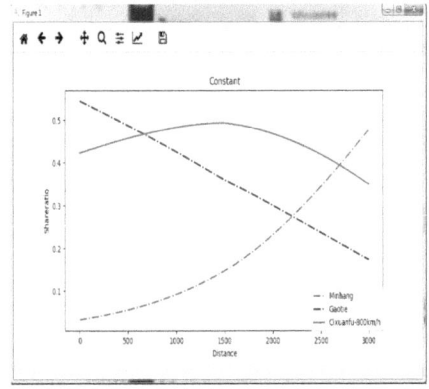

Fig. 3. Competitive Relationships in the Maglev Market (700 km/h left, 800 km/h right). Note: The original data are from the Yearbook of Civil Aviation and Railway Statistics 2012–2019

prominent. The main transportation for passenger flow between Beijing and Shanghai are aviation and high-speed railway, and the two modes of transportation together complete the passenger flow between Beijing and Shanghai is shown in Table 3.

Table 3. Passenger volume between Beijing-Shanghai corridors over the years.

Units: million	2012	2013	2014	2015	2016	2017	2018	2019
Volume of transportation	1071	1110	1217	1327	1435	1637	1787	1882

Note: The original data are from the Yearbook of Civil Aviation and Railway Statistics 2012–2019

Considering the impact of pre-demonstration and publicity of Maglev trains on fares and passenger flow, price changes due to technology development in the long term and other circumstances, the induced increase ratio after the opening of Maglev is 2.15%. It is expected that after the opening of Maglev, the total traffic capacity of Beijing-Shanghai corridor will be 32.96 million, 39.23 million and 51.77 million in 2030, 2035 and 2045 respectively (Table 4).

Table 4. Forecast of Total Passenger Transportation and Maglev Capacity of Beijing-Shanghai Corridor.

Unit: million	2030	2035	2045
Passenger volume based on the opening of the Maglev	3296	3923	5177
High-speed Maglev traffic in the corridor	1818	2164	2855

Note: The original data are from the Yearbook of Civil Aviation and Railway Statistics 2012–2019

5 Conclusions

Countries with developed transportation in the world are bravely exploring the development of maglev trains. Different countries choose different development strategies, development modes and technical routes according to their own national conditions. At present, maglev rail transit as a new type of transportation mode that enters the public's view has high investment, long cycle, relatively complex technology, and its development is constrained by a variety of factors such as technology, economy, and politics, and at the same time, it needs to face the competition of traditional wheeled rail transportation, aircraft and other transportation modes, and its development in the world is slow and full of uncertainties. The development of China's maglev rail transit has made breakthroughs in engineering technology due to strong government support. The huge demand market also brings a broad development prospect for maglev rail transportation.

Currently, long-distance transportation modes include China's high-speed railway (300 km/h) and aviation (800 km/h), while the Maglev's speed is in between the two modes of transportation. Compared to high-speed rail, high-speed Maglev greatly reduces the time needed to travel between cities by rail; compared to aviation, high-speed Maglev reduces the time needed to transfer from the airport to the destination, and it can effectively fill the high-speed transportation gap in the transportation market. The conclusion shows that: first, the 600 km/h maglev train is the most competitive maglev speed in the middle and high speed transportation market, with the 600 km/h maglev train's advantageous distance between [500,3000] kilometers; Secondly, focusing on serving middle and high-income business groups as well as tourism and leisure travelers, it mainly realizes medium- and long-distance passenger transportation between the core cities of China's economically developed 4-pole urban agglomerations, with Beijing-Shanghai, Beijing-Guangzhou, and other mainline corridors as well as point-to-point high-speed magnetic levitation trains becoming the preferred direction of development.

Acknowledgments. This study was supported by National Key R&D Program of China (Project No. 2023YFB4302500). My sincere thanks go to all the individuals and organizations that contributed to the success of this study.

. References

1. Central Committee of the CPC and the State Council: The central committee of the CPC and the state council print and issue the outlines for building a transport power. Bull. State Counc. People Repub. China **28**, 5–10 (2019)
2. Tong, Shen, Zhiwen, M.A., Xiaojie, D.U., et al.: Development status and trend analysis of high-speed maglev railways worldwide. China Railw. **11**, 94–99 (2020). https://doi.org/10. 19549/j.issn.1001-683x.2020.11.094
3. Zhang, R.H., Liu, Y.H., Xu, S.Z.: American Magplane schemes. Var. Curr. Technol. Electr. Tract. **5**, 43–46 (2005)
4. Li Yungang, Chang Wensen, Yan Yujiang: Analysis and comparison of new maglev transport technology in USA. Locomot. Electr. Drive (03): 6–9+39 (2006). https://doi.org/10.13890/j. issn.1000-128x.2006.03.002

5. Gou, J.: Review articles development status and global competition trends analysis of maglev transportation technology based on patent data. J. Pat. Data Anal. **2019** (2019). https://doi.org/10.1007/s40864-018-0087-3

6. Deng Zigang, Liu Zongxin, Li Haitao, et al.: Development status and prospect of maglev train. J. Southwest Jiaotong Univ. **57**(03): 455–474+530 (2022)

7. Xudong, Z.: China will develop a 600-kilometer per hour high-speed magnetic levitation prototype in 2020. Guangdong Transp. **1**, 46 (2018)

8. Anonymous.: 600 km/h maglev test prototype car successfully test run. Railw. Technol. Superv. **48**(08), 22 (2020)

9. CNR: Qingdao High-Speed Magnetic Levitation Transportation System [EB/OL] (2021). [Accessed 2021–11–25]. https://www.crrcgc.cc/sj/g16998/s30989/t323860.aspx

10. Yidi, Z.: 600 km/h high-speed maglev test prototype rolled off the line in Qingdao. Locomot. Electr. Drive **03**, 74 (2019)

11. Ma Guangtong, Yang Wenjiao, Wang Zhitao, et al.: Research development of superconducting maglev transportation. J. South China Univ. Technol. (Nat. Sci. Ed.) **47**(07): 68–74+82 (2019)

12. Fei, X.U., Shihui, L.U.O., Zigang, D.E.N.G.: Study on key technologies and whole speed range application of maglev rail transport. J. Railw. **41**(03), 40–49 (2019)

13. Deng Zigang, Li Haitao: Recent development of high-temperature superconducting maglev. China Mater. Prog. **36**(05): 329–334+351 (2017)

14. Anonymous: China's version of "super high-speed railway": the world's first high-temperature superconducting high-speed magnetic levitation engineered prototype test. High Technol. Ind. **27**(01), 6 (2021)

15. Fang, Lu., Liyu, Wang, Yingli, Qiao, et al.: Study on the application difficulties of ultra-high-speed magnetic levitation rail transportation system with speed over 600 km/h. Compr. Transp. **45**(05), 17–21 (2023)

16. Yu Haowei, Kou Junyu, Li Yan.: Adaptability and engineering development of 600 km/h high-speed maglev in China. J. Railw. Eng. **37**(12): 16–20+88 (2020)

17. Jiayang, X., Zigang, D.: Research progress of high-speed maglev rail transit. J. Transp. Eng. **21**(01), 177–198 (2021). https://doi.org/10.19818/j.cnki.1671-1637.2021.01.008

18. Zhang, Y.J., Wang, X.H., Fan, D.Y., et al.: Research on channel planning of ultra-high-speed pipeline maglev system. Railw. Stand. Des. **68**(02): 1–6+14 (2024). https://doi.org/10.13238/j.issn.1004-2954.202208300002

19. Zhang, Y.J., Wang, X.H., Huang, C.M., et al.: Research on the evaluation system of ultra-high-speed pipeline maglev system demonstration line. Railw. Stand. Des. **67**(12): 1–6+14 (2023). https://doi.org/10.13238/j.issn.1004-2954.202301170005

20. Gong, J. H., Xie, H. L., Yan, J. P., et al. 2020. Development and application prospect of full-speed spectrum maglev transportation technology. Urban Rail Transp. Res. **23**(09): 61–64+69. https://doi.org/10.16037/j.1007-869x.2020.09.014

21. Chaohua, W., Ji, L.S.: Experience and enlightenment of Japan's Maglev line planning and construction. Traffic Transp. **39**(3), 25–29 (2023). https://doi.org/10.3969/j.issn.1671-3400.2023.03.006

Resilience to Climate and Health Challenges

Spatial Layout Planning of Medical and Health Institutions Based on the Concept of Healthy City: A Case Study of Mianyang

Qianying Zhao[1], Yingwei Xiong[2,3(✉)], Qian Li[3], and Xinghao Cui[3(✉)]

[1] Mianyang Veterans Affairs Bureau, Mianyang 621010, China
[2] School of Urban and Rural Planning and Construction, Mianyang Teachers College, Mianyang 621000, China
375459038@qq.com
[3] Technology Research Institute of Ecology, Energy-Saving, Environmental Protection and Low-Carbon Mianyang Science & Technology City New Area, Mianyang 621010, China
cuixinghao1997@163.com

Abstract. Addressing the fundamental health needs of residents and enhancing the accessibility of healthcare services represent essential objectives in the development of healthy urban environments. Consequently, the investigation into the spatial arrangement of medical and healthcare facilities holds substantial significance. Taking Mianyang City as a case, this study investigated the spatial distribution of these facilities through spatial distribution direction analysis and kernel density analysis. The spatial distribution direction analysis showed that the layout of medical and health institutions in Mianyang extended from northwest to southeast. The kernel density analysis showed that medical facilities in Mianyang city were generally concentrated in the central urban area, with a high-density spatial distribution pattern. The spatial pattern of "single main core + multiple cores" is presented.The findings inform strategic policy recommendations aimed at optimizing the allocation and adjustment of health resources in Mianyang City, providing a basis for scientific formulation and implementation of regional health planning.

Keywords: Healthy city · Medical and health institutions · Spatial layout

1 Introduction

The notion of a 'healthy city' emerges from a comprehension of what constitutes health in the human context. This concept revolves around the creation of an urban environment through design and planning, which fosters a layout and resource distribution conducive to public health. The aim is to enhance both the physical and social environments of cities, thereby encouraging residents to adopt healthier lifestyles, ultimately advancing overall public health [1]. The genesis of modern urban planning can be traced back to addressing urban health issues. It is imperative that contemporary urban planning reorients itself towards the foundational objective of fostering urban health [2–4].

© The Author(s) 2025
B.-J. He et al. (Eds.): UCSUD 2023, LNCE 559, pp. 865–877, 2025.
https://doi.org/10.1007/978-981-97-8401-1_62

Currently, Chinese scholars have engaged in research on the theoretical and practical aspects of healthy cities. However, these studies generally require further depth, and there is a noticeable gap in providing planners and designers with effective design tools for public health interventions, particularly when compared to international research in this field [5, 6]. This calls for the prioritization of public health promotion, which should be integrated with the various challenges inherent in urban development. Besides, interdisciplinary methods should be utilized to thoroughly explore the spatial distribution of medical and healthcare facilities within healthy cities. Since the 1990s, researchers worldwide have initiated studies into the urban distribution of health-related spatial facilities. These facilities primarily encompass healthcare resources, sports and fitness centers, as well as parks and recreational areas, with a focus predominantly on their location and spatial allocation [7–10].

Fulfilling the basic health needs of residents and enhancing the accessibility of health services are crucial developmental aims for healthy cities. Thus, exploring the spatial planning of medical and healthcare institutions is significantly important. In the case of Mianyang City, there is a notable scarcity of research regarding the layout of medical facilities. This gap underscores the need for increased focus on the development of medical infrastructure in Mianyang City, which would contribute to addressing the current deficiency in this area. An optimal arrangement of medical facilities can address issues such as the uneven distribution of resources. This paper presents an analysis based on the existing layout of medical facilities in Mianyang City. Utilizing GIS spatial analysis, the study seeks to optimize the distribution of these facilities from a spatial perspective. Such optimization aims not only to improve healthcare access but also to contribute to the broader development of urban civilization.

2   Study Area and Data

2.1   Overview of the Study Area

This research focuses on Mianyang City, a significant urban area situated in the northwest of the Sichuan Basin, along the mid and upper reaches of the Fujiang River. Geographically, Mianyang City is bordered by Qingchuan County and Jian'ge County of Guangyuan City to the northeast, Nanbu County and Xichong County of Nanchong City to the east, Shehong City and Daying County of Suining City to the south, and Luojiang District, Zhongjiang County, and Mianzhu City to the west. Spanning the coordinates of $103°45'$ to $105°43'$ east longitude and $30°42'$ to $33°02'$ north latitude, Mianyang City stretches in a long, narrow belt from northwest to southeast. It measures up to 187 km in width from east to west and 256 km in length from north to south, encompassing a total area of approximately 20200 square kilometers, which constitutes about 4.2% of Sichuan Province's land area.

Within the study area, various administrative divisions are present. Fucheng District encompasses 9 subdistricts, 14 towns, and 2 townships. Youxian District has jurisdiction over 2 streets, 11 towns, and 11 townships. Anzhou District has jurisdiction over 14 towns and 4 townships. Santai County administers 41 towns and 22 townships. Yanting County has jurisdiction over 14 towns, 21 townships, and 1 nationality township. Zitong County has jurisdiction over 11 towns and 21 townships. Beichuan County has jurisdiction over

6 towns, 16 townships, and 1 nationality township. Pingwu County contains 9 towns, 3 townships, and 13 ethnic townships. Lastly, Jiangyou City comprises 4 streets, 21 towns, and 19 townships.

2.2 Study the Population Status of the Region

Within the nine counties (including cities and districts) of Mianyang City, the population distribution varies significantly. One county (city and district) has a permanent population exceeding one million people. There are three counties (cities and districts) where the population ranges between 500,000 and 1 million individuals. Additionally, three counties (cities and districts) have a population bracket of 200,000 to 500,000, while two counties (cities and districts) feature a population between 100,000 and 200,000. Notably, the combined resident population of the top three counties (or cities and districts) constitutes 61.33% of the total resident population of Mianyang City, as detailed in Table 1 (The population data utilized in this study was sourced from the "Mianyang Seventh National Population Census Bulletin" published by the Mianyang Municipal Bureau of Statistics.)

Table 1. Resident population of Mianyang city by county (city, district)

Region	Population (person)
Mianyang City	4868243
Fucheng District	1298524
Youxian District	561379
Anzhou District	372962
Santai county	955811
Yanting county	370739
Zitong county	276996
Beichuan Qiang Autonomous County	174132
Pingwu county	126357
Jiangyou City	731343

2.3 Data Source

In this study, the geographical data of medical and health resources in Mianyang City is integrated using the open API data interface of the Gaode Map Open Platform. The extraction of relevant medical and health data from Mianyang City is accomplished through the Python programming language. The search for primary medical and health services, secondary comprehensive hospitals, specialized hospitals, health centers, and clinics is conducted using the Gaode POI (Point of Interest) coding type library. This process results in the acquisition of data on medical and health resources, which are then adjusted and inputted into an Excel spreadsheet for further analysis. In total, 1775 data entries on medical and health resources were initially collected. This dataset was then cross-referenced with the "Mianyang Health and Health Industry Development Statistical Bulletin (2022)" available on the website of the Mianyang Health and Health Commission. Through this comparison, duplicate data were identified and eliminated, leading to a refined set of medical and health resource data. The attribute information of these resources includes names, grades, classifications, addresses, and geographical coordinates (longitude and latitude). Further validation and refinement of this data were achieved through field investigations and verification against the medical and health resource information published on relevant health websites of Mianyang City. This process concluded with the identification of 1378 distinct medical and health resources within the scope of the study. The detailed point attribute data of these resources were also obtained from various medical websites, encompassing information such as grade, type, longitude, latitude, address, number of beds, number of professional and technical personnel, and number of outpatients. The data collection was completed on November 13, 2023. Additionally, the population data utilized in this study was sourced from the "Mianyang Seventh National Population Census Bulletin" published by the Mianyang Municipal Bureau of Statistics.

The spatial distribution of POI points for medical and health resources in the study reveals a pattern characterized by "broad dispersion with localized concentrations." These resources are predominantly clustered in the older urban areas, with a notably higher concentration in Fucheng District and Youxian District. In contrast, the distribution of medical and health resources in peripheral urban areas like Beichuan Qiang Autonomous County, Pingwu County, and Jiangyou City is comparatively limited. This disparity in resource allocation can be attributed to the relatively recent development of the urban fringe areas. The slower pace of regional development in these outskirts has resulted in lower population density, which in turn has led to a lesser allocation of medical and health resources in each of these fringe areas (Fig. 1).

3   Research Method

The research methodology of this study is as follows: The spatial distribution of medical facilities within the designated area is analyzed using the standard deviation ellipse tool and the kernel density method. Based on this, we summarized the spatial layout characteristics of medical and health institutions in the study area. Concurrently, it identifies existing challenges within this framework and then proposes planning recommendations aimed at achieving the objectives of healthy city development.

3.1   Spatial Distribution Direction Analysis

The standard deviation ellipse method offers a robust analytical tool for assessing the layout scale of various factors across the entire study area. This method allows for the determination of the global coverage and expansion direction of these factors [11]. Starting from the mean center, the two axes of the ellipse are defined by calculating the standard deviations of the x and y coordinates. The length of the long axis is indicative

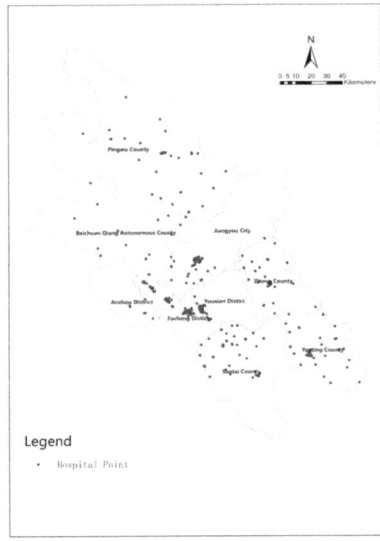

(a) Distribution map of hospital
in Mianyang city

(b) Distribution map of clinic
in Mianyang city

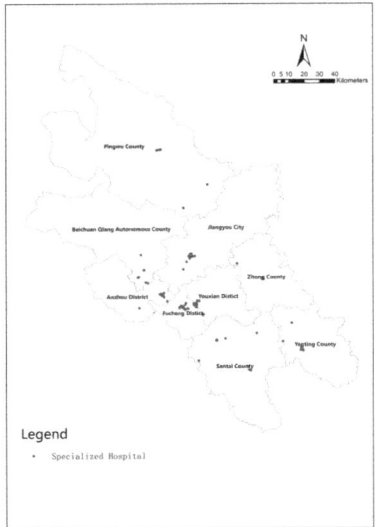

(c) Distribution map of Specialized hospital
in Mianyang city

(d) Distribution map of General
hospital in Mianyang city

Fig. 1. Distribution map of medical and health institutions in Mianyang

of the coverage area's size, while the short axis aligns with the distribution direction of the points. The area size of the resultant ellipse provides an assessment of the dispersion degree of such element points. The concept of the center of mass is pivotal in this analysis.

It represents the point at which the set of element points averages out in both the x and y directions. This point is characterized by having the shortest cumulative distance to each facility. The mathematical formulation for this calculation is as follows [12, 13]:

$$\tan\theta = \frac{\sum\limits_{i}^{n}(x_i-\bar{x})^2 - \sum\limits_{i=i}^{n}(y_i-\bar{y})^2 + \sqrt{\sum\limits_{i}^{n}(x_i-\bar{x})^2 - \sum\limits_{i=i}^{n}(y_i-\bar{y})^2 + 4[\sum\limits_{i}^{n}(x_i-\bar{x})^2 - \sum\limits_{i=i}^{n}(y_i-\bar{y})^2]}}{2\sum\limits_{i=1}^{n}\sum\limits_{i=1}^{n}(x_i-\bar{x})\sum\limits_{i=1}^{n}(y_i-\bar{y})}$$

(1)

$$\delta_x = \sqrt{\sum_{i=1}^{n}\frac{[(x_i-\bar{x})\cos\theta - (y_i-\bar{y})\sin\theta]^2}{n}}$$

(2)

$$\delta_y = \sqrt{\sum_{i=1}^{n}\frac{[(x_i-\bar{x})\cos\theta - (y_i-\bar{y})\cos\theta]^2}{n}}$$

(3)

where: δ_x is the length of the major axis of the ellipse; δ_y is the length of the minor axis of the ellipse; θ is the rotation direction Angle; n is the number of elements; (x,y) is the coordinate of the center of mass.

3.2 Spatial Distribution Density Analysis

In the realm of general geography, point model analysis primarily employs the core density method as a fundamental approach for spatial representation. This method delineates spatial density by considering the aggregation frequency, which is influenced by the spatial interactions between the focal point and other geographical locations. It adheres to the distance attenuation principle embodied in the first law of geography, wherein places in proximity to the central point exhibit higher radiation values [11]. In accordance with the first law of geography, any point within the depicted figure radiates its influence throughout the surrounding region within a distance denoted as 'h'. However, the intensity of this influence diminishes with increasing distance, eventually reaching zero attenuation within the 'h' region. The kernel density analysis method operationalizes this theory by computing density grids, with each grid cell's average value being equivalent to the sum of the average count of all points within the 'h' region of each grid. Consequently, kernel density analysis excels in capturing continuous variations in the effects emanating from POI while significantly mitigating the impact of noise interference. The specific calculation formula is presented below [12–14]:

$$f(s) = \sum_{i=1}^{n}\frac{1}{h^2}\phi(\frac{s-c_i}{h})$$

(4)

where, f(s) represents the kernel density calculation function; h represents the interval of the continuous influence range; n is the number of points within the influence range; Ci is the position of element point i; and ψ is the kernel function. The kernel function is based on the quartic kernel function, and the quartic kernel function is calculated by:

$$\phi(\frac{s-c_i}{h}) = \frac{3}{4}[1 - \frac{(s-c_i)^2}{h^2}]$$ (5)

The determination of the maximum distance width 'h' is a key step in the computation of kernel density, as this parameter directly correlates with the dispersion of the input points. A broader bandwidth is more appropriate when dealing with signals exhibiting high dispersion characteristics. For the analysis of various types of medical and health institutions within Mianyang city, an adaptive bandwidth kernel density method was employed. This approach estimates the density range surrounding each institution based on the density values of Points of Interest (POI). Subsequently, it selects an optimal search radius to generate density change maps, elucidating the spatial connections among the research subjects. It is worth noting that a higher kernel density value signifies a more densely distributed array of point elements.

4 Result Analysis

4.1 Analysis of Population Distribution in Mianyang City

Population data for Mianyang city was imported into ArcGIS, facilitating the creation of a population distribution diagram for the city, as depicted in Fig. 2. A visual examination of the figure reveals that the population predominantly clusters within the primary urban zones, namely Fucheng District, Youxian District, Jiangyou City, and the suburban regions encompassing Santai County. Conversely, Beichuan Qiang Autonomous County, Zitong County, and Pingwu County exhibit the most modest population densities.

4.2 Analysis of Spatial Layout of Medical and Health Institutions

4.2.1 Spatial Distribution Direction Analysis

The average center of medical facilities within Mianyang city is mainly located in the urban center, exhibiting a distribution pattern that extends from the northwest to the southeast. Additionally, the number of medical facilities gradually diminishes as one moves from the city center toward the periphery. This pattern is visually depicted in Fig. 3, which illustrates the predominant coverage areas through an elliptical representation. Notably, these coverage areas are primarily concentrated in the central city districts of Fucheng, Youxian, and Anzhou, as well as the suburban regions encompassing Santai County and Jiangyou City. The center of this elliptical representation corresponds to the city's geographical center. Examining the long-axis distribution direction, it becomes evident that it extends from the northwest to the southeast, characterized by a substantial long-to-short axis ratio and a discernible directional trend. This distribution pattern aligns with Mianyang city's population distribution and its urban economic development pattern.

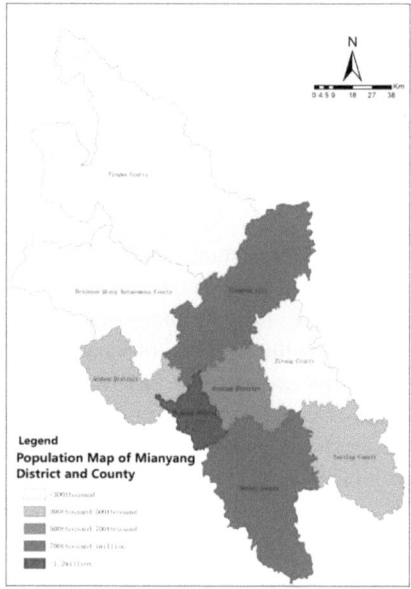

Fig. 2. Population distribution map of Mianyang

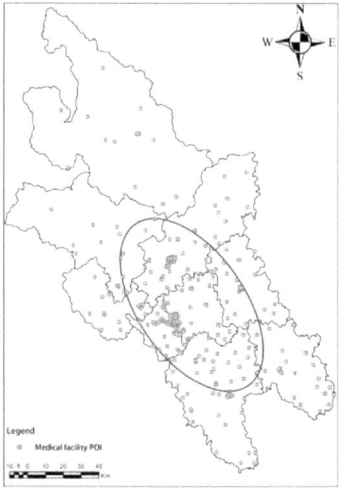

Fig. 3. Analysis on the distribution direction of medical and health institutions in Mianyang

4.2.2 Spatial Distribution Density Analysis

Kernel density analysis was performed on POI data encompassing all medical and health institutions in Mianyang City, yielding the results depicted in Fig. 4. The kernel density analysis reveals a spatial distribution pattern wherein medical facilities exhibit a general concentration within the central urban area of Mianyang City, characterized by high density. In contrast, other areas exhibit a narrower extent of concentration, thereby portraying a spatial pattern denoted as "single main core + multiple cores." The central core of this distribution resides at the intersection of Fucheng District, Youxian District, and Anzhou District in the central region of Mianyang City, boasting the highest count of medical and health institutions. Conversely, Beichuan Qiang Autonomous County and Pingwu County register the lowest numbers. The remaining districts exhibit localized concentrations, each forming its own center of medical facilities.

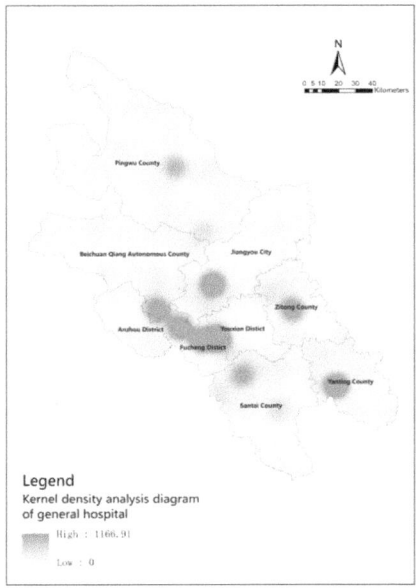

Fig. 4. The kernel density chart of medical facilities in Mianyang

Utilizing kernel density analysis, we analyzed POI data pertaining to medical and health institutions in Mianyang city, categorizing them into clinics, specialized hospitals, and general hospitals, as depicted in Fig. 5. According to the classification, the spatial distribution of various categories of medical facilities reveals distinct disparities. Clinics, being the most numerous category, exhibit a relatively uniform overall distribution. Their spatial pattern showcases a tendency towards aggregation, with clusters predominantly located in the central city area, and a scattering effect in the peripheral regions. Specialized hospitals, conversely, are the least numerous, displaying a spatial distribution characterized by a unipolar trend, with a pronounced concentration in the urban center. General hospitals, positioned between clinics and specialized hospitals in terms of quantity, showcase a multi-core spatial distribution pattern. Hotspots are discernible within Fucheng district, Youxian District, and Anzhou District, whereas peripheral areas exhibit fewer such facilities.

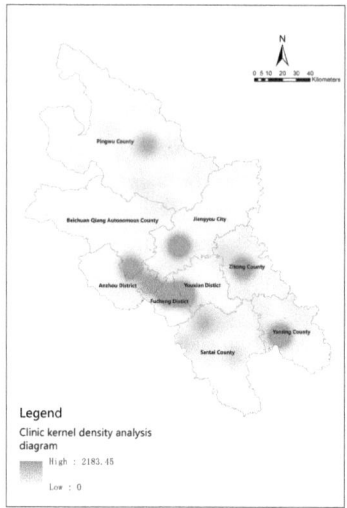

(a)The kernel density chart of Clinic

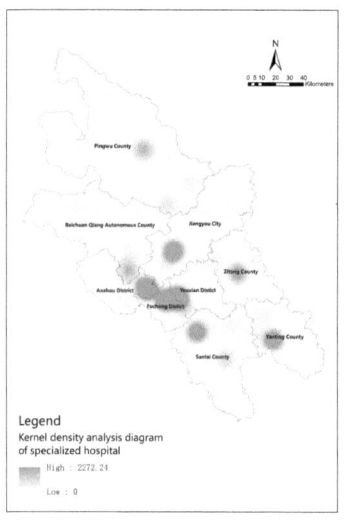

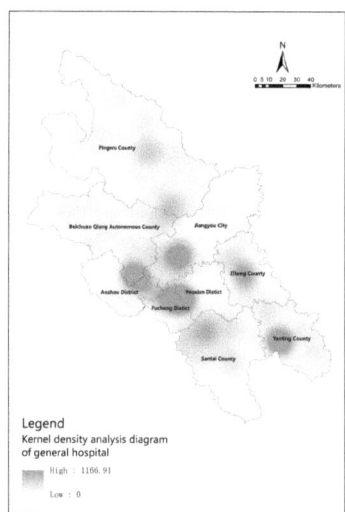

(b)The kernel density chart of Specialized hospital   (c)The kernel density chart of General hospital

Fig. 5. The kernel density chart of various medical and health institutions in Mianyang

5   Conclusion and Planning Strategy

5.1   Conclusion

The population of Mianyang city is mainly concentrated in Fucheng District, Youxian District, Jiangyou City and the surrounding suburbs of Santai County, while the population density of Beichuan Qiang Autonomous County, Zitong County and Pingwu County is relatively low. The distribution pattern of medical and health institutions extends from northwest to southeast. The urban center has a high density of medical facilities, while the medical resources in the suburbs are relatively insufficient. These analysis results provide important reference for the urban planning and health policy of Mianyang city.

5.2 Planning Strategy

Drawing upon the Mianyang City Territorial Space Planning (2020–2035) and the Mianyang City District Medical and Health Institutions Spatial Layout Plan, the Municipal Health and Health Commission, in collaboration with the Municipal Natural Resources and Planning Bureau, has undertaken a systematic organization and alignment of medical and health institutions within the municipal district. This concerted effort has taken into consideration the objectives outlined in the 14th Five-Year Plan, thereby establishing a coherent construction sequence. Guided by the overarching concept of constructing a healthy city, the following planning strategies have been articulated for the spatial layout of medical and health institutions in Mianyang city:

5.3 Strengthening the Building of Public Health Institutions

Currently, the medical and health service system in Mianyang city has largely taken shape, with a particularly well-established network of services within the older city area. However, a comprehensive analysis conducted through the fourth kernel density assessment of medical and health resource distribution in Mianyang city reveals that these resources predominantly cluster in the older city region, while their spatial distribution is notably limited in the urban periphery. The older city faces constraints related to tight urban construction land availability, leading to the partial substitution of residential and industrial land by commercial land. As a result, the medical and health resources in the older city have approached a state of saturation, making further construction impractical. Nevertheless, it's worth noting that the provision of supporting medical and health service land in the urban fringe area has not kept pace with the rapid development and construction observed in the new urban districts.

In accordance with the Mianyang City Land Use Planning for Medical and Health Services (2020–2035), strategic actions have been delineated to enhance the healthcare landscape. In the central urban area, there is a strategic imperative to reinforce general and specialized hospitals within major urban centers. Furthermore, it is essential to secure land for existing medical facilities, expedite the modernization of these facilities, and augment their comprehensive service capabilities. Simultaneously, prudent land reservation and rational allocation for medical and health services in sub-cities and new districts must be executed to align with residents' demands. In light of these objectives, the proposed optimization of the spatial layout of medical and health resources primarily centers on fine-tuning the configuration of these resources within the urban fringe. It is important to underscore that this optimization is predicated on the understanding that the overall spatial layout of medical and health service facilities will remain relatively stable [15–18]. Particularly, the well-established medical and health service network system within the older city area will remain unchanged, with a heightened focus on enhancing the spatial arrangement of medical and health resources in the urban periphery.

5.4 Healthy Urban Planning Strategies

Promote the "decentralization" of the spatial distribution of facilities, dredging public service resources in the central urban area, and promote the basic coupling of the construction scale of public service facilities at all levels with the level of economic development, the service population, and the supporting infrastructure[19] Take medical facilities as an example, a hierarchical medical facility system layout with distinct levels, clear division of labor and flexibility should be built in a healthy city, so that the classification of medical facilities is coupled with the urban space and social management level, and the tertiary hospitals are in the form of "pyramid" structure or "cabbage", so as to give play to the elastic function and sentinel role of primary medical facilities [20]. We will improve the public health system and service network based on primary medical care, and build an efficient, equitable and resilient medical and health care system.

Urban medical and health facilities at all levels shall be set up in strict accordance with the relevant national construction standards and administrative divisions. The location of a new medical and health institution should consider a location with convenient transportation, quiet environment and no pollution, and avoid adjacent places such as schools, markets and commercial districts. The layout of the general hospital should emphasize the combination with the urban public transport system and the open space system, adjacent to the adjacent bus lines or maintained within a suitable walking distance, as close as possible to parks, urban forests, waterfront green Spaces, etc. [21]. The location of infectious disease hospitals should be away from densely populated areas and public places in the suburbs of the city, away from food and feed Processing and other enterprises, while avoiding the layout near the city's water sources and above the wind direction.

References

1. Zhiming, L.: Urban planning and public health: history, theory, and practice. Planners **6**, 8 (2015)
2. Yi, W.: Healthy city oriented community planning. Planners **31**(10), 101–105 (2015)
3. Cho, C.-J.: An equity-efficiency trade-off model for the optimum location of medical care facilities. 32:99
4. Wenqi, Z., Muhammad, I.: Does healthy city construction facilitate green growth in China? Evidence from 279 cities. Environ. Sci. Pollut. Res. **30**, 102772–102789 (2023)
5. Hadorn: Racial and ethnic differences in access to and use of health care services, 1977 to 1996. Medical Care Rev. 57(1), 36–54 (2000)
6. Hurley: Measuring spatial accessibility to healthcare for populations with multiple transportation modes. Health Place 24, 115–122 (2013)
7. Zhao, S., Ren, Y., Mao, C., Yue, A.: Are cities healthy? A city health diagnose framework from the perspective of living organism. Ecol. Ind. 160, 111834 (2024)
8. Yujun, C., Bin, S.: Study on the influencing factors and optimization principles of site selection and layout of urban public sports space: Taking Yangpu District of Shanahai as an example. Sport Sci. Res. **6**, 35 (2015)
9. Mao, L., Nekorchuk, D.: Measuring spatial accessibility to healthcare for populat-ions with multiple transportation modes. Health Place **24**(115–12), 2 (2013)
10. Vojnovic I.: Building communities to promote physical activity: a multi-scale geographical analysis. Geografiska Annaler: Ser. B Human Geogr. (1) (2006)

11. Xiaomeng, W., Jin, W., Qing, Z.: Identification of hollowing phenomenon in commercial space of six district of Beiing based on checking-in data. Urban Dev. Stud. **25**(02), 77–84 (2018)
12. Wenhao, Y., Ai, T.: The visualization and analysis of POl features under network space supported by Kernel density estimation. Acta Geod. Cartogr. Sin. **44**(1), 82–90 (2015)
13. Ai, T., Wenhao, Y.: Algorithm for constructing network Voronoi diagram based on flow extension ideas. Acta Geod. Cartogr. Sin. **42**(5), 760–766 (2013)
14. Yamada, T.J.: Local Indicators of Network-constrained Clusters in Spatial Point Patterns. Geogr. Anal. **39**(3), 268–292 (2007)
15. Zhao, M., Qin, W., Zhang, S., Qi, F., Li, X., Lan, X.: Assessing the construction of a healthy city in China: a conceptual framework and evaluation index system. Public Health 220, 88–95 (2023)
16. Amri, M.: Healthy governance for cities: synergizing health in all policies (HiAP) and healthy city approaches. J. Urban Health **99**(2), 231–234 (2022)
17. Wu, L.: Planning to build a healthy city is the key to improving the livability of the city. Chin. Sci. Bull. **63**(11), 985 (2018)
18. Xie, J., Quan, M., Xie, E.: Research on health - oriented human settlements planning in the context of a healthy China: a case study of Hangzhou. City Plan. Rev. **44**(9), 48–54 (2020)
19. Rachael, M.: Urban climate science for planning healthy cities. Cities Health **8**(1), 29 (2024)
20. Ziafati Bafarasat, A., Sharifi, A.: How to achieve a healthy city: a scoping review with ten city examples. J. Urban Health 101, 120–140 (2024)
21. Barton, H., Grant, M.: Urban planning for healthy cities: a review of the progress of the European healthy cities programme. J. Urban Health **90**, 129–141 (2013)

Construction of Healthy Space Based on the Theory of Healthy City

Wei Yuan[1], Xiao-min Xia[2(✉)], Rui-li Liu[1], Yi-tian Zhou[1], and Yi Zhang[1]

[1] School of Architecture, Henan University of Technology, Zhengzhou 450001, China
[2] Institute of Civil Engineering, Zhongyuan University of Technology, Zhengzhou 450001, China

2010075@haut.edu.cn

Abstract. This article introduces the connotation and characteristics of a healthy city, puts forward the concept of building a healthy space from the perspective of land space planning and design, and explores the design path of a healthy space from the three aspects of land use complex, slow-moving system and ventilation corridor in combination with the urban physical space elements in the planning. To actively explore the construction of high-quality and healthy urban space environment.

Keywords: Healthy city · Land space planning · Health · Planning and design factors

1 Introduction

Urban planning in public health and territorial spatial planning belongs to different fields. Public health focuses on medical and health care, population health and other contents, while urban planning focuses on urban resource allocation, spatial layout and quality. However, urban planning originated from public health problems. From the second half of the 19th century to the early 20th century, western cities experienced disorderly sprawl, urban environmental deterioration and various epidemic outbreaks. To this end, a series of measures have been taken to improve the urban environment, such as determining drainage systems, urban parks, urban new housing regulations, and [1] to ensure building hygiene and fire spacing. After solving the problems of public health and disease in the early stage with urban planning, public health seems to have "withdrawn" from the field of urban planning research. In modern times, with the rapid development of urbanization, respiratory diseases caused by urban environmental pollution, obesity caused by excess nutrition, heart disease and other health problems are prominent, the research of public health has once again become the focus of urban planning research in territorial spatial planning. With the outbreak of COVID-19 in 2019 and the reality that the current urban environment has entered the post-epidemic era, how to realize sustainable urban development and create a healthy urban space environment is particularly important and necessary.

© The Author(s) 2025
B.-J. He et al. (Eds.): UCSUD 2023, LNCE 559, pp. 878–883, 2025.
https://doi.org/10.1007/978-981-97-8401-1_63

"Healthy city" concept in 1984, "healthy Toronto 2000 meeting" by the world health organization (World Health Organization), the connotation of healthy city refers to: the natural environment, social environment should be sustainable development, and constantly benign development and expand social resources, living in which people enjoy life and play their potential at the same time to support the development of the city. This shows that a healthy city not only contains the material health of people's survival and living environment, but also contains the comprehensive physical and mental health [2] of people in the city. In the early 1990s, this concept was introduced into China, and the concept of healthy people in cities gave a new and multi-dimensional perspective to the urban development in China. Professor Wang Lan of Tong ji University pointed out: "Health is not only a state of disease, but also a good physical and mental state and good social adaptability. A healthy city should be organically integrated out of a healthy population, environment and society. In order to reflect the potential of the city, a healthy city can continuously improve the environment, expand social resources, and urban residents can help each other." The characteristics of healthy cities include: 1. High quality and safe physical environment; 2. Sustainable ecosystem; 3. Stable social relations between people; 4. High participation and control of the public in health; 5. Cities meet all the needs of residents; 6. Residents obtain extensive and diverse urban experience and resources 7. Diverse and dynamic innovative spirit, cultural tradition, compatible model, and appropriate standards of health services; 8. High health level of urban people [3].

2   Healthy Space

From the 1980s, the Ottawa Charter, China's Opinions on Further Strengthening the Patriotic Health Work in the New Era (2014), Guidelines on the Construction of Healthy Cities and Villages and Towns (2016), "Healthy China 2030" Program Outline and other health intervention strategies. The theory and practice of healthy cities have had a profound impact on the work in various fields, ranging from urban management policy to health care to urban planning and design. Studying the public space that responds to public health problems from the perspective of urban planning and design has become a new trend in the development of national spatial planning discipline. Explore the design and improvement of the public space in the city, induce people to increase the opportunity of exercise behavior and improve the amount of exercise, so as to promote the overall physical and mental health of urban residents.

"Healthy space" is an expansion of the basic functions of urban public space. Through the organic integration and coordination of urban environment of land use, municipal and public service facilities, open space, residential community, street space and other urban function unit, in order to improve the urban population health, reduce carbon emissions, create high quality space environment, for the purpose of the city street space, community public space design to guide people to the activity of physical and mental health function of space. Form the design guidelines and strategies aiming at the health system space in the territorial space planning.

3   Healthy Space Design Pathways

3.1   Construction of Healthy Space

The essence of a healthy city is to guide people to carry out healthy behavior activities, so as to promote healthy and low-carbon people in urban life, work, transportation and rest. The healthy space design in urban environment is "implanted" in the public space to promote physical activities in the people, increasing the intention, opportunity and frequency of activities in the space; the healthy space can actively respond to the public health problems in cities, such as sub-health due to lack of activities. Unanticipated public health events, such as COVID-19. The Land space planning or renewal of a city needs the construction of urban structure according to certain intentions, the need to study and decide on land use, and the design of different types of urban traffic, which are three types of basic factors. The construction factors of a healthy city must also be reflected in these aspects. Therefore, the healthy space design is discussed from three aspects of land composite, slow traffic system and ventilation corridor. So as to realize the organic coordination between healthy lifestyle of urban residents and urban space environment.

3.2   Land Use and Compound

Urban and rural land use is carried out by stipulating land attribute, land type and area, land function and land distribution, which is an important basis for urban and rural planning in territorial space planning and design [4]. Land composite mainly from the land strength development and functional compound, appropriately improve the use of land development intensity and mixture of interconnected building function space, can realize the plot of traffic relationship compound, shorten people travel distance (such as transportation hub in traffic land complex, etc.), and reduce air pollution caused by motor vehicle carbon emissions, and increase the chance of people walking travel, so as to achieve the purpose of improving people's physical activity. Functional composite refers to the comprehensive development of different urban functional units in the same plot, such as urban complex development and the construction of different functional circle communities. Different functional Spaces should be integrated under a certain organizational relationship, and facilities such as rest and sports should be added to enhance the space vitality, such as roof gardens in urban complex buildings and basketball courts in outdoor squares, to form characteristic space and walking space with suitable distance. Community is one of the basic units of people feel the urban space, in the construction of different circles of community planning and design reasonable building density, convenient public transportation system, community activity center and walking space, realize the combination of mixed land and function, to create a dynamic public space, the purpose of effective use of energy and food, improve the purpose of local identification and public participation in the [5]. Functional complex can optimize the urban space environment, increase people's life vitality, promote people's positive psychology and then achieve the purpose of promoting physical and mental health.

3.3 Slow Traffic System

Research shows that some people in big cities have developed certain mental illness due to the fast pace of life, rapid information change and other reasons. Mental illness makes people in a state of physical and mental health, which is a state of sub-health. And people through slow activities in a certain space, it can promote blood circulation in the human brain, relieve the tension of muscles and nerves, can effectively relieve the production of anxiety and other negative emotions, so as to achieve the purpose of preventing depression and other psychological diseases [6]. Slow traffic refers to the traffic mode with a speed not greater than 15 km/h and mainly non-motorized traffic. Such as walking, cycling, etc. Walking is the main way of slow traffic, but also the basic form of people's physical exercise, which has the effect of improving human immunity. The slow traffic system is mainly a specific walking space in the urban public space. Through special planning and design, these walking Spaces can have a certain impact on the walking activities of people in the city, and guide people to carry out certain activities in them, so as to achieve the purpose of improving people's immunity. To create a slow traffic system with suitable scale blocks and multi-level walking space, and to have an impact on the crowd activities, so as to achieve the purpose of intervening in people's physical and mental health through walking activities. Appropriate scaling relationships in blocks can bring about better street connectivity and help encourage pedestrians to walk. The plan is to merge the original 130 m × 130 m block units into 3 × 3 grid units, and finally form a 400 m × 400 m super block. At the same time, in order to reduce the transit traffic and improve the quality of the block traffic environment, the speed of the vehicles entering the block is limited to 10 km/h, and it is stipulated that the traffic vehicles that only serve the residential area and emergency vehicles can pass [7] in the block. The plan specifies that the main function of urban streets is to create a high-quality walking space in the public space and limit the traffic space. Encourage people to take public transport in public Spaces. Multi-level walking space refers to the walking space with flat intersection, underground and elevated buildings according to the business forms and functions in the block. In leveling crossing, the comfort of walking space is realized by limiting the speed and driving route of the motor vehicle in the road (changing the linear route in the traditional road, increasing the curve route limiting the speed, etc.). Underground or elevated is the walking space design from the ground level, which is realized through the connection of the functional space of urban buildings. For example, the second floor walking system in the Hong Kong Central district separates the urban motorized traffic and pedestrian traffic through the overpass system. The first floor ensures the smooth passage of motor vehicles, and the second floor creates a pleasant walking environment and promotes the improvement of commercial vitality. Located in northern Canada, Montreal has the largest underground pedestrian system in the world. More than 1,000 shops, more than 100 restaurants, bars, residential apartments, office buildings, bus stops are connected by 30 km long pedestrian streets underground [8]. Formed a "underground block" with high strength. Through the construction of suitable pedestrian area scale and multi-level walking environment, social activities and walking behavior are generated in the urban public space, and human communication and activities are generated, so as to play a role in promoting physical and mental health.

3.4 Ventilation Corridor

Urban ventilation corridor for urban environmental problems plays an important role, in 2013 the National Development and Reform Commission and the ministry of the urban action plan to adapt to climate change and 2019 natural resources spatial planning bureau issued the cities and counties in national spatial planning guidelines, the cities and counties should be in the existing urban space planning or dredge ventilation corridor, promote the air flow in urban space, to alleviate the climate problems in the city, such as "heat island effect", "haze", etc. The construction of urban ventilation corridor is an important factor to create a healthy and comfortable urban environment, and is an important component of healthy space design. The essence of urban ventilation corridor is to flow and exchange the wind between different temperature areas in the city through the corridor, so as to realize the purpose of regulating the temperature in the city center and alleviate the problem of urban air pollution. First of all, the urban ventilation corridors and main and secondary air duct are planned according to the factors of urban prevailing monsoon conditions and natural topography, urban open space layout and road network distribution form. Secondly, control the density and height of the building layout in the ventilation corridor. Generally, the ventilation of low-density layout of high-rise buildings is the best, which is conducive to urban ventilation. The ventilation effect of high-rise buildings and low-density layout of low floors is better. The ventilation of the high-density layout of the bottom building is the weakest, which is not conducive to urban ventilation and produces heat island effect. Finally, determine the width of the ventilation corridor, the aspect ratio of the adjacent interface, and the height and density of the building layout in the corridor [9]. Finally, the urban ventilation system with the urban suburb is the compensation space for conveying low temperature air flow to the central area, and the urban and rural open public space is the air circulation guide channel.

4 Conclusion

Healthy city theory and practice is a global action strategy. It is not enough for us to advocate healthy life concept in life. We should further explore the correlation between public health and various urban material spatial elements, so as to create a healthier living environment for people. Based on the theory of healthy city, in this paper, the content of healthy space construction is explored from the perspective of territorial space planning, so as to promote the positive influence of urban space environment on public health. In the subsequent research, it is necessary to determine the index content of the elements, the applicable conditions in different geographical environments, and the setting of the health evaluation system. To improve the urban material space environment and improve the living environment.

References

1. Yang, C., Tan, S., Li, M., Dong, M.: Research on the active planning intervention pathway in healthy cities. Planner **5**, 13–14 (2022)

2. Wang, L., Liao, S., Zhao, X.: Analysis of the paths and elements of healthy urban planning. Int. Urban Plann. **4**, 4–5 (2016)
3. Xiao, M.: "Active design" creates a new perspective of urban planning and design to support a healthy lifestyle in the city of health and sports. Int. Urban Plann. **5**, 80 (2016)
4. Fan, Y., Jin, X., Xiang, X., et al.: Changes of land use function and its spatial pattern characteristics in Jiangsu Province. Geogr. Res. **38**(2), 383–398 (2019)
5. Toker, Z., Minassians, H., et al.: Good cities and healthy communities in the USA. Urban Des. Plan. **3**, 137–145 (2012)
6. Xiang, J., Li, Z., Liu, X.: Progress in walking and health. Chin. J. Sports Med. **9**(5), 575–578 (2009)
7. Cui, J., Chen, T., Zang, X.: Research on health-oriented block repair methods—take the Barcelona Super Block Project as an example. Western Habitat J. **2**:045 (2020)
8. Guo, X., Wang, D.: Interpretation of slow environment construction in Canada from the perspective of healthy city. Int. Urban Plann. **5**, 54–56 (2013)
9. Wang, J.: Research on urban ventilation corridors and their planning application. Residential Real Estate **9**, 90 (2018)

Water Ecosystem Resilience Evaluation and Blue-Green Enhancement Strategy in Tianjin from the Perspective of Social-Ecological Resilience

Yangli Li[1] and Rui Zhang[2(✉)]

[1] School of Civil Engineering and Architecture, Southwest University of Science and Technology, Mianyang 620101, China
[2] School of Architectural Engineering, Jinling Institute of Technology, Nanjing 211169, China
zhangrui1017_@tju.edu.cn

Abstract. The water ecosystem, an essential component of the United Nations Sustainable Development Goals, faces challenges such as water scarcity. Resilience is one of the central objectives of sustainable development. This study aims to investigate the water ecosystem resilience in Tianjin, providing a reference for protecting and restoring the global water ecosystem. Using the DPSIR model, an evaluation system for water ecosystem resilience was constructed. The combination weighting method was used to determine the weights of evaluation indicators. A case study on the water ecosystem resilience of Tianjin reveals spatial differences in 2018, with higher resilience in the north, and lower in the south. From a spatial planning perspective, forest coverage and wetland rate are representative evaluation indicators that significantly impact Tianjin's water ecosystem resilience. The overall connectivity index analysis method was used to analyze the representative evaluation indicators, and enhancement strategies such as building forest cities and protecting and restoring wetland systems were proposed. This study provides theoretical support for improving the water ecosystem and technical guarantees for achieving sustainable development.

Keywords: Water ecosystem resilience · DPSIR model · Combination weight · Blue-green integration · Enhancement strategies

1 Introduction

Water ecosystems are crucial for achieving the United Nations Sustainable Development Goals, such as human health and economic growth. The increasing water demand due to global population and economic growth, as well as the impact of climate change on water ecosystem quality, have put tremendous pressure on water ecosystems. The United Nations launched the "Water for Sustainable Development" international decade action plan from 2018 to 2028, to protect and restore global water ecosystems. Water ecosystem problems is mostly caused by human activities, including uncontrolled urban expansion,

© The Author(s) 2025
B.-J. He et al. (Eds.): UCSUD 2023, LNCE 559, pp. 884–902, 2025.
https://doi.org/10.1007/978-981-97-8401-1_64

extensive economic development, destructive use of ecological resources, insufficient environmental knowledge, limited environmental funding, and inefficient management systems. Although urban systems and water ecosystems belong to different disciplines and types of systems, they have a mutual influence and promotion relationship, which increases to the complexity of water ecosystem issues and creates a new demand for studying complex water ecosystem problems.

Resilience encompasses three perspectives: engineering resilience, ecological resilience, and socio-ecological resilience [1–3]. Engineering resilience emphasizes that the system has only one stable state, and the resilience of the system depends on the speed of recovery to the stable state. Ecological resilience emphasizes the maximum disturbance energy level that the system can absorb before transitioning to another equilibrium state. Socio-ecological resilience emphasizes the system's abilities to defend, adapt, and transform. Scholars researched water-related resilience, and due to different research perspectives, there are significant differences in the concept of water-related resilience. From the natural perspective, water-related resilience can be the role of maintaining the basic function of water ecosystems and maintaining the specific desired state of the social-ecological system [4–6], or it can be the ability to restore the function of water ecosystems after disturbances [7, 8]. From the artificial perspective, water-related resilience can be the ability to ensure the quality and quantity of water supply under cross-departmental cooperation strategies [9], the reliability of water supply systems during and after disaster events [10, 11], or the ability to reduce water system disasters through infrastructure construction and public participation [12]. Then, the interdisciplinary scope of research has become broader, and the concept of water-related resilience has evolved from natural or artificial ecosystems to artificial and natural ecosystems. Water-related resilience can be the ability of cities to withstand flood disasters and reorganize socio-economically [13, 14], the ability of water systems transform from risk adaption into new equilibriums, including technological, social, economic, and ecological aspects [15], or the ability of cities facing water-related pressures [16, 17]. Drawing on the definitions of water-related resilience by other scholars, a definition of water ecosystem resilience could be proposed from the perspective of social-ecological resilience [18, 19]: the ability of water ecosystems to defend, adapt, and transform in the presence of disturbances in a certain time and space, through the synergistic interaction of artificial and natural ecosystems.

The evaluation elements of the target and the connotation of resilience are two approaches to constructing a resilience evaluation system. Research on water ecosystem resilience evaluation has focused on evaluation elements of the target, mainly because the connotation of water ecosystem resilience is still being explored, making it challenging to use resilience connotation as the foundation for the evaluation system. Combining mature ecological evaluation models with the connotation of resilience provides a more effective approach to quantitative resilience analysis. Taking the Pressure-state-response (PSR), Drive force-state-response (DSR), and Drive force-pressure-state-impact-response (DPSIR) models as examples, this study explores the fundamental structure of a resilience evaluation system. The PSR model assumes that human activities hurt the ecological environment; however, the DSR model replaces pressure with driving forces, indicating that human activities have both positive and negative impacts on the ecological environment. Both PSR and DSR models only consider the impact of artificial systems on natural systems, without accounting for the potential influence of natural systems on artificial ones [20]. The DPSIR model inherits the multidimensional advantages of PSR and DSR models, compensates for their deficiencies in interconnected relationships, and can analyze ecological and environmental issues by comprehensively considering factors from both artificial and natural systems [21].

In this study, the DPSIR model and the combination of social-ecological resilience are used to construct an evaluation system for water ecosystem resilience. The improved maximum difference combination weighting method is employed to conduct an empirical study in Tianjin, identifying representative evaluation indicators. The overall connectivity index analysis method is used to analyze these indicators and propose enhancement strategies.

2 Methods

2.1 Study Area

Tianjin, with an area of approximately $11,966.45 \text{ km}^2$. With rapid economic and social development, there are prominent conflicts between Tianjin's urban development and its water ecosystem. Tianjin is considered an area with poor water security and a high risk of water scarcity on a national level [22, 23]. There are also issues such as river flow interruption, reduction of wetlands, siltation in the estuary, severe pollution discharge, insufficient water supply for the ecological environment, and serious risk of waterlogging [24, 25]. The study on water ecosystem resilience in Tianjin has a certain representativeness.

2.2 Research Method

DPSIR Model

Compared to the one-way relationships of the PSR model and DSR model [21], the Drive force-pressure-state-impact-response (DPSIR) model is an interdisciplinary and mature ecological assessment model that can consider the bidirectional relationships of both artificial and natural ecosystems [21, 26]. Due to the advantages of both the DSR and PSR models, the DPSIR model follows a logical sequence of "why-how-what can be done better," which can fully describe the stages and capabilities of resilience from the perspective of socio-ecology [27–32].

Among them, the driving force and pressure in the DPSIR model mainly address the question of "why disturbances occur." There are implicit factors of economic, social, and urban development within the driving forces that trigger explicit pressures such as water consumption and wastewater discharge. These implicit factors and explicit pressures put the system in an unfavorable position, increasing the likelihood of changes in the system's state [30]. The driving force and pressure stage represents the pre-disturbance phase of socio-ecological resilience. The state and impact in the DPSIR model mainly tackle the question of "how is resilience during disturbances." The state dimension is typically related to the condition of the ecosystem [33], where a better state indicates higher resilience of the system. The impact dimension is usually caused by driving forces and pressures, and when disturbances occur, the adverse effects on the state can reduce the system's resilience. The state and impact stage represents the disturbance phase of socio-ecological resilience. The response in the DPSIR model primarily focuses on "how to enhance resilience after disturbances." The response involves proactive measures to address the changing system. Since socio-ecological resilience follows an adaptive cycle, resilience can be enhanced through learning and implementing measures. The response stage represents the post-disturbance phase of socio-ecological resilience (Fig. 1).

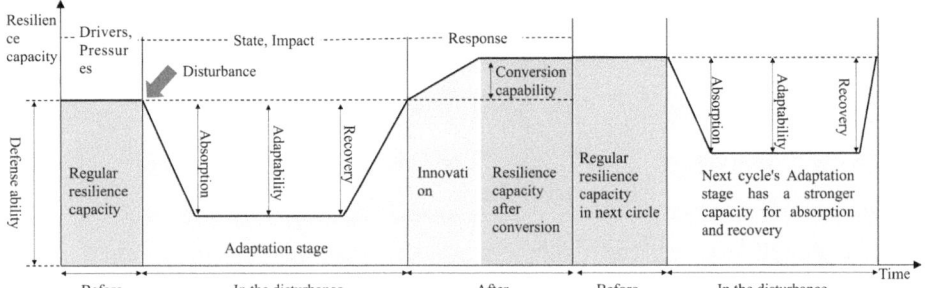

Fig. 1. Integration of the DPSIR model with resilience.

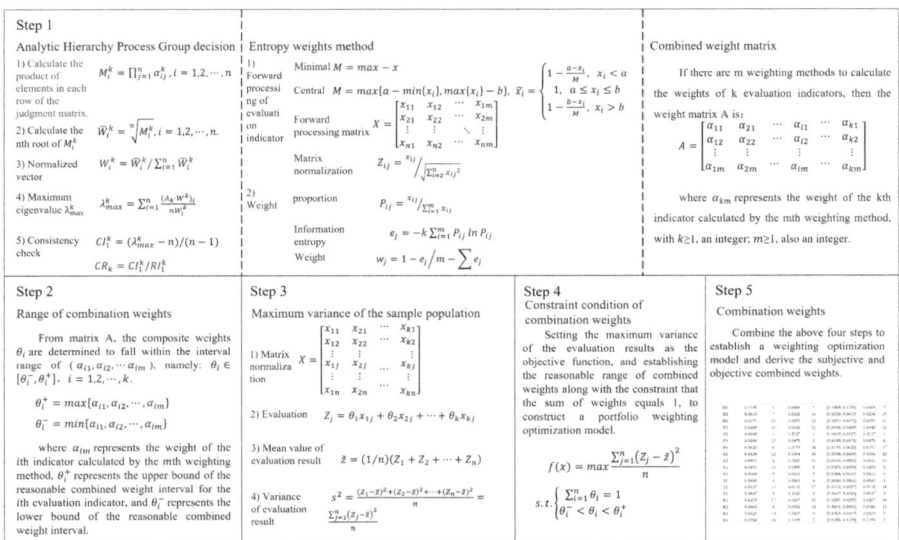

Fig. 2. Construction logic of improved level difference maximization method.

Improved Maximum Difference Combination Weighting Method

In the evaluation system, subjective weights tend to reflect the importance of evaluation indicators, while objective weights tend to reflect the current data information of evaluation indicators, which may result in significant differences in the ranking of subjective and objective weights. However, combination weights can combine the advantages of subjective and objective weights [34]. Based on the improved maximum difference combination weighting method [35], this study further reduces or eliminates the individual subjectivity of subjective weights using the group decision-making analytic hierarchy process and Delphi method, and calculates the combination weights using the improved maximum difference combination weighting method (Fig. 2).

Analysis Method of Overall Connectivity Index

From the perspective of landscape ecology, connectivity is one of the important criteria for measuring the quality of ecosystems [36]. In this study, the software for connectivity analysis [37] (Conefor Sensinode 2.6) will be used to analyze the connectivity of forests and wetlands in Tianjin.

3   Results

3.1   Evaluation of Water Ecosystem Resilience in Tianjin

Construction of Evaluation System

The selection of evaluation indicators follows nine principles: systematicity, universality, typicality, suitability, independence, quantifiability, data availability, technical operability, and foresight. Through a literature review, 18 evaluation indicators from artificial ecosystems and natural ecosystems were determined using the Delphi method. The weights of the evaluation indicators for the water ecosystem resilience in Tianjin in 2018 were determined using an improved maximum difference combination weighting method (45 iterations). Based on research findings, metric characteristics, regulations, standards, and development status and goals both domestically and internationally, evaluation metric grades are collectively defined, expanding the grading of evaluation metrics from previous domestic comparisons to global comparisons with other countries or regions. Urbanization rate and regional GDP growth rate are interval-type indicators. In this paper, the regional GDP growth rate is used as an illustrative example. The appropriate GDP growth rate, whether it should be maintained at 6%, 7%, or 8%, has long been a controversial topic. Liu Wei, the president of Renmin University of China (in 2020), suggested that China's GDP growth target for 2020 should be set at 5.5%–6% [38], arguing that anything below 5% would result in zero growth for the non-agricultural sector of the economy. However, excessively rapid GDP growth can also lead to increased emissions of major pollutants, further burdening the ecological environment. Ecological civilization construction has emerged as a significant goal in China, and the GDP growth pattern has shifted from quantity-based, extensive growth towards high-quality, stable growth. Upon reviewing GDP growth rates from 2008 to 2018, it is evident that they have consistently remained between 6.5% and 10%. Therefore, this study designates a range of 5%–10% as the optimal growth rate that can sustain economic expansion without unduly harming the ecosystem. Anything outside this range could have negative implications, while 6.5%–10% is seen as a sweet spot where relatively high economic growth can be maintained without excessively damaging the ecosystem. The water ecosystem resilience evaluation indicators were divided into five levels, with a passing score of 6, and a water ecosystem resilience evaluation system was finally constructed (Fig. 3).

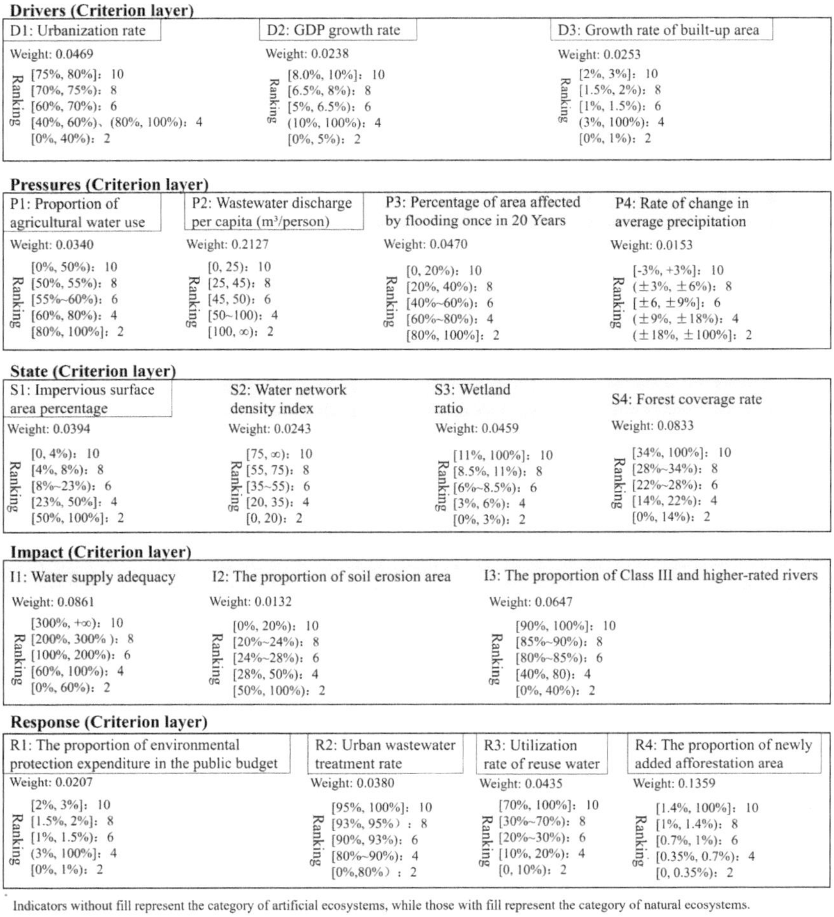

Drivers (Criterion layer)

D1: Urbanization rate	D2: GDP growth rate	D3: Growth rate of built-up area
Weight: 0.0469	Weight: 0.0238	Weight: 0.0253
Ranking: [75%, 80%]: 10 / [70%, 75%): 8 / [60%, 70%): 6 / [40%, 60%)、(80%, 100%): 4 / [0%, 40%): 2	Ranking: [8.0%, 10%]: 10 / [6.5%, 8%): 8 / [5%, 6.5%): 6 / (10%, 100%): 4 / [0%, 5%): 2	Ranking: [2%, 3%]: 10 / [1.5%, 2%): 8 / [1%, 1.5%): 6 / (3%, 100%): 4 / [0%, 1%): 2

Pressures (Criterion layer)

P1: Proportion of agricultural water use	P2: Wastewater discharge per capita (m³/person)	P3: Percentage of area affected by flooding once in 20 Years	P4: Rate of change in average precipitation
Weight: 0.0340	Weight: 0.2127	Weight: 0.0470	Weight: 0.0153
Ranking: [0%, 50%): 10 / [50%, 55%): 8 / [55%~60%): 6 / [60%, 80%): 4 / [80%, 100%]: 2	Ranking: [0, 25): 10 / [25, 45): 8 / [45, 50): 6 / [50~100): 4 / [100, ∞): 2	Ranking: [0, 20%): 10 / [20%, 40%): 8 / [40%~60%): 6 / [60%~80%): 4 / [80%, 100%]: 2	Ranking: [-3%, +3%]: 10 / (±3%, ±6%): 8 / [±6, ±9%): 6 / (±9%, ±18%): 4 / (±18%, ±100%): 2

State (Criterion layer)

S1: Impervious surface area percentage	S2: Water network density index	S3: Wetland ratio	S4: Forest coverage rate
Weight: 0.0394	Weight: 0.0243	Weight: 0.0459	Weight: 0.0833
Ranking: [0, 4%): 10 / [4%, 8%): 8 / [8%~23%): 6 / [23%, 50%): 4 / [50%, 100%]: 2	Ranking: [75%, ∞): 10 / [55, 75): 8 / [35~55): 6 / [20, 35): 4 / [0, 20): 2	Ranking: [11%, 100%]: 10 / [8.5%, 11%): 8 / [6%~8.5%): 6 / [3%, 6%): 4 / [0%, 3%): 2	Ranking: [34%, 100%]: 10 / [28%~34%): 8 / [22%~28%): 6 / [14%, 22%): 4 / [0%, 14%): 2

Impact (Criterion layer)

I1: Water supply adequacy	I2: The proportion of soil erosion area	I3: The proportion of Class III and higher-rated rivers
Weight: 0.0861	Weight: 0.0132	Weight: 0.0647
Ranking: [300%, +∞): 10 / [200%, 300%): 8 / [100%, 200%): 6 / [60%, 100%): 4 / [0%, 60%): 2	Ranking: [0%, 20%): 10 / [20%~24%): 8 / [24%~28%): 6 / [28%, 50%): 4 / [50%, 100%): 2	Ranking: [90%, 100%]: 10 / [85%~90%): 8 / [80%~85%): 6 / [40%, 80): 4 / [0%, 40%): 2

Response (Criterion layer)

R1: The proportion of environmental protection expenditure in the public budget	R2: Urban wastewater treatment rate	R3: Utilization rate of reuse water	R4: The proportion of newly added afforestation area
Weight: 0.0207	Weight: 0.0380	Weight: 0.0435	Weight: 0.1359
Ranking: [2%, 3%]: 10 / [1.5%, 2%): 8 / [1%, 1.5%): 6 / (3%, 100%]: 4 / [0%, 1%): 2	Ranking: [95%, 100%]: 10 / [93%, 95%）: 8 / [90%, 93%): 6 / [80%~90%): 4 / [0%,80%）: 2	Ranking: [70%, 100%]: 10 / [30%~70%): 8 / [20%~30%): 6 / [10%, 20%): 4 / [0, 10%): 2	Ranking: [1.4%, 100%]: 10 / [1%, 1.4%): 8 / [0.7%, 1%): 6 / [0.35%, 0.7%): 4 / [0, 0.35%): 2

Indicators without fill represent the category of artificial ecosystems, while those with fill represent the category of natural ecosystems.

Fig. 3. The water-ecosystem resilience evaluation system suitable for Tianjin in 2018.

Overall Analysis of Water Ecosystem Resilience in Tianjin

By applying the data to the water ecosystem resilience evaluation system, the classification of water ecosystem resilience levels and the comprehensive evaluation map of water ecosystem resilience in Tianjin in 2018 (Fig. 4) were obtained. Water ecosystem resilience levels in Tianjin in 2018 exhibit spatial differences with a "high in the north, low in the south, and central disparity" pattern, indicating an overall difference in water ecosystem resilience levels. The coefficient of variation, which is the ratio of the standard deviation to the mean, can measure the degree of variation and could be used for overall analysis and evaluation [39, 40].

$$C_v = \frac{\sqrt{\sum_{i=1}^{n} (S_{ij} - \overline{S_{ij}})^2 / n}}{\frac{1}{n} \sum_{i=1}^{n} S_{ij}} \tag{1}$$

where C_v represents the coefficient of variation, S_{ij} represents the water ecosystem resilience score of the j-th criterion layer in the i-th administrative region, $\overline{S_{ij}}$ represents the mean water ecosystem resilience score of all administrative regions for the j-th criterion layer, and n represents the number of administrative regions.

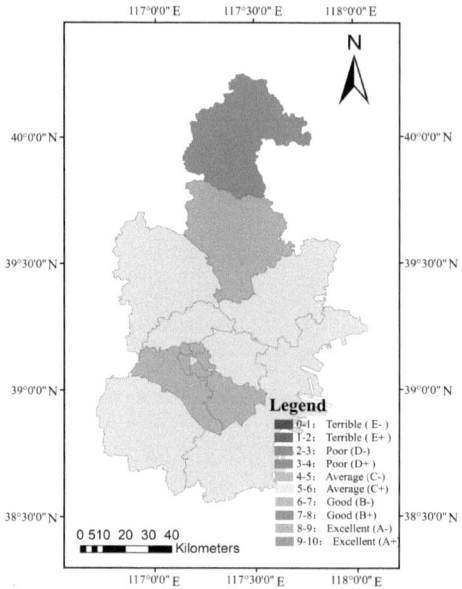

Fig. 4. Comprehensive evaluation of water-ecosystem resilience in Tianjin in 2018.

By calculating the coefficient of variation of the resilience scores of the driving forces, pressures, states, impacts, and responses of the 16 districts in Tianjin in 2018 (Eq. 1), the influence of these five aspects on the water ecosystem resilience of Tianjin could be analyzed (Fig. 5). The center point data in the figure represents the average of the coefficient of variation and mean values (0.34, 1.05). It can be seen that: ① The mean value of the water ecosystem resilience of the driving forces is the lowest, and the coefficient of variation is the highest, indicating that the urban and economic development issues in each administrative district are extremely serious and vary greatly. This is a very important reason for the low water ecosystem resilience in Tianjin. ② The mean value of the water ecosystem resilience of the pressures is the highest, and the coefficient of variation is the lowest, indicating that the water ecosystem pressures in each administrative district are generally low, which is a common reason affecting the water ecosystem resilience of Tianjin. It is necessary to continue to maintain stability to ensure the water ecosystem resilience of Tianjin. ③ The mean value of the water ecosystem resilience of the states is relatively low, and the coefficient of variation is also low, indicating that the water ecosystem states in each administrative district are generally poor and urgently need improvement. This is one of the important reasons for the low water ecosystem resilience in Tianjin. ④ The mean value of the water ecosystem resilience of the impacts is relatively low, and the coefficient of variation is relatively high, indicating that the various administrative districts are greatly affected by changes in driving forces, pressures, and states. Due to the differences in states among different administrative districts, the magnitude of the impacts also varies greatly. This is one of the important reasons for the low water ecosystem resilience in Tianjin. ⑤ The mean value of the water ecosystem resilience of the responses is relatively high, and the coefficient of variation is also high, indicating that there are significant differences among the administrative districts, which is a secondary reason for the low water ecosystem resilience in Tianjin.

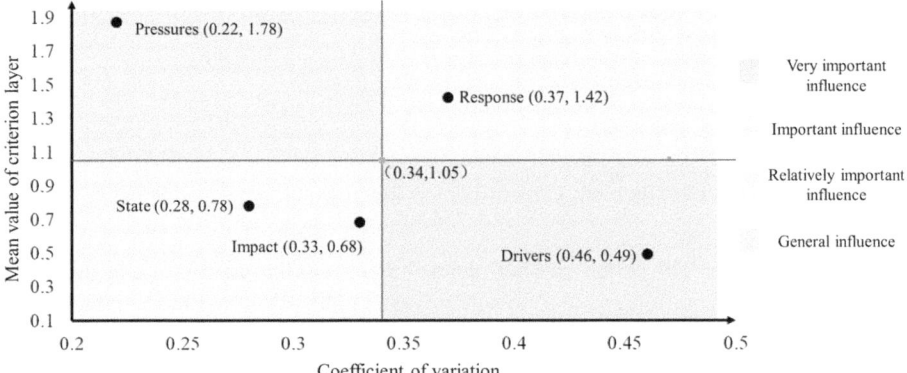

Fig. 5. Variation degree of criterion layer for water-ecosystem resilience in Tianjin.

Analysis of Representative Evaluation Indicators for Water Ecosystem Resilience

Selection of Representative Evaluation Indicators

This paper proposes enhancement strategies starting from spatial planning, using criteria such as "obvious problems," "high frequency," "high weight," and "good effect of spatial planning-related plans" to select two representative evaluation indicators: forest coverage rate and wetland rate (Fig. 6). Both indicators belong to the state criteria level and are important factors influencing water ecosystem resilience.

3.2 Analysis of Forest

Tianjin has a relatively low total forest resource and slow forest cultivation. Uneven patch sizes and low connectivity contribute to the low forest coverage rate in Tianjin.

Uneven patch size

Using ENVI software to interpret the land use of Tianjin in 2018 based on Landsat 8 satellite remote sensing images, there are a total of 23,299 forest patches in Tianjin, with a total area of approximately 1,459.90 km2, accounting for 12.20% of the total area of Tianjin, which is close to the statistical data of 12.07%. Based on the land use map of Tianjin in 2018, the patches are divided into four types according to their size. There are more small forest patches in Tianjin, and fewer super large, large, and medium-sized forest patches (Table 1). Within the scope of Tianjin, the distribution of forest patch sizes

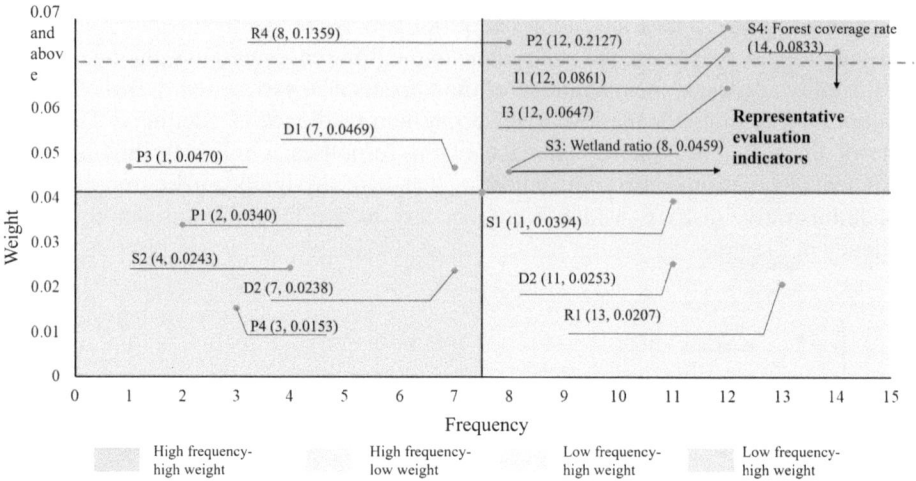

Fig. 6. Underperforming index classification and representative evaluation indicators.

is extremely uneven. The forest patches size is small, and the degree of fragmentation is high. High fragmentation can lead to a decrease in biodiversity within the patches, an increase in the risk of invasive species, a decrease in the quality of forest ecosystems, and an increase in the difficulty of forest protection and restoration.

Table 1. Types of forest patches in Tianjin.

Forest patch types	Patch type area[1]	Quantity	Percentage of total quantity	Area (km^2)	Percentage of total area
Super large-sized	10 km^2 and above	16	0.07%	535.1	36.65%
Large-sized	5 ~ 10 km^2	38	0.16%	98.98	6.78%
Medium-sized	1 ~ 5 km^2	213	0.91%	200.57	13.74%
Small-sized	0.005 ~ 1km^2	23032	98.85%	625.25	42.83%

Low Connectivity

High landscape connectivity is beneficial for maintaining ecological processes and biodiversity. The integral index of connectivity is selected to analyze the connectivity of the forest ecosystem in Tianjin. The calculation formula is as follows:

$$IIC = \frac{\sum_{i=1}^{n} \sum_{j=1}^{n} \frac{a_i * a_j}{1 + nl_{ij}}}{A_L^2}$$

where n represent the total number of ecological patches in the region, a_i and a_j respectively represent the area of the i-th and j-th patches. nl_{ij} Represents the number of routes

---

[1] 0.005km^2 is the minimum area of forest defined by the United Nations.

between patch and patch within a specified distance threshold. A_L represents the total area of ecological patches in the region, $0 \le IIC \le 1$. When $IIC = 0$, it indicates that there is no connectivity between ecological patches; when $IIC = 1$, it indicates that there is strong connectivity between all ecological patches and they can be considered as one ecological patch.

Under the integral index of connectivity, the delta integral index of connectivity can also be calculated using the following formula:

$$dIIC = 100 \frac{IIC - IIC_{remove}}{IIC} \tag{3}$$

where IIC is the integral index of connectivity, IIC_{remove} is the integral index of connectivity after removing any ecological patch.

First, four distance thresholds of 2.5, 5, 7.5, and 10 km are set. Second, starting from the edges of the patches (calculated from feature edges), the Conefor_Inputs_10 plugin in ArcGIS is used to export data on patch area, the topological distance between patches, and the number of routes between patches under the four distance thresholds. The distribution of routes between forest patches in Tianjin is exported using ArcGIS. Finally, the data is imported into the Conefor Sensinode 2.6 software to calculate the integral index of connectivity (IIC) and delta integral index of connectivity (dIIC) under the four distance thresholds. The distribution of the patch importance index of forest patches in Tianjin under different distance thresholds is exported using ArcGIS.

From the perspective of route quantity (Fig. 7), under the distance threshold of 2.5 km, there are fewer routes between forest patches in Tianjin, indicating poor connectivity and hindered biological flow between patches. As the distance threshold increases, the number of routes between forest patches in Tianjin also increases. Regardless of the distance threshold, there are no effective connecting routes between the northern and southern regions, indicating a clear forest ecosystem fragmentation between northern and southern regions. There are more routes between forest patches in northern region, indicating stronger connectivity and a healthier forest ecosystem compared to the southern region, which requires improvement.

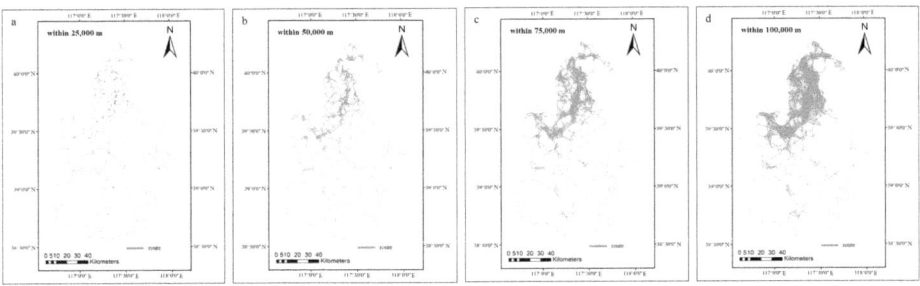

Fig. 7. Route distribution between forest patches under different distance thresholds in Tianjin.

From the perspective of the overall connectivity index (Table 2), the connectivity between forest patches in Tianjin is close to zero when the distance threshold is between

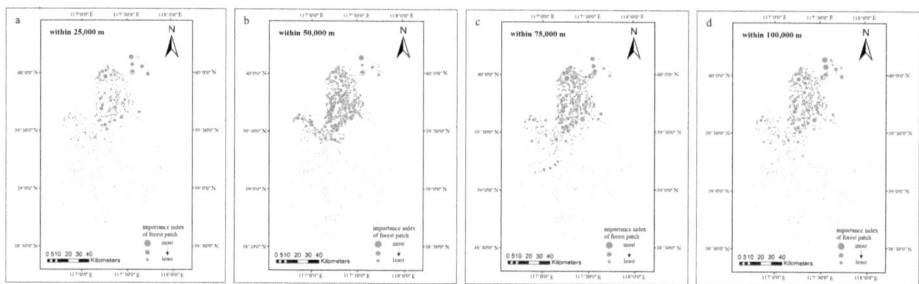

Fig. 8. Distribution of importance index for forest patch under different distance thresholds in Tianjin.

2.5 and 7.5 km. There is almost no connectivity within the forest ecosystem in Tianjin. Only when the distance threshold reaches 10 km, the IIC index is higher than 0.1, but the overall connectivity is still low. With the increase of the distance threshold, the connectivity of the forest ecosystem in Tianjin does not increase significantly, reflecting the uneven spatial distribution of forest resources and insufficient number of patches in Tianjin.

Table 2. Calculation results of connectivity indexes for forest ecosystem under different distance thresholds in Tianjin.

Indicators	2.5 km	5 km	7.5 km	10 km
Number of routes	438	1641	3319	5442
IIC	0.0219	0.0473	0.0970	0.1691
dIIC (average)	0.0886	0.2670	0.3071	0.3754

Table 3. The number of forest patches in the top 200 dIIC under different distance thresholds.

Patch Type	2.5 km	5 km	7.5 km	10 km
Super large-sized	14 (87.50%)	14 (87.50%)	14 (87.50%)	14 (87.50%)
Large-sized	27 (71.05%)	24 (63.15%)	27 (71.05%)	27 (71.05%)
Medium-sized	144 (67.60%)	116 (54.46%)	123 (57.74%)	125 (58.68%)
Small-sized	15 (0.06%)	46 (0.20%)	36 (0.15%)	34 (0.14%)

Note: The percentage represents the proportion of the top 200 dIIC patches of each type to the total number of patches in the study area

Threshold increases, the average importance index of forest patches in Tianjin shows a significant increase (Table 2), highlighting the stepping-stone function of the patches. However, it should be noted that regardless of the distance threshold, the forest patches with high importance index are concentrated in the northern region (Fig. 8), indicating strong connectivity of forest ecosystems in the northern region, while the southern region lacks important forest patches, resulting in low connectivity of forest ecosystems within the southern region and between the southern and northern regions. According to the rankings of the top 200 forest patch types based on dIIC at different distance thresholds (Table 3), almost all super large forest patches are within the top 200, many large forest patches are within the top 200, and over half of the medium-sized forest patches are within the top 200. Super large, large, and medium-sized forest patches play a crucial role in maintaining the connectivity of forest ecosystems in Tianjin and should be prioritized for protection and restoration.

3.3 Analysis of Wetland

In the early 20th century, the wetland ratio in Tianjin was approximately 45.9%. After the 1950s, it decreased to 27.3%. In the 1970s, it further declined to 8.5%. Since then, with the concept of "storage instead of drainage," reservoirs and other water facilities have been constructed, which to some extent increased the replenishment of the wetland ecological environment and steadily improved the wetland ratio. In 2018, the wetland ratio was 24.74%, an increase of 9.79 percentage points. However, there is still a significant gap between the current wetland ratio and the original state. The uneven patch sizes and low connectivity have hindered the protection and restoration of wetlands in Tianjin.

Uneven patch size

Using ENVI software to analyze the Landsat 8 satellite remote sensing images of land use in Tianjin in 2018, it was found that there are a total of 73,334 wetland patches in Tianjin, with a total area of approximately 2,768.84 km^2, accounting for 23.13% of the total area of Tianjin, which is close to the statistical data of 24.74%. Based on the land use map of Tianjin in 2018, the patches are divided into four types according to their size. There are more small wetland patches in Tianjin, and fewer super large, large, and medium-sized wetland patches (Table 4). Within the scope of Tianjin, the distribution of wetland patch sizes is extremely uneven, with small wetland patches being the dominant type. The high degree of fragmentation of wetland patches can lead to damage to the wetland ecosystem functions, a decrease in biodiversity within the patches, an increase in the risk of invasive species, a decrease in the quality of the wetland ecosystem, and an increase in the difficulty of wetland conservation and restoration.

Table 4. Types of wetland patches in Tianjin.

Wetland patch types	Patch type area	Quantity	Percentage of total quantity	Area (km^2)	Percentage of total area
Super large-sized	10 km^2 and above	29	0.04%	999.61	36.10%
Large-sized	5 ~ 10 km^2	36	0.05%	240.79	8.70%
Medium-sized	1 ~ 5 km^2	251	0.34%	526.84	19.03%
Small-sized	0.005 ~ 1 km^2	73018	99.57%	1001.59	36.17%

Low Connectivity

Based on Eqs. 2 to 3, the connectivity of wetland ecosystems is calculated at four distance thresholds: 2.5 km, 5 km, 7.5 km, and 10 km.

From the perspective of route quantity (Fig. 9), the number of routes between wetland patches in Tianjin is relatively low at a distance threshold of 2.5 km, indicating poor connectivity and limited biological exchange between wetland patches. As the distance threshold increases, the number of routes between wetland patches also increases. Regardless of the distance threshold, the distribution of route quantities between wetland patches exhibits a "multi-cluster" pattern, with super large and large wetland patches as the core, and relatively fewer connections between wetland patch clusters. The protection and restoration of the wetland ecosystem in Tianjin is currently fragmented and lacks overall coordination.

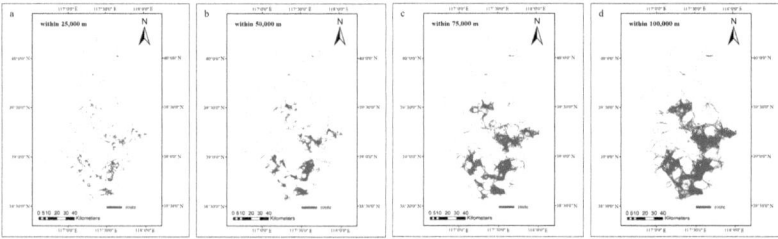

Fig. 9. Route distribution between wetland patches under different distance thresholds in Tianjin.

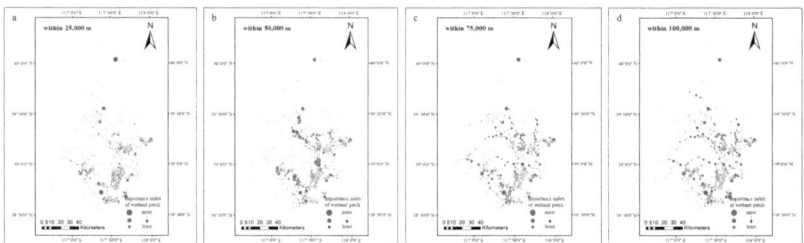

Fig. 10. Distribution of importance index for wetland patch under different distance thresholds in Tianjin.

From the perspective of the integral index of connectivity (Table 5), the connectivity between wetland patches in Tianjin is close to zero when the distance threshold is between 2.5 and 5 km. There is almost no connectivity within the wetland ecosystem in Tianjin. Only when the distance threshold reaches 7.5 km, the IIC index is higher than 0.1, but the overall connectivity is still low. This further demonstrates the weak connectivity between wetland patch clusters. With the increase of the distance threshold, the connectivity of the wetland ecosystem in Tianjin does not increase significantly, reflecting the uneven spatial distribution of wetland resources and the insufficient number of patches in Tianjin.

From the perspective of the patch importance index, as the distance threshold increases, the average importance index of wetland patches in Tianjin shows a significant

Table 5. Calculation results of connectivity indexes for wetland ecosystem under different distance thresholds in Tianjin.

Indicators	2.5 km	5 km	7.5 km	10 km
Number of routes	1699	5347	10519	16706
IIC	0.0301	0.0638	0.1056	0.1450
dIIC (average)	0.1337	0.1953	0.2671	0.3232

increase (Table 5), highlighting the stepping-stone function of the patches. However, it should be noted that regardless of the distance threshold, wetland patches with high importance index are concentrated in specific areas along the coast, rivers, and adjacent reservoirs (Fig. 10), indicating strong connectivity of wetland ecosystems in these areas, while the six districts within the city and the four surrounding districts lack important wetland patches, resulting in low connectivity of wetland ecosystems within the region. According to the types and quantities of wetland patches ranked in the top 200 based on different distance thresholds (Table 6), majority of super large and large wetland patches are within the top 200, and nearly half of the medium-sized wetland patches are within the top 200. Super large, large, and medium-sized wetland patches play a crucial role in maintaining the connectivity of the wetland ecosystem in Tianjin and require focused protection and restoration.

Table 6. The number of wetland patches in the top 200 dIIC under different distance thresholds.

Patch Type	2.5 km	5 km	7.5 km	10 km
Super large-sized	23 (79.31%)	23 (79.31%)	23 (79.31%)	23 (79.31%)
Large-sized	24 (66.67%)	21 (58.33%)	24 (66.67%)	24 (66.67%)
Medium-sized	92 (36.65%)	118 (47.01%)	133 (52.98%)	144 (57.37%)
Small-sized	61 (0.08%)	38 (0.05)	20 (0.03%)	9 (0.01%)

Note: The percentage represents the proportion of the top 200 dIIC patches of each type to the total number of patches in the study area

4   Discussion

Forests and wetlands are crucial components of blue-green systems [41, 42]. Strategies for enhancing resilience through blue-green integration include building forest cities, protecting and restoring wetland systems, and restoring blue-green ecological networks.

4.1   Building Forest Cities

To address the issues in Tianjin's forest ecosystem, forest city construction can be employed to increase forest coverage. This includes protecting existing forests and establishing urban forest parks, surrounding forest areas, and protective forest belts. Unused land identified in Tianjin's 2018 land use map can be considered as potential spaces for forest construction. The planning should align with publicized plans such as the Tianjin Urban Master Plan (2005–2020). Based on the spatial distribution of mountains, water systems, and roads, wetland spaces should be reserved for the planning of urban forest parks, surrounding forest areas, and protective forest belts. This will contribute to achieving the forest coverage target outlined in the Planning of Green Ecological Barrier Zone in the Intermediary Area between the Twin Cities of Tianjin (2018–2035).

The construction of urban forests and peri-urban forest areas is an effective means to expand the forest coverage in Tianjin. Based on the existing forest resources in Tianjin and in accordance with the requirements of the "Urban Green Space Planning Standard GB/T51346–2019" (the area of a single forest park should be above 50 hectares), this study proposes the construction of 9 urban forests and 28 peri-urban forest areas (Fig. 11.a) based on the current urban situation and natural landscape pattern, utilizing existing rivers, wetlands, and other unused land. The urban forests are all located within the outer ring road of Tianjin, increasing the forest area while providing nearby recreational spaces for residents. The peri-urban forest areas are all located outside the outer ring road of Tianjin, increasing the forest area and providing recreational spaces for outings, while also serving the function of limiting urban expansion and protecting wetland areas. The construction of protective forest belts is an effective means to enrich the linear spatial pattern of forests in Tianjin. Linear elements such as highways, expressways, and rivers have fragmented the urban space. Based on the current urban situation in Tianjin and the planning layout of peri-urban forest areas, this study suggests the construction of protective forest belts (Fig. 11.a) with the main function of protection and isolation, and with a certain width, based on road and river elements such as the S30 Beijing-Tianjin Expressway and the mainstream of the Haihe River. According to the requirements of the "Technical Specification for Urban Green Line Delineation GB/T51163-2016" and the "Greening Design Specification for Highways DB61/T 1056-2016", the width of the protective forest belts in the city and along highways should not be less than 30 m and 80 m, respectively. Referring to the width of protective green belts in Chengdu and Tianjin, it is recommended to set the width as 500 m.

4.2 Protecting and Restoring Wetland Systems

Based on the existing issues in Tianjin's wetland ecosystem, the wetland area and quantity can be increased to improve the wetland ratio in Tianjin. The focus should be on protecting the existing wetlands and adding new ones. Identify unused land in Tianjin's 2018 land use map as potential sites for wetland construction. Consider publicized planning schemes such as the Tianjin Water System Plan (2008–2020). Use the spatial distribution of lakes, ponds, rivers, and salt fields as the basic conditions for wetland expansion and wetland space reserved in forest construction strategies. This can be achieved by expanding existing wetlands and creating new ones to increase the wetland area and quantity (Fig. 11.b).

To enhance the resilience of the water ecosystem in Tianjin, it is proposed to expand the existing wetland areas, such as Beidagang Wetland, Qilihai Wetland, and Dongli Lake Wetland, by incorporating forest patches and other unused land. In addition, large wetland parks and wetland patches with stepping-stone functions should be established

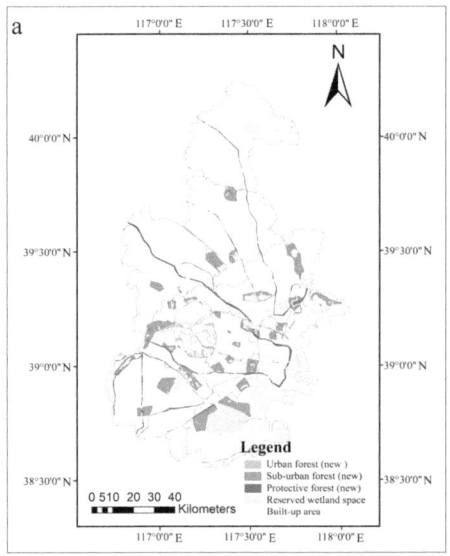
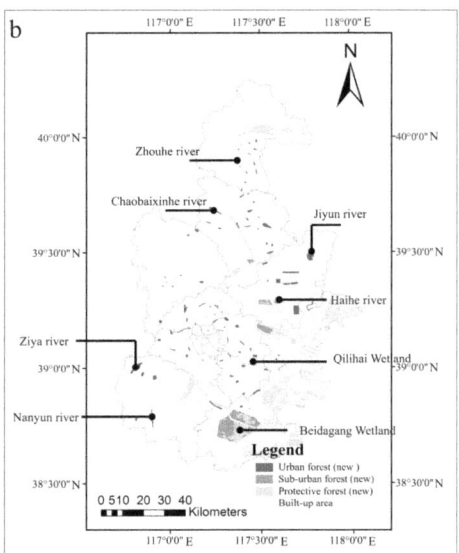

Fig. 11. The proposed forest and wetland distribution.

based on the spatial distribution of lakes, ponds, rivers, and salt fields. Examples of large wetland parks include the Ziya River, Nanyun River, and Jiyun River, which are all located upstream of rivers. These parks not only provide recreational functions but also help purify the water quality of downstream areas. Furthermore, it is suggested to designate core areas, buffer zones, and experimental areas around internationally and nationally important wetlands, such as Beidagang Wetland.

5 Conclusions

Under the national strategic background of "water-based urban development" and "resilience", a water ecosystem resilience evaluation system was constructed using the DPSIR model and an improved maximum difference combination weighting method. The case study of Tianjin was conducted to empirically investigate the system. The following conclusions were drawn:

(1) In 2018, there was a significant spatial difference in the water ecosystem resilience level in Tianjin. The overall level of water ecosystem resilience was poor, with only Ji and Bao districts achieving a good (B-) or higher level of water ecosystem resilience. After conducting an overall analysis of the water ecosystem resilience in Tianjin using the coefficient of variation, it was found that the driving criteria layer was a very important factor causing the low water ecosystem resilience in Tianjin. The state criteria layer and the impact criteria layer were important factors causing the low water ecosystem resilience in Tianjin. The response criteria layer was a secondary factor causing the low water ecosystem resilience in Tianjin. The pressure criteria layer was a general factor affecting the water ecosystem resilience

in Tianjin. The overall level of water ecosystem resilience in Tianjin is relatively poor, and Tianjin should pay attention to the optimization of driving factors and state factors for a long time.

(2) The study selected two representative evaluation indicators: forest coverage rate and wetland rate, both of which belong to the state criterion level and are important indicators that affect the resilience of water ecology. Forest coverage and wetland rate are representative evaluation indicators that affect the resilience of the water ecosystem in Tianjin. In the near future, funds should be invested to restore forests and wetlands.

Due to the length of the study, the research mainly focuses on proposing strategies for enhancing water ecosystem resilience based on representative evaluation indicators. The overall enhancement strategy was not proposed based on the overall analysis conclusion. In subsequent research, the planning strategy system can be further improved.

Funding. This research was funded by the Natural Science Foundation of Southwest University of Science and Technology, Grant agreement ID 22zx7158; Education Reform and Research Project of Southwest University of Science and Technology, Grant agreement ID 22xn0066.

References

1. Shao Yiwen, Xu Jian: Understanding urban resilience: a conceptual analysis based on Integrated international literature review. Urban Plan. Int. **30**(2), 48–54 (2015) (in Chinese)
2. Ouyang, M., Dueñas-Osorio, L., Min, X.: A three-stage resilience analysis framework for urban infrastructure systems. Struct. Saf. **36**, 23–31 (2012)
3. Li Tongyue: New progress in study on resilient cities. Urban Plan. Int. **32**(5), 15–25 (2017) (in Chinese)
4. Keys, P.W., Porkka, M., Wang-Erlandsson, L., et al.: Invisible water security: moisture recycling and water resilience. Water Secur. **8**, 100046 (2019)
5. Falkenmark, M., Wang-Erlandsson, L., Rockström, J.: Understanding of water resilience in the Anthropocene. J. Hydrol. X **2**, 100009 (2019)
6. Falkenmark, M., Wang-Erlandsson, L.: A water-function-based framework for understanding and governing water resilience in the Anthropocene. One Earth **4**(02), 213–225 (2021)
7. Chen Gang, Wang Lin, Wan Jin, et al.: Construction of water ecological resilience space based on landscape ecological pattern. Yellow River **42**(05), 87–90+96 (2020) (in Chinese)
8. Li Yuanjie, Jiang Xinhui, Chen Jun, et al.: Prediction of groundwater flow field in Linhe district of Bayan Nur City using GMS. South-to-North Water Transf. Water Sci. Technol. **14**(04), 36–41+83 (2016) (in Chinese)
9. Eriksson, M.G., Gordon, L.J., Kuylenstierna, J.: Cross-sectoral approaches help build water resilience–reflections. Aquatic Procedia **2**, 42–47 (2014)
10. Klise, K.A., Bynum, M., Moriarty, D., et al.: A software framework for assessing the resilience of drinking water systems to disasters with an example earthquake case study. Environ. Model. Softw. **95**, 420–431 (2017, Sep)
11. Amarasinghe, P., Liu, A., Egodawatta, P., et al.: Quantitative assessment of resilience of a water supply system under rainfall reduction due to climate change. J. Hydrol. **540**, 1043–1052 (2016)
12. Sweya, L.N., Wilkinson, S., Chang-Richard, A.: Understanding water systems resilience problems in Tanzania. Procedia Eng. **212**, 488–495 (2018)

13. Kasmalkar, I.G., Serafin, K.A., Miao, Y., et al.: When floods hit the road: resilience to flood-related traffic disruption in the San Francisco Bay Area and beyond. Sci. Adv. **6**(32), eaba2423 (2020)
14. Liao Guixian, Lin Hejia, Wang Yang: A theory on urban resilience to floods—a basis for alternative planning practices. Urban Plan. Int. **30**(02), 36–47 (2015) (in Chinese)
15. Rodina, L.: Defining, "water resilience": debates, concepts, approaches, and gaps. Wiley Interdiscip. Rev. Water **6**(2), e1334 (2019)
16. The City Water Resilience Approach. https://www.arup.com/perspectives/publications/research/section/the-city-water-resilience-approach, last accessed 2024/01/05
17. Bruce, A., Brown, C., Avello, P., et al.: Human dimensions of urban water resilience: perspectives from Cape Town, Kingston upon Hull, Mexico City and Miami. Water Secur. **9**, 100060 (2020)
18. Holling, C.S.: Resilience and stability of ecological systems. Annu. Rev. Ecol. Syst. **4**(1), 1–23 (1973)
19. Folke, C.: Resilience: the emergence of a perspective for social–ecological systems analyses. Glob. Environ. Chang. **16**(3), 253–267 (2006)
20. Huang Jingnan, Ao Ningqian, Xie Yuhang: Review and prospect of international common development indicator framework. Urban Plan. Int. 34(05), 94–101 (2019) (in Chinese)
21. Gari, S.R., Newton, A., Icely, J.D.: A review of the application and evolution of the DPSIR framework with an emphasis on coastal social-ecological systems. Ocean. Coast. Manag. **103**, 63–77 (2015, Jan)
22. Huang Changshuo, Geng Leihua, Wang Liqun, et al.: Evaluation on China water resources and water ecological security. Yellow River **32**(03), 14–16+140 (2010) (in Chinese)
23. Sun Caizhi, Yang Yu, Chen Xiangtao, et al.: Water risk assessment and spatial correlation patterns research in provincial scale in China. Water Resour. Prot. **31**(06), 18-26 (2015) (in Chinese)
24. Xu Xiangqin, Cai Wenqian, Wang Yan, et al.: Assessment on ecological integrity of typical lakes, reservoirs and wetlands in Tianjin, China. Chin. J. Appl. Ecol. **31**(08), 2767–2774 (2020) (in Chinese)
25. Zhou Guohua, Xu Guobin, Wu Junliang, et al.: Systematic scheme exploration for sponge city construction in Tianjin eco-city. China Water & Wastewater **36**(12), 65–69 (2020) (in Chinese)
26. Weterings, R.: Environmental indicators: typology and overview. http://www.geogr.uni-jena.de/fileadmin/Geoinformatik/projekte/brahmatwinn/Workshops/FEEM/Indicators/EEA_tech_rep_25_Env_Ind.pdf, last accessed 2024/01/05
27. Sarkki, S., Komu, T., Heikkinen, H.I., et al.: Applying a synthetic approach to the resilience of Finnish reindeer herding as a changing livelihood. Ecol. Soc. **21**(4), 14 (2016)
28. Bell, S.: DPSIR = A problem structuring method? An exploration from the "Imagine" approach. Eur. J. Oper. Res. **222**(2), 350–360 (2012)
29. Chen Danyu: Urban Resilience Assessment Based on Pressure-State-Response Model. Huazhong University of Science and Technology, Wuhan (2020) (in Chinese)
30. Norris, F.H., Stevens, S.P., Pfefferbaum, B., et al.: Community resilience as a metaphor, theory, set of capacities, and strategy for disaster readiness. Am. J. Community Psychol. **41**(1), 127–150 (2008)
31. Breton, M.: Neighborhood resiliency. J. Community Pract. **9**(1), 21–36 (2001)
32. Maguire, B., Hagan, P.: Disasters and communities: understanding social resilience. Aust. J. Emerg. Manag. **22**(2), 16–20 (2007)
33. Sparks, T.H., Butchart, S.H.M., Balmford, A., et al.: Linked indicator sets for addressing biodiversity loss. Oryx **45**(3), 411–419 (2011)
34. Li Gang, Li Jianping, Sun Xiaolei: Research on a combined method of subjective-objective weighing and the LTS rationality. Manag. Rev. **29**(12), 17–26+61 (2017) (in Chinese)

35. Li Gang: The Empirical Research of Human All-Round Development Evaluation Model Based on Scientific Outlook on Development. Dalian University of Technology, Dalian (2010) (in Chinese)
36. Zuo Shudi, Ren Yin: Study on the temporal and spatial variation of forest landscape and ecological quality during the fast urbanization process. Environ. Sci. Technol. **38**(09), 191–198+205 (2015) (in Chinese)
37. Saura, S., Torne, J.: Conefor Sensinode 2.2: a software package for quantifying the importance of habitat patches for landscape connectivity. Environ. Model. Softw. **24**(1), 135–139 (2009)
38. Liu Wei: Correctly understand the characteristics and trends of economic slowdown. Chin. Cadres Trib. (1), 6–13 (2020) (in Chinese)
39. Deng Zongbing, Su Congwen, Zong Shuwei, et al.: Measurement and analysis of China's water ecological civilization construction index. China Soft Sci. (9), 82–92 (2019) (in Chinese)
40. Deng Zongbing, He Ruofan, Chen Zheng, et al.: Study on regional differences and convergence of ecological civilization development in eight comprehensive economic areas of China. J. Quant. Technol. Econ. **37**(06), 3–25 (2020) (in Chinese)
41. Lu Min, Luo Xiaonan, Wang Yonghua, et al.: Ecological pattern of urban forest landscape of Jinan City, China. Chin. J. Appl. Ecol. **30**(12), 4117–4126 (2019) (in Chinese)
42. Liu Dongyun, Huang Xiaolei, Du Linfang, et al.: Dynamic changes of wetland landscape pattern in Tianiin, North China. Chin. J. Ecol. **31**(11), 2914–2920 (2012) (in Chinese)

Research on the Resilience Transformation Strategy of Quarantine Hotels and Concentrated Quarantine Points Under Public Health Emergencies

Chang Lin[1], Yuxi Li[1], and Haitian Huang[2(✉)]

[1] Guangzhou Academy of Fine Arts, Guangzhou 510261, China
linchang@gzarts.edu.cn
[2] Guangzhou College of Commerce, Guangzhou 511363, China
huanght0428@163.com

Abstract. This paper explores the resilience transformation of quarantine hotels and concentrated quarantine points in public health emergencies for optimized medical space usage, employing a mixed-methods approach that combines case studies and qualitative analysis. During the COVID-19 pandemic, traditional medical facilities experienced significant pressures on space and resources, underscoring the need for resilience in non-traditional medical spaces. By analyzing specific quarantine hotels and points, we identify and implement transformation measures including design adjustments, functional layout optimization, and clean/dirty zoning. These strategies significantly enhance medical resource efficiency and reduce cross-infection risks. Our findings indicate that such transformations not only improve immediate pandemic response capabilities but also bolster the overall resilience of public health systems against future crises.

Keywords: Resilient cities · Quarantine hotels · Medical space transformation

During the COVID-19 pandemic, traditional healthcare systems were confronted with unprecedented operational and capacity challenges, revealing profound gaps in preparedness for public health emergencies. This situation underscored the acute shortages of protective supplies and the urgent need for expanded isolation spaces, highlighting vulnerabilities in medical resource allocation and supply reserve systems. Despite considerable research into emergency medical response mechanisms, the role of resilience transformation of non-traditional medical spaces, such as quarantine hotels and concentrated quarantine points, in augmenting healthcare system capacities has received scant attention. This study seeks to address this research gap by presenting an in-depth analysis of resilience transformation strategies for these non-traditional medical spaces. Thus, it contributes novel insights into optimizing the usage of medical spaces and enhancing the efficiency of public health emergency responses, marking a significant advancement in the field of emergency medical planning and public health preparedness. Firstly, it is necessary to study the linkage model of the medical facility grading service system and the urban spatial structure hierarchy, and establish a medical facility layout system

B.-J. He et al. (Eds.): UCSUD 2023, LNCE 559, pp. 903–913, 2025.
https://doi.org/10.1007/978-981-97-8401-1_65

suitable for graded diagnosis and treatment [2]. Different categories of medical institutions form an orderly referral system to dynamically balance the stock at all levels; at the same time, it is required to effectively distinguish between acute and chronic diseases, avoid the treatment of emergency patients and the recovery of chronic disease patients, and also emphasize the necessity of establishing a division of labor mechanism between upper and lower level institutions [3]. Especially isolation hotels and centralized isolation points have become important auxiliary spaces for epidemic prevention and control. However, these temporary medical spaces often lack functionality, safety, and efficiency, and urgently need effective resilience transformation to meet emergency medical needs. Therefore, this study aims to explore the resilience transformation strategies of isolation hotels and centralized isolation points during public health emergencies. By deeply analyzing the structural characteristics and functional layout of existing buildings, this paper aims to propose a series of practical transformation suggestions to improve the efficiency and safety of these non-traditional medical spaces. In the context of epidemic prevention and control, improving the medical service capabilities of non-medical buildings can not only effectively alleviate the pressure on traditional hospitals but also provide valuable experience and strategies for possible future public health events.

1   Composition and Improvement of Urban Medical Space Epidemic Prevention System Facilities

After the outbreak of SARS and COVID-19, Guangzhou city has added and expanded the Guangzhou Center for Disease Control and Prevention, infectious disease hospitals, experimental research institutions, and built a series of new research institutions such as the National Respiratory Medicine Center, and improved primary healthcare institutions, health stations, and other medical command, rescue, research, and isolation facilities. It is worth noting that after the outbreak of SARS, meeting medical flow lines and medical functional requirements were once again placed at the forefront of design evaluation. Under the COVID-19 pandemic, how to enhance the overall resilience of the medical system and optimize the resilience of medical space is the focus of the design strategy research of the prevention and control system. Due to the unpredictability and uncertainty of the outbreak and spread of infectious diseases, it is not enough to rely solely on internal building modifications, but more needs to come from the systematic design and mutual support of social systems such as administrative policies, social management, medical systems, municipal support, etc. It requires the joint control and participation of the whole society to build a more flexible public health epidemic prevention system as a whole. Therefore, in addition to the resilience improvement of the medical system itself, the resilience improvement of Guangzhou's urban medical space mainly revolves around the following points:

Firstly, in terms of social management, there is a need to strengthen the hierarchical and grid-based epidemic prevention control in the central urban area. The central area of Guangzhou city has a high population density, with a mix of urban villages, residential areas, and various public buildings within the built-up area. To meet the requirements of grid-based epidemic prevention and control, it is necessary to closely track the permanent residents of the street communities. Guangzhou pioneered the mechanism of "three-person teams" for household visits in communities for epidemic prevention, which participate in epidemic prevention coordination, resident management, material allocation, etc. The screening and safeguarding aspects of the medical system's epidemic prevention work show certain management advantages. When positive cases appear, the epidemic prevention is carried out quickly by defining the control areas, management areas, and prevention areas based on their activity areas and close contacts. This minimizes the impact on social life while isolating key populations as early as possible to cut off the transmission chain. Secondly, epidemic prevention work is carried out in conjunction with traffic control measures. By controlling key places such as high-speed railway stations, subway buses, and other public transportation facilities, the frequency of travel in the hierarchical control areas is reduced, and crowd gathering is minimized to the greatest extent. It can be seen that by layering and gridding epidemic prevention management, Guangzhou city has broken down major issues into each small grid, reducing the difficulty and intensity of governance, and mobilizing grassroots personnel to achieve group prevention and control. As a mega-city with tens of millions of people, Guangzhou has effectively reduced crowd gathering and reduced the risk of epidemic spread. The "Guangzhou Experience" has become a sample for many cities in China to learn from. In terms of improving the resilience of the medical system itself, according to the three-year action plan for the construction of public health prevention and control treatment capabilities (2020–2022) jointly formulated by the Provincial Health Commission, the Provincial Development and Reform Commission, and the Provincial Traditional Chinese Medicine Bureau, the city's capacity building action plan mainly revolves around modernizing disease prevention and control, improving the urban infectious disease treatment network, upgrading provincial major epidemic treatment bases, and promoting the dual-use transformation and standardized setting of fever clinics in medical institutions. Through the three-year action plan, the four-level public health prevention and control treatment system of the province, region, city, and county is more complete, the prediction, early warning, prevention, and pathogen testing capabilities of the disease prevention and control system have reached the advanced level in China, and the major epidemic medical treatment and emergency support capabilities meet the prevention and control needs of the Guangdong-Hong Kong-Macao Greater Bay Area. Among them, the main construction content in terms of improving the urban infectious disease treatment network is the construction project of the Guangdong Provincial Public Health Medical Center (Guangdong Provincial Infectious Disease Hospital), the construction of the provincial major infectious disease comprehensive treatment center, the tropical disease and local disease comprehensive treatment center; the creation of a public health event emergency and disaster medical training base, the construction of a public health testing laboratory, a scientific and technological innovation and transformation platform, a provincial major sudden public health event emergency reserve platform (including a reserved emergency treatment site with 2000 beds). At the same time, the centralized treatment capacity of the infectious disease areas of 25 city infectious disease hospitals or general hospitals is expanded. In terms of upgrading and transforming provincial major epidemic treatment bases, the second people's hospital of Guangdong Province, the first affiliated hospital of Guangzhou University of Traditional Chinese Medicine, the first affiliated hospital of Guangzhou Medical University, and other 3 provincial major epidemic treatment bases will be constructed. The number of convertible infectious disease beds in the 3 base hospitals will increase from the current 92 (14 negative pressure) to 761 (126 negative pressure), and the number of convertible ICU beds will increase from the current 291 to 609. Secondly, in terms of the construction and promotion of dual-use transformation of urban public facilities for epidemic prevention, it is required that all cities and counties (cities, districts) thoroughly survey and sort out the situation of large public buildings such as convention and exhibition centers, sports venues, etc. in their regions. They should investigate and study the facility information that can be converted into emergency medical facilities, centralized medical isolation observation points, or makeshift hospitals through epidemic prevention conversion. They should formulate plans for expansion and the necessary conditions, forming an "Emergency Medical Facilities Distribution Map" and "Activation Sequence List". At the same time, during the expansion of such large public facilities, the needs of emergency medical care should be considered, the expansion plan should be improved, including site layout, logistics support system design, ventilation and air conditioning systems, reserving good connections with municipal facilities, information and equipment interfaces, and transformation space. This will enable rapid conversion into emergency treatment and medical isolation conditions, playing a role in "combining normal and epidemic situations". In terms of standardizing the setting up of fever clinics in medical institutions, we will upgrade and transform the fever clinics of 755 (including 21 central health clinics) medical institutions currently in operation in our province according to the standards for setting up fever clinics (except for medical institution projects included in county-level hospitals, urban infectious disease medical treatment networks, and major epidemic treatment bases). We will continue to improve the standardized construction of fever clinics in 26 central health clinics. Independent fever clinics will be set up, built according to the "three zones and two channels" standard, with at least 3 consultation rooms, treatment rooms, isolation observation wards (rooms), laboratory examination rooms, and other functional rooms and areas. On this basis, 50 new planned fever clinics will be added in traditional Chinese medicine hospitals and other specialized medical institutions across the province, bringing the total number of standardized fever clinics in the province to around 800. Different levels of medical institutions in the urban medical space system will each perform their own duties within the same system, building the resilience of the overall medical system through collaborative linkage. This paper starts with the analysis of typical cases in the subsystems of different types of urban medical spaces, including centralized isolation points and the expansion and renovation projects of fever clinics in comprehensive hospitals, and discusses the principles and strategies for resilience optimization in the normalization stage of the epidemic.

2   Centralized Isolation Points

Centralized isolation points established for epidemic prevention and control are mainly used for centralized medical isolation observation of suspected patients, key inbound populations, and close contacts of confirmed patients. Centralizing scattered isolation points can improve the concentration and efficiency of medical resources and is one of the important facilities at the screening end of overall epidemic prevention and control.

In terms of the location of centralized isolation points, large hotels, school dormitory areas, training centers in the suburbs can be utilized. At the same time, these types of hotels, school dormitory areas, and other standalone multi-story buildings can be expanded and renovated to meet isolation standards, so they have independent entrances and vertical elevators for medical staff, patients, and waste, and strict physical separation of clean and dirty areas in the layout, meeting the requirements of emergency medical spaces. Secondly, new large-scale hotel facilities can also be built according to epidemic prevention and isolation standards. They can operate and be used as normal hotels in normal times, and be converted into centralized isolation points for epidemic prevention and control when an epidemic occurs, for the isolation observation of fever patients and close contacts of confirmed patients, to improve the utilization rate of such building spaces in the switch between normal and epidemic situations.

2.1 International Health Stations

Due to the severe situation of COVID-19 prevention and control, Guangzhou, with its frequent international exchanges and business interactions, faces significant pressure in preventing imported cases of the virus. To guard against the risk of local transmission caused by imported cases and to build a solid line of defense for epidemic prevention and control, Guangzhou has pioneered the "International Health Station" model, comprehensively enhancing the safety factor of centralized quarantine facilities in terms of layout design, management model, and technical means. In addition to the International Health Station, other districts in Guangzhou are also intensifying the design and construction of similar health stations to meet future quarantine needs (Fig. 1).

The newly built Guangzhou International Health Station is a centralized quarantine facility with a high proportion of unmanned equipment applications such as centralized management through a smart platform and contactless delivery. The International Health Station is located in the Zhongluotan Mali plot in the Cloud District. Its location is a 30–40 min drive from the city center, far from densely populated residential areas and sensitive places such as schools and hospitals, and it is conveniently connected to the Guangzhou Medical University Affiliated City Eighth Hospital Jiahe District, a specialized infectious disease hospital.

The International Health Station differs from existing quarantine facilities such as hotels in two ways: First, it is larger in scale, with a total construction area of about 250,000 square meters. The International Health Station has a total of 5,074 quarantine rooms, in addition to a dining kitchen area, service center, and logistics support rooms, providing 2,000 beds for total logistics service personnel, integrating functions such as fire protection, security, kitchen and vehicle disinfection, and waste transportation. Second, it has high construction standards. As the International Health Station is located in the suburbs with fewer surrounding medical facilities, it is equipped with medical function modules such as fever clinics, general clinics, and laboratories within the park to provide medical services for those in centralized quarantine. The layout of the standard planar functional space in the centralized quarantine area of the International Health Station strictly follows the technical standards of the "three zones and two channels", namely the clean zone, semi-clean zone, and contaminated zone, and the corresponding clean and contaminated channels. Each functional module within the park is relatively

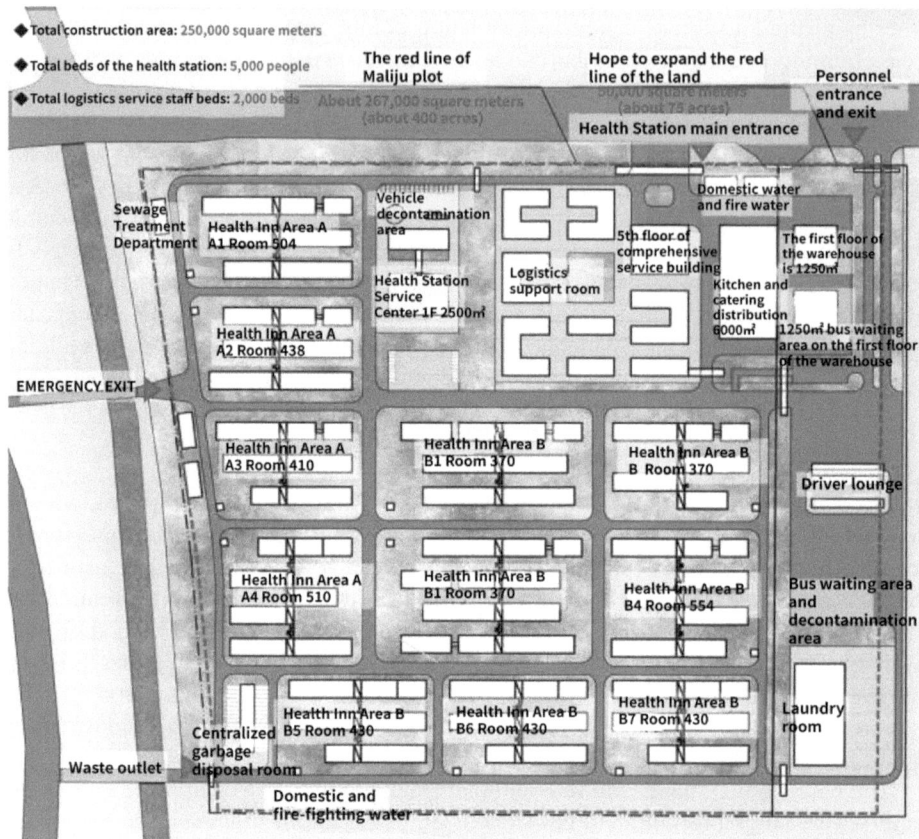

Fig. 1. The overall plan of the Guangzhou International Health Station located in the Zhongluotan Mali plot in Baiyun District, Guangzhou. (Drawn by the author)

independent, with intervals all greater than 20 m. If a positive patient appears, the area and group where they are located will be immediately closed off to cut off the transmission chain. The park staff channels, quarantine personnel channels, and material transportation channels are set up independently, with clear flow lines, separate clean and contaminated areas, and closed-loop management [5].

2.2 Renovation of Jinjiang Inn Quarantine Hotel

In addition to the large-scale centralized quarantine sites that are newly built, Guangzhou City is also renovating and expanding existing hotels and other facilities in various districts to meet the urgent needs of COVID-19 prevention and control work. This is in line with the national policy of "combining normalcy with epidemic prevention" and enhances Guangzhou's emergency medical isolation capabilities as a national central city. It is beneficial for improving and perfecting the public health prevention and control system in Guangzhou and alleviating the tense situation of public medical isolation observation hotel resources in Guangzhou.

The construction of medical isolation observation facilities in Guangzhou is mainly based on the Design Guidelines for Temporary Medical Isolation Observation Facilities (Trial) jointly issued by the National Health Commission Office and the Ministry of Housing and Urban-Rural Development Office, and the Technical Guidelines for the Second Phase of Guangzhou Health Post Construction (Draft for Comments).This study examines the renovation of the Jinjiang Inn at No. 245 Jiangyan Road, Haizhu District, to elucidate the functional classification and design considerations in transforming a conventional hotel into an effective quarantine facility. The renovation is categorized into three primary functional areas:

Supporting Service Areas: These include the hotel lobby, staff office spaces, and service facilities such as sewage treatment rooms. These areas are essential for ensuring the smooth operation of the quarantine hotel, providing necessary administrative and maintenance support.

Isolation Room Areas: The core of the quarantine hotel, these rooms are specially designed to house individuals undergoing isolation. Modifications to these rooms prioritize minimizing cross-contamination risks, incorporating features such as independent ventilation systems and easy-to-clean surfaces.

Traffic Areas: Critical for controlling the movement within the hotel, traffic areas include corridors, elevators, and emergency exits. The design emphasizes segregated pathways for different user groups (staff, isolated individuals, and service personnel) to reduce the risk of virus transmission. Special attention is given to the layout to facilitate efficient and safe movement while adhering to health protocols.

The proposed project covers an area of about 1193.17 square meters, with a total construction area of 11270.76 square meters. The building has 9 floors, divided into supporting service areas and isolation room areas.

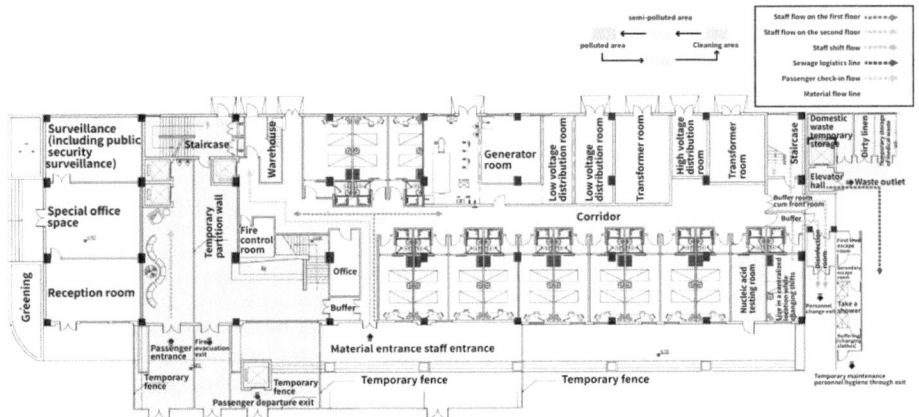

Fig. 2. Floor plan of the first floor renovation of Jinjiang Inn (Source: Drawn by the author)

The renovation of Jinjiang Inn, while successfully segregating 'clean' and 'contaminated' zones, unveils challenges inherent in retrofitting existing structures for medical use. The division of the first floor into distinct areas for specialized teams and monitoring, juxtaposed with the designated contaminated zones, exemplifies an innovative approach to space reutilization. However, this adaptation raises concerns regarding ventilation systems and the potential for airborne pathogen spread through common areas, such as elevators, which were not initially designed with medical-grade air filtration systems. Additionally, the reliance on existing elevator systems for transporting potentially infectious patients from the lobby to isolation rooms underscores a critical oversight in the design, potentially compromising the safety of both patients and staff. This case highlights the necessity for a more integrated design approach that encompasses not only spatial reconfiguration but also enhancements in building systems to ensure health and safety standards are met. The central part of the first-floor lobby uses temporary partitions to divide the fire passage and the passenger exit. Two elevators are added on the first floor for departure and waste transportation. The right area is the staff area, with added buffer rooms, shower rooms, and first and second exits for temporary maintenance personnel for sanitary passage. Buffer rooms and disinfection rooms are added between the clean area and the contaminated area passenger rooms to form a buffer space (Fig. 2).

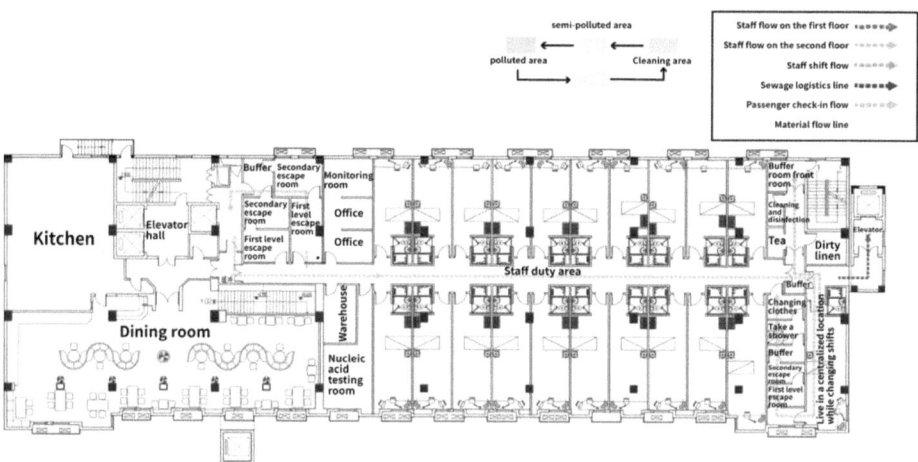

Fig. 3. Floor plan of the second floor renovation of Jinjiang Inn (Source: Drawn by the author)

The second floor of the hotel is divided into two main areas. On the left side, there is a kitchen and dining area, which is used to prepare meals for the isolation rooms on each floor, accessible via a clean staircase. The right side of the second floor is designated as the office and living area for the staff of the long-term stay isolation hotel. There are also rooms for staff to live in when they change shifts. This area is separated from the clean area by a buffer zone, which includes rooms for undressing and showering, for use by the staff before entering the clean area. Staff leaving the hotel go through a disinfection room, and the dirty staircase on the right side of the floor plan is used for collecting and transporting waste from the contaminated area. There is also a lift lobby and a corridor in the contaminated area to serve as a separation (Fig. 3).

The third floor of the hotel is a dedicated passenger isolation room area. On the left side of the floor plan, there is a pet isolation room, and on the right side, there are supervised isolation rooms and isolation rooms for people with mental abnormalities. In the middle of this floor, there is a complete sanitary pass-through room, where staff can undress, shower, and clean up after completing their service work on the third to ninth floors, before returning to the clean area. This floor also has an exit elevator in the contaminated area. Passengers who have completed their isolation period can go through an automatic disinfection room, take the exit elevator to the first floor, and leave the isolation hotel (Fig. 4).

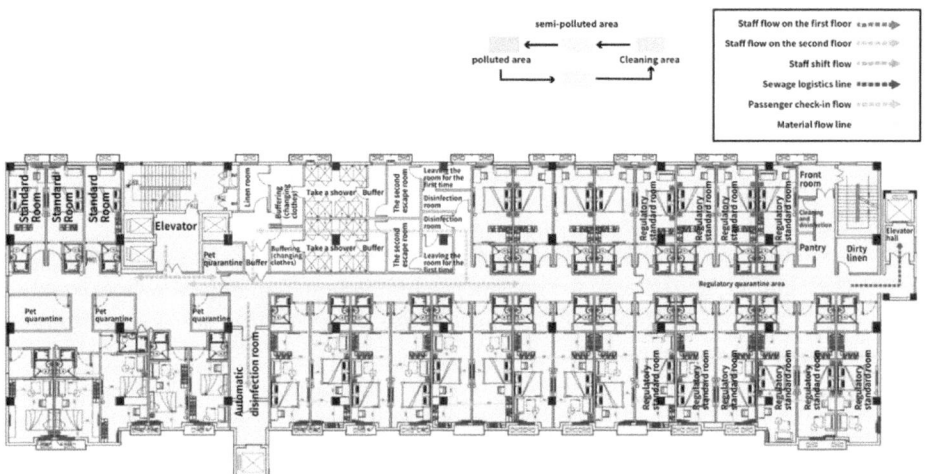

Fig. 4. Renovation floor plan of the third floor of Jinjiang Star Hotel (Source: Drawn by the author)

Secondly, in terms of ventilation and air conditioning system renovation, independent split air conditioners are installed in the guest rooms to prevent cross-infection caused by air circulation between different spaces. Vertical mechanical exhaust is set up in the bathrooms of the isolation rooms, and exhaust fans with check valves are installed on each floor. The exhaust air is discharged into the high sky through the vertical shaft. Apart from the isolation rooms, independent ventilation systems are set up in other functional areas of the floor plan. Mechanical ventilation systems are installed in the conference center, restaurant, and protective clothing dressing and undressing passages. The fresh air and exhaust air volume of each area is reasonably designed to ensure that the air flow goes from the clean area to the contaminated area. At the same time, to maintain a positive pressure environment in the corridor, a mechanical fresh air system is installed in the corridor of the isolation rooms, allowing the air to flow from the corridor to the rooms, reducing the risk of infection for service staff.

While this study concentrates on the transformation of quarantine hotels and specific quarantine points, a wider range of facilities should be explored, such as pavilions, gyms, and exhibition centers, acknowledges the breadth and depth that such investigations can add to our understanding of resilience transformation in public health crises.

Their inclusion in future research would undoubtedly enrich the discourse on pandemic preparedness and response strategies.

3 Conclusion

This study systematically investigates the resilience transformation strategies of quarantine hotels and concentrated quarantine points, offering distinct theoretical and practical insights. The research differentiates between various case scenarios, such as newly built versus renovated facilities, suburban versus urban locations, and large-scale versus smaller facilities, providing a nuanced understanding of the applicability and effectiveness of different design and planning strategies.

From a theoretical perspective, the study contributes to the literature by elucidating how resilience in public health emergencies can be achieved through strategic architectural and urban planning interventions. It demonstrates the critical role of design in enhancing the functionality and safety of non-traditional medical spaces, thereby expanding the scope of emergency medical planning theories to incorporate the physical and spatial dynamics of pandemic response.

On a practical level, the findings offer a structured approach to the transformation of quarantine facilities, underlining the importance of functional segregation, the adaptability of spaces, and the integration of health and safety protocols in design considerations. Specifically, the study highlights:

1. For newly built facilities: Emphasizes the need for flexible design that can easily adapt to changing medical needs and public health guidelines.
2. For renovated facilities: Points out the significance of retrofitting existing structures with minimal disruption, focusing on ventilation improvements, spatial reconfigurations, and enhanced sanitary measures.
3. Urban versus suburban locations: Advocates for tailored strategies that consider population density, accessibility, and potential impacts on the surrounding community.
4. Scale of facilities: Discusses the logistical and operational differences between large-scale and smaller facilities, recommending scalable solutions that maintain efficiency and infection control.

These differentiated strategies underscore the study's practical implications, providing a roadmap for policymakers, urban planners, and healthcare administrators to enhance urban resilience against future public health emergencies. By systematically concluding design and planning strategies based on case differentiation, this research affirms the potential of architectural and urban planning to significantly contribute to public health crisis management and preparedness.

Funding:. This paper is supported by individual academic enhancement project of Guangzhou Academy of Fine Arts "Research on the Resilience of Urban Medical Spaces in Guangzhou" (Grant No. 24XSC36).

References

1. Qixiang, S., Xinfan, Z.: Providing a certain system guarantee for uncertain risk events—based on the thinking of China's two major public health crises. Southeast Academic **3**, 13 (2020)
2. Lei, L., Daoyuan, C., Guoen, W., Yuanyuan, Z.: Thoughts on the planning of hierarchical medical treatment facilities in mega cities in China under public health emergencies. Modern Urban Res. **35**(10), 8 (2020)
3. Sichang, H., Daren, Z., Ruihua, Z., et al.: The current status and thoughts on the implementation of hierarchical diagnosis and treatment in China. Modern Hospital Manage. **13**(2), 3 (2015)
4. Guangdong provincial health commission. Notice on issuing the "three-year action plan for public health prevention and control capacity building in Guangdong province (2020–2022). http://wsjkw.gd.gov.cn/zwgk/content/post_3163554.html, last accessed 2023, December 26
5. Coming into use soon! Guangzhou international health station is here, a large number of exclusive images exposed. https://m.thepaper.cn/baijiahao_14523255, last accessed 2023, December 26

A Comparative Study of the Public's Risk Perception of Nuclear Energy in Different Regions

Yanling He[1(✉)], Dongqin Xia[2], Nuo Yong[2], and Huiyun Peng[1]

[1] School of Civil Engineering and Architecture, Southwest University of Science and Technology, Mianyang 621010, Sichuan, China
yanling.he@swust.edu.cn

[2] Institute of Nuclear Energy Safety Technology, Hefei Institutes of Physical Science, Chinese Academy of Sciences, Hefei 230031, Anhui, China

Abstract. In recent years, many proposed or under construction nuclear facilities projects have been canceled due to strong opposition from the local public, such as the 2013 Jiangmen anti-nuclear incident, 2016 Lianyungang anti-nuclear incident and 2018 Changsha nuclear industry relocation reconstruction incident. It can be seen that public acceptance has emerged as a bottleneck problem in the development of nuclear energy. It is very important for the sustainable development of nuclear energy to understand public risk perception and attitude of nuclear energy in different regions, especially around the nuclear power plant. Based on 184 samples of Haiyan County and 811 samples of other regions of the country, this study used descriptive statistics and other analysis to conduct a comparative empirical study on the public perception and attitude of nuclear energy between Haiyan County and other regions of the country. The study found that compared with the whole country, the people lived in Haiyan County had a slightly lower perceived risk, a higher perceived benefit and knowledge. And they placed greater trust in regulator, experts and nuclear industry. Our results could provide some implications for risk communication mechanism in the field of nuclear energy.

Keywords: Nuclear energy · Public acceptance · Risk perception

1 Introduction

The utilization of nuclear energy is widely regarded as a sustainable alternative to fossil fuels on a large scale, and it is playing an increasingly pivotal role in China's energy landscape [1, 2]. Despite the advantages of nuclear energy, such as cleanliness, safety, and economy, its use is also a "double-edged sword". Nuclear power plants have potential radiation risks, and once a nuclear accident occurs, it will lead to unpredictable consequences. The occurrence of three serious nuclear accidents worldwide (the Three Mile Island nuclear accident in 1979, the Chernobyl nuclear accident in 1986, and the Fukushima nuclear accident in 2011) has further intensified the public's fear of nuclear energy, leading to "nuclear phobia".

B.-J. He et al. (Eds.): UCSUD 2023, LNCE 559, pp. 914–923, 2025.
https://doi.org/10.1007/978-981-97-8401-1_66

In recent years, especially after the Fukushima accident, with the enhancement of public environmental protection awareness, the public's doubts and opposition to the development of nuclear power have surged, and the "NIMBY (Not In My Back Yard)" problem has become increasingly prominent. Many proposed or under-construction nuclear facility projects have been shelved due to strong opposition from the local public, such as the 2013 Jiangmen anti-nuclear incident, the 2016 Lianyungang anti-nuclear incident, and the 2018 Changsha nuclear industry relocation reconstruction incident [3, 9]. Public acceptance of nuclear power (PANE) has become a bottleneck issue in the development of nuclear energy.

Currently, a considerable body of research is dedicated to investigating the factors that influence PANE. It has been observed that both individual internal and external factors play significant roles in shaping this acceptance. Individual internal factors primarily encompass demographic variables such as gender, age, and education level [12–15], as well as social psychological aspects including trust, emotions, values, perceived risks, and perceived benefits [11, 16, 17]. External factors comprise proximity to nuclear plants, experiences with nuclear accidents, and exposure to risk information [11, 14, 16, 18]. However, there is a dearth of studies comparing public perceptions and attitudes across different regions. By comparing the different attitudes and perceptions of the public around the built nuclear power plant and those around the non-built nuclear power plants, risk communication can be made more targeted to the public in the areas where nuclear power plants are to be built, so as to increase the PANE. Thus, this article uses a questionnaire survey to study the differences in the public's perception and attitude towards nuclear power in areas with established nuclear power and other areas. The research results can provide a basis for risk communication in the field of nuclear power.

2   Research Method

2.1   Survey Samples and Data Collection

The Qinshan Nuclear Power Plant, situated in Haiyan County, Jiaxing City, Zhejiang Province, first achieved grid-connected power generation in December 1991. It is the first nuclear power plant in our country that was independently designed, built, operated, and managed. It is also currently the nuclear power base with the most nuclear power units, the richest variety of reactor types, and the largest installed capacity in the country, and is a typical representative of the nuclear power plants built in our country. At present, there are 9 operating units in the Qinshan Nuclear Power Plant, of which 7 are pressurized water reactors and 2 are heavy water reactors. The Qinshan Nuclear Power Plant has safely generated more than 600 billion kWh of electricity and has been safely operated for more than 120 reactor years.

This article takes Haiyan County, Zhejiang Province, where the Qinshan Nuclear Power Plant is located, as an example, and conducts a comparative survey of public nuclear power risk perception and attitudes in areas where nuclear power plants have been built and other areas. During December 2019, an online questionnaire survey and data collection were conducted through the professional survey tool Questionnaire Star, and a total of 995 valid questionnaires were collected, of which there were 184 valid samples in Haiyan County and 811 valid samples in other regions of the country (excluding

Haiyan). The SPSS 21.0 statistical software was used for analysis in the study. The demographic characteristics of the participants such as gender, household registration, age, and education level are shown in Table 1. The statistical results show that the survey covers a wide range of people, and the sample population is distributed in different genders, household registrations, age groups, education levels, and monthly incomes, and the sample survey has diversity and randomness.

Table 1. Demographic characteristics of survey participants.

Item	Content	Frequency	Valid percentage
Gender	Male	627	63.0%
	Female	368	37.0%
Age	Under 20	41	4.20%
	21–30	416	41.8%
	31–40	163	16.4%
	41–50	81	8.1%
	51 and above	35	3.5%
Household Registration	Urban resident	487	48.9%
	Rural resident	509	51.1%
Monthly Income	Less than ¥2000	562	56.5%
	¥2001-¥5000	179	18.0%
	¥5001-¥8000	107	10.8%
	¥8001-¥10000	56	5.6%
	More than ¥10000	91	9.1%
Degree of education	Primary school and below	8	0.8%
	Junior high	31	3.1%
	Senior high (including junior college)	107	10.7%
	Undergraduate	735	73.9%
	Postgraduate and above	114	11.5%

2.2 Variable Measurement

The main content of the questionnaire consists of three parts: The first part investigates the public's understanding of nuclear power, including views on different energy sources, acceptance of nuclear power, and knowledge of nuclear power; The second part investigates the public's trust in different organizations; The third part investigates demographic characteristics, including the respondent's gender, household registration, age, and education level, as shown in Table 1. Regarding views on different energy sources, this paper compares the public's perception differences of traditional thermal power, nuclear power, and other renewable energy sources (wind, hydro, photovoltaic energy) in established areas and other areas from the perspectives of benefits (electricity price) and risks. To ensure the high reliability and validity of this survey, all variables in this study are based on existing mature scales for formulation and adaptation. The questionnaire uses a five-star Likert scale for measurement (1 represents completely disagree, 2 represents disagree, 3 represents uncertain, 4 represents agree, 5 represents completely agree), and the variable items and sources are as shown in Table 2.

3 Data Analysis and Results

3.1 Reliability and Validity Analysis

As can be seen from Table 3, whether it is the Haiyan County sample or samples from other regions of the country, all variables have a Cronbach's α coefficient above 0.6, and the composite reliability CR is above 0.8, indicating that these variables have good reliability; all variables have factor loadings FL and average variance extracted AVE above 0.5, indicating that these samples have good validity.

3.2 Descriptive Analysis

Public Perception of Different Energy Sources. There are certain differences in the public's perception of the price and risk of different energy generation methods in the Haiyan region and other regions nationwide.

Regarding electricity prices, people in the Haiyan region believe that the price of wind power is the lowest, followed by hydropower, photovoltaic power, nuclear power, with the price of thermal power being the highest. People in other regions of the country also believe that the price of wind power is the lowest, and the price of thermal power is the highest. The difference is that the price of nuclear power is lower than that of photovoltaic power. However, according to the National Energy Administration's 2018 national electricity price situation, the price of photovoltaic power is the highest, at 859.79 yuan per megawatt-hour, followed by wind power (529.01 yuan per megawatt-hour), nuclear power (395.02 yuan per megawatt-hour), thermal power (370.52 yuan per megawatt-hour), with the lowest being hydropower, at 267.19 yuan per megawatt-hour. The survey results show that there is a significant difference between the public's perception of the prices of different energy generation methods and the reality. The reason is that renewable energy sources such as photovoltaic and wind power in our country rely on government subsidies, and the public lacks a clear understanding of the high electricity prices of renewable energy, seriously underestimating the cost of potential energy generation, which to a certain extent reduces the potential competitiveness of nuclear power in the electricity market.

Regarding the potential risks of different energy generation methods, people in the Haiyan region believe that the risk of wind power is the lowest, followed by photovoltaic power, hydropower, nuclear power, with the risk of thermal power being the highest. People in other regions of the country also believe that the risk of wind power is the

Table 2. Questionnaire design and variable setting.

Variable	Item	Scale Source
Public acceptance	I am in favor of nuclear power generation	[4]
	Nuclear power generation is an excellent way to generate electricity	
	The number of nuclear power generating plants should be increased	
Perceived Knowledge	I know quite a lot about various energy technologies	[5]
	Among my circle of friends, I'm one of the "experts" on various energy technologies	
Perceived Benefit	Nuclear energy help us to mitigate climate change	[6]
	The electricity price would become too high if nuclear energy are not expanded	
	Thanks to the expansion of nuclear energy, the energy supply will be secured in the long term	
Perceived Risk	Participants were asked to assess their perceptions of nuclear risks, including two dimensions: dread risk perception and unknown risk perception	[7]
	1. Lack of control	
	2. Dread	
	3. Catastrophic potential	
	4. Fatal consequences	
	5. Inequitable distribution of risks and benefits	
	6. Unobservable	
	7. Unknown	
	8. New	
	9. Delayed in their manifestation of harm	
Trust in regulator (Regulator trust)	1、The regulator's opinions on nuclear power are objective and reliable	[8]
	2、Current technology is sufficient to accurately assess the risks of nuclear power	
Trust in experts (Expert trust)	3. The nuclear power information provided by research institutions and industry experts is objective and reliable	
	4. Industry experts can effectively supervise and guide the construction of nuclear power plants, eliminating hidden dangers	

(*continued*)

Table 2. (*continued*)

Variable	Item	Scale Source
Trust in Nuclear Industry (Nuclear industry trust)	5. Nuclear power companies will handle nuclear accidents according to regulations	
	6. Current scientific and technological levels are sufficient to accurately assess nuclear power risks	

Table 3. Reliability and validity.

Sample	Variable	Cronbach's α	CR	Factor Loading	AVE
Haiyan County (184)	Public acceptance	0.933	0.960	0.915–0.958	0.888
	Perceived risk	0.906	0.926	0.500–0.858	0.586
	Perceived Benefit	0.731	0.851	0.764–0.865	0.657
	Perceived Knowledge	0.835	0.897	0.902	0.814
	Regulator Trust	0.872	0.940	0.942	0.887
	Trust in Experts	0.897	0.925	0.953	0.906
	Trust in Nuclear Industry	0.906	0.955	0.956	0.914
Other Regions Nationwide (811)	Public acceptance	0.911	0.946	0.888–0.948	0.854
	Perceived Risk	0.875	0.902	0.530–0.821	0.509
	Perceived Benefit	0.612	0.800	0.629–0.837	0.575
	Perceived Knowledge	0.854	0.910	0.914	0.935
	Trust in Regulator	0.699	0.871	0.878	0.771
	Trust in Experts	0.769	0.919	0.922	0.850
	Trust in Nuclear Industry	0.824	0.897	0.902	0.814

Notes: CR, composite reliability; AVE, average variance extracted

lowest, followed by photovoltaic power, hydropower, thermal power, with the highest risk being nuclear power. The survey results show that compared to other regions of the country, people in the Haiyan region have more confidence in the safety of nuclear power, which to a certain extent comes from the nearly 30 years of safe and stable operation of the Qinshan Nuclear Power Plant.

Public Attitudes and Perceptions of Nuclear Power. The survey divided the public's attitude towards nuclear power into "positive attitudes (completely agree, agree)", "uncertain", and "negative attitudes (disagree, completely disagree)". The survey results show that the proportions of people in the Haiyan region who hold positive attitudes towards the three items "I support nuclear power", "Nuclear power is a very good way of generating electricity", and "The number of nuclear power plants should increase" are 85.4%, 83.7%, and 57.6% respectively, and the proportions of those who hold negative attitudes are 9.2%, 8.2%, and 18.5% respectively, and the proportions of those who are uncertain are 5.4%, 8.1%, and 23.9% respectively. The proportions of people in other regions of the country who hold positive attitudes towards these three items are 79.6%, 70.1%, and 56.8% respectively, and the proportions of those who hold negative attitudes are 5.1%, 4.9%, and 9.7% respectively, and the proportions of those who are uncertain are 20.3%, 15.0%, and 33.5% respectively. The results show that the current PANE in our country is relatively high, and the acceptance level of nuclear power by the public in the Haiyan region is higher than in other regions of the country. However, it is worth noting that when it comes to "The number of nuclear power plants should increase", the public's attitude becomes conservative, which poses higher requirements for public communication work in new or planned areas.

Existing research shows that the public's understanding of nuclear power has a significant impact on the PANE [6]. This survey compared the public's understanding of nuclear power in the Haiyan region and other regions of the country, including perceived knowledge, perceived benefit, and perceived risk. The survey results show that there are certain differences in the public's understanding between the built-up areas and other areas. As shown in Table 3, the mean values of perceived knowledge and perceived benefit of nuclear power in the Haiyan region are higher than in other regions of the country, while the mean value of perceived risk of nuclear power in the Haiyan region is lower than in other regions of the country. This shows that people in the Haiyan region have a better understanding of nuclear power, and on the other hand, it shows that the risk communication and science popularization work of the Qinshan Nuclear Power Plant to the local people has been effective (Table 4).

Table 4. Comparison of public understanding of nuclear power.

Variable	Mean	
	Haiyan County	Nationwide
Perceived Knowledge	3.508	3.005
Perceived Benefit	3.871	3.737
Perceived Risk	2.813	3.115

Public Trust in Different Organizations. The significance of trust as a crucial determinant impacting PANE has been extensively demonstrated by numerous studies [8, 10]. This survey focused on the public's trust in different organizations, including regulator, experts, and nuclear power companies. The results show that there are significant differences in the level of public trust in different organizations between the areas where nuclear power plants have been built and other areas. Overall, people in the Haiyan County area trust regulator, experts, and nuclear power companies more. As shown in

Table 5. Comparison of Public Trust in Different Organizations.

Variable	Mean	
	Haiyan County	Nationwide
Trust in Regulator	3.919	3.654
Trust in Experts	3.927	3.739
Trust in Nuclear Industry	4.025	3.699

Table 5, people in the Haiyan County area trust nuclear power companies the most, followed by experts, and finally regulator. The reason is that nuclear power companies bear the primary responsibility for the safe operation of nuclear power plants, and the trust of the local people in Haiyan County in nuclear power companies is based on the safe operation history of the Qinshan Nuclear Power Station. However, people in other parts of the country trust experts the most, followed by nuclear power companies, and finally regulator. This shows that the public in other areas without nuclear power plants tend to trust professional, rational, and impartial experts. Therefore, public communication and science popularization work in new or planned nuclear power areas should be led by experts.

4   Conclusions and Policy Recommendations

This study, through questionnaire surveys, comparatively analyzed the public's understanding of nuclear power and trust in different organizations in the Haiyan region and other regions of the country. The following conclusions can be drawn from the research.

(1) The public generally underestimates the price of renewable energy, which to some extent reduces the potential competitiveness of nuclear power in the electricity market. Therefore, we should strengthen the knowledge popularization of electricity price in nuclear power. People in Haiyan region are more confident in the safety of nuclear power than in other regions of the country, which mainly depends on the Qinshan nuclear power plant's long history of safe operation.

(2) The residents of Haiyan exhibit a higher level of acceptance towards nuclear power compared to those in other regions, demonstrating an elevated awareness and understanding of its benefits while displaying a relatively lower perception of associated risks. This suggests that the Qinshan Nuclear Power Plant has effectively communicated potential risks of nuclear and successfully promoted scientific knowledge among the local public.

(3) Compared with other regions of the country, people in the Haiyan region have a higher degree of trust in regulator, experts, and nuclear power companies, and they trust nuclear power companies the most. The public in other areas where nuclear power plants have not been built tend to trust experts more, so the public communication and science popularization work in new or planned nuclear power areas should be led by experts.

(4) For existing nuclear power plants, tangible and effective benefits such as subsidies and job opportunities can be provided to the public around the nuclear power plants to increase the local PANE; during the site selection or construction phase of nuclear power, nuclear power knowledge can be popularized to the public through expert lectures to increase the PANE.

But there are some limitations that should be pointed out in this study. Firstly, we can further study the differences of influencing factors of PANE in different regions. Secondly, the combination of the sampling methods such as the face-to-face survey and online survey would be adopted in the future re-search in order to overcome the potential bias in the sampling.

Acknowledgement. This research was funded by the Doctoral Fund of Southwest University of Science and Technology (Grant No. 22zx7124), the National Natural Science Foundation of China (Grant No. 72204246 and 51808463), the HFIPS Director's Fund (Grant No.YZJJ2022QN38 and YZJJ2021QN35), and Sichuan Science and Technology Program (2023NSFSC1051). At the same time, the authors sincerely thank the editor and anonymous re-viewers for their insightful comments that help us improve the quality of the article.

References

1. IAEA-PRIS. Nuclear Power Reactors in the World (2023)
2. Wang, J., Li, Y., Wu, J., et al.: Environmental beliefs and public acceptance of nuclear energy in China: a moderated mediation analysis. Energy Policy **137**, 111141 (2020)
3. Ho, S.S., Kim, N., Looi, J., Leong, A.D.: Care, competency, or honesty? Framing emergency preparedness messages and risks for nuclear energy in Singapore. Energy Res. Soc. Sci. **65**, 101477 (2020)
4. Tsujikawa, N., Tsuchida, S., Shiotani, T.: Changes in the factors influencing public acceptance of nuclear power generation in Japan since the 2011 Fukushima Daiichi nuclear disaster. Risk Anal. **36**, 98–113 (2016)
5. Rijnsoever, F.J.V., Mossel, A.V., Broecks, K.P.F.: Public acceptance of energy technologies: the effects of labeling, time, and heterogeneity in a discrete choice experiment. Renew. Sustain. Energy Rev. **45**, 817–829 (2015)
6. Visschers, V.H.M., Keller, C., Siegrist, M.: Climate change benefits and energy supply benefits as determinants of acceptance of nuclear power stations: investigating an explanatory model. Energy Policy **39**(6), 3621–3629 (2011)
7. Slovic, P.: Perception of risk. Science **236**(4799), 280–285 (1987)
8. Xiao, Q.H., Liu, H.J., Feldman, M.W., et al.: How does trust affect acceptance of a nuclear power plant (NPP): a survey among people living with Qinshan NPP in China. PLoS ONE **12**(11), e0187941 (2017)
9. Zhang, T., Xia, D., Li, T., Li, Y.: The impact of public cognition on the acceptance of nuclear power. Nucl. Saf. **18**(2), 63–70 (2019)
10. Whitfield, C., Rosa, A., Dan, A., Dietz, T.: The future of nuclear power: value orientations and risk perception. Risk Anal. **29**, 425–437 (2009)
11. Vainio, A., Paloniemi, R., Varho, V.: Weighing the risks of nuclear energy and climate change: trust in different information sources, perceived risks, and willingness to pay for alternatives to nuclear power. Risk Anal. **37**, 557–569 (2017)

12. Han, Z., Lu, X., Hörhager, E.I., Yan, J.: The effects of trust in government on earthquake survivors' risk perception and preparedness in China. Nat. Hazards **86**, 437–452 (2016)
13. Trumbo, C.W., McComas, K.A.: The function of credibility in information processing for risk perception. Risk Anal. Official Publ. Soc. Risk Anal. **23**, 343–353 (2010)
14. Ryu, Y., Kim, S.: Testing the heuristic/systematic information-processing model (HSM) on the perception of risk after the Fukushima nuclear accidents. J. Risk Res., 1–20 (2016)
15. Vyncke, B., Perko, T., Van, G.B.: Information sources as explanatory variables for the Belgian health-related risk perception of the Fukushima nuclear accident. Risk Anal. **37**, 570–582 (2017)
16. McKnight, D.H., Choudhury, V., Kacmar, C.: Developing and validating trust measures for e-commerce: an integrative typology. Inf. Syst. Res. **13**, 334–359 (2002)
17. Flanagin, A.J., Metzger, M.J.: Perceptions of Internet information credibility. Journalism Mass Communication Quarterly **77**, 515–540 (2000)
18. Slovic, P.: Perception of risk. Science **236**, 280–285 (1987)

Sustainable Development of Urban Subway Tunnels Using Seismic Waves

Tong Yao[1], Xiao Luo[2], and Chen Yang[3]($\boxtimes$)

[1] School of Economics and Management, Sichuan Tourism University, Chengdu 610100, China
[2] Overseas Education College, Chengdu University, Chengdu 610106, China
[3] School of Civil Engineering, Jinjiang College, Sichuan University, Meishan 620860, China
yang7670510@163.com

Abstract. With the proposal of the low-carbon travel transportation, the subway has become the first choice for most urban residents. If it is affected by earthquakes, it will cause irreparable and significant losses. This study utilizes a three-dimensional finite element model developed in ANSYS to investigate the dynamic response of the lining of orthogonal tunnels in Chongqing's metro line 4 interval tunnel under the influence of an intensity VIII earthquake. By analysing the deformation results at various time intervals and locations, along with actual earthquake measurements, reasonable spacing and hazard site determinations for the orthogonal tunnels are obtained.

Keywords: metro,tunnel · seismic wave · earthquake

1 Introduction

The urban subway is now the preferred means of transportation for people to travel and the basis for the development of low-carbon cities. If the effects of seismic waves are encountered during the construction of subway tunnels, significant damage is often caused.

Li et al. established a numerical model of a close-fitting cross tunnel using FLAC3D finite difference software, and studied the seismic response characteristics of the close-fitting cross tunnel under strong earthquake by inputting Beijing hotel seismic waves vibrating in the horizontal direction at the model base, and derived different displacement and peak acceleration characteristics of the upper and lower tunnel under strong earthquake [1]. Based on the example of the subway shield tunnel project, Ran meticulously crafted a three-dimensional finite element model that captured the intricate soil-structure interaction by incorporating various ground shaking parameters. This approach allowed for a detailed analysis of the dynamic response of the lining of an underpass shield tunnel, taking into account diverse ground shaking effects [2]. Besides, some scholars also proposed that the transverse shear deformation occurring in any one of the overlapping tunnels under seismic waves of different propagation directions will result in a complex stress state of bending or tensile compression in the adjacent tunnel [2–7]. The transverse shear deformation will also inevitably lead to irreversible plastic deformation

B.-J. He et al. (Eds.): UCSUD 2023, LNCE 559, pp. 924–934, 2025.
https://doi.org/10.1007/978-981-97-8401-1_67

in the overlapping tunnel under the gradual increase of seismic waves [8–11]. Thus, it is necessary to study the dynamic response of the lining of orthogonal tunnels with different spacing under seismic action. In addition, the study of the seismic performance of underground orthogonal tunnel structures can summarize universal laws to help the seismic design and construction of practical projects and provide theoretical guidance for future orthogonal tunnelling projects.

Located in mountainous regions, the construction and operation of Chongqing's subway system differ significantly from those of other cities. It is not only necessary to adapt to the unique mountainous environment but also to consider the intricate layout of the lines. Consequently, a significant number of underground crossings within the subway system, where lines intersect and traverse each other, posing significant challenges to both engineering techniques and operational management. In this study, based on ANSYS 3D finite element software the orthogonal tunnel model is established with the reference of Chongqing Metro Line 4 interval tunnel under Line 3. The horizontal seismic wave is input in the horizontal direction of the model and the seismic wave vibration direction is perpendicular to the lower tunnel. By meticulously observing the fluctuations in acceleration, stress, and displacement at the orthogonal segment of the intersecting tunnel across varying spacing configurations, this study derived the dynamic response pattern of the orthogonal portion of the intersecting tunnel, thus offering insights into its seismic behavior under diverse spatial conditions. The research results can provide a theoretical reference for the design and construction of orthogonal tunnels in earthquake areas.

2 Dynamic Response Analysis

2.1 Model Building

Relying on the metro line 4 interval tunnel in Chongqing, the height, width, and lining thickness of the tunnel are 7.16 m, 7.38 m, and 0.45 m respectively, and the thickness of the overlying soil layer of the upper tunnel structure is 24 m. The upper and lower horseshoe-shaped tunnels of the double-layer orthogonal metro tunnel have the same dimensions, and the lower tunnel structure is orthogonal to the upper tunnel. To fulfill the demands of computational accuracy and considering the geological conditions of the actual project, a computational model with a side length of 100 m is chosen to establish a positive cube, and the soil model in the finite domain is used to simulate the stresses in the wireless domain. Viscoelastic artificial boundaries are applied to the front, back, left, and right sides of the model, then all degrees of freedom on the bottom surface of the model are constrained [12]. According to the geological report, the upper part of the soil uses a class V envelope, while the lower part uses a class IV envelope. After the model is established, the seismic wave excitation in the horizontal direction is used. The relevant parameters required for the model are shown in Table 1.

The Mohr-Coulomb model is used for the soil layer, and its soil-related parameters can be found in the literature [13]. To investigate the degree of influence of different spacing between upper and lower tunnels on the seismic response of the orthogonal section, this study employed four different scenarios with spacing ratios of 0.5, 1.0, 1.5, and 2.0, and analyzed the seismic dynamic response of the orthogonal tunnel under

Table 1. Physical and mechanical parameters of tunnel surrounding rock and lining

Surrounding rock parameters	Density(kg/m³)	Modulus of elasticity(Gpa)	Poisson's ratio
Level IV	2350	4.76	0.30
Level V	2050	1.31	0.34
Lining	2600	27.51	0.21

earthquake action. The computational model and spatial position of the intersecting tunnels as indicated in Fig. 1.

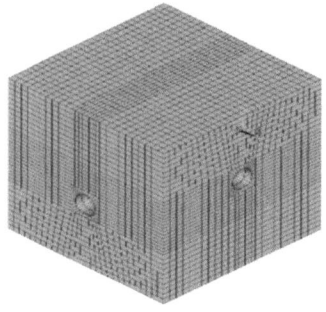

(a) Computational model   (b) Orthogonal tunnel space location

Fig. 1. Calculation model and tunnel space location

2.2 Seismic Wave Input

The input of ground shaking in this study is the acceleration input method. Since the earthquake intensity used is VIII degrees, the artificial synthetic wave with a peak acceleration is 0.20g is selected. The acceleration reflects the strongest part of the earthquake. Under the same conditions, the higher the acceleration, the stronger the response of the engineering structures. The *Code for seismic design of urban rail transit structures* stipulates the selection of seismic wave duration, and it is necessary to ensure that the selected duration should contain peak acceleration and the duration should not be less than 5–10 times the basic period of the structure, so the adjusted acceleration map as indicated in Fig. 2.

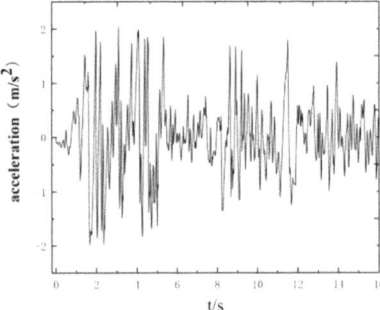

Fig. 2. Input ground vibration acceleration time curve

3 Analysis of Calculation Results

To analyze the dynamic response of the orthogonal part of the orthogonal tunnel, this paper monitors the orthogonal key parts (arch top, arch bottom, both sides of the arch waist, and two 45° lines) and some non-orthogonal parts of the orthogonal tunnel, while using different spacing (four working conditions) to compare with each other. The arrangement of each monitoring point in the orthogonal tunnel as indicated in Fig. 3.

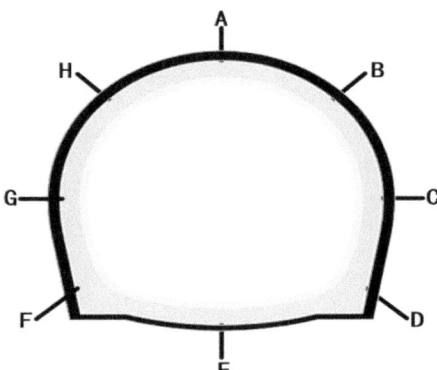

Fig. 3. Orthogonal tunnel stress and displacement monitoring points layout

3.1 Stress Analysis

Utilizing various time histories, an analysis is conducted on the principal stress profiles of the lining in orthogonally intersecting tunnels with different spacings. See Figs. 4 and 5 for illustrations (where "D" represents the spacing). The peak values of the principal stress in the lining at different spacings and various time instances are presented in Table 2.

The data from the above table gives the peak horizontal stresses of the corresponding monitoring points in the orthogonal tunnel at different spacings and different time

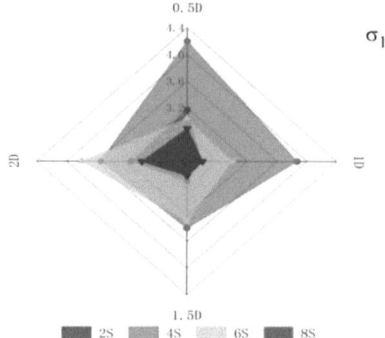

Fig. 4. Peak value of the first principal stress at different moments of the lining

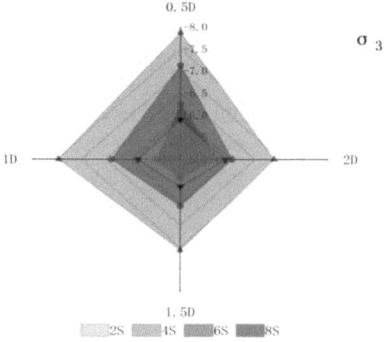

Fig. 5. Peak third principal stress diagram at different moments of lining

Table 2. Peak principal stresses in the lining at different spacing (MPa)

Time Space	2 s		4 s		6 s		8 s	
	σ_1	σ_3	σ_1	σ_3	σ_1	σ_3	σ_1	σ_3
0.5 D	3.17	-5.90	4.20	-7.87	3.05	-7.13	2.89	-6.24
1.0 D	2.59	-5.86	3.87	-7.47	3.01	-6.39	2.61	-5.47
1.5 D	2.69	-5.61	3.40	-7.06	3.31	-6.06	2.61	-5.42
2.0 D	3.14	-5.90	3.55	-6.90	3.80	-6.04	3.00	-5.76

intervals. By analyzing the data in the above table, the following conclusions can be derived.

(1) Within the range of 0.5–1.5 times the diameter of orthogonal parts in the orthogonal tunnel, the same main stress of tunnel lining decreases with the increase of spacing. Taking the time interval of 4s as an example, the tensile stress decreases from 4.2 MPa to 3.40 MPa, which is reduced by 19%. But when the spacing is 1.5 - 2.0 times the hole diameter, the opposite is true. Taking the time interval of 4s as an example

identically, the tensile stress increases from 3.40 MPa to 3.55 MPa, which increases by 1.04 times. It can be seen that the spacing between the upper and lower layers of the orthogonal tunnel is not the larger the better, and a spacing of 1.5 times the tunnel diameter is the most rational choice..

(2) As seen from the above table, the stress is higher at points H and B at the vault of the upper tunnel because of the stress concentration, followed by point A. The stress of point G and point C at the arch waist is small, while the stress at the arch foot is the largest due to the influence of the lower tunnel, which can be F>D>E. The lower tunnel is also influenced by the upper tunnel, showing that the stress at the arch top is greater than that at the foot of the arch, and the pattern of stresses at the arch waist is the same as that of the upper tunnel.

3.2 Displacement Time Course Analysis

Table 3 presents the horizontal and vertical displacements of the corresponding monitoring points at the orthogonal sections of the upper and lower tunnels under different spacing conditions.

By analyzing the data at each point in Table 3, the following conclusions can be derived.

(1) According to the analysis of the data obtained from the monitoring points of orthogonal parts under different spacing, The horizontal displacement at each monitoring point is greater than the vertical displacement, which proves that the effect of seismic waves in the horizontal direction on the displacement of the tunnel lining in the horizontal direction is larger than the effect of displacement in the vertical direction.

(2) The horizontal displacement of monitoring point A in the upper tunnel, specifically at a spacing of 0.5 times the hole diameter, significantly exceeds that of other monitoring points, reaching a magnitude of 1.021 cm. In contrast, the corresponding monitoring point in the lower tunnel exhibits a displacement of only 0.441 cm. This observation suggests that seismic waves exert a more pronounced influence on the upper tunnel compared to the lower tunnel, aligning with the findings reported in previous studies on closely spaced tunnels.The maximum horizontal displacement of the upper tunnel is 2.31 times the maximum horizontal displacement of the lower tunnel, which proves that the upper tunnel is more variable in horizontal displacement compared to the lower tunnel in the orthogonal tunnel.

(3) In the orthogonal tunnel, the horizontal displacement of each monitoring point in the upper tunnel decreases with increasing spacing, but the magnitude of horizontal displacement between different monitoring points at the same spacing shows a positive parabolic variation.

From the above conclusions, it can be seen that the displacement at point A of the upper tunnel monitoring point is the largest and the stress concentration phenomenon is also more prominent, so the displacement time interval analysis is continued for point A. The horizontal displacement time interval of point A under seismic wave as indicated in Fig. 6.

According to the conclusion of the literature [14], it is known that the arch top of the tunnel generates large tensile stresses with the increase of subsidence,When this tensile

Table 3. Peak displacement of each monitoring point at the orthogonal part of the upper and lower tunnels

Spacing between tunnels		Upper side tunnel		Lower side tunnel	
		Horizontal displacement (cm)	Vertical displacement (cm)	Horizontal displacement (cm)	Vertical displacement (cm)
A	0.5D	1.021	-0.264	0.441	-0.002
	1.0D	0.991	-0.004	0.621	0.373
	1.5D	0.911	-0.002	0.539	-0.001
	2.0D	0.862	-0.005	0.484	-0.001
B	0.5D	0.970	-0.193	0.461	-0.012
	1.0D	0.951	0.076	0.591	-0.025
	1.5D	0.871	0.065	0.510	-0.025
	2.0D	0.825	0.063	0.456	-0.024
C	0.5D	0.850	-0.154	0.441	-0.012
	1.0D	0.881	0.082	0.491	-0.017
	1.5D	0.811	0.071	0.422	-0.016
	2.0D	0.782	0.064	0.378	-0.015
D	0.5D	0.720	-0.055	0.361	-0.012
	1.0D	0.811	0.037	0.431	0.072
	1.5D	0.761	0.053	0.357	-0.018
	2.0D	0.748	0.045	0.313	-0.027
E	0.5D	0.650	-0.226	0.341	-0.001
	1.0D	0.811	-0.001	0.431	0.241
	1.5D	0.761	0.002	0.360	0.000
	2.0D	0.749	-0.0004	0.313	-0.0002
F	0.5D	0.740	-0.280	0.331	-0.014
	1.0D	0.841	-0.001	0.451	-0.020
	1.5D	0.811	0.002	0.382	-0.021
	2.0D	0.783	-0.001	0.332	-0.022
G	0.5D	0.880	-0.290	0.311	-0.012
	1.0D	0.931	-0.001	0.571	-0.012
	1.5D	0.871	0.002	0.484	-0.012
	2.0D	0.833	-0.002	0.430	-0.013
H	0.5D	0.990	-0.280	0.391	-0.012
	1.0D	0.991	-0.001	0.631	-0.020
	1.5D	0.911	0.003	0.545	-0.019
	2.0D	0.867	-0.002	0.488	-0.020

stress exceeds the tolerable range of the tunnel lining, it may lead to cracking. Therefore, the displacement change at the vault has always been a crucial focus in tunnel research. As indicated in Fig. 6, when the spacing between the upper tunnel and the lower tunnel is 0.5 times the diameter of the tunnel, the displacement is the largest, accumulating up

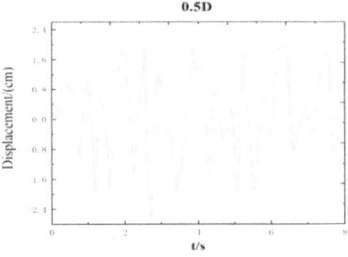

(a) 0.5 times the diameter of the tunnel

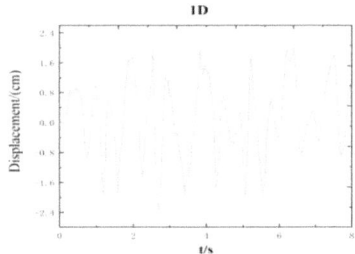

(b) 1.0 times the diameter of the tunnel

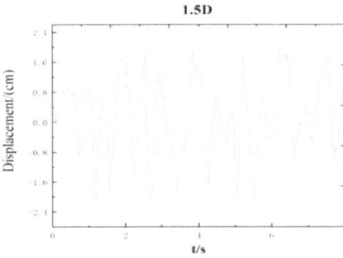

(c) 1.5 times the diameter of the tunnel

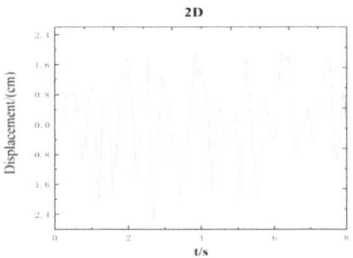

(d) 2.0 times the diameter of the tunnel

Fig. 6. Upper tunnel lining A point to the horizontal displacement map

to 2.6cm. However, the monitoring measurement in the new Austrian method stipulates that the cumulative maximum displacement of the arch top cannot exceed 2cm, so the upper tunnel has a safety hazard. According to the data of the spacing of 1.0 times the hole diameter to 2.0 times the hole diameter, it can be seen that the spacing between the two tunnels of the orthogonal tunnel is too small, which will have a greater adverse effect on the displacement of the arch top of the upper tunnel lining, especially when the spacing is 0.5 times the hole diameter.

In the study of orthogonal tunnels, it is generally recognized that the displacement at the orthogonal site is the largest and most dangerous part, but from the collected displacement data, the deformation of the lining at the upper tunnel entrance under seismic waves should not be neglected. Under the horizontal seismic wave excitation and 0.5 times the hole diameter spacing, the horizontal displacement information of the upper tunnel from 0m, 5m, 10m, 20m, and 40m is summarized as follows.

From the analysis of the data in Table 4, the following conclusions can be drawn.

The horizontal displacement of each monitoring point of the upper tunnel is in the range of 0-20m from the entrance, and the value increases with the increase of distance, and the peak position is about 20m. Within the range of 20-40m from the opening, the displacement gradually decreases. Combined with the above research results, it can be concluded that under a certain distance and horizontal seismic wave excitation the position of the maximum displacement is not in the orthogonal position, but in the range of about 20m at the entrance of the upper tunnel.

Table 4. Horizontal displacement of each monitoring point in the upper tunnel at 0.5 times the hole diameter spacing (cm)

	A	B	C	D	E	F	G	H
0m	1.021	0.977	0.850	0.716	0.656	0.739	0.823	0.994
5m	1.078	1.02	0.888	0.748	0.709	0.768	0.919	1.037
10m	1.097	1.027	0.900	0.767	0.734	0.790	0.935	1.055
15m	1.111	1.05	0.922	0.802	0.734	0.828	0.970	1.073
20m	1.110	1.040	0.926	0.813	0.787	0.840	0.968	1.074
40m	1.015	0.957	0.870	0.792	0.776	0.823	0.918	0.995

4   Conclusion

Based on the engineering background of Chongqing Metro Line 4, the dynamic effects of the orthogonal part and the upper tunnel entrance under the influence of horizontal seismic waves are studied. By analyzing the data obtained from the finite element model, the main conclusions are as follows.

(1) Under the action of horizontal seismic waves with different time intervals and different distances between the upper and lower tunnels, the lining stress changes in the orthogonal part of the orthogonal tunnel show a positive parabola phenomenon.

(2) By studying the displacement of each monitoring point under different spacing between the upper and lower tunnels of the orthogonal tunnel, it can be seen that the displacement of each monitoring point in the horizontal direction is obviously larger than that in the vertical direction under the action of seismic waves in the horizontal direction.

(4) In the orthogonal tunnel, the horizontal displacement of each monitoring point in the upper tunnel decreases with the increase of the spacing, but the magnitude of the horizontal displacement between different monitoring points at the same spacing shows a positive parabolic variation.

(5) The horizontal displacement value of each monitoring point in the upper tunnel increases with increasing distance in the range of 0-20m from the portal, and its peak value mostly appears in the distance of about 20m from the cave entrance.

Moreover, this study on the seismic performance of underground orthogonal tunnel structures can assist in the seismic design and construction of practical projects, providing theoretical guidance for future orthogonal tunnel engineering.

Acknowledgement. This paper is financially supported by the projects of Research on Transportation Optimization and Green Logistics Development Based on Digital Functional Technology by the Sichuan Tourism University (No. 2023SCTUBSSD03) and The project of A Study on the Emergency Countermeasures of Smart Tourism Scenic Spot Management in the Event of Major Natural Disasters -- A Case Study of Qingcheng Mountain Scenic Spot in Dujiangyan Irrigation Project, Sichuan by the Sichuan Tibet Smart Tourism Engineering Research Center (No. ZLGC2022B06).

References

1. Li, J., Tao, L., Wu, B., An, J., Guo, F.: Analysis of 3D dynamic response of closely overlapping tunnels during a strong earthquake. Mod. Tunnelling Technol. **51**(1), 26–31 (2014)
2. Bi, R., Xiao, T., Liu, B.: Three dimensional seismic response analysis of a shield-shaped subway tunnel. Geol. Explor. **52**(4), 712–717 (2016)
3. Lu, Z., Wang, X., Zhou, G., Feng, L., Jiang, Y.: Investigation on vibration influence law of double-shield TBM tunnel construction. Appl. Sci. **12**(15), 7727 (2022)
4. Jiang, X., Liu, W., Yang, H., Yu, L.: Study on dynamic response characteristics of slope with double-arch tunnel under seismic action. Geotech. Geol. Eng. **39**(2), 1349–1363 (2021)
5. Bauer, K., Norden, B., Ivanova, A., Stiller, M., Krawczyk, C.: Wavelet transform-based seismic facies classification and modelling: application to a geothermal target horizon in the NE German Basin. Geophys. Prospect. **68**(2), 466–482 (2020)
6. Cilingir, U., Madabhushi, S.: A model study on the effects of input motion on the seismic behaviour of tunnels. Soil Dyn. Earthq. Eng. **31**(3), 452–462 (2011)
7. Hatzigeorgiou, G., Beskos, D.: Soil–structure interaction effects on seismic inelastic analysis of 3-D tunnels. Soil Dyn. Earthq. Eng. **30**(9), 851–861 (2010)
8. Aicha, B., Mezhoud, S., Tayeb, B., Toufik, K., Abdelkader, N.: Parametric study of shallow tunnel under seismic conditions for constantine motorway tunnel, Algeria. Geotech. Geol. Eng. **40**(4), 2307–2318 (2022)
9. Guan, Z., Zhou, Y., Gou, X., Huang, H., Wu, X.: The seismic responses and seismic properties of large section mountain tunnel based on shaking table tests. Tunn. Undergr. Space Technol. **90**, 383–393 (2019)
10. Moghadam, M., Baziar, M.: Seismic ground motion amplification pattern induced by a subway tunnel: shaking table testing and numerical simulation. Soil Dyn. Earthq. Eng. **83**, 81–97 (2016)
11. Bonini, M., Barla, M., Barla, G:. FLAC applications to the analysis of swelling behavior in tunnels. In: FLAC and Numerical Modeling in Geomechanics, pp. 329–333. CRC Press (2020)
12. Jiang, X., Wang, F., Yang, H., Sun, G., Niu, J.: Dynamic response of shallow-buried small spacing tunnel with asymmetrical pressure: shaking table testing and numerical simulation. Geotech. Geol. Eng. **36**(4), 2037–2055 (2018)
13. Fu, D., Gu, Y.: Research on seismic response of mountain tunnel considering soil-structure dynamic interaction. J. Guangxi Univ. (Nat. Sci. Ed.) **44**(1), 176–182 (2019)
14. Ministry of Housing and Urban-Rural Development of the People's Republic of China. Code for seismic design of urban rail transit structures. Beijing (2010)
15. Jishnu, R., Ayothiraman, R.: Interaction of urban underground twin metro tunnels under static and earthquake loading. J. Earthq. Tsunami **14**(4), 2050019 (2020)
16. Pai, L., Wu, H.: Multi-attribute seismic data spectrum analysis of tunnel orthogonal underpass landslide

Post-pandemic Era General Hospital Fever Clinics Spatial Resilience Design Strategies

Lin Chang, Li Yuxi, and Zhong Ruizhe(✉)

Guangzhou Academy of Fine Arts, Guangzhou 510261, China
zrz@gzarts.edu.cn

Abstract. This study investigates the design challenges and opportunities for enhancing spatial resilience in general hospital fever clinics, focusing on the post-pandemic era. By integrating resilience theory with empirical case studies, we aim to identify effective design strategies that improve operational efficiency and pandemic preparedness. Our methodology combines qualitative analysis of renovation examples with a review of resilience design principles, leading to the formulation of specific spatial strategies. The key findings suggest that adaptive spatial design significantly contributes to the clinics' capacity to manage pandemic conditions effectively. This research provides a foundation for future guidelines on the spatial organization of fever clinics to better respond to public health crises.

Keywords: Post-pandemic era · Fever clinics · Spatial resilience · Pandemic transition

Since 2023, the National Health Commission has adjusted the classification of novel coronavirus infection from "Class B, Category A" to "Class B, Category B", and epidemic prevention and control has entered a new normal stage [1]. The most recent document related to the construction of fever clinics is the "Notice on Issuing a Plan for Graded Diagnosis and Treatment of COVID-19 with Medical Consortium as the Carrier" issued in December 2022, which emphasizes that community health service centers or township health clinics that meet the conditions should all set up fever clinics (outpatient clinics), fever clinics should have personnel with practicing physician qualifications, improve disinfection, inspection and testing, emergency rescue and other corresponding equipment and drug configuration, and have pre-inspection, triage, and screening functions [2]. This paper will reflect on the basic problems faced by the construction of fever clinics since the COVID-19 pandemic, introduce the concept of resilience design that is efficient in peacetime and redundant in wartime, and summarize the key points of resilience design for general hospital fever clinics in the future based on research and design examples.

In the forthcoming sections, this paperdiscuss the pressing challenges that fever clinics have faced amid the pandemic, laying the groundwork for our analysis. Subsequently, we explore resilience as a critical theoretical lens for our study, leading to the development of innovative spatial design strategies for fever clinics. Through detailed case studies, we illustrate the application of these strategies, highlighting their potential to enhance clinic adaptability and efficiency in post-pandemic healthcare settings. Our

© The Author(s) 2025
B.-J. He et al. (Eds.): UCSUD 2023, LNCE 559, pp. 935–948, 2025.
https://doi.org/10.1007/978-981-97-8401-1_68

discussion synthesizes these insights, emphasizing the paper's novel contributions to the fields of healthcare design and resilience planning.

1 Problems in the Construction of Fever Clinics

1.1 Early Construction Problems of Fever Clinics

During the epidemic, the early construction problems of fever clinics mainly focused on several levels:

First, the distribution of fever clinics is uneven at the planning level. As the "outposts" of epidemic prevention and control, grassroots fever clinics have problems such as insufficient facility density and uneven distribution. Secondly, some fever clinics are large in scale, while the density of small fever clinics in the surrounding area is small, making it difficult for fever clinics to cover the grassroots population to the maximum extent.

Second, the construction and configuration indicators of fever clinics are not entirely reasonable. The "Fever Clinic Setting Management Standards" issued in 2021 lacks a graded classification arrangement for fever clinics. Some secondary hospitals pursue the number of consultation rooms and observation rooms in fever clinics, leading to over-construction of many projects. However, some tertiary hospitals, constrained by early hospital planning and area, face problems such as insufficient space for renovation and shortage of facilities [3]. Moreover, with the issuance of fever clinic construction requirements several times, the required building area and facility equipment of fever clinics have been increasing, making their utilization rate even lower.

The primary challenge confronting fever clinics is the critical shortage of space and resources, which severely hampers their efficiency and effectiveness. This issue is exemplified by the non-standardized construction of fever clinics, leading to chaotic patient flows. A notable concern is the need for patients to travel to different departments (e.g., radiology for X-ray examination), increasing the risk of infection spread due to inadequate dedicated spaces like CT examination rooms. Such spatial and resource constraints underscore the urgency of adopting resilient design strategies to optimize clinic operations and safeguard public health. Secondly, the area of the observation room and the consultation room also has difficulty meeting the requirements. After patients enter the fever clinic, they often need to stay for 4–6 h to issue a nucleic acid report to confirm that they are not infected before they can leave. It is difficult to provide enough observation and waiting, isolation areas in the fever clinic [4].

Therefore, for the early construction problems of fever clinics and the demand changes in the post-epidemic era, we should pay attention to the construction of medical consortiums at the planning level. The construction requirements and standards of fever clinics need to be further refined, and classified construction standards should be implemented for medical space nodes of different scales and types, such as large comprehensive hospitals to grassroots clinics. In addition, in the specific process of constructing the space level of fever clinics, the concept of resilient design and epidemic combination should be integrated.

2   Resilience Concept and Key Points of Foreign Hospital Resilience Design

In order to cope with the increasing uncertainty, the concept and practice of resilience are gradually being carried out in the medical system. The application of resilience at different spatial scale levels shows different mechanisms. For the resilience concept at the level of medical facility planning, the resilience concept often serves as a conceptual framework to seek better solutions.

Foreign hospitals have conducted many studies on resilient design, such as the thematic research on the pandemic-resilient hospital: How design can help medical facilities stay operational and safe [5] jointly proposed by ARUP and HKS. In the study, a cross-disciplinary team of architects, engineers, clinicians, and medical planners discussed how hospital systems can create a flexible and resilient hospital campus. This research report aims to help medical institutions think about their current methods and planning actions' priority levels, and provides facility investment considerations that healthcare professionals can use when designing spaces to cope with the current epidemic and consider future resilience optimization. The study mentioned 7 medical design principles for epidemic resilience, namely Versatility, Surge Ready, Supports Well-being, Clean Air and Surfaces, Isolate, Contain & Separate, Flow, and Digital/Physical.

The goal of such resilient hospital design points under the influence of the resilience concept is to create a flexible and resilient campus. However, the medical design principles cannot be applied to all medical facilities to a certain extent, so it is necessary to form resilient design points at the planning and spatial level first, combined with the design practice research of various types of medical facilities, to find more specific and targeted resilient design strategies.

An integral aspect of resilience design at the spatial level is the strategic location of fever clinics within the hospital infrastructure. This strategic placement is crucial for facilitating a streamlined interface with emergency and outpatient departments, enhancing the clinic's role in triaging patients effectively and minimizing potential congestion in critical care areas. By positioning fever clinics in locations that offer direct access for ambulatory patients yet are distinct from high-traffic hospital zones, the design can significantly reduce the risk of cross-contamination. Furthermore, the operational relationship between fever clinics and other departments should be carefully managed to ensure efficient resource allocation and patient flow, especially during peak periods of infectious disease outbreaks.

3   Planning Level Fever Clinic Resilience Design Points

3.1   Reserve Development Space

From the perspective of resilience planning and design for medical facilities, it should start from the existing design standards at the beginning of the planning, and conduct research and refinement on the design standards through calculation, and adjust some of the indicators. At the same time, during the planning stage, a certain amount of long-term development space should be reserved for medical facilities, and the practice of resilience design related concepts should be attempted throughout the planning and design process. In addition, in terms of resilience planning for fever clinics, increasing the number, scale, and construction standards of fever clinics in the existing medical system, and requiring existing medical institutions of all levels and types to renovate and expand existing fever clinics through certain mandatory requirements, is one of the means to deal with sudden public health incidents. However, in normal times, the actual use efficiency of fever clinics with too large scale is not high, and the medical institutions themselves and the health planning departments at higher levels must also consider the pressure of normal operation of fever clinics, that is, to better solve the contradiction between the use efficiency of fever clinics in normal times and the redundancy in epidemic times.

3.2 Building a Regional Medical Consortium

It is worth noting that the positioning of fever clinics set up by most medical institutions is to use them as the first line of defense against epidemics. The facilities of the first line of defense must link the fever clinics with other levels of medical facilities to form a complete and comprehensive "combination punch". Therefore, at the government level, it should start from the top-level design, promote the development of smart medical care, build an information sharing platform for medical facilities at all levels, strengthen the coordination mechanism between grassroots fever clinics and regional disease control centers, city-level disease control centers and large comprehensive hospitals, and further strengthen the role of fever clinics and fever wards in future epidemic monitoring.

4 Key Points of Resilience Design for Fever Clinics at the Spatial Level

Firstly, in terms of the resilience design of the spatial level of the fever clinics in comprehensive hospitals, the proposal of its adaptability must be based on the integrity and stability of the entire hospital area, and pay attention to the synergy between normal times and epidemic times, reserve enough resilience to cope with the adjustment of long-term and short-term construction, and ensure the independence of the space after conversion and the integrity between the space function and the process. The key points of resilience design at the spatial level are mainly divided into four sub-items: function zoning conversion, function streamline conversion, buffer zone conversion, and redundant space.

4.1 Functional Zoning

In the zoning conversion, each zone should be able to independently complete its own function and have its own vertical system. For example, independent waste elevators, medical staff elevators, and patient elevators should be set up in each area. In normal design, the plan function layout should define clear function zones, at least adopt a double corridor layout mode, and the patient activity area should be strictly independent from the medical staff work area to reduce the chance of cross infection. During the transition from normal to epidemic, the medical staff work area is a clean area, including medical staff elevators, clean item storerooms, and sanitary passageways. Entrance and exit buffer rooms, item buffer rooms, etc., should be added between the clean area and the semi-contaminated area. The patient activity contaminated area should include patient elevators, patient corridors, waste packaging rooms, and other functions.

4.2 Functional Streamline

In terms of streamline transformation, the design of regular functions should set the patient elevator and the medical staff elevator independently. Based on the functional zoning, the patient entrance and the medical staff entrance should be set separately. During an epidemic, further improvements need to be made based on the regular streamline. Firstly, the core functional links such as medical staff and cleaning items should be closely connected to ensure the most basic function of the clean area. Secondly, in the three-zone two-belt three-channel nursing unit, patients usually enter each nursing unit from the patient corridor in the middle of the plane space, and the outside of the nursing unit is divided into the living balcony of the ward area by lightweight partition walls. During an epidemic, the lightweight partition walls should be removed and restored to an independent patient corridor. After the patient is admitted to the hospital, they go up to the designated floor by the independent patient elevator and enter the ward area from this patient corridor. At this time, after the medical staff pass through the sanitary pass-through links such as the first change - second change - buffer room in the clean area, they use the semi-contaminated corridor in the middle of the regular corridor, and enter each ward through the buffer room added near the semi-contaminated corridor in the middle of each nursing unit.

4.3 Buffer Zone

In terms of buffer zone transformation (Fig. 1), the dressing and shower rooms for medical staff should be set as close as possible to the entrance and exit of vertical traffic during regular layout, to meet the sanitary pass-through dressing needs during an epidemic, or to arrange spare rooms here and reserve water points. During an epidemic, firstly, an exit and entrance buffer room should be set up in the sanitary pass-through area for medical staff when setting up the clean area. That is, add partition walls and airtight doors in the entrance buffer room, and the sanitary pass-through process for medical staff is buffer room - first change - second change - buffer room. In this process, medical staff complete hand washing and disinfection - put on work clothes, hats and masks - put on protective clothing - enter the buffer room and finally enter the semi-contaminated corridor. Partition walls and airtight doors should also be added in the exit buffer room, and the sanitary pass-through process for medical staff is buffer room - first change - second change - buffer room. In this process, medical staff complete washing and removing protective clothing - removing work clothes, hats and masks - hand washing and disinfection before entering the clean corridor.

Secondly, the storeroom area can be used normally during regular times, and should be designed close to the medical staff elevator, and set double doors facing the public corridor and medical staff corridor respectively. They can be locked or opened according to specific usage conditions. In addition, during epidemic transformation, a normally

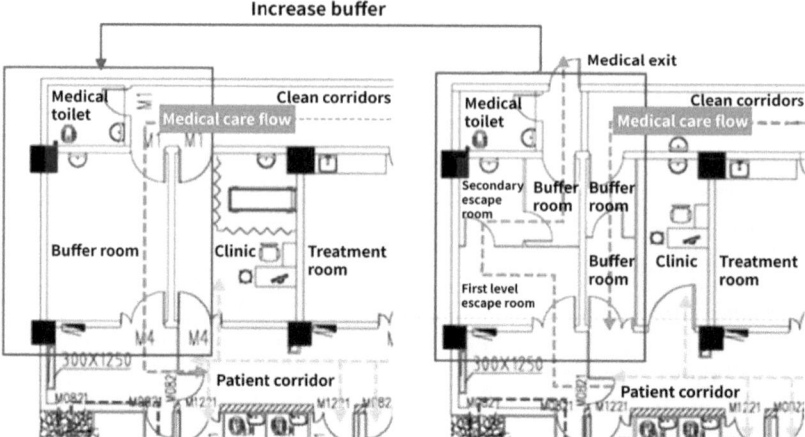

Fig. 1. Schematic diagram of buffer zone transformation during epidemic (Source: Drawn by the author)

closed airtight door and sealing measures should be added in one of the storerooms in the storeroom area to ensure tightness and form a cargo buffer room. The doors of the remaining storerooms should also add sealing measures when facing the semi-contaminated area (Fig. 2).

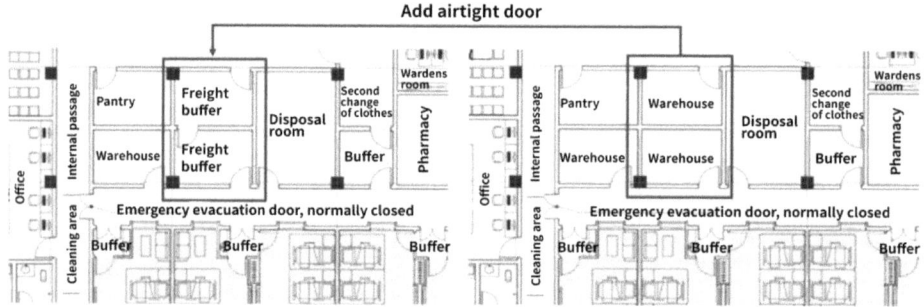

Fig. 2. Schematic diagram of semi-contaminated area storeroom area transformation during epidemic (Source: Drawn by the author)

4.4 Redundant Space

In terms of redundant space, attention should be paid to the setting of column grids during the usual layout planning to enhance the flexibility of later modifications. At the same time, important flow-through spaces should reserve a certain amount of surplus space, which is convenient for adding buffer rooms or sanitary pass-through areas later. By promoting the construction of smart hospitals, some functional rooms are replaced by self-service equipment or service technology, making the layout more flexible.

5 Design Strategies and Case Studies for the Resilience of Fever Clinic Spaces

The renovation of large comprehensive hospitals during the pandemic and the design of fever clinics is one of the focuses of medical space improvement. In terms of summarizing the design strategies of fever clinics in comprehensive hospitals and the empirical evidence of construction, this article will take the renovation of fever clinics in medical institutions such as the Third Affiliated Hospital of Guangzhou University of Chinese Medicine and the Integrated Traditional Chinese and Western Medicine Hospital of Southern Medical University as examples.

5.1 Physical Partitioning to Avoid Infection

The first type of fever clinic renovation and expansion in comprehensive hospitals, divided by area and main renovation methods, is the fever clinic co-built with other departments. The area is generally not large and is set up on the first floor or the first and second floors of the building. Since there are many departments inside the comprehensive hospital, the key point of this type of fever clinic renovation and expansion is whether it can avoid infection with other departments. The control measures for the source of infection in the fever clinics of comprehensive hospitals usually focus on the following aspects:

First is the identification of the source of infection. In the pre-examination triage area, it is necessary to screen, triage, and divert patients entering the fever clinic. At this stage, it is necessary to identify the source of infection from the outside. Therefore, before patients enter the fever clinic, they need to complete body temperature, basic situation and epidemiological investigation, nucleic acid testing, etc., which are conducive to following the consultation process for the next step of serological antibody testing, imaging examination, and routine blood tests.

Second is the physical separation of patients. After patients enter the fever clinic, in order to prevent cross-infection between infectious patients, and contact between clean area, semi-clean area medical staff and patients, it is necessary to add physical partitions and multiple channels to realize the separation of patient flow lines and medical staff flow lines, and include the air intake and exhaust in the facility space into the scope of partition consideration, and carry out more detailed division and separation of the plane area inside the fever clinic. Moreover, in areas where clean and contaminated flow lines are close, buffer rooms should be set up, and attention should be paid to the control of air pressure to reduce the possibility of cross-infection.

The renovation of the fever clinic in the third affiliated hospital of Guangzhou University of Traditional Chinese Medicine mainly utilizes the buffer space in the original building layout, dividing it into a separate sanitary passageway, which can complete the "one-off-two-off-buffer" process before entering the clean area corridor. At the same time, the original consultation room is divided by adding a partition wall, forming a buffer zone between the patient corridor and the clean area corridor, meeting the requirements of "three zones and two channels". Meanwhile, the entrance area of the fever clinic has expanded the nucleic acid sampling room and added a waiting room for nucleic acid sampling. During the renovation process, the principle of "double channels, one-way traffic, no retrograde" was followed, and buffer rooms were added as physical partitions, and induction doors were installed to reduce contact infection. However, due to the limited land use within the hospital area, the fever clinic cannot be set up as an independent building, but can only be set up as an independent treatment area with an independent entrance, and is hard-separated from other business rooms. From the outside, the fever clinic cannot be equipped with a CT device dedicated to fever patients.

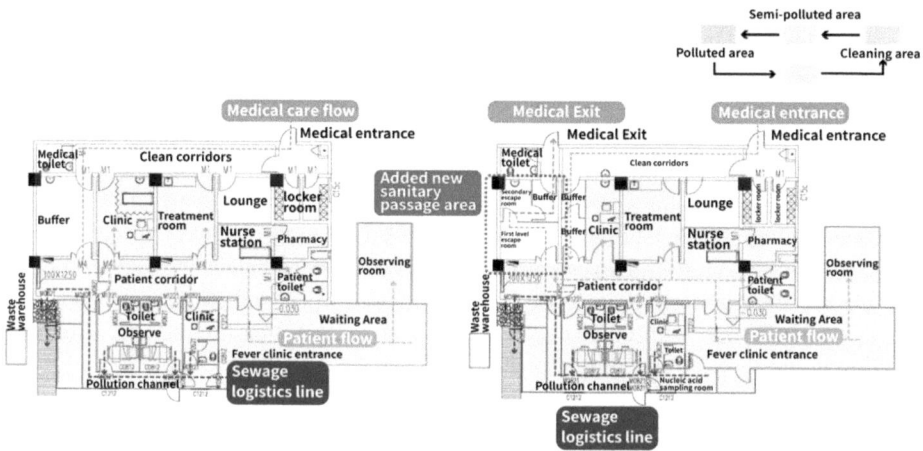

Fig. 3. Floor plan before (left) and after (right) the renovation of the fever clinic in the third affiliated hospital of Guangzhou University of Traditional Chinese Medicine (Source: Drawn by the author)

Taking the Haizhu District Traditional Chinese Medicine Hospital in Guangzhou as an example, in order to better respond to emergencies in epidemic prevention and control, the hospital has transformed part of the independent area of the medical institution into a fever clinic, with an independent entrance and exit (Fig. 3). A room in a single-sided corridor inside the hospital was added to form a building layout with the patient corridor in the middle, waste, consultation room, observation room, and blood transfusion room on the south side, and waiting room, throat swab collection room, mobile DR room, and buffer zone on the north side of the corridor. Physical partitions are added between different zones. Medical staff returning to the clean area from the patient corridor need to pass through the buffer zones such as the first and second off rooms, enter the dressing room and shower room, and then return to the clean area. At the same time, medical staff can also directly enter the patient corridor from the south side through the buffer room (Fig. 4).

5.2 Independent Zoning, Line Separation

The second type of fever clinic renovation and expansion in general hospitals, divided by area and main renovation method, is the renovation of standalone buildings into fever clinics. These fever clinics are larger in area and are generally renovated based on the existing building. The first floor of the original building is usually set up as an independent infection zone, and the second floor is used as a patient treatment area. Since the layout of the original building needs to adapt to the layout requirements of the "three zones and two channels" fever clinic, the reasonable division of the floor plan is one of the key points of the renovation.

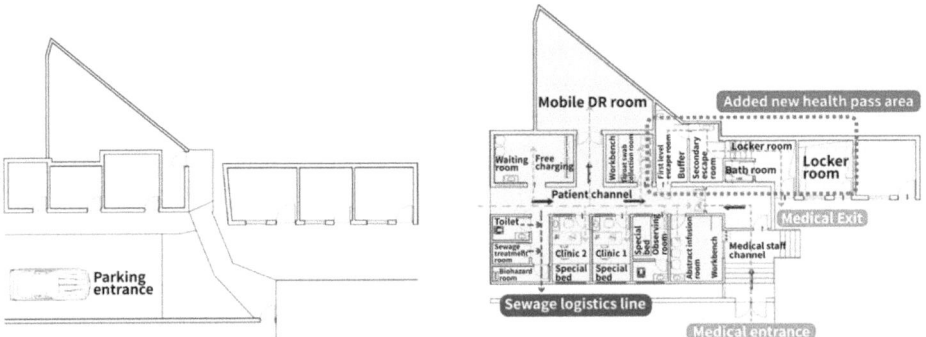

Fig. 4. Floor plan before (left) and after (right) the renovation of the fever clinic in Haizhu District Traditional Chinese Medicine Hospital, Guangzhou (Source: Drawn by the author)

Taking Southern Medical University's Integrated Traditional Chinese and Western Medicine Hospital as an example, it expanded the outpatient building within the hospital area, and built a temporary fever clinic on the basis of the original outpatient building. The fever clinic is set up as a standalone building, with a total of 2 floors, one of which serves as an independent infection area, and the second floor serves as a patient diagnosis and treatment area (Fig. 5). Based on the original outpatient clinic, clean areas, semi-contaminated areas, and contaminated areas were reasonably divided to achieve clean and dirty zoning and patient-doctor diversion, and single-person isolation observation rooms were added. The contaminated area of the temporary fever clinic built on the basis of the original outpatient building mainly consists of general fever waiting areas, special fever waiting areas, observation rooms, consultation rooms, treatment rooms, and sampling rooms, responsible for the reception, sampling, preliminary diagnosis, observation, and treatment of early patients. The renovation method of the temporary fever clinic is to divide the contaminated area, semi-contaminated area, and clean area by adding walls, and the passage also provides an additional evacuation route for medical staff. At the same time, a buffer zone is divided in the semi-contaminated area. When medical staff travel between the contaminated area and the semi-contaminated area, they follow the "exit buffer one-exit buffer two" process, implementing strict sanitary passage (Fig. 6). Due to the limited space inside the fever clinic, CT equipment and other epidemic prevention measures are used in combination with "normal epidemic". The CT room is about 20 m away from the fever clinic and has an independent passage for fever patients. If fever patients need to use CT, the CT room and passage can be completely closed and independent.

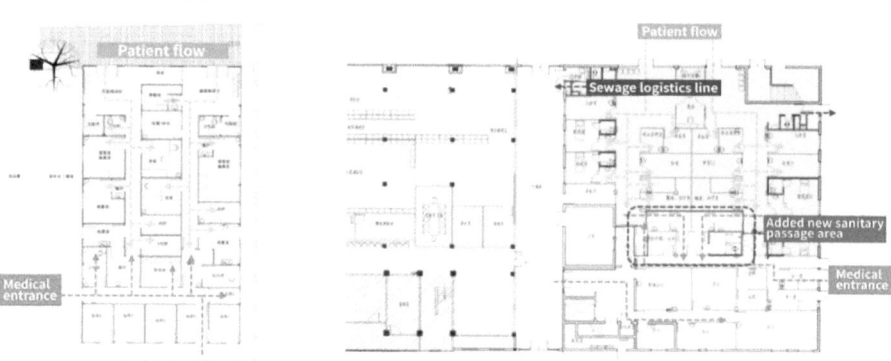

Fig. 5. Layout of the fever clinic at Southern Medical University's Integrated Traditional Chinese and Western Medicine Hospital before (left) and after (right) renovation (Source: Drawn by the author)

Fig. 6. Before (left 1) and after renovation of the fever clinic at Southern Medical University's Integrated Traditional Chinese and Western Medicine Hospital (Source: Photo by the author)

5.3 Module Construction, Normal Epidemic Transition

The third type of fever clinic renovation and expansion in general hospitals is classified according to the area and main renovation methods, which is for the emergency construction of fever clinics on reserved land. Due to the emergency construction, this type of fever clinic has fewer restrictions on the layout, generally adopting a "three-zone three-channel" layout, and separately setting up independent adult and children's fever clinics. This type of fever clinic, which is modularly constructed on reserved land, considers the sudden outbreak and initial explosive growth of cases when infectious diseases occur, often occupying a large area and scale of construction. The construction of this type of emergency fever clinic usually adopts modular construction methods such as containers, which can still be further expanded in the later stage. When used as an emergency medical facility for epidemic treatment, the timeliness of modular facility construction is highlighted. In normal times, this type of emergency medical space serves as a strategic reserve, and considering the operational issues during normal times, it can be converted into other types of medical service places. In the process of transition from epidemic to normal times, this type of fever clinic focuses on the following aspects: First, it pays attention to the coordination of recent construction and long-term planning. At the beginning of emergency construction, fully consider the possibility of long-term expansion and renovation, so that it can smoothly transition between normal times and epidemic times. Second, it can maximize the utility of emergency medical facilities during epidemics, and take into account operational pressure during normal times, fully improving the utilization rate of existing medical resources. Third, the space before and after the transition from epidemic to normal times can maintain its original independence, and the functions and processes match. Take the Zhongshan Hospital of Traditional Chinese Medicine affiliated to Guangzhou University of Traditional Chinese Medicine as an example. According to the relevant requirements for the standardized construction of fever clinics, the hospital has built a two-story independent fever clinic building with a total area of 1080 square meters and a total construction area of 2160 square meters (Fig. 6). The fever clinic is equipped with a "three-zone three-channel" standard setting, a visual intercom system, a sensor door in the removal room, 16 single-room isolation observation rooms, and independent toilets and air disinfection equipment, rescue and life support equipment, etc. The newly built fever clinic implements a full-process closed-loop management, and patients' registration, specimen collection, diagnosis and treatment, payment, testing, examination, medication and other medical activities can all be completed within the fever clinic, isolated from other medical facilities in the hospital, to avoid the risk of cross-infection in the hospital. The newly built fever clinic building of Zhongshan Hospital of Traditional Chinese Medicine affiliated to Guangzhou University of Traditional Chinese Medicine consists of general fever clinics, children's fever clinics, and special fever clinics, with two floors. The nucleic acid sampling area for yellow-coded personnel and key focus area personnel is independent of the fever clinic. Such an independently constructed fever clinic building can use the surrounding space to set up temporary waiting areas during epidemics to reduce cross-infection. The entire fever clinic building is constructed with containers, and when the number of patients visiting during an epidemic increases sharply, containers can be used again for further expansion to meet the internal space needs of the fever clinic.

Among them, the adult fever clinic is located on the first floor of the fever clinic (Fig. 7) A nucleic acid waiting area is set up next to the entrance, providing a waiting space for the external nucleic acid sampling area. After entering the adult general fever clinic from the entrance, relevant examination and testing equipment such as independent CT, DR, B-ultrasound, and blood cell analyzer are equipped in the contaminated area. There are also general fever infusion rooms, injection rooms, rescue rooms, self-service payment machines, western medicine pharmacies, 24-h self-service medicine cabinets, and two spare rooms. After the patient completes registration, sampling, diagnosis and treatment, payment, testing, examination, and medication collection, they can leave only after waiting for the result in the nucleic acid waiting area to be negative. The special fever clinic on the first floor is an independent area. After the patient's pre-examination and triage, they are guided by medical staff to enter the special fever clinic through the peripheral passage. The consultation process is basically the same as that of the general fever clinic, equipped with 2 consultation rooms, 1 treatment room, medication, blood drawing room, and 4 observation rooms, and shares the CT and DR rooms with the adult fever clinic. Medical staff working on the first floor of the fever clinic need to pass through the first removal room, second removal room, shower room, and dressing room to return to the clean area from both the special fever clinic and the adult fever clinic. The clean area is equipped with a warehouse for passing items and medicines through the observation window with the semi-contaminated area, a dining area for medical staff, and a bathroom. The transfer of contaminated items is carried out by an independent contaminated item staircase, which transfers the second-floor contaminated items to the first floor and then removes them from the fever clinic building.

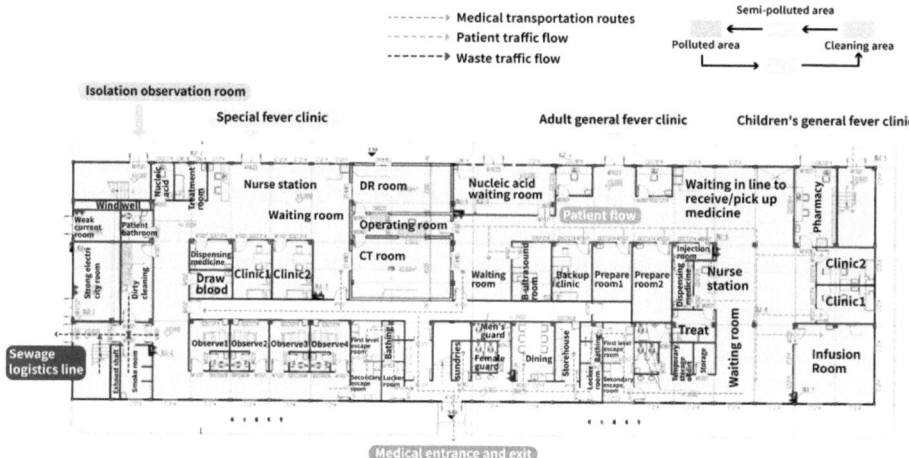

Fig. 7. Floor plan of the Fever Clinic Building of Zhongshan Hospital Affiliated to Guangzhou University of Chinese Medicine (Source: Drawn by the author)

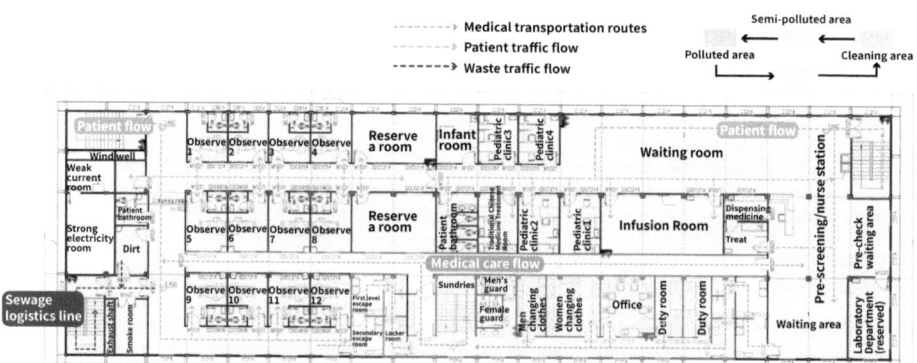

Fig. 8. Second floor plan of the Fever Clinic Building of Zhongshan Hospital Affiliated to Guangzhou University of Chinese Medicine (Source: Drawn by the author)

The second floor of the Fever Clinic Building is for the general pediatric fever clinic and 12 observation rooms. The general pediatric fever clinic enters the second floor from the independent staircase on the first floor, and has a laboratory, divided into adult-specific windows and children-specific windows, children's infusion rooms, etc. It is also specially equipped with a mother and baby room for breastfeeding mothers and babies. Medical staff can enter the second floor clean area from the independent staircase in the first floor clean area, and the process from the clean area to the contaminated area is consistent with the first floor (Fig. 8).

Through the design strategies and transformation processes of the above three types of fever clinics, it can be seen that the transformation flexibility of independently constructed fever clinics is relatively large. Buffer zones and other spaces are directly arranged according to the epidemic situation, and the flexibility and comfort of conversion are strong. Fever clinics rebuilt on the basis of existing facilities in central urban areas are usually limited by the building area and it is difficult to provide patient-specific CT examination rooms, nucleic acid testing areas, sufficient isolation rooms, etc.

6 Conclusion

In conclusion, this study on spatial resilience design strategies for post-pandemic era fever clinics contributes valuable insights into healthcare infrastructure and emergency preparedness. By synthesizing resilience theory with practical design solutions, it introduces a framework for improving the operational efficiency and adaptability of fever clinics, essential for enhancing public health resilience against future pandemics. The significance of anticipatory planning and adaptive design in healthcare facilities is highlighted, providing guidelines for the global construction and renovation of fever clinics.

Nevertheless, the research encompasses limitations, notably the focus on urban settings, which may not encapsulate the unique challenges faced by rural or less densely populated areas. Additionally, the applicability of design strategies requires consideration of local contexts and regulations, aspects not extensively covered in this analysis. Further studies are encouraged to explore these dimensions, broadening the applicability and understanding of the proposed design strategies in varied healthcare environments.

Funding:. This paper is supported by individual academic enhancement project of Guangzhou Academy of Fine Arts "Research on the Resilience of Urban Medical Spaces in Guangzhou" (Grant No. 24XSC36).

References

1. Comprehensive Group of the State Council's Joint Prevention and Control Mechanism for the Novel Coronavirus Pneumonia Epidemic: Notice on Printing and Distributing the Work Plan for Graded Diagnosis and Treatment of Novel Coronavirus Pneumonia Based on the Medical Consortium: http://www.nhc.gov.cn/yzygj/s3593g/202212/8a38e403dea2489bae9b292868073b27.shtml
2. Hao, L., Yanping, J.: Discussion on the graded planning design mode of fever clinics in China. China Hosp. Architect. Equip. **23**(11), 3–8 (2022)
3. Hao, L., Yanping, J.: Thoughts and countermeasures on the construction of fever clinics in the post-pandemic era. Contempor. Architect. **5**, 33–38 (2021)
4. HKS.: The pandemic resilient hospital: How design can help facilities stay operational and safe (July, 2021): https://www.hksinc.com/how-we-think/reports/the-pandemic-resilient-hospital-how-design-can-help-facilities-stay-operational-and-safe/

The Construction of Doctor-Patient Relationship Based on Convergence Media Under Major Public Health Events

Yanfei Ding and Yifei Zhang[✉]

Zhongnan Hospital of Wuhan University, Wuhan University, Wuhan 430071, China
zhangyf@whu.edu.cn

Abstract. As the main body of epidemic prevention and control, public hospitals are more professional, authoritative and credible. They are important participants and main parts of epidemic prevention and control propaganda. The doctor-patient relationship was once unprecedented harmonious. On this basis, the trust established is more stable, so how to build a long-term and good doctor-patient interaction based on this trust in the post-epidemic era is an important way to build the social reputation of the current public hospitals. To maintain and consolidate this harmonious relationship, the convergence media still have a great research space and broad practice platform in top-level design, agenda setting, channel development and ethical self-discipline of practitioners. Based on the interactive paradigm of convergence media and the new model of Internet medical treatment, this paper explores a new model in line with China's national conditions, namely, "filial piety relationship model", which is of great significance to the long-term maintenance of harmonious and friendly medical relations.

Keywords: major public health events · convergence media · doctor-patient relationship · filial piety and kinship relationship model

1 Introduction

"Convergence media" is a new media that makes full use of the Internet as the carrier to integrate different media, such as radio, television and newspaper, which have both common and complementary aspects, in manpower, content and publicity, so as to realize "resource integration, content integration, publicity integration and interest integration" [1]. Various media complement each other's advantages, foster strengths and circumvent weaknesses, and maximize the communication benefits. To promote the construction of a healthy China, we need to build a harmonious and friendly new doctor-patient relationship. Among them, all levels of all kinds of media to build communication channels, fair and open reporting, and create a harmonious and friendly social atmosphere play an important role [2]. Doctor-patient relationship is the main manifestation of medical ecology. The contradiction between doctors and patients has always been widespread [3]. The focus of the contradiction is the lack of effective communication and trust between doctors and patients, and some medical events amplified by the Internet have caused a wider negative public opinion.

© The Author(s) 2025
B.-J. He et al. (Eds.): UCSUD 2023, LNCE 559, pp. 949–957, 2025.
https://doi.org/10.1007/978-981-97-8401-1_69

At present, many scholars have studied the doctor-patient relationship in the new era from different aspects. Katsaliak conducted simultaneous tests of communication satisfaction among doctors, nurses and patients [4]. Duffee improved the doctor-patient relationship, accident litigation by encouraging the legislature to write informed consent into law [5]. Tang et al. improved the doctor-patient relationship through public opinion [6]. However, doctor-patient relationship is a kind of systematic engineering involving society, doctors, nurses, patients and families. It is necessary to comprehensively analyze the interaction and mutual relationship between influencing factors through top-level design and the current mainstream form of convergence media. In major public health events, medical workers have won the respect of the people across the country with the professional concept of "people first and life first" [7]. The reason for this, the doctor-patient relationship has gone beyond the simple interest relationship and sublimated into a "community of shared future" relationship, achieving an ideal harmony and unity. The medical order is gradually restored to the "normal" before the major public health event, and continuing to improve the harmonious doctor-patient relationship is a major issue. This paper discusses the role and tasks of financial media in the construction of Chinese doctor-patient relationship with "equal trust and high harmony", and puts forward the idea of a new model of doctor-patient relationship under the national conditions. The technical roadmap for this article is shown in Fig. 1.

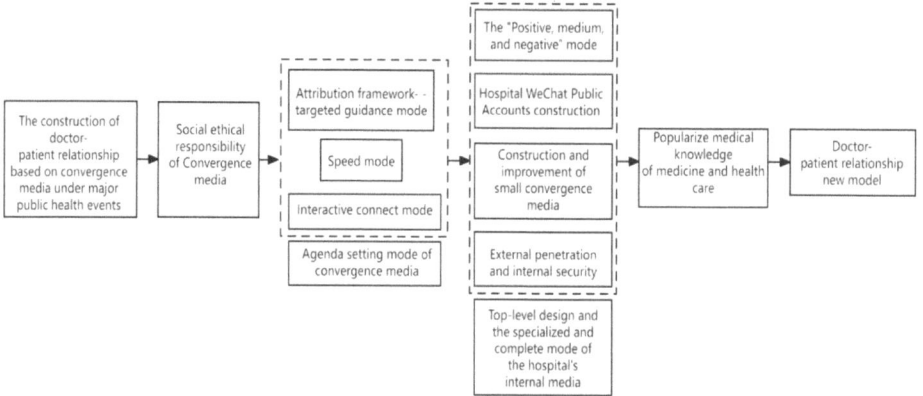

Fig. 1. Conceptual framework

2 Social Ethical Responsibility of Convergence Media

It is the inherent ethical responsibility of the media to maintain a good social and moral order. Media plays the role of "gatekeeper" in information dissemination activities, holds "abandoning power", has subjective initiative, and will integrate attitudes and views into the information to the public [8]. The good morality shown by media practitioners in the process of communication is of special significance to the formation of a good atmosphere and the improvement of the moral level of the whole society. They have the social responsibility to open the wisdom of the people and guide the people, and also have a huge influence on the social ideology. In this commercial society, people often forget some basic professional ethics. Doctors no longer give the impression that heal the wounded and saving the dying. These roles deviate from the expectations of the public. Our media has the responsibility to re-regulate the social role by reporting typical characters and events. When negative events occur, netizens will constantly search for relevant information due to their curiosity and desire to explore. In the doctor-patient relationship report, the public at least hopes to meet the following three levels of information needs through the media: one is to understand the truth and satisfy the right to know; the second is to analyze the facts and clarify their own interests; the third is to meet the desire to reveal the dark side of doctors and patients. Mass media in the role in the doctor-patient relationship, from the positive side, its comprehensive, objective and fair news, is conducive to maintain the patients and the public's right to know, to the supervision and constraints of medical institutions and medical staff behavior, so as to promote the health department of medical ethics and clean hospital construction, strengthen the consciousness of the public to maintain health rights. This study is conducive to construct the new practice of situational crisis communication theory (SCCT) in doctor-patient relationship under the background of high-quality development of public hospitals under the background of Internet [9].

3 Agenda Setting Mode of Convergence Media

3.1 Attribution Framework—Targeted Guidance Mode

According to the theory of agenda setting mode, the mass media often constructs the priority of public discussion and attention by choosing the focus of reporting. People tend to understand the issues that the media emphasizes and understand events according to the order of priority set by the mass media. This may affect the audience's perspective and position. By setting the agenda, public awareness and concern about a certain situation can be raised.

Convergence media should not be short-sighted, to establish a long-term mechanism. Design the attribution framework, grasp the key problems, conduct targeted guidance, and report should be directional. Under the new normal after the epidemic, many people believe that the root cause of the contradiction between doctors and patients is the current medical system and policy problems. While reporting the medical events objectively, the line reporter can increase the review and evaluation, and attribute the contradiction between doctors and patients according to the disadvantages of the system or policy. Such long-term attention can provide a basis for policymakers to reform, improve the medical system, and then improve the doctor-patient relationship.

3.2 Speed Mode

Reality and speed are a very important category in the study of media theory. The essence of the game between media can be summarized as: truth is the foundation, speed is the soul. In the event of a major epidemic outbreak or a serious medical event, the response speed of the media is the first requirement, so the qualified media should establish an emergency response mechanism, and configure and form a mode in terms of personnel, channels, equipment and materials. Only in this way can we take the initiative, meet the needs of the audience and defeat the opposite of the "negative state". Otherwise, when difficult to choose "silence ", "no voice", may let some of the media and the public suspicion to take the initiative, let the "clickbait" with public opinion, let the first rumors override the fact of late voice.

3.3 Interactive Connect Mode

New media has the advantage of speed, but traditional media can exist independently without attachment to new media, and it has the advantages of authority and credibility (such as newspapers). Newspaper media has a wide range of information sources, strong interview resources, high-quality news gathering and editing team and efficient production process. From the news information collection, processing integration, to the layout of the editor, until into the social circulation channels, the newspaper has formed a set of complete and efficient production line and the content of the review, filtering mechanism (including the "original information" and a large number of "derived information" editor choice), plus in the past decades and hundreds of years of accumulated credibility, these are new media cannot reach. In fact, the biggest advantage of online newspapers lies in that it can realize the necessary judgment and screening of information with the help of the newspaper media with the originality and credibility of information dissemination. Therefore, new media and traditional media should cooperate with each other, establish a cooperation model of full mutual trust and mutual assistance, the speed should be given to new media, and the tone should be set by traditional media.

4 Top-Level Design and the Specialized and Complete Mode of the Hospital's Internal Media

In a general sense, top-level design refers to the overall planning of all aspects, levels and elements of a task or a project from an overall perspective, so as to concentrate effective resources and achieve the goal efficiently and quickly [10]. The top-level design of the media mentioned here mainly refers to the overall guiding ideology, the way of purpose realization, especially the specific reporting framework design.

4.1 The "Positive, Medium, and Negative" Mode

In media reports, the proportion of "positive, medium and negative" should be appropriate (to be simple, and the patient is called "positive" and "negative"), and cases of "positive" reports are certainly the main position, but critical "negative" reports are also indispensable [11]. However, in negative reports, we should especially make a good review to find out the essence of the doctor-patient contradiction, and the purpose is to improve the doctor-patient relationship. So is this "median negative" ratio 433, or 532 or 541? How to control its cycle and rhythm? This is the topic of media science research level, and it will not be discussed here. In fact, there are many topics in the top-level design, such as the proportion design of the reporting perspective is also worth studying. We should not always look at the problem from one perspective. Adhering to the multi-perspective, multi-directional and multi-level objective reporting can continuously improve the credibility of the audience and speak out for the continuous improvement of the harmonious doctor-patient relationship. "After the middle and negative" proportion mode is determined, the assessment of media workers depends on the completion of the design proportion, rather than just the click rate or network traffic.

4.2 Hospital WeChat Public Accounts Construction

Under the background of health communication, all major hospitals attach great importance to health communication, and use new media such as WeChat Public Accounts as a carrier to produce healthy communication content, which has begun to show results in improving the health awareness of the whole people and alleviating the conflicts between doctors and patients. The health information provided by medical workers is one of the most trusted platforms by the public. According to this logic, the public account of medical institutions is naturally the most credible source of medical and health information for the public. The permanent column of the public account can be gradually promoted from less to more, from narrow to wide. If due to energy, time limits, hot columns can be rotated, can also be regularly updated.

4.3 Construction and Improvement of Small Convergence Media

Convergence media is the most effective means to promote the establishment of more and better doctor-patient communication channels. Medical institutions can make full use of the advantages of "interactive" communication on the convergence media platform, promote online Q & A and consultation, increase the production of video and audio science popularization, and try their best to change the state of information asymmetry. We should form a "small convergence media" in medical institutions. In addition to the above online Q & A video and audio production, we can also use electronic scrolling screens, knowledge leaflets, brochures (free access), so that patients can obtain more beneficial and relevant information about their illness during the long waiting time. The ward must have closed circuit television, according to the different diseases, make small short films for patients and their families to watch. The publicity board of the courtyard should be updated regularly. In addition to the publicity of chief experts and new drugs and technologies, the content of doctor-patient communication should be carefully designed.

In addition, in order to enhance the interaction between doctors and patients, we must set up a traditional "opinion book", regularly screen the problems reflected by patients in the opinion book, and give targeted replies. This kind of interaction can be presented in the official website of the hospital, and synchronized in the rolling electronic screen, publicity board and other channels.

4.4 External Penetration and Internal Security

The internal media workers of the hospital should strive to "penetrate" the influential website network platforms and win the position authority, so as to maintain the authority of the medical official media, maintain the authenticity, professionalism and seriousness of the communication content, and also play a positive role in refuting rumors and falsification, safeguarding legitimate rights and interests, and harmonious doctor-patient relationship.

The lack of departments needs to be valued urgently. "Legal Department", "hospital Department" and other departments, especially first class hospitals should be set up. With this endorsement, it is more conducive to the internal media (including the media with cooperative relations) to make reasonable, favorable and powerful reports on major medical incidents, and resolutely resist and prevent all the words and deeds that create and intensify the contradiction between doctors and patients.

5   Popularize Medical Knowledge of Medicine and Health Care

Promoting medical science knowledge is not only the social responsibility of the media, but also the responsibility of the media in medical institutions. First, the key breakthrough can choose "refuting rumors and falsification", increase the crackdown on rumors, to help the public, especially the elderly to distinguish right from wrong. Second, the young audience is more dependent on network communication, but although the network media has advantages of fast, fresh, interesting, interactive at the same time, has shortcomings of repeated content which is also easy to be forgotten, therefore, form updates, reduce content repetitive, improve topic, more impressive, is the development direction of network media. Third, attach attention to offline publicity. The media should expand the publicity channels, increase the community publicity, such as the media use its appeal to hold community health education activities, so that science popularization goes to the ordinary people, and strive to improve the health literacy of the whole people.

After over 30 years of joint efforts of the whole society, the work of law popularization in China has achieved good results. Legal knowledge is well known, among which the media has played a major role. Taking this experience, if the work of "general medicine" is carried out according to this, the communication power of today's convergence media will quickly achieve practical results and play a huge role in improving the doctor-patient relationship.

6   Doctor-Patient Relationship New Model

It is generally believed that there are three modes of doctor-patient relationship: active-passive, guiding-cooperative, and common participation. Karl Jaspers once put forward the concept model of "doctor-patient community of common destiny" as a doctor-patient relationship [8]. Doctors and patients unite as one and share weal and woe, which is also an ideal doctor-patient relationship that is widely discussed and respected. In fact, the existing various doctor-patient relationship models are not without contradictions. This paper holds that if the identity of a doctor is changed, the emotional relationship between doctors and patients is changed, and a "filial piety kinship mode" is constructed, with the assistance of media, the doctor-patient relationship can be more harmonious.

"Filial piety and foster kinship mode" is inseparable from the publicity of the convergence media. In particular, mutual trust, cost burden and other most common focus issues between doctors and patients, through the explanation and dissemination of convergence media, can strengthen the mutual trust between doctors and patients. The most important thing is to avoid doctor-patient disputes rising to the legal level. With the escort of the convergence media, a few media of the malicious hype. The construction path of "filial piety and kinship mode" is also inseparable from the role of convergence media. For example, the establishment of communication channels between doctors and patients can make full use of the convergence media to build a platform, through wechat and doctor-patient information exchange center, so that both doctors and patients can communicate with each other on the condition, treatment, psychology, emotion, etc. In the initial stage of constructing the model of filial piety and kinship, convergence media can set up a private network to introduce its characteristics and advantages, so that both doctors and patients can examine each other and choose the object with appropriate conditions. After the agreement was reached, under the continuous attention and coordination of the convergence media, the relationship model gradually entered the track of benign development. In turn, the successful pilot has provided more and better materials for the convergence media, and has continuously expanded the publicity. In this way, starting from the first case, to have a certain successful base, can formally set up the filial piety kinship model. To build the model of filial piety and kinship, it is not enough to rely on the efforts of individual doctors, but to have a perfect set of encouragement and reward mechanism to do the guarantee. For example, if a model is successful, and the patient has no adverse reactions for three months after discharge, the doctor should be given a one-time reward. The number of success can be used as an important reference for doctors' promotion and salary increase. During this period, the convergence media can follow up reports and interpret the new doctor-patient model with detailed cases, so that the harmonious doctor-patient relationship can be continuously improved. "Filial piety of kinship mode" focus on the medical side. Because the establishment of this relationship, for the patient is desired, but for the medical side is a requirement, is a duty and responsibility. The "high-quality doctor group" with profound medical humanistic quality is the basic premise of the construction of "filial piety and kinship relationship mode". This also requires the convergence media to pay special attention to this special group in terms of image publicity, use various media forms to introduce its technical characteristics, promote its virtue and benevolence, so as to improve their credibility in the eyes of patients." Over the years, in the higher medical education, we despise the medical humanities education". This is extremely detrimental to the training of a high-quality medical team. In this respect, the convergence media should also take the responsibility, use various channels in various forms, vigorously preach professional dedication, professional attitude, conscience, emotional communication skills and other people-oriented humanistic quality, to train a "benevolent teacher" to do their strength.

With the development of science and technology and the improvement of medical level, the doctor-patient relationship has not improved but continued to deteriorate. The reason is that the doctor-patient relationship which should belong to the moral trust relationship is wrongly understood as a contractual relationship or consumer service relationship. This is a very grim reality. The construction of "filial piety and kinship relationship mode" is to break through the dilemma of "performance appraisal" and "profit target", an important way to fundamentally improve the doctor-patient relationship, and also a new force for the convergence media to improve the doctor-patient relationship under the new normal. The public hospital Internet platform built in this paper based on convergence media is shown in Fig. 2.

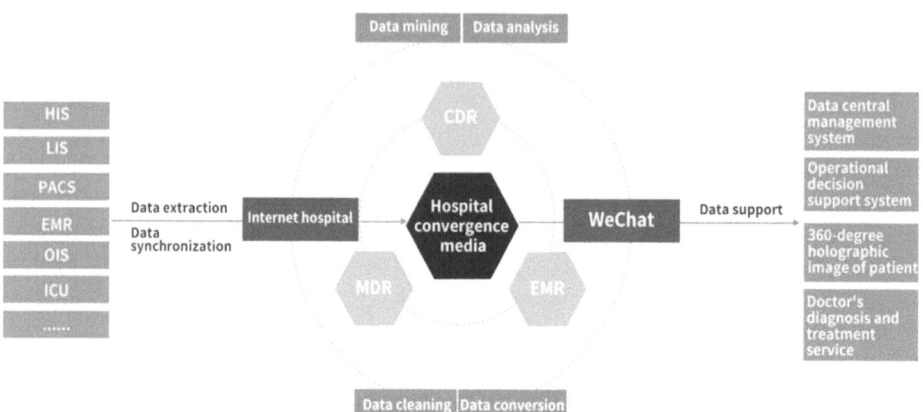

Fig. 2. Hospital convergence media

7   Conclusion

To achieve a good medical ecology of harmony and mutual assistance, "integrity, friendship" and sustainable development, although the road is blocked and long, it is incumbent to integrate media, and exploring the construction of a new doctor-patient relationship model is an important way to maintain the truly harmonious doctor-patient relationship for a long time. As long as our medical institutions join hands with the media, cooperate closely with them, strengthen their confidence and make continuous efforts, we will be able to build a new model of doctor-patient relationship with Chinese characteristics and embark on the road of happiness of "healthy China".

References

1. Triko, G., Nurfathiyah, P.: Innovations of China's Mainstream Media Convergence. Jmm-Int. J. Media Manag. **24**, 275–276 (2022)
2. Qian, Y., Zeng, G., Pan, Y., Liu, Y., Zhang, L., Li, K.: A prediction model for high risk of positive RT-PCR test results in COVID-19 patients discharged from Wuhan Leishenshan Hospital, China. Front. Public Health **9** (2021)
3. Wang, Y., Du, S.: Time to rebuild the doctor-patient relationship in China. Hepatobiliary Surg. Nutr. **12**, 235–238 (2023)
4. Katsaliaki, K.: Evaluating patient and medical staff satisfaction from doctor-patient communication. Int. J. Health Care Qual. Assur. **35**, 38–55 (2022)

5. Tang, Y., Yang, Y.T., Shao, Y.F.: Acceptance of online medical websites: an empirical study in China. Int. J. Environ. Res. Public Health **16** (2019)
6. Duffee, C.: Pathologizing pathos: suffering, technocentrism, and law in twentieth-century American medicine. J. Hist. Med. Allied Sci. (2023)
7. Liu, Y., Wang, X.J., Chen, Z.S., Zhang, Y.F., Zhao, S.H., Devici, M., Jin, L.S., Skibniewski, M.J.: Evaluating digital health services quality via social media. IEEE Transactions on Engineering Management (2023)
8. Wu, X., Chen, B., Chen, H., Feng, Z., Zhang, Y., Liu, Y.: Management of and revitalization strategy for megacities under major public health emergencies: a case study of Wuhan. Front. Public Health **9** (2022)
9. Liu, Y., Li, X., Ding, R., He, T., Wang, X.-j.: Coping ability and promotion countermeasures of medical and health institutions reputation crisis: a case study in Hubei Province. Front. Public Health **9** (2022)
10. Hu, G., Wang, Z., Jiang, S., Tian, Y., Deng, Y., Liu, Y.: Community public health safety emergency management and nursing insurance service optimization for digital healthy urban environment construction. Front. Public Health **10** (2022)
11. Guo, S., Guo, X., Zhang, X., Vogel, D.: Doctor-patient relationship strength's impact in an online healthcare community. Inf. Technol. Dev. **24**, 279–300 (2018)

Exploration and Reflection on the Construction of a Green Human Resource Management System in Public Hospitals in the Post-Pandemic Era

Yike Tian[(✉)]

Zhongnan Hospital of Wuhan University, WuhanUniversity, Wuhan, China
`yike.t@whu.edu.cn`

Abstract. For most domestic public hospitals at this stage, issues such as insufficient humanistic care, inadequate talent support policies, limited promotion paths for personnel, and inadequate assessment mechanisms are common problems that hinder talent development. To maximize the encouragement of talent innovation, it is necessary to establish and improve a training system that meets the needs of the times and the individuals themselves. This paper, from the perspective of green human resource management, delves into the effectiveness of strengthening talent team construction and establishing a green human resource management system in public hospitals in the post-epidemic era. Incorporating the developmental status of a tertiary hospital in Hubei Province, this paper ultimately summarizes the feasibility and significance of such a system.

Keywords: green human resource management · public hospital · talent team construction

1 Introduction

Since the major public health event pandemic, the world has witnessed a succession of rapidly spreading, wide-ranging, and severe epidemic diseases, including influenza, dengue fever, and whooping cough, which have continuously tested global public health and medical institutions.

As a populous nation, China's domestic medical industry has undergone significant transformations. Aligning with the national strategy of Healthy China in the new era and the health policy emphasizing prevention, it is imperative to recognize that achieving universal health and safeguarding life safety hinges on medical development, and the core of medical progress lies in the cultivation of medical talents [1–4].

Medical talents are the primary resource for the development of the health industry and an important support for the implementation of the "Healthy China" strategy. As the main battlefield for the implementation of the "Healthy China" strategy [5], the establishment of a good medical talent training system in public hospitals is more conducive to promoting the strategy. Concurrently, environmental degradation and resource

B.-J. He et al. (Eds.): UCSUD 2023, LNCE 559, pp. 958–965, 2025.
https://doi.org/10.1007/978-981-97-8401-1_70

scarcity have heightened global attention towards the concept of green management. Some studies suggest that effective green human resource management can enhance an enterprise's competitiveness and promote its sustainable development.

This paper, from the perspective of green human resource management, delves into the effectiveness of strengthening talent team construction and establishing a green human resource management system in public hospitals in the post-epidemic era. Drawing upon relevant literature and incorporating the developmental status of a tertiary hospital in Hubei Province, this paper ultimately summarizes the feasibility and significance of such a system.

2 Research Background

2.1 Overall Human Resources of the Hospital

Zhongnan Hospital of Wuhan University is a modern comprehensive tertiary hospital integrating medical treatment, teaching, scientific research, prevention, and health-care,aiming to develop into a world-class comprehensive research-oriented teaching hospital. At present, there are 3,300 beds, 7 national-level platforms, 8 national clinical key construction specialties, 53 medical science and technology innovation platforms, 4 provincial key laboratories, 4 provincial engineering (technology) research centers, and 11 provincial clinical medicine research centers, and it has ranked among the top 30 in the total technology value of hospitals in the country for 5 consecutive years. At present, there are more than 4,000 employees in the hospital, of whom 23.9% have a doctor's degree and 43.2% have a master's degree or above.

The hospital attaches great importance to the construction of talent team, constantly increases the work of "internal training and external recruitment " of talents, and builds a platform for "attracting phoenix and nesting" and a nurturing base with sufficient nutrients by providing research start-up funds, discipline construction funds, family allowances, spouse placement, scientific research performance awards and other ways. Under the strong support of the above policies, in recent years, the hospital has achieved remarkable results in the construction of talent team. At present, there are 107 doctoral supervisors, 349 master supervisors, 24 national talent titles, and 62 provincial talent titles. However, the rapid improvement of the quality of the talent team also leads to new problems and difficulties.

2.2 Main Issues

2.2.1 Weak Sense of Belonging Among Talents, and "Acclimatization Issues" for Incoming Talents

With the increasing recruitment and introduction of talents in hospitals in recent years, more and more outstanding talents at home and abroad have joined the hospitals. However, in the past few years, these foreign talents have appeared a certain degree of "acclimatization" problems, such as alienation from the superiors and subordinates of the department or colleagues, inadaptability to the work concept and methods, and lack of a strong sense of belonging and identity with the hospital, which has led to the phenomenon that imported talents leave their jobs or their personal development is slow.

2.2.2 The Work Pressure is Great, and the Physical and Mental Health is in Jeopardy.

In recent years, most of all walks of life have formed an "involution" storm As a medical staff, the pressure of clinical work is enormous, and the diagnosis and treatment work occupies almost all the time in life, and it is even more common to turn around. However, the requirements of the society for medical practitioners are still gradually improving, and the "involution" of medical staff is becoming more and more serious. The heavy pressure makes medical staff seriously lack rest. Especially in today's post-epidemic era, the continuous invasion of viruses not only increases the pressure of clinical work, but also makes many people already. In recent years, the number of employees who can't continue to be competent for clinical work due to physical health or psychological problems has increased significantly.

2.2.3 The Talent Evaluation System is Single and the Way of Personnel Development is Limited.

At present, the professional title evaluation of hospitals needs to be evaluated by the documents formulated by Wuhan University every year. The evaluation cycle is long, the standards are solidified and the promotion quota is limited. Therefore, some people who meet or even exceed the requirements of the evaluation documents fail to pass the evaluation smoothly because of the evaluation system. Especially for imported talents, due to their professional titles, they can't match the corresponding treatment and conditions in time, which brings a certain lag to their project application and research work.

3 The Concept of Green Human Resource Management

The concept of green human resource management originated from the concept of "sustainable development" at the earliest. After the concept of "sustainable development" was put forward, some scholars put forward the concept of "environmental management", which means that enterprises can reduce their impact on the environment by adjusting their enterprise structure, enterprise system and business activities. In order to cope with the change of environment, more and more enterprises introduce "environmental management" into the strategic layout of enterprises. However, "environmental management" needs the cooperation of human resources, so "environmental management" and "human resource management" are linked together, thus forming "green human resource management" [6]. (Fig. 1).

Modern Green Human Resources Management refers to the "green" concept applied to the field of human resources management to form a new management concept and management mode. Its main task is to adopt "green" management methods. To achieve the internal staff mentality harmony, human harmony and ecological harmony of the three harmonious, so as to bring the enterprise economic benefits, social benefits and ecological benefits of the unity of the comprehensive benefits, to achieve the common and sustainable development of enterprises and employees. Among them, the harmony of mentality refers to the harmony of employees themselves, including good ideological quality and professional ethics, high scientific and cultural knowledge and skills

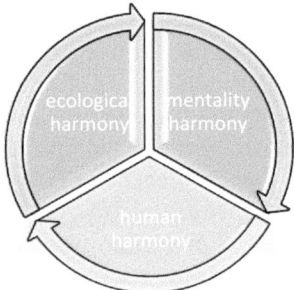

Fig. 1. Three aspects of the new management mode in modern green human resources management.

and aesthetic requirements, good self-adjustment ability, reasonable judgment of the relationship between themselves and others and nature, etc. Human harmony includes two levels of human-enterprise harmony and interpersonal harmony, namely the common development of enterprise and employees, the harmonious relationship between managers and employees, and ordinary employees; Ecological harmony refers to the harmonious coexistence between people or enterprises and nature.

4   Establish a Green Human Resource Management Model to Adapt to the Development of Hospitals

4.1   The Concept of Green Human Resource Management

The concept of green human resource management originated from the concept of "sustainable development" at the earliest. After the concept of "sustainable development" was put forward, some scholars put forward the concept of "environmental management", which means that enterprises can reduce their impact on the environment by adjusting their enterprise structure, enterprise system and business activities. In order to cope with the change of environment, more and more enterprises introduce "environmental management" into the strategic layout of enterprises. However, "environmental management" needs the cooperation of human resources, so "environmental management" and "human resource management" are linked together, thus forming "green human resource management" (citation 2).

4.2   The Necessity of Establishing a Hospital Green Human Resource Management System

According to the analysis of relevant literature, it is found that the research on green human resource management is not comprehensive at present, and most of the research focuses on the analysis of concepts and basic theories. There is not much research on how to integrate the concept of green human resource management into enterprises, and the overall analysis of the status quo, obstacles, existing problems and feasible solutions for domestic enterprises to implement green human resource management [7], especially how to scientifically establish a green human resource management model and system that adapts to the development of hospitals. As we all know, talent training is a long-term systematic project, so it is necessary to have both a scientific and reasonable training and assessment system and a perfect management and operation mechanism [8]. In view of the fact that the research proves that the practice of green human resource management can gain the competitive advantage of human resources for the sustainable development of enterprises, it is very necessary to explore how to combine ideas with practice to build a green human resource management system that conforms to the development trend of the post-epidemic era and is conducive to the construction of hospital talent team.

5   Specific Measures for Establishing a Green Human Resource Management System in Public Hospitals

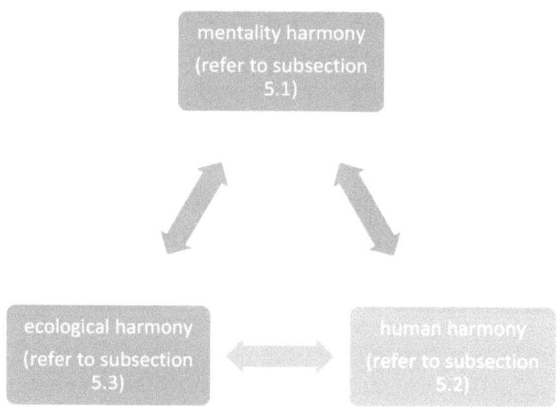

Fig. 2. Solutions to the three aspects of the new management mode in modern green human resources management.

5.1   Fostering Identity Recognition Through Internal and External Integration

Each year, approximately 150 new employees are inducted into our hospital. To ensure that these individuals gain a comprehensive understanding of the hospital's culture, operations, and career development opportunities during their initial stages, we conduct a structured pre-job training program annually in July. This training emphasizes the hospital's historical evolution, prevailing industry trends, personal development policies, welfare benefits, and other pertinent aspects. Furthermore, we establish a dedicated QQ group for new employees each year, serving as a platform for the exchange of work-related information and lifestyle tips, including rental housing options, residency training schedules, and notifications for professional title promotions. For departments that have recruited external talent, we maintain regular communication channels to actively monitor their work progress, addressing any challenges they may encounter and facilitating their smooth integration into the department. Additionally, at the conclusion of each year, we recognize and reward outstanding individuals through various awards, such as Clinical Role Model, Scientific Research Role Model, and Teaching Role Model, based on their overall performance throughout the year. This not only serves to acknowledge their contributions but also strengthens their sense of collective honor and identity within the hospital community (Fig. 2).

5.2 Maximizing Departmental Efficiency for Enhanced Humanistic Care

The hospital effectively leverages the capabilities of its administrative departments, adhering steadfastly to the working principle of "proactive and service-oriented." It carries out comprehensive visits and research in clinical front-line departments, organizing symposiums or questionnaire surveys tailored to different groups to gain a profound understanding of the ideas and needs of front-line medical staff. Based on this understanding, targeted policy documents are formulated and revised, providing meaningful assistance to address the issues and requirements faced by clinical staff. Concurrently, the hospital streamlines administrative processes, enhances work attitudes, and optimizes the work experience for clinical personnel. Annual free physical examinations are organized, psychological counseling clinics are established, and various cultural and sports activities, such as photography and marathon events, are held periodically to prioritize the physical and mental well-being of employees. Furthermore, the hospital reinforces its employee welfare system by offering birthday greetings, movie tickets, and other amenities, embodying the "humanistic Zhongnan" philosophy through these meticulous efforts.

5.3 Refining the Talent Evaluation Framework and Facilitating Individual Growth Pathways

In pursuit of broadening the talent development path, facilitating swifter achievements, and adhering to the national call for reform in scientific and technological talent development, our hospital has embarked on a comprehensive optimization and reform of the professional title evaluation system. This reform entailed the implementation of a more flexible talent evaluation mechanism, the establishment of a dual-track system for clinical and scientific research professional titles, and the expansion of talent evaluation criteria. Notably, the weight of clinical practice, teaching, and scientific research transformation capabilities was significantly increased, shifting the focus away from the sole reliance on research papers. Concurrently, we directly appointed a cohort of outstanding young talents with remarkable achievements as special deputy researchers, recognizing their potential and contributions. Since the implementation of this talent evaluation system reform, it has garnered unanimous praise from our medical staff and has significantly impacted clinical diagnosis and treatment, external collaborations, project submissions, and academic appointments. Notably, the annual average number of applicants for professional titles within the hospital has surged to over 200. This reform not only cultivates a more conducive environment for talent development within the hospital but also serves as a magnet for attracting top scientific and technological talents, effectively bolstering the core competitiveness of our hospital's talent pool.

6 Research and Reflection

In summary, green human resource management can be seamlessly integrated with hospital human resource management, thereby enhancing the scientificity, practicality, and warmth of hospital human resource management practices. This integration serves to bolster hospitals' talent competitiveness during their development process. Our research into the integration of these two concepts reveals that the exploration and application of green human resource management in public hospitals remains in its infancy. While the implemented measures have yielded certain positive outcomes, the scope and depth of these efforts are yet to be fully realized. As such, future research and work processes will prioritize the identification of synergies between the mechanisms of green human resource management and hospital development. This will involve seeking out the most appropriate strategies and methodologies for optimizing and reforming hospital human resources, while also embedding the principles of green human resource management into the routine operations of hospital human resource management. Ultimately, these efforts will contribute to the construction and nurturing of hospital talent teams.

References

1. Renwick, D.W., Redman, T., Maguire, S.: Green human resource management: A review and research agenda. Int. J. Manag. Rev. **15**(1), 1–14 (2013)
2. Arulrajah, A. A., Opatha, H.H.D.N.P., Nawaratne, N.N.J.: Green human resource management practices: A review (2015)
3. Ahmad, S.: Green human resource management: Policies and practices. Cogent Bus. & Manag. **2**(1), 1030817 (2015)
4. ALI, Md Chapol, et al.: A study of green human resources management (GHRM) and green creativity for human resources professionals. Int. J. Bus. Manag. Futur. **4**(2), 57–67 (2020)
5. Wang, D., et al.: Research on the scientific evaluation index system of medical personnel in public hospitals. Chinese Hospitals **27**(1), 39–42 (2023)
6. Zhao, S., Zhou, W.: Research status, implementation obstacles and research prospect of green human resource management in China. Leadership Sci. **10**, 104–107 (2019)
7. Guiyao Tang, Wei Sun, Jin Jia, etc. A review and prospect of Green human resource management research. Foreign Economics & Management **10** (2015)
8. An, L., Chi, B., Lun, Z.: Noteworthy problems and reflections on hospital personnel training. Chinese Hospital Management **22**(10), 34–35 (2002)

Author Index